Fish Osmoregulation

Fish Osmoregulation

Editors

Bernardo Baldisserotto
Universidade Federal de Santa Maria
Santa Maria, RS
Brazil

Juan Miguel Mancera
Universidad de Cádiz
Cádiz, Spain

B.G. Kapoor
Formerly Professor of Zoology
The University of Jodhpur
Jodhpur, India

CRC Press
Taylor & Francis Group
Boca Raton London New York

CRC Press is an imprint of the
Taylor & Francis Group, an **informa** business
A SCIENCE PUBLISHERS BOOK

First published 2007 by Science Publishers

Published 2018 by CRC Press
Taylor & Francis Group
6000 Broken Sound Parkway NW, Suite 300
Boca Raton, FL 33487-2742

ISBN-13: 978-1-57808-447 -0 (hbk)

Visit the Taylor & Francis Web site at
http://www.taylorandfrancis.com

and the CRC Press Web site at
http://www.crcpress.com

CIP data will be provided on request.

Preface

Fish lives in environments with a wide variety of chemical characteristics (fresh, brackish and seawater, acidic, alkaline, soft and hard waters). From an osmoregulatory point of view, fish have developed several mechanisms to live in these different environments. Fish osmoregulation has always attracted considerable attention and in the last years several studies have increased our knowledge of this physiological process.

In this book several specialists have analyzed and reviewed the new data published regarding fish osmoregulation. The chapters present an integrative synthesis of the different aspects of this field focusing on osmoregulation in specific environments (chapters 5 and 9) or situations (chapter 8), function of osmoregulatory organs (chapters 11, 12 and 14), general mechanisms (chapter 15) and endocrine control (chapters 2, 4, 6 and 16). In addition, interactions of osmoregulatory mechanisms with the immune system (chapter 1), diet (chapter 3) and metabolism (chapter 10) were also reviewed. Finally, new emerging techniques to study osmoregulation are analyzed (chapters 7 and 13). We hope that this book will provide a solid foundation for students and researchers and act as a guide to future perspectives in this field.

The Editors

Contents

List of Contributors

Baldisserotto Bernardo
Departamento de Fisiologia e Farmacologia, Universidade Federal de Santa Maria, 97105.900 – Santa Maria, RS, Brazil.
E-mail: bernardo@smail.ufsm.br

Bendhack Fabiano
Pontifícia Universidade Católica do Paraná. Curitiba, Paraná, Brazil.
E-mail: f.bendhack@pucpr.br

Brown J. Anne
School of Biosciences, University of Exeter, Exeter EX4 4PS, UK.
E-mail: J.A.Brown@exeter.ac.uk

Carneiro Paulo César Falanghe
Embrapa Tabuleiros Costeiros. Aracaju, Sergipe, Brazil.
E-mail: paulo@cpatc.embrapa.br

Cuesta Alberto
Fish Innate Immune System Group, Department of Cell Biology, Faculty of Biology, University of Murcia, 30100 Murcia, Spain.
E-mail: alcuesta@um.es

Esteban M. Ángeles
Fish Innate Immune System Group, Department of Cell Biology, Faculty of Biology, University of Murcia, 30100 Murcia, Spain.
E-mail: aesteban@um.es

Ferreira Francesca W.
Departamento de Biologia e Química, Universidade Regional do Noroeste do Rio Grande do Sul, 98700.000 – Ijuí, RS, Brazil.
E-mail: piscis@unijui.tche.br

Flik Gert

Department of Organismal Animal Physiology, Faculty of Science, Radboud University Nijmegen, Toernooiveld 1, 6525 ED Nijmegen, The Netherlands.
E-mail: g.flik@science.ru.nl

Freire Carolina A.

Departamento de Fisiologia, Setor de Ciências Biológicas, Universidade Federal do Paraná (UFPR), Centro Politécnico, Bairro Jardim das Américas, Curitiba, PR, CEP 81531-990, Brazil.
E-mail: cafreire@ufpr.br

Fuentes Juan

Molecular and Comparative Endocrinology, Centre of Marine Sciences, CCMAR, CIMAR Laboratório Associado, University of Algarve, Campus de Gambelas, 8005-139 Faro, Portugal.
E-mail: jfuentes@ualg.pt

Geven Edwin J.W.

Department of Organismal Animal Physiology, Faculty of Science, Radboud University Nijmegen, Toernooiveld 1, 6525 ED Nijmegen, The Netherlands.
E-mail: e.geven@science.ru.nl

Glover Chris N.

SCION, Te Papa Tipu Innovation Park, 49 Sala Street, Private Bag 3020, Rotorua, New Zealand.
E-mail: Chris.Glover@scionresearch.com

Grosell Martin

Rosenstiel School of Marine and Atmospheric Sciences, Division of Marine Biology and Fisheries, University of Miami, 4600 Rickenbacker Causeway, 33145 Miami, Florida, USA.
E-mail: mgrosell@rsmas.miami.edu

Guerreiro Pedro M.

Molecular and Comparative Endocrinology, Centre of Marine Sciences, CCMAR, CIMAR Laboratório Associado, University of Algarve, Campus de Gambelas, 8005-139 Faro, Portugal.
E-mail: pmgg@ualg.pt

Hazon Neil
School of Biology, University of St Andrews, St Andrews KY16 8LB, UK.
E-mail: nhl@st-andrews.ac.uk

Klaren Peter H.M.
Department of Organismal Animal Physiology, Faculty of Science, Radboud University Nijmegen, Toernooiveld 1, 6525 ED Nijmegen, The Netherlands.
E-mail: p.klaren@science.ru.nl

Kulczykowska Ewa
Department of Genetics and Marine Biotechnology, Institute of Oceanology of Polish Academy of Sciences, Sopot, Poland.
E-mail: ekulczykowska@iopan.gda.pl

Laiz-Carrión Raúl
Departamento de Biología, Facultad de Ciencias del Mar y Ambientales, Universidad de Cádiz, 11510 Puerto Real, Cádiz, Spain.
E-mail: raul.laiz@ca.ieo.es

Mancera Juan Miguel
Departamento de Biología, Facultad de Ciencias del Mar y Ambientales, Universidad de Cádiz, 11510 Puerto Real, Cádiz, Spain.
E-mail: juanmiguel.mancera@uca.es

Marshall William S.
Department of Biology, St. Francis Xavier University, P.O. Box 5000, Antigonish, Nova Scotia, Canada B2G 2W5.
E-mail: bmarshal@stfx.ca

McCormick Stephen D.
USGS, Conte Anadromous Fish Research Center, Turners Falls, MA, USA.
E-mail: steve_mccormick@usgs.gov

McDonald M. Danielle
Rosenstiel School of Marine and Atmospheric Science, University of Miami, Miami, Florida, 33149-1098, USA.
E-mail: mcdonald@rsmas.miami.edu

Meseguer José
Fish Innate Immune System Group, Department of Cell Biology, Faculty of Biology, University of Murcia, 30100 Murcia, Spain.
E-mail: meseguer@um.es

Parra Jorge Erick Garcia
Departamento de Ciências Agrárias, Universidade Regional Integrada do Alto Uruguai e das Missões – Campus Santiago, 97700.000 – Santiago, RS, Brazil.
E-mail: erickgarparr@yahoo.com.br

Prodocimo Viviane
Departamento de Fisiologia, Setor de Ciências Biológicas, Universidade Federal do Paraná (UFPR), Centro Politécnico, Bairro Jardim das Américas, Curitiba, PR, CEP 81531-990, Brazil.
E-mail: vprodocimo@yahoo.com.br

Sangiao-Alvarellos Susana
Dr. José L. Soengas, Laboratorio de Fisioloxía Animal, Facultade de Ciencias do Mar, Edificio de Ciencias Experimentais, Universidade de Vigo, E-36310, Vigo, Spain.
E-mail: sangiao@uvigo.es

Soengas José L.
Laboratorio de Fisioloxía Animal, Facultade de Ciencias do Mar, Edificio de Ciencias Experimentais, Universidade de Vigo, E-36310, Vigo, Spain.
E-mail: jsoengas@uvigo.es

Urbinati Elisabeth Criscuolo
Universidade Estadual Paulista. Jaboticabal, São Paulo, Brazil.
E-mail: bethurb@caunesp.unesp.br

Wilson Jonathan Mark
Laboratório de Ecofisiologia, Centro Interdisciplinar de Investigação Marinha e Ambiental, Rua dos Bragas 289, 4050-123, Porto, Portugal.
E-mail: wilson_jm@cimar.org

CHAPTER

1

Immune and Osmoregulatory System Interaction

Alberto Cuesta, José Meseguer and M. Ángeles Esteban*

INTRODUCTION

Fish, a very diverse group, were the first vertebrates to present a complete immune system about 450-500 million years ago. The innate and adaptive immune responses that they display share many similarities with the mammalian immune system. The fact that fish are poikilotherms and, therefore, subjected to environmental temperature changes, makes their adaptive responses very low and slow, which means that fish immunity is highly dependent on the innate or non-specific immune response. Therefore, study of the fish immune system is of great interest from the phylogenetical viewpoint and it is in fish that the adaptive responses first appeared. Moreover, the growth of aquaculture to provide food for the human diet has prompted researchers to investigate immunological techniques for the diagnosis and control of fish diseases, the development of vaccines being the final goal (Ellis, 1988).

Authors' address: Fish Innate Immune System Group, Department of Cell Biology, Faculty of Biology, University of Murcia, 30100 Murcia, Spain.
**Corresponding author:* E-mail: aesteban@um.es

Fish live in a changeable environment and they must adapt to these changes. As regards water salinity changes, fish are able to adapt to the environmental salinity by the mechanism known as osmoregulation. In general, fresh and marine water-living fish tend to maintain a net water influx or efflux in order to keep the plasma osmolarity constant. The organs involved in osmoregulation are the kidney, gills and intestine, which have been morpho-functionally characterized in many fish species (Meseguer *et al.*, 1981; López-Morales *et al.*, 1990; Sakamoto *et al.*, 2001; Greenwell *et al.*, 2003) and will be described in another chapter. Moreover, when the organs are engaged in osmoregulation, other functions may be affected. This happens, for example, in the case of immune functions.

The fish immune response is intended to eradicate an invading agent, the antigen. It starts with the humoral and cellular components of the innate immune system after coming into contact with structures of the pathogen known as pathogen-associated molecular patterns (PAMPs), which are common molecules not usually found in eucaryotic cells, such as viral double stranded RNA, bacterial lipopolysaccharide (LPS) and certain sugars. This response usually starts immediately and lasts several hours. The antigen is then processed and presented to the adaptive immune system components (B and T lymphocytes), which elaborate the adaptive or specific response. This entire process takes several days but, due to the lack of thermoregulation, the response achieved is never comparable in terms of effectiveness with the mammalian response. The control and integration of this immune response is carried out by cytokines, which are mainly produced by lymphocytes and monocyte/macrophages after stimulation. However, the immune response is also modulated by many other intrinsic and extrinsic factors, including environmental factors (temperature, salinity, photoperiod, etc.) and physiological status (nutrition, age, reproductive cycle, hormonal balance, stress, etc.).

Apart from the morphological features of the organs involved in osmoregulation (Meseguer *et al.*, 1981; López-Morales *et al.*, 1990), the morpho-functional properties of the teleost immune system have been characterized in our group (Esteban *et al.*, 1989, 1998, 2001; Meseguer *et al.*, 1991, 1994, 1996; Mulero *et al.*, 1994; Cuesta *et al.*, 1999, 2002, 2003, 2004; Ortuño *et al.*, 2000, 2002; Sepulcre *et al.*, 2002; Chaves-Pozo *et al.*, 2003; Rodríguez *et al.*, 2003; Salinas *et al.*, 2005;). In this chapter, we shall review the effect that salinity (as an environmental factor) may have on the fish immune responses, following by the importance and magnitude of the osmoregulatory hormones (as an intrinsic factor) to finally deal with the endocrine-osmoregulation-immunity interactions in fish whose osmotic balance has been altered.

FISH IMMUNE SYSTEM ORGANIZATION

The fish immune system is as organized and complex as it is in mammals (for reviews see Meseguer *et al.*, 1995, 1996, 2002; Zapata *et al.*, 1996; Manning, 1998; Dixon and Stet, 2001; Evans *et al.*, 2001; Magor and Magor, 2001; Secombes *et al.*, 2001). Due to variations in animal anatomy and evolutionary position of fish, morpho-functional differences exist in immune tissues and cells between fish and mammals.

Structure and Organization

The fish immune system—like that of other vertebrates—consists of physical barriers and immune organs. The first and principal barrier is the skin, which together with the gills and gut, contains large amounts of mucus. This mucus serves as an antimicrobial and antiparasitic barrier because it contains highly active immune soluble factors such as lysozyme, complement, C-reactive protein, lectins and immunoglobulins. Thus, injuries in the barriers or the lack of mucus facilitate the entry of pathogens into fish, where humoral and cellular immune effectors then begin to play their part. The most characteristic difference from mammals is the lack of bones, and therefore bone marrow, while the kidney is divided into two functional parts: the pronephros (also called anterior or head-kidney), which is the main haematopoietic organ in fish, and the opisthonephros (called posterior or trunk kidney), which is mainly dedicated to the excretory function. However, the immune functions are conserved along the entire kidney. Apart from these, there are also small batches of scattered immune cells in the gills and gut although, in general, fish leucocyte types are quite similar to their mammalian counterparts, except for granulocytes, while platelets are replaced by thrombocytes.

Innate Immune Response

Once the pathogen (bacteria, virus or parasite) has entered the fish, the host elicits an inflammatory response involving humoral (complement, lysozyme, C-reactive protein, lectins, etc.) and cellular (monocyte/

macrophages, granulocytes and lymphocytes) components of the innate immune response. Complement and lysozyme are able to kill the pathogens by puncturing their membranes. Among the cellular mechanisms, phagocytosis and cytotoxicity are the main mechanisms involved. Phagocytes (monocyte/macrophages and granulocytes) engulf the pathogen and exert their lytic function through lysosomal enzymes (peroxidases, etc.) and the production of reactive oxygen/nitrogen species (O_2^-, H_2O_2 or NO). The nonspecific cytotoxic cells (NCC) are a heterogeneous leucocyte population, functionally equivalent to the mammalian natural killer (NK) cells, which mediate the cytotoxic activity against tumor cells, virus-infected cells and parasitic protozoa. Apart from complement and lysozyme, the humoral factors include C-reactive protein, lectins, transferrin, anti-proteases, interferons and eicosanoids, which form part of the innate response and combat the pathogen by means of different mechanisms.

Adaptive Immune Response

The first functional studies carried out pointed to the presence of B and T lymphocytes in fish because of the immune responses observed, including specific cytotoxicity, antigen-specific antibody generation, delayed hypersensitivity and graft rejection. The appearance of specific antibodies directed against B or T cells and the development and application of molecular biology tools have increased our understanding of the adaptive immune responses in fish, while new findings in this area tend to confirm the similarities with the mammalian adaptive immune response, with a few exceptions. For example, the existence of rearranging genes for immunoglobulin M (IgM), T-cell receptor (TCR) and major histocompatibility (MHC) has been confirmed as has been the existence of coreceptor molecules (CD3, CD4 and CD8). Further functional studies will presumably demonstrate the great similarities existing between the mammalian and fish adaptive immune systems from a molecular and functional viewpoint.

Cytokines

Cytokines are immune system 'hormones'. They are small polypeptides or glycoproteins synthesized after leucocyte stimulation and even show pleiotropic effects. Interleukin (IL)-1, IL-2, IL-3, IL-6, interferon (IFN), tumor necrosis factor (TNF), transforming growth factor β_1 (TGB-β_1) and chemokines are the main cytokines found in fish till date. The recent availability of the cytokine gene sequences and ongoing production of recombinant cytokines will throw light on their specific functions within and outside the immune system.

Major Histocompatibility Complex (MHC)

MHCs are highly polymorphic cell surface proteins consisting of MCH class I and class II glycoproteins. They belong to the immunoglobulin superfamily of proteins and interact with the T-cell subsets through a specific TCR, initiating the adaptive immune response. They are responsible for presenting the antigen to the T lymphocytes and are considered to be the link between the innate and adaptive immune responses. Since they were first discovered by PCR techniques, the MHC from several fish species have been cloned and studied from a genetic point of view. They appear clustered in all vertebrates except for teleost fish, where they are in different chromosomes and called MH receptors. However, deeper knowledge of the involvement and functioning of the MHC in the immune response is just emerging with the use of recombinant MHC proteins and anti-MHC antibodies.

INFLUENCE OF ENVIRONMENTAL SALINITY ON FISH IMMUNE RESPONSE

Salinity is one of the most important environmental factors for aquatic organisms. In teleost fish, environmental salinity fluctuations trigger the osmoregulatory response to compensate for such changes. However, other physiological processes are also affected. For example, the immune response and fish disease resistance is modulated by salinity, as has been shown in several studies. Few experiments have examined the immunological responses after salinity disturbances in fish, the innate responses being the most analyzed thus far. The total circulating IgM levels, which reflect the immune system status without exposing the fish to a specific antigen (Yada *et al.*, 1999), has been the most examined immune parameter. On the other hand, cellular activities such as phagocytosis, respiratory burst and cytotoxicity have hardly been determined in the few investigations carried out. Future studies are needed to establish the impact of salinity on the general immunological status rather than the effect on an individual immune response.

Hyperosmotic adaptation has been mainly studied in salmonids (Table 1.1). The first studies dealt primary immune responses in coho salmon (*Oncorhynchus kisutch*), which were seen to decrease when the fish entered seawater during smoltification (Maule and Schreck, 1987). Brown trout (*Salmo trutta*) specimens transferred to seawater, on the other hand, showed increased plasmatic lysozyme activity while the phagocytic or natural cytotoxic activities of pronephric leucocytes increased or remained unchanged, respectively (Marc *et al.*, 1995). Specific antibody titres to *Yersinia ruckeri* decreased in rainbow trout (*Oncorhynchus mykiss*) 7 days after transfer to 22 ppt salinity (Betoulle *et al.*, 1995). On the other hand, the circulating IgM level of trouts was unaffected 3 days after transfer from freshwater (FW) to 12 ppt water, while the lysozyme activity was 3.5-fold increased (Yada *et al.*, 2001). The same fish were then transferred from 12 ppt to 29 ppt salinity water and 24 h later they showed the same level of IgM, while the lysozyme activity had further increased. Peripheral blood leucocyte (PBL) production of superoxide (O_2^-), measured by nitroblue tetrazolium (NBT) reduction, was greatly increased. However, the same group did not detect any change in plasmatic IgM, lysozyme activity or O_2^- production by PBLs in Mozambique tilapia (*Oreochromis mossambicus*) transferred from FW to 35 ppt salinity water for more than 1 month, although head-kidney leucocyte (HKL) production of O_2^- was increased (Yada *et al.*, 2002). Moreover, the authors conducted further research and described, for the first time, the increase of PRL-R (prolactin receptor) mRNA expression in leucocytes due to hypersaline adaptation. This PRL-R triggers a cascade into the cell, leading to the cell responses, where activation of the immune function is also produced. A recent study in Nile tilapia (*Oreochromis niloticus*) has described the lethal effect of 35 ppt environments but increased plasmatic IgM levels in specimens after 2 or 4 weeks of adaptation to 12 or 24 ppt salinity (Dominguez *et al.*, 2004). Although both tilapia species, *O. mossambicus* and *O. niloticus*, have similar life requirements, the differences observed could be due to several reasons. Apart from the different salinity conditions (time and salinity stringency), body size (50-100 or 18.2-21.7 g bw, respectively), diet ration or temperature (24 and 28°C, respectively) were also different. All these parameters influence the osmoregulatory response and also the immune response, as indicated above.

Few studies have evaluated the effects of environmental salinity changes in marine fish species (Table 1.1). In winter flounder (*Pleuronectes*

Table 1.1 Effect of salinity disturbances on fish immune responses.

Species	*Salinity acclimation*	*Immune parameter*	*Reference*
Oncorhynchus kisutch	FW to SW	↓ immune responses	Maule and Schreck (1987)
Salmo trutta	FW to SW	↑ lysozyme and phagocytosis = cytotoxicity	Marc *et al.* (1995)
Pleuronectes americanus	14 or 28 ppt	↓ blood thrombocytes in SW	Plante *et al.* (2002)
Mylio macrocephalus	33 ppt to 6 or 21ppt	↑ phagocytosis	Narnaware *et al.* (2000)
Oncorhynchus mykiss	FW to 12 or 29 ppt FW to 22 ppt	= IgM, ↑ lysozyme, ↑ O_2^- in PBLs ↓ anti-*Yersinia ruckeri* specific IgM	Yada *et al.* (2001) Betoulle *et al.* (1995)
Oreochromis mossambicus	FW to 35 ppt	= IgM and lysozyme, - O_2^- and PRL-R expression in HKLs	Yada *et al.* (2002)
Oreochromis niloticus	FW to 12 or 24 ppt	↑ IgM	Domínguez *et al.* (2004)
Sparus aurata	40 to 6 ppt 40 to 12 ppt 40 to 55 ppt	↓ peroxidases and ACH, = IgM ↓ peroxidases, ↑ ACH, = IgM ↑ IgM, = peroxidases and ACH	Cuesta *et al.* (2005a)
Epinephelus sp.	33 ppt to 20 or 40 ppt	↑ susceptibility to IPNV	Chou *et al.* (1999)
Ictalurus punctatus *Acipenser oxyrinchus desotoi* *Morone saxatilis* *Carassius auratus*	0, 1, 3 or 9 ppt	↑ resistance to *Flavobacterium columnare* with the salinity increase	Altinok and Grizzle (2001)

FW, freshwater; SW, seawater; ppt, parts per thousand; PBLs, peripheral blood leucocytes; HKLs, head-kidney leucocytes; IPNV, Infectious pancreatic necrosis virus; PRL-R, PRL receptor; ACH, alternative complement activity; ↑, increase; ↓, decrease; =, no effect.

americanus), adaptation for 2 months to seawater (SW; 28.7 ppt) or brackish water (BW; 14.7 ppt) completely abrogated the circulating thrombocytes seen in SW and increased all the stress indicators (Plante *et al.*, 2002). Two studies have also been carried out in sparids. In gilthead seabream (*Sparus aurata*), transfer from 40 ppt salinity to 55 ppt for 14 days increased the plasmatic IgM levels but did not affect the alternative complement activity or the plasmatic peroxidases content (Cuesta *et al.*, 2005a). This finding agrees with the increased IgM levels found in Nile tilapia (Dominguez *et al.*, 2004) but contrasts with those found in Mozambique tilapia and rainbow trout (Yada *et al.*, 2001, 2002). On the other hand, transfer from 40 ppt to 12 or 6 ppt salinity for 14 or 100 days decreased the peroxidase content and/or complement activity but did not influence the circulating IgM levels. In the other study, 2-5 g bw black seabream (*Mylio macrocephalus*) specimens were kept at 33, 21 or 6 ppt salinity water for 72 days (Narnaware *et al.*, 2000) and, while the phagocytic activity of pronephric leucocytes increased in those fish adapted to 6 or 21 ppt salinities compared to the fish maintained in full-seawater (33 ppt), the activity of spleenic leucocytes decreased. Moreover, the authors demonstrated that the diet ration interacted with salinity in the effect observed on the immune responses.

Many studies have demonstrated that the best culture conditions for fish, both in aquaria and fish farms, are those in which the fish species are in isoosmotic water. These conditions mean that the fish uses less energy in osmoregulation and can redirect this energy towards other physiological processes, such as growth or immune responses. In this way, the limited data related with the defence mechanisms are presented. Mortalities of 1 g bw grouper fry (*Epinephelus* sp.) specimens transferred from 33 ppt water to 20 or 40 ppt salinity water for 48 h increased (Chou *et al.*, 1999). Moreover, when they were exposed to IPNV either before or after the salinity transfer, the mortality significantly increased, reaching 100% in some cases. In another experiment, channel catfish (*Ictalurus punctatus*), goldfish (*Carassius auratus*), striped bass (*Morone saxatilis*) and gulf sturgeon (*Acipenser oxyrinchus desotoi*) were maintained in freshwater (0 ppt), 1, 3 or 9 ppt salinity (Altinok and Grizzle, 2001). After acclimation, they were exposed to an experimental infection with the bacteria *Flavobacterium columnare*. None of the gulf sturgeons died, while the mortality of the other fish species decreased with increased salinity, with no mortality observed in the fish adapted to 3 or 9 ppt salinities. However, most studies have analyzed or related salinity changes with the pathogenic potential or survival of pathogens and not with the fish defence. For example, *Ichthyophthirius multifiliis* strains isolated from rainbow trout were susceptible to more than 5 ppt salinity (Aihua and Buchmann, 2001), while the survival of the copepod *Lerneaocera branchialis*, a parasite of the aquarium cod, is salinity restricted below 16-20 ppt salinity (Knudsen and Sundnes, 1998). Apart from the direct effect of salinity on the viability of pathogens, salinity seems to affect the PAMPs because parasites incubated at different salinities change their virulence, pathogenicity and even their adherence to the fish immune system effectors (Bordas *et al.*, 1996; Altinok and Grizzle, 2001; Nitzan *et al.*, 2004; Zheng *et al.*, 2004). Results have demonstrated that salinity directly affects the pathogenicity of virus, bacteria and parasites affecting the subsequent clearance by the fish immune system.

Explanations of how the changes in osmotic pressure alter the immune function of leucocytes are not consistent. The data suggest that leucocytes, like the rest of the body cells, are affected by the osmotic pressure. However, how and why they are shifted to inhibition or activation after osmotic balance disruption remain unanswered. Although the effect of osmoregulatory hormones on these cells (see below) is supposed to be the key, some direct role must be operating in leucocyte functioning. Perhaps, alterations in the water and ionic balance are sufficient strong signals to change the immune response by themselves. Furthermore, variations in plasmatic/seric levels might be attributed to the increase/decrease of blood volume with the consequent dilution/concentration, respectively, of humoral immune mediators. However, this hypothesis cannot be supported in light of the ensuing results. These data confirm the need for more in-depth studies into the role of salinity in the immune system and disease resistance, and into the mechanisms involved.

OSMOREGULATORY HORMONES—DO THEY CONTROL THE IMMUNE SYSTEM?

It is well-known and assumed that fish present complex and bi-directional endocrine-immune interactions (Weyts *et al.*, 1999; Engelsma *et al.*, 2002). However, the mechanisms mediating such interactions are not well studied, although they are supposed to be similar to those in mammals. We shall now analyze endocrine-immune interactions, focusing on the immunomodulatory potential of those hormones that play some osmoregulatory role. The major hormones involved in fish osmoregulation, namely PRL, GH and cortisol, have been shown to act as fish immunomodulators. While PRL and GH have been found to increase immune responses, cortisol is considered a stress hormone and plays an antagonistic role. The effect of such hormones on the immune system was first defined by studies involving fish transfer to hypo- or hyperosmotic media, stressful situations and hypophysectomy and, lately confirmed by *in vitro* and *in vivo* assays conducted with purified hormones. However, more studies are needed to complete the information, regardless of their exact effect on the immune response and disease resistance. Later investigations tried to establish the precise osmoregulatory actions of several other hormones, such as corticotropin, arginine, vasotocin, epinephrine, norepinephrine, thyroid hormones (T3 and T4), estradiol, aldosterone and natriuretic peptides (see Bentley, 1998). Future research will tend to elucidate the role of the osmoregulatory hormones in the immune system and will hopefully increase our knowledge concerning the complex interactions between fish osmoregulation and immunity.

PRL and GH

These two pituitary hormones have a demonstrated immunostimulatory role in fish. First evidence pointed in this direction after the effects on the immune system in hypophysectomized fish were studied. In this sense, killifish (*Fundulus heteroclitus*) showed an important reduction in the number of circulating leucocytes (Pickford *et al.*, 1971). Removal of the pituitary in rainbow trout decreased the levels of plasmatic IgM, Ig-secreting leucocytes in head-kidney and blood, as well as O_2^- production by HKLs (Yada *et al.*, 1999; Yada and Azuma, 2002). On the other hand, lysozyme activity, the total number of leucocytes, O_2^- production by PBLs and Ig-secreting cells in thymus and spleen were unaffected. In hypophysectomized *O. mossambicus*, however, neither plasmatic IgM level nor the lysozyme activity was modified, while O_2^- production by HKLs was depressed (Yada *et al.*, 2002). These same experiments also demonstrate the reversion of the immune response caused by hypophysectomy after exogenous PRL or GH administration.

In vitro or *in vivo* treatment of fish with PRL or GH (either from fish, mammalian or recombinant source) enhances the humoral (IgM level as well as complement and lysozyme activities) and cellular (mitogenesis, phagocytosis, cytotoxicity and respiratory burst) responses of the fish immune system, as well as disease resistance (Table 1.2). They exert their

Table 1.2 Effects of principal osmoregulatory hormones (PRL, GH and cortisol) on fish immune responses.

Hormone	*Effect*	*Species*	*Reference*
PRL	↑ mitogenesis	*Oncorhynchus keta* *O. mykiss*	Sakai *et al.* (1996b) Yada *et al.* (2004a)
	↑ phagocytosis	*Sparus sarba*	Narnaware *et al.* (1998)
	↑ respiratory burst	*O. mykiss* *Oreochromis mossambicus*	Sakai *et al.* (1996c) Yada *et al.* (2002)
	↑ lysozyme activity	*O. mykiss*	Yada *et al.* (2001, 2004b)
	↑ allograft rejection	*Fundulus grandis*	Nevid and Meier (1995)
	↑ IgM levels	*O. mykiss*	Yada *et al.* (1999)
	↓ IgM levels	*Sparus aurata*	Cuesta *et al.* (2005b)
GH	↑ lymphopoiesis	*S. aurata* *S. sarba*	Calduch-Giner *et al.* (1995) Narnaware *et al.* (1997)
	↑ phagocytosis	*O. mykiss* *Oncorhynchus keta* *S. aurata*	Sakai *et al.* (1995, 1996c, 1997) Sakai *et al.* (1996b) Calduch-Giner *et al.* (1997)
	↑ mitogenesis	*O. keta* *O. mykiss*	Sakai *et al.* (1996b) Yada *et al.* (2004a)
	↑ cytotoxic activity	*O. mykiss*	Kajita *et al.* (1992)
	↑ IgM levels	*O. mykiss*	Yada *et al.* (1999)
	↑ lysozyme activity	*O. mykiss*	Yada *et al.* (2004b)
	↑ haemolytic activity	*O. mykiss*	Sakai *et al.* (1996a)
	↑ disease resistance	*O. keta*	Sakai *et al.* (1997)
	↑ respiratory burst	*O. mykiss* *O. keta* *Dicentrarchus labrax* *Oreochromis mossambicus*	Sakai *et al.* (1995, 1996c) Kitlen *et al.* (1997) Yada *et al.* (2001) Sakai *et al.* (1996b, 1997) Muñoz *et al.* (1998) Yada *et al.* (2002)
	↓ IgM levels	*Sparus aurata*	Cuesta *et al.* (2005b)
Cortisol	↓ circulating lymphocytes	*O. kisutch* *Salmo trutta* *Ictalurus punctatus* *S. salar* *Cyprinus carpio*	McLeay (1973) Pickering (1984) Ellsaesser and Clem (1987) Espelid *et al.* (1996) Wojtaszek *et al.* (2002)
	↓ leucocyte mitogenesis	*Pleuronectes platessa*	Grimm (1985)

(Table 1.2 contd.)

(*Table 1.2 contd.*)

	Oncorhynchus kisutch	Tripp *et al.* (1987)
	Ictalurus punctatus	Ellsaesser and Clem (1987)
	Salmo salar	Espelid *et al.* (1996)
	Cyprinus carpio	Weyts *et al.* (1997)
	O. mykiss	Yada *et al.* (2004)
	O. mykiss cell line RTS11	Pagniello *et al.* (2002)
↓ circulating/production IgM	*O. mykiss*	Anderson *et al.* (1982) Hou *et al.* (1999)
	C. carpio	Ruglys, (1985) Saha *et al.* (2004)
	O. kisutch	Maule *et al.* (1987) Tripp *et al.* (1987)
	Pleuronectes americanus	Carlson *et al.* (1993)
	O. tshawytscha	Milston *et al.* (2003)
	O. masou	Nagae *et al.* (1994)
↓ phagocytosis	*C. carpio*	Law *et al.* (2001)
	Oreochromis niloticus	Law *et al.* (2001)
	Sparus aurata	Esteban *et al.* (2004)
	C. carpio	Watanuki *et al.* (2002)
	C. auratus macrophage cell line	Wang and Belosevic (1995)
↓ chemotaxis	*C. auratus* macrophage cell line	Wang and Belosevic (1995)
↓ respiratory burst	*S. aurata*	Esteban *et al.* (2004)
	C. carpio	Watanuki *et al.* (2002) Kawano *et al.* (2003)
	Morone saxatilis	Stave and Roberson (1985)
↓ NO production	*C. carpio*	Watanuki *et al.* (2002)
	C. auratus macrophage cell line	Wang and Belosevic (1995)
↓ immune genes expression	*O. mykiss*	Zou *et al.* (2000)
	C. carpio	Saeij *et al.* (2003)
↑ circulating IgM	*S. aurata*	Cuesta *et al.* (2005b)
↑ apoptosis	*C. carpio*	Weyts *et al.* (1997, 1998a) Saha *et al.* (2003)
	O. mykiss	Yada *et al.* (2004)
	O. mossambicus	Bury *et al.* (1998)
↑ C-reactive protein	*P. platessa*	White and Fletcher (1985)
↑ allograft rejection	*Fundulus grandis*	Nevid and Meier (1995)

(*Table 1.2 contd.*)

(Table 1.2 contd.)

	↑ pathogen susceptibility	*O. mykiss*	Kent and Hedrick (1987)
		S. salar	Wiik *et al.* (1989) Harris *et al.* (2000)
		S. trutta	Harris *et al.* (2000)
		Salvelinus alpinus	Harris *et al.* (2000)
	Prevents apoptosis in neutrophils	*C. carpio*	Weyts *et al.* (1998b)
	Prevents stress immunodepression	*O. mykiss*	Narnaware and Baker (1996)
	= cytotoxicity	*S. aurata*	Esteban *et al.* (2004)

↑, increase; ↓, decrease; =, no effect; NO, nitric oxide.

actions after engaging their specific receptors in the cells. Both hormones belong to the cytokine/haematopoietin family, while their receptors belong to the class I superfamily of cytokine receptors (see Clevenger *et al.*, 1998; Moutoussamy *et al.*, 1998; Power, 2005). Evidence of the mRNA expression of PRL and GH, as well as their respective receptors, have been documented in lymphoid organs and isolated leucocytes in several teleosts, including tilapia, rainbow trout, gilthead seabream, orange-spotted grouper (*Epinephelus coioides*), coho salmon, goldfish, masou salmon (*Oncorhynchus masou*), japanese flounder (*Paralichtys olivaceus*) and black seabream (*Acanthopagrus schlegeli*) (Weigent *et al.*, 1988; Sandra *et al.*, 1995; Mori and Devlin, 1999; Santos *et al.*, 1999, 2001; Yang *et al.*, 1999; Prunet *et al.*, 2000; Tse *et al.*, 2000, 2003; Higashimoto *et al.*, 2001; Lee *et al.*, 2001; Yada and Azuma, 2002; Yada *et al.*, 2002; Fukada *et al.*, 2004; Zhang *et al.*, 2004; Power, 2005). In mammals, lymphocytes and macrophages are the leucocyte-types that express both hormones and hormone-receptors, and the pattern might be the same in fish. Thus, leucocyte activation may not only be due to pituitary-secreted PRL and GH but may also be caused by the self-produced hormones. In this way, both hormones could be considered as cytokines, as they are in mammals, and autocrine and paracrine actions within the immune system are actually under consideration.

Receptors for fish PRL and GH (PRL-R and GH-R respectively) show the conserved motifs of the cytokine-receptor family. Thus, fish PRL-R is only present in the long and intermediate forms with the conserved motif WSXWS (Trp-Ser-Xaa-Trp-Ser) in the extracellular domain, while in the GH-R the conserved motif found is Y/FGEFS (Tyr/Phe-Gly-Glu-Phe-Ser).

In both receptors, the single transmembrane region is followed by a cytoplasmic region containing conserved proline-rich motifs (box-1 and box-2) and phosphorylable residues. Similarly to mammals, PRL-R and GH-R binding to their respective hormones on fish leucocytes probably involves the participation of the Jak/STAT activation pathway, although there are, to date, no specific data to support this interaction in fish. However, apart from the conservation of box-1 and box-2 motifs in the receptors, the presence of the Jak/STAT pathway in fish leucocytes has been confirmed (Jaso-Friedmann *et al.*, 2001; Santos *et al.*, 2001; Fukada *et al.*, 2004; Cuesta *et al.*, 2005c). Another striking point is the cross-interactions between forms of PRL and GH-R (Auperin *et al.*, 1995; Sandra *et al.*, 1995; Shepherd *et al.*, 1997). In tilapia, the PRL_{177} is able to bind the GH-R and, could lead to a stimulation of leucocytes mimicking the effects due to either PRL or GH. These two findings are probably the most valuable for deciding future directions that should be taken. It is imperative to distinguish between the effects due to the PRL form or GH, as well as to identify which hormone receptor is responsible for the immunostimulation achieved. Molecular approaches can hopefully be conducted in order to finally clarify the molecular interactions and involvement of the Jak/STAT activation cascade in the modulation of the immune system by these pituitary hormones.

Cortisol

The principal inter-renal gland-produced hormone, cortisol, is considered the stress hormone but it is also involved in osmoregulation and the immune function. Although the immunosuppressive effects observed after stress are attributed to high levels of circulating cortisol (reviews of Balm, 1997; Wendelaar Bonga, 1997; Pickering, 1998; Harris and Bird, 2000a), we will only focus on the investigations directed at evaluating the impact of exogenous *in vitro* or *in vivo* administration of cortisol on the fish immune system (Table 1.2). In this sense, most of the studies conducted demonstrate that cortisol treatment by itself decreases the fish immune functions, as does stress. However, differences in treatment (cortisol concentration and time), fish species, leucocyte source and immune parameter measured may affect the results observed. Thus, several papers suggest that cortisol is not the mediator of the stress effects and point to the need for more and deeper studies need to be done before any general rule can be assumed. For example, Narnaware and Baker (1996) demonstrated that trout injected with cortisol recovered from the immunosuppressive effects after an acute stress. They found decreased levels of circulating lymphocytes and phagocytic activity in stressed fish. These immunological changes were abrogated and restored in those fish injected with physiological concentrations of cortisol. As a hypothesis, authors thought that cortisol might inhibit the release of catecholamines, which would be directly responsible for the stress-response in some way. Another explanation could be that cortisol mediates the expression of adhesion molecules in leucocytes and therefore their trafficking. So, as in mammals (Chung *et al.*, 1986), cortisol administration may impair lymphocyte recruitment in the lymphoid tissues, while circulating granulocyte and/or macrophage numbers may be increased (Ellsaesser and Clem, 1987; Narnaware and Baker, 1996; Ortuño *et al.*, 2001; Wojtaszek *et al.*, 2002). If these circulating phagocytes are the active cells from the spleen or pronephros, the phagocytic activity of the remaining phagocytes must be inhibited, which would agree with most studies. Weyts *et al.* (1998a,b) found more striking data. They demonstrated that cortisol did not induce apoptosis in circulating T lymphocytes and thrombocytes but did so in B lymphocytes. Moreover, circulating neutrophils treated with high cortisol levels were protected from apoptosis, making these leucocytes more able to attack the pathogens entering the body. Moreover, cortisol did not inhibit their respiratory burst, which could be essential for survival since they form part of the first line of defence. Esteban *et al.* (2004) also investigated the cortisol effect on the gilthead seabream immune response. *In vitro*, pharmacological dosages of cortisol decreased the phagocytosis of head-kidney leucocytes but unaffected the respiratory burst and cytotoxicity. On the other hand, *in vivo* administration of cortisol (reaching plasmatic levels similar to those after acute stress) increased the circulating IgM levels and left unaltered the complement activity (Cuesta *et al.*, 2005b). This variability in the data concerning the immunosuppressive effects of cortisol, as well as contrary findings, should stimulate researchers into conducting more investigations in this field to ascertain how cortisol acts and how influences the fish immune system.

To date, cortisol synthesis has only been described in the interrenal gland and not in the leucocytes. On the other hand, the expression of glucocorticoid (GR) and mineralocorticoid (MR) receptors in lymphoid organs has been mentioned in several fish species, including rainbow trout, carp, coho salmon, tilapia and *Astatotilapia burtoni* (Maule and

Schreck, 1990; Ducouret *et al.*, 1995; Tagawa *et al.*, 1997; Weyts *et al.*, 1998c; Colombe *et al.*, 2000; Bury *et al.*, 2003; Greenwood *et al.*, 2003). However, functional data support the notion that fish leucocytes contain MR and GR, as do their mammalian counterparts, although the specific cell-types expressing them are not known. Furthermore, the effects described for cortisol on the immune response are mimicked by the agonist dexamethasone and abrogated by the blocking agents cycloheximide or RU486 (Weyts *et al.*, 1998b; Law *et al.*, 2001; Pagniello *et al.*, 2002; Esteban *et al.*, 2004). Although there are evident analogies between fish and mammals as regards the receptor activation cascade and effects upon the immune related genes further studies are needed to clarify the effects of cortisol on leucocytes at molecular level.

Other Hormones

Many other fish hormones play some osmoregulatory role either by direct or indirect action. For example, they may affect the release of PRL, GH or cortisol, and modify Na^+-K^+ ATPse activity, etc. (see Bentley, 1998). However, the effects of these hormones on the fish immune system have not been studied in any depth. Thus, melanocyte-stimulating hormone (α-, β-, γ- and δ-MSH), β-endorphin (β-EP) or adrenocorticotropin hormone (ACTH) are produced in fish leucocytes and are therefore supposed to have autocrine and paracrine actions (Ottaviani *et al.*, 1995; Balm *et al.*, 1997; Amemiya *et al.*, 1999; Arnold and Rice, 2000). MSHs and β-EP are able to stimulate leucocyte proliferation and phagocyte functions, including phagocytosis, respiratory burst and the release of macrophage-stimulating factor (Harris and Bird, 1997, 1999, 2000b; Takahasi *et al.*, 1999; Watanuki *et al.*, 1999, 2000, 2003). ACTH, on the other hand, inhibits circulating leucocyte numbers and lymphocyte mitogenesis while activating phagocytosis and respiratory burst activity (McLeay, 1973; Bayne and Levy, 1991; Weyts *et al.*, 1999). Another melanotropin, the melanin-concentrating hormone (MCH), has been shown to affect fish immune responses in a similar way to the MSHs (Harris and Bird, 1997, 1999, 2000b; Watanuki *et al.*, 2003). Some sexual hormones have been found to be involved in osmoregulation and also affect the immune response. Estradiol, progesterone, testosterone or 11-ketotestosterone have been found to influence the immune response negatively, while few assays describe immunoactivation (Harris and Bird, 2000a; Law *et al.*, 2001; Watanuki *et al.*, 2002; Chaves-Pozo *et al.*, 2003;

Saha *et al.*, 2004; Cuesta *et al.*, in press). In the future, the specific role of these hormones on osmoregulation and immunity should be assayed in order to ascertain and clarify their pleiotropic functions in teleost fish and, more specifically, in osmoregulation and immunity.

ROLE OF FISH CYTOKINES IN THE ENDOCRINE SYSTEM

So far, there is no information about the effect of fish cytokines on osmoregulation. However, in mammals, bi-directional cross talk between endocrine and immune systems has been described. Mammalian pituitary cells, for example, are known to produce several cytokines (IL-1, IL-6, TNF and IFN) and respond to them by means of their specific receptors (see Thurnbull and Rivier, 1999; Engelsma *et al.*, 2002). Moreover, the administration of TNF and IL-6, but especially IL-1, stimulates the HPA-axis to produce ACTH, CRH and GC during infection, inflammation and stress in mammals. Taking into account these data and similarities between the mammalian HPA-axis with its fish HPI-axis counterpart, bi-directionality could also be assumed in fish. Although in a first step, few available data on fish confirm this parallelism and the recent availability of cytokine sequences points to promising future findings.

IL-1β gene expression is found in brain and in the pituitary of teleost fish (Engelsma *et al.*, 2001; Pelegrin *et al.*, 2001). First studies demonstrated that cortisol inhibits IL-1β mRNA levels in trout (Zou *et al.*, 2000) and carp (Engelsma *et al.*, 2001), perhaps because the hormone inactivates NF-κB, leading to no cytokine synthesis as occurs in mammals (McKay and Cidlowski, 1999). Moreover, recombinant fish IL-1β triggers the liberation of α-MSH and β-endorphin from pituitary in carp (see Engelsma *et al.*, 2002). In trout, recombinant IL-1β injection increased circulating levels of cortisol (Holland *et al.*, 2002), also demonstrating that the effect was mediated by interaction with the hypothalamus-pituitary gland. It is known that dexamethasone blocks endogenous ACTH liberation with subsequent inhibition of cortisol release. Trout treated with IL-1β and dexamethasone together did not show increased cortisol levels. These results are also in agreement with the finding of IL-1 receptor expression in brain and pituitary cells (Holland *et al.*, 2002). The scant results are promising and future studies concerning endocrine-immune system interactions, as well as with other systems, need to be conducted.

INTERACTIONS BETWEEN OSMOREGULATORY AND IMMUNE RESPONSES

Many studies are confined to describing individual effects of treatment on a specific response. However, integrative analysis of what happens throughout the animal physiology after a given treatment represents the most valuable studies but at the same time, the most difficult to achieve. Thus, information about growth, stress, metabolism, hormonal status, osmoregulation or immunity after treatment or commonly occurring situations in fish farming, such as salinity disturbance, will hopefully be of help. All these isolated data are in the process of being collated and future multidisciplinary studies will ascertain why and how they interact, as well as the consequences to the animal in terms of growth, quality, disease resistance and environmental impact.

Although effects are inter-specific, hypophysectomized fish show a lack of osmoregulation and a decreased immune response. In particular, hypophysectomized trouts and tilapia have shown reduced values of some immunological parameters (Yada *et al.*, 1999, 2002a,b) although both osmoregulatory and immune functions were restored after administration of exogenous PRL or GH, indicating their central role in both systems, although more studies should be carried to identify other potential mediators (Yada *et al.*, 1999). Many fish, including salmonids, tilapia and sparids, have shown increased pituitary expression of PRL mRNA accompanied by higher circulating levels of PRL after transfer to hypoosmotic waters (Yamauchi *et al.*, 1991; Mancera *et al.*, 1993a; Martin *et al.*, 1999; Laiz-Carrión *et al.*, 2005). However, transfer of fish from SW to lower salinity media decreased the phagocytic activity in black seabream, while in gilthead seabream the peroxidases content decreased and plasmatic IgM levels remained unaffected (Narnaware *et al.*, 2000; Cuesta *et al.*, 2005a). The complement activity of gilthead seabream was, on the other hand, differently affected and depended on the adaptation period. However, gilthead seabream is the only described case in which the increase of PRL, either by exogenous administration or as a result of transfer to hypoosmotic media, produces similar effects, that is, suppression of the immune system (Cuesta *et al.*, 2005a,b). Following with this idea, Yada *et al.* (2002) found that the hypoosmoregulatory and immunostimulant actions of PRL are drastically opposed, suggesting that the role of PRL in osmoregulation and immunity are independent. Unfortunately, there is little information about the expression of the PRL and PRL-R genes in lymphoid tissues and leucocytes and about whether they are modulated or not by plasmatic PRL levels. Yada *et al.* (2002) demonstrated that head-kidney leucocytes from tilapia increase PRL-R mRNA expression after transfer from FW to SW. This finding correlates well with the studies describing increased immune responses after hyperosmotic adaptation (see Table 1.1) and could explain part of the immunostimulation produced after hyperosmotic adaptation. Moreover, although the transfer from FW to SW decreases PRL release, favouring acclimation to saline conditions, the affinity and capacity of PRL-R is rapidly increased and maintained for several weeks (Auperin *et al.*, 1995; Sandra *et al.*, 2001). Furthermore, the expression of mRNA coding for the PRL-R gene was unaffected in head-kidney leucocytes or in the gills of hypophysectomized tilapia specimens (Auperin *et al.*, 1995; Yada *et al.*, 2002). These observations indicate that factors other than the presence and abundance of pituitary hormones might be controlling the expression of PRL-R, especially in lymphoid tissues, and, by extension, the immune function. Perhaps, paracrine actions of the leucocyte-produced PRL could be the key and need to be investigated.

GH, on the other hand, is clearly involved in hyperosmotic adaptation in salmonids but behaves differently, depending on the species and salinity in non-salmonids (Mancera and McCormick, 1998). The correlation was best observed in brown trout, which showed increased levels of plasmatic GH after transfer from FW to SW, along with increased lysozyme activity and phagocytosis (Marc *et al.*, 1995). Increased GH levels, as a result of hyperosmotic environment adaptation or exogenous administration, tend to correlate well with increased immune responses (Tables 1.1 and 1.2). However, trout exhibited lower specific antibody titres in SW than in FW (Betoulle *et al.*, 1995). The total IgM levels were unaffected or increased in several fish species adapted to hyperosmotic environments (Yada *et al.*, 2001, 2002; Dominguez *et al.*, 2004; Cuesta *et al.*, 2005a). On the other hand, seabream injected with GH showed lower values of this parameter (Cuesta *et al.*, 2005b). While the total pool of circulating IgM might be augmented by increases in salinity, the production of specific IgM is inhibited because one or more steps in the generation of specificity (antigen uptake, processing and presentation, selection of a specific IgM-producing lymphocyte B or IgM production) may be affected. Superoxide anion production was decreased in HKLs but not in PBLs after hypophysectomy, indicating differences in hormonal control in the different leucocyte sources (Yada and Azuma, 2002; Yada *et al.*, 2002). Similarly, GH injection restored IgM production in hypophysectomized trouts (Yada *et al.*, 1999). The injection of GH, together with hyperosmotic adaptation, failed to over-stimulate IgM production and lysozyme activity compared with that observed in fish only adapted to higher salinity, while superoxide production by PBLs increased (Yada *et al.*, 2001). Unfortunately, there are no studies concerning the role of osmotic change in the expression of GH-R. More and deeper analyses need to be carried out regarding GH-R expression in different physiological situations, since GH-R has been shown to interact with PRL. One form of the tilapia PRL (PRL_{177}) is structurally similar to GH and is therefore recognized by GH-R, while PRL-R does not bind GH (Auperin *et al.*, 1995; Sandra *et al.*, 1995; Shepherd *et al.*, 1997). Strikingly, this explains the increased PRL-R in SW-adapted fish and the increased immune response after GH administration or hyperosmotic adaptation. Future investigations to identify the involvement of PRL/GH-R interactions in FW or SW adaptation will be welcome.

Salinity disturbance could also be considered stressful for fish, although some data such a claim difficult to establish. Cortisol plays an important role in hyperosmotic adaptation though it can also promote adaptation to hypoosmotic environments, depending on the fish species (Mancera *et al.*, 1993b, 2002; Morgan and Iwama, 1996; Eckert *et al.*, 2001; McCormick, 2001; Laiz-Carrión *et al.*, 2003). The circulating cortisol levels reached after fish received implants of exogenous cortisol are similar to those found in fish adapted to hyperosmotic environments (Morgan and Iwama, 1996). Apart from its role in osmoregulation, cortisol is considered responsible for the inhibition of the immune system in stress situations. However, multiple interactions between endocrine-immune systems must be operating. Most of the studies based on the effect of cortisol on the immune response describe its depressive role (Table 1.2) while, experiments in which fish are adapted to hyperosmotic media and are therefore supposed to have elevated cortisol levels, generally point to activation of the immune responses (Table 1.1). Thus, there are enough data, even in the same fish species, to contradict the inhibitory hypothesis. Everything depends on the response measured and the tissue or cells used for immunologic determinations. Transfer or adaptation to hypersaline waters of coho salmon depressed the innate immune system (Maule and Schreck, 1987) while in rainbow trout the production of specific antibodies was decreased (Betoulle *et al.*, 1995). In many other studies the immune responses increased. As regards humoral factors, circulating total IgM levels are not affected in SW-adapted salmonids, which could be due to the decrease in circulating lymphocytes. However, in gilthead seabream, the IgM levels were increased both in hypersaline-adapted and cortisol-implanted specimens (Cuesta *et al.*, 2005a,c). The activity of lysozyme, which is produced and released by mature monocyte/macrophages and granulocytes, is increased after hyperosmotic adaptation. On the other hand, plasmatic cortisol impairs blood-circulating lymphocytes and their functioning (mitogenesis and the production of specific IgM) and, at the same time, they increase their susceptibility to die by apoptosis. Moreover, cortisol increases leucocyte trafficking and the number of phagocytic cells in the blood. The consequences of this cell extravasation could be an increase in lysozyme activity in the serum, the levels of free-oxygen radicals and allograft rejection due to mobilization of active leucocytes (Marc *et al.*, 1995; Nevid and Meier, 1995; Ortuño *et al.*, 2001; Yada *et al.*, 2001). Moreover, some of these data are supported by the finding that cortisol protects neutrophils against apoptosis (Weyts *et al.*, 1998b). Another consequence is the clearance of phagocytic cells from the lymphoid organs such as head-kidney and spleen. This result in myeloid precursors dividing and differentiating faster and therefore the monocyte/macrophages and granulocytes present will be more immature and, obviously, their immune responses (phagocytosis, respiratory burst, etc.) will be negatively affected. The intention behind this impairment of the defence mechanisms in organs such as the head-kidney and increase in some of the blood leucocytes is clear: the availability of active circulating phagocytes to overcome a possible pathogen invasion in altered fish homeostasis (salinity shock or other stressful situation). However, and unfortunately, the animal may not be able to overcome the pathogen as demonstrated in several studies (Kent and Hedrick, 1987; Wiik *et al.*, 1989; Chou *et al.*, 1999; Harris *et al.*, 2000). Furthermore, cortisol has been proposed as a candidate for overcoming the stress situations. Thus, trouts injected with cortisol were protected from immunosuppressive effects due to stress (Narnaware and Baker, 1996). Cortisol injection also decreased the expression of stress-related immune genes in the common carp (Kawano *et al.*, 2003). All these data suggest that cortisol plays a dual role in the immune system, as it does in the osmoregulatory response, which depends on the fish species studied and the particular parameter determined.

As commented above, other hormones, among their pleiotropic actions, may also be involved in osmoregulation and immunity. However, their effect on fish osmoregulation or immunity is not clear and hopefully will be the target of future research. The presence and expression of hormones and their receptors in endocrine and immune relevant cells as well as the mechanisms and possible interactions need to be clarified.

CONCLUDING REMARKS

Investigations demonstrate and confirm the cross-regulation and interaction between osmoregulation and immunity in teleost fish. However, the mechanisms by which salinity and osmoregulatory hormones up- or down-regulate the immune responses are not understood. The presence of hormone receptors in fish leucocytes seems to be essential but there are no data confirming this hypothesis. In this sense, experiments using receptor blockers together with osmotic shock or hormonal treatment are needed. So far, variations in hormone receptor affinity or number after hypo or hyperosmotic adaptation have been scarcely reported (Sandra *et al.*, 2001; Dean *et al.*, 2003). Moreover, the autocrine and paracrine actions of the hormones in the lymphoid tissues need to be evaluated. However, it seems evident that many other factors and interactions are also active.

Finally, the role of cytokines in osmoregulatory and endocrine organs need to be understood before we can understand these interactions. Again, the finding that pituitary cells are able to produce cytokine and their receptors opens an interesting investigation line. Obviously, the paracrine and autocrine control of the synthesis of hormones and consequently in the hormonal control of the osmoregulatory process must be determined.

References

Aihua, L. and K. Buchmann. 2001. Temperature- and salinity-dependent development of a Nordic strain of *Ichthyophthirius multifiliis* from rainbow trout. *Journal of Applied Ichthyology* 17: 273–276.

Altinok, I. and J.M. Grizzle. 2001. Effects of low salinities on *Flavobacterium columnare* infection of euryhaline and freshwater stenohaline fish. *Journal of Fish Disease* 24: 361–367.

Amemiya, Y., A. Takahashi, N. Suzuki, Y. Sasayama and H. Kawauchi. 1999. A newly characterised proopiomelanocortin in pituitaries of an elasmobranch, *Squalus acanthias*. *General and Comparative Endocrinology* 114: 387–395.

Anderson, D.P., B.S. Roberson and O.W. Dixon. 1982. Immunosuppression induced by a corticosteroid or an alkylating agent in rainbow trout (*Salmo gairdneri*) administered a *Yersinia ruckeri* bacterin. *Developmental and Comparative Immunology* S2: 197–204.

Arnold, R.E. and C.D. Rice. 2000. Channel catfish, *Ictalurus punctatus*, leucocytes secrete immunoreactive adrenal corticotropin hormone (ACTH). *Fish Physiology and Biochemistry* 22: 303–310.

Aurperin, B., F. Rentier-Delrue, J.A. Martial and P. Prunet. 1995. Regulation of gill prolactin receptors in tilapia (*Oreochromis niloticus*) after a change in salinity or hypophysectomy. *Journal of Endocrinology* 145: 213–220.

Bayne, C.J. and S. Levy. 1991. The respiratory burst response of rainbow trout *Oncorhynchus mykiss* (Waldbaum), phagocytes is modulated by sympathetic neurotransmitters and the 'neuro' peptide ACTH. *Journal of Fish Biology* 38: 609–619.

Bentley, P.J. 1998. Hormones and osmoregulation. In: *Comparative Vertebrate Endocrinology*, P.J. Bentley (ed.). Cambridge University Press, Cambridge, pp. 337–378.

Betoulle, S., D. Troutaud, N. Khan and P. Deschaux. 1995. Antibody response, cortisolemia and prolactinemia in rainbow trouts. *Comptes Rendus Academie des Sciences Paris* 318: 677–681.

Bordas, M.A., M.C. Balebona, I. Zorrilla, J.J. Borrego and M.A. Moriñigo. 1996. Kinetics of the adhesion of selected fish-pathogenic *Vibrio* strains to skin mucus of gilt-head sea bream. *Applied and Environmental Microbiology* 62: 3650–3654.

Bury, N.R., L. Jie, G. Flik, R.A.C. Lock and S.E. Wendelaar Bonga. 1998. Cortisol protects against copper induced necrosis and promotes apoptosis in fish gill chloride cells *in vitro*. *Aquatic Toxicology* 40: 193–202.

Bury, N.R., A. Sturm, P. Le Rouzic, C. Tethimonier, B. Ducouret, Y. Guiguen, M. Robinson-Rechavi, V. Laudet, M.E. Rafestin-Oblin and P. Prunet. 2003. Evidence for two distinct functional glucocorticoid receptors in teleost fish. *Journal of Molecular Endocrinology* 31: 141–156.

Calduch-Giner, J.A., A. Sitja-Bobadilla, P. Alvarez-Pellitero and J. Pérez-Sánchez. 1995. Evidence for a direct action of GH on haemopoietic cells of a marine fish, the gilthead sea bream (*Sparus aurata*). *Journal of Endocrinology* 146: 459–467.

Calduch-Giner, J.A., A. Sitja-Bobadilla, P. Alvarez-Pellitero and J. Pérez-Sánchez. 1997. Growth hormone as an *in vitro* phagocyte-activating factor in the gilthead sea bream (*Sparus aurata*). *Cell and Tissue Research* 287: 535–540.

Carlson, R.E., D.P. Anderson and J.E. Bodammer. 1993. *In vivo* cortisol administration suppresses the *in vitro* primary immune response of winter flounder lymphocytes. *Fish and Shellfish Immunology* 3: 299–312.

Chaves-Pozo, E., P. Pelegrín, V. Mulero, J. Meseguer and A. García-Ayala. 2003. A role for acidophilic granulocytes in the testis of the gilthead seabream (*Sparus aurata* L., Teleostei). *Journal of Endocrinology* 179: 165–174.

Chou, H.Y., T.Y. Peng, S.J. Chang, Y.L. Hsu and J.L. Wu. 1999. Effect of heavy metal stressors and salinity shock on the susceptibility of grouper (*Epinephelus* sp.) to infectious pancreatic necrosis virus. *Virus Research* 63: 121–129.

Chung, H-T., W.E. Samlowski and R.A. Daynes. 1986. Modification of the murine immune system by glucocorticosteroids: Alteration of the tissue localization properties of circulating lymphocytes. *Cellular Immunology* 101: 571–589.

Clevenger, C.V., D.O. Freier and J.B. Kline. 1998. Prolactin receptor signal transduction in cells of the immune system. *Journal of Endocrinology* 157: 187–197.

Colombe, L., A. Fostier, N. Bury, F. Pakdel and Y. Guiguen. 2000. A mineralocorticoid receptor in rainbow trout, *Oncorhynchus mykiss*: cloning and characterization of its steroid binding domain. *Steroids* 65: 319–328.

Cuesta, A., M.A. Esteban and J. Meseguer. 1999. Natural cytotoxic activity of gilthead seabream (*Sparus aurata* L.) leucocytes. Assessment by flow cytometry and microscopy. *Veterinary Immunology and Immunopathology* 71: 161–171.

Cuesta, A., J. Ortuño, A. Rodríguez, M.A. Esteban and J. Meseguer. 2002. Changes in some innate defence parameters of seabream (*Sparus aurata* L.) induced by retinol acetate. *Fish and Shellfish Immunology* 13: 279–291.

Cuesta, A., M.A. Esteban and J. Meseguer. 2003. *In vitro* effect of chitin particles on the innate cellular immune system of gilthead seabream (*Sparus aurata* L.). *Fish and Shellfish Immunology* 15: 1–11.

Cuesta, A., J. Meseguer and M.A. Esteban. 2004. Total serum immunoglobulin M levels are affected by immunomodulators in seabream (*Sparus aurata* L.). *Veterinary Immunology and Immunopathology* 101: 203–210.

Cuesta, A., R. Laiz-Carrión, M.P. Martín del Río, J. Meseguer, J.M. Mancera and M.A. Esteban. 2005a. Salinity influences the humoral immune parameters of gilthead seabream (*Sparus aurata* L.). *Fish and Shellfish Immunology* 18: 255–261.

Cuesta, A., F. Arjona, R. Laiz-Carrión, M.P. Martín del Río, J. Meseguer, J.M. Mancera and M.A. Esteban. 2005b. Effect of PRL, GH and cortisol on the serum complement and IgM levels in gilthead seabream (*Sparus aurata* L.). *Fish and Shellfish Immunology* (In press).

Cuesta, A., M.A. Esteban and J. Meseguer. 2005c. Molecular characterization of the nonspecific cytotoxic cell receptor (NCCRP-1) demonstrates gilthead seabream NCC heterogeneity. *Developmental and Comparative Immunology* 29: 637–650.

Cuesta, A., L. Vargas-Chawff, A. García-López, F.J. Arjona, G. Martinez-Rodríguez, J. Meseguer, J.M. Maucera and M.A. Estebau. 2007. Effect of sex-steroid hormones, testosterone and estradiol, on humoral immune parameters of gilthead seabream. *Fish and Shellfish Immunology* doi: 10.1016/j.fsi.2007.01.015.

Dean, D.B., Z.W. Whitlow and R.J. Borski. 2003. Glucocorticoid receptor upregulation during seawater adaptation in a euryhaline teleost, the tilapia (*Oreochromis mossambicus*). *General and Comparative Endocrinology* 132: 112–118.

Dixon, B. and R.J.M. Stet. 2001. The relationship between major histocompatibility receptors and innate immunity in teleost fish. *Developmental and Comparative Immunology* 25: 683–699.

Ducouret, B., M. Tujague, J. Ashraf, N. Mouchel, N. Servel, Y. Valotaire and E.B. Thompson. 1995. Cloning of a fish glucocorticoid receptor shows that it contains a deoxyribonucleic acid-binding domain different from that of mammals. *Endocrinology* 136: 3774–3783.

Domínguez, M., A. Takemura, M. Tsuchiya and S. Nakamura. 2004. Impact of different environmental factors on the circulating immunoglobulin levels in the Nile tilapia, *Oreochromis niloticus*. *Aquaculture* 241: 491–500.

Eckert, S.M., T. Yada, B.S. Shepherd, M.H. Stetson, T. Hirano and G. Grau. 2001. Hormonal control of osmoregulation in the channel catfish *Ictalurus punctatus*. *General and Comparative Endocrinology* 122: 270–286.

Ellis, A.E. 1988. General principles of fish vaccination. In: *Fish Vaccination*, A.E. Ellis (ed.). Academic Press, London, pp. 1–19.

Ellsaesser, C.F. and L.W. Clem. 1987. Cortisol-induced hematologic and immunologic changes in channel catfish (*Ictalurus punctatus*). *Comparative Biochemistry and Physiology* A 87: 405–408.

Engelsma, M.Y., R.J.M. Stet, H. Schipper and B.M.L. Verburg-van Kemenade. 2001. Regulation of interleukin 1 beta RNA expression in the common carp *Cyprinus carpio* L. *Developmental and Comparative Immunology* 25: 195–203.

Engelsma, M.Y., M.O. Huising, W.B. van Muiswinkel, G. Flik, J. Kwang, H.F.J. Savelkoul and B.M.L. Verburg-van Kemenade. 2002. Neuroendocrine-immune interactions in fish: a role for interleukin-1. *Veterinary Immunology and Immunopathology* 87: 467–479.

Espelid, S., G.B. Løkken, K. Steiro and J. Bgwald. 1996. Effects of cortisol and stress on the immune system in Atlantic salmon (*Salmo salar* L.). *Fish and Shellfish Immunology* 6: 95–110.

Esteban, M.A., J. Meseguer, A. García-Ayala and B. Agulleiro. 1989. Erythropoiesis and thrombopoiesis in the head-kidney of the sea bass (*Dicentrarchus labrax* L.): An ultrastructural study. *Archives of Histology and Cytology* 52: 407–419.

Esteban, M.A., V. Mulero, J. Muñoz and J. Meseguer. 1998. Methodological aspects of assessing phagocytosis of *Vibrio anguillarum* by leucocytes of gilthead seabream (*Sparus aurata* L.) by flow cytometry and electron microscopy. *Cell and Tissue Research* 293: 133–141.

Esteban, M.A., A. Cuesta, J. Ortuño and J. Meseguer. 2001. Immunomodulatory effects of dietary intake of chitin on gilthead seabream (*Sparus aurata* L.) innate immune system. *Fish and Shellfish Immunology* 11: 303–315.

Esteban, M.A., A. Rodríguez, A. García-Ayala and J. Meseguer. 2004. Effects of high doses of cortisol on innate cellular immune response of seabream (*Sparus aurata* L.). *General and Comparative Endocrinology* 137: 89–98.

Evans, D.L., J.H. Leary III and L. Jaso-Friedmann. 2001. Nonspecific cytotoxic cells and innate immunity: regulation by programmed cell death. *Developmental and Comparative Immunology* 25: 791–805.

Fukada, H., Y. Ozaki, A.L. Pierce, S. Adachi, K. Yamauchi, A. Hara, P. Swanson and W.W. Dickhoff. 2004. Salmon growth hormone receptor: molecular cloning, ligand specificity, and response to fasting. *General and Comparative Endocrinology* 139: 61–71.

Greenwell, M.G., J. Sherrill and L.A. Clayton. 2003. Osmoregulation in fish. Mechanisms and clinical implications. The veterinary clinics of North America. *Exotic Animal Practice* 6: 169–189.

Greenwood, A.K., P.C. Butler, R.B. White, U. DeMarco, D. Pearce and R.D. Fernald. 2003. Multiple corticosteroid receptors in a teleost fish: Distinct sequences, expression patterns, and transcriptional activities. *Endocrinology* 144: 4226–4236.

Grimm, A.S. 1985. Suppression by cortisol of the mitogen-induced proliferation of peripheral blood leucocytes from plaice, *Pleuronectes platessa* L. In: *Fish Immunology*, M.J. Manning and M.F. Tatner (eds.). Academic Press, London, pp. 263–271.

Harris, J. and D.J. Bird. 1997. The effects of α-MSH and MCH on the proliferation of rainbow trout (*Oncorhynchus mykiss*) lymphocytes *in vitro*. In: *Advances in Comparative Endocrinology*, S. Kawashima and S. Kikuyama (eds.). Monduzzi Editoire, Bologna, pp. 1023–1026.

Harris, J. and D.J. Bird. 1999. Effects of melanin-concentrating hormone and α-melanocyte stimulating hormone on immune responses of rainbow trout. In: *Recent Developments in Comparative Endocrinology and Neurobiology*, E.W. Roubos, S.E. Wendelaar Bonga, H. Vaudry and A. De Loof (eds.). Shaker, Maastricht, pp. 343–345.

Harris, J. and D.J. Bird. 2000a. Modulation of the fish immune system by hormones. *Veterinary Immunology and Immunopathology* 77: 163–176.

Harris, J. and D.J. Bird. 2000b. Supernatants from leucocytes treated with melanin-concentrating hormone (MCH) and α-melanocyte stimulating hormone (α-MSH) have a stimulatory effect on rainbow trout (*Oncorhynchus mykiss*) head kidney phagocytes *in vitro*. *Veterinary Immunology and Immunopathology* 76: 117–124.

Harris, P.D., A. Soleng and T.A. Bakke. 2000. Increased susceptibility of salmonids to the monogenean *Gyrodactylus salaries* following administration of hydrocortisone acetate. *Parasitology* 120: 57–64.

Higashimoto, Y., N. Nakao, T. Ohkubo, M. Tanaka and K. Nakashima. 2001. Structure and tissue distribution of prolactin receptor mRNA in Japanese flounder (*Paralichtys olivaceus*): conserved and preferential expression in osmoregulatory organs. *General and Comparative Endocrinology* 123: 170–179.

Holland, J.W., T.G. Pottinger and C.J. Secombes. 2002. Recombinant interleukin-1β activates the hypothalamic-pituitary-interrenal axis in rainbow trout, *Oncorhynchus mykiss*. *Journal of Endocrinology* 175: 261–267.

Hou, Y., Y. Suzuki and K. Aida. 1999. Effects of steroids on the antibody producing activity of lymphocytes in rainbow trout. *Fisheries Science* 65: 850–855.

Jaso-Friedmann, L., J.H. Leary III and D.L. Evans. 2001. The nonspecific cytotoxic cell receptor (NCCRP-1): molecular organization and signaling properties. *Developmental and Comparative Immunology* 25: 701–711.

Kajita, Y., M. Sakai, M. Kobayashi and H. Kawauchi. 1992. Enhancement of non-specific cytotoxic activity of leucocytes in rainbow trout *Oncorhynchus mykiss* injected with growth hormone. *Fish and Shellfish Immunology* 2: 155–157.

Kawano, H., T. Kono, H. Watanuki, R. Savan and M. Sakai. 2003. Analysis of genes expressed in head kidney of common carp *Cyprinus carpio* L. treated with cortisol. *Comparative Biochemistry and Physiology* B 136: 875–886.

Kent, M.L. and R.P. Hedrick. 1987. Effects of cortisol implants on the PKX myxosporean causing proliferative kidney disease in rainbow trout, *Salmo gairdneri*. *Journal of Parasitology* 73: 455–461.

Kitlen, J.W., E.K. Hejbol, T. Zinck, K. Varming, J.C. Byatt and E. McLean. 1997. Growth performance and respiratory burst activity in rainbow trout treated with growth hormone and vaccine. *Fish and Shellfish Immunology* 7: 297–304.

Knudsen, K.K. and G. Sundnes. 1998. Effects of salinity on infection with *Lerneaocera branchialis* (L.) (Copepoda: Pennelidae). *Journal of Parasitology* 84: 700–704.

Laiz-Carrión, R., M.P. Martín del Río, J.M. Míguez, J.M. Mancera and J.L. Soengas. 2003. Influence of cortisol on osmoregulation and energy metabolism in gilthead seabream *Sparus aurata*. *Journal of Experimental Zoology* A298: 105–118.

Laiz-Carrión, R., B. Bredruel, J. Fuentes, J.M. Guzmán, M.P. Martín del Río, D.M. Power and J.M. Mancera. 2005. Prolactin, growth hormone and somatolactin pituitary expression in gilthead sea bream *Sparus auratus* in different osmotic conditions. (Submitted).

Law, W.Y., W.H. Chen, Y.L. Song, S. Dufour and C.F. Chang. 2001. Differential *in vitro* suppressive effects of steroids on leucocyte phagocytosis in two teleosts, tilapia and common carp. *General and Comparative Endocrinology* 121: 163–172.

Lee, L.T.O., G. Nong, D.L.Y. Tse and C.H.K. Cheng. 2001. Molecular cloning of a teleost growth hormone receptor and its functional interaction with human growth hormone. *Gene* 270: 121–129.

López-Morales, E., J. Meseguer, M.T. Lozano and B. Agulleiro. 1990. Ultrastructure of the nephron of grey mullets (*Mugil cephalus* L. and *Liza saliens* Risso 1810). *Anatomischer Anzeiger* 170: 49–61.

Magor, B.G. and K.E. Magor. 2001. Evolution of effectors and receptors of innate immunity. *Developmental and Comparative Immunology* 25: 651–682.

Mancera, J.M. and S.D. McCormick. 1998. Osmoregulatory actions of the GH/IGF axis in non-salmonids teleosts. *Comparative Biochemistry and Physiology* B121: 43–48.

Mancera, J.M., P. Fernández-Llebrez, J.M Grondona and J.M. Pérez-Fígares. 1993a. Influence of environmental salinity on prolactin and corticotropic cells in the euryhaline gilthead sea bream (*Sparus aurata* L.). *General and Comparative Endocrinology* 190: 220–231.

Mancera, J.M., J.M. Pérez-Fígares and P. Fernández-Llebrez. 1993b. Effect of cortisol on brackish water adaptation in the euryhaline gilthead sea bream (*Sparus aurata*). *Comparative Biochemistry and Physiology* A107: 397–402.

Mancera, J.M., R. Laiz-Carrión and M.P. Martín del Río. 2002. Osmoregulatory action of PRL, GH, and cortisol in the gilthead seabream (*Sparus aurata* L.). *General and Comparative Endocrinology* 129: 95–103.

Manning, M.J. 1998. Immune defence systems. In: *Biology of Farmed Fish*, K.D. Black and A.D. Pickering (eds.). Sheffield Academic Press, Sheffield, pp. 180–221.

Marc, A.M., C. Quentel, A. Severe, P.Y. Le Bail and G. Boeuf. 1995. Changes in some endocrinological and non-specific immunological parameters during seawater exposure in the brown trout. *Journal of Fish Biology* 46: 1065–1081.

Martin, S.A.M., A.F. Youngson and A. Ferguson. 1999. Atlantic salmon (*Salmo salar*) prolactin cDNA sequence and its mRNA expression after transfer of fish between salinities. *Fish Physiology and Biochemistry* 20: 351–359.

Maule, A.G. and C.B. Schreck. 1987. Changes in the immune system of coho salmon (*Oncorhynchus kisutch*) during the parr-to-smolt transformation and after implantation of cortisol. *Canadian Journal of Fisheries and Aquatic Sciences* 44: 161–166.

Maule, A.G. and C.B. Schreck. 1990. Glucocorticoid receptors in leucocytes and gill of juvenile coho salmon (*Oncorhynchus kisutch*). *General and Comparative Endocrinology* 77: 448–455.

McCormick, S.D. 2001. Endocrine control of osmoregulation in fish. *American Zoologist* 282: 290–300.

McKay, L.I. and J.A. Cidlowski. 1999. Molecular control of immune/inflammatory responses: interactions between nuclear factor-κB and steroid receptor-signalling pathways. *Endocrine Reviews* 20: 435–459.

McLeay, D.J. 1973. Effects of cortisol and dexamethasone on the pituitary-interrenal axis and abundance of white blood cell types in juvenile coho salmon, *Oncorhynchus kisutch*. *General and Comparative Endocrinology* 21: 441–450.

Meseguer, J., B. Agulleiro and F. Hernández. 1981. Fine structure of chloride cells front *Sparus auratus* gills (Teleost). *Morfología Normal y Patológica* 6: 139–152.

Meseguer, J., M.A. Esteban and B. Agulleiro. 1991. Stromal cells, macrophages and lymphoid cells in the head-kidney of sea bass (*Dicentrarchus labrax* L.). An ultrastructural study. *Archives of Histology and Cytology* 54: 299–309.

Meseguer, J., A. López-Ruiz and M.A. Esteban. 1994. Cytochemical characterization of leucocytes from the seawater teleost, gilthead seabream (*Sparus aurata* L.). *Histochemistry* 102: 37–44.

Meseguer, J., A. López-Ruiz and A. García-Ayala. 1995. Reticulo-endothelial stroma of the head-kidney from the seawater teleost gilthead seabream (*Sparus aurata* l.): An ultrastructural and Cytochemical study. *Anatomical Record* 241: 303–309.

Meseguer, J., M.A. Esteban and V. Mulero. 1996. Natural cytotoxic response in fish: A review. In: *Ichthyology. Recent Research Advances*, D.N. Saksena (ed.). Oxford & IBH Publishing Co. Pvt. Ltd., New Delhi, pp. 195–208.

Meseguer, J., M.A. Esteban and A. Rodríguez. 2002. Are thrombocytes and platelets true phagocytes? *Microscopy Research and Technique* 57: 491–497.

Milston, R.H., A.T. Vella, T.L., Crippen, M.S. Fitzpatrick, J.A. Leong and C.B. Schreck. 2003. *In vitro* detection of functional humoral immunocompetence in juvenile chinook salmon (*Oncorhynchus tshawytscha*) using flow cytometry. *Fish and Shellfish Immunology* 15: 145–158.

Morgand, J.D. and G.K. Iwama. 1996. Cortisol induced changes in oxygen consumption and ionic regulation in coastral cutthroat trout parr. *Fish Physiology and Biochemistry* 15: 385–394.

Mori, T. and R.H. Devlin. 1999. Transgene and host growth hormone gene expression in pituitary and nonpituitary tissues of normal and growth hormone transgenic salmon. *Molecular and Cellular Endocrinology* 149: 129–139.

Moutoussamy, S., P.A. Kelly and J. Finidori. 1998. Growth-hormone-receptor and cytokine-family signalling. *European Journal of Biochemistry* 255: 1–11.

Mulero, V., M.A. Esteban, J. Muñoz and J. Meseguer. 1994. Non-specific cytotoxic response against tumor target cells mediated by leucocytes from seawater teleosts, *Sparus aurata* and *Dicentrarchus labrax*: an ultrastructural study. *Archives of Histology and Cytology* 57: 351–358.

Muñoz, P., J.A. Calduch-Giner, A. Sitjà-Bobadilla, P. Álvarez-Pellitero and J. Pérez-Sánchez. 1998. Modulation of the respiratory burst activity of Mediterranean sea bass (*Dicentrarchus labrax*) phagocytes by growth hormone and parasitic status. *Fish and Shellfish Immunology* 8: 25–36.

Nagae, M., H. Fuda, K. Ura, H. Kawamura, S. Adachi, A. Hara and K. Yamauchi. 1994. The effect of cortisol administration on blood plasma immunoglobulin M (IgM) concentrations in masu salmon (*Oncorhynchus masou*). *Fish Physiology and Biochemistry* 13: 41–48.

Narnaware, Y.K. and B.I. Baker. 1996. Evidence that cortisol may protect against the immediate effects of stress on circulating leucocytes in the trout. *General and Comparative Endocrinology* 103: 359–366.

Narnaware, Y.K., S.P. Kelly and N.Y.S. Woo. 1997. Effect of injected growth hormone on phagocytosis in silver sea bream (*Sparus sarba*) adapted to hyper- and hypo-osmotic salinities. *Fish and Shellfish Immunology* 7: 515–517.

Narnaware, Y.K., S.P. Kelly and N.Y.S. Woo. 1998. Stimulation of macrophage phagocytosis and lymphocyte count by exogenous prolactin administration in silver sea bream (*Sparus sarba*) adapted to hyper- and hypo-osmotic salinities. *Veterinary Immunology and Immunopathology* 61: 387–391.

Narnaware, Y.K., S.P. Kelly and N.Y. Woo. 2000. Effect of salinity and ration size on macrophage phagocytosis in juvenile black seabream (*Mylio macrocephalus*). *Journal of Applied Ichthyology* 16: 86–88.

Nevid, N.J. and A.H. Meier. 1995. Timed daily administrations of hormones and antagonists of neuroendocrine receptors after day-night rhythms of allograft rejection in the gulf killifish, *Fundulus grandis*. *General and Comparative Endocrinology* 97: 327–339.

Nitzan, S., B. Schwartsburg and E.D. Heller. 2004. The effect of growth medium salinity of *Photobacterium damselae* subsp. *piscicida* on the immune response of hybrid bass (*Morone saxatilis* × *M. chrysops*). *Fish and Shellfish Immunology* 16: 107–116.

Ortuño, J., M.A. Esteban and J. Meseguer. 2000. Kinetics of hydrogen peroxide production during *in vitro* respiratory burst of seabream (*Sparus aurata* L.) head-kidney leucocytes, as measured by a flow cytometric method. *Fish and Shellfish Immunology* 10: 725–729.

Ortuño, J., M.A. Esteban and J. Meseguer. 2001. Effects of short-term crowding stress on the gilthead seabream (*Sparus aurata* L.) innate immune response. *Fish and Shellfish Immunology* 11: 187–197.

Ortuño, J., M.A. Esteban and J. Meseguer. 2002. Effects of four anaesthetics on the innate immune response of gilthead seabream (*Sparus aurata* L.). *Fish and Shellfish Immunology* 12: 49–59.

Ottaviani, E., A. Franchini and C. Francheschi. 1995. Evidence for the presence of immunoreactive POMC-derived peptides and cytokines in the thymus of the goldfish (*Carassius auratus*). *Histochemistry Journal* 27: 597–601.

Pagniello, K.B., N.C. Bols and L.E. Lee. 2002. Effect of corticosteroids on viability and proliferation of the rainbow trout monocyte/macrophage cell line, RTS11. *Fish and Shellfish Immunology* 13: 199–214.

Pelegrín, P., J. García-Castillo, V. Mulero and J. Meseguer. 2001. Interleukin-1β isolated from a marine fish reveals up-regulated expression in macrophages following activation with lipopolysaccharide and lymphokines. *Cytokine* 16: 67–72.

Pickering, A.D. 1984. Cortisol-induced lymphocytopenia in brown trout, *Salmo trutta* L. *Freshwater Biology* 21: 47–55.

Pickering, A.D. 1998. Stress responses of farmed fish. In: *Biology of Farmed Fish*, K.D. Black and A.D. Pickering (eds.). Sheffield Academic Press, Sheffield, pp. 222–255.

Pickford, G.E., A.K. Srivastava, A.M. Slicher and P.K.T. Pang. 1971. The stress response in the abundance of circulating leucocytes in the killifish, *Fundulus heteroclitus*. I. The cold-shock sequence and the effects of hypophysectomy. *Journal of Experimental Zoology* 177: 89–96.

Plante, S., C. Audet, Y. Lambert and J. de la Noüe. 2002. The effects of two rearing salinities on survival and stress of winter flounder broodstock. *Journal of Aquatic Animal Health* 14: 281–287.

Power, D.M. 2005. Developmental ontogeny of prolactin and its receptor in fish. *General and Comparative Endocrinology* 142: 25–33.

Prunet, P., O. Sandra, P. Le Rouzic, O. Marchand and V. Laudet. 2000. Molecular characterization of the prolactin receptor in two fish species, tilapia *Oreochromis niloticus* and rainbow trout, *Oncorhynchus mykiss*: A comparative approach. *Canadian Journal of Physiology and Pharmacology* 78: 1086–1096.

Rodríguez, A., M.A. Esteban and J. Meseguer. 2003. Phagocytosis and peroxidase release by seabream (*Sparus aurata* L.) leucocytes in response to yeast cells. *Anatomical Record* 272: 415–423.

Ruglys, M.P. 1985. The secondary immune response of young carp, *Cyprinus carpio* L., following injection of cortisol. *Journal of Fish Biology* 26: 429–434.

Saeij, J.P., B.M.L. Verburg-van Kemenade, W.B. van Muiswinkel and G.F. Wiegertjes. 2003. Daily handling stress reduces resistance of carp to *Trypanoplasma borreli*: *in vitro* modulatory effects of cortisol on leucocyte function and apoptosis. *Developmental and Comparative Immunology* 27: 233–245.

Saha, N.R., T. Usami and Y. Suzuki. 2003. A double staining flow cytometric assay for the detection of steroid induced apoptotic leucocytes in common carp (*Cyprinus carpio*). *Developmental and Comparative Immunology* 27: 351–363.

Saha, N.R., T. Usami and Y. Suzuki. 2004. *In vitro* effects of steroid hormones on IgM-secreting cells and IgM secretion in common carp (*Cyprinus carpio*). *Fish and Shellfish Immunology* 17: 149–158.

Sakai, M., M. Kobayashi and H. Kawauchi. 1995. Enhancement of chemiluminescent responses of phagocytic cell from rainbow trout, *Oncorhynchus mykiss*, by injection of growth hormone. *Fish and Shellfish Immunology* 5: 375–379.

Sakai, M., M. Kobayashi and H. Kawauchi. 1996a. *In vitro* activation of fish phagocytic cells by GH, prolactin and somatostatin. *Journal of Endocrinology* 151: 113–118.

Sakai, M., M. Kobayashi and H. Kawauchi. 1996b. Mitogenic effect of growth hormone and prolactin on chum salmon *Oncorhynchus keta* leucocytes *in vitro*. *Journal of Endocrinology* 53: 185–189.

Sakai, M., Y. Kajita, M. Kobayashi and H. Kawauchi. 1996c. Increase in haemolytic activity of serum from rainbow trout *Oncorhynchus mykiss* injected with exogenous growth hormone. *Fish and Shellfish Immunology* 6: 615–617.

Sakai, M., Y. Kajita, M. Kobayashi and H. Kawauchi. 1997. Immunostimulating effect of growth hormone: *in vivo* administration of growth hormone in rainbow trout enhances resistance to *Vibrio anguillarum* infection. *Veterinary Immunology and Immunopathology* 57: 1–6.

Sakamoto, T., K. Uchida and S. Yokota. 2001. Regulation of the ion-transporting mitochondrion-rich cell during adaptation of teleost fishes to different salinities. *Zoological Science* 18: 1163–1174.

Salinas, I., A. Cuesta, M.A. Esteban and J. Meseguer. 2005. Dietary administration of *Lactobacillus delbrueckii* and *Bacillus subtilis*, single or combined, on gilthead seabream cellular innate immune responses. *Fish and Shellfish Immunology* 19: 67–77.

Sandra, O., F. Sohm, A. de Luze, P. Prunet, M. Edery and P.A. Kelly. 1995. Expression cloning of a cDNA encoding a fish prolactin receptor. *Proceedings of the National Academy of Sciences of the United States of America* 92: 6037–6041.

Sandra, O., P. Le Rouzic, F. Rentier-Delrue and P. Prunet. 2001. Transfer of tilapia (*Oreochromis niloticus*) to a hyperosmotic environment is associated with sustained expression of prolactin receptor in intestine, gill, and kidney. *General and Comparative Endocrinology* 123: 295–307.

Santos, C.R.A., L. Brinca, P.M. Ingleton and D.M. Power. 1999. Cloning, expression, and tissue distribution of prolactin in adult sea bream (*Sparus aurata*). *General and Comparative Endocrinology* 114: 57–66.

Santos, C.R.A., P.M. Ingleton, J.E.B. Cavaco, P.A. Kelly, M. Edery and D.M. Power. 2001. Cloning, characterization, and tissue distribution of prolactin receptor in the sea bream (*Sparus aurata*). *General and Comparative Endocrinology* 121: 32–47.

Secombes, C.J., T. Wang, S. Hong, S. Peddie, M. Crampe, K.J. Laing, C. Cunningham and J. Zou. 2001. Cytokines and innate immunity of fish. *Developmental and Comparative Immunology* 25: 713–723.

Sepulcre, M.P., P. Pelegrin, V. Mulero and J. Meseguer. 2002. Characterisation of gilthead seabream acidophilic granulocytes by a monoclonal antibody unequivocally points to their involvement in fish phagocytic response. *Cell and Tissue Research* 308: 97–102.

Shepherd, B.S., T. Sakamoto, R.S. Nishioka, N.H. Richman 3rd, I. Mori, S.S Madsen, T.T. Chen, T. Hirano, H.A. Bern and E.G. Grau. 1997. Somatotropic actions of the homologous growth hormone and prolactins in the euryhaline teleost, the tilapia, *Oreochromis mossambicus*. *Proceedings of the National Academy of Sciences of the United States of America* 94: 2068–2072.

Stave, J.W. and B.S. Roberson. 1985. Hydrocortisone suppresses the chemiluminescent response of stripped bass phagocytes. *Developmental and Comparative Immunology* 9: 77–84.

Tagawa, M., H. Hagiwara, A. Takemura, S. Hirose and T. Hirano. 1997. Partial cloning of the hormone-binding domain of the cortisol receptor in tilapia, *Oreochromis mossambicus*, and changes in the mRNA level during embryonic development. *General and Comparative Endocrinology* 108: 132–140.

Takahashi, A., Y. Amemiya, M. Sakai, A. Yasuda, N. Suzuki, Y. Sasayama and H. Kawauchi. 1999. Occurrence of four MSHs in dogfish POMC and their immunomodulating effects. *Annals of the New York Academy of Sciences* 885: 459–463.

Tripp, R.A., A.G. Maule, C.B. Schreck and S.L. Kaattari. 1987. Cortisol mediated suppression of salmonids lymphocyte responses *in vitro*. *Developmental and Comparative Endocrinology* 11: 565–576.

Tse, D.L.Y., B.K.C. Chow, C.B. Chan, L.T.O. Lee and C.H.K. Cheng. 2000. Molecular cloning and expression studies of a prolactin receptor in goldfish (*Carassius auratus*). *Life Sciences* 66: 593–605.

Tse, D.L.Y., M.C.L. Tse, C.B. Chan, W.M. Zhang, H.R. Lin and C.H.K. Cheng. 2003. Seabream growth hormone receptor: molecular cloning and functional studies of the full-length cDNA, and tissue expression of two alternatively spliced forms. *Biochimica et Biophysica Acta* 1625: 64–76.

Turnbull, A.V. and C. Rivier. 1999. Regulation of hypothalamic-pituitary-adrenal axis by cytokines: actions and mechanisms of action. *Physiological Reviews* 79: 1–71.

Wang, R. and M. Belosevic. 1995. The *in vitro* effects of estradiol and cortisol on the function of a long-term goldfish macrophage cell line. *Developmental and Comparative Immunology* 19: 327–336.

Watanuki, H., A. Takahashi, A. Yasuda and M. Sakai. 1999. Kidney leucocytes of rainbow trout, *Oncorhynchus mykiss*, are activated by intraperitoneal injection of β-endorphin. *Veterinary Immunology and Immunopathology* 71: 89–97.

Watanuki, H., Y. Gushiken, A. Takahashi, A. Yasuda and M. Sakai. 2000. *In vitro* modulation of fish phagocytic cells by β-endorphin. *Fish and Shellfish Immunology* 10: 203–212.

Watanuki, H., T. Yamaguchi and M. Sakai. 2002. Suppression in function of phagocytic cells in common carp *Cyprinus carpio* L. injected with estradiol, progesterone or 11-ketotestosterone. *Comparative Biochemistry and Physiology* C 132: 407–413.

Watanuki, H., M. Sakai and A. Takahashi. 2003. Immunomodulatory effects of alpha melanocyte stimulating hormone on common carp (*Cyprinus carpio* L.). *Veterinary Immunology and Immunopathology* 91: 135–140.

Weigent, D.A., J.B. Baxter, W.E. Wear, L.R. Smith, K.L. Bost and J.E. Blalock. 1988. Production of immunoreactive growth hormone by mononuclear leucocytes. *FASEB Journal* 2: 2812–2818.

Wendelaar Bonga, S.E. 1997. The stress response in fish. *Physiological Reviews* 77: 591–625.

Weyts, F.A., B.M.L. Verburg-van Kemenade, G. Flik, J.G. Lambert and S.E. Wendelaar Bonga. 1997. Conservation of apoptosis as an immune regulatory mechanism: effects of cortisol and cortisone on carp lymphocytes. *Brain Behaviour and Immunity* 11: 95–105.

Weyts, F.A., G. Flik, J.H. Rombout and B.M.L. Verburg-van Kemenade. 1998a. Cortisol induces apoptosis in activated B cells, not in other lymphoid cells of the common carp, *Cyprinus carpio* L. *Developmental and Comparative Immunology* 22: 551–562.

Weyts, F.A., G. Flik and B.M.L. Verburg-van Kemenade. 1998b. Cortisol inhibits apoptosis in carp neutrophilic granulocytes. *Developmental and Comparative Immunology* 22: 563–572.

Weyts, F.A., B.M.L. Verburg-van Kemenade and G. Flik. 1998c. Characterisation of glucocorticoid receptors in peripheral blood leucocytes of carp, *Cyprinus carpio* L. *General and Comparative Endocrinology* 111: 1–8.

Weyts, F.A., N. Cohen, G. Flik and B.M.L Verburg-van Kemenade. 1999. Interactions between the immune and endocrine system and the hypothalamo-pituitary-interrenal axis in fish. *Fish and Shellfish Immunology* 9: 1–20.

White, A. and T.C. Fletcher. 1985. The influence of hormones and inflammatory agents on C-reactive protein, cortisol and alanine aminotransferase in the plaice (*Pleuronectes platessa* L.). *Comparative Biochemistry and Physiology* C 80: 99–104.

Wiik, R., K. Andersen, I. Uglenes and E. Egidius. 1989. Cortisol-induced increase in susceptibility of Atlantic salmon, *Salmo salar*, to *Vibrio salmonicida*, together with effects on the blood pattern. *Aquaculture* 83: 201–215.

Wojtaszek, J., D. Dziewulska-Szwajkowska, M. Lozinska-Gabska, A. Adamowicz and A. Dzugaj. 2002. Hematological effects of high dose of cortisol on the carp (*Cyprinus carpio* L.): cortisol effect on the carp blood. *General and Comparative Endocrinology* 125: 176–183.

Yada, T. and T. Azuma. 2002. Hypophysectomy depresses immune functions in rainbow trout. *Comparative Biochemistry and Physiology* C 131: 93–100.

Yada, T., M. Nagae, S. Moriyuma and T. Azuma. 1999. Effects of prolactin and growth hormone on plasma immunoglobulin M levels of hypophysectomized rainbow trout, *Oncorhynchus mykiss*. *General and Comparative Endocrinology* 115: 46–52.

Yada, T., T. Azuma and Y. Takagi. 2001. Stimulation of non-specific immune functions in seawater-acclimated rainbow trout, *Oncorhynchus mykiss*, with reference to the role of growth hormone. *Comparative Biochemistry and Physiology* B 129: 695–701.

Yada, T., K. Uchida, S. Kajimura, T. Azuma, T. Hirano and E.G. Grau. 2002. Immunomodulatory effects of prolactin and growth hormone in the tilapia, *Oreochromis mossambicus*. *Journal of Endocrinology* 173: 483–492.

Yada, T., I. Misumi, K. Muto, T. Azuma and C.B. Schreck. 2004a. Effects of prolactin and growth hormone on proliferation and survival of cultured trout leucocytes. *General and Comparative Endocrinology* 136: 298–306.

Yada, T., K. Muto, T. Azuma and K. Ikuta. 2004b. Effects of prolactin and growth hormone on plasma levels of lysozyme and ceruloplasmin in rainbow trout. *Comparative Biochemistry and Physiology* C 139: 57–63.

Yamauchi, K., R.S. Nishioka, G. Young, T. Ogasawara, T. Hirano and H.A. Bern. 1991. Osmoregulation and circulating growth hormone and prolactin in hypophysectomized coho salmon (*Oncorhynchus kisutch*) after transfer to freshwater and seawater. *Aquaculture* 92: 33–42.

Yang, B-Y., M. Greene and T.T. Chen. 1999. Early embryonic expression of the growth hormone family protein genes in the developing rainbow trout, *Oncorhynchus mykiss*. *Molecular Reproduction and Development* 53: 127–134.

Zapata, A.G., A. Chibá and A. Varas. 1996. Cells and tissues of the immune system of fish. In: *Fish Physiology: The Fish Immune System. Organism, Pathogen, and Environment*, G.K. Iwama and T. Nakanishi (eds.). Academic Press, San Diego, vol. 15, pp. 1–62.

Zhang, W., J. Tian, L. Zhang, Y. Zhang, X. Li and H. Lin. 2004. cDNA sequence and spatio-temporal expression of prolactin in the orange-spotted grouper, *Epinephelus coioides*. *General and Comparative Endocrinology* 136: 134–142.

Zheng, D., K. Mai, S. Liu, L. Cao, Z. Liufu, W. Xu, B. Tan and W. Zhang. 2004. Effect of temperature and salinity on virulence of *Edwardsiella tarda* to Japanese flounder, *Paralichthys olivaceus* (Temminck et Schlegel). *Aquaculture Research* 35: 494–500.

Zou, J., J. Holland, O. Pleguezuelos, C. Cunningham and C.J. Secombes. 2000. Factors influencing the expression of interleukin-1 beta in cultured rainbow trout (*Oncorhynchus mykiss*) leucocytes. *Developmental and Comparative Immunology* 24: 575–582.

CHAPTER

2

The Involvement of the Thyroid Gland in Teleost Osmoregulation

Peter H.M. Klaren*, Edwin J.W. Geven and Gert Flik#

INTRODUCTION

It is not our goal—nor is it desirable—to provide a review of fish thyroid physiology here. Indeed, others have comprehensively and authoritatively treated the physiology of the piscine thyroid gland and thyroid hormones (Eales and Brown, 1993; Leatherland, 1994). We have chosen to describe some thyroidological aspects concisely, aiming to identify less well-investigated areas of piscine thyroidology. Specifically, we wish to focus briefly on the teleost hypothalamus-pituitary gland-thyroid axis, the regulation of which allows bidirectional communication with the teleost stress axis. We shall also discuss the presence of heterotopic thyroid follicles in osmoregulatory organs. With this contribution, we wish to

Authors' address: Department of Organismal Animal Physiology, Faculty of Science, Radboud University Nijmegen, Toernooiveld 1, 6525 ED Nijmegen, The Netherlands.
Corresponding authors: E-mail: *p.klaren@science.ru.nl; #g.flik@science.ru.nl

suggest the use of parameters other than thyroid gland morphology and plasma thyroid hormone concentrations, and to further the investigation of the involvement of thyroid hormones in osmoregulation and other aspects of fish physiology.

THYROID HORMONE BIOSYNTHESIS AND PLASMA TRANSPORT

Biosynthesis

Transcellular iodide transport by the thyrocyte is established by a concerted action of basolaterally and apically located transporters. A Na^+/I^- symporter (NIS) (Dai *et al.*, 1996) located in the basolateral membrane allows the thyrocyte to load systemic iodide from the circulation. NIS activity is inhibited directly by thiocyanate and perchlorate (Van Sande *et al.*, 2003), and indirectly by ouabain (Ajjan *et al.*, 1998) through the inhibition of Na^+, K^+-ATPase and the subsequent collapse of the transmembrane Na^+ gradient which drives iodide transport. The novel Cl^-/anion exchanger pendrin (Scott *et al.*, 1999) is believed to constitute the apical iodide extrusion pathway (Bidart *et al.*, 2000; Royaux *et al.*, 2000; Yoshida *et al.*, 2002). Recently, a human, perchlorate-sensitive, apical iodide transporter (hAIT), with high homology to hNIS, has been proposed as an alternative transport mechanism (Rodriguez *et al.*, 2002) (reviews on thyroid gland iodine metabolism: Spitzweg *et al.*, 2000; Dunn and Dunn, 2001).

To date, no piscine homologues of the transporters involved in thyrocyte transcellular iodide movement have been identified. Even so, the presence of NIS in teleosts can be inferred from the reduced accumulation of radioiodide by zebrafish (Brown, 1997) and Mozambique tilapia (*Oreochromis mossambicus*) (our unpublished results) after treatment with the goitrogen perchlorate, and, in agnathans, from the drastically decreased thyroidal iodide uptake and plasma thyroid hormone levels in larval lampreys treated with different goitrogens (Manzon *et al.*, 2001; Manzon and Youson, 2002).

Thyroglobulin is a large (ca. 660 kDa) homodimeric glycosylated protein synthesized by the thyrocyte and secreted into the follicular lumen where it comprises a major component of the colloid. Thyroglobulins were identified in cyclostomes and elasmobranchs (Suzuki *et al.*, 1975; Monaco *et al.*, 1976, 1978) and teleosts (Kim *et al.*, 1984; Baumeister and Herzog, 1988). In the afollicular endostyle of larval cyclostomes, thyroglobulin was found to be localized in the cytoplasm and associated with the apical membrane of a subpopulation of cells (Wright *et al.*, 1978a,b).

Thyroid peroxidase (TPO) is an integral protein of the thyrocyte apical membrane. The enzyme's catalytic site is located extracellularly and faces the follicular colloid where it catalyzes the oxidation of iodide (I^-) to iodonium (I^+). TPO further catalyzes iodine organification, which involves the substitution of hydrogen atoms at the 3- and 5-positions of the phenolic ring of tyrosine residues in thyroglobulin with iodonium. This results in the formation of mono- (MIT) and diiodotyrosines (DIT). TPO also catalyzes the coupling of iodotyrosine residues to form the iodothyronines thyroxine, T4 (3,5,3′5′-tetraiodothyronine), by the coupling of two DIT molecules, and some 3,5,3′-triiodothyronine, T3 (MIT + DIT). Organification and iodothyronine formation are inhibited by the TPO-inhibitors 6-*n*-propyl-2-thiouracil (PTU) and methimazole (MMI), which are clinically used as thyrostatics to treat hyperthyroidism. No piscine TPO homologues have been identified so far, but treatment of fishes with PTU or MMI successfully induces hypothyroidism (De *et al.*, 1989; Van der Geyten *et al.*, 2001; Varghese *et al.*, 2001; Elsalini and Rohr, 2003) from which the presence of TPO can be inferred.

Thyroglobulin is stored in the follicular lumen where it forms the major constituent of the colloid. Micropinocytosis and colloid resorption produce endosomes that fuse with primary lysosomes to form fagosomes. Endo- and exopeptidase activities hydrolytically digest thyroglobulin to smaller dipeptide fragments with the concomitant release of iodothyronines. The thyroid hormones are secreted across the basolateral membrane of the thyrocyte through an as yet unknown mechanism, but which would most likely include a membrane transport protein.

PLASMA TRANSPORT AND CELLULAR UPTAKE OF THYROID HORMONES

Native iodothyronines are lipophilic, and binding proteins facilitate convective plasma transport of thyroid hormones. In mammals, thyroid hormones are bound to (in the order of decreasing T4-binding affinities): thyroxine-binding globuline (TBG), transthyretin (TTR, previously designated thyroxine-binding prealbumin or TBPA) and albumin

(Schreiber and Richardson, 1997; Schussler, 2000). Typical values for free T4 (fT4) and free T3 (fT3) fractions in mammalian plasma are 0.02–0.05 and 0.2–0.5% of the total T4 and T3 concentrations, respectively. In fish, the fT4 fraction (ranging from 0.15 to 0.4% in salmonids) is generally higher than in mammals, and, in contrast to mammals, exceed fT3 fractions (ranging from 0.1 to 0.2% in salmonids) (Eales *et al.*, 1983; Eales and Shostak, 1985).

In fishes, albumin is a common protein in plasma which can bind T4 (Richardson *et al.*, 1994), but much less is known about other thyroid hormone binding proteins. Cyr and Eales (1992) suggested that changes in plasma free T4 concentrations in estradiol-treated rainbow trout were mediated through lipoproteins and vitellogenin. Their observations were confirmed by experimental results obtained on rainbow trout plasma lipoproteins (Babin, 1992) and vitellogenin in killifish (*Fundulus heteroclitus*) (Monteverdi and Di Giulio, 2000) and gilthead seabream (*Sparus auratus*) (Funkenstein *et al.*, 2000). Indeed, lipoproteins are considered to be a primitive plasma hormone transport modality (Benvenga, 1997). Only fairly recently, a full length cDNA encoding a TTR protein was isolated from seabream liver (Santos and Power, 1999; Santos *et al.*, 2002). It is biochemically distinct from TTR of higher vertebrates, i.e., it preferentially binds T3 over T4, and it does not form dimers with retinol-binding protein as it does in mammals (Santos and Power, 1999; Folli *et al.*, 2003). It could well be that the relatively high fT4 fraction in rainbow trout (Cyr and Eales, 1992) results from the binding properties of plasma proteins, rather than from a high secretion rate of the thyroid gland.

Total plasma thyroid hormone concentrations are, thus, greatly determined by the spectrum and concentrations of proteins in the plasma, which, in turn, are determined by physiological and pathological factors such as nutritional state, reproduction, disease, and developmental state (Richardson *et al.*, 2005), and, indeed, osmoregulatory activity of fishes (Sangiao-Alvarellos *et al.*, 2003). It is generally assumed that target cells can only take up the free forms of thyroid hormones. Free T4 and fT3 concentrations are, therefore, more relevant to thyroid status than are total hormone concentrations, as it is in (human) clinical diagnostics (Midgley, 2001). We have measured increased fT4 levels, with fT3 levels unchanged, in gilthead seabream that were adapted to low salinity water (Klaren *et al.*, 2007), indicating that the free thyroid hormone level is responsive to an osmotic challenge. Unfortunately, many research papers often do not report whether total or free thyroid hormone concentrations were measured in fish plasma. Considering the (nanomolar) hormone concentrations reported, we assume measurements mostly represent total thyroid hormone levels.

It has long been assumed that, due to their lipophilic nature, iodothyronines cross the plasma membrane by diffusion only. Only recently several carrier proteins for thyroid hormones have been proposed. They include members of the organic anion transporter family (Abe *et al.*, 1998; Friesema *et al.*, 1999; Fujiwara *et al.*, 2001) and amino acid transporters (Friesema *et al.*, 2001, 2003). No piscine orthologues have been identified to date. The thyroid hormone carriers identified thus far display different preferences for the transport of T4 and T3 (for a review see Jansen *et al.*, 2005). It follows that the repertoire of carriers expressed by a cell or tissue determines the bioactivity of T4 and T3, and, hence, *in vivo* or *in vitro* treatments with thyroid hormones do not necessarily have to result in an intracellular hyperthyroidism.

PERIPHERAL METABOLISM

Once secreted into the circulation, thyroid hormones are subject to a series of metabolizing pathways, which lead to major and minor iodothyronine metabolites (Fig. 2.1). Acknowledgement of these is not trivial, since they possess highly different reactivities towards metabolizing enzymes and receptors. The extensive peripheral metabolism of thyroid hormones bears an analogy to the complex posttranslational processing seen for some peptide hormones.

Metabolic pathways other than deiodination (which is treated in more detail in the next section) involve sulfation (catalyzed by sulfotransferases) and glucuronidation (UDP-glucuronyltransferases) to yield conjugated thyroid hormones in mammals (Visser, 1996) and teleosts (Sinclair and Eales, 1972; Finnson and Eales, 1996, 1997, 1998). Conjugated iodothyronines are considered to be biologically inactive, and the increased water solubility to facilitate urinary and biliary excretion. The presence of glucuronidated and sulfated iodothyronines in fish bile and urine (Parry *et al.*, 1994; Finnson and Eales, 1996) corroborates the role of hepatic conjugation as a clearance pathway. Interestingly, in healthy, fasted Mozambique tilapia a substantial fraction (ca. 8%) of the total plasma T3 pool was found to be glucuronidated (DiStefano *et al.*, 1998).

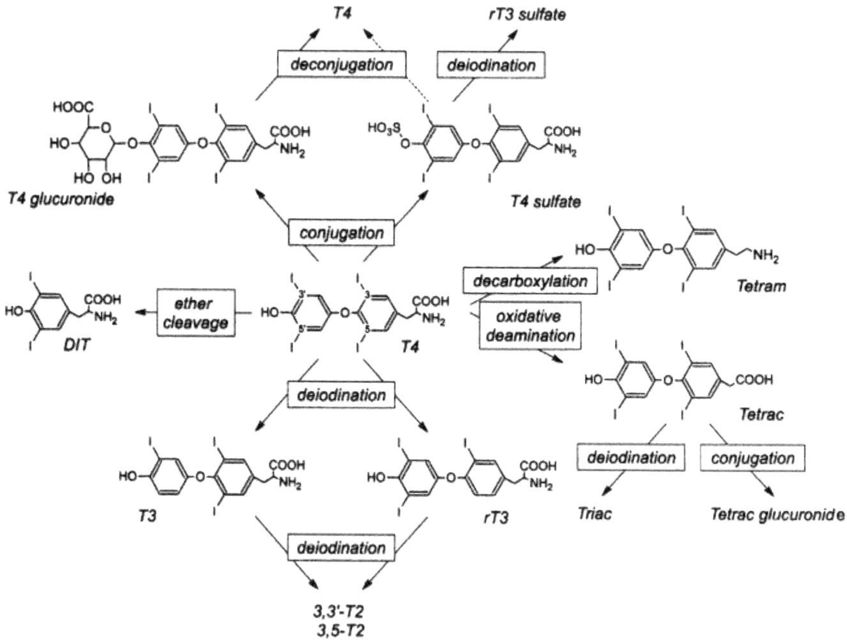

Fig. 2.1 Pathways of thyroid hormone metabolism (adapted from Köhrle *et al.*, 1987). Here, T4 is chosen as the central metabolite, but most reactions are applicable to, respectively, T3 and T2s as well. Note: T4 sulfate is not susceptible to deconjugation by sulfatase activity (as indicated by the dashed arrow) or outer ring deiodination by D1, but T3 sulfate is. Abbreviations: DIT, diiodotyrosine; Tetrac, tetraiodoacetic acid; Tetram, tetraiodothyronine; Triac, triiodotetraacetic acid.

This and other observations in mammals (van der Heide *et al.*, 2002, 2004) hint at a role of thyroid hormone conjugation other than the facilitation of excretion. Indeed, sulfation and glucuronidation greatly affect the reactivity of iodothyronines towards deiodinases, receptors, binding proteins, and cellular uptake. (Hays and Hsu, 1988; Hays and Cavalieri, 1992; Visser, 1994, 1996; van der Heide *et al.*, 2007). When we transferred gilthead seabream from seawater to low salinity water (1 ppt salinity), we not only measured increased plasma fT4 levels and decreased branchial outer ring deiodination activities, but also differential responses of enzyme activities putatively involved in the conjugation and deconjugating pathways of peripheral thyroid hormone metabolism (Klaren *et al.*, 2007).

The total potential effect of secreted T4, of which the thyroid is the only source, is very likely to be much more than the added effects of T3 and T4. Iodothyronine metabolites could well play subtle but important roles—locally and systemically—in organismal physiology.

Deiodination

Deiodination involves the enzymatic removal of an iodine atom from the outer (phenolic) ring and/or inner (tyrosyl) ring of the iodothyronine molecule. Outer ring deiodination of T4 is required to yield the potent bioactive hormone 3,5,3′-triiodothyronine (T3). Three mammalian iodothyronine deiodinases (D1, D2, D3) have been characterized, and all three are selenoenzymes with a selenocysteine in the catalytic centre, a specific iodothyronine substrate affinity and tissue distribution, and preference for inner or outer ring deiodination (Fig. 2.2). Only the mammalian D1 isozyme is sensitive to inhibition by the thyrostatic PTU (see reviews: Köhrle, 1999; Bianco *et al.*, 2002; Kuiper *et al.*, 2005).

Teleost deiodinases resemble their mammalian counterparts in their primary structure, but, although it has been suggested that they are more

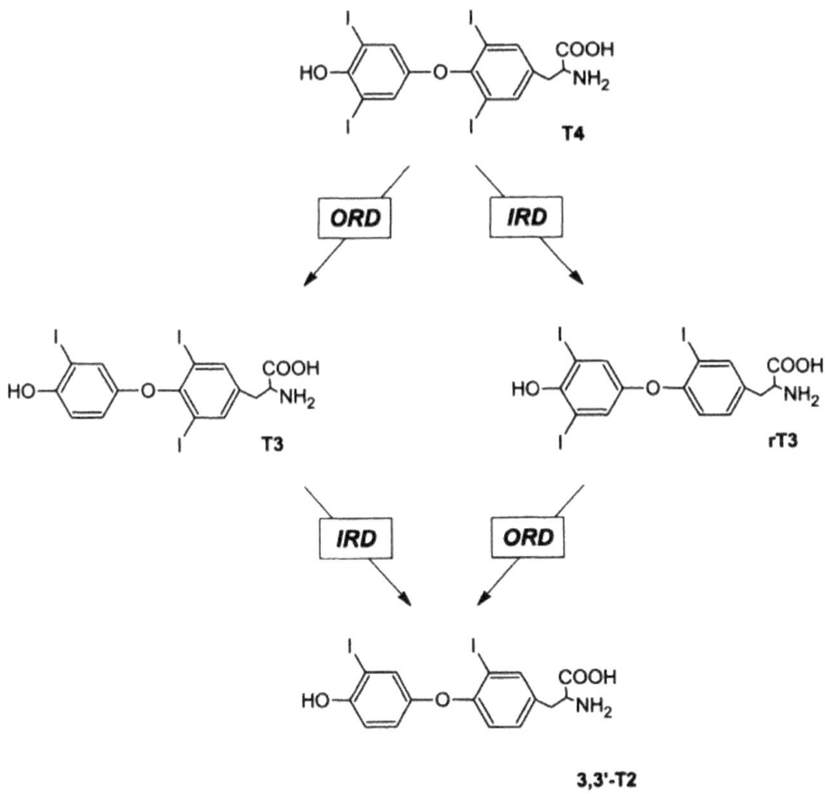

Fig. 2.2 Pathways for inner and outer ring deiodination.

similar to human orthologues than is generally accepted (Mol *et al.*, 1998), some peculiar biochemical differences certainly exist (see review: Orozco and Valverde-R, 2005). Teleost D1 is relatively insensitive to PTU inhibition (Sanders *et al.*, 1997; Orozco *et al.*, 2003; Klaren *et al.*, 2005b), as opposed to mammalian D1. In a number of teleost species (e.g., gilthead seabream, Senegalese sole (*Solea senegalensis*)), D1 activity is inhibited by dithiothreitol, the thiol cofactor that, *in vitro*, potently activates mammalian deiodinases (Mol, 1996; Klaren *et al.*, 2005b). Moreover, whereas the inner ring deiodination rates of sulfated T4 and T3 by rat D1 is greatly enhanced over that of native iodothyronines, no evidence was found for the deiodination of thyroid hormone sulfates in rainbow trout liver (Finnson *et al.*, 1999).

Deiodinases are sensitive to the thyroid status of an animal, i.e., they are regulated by the very substrates and products that these enzymes use and produce. Indeed, Eales and colleagues (Fok and Eales, 1984; Eales and Finnson, 1991) found that outer ring deiodination of T4 in the liver of rainbow trout was reduced upon treatment with T4 or T3. Evidence for the direct involvement of deiodinases was obtained in Nile tilapia, where hepatic D1 and D2 mRNA levels and deiodinating activities were upregulated, but D3 activities in brain, gills and liver were reduced upon methimazole-induced hypothyroidism (Mol *et al.*, 1999; Van der Geyten *et al.*, 2001). The highest D2 activities are detected in the teleost liver, and experimental hyperthyroidism decreases, and hypothyroidism increases hepatic D2 gene expression and enzyme activity (Mol *et al.*, 1999; Van der Geyten *et al.*, 2001; García *et al.*, 2004). Hepatic and renal D1 enzyme activities appear to be insensitive to experimental hyperthyroidism, induced by treatment with T3, in several teleost species (Finnson and Eales, 1999; Mol *et al.*, 1999; García *et al.*, 2004), although in killifish hepatic D1 mRNA levels were reduced upon 12-24 h treatment with T4, T3, or, surprisingly, 3,5-T2 (García *et al.*, 2004). Interestingly, the choice of treatment to induce hyperthyroidism is relevant here. Van der Geyten *et al.* (2005) have recently shown that treatment of Nile tilapia with T3 does not affect hepatic and renal D1 activities, in accordance with the results mentioned above, whereas treatment with T4 increased hepatic D1 activity. Not only because of the differential regulation of deiodinases by thyroid hormone, but also as a result of the specific preferences for the transport of thyroid hormones by carrier proteins (see previous chapter), experimental treatments with T4 or T3 are clearly not equivalent, and will result in physiologically different hyperthyroid states. This should be an important consideration in the design of physiological experiments.

D1 and D2 possess outer ring deiodination activities, and are thus involved in the activation of the thyroid prohormone T4 to T3. On the contrary, D3 possesses an inner ring deiodination activity only, and its obvious role in thyroid hormone metabolism is the irreversible inactivation of T4 and T3. It should be appreciated that the regulation of deiodinases by iodothyronines can confound experimental results obtained on fishes with experimentally induced hyper- or hypothyroidism. Experimental treatments are mostly aimed at attaining elevated or decreased systemic plasma thyroid hormone levels, but hyper- or hypothyroidism thus induced is not always reflected locally, at the organ or cellular level. Indeed, experimentally elevated plasma or tissue T4 levels in fish do not always result in elevated T3 concentrations (Blaschuk *et al.*, 1982; Fok and Eales, 1984; Inui *et al.*, 1989; Peter *et al.*, 2000). Deiodinases are important determinants of the thyroid status of an animal, and could be useful parameters in assessing the activity of the thyroid peripheral system. Indeed, in rainbow trout and gilthead seabream, osmotic challenge resulted in altered outer ring deiodination activities in gills, liver and kidney (Orozco *et al.*, 2002; Klaren *et al.*, 2007).

THYROID HORMONE TARGETS

Although (non-genomic) actions of T4 and T3 have been described in mammalian cells (see review: Davis *et al.*, 2002), T4 is still traditionally considered to be a prohormone with few or no biological activities. The most biologically potent iodothyronine, T3, binds intracellularly to a thyroid hormone receptor (TR), a type-2 nuclear receptor of which four isoforms ($\alpha 1$, $\alpha 2$, $\beta 1$, $\beta 2$) are currently known. The T3-TR ligand-receptor complex heterodimerizes with a retinoic X receptor (RXR), and can then bind to a thyroid response element (TRE) in or near the promoter region of thyroid hormone responsive genes, leading to the activation or suppression of gene expression (see review: Yen, 2001).

Thyroid Hormone and Na,K-ATPase

The actions of T3 are truly pleiotropic, as exemplified by DNA-microarray analyses of hepatic gene expression in hyper- and hypothyroid mice (Feng *et al.*, 2000). From a fish' osmoregulatory perspective, however, the most important target of T3 action is likely to be the Na,K-ATPase sodium pump. The Na,K-ATPase holoprotein is a tetramer consisting of two α- and two β-subunits. In mammals, both subunits are sensitive to T3 in an organ-specific manner (Horowitz *et al.*, 1990; Bajpai and Chaudhury, 1999; Yalcin *et al.*, 1999). Several studies point to the regulatory actions of T3 treatment on *in vitro* teleost branchial, renal and hepatic Na,K-ATPase activities (De *et al.*, 1987; Peter *et al.*, 2000; Shameena *et al.*, 2000), indicating that in fishes (subunits of) the sodium pump also are sensitive to T3. Na,K-ATPase is a pivotal enzyme involved in ion transport and osmoregulation, and its activity in branchial and renal epithelia is increased upon seawater acclimation of fish (Madsen *et al.*, 1995; Seidelin *et al.*, 2000). This would make thyroid hormones *a priori* important determinants of osmoregulatory capacity in teleosts.

Considering the presence of positive thyroid response elements in the Na,K-ATPase subunit gene promoter regions, one would predict increased branchial and renal enzyme activities upon thyroid hormone treatment. However, evidence for this relation is still inconclusive. Indeed, in cultured pavement cells from rainbow trout treatment with T3 resulted in a ca. 35% increase in Na,K-ATPase activity (Kelly and Wood, 2001). However, this effect was not T3-dose dependent and was not correlated with net transepithelial Na^+ fluxes. In *in vivo* studies in salmonids, plasma T4 concentrations were found to be correlated with branchial Na,K-ATPase activity (measured *in vitro* as the rate of ouabain-sensitive ATP hydrolysis) (Folmar and Dickhoff, 1979; Virtanen and Soivio, 1985). Contrary, Na,K-ATPase activities were unaffected (Saunders *et al.*, 1985; Dangé, 1986; Madsen, 1989, 1990; Shrimpton and McCormick, 1998; Mancera and McCormick, 1999) or even reduced (Omeljaniuk and Eales, 1986) by treatment with T4 or T3 in a number of teleost species. Peter *et al.* (2000) measured a tissue-specific response of Na,K-ATPase to treatments with T4 or T3 in freshwater Mozambique tilapia; at low doses branchial sodium pump activities were increased by 50 to 70%, whereas those in kidney were decreased. At higher hormone doses, these effects disappeared. We have as yet found no satisfying explanation for this result, and, indeed, for the lack of effect of thyroid hormone treatment on *in vitro* Na,K-ATPase activities. We have to consider the methodology of assaying Na,K-ATPase activities *in vitro*, which is mostly performed on homogenates or partially purified membrane preparations. This does not allow a discrimination between enzyme activities in (sub)cellular fractions involved in osmoregulatory processes (i.e., branchial chloride cells, basolateral membranes) and in fractions that are not (i.e., branchial pavement cells, Na,K-ATPase in intracellular compartments). The measurement of [^{3}H]-ouabain binding sites (Clausen, 1996) in teleost branchial and renal tissues could be very fruitful to assess the effects of thyroid hormones on Na,K-ATPase.

THYROID GLAND REGULATION

Pituitary Gland—TSH

Thyroid-stimulating hormone (TSH, or thyrotropin) is the key regulating factor of thyroid function. Treatment with heterologous TSH results in elevated plasma T4 levels in a number of teleost species *in vivo* (Grau and Stetson, 1977; Brown and Stetson, 1983; Specker and Richman, 1984; Brown *et al.*, 1985; Leatherland, 1987; Inui *et al.*, 1989; Byamungu *et al.*, 1990; Bandyopadhyay and Bhattacharya, 1993) and an increased release of T4 from Hawaiian parrotfish (*Scarus dubius*) thyroid *in vitro* (Grau *et al.*, 1986; Swanson *et al.*, 1988). The parrotfish' thyroid, which consists of two distinct lobes—and this is an exceptional organization in fish, as the thyroid gland is diffusely organised in other species—is particularly suitable for static or perifusion incubations. *In vitro* studies conducted on parrotfish performed by Grau *et al.* (1986) and Swanson *et al.* (1988) are unique and have allowed a direct assessment of thyroid gland output and other aspects of thyroid gland physiology. Still, with the help of detailed anatomical knowledge, careful dissection and perhaps a mild digestion of surrounding tissues, it could well be feasible to subject thyroid glands from other teleost species to *in-vitro* examinations, as was performed by Bhattacharya *et al.* (1976a). In this respect, the technique developed by Toda *et al.* (2002) to maintain cultured porcine thyroid follicles in a three-dimensional extracellular matrix environment could be promising.

A negative feedback exists between plasma thyroid hormones and TSH secretion by pituitary thyrotrophs. Specifically, in several teleost species T4 and T3 down-regulate the pituitary TSH β-subunit mRNA content (Larsen *et al.*, 1997; Pradet-Balade *et al.*, 1997, 1999; Sohn *et al.*, 1999; Yoshiura *et al.*, 1999; Chatterjee *et al.*, 2001; Geven *et al.*, 2006).

Hypothalamus—TRH and CRH

Thyrotropin-releasing hormone (TRH), a hypothalamic tripeptide (pGlu-His-Pro-NH_2) controls pituitary TSH cells and also functions as a neurotransmitter throughout the central nervous system. It has a widespread distribution in the teleost brain (Kreider *et al.*, 1988; Batten *et al.*, 1990a, b; Hamano *et al.*, 1990; Matz and Takahashi, 1994; Matz and Hofeldt, 1999; Diaz *et al.*, 2001, 2002). In the brain and pituitary gland of white sucker (*Catostomus commersonii*) and salmonids two TRH-R subtypes have been identified (Schwartzentruber and Omeljaniuk, 1995; Ohide *et al.*, 1996; Harder *et al.*, 2001; Kumar and Trant, 2001). Although TRH circulates peripherally in Mozambique tilapia (Lamers *et al.*, 1994), no TRH-binding sites outside the fish brain are known (for reviews see Burt and Ajah, 1984; Kumar and Trant, 2001).

Recently, it has been shown that TRH upregulates TSH β-subunit gene expression in a teleost (bighead carp, *Aristichtys nobilis*) pituitary (Chatterjee *et al.*, 2001). However, the hypophysiotropic and thyrotropic effects of TRH differ greatly among species. In arctic charr (*Salvelinus alpinus*) low doses (≤0.1 μg/g body weight) of TRH elevate plasma T4 levels, but rainbow trout responded only to higher doses (1 μg/g) (Eales and Himick, 1988). In many other fishes, TRH has been reported to be without effect on TSH release from pituitary cells or on the thyroid status of the animal, whereas the release of growth hormone, prolactin or αMSH was stimulated (Gorbman and Hyder, 1973; Dickhoff *et al.*, 1978; Lamers *et al.*, 1994; Melamed *et al.*, 1995; Kagabu *et al.*, 1998; Larsen *et al.*, 1998). TRH clearly is a misnomer for the piscine tripeptide pGlu-His-ProNH$_2$.

Corticotropin-releasing hormone (CRH) is abundantly expressed in the teleost brain (Pepels *et al.*, 2002a). Although in Mozambique tilapia around 80% of the total brain immunoreactive CRH content is localized outside the hypothalamus (Pepels *et al.*, 2002b), the presence of CRH in neurons of the hypothalamic preoptic area (nucleus preopticus, NPO), which project to the pituitary, is of special relevance for pituitary gland regulation (see review: Meek and Nieuwenhuys, 1998). The 'classical' action of hypothalamic CRH in all vertebrates including teleosts is the regulation of the release of adrenocorticotropic hormone (ACTH) from the pituitary pars distalis which, in turn, regulates the secretion of cortisol from the head kidney's interrenal cells during a stress response (Ando *et al.*, 1999; Huising *et al.*, 2004; Metz *et al.*, 2004; Flik *et al.*, 2005). The involvement of CRH in the regulation of the thyroid gland can be inferred from the observation that lesions of the NPO in tilapia resulted, after 10 days, in increases of plasma T4 and rT3, but not T3 (Sukumar *et al.*, 1997). Lesions in other parts of the hypothalamus, i.e., the anterior and posterior nucleus lateralis tuberalis, had no effect, and the pituitary contents of growth hormone and prolactin remained unchanged by any of these lesions (Sukumar *et al.*, 1997), indicating a specific thyrotropic action of a specific cell population in the NPO. Reversely, treatment of the catfish *Clarius batrachus* with thyrostatics resulted in a decrease of nuclear dimensions of NPO cells (Dixit, 1976). Heterologous CRH and other members of the CRH family, e.g., urotensin I and sauvagine, were found to be potent stimulators of TSH release from cultured pituitary cells from coho salmon (Larsen *et al.*, 1998). Intracerebroventricular administration of ovine CRH (oCRH) in fed goldfish (*Carassius auratus*) decreased total thyroid T4 and T3 content. In fasted goldfish, oCRH treatment increased the free T4 and decreased free T3 contents of the thyroid (De Pedro *et al.*, 1995).

It appears that through an involvement of the shared signal molecule CRH, the corticotrope and thyrotrope axes in fishes are intertwined [(see Kühn *et al.* (1998) for other vertebrates)], and this would open an interesting field of research. We have hypothesized that thyroid hormones, the release of which is regulated by CRH, feed back on the NPO to modulate the CRH activity which would have a concomitant effect on the hypothalamus-pituitary-interrenal axis. Interesting results were obtained from studies where T4 treatment or thyroidectomy of rats resulted in reciprocal changes in the expression of CRH mRNA and other hypothalamic peptides as well in the paraventricular nucleus (homologous to the teleost NPO) (Ceccatelli *et al.*, 1992; Dakine *et al.*, 2000). We have also found that carp with experimentally induced hyperthyroidism displayed a marked hypocortisolinemia, and this correlated with increased mRNA-levels of CRH binding protein, a primitive endogenous CRH antagonist (Huising and Flik, 2005; Geven *et al.*, 2005). We interpret this as a proof of principle of the hypothesis that thyroid hormones affect the activity of the hypothalamus-pituitary-interrenal axis at a central, hypothalamic level. This could also form the molecular basis of the interactions between thyroid hormones and cortisol in fish. It has been observed that cortisol treatment reduces plasma T4 in European eel (*Anguilla anguilla*) and coho salmon (Redding *et al.*, 1984, 1986), but negative effects have also been observed (Leatherland, 1987; Redding *et al.*, 1991). Direct proof for the synergistic effect of thyroid hormone and cortisol on osmoregulatory capacity comes from cultured pavement cell epithelia from rainbow trout where, compared with T3 alone, a co-incubation of T3 and cortisol decreased unidirectional Na^+- and Cl^--fluxes under symmetrical culture conditions (Kelly and Wood, 2001).

Hypothalamic actions of thyroid hormones could well be involved in the synergistic or additive effects of thyroid hormones and cortisol in teleost osmoregulation (for review see McCormick, 2001).

OSMOREGULATORY ASPECTS OF THYROID HORMONES

Ontogeny and Development of Osmoregulatory Capacity

Early experiments by the thyroidology pioneer J. Friedrich Gudernatsch clearly indicated the determining role of the thyroid gland in (amphibian) metamorphosis. The involvement of the thyroid in teleost metamorphosis and ontogeny is exemplified by the observation that exogenous administration of T4 induced metamorphosis in flatfish whereas the thyrostatic thiourea inhibited it (Solbakken *et al.*, 1999; Stickney and Liu, 1999). Reversely, T4 rescued zebrafish (*Danio rerio*) from developmental arrest induced by thyrostatics (Brown, 1997). Thyroid hormones are involved in many ontogenetic events, e.g., the development of the olfactory region of the brain, and of ultraviolet photosensitivity of the retina during parr-smolt transformation of salmonids (Browman and Hawryshyn, 1994; Morin *et al.*, 1997; Alexander *et al.*, 1998). The pervasive effects of thyroid hormones on ontogenetic development are perhaps best illustrated in the teleost ice goby (*Leucopsarion petersii*) in which an inactive thyroid gland is considered to be causal to the neotenic phenotype of the adult animal (Harada *et al.*, 2003). The developmental effects of thyroid hormones are regulated through the differential expression of thyroid hormone receptor subtypes, which are expressed already before midblastula stage in zebrafish embryos (Liu *et al.*, 2000) and which are temporally and regionally differentially expressed in several teleost species (Yamano and Miwa, 1998; Power *et al.*, 2001; Marchand *et al.*, 2004).

Anadromic and catadromic species, in particular, encounter greatly varying environments during their natural life span. Plasma thyroid hormone levels vary with the migration of salmonids (Eales, 1965; Sower and Schreck, 1982; Cyr *et al.*, 1988; Youngson and Webb, 1993; Iwata, 1995; Persson *et al.*, 1998), indicating that they are possibly responsive to environmental salinity. However, it cannot be excluded that the observed changes in plasma thyroid levels are involved in the adaptation to other environmental factors than salinity, or aspects other than strictly osmoregulation, e.g., parr-smolt transformation and spawning. More direct evidence for the involvement of thyroid hormones in osmoregulation comes from studies where fishes, *ceteris paribus*, were transferred to different salinities and where changes in thyroid gland morphology, and plasma levels of T4 and/or T3 were measured (Olivereau *et al.*, 1977; Folmar and Dickhoff, 1979; Redding *et al.*, 1984, 1991; Grau *et al.*, 1985; Klaren *et al.*, 2007). Yet, in this context, there have been reports of negative results, i.e., no changes in thyroid hormone plasma levels (McCormick and Saunders, 1990; Redding *et al.*, 1991).

Treatment of animals with the thyrostatic thiourea apparently reduces their osmoregulatory capacity (Knoeppel *et al.*, 1982; Madsen, 1989; Schreiber and Specker, 1999). Results from these studies are equivocal, e.g., the treatment with T4 not always rescues from the treatment with the thyrostatic, indicating that the actions of thiourea in fish are not always specifically targeted at the thyroid gland.

HETEROTOPIC THYROID FOLLICLES IN OSMOREGULATORY ORGANS

In contrast to the compact thyroid gland found in higher vertebrates, in virtually all teleosts and adult Cyclostomata the thyroid gland consists of rather diffusely scattered follicles (single or in small groups) in the region ventral to the pharynx, along the ventral aorta and where the branchial arteries branch off. Almost a century ago, Gudernatsch (1911) reported on the dispersal of thyroid follicles over larger areas in the subpharyngeal region. His early observations coincided with reports on 'so-called' thyroid carcinomas in brook trout (*Salvelinus fontinalis*) (Marine and Lenhart, 1910, 1911). Baker-Cohen (1959), who summarized the pertinent literature on extrapharyngeal thyroid tissue, concluded that in several teleost species the kidney was the most common location of heterotopic thyroid follicles. Ever since, reports have appeared on heterotopic thyroid follicles mostly in the head-kidney (pronephros) and trunk-kidney (opistonephros) area in a large number of species, with a surge of publications some three decades ago (Chavin, 1956; Olivereau, 1960; Lysak, 1964; Frisén and Frisén, 1967; Srivastava and Sathyanesan, 1967, 1971a, b; Peter, 1970; Qureshi, 1975; Bhattacharya *et al.*, 1976a, b; Joshi and Sathyanesan, 1976; Qureshi and Sultan, 1976; Qureshi *et al.*, 1978; Agrawala and Dixit, 1979; Sharma and Kumar, 1982). (See Figure 2.3 with

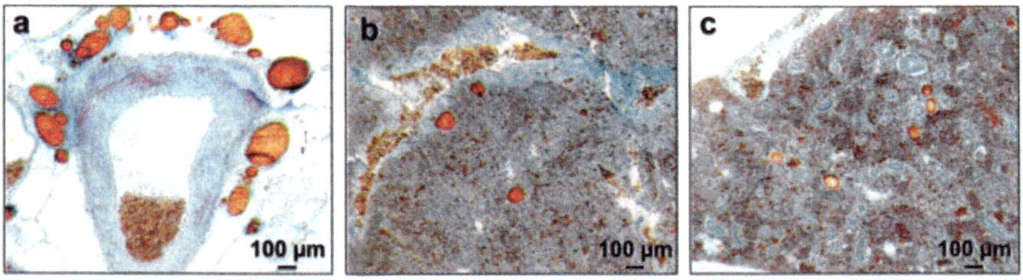

Fig. 2.3 Thyroid follicles in tissues of common carp (*Cyprinus carpio*) treated with the trichrome Light Green-Orange G-fuchsin which stains the follicular colloid bright red. a. Subpharyngeal region where thyroid follicles are arranged around the ventral aorta. b. Head-kidney tissue. c. Trunk-kidney tissue.

our unpublished results on common carp, *Cyprinus carpio*.) Despite the obvious involvement of head-kidney and kidney in fish osmoregulation, the subject of renal heterotopic thyroid follicles has received little attention since.

The head-kidney, an organ analogous to the adrenal glands in mammals, is unique to teleost fish. Chromaffin cells and interrenal cells produce and secrete catecholamines and cortisol, respectively. Heterotopic thyroid follicles, when present, are functional as evidenced by the accumulation of iodine and synthesis of the thyroxine precursors MIT and DIT, and T4, processes that are sensitive to TSH and thyrostatics. In some species head-kidney iodine accumulation exceeds that in the pharyngeal region (our unpublished results on common carp, *Cyprinus carpio*) and shows a seasonal variation (Chavin and Bouwman, 1965; Srivastava and Sathyanesan, 1971a; Bhattacharya *et al.*, 1976a). Thus, heterotopic thyroid follicles must be active endocrine tissues, sensitive to physiological regulators. It has been suggested that development of heterotopic thyroid tissue may reflect a compensatory mechanism to iodine deficiency, as iodide supplementation prevented but not reversed the development of heterotopic thyroid tissue (Baker-Cohen, 1959). However, ca. 40% of the cases of renal heterotopic thyroid reviewed by Baker-Cohen were animals with normal, non-goitrous pharyngeal thyroid tissue that probably did not experience an iodine deficiency. It may be that heterotopic follicles concern metastatic thyroid carcinoma, as teleostean thyroid tissue appears particularly sensitive to carcinogens and high-energy radiation (Woodhead and Scully, 1977; Blasiola *et al.*, 1981; Hoover, 1984; Woodhead *et al.*, 1984; Bunton and Wolfe, 1996; Chen *et al.*, 1996; Toussaint *et al.*, 1999). Indeed, a major part of Baker-Cohen's observations were performed on a platyfish strain (*BH* strain) which has a very high thyroid tumour incidence compared to other strains. Moreover, heterotopic renal thyroid follicles occur not only in feral fish, but also in fish kept or bred under laboratory conditions (Baker-Cohen, 1959; Frisén and Frisén, 1967; Qureshi, 1975). Interestingly, Baker-Cohen (1959) observed normal thyroid tissue in head-kidney of platyfish of strain *30*, a strain that does not develop thyroid tumours while the *BH* strain does. We observed thyroid follicles in head-kidney and renal tissue of our laboratory-bred and -raised carp and tilapia (unpublished results). The presence of heterotopic thyroid follicles is most likely a normal anatomical feature in healthy animals. It is interesting to note that in fish the location of the majority of thyroid follicles, i.e., near the branchial efferents of the ventral aorta, kidney and head-kidney, always is in a well-innervated area and associated with a portal system. This could have a physiological significance for systemic thyroid hormone homeostasis.

The preference of heterotopic thyroid follicles for residence in head-kidney tissue, the close juxtaposition of iodothyronine-producing cells with interrenal (cortisol-producing) cells, chromaffin (catecholamine-producing) cells, and haematopoietic cells strongly hints at some paracrine relationship between thyroid and interrenal tissue. Initial experiments in our department failed to establish an effect of T4 or T3 on the *in vitro* cortisol release from tilapia head-kidney. However, the putative relation between the head-kidney and its thyroid tissue could well be more intricate, and deserves further and more detailed investigation.

CONCLUSION

The best known *in vivo* effect of thyroid hormones is the stimulation of the basal metabolic rate, and from this alone the involvement of thyroid hormones in osmoregulatory processes can be inferred. The multiple effects of thyroid hormone and probably its metabolites as well on, e.g., development (of osmoregulatory capacity, *sensu stricto*) and Na,K-ATPase activities strengthen the important role of iodothyronines in teleost osmoregulation. However, a generalized mode of action does not, as yet, emerge, not only because of the relative paucity in physiological experimentation in teleost thyroidology, but also because of the synergistic/additive effects of thyroid hormones with cortisol (and growth hormone) in osmoregulation.

References

Abe, T., M. Kakyo, H. Sakagami, T. Tokui, T. Nishio, M. Tanemoto, H. Nomura, S.C. Hebert, S. Matsuno, H. Kondo and H. Yawo. 1998. Molecular characterization and tissue distribution of a new organic anion transporter subtype (oatp3) that transports thyroid hormones and taurocholate and comparison with oatp2. *Journal of Biological Chemistry* 273: 22395–22401.

Agrawala, N. and R.K. Dixit. 1979. Seasonal variations in the pharyngeal and pronephric thyroid tissues of the fresh water teleost *Puntius sophore* (Ham). *Zeitschrift für Zellforschung und Mikroskopische Anatomie* 93: 138–146.

Ajjan, R.A., C. Findlay, R.A. Metcalfe, P.F. Watson, M. Crisp, M. Ludgate and A.P. Weetman. 1998. The modulation of the human sodium iodide symporter activity by Graves' disease sera. *Journal of Clinical Endocrinology and Metabolism* 83: 1217–1221.

Alexander, G., R. Sweeting and B. McKeown. 1998. The effect of thyroid hormone and thyroid hormone blocker on visual pigment shifting in juvenile coho salmon (*Oncorhynchus kisutch*). *Aquaculture* 168: 157–168.

Ando, H., M. Hasegawa, J. Ando and A. Urano. 1999. Expression of salmon corticotropin-releasing hormone precursor gene in the preoptic nucleus in stressed rainbow trout. *General and Comparative Endocrinology* 113: 87–95.

Babin, P.J. 1992. Binding of thyroxine and 3,5,3'-triiodothyronine to trout plasma lipoproteins. *American Journal of Physiology* 262: E712–E720.

Bajpai, M. and S. Chaudhury. 1999. Transcriptional and post-transcriptional regulation of Na^+,K^+-ATPase α isoforms by thyroid hormone in the developing rat brain. *Neuroreport* 10: 2325–2328.

Baker-Cohen, K.F. 1959. Renal and other heterotopic thyroid tissue in fishes. In: *Proceedings of the Columbia University Symposium on Comparative Endocrinology*. Cold Spring Harbor, New York, May 25-29, 1958, A. Gorbman (ed.), John Wiley & Sons, New York, pp. 283–301.

Bandyopadhyay, S. and S. Bhattacharya. 1993. Purification and properties of an Indian major carp (*Cirrhinus mrigala*, Ham.) pituitary thyrotropin. *General and Comparative Endocrinology* 90: 192–204.

Batten, T.F., M.L. Cambre, L. Moons and F. Vandesande. 1990a. Comparative distribution of neuropeptide-immunoreactive systems in the brain of the green molly, *Poecilia latipinna*. *Journal of Comparative Neurology* 302: 893–919.

Batten, T.F., L. Moons, M.L. Cambre, F. Vandesande, T. Seki and M. Suzuki. 1990b. Thyrotropin-releasing hormone-immunoreactive system in the brain and pituitary gland of the sea bass (*Dicentrarchus labrax*, Teleostei). *General and Comparative Endocrinology* 79: 385–392.

Baumeister, F.A. and V. Herzog. 1988. Sulfation of thyroglobulin: a ubiquitous modification in vertebrates. *Cell and Tissue Research* 252: 349–358.

Benvenga, S. 1997. A thyroid hormone binding motif is evolutionarily conserved in apolipoproteins. *Thyroid* 7: 605–611.

Bhattacharya, S., R.H. Das and A.G. Datta. 1976a. Iodine metabolism in dispersed pharyngeal and head kidney teleostean thyroid cells obtained by continuous trypsinization. *General and Comparative Endocrinology* 30: 128–130.

Bhattacharya, S., P. Dasgupta and D. Kumar. 1976b. Thyroid hormone synthesis by pharyngeal and head-kidney cell-free preparation from a teleost fish *Clarias batrachus* L. *Indian Journal of Experimental Biology* 14: 227–231.

Bianco, A.C., D. Salvatore, B. Gereben, M.J. Berry and P.R. Larsen. 2002. Biochemistry, cellular and molecular biology, and physiological roles of the iodothyronine selenodeiodinases. *Endocrine Reviews* 23: 38–89.

Bidart, J.M., C. Mian, V. Lazar, D. Russo, S. Filetti, B. Caillou and M. Schlumberger. 2000. Expression of pendrin and the Pendred syndrome (PDS) gene in human thyroid tissues. *Journal of Clinical Endocrinology and Metabolism* 85: 2028–2033.

Blaschuk, O.W., J.C. Jamieson and J.G. Eales. 1982. Properties of hexosaminidases in cell-free extracts of rainbow trout livers and effects of thyroid hormones. *Comparative Biochemistry and Physiology B - Biochemistry and Molecular Biology* 73: 729–734.

Blasiola, G.C., Jr., J.C. Turnier and E.E. Hurst. 1981. Metastatic thyroid adenocarcinomas in a captive population of kelp bass, *Paralabrax clathratus*. *Journal of the National Cancer Institute* 66: 51–59.

Browman, H.I. and C.W. Hawryshyn. 1994. The developmental trajectory of ultraviolet photosensitivity in rainbow trout is altered by thyroxine. *Vision Research* 34: 1397–1406.

Brown, C.L., E.G. Grau and M.H. Stetson. 1985. Functional specificity of gonadotropin and thyrotropin in *Fundulus heteroclitus*. *General and Comparative Endocrinology* 58: 252–258.

Brown, C.L. and M.H. Stetson. 1983. Prolactin-thyroid interaction in *Fundulus heteroclitus*. *General and Comparative Endocrinology* 50: 167–171.

Brown, D.D. 1997. The role of thyroid hormone in zebrafish and axolotl development. *Proceedings of the National Academy of Sciences of the United States of America* 94: 13011–13016.

Bunton, T.E. and M.J. Wolfe. 1996. *N*-methyl-*N'*-nitro-*N*-nitrosoguanidine-induced neoplasms in medaka (*Oryzias latipes*). *Toxicologic Pathology* 24: 323–330.

Burt, D.R. and M.A. Ajah. 1984. TRH receptors in fish. *General and Comparative Endocrinology* 53: 135–142.

Byamungu, N., S. Corneillie, K. Mol, V. Darras and E.R. Kühn. 1990. Stimulation of thyroid function by several pituitary hormones results in an increase in plasma thyroxine and reverse triiodothyronine in tilapia (*Tilapia nilotica*). *General and Comparative Endocrinology* 80: 33–40.

Ceccatelli, S., L. Giardino and L. Calza. 1992. Response of hypothalamic peptide mRNAs to thyroidectomy. *Neuroendocrinology* 56: 694–703.

Chatterjee, A., Y.-L. Hsieh and J.Y.L. Yu. 2001. Molecular cloning of cDNA encoding thyroid stimulating hormone β subunit of bighead carp *Aristichtys nobilis* and regulation of its gene expression. *Molecular and Cellular Endocrinology* 174: 1–9.

Chavin, W. 1956. Thyroid follicles in the head-kidney of the goldfish, *Carassius auratus* (Linnaeus). *Zoologica* 41: 101–104.

Chavin, W. and B.N. Bouwman. 1965. Metabolism of iodine and thyroid hormone synthesis in the goldfish, *Carassius auratus* L. *General and Comparative Endocrinology* 5: 493–503.

Chen, H.C., I.J. Pan, W.J. Tu, W.H. Lin, C.C. Hong and M.R. Brittelli. 1996. Neoplastic response in Japanese medaka and channel catfish exposed to *N*-methyl-*N'*-nitro-*N*-nitrosoguanidine. *Toxicologic Pathology* 24: 696–706.

Clausen, T. 1996. The Na^+,K^+ pump in skeletal muscle: quantification, regulation and functional significance. *Acta Physiologica Scandinavica* 156: 227–235.

Cyr, D.G. and J.G. Eales. 1992. Effects of short-term 17β-estradiol treatment on the properties of T4-binding proteins in the plasma of immature rainbow trout, *Oncorhynchus mykiss*. *Journal of Experimental Zoology* 262: 414–419.

Cyr, D.G., N.R. Bromage, J. Duston and J.G. Eales. 1988. Seasonal patterns in serum levels of thyroid hormones and sex steroids in relation to photoperiod-induced changes in spawning time in rainbow trout, *Salmo gairdneri*. *General and Comparative Endocrinology* 69: 217–225.

Dai, G., O. Levy and N. Carrasco. 1996. Cloning and characterization of the thyroid iodide transporter. *Nature (Lond.)* 379: 458–460.

Dakine, N., C. Oliver and M. Grino. 2000. Thyroxine modulates corticotropin-releasing factor but not arginine vasopressin gene expression in the hypothalamic paraventricular nucleus of the developing rat. *Journal of Neuroendocrinology* 12: 774–783.

Dangé, A.D. 1986. Branchial Na^+-K^+-ATPase activity in freshwater or saltwater acclimated tilapia, *Oreochromis (Sarotherodon) mossambicus*: effects of cortisol and thyroxine. *General and Comparative Endocrinology* 62: 341–343.

Davis, P.J., H.C. Tillmann, F.B. Davis and M. Wehling. 2002. Comparison of the mechanisms of nongenomic actions of thyroid hormone and steroid hormones. *Journal of Endocrinological Investigation* 25: 377–388.

De Pedro, N., B. Gancedo, A.L. Alonso-Gomez, M.J. Delgado and M. Alonso-Bedate. 1995. CRF effect on thyroid function is not mediated by feeding behavior in goldfish. *Pharmacology, Biochemistry and Behavior* 51: 885–890.

De, S., A.K. Ray and A.K. Medda. 1987. Nuclear activation by thyroid hormone in liver of Singi fish: changes in different ion-specific adenosine triphosphatases activities. *Hormone and Metabolic Research* 19: 367–370.

De, S., A.K. Ray and A.K. Medda. 1989. Effects of L-thyroxine and L-triiodothyronine on protein and nucleic acid contents of liver of 6N-2-propylthiouracil treated hypothyroid singi fish, *Heteropneustes fossilis* Bloch. *Hormone and Metabolic Research* 21: 416–420.

Diaz, M.L., M. Becerra, M.J. Manso and R. Anadon. 2001. Development of thyrotropin-releasing hormone immunoreactivity in the brain of the brown trout *Salmo trutta fario*. *Journal of Comparative Neurology* 429: 299–320.

Diaz, M.L., M. Becerra, M.J. Manso and R. Anadon. 2002. Distribution of thyrotropin-releasing hormone (TRH) immunoreactivity in the brain of the zebrafish (*Danio rerio*). *Journal of Comparative Neurology* 450: 45–60.

Dickhoff, W.W., J.W. Crim and A. Gorbman. 1978. Lack of effect of synthetic thyrotropin releasing hormone on Pacific hagfish (*Eptatretus stouti*) pituitary-thyroid tissues *in vitro*. *General and Comparative Endocrinology* 35: 96–98.

DiStefano, J.J., III, B. Ron, T.T. Nguyen, G.M. Weber and E.G. Grau. 1998. 3,5,3'-triiodothyronine (T_3) clearance and T3-glucuronide (T_3G) appearance kinetics in plasma of freshwater-reared male tilapia, *Oreochromis mossambicus*. *General and Comparative Endocrinology* 111: 123–140.

Dixit, V.P. 1976. Karyometric studies of the nucleus preopticus in fish (*Clarias batrachus*) and toad (*Bufo andersonii*). *Acta Anatomica* 95: 182–189.

Dunn, J.T. and A.D. Dunn. 2001. Update on intrathyroidal iodine metabolism. *Thyroid* 11: 407–414.

Eales, J.G. 1965. Factors influencing seasonal changes in thyroid activity in juvenile steelhead trout, *Salmo gairdneri*. *Canadian Journal of Zoology* 43: 719–729.

Eales, J.G. and S.B. Brown. 1993. Measurement and regulation of thyroidal status in teleost fish. *Reviews in Fish Biology and Fisheries* 3: 299–347.

Eales, J.G. and K.W. Finnson. 1991. Response of hepatic thyroxine 5'-deiodinase of rainbow trout, *Oncorhynchus mykiss*, to chronic ingestion of 3,5,3'-triiodo-L-thyronine. *Journal of Experimental Zoology* 257: 230–235.

Eales, J.G. and B.A. Himick. 1988. The effects of TRH on plasma thyroid hormone levels of rainbow trout (*Salmo gairdneri*) and arctic charr (*Salvelinus alpinus*). *General and Comparative Endocrinology* 72: 333–339.

Eales, J.G. and S. Shostak. 1985. Free T_4 and T_3 in relation to total hormone, free hormone indices, and protein in plasma of rainbow trout and arctic charr. *General and Comparative Endocrinology* 58: 291–302.

Eales, J.G., R.J. Omeljaniuk and S. Shostak. 1983. Reverse T_3 in rainbow trout, *Salmo gairdneri*. *General and Comparative Endocrinology* 50: 395–406.

Elsalini, O.A. and K.B. Rohr. 2003. Phenylthiourea disrupts thyroid function in developing zebrafish. *Development Genes and Evolution* 212: 593–598.

Feng, X., Y. Jiang, P. Meltzer and P.M. Yen. 2000. Thyroid hormone regulation of hepatic genes *in vivo* detected by complementary DNA microarray. *Molecular Endocrinology* 14: 947–955.

Finnson, K.W. and J.G. Eales. 1996. Identification of thyroid hormone conjugates produced by isolated hepatocytes and excreted in bile of rainbow trout, *Oncorhynchus mykiss*. *General and Comparative Endocrinology* 101: 145–154.

Finnson, K.W. and J.G. Eales. 1997. Glucuronidation of thyroxine and 3,5,3'-triiodothyronine by hepatic microsomes in rainbow trout, *Oncorhynchus mykiss*. *Comparative Biochemistry and Physiology* C 117: 193–199.

Finnson, K.W. and J.G. Eales. 1998. Sulfation of thyroid hormones by liver of rainbow trout, *Oncorhynchus mykiss*. *Comparative Biochemistry and Physiology* C 120: 415–420.

Finnson, K.W. and J.G. Eales. 1999. Effect of T_3 treatment and food ration on hepatic deiodination and conjugation of thyroid hormones in rainbow trout, *Oncorhynchus mykiss*. *General and Comparative Endocrinology* 115: 379–386.

Finnson, K.W., J.M. McLeese and J.G. Eales. 1999. Deiodination and deconjugation of thyroid hormone conjugates and type I deiodination in liver of rainbow trout, *Oncorhynchus mykiss*. *General and Comparative Endocrinology* 115: 387–397.

Flik, G., P.H.M. Klaren, E.H. van den Burg, J.R. Metz and M.O. Huising. 2006. CRF and stress in fish. *General and Comparative Endocrinology* 146: 36–44.

Fok, P. and J.G. Eales. 1984. Regulation of plasma T3 levels in T4-challenged rainbow trout, *Salmo gairdneri*. *General and Comparative Endocrinology* 53: 197–202.

Folli, C., N. Pasquato, I. Ramazzina, R. Battistutta, G. Zanotti and R. Berni. 2003. Distinctive binding and structural properties of piscine transthyretin. *FEBS Letters* 555: 279–284.

Folmar, L.C. and W.W. Dickhoff. 1979. Plasma thyroxine and gill Na^+-K^+ ATPase changes during seawater acclimation of coho salmon, *Oncorhynchus kisutch*. *Comparative Biochemistry and Physiology* A63: 329–332.

Friesema, E.C.H., R. Docter, E.P. Moerings, F. Verrey, E.P. Krenning, G. Hennemann and T.J. Visser. 2001. Thyroid hormone transport by the heterodimeric human system L amino acid transporter. *Endocrinology* 142: 4339–4348.

Friesema, E.C.H., R. Docter, E.P.C.M. Moerings, B. Stieger, B. Hagenbuch, P.J. Meier, E.P. Krenning, G. Hennemann and T.J. Visser. 1999. Identification of thyroid hormone transporters. *Biochemical and Biophysical Research Communications* 254: 497–501.

Friesema, E.C.H., S. Ganguly, A. Abdalla, J.E. Manning Fox, A.P. Halestrap and T.J. Visser. 2003. Identification of monocarboxylate transporter 8 as a specific thyroid hormone transporter. *Journal of Biological Chemistry* 278: 40128–40135.

Frisén, L. and M. Frisén. 1967. Analysis of the topographic distribution of thyroid activity in a teleost fish: *Carassius carassius* L. *Acta Endocrinologica* 56: 533–546.

Fujiwara, K., H. Adachi, T. Nishio, M. Unno, T. Tokui, M. Okabe, T. Onogawa, T. Suzuki, N. Asano, M. Tanemoto, M. Seki, K. Shiiba, M. Suzuki, Y. Kondo, K. Nunoki, T. Shimosegawa, K. Iinuma, S. Ito, S. Matsuno and T. Abe. 2001. Identification of thyroid hormone transporters in humans: different molecules are involved in a tissue-specific manner. *Endocrinology* 142: 2005–2012.

Funkenstein, B., C.J. Bowman, N.D. Denslow, M. Cardinali and O. Carnevali. 2000. Contrasting effects of estrogen on transthyretin and vitellogenin expression in males of the marine fish, *Sparus aurata*. *Molecular and Cellular Endocrinology* 167: 33–41.

García, G.C., M.C. Jeziorski, C. Valverde-R and A. Orozco. 2004. Effects of iodothyronines on the hepatic outer-ring deiodinating pathway in killifish. *General and Comparative Endocrinology* 135: 201–209.

Geven, E.J.W., F. Verkaar, G. Flik and P.H.M. Klaren. 2006. Experimental hyperthyroidism and central mediators of stress axis and thyroid axis activity in common carp (*Cyprinus carpio* L.). *Journal of Molecular Endocrinology* 37: 443–452.

Gorbman, A. and M. Hyder. 1973. Failure of mammalian TRH to stimulate thyroid function in the lungfish. *General and Comparative Endocrinology* 20: 588–589.

Grau, E.G. and M.H. Stetson. 1977. The effects of prolactin and TSH on thyroid function in *Fundulus heteroclitus*. *General and Comparative Endocrinology* 33: 329–335.

Grau, E.G., A.W. Fast, R.S. Nishioka and H.A. Bern. 1985. The effects of transfer to novel conditions on blood thyroxine in coho salmon. *Aquaculture* 45: 377.

Grau, E.G., L.M.H. Helms, S.K. Shimoda, C.-A. Ford, J. LeGrand and K. Yamauchi. 1986. The thyroid gland of the Hawaiian parrotfish and its use as an *in vitro* model system. *General and Comparative Endocrinology* 61: 100–108.

Gudernatsch, J.F. 1911. The thyroid gland of teleosts. *Journal of Morphology* 21: 709–782.

Hamano, K., K. Inoue and T. Yanagisawa. 1990. Immunohistochemical localization of thyrotropin-releasing hormone in the brain of carp, *Cyprinus carpio*. *General and Comparative Endocrinology* 80: 85–92.

Harada, Y., S. Harada, I. Kinoshita, M. Tanaka and M. Tagawa. 2003. Thyroid gland development in a neotenic goby (ice goby, *Leucopsarion petersii*) and a common goby (ukigori, *Gymnogobius urotaenia*) during early life stages. *Zoological Science* 20: 883–888.

Harder, S., O. Dammann, F. Buck, H. Zwiers, K. Lederis, D. Richter and T.O. Bruhn. 2001. Cloning of two thyrotropin-releasing hormone receptor subtypes from a lower vertebrate (*Catostomus commersoni*): functional expression, gene structure, and evolution. *General and Comparative Endocrinology* 124: 236–245.

Hays, M.T. and R.R. Cavalieri. 1992. Deiodination and deconjugation of the glucuronide conjugates of the thyroid hormones by rat liver and brain microsomes. *Metabolism: Clinical and Experimental* 41: 494–497.

Hays, M.T. and L. Hsu. 1988. Equilibrium dialysis studies of plasma binding of thyroxine, triiodothyronine and their glucuronide and sulfate conjugates in human and cat plasma. *Endocrine Research* 14: 51–58.

Hoover, K.L. 1984. Hyperplastic thyroid lesions in fish. *National Cancer Institute Monographs* 65: 275–289.

Horowitz, B., C.B. Hensley, M. Quintero, K.K. Azuma, D. Putnam and A.A. McDonough. 1990. Differential regulation of Na,K-ATPase $\alpha 1$, $\alpha 2$, and β subunit mRNA and protein levels by thyroid hormone. *Journal of Biological Chemistry* 265: 14308–14314.

Huising, M.O. and G. Flik. 2005. The remarkable conservation of corticotropin-releasing hormone-binding protein (CRH-BP) in the honeybee (*Apis mellifera*) dates the CRH system to a common ancestor of insects and vertebrates. *Endocrinology* 146: 2165–2170.

Huising, M.O., J.R. Metz, C. van Schooten, A.J. Taverne-Thiele, T. Hermsen, B.M.L. Verburg-van Kemenade and G. Flik. 2004. Structural characterisation of a cyprinid (*Cyprinus carpio* L.) CRH, CRH-BP and CRH-R1, and the role of these proteins in the acute stress response. *Journal of Molecular Endocrinology* 32: 627–648.

Inui, Y., M. Tagawa, S. Miwa and T. Hirano. 1989. Effects of bovine TSH on the tissue thyroxine level and metamorphosis in prometamorphic flounder larvae. *General and Comparative Endocrinology* 74: 406–410.

Iwata, M. 1995. Downstream migratory behavior of salmonids and its relationship with cortisol and thyroid hormones: A review. *Aquaculture* 135: 131–139.

Jansen, J., E.C.H. Friesema, C. Milici and T.J. Visser. 2005. Thyroid hormone transporters in health and disease. *Thyroid* 15: 757–768.

Joshi, B.N. and A.G. Sathyanesan. 1976. Presence of functional renal thyroid in the teleost *Cirrhinus mrigala* (Ham). *Indian Journal of Experimental Biology* 14: 700–701.

Kagabu, Y., T. Mishiba, T. Okino and T. Yanagisawa. 1998. Effects of thyrotropin-releasing hormone and its metabolites, cyclo(His-Pro) and TRH-OH, on growth hormone and prolactin synthesis in primary cultured pituitary cells of the common carp, *Cyprinus carpio*. *General and Comparative Endocrinology* 111: 395–403.

Kelly, S.P. and C.M. Wood. 2001. The physiological effects of 3,5',3'-triiodo-L-thyronine alone or combined with cortisol on cultured pavement cell epithelia from freshwater rainbow trout gills. *General and Comparative Endocrinology* 123: 280–294.

Kim, P.S., J.T. Dunn and D.L. Kaiser. 1984. Similar hormone-rich peptides from thyroglobulins of five vertebrate classes. *Endocrinology* 114: 369–374.

Klaren, P.H.M., J.M. Guzmán, S.J. Reutelingsperger, J.M. Mancera and G. Flik. 2007. Low salinity acclimation and thyroid hormone metabolizing enzymes in gilthead seabream (*Sparus auratus*). *General and Comparative Endocrinology*: doi:10.1016/j.ygcen.2007.02.010.

Klaren, P.H.M., R. Haasdijk, J.R. Metz, L.M.C. Nitsch, V.M. Darras, S. Van der Geyten and G. Flik. 2005b. Characterization of an iodothyronine 5'-deiodinase in gilthead seabream (*Sparus auratus*) that is inhibited by dithiothreitol. *Endocrinology* 146: 5621–5630.

Knoeppel, S.J., D.L. Atkins and R.K. Packer. 1982. The role of the thyroid gland in osmotic and ionic regulation in *Fundulus heteroclitus* acclimated to freshwater and seawater. *Comparative Biochemistry and Physiology* A73: 25–29.

Köhrle, J. 1999. Local activation and inactivation of thyroid hormones: the deiodinase family. *Molecular and Cellular Endocrinology* 151: 103–119.

Köhrle, J., G. Brabant and R.D. Hesch. 1987. Metabolism of the thyroid hormones. *Hormone Research* 26: 58–78.

Kreider, M.S., A. Winokur, S. Manaker, A.I. Pack and A.P. Fishman. 1988. Characterization of thyrotropin-releasing hormone in the central nervous system of African lungfish. *General and Comparative Endocrinology* 72: 115–122.

Kühn, E.R., K.L. Geris, S. Van der Geyten, K.A. Mol and V.M. Darras. 1998. Inhibition and activation of the thyroidal axis by the adrenal axis in vertebrates. *Comparative Biochemistry and Physiology* A 120: 169–174.

Kuiper, G.G.J.M., M.H.A. Kester, R.P. Peeters and T.J. Visser. 2005. Biochemical mechanisms of thyroid hormone deiodination. *Thyroid* 15: 787–798.

Kumar, R.S. and J.M. Trant. 2001. Piscine glycoprotein hormone (gonadotropin and thyrotropin) receptors: A review of recent developments. *Comparative Biochemistry and Physiology* B 129: 347–355.

Lamers, A.E., G. Flik and S.E. Wendelaar Bonga. 1994. A specific role for TRH in release of diacetyl α-MSH in tilapia stressed by acid water. *American Journal of Physiology* 267: R1302–R1308.

Larsen, D.A., J.T. Dickey and W.W. Dickhoff. 1997. Quantification of salmon α- and thyrotropin (TSH) β-subunit messenger RNA by an RNase protection assay: Regulation by thyroid hormones. *General and Comparative Endocrinology* 107: 98–108.

Larsen, D.A., P. Swanson, J.T. Dickey, J. Rivier and W.W. Dickhoff. 1998. *In vitro* thyrotropin-releasing activity of corticotropin-releasing hormone-family peptides in coho salmon, *Oncorhynchus kisutch*. *General and Comparative Endocrinology* 109: 276–285.

Leatherland, J.F. 1987. Thyroid response to ovine thyrotropin challenge in cortisol- and dexamethasone-treated rainbow trout, *Salmo gairdneri*. *Comparative Biochemistry and Physiology* A 86: 383–387.

Leatherland, J.F. 1994. Reflections on the thyroidology of fishes: from molecules to humankind. *Guelph Ichthyology Reviews* 2: 1–67.

Liu, Y.W., L.J. Lo and W.K. Chan. 2000. Temporal expression and T3 induction of thyroid hormone receptors α1 and β1 during early embryonic and larval development in zebrafish, *Danio rerio*. *Molecular and Cellular Endocrinology* 159: 187–195.

Lysak, A. 1964. Thyroid centres in carp and in some other teleost fishes revealed by iodine I^{131}. *Acta Biologica Cracoviensa Series Zoologia* 7: 21–46.

Madsen, S.S. 1989. Extrathyroidal effects of thiourea treatment in rainbow trout (*Salmo gairdneri*) rapidly transferred from fresh water to dilute sea-water. *Comparative Biochemistry and Physiology* A 94: 277–282.

Madsen, S.S. 1990. Effect of repetitive cortisol and thyroxine injections on chloride cell number and Na^+/K^+-ATPase activity in gills of freshwater acclimated rainbow trout, *Salmo gairdneri*. *Comparative Biochemistry and Physiology* A 95: 171–175.

Madsen, S.S., M.K. Jensen, J. Nohr and K. Kristiansen. 1995. Expression of Na^+-K^+-ATPase in the brown trout, *Salmo trutta: in vivo* modulation by hormones and seawater. *American Journal of Physiology* 269: R1339–R1345.

Mancera, J.M. and S.D. McCormick. 1999. Influence of cortisol, growth hormone, insulin-like growth factor I and 3,3',5-triiodo-L-thyronine on hypoosmoregulatory ability in the euryhaline teleost *Fundulus heteroclitus*. *Fish Physiology and Biochemistry* 21: 25–33.

Manzon, R.G. and J.H. Youson. 2002. $KClO_4$ inhibits thyroidal activity in the larval lamprey endostyle in vitro. *General and Comparative Endocrinology* 128: 214–223.

Manzon, R.G., J.A. Holmes and J.H. Youson. 2001. Variable effects of goitrogens in inducing precocious metamorphosis in sea lampreys (*Petromyzon marinus*). *Journal of Experimental Zoology* 289: 290–303.

Marchand, O., M. Duffraisse, G. Triqueneaux, R. Safi and V. Laudet. 2004. Molecular cloning and developmental expression patterns of thyroid hormone receptors and T3 target genes in the turbot (*Scophtalmus maximus*) during post-embryonic development. *General and Comparative Endocrinology* 135: 345–357.

Marine, D. and C.H. Lenhart. 1910. Observations and experiments on the so-called thyroid carcinoma of brook trout (*Salvelinus fontinalis*) and its relation to endemic goitre. *Journal of Experimental Medicine* 12: 311–337.

Marine, D. and C.H. Lenhart. 1911. Further observations and experiments on the so-called thyroid carcinoma of brook trout and its relation to endemic goitre. *Journal of Experimental Medicine* 13: 455–475.

Matz, S.P. and G.T. Hofeldt. 1999. Immunohistochemical localization of corticotropin-releasing factor in the brain and corticotropin-releasing factor and thyrotropin-stimulating hormone in the pituitary of chinook salmon (*Oncorhynchus tshawytscha*). *General and Comparative Endocrinology* 114: 151–160.

Matz, S.P. and T.T. Takahashi. 1994. Immunohistochemical localization of thyrotropin-releasing hormone in the brain of chinook salmon (*Oncorhynchus tshawytscha*). *Journal of Comparative Neurology* 345: 214–223.

McCormick, S.D. 2001. Endocrine control of osmoregulation in teleost fish. *Integrative and Comparative Biology* 41: 781–794.

McCormick, S.D. and R.L. Saunders. 1990. Influence of ration level and salinity on circulating thyroid hormones in juvenile Atlantic salmon (*Salmo salar*). *General and Comparative Endocrinology* 78: 224–230.

Meek, J. and R. Nieuwenhuys. 1998. Holosteans and teleosts. In: *The Central Nervous System of Vertebrates*, R. Nieuwenhuys, H.J. ten Donkelaar and C. Nicholson. Springer-Verlag, Berlin, pp. 759–937.

Melamed, P., N. Eliahu, B. Levavi-Sivan, M. Ofir, O. Farchi-Pisanty, F. Rentier-Delrue, J. Smal, Z. Yaron and Z. Naor. 1995. Hypothalamic and thyroidal regulation of growth hormone in tilapia. *General and Comparative Endocrinology* 97: 13–30.

Metz, J.R., M.O. Huising, J. Meek, A.J. Taverne-Thiele, S.E. Wendelaar Bonga and G. Flik. 2004. Localisation, expression and control of adrenocorticotropic hormone in the nucleus preopticus and pituitary gland of common carp (*Cyprinus carpio* L.). *Journal of Endocrinology* 182: 23–31.

Midgley, J.E. 2001. Direct and indirect free thyroxine assay methods: theory and practice. *Clinical Chemistry* 47: 1353–1363.

Mol, K.A. 1996. *A Study on Peripheral Deiodination of Thyroid Hormones in Fish*. Ph.D. Thesis, Katholieke Universiteit Leuven, Belgium.

Mol, K.A., S. Van der Geyten, C. Burel, E.R. Kühn, T. Boujard and V.M. Darras. 1998. Comparative study of iodothyronine outer ring and inner ring deiodinase activities in five teleostean fishes. *Fish Physiology and Biochemistry* 18: 253–266.

Mol, K.A., S. Van der Geyten, E.R. Kühn and V.M. Darras. 1999. Effects of experimental hypo- and hyperthyroidism on iodothyronine deiodinases in Nile tilapia, *Oreochromis niloticus*. *Fish Physiology and Biochemistry* 20: 201–207.

Monaco, F., M. Andreoli, S. Cataudella and J. Roche. 1976. Biosynthesis of thyroglobulin in an adult lamprey, *Lampetra planeri* (Bloch). *Comptes Rendus des Seances de la Societe de Biologie et de ses Filiales* 170: 59–64.

Monaco, F., M. Andreoli, A. La Posta and J. Roche. 1978. Thyroglobulin biosynthesis in a larval (ammocoete) and adult freshwater lamprey (*Lampetra planeri* Bl.). *Comparative Biochemistry and Physiology* B 60: 87–91.

Monteverdi, G.H. and R.T. Di Giulio. 2000. Vitellogenin association and oocytic accumulation of thyroxine and 3,5,3'-triiodothyronine in gravid *Fundulus heteroclitus*. *General and Comparative Endocrinology* 120: 198–211.

Morin, P.P., S. Winberg, G.E. Nilsson, T.J. Hara and J.G. Eales. 1997. Effects of L-thyroxine on brain monoamines during parr-smolt transformation of Atlantic salmon (*Salmo salar* L.). *Neuroscience Letters* 224: 216–218.

Ohide, A., H. Ando, T. Yanagisawa and A. Urano. 1996. Hydropathy profiles of predicted thyrotropin-releasing hormone precursors are highly conserved despite low similarity of primary structures. *Journal of Neuroendocrinology* 8: 695–701.

Olivereau, M. 1960. Hyperplasie thyroidienne et présence de follicules thyroidiens intrarénaux, chez un exemplaire de *Typhlogarra widdowsoni* Trewavas, poisson aveugle et cavernicole de l'Irak. *Annales Royale Sociètè Zoologicae Belgique* 90: 117–125.

Olivereau, M., J.M. Olivereau and C. Aimar. 1977. Modifications hypophysaires et thyroidiennes chez le pleurodèle en milten salin. *General and Comparative Endocrinology* 32: 195–204.

Omeljaniuk, R.J. and J.G. Eales. 1986. The effect of 3,5,3'-triiodo-L-thyronine on gill Na^+/K^+-ATPase of rainbow trout *Salmo gairdneri*, in fresh water. *Comparative Biochemistry and Physiology* B 84: 427–429.

Orozco, A. and C. Valverde-R. 2005. Thyroid hormone deiodination in fish. *Thyroid* 15: 799–813.

Orozco, A., P. Villalobos and C. Valverde-R. 2002. Environmental salinity selectively modifies the outer-ring deiodinating activity of liver, kidney and gill in the rainbow trout. *Comparative Biochemistry and Physiology* A 131: 387–395.

Orozco, A., P. Villalobos, M.C. Jeziorski and C. Valverde-R. 2003. The liver of *Fundulus heteroclitus* expresses deiodinase type 1 mRNA. *General and Comparative Endocrinology* 130: 84–91.

Parry, J.E., C. Zhang and J.G. Eales. 1994. Urinary excretion of thyroid hormones in rainbow trout, *Oncorhynchus mykiss*. *General and Comparative Endocrinology* 95: 310–319.

Pepels, P.P.L.M., J. Meek, S.E. Wendelaar Bonga and P.H.M. Balm. 2002a. Distribution and quantification of corticotropin-releasing hormone (CRH) in the brain of the teleost fish *Oreochromis mossambicus* (tilapia). *Journal of Comparative Neurology* 453: 247–268.

Pepels, P.P.L.M., G. Pesman, H. Korsten, S.E. Wendelaar Bonga and P.H.M. Balm. 2002b. Corticotropin-releasing hormone (CRH) in the teleost fish *Oreochromis mossambicus* (tilapia): *in vitro* release and brain distribution determined by a novel radioimmunoassay. *Peptides* 23: 1053–1062.

Persson, P., K. Sundell, B.T. Björnsson and H. Lundqvist. 1998. Calcium metabolism and osmoregulation during sexual maturation of river running Atlantic salmon. *Journal of Fish Biology* 52: 334–349.

Peter, R.E. 1970. Comparison of the activity of the pronephric thyroid and the pharyngeal thyroid of the goldfish, *Carassius auratus*. *General and Comparative Endocrinology* 15: 88–94.

Peter, M.C.S., R.A.C. Lock and S.E. Wendelaar Bonga. 2000. Evidence for an osmoregulatory role of thyroid hormones in the freshwater Mozambique tilapia *Oreochromis mossambicus*. *General and Comparative Endocrinology* 120: 157–167.

Power, D.M., L. Llewellyn, M. Faustino, M.A. Nowell, B.T. Björnsson, I.E. Einarsdottir, A.V.M. Canario and G.E. Sweeney. 2001. Thyroid hormones in growth and development of fish. *Comparative Biochemistry and Physiology* C 130: 447–459.

Pradet-Balade, B., C. Burel, S. Dufour, T. Boujard, S.J. Kaushik, B. Querat and G. Boeuf. 1999. Thyroid hormones down-regulate thyrotropin beta mRNA level *in vivo* in the turbot (*Psetta maxima*). *Fish Physiology and Biochemistry* 20: 193–199.

Pradet-Balade, B., M. Schmitz, C. Salmon, S. Dufour and B. Querat. 1997. Down-regulation of TSH subunit mRNA levels by thyroid hormones in the European eel. *General and Comparative Endocrinology* 108: 191–198.

Qureshi, T.A. 1975. Heterotopic thyroid follicles in the accessory mesonephric lobes of *Heteropneustes fossilis* (Bloch). *Acta Anatomica* 93: 506–511.

Qureshi, T.A. and R. Sultan. 1976. Thyroid follicles in the head-kidney of teleosts. *Anatomische Anzeiger* 139: 332–336.

Qureshi, T.A., D.K. Belsare and R. Sultan. 1978. Head-kidney thyroid in some Indian teleosts. *Zeitschrift für Mikroskopische und Anatomische Forschung* 92: 352–358.

Redding, J.M., C.B. Schreck, E.K. Birks and R.D. Ewing. 1984. Cortisol and its effects on plasma thyroid hormone and electrolyte concentrations in fresh water and during seawater acclimation in yearling coho salmon, *Oncorhynchus kisutch*. *General and Comparative Endocrinology* 56: 146–155.

Redding, J.M., A. deLuze, J. Leloup-Hatey and J. Leloup. 1986. Suppression of plasma thyroid hormone concentrations by cortisol in the European eel *Anguilla anguilla*. *Comparative Biochemistry and Physiology* A 83: 409–413.

Redding, J.M., R. Patiño and C.B. Schreck. 1991. Cortisol effects on plasma electrolytes and thyroid hormones during smoltification in coho salmon *Oncorhynchus kisutch*. *General and Comparative Endocrinology* 81: 373–382.

Richardson, S.J., A.J. Bradley, W. Duan, R.E.H. Wettenhall, P.J. Harms, J.J. Babon, B.R. Southwell, S. Nicol, S.C. Donnellan and G. Schreiber. 1994. Evolution of marsupial and other vertebrate thyroxine-binding plasma proteins. *American Journal of Physiology - Regulatory, Integrative and Comparative Physiology* 266: R1359–R1370.

Richardson, S.J., J.A. Monk, C.A. Shepherdley, L.O.E. Ebbesson, F. Sin, D.M. Power, P.B. Frappell, J. Köhrle and M.B. Renfree. 2005. Developmentally regulated thyroid hormone distributor proteins in marsupials, a reptile, and fish. *American Journal of Physiology - Regulatory, Integrative and Comparative Physiology* 288: R1264–R1272.

Rodriguez, A.-M., B. Perron, L. Lacroix, B. Caillou, G. Leblanc, M. Schlumberger, J.-M. Bidart and T. Pourcher. 2002. Identification and characterization of a putative human iodide transporter located at the apical membrane of thyrocytes. *Journal of Clinical Endocrinology and Metabolism* 87: 3500–3503.

Royaux, I.E., K. Suzuki, A. Mori, R. Katoh, L.A. Everett, L.D. Kohn and E.D. Green. 2000. Pendrin, the protein encoded by the Pendred syndrome gene (*PDS*), is an apical porter of iodide in the thyroid and is regulated by thyroglobulin in FRTL-5 cells. *Endocrinology* 141: 839–845.

Sanders, J.P., S. Van der Geyten, E. Kaptein, V.M. Darras, E.R. Kühn, J.L. Leonard and T.J. Visser. 1997. Characterization of a propylthiouracil-insensitive type I iodothyronine deiodinase. *Endocrinology* 138: 5153–5160.

Sangiao-Alvarellos, S., R. Laiz-Carrión, J.M. Guzmán, M.P. Martín del Río, J.M. Miguez, J.M. Mancera and J.L. Soengas. 2003. Acclimation of *S. aurata* to various salinities alters energy metabolism of osmoregulatory and nonosmoregulatory organs. *American Journal of Physiology - Regulatory, Integrative and Comparative Physiology* 285: R897–R907.

Santos, C.R. and D.M. Power. 1999. Identification of transthyretin in fish (*Sparus aurata*): cDNA cloning and characterisation. *Endocrinology* 140: 2430–2433.

Santos, C.R., L. Anjos and D.M. Power. 2002. Transthyretin in fish: state of the art. *Clinical Chemistry and Laboratory Medicine* 40: 1244–1249.

Saunders, R.L., S.D. McCormick, E.B. Henderson, J.G. Eales and C.E. Johnston. 1985. The effect of orally administered 3,5,3'-triiodo-L-thyronine on growth and salinity tolerance of Atlantic salmon (*Salmo salar* L.). *Aquaculture* 45: 143–156.

Schreiber, G. and S.J. Richardson. 1997. The evolution of gene expression, structure and function of transthyretin. *Comparative Biochemistry and Physiology* B116: 137–160.

Schreiber, A.M. and J.L. Specker. 1999. Metamorphosis in the summer flounder, *Paralichthys dentatus*: thyroidal status influences salinity tolerance. *Journal of Experimental Zoology* 284: 414–424.

Schussler, G.C. 2000. The thyroxine-binding proteins. *Thyroid* 10: 141–149.

Schwartzentruber, R.S. and R.J. Omeljaniuk. 1995. Thyrotropin-releasing hormone receptors in the pituitary of rainbow trout (*Oncorhynchus mykiss*). *General and Comparative Endocrinology* 97: 209–219.

Scott, D.A., R. Wang, T.M. Kreman, V.C. Sheffield and L.P. Karniski. 1999. The Pendred syndrome gene encodes a chloride-iodide transport protein. *Nature Genetics* 21: 440–443.

Seidelin, M., S.S. Madsen, H. Blenstrup and C.K. Tipsmark. 2000. Time-course changes in the expression of Na^+, K^+-ATPase in gills and pyloric caeca of brown trout (*Salmo trutta*) during acclimation to seawater. *Physiological and Biochemical Zoology* 73: 446–453.

Shameena, B., S. Varghese, S. Leena and O.V. Oommen. 2000. 3,5,3'-triiodothyronine (T_3) and 3',5'-diiodothyrone (T_2) have short-term effects on lipid metabolism in a teleost *Anabas testudineus* (Bloch): evidence from enzyme activities. *Endocrine Research* 26: 431–444.

Sharma, R. and S. Kumar. 1982. Distribution of thyroid follicles and nerves in the kidney of a teleost, *Clarias batrachus* (Linn.). *Zeitschrift für Mikroskopische und Anatomische Forschung* 96: 1069–1077.

Shrimpton, J.M. and S.D. McCormick. 1998. Regulation of gill cytosolic corticosteroid receptors in juvenile Atlantic salmon. Interaction effects of growth hormone with prolactin and triiodothyronine. *General and Comparative Endocrinology* 112: 262–274.

Sinclair, D.A. and J.G. Eales. 1972. Iodothyronine-glucuronide conjugates in the bile of brook trout, *Salvelinus fontinalis* (Mitchill) and other freshwater teleosts. *General and Comparative Endocrinology* 19: 552–598.

Sohn, Y.C., Y. Yoshiura, H. Suetake, M. Kobayashi and K. Aida. 1999. Isolation and characterization of the goldfish thyrotropin β subunit gene including the 5'-flanking region. *General and Comparative Endocrinology* 115: 463–473.

Solbakken, J.S., B. Norberg, K. Watanabe and K. Pittman. 1999. Thyroxine as a mediator of metamorphosis of Atlantic halibut, *Hippoglossus hippoglossus*. *Environmental Biology of Fishes* 56: 53–65.

Sower, S.A. and C.B. Schreck. 1982. Steroid and thyroid hormones during sexual maturation of coho salmon (*Oncorhynchus kisutch*) in seawater or fresh water. *General and Comparative Endocrinology* 47: 42–53.

Specker, J.L. and N.H. Richman, 3rd. 1984. Environmental salinity and the thyroidal response to thyrotropin in juvenile coho salmon (*Oncorhynchus kisutch*). *Journal of Experimental Zoology* 230: 329–333.

Spitzweg, C., A.E. Heufelder and J.C. Morris. 2000. Thyroid iodine transport. *Thyroid* 10: 321–330.

Srivastava, S.S. and A.G. Sathyanesan. 1967. Presence of functional renal thyroid follicles in the Indian mud eel *Amphipnous cuchia* (Ham.). *Naturwissenschaften* 54: 146.

Srivastava, S.S. and A.G. Sathyanesan. 1971a. Studies on the histophysiology of the pharyngeal and heterotopic renal thyroid in the freshwater teleost *Puntius sophore* (Ham.). *Zeitschrift für Mikroskopische und Anatomische Forschung* 83: 145–165.

Srivastava, S.S. and A.G. Sathyanesan. 1971b. Histophysiological studies on the pharyngeal and ectopic renal thyroid of the Indian mud-eel *Amphipnous cuchia* (Ham.). *Endokrinologie* 57: 260–269.

Stickney, R.R. and H.W. Liu. 1999. Maintenance of broodstock, spawning, and early larval rearing of Pacific halibut, *Hippoglossus stenolepis*. *Aquaculture* 176: 75–86.

Sukumar, P., A.D. Munro, E.Y.M. Mok, S. Subburaju and T.J. Lam. 1997. Hypothalamic regulation of the pituitary-thyroid axis in the tilapia *Oreochromis mossambicus*. *General and Comparative Endocrinology* 106: 73–84.

Suzuki, S., A. Gorbman, M. Rolland, M.F. Montfort and S. Lissitzky. 1975. Thyroglobulins of cyclostomes and an elasmobranch. *General and Comparative Endocrinology* 26: 56–69.

Swanson, P., E.G. Grau, L.M. Helms and W.W. Dickhoff. 1988. Thyrotropic activity of salmon pituitary glycoprotein hormones in the Hawaiian parrotfish thyroid *in vitro*. *Journal of Experimental Zoology* 245: 194–199.

Toda, S., K. Watanabe, F. Yokoi, S. Matsumura, K. Suzuki, A. Ootani, S. Aoki, N. Koike and H. Sugihara. 2002. A new organotypic culture of thyroid tissue maintains three-dimensional follicles with C cells for a long term. *Biochemical and Biophysical Research Communications* 294: 906–911.

Toussaint, M.W., M.J. Wolfe, D.T. Burton, F.J. Hoffmann, T.R. Shedd and H.S. Gardner, Jr. 1999. Histopathology of Japanese medaka (*Oryzias latipes*) chronically exposed to a complex environmental mixture. *Toxicologic Pathology* 27: 652–663.

Van der Geyten, S., A. Toguyeni, J.F. Baroiller, B. Fauconneau, A. Fostier, J.P. Sanders, T.J. Visser, E.R. Kühn and V.M. Darras. 2001. Hypothyroidism induces type I iodothyronine deiodinase expression in tilapia liver. *General and Comparative Endocrinology* 124: 333–342.

Van der Geyten, S., N. Byamungu, G.E. Reyns, E.R. Kühn and V.M. Darras. 2005. Iodothyronine deiodinases and the control of plasma and tissue thyroid hormone levels in hyperthyroid tilapia (*Oreochromis niloticus*). *Journal of Endocrinology* 184: 467–479.

van der Heide, S.M., T.J. Visser, M.E. Everts and P.H.M. Klaren. 2002. Metabolism of thyroid hormones in cultured cardiac fibroblasts of neonatal rats. *Journal of Endocrinology* 172: 111–119.

van der Heide, S.M., B.J.L.J. Joosten, M.E. Everts and P.H.M. Klaren. 2004. Activities of UDP-glucuronyltransferase, β-glucuronidase and deiodinase types I and II in hyper- and hypothyroid rats. *Journal of Endocrinology* 181: 393–400.

van der Heide, S.M., B.J.L.J. Joosten, B.S. Dragt, M.E. Everts and P.H.M. Klaren. 2007. A physiological role for glucuronidated thyroid hormones: preferential uptake by H9c2(2-1) myotubes. *Molecular and Cellular Endocrinology* 264: 109–117.

Van Sande, J., C. Massart, R. Beauwens, A. Schoutens, S. Costagliola, J.E. Dumont and J. Wolff. 2003. Anion selectivity by the sodium iodide symporter. *Endocrinology* 144: 247–252.

Varghese, S., B. Shameena and O.V. Oommen. 2001. Thyroid hormones regulate lipid peroxidation and antioxidant enzyme activities in *Anabas testudineus* (Bloch). *Comparative Biochemistry and Physiology B* 128: 165–171.

Virtanen, E. and A. Soivio. 1985. The patterns of T3, T4, cortisol and Na^+-K^+-ATPase during smoltification of hatchery-reared *Salmo salar* and comparison with wild smolts. *Aquaculture* 45: 97–109.

Visser, T.J. 1994. Role of sulfation in thyroid hormone metabolism. *Chemico-Biological Interactions* 92: 293–303.

Visser, T.J. 1996. Pathways of thyroid hormone metabolism. *Acta Medica Austriaca* 23: 10–16.

Woodhead, A.D. and P.M. Scully. 1977. A comparative study of the pretumorous thyroid gland of the gynogenetic teleost, *Poecilia formosa*, and that of other poeciliid fishes. *Cancer Research* 37: 3751–3755.

Woodhead, A.D., R.B. Setlow and V. Pond. 1984. The Amazon molly, *Poecilia formosa*, as a test animal in carcinogenicity studies: chronic exposures to physical agents. *National Cancer Institute Monographs* 65: 45–52.

Wright, G.M., M.F. Filosa and J.H. Youson. 1978a. Light and electron microscopic immunocytochemical localization of thyroglobulin in the thyroid gland of the anadromous sea lamprey, *Petromyzon marinus* L., during its upstream migration. *Cell and Tissue Research* 187: 473–478.

Wright, G.M., M.F. Filosa and J.H. Youson. 1978b. Immunocytochemical localization of thyroglobulin in the endostyle of the anadromous sea lamprey, *Petromyzon marinus* L. *American Journal of Anatomy* 152: 263–268.

Yalcin, Y., D. Carman, Y. Shao, F. Ismail-Beigi, I. Klein and K. Ojamaa. 1999. Regulation of Na/K-ATPase gene expression by thyroid hormone and hyperkalemia in the heart. *Thyroid* 9: 53–59.

Yamano, K. and S. Miwa. 1998. Differential gene expression of thyroid hormone receptor α and β in fish development. *General and Comparative Endocrinology* 109: 75–85.

Yen, P.M. 2001. Physiological and molecular basis of thyroid hormone action. *Physiological Reviews* 81: 1097–1142.

Yoshida, A., S. Taniguchi, I. Hisatome, I.E. Royaux, E.D. Green, L.D. Kohn and K. Suzuki. 2002. Pendrin is an iodide-specific apical porter responsible for iodide efflux from thyroid cells. *Journal of Clinical Endocrinology and Metabolism* 87: 3356–3361.

Yoshiura, Y., Y.C. Sohn, A. Munakata, M. Kobayashi and K. Aida. 1999. Molecular cloning of the cDNA encoding the b subunit of thyrotropin and regulation of its gene expression by thyroid hormones in the goldfish, *Carassius auratus*. *Fish Physiology and Biochemistry* 21: 201–210.

Youngson, A.F. and J.H. Webb. 1993. Thyroid hormone levels in Atlantic salmon (*Salmo salar*) during the return migration from the ocean to spawn. *Journal of Fish Biology* 42: 293–300.

CHAPTER

3

Diet and Osmoregulation

Francesca W. Ferreira[1] and Bernardo Baldisserotto[2,*]

INTRODUCTION

Fish adapted to freshwater and waters with low salinity present a diffusive ion loss to the environment via gills and skin, as well as by feces and urine. This ion loss must be compensated by an active influx from the water by the gills (Wood, 2001), from the diet by the intestine (Dabrowski *et al.*, 1986; Buddington and Diamond, 1987; Bogé *et al.*, 1988; Baldisserotto *et al.*, 1993; Kerstetter and White, 1994; Baldisserotto and Mimura, 1995; Bijvelds *et al.*, 1998), and in some species as the swamp eel, *Synbranchus marmoratus*, it might be also complemented by the skin (Stiffler *et al.*, 1986). Another complicating factor in freshwater fishes is that most studies of *in vitro* intestinal absorption/transporters were done with fasting fishes. Feeding drastically changes the ionic situation of rainbow trout, *Oncorhynchus mykiss*, intestine (Dabrowski *et al.*, 1986), and the addition

Authors' addresses: [1]Departamento de Biologia e Química, Universidade Regional do Noroeste do Rio Grande do Sul, 98700.000 – Ijuí, RS, Brazil.

[2]Departamento de Fisiologia e Farmacologia, Universidade Federal de Santa Maria, 97105.900 – Santa Maria, RS, Brazil.

**Corresponding author*: E-mail: bernardo@smail.ufsm.br

of several amino acids or glucose to the mucosal side of the medium bathing to intestines *in vitro* increases the flow of Na^+ toward the serosal side in various teleost species (Ferraris and Ahearn, 1984; Vilella *et al.*, 1988, 1989; Bogé and Péres, 1990). In addition, intestinal Ca^{2+} absorption is affected by the diet (Baldisserotto *et al.*, 2006).

The drinking rate of freshwater teleosts is low (370-1400 $\mu l \cdot h^{-1} \cdot kg^{-1}$) (see Flik *et al.*, 1985), but the intestine (or the pyloric ceca, when present) can absorb Na^+, Cl^-, Ca^{2+}, and Mg^{2+} (and probably other ions) provided by feeding (Dabrowski *et al.*, 1986; Buddington and Diamond, 1987; Bogé *et al.*, 1988; Baldisserotto *et al.*, 1993; Kerstetter and White, 1994; Baldisserotto and Mimura, 1995; Bijvelds *et al.*, 1998). Therefore, diet can be an important ion source for osmoregulatory needs of fish living in hyposaline environments. Dietary salt supplementation can also decrease energy spent on osmoregulation and consequently more will be available for growth (Gatlin *et al.*, 1992; D'Cruz and Wood, 1998).

On the other hand, fish that live in waters with high salinity have problems of excessive ion influx, which must be eliminated by the gills and urine. The digestive tract absorbs ions, but the aim of this process is to provide absorption of the ingested water and not ions from the diet (Kirsch *et al.*, 1984). Therefore, present chapter will deal mainly with the contribution of the diet to osmoregulation of fish adapted to low salinities. Moreover, emphasis will be on the direct effects of dietary composition on osmoregulation and not indirect effects (morphological changes) due to lack of specific nutrients as vitamins, for example.

DIETARY Na^+ AND Cl^-

Rainbow trout can survive for long fasting periods without a significant decrease on blood Na^+ concentration (Heming and Paleczny, 1987). Consequently, Na^+ branchial influx is adequate for maintaining ionic balance in fish even when food consumption (and, consequently, dietary Na^+ intake) is low, as in winter (Smith *et al.*, 1989). However, intestinal influx of dietary Na^+ in rainbow trout collected from the wild in summer is similar to branchial influx (Smith *et al.*, 1989). Apparently, most dietary Na^+ ingested is absorbed by the gut (Salman, 1987; Smith *et al.*, 1995), because feces have a low amount of salts even in rainbow trout fed with high salt content in the diet (Salman and Eddy, 1988). Rainbow trout fed high NaCl diets (1.8 and 3% Na^+) showed a decrease of 40.8 and 44.0% on waterborne Na^+ whole body uptake rates relative to controls (diet with

0.6% Na^+). Moreover, Na^+ efflux was 12% and 38% higher in fish fed 1.8% and 3% sodium-enriched diets, respectively. The increase of plasma Na^+ concentration due to high dietary Na^+ (38.1% in fish fed with 3% sodium-enriched diet) (Fig. 3.1) probably causes a downregulation of a branchial uptake route through an apical sodium channel, which reduces waterborne Na^+ uptake. Fish fed high-sodium diets (3%) also drank 58% more water than controls (Pyle *et al.*, 2003). This increase on drinking rate is needed to counterbalance the increase on plasma Na^+ concentration (Salman and Eddy, 1988).

An increase of dietary NaCl from 2 to 12% in rainbow trout promoted a two-fold increase of the number of chloride cells in the gills and gill Na^+/K^+- ATPase activity (Salman and Eddy, 1987). Na^+/K^+- ATPase activity in the proximal intestine (pyloric ceca and anterior intestine) was also stimulated by Na^+-supplemented diets in rainbow trout (Pyle *et al.*, 2003), but not in bluegill, *Lepomis macrochirus* (Musselman *et al.*, 1995). The proportion of chloride cells related to total branchial cells also increased from 8% in rainbow trout fed with 1.3% dietary salt, to 10.5% in fish fed with 12% dietary NaCl (Salman and Eddy, 1987). High dietary NaCl (11.6%) did not alter the typical freshwater renal mechanism in rainbow

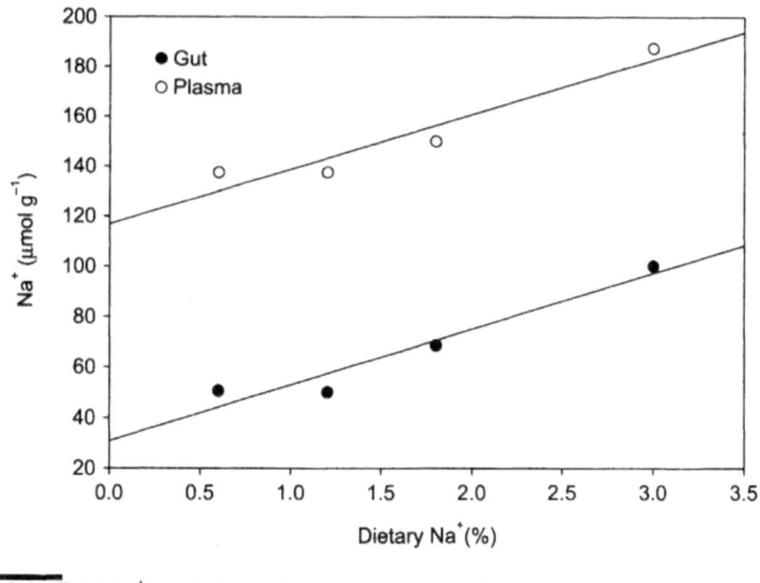

Fig. 3.1 Total Na^+ levels in gut tissue and plasma of rainbow trout fed for 7 days with different Na^+ levels in the diet (and also exposed to 20 µg L^{-1} waterborne Cu for 6 h). Adapted from Pyle *et al.* (2003).

trout, where the majority of filtered ions are reabsorbed to produce a relatively large volume of dilute urine. However, there was an increase on glomerular filtration rate and urinary flow rate, which approximately doubled the Na^+ and Cl^- urinary excretion rate. Even with this increase of salt urinary excretion rate, salt renal excretion represented only 10% of the total body loss and is similar to fecal salt loss (Salman and Eddy, 1988). Increase of drinking rate (Pyle *et al.*, 2003), gill water permeability and intestinal water absorption would provide the water needed for this increase of urinary flow rate (Salman and Eddy, 1988).

The principal mechanism for Na^+ homeostasis is the variation on branchial Na^+ influx and efflux rates, and the main way of excreting large salt loads is to increase Na^+ branchial efflux (Smith *et al.*, 1995; Pyle *et al.*, 2003). In rainbow trout fed on freshwater shrimp *Gammarus pulex* 60% of the ingested Na^+ was absorbed within 5 h (Smith *et al.*, 1989). Similarly, rainbow trout receiving a commercial diet supplemented with 12% NaCl constituted a mean Na^+ load of 36.2 mmol·kg fish^{-1}, of which around 85% was absorbed within 7 h (maximum time of the experiment). Absorption from the gut increased the Na^+ plasma levels when compared with levels in unfed fish, and within 1 h, branchial Na^+ efflux increased and remained high for 7 h, indicating that excretion of excess Na^+ was incomplete at the end of this period (Smith *et al.*, 1995). Blood Cl^- levels were unchanged in brook trout (*Salvelinus fontinalis*) fed a NaCl load of 15.3 mmol·kg fish^{-1}, but ingestion of 46 mmol.kg fish^{-1} increased blood Cl^- levels up to 40% above control values 7 h after feeding. Blood Cl^- levels returned to control values after 24 h, but ingestion of higher NaCl load (77 mmol·kg fish^{-1}) led to a prolonged increase in blood Cl^- levels and in many cases, death (Phillips, 1944). Plasma Cl^- levels in bluegill maintained in freshwater and fed diet supplemented with 2 or 4% NaCl were also higher than in fish kept in freshwater and fed a diet without NaCl supplementation (Musselman *et al.*, 1995).

In acidic water, excess H^+ can inhibit the Na^+/H^+ exchanger (Potts, 1994) and create a gradient too steep for further extrusion of protons (Lin and Randall, 1991), reducing Na^+ uptake by the gills. Moreover, high H^+ concentrations disrupt the tight junctions of gill epithelia, increasing ion loss by a paracellular route, leading to whole body ion loss, as observed in rainbow trout (McDonald and Wood, 1981) and silver catfish, *Rhamdia quelen* (Zaions and Baldisserotto, 2000). Under these conditions, dietary salts may become very important in maintaining body ion levels during acid stress (D'Cruz and Wood, 1998). Starved fish (or fed with a very limited diet) showed ionoregulatory changes during exposure to acidic environment (D'Cruz *et al.*, 1998), but when they were fed with adequate amount of salts the effect of low pH was reduced or did not occur (Dockrey *et al.*, 1996; D'Cruz *et al.*, 1998). Therefore, dietary salt can replace branchial ion loss (D'Cruz and Wood, 1998).

Some studies proposed that acidic pH may impair growth in rainbow trout due to a decrease on food consumption (for review see D'Cruz and Wood, 1998), as was observed in silver catfish (Copatti *et al.*, 2005). However, Dockray *et al.* (1996), Reid *et al.* (1996, 1997) and D'Cruz *et al.* (1998) verified that chronic exposure of rainbow trout to low pH seemed to stimulate appetite. Rainbow trout exposed to acidic pH for 28 days and starved showed significantly lower plasma Na^+ (but not whole body Na^+, Cl^- and K^+) than before the acid challenge. Those fed with a low NaCl diet (0.1–0.18%), independently of energy content, also presented a decrease on plasma Na^+ and whole body Na^+ and Cl^- (the last, only the low energy diet), but fish fed with 0.6% NaCl did not show any ionic imbalance. Therefore, is the salt content of the food rather than the energy content that is critical in protecting against the effect of acidic pH (D'Cruz and Wood, 1998)?

An adequate dietary Na^+ level could have lower metabolic cost when associated with active branchial ion transport, and the saved energy could be used for growth (Smith *et al.*, 1989). Atlantic salmon (*Salmo salar*) has a whole body Na^+ content of 25 mmol/kg (Talbot *et al.*, 1986), so Smith *et al.* (1989) estimated that a 10 g fish (whole body Na^+ content of 0.25 mmol) doubling in weight over a year would need 0.25 mmol Na^+. According to the same authors, this amount is easily obtained by feeding and branchial uptake because total Na^+ influx in rainbow trout in June is over 1000 times greater. It must also be considered that branchial Na^+ fluxes may be rapidly adjusted to diet changes (Smith *et al.*, 1995), and therefore, a high Na^+ diet might not improve growth in rainbow trout in optimum water conditions. However, when fish are exposed to acidic pH branchial ion influx is lower and the efflux is higher than in neutral waters, and dietary salt supplementation may help to maintain ionic balance (D'Cruz and Wood, 1998).

Salinity has variable effects on growth of euryhaline species, and growth is not necessarily maximal at isosmotic conditions (Brett, 1979; Musselman *et al.*, 1995; Likongwe *et al.*, 1998). Red drum (*Sciaenops ocellatus*) is commonly found in waters with 20–40‰, and juveniles growth is improved in freshwater with a diet supplemented with 2% NaCl or 2% NaCl + 2% KCl. However, NaCl dietary supplementation did not affect any change in growth of juveniles exposed to brackish (6‰) and seawater (35‰); neither was blood osmolality of fish maintained in fresh or brackish water and transferred to seawater. These results suggest that salinity of 6‰ may be close to the threshold at which dietary salt supplementation promotes growth in red drum (Gatlin III *et al.*, 1992). Chinook salmon (*Oncorhynchus tshawytscha*) fed diets supplemented with 7% NaCl or 5% NaCl + 2% KCl showed higher tolerance to seawater transfer (Zaugg *et al.*, 1983). Diets supplemented with 10% NaCl also improved survival to seawater transfer of two tilapia species (*Oreochromis mossambicus* and *O. spilurus*) and the hybrid *O. aureus* × *O. niloticus* (Al-Amoundi, 1987), as well as brown trout (*Salmo trutta*) (Arzel *et al.*, 1993). Nile tilapia (*O. niloticus*) maintained in freshwater and fed diet supplemented with 8% NaCl for 30 days showed a higher growth rate than those fed diet without NaCl supplementation, while dietary NaCl did not change significantly growth rate in fish kept in brackish water (10 and 20‰) (Fontaínhas-Fernandes *et al.*, 2002). However, Nile tilapia fed on a diet of 8% NaCl presented significantly lower plasma Cl^- and osmolarity after transference to brackish water than those fed diet without NaCl supplementation, indicating a reduction of osmotic imbalance (Fontaínhas-Fernandes *et al.*, 2001).

DIETARY Ca^{2+}

Fish take up calcium for growth and homeostasis predominantly via the gills, directly from the water. This branchial Ca^{2+} uptake is an active and a more or less continuous process and largely independent of waterborne Ca^{2+} (Flik, 1996). Fish also take up Ca^{2+} from food or water drunk by the intestine (Flik *et al.*, 1993a, b; Flik and Verbost, 1995), but under normal, no stressed conditions, the drinking rate of freshwater fish is very low, and the contribution of the intestine to Ca^{2+} uptake is restricted to dietary calcium (Flik, 1995). The contribution of gills and intestine to Ca^{2+} uptake is variable and depends on waterborne and dietary Ca^{2+} concentration. In fish exposed to low waterborne Ca^{2+} the relative contribution of the food increases, whereas feeding low-Ca^{2+} diets stimulates the branchial uptake. There is also evidence that fish rely on Ca^{2+} intestinal uptake when extensive amounts of Ca^{2+} are required for gonadal maturation (Flik *et al.*, 1995). A total lack of dietary Ca^{2+} can be completely compensated by branchial uptake, but very low waterborne Ca^{2+} induces hypocalcemia and impairs growth (Shoenmakers *et al.*, 1993; Flik *et al.*, 1996).

Low waterborne (0.125 mmol) and dietary Ca^{2+} reduced growth rate in brook trout (*Salvelinus fontinalis*) (Rodgers, 1984), demonstrating that a minimum Ca^{2+} uptake by gills and/or intestine is needed for normal fish growth. Channel catfish (*Ictalurus punctatus*) reared in low waterborne Ca^{2+} (< 0.25 mmol) required 4.5 mg $Ca^{2+}g^{-1}$ food for normal growth and tissue mineralization (Robinson *et al.*, 1986), while blue tilapia (*Oreochromis aureus*) reared in similar conditions fed with 7.5 mg $Ca^{2+}g^{-1}$ food showed higher bone and scale Ca^{2+} concentration (but only higher scale Mg^{2+} concentration and bone phosphorus concentration) than those fed with diet deprived of Ca^{2+} (O'Connell and Gatlin, 1994) (Fig. 3.2). Striped bass (*Morone saxatilis*) juveniles maintained at 0.68 mmol Ca^{2+} presented high whole body waterborne Ca^{2+} uptake compared to other teleosts, but still much lower than the rate of Ca^{2+} assimilation necessary for optimum growth of this species (Grizzle *et al.*, 1993). However, dietary Ca^{2+} was dispensable for rainbow trout when waterborne Ca^{2+} was above 0.6 mmol (Ogino and Takeda, 1978), and Ca^{2+}-supplemented diets from 3.6 to 11 mg $Ca^{2+}g^{-1}$ food did not change growth of this species when reared at water with 0.75 mmol Ca^{2+} (Barnett *et al.*, 1979).

Rainbow trout maintained at waterborne Ca^{2+} 1 mmol fed on high dietary Ca^{2+} levels (60 mg $Ca^{2+}g^{-1}$ food) showed 52–64% lower whole body waterborne Ca^{2+} uptake compared to fish fed with lower dietary Ca^{2+} levels (20 mg $Ca^{2+}g^{-1}$ food) (Baldisserotto *et al.*, 2004 a, b) (Fig. 3.3). A diet supplemented with $CaCl_2$ to yield 30 mg $Ca^{2+}g^{-1}$ food did not change rainbow trout growth, but a higher dietary level of $CaCl_2$ (60 mg $Ca^{2+}g^{-1}$ food) led to 21.6% mortality and decreased weight gain. The deaths observed in the treatment with a high amount of $CaCl_2$ probably were due to metabolic acidosis and/or to a sharp increase on Ca^{2+} plasma levels seen after the first feeding with this diet (Baldisserotto *et al.*, 2004a). However, in rainbow trout fed the same dietary Ca^{2+} levels but supplemented with $CaCO_3$, mortality was not observed (Baldisserotto *et al.*, 2004b). Therefore, supplementation with $CaCO_3$ seems to be safer than with $CaCl_2$.

Fishes adapted to seawater drink water with a high Ca^{2+} content (approximately 10 mmol/L), and do not decrease branchial Ca^{2+} uptake,

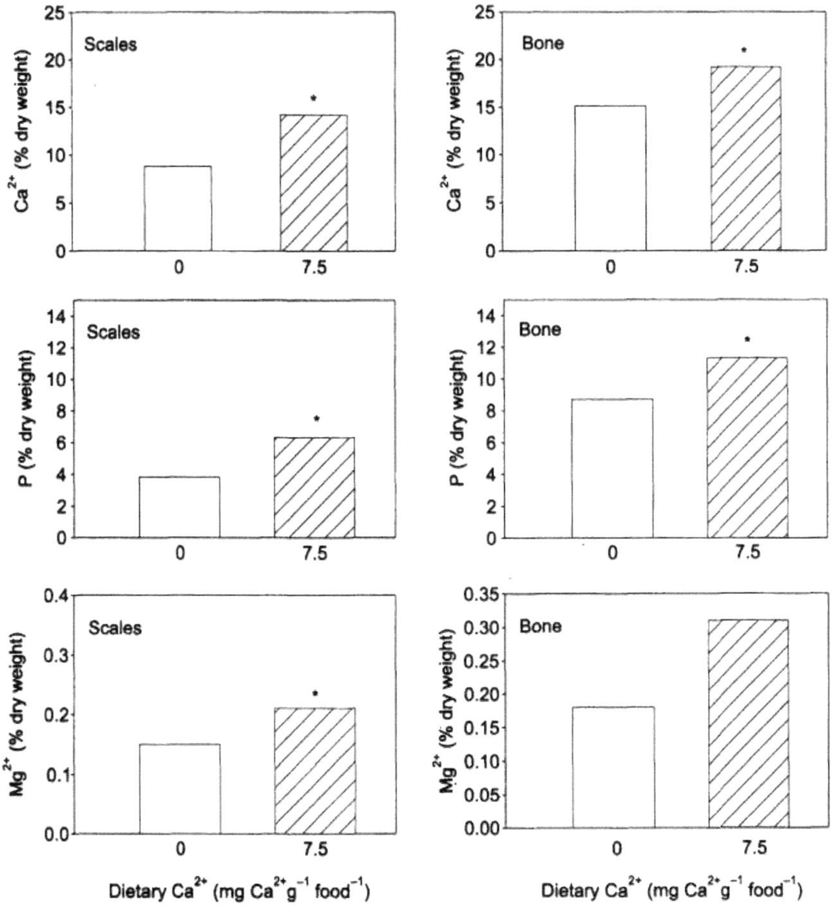

Fig. 3.2 Calcium, Mg^{2+} and P concentrations in scales and bone of blue tilapia fed with diets with different Ca^{2+} levels for 24 weeks. Data from O'Connell and Gatlin, 1994. *significantly different from 0 mg Ca^{2+} g^{-1} $food^{-1}$ by ANOVA ($P<0.05$)

which suffices for growth and homeostasis. In addition, Ca^{2+} intestinal absorption is greatly reduced when compared to freshwater fishes, and this reduction is correlated with a decrease in the activity of Ca^{2+} transporters in the enterocyte plasma membrane (Flik and Verbost, 1993).

DIETARY PHOSPHORUS

In addition to being a major constituent of structural components of skeletal tissues, phosphorus is located in every cell of the body. It is an

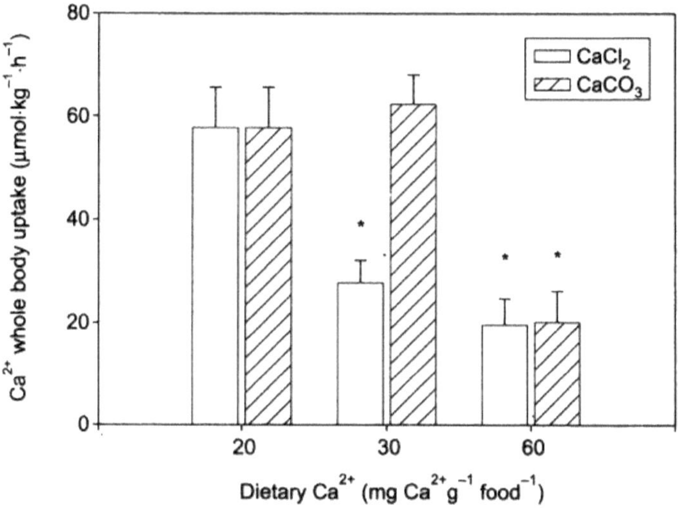

Fig. 3.3 Ca^{2+} (A) whole body uptake of rainbow trout exposed to diets with different Ca^{2+} concentrations. Means ± 1 SEM (N = 8–9). Adapted from Baldisserotto *et al.* (2004a, b). *significantly different from 20 mg/g Ca^{2+} by One-way ANOVA and Tukey test ($P<0.05$)

important constituent of nucleic acids and cells membranes and is directly involved in all the energy producing reactions of the cell. It plays an important role on carbohydrate, lipid and amino acid metabolism and in muscle and nervous tissues metabolism, as well as on various metabolic process involving buffers body fluids. The phosphate metabolism of fish has not been as well studied as that of Ca^{2+}. Food is the main source of phosphorus because phosphate concentrations are low in both freshwater and seawater (approximately 0.02 mg L^{-1}). The uptake of phosphorus from water has been repeatedly demonstrated (Lall, 2002). The absorption of dietary phosphorus is affected by the level of phosphate in the blood. Phosphorus accumulates mainly in soft tissues (heart, liver, kidney, and blood) and, to a limited extent, in skeletal tissues (Mol *et al.*, 1999). Numerous studies with monogastric animals have shown that an optimum dietary Ca^{2+}/phosphorus ratio is important: an increase of dietary Ca^{2+}/ phosphorus ratio interferes with phosphorus absorption, and conversely, a high phosphorus/Ca^{2+} ratio may restrict Ca^{2+} absorption. Although the magnitude of effect changes with species and forms of Ca^{2+} and phosphorus present in the diet, such studies on Ca^{2+}/ phosphorus ratio in fish diet are limited (Lall, 2002).

DIETARY Mg^{2+}

In fish, as in all vertebrates, Mg^{2+} is found mineralized in bony tissues as an ionized form (Mg^{2+}) or complexed with proteins in all the tissues (Bijvelds *et al.*, 1998). The magnesium pool of bones and scales may be used as a reserve to maintain normal Mg^{2+} levels in soft tissues when Mg^{2+} intake is low (Bijvelds *et al.*, 1996). Of the remaining Mg^{2+} pool in the soft tissues, only a small percentage is found in the extracellular fluid. The total plasma Mg^{2+} concentration in most cases does not exceed 2 mmol L^{-1}, and the ionic concentration is normally less than 1 mmol L^{-1} (Bijvelds *et al.*, 1997). The ionic Mg^{2+} level in the cytoplasm is kept relatively low, i.e. in the submillimolar range, typically representing less than 10% of total Mg^{2+} content to the cell (Bijvelds *et al.*, 1998).

As Mg^{2+} plays an important role in cells, and intracellular and extracellular Mg^{2+} levels are maintained within narrow limits, the vertebrates must have developed effective mechanisms by which Mg^{2+} is transported, stored and is concentration regulated. However, transport of Mg^{2+} across intestinal and kidney epithelia, key process to the understanding of Mg^{2+} regulation, is still poorly understood. The mechanism of branchial Mg^{2+} uptake has not been demonstrated (Bijvelds *et al.*, 1998). Freshwater fish are threatened by diffusive losses of Mg^{2+} across the body surfaces because Mg^{2+} concentration in freshwater is typically well below 0.5 mmol L^{-1}. These losses have to be compensated for by Mg^{2+} uptake from the diet and from the water. Renal excretion must be limited by reabsorption of filtered Mg^{2+} (Bijvelds *et al.*, 1998).

The hydrolytic activity of ATPase is described as Mg^{2+} dependent. ATPase affects the utilization of stored energy from ATP which is required for excretion or intake of ions across the gill membranes against a concentration gradient. The increased branchial ATPase activity during smoltification has made this an indicator enzyme in monitoring the hypo-osmoregulation ability in salmonids. The activity of Mg^{2+} ATPase in renal tissue displays positive correlation with branchial ATPase activity during smoltification and may be involved on Mg^{2+} excretion. The concentration of Mg^{2+} in freshwater differs widely from that in sea water, and dietary Mg^{2+} fed in the freshwater phase possibly affects the regulatory ability of Mg^{2+} in salmon transferred to seawater (El-Mowafi, 1997).

In freshwater fish dependent primarily on diet for their Mg^{2+} uptake, optimal growth is usually achieved when dietary Mg^{2+} is 15–20 mmol kg^{-1}. Prolonged feeding with lower dietary Mg^{2+} content may lead to a decrease on growth rate, Mg^{2+} depletion of the tissues, muscle dysfunction, neurological disorders and high mortality (Bijvelds *et al.*, 1996). A low dietary Mg^{2+} intake induced high body Ca^{2+} levels in rainbow trout (Cowey *et al.*, 1977), tilapia (*O. niloticus* and *O. mossambicus*) (Dabrowski *et al.*, 1989; Bijvelds *et al.*, 1997a) and guppy (*Poecilia reticulata*) (Shim and Ng, 1988). Magnesium affects the permeability of the intestinal epithelium to ions (Fordtran *et al.*, 1985) and low luminal Mg^{2+} concentration, therefore, increase epithelial permeability to ions, possibly stimulating paracellular Ca^{2+} absorption (Karbach and Feldmeier, 1991). Minutes perturbations to cellular Mg^{2+} homeostasis affects the acid-base regulation of the cells involved in bone formation, and this may lead to supersaturation of plasma Ca^{2+}, resulting in spontaneous calcification process in soft tissues (Driessens *et al.*, 1987). In line with this hypothesis, there was calcification of renal tissue in Mg^{2+}-deficient rainbow trout (Cowey *et al.*, 1977).

It has been suggested that in Mg^{2+}-deficient rainbow trout, the increase in muscle Na^+ content was due to a decrease in cell water content coupled with an increase in extracellular volume (Cowey *et al.*, 1977; Bijvelds *et al.*, 1998). Such changes in the mineral status and water content of tissues are suggestive of changes in cell membrane permeability that could cause an increased turnover. The action of Mg^{2+} on the fluidity of cellular membranes may underlie this phenomenon. The permeability of osmoregulatory epithelia may also be affected since it has been demonstrated that external Mg^{2+} and Ca^{2+} levels influence gill permeability to both water and ions (Wendelaar Bonga *et al.*, 1993; Bijvelds *et al.*, 1996a, 1998). It is also plausible that this relationship reflects the dependence of cellular ion transport mechanisms, such as Na^+/K^+-ATPase (Rude, 1989), the $Na^+/K^+/Cl^-$ symporters (Flatman, 1988) and cations channels (Horie *et al.*, 1987; Dorop and Clausen, 1993; Bijvelds, 1998) on Mg^{2+}. Moreover, Mg^{2+} may influence ion movement across cellular membranes through its action on membrane permeability (Bijvelds *et al.*, 1998).

Internally, Mg^{2+} may have similar actions on membrane permeability and ion turnover in osmoregulatory organs such as the gills. For instance, in Mozambique tilapia, low dietary Mg^{2+} caused proliferation of branchial chloride cells (Bijvelds *et al.*, 1996a), and a decrease in the Na^+ influx across the gills (Van der Velden *et al.*, 1992b). Such changes are indicative of an increased turnover of these cells in the gill epithelium. Renewal branchial epithelium may be a response to disturbance in ion transport across the gills, since both the epithelial permeability to ions and water (Wendelaar Bonga *et al.*, 1993) and the activity of cellular ions transporters (Flatman, 1993) are controlled by Mg^{2+} (Bijvelds *et al.*, 1997). Furthermore, in common carp (*Cyprinus carpio*), Mg^{2+} deficiency is associated with changes in branchial ion regulation (an increase in opercular chloride cell density and a decrease in branchial Na^+/K^+ATPase activity) that coincide with an increased bone Na^+ content (Van der Velden *et al.*, 1992a).

Freshwater fish depend on Mg^{2+} absorption from the intestinal tract to meet most of their Mg^{2+} requirement (Van der Velden *et al.*, 1992a, b). It is widely recognized that the intestine is the most important route for Mg^{2+} uptake in freshwater fish (Gatlin *et al.*, 1992; Shearer and Asgard, 1992). At low dietary Mg^{2+} levels, absorption of this ion is highly efficient, suggesting a regulated intestinal Mg^{2+} transport route. In this condition, urinary Mg^{2+} excretion decreased, and this suggests that tubular reabsorption is increased. As renal and intestinal epithelial cells maintain a large potential difference (inside negative) across the plasma membrane, an active extrusion mechanism is indicated, both to maintain the low intracellular Mg^{2+} concentration and to allow transcellular Mg^{2+} transport (Bijvelds *et al.*, 1996a).

Acknowledgements

B. Baldisserotto received a CNPq (Conselho Nacional de Desenvolvimento Tecnológico-Brazil) research grant.

References

Al-Amoudi, M.M. 1987. The effects of high salt diet on the direct transfer of *Oreochromis mossambicus*, *O. siplurus*, *O. niloticus* hybrids to sea water. *Aquaculture* 64: 333–338.

Arzel, J., R. Metailler, G. Boeuf, F. Baudin-Lauencin, H. Barone and J. Guillaume. 1993. Effect of high extra dietary sodium chloride in *Salmo trutta* on transfer to seawater. In: *Fish Nutrition in Practice*. S.J. Kaushik and P. Luquet (eds.), IV Int. Symposium Fish Nutrition and Feeding, Biarritz, 24-27 June 1991, INRA, pp. 903–906.

Baldisserotto, B. 2003. Osmoregulatory adaptations of freshwater teleosts. In: *Fish Adaptations*. A.L. Val and B.G. Kapoor (eds.). Science Publishers, Inc, Enfield (NH), USA, pp 179–201.

Baldisserotto, B. and Olga M. Mimura. 1995. Ion and water transport in the gut on the freshwater teleost *Prochilodus scrofa*. *Ciência e Cultura* 47: 83–85.

Baldisserotto, B., O.M. Mimura and L.C. Salomão. 1993. Effect of pH on ion and water transport in the gut of freshwater teleost *Synbranchus marmoratus*. *Ciência e Cultura* 45: 396–398.

Baldisserotto, B., C. Kamunde, A.Y.O. Matsuo and C.M. Wood. 2004a. A protective effect of dietary calcium against acute waterborne cadmium uptake in rainbow trout. *Aquatic Toxicology* 67: 57–73.

Baldisserotto, B., C. Kamunde, A.Y.O. Matsuo and C.M. Wood. 2004b. Acute waterborne cadmium uptake in rainbow trout is reduced by dietary calcium carbonate. *Comparative Biochemistry and Physiology* C137: 363–372.

Baldisserotto, B., M.J. Chowdhury and C.M. Wood. 2006. *In vitro* analysis of intestinal absorption of cadmium and calcium in rainbow trout fed with calcium- and cadmium-supplemented diets. *Journal of Fish Biology* 69: 658–667.

Bijvelds, M.J.C., G. Flik, Z. Kolar, Z. and S.E. Wanderlaar Bonga. 1996a. Uptake, distribution and excretion of magnesium in *Oreochromis mossambicus* dependence on magnesium in diet and water. *Fish Physiology and Biochemistry* 15: 287–298.

Bijvelds, M.J.C., Z. Kolar, S.E. Wendelaar Bonga and G. Flik. 1996b. Magnesium transport across the basolateral plasma membrane of fish enterocyte. *Journal of Membrane Biology* 154: 217–225.

Bijvelds, M.J.C., Z. Kolar and S.E. Wenderlaar Bonga. 1997a. Mineral balance in *Oreochromis mossambicus*: Dependence on magnesium in diet and water. *Fish Physiology and Biochemistry* 16: 323–331.

Bijvelds, M.J.C., Z. Kolar, S.E. Wenderlaar Bonga and G. Flik. 1997b. Magnesium transport in plasma membrane vesicles of renal epithelium of the Mozambique tilapia (*Oreochromis mossambicus*). *Journal of Experimental Biology* 200: 1931–1939.

Bijvelds, M.J.C., J.A. Van Der Velden, Z. Kolar and G. Flik. 1998. Magnesium transport in freshwater teleosts. *Journal of Experimental Biology* 201: 1981–1990.

Bogè, G. and G. Pérès. 1990. Chloride requirements of sodium cotransporter systems. In: *Comparative Physiology: Comparative Aspects of Sodium Cotransport Systems*. R.K.H. Kinne (ed.). Karger, Basel, Vol. 7, pp. 186–215.

Bogè, G., L. Lopes and G. Pérès. 1988. An *in vivo* study of the role of pyloric caeca in water absorption in rainbow trout (*Salmo gairdneri*). *Comparative Biochemistry and Physiology* 91: 9–13.

Brett, J.R. 1979. Environmental factors and growth. In: *Fish Physiology*, W.S. Hoar, D.J. Randall and J.R. Brett (eds.). Academic Press, New York, Vol. 8: Bioenergetics and Growth, pp. 599–675.

Buddigton, R.K., J.W. Chen and J. Diamond. 1987. Genetic and phenotypic adaptation of intestinal nutrient transport to diet in fish. *Journal of Physiology* 393: 261–281.

Copatti, C.E , I.J. Coldebella, J. Radünz Neto, L.O. Garcia, M.C. Rocha, M.C. da and B. Baldisserotto. 2005. Effect of dietary calcium on growth and survival of silver catfish fingerlings, *Rhamdia quelen* (Heptapteridae), exposed to different water pH. *Aquaculture Nutrition* 11: 345–350.

Cowey, C.B., D. Knox, J.W. Adron, S. George and B. Pirie. 1977. The production of renal calcinosis by magnesium deficiency in rainbow trout (*Salmo gairdneri*). *British Journal of Nutrition* 38: 127–135.

Dabrowski, K., C. Leray, G. Nonnotte and D.A. Colin. 1986. Protein digestion and ion concentration in rainbow trout (*Salmo gairdneri* Rich.) digestive tract in sea and freshwater. *Comparative Biochemistry and Physiology* A83: 27–39.

Dabrowski, H., K.H. Meyer-Burgdorff and K.D. Günther. 1991. Magnesium status in freshwater fish, common carp (*Cyprinus carpio* L.) and dietary protein-magnesium interaction. *Fish Physiology and Biochemistry* 9: 165–172.

D'Cruz, L.M. and C.M. Wood. 1998. The influence of dietary salt and energy on the response to low pH in juvenile rainbow trout. *Physiological Zoology* 71: 642–657.

D'Cruz, L.M., J.J. Dockray, I.J. Morgan and C.M. Wood. 1998. Physiological effects of sublethal acid exposure in juvenile rainbow trout on a limited ration during a simulated global warming scenario. *Physiological Zoology* 71: 359–376.

Dockray, J.J., S.D. Reid and C.M. Wood. 1996. Effects of elevated summer temperatures and reduces pH on metabolism and growth of juvenile rainbow trout (*Oncorhynchus mykiss)* on unlimited ration. *Canadian Journal of Fisheries and Aquatic Sciences* 25: 2752–2763.

Driessens, F.C.M., R.M.H. Verbeck, J.W.E. Van Dijk and J.M.P.M. Borggreven. 1987. Response of plasma calcium and phosphate to magnesium depletion. A review and its physiological interpretation. *Magnesium Bulletin* 9: 193–201.

Dorop, I. and T. Clausen. 1993. Correlation between magnesium and potassium contents in muscle: role of Na^+-K^+ pump. *American Journal of Physiology* 204: C457–C463.

El-Mowafi, A.F.A., R. Waagbo and A. Maage. 1997. Effect of low magnesium on immune response and osmoregulation in Atlantic salmon. *Journal of Aquatic Animal Health* 9: 8–17.

Ferraris, R. and G.A. Ahearn. 1984. Sugar and aminoacid transport in fish intestine. *Comparative Biochemistry and Physiology* A77 :397–413.

Flatman. P.W. 1993. The role of magnesium in relating ion transport. In: *Magnesium and the Cell*, N.J. Buich (ed.). Academic Press, London, pp. 197–216.

Flik, G. and P.M. Verbost. 1995. Cellular mechanisms in calcium transport and homeostasis in fish. In: *Biochemistry and Molecular Biology of Fishes*, P.W. Hochachka and T.P. Mommsen (eds.). Elsevier, Amsterdam, Vol. 5, pp. 251–263.

Flik, G., J.H. Van Rijs and S.E. Wendelaar Bonga. 1985. Evidence for high-affinity Ca^{2+}-ATPase activity and ATP-driven Ca^{2+} transport in membrane preparations of the gill epithelium of the cichlid fish *Oreochromis mossambicus*. *Journal of Experimental Biology* 119: 335–347.

Flik, G., J.A. Van Der Velden, K.J. Dechering, J.M. Verbost, J.M.T.H. Schoenmakers, Z.I. Kolar, and S.E. Wendelaar Bonga. 1993a. Ca^{2+} and Mg^{2+} transport in gills and gut of tilapia, *Oreochromis mossambicus*: A review. *Journal of Experimental Zoology* 265: 356–366.

Flik, G., W. Atsma, J.C. Fenwick, F. Rentier-Delrue, J. Smal and S.E. Wendelaar Bonga. 1993b. Homologous recombinant growth hormone and calcium metabolism in the tilapia, *Oreochromis mossambicus*, adapted to fresh water. *Journal of Experimental Biology* 185: 107–119.

Flik, G., P.M. Verbost and S.E. Wendelaar Bonga. 1995. Calcium transport processes in fishes. In: *Fish Physiology*, C.M. Wood (ed.). Istuttle North, T.J. CED, San Diego, Vol. 14, pp. 317–341.

Flik, G., P.H.M. Klaren, T.J.M. Schoenmakers, M.J.C. Bijvelds, P.M. Verbost and S.E. Wendelaar Bonga. 1996. Cellular calcium transport in fish: Unique and universal mechanisms. *Physiological Zoology* 69: 403–417.

Fontaínhas-Fernandes, A.A., F. Russell-Pinto, E. Gomes, M.A. Reis-Henriques and J. Coimbra. 2001. The effect of dietary sodium chloride on some osmoregulatory parameters of the teleost, *Oreochromis niloticus*, after transfer from freshwater to seawater. *Fish Physiology and Biochemistry* 23: 307–316.

Fontaínhas-Fernandes, A.A., E. Gomes, M.A. Reis-Henriques, and J. Coimbra. 2002. Efeito da suplementação da dieta com NaCl no crescimento de tilápia *Oreochromis niloticus* cultivada em diferentes salinidades. *Arquivo Brasileiro de Medicina Veterinária e Zootecnia* 54: 204–211.

Fordtran, J.S., S.G. Morawski and C.A. Santa Ana. 1985. Effect of magnesium an active and passive sodium transport in the human ileum. *Gastroenterology* 89: 1050–1053.

Gatlin III, D.M., D.S. MacKenzie, S.R. Craig and W.H. Neill. 1992. Effects of dietary sodium chloride on red drum juveniles in waters of various salinities. *The Progressive Fish-Culturist* 54: 220–227.

Grizzie, J.M., K.A. Cummins and C.J. Ashfield. 1993. Effects of environments concentrations of calcium and sodium on the calcium flux in stresses 34-day-old striped bass. *Canadian Journal of Zoology* 71: 1379–1384.

Grosell, M. and C.M. Wood. 2002. Copper uptake across rainbow trout gills: Mechanisms of apical entry. *Journal of Experimental Biology* 205: 1179–1188.

Horie, M., H. Irisawa and A. Noma. 1987. Voltage-dependent magnesium block of adenosine-triphosphate-sensitive potassium channel in guinea-pig ventricular cells. *Journal of Physiology* 387: 251–272.

Karbach, U. and H. Feldmeier. 1991. New clinical and experimental aspects of intestinal magnesium transport. *Magnesium Research* 4: 9–22.

Kerstetter, T.H. and R.J. White. 1994. Changes in intestinal water absorption in coho salmon during short-term seawater adaptation: A developmental study. *Aquaculture* 121: 171–180.

Kleinow, K.M. and M.O. James. 2001. Response of teleost gastrointestinal system to xenobiotics. In: *Target Organ Toxicity in Marine and Freshwater Teleosts*, D. Schlenck and W.H. Benson (eds). Taylor & Francis, London, Vol. 1: Organs, pp. 269–339.

Lall, S.P. 2002. The minerals. In: *Fish Nutrition*, J.E. Halver and R.W. Hardy (eds.). Academic Press, San Diego, pp. 259–308.

Likongwe, J.S., T.D. Stecko, J.R. Stauffer Jr. and R.F. Carline. 1998. Combined effects of water temperature and salinity on growth and feed utilization of juvenile Nile tilapia *Oreochromis niloticus* (Linnaeus). *Aquaculture* 146: 37–46.

Lin, H. and D.J. Randall. 1991. Evidence for the presence of an electrogenic proton pump in the trout gill epithelium. *Journal of Experimental Biology* 161: 119–134.

Kirsch, R., W. Humbert and J.L. Rodeau. 1984. Control of the blood osmolarity in fishes with references to the functional anatomy of the gut. In: *Osmoregulation in Estuarine and Marine Animals*, A. Pequex, R. Gilles and L. Bolis (eds.). Springer-Verlag, Berlin, pp. 68–89.

McDonald, D.G. and C.M. Wood. 1981. Branchial and renal acid and ion fluxes in the rainbow trout, *Salmo gairdneri*, at low environmental pH. *Journal of Experimental Biology* 93: 101–118.

Mol, J.H., W. Astma, G. Flik, H. Bouwmeester and J.W.M. Osse. 1999. Effect of low ambient mineral concentrations on the accumulation of calcium, magnesium and phosphorus by the early stages of the air-breathing armoured catfish *Megalechis personata* (Siluriformes: Callichthyidae). *Journal of Experimental Biology* 202: 2121–2129.

Musselman, N.J., M.S. Peterson and W.J. Diehl. 1995. The influence of salinity and prey content on growth and intestinal Na^+/K^+-ATPase activity of juvenile bluegill, *Lepomis macrochirus*. *Environmental Biology of Fish* 42: 303–311.

O'Connell, J.P. and D.M. Gatlin III. 1994. Effects of dietary calcium and vitamin D_3 on weight gain and mineral composition of blue tilapia (*Oreochromis aureus*) in low-calcium water. *Aquaculture* 125: 107–117.

Phillips, A.M. 1944. The physiological effect of sodium chloride upon brook trout. *Transactions of American Fisheries Society* 74: 297–304.

Pyle, G.G., C. Kamunde, C.M. Wood and D.G. McDonald. 2003. Dietary sodium inhibits aqueous copper uptake in rainbow trout (*Oncorhynchus mykiss*). *Journal of Experimental Biology* 206: 609–618.

Reid, S.D., D.G. McDonald and C.M. Wood. 1996. Interactive effects of temperature and pollutant stress. In: *Global Warming: Implications for Freshwater and Marine Fish*, C.M. Wood and D.G. McDonald (eds.). Cambridge University Press, Cambridge, pp. 325–349.

Reid, S.D., J.J. Dockray, T.K. Linton, D.G. McDonald and C.M. Wood. 1997. Effect of chronic environmental acidification and summer global warming scenario: Protein synthesis in juvenile rainbow trout (*Oncorhynchus mykiss*). *Canadian Journal of Fisheries and Aquatic Science* 54: 2014–2024.

Robinson, E.H., S.D. Rawles, P.B. Brown, H.E. Yete and L.H. Greene. 1986. Dietary calcium requirements of channel catfish *Ictalurus punctatus* reared in calcium-free water. *Aquaculture* 53: 263–270.

Robinson, E.H., D. Labomascus, P.B. Brown and T.L. Linton. 1987. Dietary calcium and phosphorus requirements of *Oreochromis aureus* reared in calcium-free water. *Aquaculture* 64: 267–276.

Roy, P.K., P.E. Witten, B.K. Hall and S.P. Lall. 2002. Effects of dietary phosphorus on bone growth and mineralization of vertebrae in haddock (*Melanogrammus aeglefinus* L.). *Fish Physiology and Biochemistry* 27: 35–48.

Rude, R.K. 1989. Physiology of magnesium metabolism and the important role of magnesium in potassium deficiency. *American Journal of Cardiology* 63: G31–G34.

Salman, N.A. and F.B. Eddy. 1987. Response of chloride cell numbers and gill Na^+/K^+ ATPase activity of freshwater rainbow trout (*Salmo gairdneri* Richardson) to salt feeding. *Aquaculture* 61: 41–48.

Salman, N.A. and F.B. Eddy. 1988. Kidney function in response to salt feeding in rainbow trout (*Salmo gairdneri* Richardson). *Comparative Biochemistry and Physiology* A89: 535–539.

Shearer, K.D. and T. Asgard. 1992. The effect of waterborne magnesium on the dietary magnesium requirement of the rainbow trout (*Oncorhynchus mykiss*). *Fish Physiology and Biochemistry* 9: 387–392.

Schoenmakers, T.J.M., P.M. Verbost, G. Flik and S.E. Wendelaar Bonga. 1993. Transcellular intestinal calcium transport in freshwater and seawater fish and its dependence on sodium/calcium exchange. *Journal of Experimental Biology* 176: 195–206.

Shim, K.F., and S.H. Ng. 1988. Magnesium requirement of the *Poecilia reticulata* Peters. *Aquaculture* 73: 131–141.

Smith, N.F., C. Talbot and F.B. Eddy. 1989. Dietary salt intake its relevance to ionic regulation in freshwater salmonids. *Journal of Fish Biology* 35: 749–753.

Smith, N.F., F.B. Eddy and C. Talbot. 1995. Effect of dietary salt load on transepithelial Na^+ exchange in freshwater rainbow trout (*Oncorhynchus mykiss*). *Journal of Experimental Biology* 198: 2359–2364.

Stiffler, D.F., J.B. Graham, K.A. Dickson and W. Stockmann. 1986. Cutaneous ion transport in the freshwater teleost *Synbranchus marmoratus*. *Physiological Zoology* 59: 406–418.

Talbot,C., T. Preston and B.W. East. 1986. Body composition of Atlantic salmon (*Salmo salar* L.) studied by neutron activation analysis. *Comparative Biochemistry and Physiology* A85: 445–450.

Van Der Velden, J.A., G. Flik, F.A. Spanings, T.G. Verburg, Z.I. Kolar and S.E. Wendelaar Bonga. 1992a. Physiological effects of low-magnesium feeding in the common carp, *Cyprinus carpio*. *Journal of Experimental Zoology* 264: 237–244.

Van Der Velden, J.A., G. Flik and S.E. Wendelaar Bonga. 1992b. Prolactin cell activity and ion regulation in tilapia, *Oreochromis mossambicus* (Peters): Effects of low magnesium diet. *Journal of Fish Biology* 40: 875–885.

Vilella, S., G.A. Ahearn, G. Cassano, G. and C. Storelli. 1988. Na^+-dependent L-proline transport by eel intestinal brush-border membrane vesicles. *American Physiological Society*, Vol. 1, pp. 648–653.

Vilella, S., G. Cassano and C. Storelli. 1989. Haw many Na^+-dependent carriers for L-alanine and L-proline in the eel intestine? Studies with brush-border membrane vesicles. *Biochimica et Biophysica Acta* 984: 188–192.

Wendelaar Bonga, S.E. 1993. Endocrinology. In: *The Physiology of Fishes*, D.H. Evans (ed.). CRC Press, Boca Raton, pp. 469–502.

Wilkie, M.P. and C.M. Wood. 1996. The adaptations of fish to extremely alkaline environments. *Comparative Biochemistry and Physiology* B113: 665–673.

Zaugg, W.S., D.D. Roley. E.F. Prentice, K.X. Gores and F.W. Waknitz. 1983. Increase sea water survival and contribution to the fishery of Chinook salmon *Oncorhynchus tshawytscha* by supplemental dietary salt. *Aquaculture* 32: 183–188.

Zohouri, M.A., G.G. Pyle and C.M. Wood. 2001. Dietary Ca inhibits waterborne Cd uptake in Cd-exposed rainbow trout, *Oncorhynchus mykiss*. *Comparative Biochemistry and Physiology* C130: 347–356.

CHAPTER

4

The Renin-Angiotensin Systems of Fish and their Roles in Osmoregulation

J. Anne Brown[1,*] and Neil Hazon[2]

The renin-angiotensin system (RAS) plays an important role in the control of salt and water balance of most vertebrates. Our understanding of this system, and its many actions, is most detailed in mammals (Bader *et al.*, 2001; Gociman *et al.*, 2004; Navar and Nishiyama, 2004). Here, a role in the pathogenesis of hypertension has led to pharmaceutical targeting of the system and its receptors to manipulate vascular tone (Croom *et al.*, 2004). It is amongst the fish, however, where the actions of the vertebrate (and pre-vertebrate) RAS developed, that we may perhaps identify the fundamental (primitive) actions of the vertebrate RAS. Research on fish has revealed both direct and indirect vasoconstrictor actions of angiotensin (Olson, 1992; Bernier and Perry, 1999; Rankin *et al.*,

Authors' addresses: [1]School of Biosciences, University of Exeter, Exeter EX4 4PS, UK.
[2]School of Biology, University of St Andrews, St Andrews KY16 8LB, UK.
E-mail: nhl@st-andrews.ac.uk
**Corresponding author*: E-mail: J.A.Brown@exeter.ac.uk

2004) similar to those initially seen when extracts of mammalian kidneys were first injected into the vasculature resulting in angiotensin formation *in vivo*. But amongst fish, the osmoregulatory actions of angiotensin also play a significant part in dealing with some of the osmotic challenges they face. In this chapter we will review our current knowledge of the renin-angiotensin systems of fish, and our understanding of the roles of angiotensin in their osmoregulatory processes, discussing both the circulating RAS and the recent evidence of tissue-specific paracrine systems.

RAS: THE BIOCHEMICAL CASCADE

The series of biochemical events that generate the physiologically active peptides, angiotensin II (Ang II) and angiotensin III (Ang III), as well as other peptide fragments in circulation are outlined in Fig. 4.1. The initial event is the synthesis and release of the proteolytic enzyme renin that cleaves angiotensinogen. In mammals, circulating levels of angiotensinogen, constitutively secreted by the liver are reported to be one thousand times or more greater than the normal concentrations of circulating Ang II, the main physiologically active component of the system (e.g., Klett and Granger, 2001; Lantelme *et al.*, 2002), which has been taken to suggest that hepatic angiotensinogen synthesis is not the major rate-limiting step in determining circulating Ang II. Nevertheless, plasma concentrations of angiotensinogen do appear to influence the activity of the mammalian renin-angiotensin systems and physiological results such as blood pressure elevation. This has been argued from a biochemical point of view to reflect the fact that plasma concentrations of angiotensinogen are within the concentration range of the K_m value for the enzyme substrate (renin-angiotensinogen) reaction (Klett and Granger, 2001). Liver mRNA angiotensinogen and plasma angiotensinogen have been reported to influence maximal angiotensin formation (Brasier and Li, 1996; Klett and Granger, 2001), and Ang II itself can regulate hepatic angiotensinogen mRNA, as well as renin mRNA (Nakamura *et al.*, 1990). Recent studies suggest significant regulation of mRNA for angiotensinogen in the trout. After transfer of rainbow trout to a hyperosmotic environment, hepatic angiotensinogen mRNA was increased in parallel with rising plasma osmolality (Paley *et al.*, 1996; Aust, 2002). These studies only became feasible after the cloning and sequencing of angiotensinogen cDNA in the trout (Paley *et al.*, 2003). The

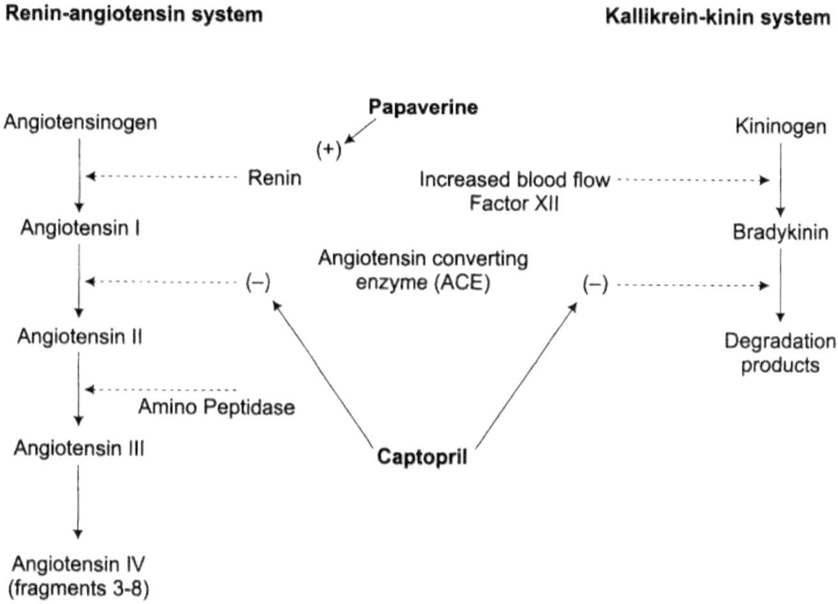

Fig. 4.1 Renin-angiotensin and kallikrein-kinin systems showing simplified biochemical cascades in circulation and the effects of enzymes, the ACE inhibitor, captopril, and the vasodilator, papaverine. This diagram does not show the alternative pathways for possible processing of angiotensinogen by angiotensinogenases and aminopeptidases that may occur in tissues.

angiotensinogen gene (and protein) sequence has only been identified in three teleostean fish species: the zebrafish, *Danio rerio* (GenBank Data Base: AL772289; BC095585; NB198063), the pufferfish, *Takifugu rubripes* (GenBank Data Base: BK001021) and the rainbow trout, *Oncorhynchus mykiss* (GenBank Data Base: AJ579373). The rainbow trout and zebrafish sequences show 64% identity, but have a very low homology with mammalian angiotensinogens, except in the region encoding angiotensin I, probably because of the redundancy of the majority of the molecule with little need to conserve primary structure; the most important functional parts of the molecule are the renin cleavage site and the angiotensin I-containing region.

Renin is synthesized in differentiated smooth muscle cells of vasculature supplying the renal glomerulus. This site of renin production has long been considered the major source of circulating renin for the systemic RAS. The classical mammalian picture of renin-producing cells within the wall of the afferent glomerular arteriole has led to their description as juxtaglomerular cells, but in fish these cells often occur well away from the glomerulus (Kobayashi and Takei, 1996). In marine fish not possessing glomeruli, the aglomerular teleosts, juxtaglomerular cells would clearly be a misnomer and 'granular epithelioid cells' is a preferable and more accurate term.

Identification of the cells producing renin in fish has often proved more difficult than in mammals. Initial studies relied on complex histological staining of the tiny renin granules. In mammals, these approaches have now largely been overridden by immunohistochemistry using specific renin antisera (e.g., Gomez *et al.*, 1990). However, mammalian antibodies rarely bind to fish renins and specific antisera are still needed for research on fish. The sequencing of mammalian renin genes has enabled the development of molecular probes to investigate renin mRNA (Hackenthal *et al.*, 1990), but amongst fish the renin gene has as yet only been identified in the zebrafish, *D. rerio* and pufferfish, *Takifugu rubripes* (Liang *et al.*, 2004), hence impeding progress in investigations of expression of the renin gene in most fish. Intriguingly, abundant expression of renin mRNA in early development of zebra fish has been demonstrated but the functional significance of this is as yet unclear (Liang *et al.*, 2004) and the specific cells expressing the renin gene have not yet been identified.

Despite the lack of an ability to investigate renin expression in fish other than the zebrafish and pufferfish, the kidney seems very likely to be the major source of renin—as in other vertebrates—but the existence of additional sites of renin production, as part of tissue specific RAS, has been known for many years in mammals (Hackenthal *et al.*, 1990; Yanegawa *et al.*, 1991; Bader *et al.*, 2001). In fish, there is less evidence as yet of tissue-specific RASs, but this should not be taken as indicating that they do not exist; it is more likely to be merely a reflection of the lack of the appropriate studies, rather than a lack of paracrine RASs. The first extra-renal tissue that was revealed to synthesize renin in teleost fish was the corpuscles of Stannius (CS), the small endocrine glands that lie on the surface of the kidney or embedded within the kidney (Hasegawa *et al.*, 1984). The CS have been argued to contain a calcium-regulating hormone, although in eel, renin secreted by the CS has been suggested to play a role in regulation of cardiovascular function, rather than in calcium regulation (Butler and Zhang, 2001; Butler *et al.*, 2003). Renin, either in circulation or in tissue-specific systems, acts on the substrate, angiotensinogen, in order to form the decapeptide, angiotensin I (Ang I), as shown in Fig. 4.1. It is, therefore, important that recent studies have also identified angiotensinogen mRNA in the CS (Aust, 2002), suggesting that they contain a complete RAS. The role of this paracrine system is as yet unclear, but its involvement in osmoregulation seems likely. Our recent studies of angiotensinogen gene expression in trout also provide strong support for the occurrence of a paracrine RAS in the kidney (Paley *et al.*, 1996; Brown *et al.*, 2000) and brain (Aust, 2002). Once again, the roles of these systems are as yet uncertain, but likely interaction with osmoregulatory processes will be discussed in later sections of this chapter.

An important step in the systemic RAS cascade (Fig. 4.1) is the conversion of Ang I to the octapeptide, Ang II, by angiotensin converting enzyme (ACE), although in mammalian paracrine RASs such as that in the heart, Ang II may be generated directly from angiotensinogen, through the actions of cathepsin G, bypassing Ang I, and chymase may act as the enzyme forming Ang II from Ang I (Nishimura, 2001). Angiotensin converting enzyme, nevertheless, has a major role to play in converting circulating Ang I to Ang II. ACE activity can be measured using the tri-peptide substrate, hippuryl-histidyl-leucine, which is cleaved by ACE to form His-Leu and hippuric acid. Most studies of ACE activity in fish tissues have measured hippuric acid spectrophotometrically, but fluorimetric assay of His-Leu offers both higher sensitivity and reliability (Casarini *et al.*, 1997; Cobb *et al.*, 2002a, 2004). ACE-like activity has been measured in all of the major groups of fish and for an extensive range of tissue types, including kidney, heart, intestine, gill and brain (Lipke and Olson, 1988; Uva *et al.*, 1992; Cobb *et al.*, 2002a, 2004). Much of the ACE is membrane bound in vascular endothelial plasma membranes where the long extracellular part of the molecule contains the active site, although cellular mechanisms also regulate the secretion of soluble ACE into a range of body fluids, including serum (Erdos, 1990; Ramchandran *et al.*, 1994; Casarini *et al.*, 1997). Particularly high levels of ACE activity are present in tissues with an extensive vascular endothelium, such as mammalian, reptilian and amphibian lung (Lipke and Olson, 1988; Bramucci *et al.*, 2004), a feature likely to explain the high rate of conversion of Ang I to Ang II as it passes through the pulmonary circulation. In fish, the functional homologue of the mammalian lung, the gills, has a similarly extensive vascular surface area and, in many species, high ACE activity (see Cobb *et al.*, 2004 for a review of the measurements).

The respiratory lamellar pathway through the gills is a major site for Ang I to Ang II conversion (Olson *et al.*, 1989). However, other tissues, in particular, the kidney and brain also have significant ACE activity (Polanco *et al.*, 1990; Uva *et al.*, 1992; Cobb *et al.*, 2002a).

The key role of ACE in the RAS has led to the development of pharmaceutical inhibitors and allowed their use in physiological studies aimed at investigating the actions of Ang II in fish. However, ACE inhibitors such as captopril result in both a decline in Ang II and an increase in circulating and tissue kinins, as ACE is identical to kininase II which deactivates bradykinin (see Fig. 4.1), as well as cleaving other peptides (Erdos, 1990). So, the action of kinins also needs careful consideration. Nevertheless, ACE inhibitors have proved to be extremely important tools in exploring the physiological role of the RAS of fish, as we shall see in later sections of this chapter.

Not all of ACE is endothelial. For example, the brush border of the renal proximal tubule epithelium contains high levels of ACE (Yanegawa *et al.*, 1991) that is likely to be linked to a paracrine RAS generating nanomolar concentrations of intratubular Ang II (Navar and Nishiyama, 2004). Recent measurements of angiotensins in urine excreted from the perfused kidney, where a hepatic supply of angiotensinogen does not exist, suggests a similar renal RAS exists in teleosts (Aust, 2002).

ANGIOTENSIN SEQUENCES IN FISH

Over the past 25 years or so, angiotensins have been sequenced in all major groups of fish, with the only exception now being the jawless hagfish (Fig. 4.2). Until relatively recently, angiotensin had remained unsequenced in lungfish and elusive in both cartilaginous elasmobranchs and the jawless agnathan fish, but homologous incubations of kidneys and plasma finally resolved the long-standing controversy as to the existence of the renin-angiotensin system in elasmobranchs and lampreys (Nishimura *et al.*, 1970; Henderson *et al.*, 1981; Takei *et al.*, 1993, 2004; Rankin *et al.*, 2004). Angiotensin II in lampreys has proved to be identical to that in most teleost fish, suggesting an early origin of this peptide structure. Indeed, the structure of the physiologically active octapeptide, Ang II is largely conserved throughout vertebrate groups, but with interchange of the asparagine and aspartic acid at position 1 and valine and isoleucine at position 5 in the agnathan lampreys, most teleosts and the lungfish compared to mammalian species (Fig. 4.2). Recent searching

	1	2	3	4	5	6	7	8	9	10
Mammals										
human, horse, dog, sheep, pig, rat, rabbit[7]	**Asp**	Arg	Val	Tyr	**Ile**	His	Pro	Phe	**His**	Leu
Agnathans										
River lamprey (*Lampetra fluviatilis*)[11]	**Asn**	Arg	Val	Tyr	**Val**	His	Pro	Phe	**Thr**	Leu
Sea lamprey (*Petromyzon marinus*)[12]	**Asn**	Arg	Val	Tyr	**Val**	His	Pro	Phe	**Thr**	Leu
Holosteans										
Bowfin (*Amia calva*)[8]	**Asp**	Arg	Val	Tyr	**Val**	His	Pro	Phe	**Asn**	Leu
Teleosts										
Japanese goosefish (*Lophius litulon*)[3]	**Asn**	Arg	Val	Tyr	**Val**	His	Pro	Phe	**His**	Leu
Chum salmon (*Oncorhynchus keta*)[2*]	**Asn**	Arg	Val	Tyr	**Val**	His	Pro	Phe	**Asn**	Leu
Rainbow trout (*Oncorhynchus mykiss*)[6]	**Asn**	Arg	Val	Tyr	**Val**	His	Pro	Phe	**Asn**	Leu
Flounder (*Platichthys flesus*)[10]	**Asn**	Arg	Val	Tyr	**Ile**	His	Pro	Phe	**Thr**	Leu
Eel (*Anguilla japonica*)[1,] (*A. rostrata*) [4*]	**Asn**	Arg	Val	Tyr	**Val**	His	Pro	Phe	**Gly**	Leu
Elasmobranchs										
Dogfish (*Triakis scyllia*)[5]	**Asn**	Arg	Pro	Tyr	**Ile**	His	Pro	Phe	**Gln**	Leu
Lungfish										
(*Neoceratodus forsteri*)[9]	**Asn**	Arg	Val	Tyr	**Val**	His	Pro	Phe	**Thr**	Leu

References: [1]Hasegawa *et al.* (1983); [2]Takemoto *et al.* (1983); [3]Hasegawa *et al.* (1984); [4]Khosla *et al.* (1985); [5]Takei *et al.* (1993); [6]Conlon *et al.* (1996); [7]see Kobayashi and Takei (1996); [8]Takei *et al.* (1998); [9]Joss *et al.* (1999); [10]Balment *et al.* (2003); [11]Rankin *et al.* (2004); [12]Takei *et al.* (2004). [*]Asp[1], Val[5]-Ang also identified but unlikely to be naturally present; Asp[1] and Asn[1] interconvertible during incubation.

Fig. 4.2 Angiotensin I and Angiotensin II amino acid sequences of peptides isolated from fish species, compared to the peptide sequence in most mammals. Amino acids at positions 3, 5 and 9 (Ang I only) that differ between species are highlighted. Further differences occur in the dogfish and flounder.

of angiotensinogen sequences in genome and EST databases has also identified that Ang II in fugu, stickleback, medaka, carp and zebrafish are identical to the typical (not flounder) teleostean sequence (Takei *et al.*, 2004). Intriguingly, elasmobranch angiotensin was found to be more unusual, with asparagine at position 1 (as in teleost fish), isoleucine at position 5 (as in mammals) and a unique proline substitution at position 3 (Fig. 4.2). Due to the large differences in the aliphatic side chains of proline and valine, and the influence this has on protein architecture, proline^3 will profoundly affect the tertiary structure of Ang II (Creighton, 1993). This may well explain why some antibodies generated against mammalian Ang II structure fail to recognize elasmobranch Ang II. However, a homologous radioimmunoassay (RIA) for elasmobranch Ang II has been established (Tierney *et al.*, 1998) and basal circulating concentrations in the European spotted dogfish, *Scyliorhinus canicula* shown to be in the range of 100-150 pM, although these values may rise up to ten fold during experimental manipulations (Anderson *et al.*, 2002a).

Only in hagfish is an Ang II yet to be isolated. Lampreys and hagfish are the sole survivors of a once flourishing radiation of jawless fishes. Although they are still grouped together as cyclostomes, molecular phylogenetic analyses suggest a diphyletic origin, with lampreys more closely related to elasmobranchs and other jawed vertebrates, while hagfish appear to be the survivors of a group that diverged from the main craniate lineage at least 500 million years ago (Janvier, 1999). Granules of renin could not be identified in the renal vasculature of the Japanese hagfish (*Paramyxine atami*) or the Atlantic hagfish (*Myxine glutinosa*) and incubation of renal extracts of *P. atami* with homologous or heterologous plasma did not yield a pressor substance (Nishimura *et al.*, 1970; Nishimura, 1985), suggesting a lack of a RAS. However, recent radioimmunoassay has detected 'immunoreactive angiotensins' in hagfish plasma, at around 250 pM (Cobb *et al.*, 2004). The antisera employed has a high binding affinity for Ang II and low cross reactivity with Ang I, but high cross reactivity to Ang III and IV, so measurement of Ang II, III and IV might be predicted. The concentration measured is in the same order of magnitude as angiotensin concentrations reported in fish and amphibians and the parallelism of serially diluted plasma with angiotensin standards supports the existence of circulating angiotensin in hagfish, although the sequence of this angiotensin clearly is yet to be determined.

ANGIOTENSIN RECEPTORS

Amongst teleost fish, angiotensin receptors have been identified in many tissues (see Table 4.1) giving us clues as to probable sites of action, including actions on osmoregulatory tissues. For example, binding of angiotensin to the chloride cells of eel gills (Marsigliante *et al.*, 1996, 1997a) led to research that has identified rapid actions on gill Na^+,K^+-ATPase and a highly probable role in determining sodium balance. In teleostean kidneys, Ang II binding sites have been identified in the renal tubules of several species (Cobb and Brown, 1992; Marsigliante *et al.*, 1996, 1997b), suggesting actions on renal transport processes, but fewer studies have identified glomerular receptors. No Ang II binding to the eel glomerulus was detected by immunocytochemistry (Marsigliante *et al.*, 1996) although this may simply reflect failure of the monoclonal antibody (selective to mammalian AT_1 receptors) to recognize teleostean Ang receptors. Radioligand-binding techniques have shown the occurrence of an Ang II binding site in the renal glomerulus of the trout and displacement studies suggest the receptor has some similarities to the mammalian AT_1 sub-type, although significant differences were revealed in signalling studies (Cobb and Brown, 1993; Brown *et al.*, 1997; Cobb *et al.*, 1999). The glomerular receptor of trout has been linked to renal antidiuretic actions and this action will be discussed more fully later when reviewing the renal actions of Ang II in fish.

The apparent cloning of a fragment of an Ang receptor gene from the rainbow trout (Parkyn *et al.*, 1997) was later revealed to encode a protein with a higher homology to mammalian and *Xenopus* angiotensin receptor-like proteins (Aust, 2002), that do not bind Ang II, and despite further cloning attempts the trout receptor remains elusive. However, cloning of a cDNA encoding the entire angiotensin receptor in eel liver (Tran van Chuoi *et al.*, 1999) confirms a receptor with a higher homology to AT_1 than AT_2 receptors, although phylogenetic analysis separated the eel angiotensin receptor from all mammalian AT_1 receptors and AT_2 receptors and the angiotensin receptors of the African clawed toad, *Xenopus* (Sandberg, 1994).

In elasmobranch fish, autoradiographic and ligand-binding techniques have indicated the presence of Ang II receptors in a range of tissues (see Table 4.1) including brain, heart, intestine, gill, kidney rectal gland and liver (Hazon *et al.*, 1997a; Tierney *et al.*, 1997a; Cerra *et al.*, 2001), although these studies were unable to determine receptor sub-types. More

Table 4.1 Angiotensin receptors or binding sites identified in fish tissues. The techniques employed in each study are listed here.

Fish group species	*Tissues with identified receptors*	*Technique and receptor affinity (K_d or IC_{50})*	*Reference*
Teleosts			
Rainbow trout (*Oncorhynchus mykiss*)	Kidney: glomerulus and its vascular pole, proximal tubule; urinary bladder; dorsal and ventral aorta; intestine; oesophagus; liver; skin; heart; interrenal tissue; gill; brain	Autoradiography: ^{125}I-Asn^1, Val^5-Ang II	Brown *et al.* (1990) Cobb and Brown (1992)
(*O. mykiss*)	Glomerulus	Radioligand-binding Asn^1, Val^5-Ang II: K_d = 0.32-0.39 nM; Losartan (mammalian AT_1 subtype inhibitor) IC_{50}: 27 nM	Cobb and Brown (1993) Brown *et al.* (1997)
(*O. mykiss*)	Glomerulus	Ca^{2+} signalling: calcium green fluorescence	Brown *et al.* (1997) Cobb *et al.* (1999)
Eel			
(*A. anguilla*)	Intestine: brush border membrane liver kidney tubules: brush border membrane	Radioligand binding and isoelectric focusing: Ang II K_d intestine—3.4 nM	Marsigliante *et al.* (1994)
(*A. anguilla*)	Gill: chloride cells, cytoplasm, membranes	Immunocytochemistry and isoelectric focussing gill membranes	Marsigliante *et al.* (1996, 1997a)
(*A. anguilla*)	Kidney tubules	Ca^{2+} signalling: fura 2; inositol phosphate generation	Marsigliante *et al.* (2000)
(*A. anguilla*)		Cloning cDNA	Tran van Chuoi *et al.* (1998)
Toadfish (*Opsanus tau*)	Vascular smooth muscle: ventral and dorsal aortae, branchial arteries, coeliac artery	Radioligand binding: Asn^1, Val^5-Ang II K_d: 1 nM; Ca^{2+} signalling: fura 2	Qin *et al.* (1999)
Icefish (*Chionodraco hamatus*)	Intestine; gill (chloride cells and pavement cells); kidney tubules basolateral cytoplasm; weaker on apical membranes	Immunohistochemistry	Marsigliante *et al.* (1997b)

(*Table 4.1 contd.*)

(Table 4.1 contd.)

Elasmobranchs			
Japanese dogfish (*Triakis scyllia*)	Interrenal; gills; rectal gland; kidney; brain; heart; intestine; liver	*In vitro* autoradiography and radioligand-binding ^{125}I-Asn^{1}, Pro^{3}, Ile^{5}-Ang II. Gill cells: 2 sites—K_d: 0.11 nM and 1.75 nM	Tierney *et al.* (1997a) Hazon *et al.* (1997a)
European spotted dogfish (*Scyliorhinus canicula*)	Heart	*In vitro* autoradiography ^{125}I-Asn^{1}, Pro^{3}, Ile^{5}-Ang II binding sites in ventricle, atrium and conus arteriosus; Ventricle: 2 sites—K_d: 0.05 nM and 1.3 nM	Cerra *et al.* (2001)
(*S. canicula*)	Rectal gland, interrenal gland	*In vitro* autoradiography ^{125}I-Asn^{1}, Pro^{3}, Ile^{5}-Ang II. Rectal gland sub-caspule Ang II K_d: 0.41 nM; Interrenal: 2 sites—K_d: 0.25 nM and 7 nM	Hazon *et al.* (1997a)
(*S. canicula*)	Kidney, rectal gland, interrenal tissue, gill brain, liver, spleen, heart, intestine, muscle	Cloning cDNA: homology to mammalian AT_{1A} receptor (50%) and AT_2 receptors (38%). mRNA angiotensin receptor expression	Aust (2002)

recently, cloning of a full-length cDNA for an angiotensin receptor in the European lesser spotted dogfish, *S. canicula*, an elasmobranch fish, again shows greater similarity to mammalian AT_1 receptors (~50% identity) than mammalian AT_2 receptors (~38% identity) and phylogenetic analysis groups the dogfish angiotensin receptor with AT_1 receptors (Aust, 2002). Seventeen of the 20 amino acid residues identified as essential for structural integrity, angiotensin binding and receptor internalization of mammalian AT_1 receptors are conserved in the dogfish receptor. However, low conservation of extracellular loop 2, and non-conservative substitution of the amino acid predicted to bind to the unique Pro^3 of dogfish angiotensin, provide evidence for a predictable co-evolution of the peptide and its receptor.

Expression of angiotensin receptor mRNA in dogfish tissues and ligand-binding studies in teleosts show the occurrence of angiotensin receptors in brain, liver, spleen, heart, rectal gland (elasmobranchs only), intestine, kidney, interrenal tissue, gill and muscle tissue, giving evidence for a diverse range of physiological actions, including roles in osmoregulation that will be examined later in this chapter.

EURYHALINITY AND ACTIVATION OF THE RENIN-ANGIOTENSIN SYSTEM

There is evidence that the RAS of fish is activated in association with exposure to higher salinities. Amongst the euryhaline teleosts, the species most studied with regard to the RAS are the eel (European and Japanese eel), flounder and rainbow trout. For example, circulating levels of Ang II are three times higher in seawater (SW)-adapted European eel, *Anguilla anguilla* (~33 fmol/ml) than in freshwater (FW) eels (Tierney *et al.*, 1995a). In the Japanese eel too, circulating angiotensin increased transiently after transfer to SW (Okawara *et al.*, 1987), but then returned to basal levels (Okawara *et al.*, 1987; Tsuchida and Takei, 1998), indicating a lack of a persistent elevation and suggesting roles in short-term adaptive function of osmoregulatory tissues. This pattern of response does not seem to apply in the jawless lampreys.

Our understanding of the RAS of jawless cyclostomes and its potential links to euryhalinity is mostly derived from work on the river lamprey, *Lampetra fluviatilis*. Capture of lampreys in the Ringkøbing Fjord, Denmark at the start of their upstream migration, has allowed measurements of Ang in lampreys acclimated for several weeks to a range of salinities, and after transfer from one salinity to another (Rankin *et al.*, 2001; Brown *et al.*, 2005). The first measurements of plasma Ang concentrations in river lampreys separated Ang II and Ang III (see Fig. 4.1) by high-pressure liquid chromatography (HPLC) (Rankin *et al.*, 2001). Both Ang II and Ang III were found to be significantly higher in SW-acclimated rather than in FW-acclimated river lampreys. These first studies also suggested a possible link between circulating Ang and rising plasma osmolality, and recent investigations have shown changes in plasma Ang concentrations after rapid changes in environmental salinity. Figure 4.3 shows a significant decrease in plasma Ang after the decrease in environmental salinity, alongside declining plasma osmolality and haematocrit, an index of plasma volume in lampreys that lack a splenic source of stored red blood cells or the secondary circulation that occurs in teleost fish (Brown *et al.*, 2005). Reduced RAS activity due to volume expansion in lampreys experiencing an acute decrease in the external osmolality is in keeping with the rapid rise in plasma Ang that was shown to occur after blood volume depletion (Brown *et al.*, 2005). Volume receptors are, therefore, likely to play a major role in regulating circulating Ang in lampreys during their exposure to different salinities.

In addition to volume receptors, osmo (sodium) receptors also appear to play a role in regulating the RAS of river lampreys. The evidence for this hypothesis comes from more than one experimental approach. Firstly, after an acute increase in external salinity, rising plasma Ang occurred but haematocrit was unaffected, indicating little change in plasma volume (Fig. 4.3). Secondly, intraperitoneal injections of known volumes of saline influence plasma concentrations of Ang in ways that suggest both volume and osmo/salt receptors. Intraperitoneal injection of 1% by volume of body mass of isosmotic saline led to a rapid reduction in circulating Ang that was not apparent after injection of the same volume of hyperosmotic saline, even though similar volume expansion was implied by the declining blood haematocrit (Fig. 4.4). The increased plasma osmolality after hyperosmotic injection suggests osmo/salt receptors interact with volume receptors in regulating the lamprey RAS. In placing these signals in context, with regard to the euryhalinity of lampreys, changes in plasma volume and electrolytes when anadromous lampreys migrate between FW and SW will act in a complementary fashion; high plasma volume and lowering of plasma osmolality in FW would both act to depress the RAS while reduced plasma volume and increased plasma osmolality would both

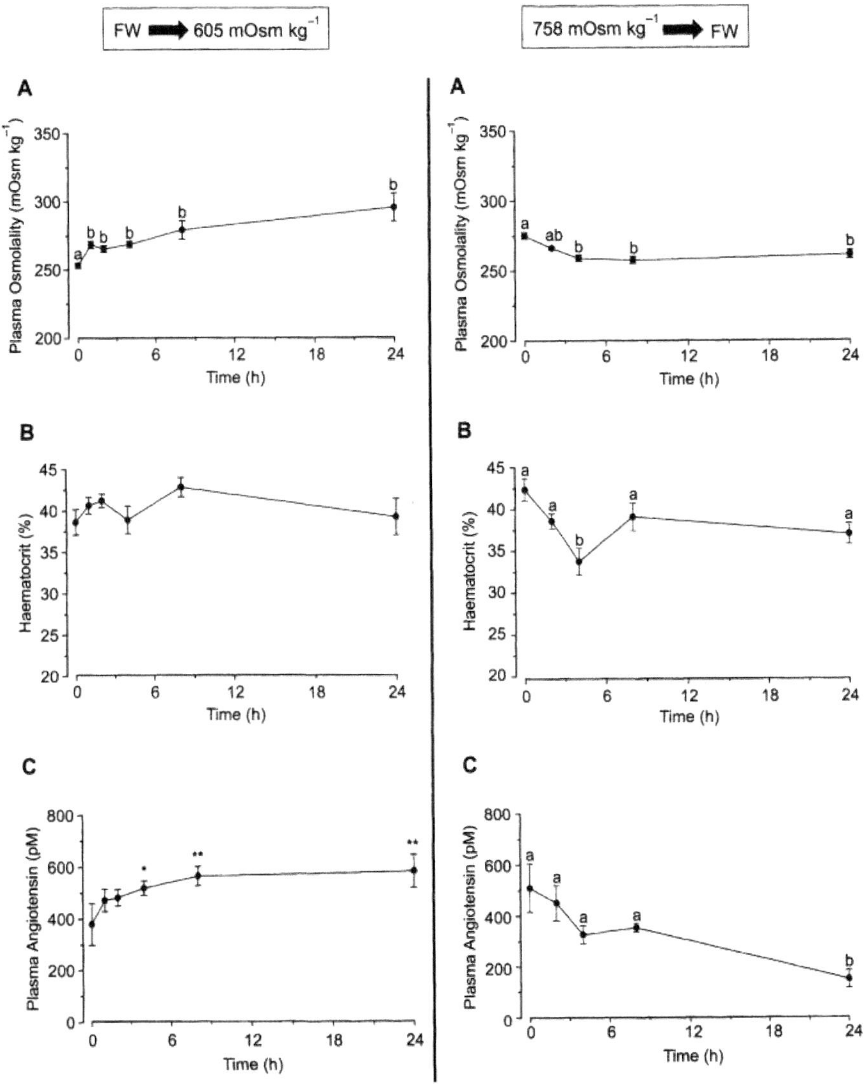

Fig. 4.3 Effects of a rapid increase in environmental salinity (FW to 21 ppt; 605 mOsm kg^{-1}) and a rapid decrease in environmental salinity (758 mOsm kg^{-1} to 22 mOsm kg^{-1}) in the river lamprey. Changes in (A) plasma osmolality (mOsm kg^{-1}), (B) blood haematocrit (%), indicative of changes in plasma volume in lampreys, and (C) plasma angiotensin concentrations (pM) are shown (n=7 to 9 at different time points). Values are means ± SEM. Groups with different letters differ significantly (ANOVA and *post hoc* multiple comparison tests). Asterisks indicate groups that differ significantly from time 0 h (* $p<0.05$, ** $p<0.01$, ANOVA and linear contrast analyses). Data presented in Brown *et al.* (2005).

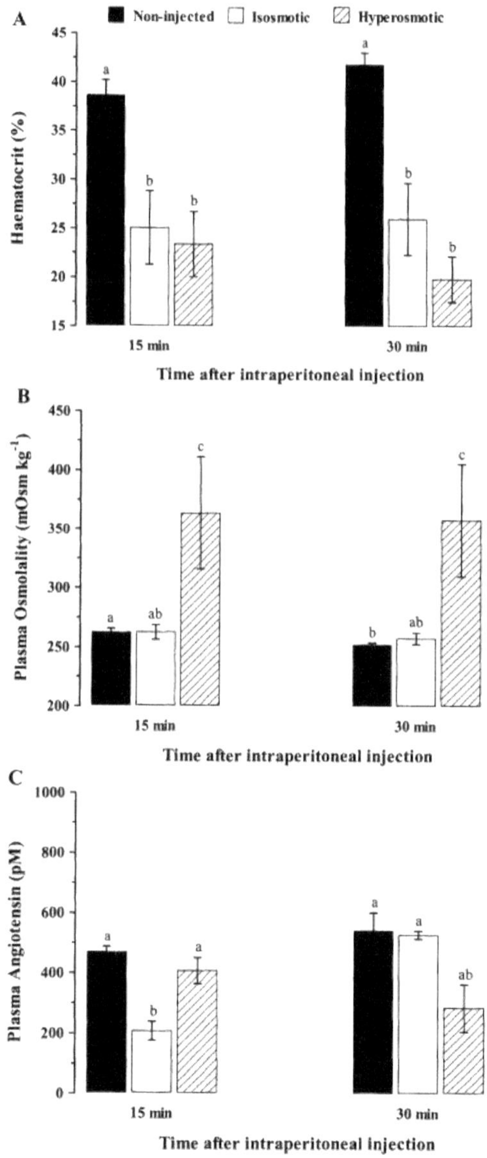

Fig. 4.4 Plasma angiotensin (pM) in river lampreys held in freshwater (Panel C), 15 min and 30 min after intraperitoneal injection of 1% body mass by volume of isosmotic saline (120 mM NaCl; 233 mOsm kg^{-1}; n=5 at each time point) or hyperosmotic saline (4M NaCl; n=5 at 30 min; n=8 at 30 min). Control lampreys (non-injected) were held in light anaesthesia for 15 min (n=8) and 30 min (n=4) in the absence of further manipulation. Panel (A) shows haematocrit (%) and panel (B) shows plasma osmolality (mOsm kg^{-1}) of these fish. Values are means ± SEM. Different letters above error bars signify groups that differ significantly (ANOVA and post hoc multiple comparison tests). Data presented in Brown *et al.* (2005).

activate the RAS when meeting hyperosmotic environments. However, lampreys feed on fish blood and tissues during the marine phase of their life cycle, and this could also have major impacts that are not necessarily complementary. Teleost blood would expand extracellular fluid volume with an approximately isosmotic load, reducing plasma Ang, but sea lampreys feed on sharks and this causes simultaneous volume and hyperosmotic challenges (Wilkie *et al.*, 2004). The implications of these differences at a functional level are fascinating but as yet unexplored.

Although many elasmobranch fish only inhabit the marine environment, there are true euryhaline elasmobranch fish that inhabit both full FW and SW environments and many species exhibit some capacity to acclimate to dilute SW, at least under laboratory conditions, and may be regarded as partially euryhaline (Hazon *et al.*, 2003). In SW, fully euryhaline elasmobranch fish osmoregulate as stenohaline marine fish by maintaining the body fluid osmolality slightly hyperosmotic to SW primarily due to the hepatic and muscle (Kajimura *et al.*, 2006) synthesis (via the ornithine–urea cycle) and renal retention of the nitrogenous compound urea (Thorson *et al.*, 1973; Piermarini and Evans, 1998). In FW, body fluid osmolality is maintained hyperosmotic to the environment and increased in comparison to teleosts (greater than twofold). This hyper/hyper osmoregulatory strategy in FW/SW for euryhaline elasmobranch fish is distinctly different to that employed by euryhaline teleosts. Until recently, our knowledge of the osmoregulatory strategy of fully euryhaline elasmobranchs was based largely on work carried out on the bullshark, *Carcharhinus leucas* (Thorson, 1962, 1967; Thorson and Gerst, 1972; Thorson *et al.*, 1973). Studies have now been extended to include the euryhaline Atlantic stingray, *Dasyatis sabina* (Piermarini and Evans, 1998; Piermarini and Evans, 2000, 2001; Janech and Piermarini, 2002) together with further studies of *C. leucas* (Pillans and Franklin, 2004; Pillans *et al.*, 2005). It is evident that there is a marked shift in the percentage contribution that sodium, chloride and urea provide to plasma osmolality in FW and SW environments. It appears that the euryhaline species, *D. sabina* and *C. leucas*, although displaying reduced plasma osmolality in FW compared to SW, maintain the capacity to retain urea as they acclimate to reduced environmental salinities (Hazon *et al.*, 2003). However, when both fully and partially euryhaline elasmobranchs acclimate to environments of higher salinity, plasma osmolality must increase rapidly so that plasma osmolality is maintained iso- or slightly hyperosmotic to the new environment. In this case plasma concentrations of sodium and chloride increase before urea (Anderson *et al.*, 2002b, 2006). The implication of these findings will be discussed later in this chapter (see text on p-109, **Drinking**).

There is now evidence that, as in teleost fish and lampreys, the RAS in elasmobranchs is activated in association with exposure to higher environmental salinities. In the partially euryhaline lesser spotted dogfish, *S. canicula*, Ang II concentration increased after acute transfer from 80% to 100% SW (Fig. 4.5; Anderson *et al.*, 2002a). In the fully euryhaline bullshark *C. leucas*, Ang II concentrations were also elevated during a more long-term transfer from FW to SW that occurred over a period of days (Anderson *et al.*, 2005).

ROLES OF ANGIOTENSIN IN OSMOREGULATION

Angiotensin has a wide range of biological actions in different tissues. Osmoregulatory actions generally appear to be fast acting and of relatively short duration. However, the continual presence of significant circulating levels of angiotensin in fish may result in longer-term regulatory effects. Also, the stimulation of the release of longer-acting steroid hormones may have prolonged effects on osmoregulatory processes. It is clear that in

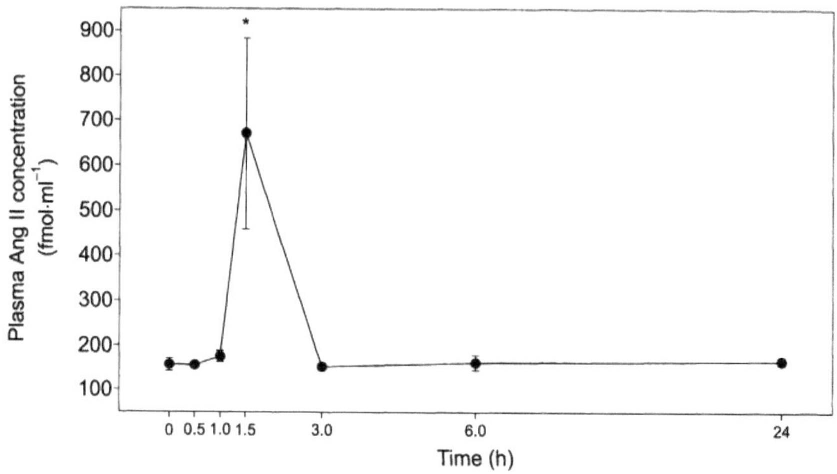

Fig. 4.5 Plasma levels of Ang II (fmol·ml^{-1}) in *S. canicula* following rapid transfer from 80% SW to 100% SW (n=7). Statistical comparison was made between values taken at time 0 h and immediately after transfer. Ang II was significantly increased within 90 minutes of transfer. Values are means ± S.E.M; *p<0.05 (Students t-test). Taken from Anderson *et al.* (2001a) (with permission).

mammals both a systemic circulating RAS and local paracrine systems exist and recent studies in fish are beginning to reveal similar complexities. In the following sections, we will review the actions of Ang II in the peripheral system and in paracrine systems and their regulation of drinking, renal function, rectal gland (in elasmobranchs), gill and interrenal function.

DRINKING

Amongst teleost fish, drinking is an important component of volume homeostasis in a hyperosmotic environment where loss of water across the gills is balanced by copious drinking of the environmental medium and absorption of a large proportion of the water passing through the intestine. Freshwater teleosts, already facing a continual osmotic water influx, were initially thought to drink none of the surrounding freshwater, but a closer examination showed this to be untrue (see Table 4.2: Kobayashi *et al.*, 1983; Fuentes and Eddy, 1997; Tsuchida and Takei, 1999), although drinking of FW may normally be linked to food digestion rather than any osmoregulatory requirement for water (Rankin, 2002).

The renin-angiotensin system was first revealed to play a prominent role in the regulation of drinking in studies of rats in which ligature of the vena cava (reducing cardiac output) stimulated drinking, but nephrectomized rats showed no drinking response indicating that a renal factor, believed to be renin, was necessary for the drinking response (Fitzsimons, 1964). Further studies led to the idea that angiotensin is an important physiological regulator of drinking (Fitzsimons and Simons, 1969; Fitzsimons, 1998). Angiotensin II induced drinking has now been demonstrated in all vertebrate groups, with the only apparent exception so far being the jawless fish.

Thirst and physiological drinking in mammals, birds and reptiles is determined by three main stimuli: increasing plasma osmolality (and linked cellular dehydration), decreased blood volume (= extracellular dehydration), and an increase in Ang II (Fitzsimons, 1998). Drinking rates in teleosts seem to be regulated primarily by extracellular dehydration (blood volume) and Ang II, which in themselves can be linked (Nishimura *et al.*, 1979; Olson, 1992), with little evidence or no evidence of stimulation of drinking by cellular dehydration after injections of hypertonic saline (Beasley *et al.*, 1986; Takei *et al.*, 1988). A dipsogenic action of Ang II was first demonstrated in two euyhaline teleosts, the eel and the killifish (Hirano *et al.*, 1978; Malvin *et al.*, 1980). Since then, dispogenic responses to Ang I (subsequently converted to Ang II) and/or Ang II have been shown in many teleosts (Table 4.2).

An early study compared the effects of intraperitoneal injection of Ang II on drinking in 37 species of teleosts, some considered stenohaline FW species, some euryhaline species and a group of stenohaline marine species, after adding phenol red to the environmental medium, as a marker for drinking (Kobayashi *et al.*, 1983). Many of the 'freshwater' species did not respond to Ang II, even though the circulating concentrations reached are likely to have been at pharmacological levels, at least transiently. Unresponsive species were *Pseudorasbora parva* (topmouth gudgeon), *Rhodeus ocellatus* (rose bitterling), *R. lanceolatus* (slender bitterling), *Ctenopharyngodon idealla* (grass carp), *Cobitis anguillicaudatus* (Asian pond loach), *Salvelinus leucomaenis* (Japanese char), *Pungitis sinensis* (Chinese stickleback), *Oreochromis mossambicus* (Mozambique tilapia) and *Cyprinus carpio* (carp) and this list was later extended when a lack of a drinking response to Ang II was reported in *Natropis cornutus* (common shiner), *Cottus bairdi* (mottled sculpin) and *Carassius auratus* (goldfish) (Beasley *et al.*, 1986). The goldfish had showed a drinking response to Ang II in Kobayashi's studies (Table 4.2) which raises certain questions relating to experimental protocols. The sequence of the Ang II employed is particularly important. Asp^1, Ile^5-Ang II was used by Beasley *et al.* but is unlikely to be the endogenous sequence (see Fig. 4.2). However, it has also been argued that the strain of goldfish used by Kobayashi *et al.* (1996), (that showed a drinking response to Ang II) was particularly saline tolerant (Kobayashi and Takei, 1996). Since these studies, intramuscular Ang I (which would be converted *in vivo* to Ang II) has also been shown to induce a dispogenic response in carp held in freshwater (Perrott *et al.*, 1992; see Table 4.2), but the carp was a non-responder to Ang II in Kobayashi's studies, and this too may reflect differences in salinity tolerance of the strains of carp employed. However, despite the drinking response to Ang I in the carp held in freshwater, and the evidence that inhibition of Ang I conversion to Ang II inhibited the drinking response, an effect of angiotensin on drinking in carp could not be demonstrated when carp were held in hyperosmotic brackish water (40% SW) (Perrott *et al.*, 1992). In these conditions, the carp increased their plasma osmolality by almost 50%, but did remain hyposmotic to the environment. This draws our attention to the importance of saline tolerance in any

Table 4.2 Teleost and elasmobranch fish: Ang-II stimulated drinking in fish held in freshwater (FW) or seawater (SW).

Species	*Environment*	*Ang II dose (per g body wt)*	*Peptide sequence*
Freshwater Teleosts			
Carassius auratus[3]	FW	ip 100 ng	Asn^1,Val^5-Ang II
C. carassius[3]	FW	ip 100 ng	Asn^1,Val^5-Ang II
Leuciscus hakonensis[3]	FW	ip 10 ng	Asn^1,Val^5-Ang II
Gambusia affinis[3]	FW	ip 100 ng	Asn^1,Val^5-Ang II
Gyrinocheilus anymonieri[3]	FW	ip 10 ng	Asn^1,Val^5-Ang II
Parasilurus asotus[3]	FW	ip 200 ng	Asn^1,Val^5-Ang II
Tridentiger obscurus[3]	FW	ip 100 ng	Asn^1,Val^5-Ang II
Cyprinus carpio[8]	FW	im 0.3 nmol	Val^5-Ang I
Euryhaline Teleosts			
Anguilla anguilla[8]	FW & SW	im 0.3 nmol	Val^5-Ang II
A. japonica[3]	FW	ip 10 ng	Asn^1,Val^5-Ang II
A. japonica[1, 5]	FW	min. ia 0.5 ng	Asp^1 or Asn^1, Ile-Ang II
A. japonica[1]	SW	min. ia 0.05 ng	Asn^1,Ang II
A. japonica[11]	FW	vascular; 0.1-10 fmol/min	Asn^1,Val^5-Ang II
A. japonica[13]	SW	intracranial 0.25-25 pmol	Asn^1,Val^5-Ang II
Oryzias latipes[3]	FW	ip 10 ng	Asn^1,Val^5-Ang II
Platichthys flesus[8]	FW	im 0.3 nmol	Val^5 - Ang II
P. flesus[4,6]	FW/SW	iv 6, 150 ng/min	Asn^1,Val^5 -Ang II
Salmo salar[10]	FW	im 0.4 nmol	Asn^1,Val^5-Ang II
Oncorhynchus mykiss[9]	FW	im 0.2 - 2 nmol	Asn^1,Val^5-Ang II
Seawater Teleosts			
Glossobius giuris fasciatopunctatus[3]	SW	ip 50 ng	Asn^1,Val^5-Ang II
Callionymus richardsoni[3]	SW	ip 1 ng	Asn^1,Val^5-Ang II
Hypodytes rubripinnis[3]	SW	ip 5 ng	Asn^1,Val^5-Ang II
Chasmichthys gulosus[3]	SW	ip 20 ng	Asn^1,Val^5-Ang II
Sillago japonica[3]	SW	ip 100 ng	Asn^1,Val^5-Ang II
Mugil cephalus[3]	SW	ip 500 ng	Asn^1,Val^5-Ang II
Pseudopleuronectes americanus[7]	SW	im 13 ng	Asp^1,Ile^5-Ang II
Pleuronectes platessa[8]	SW	im 0.3 nmol	Val^5-Ang II
Myxocephalus octodecemspinosus[7]	SW	im 13 ng	Asp^1,Ile^5-Ang II
Myxocephalus scorpius[8]	SW	im 0.3 nmol	Val^5-Ang II
Limanda limanda[8]	SW	im 0.3 nmol	Val^5-Ang II
Merlangius merlangus[8]	SW	im 0.3 nmol	Val^5-Ang II

(Table 4.2 contd.)

(Table 4.2 contd.)

Seawater Elasmobranch			
Scyliorhinus canicula[12]	SW	ia 0.01 - 2 nmol/kg	Asn[1],Pro[3],Ile[5]-Ang II
Triakis scyllia[12]	SW	ia 0.001 - 2 nmol/kg	Asn[1],Pro[3],Ile[5]-Ang II

ip = intraperitoneal; im = intramuscular; ia = intraarterial; FW = freshwater; SW = seawater

References: [1]Hirano *et al.* (1978); [2]Malvin *et al.* (1980); [3]Kobayashi *et al.* (1983); [4]Carrick and Balment (1983); [5]Hirano and Hasegawa (1984); [6]Balment and Carrick (1985); [7]Beasley *et al.* (1986); [8]Perrott *et al.* (1992); [9]Fuentes and Eddy (1996); [10]Fuentes and Eddy (1997); [11]Tsuchida and Takei (1999); [12]Anderson *et al.* (2001a); [13]Kozaka *et al.* (2003).

separation of freshwater, marine and euryhaline species when comparing Ang-II sensitive and Ang-II insensitive species, but the salinity tolerances and physiological adaptability to saline waters of many, so-called 'freshwater' teleosts are not established. The fundamental separation should lie in whether a teleost fish species can achieve hyposmoregulation in saline waters—which can only occur if a drinking response is attuned to the volume loss resulting from the hyposmoregulation—along with appropriate changes in intestinal, gill and kidney function to absorb the gut fluid, reduce urinary loss of fluid and excrete excess salts. These physiological processes occur in all marine teleosts. But, in a constant marine environment, there will be a stable water loss. Kobayashi *et al.* (1983) suggested that stenohaline marine teleosts do not show a drinking response to Ang II after showing that several of the exclusively marine teleosts: *Sardinops melanosticata* (Japanese sardine), *Trachurus japonicus* (Japanese horse mackerel), *Sebastes inermis* (black rockfish) and *Rudarius ercodes* (Japanese file fish), did not show dipsogenic responses to intraperitoneal injections of Ang II. However, several species that are usually considered stenohaline marine species: *Myxocephalus octodecemspinosus* (long-horned sculpin), *Pseudopleuronectes americanus* (winter flounder), *Pleuronectes platessa* (plaice), *Limanda limanda* (dab) and *Merlangius merlangus* (whiting), have subsequently shown drinking responses to Ang II injections (Beasley *et al.*, 1986; Perrott *et al.*, 1992; see Table 4.2). Nevertheless, Ang II effects on drinking are likely to be most important in responding to variable salinities and a need to continually adjust drinking rates in euryhaline fish.

The drinking response to Ang II has been most thoroughly investigated in the euryhaline flounder, *Platichthys flesus*, and in Japanese and European eels (Hirano *et al.*, 1978; Carrick and Balment, 1983; Kobayashi *et al.*, 1983; Balment and Carrick, 1985; Perrott *et al.*, 1992; Tierney *et al.*, 1995a,b; Takei and Tsuchida, 2000; Kozaka *et al.*, 2003), although many of these studies have been pharmacological in nature, using high levels of angiotensin. To examine the effects of physiological concentration of Ang II, intravenous infusions of low levels would be more appropriate, but few studies of this nature have as yet been undertaken. One such study infused 0.1 to 10 pmol/kg/min into volume replete, FW-acclimated, Japanese eels (with low basal drinking rates) and showed that even in FW eels, exogenous Ang II has a potent dipsogenic action in this euryhaline species (Tsuchida and Takei, 1999).

An alternative, powerful approach that has been employed to investigate the regulation of drinking in fish, rather than simply administering angiotensin (or angiotensin receptor antagonists/agonists), is the pharmacological manipulation of the endogenous RAS. The RAS of teleosts is activated in response to blood volume depletion (Nishimura *et al.*, 1979; Olson, 1992), so one favoured approach has been the use of papaverine, a smooth muscle relaxant (see Fig. 4.1). In eels, papaverine treatment results in a prolonged lowering of blood pressure (Tierney *et al.*, 1995a). Papaverine treatment of both FW- and SW-acclimated eels was found to increase circulating Ang II in the eel, presumably in response to the observed hypotension, and stimulated drinking, but the drinking response could be blocked by treatment with captopril, an inhibitor of Ang I to Ang II conversion (Tierney *et al.*, 1995 a,b). Similarly, administration of sodium nitroprusside (a nitric oxide releasing agent that induces hypotension in trout) stimulated drinking in Atlantic salmon fry held in FW, and this drinking was abolished by the angiotensin converting enzyme inhibitor, enalapril (Fuentes *et al.*, 1996).

Inhibition of Ang I conversion to Ang II using ACE inhibitors alone has had varied effects on drinking rates. Large doses of captopril (72 mg/kg injected intramuscularly or intravenously) inhibited drinking in the euryhaline SW-acclimated European eel (Perrott *et al.*, 1992; Tierney *et al.*, 1995a) but in two marine species that do not enter FW, winter flounder and longhorn sculpin, captopril had no effect on basal drinking rates, even though Ang II stimulated drinking by these two species (Beasley *et al.*, 1986). Physiological manipulation of the RAS to regulate drinking rates in stenohaline marine species, therefore, appears to have less prominent effects than in euryhaline species.

In a study of the copious drinking of SW-acclimated Japanese eels, slow infusion of captopril at 0.01 or 1 mg/kg/min significantly reduced drinking rates, even though Ang II was only significantly reduced at the highest dose (Takei and Tsuchida, 2000). This could mean that captopril at low doses achieves its effects through action on a brain RAS, and further work is needed to clarify whether a brain RAS exists in the eel or not. However, effects of captopril administration may be confounded by the actions of bradykinin that we might expect to increase during ACE inhibition, due to inhibition of bradykinin breakdown too (see Fig. 4.1). Bradykinin has been shown to inhibit drinking in SW-acclimated eels, independent of any change in blood pressure (Takei *et al.*, 2001), so depressed drinking rates after capropril treatment could simply reflect the actions of elevated bradykinin.

Although experiments with captopril do not enable a clear distinction of the effects of any brain RAS from those of peripheral angiotensin, an alternative experimental strategy, the infusion of an antiserum to Ang II into seawater-adapted eels, neutralizing the effects of free circulating Ang II, has added to the picture (Takei and Tsuchida, 2000). While plasma Ang II was dramatically reduced by this procedure, declining from more than 60 fmol/ml to undetectable levels, drinking rates did not decrease significantly. This suggested that the normal copious drinking of the SW-acclimated eel does not rely simply on circulating angiotensin, even though intravascular administration of angiotensin can exert a potent dipsogenic action in the eel (Tsuchida and Takei, 1999).

The dipsogenic action of Ang II in teleosts raises questions as to the localisation of Ang II receptors that result in drinking. In mammals, three circumventricular organs (CVOs), the subfornical organ (SFO), the organ vasculosum of the lamina terminalis (OVLT) and the area postrema (AP), highly vascularized brain centres that lie outside the blood brain barrier, have each been implicated in responding to circulating Ang II (Fitzsimons, 1998). For example, in rat the SFO responds to femtomole amounts of Ang II injected directly into the tissue, while destruction of this tissue blocks drinking responses to intravenous Ang II. However, in teleosts brain lesions appeared to have relatively little effect on Ang II-induced drinking responses (Hirano *et al.*, 1978) and led to the suggestion that Ang II in teleosts acts at the level of the medulla oblongata to induce swallowing (Takei *et al.*, 1985). In keeping with this idea, specific Ang II binding sites have been identified in the medulla oblongata of the rainbow trout, although further binding sites were also identified in the cerebellum (Cobb and Brown, 1992). In this autoradiographic study, estimates of receptor density were similar for trout acclimated to FW and SW, but as yet there has been no characterization of these receptors, which may differ in binding affinities. In fish species that are restricted to FW over their entire life span, it is feasible that on an evolutionary timescale receptors responsible for inducing the drinking response have been lost, and that this accounts for the apparent lack of a dipsogenic response to Ang II in many FW species.

In mammals, distinct Ang II receptors have also been identified within the blood brain barrier, and argued to be inaccessible or less accessible to circulating Ang II, instead forming a component of an exclusively brain RAS. The brain RAS is also thought to participate in controlling water intake, as well as other physiological actions of Ang II such as blood pressure regulation (Bader *et al.*, 2001). Angiotensin II immunoreactivity and binding sites are widely distributed in the mammalian central nervous system and their actions may include brain remodelling and cell differentiation that could include long-term changes in neuronal organization affecting such aspects as sodium appetite (Fitzsimons, 1998).

Specific receptors for Angiotensin in the brain have been indicated by recent studies of SW-acclimated eels in which angiotensin was injected via a cerebral cannula inserted into the fourth ventricle. Intracranial Ang II rapidly stimulated drinking (Kozaka *et al.*, 2003). Intravenously injected Ang II that might only have access to brain areas lacking a blood brain barrier (believed to include the magnocellular preoptic nucleus and anterior tuberal nucleus of the hypothalamus and the area postrema of the medulla oblongata) also enhanced water intake in eels (Tsuchida and Takei, 1999; Ando *et al.*, 2000; Kozaka *et al.*, 2003). However, comparison of the time course of the effects of intracranial and intravascular Ang II on drinking showed that intracranial Ang II exerts a more persistent effect than circulating Ang II (Kozaka *et al.*, 2003). The authors argued that this might reflect the release of atrial natriuretic peptide in response to circulating angiotensin that inhibits drinking in the Japanese eel (Takei, 2000). Thus, endogenous stimulation of drinking in euryhaline teleosts is perhaps more likely to involve local release of Ang II from a brain RAS. Although there is no direct evidence of a brain RAS as yet in the eel, the evidence of a brain RAS in the rainbow trout does support this idea.

Until fairly recently, it was thought that elasmobranch fish did not drink (Kobayashi *et al.*, 1983). However, this assumption was based largely on elasmobranchs acclimated only to full SW where plasma osmolality is maintained iso- or slightly hyper-osmotic to the surrounding SW and a small but constant influx of water means there is no physiological requirement to drink. Furthermore, these early studies employed phenol red as the marker for drinking rate and this is now regarded as a relatively insensitive technique. For example, Kobayashi *et al.* (1983) reported a lack of drinking response in the Japanese dogfish, *Triakis scyllia*. However, using the sensitive oesophageal cannulation technique Anderson *et al.* (2001a) demonstrated that *T. scyllia* does indeed drink and drinking rate increases in response to transfer to higher salinity environments (Anderson *et al.*, 2002c). On entering environments of higher salinity, plasma osmolality of both fully and partially euryhaline elasmobranchs must increase rapidly as plasma osmolality is maintained iso- or slightly hyper-osmotic to the new environment. The first report of drinking in elasmobranch fish during acclimation to increased environmental salinity was in the European spotted dogfish, *S. canicula* (Hazon *et al.*, 1997b). Subsequently, Anderson *et al.* (2002c) demonstrated that this drinking response was maximal approximately 2 hours after transfer from 80 to 100% seawater and resulted from extracellular dehydration (Anderson *et al.*, 2002b), as in teleost fish (Takei *et al.*, 1988) and not cellular dehydration, as occurs in mammals (Fitzimons, 1998). Indeed, cellular dehydration induced by injection of hyper-osmotic sodium chloride-rich Ringer caused an inhibition in the drinking response in *S. canicula* (Anderson *et al.*, 2002b). As discussed earlier, the rapid increase in plasma concentrations of sodium and chloride before plasma urea concentrations increase suggests that imbibed fluid is rapidly reabsorbed in the gastointestinal tract in order to raise plasma osmolality. This response maintains the iso-hyper osmoregulatory strategy as part of the acclimation response to entering environments of higher salinity. However, to date, there is no direct information of gut absorption of imbibed fluid in elasmobranchs. It is evident, therefore, that elasmobranchs do have the physiological capacity to drink and will do so when there is a requirement to rapidly increase plasma osmolality following transfer to a hyper-osmotic medium.

The first evidence that the RAS was involved in the control of the dipsogenic response of elasmobranchs was the demonstration that administration of papaverine, a smooth muscle relaxant that stimulates the endogenous RAS (Fig. 4.1), caused a significant increase in the drinking rate in *S. canicula* (Hazon *et al.*, 1989). Furthermore, the ACE inhibitor captopril, when co-administered with papaverine, reduced the papaverine-induced drinking response to basal levels. However, heterologous Ang II administration did not alter basal drinking rates. Recently, homologous Ang II has been shown to induce a dose-dependent increase in drinking rates in both *S. canicula* and in the Japanese dogfish *T. scyllia* (Table 4.2; Anderson *et al.*, 2001a). The drinking response was further investigated in these two species of fish after acclimation to 80% SW and then rapid transfer to 100% SW (Anderson *et al.*, 2002a). Plasma sodium and chloride concentrations increased rapidly within 6 hours although plasma urea concentrations remained unaltered. The drinking rate, determined by oesophageal cannulation (Anderson *et al.*, 2001a), was significantly increased, reaching maximal rates within 2-3 hours post-transfer. Plasma concentrations of Ang II increased rapidly, reaching a maximum concentration 90 minutes after transfer to 100% SW (Fig. 4.6). The time course of the Ang II response was similar to, but just preceded, the increase in drinking rate (Anderson *et al.*, 2002c). It is clear,

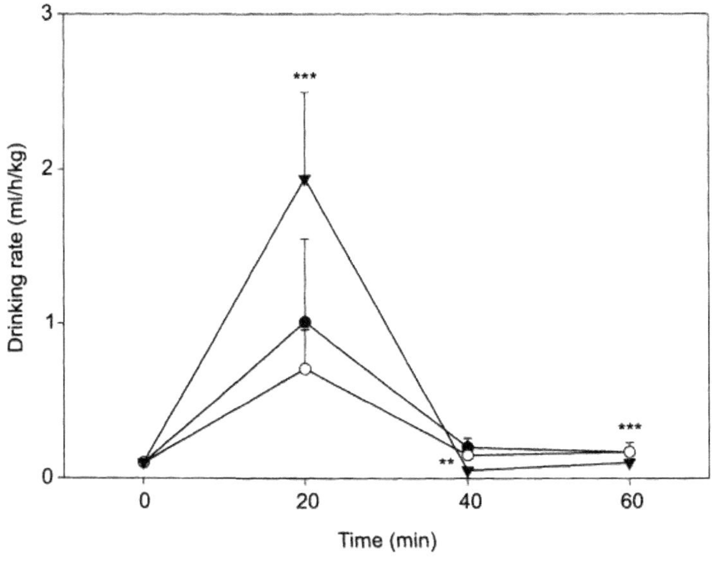

Fig. 4.6 Short term assessment of drinking rate in *S. canicula* following administration of 2 nmol/kg homologous Ang I (○), Ang II (●) and papaverine 10 mg/kg (▼). In the first 20 min post-injection drinking rates in all 3 treatments were significantly higher than basal rates ***$p<0.001$. Drinking rates after Ang I and Ang II treatment remained significantly higher than basal for 60 min post-injection, ***$p<0.001$ ($n>6$). Between 20 and 40 min, drinking rates after papaverine treatment were significantly lower than basal levels **$p<0.01$ (n = 9, 7 and 8 for the respective treatments). Values are means ± SEM. From Anderson *et al.* (2001a) (with permission).

therefore, that the RAS is closely involved in the drinking response of elasmobranch fish.

Drinking has also been demonstrated in lampreys—both the sea lamprey, *Petromyzon marinus* and the river lamprey, *Lampetra fluviatilis* (Rankin, 2002). *L. fluviatilis* acclimated to ocean strength seawater (1015 mOsm kg^{-1}) drank at 7.4 ± 0.9 ml kg^{-1} h^{-1} (n=6) while FW-acclimated river lampreys, that experience osmotic water influx, only drank 0.08 ± 0.04 ml kg^{-1} h^{-1}. Drinking would, therefore, appear to be an essential component in the volume homeostasis of lampreys. Drinking in SW-acclimated lampreys was coupled to absorption of more that 99% of the sodium in the intestine (Rankin, 1997) and is likely to drive water absorption by processes similar to those now well known in teleosts. However, unlike those in teleosts, studies so far have failed to demonstrate any effects of the RAS on drinking in lampreys. Kobayashi *et al.* (1983) initially found no increase in drinking by FW-acclimated lampreys after Ang II injections. This could reflect an inability to reinstate marine osmoregulatory mechanisms after migration to FW where progressive gut degeneration occurs, but further studies involving lampreys caught early in their upstream migration and transferred back to hyperosmotic brackish water so as to reinitiate drinking, still failed to reveal any evidence of angiotensin effects on drinking (Rankin *et al.*, 2001). The circulating concentrations of angiotensin achieved in these studies were not measured, but the results do suggest limited or no effect of angiotensin on drinking. Papaverine, the smooth muscle relaxant that has been used so effectively to demonstrate drinking linked to volume regulation in elasmobranchs and teleosts (Hazon *et al.*, 1989; Perrott *et al.*, 1992; Tierney *et al.*, 1995b) did not stimulate drinking in lampreys acclimated to 50% SW, and in fact appeared to inhibit drinking (Rankin *et al.*, 2001). The effects on blood pressure were not explored; so papaverine may not have worked effectively in this experiment. However, injection of the ACE inhibitor, captopril, that inhibits conversion of Ang I to Ang II, did reduce drinking rates of lampreys acclimated to 50% SW (Rankin *et al.*, 2001). Although the action of captopril on bradykinin formation and its inhibition of drinking in teleosts (Takei *et al.*, 2001) must be borne in mind, further studies are warranted. As things stand, the evidence for angiotensin action on drinking in lampreys is weak, but further work is warranted.

Drinking, as a component of the osmoregulatory mechanisms to achieve volume homeostasis, is clearly of the greatest importance in fish that suffer osmotic water loss, and for this reason, might be considered of minimal importance in hagfish that are virtually isosomotic to a marine environment. Early studies suggested that hagfish do drink their environmental medium, but found no evidence for water absorption from the gut (Morris, 1960), suggesting that drinking, plays a minimal role in fluid balance in hagfish. However, even in this environment, fluid loss in the small amounts of urine that are excreted does need replacement, and drinking (as seen in teleosts and elasmobranchs) would be one way to achieve this. Further research has confirmed that hagfish (inshore hagfish, *Eptatretus burgeri*) drink their environmental medium, but found no evidence of any action of angiotensin on drinking rates (Kobayashi *et al.*, 1983).

To be effective in acquiring water, drinking responses need to be accompanied by functional changes in the gut to achieve solute-linked water absorption. Recent study indicates that bicarbonate secretion in the marine teleost intestine also plays an important role in facilitating water absorption (Wilson *et al.*, 2002). Angiotensin II ligand-binding sites appear to be present in the teleost intestine (Cobb and Brown, 1992), so Ang II may play a role in regulation of transport processes to maximize water absorption. This idea has yet to be examined in fish, but is in keeping with the evidence that Ang II increases sodium and water transport across all regions of the mammalian intestine (Jin *et al.*, 1998) and stimulates an alkaline secretion in the duodenum (Johansson *et al.*, 2001). If similar actions exist in teleost fish, they could play an important part in SW acclimation.

RENAL FUNCTION

The primary role of the kidneys in freshwater teleosts is to excrete the excess water resulting from osmotic water influx while retaining solutes. Glomerular filtration rates (GFRs) in FW teleosts are, therefore, typically high, and only a low proportion of the filtered water (usually less than half, but sometimes as little as a few %) is absorbed in the renal tubules (Rankin *et al.*, 1985; Brown *et al.*, 1993). In contrast, GFR in SW teleosts is typically much lower than that of FW teleosts and filtration is followed by an increased reabsorption of water with tubular secretion of calcium, magnesium, sulphate and phosphate (Brown *et al.*, 1993; Pelis and Renfro, 2004) to produce a more concentrated urine, usually approaching the overall osmotic concentration of plasma and rich in divalent ions.

The first evidence that angiotensin plays a role in regulation of renal functional changes in euryhaline teleosts was obtained from experiments involving intravenous infusion of Ang II into anaesthetized rainbow trout (Brown *et al.*, 1980). The profound antidiuretic action of Ang II in these experiments was found to be primarily due to a reduction in GFR. Specific binding sites for Ang II in the trout glomerulus and associated with the vascular pole were subsequently demonstrated (Brown *et al.*, 1990; Cobb and Brown, 1993).

The reduced GFR in trout receiving exogenous Ang II could in theory reflect reduced individual (single nephron) filtration rates (SNGFRs) or a reduction in the number of actively filtering glomeruli, although the latter are the most usual cause of changes in GFR in teleosts (Brown *et al.*, 1978, 1993; Rankin *et al.*, 1985). Changes in glomerular ultrastructure occur after Asn^1, Val^5-Ang II infusion of freshwater trout, with a reduced interdigitation of epithelial pedicels and flattening of the epithelial podocytes (Gray and Brown, 1987) and the resultant ultrastructure is reminiscent of the appearance of glomeruli in SW-acclimated trout (Brown *et al.*, 1983). However, SNGFRs appear to be higher in SW-acclimated trout than FW-acclimated trout, and angiotensin infusion had no effect on the SNGFRs of FW-acclimated trout, even though overall GFR was reduced by 50% (Brown *et al.*, 1980), so structural changes do not seem to result in a reduced ultrafiltration coefficient, thereby decreasing SNGFR. Instead, Ang II treatment increases the proportion of non-filtering glomeruli (Brown *et al.*, 1980, 1993). Direct evidence of this action was obtained through the use of sodium ferrocyanide (a marker for GFR that can be visualized after formation of Prussian blue, revealing the filtering nephrons) by a method originally devised by Hanssen in 1958, but that gained wide use in renal studies after Baines and de Rouffignac developed the technique in the early 1960s so as to obtain measurements for individual (single nephron) filtration rates. However, the original, non-quantitative Hanssen approach has been the most widely used method (Aukland, 2001). In the rainbow trout, this technique provided direct evidence that Ang II reduces the filtering population of nephrons and comparison of the proportion of filtering nephrons in FW- and SW-acclimated fish suggested that Ang II is important in regulation of glomerular intermittency and renal responses to water availability. Experiments using ferrocyanide also suggested reduced blood flow to individual glomeruli (assessed qualitatively by the intensity of Prussian blue precipitation after ferrocyanide infusion), presumably, in the extreme, leading to the cessation of filtration in many of the nephrons (Brown *et al.*, 1980, 1993).

In considering the studies that have employed infusions of exogenous peptides to examine the renal action of Ang II in teleosts, it is essential to appreciate the complications that may arise. Angiotensin is a powerful vasoconstrictor in all vertebrates (Kobayashi and Takei, 1996) and this action may confuse interpretation of the renal effects of Ang II, particularly as in fish, renal function is not autoregulated, in contrast to the position in mammals (Brown *et al.*, 1993). Therefore, in teleosts, changes in blood pressure often (but not always) result in parallel changes in GFR and urine output (Nishimura and Sawyer, 1976; Brown *et al.*, 1993). As a result, prolonged infusion of Ang II in trout resulted in an initial antidiuresis that was then abated by the coincident pressor action (Gray and Brown, 1985). This may explain why some of the early studies of the renal effects of Asn^1, Val^5-Ang II in both the lungfish, *Neoceratodus forsteri*, and the American eel, *Anguilla rostrata* suggested a diuretic action; an intraarterial infusion of exogenous Ang II was reported to result in a diuresis in the American eel as a result of increased GFR but effects only occurred after clear vasopressor responses to 10 and 100 ng/kg/min (Nishimura and Sawyer, 1976).

An additional complication in interpreting results obtained so far in whole animal studies is that circulating concentrations of Ang II during intravenous infusions of Ang II or after injections of Ang II have not been determined, so it is not possible to be certain that physiological effects have been examined. However, use of an alternative *in vitro* approach, the *in situ* perfused kidney that was developed in the early 1990s by Dunne and Rankin, has added to the picture, and enabled the effects of physiological concentrations of Ang II to be investigated. Using this preparation, physiological concentrations of Ang II (Dunne and Rankin, 1992) and the heptapaptide, (angiotensin 2-8) Ang III (Brown and Balment, 1997; Pope, 2002) have been shown to be antidiuretic. Ang III at 10^{-11} M decreased urine production by 22% while 10^{-9} M decreased urine output by 74 %. These studies support the hypothesis that Ang II exerts an antidiuretic action on the teleost kidney and may play an important role in the antidiuresis that occurs in acclimation to hyperosmotic environments. In addition to an antidiuretic action of circulating Ang II, studies with the perfused kidney indicated stimulation of magnesium secretion (Dunne and Rankin, 1992).

Tubular actions of Ang II were also suggested by studies in aglomerular teleosts. In the aglomerular goosefish, *Lophius americanus*, Ang II was reported to be diuretic and natriuretic (Churchill *et al.*, 1979) and these effects cannot, in aglomerular fish, reflect a direct link between blood pressure and glomerular filtration! It was suggested that this action is likely to reflect inhibition by Ang II of the reabsorption of sodium chloride and the osmotic water reabsorption. This is contrary to the significant increase in water reabsorption that accompanies the glomerular antidiuretic action of Ang II in the glomerular trout kidney (Brown *et al.*, 1980). In further studies conducted on the aglomerular goosefish, inhibition of ACE to lower circulating Ang II resulted in a decrease in urine output and increased excretion of potassium, magnesium and calcium without affecting blood pressure (Churchill *et al.*, 1985). This supports the concept that endogenous Ang II is diuretic in aglomerular fish independent of actions on blood pressure, although changes in intrarenal blood flow and its distribution cannot be ruled out. Actions of Ang II on tubular transport processes are a possible explanation, but direct studies of the actions of Ang II on renal tubular transport processes in either glomerular or aglomerular teleost nephron have yet to be performed.

In the past 5 years, increasing evidence for a complete renal RAS that has the potential to operate independently of the systemic system has been obtained in trout. Parts of a renal RAS in fish had long been identified within the kidney, as this organ is the established prime site for renin synthesis. Added to this, many studies have reported angiotensin converting enzyme activity in teleost kidneys (reviewed in Cobb *et al.*, 2004) and our recent research has demonstrated expression of the ACE gene in the trout kidney (Paley *et al.*, 1998). Demonstration of angiotensinogen mRNA in kidney tissue provided the final element for a complete renal RAS (Brown *et al.*, 1995, 2000; Aust, 2002). More recently, extraction of angiotensin from kidney homogenate after *in vitro* perfusion of the trout trunk has added weight to the hypothesis that an active intrarenal RAS occurs in trout (Brown, Aust and Frankling, unpublished).

The first studies aimed at investigating the actions of the renal RAS of trout, used the *in situ* perfused trout kidney preparation treated with captopril (Brown *et al.*, 2000). This preparation excludes the liver and so precludes existence of hepatic angiotensinogen and the systemic RAS. Captopril treatment (to inhibit Ang II formation) in preparations perfused at a normal pressure head (38 cm H_2O) had no effect on renal function, but when preparations were perfused at a sub-physiological pressure of 25 cm H_2O, which reduced perfusate delivery rates, captopril treatment resulted in a glomerular diuresis. These results suggest firstly that the intrarenal RAS in teleost fish is activated by low vascular pressure and/or blood flow rates, and secondly that the angiotensin formed by the intrarenal RAS plays an important role in regulating urine production. The increased GFR during captopril treatment implies possible vascular actions of a renal RAS that would *in vivo* presumably work hand in hand with the established vascular actions of circulating Ang II. However, further work is essential before the role and action of the renal RAS in fish and how it operates alongside the systemic RAS can be fully appreciated.

Further work gives several lines of evidence that suggest renal tubular actions of an intrarenal RAS in teleosts and it may be here that the distinction from the actions of circulating Ang II appear. Firstly, *in situ* hybridization has localized angiotensinogen mRNA in the proximal tubules of the rainbow trout, suggesting that this is an important site for Ang II formation. Secondly, measurable amounts of angiotensin have been collected in urine produced by *in situ* perfused trout trunks (Aust, 2002). Since proteases within the renal tubule and collecting ducts will degrade the majority of intratubular angiotensin, measurable urinary concentrations suggest that high concentrations are likely to occur in tubular fluid *in vivo*, and act on the apical membrane of the tubular epithelium. In the mammalian kidney, nanomolar concentrations of Ang II occur in tubular fluid (Navar and Nishiyama, 2004) compared to pM levels in circulation. Current evidence from the trout leads us to believe that there was an early evolution of this control system, with teleosts possessing an intratubular RAS. Binding of ^{125}I-Ang II to trout tissues suggests Ang II binding sites in the renal tubules (Cobb and Brown, 1992) and, in the eel, a renal tubular apical binding site specific for Ang II was reported (Marsigliante *et al.*, 1994). At the time of these studies, it was envisaged that these receptors would bind circulating Ang-II after filtration, but in the light of the recent demonstration of an intrarenal RAS it seems more likely that Ang II generated within the tubular fluid will act on the apical membrane ligand binding sites.

In early studies, intravenous Ang II infusion increased the renal tubular water reabsorption of FW-acclimated trout, increasing the urine to plasma concentration ratio of iothalamate, the GFR marker employed

(Brown *et al.*, 1980). Whether this action reflected the effects on the basal or apical epithelium could not be determined from the whole animal studies. To obtain a clear picture of renal tubular actions of Ang II in teleosts we now require more detailed studies such as microperfusion of teleost nephrons to specifically examine the actions of Ang II and Ang III on epithelial transport processes that may result *in vivo* from circulating Ang II, or Ang II generated within the kidney, including within the renal tubules. Of the few studies so far, one has examined the effects of Ang II on Na^+, K^+-ATPase activity after administration of Ang II to perfused kidneys and after exposure of isolated renal cells to Ang II (Marsigliante *et al.*, 2000). Both approaches showed that Ang II stimulates Na^+, K^+-ATPase in FW-acclimated eels, in a dose-dependent way, with maximal 1.8 × increase in the presence of 100 nM Ang II, but with no effect in SW-acclimated eels, where Na^+, K^+-ATPase activity was already elevated. Renal tubular Na^+, K^+-ATPase is involved in sodium transport across the basolateral border of the renal tubules and in SW-acclimated fish leads to water reabsorption (with sodium being regulated by the gills). Hence, the effects on Na^+, K^+-ATPase activity reported in eels may explain the earlier observations of increased renal tubular water absorption in the trout.

Marine elasmobranch fish produce urine that is hypoosmotic to blood plasma and the kidney is not regarded as a site for net excretion of sodium and chloride: this is the role of the rectal gland. The elasmobranch kidney consists of a series of nephrons that are both structurally and functionally highly complex (Lacy and Reale, 1991a, b, 1995; Hentschel *et al.*, 1993; Lacy *et al.*, 1995). Marine elasmobranchs maintain their plasma osmolality slightly higher than that of the surrounding environment primarily due to the hepatic and muscle synthesis and renal retention of the nitrogenous compound urea. As more than 90% of filtered urea is reabsorbed, the major role of the elasmobranch nephron appears to be urea retentive, although much further work is required to elucidate the full functions of the nephron. One of the major reasons for this lack of knowledge is that renal studies in marine elasmobranch fish are technically difficult to perform, at least in part, due to the very long and convoluted urinary sinuses. Establishing basal urine flow rates—a prerequisite for *in vivo* renal studies—is almost impossible without extremely long periods of urine collection. In order to avoid some of these complications and to determine the actions of individual peptides on kidney function, a perfused trunk preparation with the kidney *in situ* was developed for elasmobranch fish (Wells *et al.*, 2002) based on a technique previously described in teleosts (Dunne and Rankin, 1992; Amer and Brown, 1995). To investigate the possibility that the RAS in elasmobranchs may be involved in control of GFR, the *in situ* trunk preparation in *S. canicula* was perfused with 10^{-9} M Ang II and resulted in a significant decrease in urine flow rate and GFR. This was associated with a decrease in the filtering population of nephrons as determined by the ferrocyanide technique (Wells *et al.*, 2005). The renal effects of Ang II did indeed appear to be glomerular as no significant changes were observed in any of the tubular parameters measured (Wells *et al.*, 2005). These results suggest that RAS is important in the control of GFR as previously proposed for teleosts (Brown *et al.*, 1980).

Ang II has been shown to have profound vasoconstrictor effects in elasmobranchs (Tierney *et al.*, 1997b; Hamano *et al.*, 1998) and this may, at least in part, account for the effects in the kidney. Glomerular bypass shunts have been identified in *S. canicula* (Brown and Green, 1992) and it is possible that Ang II may act at these shunts to vary the degree of glomerular perfusion and control the proportion of active glomeruli. Tierney *et al.* (1997a) reported the presence of Ang II receptors in the elasmobranch kidney as has been previously reported for teleost fishes (Cobb and Brown, 1993), although the precise location of these receptors remains to be established. Preliminary evidence has also been presented for an intrarenal RAS in the dogfish, *S. canicula* (Wells *et al.*, 2003) that is likely to play a role in regulation of renal function.

As in teleosts and elasmobranchs, Ang II appears to exert an antidiuretic action in the lampreys, although the effects appear less pronounced than in the teleosts (trout) or elasmobranchs (dogfish). Indeed, initial studies reported a lack of any renal effect of Ang II injected into the caudal vein of river lampreys adapted to a brackish medium (Rankin *et al.*, 1985), but this may have been because urine flow rates were already very low, and further reduction hard to discern. Later studies in freshwater-acclimated lampreys indicated an antidiuretic action (Rankin, 1997). More recent research has suggested that the pressor action of exogenous Ang II (Rankin *et al.*, 2004) may compromise the detection of an antidiuretic action since lampreys, like teleosts, cannot autoregulate their GFR. Increased blood pressure results in an increased GFR, but in lampreys the glomerular diuresis results from an elevation in individual nephron filtration rates, rather than an increase in the filtering population of glomeruli, as there is a lack of glomerular intermittency (Brown and Rankin, 1999). In both sea lamprey, *Petromyzon marinus* and the river lamprey, *Lampetra fluviatilis*, acclimated to FW after capture during upstream migration, only high doses of Ang II (10^{-10} to 4×10^{-9} moles min^{-1} kg body wt^{-1} infused intravenously) were antidiuretic (Cobb *et al.*, 2002b), and no changes in urine composition were identified. The circulating concentrations achieved in these experiments would be likely to have exceeded physiological levels, so the significance of Ang II in regulating renal function in river lampreys remains unclear. An antidiuretic action would be most appropriate in marine lampreys where renal water conservation plays an important role in osmoregulation (Rankin, 1997), but experiments that manipulate the RAS and explore renal function in lampreys acclimated to SW have yet to be performed.

RECTAL GLAND

Elasmobranch fish possess a unique gland, the rectal gland that secretes a fluid isosmotic to blood, but essentially composed of sodium and chloride (Shuttleworth, 1988). The rectal gland is highly vascularized and secretion rates of the gland are highly dependent on blood flow to the secretory epithelia (Anderson *et al.*, 2002c) with reduced blood flow through the gland demonstrated at minimal secretion rates (Kent and Olson, 1982). The intermittent nature of rectal gland secretion suggests that it may be regulated by a series of endocrine/neuroendocrine factors acting to regulate blood flow. As Ang II is a potent vasoconstrictor in elasmobranchs (Tierney *et al.*, 1997b; Hamano *et al.*, 1998) and Ang II-like binding sites and angiotensin converting enzyme like activity have been reported in the rectal gland of *S. canicula* (Masini *et al.*, 1994; Hazon *et al.*, 1997a), Ang II seemed a likely candidate to influence rectal gland function. However, perfusion of the isolated rectal gland of *S. canicula* with 10^{-9} M Ang II did not affect either secretion rate or vascular perfusion of the secretory epithelium (Anderson *et al.*, 2002c). Several factors have been reported to both stimulate (CNP, Solomon *et al.*, 1992; VIP, Stoff *et al.*, 1977; rectin, Shuttleworth, 1988; scyliorhinin II, Anderson *et al.*, 1995a) and inhibit (NPY, Silva *et al.*, 1993; somatostatin, Silva *et al.*, 1985) rectal gland function, and it may be that the actions of Ang II involve interactions with some of these factors (Anderson *et al.*, 1995b), although it should also be considered that some of the reported stimulatory effects may be species dependent.

GILL FUNCTION

Gills play an important role in salt balance in teleost fish and autoradiographic studies suggests the existence of Ang II binding sites (Table 4.1; Cobb and Brown, 1992) that may have a significant role in osmoregulation. One of the few studies of the effects of Ang II on gill function showed a reduced transepithelial potential in the isolated perfused gill of flounders, suggesting reduced chloride transport (Lyndon, 1993). The idea of Ang II actions on branchial transport processes has since been supported by the identification of Angiotensin receptors on chloride cells of the eel (Marsigliante *et al.*, 1996, 1997a). Adaptation to SW increases the number of chloride cells (and associated Na^+, K^+-ATPase activity) linked to the role of the gills in sodium efflux, so it is perhaps not surprising that Ang II binding sites in the gills were shown to increase after acclimation to SW, but even after normalization to the number of chloride cells present a three fold increase in receptor density was reported (Marsigliante *et al.*, 1997a), supporting the idea that Ang II regulates gill function. There are many hormones that influence gill Na^+, K^+-ATPase activity (e.g. cortisol, growth hormone, insulin growth factor) and to the list we must add Ang II. The action of Ang II on gill function is rapid with short-term exposure of gill homogenates to Ang II or a 30 min perfusion of gills with Ang II resulting in increased Na^+, K^+-ATPase activity in FW-acclimated eels (Marsigliante *et al.*, 1997a). These studies also showed the existence of two Angiotensin receptor isoforms in the gills with both present in SW-acclimated eels, but one was lost in the FW-acclimated eels, but these complexities that have yet to be fully understood.

In contrast to teleost fish, the role of the gill in elasmobranch osmoregulation is poorly understood. Chloride cells have been identified in elasmobranch gills (Wright, 1973; Laurent and Dunel, 1980), but elasmobranch branchial Na^+-K^+-ATPase activity was reported as 10-15 times less than that found in teleosts (Jampol and Epstein, 1970). Investigation of Na^+, K^+-ATPase activity and abundance in *D. sabina* (Piermarini and Evans, 2000) demonstrated that SW acclimation induced a decrease in the activity and relative abundance of Na^+, K^+-ATPase in the gills. In FW, *D. sabina* were shown to have a higher number of Na^+, K^+-ATPase-rich cells localized primarily on the gill lamellae. The activity, location and abundance of Na^+, K^+-ATPase in FW-acclimated *D. sabina* suggests a role in ion-uptake for Na^+, K^+-ATPase-rich cells (Piermarini and Evans, 2000). Following the identification and localization of the vacuolar proton-ATPase (V-H^+-ATPase) in the gills of *D. sabina,* a dual role was proposed for V-H^+-ATPase and Na^+, K^+-ATPase cell types in combined regulation of sodium and chloride with acid base balance in both FW and SW (Piermarini and Evans, 2001).

In terms of RAS control of gill function in elasmobranchs, the only evidence available to date is indirect. Receptors for Ang II have been identified in membrane fractions prepared from gill cells of the Japanese dogfish, *T. scyllia* (Tierney *et al.*, 1997a). However, it is not known whether Ang II acts on the gill epithelia in a similar way to teleost fish although it is highly probable that Ang II will influence the perfusion of blood through the gills due to its vasopressor action on elasmobranchs vasculature (Hazon *et al.*, 1999).

STEROIDOGENIC ACTIONS OF ANGIOTENSIN AND LINKS TO OSMOREGULATION

In higher vertebrates, a major role for Ang II is in control of sodium retention through the stimulation of adrenal steroid release (Kobayashi and Takei, 1996). There is considerable evidence that a similar action occurs in teleost fish. For example, Ang II has been shown to stimulate secretion of cortisol, the major interrenal corticosteroid, after intravenous administration to flounder *in vivo*. Plasma cortisol increased over a more prolonged time course than the elevation in blood pressure (Perrott and Balment, 1990). Although the dose employed was high (2.5 μg/kg body wt), the Ang II sequence used in this study (Asp^1, Val^5-Ang II) is now known to not be the endogenous sequence and this may have reduced sensitivity. In trout, Ang II has been shown to stimulate cortisol secretion *in vitro* from perfused interrenal tissue acting either synergistically or additively with adrenocorticotrophin (ACTH) (Decourt and Lahlou, 1987; Arnold Reed *et al.*, 1994). In teleosts, cortisol has well-established osmoregulatory actions particularly on the gills and intestine, so corticosteroidogenic actions of Ang II in teleosts could indirectly have profound osmoregulatory effects.

In the elasmobranch fish, *T. scyllia*, the greatest specific binding of Ang II occurred in the interrenal gland (Tierney *et al.*, 1997a) and both homologous (Anderson *et al.*, 2001b) and heterologous (O'Toole *et al.*,

1990) Ang II have been shown to be potently steroidogenic in *S. canicula*. The major corticosteroid produced by the elasmobranch interrenal gland is 1α-hydroxycorticosterone (1α-OH-B) (Idler and Truscott, 1966) and the occurrence of receptors for 1α-OH-B in the rectal gland, gill and kidney of *Raja ocellata* (Idler and Kane, 1980) suggests osmoregulatory roles, but there is much less information on the actions of corticosteroids in non-teleostean fish. Armour *et al.* (1993) showed that 1α-OH-B acts on renal and extra-renal sites to reduce sodium loss, so Ang II can be predicted to indirectly influence sodium balance. Although the exact osmoregulatory roles of 1α-OH-B has yet to be determined, it is clear that the steroid may influence salt and/or urea flux.

In the lungfish, the most closely related of the fishes to the tetrapods, corticosterone, aldosterone and cortisol have all been reported to be present in the plasma with increased levels of corticosterone or aldosterone resulting from injection of Ang II (Joss *et al.*, 1994), but the effects varied depending on the amino acid sequence of the peptide. Since this study, native Ang II has been identified in the Australian lungfish (Fig. 4.2; Joss *et al.*, 1999) and this allows comparison of the effects of native and non-native sequences. Intramuscular injection of Asn^1 Val^5-Ang II, the endogenous peptide, increased circulating levels of aldosterone, as did Asp Ile^5-Ang II (Joss *et al.*, 1994). More recently, further studies have confirmed the release of aldosterone *in vitro* by perfused interrenal tissue stimulated with lungfish Ang II, but alteration of the first amino acid from asparagine to aspartic acid was sufficient to block the release of aldosterone, whereas isoleucine substitution for valine as the 5th amino acid restored selectivity, but with a reduced sensitivity (Joss *et al.*, 1999). Studies to investigate whether aldosterone in lungfish acts on sodium regulation, as in tetrapods are now needed.

SUMMARY AND FUTURE DIRECTIONS

This chapter has reviewed our understanding of the integrated actions of the systemic RAS of fish groups and the recent evidence of local paracrine systems. It is as yet unclear exactly how systemic and local systems operate alongside each other. Differences in receptor affinities as well as local concentrations of Ang II or other angiotensin fragments may ultimately explain how the two systems operate in tandem, but further work is required to complete the jigsaw. Their roles in regulating drinking, gill and kidney function and stimulating longer-acting steroidogenesis with further osmoregulatory actions form important elements in contributing to the achievement of body fluid homeostasis, but much remains to be explored. Our current knowledge is perhaps greatest amongst the teleosts but considering that the elasmobranch RAS has only been discovered relatively recently, much progress has been made in elucidating physiological roles in the osmoregulatory processes of marine elasmobranchs. However, there are considerable gaps in our knowledge and, in particular, relatively little is known of RAS function in the small number of fully euryhaline elasmobranch fish that inhabit freshwater and seawater environments, and nothing is known for the members of the ray family Potamotrygonidae (e.g., *Potamotrygon)* that are considered to be permanently resident in freshwaters in the Amazon and Orinoco river systems, as much as 4500 km from the sea. Although there is evidence of a RAS in the agnathan lampreys and this system has been shown to be activated by volume depletion with an apparent interactive effect of sodium/osmoreceptors, and after an increase in the external osmolality, as yet our understanding of the role of the RAS in these fish (or whether they too possess paracrine RASs) is relatively poor. A vasoconstrictor response is clear and this may represent the fundamental (primitive) action of the agnathan RAS as a distinct role in stimulation of drinking is not apparent and the action on the kidney relatively weak, potentially compromised by vasopressor actions.

References

Amer, S. and J.A. Brown. 1995. Glomerular actions of arginine vasotocin in the *in situ* perfused trout kidney. *American Journal of Physiology* 269: R775–R780.

Anderson, W.G., J.M. Conlon and N. Hazon. 1995a. Characterization of the endogenous intestinal peptide that stimulates the rectal gland of *Scyliorhinus canicula. American Journal of Physiology* 37: R1359–R1364.

Anderson, W.G., M.L. Tierney, Y. Takei and N. Hazon. 1995b. Natriuretic hormones in elasmobranch fish; possible interactions with other endocrine systems. *Physiological Zoology* 68: 182.

Anderson, W.G., Y. Takei and N. Hazon. 2001a. The dipsogenic effect of the renin angiotensin system in elasmobranch fish. *General and Comparative Endocrinology* 124: 300–307.

Anderson, W.G., M.C. Cerra, A. Wells, M.L. Tierney, B. Tota, Y. Takei and N. Hazon. 2001b. Angiotensin and angiotensin receptors in cartilaginous fishes: a review. *Comparative Biochemistry and Physiology* 128: 31–40.

Anderson, W.G., A. Wells, Y. Takei and N. Hazon. 2002a. The control of drinking in elasmobranch fish with special reference to the renin angiotensin system. In:

Osmoregulation and Drinking in Vertebrates, N. Hazon and G. Flik (eds.). BIOS Scientific, Oxford, pp. 19–30.

Anderson, W.G., Y. Takei and N. Hazon. 2002b. Osmotic and volaemic effects on drinking rate in elasmobranch fish. *Journal of Experimental Biology* 205: 1115–1122.

Anderson, W.G., J.P. Good and N. Hazon. 2002c. Changes in secretion rate and vascular perfusion in the rectal gland of the European lesser spotted dogfish (*Scyliorhinus canicula* L.) in response to environmental and hormonal stimuli. *Journal Fish Biology* 60: 1580–1590.

Anderson, W.G., S. Hyodo, T. Tsukada, L. Meischke, R.D. Pillans, J.P. Good, Y. Takei, G. Cramb, C.E. Franklin and N. Hazon. 2005. Sequence, circulating levels and expression of C-type natriuretic peptide in a euryhaline elasmobranch, *Carcharhinus leucas*. *General and Comparative Endocrinology* 144: 90–98.

Ando, M., Y. Fujii, T. Kadota, T. Kozaka, T. Mukuda, I. Takase and A. Kawahara. 2000. Some factors affecting drinking behavior and their interactions in seawater-acclimated eels, *Anguilla japonica*. *Zoological Science* 17: 171–178

Armour, K.J., L.B. O'Toole and N. Hazon. 1993. The effect of dietary protein restriction on the secretory dynamics of 1α-hydroxycorticosterone and urea in the dogfish, *Scyliorhinus canicula*: A possible role for 1α-hydroxycorticosterone in sodium retention. *Journal of Endocrinology* 138: 275–282.

Arnold-Reed, D.E. and R.J. Balment. 1994. Peptide hormones influence *in vitro* interrenal secretion of cortisol in the trout, *Oncorhynchus mykiss*. *General and Comparative Endocrinology* 96: 85–91.

Aukland, K. 2001. Odd E. Hanssen and the Hanssen method for measurement of single nephron glomerular filtration rate. *American Journal of Physiology* 281: F407–413.

Aust, J.G. 2002. *Molecular and Physiological Investigations of Fish Renin Angiotensin Systems*. Ph.D. Thesis, University of Exeter, Exeter, UK.

Bader, M., J. Peters, O. Baltatu, D.N. Muller, F.C. Luft and D. Ganten. 2001. Tissue renin-angiotensin systems: new insights from experimental animal models in hypertension research. *Journal of Molecular Medicine* 79: 76–102.

Balment, R.J. and S. Carrick. 1985. Endogenous renin-angiotensin system and drinking behaviour in flounder. *American Journal of Physiology* 248: R157–R160.

Balment, R.J., J.M. Warne and Y. Takei. 2003. Isolation, synthesis, and biological activity of flounder [Asn^1, Ile^5, Thr^9] angiotensin I. *General and Comparative Endocrinology* 130: 92–98.

Beasley, D., D.N. Shier, R.L. Malvin and G. Smith. 1986. Angiotensin-stimulated drinking in marine fish. *American Journal of Physiology* 250: R1034–1038.

Bernier, N.J. and S.F. Perry. 1999. Cardiovascular effects of angiotensin-II-mediated adrenaline release in rainbow trout *Oncorhynchus mykiss*. *Journal of Experimental Biology* 202: 55–66.

Bramucci, M., L. Quassinti, E. Maccari, O. Murri and D. Amici. 2004. Seasonal change in angiotensin converting enzyme activity in male and female frogs (*Rana esculenta*). *Comparative Biochemistry and Physiology* A137: 605–610.

Brasier, A.R. and J. Li. 1996. Mechanisms for inducible control of angiotensinogen gene transcription. *Hypertension* 27: 465–475.

Brown, J.A. and R.J. Balment. 1997. Teleost renal function: regulation by arginine vasotocin and by angiotensins. In: *Ionic Regulation in Animals*, N. Hazon, F. B. Eddy and G. Flik (eds.). Springer-Verlag, Heidelberg, pp. 150–164.

Brown, J.A. and C. Green. 1992. Glomerular bypass shunts and distribution of glomeruli in the kidney of the lesser spotted dogfish *Scyliorhinus canicula*. *Cell and Tissue Research* 269: 299–304.

Brown, J.A. and J.C. Rankin. 1999. Lack of glomerular intermittency in the river lamprey, *Lampetra fluviatilis* acclimated to sea water and following acute transfer to iso-osmotic brackish water. *Journal of Experimental Biology* 202: 939–946.

Brown, J.A., B.A. Jackson, J.A. Oliver and I.W. Henderson. 1978. Single nephron glomerular filtration rates (SNGFR) in the trout, *Salmo gairdneri*. Validation of the use of ferrocyanide and the effects of environmental salinity. *Pflugers Archives* 377: 101–108.

Brown, J.A., J.A. Oliver, I.W. Henderson and B.A. Jackson. 1980. Angiotensin and single nephron glomerular function in the rainbow trout *Salmo gairdneri*. *American Journal of Physiology* 239: R509–R514.

Brown, J.A., S.M. Taylor and J.C. Gray. 1983. Glomerular ultrastructure of the trout, *Salmo gairdneri*. Glomerular capillary epithelium and the effects of environmental salinity. *Cell and Tissue Research* 230: 205–218.

Brown, J.A., S.M. Taylor and J.C. Gray. 1990. Glomerular receptors for angiotensin II in the rainbow trout, *Salmo gairdneri*. *Cell and Tissue Research* 259: 479–482.

Brown, J.A., J.C. Rankin and S.D. Yokota. 1993. Glomerular haemodynamics of filtration in single nephrons of non-mammalian vertebrates. In: *New Insights in Vertebrate Kidney Function*, J.A. Brown, R.J. Balment and J.C. Rankin (eds.). Cambridge University Press, Cambridge, pp. 1–44.

Brown, J.A., R.K. Paley, S. Amer and S.J. Aves. 1995. Evidence for an intrarenal renin-angiotensin system in the rainbow trout. *Journal of Endocrinology* 147: (supplement) P79.

Brown, J.A., S.K. Pope, S. Amer, C.S. Cobb and R. Williamson. 1997. Angiotensin receptors in teleost fish glomeruli. In: *Advances in Comparative Endocrinology*, S. Kawashima and S. Kikuyama (eds.). Monduzzi Editore, Bologna, Vol. 2, pp. 1313–1319.

Brown, J.A., R.K Paley, S. Amer and S.J. Aves. 2000. Evidence for an intrarenal renin-angiotensin system in the rainbow trout, *Oncorhynchus mykiss*. *American Journal of Physiology* 278: R1685–R1691.

Brown, J.A., C.S. Cobb, S.C. Frankling and J.C. Rankin. 2005. Activation of the newly-discovered cyclostome renin-angiotensin system of the river lamprey *Lampetra fluviatilis*. *Journal of Experimental Biology* 208: 223–232.

Butler, D.G. and D.H. Zhang. 2001. Corpuscles of Stannius secrete renin or an isorenin that regulates cardiovascular function in freshwater North American eel, *Anguilla anguilla* LeSueur. *General and Comparative Endocrinology* 124: 199–217.

Butler, D.G., D.H. Zhang, R. Villadiego, G.Y. Oudit, J.H. Youson and M.Z.A. Cadinouche. 2003. Responses by the corpuscles of Stannius to hypotensive stimuli in three divergent ray-finned fishes (*Amia calva, Anguilla rostrata* and *Catastomus commersoni*): cardiovascular and morphological changes. *General and Comparative Endocrinology* 132: 198–208.

Carrick, S. and R.J. Balment. 1983. The renin-angiotensin system and drinking in the euryhaline flounder, *Platichthys flesus*. *General and Comparative Endocrinology* 51: 423–433.

Casarini, D.E., M.A. Boim, R.C.R. Stella, M.H. Kreiger-Azzolini, J.E. Kreiger and N. Schor. 1997. Angiotensin I converting enzyme activity in tubular fluid along the rat nephron. *American Journal of Physiology* 272: F405–F409.

Cerra, M.C., M.L. Tierney, Y. Takei, N. Hazon and B. Tota. 2001. Localisation and partial characterisation of angiotensin II receptors in the heart of *Scyliorhinus canicula*. *General and Comparative Endocrinology* 121: 126–134.

Churchill, P.C., R.C. Malvin, M.C. Churchill and F.D. McDonald. 1979. Renal function in *Lophius americanus*: Effects of angiotensin II. *American Journal of Physiology* 236: R297–301.

Churchill, P., R. Malvin, M. Churchill, D. Beasley and D. Shier. 1985. Antidiuretic effect of [Sar^1, Val^5, Ala^8] angiotensin II in *Lophius americanus*. *Journal of Experimental Zoology* 233: 15–20.

Cobb, C.S. and J.A. Brown. 1992. Localization of angiotensin II binding to tissues of the rainbow trout, *Oncorhynchus mykiss* adapted to freshwater and seawater: an autoradiographic study. *Journal of Comparative Physiology* B 162: 197–202.

Cobb, C.S. and J.A. Brown. 1993. Characterization of putative glomerular receptors for angiotensin II in the rainbow trout *Oncorhynchus mykiss* using the antagonists Losartan, PD 123177 and saralasin. *General and Comparative Endocrinology* 92: 123–131.

Cobb, C.S., R. Williamson and J.A. Brown. 1999. Angiotensin II-induced calcium signalling in isolated glomeruli from fish kidneys (*Oncorhynchus mykiss*) and effects of losartan. *General and Comparative Endocrinology* 113: 312–321.

Cobb, C.S., S.C. Frankling, J.C. Rankin and J.A. Brown. 2002a. Angiotensin converting enzyme-like activity in tissues from the river lamprey or lampern, *Lampetra fluviatilis*, acclimated to freshwater and seawater. *General and Comparative Endocrinology* 127: 8–15.

Cobb, C.S., S.C. Frankling and J.A. Brown 2002b. The renin-angiotensin system of agnathan fish. *Comparative Biochemistry and Physiology* Part A 132 Supplement: S43 A6.7.

Cobb, C.S., S.C. Frankling, M.C. Thorndyke, F.B. Jensen, J.C. Rankin and J.A. Brown. 2004. Angiotensin I-converting enzyme-like activity in tissues from the Atlantic hagfish (*Myxine glutinosa*) and detection of immunoreactive plasma angiotensins. *Comparative Biochemistry and Physiology* B138: 357–364.

Conlon, J.M., K.J. Yano and K.R. Olson. 1996. Production of [Asn^1,Val^5] angiotensin II and [Asp^1,Val^5] angiotensin II in kallikrein-treated trout plasma (T60K). *Peptides* 17: 527–530.

Creighton, T.E. 1993. *Proteins, Structures and Molecular Properties*, 2nd Edition. W.H. Freeman and Co, New York.

Croom, K.F., M.P. Curran, K.L. Goa and C.M. Perry. 2004. Irbesartan—A review of its use in hypertension and in the management of diabetic nephropathy. *Drugs* 64: 999–1028.

Decourt, C. and B. Lahlou. 1987. Evidence for the direct intervention of angiotensin-II in the release of cortisol in teleost fishes. *Life Sciences* 41: 1517–1524.

Dunne, J.B. and J.C. Rankin. 1992. Effect of atrial natriuretic peptide and angiotensin II on salt and water excretion by the perfused rainbow trout kidney. *Journal of Physiology* 446: 92P.

Erdos, E.G. 1990. Angiotensin I converting enzyme and the changes in our concepts through the years. *Hypertension* 16: 363–370.

Fitzsimons, J.T. 1964. Drinking caused by constriction of the inferior vena cava in the rat. *Nature (Lond.)* 204: 479–480.

Fitzsimons, J.T. 1998. Angiotensin, thirst, and sodium appetite. *Physiological Reviews* 78: 583–686.

Fitzsimons, J.T. and B.J. Simons. 1969. The effect of drinking in the rat of intravenous infusion of angiotensin, given alone or in combination with other stimuli of thirst. *Journal of Physiology* (London) 203: 45–57.

Fuentes, J. and F.B. Eddy. 1996. Drinking in freshwater-adapted rainbow trout fry, *Oncorhynchus mykiss* (Walbaum), in response to angiotensin I, angiotensin II, angiotensin converting enzyme inhibition, and receptor blockade. *Physiological Zoology* 69: 1555–1569.

Fuentes, J. and F.B. Eddy. 1997. Effect of manipulation of the renin-angiotensin system in control of drinking in Atlantic salmon (*Salmo salar* L.) in freshwater and after transfer to seawater. *Journal of Comparative Physiology* 167: 438-443.

Fuentes, J., J.C. McGeer and F.B. Eddy. 1996. Drinking rate in juvenile Atlantic salmon, *Salmo salar* fry in response to a nitric oxide donor, sodium nitroprusside and an inhibitor of angiotensin converting enzyme, enalapril. *Fish Physiology and Biochemistry* 15: 65–69.

Gociman, B., A. Rohrwasser, P. Lantelme, T. Cheng, G. Hunter, S. Monson, J. Hunter, E. Hillas, P. Lott, T. Ishigami and J.M. Lalouel. 2004. Expression of angiotensinogen in proximal tubule as a function of glomerular filtration rate. *Kidney International* 65: 2153–2160.

Gomez, R.A., R.L. Chevalier, A.D. Everett, J.P. Elwood, M.J. Peach, K.R. Lynch and R.M. Carey. 1990. Recruitment of renin gene-expressing cells in adult-rat kidneys. *American Journal of Physiology* 259: F660–F665.

Gray, J.C. and J.A. Brown. 1985. Renal and cardiovascular effects of angiotensin II in the rainbow trout, *Salmo gairdneri*. *General and Comparative Endocrinology* 59: 375–381.

Gray, C.J. and J.A. Brown. 1987. Glomerular ultrastructure of the trout, *Salmo gairdneri*. Effects of angiotensin II and adaptation to seawater. *Cell and Tissue Research* 249: 437–442.

Hackenthal, E., M. Paul, D. Ganten and R. Taugner. 1990. Morphology, physiology, and molecular biology of renin secretion. *Physiological Reviews* 70: 1067–1116.

Hasegawa, Y., T. Nakajima and H. Sokabe. 1983. Chemical structure of angiotensin formed with kidney renin in the Japanese eel, *Anguilla japonica*. *Biomedical Research* 4: 417–420.

Hasegawa, Y., T.X. Watanabe, T. Nakajima and H. Sokabe. 1984. Chemical structure of angiotensin formed by incubating plasma with corpuscles of Stannius in the Japanese goosefish *Lophius litulon*. *General and Comparative Endocrinology* 54: 264–269.

Hamano, K., M.L. Tierney, K. Ashida, Y. Takei and N. Hazon. 1998. Direct vasoconstrictor action of homologous angiotensin II on isolated arterial ring preparations in an elasmobranch. *Journal of Endocrinology* 158: 419–423.

Hazon, N., R.J. Balment, M. Perrott and L.B. O'Toole. 1989. The renin-angiotensin system and vascular and dipsogenic regulation in elasmobranchs. *General and Comparative Endocrinology* 74: 230–236.

Hazon, N., M.C. Cerra, M.L. Tierney, B. Tota and Y. Takei. 1997a. Elasmobranch renin angiotensin system and the angiotensin receptor. In: *Advances in Comparative Endocrinology*, S. Kawashima and S. Kikuyama (eds.). Monduzzi Editore. Bologna, Vol. 2, pp. 1313–1319.

Hazon, N., M.L. Tierney, W.G. Anderson, S. MacKenzie, C. Cutler and G. Cramb. 1997b. Ion and water balance in elasmobranch fishes. In: *Ionic Regulation in Animals*, N. Hazon, B. Eddy and G. Flik (eds.), Springer-Verlag. Berlin, pp. 70–86.

Hazon, N., M.L. Tierney and Y. Takei. 1999. Renin-angiotensin system in elasmobranch fish: A review. *Journal of Experimental Zoology* 284: 526–534.

Hazon, N., A. Wells, R.D. Pillans, J.P. Good, W.G. Anderson and C.E. Franklin. 2003. Urea-based osmoregulation and endocrine control in elasmobranch fish with special reference to euryhalinity. *Comparative Biochemistry and Physiology* 136: 685–700.

Henderson, I.W., J.A. Oliver, A. Mckeever and N. Hazon. 1981. Phylogenetic aspects of the renin-angiotensin system. In: *Advances in Physiological Sciences: Advances in Animal and Comparative Physiology*. G. Pethes and V.L. Frenyo (eds.). Pergamon Press, Budapest, Vol. 20, pp. 355–363.

Hentschel, H., S. Mahler, P. Herter and M. Elger. 1993. Renal tubule of dogfish, *Scyliorhinus caniculus*—a comprehensive study of structure with emphasis on intramembrane particles and immunoreactivity for H^+-K^+-adenosine triphosphatase. *Anatomical Record* 235: 511–532.

Hirano, T., Y. Takei and H. Kobayashi. 1978. Angiotensin and drinking in the eel and the frog. In: *Osmotic and Volume Regulation. Alfred Benzon Symposium*. Barker Jorgensen and E. Skadhauge (eds.), Copenhagen. XI. C. pp 123–128.

Hirano, T. and S. Hasegawa. 1984. Effects of angiotensins and other vasoactive substances on drinking in the eel *Anguilla japonica*. *Zoological Science* 1: 106–113.

Idler, D.R. and K.M. Kane. 1980. Cytosol receptor glycoprotein for 1α-hydroxycorticosterone in tissues of an elasmobranch fish (*Raja ocellata*). *General and Comparative Endocrinology* 42: 259–266.

Idler, D.R. and B. Truscott. 1966. 1α-Hydroxycorticosterone from cartilagenous fish: a new adrenal steroid in blood. *Journal of Fisheries Research Board Canada* 23: 615–619.

Jampol, L.M. and F.M. Epstein. 1970. Sodium-potassium-activated adenosinetriphosphate and osmotic regulation by fishes. *American Journal of Physiology* 218: 607–611.

Janech, M.G. and P.M. Piermarini. 2002. Renal water and solute excretion in the Atlantic stingray in fresh water. *Journal of Fish Biology* 61: 1053–1057.

Janvier, P. 1999. Catching the first fish. *Nature (Lond.)* 402: 21–22.

Jin, X.H., Z.Q. Wang, H.M. Siragy, R.L. Guerrant and R.M. Carey. 1998. Regulation of jejunal sodium and water absorption by angiotensin subtype receptors. *American Journal of Physiology* 275: R515–R523.

Johansson, B., M. Holm, S. Ewert, A. Casselbrant, A. Pettersson and L. Fandriks. 2001. Angiotensin II type 2 receptor-mediated duodenal mucosal alkaline secretion in the rat. *American Journal of Physiology* 280: G1251–G1260.

Joss, J.M.P., D.E. Arnold-Reed and R.J. Balment. 1994. The steroidogenic response to angiotensin in the Australian lungfish, *Neoceratodus forsteri*. *Journal of Comparative Physiology B* 164: 378–382.

Joss, J.M.P., Y. Itahara, T.X. Watanabe, K. Nakajima and Y. Takei. 1999. Teleost-type angiotensin is present in Australian lungfish, *Neoceratodus forsteri*. *General and Comparative Endocrinology* 114: 206–212.

Kajimura, M., P.J. Walsh, T.P. Mommsen, C.M. Wood. 2006. The dogfish shark (*Squalus acanthias*) increases both hepatic and extrahepatic ornithine urea cycle enzyme activities for nitrogen conservation after feeding. *Physiological and Biochemical Zoology* 79: 602–613.

Kent, B. and K.R. Olson. 1982. Blood flow in the rectal gland of *Squalus acanthias*. *American Journal of Physiology* 243: 296–303.

Khosla, M.C., H. Nishimura, Y. Hasegawa and F. M. Bumpus. 1985. Identification and synthesis of [1-asparagine, 5-valine, 9-glycine] angiotensin I produced from plasma of American eel *Anguilla rostrata*. *General and Comparative Endocrinology* 57: 223–233.

Klett, C.P.R. and J.P. Granger. 2001. Physiological elevation in plasma angiotensinogen increases blood pressure. *American Journal of Physiology* 281: R1437–R1441.

Kobayashi, H. and Y. Takei. 1996. *The Renin-Angiotensin System—Comparative Aspects*, Springer-Verlag, Berlin.

Kobayashi, H., H.Y. Uemura, Y. Takei, N. Itasu, M. Ozawa and K. Ichinohe. 1993. Drinking induced by angiotensin II in fishes. *General and Comparative Endocrinology* 49: 295–306.

Kozaka, T., Y. Fujii and M. Ando. 2003. Central effects of various ligands on drinking behavior in eels acclimated to seawater. *Journal of Experimental Biology* 206: 687–692.

Lacy, E.R. and E. Reale. 1991a. Fine structure of the elasmobranch renal tubule: intermediate, distal and collecting duct segments of the little skate. *American Journal of Anatomy* 192: 478–497.

Lacy, E.R. and E. Reale. 1991b. Fine structure of the elasmobranch renal tubule: neck and proximal segments of the little skate. *American Journal of Anatomy* 190: 118–132.

Lacy, E.R. and E. Reale. 1995. Functional morphology of the elasmobranch nephron and retention of urea. In: *Cellular and Molecular Approaches to Fish Ionic Regulation*, C.M. Wood and T.J. Shuttleworth (eds.). Academic Press, San Diego, Vol. 14, pp. 107–146.

Lantelme, P., A. Rohrwasser, B. Gociman, E. Hillas, T. Cheng, G. Petty, J. Thomas, S. Xiao, T. Ishigami, T. Herrmann, D.A. Terreros, K. Ward and J-M. Lalouel. 2002. Effects of dietary sodium and genetic background on angiotensinogen and renin in mouse. *Hypertension* 39: 1007–1014.

Liang, P., C.A. Jones, B.W. Bisgrove, L. Song, S.T. Glenn, H.J. Yost and K.W.Gross. 2004. Genomic characterization and expression analysis of the first nonmammalian renin genes from zebrafish and pufferfish. *Physiological Genomics* 16: 314–322.

Laurent, P. and S. Dunel. 1980. Morphology of gill epithelia in fish. *American Journal of Physiology* 238: R147–R159.

Lipke, D.W. and K.R. Olson 1988. Distribution of angiotensin-converting enzyme-like activity in vertebrate tissues. *Physiological Zoology* 61: 420–428.

Lyndon, A.R. 1993. Effect of angiotensin II on transepithelial potential in the isolated perfused flounder (*Platichthys flesus*) gill. *Journal of Fish Biology* 42: 609–610.

Malvin, R.L., D. Schiff and S. Eiger. 1980. Angiotensin and drinking rates in euryhaline killifish. *American Journal of Physiology* 239: R31–R34.

Marsigliante, S., T. Verri, S. Barker, E. Jimenez, G.P. Vinson and C. Storelli. 1994. Angiotensin II receptor subtypes in eel (*Anguilla anguilla*). *Journal of Molecular Endocrinology* 12: 61–69.

Marsigliante, S., A. Muscella, S. Vilella, G. Nicolardi, L. Ingrosso, V. Ciardo, V. Zonno, G.P. Vinson, M.M. Ho and C. Storelli. 1996. A monoclonal antibody to mammalian angiotensin II AT_1 receptor recognizes one of the angiotensin II receptor isoforms expressed by the eel (*Anguilla anguilla*). *Journal of Molecular Endocrinology* 16: 45–56.

Marsigliante, S., A. Muscella, G.P. Vinson and C. Storelli. 1997a. Angiotensin II receptors in the gill of sea water- and freshwater-adapted eel. *Journal of Molecular Endocrinology* 18: 67–76.

Marsigliante, S., R. Acierno, M. Maffia, A. Muscella, G.P. Vinson and C. Storelli. 1997b. Immunolocalisation of angiotensin II receptors in icefish (*Chionodraco hamatus*) tissues. *Journal of Endocrinology* 154: 193–200.

Marsigliante, S., A. Muscella, S. Barker and C. Storelli. 2000. Angiotensin II modulates the activity of the Na^+/K^+ ATPase in eel kidney. *Journal of Endocrinology* 165: 147–156.

Masini, M.A., B. Uva, M. Devecchi and L. Napoli. 1994. Renin-like activity, angiotensin I-converting enzyme-like activity, and osmoregulatory peptides in the dogfish rectal gland. *General and Comparative Endocrinology* 93: 246–254.

Morris R. 1960. General problems of osmoregulation with special reference to cyclostomes. In: *Hormones in Fish, I.C. Jones* (ed.). *Symposium of Zoological Society of London* No. 1, pp. 1–16.

Nakamura, A., H. Iwao, K. Kukui, S. Kimura, T. Tamaki, S. Nakanishi and Y. Abe. 1990. Regulation of liver angiotensinogen and kidney renin messenger-RNA levels by angiotensin-II. *American Journal of Physiology* 258: E1–E6.

Navar, L.G. and A. Nishiyama. 2004. Why are angiotensin concentrations so high in the kidney? *Current Opinion in Nephrology and Hypertension* 13: 107–115.

Nishimura, H. 1985. Evolution of the renin-angiotensin system and its role in control of cardiovascular function in fishes. In: *Evolutionary Biology of Primitive Fishes*, R.E. Foreman, J.M. Gorbman, J.M. Dodd and R. Olsson (eds). Plenum Press, New York, pp. 275–293.

Nishimura, H. 2001. Angiotensin receptors—Evolutionary overview and perspectives. *Comparative Biochemistry and Physiology* A 128: 11–30.

Nishimura, H. and W.B. Sawyer. 1976. Vasopressor, diuretic and natriuretic responses to angiotensins by the American eel, *Anguilla rostrata*. *General and Comparative Endocrinology* 29: 337–348.

Nishimura, H., M. Oguri, M. Ogawa, H. Sokabe and I. Imai. 1970. Absence of renin in kidneys of elasmobranchs and cyclostomes. *American Journal of Physiology* 218: 911–915.

Nishimura, H., L.G. Lunde and A. Zucker. 1979. Renin response to hemorrhage and hypotension in the aglomerular toadfish *Opsanus tau*. *American Journal of Physiology* 237: H105–111.

Olson, K.R. 1992. Blood and extracellular fluid volume regulation: role of the renin-angiotensin, kallikrein-kinin system and atrial peptides. In: *Fish Physiology*. W. S. Hoar, D. J. Randall and A. P. Farrell (eds), Academic Press, London, Vol. 12B, pp. 135–254.

Olson, K.R., D.W. Lipke, D. Kullman, A.P. Evans and J.W. Ryan. 1989. Localisation of angiotensin-converting enzyme in the trout gill. *Journal of Experimental Biology* 250: 109–115.

Okawara Y., T. Karakida, M. Aihara, K. Yamaguchi and H. Kobayashi. 1987. Involvement of angiotensin II in water intake in the Japanese eel, *Anguilla japonica*. *Zoological Science* 4: 523–528.

O'Toole, L.B., K.J. Armour, C. Decourt, N. Hazon, B. Lahlou and I.W. Henderson. 1990. Secretory patterns of 1α-hydroxycorticosterone in the isolated perifused interrenal gland of the dogfish, *Scyliorhinus canicula*. *Journal of Molecular Endocrinology* 5: 55–60.

Paley, R.K., J.A. Brown and S.J. Aves. 1996. The intrarenal renin-angiotensin system of the euryhaline rainbow trout. *Annales d'Endocrinologie* 57: S4 P145.

Paley, R.K., J.A. Brown and S.J. Aves. 1998. Cloning, sequence and expression analysis of a rainbow trout angiotensin converting enzyme cDNA. *Pflugers Archiv*, 435S: P49.

Paley, R.K., J.G. Aust, S.J. Aves and J.A. Brown. 2003. Angiotensinogen gene expression in the rainbow trout: mRNA localisation and effects of saline exposure on renal and hepatic expression. *GenBank Data Base* AJ579373.

Parkyn, G., S.J. Aves and J.A. Brown. 1997. Cloning and characterisation of the rainbow trout angiotensin receptors. In: *Advances in Comparative Endocrinology*, S. Kawashima and S. Kikuyama (eds.), Monduzzi Editore, Bologna, Vol. 2, pp. 1329–1332.

Pelis, R.M. and J.L Renfro. 2004. Role of tubular secretion and carbonic anhydrase in vertebrate renal sulfate excretion. *American Journal of Physiology* 287: R491–R501.

Perrott, M.N. and R.J. Balment. 1990. The renin-angiotensin system and the regulation of plasma cortisol in the flounder, *Platichthys flesus*. *General and Comparative Endocrinology* 78: 414–420.

Perrott, M.N., C.E. Grierson, N. Hazon and R. Balment. 1992. Drinking behaviour in sea water and fresh water teleosts, the role of the renin-angiotensin system. *Fish Physiology and Biochemistry* 10: 161–168.

Piermarini, P.M. and D.H. Evans. 1998. Osmoregulation of the Atlantic stingray (*Dasyatis sabina*) from the freshwater lake Jesup of the St. Johns River, Florida. *Physiological Zoology* 71: 553–560.

Piermarini, P.M. and D.H. Evans. 2000. Effects of environmental salinity on Na^+/K^+-ATPase in the gills and rectal gland of a euryhaline elasmobranch (*Dasyatis sabina*). *Journal of Experimental Biology* 203: 2957–2966.

Piermarini, P.M. and D.H. Evans. 2001. Immunochemical analysis of the vacuolar proton-ATPase β-subunit in the gills of a euryhaline stingray (*Dasyatis sabina*): Effects of salinity and relation to Na^+/K^+-ATPase. *Journal of Experimental Biology* 204: 3251–3259.

Pillans, R.D. and C.E. Franklin. 2004. Plasma osmolyte concentrations of bull sharks *Carcharhinus leucas*, captured along a salinity gradient. *Journal of Comparative Physiology* B 138: 363–371.

Pillans, R.D., J.P Good, W.G Anderson, N. Hazon and C.E. Franklin. 2005. Freshwater to seawater acclimation of juvenile bull sharks (*Carcharhinus leucas*): plasma osmolytes and Na^+/K^+ATPase activity in gill, rectal gland, kidney and intestine. *Journal of Comparative Physiology* B 175: 37–44.

Polanco, M.J., M.I. Mata, M.T. Agapito and J.M. Recio. 1990. Angiotensin-converting enzyme distribution and hypoxia response in mammal, bird, and fish. *General and Comparative Endocrinology* 79: 240–245.

Pope, S.K. 2002. *Control of Renal Function by Peptide Hormones in the Rainbow Trout, Oncorhynchus mykiss*. Ph.D. Thesis, University of Exeter, Exeter, UK.

Qin, Z., H-Q. Yan and H. Nishimura. 1999. Vascular angiotensin II receptor and calcium signalling. *General and Comparative Endocrinology* 115: 122–131.

Ramchandran, R., G.C. Sen, K. Misono and I. Sen. 1994. Regulated cleavage secretion of the membrane-bound angiotensin-converting enzyme. *Journal of Biological Chemistry* 269: 2125–2130.

Rankin, J.C., V. Griffiths, A.J. McVicar and I.D. Gilham. 1985. Control of kidney function in the river lamprey, *Lampetra fluviatilis*. In: *Current Trends in Comparative Endocrinology*, B. Lofts and W. N. Holmes (eds.). Hong Kong University Press, Hong Kong.

Rankin, J.C., 1997. Osmotic and ionic regulation in cyclostomes. In: *Ionic Regulation in Animals*, N. Hazon, F. B. Eddy and G. Flik (eds.), Springer-Verlag, Berlin, pp. 50–69.

Rankin, J.C. 2002. Drinking in hagfishes and lampreys. In: *Osmoregulation and Drinking in Vertebrates*, N. Hazon and G. Flik (eds.). BIOS, Oxford, pp. 1–17.

Rankin, J.C., C.S. Cobb, S.C. Frankling and J.A. Brown. 2001. Circulating angiotensins in the river lamprey, *Lampetra fluviatilis*, acclimated to freshwater and seawater: possible involvement in the regulation of drinking. *Comparative Biochemistry and Physiology* B 129: 311–318.

Rankin, J.C., T.X. Watanabe, K. Nakajima, C. Broadhead and Y. Takei. 2004. Identification of angiotensin I in a cyclostome, *Lampetra fluviatilis*. *Zoological Science* 21: 173–179.

Sandberg, K. 1994. Structural analysis and regulation of angiotenin II receptors. *Trends in Endocrinology and Metabolism* 5: 28–35.

Shuttleworth, T.J. 1988. Salt and water balance—extrarenal mechanisms. In: *Physiology of Elasmobranch Fishes*, T. J. Shuttleworth (ed.). Springer-Verlag, Berlin, pp. 171–199.

Silva, P., J.S. Stoff, D.R. Leone and F.H. Epstein. 1985. Mode of action of somatostatin to inhibit secretion by shark rectal gland. *American Journal of Physiology* 249: R329–R334.

Silva, P., F.H. Epstein, K.J. Karnaky, S. Reichlin and J.N. Forrest. 1993. Neuropeptide Y inhibits chloride secretion in the shark rectal gland. *American Journal of Physiology* 265: R439–R446.

Solomon, R., A. Protter, G. McEnroe, J.G. Porter and P. Silva. 1992. C-type natriuretic peptides stimulate chloride secretion in the rectal gland of *Squalus acanthias*. *American Journal of Physiology* 262: R707–R711.

Stoff, J.S., R. Hallac, R. Rosa, P. Silva, J. Fischer and F.H. Epstein. 1977. The role of vasoactive intestinal peptide (VIP) in the regulation of active chloride secretion in the rectal gland of *Squalus acanthias*. *Bulletin of Mount Desert Island Biological Laboratory* 17: 66.

Takei, Y. 2000. Comparative physiology of body fluid regulation in vertebrates with special reference to thirst regulation. *Japanese Journal of Physiology* 50: 171–186.

Takei, Y. and T. Tsuchida. 2000. Role of the renin-angiotensin system in drinking of seawater-adapted eels *Anguilla japonica*: a reevaluation. *American Journal of Physiology* 279: R1105–R1111.

Takei, Y., H. Uemura and H. Kobayashi. 1985. Angiotensin and hydromineral balance: with special reference to induction of drinking behaviour. In: *Current Trends in Comparative Endocrinology*, B. Lofts and W. N. Holmes (eds.). Hong Kong University Press, Hong Kong, pp. 933–936.

Takei, Y., J. Okubo and K. Yamaguchi. 1988. Effects of cellular dehydration on drinking and plasma angiotensin II level in the eel, *Anguilla japonica*. *Zoological Science* 5: 43–51.

Takei, Y., Y. Hasegawa, T.X. Watanabe, K. Nakajima and N. Hazon. 1993. A novel angiotensin I isolated from an elasmobranch fish. *Journal of Endocrinology* 139: 281–285.

Takei, Y., Y. Itahara, D.G. Butler, T.X. Watanabe and G.Y. Oudit. 1998. Tetrapod-type angiotensin is present in a holostean fish, *Amia calva*. *General and Comparative Endocrinology* 110: 140–146.

Takei,Y., T. Tsuchida, L. Zhihong and M. Conlon. 2001. Antidipsogenic effects of eel bradykinins in the eel *Anguilla japonica*. *American Journal of Physiology* 281: R1090–R1096.

Takei, Y., J.M.P Joss, W. Kloas and J.C. Rankin. 2004. Identification of angiotensin I in several vertebrate species: its structural and functional evolution. *General and Comparative Endocrinology* 135: 286–292.

Takemoto, Y., T. Nakajima, Y. Hasegawa, T.X. Watanabe, H. Sokabe, S. Kumagae and S. Sakakibara. 1983. Chemical structures of angiotensins formed by incubating plasma with the kidney and the corpuscles of Stannius in the chum salmon, *Oncorhynchus keta*. *General and Comparative Endocrinology* 51: 219–227.

Thorson, T.B. 1962. Partitioning of body fluids in the Lake Nicaraguan shark and three marine sharks. *Science* 138: 688–690.

Thorson, T.B. 1967. Osmoregulation in freshwater elasmobranchs. In: *Sharks, Skates and Rays*, P.W. Gilbert R.F. Mathewson and D.P. Rall (eds.), John Hopkins University Press, Baltimore, pp. 265–270.

Thorson, T. B. and J.W. Gerst. 1972. Comparison of some parameters of serum and uterine fluid of pregnant, viviparous sharks (*Carcharinus leucas*) and serum of their near-term young. *Comparative Biochemistry and Physiology* 42A: 33–40.

Thorson, T.B., C.M. Cowan and D.E. Watson. 1973. Body fluid solutes of juveniles and adults of the euryhaline bull shark *Carcharinus leucas* from freshwater and saline environments. *Physiological Zoology* 46: 29–42.

Tierney, M.L., G. Luke, G. Cramb and N. Hazon. 1995a. The role of the renin-angiotensin system in the control of blood pressure and drinking in the European eel, *Anguilla anguilla*. *General and Comparative Endocrinology* 100: 39–48.

Tierney, M.L., G. Cramb and N. Hazon. 1995b. Stimulation of the renin-angiotensin and drinking by papaverine in the seawater eel, *Anguilla anguilla*. *Journal of Fish Biology* 46: 721–724.

Tierney, M.L., Y. Takei and N. Hazon. 1997a. The presence of angiotensin II receptors in elasmobranchs. *General and Comparative Endocrinology* 105: 9–17.

Tierney, M.L., K. Hamano, W.G. Anderson, Y. Takei, K. Ashida and N. Hazon. 1997b. Interactions between the renin-angiotensin system and catecholamines on the cardiovascular system of elasmobranchs. *Fish Physiology and Biochemistry* 17: 333–337.

Tierney, M.L., Y. Takei and N. Hazon. 1998. A radioimmumoassay for the determination of angiotensin II in elasmobranch fish. *General and Comparative Endocrinology* 111: 299–305.

Tran van Chuoi, M., C.T. Dolphin, S. Barker, A.J. Clark and G.P. Vinson. 1998. Molecular cloning and characterization of the cDNA encoding the angiotensin II receptor of European eel *Anguilla anguilla*. GenBank Data Base AJ005132.

Tsuchida T. and Y. Takei. 1998. Effects of homologous atrial natriuretic peptide on drinking and plasma ANG II level in eels. *American Journal of Physiology* 275: R1605–R1610.

Tsuchida, T. and Y. Takei. 1999. A potent dipsogenic action of homologous angiotensin II infused at physiological doses in eels. *Zoological Science* 16: 479–483.

Uva, B.M., M.A. Masini, N. Hazon, L.B. O'Toole, I.W. Henderson and P. Ghiani. 1992. Renin and angiotensin converting enzyme in elasmobranchs. *General and Comparative Endocrinology* 86: 407–412.

Wilkie, M.P., S. Turnbull, J. Bird, J.Y.S. Wang, J.F. Claude and J.H. Youson. 2004. Lamprey parasitism of sharks and teleosts: high capacity urea excretion in an extant vertebrate relic. *Comparative Biochemistry and Physiology* A138: 485–492.

Wilson, R.W., J.M. Wilson and M. Grosell. 2002. Intestinal bicarbonate secretion by marine teleost fish—why and how? *Biochimica et Biophysica Acta* 1566: 182–193.

Wells, A., W.G. Anderson and N. Hazon. 2002. Development of an *in situ* perfused kidney preparation for elasmobranch fish: Action of arginine vasotocin. *American Journal of Physiology* 282: R1636–R1642.

Wells, A., W.G. Anderson and N. Hazon. 2003. Evidence for an intrarenal renin-angiotensin system in the European lesser-spotted dogfish. *Journal of Fish Biology* 63: 1337–1340.

Wells, A., W.G. Anderson, J.E. Cains, M.W. Cooper and N. Hazon. 2006. Effects of angiotensin II and C-type natriuretic peptide on the *in situ* perfused trunk preparation of the dogfish. *General and Comparative Endocrinology* 145: 109–115.

Wright, D.E. 1973. The structure of the gills of the elasmobranch *Scyliorhinus canicula*. *Zeitschrift fur Zellforschung* 144: 489–509.

Yanegawa, N., A.W. Capparelli, O.D Jo, A. Friedal, J.D. Barrett and P. Eggena. 1991. Production of angiotensinogen and renin-like activity by rabbit proximal tubular cells in culture. *Kidney International* 39: 938–941.

CHAPTER

5

Effect of Water pH and Hardness on Survival and Growth of Freshwater Teleosts

Jorge Erick Garcia Parra[1] and Bernardo Baldisserotto[2,*]

INTRODUCTION

The freshwater contains a variable amount of dissolved substances (salts and organic compounds) (Table 5.1), depending on the soil in which it occurs. Water acidification may occur in places where the soil contains acidic cations, as Al^{3+}, or iron pyrite, which under oxygenating conditions, forms sulfuric acid (Zweig *et al.*, 1999). Some lakes are acidified by streams with sulfuric and hydrochloric acid of volcanic origin, as Lake Usoriko (Japan) (Takatsu *et al.*, 2000). The presence of humic and fulvic acids, formed in the soil through the decomposition of organic matter, can

Authors' addresses: [1]Departamento de Ciências Agrárias, Universidade Regional Integrada do Alto Uruguai e das Missões – Campus Santiago, 97700.000 – Santiago, RS, Brazil.
[2]Departamento de Fisiologia e Farmacologia, Universidade Federal de Santa Maria, 97105.900 – Santa Maria, RS, Brazil.
**Corresponding author:* E-mail: bernardo@smail.ufsm.br

Table 5.1 Ion concentration (mmol) in some freshwater environments.

	Rio Negro, Brazil[1]	Nijmejen, The Netherlands[2]	Pantanal, Brazil[3]	Lake Usoriko, Japan[4]	Lake Van, Turkey[5]
pH	3.9 -5.0	7.6	10.2	3.0-3.6	9.8
Na^+	0.039	5.0	11.1	1.07	336.9
K^+	0.006	0.06	3.0	0.07	13.0
Ca^{2+}	0.007	0.8	0.23	0.27	0.11
Mg^{2+}	0.004	0.2	0.04	0.08	3.9
Cl^-	0.014	4.2	-	1.35	153.7
SO_4^-	0.003	0.5	-	1.34	24.3
HCO_3^-	0.018	-	-	-	-
Hardness	0.95	98.7	27.0	15.7	366.5

[1]Mortatti and Probst (2003); [2]Li *et al.* (1995), [3]Galvão *et al.* (2003); [4]Takatsu *et al.* (2000); [5]Danulat and Kempe (1992)

also reduce water pH down to 3.5, as observed in Amazonian blackwaters (Matsuo and Val, 2003) and some densely vegetated swamps and bogs of North America (Patrick *et al.*, 1981; Gonzalez, 1996). Alkaline waters may be the consequence of phytoplankton or aquatic plants blooms, which decrease the CO_2 available in the water during daylight (Wood, 2001). There are also natural alkaline lakes, some of them with high concentration of ions (Table 5.1). Soft waters have a low content of salts, mainly Ca^{2+} and Mg^{2+}, but if the soil contains limestone, water can dissolve large amounts of Ca^{2+} and Mg^{2+} salts and is then termed hard water (Baldisserotto, 2003). In some places, water hardness can also be increased by the presence of Ba^{2+} and Fe^{2+}.

SURVIVAL IN ACIDIC AND ALKALINE WATERS

Most teleosts species survive to acute pH changes down to water pH 4.0–5.0 or up to 9.0-10.0 (Table 5.2) (Alabaster and Lloyd, 1982), but exposure to more acidic or alkaline waters is lethal within a few hours (Figs. 5.1 and 5.2). However, species that inhabit acidic waters, as cardinal tetra, *Paracheirodon axelrodi*, and banded sunfish, *Enneacanthus obesus*, can withstand water pH 3.5 indefinitely (Gonzalez, 1996; Matsuo and Val, 2003). Juveniles and adults of *Tribolodon hakonensis* survive in the acidic waters of Lake Usoriko, but must migrate to neutral waters to breed (Satake *et al.*, 1995). Some species, as the cyprinid *Chalcarburnus tarichi*, the Lahontan cutthroat trout (*Oncorhynchus clarki henshawi*), and the scale-less carp *Gymnocypris przewalskii* live in alkaline waters (9.4-9.8), but

Table 5.2 Water pH survival range (100% survival) of some freshwater teleosts species. nd – non-determined.

Species	*lower pH*	*Higher pH*	*Exposure time (days)*	*Reference*
Enneacanthus obesus	3.5	nd	21	Gonzalez (1996)
Paracheirodon axelrodi	3.5	nd	5	Gonzalez *et al.* (1998)
Tribolodon hakonensis	3.6	nd	3	Kaneko *et al.* (1999)
Odontesthes bonariensis	4.9	10.4	4	Gómez (1998)
Prochilodus lineatus	4.0	9.5	5	Zaniboni-Filho *et al.* (2002)
Rhamdia quelen	4.0	9.0	4	Zaions and Baldisserotto (2000)
Oncorhynchus mykiss	4.4	9.2	15	see Alabaster and Lloyd (1982)
Alcolapia grahami	5.0	11.0	5	Reite *et al.* (1974)

perform annual spawning migrations into rivers with lower water pH (8.2-8.5) (Danulat and Selcuk, 1992; Wilkie *et al.*, 1994; Wang *et al.*, 2003). A species that lives and reproduces in a very alkaline water pH (10.0) is the Lake Magadi tilapia, *Alcolapia grahami* (Wilson *et al.*, 2004).

GENERAL ASPECTS OF OSMOREGULATION IN ACIDIC AND ALKALINE SOFT WATERS

Mortality of fishes exposed to acidic soft water seems related to a decrease of around 50% of plasma ions, mainly Na^+ and Cl^- (Freda and McDonald, 1988). The rapid ions loss during acid exposure entrain hematological (increase of hematocrit, hemoglobin and plasma protein) and fluid volumes disturbances which kill the fish through circulatory failure (Wood, 1989). Decrease on plasma Ca^{2+} and Mg^{2+} was observed in rainbow trout exposed to water pH 4.0 for 22 days (Giles *et al.*, 1984), but not in the same species, common shiners (*Notropis cornutus*), *and* yellow perch (*Perca flavescens*) after 14 days (Freda and McDonald, 1988). In low water pH, acid load through the gills is the source of acid-base disturbance, and there is an increase of H^+ and NH_4^+ excretion by the urine to compensate this problem (Wood *et al.*, 1999).

Exposure to low water pH increases branchial Na^+ efflux due to an opening of tight junctions of gill epithelia, increasing ion loss by a paracellular route (Gonzalez, 1996; Wood, 2001). There is also a decrease on Na^+ influx when fishes are exposed to acidic waters, with a blockade of almost 100% at pH 4.0 in rainbow trout. The Na^+ uptake blockade is caused by inhibition of the apical Na^+/H^+, NH_4^+ exchangers by external

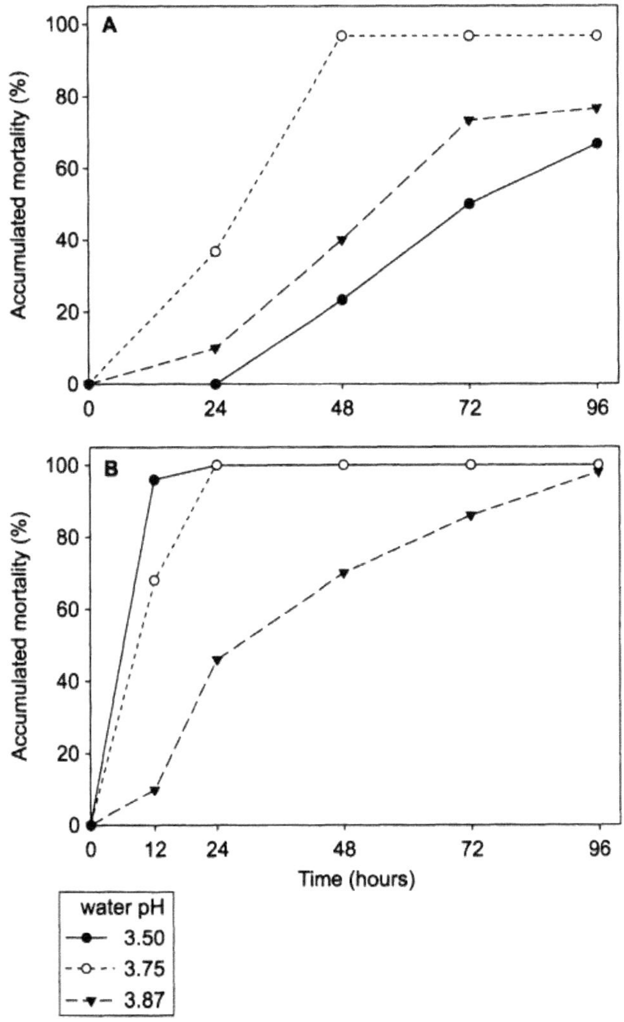

Fig. 5.1 Accumulated mortality as a function of time in *Prochilodus lineatus* (A) and *Rhamdia quelen* (B) exposed to acidic water pH and soft water (20-30 mg L^{-1} $CaCO_3$). Adapted from Zaions and Baldisserotto (2000) and Zaniboni *et al.* (2002).

H^+, or by creating a gradient too step for further extrusion of protons (Wood, 2001). However, as species that live in very acidic waters are able to take up salts at water pH 3.5-4.0, it seems that the model proposed for rainbow trout is not applicable to these species (Gonzalez *et al.*, 1998, 2002).

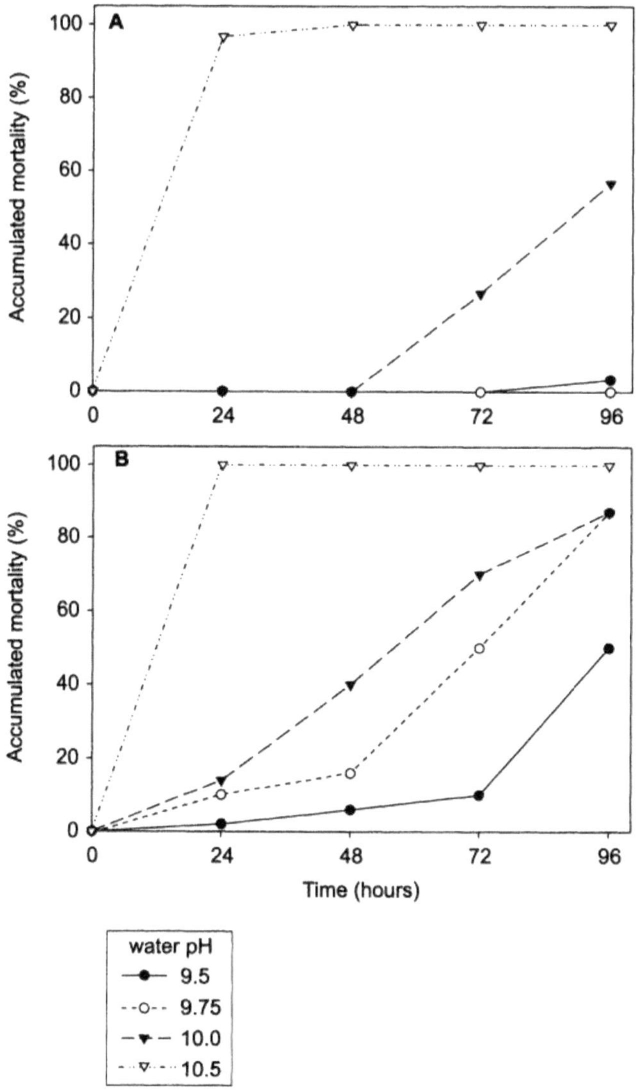

Fig. 5.2 Accumulated mortality as a function of time in *Prochilodus lineatus* (A) and *Rhamdia quelen* (B) exposed to alkaline water pH and soft water (20-30 mg L^{-1} $CaCO_3$). Adapted from Zaions and Baldisserotto (2000) and Zaniboni *et al.* (2002).

The main problems in alkaline waters are the inhibition of ammonia excretion and increase of CO_2 excretion (Wood, 2001). At circumneutral water pH ammonia leaves the gills by diffusion in the form of NH_3, and is converted to NH_4^+ in the water, maintaining a favorable gradient for NH_3

diffusion. When the water is alkaline, there is less H^+ available to transform NH_3 into NH_4^+, and the NH_3 gradient blood–water decreases. The corresponding decrease in water CO_2 creates a higher blood–water gradient, which promotes branchial CO_2 losses (Wilkie and Wood, 1996). The resultant respiratory alkalosis increases plasma pH levels, as observed in rainbow trout exposed to water pH 10.1 (Yesaki and Iwama, 1992). In addition, high water pH also inhibits branchial Na^+/NH_4^+ and Cl^-/HCO_3^- exchangers (Wilkie and Wood, 1996), which explains reductions of plasma Na^+ and Cl^- in rainbow trout (Yesaki and Iwama, 1992).

Species adapted to very alkaline waters developed mechanisms to overcome these problems. Lake Magadi tilapia excretes virtually all of its waste nitrogen as urea (Wilkie and Wood, 1996), and base excretion is possible through a modified 'seawater type' chloride cell (Laurent *et al.*, 1995). *C. tarichi*, from Lake Van, also excretes urea, but NH_3 excretion is maintained, either by the gills due to the specific composition of Lake Van, which allows maintenance of the gills-water NH_3 gradient or by an increase of renal excretion (Danulat and Kempe, 1992). On the other hand, Lahontan cutthroat trout survives in alkaline waters because of its ability to reduce ammonia production and possibly stimulation of gill Cl^-/HCO_3^- exchange by branchial cell chloride proliferation (Wilkie *et al.*, 1994). The scale-less carp *G. przewalskii* also reduces NH_3 production, accumulates this metabolite in the muscle to attenuate levels in other tissues, and possibly incorporates NH_3 into amino acids (but not urea) (Wang *et al.*, 2003).

EFFECT OF WATER HARDNESS ON SURVIVAL AND OSMOREGULATION

Rainbow trout maintain constant plasma Ca^{2+} irrespective of waterborne Ca^{2+}. Specimens of this species exposed to low waterborne Ca^{2+} (2.5 mg L^{-1} $CaCO_3$) showed an increase in number of chloride cells on lamellae and large apical surfaces to increase ion uptake (Perry and Wood, 1985). The armored catfish, *Hypostomus tietensis*, traíra, *Hoplias malabaricus*, and jejú, *Hoplerythrinus unitaeniatus*, when transferred from 1.7 mg L^{-1} $CaCO_3$ to distilled or deionized water also showed high proliferation of gill chloride cells, but the apical surface in direct contact with the external medium increased only in traíra in the first days after transference (Fernandes and Perna-Martins, 2002; Moron *et al.*, 2003). In addition, in armored catfish kept in distilled water, the chloride cells were buried and recessed under adjacent pavement cells and the apical surface was arranged in a sponge-like structure, developing an apical crypt. These modifications may be an adaptation to provide a microenvironment inside the crypt, preventing ion loss and favoring ion uptake in a much diluted environment (Fernandes and Perna-Martins, 2002). Transference of traíra from 1.7 to 85 mg L^{-1} $CaCO_3$ also induced a transient chloride cell proliferation in the lamellar epithelia, but the same was not observed in jejú (Moron *et al.*, 2003).

Channel catfish, reared at 407 mg L^{-1} $CaCO_3$, accumulates taurine and other amino acids in the muscle-free pool as a strategy for elevating osmolality in hard water without raising the ionic concentration of the body fluids. This strategy is analogous to the accumulation of urea in other vertebrates, but provides an additional advantage: taurine accumulation is metabolically safe. However, as the absolute changes in free amino acids in muscle and plasma were small (compared to fish exposed to 17.9 mg L^{-1} $CaCO_3$), they might play a limited role in the overall osmoregulation of channel catfish (Buentello and Gatlin, 2002).

Most Mg^{2+} uptake is by food intake, but if the water presents an adequate amount of Mg^{2+}, branchial uptake may be enough to compensate a low-Mg^{2+} diet in some species (Bijvelds *et al.*, 1998). However, in Mozambique tilapia, *Oreochromis mossambicus*, Mg^{2+} intake from the water—either via the integument or drinking—did not increase in low-Mg^{2+} fed fish, despite an increased opercular chloride cell density (Bijvelds *et al.*, 1996). High waterborne Mg^{2+} (up to 50 mmol) did not affect *Gasterosteus aculeatus*, Mozambique tilapia and goldfish, *Carassius auratus* (Bijvelds *et al.*, 1998).

Most teleosts exposed to acidic or alkaline waters showed a higher survival in hard rather than in soft waters (Freda and McDonald, 1988; Yesaki and Iwama, 1992; Laitinen and Karttunen, 1994; Townsend and Baldisserotto, 2001). Rainbow trout exposed to water pH 3.0 or 3.2 presented higher survival at water hardness 165 mg L^{-1} $CaCO_3$ than those exposed to the same pH and water hardness 10 mg L^{-1} $CaCO_3$ (McDonald *et al.*, 1980). Specimens of this species transferred from water pH 6.8 to 10.1 showed decreased Na^+ and Cl^- plasma levels at water hardness 4.0 mg L^{-1} $CaCO_3$, but when maintained at water hardness 320 mg L^{-1} $CaCO_3$, did not change ion levels in the plasma and also showed higher survival at water pH 10.1 (Yesaki and Iwama, 1992). In addition, survival was higher in recently hatched rainbow trout larvae exposed to pH 4.7 and water hardness of 90 mg L^{-1} $CaCO_3$ (increased with addition of Ca^{2+} and Mg^{2+}) than in those exposed to the same pH and water hardness of 12 mg L^{-1} $CaCO_3$ (Laitinen and Karttunen, 1994). The increase of water hardness from 20 up to 300 mg L^{-1} $CaCO_3$ (using $CaCl_2$) increased the survival rate of silver catfish juveniles exposed water pH 3.75, 10.0 and 10.5, but did not affect significantly survival (which was 100%) at less extreme pH even at 600 mg L^{-1} $CaCO_3$ (higher values were not tested) (Townsend and Baldisserotto, 2001).

The protective effect of water hardness against low pH is dependent of the affinity of branchial tight junctions for Ca^{2+}. This ion is important at stabilizing these tight junctions and, consequently, decreasing gill ion loss (Wood, 2001). Apparently, one of the mechanisms for survival of species that live in very soft waters, as those from the Rio Negro and the banded sunfish, is the high branchial affinity for Ca^{2+}, because even water hardness of 0.4-2.0 mg L^{-1} $CaCO_3$ provides enough Ca^{2+} to saturate tight junctions binding sites (Gonzalez *et al.*, 1998). However, dissolved organic carbon present in high amounts in these blackwaters also protects in some way against the deleterious effect of low water pH (Gonzalez *et al.*, 2002). Another strategy is used by yellow perch, because waterborne Ca^{2+} levels do not affect Na^+ uptake in this species. It seems that this species has transporters with high affinity by Na^+ and, therefore, Na^+ influx is maintained even at water pH 4.0 (Freda and McDonald, 1988). However, for most of the species studied, the increase of water hardness (waterborne Ca^{2+}) decreases ion loss in acidic pH (Fig. 5.3).

EFFECT OF WATER PH ON HATCHING AND GROWTH

The 6.5-9.0 water pH range is usually considered best for teleosts reproduction and growth (Zweig *et al.*, 1999), and any increase of acidity of the water impairs hatching and growth. Gametogenesis is impaired and 11% of the fertilized embryos showed deformation in rainbow trout exposed to pH 4.5. Hatching does not occur in grumatã, *Prochilodus lineatus* (Reynalte, 2000), and mortality is high in common carp, *Cyprinus carpio* (Jezierka and Witeska, 1995), when fertilization is done at pH 5.0, and is affected in *Perca fluviatilis* exposed to pH 5.5 or lower (Runn *et al.*, 1977). Exposure to low pH (5.5-6.0) also reduced length and weight of silver catfish larvae compared to those maintained at pH 8.0-8.5 (Lopes *et al.*, 2001). Similar results were obtained with the flagfish, *Jordanella floridae*, whose larvae showed reduced growth when exposed to pH 5.5

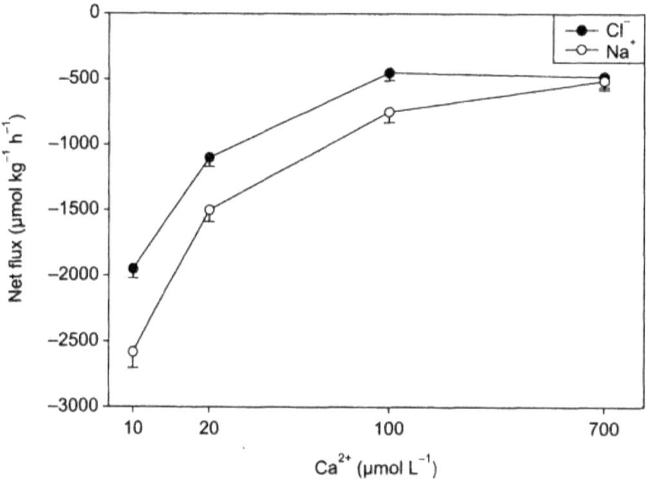

Fig. 5.3 Whole body ion net fluxes in tambaqui, *Colossoma macropomum*, exposed to water pH 3.5 and different waterborne Ca^{2+} levels. Data from Gonzalez *et al.* (1998) and Wood *et al.* (1998).

compared to larvae maintained in the 6.0 - 6.8 range. In addition, exposure of larvae of this species to pH 5.0 impaired growth and reduced survival compared to pH 5.5 (Graig and Baksi, 1977).

Brook trout fingerlings presented lower growth at water pH 5.5, 6.0 and 6.5 than at pH 7.1 (Menendez, 1976), and at water pH 4.2-5.0 than 5.2-6.5 (Norrgren and Degerman, 1992). Rainbow trout showed better growth at neutral pH (7.2) than at acidic pH (4.4) (Nelson, 1982). Exposure of silver catfish fingerlings to alkaline (9.0) or acidic water (5.5) also reduced growth compared to neutral pH (7.5) (Copatti *et al.*, 2005). However, post larvae of grumatã showed better development and survival at waters pH 6 than at pH 7 and 8 (Reynalte *et al.*, 2000), and exposure of brown trout, *Salmo trutta*, to water pH 5.0, 5.44 and 6.0 did not influence standard growth rate (Norrgren and Degerman, 1992).

Additional studies regarding the growth of fish whose natural environment is acidic or alkaline (as those described in the previous sections of this chapter) are still lacking. Probably, their best pH range for growth is not circumneutral, because the freshwater angelfish, *Pterophyllum scalare*, from Rio Negro, showed better growth in the 5-6.9 pH range (Chellappa, 2005).

EFFECT OF WATER HARDNESS ON HATCHING AND GROWTH

Water hardening (swelling process of flaccid newly shed eggs when they first contact water and absorb water) of fertilized eggs varies according to species and water hardness. When hardness is low, increase in egg diameter is greater (Gonzal *et al.*, 1987; Spade and Bristow, 1999). Survival of eye-up eggs of rainbow trout, Atlantic salmon and brook trout is higher when eggs are hatched at 139-230 mg·L^{-1} $CaCO_3$ (Ketola *et al.*, 1988). Hatching rate of stripped bass, *Morone saxatilis*, is higher at 200 mg·L^{-1} $CaCO_3$ (Spade and Bristow, 1999), and the recommended range for hatching of silver carp, *Hypophthalmichthys molitrix*, is 300–500 mg·L^{-1} $CaCO_3$ (Gonzal *et al.*, 1987). However, higher larval survival and growth in African catfish, *Clarias gariepinus*, and silver catfish is in the 60-70 mg·L^{-1} $CaCO_3$ water hardness range (Molokwu and Okpokwasili, 2002; Silva *et al.*, 2003, 2005; Townsend *et al.*, 2003) (Fig. 5.4). Channel catfish swim-up fry exposed to 0, 0.4, 2, 4, or 40 mg·L^{-1} Ca^{2+} (0, 1, 5, 10, and 100 mg·L^{-1} $CaCO_3$) showed best growth at 4 and 40 mg·L^{-1} Ca^{2+} (10 and 100 mg·L^{-1} $CaCO_3$) (Tucker and Steeby, 1993). The same authors observed an abnormal behavior (fry appeared lethargic and were spread out over the bottom) in water with low Ca^{2+} concentration (below 2 mg·L^{-1} Ca^{2+}). At least in case of silver catfish larvae, the increase of water hardness from 20 to 70 mg·L^{-1} $CaCO_3$ using either Ca^{2+} or Mg^{2+} improved hatch rate, but increase of waterborne Ca^{+2} above 20 mg·L^{-1}, irrespective of water hardness, is not recommend for incubation of silver catfish eggs because it reduced post-hatch (2 days after hatching) survival and larval weight and length after 21 days (Silva *et al.*, 2003, 2005). On the other hand, 30-40-day-old striped bass fingerlings presented higher survival at 278 mg·L^{-1} $CaCO_3$ than at 8-10 mg·L^{-1} $CaCO_3$, but the increase of water hardness must be done with $CaCl_2$ and not $MgCl_2$, showing the importance of Ca^{2+} in this period (Grizzle *et al.*, 1990). However, survival of recently hatched larvae (still with yolk sac) of this species were not affected by water hardness of 3-250 mg·L^{-1} of $CaCO_3$, indicating that the mechanisms for osmoregulation are different than in older striped bass (Grizzle *et al.*, 1992).

White bass, *Morone chrysops*, and sunshine bass, *M. chrysops* female × *Morone saxatilis* male (23-30-day-old) died within a few hours in water with 5-6 mg·L^{-1} $CaCO_3$. Addition of NaCl (0.7-5.0 g·L^{-1}) alone did not increase the survival rate in white bass, and only 25% sunshine bass

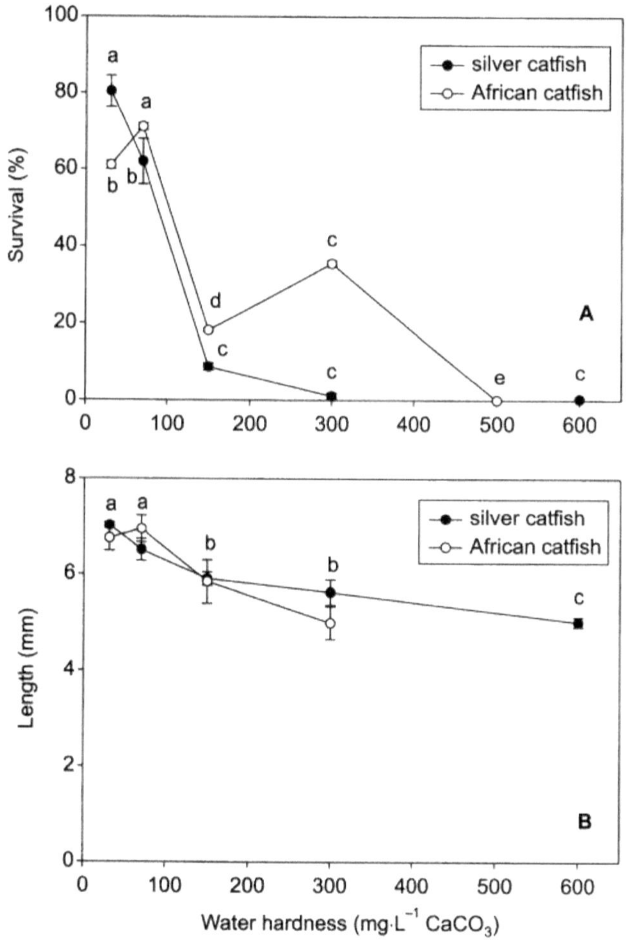

Fig. 5.4 Survival (A) and length (B) of African catfish (second day after hatching) and silver catfish (9 days after hatching) larvae to different water hardness levels. Data from Molokwu and Okpokwasili (2002) and Townsend *et al.* (2003). Different letters indicate significant difference among different water hardness in the same species.

survived. Survival rate of white bass increased with 5.0 g·NaCl L^{-1} and increasing water hardness up to 85.7% at 35 mg·L^{-1} $CaCO_3$, and a higher concentration of Ca^{2+} (up to 50 mg·L^{-1} $CaCO_3$) did not increase survival. Addition of Ca^{2+} to increase water hardness to 210 mg·L^{-1} $CaCO_3$ increased sunshine bass survival to 64% (Grizzle and Mauldin, 1999). However, neither the hatching rate nor the growth of Mozambique tilapia larvae were affected by exposure to waters with 2-3 or 88-96 mg·L^{-1} $CaCO_3$ (using $CaCl_2$ or $CaSO_4$ to increase water hardness) (Hwang *et al.*, 1996).

Water hardness from 12.5 to 200 mg·L^{-1} $CaCO_3$ (increased with $CaCl_2$) did not significantly affect the final weight, food conversion, condition factor, and plasma Ca^{2+} levels of sunshine bass after 42 days (Seals *et al.*, 1994). However, water hardness 200 mg·L^{-1} $CaCO_3$ improved sunshine bass post-harvest survival compared to lower water hardness (Grizzle *et al.*, 1985). The increase of water hardness with $MgSO_4$ up to 400 mg·L^{-1} $CaCO_3$ (with $MgSO_4$) reduced the survival rate of channel catfish juveniles to 0%, but when $CaCO_3$ was used to increase water hardness survival was higher (95%) (Perschbacher and Wurts, 1999).

Juvenile brook trout reared at water pH 6.5 and water hardness 100 mg·L^{-1} $CaCO_3$ presented a higher growth rate than those reared at the same pH and water hardness 12.5 mg·L^{-1} $CaCO_3$. Exposure of juveniles of this species to water pH 5.3 reduced growth around 33-38% at both water hardness (Rodgers, 1984). Apparently, the effect of water hardness on growth varies according to species and water quality. For some species (those which in the natural environment are found in hard or moderately hard water) hard water is needed for good development, while in others it might ameliorate the deleterious effect of non-optimal conditions.

Acknowledgements

B. Baldisserotto received a CNPq (Conselho Nacional de Desenvolvimento Tecnológico—Brazil) research grant.

References

Alabaster, J.S. and R. Lloyd. 1982. *Water Quality Criteria for Freshwater Fish.* 2nd Edition. FAO, Cambridge.

Baldisserotto, B. 2003. Osmoregulatory adaptations of freshwater teleosts. In: *Fish Adaptations*, A.L. Val and B.G. Kapoor (eds.). Science Publishers, Inc., Enfield (NH), USA, pp. 179–201.

Bijvelds, M.J.C., G. Flik, Z.I. Kolar and S.E.Wendelaar Bonga. 1996. Uptake, distribution and excretion of magnesium in *Oreochromis mossambicus*: dependence on magnesium in diet and water. *Fish Physiology and Biochemistry* 15: 287–298.

Bijvelds, M.J.C., J.A. Van der Velden, Z.I. Kolar and G. Flik. 1998. Magnesium transport in freshwater teleosts. *Journal of Experimental Biology* 201: 1981–1990.

Buentello, J.A. and D.M. Gatlin. 2002. Preliminary observations on the effects of water hardness on free taurine and other amino acids in plasma and muscle of channel catfish. *North American Journal of Aquaculture* 64: 95–102.

Chellapa, S. 2005. Acará bandeira, *Pterophyllum scalare*. In: *Espécies Nativas para Piscicultura no Brasil*. Baldisserotto, B. and Gomes, L.C. (eds.). Editora da Universidade Federal de Santa Maria. Santa Maria, pp. 393–402.

Danulat, E. and S. Kempe. 1992. Nitrogenous waste excretion and accumulation of urea and ammonia in *Chalcalburnus tarichi* (Cyprinidae), endemic to the extremely alkaline Lake Van (Eastern Turkey). *Fish Physiology and Biochemistry* 9: 377–386.

Danulat, E. and B. Selcuk. 1992. Life history and environmental conditions of the anadromous *Chalcalburnus tarichi* (Cyprinidae) in the highly alkaline Lake Van, Eastern Anatolia, Turkey. *Archives Fur Hydrobiologie* 126: 105–125.

Fernandes, M.N. and S.A. Perna-Martins. 2002. Chloride cell responses to long-term exposure to distilled and hard water in the gill of the armored catfish, *Hypostomus tietensis* (Loricariidae). *Acta Zoologica* 83: 321–328.

Freda, J. and D.G. McDonald. 1988. Physiological correlates of interspecific variation in acid tolerance in fish. *Journal of Experimental Biology* 136: 243–258.

Galvão, L., W. Pereira Filho, M. Abdon, E. Novo, J. Silva and F. Ponzoni. 2003. Spectral reflectance characterization of shallow lakes of the Brazilian Pantanal wetlands with field and airborne hyperspectral. *International Journal of Remote Sensing* 24: 4093–4112.

Giles, M.A., H.S. Majewski and B. Hobden. 1984. Osmoregulatory and hematological responses of rainbow trout (*Salmo gairdneri*) to extended environmental acidification. *Canadian Journal of Fisheries and Aquatic Sciences* 41: 1686–1694.

Gómez, S.E. 1998. Niveles letales de pH en *Odontesthes bonariensis* (Atheriniformes, Atherinidae). *Iheringia*, Série Zoologia 85: 101–108.

Gonzal, A.C., E.V. Aralar and J.M.F. Pavico. 1987. The effects of water hardness on the hatching and viability of silver carp *Hypophthalmichthys molitrix* eggs. *Aquaculture* 64: 111–118.

Gonzalez, R.J. 1996. Ion regulation in ion poor waters of low pH. In: *Physiology and Biochemistry of the Fishes of the Amazon*, A.L. Val, V.M.F. Almeida-Val and D.J. Randall (eds.). Instituto Nacional de Pesquisas da Amazônia (INPA), Manaus, pp. 111–121.

Gonzalez, R.J., R.W. Wilson, C.M. Wood, M.L. Patrick and A.L. Val. 2002. Diverse strategies for ion regulation in fish collected from the ion-poor, acidic Rio Negro. *Physiological and Biochemical Zoology* 75: 37–47.

Gonzalez, R.J., C.M. Wood, R.W. Wilson, M.L. Patrick, H.L. Bergman, A. Narahara and A.L. Val. 1998. Effects of water pH and calcium concentration on ion balance in fish of the Rio Negro, Amazon. *Physiological Zoology* 71: 15–22.

Graig, G.R. and W.F. Baksi. 1977. The effects of depressed pH on flagfish reproduction, growth and survival. *Water Research* 11: 621–626.

Grizzle, J.M. and A.C. Mauldin. 1999. Increased post-harvest survival of young white bass and sunshine bass by addition of calcium and sodium chloride to soft water. *North American Journal of Aquaculture* 61: 146–149.

Grizzle, J.M., A.C. Mauldin, D. Young and E. Henderson. 1985. Survival of juvenile striped bass (*Morone saxatilis*) and *Morone* hybrid bass (*Morone chrysops* × *Morone saxatilis*) increased by addition of calcium to soft water. *Aquaculture* 46: 167–171.

Grizzle, J.M., A.C. Mauldin II, D. Young and E. Henderson. 1990. Effects of environmental calcium and sodium on post-harvest survival of juvenile striped bass. *Journal of Aquatic Animal Health* 2: 104–108.

Grizzle, J.M., A.C. Mauldin and C.J. Ashfield. 1992. Effects of sodium chloride and calcium chloride on survival of larval striped bass. *Journal of Aquatic Animal Health* 4: 281–285.

Hwang, P.P., Y.C. Tung and M.H. Chang. 1996. Effect of environmental calcium levels on calcium uptake in tilapia larvae (*Oreochromis mossambicus*). *Fish Physiology and Biochemistry* 15: 363–370.

Jezierka, B. and M. Witeska. 1995. The influence of pH on embryonic development of common carp (*Cyprinus carpio*). *Archives of Polish Fisheries* 3: 85–94.

Kaneko, T., S. Hasegawa, K. Uchida, T. Ogasawara, A. Oyagi and T. Hirano. 1999. Acid tolerance of Japanese dace (a cyprinid teleost) in Lake Osorezan, a remarkable acid lake. *Zoological Science* 16: 871–877.

Ketola, H.G., D. Longacre, A. Greulich, L. Phetterplace and R. Lashomb. 1988. High calcium concentration in water increases mortality of salmon and trout eggs. *Progressive Fish-Culturist* 50: 129–135.

Laitinen, M. and M. Karttunen. 1994. Effects and calcium and magnesium in acid water on the ion balance of eggs and alevins of rainbow trout (*Oncorhynchus mykiss*). In: *Chronic Effect of Pollutants on Freshwater Fish*, R. Müller and R. Lloyd (eds.). FAO Fishing New Books, Cambridge, pp. 262–272.

Laurent, P., J.N. Maina, H.L. Bergman, A. Narahara, P. J. Walsh and C.M. Wood. 1995. Gill structure of a fish from an alkaline lake—effect of short-term exposure to neutral conditions. *Canadian Journal of Zoology* 73: 1170–1181.

Li, J., J. Eygensteyn, R.A.C. Lock, P.M. Verbost, A.J.H. Van Der Heijden, S.E. Wendelaar Bonga and G. Flik. 1995. Branchial chloride cells in larvae and juveniles of freshwater tilapia *Oreochromis mossambicus*. *Journal of Experimental Biology* 198: 2177–2184.

Lopes, J.M., L.V.F. Silva and B. Baldisserotto. 2001. Survival and growth of silver catfish larvae exposed to different water pH. *Aquaculture International* 9: 73–80.

Matsuo, A.Y.O. and A.L. Val. 2003. Fish adaptations to Amazonian blackwaters. In: *Fish Adaptations*, A.L. Val and B.G. Kapoor (eds.). Science Publishers, Inc., Enfield (NH), USA. pp. 1–36.

McDonaıd, D.G., H. Hōbe and C.M. Wood. 1980. The influence of calcium on the physiological responses of the rainbow trout (*Salmo gairdneri*) to low environmental pH. *Journal of Experimental Biology* 88: 109–131.

Molokwu, C.N. and G.C. Okpokwasili. 2002. Effect of water hardness on egg hatchability and larval viability of *Clarias gariepinus*. *Aquaculture International* 10: 57–64.

Moron, S.E., E.T. Oba, C.A. Andrade and M.N. Fernandes. 2003. Chloride cell responses to ion challenge in two tropical freshwater fish, the erythrinids *Hoplias malabaricus* and *Hoplerythrinus unitaeniatus*. *Journal of Experimental Zoology* A 298: 93–104.

Mortatti, J. and J.-L. Probst. 2003. Silicate rock weathering and atmospheric/soil CO_2 uptake in the Amazon basin estimated from river geochemistry: Seasonal and spatial variations. *Chemical Geology* 197: 177–196.

Nelson, J.A. 1982. Physiological observations on developing rainbow trout, *Salmo gairdnieri* (Richardson), exposed to low pH and varied calcium ion concentrations. *Journal of Fish Biology* 20: 359–372.

Norrgren, L. and E. Degerman. 1992. The influence of liming on embryos and yolk sac fry of Atlantic salmon and brown trout in acidified river in Sweden. *Journal of Fish Biology* 23: 567–576.

Patrick, R., V.P. Binetti and S.G. Halterman. 1981. Acid lakes from natural and anthropogenic causes. *Science* 211: 446–448.

Perry, S.F. and C.M. Wood. 1985. Kinetics of branchial calcium uptake in rainbow trout: Effects of acclimation to various external calcium levels. *Journal of Experimental Biology* 116: 411–433.

Perschbacher, P.W. and W.A. Wurts. 1999. Effects of calcium and magnesium hardness on acute copper toxicity to juvenile channel catfish *Ictalurus punctatus*. *Aquaculture* 172: 275–280.

Reite, O.B., G.M.O. Maloiy and B. Aasehaug. 1974. pH, salinity and temperature tolerance of Lake Magadi *Tilapia*. *Nature (Lond.)* 274: 315.

Reynalte Tataje, D.A. 2000. Efeito do pH da água na incubação e larvicultura do curimbatá *Prochilodus lineatus* Valenciennes, 1847 (Characiformes, Prochilodontidae). M.Sc. Thesis in Aquaculture – Universidade Federal de Santa Catarina, Florianópolis, Brazil.

Reynalte, D.A., R.L. Serafini and E. Zaniboni-Filho, 2000. Influência do pH da água na sobrevivência e crescimento das pós-larvas de curimbata *Prochilodus lineatus* Valenciennes, 1847 (Characiformes, Prochilodontidae). In: Simpósio Brasileiro de Aquicultura, 2000, Florianópolis, Brazil. *Anais*, CD-ROM.

Rodgers, D.W. 1984. Ambient pH and calcium concentrations as modifiers of growth and calcium dynamics of brook trout, *Salvelinus fontinalis*. *Canadian Journal of Fisheries and Aquatic Sciences* 41: 1774–1780.

Runn, P., N. Johansson and G. Milbrink. 1977. Some effects of low pH on the hatchability of eggs of perch, *Perca fluviatilis* L. *Zoon* 5: 115–125.

Satake, K., A. Oyagi and Y. Iwao. 1995. Natural acidification of lakes and rivers in Japan: The ecosystem of Lake Usoriko (pH 3.4-3.8). *Water, Air and Soil Pollution* 85: 511–516.

Seals, C., C.J. Kempton, J.R. Tomasso and T.I.J. Smith. 1994. Environmental calcium does not affect production or selected blood characteristics of sunshine bass reared under normal culture conditions. *Progressive Fish-Culturist* 56: 269–272.

Silva, L.V.F., J.I. Golombieski and B. Baldisserotto, 2003. Incubation of silver catfish, *Rhamdia quelen* (Pimelodidae), eggs at different calcium and magnesium concentrations. *Aquaculture* 228: 279–287.

Silva, L.V.F., J.I. Golombieski and B. Baldisserotto. 2005. Growth and survival of silver catfish larvae, *Rhamdia quelen* (Heptapteridae), at different calcium and magnesium concentrations. *Neotropical Ichthyology* 3: 299–304.

Spade, S. and B. Bristow. 1999. Effects of increasing water hardness on egg diameter and hatching rates of striped bass eggs. *North American Journal of Aquaculture* 61: 263–265.

Takatsu, A., Y. Ezoe, S. Eyama, A. Uchiumi, K. Tsunoda and K. Satake. 2000. Aluminum in lake water and organs of a fish *Tribolodon hakonensis* in strongly acidic lakes with a high aluminum concentration. *Limnology* 1: 185–189.

Townsend, C.R. and B. Baldisserotto. 2001. Survival of silver catfish fingerlings exposed to acute changes of water pH and hardness. *Aquaculture International* 9: 413–419.

Townsend, C.R., L.V.F. Silva and B. Baldisserotto. 2003. Growth and survival of *Rhamdia quelen* (Siluriformes, Pimelodidae) larvae exposed to different levels of water hardness. *Aquaculture* 215: 103–108.

Tucker, C.S. and J.A. Steeby. 1993. A practical calcium hardness criterion for channel catfish hatchery water supplies. *Journal of the World Aquaculture Society* 24: 396–401.

Wang, Y.X.S., R.J. Gonzalez, M.L. Patrick, M. Grosell, C.G. Zhang, Q.A. Feng, J.Z. Du, P.J. Walsh and C.M. Wood. 2003. Unusual physiology of scale-less carp, *Gymnocypris przewalskii*, in Lake Qinghai: a high altitude alkaline saline lake. *Comparative Biochemistry and Physiology* A134: 409–421.

Wilkie, M.P. and C.M. Wood. 1996. The adaptations of fish to extremely alkaline environments. *Comparative Biochemistry and Physiology* B113: 665–673.

Wilkie, M.P., P.A. Wright, G.K. Iwama and C.M. Wood. 1994. The physiological adaptations of the Lahontan cutthroat trout (*Oncorhynchus clarki henshawi*) following transfer from well water to the highly alkaline waters of Pyramid Lake, Nevada (pH 9.4). *Physiological Zoology* 67: 355–380.

Wilson, P.J., C.M. Wood., P.J. Walsh, A.N. Bergman, H.L. Bergman, P. Laurent and B.N. White. 2004. Discordance between genetic structure and morphological, ecological, and physiological adaptation in Lake Magadi tilapia. *Physiological and Biochemical Zoology* 77: 537–555.

Wood, C.M. 1989. The physiological problems of fish in acid waters. In: *Acid Toxicity and Aquatic Animals*, R. Morris, E.W. Taylor, D.J.A. Brown and J.A. Brown (eds.). Cambridge University Press, Cambridge, pp. 125–152.

Wood, C.M. 2001. Toxic response of the gill. In: *Target Organ Toxicity in Marine and Freshwater Teleosts*, D. Schlenk and W.H. Benson (eds.). Taylor & Francis, London, pp. 1–89.

Wood, C.M., R.W. Wilson, R.J. Gonzalez, M.L. Patrick, H.L. Bergman, A. Narahara and A.L. Val. 1998. Responses of an Amazonian teleost, the tambaqui (*Colossoma macropomum*), to low pH in extremely soft water. *Physiological Zoology* 71: 658–670.

Wood, C.M., C.L. Milligan and P.J. Walsh. 1999. Renal responses of trout to chronic respiratory and metabolic acidosis and metabolic alkalosis. *American Journal of Physiology* 277: R482–R492.

Yesaki, T.Y. and G.K. Iwama. 1992. Survival, acid-base regulation, ion regulation, and ammonia excretion in rainbow trout in highly alkaline hard water. *Physiological Zoology* 65: 763–787.

Zaions, M.I. and B. Baldisserotto. 2000. Na^+ and K^+ body levels and survival of fingerlings of *Rhamdia quelen* (Siluriformes: Pimelodidae) exposed to acute changes of water pH. *Ciência Rural* 30: 1041–1045.

Zaniboni-Filho, E., S. Meurer, J.I. Golombieski, L.V.F. Silva and B. Baldisserotto. 2002. Survival of *Prochilodus lineatus* (Valenciennes) fingerlings exposed to acute pH changes. *Acta Scientiarum* 24: 917–920.

Zweig, R.D., J.D. Morton and M.M. Stewart. 1999. *Source Water Quality for Aquaculture*. World Bank, Washington.

CHAPTER

6

Arginine Vasotocin and Isotocin: Towards their Role in Fish Osmoregulation

Ewa Kulczykowska

NEUROPEPTIDES ARGININE VASOTOCIN AND ISOTOCIN—GENERAL VIEW

Neuropeptides are defined as peptides synthesized in neurons that play an important role in transmitting information in the nervous system. The mechanism of known neuropeptides biosynthesis is similar in general: the gene of the peptide precursor is first transcribed into m-RNA, which is then translated into the propeptide. After various modifications (i.e., proteolysis of the propeptide), the mature peptide packed into secretory granules is transported via axonal flow to the nerve terminal where it is stored and released in response to appropriate stimulus.

That view does fully apply to arginine vasotocin (AVT) and isotocin (IT), fish neuropeptides synthesized in the hypothalamic magnocellular

Author's address: Department of Genetics and Marine Biotechnology, Institute of Oceanology of Polish Academy of Sciences, Sopot, Poland.
E-mail: ekulczykowska@iopan.gda.pl

neurons of the NPO (nucleus preopticus) from where they are transported to the neurohypophysis for storage and release. The identification of the neuronal origin of both neurohypophysial peptides and evidence for the presence of separate hypothalamic neurosecretory neurons producing AVT and IT in fish have been provided by various methods. These procedures include immunocytochemistry either alone or combined with carbocyanine tract tracing and confocal double-color immunofluorescent microscopy (Goossens *et al.*, 1977; Van den Dungen *et al.*, 1982; Holmquist and Ekström, 1995; Saito *et al.*, 2004). The peptides, closely related to mammalian vasopressin and oxytocin, containing nine amino acids residues, have been identified in the hypothalamo-neurohypophysial system of teleosts by Acher *et al.* (1961, 1962). As is the case with other neuropeptides, both AVT and IT are produced as a part of larger precursor molecule. The vasotocin and isotocin precursor sequences consist of a signal peptide, hormone and neurophysin (Fig. 6.1). Both pro-vasotocin and pro-isotocin have elongated carboxyl-terminals with a leucin-rich segment similar to copeptine-like sequence of vasopressin precursor, but its glycosylation does not appear to be possible. The polypeptide neurophysin, cysteine rich and capable of binding the neurohypophysial hormone, has been shown for the first time in fish by Pickering (1968). Since the early nineties, the structural organization of pro-vasotocin and pro-isotocin genes has been described in several fish species: *Catostomus commersoni, Oncorhynchus keta, O. masou, Platichthys flesus, Triakis scyllium, Neoceratodus forsteri, Danio rerio* (Heierhorst *et al.*, 1989, 1990; Suzuki *et al.*, 1992; Hyodo *et al.*, 1997, 2004; Warne *et al.*, 2000; Unger and Glasgow, 2003). Prior to secretion, each hormone is stored in secretory granules in the form of a non-covalent complex with its associated neurophysin, which is important during formation of the mature nonapeptide hormone. Release of the complex into the blood by exocytosis leads to its spontaneous dissociation. In many teleost species, especially salmonids, a duplication of the nonapeptides genes possessing the different expression level has been presented (Hiraoka *et al.*, 1993).

The homologous neurohypophysial hormones have been identified in representatives of all vertebrates systematic groups (Acher, 1995; Bentley, 1998). In general, they are arranged in a five-amino acid ring, joined by a disulfide bridge and a side chain of three amino acids. So far, at least 12 homologous nonapeptides have been identified among the vertebrates (Fig. 6.2). Amino acid substitution occurs at the second, third, fourth, or

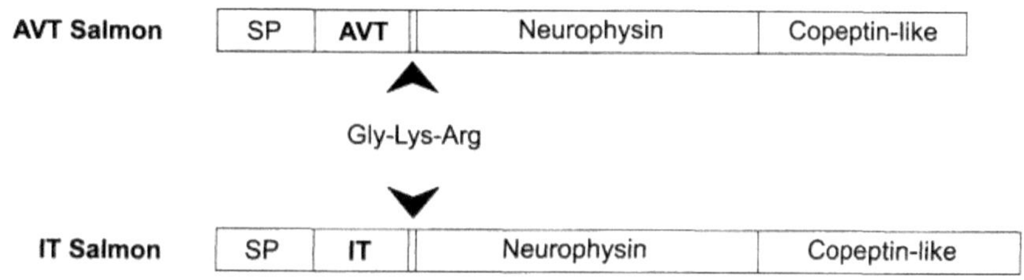

Fig. 6.1 Structure of pro-hormones for pro-vasotocin and pro-isotocin in fish.

eighth position in the molecule. The distribution of these natural analogs is well-defined: in mammals vasopressin and oxytocin are the main neurohypophysial hormones, in non-mammalian vertebrates arginine vasotocin (vasopressin-like peptide) is present together with usually one of the eight already identified variants of oxytocin-like peptide. There is a great deal of variability in nonapeptides variants between fish (Acher, 1993; Acher and Chauvet, 1995). Both nonapeptides vasotocin and isotocin are found in all bony fish except for lungfish possessing mesotocin instead of isotocin and chondrosteans in which glumitocin, aspargtocin, asvatocin, phasvatocin or valitocin are recognized besides vasotocin. In cyclostome fish, however, vasotocin is a sole neurohypophysial nonapeptide. Since vasotocin is present in all vertebrates, it has been considered as a precursor molecule for the neurohypophysial hormones family (Bentley, 1998). It is worth emphasizing that the neurohypophysial hormones present in one species differ by two amino acid substitutions at the most, but their biological activities are considerable distinct, i.e., vasopressin—mammalian well-known antidiuretic hormone and oxytocin—hormone involved in parturition and lactation (Bentley, 1998).

Neuropeptides action is generally determined by their binding to the specific receptors in the central nervous system (CNS), or in other sites of the body. In the CNS, neuropeptides play a role of neurotransmitters and/or neuromodulators, while distributed with the circulatory system they act as hormones. Just the same, AVT and IT may act, either locally in the CNS or in the peripheral target organs (Goossens *et al.*, 1977; Van den Dungen *et al.*, 1982; Acher, 1993; Acher and Chauvet, 1995; Goodson and Bass, 2000).

It is a well-established fact that the neurohypophysial hormones receptors belong to the superfamily of guanine nucleotide binding protein (G-protein)-coupled receptors. The topography of the receptors is typical

ARGININE VASOTOCIN

Cys - *Tyr - Ile - Gln* - Asn - Cys - Pro - *Arg* - Gly - NH_2
1 2 3 4 5 6 7 8 9

ARGININE VASOPRESSIN

Cys - *Tyr - Phe - Gln* - Asn - Cys - Pro - *Arg* - Gly - NH_2

LYSINE VASOPRESSIN

Cys - *Tyr - Phe - Gln* - Asn - Cys - Pro - *Lys* - Gly - NH_2

PHENYPRESSIN

Cys - *Phe - Phe - Gln* - Asn - Cys - Pro - *Arg* - Gly - NH_2

ISOTOCIN

Cys - *Tyr - Ile - Ser* - Asn - Cys - Pro - *Ile* - Gly - NH_2

OXYTOCIN

Cys - *Tyr - Ile - Gln* - Asn - Cys - Pro - *Leu* - Gly - NH_2

MESOTOCIN

Cys - *Tyr - Ile - Gln* - Asn - Cys - Pro - *Ile* - Gly - NH_2

VALITOCIN

Cys - *Tyr - Ile - Gln* - Asn - Cys - Pro - *Val* - Gly - NH_2

GLUMITOCIN

Cys - *Tyr - Ile - Ser* - Asn - Cys - Pro - *Gln* - Gly - NH_2

ASPARGTOCIN

Cys - *Tyr - Ile - Asn* - Asn - Cys - Pro - *Leu* - Gly - NH_2

ASVATOCIN

Cys - *Tyr - Ile - Asn* - Asn - Cys - Pro - *Val* - Gly - NH_2

PHASVATOCIN

Cys - *Tyr - Phe - Asn* - Asn - Cys - Pro - *Val* - Gly - NH_2

Fig. 6.2 Amino acid sequences of neurohypophysial peptides in vertebrates.

with seven helical transmembrane-spanning domains, four extracellular domains and three intracellular domains (Bentley, 1998) (Fig. 6.3). Vasotocin and isotocin receptor transcripts are widely distributed in different fish organs: brain, pituitary, spleen, lateral line, ovary, bladder, intestine, liver, heart, gills, kidney and skeletal and smooth muscle, suggesting a function of nonapeptides there (Mahlmann *et al.*, 1994; Hausmann *et al.*, 1995). So far, to this author's knowledge, AVT and IT receptors have been cloned in two fish species, *C. commersoni* and flounder (*Platichthys flesus*) (Mahlmann *et al.*, 1994; Hausmann *et al.*, 1995; Warne, 2001). Cloning of the vasotocin and isotocin receptor genes has revealed the presence of residues that are conserved among nonapeptide receptors and have been suggested to contribute to the ligand binding domain (Hausmann *et al.*, 1996) (Fig. 6.3). There is a growing body of evidence

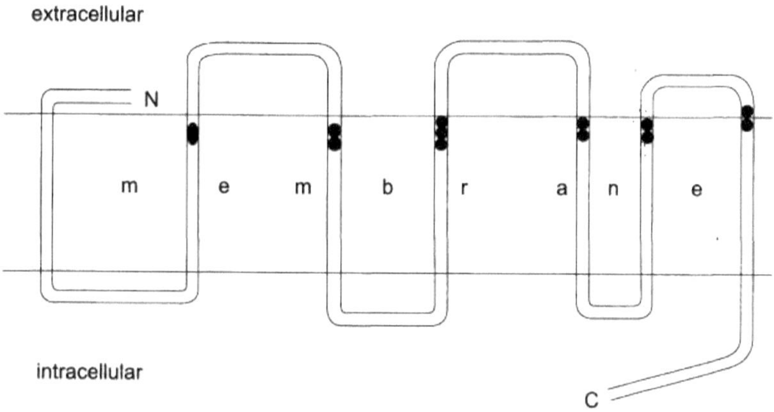

Fig. 6.3 Topography of the arginine vasopressin and isotocin receptors. The receptors are the members of the G protein-coupled receptor superfamily. The conservative regions marked in black are probably the sites of interaction with the hormone.

that AVT, which binds to the receptors on the pre-synaptic or post-synaptic membranes in the central nervous system of fish, acts as neurotransmitter and/or neuromodulator and influences reproductive physiology and related social behavior (Foran and Bass, 1998; Goodson and Bass, 2000). Recently, the target neurons for brain action of vasotocin were identified in newts (Lewis *et al.*, 2004). On the other hand, a number of studies in fish have demonstrated a role of AVT, when distributed with circulation, in maintenance of water and electrolytes balance, cardiovascular activity and regulation of endocrine secretion (Babiker and Rankin, 1978, 1979; Fryer and Leung, 1982; Acher, 1993; Conklin *et al.*, 1997).

SIGNALS FOR AVT/IT SYNTHESIS AND RELEASE

Several studies have pointed out that synthesis of teleosts nonapeptides in hypothalamus and their secretion into circulation in the neurohypophysis have changed in response to environmental salinity (Maetz and Lahlou, 1974; Haruta *et al.*, 1991; Hyodo and Urano, 1991; Perrott *et al.*, 1991). In early sixties, it was already observable that neurosecretory material in hypothalamus and pituitary was reduced after transfer of rainbow trout (*Oncorhynchus mykiss*), eel (*Anguilla anguilla*) and stickleback (*Gasterosteus aculeatus*) to a hyperosmotic medium (Fridberg and Olsson, 1959; Holmes and McBean, 1963; Sharratt *et al.*, 1964). Moreover, in rainbow trout transferred from FW to SW, Carlson and Holmes (1962) and Elders (1964) demonstrated the transitory decrease in antidiuretic and pressor activity of pituitary, which recovered after 6 hours. Novel studies by Hyodo and Urano (1991) applying an *in situ* hybridization method have coincided well with those early results. They have shown in rainbow trout, that proAVT mRNA level is markedly decreased by transfer of fish from FW to 80% SW and remains consistently low by day 14. Just after retransfer of fish to FW, proAVT mRNA level rises to the initial FW value. A significant decrease in proIT mRNA levels in SW salmonid fish is also observed but only on day 1 after transfer. The results have suggested that the acute change of water salinity is an important factor influencing neurohypophysial hormones synthesis and/or release (Hyodo and Urano, 1991; Urano *et al.*, 1994).

The changes in the pattern of secretion from pituitary and resulting plasma concentrations of the AVT evoked by alterations in environmental tonicity was exhibited in the flounder, rainbow trout and carp (*Cyprinus carpio*) by Perrott *et al.* (1991). In the euryhaline flounder and trout, the higher pituitary AVT in FW-adapted fish was associated with a higher plasma concentration of the hormone compared with SW-acclimated fish. However, in the stenohaline carp, there was no difference in plasma AVT between fish adapted to either FW or 40% SW. On the other hand, the initial response of the euryhaline flounder to hypotonic challenge involved a decrease in plasma AVT concentration (Bond *et al.*, 2002) similarly to that observed in rainbow trout by Kulczykowska (1997). In other euryhaline fish medaka (*Oryzias latipes*), the AVT content in the pituitary temporarily decreased after transfer to SW, thus, suggesting an increase of AVT release during the first hours of SW adaptation. During readaptation to FW, however, pituitary content of AVT was elevated within the first 2 hours after transfer, again indicating the pronounced storage of the hormone (Haruta *et al.*, 1991). In dogfish (*Triakis scyllium*), marine elasmobranch utilizing a unique strategy for adaptation to elevated salinity by accumulation of urea, AVT secretory activity was also enhanced by transfer to hyperosmotic environment (Hyodo *et al.*, 2004). A teleost species *Rivulus marmoratus*, dwellers of brackish waters of tropical mangrove forests, while exposed to different salinities, reacted by a varying pattern of vasotocin immunoreactivity in the pituitary and in the preoptic nucleus (Nürnberger *et al.*, 1996). The response of circulating AVT and IT to osmotic challenge was also reported in rainbow trout transferred from

FW to brackish water (BW): the AVT and IT levels increased significantly 2 hr and 24 hr after transfer, respectively. In FW-transferred fish, both hormones decreased steadily, achieving the lowest values 2 and 24 hr after transfer. However, after 10 days of fish adaptation to BW plasma, AVT concentration decreased below FW value, whereas after 10 days of acclimatization in FW that increased above BW value (Kulczykowska, 1997). In rainbow trout subjected to acute osmotic stress, a significant increase in plasma AVT, but not in plasma IT level was observed (Kulczykowska, 2001). Although it was not clear whether plasma AVT/IT variations reflect changes in AVT/IT production or secretion rates, or both, but the hormonal responses to the acute and prolonged changes of water salinity were evident. Yet both hormones synthesis and release seemed to be differentially sensitive to plasma osmolality changes (Kulczykowska, 1997, 2001).

Taken together, it appears that in the euryhaline fish species, the synthesis, storage and secretion of AVT are sensitive to environmental stimuli such as exposure to extreme salinities. A question here arises: what is an internal signal for the neurohypophysial hormone release?

It is a well-established fact in mammals that an increase in plasma osmolality is the principal physiological stimulus for vasopressin secretion. Also, severe hypovolemia and/or hypotension are strong signals for this nonapeptide release (Szczepańska-Sadowska *et al.*, 1983). It is not clear, however, if similar signals are also engaged in vasotocin and isotocin secretion in fish. A consistent relationship between plasma AVT concentration and plasma Na^+, Cl^-, and osmolality was demonstrated in seawater-adapted flounder (Warne and Balment, 1995). In rainbow trout transferred between FW and BW, plasma vasotocin concentration also correlated with plasma osmolality (Kulczykowska, 1997). On the other hand, although an increase in plasma AVT concentration with increasing plasma osmolality or sodium concentration was apparent for SW-adapted flounder, this was not the case in FW-adapted fish (Perrott *et al.*, 1991; Balment *et al.*, 1993; Harding *et al.*, 1997). Thus, the other factor involved in controlling of vasotocin secretion, i.e., volemic status of the animal, was considered. However, the volemic expansion or hemorrhage protocols applied in chronically cannulated flounder studies failed to demonstrate that plasma AVT level may be sensitive to volemic status (Warne and Balment, 1995). Therefore, a rapid increase in plasma osmolality seems to be a major factor to control circulating AVT level in fish.

ARGININE VASOTOCIN/ISOTOCIN: TARGETS OF ACTION AND MECHANISMS

Overview

To have a clearer view of the role of neurohormones in osmoregulation in fish, it is essential herein to remind the general rules of that process. The biological structures and regulatory mechanisms participating in osmoregulation in aquatic and terrestrial vertebrates are different, but they have also many common features, i.e., neurohypophysial hormones analogs playing similar regulatory roles (Bentley, 1998). Terrestrial vertebrates are equipped with kidneys, which are highly specialized in preserving water and ions. Fish, in general, posses less complicated mechanism for regulation of ions and water balance, because of easier access to these elements in water. However, water and ions movements in freshwater and seawater fish require the regulation in the opposite direction: marine fish must secrete NaCl and conserve water and freshwater fish must accumulate salt from environment and eliminate excess water. Many species use both strategies to maintain electrolytes balance during a lifetime, i.e., migrating salmon, eel, etc. (Bentley, 1971).

There are three main organs engaged in osmoregulation in fish (in order of importance): gill, gastrointestinal tract (GIT) and kidney. The responses of fish to osmotic perturbation are summarized in Fig. 6.4.

Osmoregulatory tissues possess a high level of membrane-bound enzyme sodium ion-potassium ion-adenosinetriphosphatase activity, i.e., Na^{+}- K^{+}- ATPase, 'sodium pump'. The gill is the primary site of active ion transport responsible for body electrolyte homeostasis in teleosts. In freshwater teleosts, the gills actively absorb salt and passively intake water due to the osmotic influx, whereas seawater teleosts actively excrete NaCl. The branchial chloride cells play an essential role in the seawater acclimation of euryhaline teleost being responsible for Cl^{-} secretion (Bentley, 1971). The second important organ for water and salt exchange in fish is the gastrointestinal tract. The GIT's role differs in various fish species. Fish living in fresh water scarcely drink, but the intestine actively transports sodium from the lumen to the blood. Conversely, drinking and subsequent absorption of water by the intestine is essential for seawater fish (Bentley, 1971). The kidney does not have a pivotal role in fish osmoregulation, but its significance should not be underestimated. Fish are the only vertebrates with kidneys able to produce urine by glomerular and

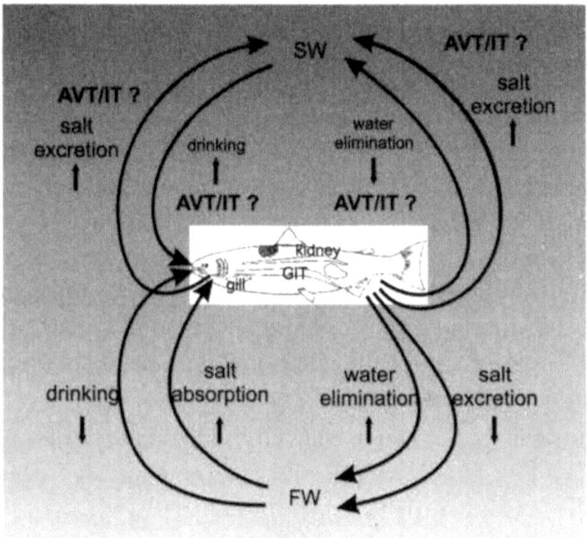

Fig. 6.4 Physiological responses of fish to different water salinities: potential involvement of the neurohypophysial neuropeptides. Modified from: Kulczykowska, E. (2002). A review of multifunctional hormone melatonin and a new hypothesis involving osmoregulation. *Reviews in Fish Biology and Fisheries* 11, 321-330.

aglomerular mechanisms defined by Beyenbach (1995) and Renfro (1999). Glomeruli appear to be particularly suited to the excretion of the large volumes of water accumulated by fish living in fresh water. The renal tubules of fish are differentiated from one or two tubular segments to the complement structure of the vertebrate nephron. Thus, the contribution of the kidney to extracellular fluid homeostasis is not universal in fish, and renal function spans the whole spectrum from glomerular filtration to tubular secretion (for review see: Beyenbach, 1995; Renfro, 1999; Dantzler, 2003; Nishimura and Fan, 2003). The main function of the proximal tubule of glomerular kidney in higher vertebrates is to reabsorb fluid. Surprisingly, the renal proximal tubules of glomerular kidneys in marine and euryhaline fish, e.g., winter flounder (*Pseudopleuronectes americanus*), dogfish shark (*Squalus acanthias*) and killifish (*Fundulus heteroclitus*)) appear to secrete fluids containing Na^+ and Cl^-. Interestingly, the same phenomenon is observed in the kidney of aglomerular toadfish (*Opsanus tau*). In elasmobranchs, the kidney is a major site of urea retention.

It is a well-established fact in mammals that neurohypophysial hormone arginine vasopressin regulates water and ions transport by epithelia (kidney tubules, skin, bladder) by stimulating adenylate cyclase via V2-type receptors. Other effects of that hormone (vasoconstriction, glycogenolysis, platelet aggregation) are mediated by stimulation of phosphoinositide breakdown and/or calcium mobilization via V1-type receptors (Bentley, 1998). In non-mammalian vertebrates—among them fish—arginine vasopressin is replaced by arginine vasotocin. If vasotocin triggers both signaling pathways and fulfils essentially similar functions in fish as vasopressin does so in mammals is not clear and requires elucidation. A role of the second neurohypophysial hormone in teleost, i.e., oxytocin-like isotocin is even more enigmatic.

Are osmoregulatory organs: gill, GIT and kidney the physiological goals for AVT/IT action? This question will be addressed below.

Gill

The first suggestions that the gill may be a target organ for AVT/IT action appeared as early as in the sixties. The gills of fish living in seawater are the site of considerable sodium exchange, with the outflux exceeding the influx by an amount which is about equivalent to that gained through the gut (urinary loss is insignificant). In seawater fish, chloride cells secrete chloride actively into the external water. It was proved that injections of neurohypophysial peptides—vasotocin and oxytocin—facilitated the branchial outflux of sodium in flounder transferred from fresh water to seawater (Motais and Maetz, 1967). In fish living in fresh water, however, sodium and chloride depletion stimulates the accumulation of sodium by the gills. It was demonstrated in goldfish that isotocin and, to a lesser extent vasotocin, while administered, increased the rate of sodium uptake across the gills (Maetz *et al.*, 1964).

A more updated data will evidently broaden this view. There is no doubt that an adenylate cyclase in plasma membranes from the teleost gill is sensitive to fish neurohypophysial peptides. In rainbow trout, AVT and IT, when administered in concentrations between 1 and 100 pM, have inhibited both basal and glucagon stimulated activity of the enzyme (Guibbolini and Lahlou, 1987a). The ability of the inhibition appeared to be sensitive to the environmental salinity, being especially expressed in high-salt media (Guibbolini and Lahlou, 1987a). The first evidence for a saturable, reversible and high-affinity specific binding of AVT by cells isolated from the gills of eels adapted to freshwater (FW) and seawater (SW) have been provided by means of ^{125}I-labelled AVT binding (Guibbolini *et al.*, 1988). Further studies conducted on rainbow trout gill epithelium, using specific neurohypophysial analogues, strongly suggested the presence of a new type of vasotocin and isotocin receptors designated as NH_f (Guibbolini and Lahlou, 1990), closer to the V1 than to the V2-type, with reference to the mammalian model. In experiments with the V1 receptor agonists in rainbow trout gill epithelium, reduction of adenylate cyclase activity have been more noticeable in SW rather than in FW (Guibbolini and Lahlou, 1990). Then it has been confirmed that the effects of fish nonapeptides are mediated by a G_i protein sensitive to pertussis toxin and guanine nucleotides (Guibbolini and Lahlou, 1992). Shortly afterwards, the molecular structures of AVT and IT receptors has been established by Mahlmann *et al.* (1994) and Hausmann *et al.* (1995) in teleost fish *C. commersoni*. The sequence of AVT and IT receptor has displayed the similarity to the mammalian V1-type receptor and the oxytocin receptor, respectively. Mutational analysis has shown that the different regions of the vasotocin receptor participate in hormone binding (Hausmann *et al.*, 1996) (Fig. 6.3). Subsequently, an AVT receptor from the euryhaline flounder has been cloned and its specificity for AVT has been documented (Warne, 2001). It has been demonstrated that activation of the receptor is coupled to phospholipase C and the inositolphosphate/calcium pathway, similarly to that presented in *C. commersoni*. Moreover, the flounder receptor has been shown to be more similar to the mammalian V1 than to the V2-type, in terms of its potent activation by AVP agonists for V1 receptor subtype and its higher homology with V1-type receptors (Warne, 2001). The AVT receptor mRNA expression studies confirm gills as a site of the hormone action in bony fish (Mahlmann *et al.*, 1994; Warne, 2001).

Although the results pointed out the fish gill as a physiological target organ for neurohypophysial hormones, the heterogeneity of the branchial epithelium comprising gill chloride, respiratory, pillar and mucous cells made it difficult to identify the exact target cells. Only when a detailed study on primary cultures of gill cells from a marine fish sea bass has been performed, the respiratory like cells have been shown to be an effective goals for both neurohypophysial peptides (Guibbolini and Avella, 2003). It has been demonstrated that both AVT and IT induce a dose-dependent stimulation of Cl^- secretion through the epithelium. It is consistent with the phenomenon of ions excretion through the gills observed in fish living in seawater. It has been suggested that the physiological effects of AVT and IT are mediated via two pharmacologically similar, V1-like receptors, located in gill respiratory-like cells. Probably V1-type receptors are also present in the chloride cells. Early experiments with isolated gill cells from SW-adapted eels, showed that AVT is a potent regulator of Cl^-secretion in the chloride cells (Guibbolini *et al.*, 1988). Moreover, in rainbow trout adapted to SW, an increase of gill chloride cells number and a more pronounced inhibition of adenylate cyclase activity by AVT and IT was observed, which may indicate the multiplication of the AVT/IT receptors sites located on the chloride cells during SW adaptation of euryhaline fish (Guibbolini and Lahlou, 1987a, b). Further, the morphological changes in the chloride cells of the epithelium in respect to the external salinity, which correlate strikingly with the AVT-binding parameters, point out to these cells as the site of AVT action (Guibbolini *et al.*, 1988). Also, recent studies of the estuarine teleost killifish (*Fundulus heteroclitus*) suggest that sodium chloride secretion upregulated on return of the fish to full seawater, is mediated via arginine vasotocin receptors present in basolateral membrane of mitochondria-rich cell in gill epithelium (Marshall, 2003).

Considering the fish gill as a target organ for neurohypophysial hormones, also the branchial blood vessels should be taken into account as potential effectors of the hormone action. AVT is known to be a vasopressor in all major vertebrate groups, among them in fish (Le Mevel *et al.*, 1993; Bentley, 1998). The vascular action of both neurohypophysial hormones in the branchial vasculature has been shown in many experiments (Bennett and Rankin, 1986; Oudit and Butler, 1995; Conklin *et al.*, 1996, 1997, 1999). In the isolated perfused gills of the European eel, Bennett and Rankin (1986) demonstrated a vasoconstriction of the arterio-arterial pathway and a decrease in arterio-arterial flow with no effect on the arterio-venous component of branchial flow. More recently, Oudit and Butler (1995) have postulated that AVT-induced vasoconstriction of arterio-arterial pathway is responsible for increased branchial shunting in freshwater eel. AVT in physiological concentrations produced contraction in afferent branchial arteries in holostean gar (*Lepisosteous* spp.) (Conklin *et al.*, 1996). In rainbow trout, AVT injection had a greater impact on branchial resistance than it did on systemic resistance, but in the isolated perfused gill, the hormone affected both the arterio-arterial and arterio-venous pathways (Conklin *et al.*, 1997). In the same study, the AVT at low concentrations stimulated constriction in the arterio-venous pathway and thus increased flow through the alternative arterio-arterial pathway. On the contrary, the AVT at elevated levels decreased arterio-arterial flow while increased arterio-venous flow (Conklin *et al.*, 1997). The alterations in redistribution of blood in gills may change AVT and IT delivery to the sites of ion exchange in gill epithelium and influence the ions and water transport via epithelium, as it was suggested by Maetz and Lahlou (1974).

The magnitude of AVT-stimulated constriction of the isolated efferent branchial artery in rainbow trout—high above that of other well-known vasoconstrictors—indicates a significance of AVT in regulation of gill vascular resistance, at least in this species. Studies in rainbow trout and eel have shown that the gill vasculature is far less sensitive to IT than to AVT (Bennett and Rankin, 1986; Conklin *et al.*, 1999). Vascular actions of neurohypophysial peptides have also been studied in free-swimming, chronically cannulated flounder by Warne and Balment (1997a, b). The initial fall in dorsal aortic blood pressure observed immediately after AVT and IT injections have been proposed to be due to AVT/IT dependent branchial vasoconstriction. It has been suggested that the AVT primary effect is a constriction of the arterio-arterial pathway (Warne and Balment, 1997a, b), as it has been reported in the *in vitro* perfused gill preparation by Bennett and Rankin (1986). A subsequent rise in the post-branchial blood pressure in response to AVT and lack of this effect following IT injection would suggest the presence of different types of hormones receptors (Warne and Balment, 1997a, b). However, it should be stressed here that the most of vascular effects in gill have been observed at hormones doses high above those considered as physiologically relevant. Also in the study of conscious, chronically cannulated Atlantic cod (*Gadus morhua*), the biphasic vascular reaction to administered AVT was demonstrated, but the dose of the hormone can be considered evidently as pharmacological. The absence of any response to IT injection was obvious in this study (Kulczykowska, 1998).

Taking together, a physiological action of both peptides in the regional blood flow distribution in fish gill and significance of this mechanism in osmoregulatory process needs to be reconsidered.

Gastrointestinal Tract (GIT)

It is well known that fish living in freshwater drink little water, but teleosts in the sea must drink salt water permanently in order to prevent dehydration. The intestine of freshwater fish actively absorbs sodium from the food, while fasting the fish gain little sodium from the water they drink. In seawater, a salt absorption and coupled to this water absorption take place in teleost fish, but the cartilaginous fish have developed an original urea-based osmoregulation model and use a rectal salt gland for sodium chloride excretion (Bentley, 1971).

There are no existing data—to this author's knowledge—on the potential physiological action of neurohypophysial hormones in GIT of fish, although the IT receptor transcripts in the intestine of C. *commersoni* (Hausmann *et al.*, 1995) and AVT receptors in the GIT's smooth muscle of rainbow trout (Conklin *et al.*, 1999) were reported. AVT has been shown to contract gastrointestinal tissue, but at concentrations 10-100 times higher than those that contract blood vessels (Conklin *et al.*, 1999). It may mean that GIT is not a natural site of AVT action or else a local supplementing production of AVT takes place here, as has been shown for mammalian AVP (Friedmann *et al.*, 1993). An immuno-histochemical localization of AVP in cells of mucosal epithelium and in fibers near the capillaries situated along the basal side of the epithelium cells suggests an action of this nonapeptide in mammalian GIT (Sanchez-Franco *et al.*, 1986). Whether there is the case of AVT in fish, is a matter of speculation. In marine cartilaginous fish, the rectal salt gland devoted to sodium chloride excretion is considered as a potential site of AVT and oxytocin-like peptides action (Acher *et al.*, 1999), but so far, to the author's knowledge, no investigations have been carried out.

Kidney

A renal function in fish depends on the external environment and can exemplify hyperosmoregulation in FW medium or hypoosmoregulation in SW medium. In freshwater teleosts, the major role of the kidney is to eliminate excess water and to retain electrolytes. In the distal tubule, NaCl is actively reabsorbed without the accompaniment of water. In marine teleost, however, the role of the kidney is different, i.e., to conserve water. The proximal tubules reabsorb NaCl to restore water, but the distal tubules are degenerated and a considerable volume of water is lost (Bentley, 1971).

Adaptations of euryhaline teleosts to both hypo- and hyperosmotic media probably include changes of both: the glomerular filtration rate and the permeability of the renal tubules. However, in contrast to well-documented AVP renal effects in mammals, our current knowledge of AVT actions in fish kidney are far behind.

In the early studies in chronically cannulated eels, AVT administration was shown to induce dose dependent changes in urine production (Chester Jones *et al.*, 1969; Henderson and Wales, 1974; Babiker and Rankin, 1978). In the freshwater eel, doses less than 0.1 ng/kg body weight were antidiuretic while doses greater than 1 ng/kg body weight were diuretic (Henderson and Wales, 1974). Similar results were obtained by Babiker and Rankin (1978) with low doses of AVT and IT (1 pg - 1 ng/kg body weight) reducing urine production in eels adapted to FW (but not in SW fish), and high doses (more than 10 ng/kg body weight) resulting always in diuresis. In the light of more recently developed AVT/IT assays (Warne *et al.*, 1994; Pierson *et al.*, 1995; Gozdowska and Kulczykowska, 2004), sensitive enough to measure the circulating hormones levels, it appears that the diuretic actions reported in earlier studies were present only at high, pharmacological hormones concentrations. Smaller doses of the hormones, which gave plasma AVT/IT concentrations within physiological range, were antidiuretic.

By analogy with mammalian AVP, two components of AVT/IT action in fish kidney, i.e., vascular and tubular should be considered. The diuresis observed after pharmacological AVT doses (100 and more times AVT physiological levels) was accompanied by a significant increase in systemic blood pressure (Henderson and Wales, 1974). In many teleost fish—in contrast to mammals—glomerulotubular balance seems to be poorly developed, and GFR is readily increased by an increase in renal perfusion pressure (Nishimura and Bailey, 1982). Therefore, the changes in systemic blood pressure affect glomerular filtration in fish kidney. Thus, AVT and IT inducing hypertension may be responsible for an increase in GFR, and in this way affect the water/ions balance in fish. The model of *in situ* trout kidney perfused under constant pressure was employed by Amer and Brown (1995) to evaluate the effect of 10^{-9} and 10^{-11} M AVT on glomerular function. The investigations have clearly demonstrated that both concentrations of AVT have a potent glomerular antidiuretic action resulting from the decrease in the filtering population of glomeruli. The AVT induced antidiuresis accompanied by a fall in the number of filtering nephrons has also been shown in freshwater eel (Henderson and Wales, 1974). Moreover, in a model of trout trunk preparation, the renal response to AVT would appear to be a result of glomerular action of the hormone rather than tubular one (Warne *et al.*, 2002). It has been proposed by Warne *et al.* (2002) that the mechanism of that involved smooth muscle constriction of the renal arterioles regulating blood flow to the glomeruli. Whether it would be V1 type AVT receptor, as suggested by Pang *et al.* (1983) using mammalian V1 receptor antagonists in the trout trunk model, still remains to be elucidated. A decrease in GFR, population of filtering nephrons and urine flow rate induced by AVT has also been shown in a perfused trunk preparation of *Scyliorhinus canicula* by Wells *et al.* (2002).

In addition to the glomerular effect of the hormones, their tubular action has also been taken into account. In isolated nephron of the trout, a dose-dependent rise in cAMP production as a result of the administration of 10^{-5} - 10^{-11} M AVT was demonstrated by Perrott *et al.* (1993) and the presence of a receptor similar to mammalian V2-type receptor was suggested (Balment *et al.*, 1993). More recently, significant stimulation of intracellular accumulation of cAMP by AVT at lower concentrations (10^{-12} M) has been observed in the case of *in vitro* preparation of trout kidney tubules (Warne *et al.*, 2002). Amer and Brown (1995) in their trout trunk studies have presented an increase of tubular reabsorption of water after small dose of AVT (10^{-11} M) and a decrease of that after the higher dose of the hormone (10^{-9} M). On the other hand, Nishimura *et al.* (1983) in early studies of the isolated renal tubules in freshwater trout suggested that in the distal tubule water flux and osmotic water permeability were not affected by neurohypophysial peptides. As yet, there has been no clear picture of either a site of AVT/IT action or a presence of well-defined type AVT/IT receptors in fish renal tubules. The fish nephron lacks the loop of Henle and cannot produce concentrated urine by way of water withdrawal in the collecting duct. Thus, the tubular action of AVT, if any, is probably different from that of AVP in mammalian kidney and any analogies between both nonapeptides should be drawn carefully.

The fluid produced by the kidney passes to a urinary bladder, organ of a high permeability to water and solute, where the composition of renal output can be modified probably by the hormones. In the 1990s, the AVT and IT receptors have been identified in urinary bladder in *C. commersoni* by Mahlmann *et al.* (1994) and Hausmann *et al.* (1995), but a direct role of nonapeptides remains to be determined.

INTERACTIONS WITH OTHER HORMONES

The hormonal regulation of water and ion homeostasis requires participation and interaction of many endocrine systems at many functional levels in the organism. Hence, the role of potential relationships between AVT/IT and other hormones systems contributing to osmoregulation in fish is considered. However, such interactions, with few exceptions, have not been studied to date in fish and, therefore, data presented herein are scarce.

Cortisol, prolactin and growth hormone (GH) have been well documented to be involved in osmoregulation in fish since the sixties (Bentley, 1971). The investigations undertaken in goldfish demonstrated that both AVT and IT can stimulate cortisol secretion in this species and suggested the corticotropin-releasing factor (CRF) activity of these hormones (Fryer and Leung, 1982). Moreover, Pierson *et al.* (1995) showed that both neurohypophysial hormones induced in a dose-dependent way the corticotropin (ACTH) release from trout pituitary. On the other hand, in view of papers by Fryer *et al.* (1985) and Olivereau and Olivereau (1990), the participation of AVT in stimulation of ACTH release in teleosts seems to be less clear, but can not be excluded. In terms of other osmoregulatory hormones, i.e., GH and prolactin, there is a lack of satisfactory data on the relationship with neurohypophysial nonapeptides, although the binding sites for AVT found in GH-producing cells in the pituitary may suggest an involvement of AVT in GH release (Moons *et al.*, 1989).

The renin-angiotensin system (RAS) appears to be an important factor in the regulation of blood pressure and drinking, and thus in the control of body fluid homeostasis in teleost fish (Russell *et al.*, 2001). Hence, a possible interaction between angiotensin II, a potent dipsogenic hormone, and neurohypophysial hormones in fish has also been considered. In mammals, administration of angiotensin II stimulates vasopressin secretion from pituitary (Yamaguchi *et al.*, 1985). In the chronically cannulated flounder model, however, an inhibitory effect of angiotensin II on AVT release has been observed (Balment *et al.*, 2003). It is not yet clear whether angiotensin II is a factor engaged in physiological control of AVT release in fish, but a potential relationship between hormones would be of special significance in species migrating from fresh- to seawater, when angiotensin II stimulate drinking (Takei, 2000).

ANALOGY BETWEEN NEUROHYPOPHYSIAL HORMONES IN FISH AND MAMMALS: A USEFUL PARADIGM?

The advanced knowledge of AVP engagement in mammalian osmoregulation and the absence of equivalent data on AVT and IT in fish are templates for drawing analogies. Is that legitimate?

In mammals, the kidney is a sole important osmoregulatory organ and vasopressin is a main antidiuretic hormone acting here. On the basis of functional and pharmacological criteria, two types of renal vasopressin receptors have been distinguished: V1 receptors which activation is associated with mobilizing intracellular calcium or stimulating phosphoinositide breakdown and V2 receptors leading to the activation of adenylate cyclase. Generally, V1 type receptor is associated with the vascular action of AVP and V2 type receptor is linked to the tubular antidiuretic action of the hormone in the distal parts of nephron.

In fish, on the other hand, the many structures such as gill, GIT, salt gland and urinary bladder act in concert with the kidney so as to maintain homeostasis of body fluids and electrolytes. Moreover, the kidney seems to be less important organ in fish osmoregulation process. The precise characterization of AVT and IT receptors and their distribution in osmoregulatory organs is far from being established. In fish species studied to date, i.e., flounder and *C. commersoni*, AVT receptors have a higher homology with vasopressin V1 type than with the V2 type and are linked to the phospholipase C-phosphatidylinositol signaling pathway. On the other hand, the isotocin receptors are closely related to mammalian oxytocin receptors and vasopressin V1 type. Also, V2 type receptors linked to adenylate cyclase seem to be present in fish kidney. A dose-dependent increase of cAMP production after AVT administration was shown in trout renal tubules *in vitro* (Perrott *et al.*, 1993). Moreover, a significant accumulation of cAMP by AVT within the range of physiological concentration was presented in trout nephron suspension (Warne *et al.*, 2002). The occurrence of V2 type receptors was also suggested in other transport epithelia in fish. Urea excretion in marine toadfish (*Opsanus beta*) gill was proposed to involve a specific transport mechanism analogous to the vasopressin-sensitive renal urea transporter of mammals linked to V2 type receptors (Walsh, 1997). It was shown in this aglomerular fish, that pulsatile branchial urea excretion was under the control of arginine vasotocin (Perry *et al.*, 1998). However, the recent experimental data by Wood *et al.* (2001) do not confirm this hypothesis. Therefore, further studies are essential to examine whether in fish nephron the AVT receptors related to mammalian V2 type are present. The character of the tubular action of AVT in fish, if any, would be probably different from that of AVP in mammals, because the fish nephron lacks the loop of Henle and cannot produce concentrated urine by water withdrawal in the collecting duct.

Thus, the mammalian paradigm offered to elucidate the mode of involvement of AVT/IT in osmoregulation in fish, although essential considering a lack of satisfactory fish data, can be applied only within strict limit.

SUMMARY. ARE THERE CONVINCING EVIDENCES ON A ROLE OF AVT/IT IN FISH OSMOREGULATION?

In this chapter, the current state of the knowledge of the role of neurohypophysial hormones arginine vasotocin and isotocin in fish osmoregulation has been summarized.

It has been shown that the synthesis of nonapeptides in hypothalamus and their secretion into circulation from the neurohypophysis are sensitive to an important environmental factor: salinity. Rapid transfer of teleost fish between fresh and seawaters results in altered mRNA expression of AVT precursors in hypothalamic neurons and consequently in altered content of AVT in pituitary. The patterns of plasma AVT and IT concentrations in response to external salinity changes suggest the role of the hormones in the mechanism of fast adaptation. AVT and IT releases appear to be controlled independently. There are three main osmoregulatory organs in fish: gill, GIT and kidney, which are considered as potential goals for neurohypophysial nonapetides action. Physiological responses of fish to different water salinities and suggested involvement of AVT/IT are summarized in Fig. 6.4.

There are clearly many unresolved questions to be explored. Data on the osmoregulatory role of AVT in fish still remains fragmentary. Implication of isotocin in this process is even less clear. Moreover, the physiological role of neurohypophysial hormones in the maintenance of water/ions homeostasis in fish does not seem to be uniform among species, in contrast to their antidiuretic and sodium-retaining function in tetrapods. A mammalian paradigm, although helpful in interpretation of fish data, needs to be verified by studies in fish. The new experimental approaches and modern techniques addressed to arginine vasotocin and isotocin gives a promise to elucidate a role of both nonapeptides in fish osmoregulation.

References

Acher, R. 1993. Neurohypophysial peptide systems: processing machinery, hydro-osmotic regulation, adaptation and evolution. *Regulatory Peptides* 45: 1–13.

Acher, R. 1995. Evolution of neurohypophysial control of water homeostasis: Integrative biology of molecular, cellular and organismal aspects. In: *Recent Progress of Vasopressin and Oxytocin Research*, T. Saito, K. Kurosawa and S. Yoshida (eds.) Neurohypophysis, Elsevier Science, Amsterdam, pp. 39–54.

Acher, R. and J. Chauvet. 1995. The neurohypophysial endocrine regulatory cascade: Precursors, mediators, receptors, and effectors. *Frontiers in Neuroendocrinology* 16, 237–289.

Acher, R., J. Chauvet, M.T. Chauvet and D. Crepy. 1961. Les hormones neurohypophysaires des poissons: Isolement d'une vasotocine du tacaud (*Gadus luscus* L.). *Biochimica et Biophysica Acta* 51: 419–420.

Acher, R., J. Chauvet, M.T. Chauvet and D. Crepy. 1962. Isolement d'une nouvelle hormone neurohypophysaire, l'isotocine presente chez les poissons osseux. *Biochimica et Biophysica Acta* 58: 624–625.

Acher, R., J. Chauvet, M.T. Chauvet and Y. Rouille. 1999. Unique evolution of neurohypophysial hormones in cartilaginous fishes: Possible implication for urea-based osmoregulation. *Journal of Experimental Zoology* 284: 475–484.

Amer, S. and A. Brown. 1995. Glomerular actions of arginine vasotocin in the *in situ* perfused trout kidney. *American Journal of Physiology* 269: R775–R780.

Babiker, M.M. and J.C. Rankin. 1978. Neurohypophysial hormonal control of kidney function in the European eel (*Anguilla anguilla* L.) adapted to sea-water or fresh water. *Journal of Endocrinology* 76: 347–358.

Babiker, M.M. and J.C. Rankin. 1979. Renal and vascular effects of neurohypophysial hormones in the African lungfish *Protopterus annectens* (Owen). *General and Comparative Endocrinology* 37: 26–34.

Balment, R.J., J.M. Warne, M. Tierney and N. Hazon. 1993. Arginine vasotocin (AVT) and fish osmoregulation. *Fish Physiology and Biochemistry* 11: 189–194.

Balment, R.J., J.M. Warne and Y. Takei. 2003. Isolation, synthesis, and biological activity of flounder [Asn^1, Ile^5, Thr^9] angiotensin I. *General and Comparative Endocrinology* 130: 92–98.

Bennett, M.B. and J.C. Rankin. 1986. The effect of neurohypophysial hormones on the vascular resistance of the isolated perfused gill of the European eel *Anguilla anguilla* L. *General and Comparative Endocrinology* 64: 60–66.

Bentley, P.J. 1971. Endocrines and osmoregulation. A comparative account of the regulation of water and salt in vertebrates. In: *Zoophysiology and Ecology*. W.S. Hoar,

J. Jacobs, H. Langer and M. Lindauer (eds.). Springer-Verlag, New York: Vol. 1. pp. 1–37, 199–254.

Bentley, P.J. 1998. *Comparative Vertebrates Endocrinology*. 3rd Edition. Cambridge University Press, Cambridge, pp. 67–176.

Beyenbach, K. 1995. Secretory electrolyte transport in renal proximal tubules of fish. In: *Cellular and Molecular Approaches to Fish Ionic Regulation*, C.M. Wood and T.J. Shuttleworth (eds.). Academic Press, San Diego, pp. 85–105.

Bond, H., M.J. Winter, J.M. Warne, C.R. McCrohan and R.J. Balment. 2002. Plasma concentrations of arginine vasotocin and urotensin II are reduced following transfer of the euryhaline flounder (*Platichthys flesus*) from seawater to fresh water. *General and Comparative Endocrinology* 125: 113–120.

Carlson, I.H. and W.N. Holmes. 1962. Changes in the hormone content of the hypothalamo-hypophysial system of the rainbow trout, *Salmo gairdneri*. *Journal of Endocrinology* 24: 23–32.

Chester Jones, I., D.K.O. Chan and J.C. Rankin. 1969. Renal function in the European eel (*Anguilla anguilla* L.): Effects of the caudal neurosecretory system, corpuscles of Stannius, neurohypophysial peptides and vasoactive substances. *Journal of Endocrinology* 43: 21–31.

Conklin, D.J., N.W. Mick and K.R. Olson. 1996. Arginine vasotocin relaxation of gar (*Lepisosteous* spp.) hepatic vein *in vitro*. *General and Comparative Endocrinology* 104: 52–60.

Conklin, D.J., A. Chavas, D.W. Duff, L. Weaver, Y. Zhang and K.R. Olson. 1997. Cardiovascular effects of arginine vasotocin in the rainbow trout *Oncorhynchus mykiss*. *Journal of Experimental Biology* 200: 2821–2832.

Conklin, D.J., M.P. Smith and K.R. Olson. 1999. Pharmacological characterization of arginine vasotocin vascular smooth muscle receptors in the trout (*Oncorhynchus mykiss*) *in vitro*. *General and Comparative Endocrinology* 114: 36–46.

Danzler, W.H. (2003). Regulation of renal proximal and distal tubule transport: sodium, chloride and organic anions. *Comparative Biochemistry and Physiology* A 136: 453–478.

Foran, C.M. and A.H. Bass. 1998. Preoptic AVT immunoreactive neuron of a teleost fish with alternative reproductive tactics. *General and Comparative Endocrinology* 111: 271–282.

Friedmann, A.S., V.A. Memoli, X.M. Yu and W.G. North. 1993. Biosynthesis of vasopressin by gastrointestinal cells of Brattleboro and Long-Evans rats. *Peptides* 14: 607–612.

Fridberg, G., and R. Olsson. 1959. The preopticohypophysial system, nucleus tuberis lateralis and subcommisural organ of *Gasterosteus aculeatus* after changes in osmotic stimuli. *Zeitschrift fur Zellforschung und Mikroscopist Anatomie* 49: 531–540.

Fryer, J.N., and E. Leung. 1982. Neurohypophysial hormonal control of cortisol secretion in the teleost *Carassius auratus*. *General and Comparative Endocrinology* 48: 425–431.

Fryer, J., K. Lederis and J. Rivier. 1985. ACTH-releasing activity of urotensin I and ovine CRF: Interactions with arginine vasotocin, isotocin and arginine vasopressin. *Regulatory Peptides* 11: 11–15.

Goodson, J.L. and A.H. Bass. 2000. Forebrain peptides modulate sexually polymorphic vocal circuitry. *Nature (Lond.)* 403: 769–772.

Goossens, N., K. Dierickx and F. Vandesande. 1977. Immunocytochemical localization of vasotocin and isotocin in the preopticohypophysial neurosecretory system of teleosts. *General and Comparative Endocrinology* 32: 371–375.

Gozdowska, M. and E. Kulczykowska. 2004. Determination of arginine-vasotocin and isotocin in fish plasma with solid-phase extraction and fluorescence derivatization followed by high-performance liquid chromatography. *Journal of Chromatography* B 807: 229–233.

Guibbolini, M.E. and M. Avella. 2003. Neurohypophysial hormone regulation of Cl- secretion: Physiological evidence for V1-type receptors in sea bass gill respiratory cells in culture. *Journal of Endocrinology* 176: 111–119.

Guibbolini, M.E. and B. Lahlou. 1987a. Neurohypophysial peptide inhibition of adenylate cyclase activity in fish gills. The effect of environmental salinity. *FEBS Letters* 220: 98–102.

Guibbolini, M.E. and B. Lahlou. 1987b. Adenylate cyclase activity in fish gills in relation to salt adaptation. *Life Sciences* 41: 71–78.

Guibbolini, M.E. and B. Lahlou. 1990. Evidence for presence of a new type of neurohypophysial hormone receptor in fish gill epithelium. *American Journal of Physiology* 258: R3–R9.

Guibbolini, M.E. and B. Lahlou. 1992. G_i protein mediates adenylate cyclase inhibition by neurohypophyseal hormones in fish gill. *Peptides* 13: 865–871.

Guibbolini, M.E., I.W. Henderson, W. Mosley and B. Lahlou. 1988. Arginine vasotocin binding to isolated branchial cells of the eel: effect of salinity. *Journal of Molecular Endocrinology* 1: 125–130.

Harding, K.E., J.M. Warne, S. Hyodo and R.J. Balment. 1997. Pituitary and plasma AVT content in the flounder (*Platichthys flesus*). *Fish Physiology and Biochemistry* 17: 357–362.

Haruta, K., T. Yamashita and S. Kawashima. 1991. Changes in arginine vasotocin content in the pituitary of the medaka (*Oryzias latipes*) during osmotic stress. *General and Comparative Endocrinology* 83: 327–336.

Hausmann, H., W. Meyerhof, H. Zwiers, K. Lederis and D. Richter. 1995. Teleost isotocin receptor: Structure, functional expression, mRNA distribution and phylogeny. *FEBS Letters* 370: 227–230.

Hausmann, H., A. Richters, H.-J. Kreienkamp, W. Meyerhof, H. Mattes, K. Lederis, H. Zwiers and D. Richter. 1996. Mutational analysis and molecular modeling of the nonapeptide hormone binding domains of the [Arg^8] vasotocin receptor. *Proceedings of the National Academy of Sciences of the United States of America* 93: 6907–6912.

Heierhorst, J., S.D. Morley, J. Figureoa, C. Krentler, K. Lederis and D. Richter. 1989. Vasotocin and isotocin precursors from the white sucker, *Catostomus commersoni*: Cloning and sequence analysis of the cDNAs. *Proceedings of the National Academy of Sciences of the United States of America* 86: 5242–5246.

Heierhorst, J., S. Mahlmann, S.D. Morley, I.R. Coe, N.M. Sherwood and D. Richter. 1990. Molecular cloning of two distinct vasotocin precursor cDNAs from chum salmon, *Oncorhynchus keta* suggests an ancient gene duplication. *FEBS Letters* 260: 301–304.

Henderson, I.W. and N.A.M. Wales. 1974. Renal diuresis and antidiuresis after injections of arginine vasotocin in the freshwater eel (*Anguilla anguilla* L.). *Journal of Endocrinology* 61: 487–500.

Hiraoka, S., M. Suzuki, T. Yanagisawa, M. Iwata and A. Urano. 1993. Divergence of gene expression in neurohypophysial hormone precursors among salmonids. *General and Comparative Endocrinology* 92: 292–301.

Holmes, W.N. and R.L. McBean. 1963. Studies on the glomerular filtration rate of rainbow trout (*Salmo gairdneri*). *Journal of Experimental Biology* 40: 335–341.

Holmqvist, B.I. and P. Ekström. 1995. Hypophysiotrophic systems in the brain of the Atlantic salmon. Neuronal innervation of the pituitary and the origin of pituitary dopamine and nonapeptides identified by means of combined carbocyanine tract tracing and immunocytochemistry. *Journal of Chemical Neuroanatomy* 8: 125–145.

Hyodo, S., Y. Kato, M. Ono and A. Urano. 1991. Cloning and sequence analyses of cDNA encoding vasotocin and isotocin precursors of chum salmon, *Oncorhynchus keta*: Evolutionary relationships of neurohypophysial hormone precursors. *Journal of Comparative Physiology* B 160: 601–608.

Hyodo, S. and A. Urano. 1991. Changes in expression of provasotocin and proisotocin genes during adaptation to hyper- and hypo-osmotic environments in rainbow trout. *Journal of Comparative Physiology* B 161: 549–556.

Hyodo, S., S. Ishii and J.M.P. Joss. 1997. Australian lungfish neurohypophysial hormone genes encode vasotocin and [Phe^2]mesotocin precursors homologous to tetrapod-type precursors. *Proceedings of the National Academy of Sciences of the United States of America* 94: 13339–13344.

Hyodo, S., T. Tsukada and Y. Takei 2004. Neurohypophysial hormones of dogfish, *Triakis scyllium*: structures and salinity-dependent secretion. *General and Comparative Endocrinology* 138: 97–104.

Kulczykowska, E. 1997. Response of circulating arginine vasotocin and isotocin to rapid osmotic challenge in rainbow trout. *Comparative Biochemistry and Physiology* A118: 772–778.

Kulczykowska, E. 1998. Effects of arginine vasotocin, isotocin and melatonin on blood pressure in the conscious Atlantic cod (*Gadus morhua*): hormonal interactions? *Experimental Physiology* 83: 809–820.

Kulczykowska, E. 2001. Responses of circulating arginine vasotocin, isotocin, and melatonin to osmotic and disturbance stress in rainbow trout (*Oncorhynchus mykiss*). *Fish Physiology and Biochemistry* 24: 201–206.

Kulczykowska, E. 2002. A review of the multifunctional hormone melatonin and a new hypothesis involving osmoregulation. *Reviews in Fish Biology and Fisheries* 11: 321–330.

Lederis, K. 1964. Fine structure and hormone content of the hypothalamo-neurohypophysial system of the rainbow trout (*Salmo irideus*) exposed to seawater. *General and Comparative Endocrinology* 4: 638–661.

Le Mevel, J.C., T.-F. Pomantung, D. Mabin and H. Vaudry. 1993. Effects of central and peripheral administration of arginine vasotocin and related neuropeptides on blood pressure and heart rate in the conscious trout. *Brain Research* 610: 82–89.

Lewis, C.M., E.K. Dolence, Z. Zhang and J.D. Rose. 2004. Fluorescent vasotocin conjugate for identification of the target cells for brain actions of vasotocin. *Bioconjugate Chemistry* 15: 909–914.

Maetz, J. and B. Lahlou. 1974. Actions of neurohypophysial hormones in fishes. In: *Handbook of Physiology-Endocrinology*, R.O. Greep and E.B. Astwood (eds.). American Physiological Society, Baltimore, Vol. 4, Part 1, pp. 521–544.

Maetz, J., J. Bourguet, B. Lahlou and J. Hourdry. 1964. Peptides neurohypophysaires et osmoregulation chez *Carassius auratus*. *General and Comparative Endocrinology* 4: 508–522.

Mahlmann, S., W. Meyerhof, H. Hausmann, J. Heierhorst, C. Schönrock, H. Zwiers, K. Lederis and D. Richter. 1994. Structure, function, and phylogeny of [Arg^8]vasotocin receptors from teleost fish and toad. *Proceedings of the National Academy of Sciences of the United States of America* 91: 1342–1345.

Marshall, W.S. 2003. Rapid regulation of NaCl secretion by estuarine teleost fish: coping strategies for short-duration freshwater exposures. *Biochimica et Biophysica Acta* 1618: 95–105.

Moons, L., M. Cambre, T.F.C. Batten and F. Vandesande. 1989. Autoradiographic localization of binding sites for vasotocin in the brain and pituitary of the sea bass (*Dicentrarchus labrax*). *Neuroscience Letters* 100: 11–16.

Motais, R. and J. Maetz. 1967. Arginine vasotocine et evolution de la permeabilite branchiale au sodium au cours du passage d'eau douce en eau de mer chez le Flet. *Journal of Physiology* (Paris) 59: 271.

Nishimura, H. and J.R. Bailey. 1982. Intrarenal renin-angiotensin system in primitive vertebrates. *Kidney International* 12: S185–192.

Nishimura, H. and Z. Fan. 2003. Regulation of water movement across vertebrate renal tubules. Comparative Biochemistry and Physiology A 136: 479–498.

Nishimura, H., M. Imai and M. Ogawa. 1983. Sodium chloride and water transport in the renal distal tubule of the rainbow trout. *American Journal of Physiology* 244: F247–F254.

Nürnberger, F., H.-W. Korf, M.A. Ali and D.L.G. Nokes. 1996. Salinity and vasotocin immunoreactivity in the brain of *Rivulus marmoratus* (Teleostei). *Naturwissenchaften* 83: 326–328.

Olivereau, M. and J. Olivereau 1990. Effect of pharmacological adrenalectomy on corticotropin-releasing factor-like and arginine vasotocin immunoreactivities in the brain and pituitary of the eel: immunocytochemical study. *General and Comparative Endocrinology* 80: 199–215.

Oudit, G.Y. and D.G. Butler. 1995. Cardiovascular effects of arginine vasotocin, atrial natriuretic peptide, and epinephrine in freshwater eels. *American Journal of Physiology* 268: R1273–R1280.

Pang, P.K.T., P.B. Furspan and W.H. Sawyer. 1983. Evolution of neurohypophyseal hormone actions in vertebrates. *American Zoologist* 23: 655–662.

Perrott, M.N., S. Carrick and R.J. Balment. 1991. Pituitary and plasma arginine vasotocin levels in teleost fish. *General and Comparative Endocrinology* 83: 68–74.

Perrott, M.N., R.J. Sainsbury and R.J. Balment. 1993. Peptide hormone-stimulated second messenger production in the teleostean nephron. *General and Comparative Endocrinology* 89: 387–395.

Perry, S.F., K.M. Gilmour, C.M. Wood, P. Part, P. Laurent and P.J. Walsh. 1998. The effect of arginine vasotocin and catecholamines on nitrogen excretion and the cardio-respiratory physiology of the gulf toadfish, *Opsanus beta*. *Journal of Comparative Physiology*B 168: 461–472.

Pickering, B.T. 1968. A neurophysin from cod (*Gadus morhua*) pituitary glands: isolation and properties. *Journal of Endocrinology* 42: 143–152.

Pierson, P.M., M.E. Guibbolini, N. Mayer-Gostan and B. Lahlou. 1995. ELISA measurements of vasotocin and isotocin in plasma and pituitary of the rainbow trout: Effect of salinity. *Peptides* 16: 859–865.

Renfro, J.L. 1999. Recent developments in teleost renal transport. *Journal of Experimental Zoology* 283: 653–661.

Russell, M.J., A.M. Klemmer and K.R. Olson. 2001. Angiotensin signaling and receptor types in teleost fish. *Comparative Biochemistry and Physiology* A 128: 41–51.

Saito, D., M. Komatsuda and A. Urano. 2004. Functional organization of preoptic vasotocin and isotocin neurons in the brain of rainbow trout: central and neurohypophysial projections of single neurons. *Neuroscience* 124: 973–984.

Sanchez-Franco, F., L. Cacicedo, J.L. Vasallo, J.L. Blazquez and L. Munoz Baragan. 1986. Arginine-vasopressin immunoreactive material in the gastrointestinal tract. *Histochemistry* 85: 419–422.

Sharratt, B.M., D. Bellamy and I. Chester Jones. 1964 Adaptation of the silver eel (*Anguilla anguilla*) to seawater and to artificial media together with observations on the role of the gut. *Comparative Biochemistry and Physiology* 11: 19–30.

Suzuki, M., S. Hyodo, A. Urano. 1992. Cloning and sequence analyses of vasotocin and isotocin precursor cDNA in the masu salmon, *Oncorhynchus masou*: evolution of neurohypophysial hormone precursors. *Zoological Science* 9: 157–167.

Szczepańska-Sadowska, E., D. Gray and C. Simon-Oppermann. 1983. Vasopressin in blood and third ventricle CSF during dehydration, thirst, and hemorrhage. *American Journal of Physiology* 245: R549–R555.

Takei, Y. 2000. Comparative physiology of body fluid regulation in vertebrates with special reference to thirst regulation. *Japanese Journal of Physiology* 50: 171–186.

Unger, J.L. and E. Glasgow. 2003. Expression of isotocin-neurophysin mRNA in developing zebrafish. *Gene Expression Patterns* 3: 105–108.

Urano, A., K. Kubokawa and S. Hiraoka. 1994. Expression of the vasotocin and isotocin gene family in fish. In: *Fish Physiology*, N.M. Sherwood and C.L. Hew (eds.). Academic Press, New York, Vol. 13, pp. 101–132.

Van den Dungen, H.M., R.M. Buijs, C.W. Pool and M. Terlou. 1982. The distribution of vasotocin and isotocin in the brain of the rainbow trout. *Journal of Comparative Neurology* 212: 146–157.

Walsh, P.J. 1997. Evolution and regulation of urea synthesis and ureotely in (batrachoidid) fishes. *Annual Review of Physiology* 59: 299–323.

Warne, J.M. 2001. Cloning and characterization of an arginine vasotocin receptor from the euryhaline flounder *Platichthys flesus*. *General and Comparative Endocrinology* 122: 312–319.

Warne, J.M. and R.J. Balment. 1995. Effect of acute manipulation of blood volume and osmolality on plasma [AVT] in seawater flounder. *American Journal of Physiology* 269: R1107–R1112.

Warne, J.M. and R.J. Balment. 1997a. Changes in plasma arginine vasotocin (AVT) concentration and dorsal aortic blood pressure following AVT injection in the teleost *Platichthys flesus*. *General and Comparative Endocrinology* 105: 358–364.

Warne, J.M., and R.J. Balment. 1997b. Vascular actions of neurohypophysial peptides in the flounder. *Fish Physiology and Biochemistry* 17: 313–318.

Warne, J.M., N. Hazon, J.C. Rankin and R.J. Balment. 1994. A radioimmunoassay for the determination of arginine vasotocin (AVT): plasma and pituitary concentrations in fresh- and seawater fish. *General and Comparative Endocrinology* 96: 438–444.

Warne, J.M., S. Hyodo, K. Harding and R.J. Balment. 2000. Cloning of pro-vasotocin and pro-isotocin cDNAs from the flounder *Platichthys flesus*; levels of hypothalamic mRNA following acute osmotic challenge. *General and Comparative Endocrinology* 119: 77–84.

Warne, J.M., K.E. Harding and R.J. Balment. 2002. Neurohypophysial hormones and renal function in fish and mammals. *Comparative Biochemistry and Physiology* B 132: 231–237.

Wells, A., W.G. Anderson and N. Hazon. 2002. Development of an *in situ* perfused kidney preparation for elasmobranch fish: action of arginine vasotocin. *American Journal of Physiology* 282: R1636–R1642.

Wood, C.M., J.M. Warne, Y. Wang, M.D. McDonald, R.J. Balment, P. Laurent and P.J. Walsh. 2001. Do circulating plasma AVT and/or cortisol levels control pulsatile urea excretion in the gulf toadfish (*Opsanus beta*)? *Comparative Biochemistry and Physiology* A 129: 859–872.

Yamaguchi, K., M. Kioke and H. Hama. 1985. Plasma vasopressin response to peripheral administration of angiotensin in conscious rats. *American Journal of Physiology* 248: R249–R256.

CHAPTER

7

Cellular and Molecular Approaches to the Investigation of Piscine Osmoregulation: Current and Future Perspectives

Chris N. Glover

The maintenance of salt and water balance in fish is the consequence of a tightly regulated, integrated network of molecular and cellular processes operating within a variety of cell types and across a range of tissues. In essence, these same cells and tissues, often utilizing the same molecular and cellular pathways, are capable of ionic and osmotic homeostasis in environments as disparate as dilute freshwater, where the fish is faced with salt loss and water gain, and seawater where water loss and salt loading are the major challenges. Cellular and molecular components of these

Author's address: SCION, Te Papa Tipu Innovation Park, 49 Sala Street, Private Bag 3020, Rotorua, New Zealand.

E-mail: Chris.Glover@scionresearch.com

homeostatic pathways are under the control of a complex assortment of endocrine factors, the presence of which depends on environmental, developmental and physiological cues. Osmoregulatory hormones may have distinct effects, depending on circulating concentrations, and their impact may vary with the presence of other circulating mediators, as well as the molecular and cellular wiring of the target cell. This wiring itself can be determined by genetic constitution, life history and environmental factors. Consequently, even closely related species will have a different assortment and/or arrangement of osmoregulating machinery and mediators. Given its complexity, our understanding of exactly how the processes underlying the ability of fish to maintain homeostasis are initiated, regulated and integrated is in its infancy. Recent developments in cellular and molecular techniques will provide the impetus for many of the advances in this field in future years.

This chapter seeks to provide insight into how advances in cellular and molecular techniques have been, and can be, applied to further our understanding of fish osmoregulation. It is not intended to act as a practical guide to these techniques; nor is it the intent to provide a comprehensive summary of the progress that has been made. The emphasis of this review will be on developments in the last decade. Thus, to a limited degree, it may act as an update to the monograph on cellular and molecular approaches in fish osmoregulation that appeared ten years ago (Wood and Shuttleworth, 1995). Many of the techniques and their applications have changed little over this period and consequently will not be detailed in this synopsis. This extends to preparations such as the killifish opercular epithelium, and the accompanying suite of electrophysiological techniques, which continue to provide important insights into the passage of ions across osmoregulatory epithelia (e.g., Evans *et al.*, 2004; Marshall *et al.*, 2005), but which, in terms of utility and applicability, remain relatively unchanged. Devoted discussions of the elasmobranch rectal gland (Silva *et al.*, 1990), and fish urinary bladder (Marshall, 1995), have been excluded for similar reasons. As a guide, the material herein will be aimed at the reader with sound background knowledge of fish osmoregulation, and an understanding of the basic concepts of cellular and molecular biology.

UTILIZATION AND APPLICATION OF CELLULAR TECHNIQUES IN FISH OSMOREGULATION

As the biological barrier between the internal and external milieux, epithelia are critical for the maintenance of salt and water balance. In addition to functions related to ion and water transport, they may also be involved in essential biological processes such as digestion, gas exchange, and excretion. By virtue of their tightly regulated multifunctional roles, epithelia are characterized by both the heterogeneity of their cellular composition, and their responsiveness to endocrine and neural control. For discerning mechanisms of physiological action these characteristics are disadvantageous. This has led to the development of models where physiological function can be assessed in relatively homogenous cell or membrane populations free from the confounding actions of central mediators. The extrication of cells from an environment where they are exposed to a number of simultaneous inputs, and their subsequent placement in conditions where they can respond normally but are under the strict control of the experimenter, permits the detailed elucidation of molecular, biochemical and physiological pathways. Delineating mechanisms of ionic and osmotic homeostasis in a relatively simplistic setting allows the development of hypotheses that can subsequently be tested in more biologically complicated systems, and across a range of cell types, tissues and species.

Cellular approaches not only offer scientific advantages, but also technical, ethical and economic benefits over the use of whole animal physiological investigations (Castaño *et al.*, 2003). Many species of interest are not easily maintained in laboratory settings, and therefore are not readily accessible for experimentation. Access to cell lines, for example, can overcome this problem. The development and use of suitable fish cellular models will reduce the dependence on whole animal studies, of obvious benefit from an animal welfare perspective. The possibility of minimizing expensive animal husbandry, and the ability to utilize a single animal for multiple manipulations, is of obvious financial benefit.

Cellular techniques utilized in fish biology—and that will be discussed in the following chapter—include reconstituted epithelia, single cell preparations, and isolated cellular components. In all cases, these techniques facilitate the study of fish osmoregulation by improving manipulability, homogeneity, and accessibility to the epithelial cells of greatest interest to the researcher. The utility and applicability of these methods will be critically assessed with particular regard to fish osmoregulation by highlighting the progress made using these techniques to date, and speculating on how they may advance the field in the future. Methods often associated with cellular models (i.e., pharmacological agents) will also be considered.

'Yolk-balls'—A Novel Surrogate Preparation for Chloride Cell Investigation

The osmotic challenges faced by early life stages of teleost fish are similar to those of juvenile and adult forms. However, unlike in their more developed brethren, the structures primarily associated with osmotic and ionic control (gill, gut and kidney), are not fully functional (Alderice, 1988; Varsamos *et al.*, 2005). The timing of gill, kidney and gut development varies among species. However, the mechanism used by embryonic and larval forms to compensate for passive ion and water flux appears to be highly conserved. The occurrence of chloride cells on the integument has been noted in all studied teleost species (e.g., Hwang and Hirano, 1985, Katoh *et al.*, 2000, Varsamos *et al.*, 2002), and are believed to be critical for osmoregulation in developing fish. These mitochondria-rich cells, characterized by extensive basolateral invaginations, a high metabolic activity, and a morphology that can change considerably with osmotic environment, are believed to be important osmoregulatory loci (Evans *et al.*, 1999). In most developing species, integument chloride cell clusters are primarily associated with the yolk-sac membrane.

Recently, a method has been developed that permits the separation of the yolk-sac from the developing embryo, and which may provide an *in vitro* model of considerable utility (Shiraishi *et al.*, 2001; Kaneko *et al.*, 2002). This technique involves the surgical removal of the yolk sac from the developing fish, in this case tilapia (*Oreochromis mossambicus*). The incision wound heals rapidly such that three hours after the surgery, the yolk sac membrane has completely resealed and encapsulates the yolk (Shiraishi *et al.*, 2001). This 'yolk-ball' preparation offers many potential advantages as a tool for examining osmoregulatory function in fish. A major benefit is that the yolk-sac membrane chloride cells appear to be similar to those that later develop in the adult gill (Kaneko *et al.*, 2002). They are mitochondria-rich and immunoreactive to the basolateral sodium pump (Na^+/K^+-ATPase) thought to drive ion uptake (Kaneko *et al.*, 2002). Morphologically, the extensive tubular elongations of branchial chloride cells are replaced by a more compact, flattened appearance in the yolk sac, a limitation likely imposed by the reduced thickness of the epidermal layer (Kaneko *et al.*, 2002). Functionally, these cells appear to respond to changes in environmental salinity in a similar manner to the adult forms, even in this liberated state. For example, in response to seawater, cells increase in size and develop characteristic membrane elaborations (Kaneko *et al.*, 2002). This suggests that the yolk ball has an inherent mechanism for both sensing and responding to external salinity. Structural confirmation of this ability was provided by the demonstration of functional receptors for the important osmoregulatory hormones prolactin and cortisol in yolk-sac membranes (Shiraishi *et al.*, 1999).

The yolk ball offers several advantages over the isolated yolk sac membrane in terms of its usefulness as a model system for the examination of osmoregulation. In order to maintain a chloride cell population, yolk-sac membranes required the application of exogenous cortisol, the hormone closely associated with seawater acclimation (Ayson *et al.*, 1995). Exogenous application of cortisol had little effect on the responses of freshwater- or seawater-exposed yolk-balls (Shiraishi *et al.*, 2001). However, on exposure to seawater, the yolk ball chloride cells responded both morphologically and physiologically, in a manner similar to that observed on seawater exposure *in vivo* (Shiraishi *et al.*, 2001; Hiroi *et al.*, 2005). Such a response is associated with a cortisol surge (McCormick, 2001). This is an important observation. First, it indicates that the yolk-ball preparation is capable of sensing osmotic changes in the environment, independent of higher control. Second, it suggests that the yolk itself may play a vital role in the osmoregulatory capacity of the preparation. The ability to respond to seawater exposure is likely to be the consequence of maternally supplied cortisol present in the yolk (Tagawa, 1996). This is of particular interest in that it indicates the yolk may play a role analogous to the circulation in intact animals. Consequently, from an experimental perspective, it provides a situation whereby the yolk contents may be manipulated in order to investigate aspects of osmoregulatory control of chloride cell function. Due to the rapid healing of the membrane following surgery, it may be possible to introduce salinity changes and endocrine factors into the yolk, equivalent to infusion or injection *in vivo*, but without the side effects of higher control (e.g., alterations in blood perfusion, release of endocrine regulators). The yolk-ball preparation, therefore, allows chloride cell structure and function to be examined under conditions where both apical and basolateral influences can be monitored either independently or simultaneously.

Many of the advantages of this preparation are similar to those offered by double-seeded primary gill cell culture techniques (see below). Both include populations of chloride cells, and the capacity to control apical and basolateral exposure conditions. A distinct benefit of the yolk-ball technique over primary cell cultures as a model system for the examination of chloride cell function is its relative ease. They are relatively simple to acquire, maintain viability with a minimum of effort, and can be easily experimentally manipulated. In contrast, cell cultures are an often laborious and expensive method.

The yolk-ball preparation is a relatively new development, and there is much scope for further investigation. To date, the technique has only been utilized in tilapia. Expansion to other fish species should be amenable. The manipulability, viability and *in vitro* nature of the yolk-ball preparation are of significant advantage in that osmotic environment can be altered, while systemic concerns are largely eliminated. Thus, the comparison of yolk-balls between stenohaline freshwater and marine species, along with euryhaline teleosts, may offer insight into the 'wiring' of homeostatic control. It is also a method applicable to the investigation of osmoregulatory development. Yolk balls of different stages may offer perspectives into the ontogeny of important cellular sensors of, and responses to, osmotic and ionic imbalance.

Cultured Fish Cells—Reconstituting Epithelia for Transport Studies

Branchial cell culture

Investigations of ionic and osmotic regulation in fish generally focus on the gill (see Evans *et al.*, 1999; Marshall, 2002 for reviews). Its elegant system of arches, filaments and lamellae result in a branchial surface area that can represent more than half of the total body surface area. This large surface is well vascularized with a short diffusional distance between blood and water. These properties, in association with the often large osmotic and ionic gradients between the animal and the external milieu, render the gill a critical osmoregulatory organ. The many other functions of branchial tissue, including gas exchange and excretion, are also favored by this morphology, and must be balanced with osmoregulatory demands. The intricate architecture of the gill, combined with its diverse physiological roles, has been a limiting factor in determining aspects of ion and water passage across this epithelial surface. The desire for a cellular method that permits the maintenance of gill properties and function, while offering a lamellar surface amenable to apical and basolateral manipulation, has resulted in the development of branchial cell culture methods (Leguen *et al.*, 2001; Wood *et al.*, 2002). Opercular epithelial preparations of the killifish incorporate many of these advantages (see Marshall, 1995), but to date these methods are most useful only as a model for ion transport in seawater fish, especially with regard to chloride secretion and calcium uptake (Zadunaisky, 1984; Marshall, 1995; Wood *et al.*, 2002).

Following the successful culture of branchial cells in flasks (Avella *et al.*, 1994), methods were developed—in both marine and freshwater fish species—to allow growth of branchial cells on permeable filters (Avella and Ehrenfeld, 1997; Wood and Pärt, 1997). These techniques resulted in homogenous cultures of pavement cells that formed tight epithelia and developed apical and basolateral morphologies. Pavement cells are the less differentiated branchial epithelial cells with dual roles in gas exchange and ionic regulation (Evans *et al.*, 1999). Mucus-secreting cells and chloride cells were virtually absent. These cultures permitted analysis of ion transport characteristics of pavement cells in both marine and freshwater gills. For example, electrogenic chloride movement was determined using Ussing techniques in the sea bass, *Dicentrarchus labrax* (Avella *et al.*, 1999). The epithelial integrity of these preparations is such that the apical bathing medium can be replaced with freshwater to recreate conditions that may exist *in vivo*.

These 'single-seeded' preparations lack chloride cells, and thus are not good surrogate models of intact gill. This led to the development of 'double-seeded' techniques for freshwater rainbow trout. These utilize an initially cultured layer of pavement cells as a seeding platform for a second application of gill cells and result in a culture containing chloride cells (Fletcher *et al.*, 2000). In fact, the relative proportions of cells in these double-seeded reconstituted cultures resemble those found in freshwater *in vivo* (~85% pavement cells, ~15% chloride cells; Perry and Walsh, 1989). These preparations form very tight epithelia and have been used to reinvestigate the mechanisms of ammonia excretion across branchial surfaces (Wood *et al.*, 2002), for example. This technique has yet to be extended to marine fish.

There is a major caveat regarding the usefulness of these preparations for studies investigating piscine osmoregulation. To date, these cultures display somewhat anomalous active ion transport properties in freshwater. Diffusive ion fluxes appeared to closely resemble those of intact fish (Wood and Pärt, 1997). However, in single-seeded preparations lacking chloride cells, a small active chloride influx was present under asymmetric culture conditions (basolateral culture medium/apical freshwater). This is contrary to the established ion transport dogma, which localizes this process to chloride cells. The active transport component persisted in double-seeded cultures (with chloride cells), but at a magnitude that was considerably less than expected, and which represented only a small proportion of diffusive ion movement (Wood *et al.*, 2002). In double-seeded cultures, active calcium uptake was present and quantitatively resembles that *in vivo*, but was accompanied by an unanticipated active calcium efflux (Wood *et al.*, 2002).

A recent study suggested a possible reason for observed discrepancies between cultured cells and the *in vivo* condition. To mimic circulating cortisol levels, this hormone was applied to the serosal surface of double-seeded rainbow trout cultures exposed to apical freshwater (Zhou *et al.*, 2003). This resulted in the development of a small active influx component for both sodium and chloride. Although the transport rates were small, this represented the first demonstration of active sodium transport by a branchial gill culture. This suggests that hormonal supplementation of media may provide the key to establishing a functional, ion transporting cultured gill epithelium. Until active transport can be established in these models, they have limited utility for many applications of relevance to the study of fish osmoregulation.

Depending on the nature of the experiment, culturing cells on supports may offer little advantage over suspended cell techniques (Sandbacka *et al.*, 1999), or simple gill filament culture systems (Bury *et al.*, 1998; Mazon *et al.*, 2004). Although lacking cellular polarity, and/or exhibiting reduced viability, these latter methods are advantageous in terms of cost and labor. For example, results from studies examining the impact of hypotonic shock on cells in culture resembled those conducted with freshly suspended cells (Leguen and Prunet, 2004). This highlights the integrity of the branchial culture technique for studies of relevance to fish osmoregulation, but also its potential redundancy for certain experimental manipulations.

Renal cell culture

The kidney of fish has two main roles in terms of osmotic and ionic homeostasis. In freshwater, it forms a hypotonic urine as a means of conserving ions and ridding the animal of osmotic water influx. In marine environments, the kidney is primarily concerned with ion excretion and water conservation. A number of cellular methods exist for examining the function of renal cells from fish. Enriched membrane vesicle approaches (e.g., Pane *et al.*, 2006), and tubular perfusion techniques (see Beyenbach and Dantzler, 1990) for example, have proven useful for elucidating aspects of kidney function in fish. These preparations may have limited applicability and utility for the study of piscine ionic and osmotic homeostasis, depending on the aspect of osmoregulation under consideration. Culture techniques also exist for renal proximal tubule cells of the winter flounder (*Pleuronectes americanus*), a marine teleost.

Renal cell cultures of fish were developed primarily for the purpose of investigating ion transport, and as such the utility of these preparations has been well described (Dickman and Renfro, 1986; Renfro, 1995). This method involves culturing on contractible collagen gels, which act to enhance differentiation, while permitting experimental access to both peritubular and luminal surfaces. The cells have established apical/basolateral polarity and, when examined using Ussing chambers, are observed to replicate transepithelial solute transfer in a manner similar to that thought to occur *in vivo* (Dickman and Renfro, 1986; Dudas and Renfro, 2001).

Recent investigations using this technique have yielded information regarding the renal secretion of taurine, the important cell volume regulator. Combining cell culture with electrophysiological methods taurine secretion was found to be closely coupled to both extracellular osmolarity and taurine levels (Benyajati and Renfro, 2000). Further recent studies have investigated the roles of cortisol, Na^+/H^+ exchange and carbonic anhydrase in renal sulfate excretion. By incorporating renal cell culture with immunoblotting and molecular cloning, putative functions of these entities in marine teleost proximal tubule cells were elucidated (Pelis *et al.*, 2003, 2005).

This technique has, therefore, proven to be a highly useful *in vitro* model for examining renal function in marine teleosts. The diversity of piscine kidney structure and function suggests the extension of this preparation to other marine species, and, if applicable, to freshwater fish, will provide new insights into renal function.

Intestinal cell/tissue culture

The intestine performs vital absorptive and excretory roles that act to maintain osmotic balance. This is especially the case in marine fish where, by virtue of the drinking response that compensates for diffusive ion loss, the intestinal lumen is directly exposed to an ion-rich medium. A number of *in vitro* techniques exist for examining the physiology of the piscine gut, including intestinal perfusion, gut sacs, enriched membrane vesicles, and isolated intestinal cells (see Glover *et al.*, 2003a, b; Burke and Handy, 2005). Insofar as osmoregulatory investigations are concerned, these methods may suffer from their lack of polarity, viability, and/or homogeneity. Consequently, a piscine enterocyte cell culture model, which overcomes many of these drawbacks, is of considerable utility.

There are literature reports of intestinal cell cultures from lamprey (Ma and Collodi, 1996, 1999), but these remain almost completely uncharacterized. Further attempts to culture intestinal cells from fish have not yet yielded publishable results. Difficulties in intestinal cell culture are not unique to fish, and may relate to the requirement of adequate feeder layers to provide the necessary signals for cellular differentiation and maintenance (Ali and Reynolds, 1996; Sambruy *et al.*, 2001).

Recently, an intestinal tissue explant technique was reported for sockeye salmon (*Oncorhynchus nerka*; Veillette and Young, 2005). These authors demonstrated that explants from pyloric caecae and the posterior intestine were responsive to cortisol, with effects on Na^+/K^+-ATPase similar to those reported in other preparations at similar hormone levels. However, the activity of Na^+/K^+-ATPase deteriorated over time, in association with degradation of the basolateral membrane domain. This is a novel method, and with further refinement and optimization of culture conditions, it may offer considerable scope as a tool for investigating osmoregulatory function in fish intestine.

There is significant heterogeneity between fish species, and indeed along the gastrointestinal tract of fish, in terms of intestinal architecture (see Ferraris and Ahearn, 1984). Species variation is primarily governed by diet, and may be reflected in differences in epithelial folding and mucus-secreting goblet cells for example. It remains possible that different fish species, or explants from different sections of the digestive tract, may prove more amenable to culture. Ideally, an intestinal epithelial culture technique will be developed in the future that allows culturing of a lamellar epithelial surface to facilitate transport studies by permitting manipulation of apical and/or serosal media in confluent cells with polarity.

Skin cell/tissue culture

The role of the integument in salt and water balance is mostly overlooked in favor of the relatively greater roles performed by the gills, gut and kidney. The skin of fish can, however, play a significant role in addressing the osmotic imbalance that exists between the animal and the external milieu. This role extends from ion transport (e.g., Marshall *et al.*, 1992; McCormick *et al.*, 1992) to the secretion of mucus that may act as a barrier to diffusive ion and water movement across the body surface (Shephard, 1994). Whole skin preparations from frog have been essential in the development of much of the current osmotic theory (e.g., Ussing and Zerahn, 1951), and have been applied to great effect in fish (e.g., Marshall *et al.*, 1992). More recently, several cell culture techniques have been developed for the examination of skin cell properties. These techniques consist of the culture of skin cell explants (Mothersill *et al.*, 1995), and a primary skin cell culture (Lamche *et al.*, 1998).

These preparations have been principally developed as ecotoxicological tools, where the skin—as a first line of defence—is a toxicologically important barrier. Their success as *in vitro* tools may not, therefore, necessarily depend upon the maintenance of functional properties. As such, their value as osmoregulatory models is largely undescribed. The explant technique is known to produce cells that respond to cortisol (van der Salm *et al.*, 2002). This indicates that endocrine responsiveness remains intact, an important tenet of an *in vitro* model. However, explants are unable to be cultured in lamellar sheets, reducing their utility. Like early gill primary culture models, skin cell cultures currently lack chloride cells, reducing their usefulness for osmoregulation investigations. This may be overcome by applying 'double-seeding' techniques that have recently been established for chloride cell incorporation in fish gills (see above).

To increase their utility, it may be possible to culture skin from specialized skin regions. In the rainbow trout, for example, the skin on the cleithrum bone appears to be relatively rich in chloride cells, and is likely to perform a role in osmoregulation (Marshall *et al.*, 1992). A further example is the pectoral skin of mudskippers, which is reported as having a high density of chloride cells that appear to function in chloride secretion in seawater (Yokota *et al.*, 1997). Difference in ion uptake between chloride cells of the gill and skin are apparent (Flik and Verbost, 1993). It is possible, therefore, that the greatest utility for skin cell culture preparations lies in exploring these differences.

Primary cultures versus *cell lines*

In fish osmoregulation, cell lines based on piscine epithelia are less commonly exploited than are primary cultures. They have, however, found favor in fish toxicological and disease research (e.g., Castaño *et al.*, 2003; Butler and Nowak, 2004). Cell lines are easier to maintain than primary cultures, and thus offer benefits in terms of experimental and labor costs. Additionally, a considerable advantage of cell lines is that they offer a reproducible cellular phenotype. In contrast, freshly isolated cells are prone to heterogeneity associated with cell culturing technique and individual physiological condition (Castaño *et al.*, 2003). However, it is generally considered that cell lines are less differentiated than primary cultures, and thus may be poorly representative of the cells from which they are derived.

In the few studies where piscine cell lines have been utilized to investigate aspects of cellular ion and water transport, further potential difficulties have emerged. In a comparative study, a rainbow trout gill epithelial cell line was found to be less capable of generating eicosanoids than either intact gill or freshly isolated cells (Holland *et al.*, 1999). Eicosanoids are important cellular signalling molecules with critical roles in fish osmoregulation (Van Praag *et al.*, 1987). Using a cyprinid epithelial cell line (*Epithelioma Papulosum Cyprini*; EPC), standard culture conditions were shown to result in vitamin E deficient cells (George *et al.*, 2000). This resulted in a reduced oxidative defence status in these cells, and increased their susceptibility to toxicants. Deviations in other metabolites, of critical importance to osmoregulatory function, may also be induced by poor culture media. While media deficiencies are likely to be deleterious to assessing physiologically relevant functions in both primary culture and cell lines, the effects are likely to be more significant in cell lines due to their extended culture histories. Further problems may also exist in that raising medium osmolarity of the EPC cell line can result in enhanced apoptosis and cell death (Hashimoto *et al.*, 1998). This indicates that these cultures may be unsuitable for many of the manipulations that may be of relevance to researchers of fish osmoregulation. This finding is also in contrast to recent studies that show primary gill cell cultures maintain physiological integrity in response to apical dilution (Zhou *et al.*, 2004), although their response to hypertonicity has not been investigated.

Studies that depend more stringently on physiological parameters that resemble those in intact organisms and tissues are likely to be better served by primary cultures. However the genetic homogeneity of cell lines is likely to offer some advantages over primary cultures in terms of molecular endpoints. The EPC cell line has, for example, been successfully used to isolate a taurine transporter that is likely to perform a role in cell volume regulation (Takeuchi *et al.*, 2000).

Purifying Epithelial Cell Types—Techniques for Assigning Osmoregulatory Function

Considerable difficulties exist with the assignation of physiological function to specific cell types within multifunctional epithelia of mixed cell populations. Consequently, in the fish gill, several controversies pertaining to ion transport exist. With regard to sodium transport, for example, both the nature of the transporters involved (see 'Cloning of transporters'; Fig. 7.1) and the roles of pavement and chloride cells in this process, are unresolved. To address these issues, methods are required that permit the examination of individual cell types. Based on cell isolation methods in mammalian epithelia, techniques have recently been developed in fish that are capable of isolating cell types from the gill. Following isolation these cells remain viable for molecular, biochemical and physiological investigation. These tools have been integral to studies that have ascribed biological function to cell type, and to research that has challenged traditional branchial cell type classification.

The experimental utility of relatively pure chloride cell populations has been recognized for some time. With this goal in mind, a number of different isolation techniques have been used (e.g., Sargent *et al.*, 1975; Naon and Mayer-Gostan, 1983; Perry and Walsh, 1989). These methods suffered from either one or a combination of difficulties relating to low purity, reduced viability, and poor applicability to both freshwater and marine fish. Recently, a Percoll density gradient centrifugation technique

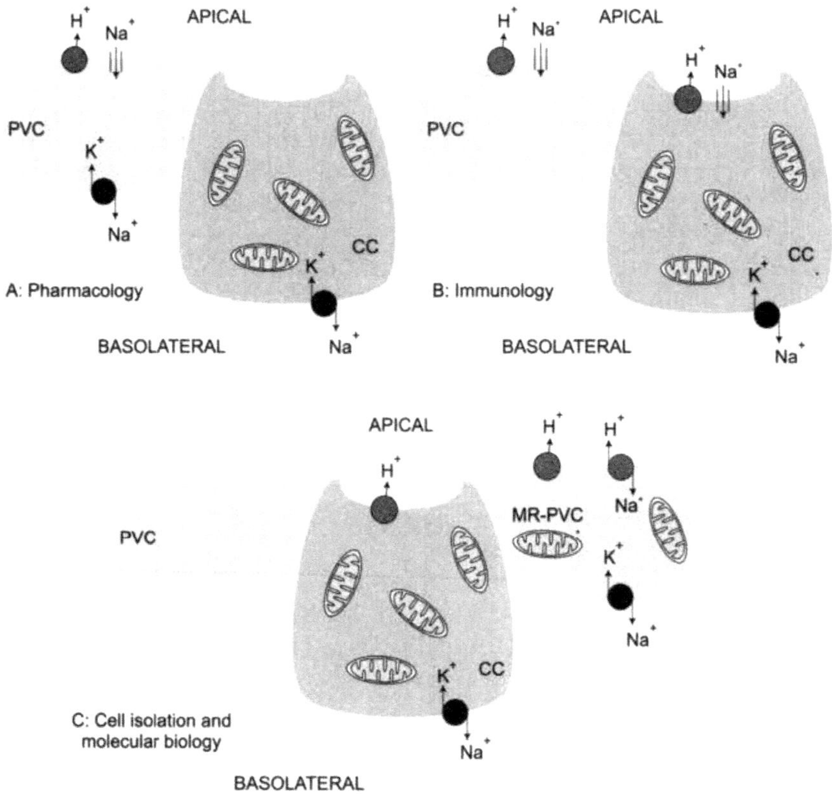

Fig. 7.1 Schematic diagram illustrating how molecular and cellular advances have altered our understanding of sodium transport in freshwater rainbow trout. Pharmacological (A) and immunological (B) studies may produce conflicting results, and have often been unable to attribute a transport function to a particular cell type (see text for discussion). The models demonstrated here are primarily based on data from Lin and Randall (1993) and Wilson *et al.* (2000a). The development of cell isolation protocols and molecular cloning of transport proteins has considerably altered these models (C). This hypothesis of sodium transport awaits further molecular and immunolocalization confirmation. See 'Molecular cloning of transporters', 'Purifying epithelial cell types' and 'Pharmacology of ion transport' for details.

was developed that resulted in highly pure (89-98%) populations of pavement and chloride cells from eel gill (Wong and Chan, 1999b). The low cytotoxicity and adjustable osmolarity of Percoll as a substrate resulted in viable cells that displayed biochemical and morphological properties resembling those *in vivo* (Wong and Chan, 1999b). This technique has been applied with success to eel and rainbow trout from both freshwater and marine environments (Wong and Chan, 1999b; Goss *et al.*, 2001; Hawkings *et al.*, 2004).

A significant finding from these initial studies of isolated gill cells was the elucidation of two mitochondria-rich cell types (Wong and Chan, 1999b; Goss *et al.*, 2001), confirming earlier morphological observations (e.g., Pisam and Rambourg, 1991). The nature, and scientific implication, of mitochondria-rich subtypes will be discussed below. Methodologically, the presence of two mitochondria-rich cells required an additional isolation step. In eel, gill cell subtypes were further separated by flow cytometric techniques. Variation in cell properties such as size, granularity, and the autofluorescence of mitochondria-associated metabolites, permitted these cells to be adequately sorted into homogenous populations. Furthermore, changes in these properties were used to monitor the alterations that occurred in these cell populations as the animal were acclimated to seawater or exposed to cortisol (Wong and Chan, 1999a, 2001). Laser scanning cytometry, a related but more sophisticated technique, has also been applied to separate and assess the effects of salinity changes on specific cell types of the killifish gill (Lima and Kültz, 2004).

In rainbow trout, flow cytometry was not a viable method of isolating these cell types (Galvez *et al.*, 2002), suggesting that species differences are likely to exist in cellular properties amenable to flow cytometry. Instead, it was found that the two trout mitochondria-rich cell subtypes could be distinguished on the basis of their affinity for peanut lectin agglutinin (PNA; Goss *et al.*, 2001). One population of mitochondria-rich cells bound PNA, the other did not. The mitochondria-poor cells (pavement cells) already separated by centrifugation, also failed to exhibit PNA binding. PNA specifically binds to glycosylated membrane proteins with a terminal D-galactosyl residue, and this property has previously been used to distinguish between intercalated cell subtypes in mammalian kidney (LeHir, 1982). Fluorescein isothiocyanate (FITC) conjugated-PNA, and anti-FITC antibodies conjugated to iron particles were then used in combination with magnetic separation techniques to isolate the two mitochondria-rich cell subtypes in trout (Galvez *et al.*, 2002). Cells were initially incubated in PNA, then in the antibody, before being passed through a magnetic field. Those cells without PNA affinity (non-magnetic) were flushed through, while PNA-positive cells (with iron particle) were trapped, and later eluted. This technique resulted in viable, homogenous cell populations that can be utilized for subsequent analyses. This isolation protocol has been applied, for example, to investigate the relative changes in distinct cell population enzyme activities that occur with seawater acclimation (Hawkings *et al.*, 2004).

From a functional perspective, gill cells have been traditionally considered as either pavement or chloride cells. Categorization is usually made by virtue of simple bioassays based on the higher mitochondrial content of the latter cell type. The presence of two mitochondria-rich cell types, with properties distinct enough to permit their successful separation, suggests that the situation is more complicated. Supported by mitochondrial density data and the nature of cellular changes associated with seawater exposure, Wong and Chan (1999a) concluded that the additional mitochondria-rich cell type in eel was a chloride cell. Contrastingly, based on transmission electron microscopy and differential PNA binding, this additional subtype was classified as a mitochondria-rich pavement cell (Goss *et al.*, 2001). A number of possibilities for such a discrepancy exist. Some assays used for characterizing cell types are more robust than others and this could have influenced the findings. Alternatively, the different species used may account for these differences. The possibility of undiscovered mitochondria-rich pavement cells in eel, and chloride cells in trout, would also influence the conclusions drawn. Irrespective, it is clear that cell subtypes exist in fish gill.

Isolation of cell subtypes has significant implications for many existing controversies in fish osmoregulation. Of particular relevance is the reassessment of branchial sodium transport in freshwater fish (see Fig. 7.1). One theory states that sodium uptake is achieved via an epithelial sodium channel (ENaC), and driven by an apical proton pump (H^+-ATPase; see also 'Cloning of transporters'). Immunological evidence has demonstrated the presence of a V-type H^+-ATPase either exclusively in pavement cells (Sullivan *et al.*, 1995), or in both pavement and chloride cells (Wilson *et al.*, 2000a, b). Galvez and colleagues (2002) attempted to address this incongruity using their isolated cell technique in rainbow trout. Following cell separation, mitochondria-rich cells that were PNA negative (i.e., mitochondrial-rich pavement cells) exhibited protein expression levels of H^+-ATPase that were two-fold greater than those of chloride cells. These mitochondria-rich pavement cells also demonstrated a relative increase in proton pump expression under conditions that promote acid excretion (hypercapnia). In an additional study, cell isolation was used in combination with pharmacological evidence to suggest the presence of an ENaC in the same cell type (Reid *et al.*, 2003). These results indicate that the mitochondria-rich pavement cell may be the branchial location for acid-mediated sodium uptake. It is interesting to note that in the mammalian kidney, of the two mitochondria-rich cell subtypes, the one that is PNA-negative is also thought responsible for acid secretion, analogous to the scenario in trout gill (Reid *et al.*, 2003). This finding appears to resolve the seemingly conflicting immunological evidence. If the mitochondria-rich pavement cell is the primary locus for branchial sodium uptake, then under more simplistic categorization it could be classified as either a pavement cell (based on morphology) or a chloride cell (based on mitochondrial content), leading to either of these cell types being attributed with sodium transport properties.

Cell isolation protocols do have some significant drawbacks. Isolated cells tend to lose their apical/basolateral polarity (Galvez *et al.*, 2002; Reid *et al.*, 2003). This may result in the redistribution of ion transport structures and functions, and will limit the usefulness of these techniques. Furthermore, the passage of ions across epithelia is often dependent upon cooperation between epithelial cell types (Do and Civan, 2004). As such, transport may not function in isolated cells, as it does in epithelia. Nevertheless, such methods offer considerable promise, especially in conjunction with other molecular and cellular techniques. The custom seeding of primary gill cell cultures with specific isolated cell types, for example, will overcome polarity problems, and could be a tool of some significance for examining transport function. In addition, this preparation has great potential for molecular cloning studies, where entities involved in ion transport could be assigned to particular cell populations. Advances in RNA isolation and amplification technology (Ginsberg, 2005), are likely to overcome many of the technical difficulties associated with such methods.

Enriched Membrane Vesicles—Mechanistic and Structural Analysis of Transport

The ability to preferentially enrich and encapsulate a specific membrane component of an epithelium is of considerable value. It enables both the localization of specific functions to a membrane surface, and the mechanistic characterization of epithelial processes (Berteloot and Semenza, 1990). Enriched membrane vesicles permit the delineation of osmoregulatory processes in a system relatively free from a number of confounding impacts. These include factors such as the unstirred water layer of epithelial surfaces, transport systems on other membranes, and systemic feedback. While these are physiologically important features of epithelia, they can preclude measurement of transport mechanism. Membrane vesicles are ideally suited for kinetic analyses. They also offer the advantage of being able to manipulate 'intracellular' and 'extracellular' constitution, in terms of both ion composition and the application of pharmacological agents. Of particular benefit is the capacity to examine function in the native membrane.

These advantages have been extensively capitalized upon in studies of both mammalian and piscine epithelia. Enriched membrane techniques have been applied to gill, kidney and intestine of fish, with a particular focus the characterization of divalent ion transport (Bijvelds *et al.*, 1996; Flik *et al.*, 1996). In contrast, examination of the 'strong' osmoregulatory ions, sodium, potassium and chloride across isolated membrane preparations, has been comparatively neglected (although see De Giorgi *et al.*, 1992).

Methodological considerations are likely to explain this lack of research activity. The osmotic reactivity of these sealed vesicle preparations, for example, may be mitigating. The investigation of transport processes may require manipulation of electrochemical gradients, which can result in altered intravesicular space as the vesicles shrink or expand. Furthermore, radiotracers or ion-specific dyes used to monitor ion movement rapidly equilibrate owing to the small intravesicular volume. Consequently, it may be difficult to monitor the initial rate of ion transfer, a measurement that is favored owing to its independence from properties such as vesicle size (Berteloot and Semenza, 1990). However, these problems are not significant and can often be overcome by experimental advances such as stop-flow fluorimetry, which permits assessment of transport over fractions of a second (e.g., Rivers *et al.*, 1998). In other comparative models, membrane vesicles have offered key insights into strong ion transport (e.g., Towle, 1993; Dubinsky *et al.*, 2000). These preparations are, therefore, still likely to be useful for the investigation of epithelial ion movement in fish. It is important to note that the ability to isolate enriched and sealed vesicles of suitable purity appears to be technique- and investigator-dependent, and that certain membranes and epithelia are more suited to this application than others. Apical (brush-border) enrichment of the intestine and basolateral enrichment of the gill, for example, appear to be the most robust preparations.

Enriched membrane studies may also have significant scope for the investigation of membrane lipids and their roles in osmoregulation. Basolateral membrane isolation techniques have been used to investigate the role of specific membrane phospholipid microdomains on the activity of vital osmoregulatory enzymes in fish gill (Lingwood *et al.*, 2004, 2005). Findings from these studies suggest that migration of Na^+/K^+-ATPase to sulfogalactosylceramide-rich microdomains is responsible for the upregulation of this enzyme during acclimation to seawater (Lingwood *et al.*, 2005). The importance of lipids in osmoregulation has been suggested by studies that show lipid composition changes with salinity acclimation (Hansen *et al.*, 1999; Hansen and Grosell, 2004), but until the utilization of enriched membrane techniques the significance of this was not well understood.

Membrane isolation also facilitated the discovery of the unusually high cholesterol to phospholipid ratio in the basolateral membrane of a marine elasmobranch gill (Fines *et al.*, 2001). This finding may explain the low urea permeability of the elasmobranch gill, which is partially responsible for the ability of these animals to maintain this important osmolyte against a strong diffusive gradient for urea loss. Recent investigations, utilizing membrane vesicles from both teleost and elasmobranch gill (Hill *et al.*, 2004), and from shark rectal gland (Zeidel *et al.*, 2005), are somewhat contradictory. These studies suggest that the low permeability to urea and water it is not due to lipid composition. Instead, it is proposed that factors such as vascular perfusion, unstirred water layers (Hill *et al.*, 2004), or specific protein components (Zeidel *et al.*, 2005) may play an essential role. Studies have also demonstrated the presence of specific urea transporters in the branchial basolateral membrane vesicles of both sharks and teleosts (Fines *et al.*, 2001; Walsh *et al.*, 2001; McDonald and Wood, 2004). These transporters would act to pump urea from the gill back into the circulation, thus reducing apparent urea permeability. Irrespective of the factors involved, the efficacy of membrane vesicles in the investigation of urea transport is clear.

The examples described above illustrate the utility of membrane vesicle preparations for the analysis of aspects of salt and water balance in fish. As with most techniques, however, there are important experimental caveats that need to be overcome. The purity of the vesicle population, the orientation of the membrane (right-side-out *versus* inside-out), and confirmation of osmotically active vesicles are important details, that are all too frequently unreported. Appropriate marker enzymes can be used to confirm purity and orientation. However, regional heterogeneity of enzyme activity and content (Almansa *et al.*, 2001; Dopido *et al.*, 2004) may influence the conclusions drawn. For this reason, it may be necessary to carefully characterize the marker enzymes of choice. A recent study labelled the apical surface of winter flounder gills with a biotin derivative (Sáenz *et al.*, 2003). After employing the membrane isolation protocol, the putative apical membrane fraction was incubated with avidin, and biotin content was determined via the peroxidase assay. The simultaneous measurement of the proposed apical membrane marker ADPase correlated well with the biotin-avidin results. This established the suitability of using ADPase activity as a marker of branchial apical membranes in this species.

With the expanding array of molecular and cellular techniques, the use of enriched membrane vesicles is perhaps an overlooked method. Data from several recent studies indicate their utility. Membrane vesicle preparations may offer an important bridge between the structural information resulting from molecular cloning of transporters, and their physiological roles in osmoregulating epithelia.

Pharmacology of Ion Transport—Caveats for Comparative Biologists

There are hundreds of pharmacological agents that can be used to block aspects of ion and water transport. These range from natural toxins to synthetic chemicals, with actions varying from those on second messenger systems to those affecting the transporters themselves. Many of these agents are targeted as therapeutics for human disease, and although a widely utilized and valuable tool for elucidating transport mechanisms in comparative physiology, their applicability to such models is rarely questioned.

The mechanism of sodium transport across the fish gill is an ongoing controversy. Based partly on pharmacological evidence the favored model, at least during the 1990s, stated that uptake was achieved through an epithelial sodium channel (ENaC), powered by an apical proton pump (V-type H^+-ATPase; Avella and Bornacin, 1989; Lin and Randall, 1991, 1993; see also 'Cloning of transporters' below; Fig. 7.1). More recently, other evidence has supported the alternative hypothesis, that of an apical sodium-proton exchanger (NHE). The ongoing controversy perhaps warrants a critical re-examination of the foundation for pharmacological transporter blockade.

Amiloride is a widely used pharmacological agent that inhibits the function of the NHE. Some NHE isoforms, including the major apical branchial NHE found in fish (NHE3; Choe *et al.*, 2002, 2005; Hirata *et al.*, 2003), remain relatively resistant to blockade (Masareel *et al.*, 2003). Amiloride is somewhat promiscuous, and will also block both ENaC and the sodium/calcium exchanger (Masareel *et al.*, 2003), and may disrupt cellular signalling pathways (Holland *et al.*, 1983). Furthermore, amiloride is known to promote the binding of atrial natriuretic peptide to its receptor (De Lean, 1986). This effect could modify sodium and water transport independently of direct blocking effects on transport moieties in systemic preparations. As a result, the interpretation of amiloride actions cannot necessarily be restricted to consideration of its impact on NHE.

There are still important roles for pharmacological agents in the study of fish osmoregulation. To a large degree, the promiscuity of amiloride can be minimized by the careful choice of dose. Furthermore, the development of more specific inhibitors (e.g., phenamil, HOE-694) has eliminated many of the known non-specific actions. Despite these measures, precautions must be taken to ensure the suitability of pharmacological studies. A major challenge is the heterogeneity of the biological material available for osmoregulatory investigation. This can display significant variation in properties depending on fish species, osmoregulatory tissue, and the physiological state of the source animal. In addition, it may be simplistic to expect that pharmacological agents applied to mammalian preparations will have identical actions in fish. It is known, for example, that some fish NHE's have unique phenotypes (Nickerson *et al.*, 2003). This may lend to them properties that alter the efficacy of pharmacological attack. Subsequently, considerable investigation of inhibitor activity is required prior to use, a step rarely undertaken (cf., Reid *et al.*, 2003).

Amiloride is one of the better characterized pharmacological agents used in fish studies. It may be that its wide array of non-specific cellular actions is a case of 'familiarity breeds contempt'. The more that is known about a blocker, the more likely it is that promiscuous actions can be ascribed. Thus the new generation of pharmacological agents, while appearing to exhibit fewer non-specific actions, may have effects on non-target pathways that remain undiscovered. It is also worthwhile highlighting the fact that variations in isoforms may also continue to influence effects of these new blocking agents. HOE-694, for example, is touted as a useful agent against NHE, and has a high specificity for this transporter. However, it is only a poor inhibitor of NHE3 (Masareel *et al.*, 2003), the main branchial isoform in fish (Choe *et al.*, 2002, 2005; Hirata *et al.*, 2003).

Ouabain, a plant alkaloid, is another commonly exploited pharmacological tool in fish osmoregulation. It is a potent inhibitor of the basolateral Na^+/K^+-ATPase that is responsible for driving transepithelial ion transport. Following the discovery of ouabain as a circulating hormone in mammals (e.g., Mathews *et al.*, 1991), endogenous immunoreactive ouabain was subsequently identified in fish plasma, where its levels fluctuate in response to salinity changes (Kajimura *et al.*, 2004). The primary cellular target of ouabain appears to be Na^+/K^+-ATPase. However, effects of this hormone on the release of the important osmoregulatory hormone prolactin have been observed at levels less than those required for effects on the sodium pump (Kajimura *et al.*, 2005). Ouabain may also have direct impacts on cellular signalling pathways, at least in mammals (Harwood and Yaqoob, 2005). This suggests that ouabain could exert biologically relevant impacts on endogenous pathways, beyond those previously attributed to effects arising from inhibition of Na^+/K^+-ATPase. This may have significant consequences for the use of ouabain in fish osmoregulation, and again suggests that results arising from the use of pharmacological agents in comparative models have to be carefully interpreted.

MOLECULAR BIOLOGY OF FISH OSMOTIC AND IONIC REGULATION

Fish are the largest, and arguably the most successful, extant group of vertebrates. The diversity of fish species, and the environments they inhabit, provides an ideal opportunity for examining the mechanisms of ionic and osmotic homeostasis. In particular, it permits assessment of how alterations in the osmoregulatory apparatus have contributed to this vast radiation. Their ability to thrive in highly saline, extremely dilute, acid-rich or strongly alkaline waters depends, to a large extent, on the adaptability of the osmoregulatory apparatus. The genetic material encoding the ion transporters, hormones, endocrine receptors and signal transducers can differ considerably between even closely related fish species. Differences in the number, the nature, and the regulation of these moieties will have significant impact upon the processes that enable salt and water balance, and may explain how fish have been able to adapt to such a wide range of osmotic environments.

Our knowledge of the differences in genomes, genes, proteins and metabolites between fish species is poor. Our understanding of how these differences may influence fish osmoregulation is even less well developed. This ignorance is due in part to a lack of bioinformatics information related to fish. Progress is being made on a selected few species (see Table 7.1), with detailed sequence knowledge of the zebrafish and pufferfish genomes representing a significant advance. However, the phylogenetic distance between these species and those that are often considered of greater relevance to the fish physiologist is often considerable. Theoretically, the tremendous diversity in form, habitat, physiology and genetics of fish, is of great benefit for examination of osmoregulatory structures and processes. Practically, however, it may represent an important obstacle to the elucidation of molecular processes.

Despite the difficulties associated with utilizing molecular tools in fish osmoregulation, considerable progress has been made. The use of cloning techniques has provided structural support for the ion-transporting theories established from physiological data. It has also offered insight into

Table 7.1 Genetic resources for the piscine molecular biologist.

Sequencing projects	*Website(s)*
Zebrafish (*Danio rerio*)	http://www.sanger.ac.uk/Projects/D_rerio/
Japanese pufferfish (*Fugu*)	http://genome.jgi-psf.org/Takru4/Takru4.home.html http://fugu-sg.org/
Green-spotted pufferfish (*Tetraodon*)	http://www.genoscope.cns.fr/externe/tetranew/
Japanese medaka (*Oryzias latipes*)	http://dolphin.lab.nig.ac.jp/medaka/
Atlantic salmon (*Salmo salar*)	http://salmongenome.no http://web.uvic.ca/cbr/grasp
Threespine stickleback (*Gasterosteus aculeatus*)	http://www.genome.gov/12512292
Rainbow trout (*Oncorhynchus mykiss*)	http://dga.jouy.inra.fr/cgi-bin/lgbc/main/.pl?base=rainbow
Other resources	
Genbank (Nucleotide database for all taxa)	http://www.ncbi.nlm.nih.gov
Swiss-Prot (Proteomic database for all taxa)	http://www.expasy.ch/sprot
Zebrafish gene resources and links	http://zfin.org
Medaka resources and links	http://biol1.bio.nagoya-u.ac.jp:8000/

unforeseen complexities in endocrine regulation of fish salt and water balance. Examples of these advances will be discussed below. The potential use of methods for manipulating and monitoring cellular expression will also be addressed. The utilization of molecular biology in fish osmoregulation has significant pitfalls, but the opportunities it offers in terms of new discoveries are unrivalled.

Physiological *versus* Genomic Models—As the Krogh Flies...

The choice of fish species in studies of piscine salt and water balance has been driven extensively by Krogh's principle. This suggests that for any biological question, nature provides an experimental system best suited for its investigation (Krebs, 1975). Osmoregulation research has generally focussed on those animals with the ability to acclimate to altered environmental osmolarity, or which undergo life history- or migration-related changes exposing them to vastly different ionic and osmotic challenges. The changes in epithelial structure and physiology that accompany this acclimation have been invaluable for our understanding of the mechanisms that are responsible for, and which regulate, osmotic and ionic homeostasis (Wheatly and Gao, 2004). More practical concerns ensure those species that are easily cultured and of economic importance are especially favored.

The application of Krogh's principle in fish genetics has seen the development of model organisms such as the Japanese pufferfish (*Takifugu rubripes*; *Fugu*), a fish with a compact genome densely populated with coding sequences (Brenner *et al.*, 1993). The relatively small genome size of *Fugu* has made it a target for full genome sequencing (Aparicio *et al.*, 2002). These attributes facilitate studies examining a wide range of biological phenomena, including the importance of non-coding DNA (Woolfe *et al.*, 2005), gene regulation (Goode *et al.*, 2005), and even investigation of human genetic disease (Yu *et al.*, 2005). By virtue of the genetic resources available, this fish is a model species of considerable importance. However, this species is not readily available for laboratory culture in many areas of the world. Its stenohaline nature also renders it inappropriate for many of the investigations of greatest interest to fish osmoregulation. The genome sequence of the zebrafish is also near completion, an effort driven mainly by the value of this fish as a developmental biology model. While this is a more accessible species, it too lacks applicability for investigating salt and water balance.

There are, however, increasing amounts of gene information available for other fish species, in the form of expressed sequence tag and cDNA sequencing projects (see Table 7.1). Consequently, with time, the available sequence information for physiological Krogh's models will increase. This will significantly increase the applicability of molecular techniques to investigators of fish osmoregulation. Even the genomes that have already been sequenced are of some value. These can provide, in some cases, an excellent starting point for homology-based cloning or transcriptomic studies, and for bioinformatic database mining approaches (see below, also Wheatly and Gao, 2004).

It has been suggested that the Japanese medaka (*Oryzias latipes*) may offer greater utility as both a physiological and genetic model (Inoue and Takei, 2003). This is a well-established transgenic model, and considerable sequence information is available (Tanaka and Kinoshita, 2001; Inoue and Takei, 2003). Essentially a freshwater species, *O. latipes* can be acclimated to different salinities. It will, for example, spawn, hatch and undergo embryonic development in seawater. However, adult survival is relatively poor, and is only achieved by careful acclimation (Inoue and Takei, 2002). This suggests changes that may be interpreted as osmotic acclimation may better represent generalized stress effects. Of greater applicability is the presence of closely related species that naturally occur in varying salinities (Inoue and Takei, 2003). Comparison of molecular responses and expression between species with closely related phylogenies may yield insights into the mechanisms and fundamental processes linked to the ability to adapt to different salinities (Inoue and Takei, 2003).

Currently, a distinction exists between the organisms of greatest interest to the molecular biologist and those with greatest applicability to the physiologist. This is a disparity that is rapidly diminishing as sequence information in species with greater utility for the examination of fish osmoregulation increases. Future sequencing efforts will be further aided by the continuing advances in tools that improve speed, cost and accuracy of genome sequencing (Bonetta, 2006). Even with the existing datasets, the scope for molecular investigations of fish osmoregulation has increased significantly over the previous decade. In another decade from now, many of the arguments regarding the lack of genetically characterized models of relevance to piscine salt and water balance will very likely be redundant such is the rapid advance of this field.

Gene Duplication and Isoform Switching—The Unravelling Complexity of Fish Biology

Gene duplication represents a significant challenge for the piscine molecular biologist. There is considerable support for a whole genome duplication event that occurred early in the teleost lineage (Christoffels *et al.*, 2004; Jaillon *et al.*, 2004). The resulting multiple gene copies are believed to be partially responsible for an enhanced adaptive capacity and, consequently, may have promoted the radiation of teleost fish (Venkatesh, 2003). The genome of fish remains relatively plastic, and evidence of localized duplication events effecting osmoregulatory genes exists (e.g., Rajarao *et al.*, 2001; Inoue *et al.*, 2003). These localized gene duplications may be restricted to species or certain lineages of teleosts. Additionally, some fish have lost gene copies arising from duplication events, further contributing to the molecular complexity of osmoregulation in fish. The result of this genetic flux is a gene- and species-specific distribution of gene duplicates.

A number of osmoregulatory moieties with gene duplicates have been identified in fish. The V-type H^+-ATPase B subunit (Niederstatter and Pelster, 2000) and an aquaporin isoform 1 (Cutler and Cramb, 2001) in eel are represented by duplicated coding sequences. The multiple copies of Na^+/K^+-ATPase α and β subunit isoforms in zebrafish (Rajarao *et al.*, 2001), rainbow trout (Richards *et al.*, 2003) and Atlantic salmon (Gharbi *et al.*, 2005) are likely to have arisen by gene duplication. The complicated natriuretic peptide system in medaka and pufferfish is proposed to have resulted from both whole genome duplication and local tandem duplication events (Inoue *et al.*, 2003).

The physiological implications of possessing an assortment of closely related genes are many and varied. The presence of duplicate coding sequences may spawn new functions for an otherwise superfluous gene. The duplicate copies of the midkine growth factor genes in zebrafish, for example, have different functional properties than their mammalian counterpart (Winkler *et al.*, 2003). Additionally, fish antifreeze proteins are believed to be derived from a gene duplicate originally coding for the protease precursor trypsinogen (Chen *et al.*, 1997).

There is also evidence that gene duplicates of ion transport proteins can be independently regulated. Richards and colleagues (2003) identified three Na^+/K^+-ATPase $\alpha 1$ subunit copies in rainbow trout, denoted as $\alpha 1a$, $\alpha 1b$ and $\alpha 1c$. The tissue distribution of $\alpha 1a$ and $\alpha 1b$ was similar, indicating a conserved functional role. Upon exposure of trout to seawater, these two transcripts were differentially regulated. While α1a expression decreased in the gill, α1b expression increased (Richards *et al.*, 2003). This isoform switching is likely to be controlled hormonally, as it accompanied the cortisol surge observed after seawater transfer (Richards *et al.*, 2003). The opposite pattern (an increase in α1a expression) was observed in sockeye salmon that migrated from seawater to freshwater (Shrimpton *et al.*, 2005). These studies indicate the presence of differential osmoregulatory gene duplicate regulation and also suggest that subtle differences between these duplicated genes could offer functional advantages, depending on osmotic environment.

In mammals, four distinct α, and three β subunits exist, that can combine in order to generate a functional sodium pump (Blanco, 2005). Orthologs for some, but not all, of these subunit isoforms have been discovered in fish (see Gharbi *et al.*, 2005). The functional Na^+/K^+-ATPase is a heterodimer, consisting of a single α and a single β subunit, with other regulatory elements playing an important role (e.g., Blanco, 2005). The variety of subunit isoforms and the gene duplication of a number of these isoforms indicates the potential for a vast assortment of different heterodimers in fish. These are likely to be expressed in a cell- and tissue-dependent manner. The possibility that each heterodimer has distinct functional and regulatory properties (see Richards *et al.*, 2003; Deane and Woo, 2005), hints at the previously unrealized complexity of this protein in fish osmoregulatory epithelia.

Independent regulation and isoform switching of Na^+/K^+-ATPase has important consequences for the monitoring of this enzyme in osmoregulation studies. Commonly used assays may greatly oversimplify or even misrepresent experimental changes in this protein (Richards *et al.*, 2003). For example, the measurement of Na^+/K^+-ATPase activity in a tissue homogenate may represent the summed activities of a number of different heterodimers. The overall effect may, therefore, mask the impact on individual heterodimers. Gene expression assays are, thus, likely to offer a more sensitive measure of cellular Na^+/K^+-ATPase dynamics (e.g., Scott *et al.*, 2004).

Na^+/K^+-ATPase is a key enzyme in fish osmoregulation, but represents only a single point in a single pathway influencing ion and water movement across osmoregulatory epithelia in fish. It is possible that other components of ion transport pathways are represented by differentially regulated duplicate genes and also exhibit isoform switching. Given all the elements that may combine to ensure ionic homeostasis, the complexity of these processes is likely to be extensive.

Cloning of Transporters—Molecular Support for Physiological Theories

Physiological studies can provide only circumstantial evidence for the existence of specific transporters and ion uptake pathways. Recent efforts have utilized molecular cloning techniques to establish a structural basis for ion transport in fish epithelia. Over the past decade the presence of ion channels (e.g., cystic fibrosis transmembrane conductance regulator), bicarbonate transporters, sodium/proton exchangers, chloride cotransporters, proton pumps (V-type H^+-ATPase), sodium pumps (Na^+/K^+-ATPase) and water channels (aquaporins), has been elucidated (see Cutler and Cramb, 2001; Hirose *et al.*, 2003; Perry *et al.*, 2003 for reviews).

Structural evidence has permitted reassessment of many of the ion transport hypotheses established by previous physiological investigations. For example, evidence regarding the nature of proton-facilitated sodium transport in the freshwater fish gill is contradictory. Uncertainty extends to both the likely cellular location of this transport process (see 'Purifying epithelial cell types' above), and the structural entities involved (see Kirschner, 2004). Two physiological models have been proposed. In one, sodium exchange for a proton is mediated by an apical sodium/proton exchanger (NHE). This model relies on the creation of a favorable electrochemical gradient for sodium influx, likely generated by the actions of the basolateral Na^+/K^+-ATPase, and the efflux of protons down a concentration gradient. Calculations based on the few existing measures of intracellular sodium and pH suggest this model may not viable, at least in rainbow trout (Avella and Bornancin, 1989). An alternative model states that an apical proton pump (H^+-ATPase), in concert with the basolateral Na^+/K^+-ATPase, generates an electrochemical gradient that favors the inward passage of sodium via an epithelial sodium channel (ENaC). This model is supported on the basis of physiological and pharmacological evidence (Lin and Randall, 1991, 1993).

Immunological studies aimed at investigating ion transporters in gill epithelia provided ambiguous data. Supporting the favored model was evidence demonstrating the presence of proton pumps on apical branchial surfaces, and sodium pumps on the basolateral surface of branchial epithelial cells (e.g., Wilson *et al.*, 2000a, b). In some species, however, these activities were not located to the same cell types (Piermarini and Evans, 2001; Tresguerres *et al.*, 2005), contradicting the model. Conversely, in tilapia and rainbow trout, a putative ENaC was discovered in the same cell type as the proton pump (Wilson *et al.*, 2000a), favoring the pump/channel coupling hypothesis. The presence of an apical NHE has been discerned in some species (Wilson *et al.*, 2000b; Edwards *et al.*, 2002; Choe *et al.*, 2005). While this does not necessarily contradict the model, it does suggest the alternative hypothesis involving an NHE should bear further consideration. In general, the findings from immunological studies were somewhat confusing, and suggested that transport mechanism could be species dependent.

Pharmacological and immunological evidence is subject to several pitfalls. Some of these have been detailed above (see 'Pharmacology of ion transport'). In immunological studies, antibodies raised to different subunits or regions of the protein of interest can give different results. For instance, the presence of the ENaC β-subunit, but not the α-subunit, can be detected in fish gill (Wilson *et al.*, 2000a). Furthermore, the use of heterologous antibodies may cause misleading conclusions. Katoh and colleagues (2003), for example, showed a basolateral locus for the proton pump using a homologous antibody in killifish, in disagreement with apical locations from studies utilizing heterologous antibodies (e.g., Wilson *et al.*, 2000a, b). Molecular cloning of transport proteins provides a structural framework in which to interpret the results of these studies.

Molecular cloning complements physiological, pharmacological and immunological approaches, and can be used as a tool to explain some of the apparent experimental contradictions that may arise from these other techniques. A vacuolar-type H^+-ATPase has been cloned from the gill of a number of fish species (Perry *et al.*, 2000; Boesch *et al.*, 2003; Hirata *et al.*, 2003; Katoh *et al.*, 2003), lending support to the physiological, pharmacological, and immunological evidence for its existence in this tissue. Likewise, a number of sequences now exist for NHE isoforms in fish (e.g., Choe *et al.*, 2002, 2005; Hirata *et al.*, 2003). Of particular interest is the NHE3 isoform which appears to be the major apical form in fish gills (Choe *et al.*, 2002, 2005; Hirata *et al.*, 2003). Comparisons between the stingray and zebrafish NHE3 show only a 53% amino acid identity between the predicted protein sequences of these two species. Such divergence may explain the core difficulty with the use of non-homologous antibodies (non-homologous antibody recognition sites).

The fact that sequences exist in fish species for components of both of the competing models in freshwater fish gill provides no clue as to which is most likely to explain sodium transport. Of more interest is the evidential failure to clone a piscine ENaC. No sequences from the completed pufferfish genome share homology with cloned ENaCs from mammalian species. Even though pharmacological and immunological evidence has suggested the presence of ENaC in fish gill, it is difficult to support the existence of a proton pump/sodium channel coupled model without the structural basis for these moieties in the genome of fish. It is possible that attempts to clone ENaC may be successful in other species, but there is no molecular support for its occurrence in any fish examined to date.

Cloning studies offer important supporting evidence for the presence of ion transporters in osmoregulating epithelia. As such they provide the basis for investigations exploring both the cellular localization and physiological roles of the cloned gene products. Integration of cloning studies with other molecular and cellular methods is, therefore, likely to significantly enhance our existing knowledge of fish osmoregulation.

Corticosteroid Hormone Receptors—Molecular Insight into Endocrine Regulation

The hormonal control of salt and water balance in fish is complex. A wide variety of endocrine, paracrine and autocrine factors have been identified over recent years which can act as effector molecules on various aspects of osmotic regulation. Using homology-based cloning, many hormones, their synthesizing enzymes, and their receptors have now been sequenced in fish, and homologous assays have been established to accurately determine the circulating levels of endocrine factors in fish blood and tissues. These molecular investigations have often yielded new information regarding the ancestral role of these hormones, and also provided insight into aspects of ion regulation in other vertebrate systems. It is not the scope of the current chapter to summarize these findings as this has been the focus of a number of recent reviews (e.g., Mommsen *et al.*, 1999; McCormick, 2001; Sakamoto *et al.*, 2001; Evans, 2002; Manzon, 2002; Takei and Hirose, 2002). This section will, instead, concentrate on the intracellular steroid receptors of the corticoid hormones. In this area, recent advances in molecular techniques have produced surprising findings that highlight the complexity of osmoregulation in teleost fish.

In mammals, the two major circulating corticosteroids are cortisol and aldosterone. Cortisol is primarily associated with energy metabolism, whereas aldosterone is mainly involved with the modulation of ion metabolism. These effects are mediated by two hormone receptors, the glucocorticoid receptor (GR) that facilitates cortisol action, and the mineralocorticoid receptor (MR) that promotes aldosterone activity. Interestingly, the MR has a similar affinity for both cortisol and aldosterone (Arriza *et al.*, 1987). In fact, given the relative circulating levels of these hormones, the MR should be bound almost exclusively by cortisol (Arriza *et al.*, 1987). Specificity of aldosterone response, however, is governed by the presence of 11β-hydroxysteroid dehydrogenase type 2 (11HSD2), an enzyme that degrades cortisol in aldosterone-responsive tissues (Funder *et al.*, 1988).

Biochemical and physiological investigations demonstrated what appeared to be a less sophisticated mechanism of corticoid action in fish. It is generally assumed that there is no circulating aldosterone in fish (e.g., Balm *et al.*, 1989; Sangalang and Uthe, 1994). Instead, it appears that both glucocorticoid and mineralocorticoid functions in fish are achieved by cortisol, albeit in conjunction with other endocrine factors (Evans, 2002).

Steroid-binding assays have demonstrated the presence of GR in fish tissues (Sandor *et al.*, 1984), but molecular evidence was first provided by Ducouret and colleagues (1995), who cloned the first piscine GR from rainbow trout. Based on a full-length cloned GR sequence, *in vitro* transcription and translation techniques were used to generate the GR. Results from steroid-binding assays demonstrated a considerably higher binding affinity for cortisol than aldosterone, functionally characterizing the receptor as a GR. Furthermore, this receptor was transcriptionally active in a transfection assay, promoting the expression of a reporter gene in the presence of a glucocorticoid stimulus. Northern analysis was then performed that localized the receptor to osmoregulatory tissues. Subsequently, GR have been fully or partially cloned in other fish species (Tagawa *et al.*, 1997; Tokuda *et al.*, 1999; Greenwood *et al.*, 2003; Terova *et al.*, 2005).

Following the initial report, a second group discerned a related GR in rainbow trout that was a splice variant of the first (rtGR1b; Takeo *et al.*, 1996). The most significant difference between the initially cloned GR (rtGR1a) and mammalian GR forms was the presence of an additional 9 amino acid sequence located between the highly conserved zinc fingers constituting the DNA-binding domain (Ducouret *et al.*, 1995; Lethimonier *et al.*, 2002). rtGR1b lacked this insert, and thus resembled mammalian GR. Gene fusion proteins with the native form (rtGR1a) or with the 9 amino acids deleted (i.e., similar to rtGR1b) were constructed. A higher constitutive (hormone-independent) expression of the reporter gene driven by the native GR form was noted. This was attributed to a higher binding affinity, relative to the insert-deleted GR (Lethimonier *et al.*, 2002). This difference in activity was noted only when the fusion proteins were controlled by a single copy of the glucocorticoid responsive element (GRE) in the promoter region. When constitutive expression was driven by multiple GREs, there were no differences between the two forms (Lethimonier *et al.*, 2002). There were also no differences between the two forms in the presence of hormonal stimulus. This is in contrast to the findings of Greenwood *et al.* (2003). These authors reported enhanced activity of the non-insert form of cichlid GR (i.e., equivalent to rtGR1b). These differences may be species-dependent. It has been suggested that this splice variation could be one means of differentially regulating cortisol function. If GR splice variants activate genes controlled by promoters with a varying number of GREs in a differential manner, then this could explain how cortisol specificity is maintained in light of its diverse, pluripotent actions (Greenwood *et al.*, 2003). These results suggest that molecular structure is likely to play a significant role in the regulation of cellular processes under the control of cortisol. This finding may also help explain why there is still controversy regarding the roles of cortisol in different fish species (Mommsen *et al.*, 1999).

Given teleost genome duplication, it was of little surprise that a second GR was cloned from rainbow trout. Using similar methods to those described above, Bury and colleagues (2003) were able to isolate and functionally describe rtGR2. This receptor was found to be stimulated by lower levels of cortisol than rtGR1 and, like the first receptor, no biologically relevant activity for aldosterone was discerned. The study of Greenwood and colleagues (2003) identified a homologue to rtGR2 in cichlid fish. The rtGR2 contains a 4 amino acid insert in the DNA binding domain, while the cichlid equivalent resembled the mammalian form (no insert). The different cichlid GRs (the two GR1 splice variants, and the GR2) were characterized by diverse tissue distributions and activities in the presence of cortisol (Greenwood *et al.*, 2003).

The one hormone-one receptor model of cortisol action was further revised by the cloning of two receptors from rainbow trout with homology to mammalian mineralocorticoid receptors (Colombe *et al.*, 2001; Sturm *et al.*, 2005). These receptors, termed rtMRa and rtMRb, are likely to have resulted from gene duplication (Sturm *et al.*, 2005). A cichlid MR has also been cloned (Greenwood *et al.*, 2003). These findings are particularly curious given the lack of circulating aldosterone in fish. Using steroid-binding assays, the cichlid MR was found to have a high affinity for both cortisol and aldosterone (Greenwood *et al.*, 2003). In fact, this receptor had an affinity for cortisol that was 100-fold greater than the GRs, and was 2.5 fold higher that the MR affinity for aldosterone. This is a scenario similar to that found in mammals (Arriza *et al.*, 1987). In rainbow trout, aldosterone was favored over cortisol as a ligand (Sturm *et al.*, 2005), again suggesting species-specificity.

These studies raise significant questions as to the various roles and regulation of these receptors in fish (see Gilmour, 2005). They also generate renewed interest into the possible existence of piscine aldosterone. It is, however, important to note that piscine MRs have yet to be localized to tissues directly responsible for osmoregulation. Instead, they appear to be primarily localized to the brain (Greenwood *et al.*, 2003; Sturm *et al.*, 2005). It is possible that aldosterone may be a local effector with roles in mediating osmoregulation via modification of central control. A localized distribution, coupled with the high affinity of the MR for aldosterone may mean that effective levels of aldosterone elude detection via common assays. Although bioinformatic searches have failed to detect an aldosterone synthase in *Fugu* (Baker, 2003), there appears to be sufficient evidence to reinvestigate the presence of aldosterone in piscine tissues. It is also possible that the MR is a ligand for the corticosteroid precursor 11-deoxycorticosterone, as this molecule also binds rtMR with high affinity (Sturm *et al.*, 2005). It remains to be determined if this is a physiological ligand with biologically relevant effects.

The ability of a non-cortisol ligand to mediate its actions via an MR will likely require a mechanism ensuring the specificity of that ligand. As described above, in mammalian systems, enzymatic control ensures the specificity of aldosterone action (Funder *et al.*, 1988). Recent reports have demonstrated the existence of this enzyme (11HSD2) in fish (Jiang *et al.*, 2003; Kusakabe *et al.*, 2003). In testes 11HSD2 is responsible for both the degradation of cortisol and the formation of 11 keto-testosterone, an important piscine androgen (Kusakabe *et al.*, 2003). 11HSD2 is expressed in osmoregulatory tissues (Kusakabe *et al.*, 2003). The evidence for a role of this enzyme in regulation of reproduction is strong. A functional role in regulating osmoregulation has not yet been explored.

The pluripotent effects of cortisol are likely to be explained by differential regulation of multiple receptor forms in a species-, tissue- and cell-specific manner. This highlights the immense underlying complexity of what was considered a relatively simple system (see Fig. 7.2). Molecular evidence has begun to unravel the intricate regulation of cortisol functions in fish, albeit in only a few species and tissues. Nevertheless, these pioneering studies should provide the structural basis for the expansion of research into other species.

These recent studies exploring the corticosteroid system in fish have thus raised more questions than they have provided answers. These questions have specific relevance for the functioning of osmoregulation in fish, but also have considerable implications for the evolution of this system. The ancestral roles of steroid hormones, their regulation, and their receptors require re-evaluation.

Transgenics and Antisense Applications—Manipulating Gene Expression

Techniques that are capable of modifying the expression of genes are of considerable value in fish biology. Expression-modifying methods include stable, germline manipulations that are inheritable, such as transgenic and knockout models. Additionally, methods that target a specific gene or subset of genes and that are only effective over limited time spans also exist. As an example of the former category, transgenic growth-enhanced and freeze-resistant aquacultural species offer considerable potential for industry (Zbikowska, 2003), while transgenic and 'knock-out' zebrafish are currently being utilized in developmental studies to great effect (Udvadia and Linney, 2003). While such inheritable, germline-modified models offer insight into physiological processes, they currently have limited applicability and utility for the study of fish osmoregulation. This is due in part to the likely lethality of gene-knockouts of the well-known osmoregulatory mediators such as the ion transporters. Furthermore, because relatively little is known regarding the genomics of other entities active in osmoregulation, such techniques are unlikely to be cost-effective

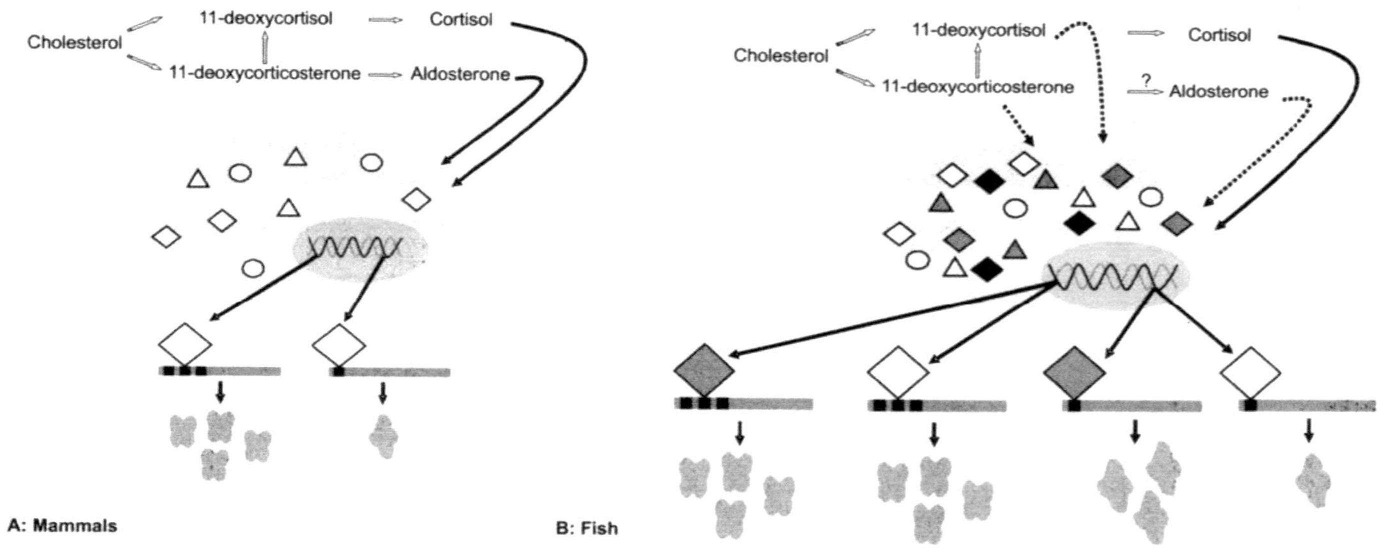

Fig 7.2 Schematic diagram illustrating the complexity of corticosteroid regulation of ion transport in teleost fish as elucidated by cellular and molecular approaches. In mammals (A) a single receptor exists for each of the two circulating hormones. Cortisol will bind to its receptor (*white diamond*), and aldosterone to its receptor (*white triangle*). Specificity is ensured by the presence of the aldosterone-degrading enzyme 11β-hydroxysteroid dehydrogenase type 2 (*white circle*). Cortisol will drive the expression of genes with glucocorticoid responsive elements (GREs). In fish (B), there is no convincing evidence for either aldosterone or aldosterone synthase. Multiple isoforms of both glucocorticoid receptors (*white, grey and black diamonds*), and mineralocorticoid receptors exist (*white and grey triangles*), while 11β-hydroxysteroid dehydrogenase type 2 (*white circle*) has also been described. Given the lack of aldosterone, it has been suggested that 11-deoxycorticosterone and 11-deoxycortisol may be relevant mineralocorticoid ligands. Constitutive expression of GRE-driven genes differs depending on the glucocorticoid receptor that binds, and the number of GREs present. See text for more details.

until suitable targets are identified and sufficient knowledge is acquired regarding their roles in osmoregulatory processes.

A number of techniques exist for the transient modification of the expression of single genes. Techniques such as antisense morpholino oligonucleotides, small interfering RNA (siRNA), and ribozymes are commonly used to block expression of specific genes in mammals (see reviews by Kurreck, 2003; Bantounas *et al.*, 2004; Scanlon, 2004), and are equally applicable in piscine systems. This is a rapidly evolving area of molecular biology, and ongoing refinements of these methods continue to improve the specificity and magnitude of gene silencing. One such modification, with demonstrated application in fish models, is the ability to temporally control gene inactivation. Caged-RNA can be used to selectively silence zebrafish genes. Exposure to ultraviolet light, however, frees these genes for expression, permitting the investigator to switch on transcription (Ando *et al.*, 2001). This technique could have significant application for the examination of salinity acclimation, for example. The capacity to regulate specific genes at specific points of acclimation would provide detailed knowledge regarding their roles in this process.

Tools capable of modifying expression on a transient basis are advantageous over germ-line mutations in that they can be performed in cell culture, and are more cost-effective. Often, however, gene expression can be controlled only for a limited time, and these techniques are often most successful in early life stages, when the study of osmoregulatory function may be of less relevance. These techniques depend on pre-existing genetic knowledge, and as such, they are currently favored for use in fish for which adequate genetic information is available.

Antisense manipulations have been applied to examine specific cellular functions of developmental genes in the rainbow trout (Boonanuntanasarn *et al.*, 2002, 2005), zebrafish (Chen and Ekker, 2005), and medaka (Yamamoto and Suzuki, 2005). The application of such methods may extend to the control of aquacultural viruses (Alonso *et al.*, 2005). To date, however, such technologies have not been specifically applied to examine fish osmoregulation, although as genetic knowledge regarding the moieties involved expands, so too does the potential for the use of such gene expression modifications.

'-omics'—Monitoring Global Changes in Cellular Expression

Global analysis of cellular expression is a powerful molecular tool. The technological advances in microarray development and analysis of cellular protein and metabolite profiles have facilitated a recent explosion in the use of '-omic' approaches to biological problems. Examination of the simultaneous expression of thousands of genes, proteins or metabolites offers obvious benefits over more conventional analyses. We are only beginning to understand the complicated cellular cascades that enable fish and other aquatic animals to cope with life in water. Until now, the cellular and molecular intricacies of osmoregulation have been revealed by studies concentrating on the measurement of individual enzymes, signalling molecules, receptors, or a small assemblage of these important entities (Deane and Woo, 2004). By providing a snapshot of the cellular pathways that may be activated or deactivated under a given condition, the use of global expression techniques permits the complex intricacies of osmoregulation to be elucidated in a holistic manner, potentially defining novel homeostatic mechanisms. As outlined below, these techniques have already facilitated several advances in the field of piscine salt and water balance.

Transcriptomics

The use of physiological model organisms in transcriptomic studies is restricted by the lack of bioinformatic information for these species. Thus, while changes in expression profiles of proteins and genes can be monitored, the identification of these entities is often limiting. There are, however, transcriptomic approaches which remain amenable. Recent evidence from ecotoxicological studies in fish employing a model-hopping approach has provided insight into changes in cellular physiology (Hogstrand *et al.*, 2002). In these preparations, cDNA derived from the species of interest—in this case rainbow trout—was arrayed against a cDNA array from a better-characterized species, *Fugu*. These approaches are most robust when species are closely related, but can withstand a certain degree of phylogenetic distance (Renn *et al.*, 2004).

Differential display polymerase chain reaction (PCR) is another technique that avoids some of the pitfalls associated with a poorly genetically characterized species. This technique involves the arbitrary priming of mRNA from both control and treatment groups, with the subsequent PCR amplification highlighting the presence of unique or differentially-regulated transcripts (McClelland *et al.*, 1995). This approach has been the subject of several investigations on fish osmoregulation. Screening of gill transcripts of eels acclimated to differing salinities highlighted the presence of an expressed sequence, that on full cDNA cloning and expression in a cell line, was identified as an inwardly rectifying potassium channel (Suzuki *et al.*, 1999). Subsequent immunohistochemical localization to osmoregulatory tissues indicated that this channel plays a role in seawater acclimation in the eel. Further studies in a similar vein have also elucidated the importance of cellular structure changes in euryhalinity, by demonstrating the altered expression of genes associated with the actin scaffold of chloride cells (Suzuki *et al.*, 1999; Sakamoto *et al.*, 2002). The discovery of these actin-associated proteins has since led to significant advances in the cloning and understanding of factors involved with cell morphology changes during osmotic challenges (Mistry *et al.*, 2004). These are amongst a number of transcripts described from such differential display methods that exhibit differential expression based on salinity (Suzuki *et al.*, 1999; Sakamoto *et al.*, 2000, 2002; Takeuchi *et al.*, 2000). The role of most of these proteins is not yet well understood.

Another means of probing global expression differences in non-model species is suppressive subtractive hybridization (Diatchenko *et al.*, 1996). This technique involves the annealing of cDNA derived from control transcripts to that derived from experimental transcripts. Transcripts that are shared between control and experimental treatments are not amplified by subsequent PCR amplification. This method has been applied to examine factors responsible for the acclimation of freshwater tilapia to seawater (Fiol and Kültz, 2005). Two transcription factors were identified that were upregulated within hours of transfer. Follow-up investigations showed protein levels of these factors returned to basal levels within 24 hours and that induction was saline-dependent (Fiol and Kültz, 2005). These transcription factors are likely to be key regulatory elements in seawater acclimation.

Perhaps the technique that offers the greatest scope for the discovery of transcripts involved in osmoregulation is the homologous microarray. Microarray studies investigating aspects of osmoregulation have been conducted in plants (*Arabidopsis*, Satoh *et al.*, 2002) and in the nematode (Lamitina and Strange, 2003), and have yielded novel information regarding osmoregulation in these species. For example, the former study discovered the presence of a proline responsive element, by examination of sequence information derived from proline-inducible transcripts. The potential utility of such studies suggests that they can make a significant contribution to investigations of salt and water balance in fish.

The bioinformatics data that facilitates microarray studies is still generally lacking. However, this information is growing rapidly, and has advanced to the stage where tools are becoming available for the investigation of genomics in more physiologically interesting species. For example, there are now commercially available microarray chips for Atlantic salmon (Rise *et al.*, 2004). This technology, coupled with the expansion of bioinformatic data that accompanies such advances, promises significant potential for the elucidation of osmoregulatory processes in this, and closely related species.

Methods for exploring changes in gene transcription are all applicable for examination of aspects of osmotic regulation in fish. In fact, many have already yielded valuable new information regarding the cellular changes associated with altered environmental salinity.

Proteomics

Proteomic approaches are perhaps more amenable than transcriptomics for examining aspects of salt and water balance in fish. This partly reflects the fact that protein expression levels are a more accurate portrayal of cellular physiology than are transcript levels. In mammalian liver, for example, the correlation between mRNA and the corresponding protein levels was less than 0.5 (Anderson and Seilhamer, 1997). In rainbow trout hepatocytes, cortisol causes glucocorticoid receptor mRNA levels to increase, but protein expression levels of glucocorticoid receptor actually decrease (Sathiyaa and Vijayan, 2003). Similarly, increases in expression levels of the killifish chloride channel CFTR on seawater exposure are not matched by increases in CFTR protein (Scott *et al.*, 2004). Consequently, monitoring changes in protein profile may provide greater functional insight than genomic approaches.

The technology available for conducting proteomic experiments is appropriate for research in non-model organisms. Unlike the most robust genomic methods that require some degree of *a priori* genetic knowledge, two-dimensional polyacrylamide gel electrophoresis (2D-PAGE) or surface enhanced laser desorption/ionization (SELDI) can be performed without extensive knowledge of the fish proteome. 2D-PAGE is the most common proteomic approach, and separates proteins in a gel on the basis of size and isoelectric point. It is the more standard application, and offers greater scope for determining the identity of unknown proteins. SELDI is a more recent technological advance, employing time of flight mass spectrometry with specific assay chip matrices to derive information regarding a proteins size and some information regarding its physicochemical properties (Merchant and Weinberger, 2000). It offers high sensitivity, is perhaps more user-friendly, and has the added advantage that membrane based proteins are more easily analyzed, thus offering a greater potential for studies of transport proteins (Jenkins and Pennington, 2001).

The biggest drawback to using proteomic approaches is that there is even less bioinformatic data available regarding fish proteins than fish genes. In the field of fish ecotoxicology, however, it is often enough to simply state the physical characteristics of unique proteins in a given exposure condition and therefore proteomic studies are proving of value for the identification of biomarkers in this field (Hogstrand *et al.*, 2002). Gel-based proteomic approaches do, however, permit the investigator to pick proteins of interest and have these sequenced, yielding additional information regarding the possible identity of the protein. A recent 2D-PAGE study in rainbow trout was able to identify nineteen of 23 proteins excized from the gel, exhibiting the utility of this technique, even in relatively poorly characterized species (Martin *et al.*, 2003). Advanced SELDI applications may also include fragmentation analysis, permitting additional information to be derived from proteins of interest. Both these technologies hold considerable promise for studies of piscine osmoregulation in the future.

There are, however, examples of how these technologies are being currently applied to investigate cellular changes that result from osmotic changes in the environment. Using 2D-PAGE, Kültz (1996) identified 21 proteins that were induced in isolated gill cells of *Gillichthys mirabilis* upon hyperosmotic shock and a further 14 that were induced upon hypoosmotic shock. Nine of these proteins were induced in both conditions, but only three hyperosmotically-induced and five hypoosmotically-induced proteins were also induced following thermal shock. The changes observed were, thus, specific to the type of the challenge, suggesting branchial cells can detect not only changes in the environment but the nature of these changes, and can respond in an appropriate manner, independent of higher control.

In a similar study, 2D-PAGE analysis identified nine proteins that were induced by hypoosmotic shock in gill cells of *Gillichthys*, but none of these appeared to share significant homology to any osmotically differentiated proteins determined in the first study (Kültz and Somero, 1996). This indicates that important differences may exist between preparations, even those from the same species. This may result from subtle changes in experimental methodology, or the physiological state of the animals themselves. Consequently, the high sensitivity of such techniques may be both a blessing and a curse, permitting the detection of relatively minor changes in osmoregulatory entities, but also potentially delineating changes that are the result of inherent individual variation. Recent studies have, however, suggested that in general '-omic' techniques are robust and offer high reproducibility (Larkin *et al.*, 2005). Nevertheless, careful consideration of experimental parameters such as sample size is critical (Wei *et al.*, 2004).

Proteomic approaches can also be adapted for a more targeted analysis. This was highlighted by a recent study investigating the killifish cystic fibrosis transmembrane regulator (CFTR) chloride channel (Bridges *et al.*, 2003). Protein constructs of the C-terminal region of the CFTR protein were conjugated in a gel column. A lysate of killifish operculum was passed through the column, and following elution, proteins bound to the conjugated construct were separated and analyzed via MALDI-TOF. This elegant approach permitted identification of important accessory proteins involved in CFTR regulation, and thus discerned the role of actin organization in this process.

Metabolomics

Metabolomics is, in essence, the assessment of the metabolic substrates and metabolites that are responsible for, or result from, normal cellular function (Goodacre *et al.*, 2004). This approach is fast finding favor as a complementary technique to transcriptomics and proteomics, as it captures cellular expression at its functional endpoint. Metabolomics is advantageous from a comparative biology perspective in the sense that the chemicals analyzed in the metabolome are conserved across all phyla, meaning that species-specific genetic knowledge is not required for either conducting or analysing experiments. This implies that, in the short term, metabolomics has greater application to non-model species, especially those for which very little genetic knowledge exists, yet are relevant for the examination of osmoregulation. Furthermore, while the platform for conducting metabolomic studies is expensive, the 'per sample' running costs are likely to be less than those for the other global expression analyses (Goodacre *et al.*, 2004).

A number of different metabolomic platforms exist. Those most commonly used are nuclear magnetic resonance spectroscopy or mass spectroscopy, often in combination with a chromatographic technique. Each platform offers different advantages depending on the number and nature of metabolites of interest and the required sensitivity (Weckwerth and Morgenthal, 2005). The ability to detect a given compound depends, in part, on its standard having being characterized under analytical conditions equivalent to those under use.

No studies currently exist that have examined the effect of osmoregulation on the metabolome. The technique has, however, been applied with success in fish. Viant (2003) demonstrated specific changes in cellular composition during embryogenesis in the Japanese medaka. Metabolomic approaches will be best suited to specific subsets of fish osmoregulation studies. In many cases, the entities of greatest interest in osmoregulation are the ions, the transporters that facilitate their movement across epithelia, and the regulatory molecules that control these processes. Metabolomic approaches are not appropriate for the examination of these cellular constituents. However, investigation of the effect of osmoregulation on energy metabolism (Sangiao-Alvarellos *et al.*, 2005), for example, would benefit significantly from a metabolomic approach.

Perspectives

The past ten years have witnessed considerable advances in our understanding of osmoregulation in fish. This has been driven to a large extent by the development of refined cellular models, and the indoctrination of molecular techniques. The utilization of these tools has challenged traditional osmoregulatory dogmas, and has also elucidated the significant complexity of ionic and osmotic homeostasis. The creation of novel information has also provided new challenges for researchers in this discipline.

The nature of many of the biological techniques described in this review is the vast quantities of information they can provide. In particular, large-scale expression studies and genomic sequencing projects generate extensive datasets. Such investigations, however, provide only limited information of gene function. A lesson from the studies already conducted is that ascribed roles and regulation of osmoregulatory entities in mammalian systems may not necessarily hold in fish. Consequently, there is significant scope for the elucidation of functional actions for many of the structural elements with putative roles in osmoregulation. Cellular models will provide the impetus for this characterization.

By breaking ionic and osmotic homeostasis down to its basic constituents and observing them in relatively simplistic model systems, significant insight into regulatory and functional mechanisms can be discerned. Effects at this level must, however, be integrated with studies that examine osmoregulatory processes in intact animals, where neural and endocrine feedback, and interactions between osmoregulatory epithelia will alter these processes. Extracellular influences will also be important. Cardiovascular modification of osmoregulatory surface perfusion and the roles of mucus as an unstirred water layer, are examples of tissue-level processes that may have significant influences on ionic status in fish. These may not be accounted for in cellular and molecular studies. Equally, consideration should also exist for the influence of organism-level factors that may impact upon the fish, its cells and molecules. Behavioral osmoregulation (Lyse *et al.*, 1998), the presence of ion-disrupting toxicants (Wood, 2001), and switching of energy resources to reproduction (Le Francois and Blier, 2003), for example, will all have significant influences.

The cellular tools described herein are well suited for mechanistic approaches to fish osmoregulation. In association with the structural information generated by the advances in molecular methods and physiological data, models of osmoregulatory function are becoming increasingly refined. The vast diversity of fish, in terms of their genetic composition, life histories, and habitats is poorly represented by the relatively few species examined to date. Knowledge of the key osmoregulatory structures and mechanisms permits testing of osmoregulatory theories across a far greater range of species, without requiring access to the whole animal. Molecular and cellular approaches permit an increasingly developed understanding of the osmoregulatory minutiae. At the same time, they also facilitate the expansion of investigations to organisms previously considered impractical for research of this nature. These advances have resulted in significant improvements in our knowledge of ionic and osmotic homeostasis in fish, and will continue to stimulate and provoke future insights.

Acknowledgements

Drs. Fernando Galvez, Chris Wood and Pål Olsvik are thanked for their various important contributions to this work.

References

Alderice, D.F. 1988. Osmotic and ionic regulation in teleost eggs and larvae. In: *Fish Physiology*, W.S. Hoar and D.J. Randall (eds.). Academic Press, San Diego, Vol. 11A, pp. 163–251.

Ali, A. and D.L. Reynolds. 1996. Primary cell culture of turkey intestinal epithelial cells. *Avian Diseases* 40: 103–108.

Almansa, E., J.J. Sanchez, S. Cozzi, M. Casariego, J. Cejas and M. Diaz. 2001. Segmental heterogeneity in the biochemical properties of the Na^+-K^+-ATPase along the intestine of the gilthead seabream (*Sparus aurata* L.). *Journal of Comparative Physiology* B 171: 557–567.

Alonso, M., D.A. Stein, E. Thomann, H.M. Moulton, J.C. Leong, P. Iversen and D.V. Mourich. 2005. Inhibition of infectious haematopoietic necrosis virus in cell cultures with peptide-conjugated morpholino oligomers. *Journal of Fish Diseases* 28: 399–410.

Anderson, L. and J. Seilhamer. 1997. A comparison of selected mRNA and protein abundances in human liver. *Electrophoresis* 18: 533–537.

Ando, H., T. Furuta, R.Y. Tsien and H. Okamoto. 2001. Photo-mediated gene activation using caged RNA/DNA in zebrafish embryos. *Nature Genetics* 28: 317–325.

Aparicio, S., J. Chapman, E. Stupka, N. Putnam, J. Chia, P. Dehal, A. Christoffels, S. Rash, S. Hoon, A. Smit, M.D.S. Gelpke, J. Roach, T. Oh, I.Y. Ho, M. Wong, C. Detter, F. Verhoef, P. Predki, A. Tay, S. Lucas, P. Richardson, S.F. Smith, M.S. Clark, Y.J.K. Edwards, N. Doggett, A. Zharkikh, S.V. Tavtigian, D. Pruss, M. Barnstead, C. Evans, H. Baden, J. Powell, G. Glusman, L. Rowen, L. Hood, Y.H. Tan, G. Elgar, T. Hawkins, B. Venkatesh, D. Rokhsar and S. Brenner. 2002. Whole-genome shotgun assembly and analysis of the genome of *Fugu rubripes*. *Science* 297: 1301–1310.

Arriza, J.L., C. Weinberger, G. Cerelli, T.M. Glaser, B.I. Handelin, D.E. Housman and R.M. Evans. 1987. Cloning of human mineralocorticoid receptor complementary DNA: structural and functional kinship with the glucocorticoid receptor. *Science* 237: 268–275.

Avella, M. and M. Bornancin. 1989. A new analysis of ammonia and sodium transport through the gills of the freshwater rainbow trout (*Salmo gairdneri*). *Journal of Experimental Biology* 142: 155–175.

Avella, M. and J. Ehrenfeld. 1997. Fish gill respiratory cells in culture: a new model for Cl^- secreting epithelia. *Journal of Membrane Biology* 156: 87–97.

Avella, M., J. Berhaut and P. Payan. 1994. Primary culture of gill epithelial cells from the sea bass *Dicentrarchus labrax in vitro*. *Cellular and Developmental Biology* A30: 41–49.

Avella, M., P. Pärt and J. Ehrenfeld. 1999. Regulation of Cl^- secretion in seawater fish (*Dicentrarchus labrax*) gill respiratory cells in primary culture. *Journal of Physiology* 516: 353–363.

Ayson, F.G., T. Kaneko, S. Hasegawa and T. Hirano. 1995. Cortisol stimulates the size and number of mitochondrion-rich cells in the yolk-sac membrane of embryos and larvae of tilapia, *Oreochromis mossambicus, in vitro* and *in vivo*. *Journal of Experimental Zoology* 272: 419–425.

Baker, M.E. 2003. Evolution of glucocorticoid and mineralocorticoid responses: Go fish. *Endocrinology* 144: 4223–4225.

Balm, P.H.M., J.D.G. Lambert and S.E. Wendelaar Bonga. 1989. Corticosteroid biosynthesis in the inter-renal cells of teleost fish, *Oreochromis mossambicus*. *General and Comparative Endocrinology* 76: 53–62.

Bantounas, I., L.A. Phylactou and J.B. Uney. 2004. RNA interference and the use of small interfering RNA to study gene function in mammalian systems. *Journal of Molecular Endocrinology* 33: 545–557.

Benyajati, S. and J.L. Renfro. 2000. Taurine secretion in primary monolayer cultures of flounder renal epithelium: Stimulation by low osmolality. *American Journal of Physiology* 279: R704–R712.

Berteloot, A and G. Semenza. 1990. Advantages and limitations of vesicles for the characterization and the kinetic analysis of transport systems. *Methods in Enzymology* 192: 409–437.

Beyenbach, K.W. and W.H. Dantzler. 1990. Comparative kidney tubule sources, isolation, perfusion, and function. *Methods in Enzymology* 191: 167–226.

Bijvelds, M.J.C., Z.I. Kolar, S.E. Wendelaar Bonga and G. Flik. 1996. Magnesium transport across the basolateral plasma membrane of the fish enterocyte. *Journal of Membrane Biology* 154: 217–225.

Blanco, G. 2005. Na^+,K^+-ATPase subunit heterogeneity as a mechanism for tissue-specific ion regulation. *Seminars in Nephrology* 25: 292–303.

Boesch, S.T., B. Eller and B. Pelster. 2003. Expression of two isoforms of the vacuolar-type ATPase subunit B in the zebrafish *Danio rerio*. *Journal of Experimental Biology* 206: 1907–1915.

Bonetta, L. 2006. Genome sequencing in the fast lane. *Nature Methods* 3: 141–147.

Boonanuntanasarn, S., G. Yoshizaki and T. Takeuchi. 2002. Specific gene silencing using small interfering RNAs in fish embryos. *Biochemical and Biophysical Research Communications* 310: 1089–1095.

Boonanuntanasarn, S., T. Takeuchi and G. Yoshizaki. 2005. High-efficiency gene knockdown using chimeric ribozymes in fish embryos. *Biochemical and Biophysical Research Communications* 336: 438–443.

Brenner, S., G. Elgar, R. Sandford, A. Macrae, B. Venkatesh and S. Aparicio. 1993. Characterization of the pufferfish (*Fugu*) genome as a compact model vertebrate genome. *Nature (Lond.)* 366: 265–268.

Bridges, J.T., C.R. Stanton and J.E. Mickle. 2003. CFTR accessory protein identification via comparative proteomics. *Mount Desert Island Biological Laboratory Bulletin* 42: 91–92.

Burke, J. and R.D. Handy. 2005. Sodium-sensitive and -insensitive copper accumulation by isolated intestinal cells of rainbow trout, *Oncorhynchus mykiss*. *Journal of Experimental Biology* 208: 391–407.

Bury, N.R., J. Li, R.A.C. Lock and S.E. Wendelaar Bonga. 1998. Cortisol protects against copper-induced necrosis and promotes apoptosis in fish gill chloride cells *in vitro*. *Aquatic Toxicology* 40: 193–202.

Bury, N.R., A. Sturm, P. Le Rouzic, C. Lethimonier, B. Ducouret, Y. Guiguen, M. Robinson-Rechavi, V. Laudet, M.E. Rafestin-Oblin and P. Prunet. 2003. Evidence for two distinct functional glucocorticoid receptors in teleost fish. *Journal of Molecular Endocrinology* 31: 141–156.

Butler, R. and B.F. Nowak. 2004. *In vitro* interactions between *Neoparamoeba* sp. and Atlantic salmon epithelial cells. *Journal of Fish Diseases* 27: 343–349.

Castaño, A., N. Bols, T. Braunbeck, P. Dierickx, M. Halder, B. Isomaa, K. Kawahara, L.E.J. Lee, C. Mothersill, P. Pärt, G. Repetto, J.R. Sintes, H. Rufli, R. Smith, C.M. Wood and H. Segner. 2003. The use of fish cells in ecotoxicology. *Alternatives to Laboratory Testing in Animals* 31: 317–353.

Chen, E. and S.C. Ekker. 2005. Zebrafish as a genomic research model. *Current Pharmaceutical Biotechnology* 5: 409–413.

Chen, L.B., A.L. DeVries and C.H.C. Cheng. 1997. Evolution of antifreeze glycoprotein gene from a trypsinogen gene in Antarctic notothenioid fish. *Proceedings of the National Academy of Sciences of the United States of America* 94: 3811–3816.

Choe, K.P., A.I. Morrison-Shetlar, B.P. Wall and J.B. Claiborne. 2002. Immunological detection of Na^+/H^+ exchangers in the gills of a hagfish *Myxine glutinosa*, an elasmobranch, *Raja erinacea*, and a teleost, *Fundulus heteroclitus*. *Comparative Biochemistry and Physiology* A131: 375–385.

Choe, K.P., A. Kato, S. Hirose, C. Plata, A. Sindić, M.F. Romero, J.B. Claiborne and D.H. Evans. 2005. NHE3 in an ancestral vertebrate: Primary sequence, distribution, localization, and function in gills. *American Journal of Physiology* 229: R1520–R1534.

Christoffels, A., E.G.L. Koh, J.-M. Chia, S. Brenner, S. Aparicio and B. Venkatesh. 2004. *Fugu* genome analysis provides evidence for a whole-genome duplication early during the evolution of ray-finned fishes. *Molecular Biology and Evolution* 21: 1146–1151.

Colombe, L., A. Fostier, N. Bury, F. Pakdal and Y. Guiguen. 2000. A mineralocorticoid-like receptor in the rainbow trout, *Oncorhynchus mykiss*: Cloning and characterization of its steroid binding domain. *Steroids* 65: 319–328.

Cutler, C.P. and G. Cramb. 2001. Molecular physiology of osmoregulation in eels and other teleosts: The role of transporter isoforms and gene duplication. *Comparative Biochemistry and Physiology* A130: 551–564.

De Giorgi, A., L. Carnimeo and A. Corcelli. 1992. Mechanism of Cl^- transport in eel intestinal brush-border membrane vesicles. *European Journal of Physiology* 420: 551–558.

De Lean, A. 1986. Amiloride potentiates atrial natriuretic factor inhibitory action by increasing receptor binding in bovine adrenal zona glomerulosa. *Life Science* 39: 1109–1116.

Deane, E.E. and N.Y.S. Woo. 2004. Differential gene expression associated with euryhalinity in sea bream (*Sparus sarba*). *American Journal of Physiology* 287: R1054–R1063.

Deane, E.E. and N.Y.S. Woo. 2005. Cloning and characterization of sea bream Na^+-K^+-ATPase α and β subunit genes: *In vitro* effects of hormones on transcriptional and translational expression. *Biochemical and Biophysical Research Communications* 331: 1229–1238.

Diatchenko, L., Y.-F. C. Lau, A.P. Campbell, A. Chenchik, F. Moqadam, B. Huang, S. Lukyanov, K. Lukyanov, N. Gurskaya, E.D. Sverdlov and P.D. Siebert. 1996. Suppression subtractive hybridization: A method for generating differentially regulated or tissue-specific cDNA probes and libraries. *Proceedings of the National Academy of Sciences of the United States of America* 93: 6025–6030.

Dickman, K.G. and J.L. Renfro. 1986. Primary culture of flounder renal tubule cells: transepithelial transport. *American Journal of Physiology* 251: F424–F432.

Do, C.W. and M.M. Civan. 2004. Basis of chloride transport in ciliary epithelium. *Journal of Membrane Biology* 200: 1–13.

Dopido, R., C. Rodriguez, T. Gomez, N.G. Acosta and M. Diaz. 2004. Isolation and characterization of enterocytes along the intestinal tract of the gilthead seabream (*Sparus aurata* L.). *Comparative Physiology and Biochemistry* A139: 21–31.

Dubinsky, W.P., O. Mayorga-Wark and S.G. Schultz. 2000. Potassium channels in basolateral membrane vesicles from *Necturus* enterocytes: Stretch and ATP sensitivity. *American Journal of Physiology* 279: C634–C638.

Ducouret, B., M. Tujague, J. Ashraf, N. Mouchel, N. Serval, Y. Valotaire and E.B. Thompson. 1995. Cloning of a teleost fish glucocorticoid receptor shows that it contains a deoxyribonucleic acid-binding domain different to that of mammals. *Endocrinology* 136: 3774–3783.

Dudas, P.L. and J.L. Renfro. 2001. Assessment of tissue-level kidney functions with primary cultures. *Comparative Biochemistry and Physiology* A128: 199–206.

Edwards, S.L., J.A. Donald, T. Toop, M. Donowitz and C.-M. Tse. 2002. Immunolocalization of sodium/proton exchanger-like proteins in the gills of elasmobranchs. *Comparative Biochemistry and Physiology* A131: 257–265.

Evans, D.H. 2002. Cell signaling and ion transport across the fish gill epithelium. *Journal of Experimental Zoology* 293: 336–347.

Evans, D.H., P.M. Piermarini and W.T.W. Potts. 1999. Ionic transport in the fish gill epithelium. *Journal of Experimental Zoology* 283: 641–652.

Evans, D.H., R.E. Rose, J.M. Roeser and J.D. Stidham. 2004. NaCl transport across the opercular epithelium of *Fundulus heteroclitus* is inhibited by an endothelin to NO, superoxide, and prostanoid signaling axis. *American Journal of Physiology* 286: R560–R568.

Ferraris, R.P. and G.A. Ahearn. 1984. Sugar and amino acid transport in fish intestine. *Comparative Biochemistry and Physiology* A77: 397–413.

Fines, G.A., J.S. Ballantyne and P.A. Wright. 2001. Active urea transport and an unusual basolateral membrane composition in the gills of a marine elasmobranch. *American Journal of Physiology* 280: R16–R24.

Fiol, D.F. and D. Kültz. 2005. Rapid hyperosmotic coinduction of two tilapia (*Oreochromis mossambicus*) transcription factors in gill cells. *Proceedings of the National Academy of Sciences of the United States of America* 102: 927–932.

Fletcher, M., S.P. Kelly, P. Pärt, M.J. O'Donnell and C.M. Wood. 2000. Transport properties of cultured branchial epithelia from rainbow trout: A novel preparation with mitochondria-rich cells. *Journal of Experimental Biology* 203: 1523–1537.

Flik, G. and P.M. Verbost. 1993. Calcium transport in fish gills and intestine. *Journal of Experimental Biology* 184: 17–29.

Flik, G., P.H.M. Klaren, T.J.M. Schoenmakers, M.J.C. Bijvelds, P.M. Verbost and S.E. Wendelaar Bonga. 1996. Cellular calcium transport in fish: Unique and universal mechanisms. *Physiological Zoology* 69: 403–417.

Funder, J.W., P.T. Pearce, R. Smith and I. Smith. 1988. Mineralocorticoid action: target tissue specificity is enzyme, not receptor, mediated. *Science* 242: 583–585.

Galvez, F., S.D. Reid, G. Hawkings and G.G. Goss. 2002. Isolation and characterization of mitochondria-rich cell types from the gill of freshwater rainbow trout. *American Journal of Physiology* 282: R658–R668.

George, S., C. Riley, J. McEvoy and J. Wright. 2000. Development of a fish *in vitro* cell culture model to investigate oxidative stress and its modulation by dietary vitamin E. *Marine Environmental Research* 50: 541–544.

Gharbi, K., M.M. Ferguson and R.G. Danzmann. 2005. Characterization of Na, K-ATPase genes in Atlantic salmon (*Salmo salar*) and comparative genomic organization with rainbow trout (*Oncorhynchus mykiss*). *Molecular Genetics and Genomics* 273: 474–483.

Gilmour, K.M. 2005. Mineralocorticoid receptors and hormones: Fishing for answers. *Endocrinology* 146: 44–46.

Ginsberg, S.D. 2005. RNA amplification strategies for small sample populations. *Methods* 37: 229–237.

Glover, C.N., S. Balesaria, G.D. Mayer, E.D. Thompson, P.J. Walsh and C. Hogstrand. 2003a. Intestinal zinc uptake in two marine teleosts, squirrelfish (*Holocentrus adscensionis*) and Gulf toadfish (*Opsanus beta*). *Physiological and Biochemical Zoology* 76: 321–330.

Glover, C.N., N.R. Bury and C. Hogstrand. 2003b. Zinc uptake across the apical membrane of freshwater rainbow trout intestine is mediated by high affinity, low affinity and histidine-facilitated pathways. *Biochimica et Biophysica Acta-Biomembranes* 1614: 211–219.

Goodacre, R., S. Vaidyanathan, W.B. Dunn, G.G. Harrigan and D.B. Kell. 2004. Metabolomics by numbers: Acquiring and understanding global metabolite data. *Trends in Biotechnology* 22: 245–252.

Goode, D.K., P. Snell, S.F. Smith, J.E. Cooke and G. Elgar. 2005. Highly conserved regulatory elements around the SHH gene may contribute to the maintenance of conserved synteny across human chromosome 7q36.3. *Genomics* 86: 172–181.

Goss, G.G., S. Adamia and F. Galvez. 2001. Peanut lectin binds to a subpopulation of mitochondria-rich cells in the rainbow trout gill epithelium. *American Journal of Physiology* 281: R1718–R1725.

Greenwood, A.K., P.C. Butler, R.B. White, U. De Marco, D. Pearce and R.D. Fernald. 2003. Multiple corticosteroid receptors in teleost fish: Distinct sequences, expression patterns, and transcriptional activities. *Endocrinology* 144: 4226–4236.

Hansen, H.J.M. and M. Grosell. 2004. Are membrane lipids involved in osmoregulation? Studies *in vivo* on the European eel, *Anguilla anguilla*, after reduced ambient salinity. *Environmental Biology of Fishes* 70: 57–65.

Hansen, H.J.M., M.H. Grosell and P. Rosenkilde. 1999. Gill lipid metabolism and unidirectional Na^+ flux in the European eel (*Anguilla anguilla*) after transfer to dilute media: the formation of wax alcohols as a primary response. *Aquaculture* 177: 277–283.

Harwood, S. and M.M. Yaqoob. 2005. Ouabain-induced cell signaling. *Frontiers in Bioscience* 10: 2011–2017.

Hashimoto, K., Y. Matsuo, Y. Yokohama, H. Toyohara and M. Sakaguchi. 1998. Induction of apoptosis in fish cells by hypertonic stress. *Fisheries Science* 64: 820–825.

Hawkings, G., F. Galvez and G.G. Goss. 2004. Seawater acclimation causes independent alterations in Na^+/K^+- and H^+-ATPase activity in isolated mitochondria-rich cell subtypes of the rainbow trout gill. *Journal of Experimental Biology* 207: 905–912.

Hill, W.G., J.C. Mathai, R.H. Gensure, J.D. Zeidel, G. Apodaca, J.P. Saenz, E. Kinne-Saffran, R. Kinne and M.L. Zeidel. 2004. Permeabilities of teleost and elasmobranch gill apical membranes: evidence that lipid bilayers alone do not account for barrier function. *American Journal of Physiology* 287: C235–C242.

Hirata, T., T. Kaneko, T. Ono, T. Nakazato, N. Furukawa, S. Hasegawa, S. Wakabayashi, M. Shigekawa, M.H. Chang, M.F. Romero and S. Hirose. 2003. Mechanism of acid adaptation in a fish living in a pH 3.5 lake. *American Journal of Physiology* 284: R1199–R1212.

Hiroi, J., H. Miyazaki, F. Katoh, R. Ohtani-Kaneko and T. Kaneko. 2005. Chloride turnover and ion-transporting activities of yolk-sac preparations (yolk balls) separated from Mozambique tilapia embryos and incubated in freshwater and seawater. *Journal of Experimental Biology* 208: 3851–3858.

Hirose, S., T. Kaneko, N. Naito and Y. Takei. 2003. Molecular biology of major components of chloride cells. *Comparative Biochemistry and Physiology* B136: 593–620.

Hogstrand, C., S. Balesaria and C.N. Glover. 2002. Application of genomics and proteomics for study of the integrated response to zinc exposure in a non-model fish species, the rainbow trout. *Comparative Biochemistry and Physiology* B133: 523–535.

Holland, J.W., G.W. Taylor and A.F. Rowley. 1999. The eicosanoid generating capacity of isolated cell populations from the gills of the rainbow trout, *Oncorhynchus mykiss*. *Comparative Biochemistry and Physiology* C122: 297–306.

Holland, R., J.R. Woodgett and D.G. Hardie. 1983. Evidence that amiloride antagonises insulin-stimulated protein phosphorylation by inhibiting protein kinase activity. *FEBS Letters* 154: 269–273.

Hwang, P.P. and R. Hirano. 1985. Effects of environmental salinity on intercellular organization and junctional structure of chloride cells in early teleost development. *Journal of Experimental Zoology* 236: 115–126.

Inoue, K. and Y. Takei. 2002. Diverse adaptability in *Oryzias* species to high environmental salinity. *Zoological Science* 19: 727–734.

Inoue, K. and Y. Takei. 2003. Asian medaka fishes offer new models for studying the mechanisms of seawater adaptation. *Comparative Biochemistry and Physiology* B136: 635–645.

Inoue, K., K. Naruse, S. Yamagami, H. Mitani, N. Suzuki and Y. Takei. 2003. Four functionally distinct C-type natriuretic peptides found in fish reveal evolutionary history of the natriuretic peptide system. *Proceedings of the National Academy of Sciences of the United States of America* 100: 10079–10084.

Jaillon, O., J.M. Aury, F. Brunet, J.L. Petit, N. Stange-Thomann, E. Mauceli, L. Bouneau, C. Fischer, C. Ozouf-Costaz, A. Bernot, S. Nicaud, D. Jaffe, S. Fisher, G. Lutfalla, C. Dossat, B. Segurens, C. Dasilva, M. Salanoubat, M. Levy, N. Boudet, S. Castellano, R. Anthouard, C. Jubin, V. Castelli, M. Katinka, B. Vacherie, C. Biemont, Z. Skalli, L. Cattolico, J. Poulain, V. de Berardinis, C. Cruaud, S. Duprat, P. Brottier, J.P. Coutanceau, J. Gouzy, G. Parra, G. Lardier, C. Chapple, K.J. McKernan, P. McEwan, S. Bosak, M. Kellis, J.N. Volff, R. Guigo, M.C. Zody, J. Mesirov, K. Lindblad-Toh,

B. Birren, C. Nusbaum, D. Kahn, M. Robinson-Rechavi, V. Laudet, V. Schachter, F. Quetier, W. Saurin, C. Scarpelli, P. Wincker, E.S. Lander, J. Weissenbach and H.R. Crollius. 2004. Genome duplication in the teleost fish *Tetraodon nigroviridis* reveals the early vertebrate proto-karyotype. *Nature (London)* 431: 946–957.

Jenkins, R.E. and S.R. Pennington. 2001. Arrays for protein expression profiling: Towards a viable alternative to two-dimensional gel electrophoresis. *Proteomics* 1: 13–29.

Jiang, J.Q., D.S. Wang, B. Senthilkumaran, T. Kobayashi, H.K. Kobayashi, A. Yamaguchi, W. Ge, G. Young and Y. Nagahama. 2003. Isolation, characterization and expression of 11β-hydroxysteroid dehydrogenase type 2 cDNAs from the testes of Japanese eel (*Anguilla japonica*) and Nile tilapia (*Oreochromis niloticus*). *Journal of Molecular Endocrinology* 31: 305–315.

Kajimura, S., T. Hirano, S. Moriyama, O. Vakkuri, J. Leepaluoto and E.G. Grau. 2004. Changes in plasma concentrations of immunoreactive ouabain in the tilapia in response to changing salinity: Is ouabain a hormone in fish? *General and Comparative Endocrinology* 135: 90–99.

Kajimura, S., A.P. Seale, T. Hirano, I.M. Cooke and E.G. Grau. 2005. Physiological concentrations of ouabain rapidly inhibit prolactin release from the tilapia pituitary. *General and Comparative Endocrinology* 143: 240–250.

Kaneko, T., K. Shiraishi, F. Katoh, S. Hasegawa and J. Hiroi. 2002. Chloride cells during early life stages of fish and their functional differentiation. *Fisheries Science* 68: 1–9.

Katoh, F., A. Shimizu, K. Uchida and T. Kaneko. 2000. Shift of chloride cell distribution during early life stages in seawater-adapted killifish, *Fundulus heteroclitus*. *Zoological Science* 17: 11–18.

Katoh, F., S. Hyodo and T. Kaneko. 2003. Vacuolar-type proton pump in the basolateral plasma membrane energises ion uptake in branchial mitochondria-rich cells of killifish *Fundulus heteroclitus*, adapted to a low ion environment. *Journal of Experimental Biology* 206: 793–803.

Kirschner, L.B. 2004. The mechanism of sodium chloride uptake in hyperregulating aquatic animals. *Journal of Experimental Biology* 207: 1439–1452.

Krebs, H.A. 1975. The August Krogh principle: 'For many problems there is an animal on which it can be most conveniently studied.' *Journal of Experimental Zoology* 194: 309–344.

Kurreck, J. 2003. Antisense technologies: improvement through novel chemical modifications. *European Journal of Biochemistry* 270: 1628–1644.

Kültz, D. 1996. Plasticity and stressor specificity of osmotic and heat shock responses of *Gillichthys mirabilis* gill cells. *American Journal of Physiology* 271: C1181–C1193.

Kültz, D. and G.N. Somero. 1996. Differences in protein patterns of gill epithelial cells of the fish *Gillichthys mirabilis* after osmotic and thermal acclimation. *Journal of Comparative Physiology* B 166: 88–100.

Kusakabe, M., I. Nakamura and G. Young. 2003. 11β-hydroxysteroid dehydrogenase complementary deoxyribonucleic acid in rainbow trout: cloning, sites of expression, and seasonal changes in gonads. *Endocrinology* 144: 2534–2545.

Lamche, G., W. Meier, M. Suter and P. Burkhardt-Holm. 1998. Primary culture of dispersed skin epidermal cells of rainbow trout *Oncorhynchus mykiss* Walbaum. *Cellular and Molecular Life Sciences* 54: 1042–1051.

Lamitina, S.T. and K. Strange. 2003. Whole-genome microarray analysis of cellular osmoregulation in the nematode *C. elegans*. *FASEB Journal* 17: A899.

Larkin, J.E., B.C. Frank, H. Gavras, R. Sultana and J. Quackenbush. 2005. Independence and reproducibility across microarray platforms. *Nature Methods* 2: 337–356.

Le Francois, N.R. and P.U. Blier. 2003. Reproductive events and associated reduction in the seawater adaptability of brook charr (*Salvelinus fontinalis*): Evaluation of gill metabolic adjustments. *Aquatic Living Resources* 16: 69–76.

Leguen, I. and P. Prunet. 2004. Effect of hypotonic shock on cultured pavement cells from freshwater or seawater rainbow trout gills. *Comparative Biochemistry and Physiology* A137: 259–269.

Leguen, I., J.P. Cravedi, M. Pisam and P. Prunet. 2001. Biological functions of trout pavement-like gill cells in primary culture on solid support: pH_i regulation, cell volume regulation and xenobiotic biotransformation. *Comparative Biochemistry and Physiology* A128: 207–222.

LeHir, M., B. Kaissling, B.M. Koeppen and J.B. Wade. 1982. Binding of peanut lectin to specific epithelial cell types in kidney. *American Journal of Physiology* 242: C117–C120.

Lethimonier, C., M. Tujague, L. Kern and B. Ducouret. 2002. Peptide insertion in the DNA-binding domain of fish glucocorticoid receptor is encoded by an additional exon and confers particular functional properties. *Molecular and Cellular Endocrinology* 194: 107–116.

Lima, R. and D. Kültz. 2004. Laser scanning cytometry and tissue microarray analysis of salinity effects on killifish chloride cells. *Journal of Experimental Biology* 207: 1729–1739.

Lin, H. and D.J. Randall. 1991. Evidence for the presence of an electrogenic proton pump on the trout gill epithelium. *Journal of Experimental Biology* 161: 119–134.

Lin, H. and D.J. Randall. 1993. H^+-ATPase activity in crude homogenates of fish gill tissue: Inhibitor sensitivity and environmental and hormonal regulation. *Journal of Experimental Biology* 180: 163–174.

Lingwood, D., L.J. Fisher, J.W. Callahan and J.S. Ballantyne. 2004. Sulfatide and Na^+-K^+-ATPase: A salinity-sensitive relationship in the gill basolateral membrane of rainbow trout. *Journal of Membrane Biology* 201: 77–84.

Lingwood, D., G. Harauz and J.S. Ballantyne. 2005. Regulation of fish gill Na^+-K^+-ATPase by selective sulfatide-enriched raft partitioning during seawater adaptation. *Journal of Biological Chemistry* 280: 36545–36550.

Lyse, A.A., Steffansson, S.O. and A. Ferno. 1998. Behaviour and diet of sea trout post-smolts in a Norwegian fjord system. *Journal of Fish Biology* 52: 923–936.

Ma, C. and P. Collodi. 1996. Culture of cells from tissues of adult and larval sea lamprey. *Cytotechnology* 21: 195–203.

Ma, C. and P. Collodi. 1999. Preparation of primary cell cultures from lamprey. *Methods in Cell Science* 21: 39–46.

Manzon, L.A. 2002. The role of prolactin in fish osmoregulation: A review. *General and Comparative Endocrinology* 125: 291–310.

Marshall, W.S. 1995. Transport processes in isolated teleost epithelia: opercular epithelium and urinary bladder. In: *Cellular and Molecular Approaches to Fish Ionic Regulation*, C.M. Wood and T.J. Shuttleworth (eds.). Academic Press, New York, pp. 1–23.

Marshall, W.S. 2002. Na^+, Cl^-, Ca^{2+} and Zn^{2+} transport by fish gills: Retrospective review and prospective synthesis. *Journal of Experimental Zoology* 293: 264–283.

Marshall, W.S., S.E. Bryson and C.M. Wood. 1992. Calcium transport by isolated skin of rainbow trout. *Journal of Experimental Biology* 166: 297–316.

Marshall, W.S., C.G. Ossum and E.K. Hoffmann. 2005. Hypotonic shock mediation by p38 MAPK, JNK, PKC, FAK, OSR1 and SPAK in osmosensing chloride secreting cells of killifish opercular epithelium. *Journal of Experimental Biology* 208: 1063–1077.

Martin, S.A.M., O. Vilhelmsson, F. Médale, P. Watt, S. Kaushik and D.F. Houlihan. 2003. Proteomic sensitivity to dietary manipulations in rainbow trout. *Biochimica et Biophysica Acta* 1651: 17–29.

Masereel, B.A., L. Pochet and D. Laeckmann. 2003. An overview of inhibitors of Na^+/H^+ exchanger. *European Journal of Medicinal Chemistry* 38: 547–554.

Mathews, W.R., D.W. Ducharme, J.M. Hamlyn, D.W. Harris, F. Mandel, M.A. Clark and J.H. Ludens. 1991. Mass-spectral characterization of an endogenous digitalis-like factor from human plasma. *Hypertension* 17: 930–935.

Mazon, A.F., D.T. Nolan, R.A.C. Lock, M.N. Fernandes and S.E. Wendelaar Bonga. 2004. A short-term *in vitro* gill culture system to study the effects of toxic (copper) and non-toxic (cortisol) stressors on the rainbow trout, *Oncorhynchus mykiss* (Walbaum). *Toxicology in Vitro* 18: 691–701.

McClelland, M., F. Mathieu-Daude and J. Welsh. 1995. RNA fingerprinting and differential display using arbitrarily primed PCR. *Trends in Genetics* 11: 242–247.

McCormick, S.D. 2001. Endocrine control of osmoregulation in teleost fish. *American Zoologist* 41: 781–794.

McCormick, S.D., S. Hasegawa and T. Hirano. 1992. Calcium uptake in the skin of a freshwater teleost. *Proceedings of the National Academy of Sciences of the United States of America* 89: 3635–3638.

McDonald, M.D. and C.M. Wood. 2004. Evidence for facilitated diffusion of urea across the gill basolateral membrane of the rainbow trout (*Oncorhynchus mykiss*). *Biochimica et Biophysica Acta- Biomembranes* 1663: 89–96.

Merchant, M. and S.R. Weinberger. 2000. Recent advancements in surface-enhanced laser desorption/ionization time-of-flight mass spectrometry. *Electrophoresis* 21: 1164–1177.

Mistry, A.C., A. Kato, Y.H. Tran, S. Honda, T. Tsukada, Y. Takei and S. Hirose. 2004. FHL5, a novel actin-binding protein, is highly expressed in eel pillar cells and responds to wall tension. *American Journal of Physiology* 287: R1141–R1154.

Mommsen, T.P., M.M. Vijayan and T.W. Moon. 1999. Cortisol in teleosts: Dynamics, mechanisms of action, and metabolic regulation. *Reviews in Fish Biology and Fisheries* 9: 211–268.

Mothersill, C., F. Lyng, M. Lyons and D. Cottell. 1995. Growth and differentiation of epidermal cells of the rainbow trout established as explants and maintained in various media. *Journal of Fish Biology* 46: 1011–1025.

Naon, R. and N. Mayer-Gostan. 1983. Separation by velocity sedimentation of the gill epithelial cells and their ATPase activities in the seawater-adapted eel *Anguilla anguilla* L. *Comparative Biochemistry and Physiology* A75: 541–547.

Nickerson, J.G., S.G. Dugan, G. Drouin, S.F. Perry and T.W. Moon. 2003. Activity of the unique β-adrenergic Na^+/H^+ exchanger in trout erythrocytes is controlled by a novel β_3-AR subtype. *American Journal of Physiology* 285: R526–R535.

Niederstatter, H. and B. Pelster. 2000. Expression of two vacuolar-type ATPase B subunit isoforms in swimbladder gas gland cells of the European eel: Nucleotide sequences and deduced amino acid sequences. *Biochimica et Biophysica Acta* 1491: 133–142.

Pane, E.F., C.N. Glover, M. Patel and C.M. Wood. 2006. Characterization of nickel transport in renal brush border membrane vesicles isolated from the freshwater rainbow trout (*Oncorhynchus mykiss*). *Biochimica et Biophysica Acta* 1758: 74–84.

Pelis, R.M., J.E. Goldmeyer, J. Crivello and J.L. Renfro. 2003. Cortisol alters carbonic anhydrase-mediated renal sulphate secretion. *American Journal of Physiology* 285: R1430–R1438.

Pelis, R.M., S.L. Edwards, S.C. Kunigelis, J.B. Claiborne and J.L. Renfro. 2005. Stimulation of renal sulphate excretion by metabolic acidosis requires Na^+/H^+ exchange induction and carbonic anhydrase. *American Journal of Physiology* 289: F208–F216.

Perry, S.F. and P.J. Walsh. 1989. Metabolism of isolated fish gill cells: contribution of epithelial chloride cells. *Journal of Experimental Biology* 144: 507–520.

Perry, S.F., M.L. Byers and D.A. Johnson. 2000. Cloning and molecular characterization of the trout (*Oncorhynchus mykiss*) vacuolar H^+-ATPase B-subunit. *Journal of Experimental Biology* 203: 459–470.

Perry, S.F., A. Shahsavarani, T. Georgalis, M. Bayaa, M. Furmisky and S.L.Y. Thomas. 2003. Channels, pumps, and exchangers in the gill and kidney of freshwater fishes: Their role in ionic and acid-base regulation. *Journal of Experimental Zoology* A300: 53–62.

Piermarini, P.M. and D.H. Evans. 2001. Immunochemical analysis of the vacuolar proton-ATPase B-subunit in the gills of a euryhaline stingray (*Dasyatis sabina*): Effects of salinity and relation to Na^+/K^+-ATPase. *Journal of Experimental Biology* 204: 3251–3259.

Pisam, M. and A. Rambourg. 1991. Mitochondria-rich cells in the gill epithelium of teleost fish—an ultrastructural approach. *International Review of Cytology* 130: 191–232.

Rajarao, S.J., V.A. Canfield, M.A. Mohideen, Y.L. Yan, J.H. Postlethwait, K.C. Cheng and R. Levenson. 2001. The repertoire of Na,K-ATPase α and β subunit genes expressed in the zebrafish, *Danio rerio*. *Genome Research* 11: 1211–1220.

Reid, S.D., G.S. Hawkings, F. Galvez and G.G. Goss. 2003. Localization and characterization of phenamil-sensitive Na^+ influx in isolated rainbow trout gill epithelial cells. *Journal of Experimental Biology* 206: 551-559.

Renfro, J.L. 1995. Solute transport by flounder renal cells in primary culture. In: *Cellular and Molecular Approaches to Fish Ionic Regulation*, C.M. Wood and T.J. Shuttleworth (eds.). Academic Press, New York, pp. 147–171.

Renn, S.C.P., N. Aubin-Horth and H.A. Hofmann. 2004. Biologically meaningful expression profiling across species using heterologous hybridization to a cDNA microarray. *BMC Genomics* 5: 42.

Richards, J.G., J.W. Semple, J. S. Bystriansky and P.M. Schulte. 2003. Na^+/K^+-ATPase α-isoform switching in gills of rainbow trout (*Oncorhynchus mykiss*) during salinity transfer. *Journal of Experimental Biology* 206: 4475–4486.

Rise, M.L., K.R. von Schalburg, G.D. Brown, M.A. Mawer, R.H. Devlin, N. Kuipers, M. Busby, M. Beetz-Sargent, R. Alberto, A.R. Gibbs, P. Hunt, R. Shukin, J.A. Zeznik,

C. Nelson, S.R.M. Jones, D.E. Smailus, S.J.M. Jones, J.E. Schein, M.A. Marra, Y.S.N. Butterfield, J.M. Stott, S.H.S. Ng, W.S. Davidson and B.F. Koop. 2004. Development and application of a salmonid EST database and cDNA microarray: Data mining and interspecific hybridization characteristics. *Genome Research* 14: 478–490.

Rivers, R., A. Blanchard, D. Eladari, F. Leviel, M. Palliard, R.A. Podevin and M.L. Zeidel. 1998. Water and solute permeabilities of medullary thick ascending limb apical and basolateral membranes. *American Journal of Physiology* 274: F453–F462.

Sáenz, J.P., R. Kinne and E. Kinne-Saffran. 2003. Determination of the purity of membrane fractions isolated from flounder (*Pleuronectes americanus*) gills by a selective biotin exposure technique. *Mount Desert Island Biological Laboratory Bulletin* 42: 85–88.

Sakamoto, T., N. Ojima and M. Yamashita. 2000. Induction of mRNAs in response to acclimation of trout cells to different osmolalities. *Fish Physiology and Biochemistry* 22: 255–262.

Sakamoto, T., K. Uchida and S. Yokota. 2001. Regulation of the ion-transporting mitochondrion-rich cell during adaptation of teleost fishes to different salinities. *Zoological Science* 18: 1163–1179.

Sakamoto, T., H. Yasunaga, S. Yokota and M. Ando. 2002. Differential display of skin mRNAs regulated under varying environmental conditions in a mudskipper. *Journal of Comparative Physiology* B 172: 447–453.

Sambruy, Y., S. Ferruza, G. Ranaldi and I. De Angelis. 2001. Intestinal cell culture models. *Cell Biology and Toxicology* 17: 301–317.

Sandbacka, M., P. Pärt and B. Isomaa. 1999. Gill epithelial cells as tools for toxicity screening—comparison between primary cultures, cells in suspension and epithelia on filters. *Aquatic Toxicology* 46: 23–32.

Sandor, T., J.A. Di Battista and A.Z. Mehdi. 1984. Glucocorticoid receptors in the gill tissue of fish. *General and Comparative Endocrinology* 53: 353–364.

Sangalang, G.B. and J.F. Uthe. 1994. Corticosteroid activity, *in vitro*, in interrenal tissue of Atlantic salmon (*Salmo salar*) parr. *General and Comparative Endocrinology* 95: 273–285.

Sangiao-Alvarellos, S., F.J. Arjona, M.P.M. Del Rio, J.M. Miguez, J.M. Mancera and J.L. Soengas. 2005. Time course of osmoregulatory and metabolic changes during osmotic acclimation in *Sparus auratus*. *Journal of Experimental Biology* 208: 4291–4304.

Sargent, J.R., A.J. Thomson and M. Bornancin. 1975. Activities and localization of succinic dehydrogenase and Na^+/K^+-activated adenosine triphosphatase in the gills of fresh water and sea water eels *Anguilla anguilla*. *Comparative Biochemistry and Physiology* B51: 75–79.

Sathiyaa, R. and M.M. Vijayan. 2003. Autoregulation of glucocorticoid receptor by cortisol in rainbow trout hepatocytes. *American Journal of Physiology* 284: C1508–C1515.

Satoh, R., K. Nakashima, M. Seki, K. Shinozaki and K. Yamaguchi-Shinozaki. 2002. ACTCAT, a novel cis-acting element for proline- and hypoosmolarity-responsive expression of the ProDH gene encoding proline dehydrogenase in *Arabidopsis*. *Plant Physiology* 130: 709–719.

Scanlon, K.J. 2004. Anti-genes: siRNA, ribozymes and antisense. *Current Pharmaceutical Biotechnology* 5: 415–420.

Scott, G.R., J.G. Richards, B. Forbush, P. Isenring and P.M. Schulte. 2004. Changes in gene expression in gills of the euryhaline killifish *Fundulus heteroclitus* after abrupt salinity transfer. *American Journal of Physiology* 287: C300–C309.

Shephard, K.L. 1994. Functions for fish mucus. *Reviews in Fish Biology and Fisheries* 4: 401–429.

Shiraishi, K., M. Matsuda, T. Mori and T. Hirano. 1999. Changes in the expression of prolactin- and cortisol-receptor genes during early life stages of euryhaline tilapia (*Oreochromis mossambicus*) in freshwater and seawater. *Zoological Science* 16: 139–146.

Shiraishi, K., J. Hiroi, T. Kaneko, M. Matsuda, T. Hirano and T. Mori. 2001. *In vitro* effects of environmental salinity and cortisol on chloride cell differentiation in embryos of Mozambique tilapia, *Oreochromis mossambicus*, measured using a newly developed 'yolk-ball' incubation system. *Journal of Experimental Biology* 204: 1883–1888.

Shrimpton, J.M., D.A. Patterson, J.G. Richards, S.J. Cooke, P.M. Schulte, S.G. Hinch and A.P. Farrell. 2005. Ionoregulatory changes in different populations of maturing sockeye salmon *Oncorhynchus nerka* during ocean and river migration. *Journal of Experimental Biology* 208: 4069–4078.

Silva, P., R.J. Solomon and F.H. Epstein. 1990. Shark rectal gland. *Methods in Enzymology* 192: 754–766.

Sturm, A., N. Bury, L. Dengreville, J. Fagart, G. Flouriot, M.E. Rafestin-Oblin and P. Prunet. 2005. 11-deoxycorticosterone is a potent agonist of the rainbow trout (*Oncorhynchus mykiss*) mineralocorticoid receptor. *Endocrinology* 146: 47–55.

Sullivan, G.V., J.N. Fryer and S.F. Perry. 1995. Immunolocalization of proton pumps (H^+-ATPase) in pavement cells in rainbow trout gill. *Journal of Experimental Biology* 198: 2619–2629.

Suzuki, Y., M. Itakura, M. Kashiwagi, N. Nakamura, T. Matsuki, H. Sakuta, N. Naito, K. Takano, T. Fujita and S. Hirose. 1999. Identification by differential display of a hypertonicity-inducible inward rectifier potassium channel highly expressed in chloride cells. *Journal of Biological Chemistry* 274: 11376–11382.

Tagawa, M. 1996. Current understanding of the presence of hormones in fish eggs. In: *Survival Strategies in Early Life Stages of Marine Resources*, Y. Watanabe, Y. Yamashita and Y. Oozeki (eds.). A.A. Balkema, Rotterdam, pp. 27–38.

Tagawa, M., H. Hagiwara, A. Takemura, S. Hirose and T. Hirano. 1997. Partial cloning of the hormone-binding domain of the cortisol receptor in tilapia, *Oreochromis mossambicus*, and changes in the mRNA level during embryonic development. *General and Comparative Endocrinology* 108: 132–140.

Takei, Y. and S. Hirose. 2002. The natriuretic peptide system in eels: a key endocrine system for euryhalinity. *American Journal of Physiology* 282: R940–R951.

Takeo, J., J.I. Hata, C. Segawa, H. Toyohara and S. Yamashita. 1996. Fish glucocorticoid receptor with splicing variants in the DNA binding domain. *FEBS Letters* 389: 244–248.

Takeuchi, K., H. Toyohara and M. Sakaguchi. 2000. A hyperosmotic stress-induced mRNA of carp cell encodes Na^+- and Cl^--dependent high affinity taurine transporter. *Biochimica et Biophysica Acta* 1464: 219–230.

Tanaka, M. and M. Kinoshita. 2001. Recent progress in the generation of transgenic medaka (*Oryzias latipes*). *Zoological Science* 18: 615–622.

Terova, G., R. Gornati, S. Rimoldi, G. Bernardini and M. Saroglia. 2005. Quantification of a glucocorticoid receptor in sea bass (*Dicentrarchus labrax* L.) reared at high stocking density. *Gene* 357: 144–151.

Tokuda, Y., K. Touhata, M. Kinoshita, H. Toyohara, M. Sakaguchi, Y. Yokoyama, T. Ichikawa and S. Yamashita. 1999. Sequence and expression of a cDNA encoding Japanese flounder glucocorticoid receptor. *Fisheries Science* 65: 46–471.

Towle, D.W. 1993. Ion transport systems in membrane vesicles isolated from crustacean tissues. *Journal of Experimental Zoology* 265: 387–396.

Tresguerres, M., F. Katoh, H. Fenton, E. Jasinska and G.G. Goss. 2000. Regulation of branchial V-H^+-ATPase, Na^+/K^+-ATPase and NHE2 in response to acid and base infusions in the Pacific spiny dogfish (*Squalus acanthias*). *Journal of Experimental Biology* 208: 345–354.

Udvadia, A.J. and E. Linney. 2003. Windows into development: historic, current, and future perspectives on transgenic zebrafish. *Developmental Biology* 256: 1–17.

Ussing, H.H. and K. Zerahn. 1951. Active transport of sodium as the source of electric current in the short-circuited isolated frog skin. *Acta Physiologica Scandanavia* 23: 110–127.

van der Salm, A.L., D.T. Nolan and S.E. Wendelaar Bonga. 2002. *In vitro* evidence that cortisol directly modulates stress-related responses in the skin epidermis of the rainbow trout (*Oncorhynchus mykiss* Walbaum). *Fish Physiology and Biochemistry* 27: 9–18.

Van Praag, D., S.J. Farber, E. Minkin and N. Primor. 1987. Production of eicosanoids by the killifish gills and opercular epithelia and their effect on active transport of ions. *General and Comparative Endocrinology* 67: 50–57.

Varsamos, S., J.P. Diaz, G. Charmantier, C. Blasco, R. Connes and G. Flik. 2002. Location and morphology of chloride cells during the post-embryonic development of the European sea bass, *Dicentrarchus labrax*. *Anatomy and Embryology* 205: 203–213.

Varsamos, S., C. Nebel and G. Charmantier. 2005. Ontogeny of osmoregulation in post-embryonic fish: A review. *Comparative Biochemistry and Physiology* A141: 401–429.

Veillette, P.A. and G. Young. 2005. Tissue culture of sockeye salmon intestine: Functional response of Na^+-K^+-ATPase to cortisol. *American Journal of Physiology* 288: R1598–R1605.

Venkatesh, B. 2003. Evolution and diversity of fish genomes. *Current Opinion in Genetics and Development* 13: 588–592.

Viant, M.R. 2003. Improved methods for the acquisition and interpretation of NMR metabolomic data. *Biochemical and Biophysical Research Communications* 310: 943–948.

Walsh, P.J., Y. Wang, C.E. Campbell, G. De Boeck and C.M. Wood. 2001. Patterns of nitrogenous waste excretion and gill urea transporter mRNA expression in several species of marine fish. *Marine Biology* 139: 839–844.

Weckwerth, W. and K. Morgenthal. 2005. Metabolomics: From pattern recognition to biological interpretation. *Drug Discovery Today* 10: 1551–1558.

Wei, C., J. Li and R. Bumgarner. 2004. Sample size for detecting differentially expressed genes in microarray experiments. *BMC Genomics* 5: 87.

Wheatly, M.G. and Y. Gao. 2004. Molecular biology of ion motive proteins in comparative models. *Journal of Experimental Biology* 207: 3253–3263.

Wilson, J.M., P. Laurent, B.L. Tufts, D.J. Benos, M. Donowitz, A.W. Vogl and D.J. Randall. 2000a. NaCl uptake by the branchial epithelium in freshwater teleost fish: An immunological approach to ion transport protein localization. *Journal of Experimental Biology* 203: 2279–2296.

Wilson, J.M., D.J. Randall, M. Donowitz, A.W. Vogl and A.K.-Y. Ip. 2000b. Immunolocalizaton of ion transport proteins to branchial epithelium mitochondria-rich cells in the mudskipper (*Periophthalmodon schlosseri*). *Journal of Experimental Biology* 203: 2297–2310.

Winkler, C., M. Shafer, J. Duschl, M. Schartl and J.N. Volff. 2003. Functional divergence of two zebrafish midkine growth factors following fish-specific duplication. *Genome Research* 13: 1067–1081.

Wong, C.K.C. and D.K.O. Chan. 1999a. Chloride cell subtypes in the gill epithelium of Japanese eel *Anguilla japonica*. *American Journal of Physiology* 276: R517–R522.

Wong, C.K.C. and D.K.O. Chan. 1999b. Isolation of viable cell types from the gill epithelium of Japanese *eel Anguilla japonica*. *American Journal of Physiology* 276: R363–R372.

Wong, C.K.C. and D.K.O. Chan. 2001. Effects of cortisol on chloride cells in the gill epithelium of Japanese eel, *Anguilla japonica*. *Journal of Endocrinology* 168: 185–192.

Wood, C.M. 2001. Toxic responses of the gill. In: *Target Organ Toxicity in Marine and Freshwater Teleosts*, Vol. 1, *Organs*, D. Schlenk and W.H. Benson (eds.). Taylor and Francis, London, pp. 1–89.

Wood, C.M. and P. Pärt. 1997. Cultured branchial epithelia from freshwater fish gills. *Journal of Experimental Biology* 204: 1047–1059.

Wood, C.M. and T.J. Shuttleworth (eds.). 1995. *Cellular and Molecular Approaches to Fish Ionic Regulation*. Academic Press, New York.

Wood, C.M., S.P. Kelly, B. Zhou, M. Fletcher, M. O'Donnell, B. Eletti and P. Pärt. 2002. Cultured gill epithelia as models for the freshwater fish gill. *Biochimica et Biophysica Acta* 1566: 72–83.

Woodruff, T.K. 2004. Cellular localization of mRNA and protein: *In situ* hybridization histochemistry and *in situ* ligand binding. *Methods in Cell Biology* 57: 333–351.

Woolfe, A., M. Goodson, D.K. Goode, P. Snell, G.K. McEwen, T. Vavouri, S.F. Smith, P. North, H. Callaway, K. Kelly, K. Walter, I. Abnizova, W. Gilks, Y.J.K. Edwards, J.E. Cooke and G. Elgar. 2005. Highly conserved non-coding sequences are associated with vertebrate development. *PLoS Biology* 3: 116–130.

Yamamoto, T. and N. Suzuki. 2005. Expression and function of cGMP-dependent protein kinase type I during medaka fish embryogenesis. *Journal of Biological Chemistry* 280: 16979–16986.

Yokota, S., K. Iwata, Y. Fujii and M. Ando. 1997. Ion transport across the skin of the mudskipper *Periophthalmus modestus*. *Comparative Physiology and Biochemistry* A118: 903–910.

Yu, W.P., J.M.M. Tan, K.C.M. Chewe, T. Oh, P. Kolatkar, B. Venkatesh, T.M. Dawson and K.L. Lim. 2005. The 350-fold compacted *Fugu* parkin gene is structurally and functionally similar to human Parkin. *Gene* 346: 97–104.

Zadunaisky, J.A. 1984. The chloride cell: the active transport of chloride and the paracellular pathways. In: *Fish Physiology*, W.S. Hoar and D.J. Randall (eds.). Academic Press, San Diego, Vol. 10 B, pp. 129–176.

Zbikowska, H.B. 2003. Fish can be first—advances in fish transgenesis for commercial applications. *Transgenic Research* 12: 379–389.

Zeidel, J.D., J.C. Mathai, J.D. Campbell, W.G. Ruiz, G.L. Apodaca, J. Riordan, M.L. Zeidel. 2005. Selective permeability barrier to urea in shark rectal gland. *American Journal of Physiology* 289: F83–F89.

Zhou, B.S., S.P. Kelly, J.P. Ianowski and C.M. Wood. 2003. Effects of cortisol and prolactin on Na^+ and Cl^- transport in cultured branchial epithelium from FW rainbow trout. *American Journal of Physiology* 285: R1305–R1316.

Zhou, B.S., S.P. Kelly and C.M. Wood. 2004. Response of developing cultured freshwater gill epithelia to gradual apical media dilution and hormone supplementation. *Journal of Experimental Zoology* A301: 867–881.

CHAPTER

8

Osmoregulation and Fish Transportation

Paulo César Falanghe Carneiro[1], Elisabeth Criscuolo Urbinati[2] and Fabiano Bendhack[3,*]

OSMOREGULATION

Water is the most abundant constituent of all cells with the exception of the anhydrobiotic states of extremely few, highly specialized cells. Most cells have electrolyte (inorganic ion) concentrations accounting for an osmolality of 200–400 mosmol/kg (Wood and Shuttleworth, 1995).

Cellular osmoregulation constitutes a phylogenetically conserved set of highly complex responses to changes in external osmolality/tonicity to maintain cell volume, intracellular concentrations of macro- and micromolecules, protein structure and function, and genomic integrity. The ubiquitous importance of cellular osmoregulation suggests that a high and variable environmental salt concentration represents a severe threat

Authors' addresses: [1]Embrapa Tabuleiros Costeiros. Aracaju, Sergipe, Brazil.
E-mail: paulo@cpatc.embrapa.br
[2]Universidade Estadual Paulista. Jaboticabal, São Paulo, Brazil.
E-mail: bethurb@caunesp.unesp.br
[3]Pontifícia Universidade Católica do Paraná. Curitiba, Paraná, Brazil.
**Corresponding author:* E-mail: f.bendhack@pucpr.br

to the cells. To understand the nature of such a threat, we have to consider how exactly these cells function. Cell function depends on tightly controlled ionic and electrostatic interactions between macromolecules, mainly DNA, RNA, and proteins (Kültz, 2000). In addition, the structure and function of proteins further depends on water activity and electrolyte concentration (Hochachka and Somero, 2002). Cell metabolism has been optimized, during the course of evolution, to function at particular concentrations of Na^+, Cl^-, K^+, Mg_2^+, Ca_2^+, Zn_2^+, and other electrolytes that define intracellular ionic strength. Consequently, any uncontrolled change in the concentration of these cations interferes with cell metabolism and function.

The successful establishment of the fish species in different habitats and environments depends on its ability to cope with salinity differences between internal (plasma) and external (water) environments through osmoregulation. Thus, salinity is one of the main environmental factors exerting a selective pressure on aquatic organisms. The mechanisms involved result in the maintenance of almost constant or only slightly variable blood osmolality, over a species-dependent range of salinity. Below this salinity range, and especially in freshwater, a fish hyper-osmoregulates and it is submitted to passive osmotic influx of water and diffusive loss of ions, mainly Na^+ and Cl^-. Above the isosmotic salinity, the fish hypoosmoregulates and is exposed to ion invasion and dehydration. Limiting and compensatory mechanisms include low integument permeability, active ion uptake in the branchial chambers (mainly gills), low drinking rate, and production of a high volume of hypotonic urine (Varsamos *et al.*, 2005).

GILL STRUCTURE AND FUNCTION

Although osmoregulation in fishes is mediated by a suite of structures, including the gastrointestinal epithelium and kidney, the gill is the major site of ion movements that balance diffusional gains or losses. The fish gill is a morphologically and functionally complex tissue that is the site of numerous, interconnected physiological processes, which are vital to maintaining systemic homeostasis in the face of changing internal and external conditions. The features of the gills that enhance gas exchange also make the gills susceptible to osmotic and ionic movements between the environment and extracellular fluids, and this condition necessitates osmoregulation. The branchial epithelium is the primary site of transport processes that counter the effects of osmotic and ionic gradients, as well as the principal site for pH regulation and nitrogenous waste excretion. Thus, the branchial epithelium in fishes is a multipurpose organ that plays a central role in a suite of physiological responses to environmental and internal changes. The structure and function of the branchial epithelium as a site of gas exchange dictates that it is also the site of osmosis and ionic diffusion if gradients exist, and the maintenance of blood and intracellular tonicity in the face of these osmotic and ionic gradients across the fish branchial epithelium takes energy (Varsamos *et al.*, 2005). The fish gill epithelium is composed of several distinct cell types: pavement cells (> 90%); mitochondrion-rich cells, also called chloride cells (CCs); and accessory cells, in addition to mucous cells that are usually on the leading or trailing edge of the filament (Laurent *et al.*, 1994).

FRESHWATER AND SALTWATER FISHES

Modern freshwater (FW) fishes balance the osmotic uptake of water through substantial glomerular filtration rates and urine flows and minimizing renal salt loss by a significant tubular reabsorption of necessary ions. Net ionic losses in the urine and diffusional outflux across the gill are balanced by active uptake mechanisms in the gill epithelium plus any ionic gain from food. Seawater (SW) fishes face the reverse osmotic and ionic gradients across the branchial epithelium to those found in the freshwater environment. In teleost saltwater species—which are distinctly hypo-osmotic to seawater—the osmotic loss of water is balanced by ingestion of seawater and subsequent intestinal uptake of NaCl to withdraw needed water from the lumen contents. The resulting salt load adds to that produced by the diffusional uptake of NaCl across the gills. The sum is balanced by gill epithelial NaCl extrusion, because the lack of a loop of Henle precludes the production of urine that is hyperosmotic to the plasma (Evans *et al.*, 2005).

The key enzyme to transport processes in the gill and intestine is the membrane-spanning protein Na^+, K^+-ATPase. Therefore, the regulation of Na^+, K^+-ATPase expression in these organs is of major importance to fish during SW acclimation. In the opercular membrane of SW-acclimated fish, chloride cells (CCs) comprise the cellular sites for the excretion of monovalent ions (Foskett and Scheffey, 1982) and these cells increase in size and number during the SW acclimation (Laurent and Dunel, 1980). Even though this cellular function has never been directly demonstrated in the gill, there is ample evidence to suggest that CCs have a similar function in the gills of SW-acclimated fish. On the other hand, the role of CCs in freshwater FW-acclimated fish is currently being debated (Perry, 1997).

STRESS AND OSMOREGULATION

Stress in fish is a state caused by a stressor that deviates from a normal resting state. Stress responses are considered a disturbance in the physiological condition of fish altering their homeostasis and health (Iwama *et al.*, 1997a).

The primary stress response of fish results in the activation of the brain-sympathetic-chromafin cells axis (Perry and Reid, 1993) and the brain-pituitary interrenal axis (Wendelaar Bonga, 1997), leading to the rapid release of catecholamines and cortisol into the blood stream. The former event contributes to the initial hyperglycemia and increased permeability of the surface epithelia, while the latter combines glucocorticoid and mineralocorticoid actions (Wendelaar Bonga, 1997). The effects at the level of the gill are particularly important because this is the site of active ion uptake in freshwater and extrusion in seawater (Perry and Laurent, 1993), as well as ammonia excretion and gas exchange (Wilkie, 1997). Generally, when stress disturbs the ability of a fish to osmoregulate, the ionic composition of its blood comes to resemble that of the surrounding medium (Pickering, 1993). Stress increases gill permeability to water and leads to plasma electrolyte changes in a hyperosmotic or hypoosmotic environment (Cech *et al.*, 1996). Each fish species regulates the blood fluid concentration within certain limits and departure from this range is likely to be the result of stress. The point at which the effect of such influences ceases to be normal and becomes stressful is difficult to be defined but it is easy to observe the gross effects of stress. However, by the time these effects on osmotic and ionic regulation are apparent, the fish may be severely damaged (Eddy, 1981; McDonald and Milligan, 1997). The osmotic regulation under stress conditions is extremely expensive and substantial body energy is used to restore the normal ionic status of the fish. It is widely accepted to possess an energy-saving effect to rear fish in salinities near their isoosmotic point (Febry and Lutz, 1987; Gaumet *et al.*, 1995). It was estimated that the energy required for ion regulation in FW trout gills to be ~1.6% of the resting metabolic rate and the energy required for ion regulation to be higher in SW gills (5.7%) (Kirschner, 1995). An isolated, saline-perfused gill arch preparation was used to compare gill energetics in FW- and SW-adapted trout. The consumption of FW gills was 33% higher than SW gills. On a whole animal basis, total gill oxygen consumption in FW and SW trout accounted for 3.9 and 2.4% of resting metabolic rate, respectively (Morgan and Iwama, 1999).

STRESS DURING FISH TRANSPORTATION

Stress is present in several steps of fish production, including capture and transport. Transporting fish is an extremely important part of fish culture. Live harvested fish may be transported to the market, either to a fish-processing plant or a fee fishing operation. In both cases, fish must arrive in good physiological condition and satisfy the criteria of the purchaser. Especially in the latter case, fish must be able to hook in a very short period of time after released into the pond (Wurts, 1995). Transportation is a traumatic procedure that consists of a succession of adverse stimuli, including initial capture, loading into the transport containers, the actual transport, unloading and stocking into the new environment (Robertson *et al.*, 1988). Transport, as any other stressor, may alter the normal homeostasis of fish and influence the osmoregulation process negatively (Carneiro and Urbinati, 2001).

Several substances have been used to minimize transport stress as sodium chloride, calcium sulphate, anesthetics, and others. Sodium chloride has been used as a mean of stress reduction and increases survival during transportation of freshwater fish in order to balance with water gain and electrolytes losses (Carmichael *et al.*, 1984; Carmichael and Tomasso, 1998). The addition of salt to transport water is known to reduce osmoregulatory disturbances and other physiological responses to stress factors, besides reducing mortality (Tomasso *et al.*, 1980; Johnson and Metcalf, 1982; McDonald and Milligan, 1997).

To evaluate the efficiency of salt as a stress reductor, juveniles of tambaqui *Colossoma macropomum*, a freshwater species of the Amazon Basin, were transported in customized plastic boxes in water containing 0, 2, 5 and 8 g of salt/L (Gomes *et al.*, 2003). Plasma cortisol presented a significant increase after transportation in water either without salt or with 2 g of salt/L, returning to normal levels after 96 hours. The fish exposed to all salt concentrations had plasma glucose increased after transportation, except the treatment with 8 g of salt/L of water, returning to normal levels within 24 hours. Our studies with matrinxã *Brycon cephalus*, another Amazonian stenohaline freshwater fish, showed positive results using common salt during a 4-hour transport (Carneiro and Urbinati, 2001). Fish transported in water without added salt (control) showed reduction of the serum sodium levels after transport. Plasma chloride levels decreased upon arrival in control fish as well as in those transported in water with 0.1 and 0.3% of added NaCl but not in fish transported in water with 0.6% of NaCl (Fig. 8.1). During stressful situations, fish present an increased blood flow and permeability of gills caused mainly by the action of the catecholamines which facilitate the exchange of carbonic dioxide for oxygen and improve the supply of the latter one to support the higher demand for this gas. Nevertheless, any change in the tegument permeability allows electrolyte loss, predominantly sodium and chloride, and the water influx, causing serious osmoregulatory and electrolytic disturbances in freshwater fish. Besides, sodium chloride has shown positive effect when used during transportation of several fishes, some species present higher sensibility to this substance. Gomes *et al.* (1999) observed increasing mortality and Na^+ body levels of jundia fingerlings transported in water containing salt at 0, 1, 3 and 6 g/L.

TRANSPORT WATER

Water quality is also an important factor in determining the success during fish transportation. Transport of live fish in closed systems results in significant degradation of water quality throughout the transport period. Excretory products, mucus and regurgitated food degrade the water quality and also stress the fish (Berka, 1986).

Total ammonia (NH_3 plus NH_4^+) is the primary waste product from the protein metabolism in fish, with un-ionized ammonia being the most toxic in the hauling water. Within the transportation unit, the main sources of ammonia are the excretion by fish as a normal part of their metabolism. As the total ammonia is released into the water, it takes the form of un-ionized ammonia (NH_3), which is highly toxic for fish, and that of ionized ammonia (NH_4^+). The main environmental factors affecting the proportional amount of these two forms, and thus, ammonia toxicity are the ambient temperature and the pH (Fivelstad *et al.*, 1993).

The increase in carbon dioxide, normally observed during transportation, causes water pH to decrease. Low pH increases the proportion of the toxic form of carbon dioxide (CO_2), but decreases the

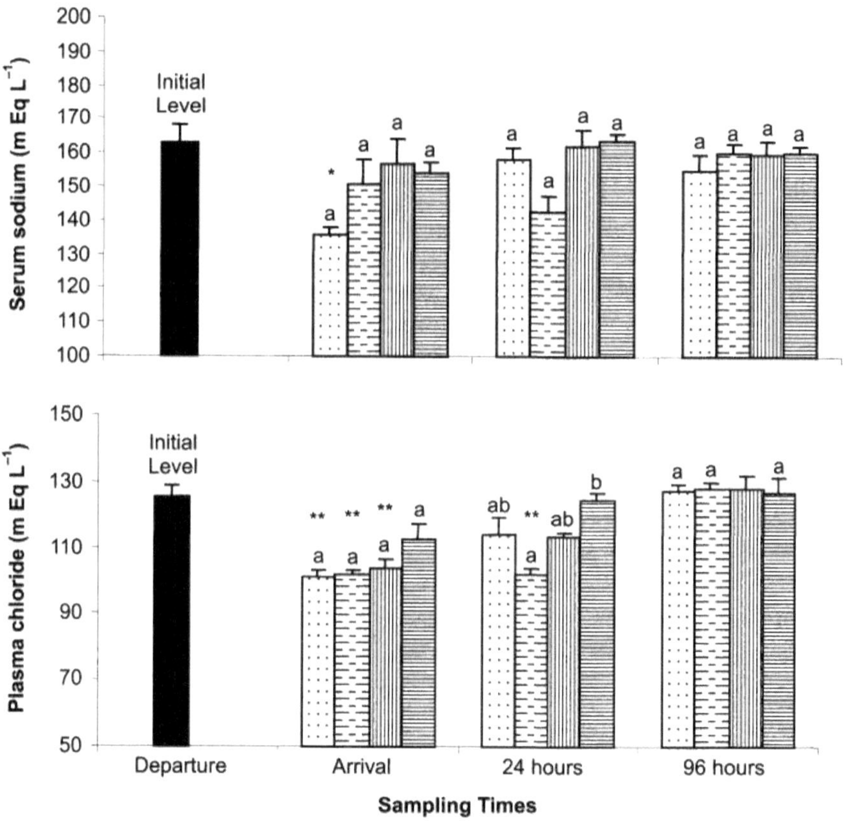

Fig. 8.1 Serum sodium (mEq L^{-1}) and plasma chloride (mEq L^{-1}) of matrinxã *Brycon cephalus* submitted to transport stress in the presence of added NaCl at the concentrations 0.0 (dotted bars), 0.1 (Dashed bars), 0.3 (vertical lined bars), and 0.6 % (horizontal lined bars). Different letters indicate differences ($P<0.05$) among treatments at the same sampling time. Single and double asterisks indicate differences ($P<0.05$ and $P<0.01$, respectively) between treatments and initial level (shaded bar); all treatments share the same initial level. Vertical bars represent SEM (N=5). (Carneiro and Urbinati, 2001).

proportion of the toxic form of ammonia (NH_3) (Amend *et al.*, 1982). Carbon dioxide gas behaves as an acid when dissolved in water. It readily enters the blood plasma as nontoxic bicarbonate ions and renders the plasma more acidic.

Accumulation of acid in the plasma can have detrimental effects on the fish osmoregulatory process (Berka, 1986). Low pH impairs the oxygen-carrying function of haemoglobin. The red blood cells defend their internal pH against a fall in blood plasma pH using ionic 'pumps' built into their cell envelopes. The acid (H^+) is removed from the cells in exchange for Na^+. The osmotic pressure rises within and water follows the salt, causing the red blood cells to swell. This response can be shown clinically by a rise in the volume of red cells in the blood (measured as the haematocrit) without a rise in haemoglobin concentration (Fievet *et al.*, 1988).

Stressful conditions cause passive ion loss and water influx in fish maintained in hypoosmotic environments, thus negatively affecting the function of active exchange mechanisms (Wendelaar Bonga, 1997). The accumulation of ammonia usually observed in transport water may cause serious problems to the fish, such as increased oxygen consumption and heart rate, decreased plasma sodium and alteration of the acid-base balance (Eddy, 1981; Tomasso, 1994; El-Shafey, 1998). Fasting prior to transport diminishes the amount of food in the digestive tract and reduces ammonia excretion, being a common practice before transport procedures of virtually all species or sizes. In spite of NH_3 diffusion being the preferential mechanism of ammonia excretion in teleost, there is a branchial exchange system between NH_4^+ and Na^+ ions that may become more important in an environment with sodium ions and high levels of NH_3 (Weirich *et al.*, 1993; Tomasso, 1994). Carneiro and Urbinati (2001) demonstrated that matrinxã transported for 4 hours in water containing 0.6% of added NaCl presented plasma levels of ammonia lower than fish transported in water containing 0.0 or 0.1% (Fig. 8.2). In fact, fish submitted to the highest salt concentration first recovered the initial ammonia level, suggesting the presence of the NH_4^+/Na^+ exchange system in matrinxã.

Calcium is another ion that may be present in water, being essential to biological processes in fish and linked to osmoregulation, mainly to the exchange Ca^{+2}/Na^+ and cellular permeability (Flik and Verbost, 1993). The calcium is a low-cost salt, of frequent use in fish transportation, although its efficiency is yet to be confirmed. The intracellular calcium concentration of gill cells of a majority of fish species is less than 1 μmol. Nevertheless, even in soft fresh waters with less than 10 μmol of this ion, there is diffusional uptake of Ca^{+2}. The calcium enters to the chloride cells (CCs) across apical membrane, is transported via cytoplasm and extruded across the basolateral membrane via Ca^{2+} ATPase or a Na^+/Ca^{2+} exchanger (Flik *et al.*, 1995; Evans *et al.*, 2005) (Fig. 8.3).

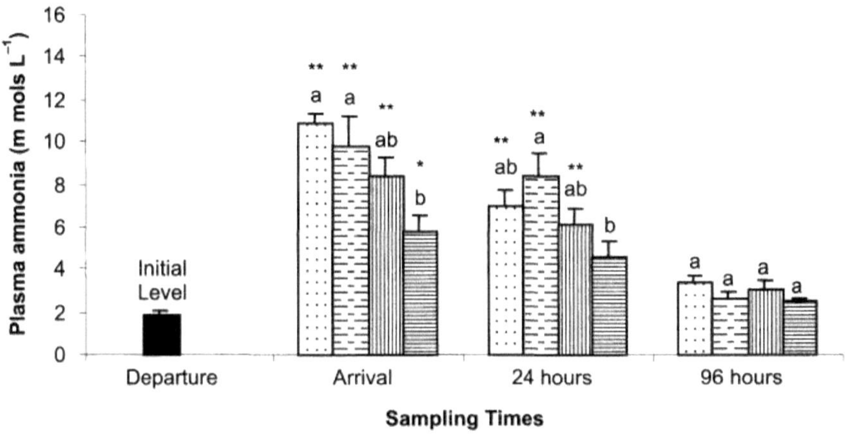

Fig. 8.2 Plasma ammonia (mmol L^{-1}) of matrinxã submitted to transport stress in the presence of added NaCl at the concentrations 0.0 (dotted bars), 0.1 (Dashed bars), 0.3 (vertical lined bars), and 0.6% (horizontal lined bars). See Fig. 8.1 for explanation of letters and asterisks accompanying the value bars (Carneiro and Urbinati, 2001).

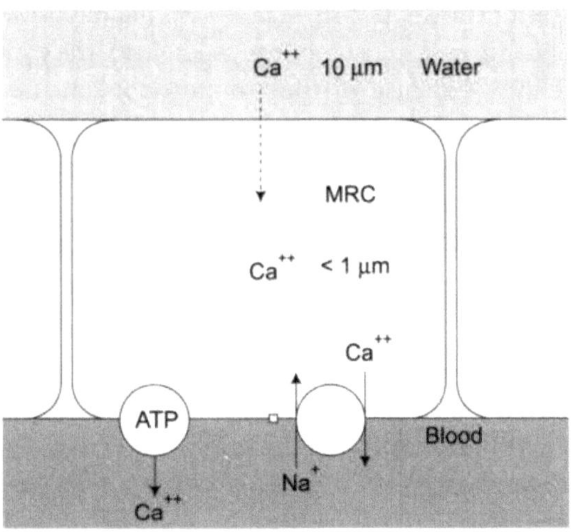

Fig. 8.3 Working model for the calcium uptake mechanisms across the freshwater teleost gill epithelium. (Modified from Evans *et al.*, 2005.)

Teleosts regulate the calcium uptake through the corpuscles of Stannius that produces stanniocalcin, a hypocalcaemic hormone that plays an active role in reducing the uptake of this ion in hypercalcic water (Flik *et al.*, 1995). Besides the prolactin and somatolactin (Kaneko and

Hirano, 1993), cortisol also influences the regulation of the calcium uptake, promoting hypercalcaemic effects in fish (Flik and Verbost, 1993).

The ion calcium on water, through anionic reactions, may alter the hydration of organic structures on gills and thereby reduce their permeability to both sodium and chloride ions (Potts, 1984), minimizing ions losses caused by catecholamines action (Wendelaar Bonga, 1997), normally released into the blood stream during transportation. The ion calcium presumably also acts on the branchial process of ammonia excretion by increasing the thickness of mucus layer at the gill surface, providing a more stable acidic boundary layer in which conversion of NH_3 to NH_4 would be enhanced, and in which a higher blood to water NH_3 gradient maintained (Iwama *et al.*, 1997b). The higher water H^+ concentration provided by the mucus layer near the gill epithelium rapidly ionizes ammonia excreted to NH_4^+, contributing to maintain the gradient difference and the excretion process. In our recent research, the circulating levels of cortisol increased four times in the control matrinxã and six times in the $CaSO_4$ exposed fish after the packing in the bags. On the other hand, after transport, fish exposed to calcium presented reduced levels of cortisol compared to fish not exposed. Also, the excess of calcium on water induced a decrease of the calcium concentration on blood. During the recovery period, the lower water calcium concentrations and the cortisol effect may have stimulated the uptake of this ion, being that calcitropic action of cortisol become noticeable only in the long term, approximately 4 days after.

The ions losses and the tendency to water influx are electrolytic disturbances that can be reduced in freshwater fish transportation (Carneiro and Urbinati, 2001). Also, in our recent works, the chloride levels were maintained in calcium-exposed matrinxã. Besides, sodium decreased after transportation (Fig. 8.4). Sodium levels also decreased in striped bass (*Morone saxatilis*) transported in water with addition of 100 mg/L of $CaCl_2$ (Mazik *et al.*, 1991).

CONCLUSION

Transport, like any other stressor among the fish culture procedures, provokes osmoregulatory alterations in fish. Physiological changes might alter among fish species, but generally all fishes exhibit a similar pattern concerning hormones releases and electrolyte disturbances, mainly Na^+, Cl^- and Ca^{2+}. Understanding and previewing these osmoregulatory

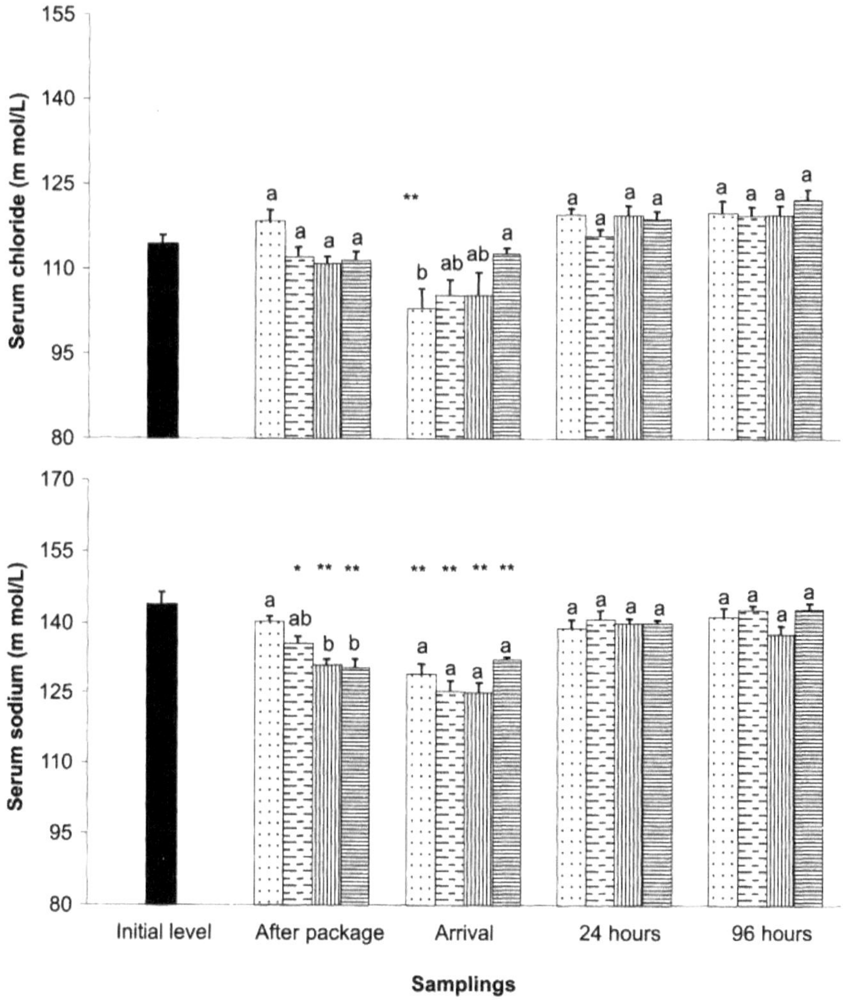

Fig. 8.4 Serum chloride (mmol/L), sodium (mmol/L) of matrinxã transported in water containing $CaSO_4$ at concentrations of 0 mg/L (dotted bars), 75 mg/L (dashed bars), 150 mg/L (vertical striped bars), 300 mg/L (horizontal striped bars). See Figure 8.1 for explanation of letters and asterisks accompanying the value bars. Vertical bars represent SEM (n=8). (Unpublished data).

changes during transportation can help a fish farmer diminish mortality and loss, as well as increase profit. The use of substances as common salt and calcium sulphate in the transport water, as well as adopt practices like handling fish carefully and fasting them before transport are part of the normal procedures that help fish farmers to cope with osmoregulatory disturbances caused by transportation.

References

Amend, D.F., T.R. Croy, B.A. Goven, K.A. Johnson and D.H. McCarthy. 1982. Transportation of fish in closed systems: Methods to control ammonia, carbon dioxide, pH, and bacterial growth. *Transactions of the American Fisheries Society* 111: 603–611.

Berka, R. 1986. The transport of live fish. A review. *EIFAC Technical Report* 48, FAO, Rome.

Carmichael, G.J. and J.R. Tomasso. 1998. Survey of fish transportation equipment and techniques. *Progressive Fish-Culturist* 50: 155–159.

Carmichael, G.J., J.R. Tomasso, B.A. Simco and K.B. Davis. 1984. Characterization and alleviation of stress associated with hauling largemouth bass. *Transactions of the American Fisheries Society* 113: 778–785.

Carneiro, P.C.F. and E.C. Urbinati. 2001. Salt as a stress response mitigator of matrinxã *Brycon cephalus* (Günther) during transport. *Aquaculture Research* 32: 297–304.

Cech, J.J.R., S.D. Bartholow, P.S. Young and T.E Hopkins. 1996. Striped bass exercise and handling stress in freshwater: Physiological responses to recovery environment. *Transactions of the American Fisheries Society* 125: 308–320.

Eddy, F.B. 1981. Effects of stress on osmotic and ionic regulation in fish. In: *Stress and Fish*, A.D. Pickering (ed.). Academic Press, London, pp. 77–102.

El-Shafey, A.A.M. 1998. Effect of ammonia on respiratory functions of blood of *Tilapia zilli*. *Comparative Biochemistry and Physiology* A121: 305–313.

Evans, D.H., P.M. Piermarini and K.P. Choe. 2005. The multifunctional fish gill: dominant site of gas exchange, osmoregulation, acid-base regulation, and excretion of nitrogenous waste. *Physiology Reviews* 85: 97–177.

Febry, R. and P. Lutz. 1987. Energy partitioning in fish: the activity-related cost of osmoregulation in euryhaline cichlid. *Journal of Experimental Biology* 128: 63–85.

Fievet, B., G. Claireaux, S. Thomas and R. Motais. 1988. Adaptive respiratory responses of trout to acute hypoxia: III. Ion movements and pH changes in the red blood cell. *Respiration Physiology* 74: 99–114.

Fivelstad, S., H. Kallevik, H.M. Iversen, T. Møretrø, K. Vage and M. Binde. 1993. Sublethal effects of ammonia in soft water on Atlantic salmon smolts at a low temperature. *Aquaculture International* 1: 157–169.

Flik, G. and P.M. Verbost. 1993. Calcium transport in fish gills and intestine. *Journal of the Experimental Biology* 184: 17–29.

Flik, G., P.M. Verbost and S.E. Wendelaar-Bonga. 1995. Calcium transport process in fish. In: *Cellular and Molecular Approaches to Fish Ionic Regulation*, C.M. Wood and T.J. Shuttleworth (eds.). Academic Press, San Diego, pp. 317–342.

Foskett, J.K. and C. Scheffey. 1982. The chloride cell: definitive identification as the salt-secretory cell in teleosts. *Science* 215: 164–166.

Gaumet, F., G. Bouef, A. Severe, A. LeRoux and N. Mayer-Gostan. 1995. Effects of salinity on the ionic balance and growth of juvenile turbot. *Journal of Fish Biology* 47: 865–876.

Gomes, L.C., J.I. Golombieski, A.R. Chippari-Gomes and B. Baldisserotto. 1999. Effect of salt in the water for transport on survival and on Na^+ and K^+ body levels of silver catfish, *Rhamdia quelen*, fingerlings. *Journal of Applied Aquaculture* 9: 1–9.

Gomes, L.C., C.A.R.M. Araújo-Lima, R. Roubach and E.C. Urbinati. 2003. Avaliação dos efeitos da adição de sal e da densidade no transporte de tambaqui (*Colossoma macropomum*). *Pesquisa Agropecuária Brasileira* 38: 283–290.

Hochachka, P.W. and G.N. Somero. 2002. *Biochemical Adaptation: Mechanism and Process in Physiological Evolution*. Oxford University Press, New York.

Iwama, G.K., A.D. Pickering, J.P. Sumpter and C.B. Schreck (eds.). 1997a. *Fish Stress and Health in Aquaculture*. Cambridge University Press, New York.

Iwama, G.K., J.C. McGeer, P.A. Wright, M.P. Wilkie and C.M. Wood. 1997b. Divalent cations enhance ammonia excretion in Lahontan cutthroat trout in highly alkaline water. *Journal of Fish Biology* 50: 1061–1073.

Johnson, D.L. and M.T. Metcalf. 1982. Causes and controls of freshwater drum mortality during transportation. *Transactions of the American Fisheries Society* 111: 58–62.

Kaneko, T. and T. Hirano. 1993. Role of prolactin and somatolactin in calcium regulation in fish. *Journal of the Experimental Biology* 184: 31–45.

Kirschner, L.B. 1995. Energetics of osmoregulation in freshwater vertebrates. *Journal of Experimental Zoology* 271: 243–252.

Kültz, D. 2000. Osmotic regulation of DNA activity and the cell cycle. In: *Cell and Molecular Responses to Stress*, K. Storey and J. Storey (eds.). Elsevier, Amsterdam, pp. 157–179.

Laurent, P. and S. Dunel. 1980. Morphology of gill epithelia in fish. *American Journal of Physiology* 238: 147–159.

Laurent, P., S. Dunel-Erb, C. Chevalier and J. Lignon. 1994. Gill epithelial kinetics in a freshwater teleost, *Oncorhynchus mykiss* during adaptation to ion-poor water and hormonal treatments. *Fish Physiology and Biochemistry* 5: 353–370.

McDonald, G. and L. Milligan.1997. Ionic, osmotic and acid-base regulation in stress. In: *Fish Stress and Health in Aquaculture*, G.W. Iwama, A.D. Pickering, J.P. Sumpter, C.B. Schreck (eds.). Cambridge University Press, Cambridge, pp. 119–144.

Morgan, J.D. and G.K. Iwama. 1999. Energy cost of NaCl transport in isolated gills of cutthroat trout. *American Journal of Physiology* 277: 631–639.

Mazik, P.M., B.A. Simco and N.C. Parker. 1991. Influence of water hardness and salts on survival and physiological characteristics of striped bass during and after transport. *Transactions of the American Fisheries Society* 120: 121–126.

Perry, S.F. 1997. The chloride cell structure and function in the gills of freshwater fishes. *Annual Review of Physiology* 59: 325–347.

Perry, S.F. and P. Laurent. 1993. Environmental effects on fish gill structure and function. In: *Fish Ecophysiology*, J.C. Rankin and F.B. Jensen (eds.). Chapman & Hall, London, pp. 231–264.

Perry, S.F. and S.D. Reid. 1993. β-adrenergic signal transduction in fish: interactive effects of catecholamines and cortisol. *Fish Physiology and Biochemistry* 11: 195–203.

Pickering, A.D. 1993. Stress and adaptation: A. Husbandry and stress. In: *Recent Advances in Aquaculture*, J.R. Muir and R.J. Roberts (eds.). Blackwell, Oxford, pp. 155–169.

Potts, W.T.W. 1984. Transepithelial potentials in fish gills. In: *Ion and Water Transfer*, W.S. Hoar, D.J. Randall and J.R. Brett (eds.). Academic Press, Orlando, pp. 105–128.

Robertson, L., P. Thomas and C.R. Arnold. 1988. Plasma cortisol and secondary stress responses of cultured red drum (*Sciaenops ocellatus*) to several transportation procedures. *Aquaculture* 68: 115–130.

Tomasso, J.R., K.B. Davis and N.C. Parker. 1980. Plasma corticosteroid and electrolyte dynamics of hybrid striped bass (*White bass* × *Striped bass*) during netting and hauling. *Proceedings of the World Mariculture Society* 11: 303–310.

Tomasso, J.R. 1994. Toxicity of nitrogenous wastes to aquaculture animals. *Reviews in Fisheries Science* 2: 291–314.

Varsamos, S., C. Nebel and G.T. Charmantier. 2005. Ontogeny of osmoregulation in postembryonic fish: A review. *Comparative Biochemistry and Physiology* A141: 401–429.

Weirich, C.R., J.R. Tomasso and T.I.J. Smith. 1993. Toxicity of ammonia and nitrite to sunshine bass in selected environments. *Journal of Aquatic Animal Health* 5: 64–72.

Wendelaar Bonga, S.E. 1997. The stress response in fish. *Physiological Reviews* 7: 591–625.

Wilkie, M.P. 1997. Mechanisms of ammonia excretion across fish gills. *Comparative Biochemistry and Physiology* A118: 39–50.

Wood, C.M. and T.J. Shuttleworth (eds.). 1995. *Cellular and Molecular Approaches to Fish Ionic Regulation*. Academic Press, San Diego.

Wurts, W.A. 1995. Using salt to reduce handling stress in channel catfish. *World Aquaculture* 26: 80–81.

CHAPTER

9

Special Challenges to Teleost Fish Osmoregulation in Environmentally Extreme or Unstable Habitats

Carolina A. Freire* and Viviane Prodocimo

INTRODUCTION

The sea, beyond a certain minimal depth and distance from land, is actually a very stable environment. However, even if it is environmentally stable, certain oceanic habitats offer special physiological challenges such as freezing cold water in polar seas or extremely high pressure in deep seas, which end up interfering with blood hypo-regulation of marine teleost fish. Some habitats, however, display a wide variation in

Authors' address: Departamento de Fisiologia, Setor de Ciências Biológicas, Universidade Federal do Paraná (UFPR), Centro Politécnico, Bairro Jardim das Américas, Curitiba, PR, CEP 81531-990, Brazil.
**Corresponding author:* E-mail: cafreire@ufpr.br

This chapter is dedicated to the memory of one of my Ph.D. advisors, Frau Dr Evamaria Kinne-Saffran. I owe much to her (CAF).

environmental parameters on a short time scale, imposing the need for significant and rapid modulation in osmotic homeostasis processes of their inhabitants. These variable habitats are mainly the estuarine and intertidal ecosystems, respectively, interfaces between seawater and fresh water, or between seawater and coastal land. Strictly fresh water bodies may be more or less stable, depending on their volume and latitudinal location. However, in temperate latitudes, some may partially freeze in winter, and in tropical low latitudes some may be either extremely hot or they may evaporate or present strong acid-based challenges due to intense organic matter decay.

Approximately 40% of all extant bony fishes (Ostheichthyes) live in freshwater lakes and rivers, water bodies that cover ~1% of the surface of the earth, encompassing only ~0.01% of the planet's volume of water. The other 60% of the species live in seawater. Of these, 78% are coastal or littoral (Nelson, 1994). The oceans cover 70% of the surface of the earth, and hold 97% of the water volume (Nelson, 1994; Pough *et al.*, 1996). These numbers reflect the great freshwater diversity of bony fishes. In addition, both in freshwater and seawater, the largest number of species is tropical (Nelson, 1994). The bony fishes (the largest taxon of vertebrates) include the Dipneusti (lungfishes) and Crossopterygii (coelacanth), and the Brachiopterygii (bichirs), besides the Actinopterygii (ray-finned fishes). The Actinopterygii are the Chondrostei (sturgeons and paddlefishes) and the Neopterygii; the latter includes the Lepisosteiformes (gars), the Amiiformes (bowfins), and the teleosts (Nelson, 1994; Pough *et al.*, 1996). Teleostean evolution occurred largely in the Mesozoic period, with all phyletic lineages established in the Cretaceous (Gilbert, 1993). This chapter will deal only with teleosts, which comprise the vast majority of bony fish nowadays occupying a wide variety of habitats, and will focus on showing how some environments challenge fish osmoregulation by offering extreme values of temperature, pressure, or salinity, or either fast/steep changes in those parameters, along tidal cycles.

EXTREMELY LOW TEMPERATURES

The special challenges to fish osmoregulation presented by extremely low temperatures find a 'natural laboratory' in Antarctica. This term has been used for the intertidal habitat due to its suitability for investigations of the effect of physical factors on animal physiology and natural distribution patterns (Hofmann *et al.*, 2002; Tomanek, 2002), but also seems to apply for Antarctica. Seawater in Antarctica displays an average temperature of −1.9°C, and it has been that cold for over 10 million years (Eastman, 1993; Fields and Somero, 1998). The upper 30 m of seawater in Antarctica contains small ice crystals, and fishes that dwell in superficial waters could not exist in a super cooled state, due to the abundance of seeding ice crystals (Eastman, 1993). They must possess an antifreeze protection to prevent body fluids from freezing. Antifreezers in Antarctic notothenioids are synthesized in the liver, of molecular weight ranging from 2.6 to 33.7 kDa, consisting of a peptide sequence of variable length with a repeating unit of the tripeptide Ala-Ala-Thr, with the threonine residue glycosilated (Eastman, 1993). These antifreeze glycopeptides (AFGP) are present in the blood of Antarctic notothenioids in higher concentration (35 mg/ml) than in other body fluids (Eastman, 1993). AFGP exist in higher concentrations in fishes that remain in shallower ice-laden waters. The concentration of AFGP decreases with the increase in the depth inhabited by the fish, as a result of the pressure effect reducing the freezing point of the water and the reduction until a virtual absence of ice crystals in very deep waters (Gordon *et al.*, 1962; Eastman, 1993). In fish inhabiting these deep waters, the amount of AFGP is also dependent on their mode of life: active species display an even lower content of AFGP than sluggish species (Wöhrmann, 1998).

There is a strong association between the presence of AFGP in the fluids of notothenioids, and the absence of glomeruli in their kidneys. The secondary loss of glomeruli has been selected as it prevents the loss of AFGP in the urine; molecules below ~68 kDa pass through the glomerular barrier (Eastman, 1993). Or, alternatively, the energetic costs of reabsorption or resynthesis of AFGPs would be huge. Aglomerularism is of significant adaptive value in the Antarctic environment, and is apomorphic in notothenioids, as notothenioids of lower latitudes (New Zealand, Southern South America) lack AFGP and display glomerular kidneys (Eastman and DeVries, 1986; Eastman, 1993). Interestingly, the prevention of urinary loss of AFGP is not the sole answer for the condition of aglomerularism, as it is found in temperate or tropical species, even freshwater species or marine species that tolerate seawater dilution (Eastman and DeVries, 1986; Beyenbach and Baustian, 1989; Beyenbach, 1995, 2004; Bone *et al.*, 1995; Baustian *et al.*, 1997). Aglomerular nephrons produce urine basically by proximal tubular secretion of NaCl, Mg, and sulphate (Beyenbach and Baustian, 1989; Beyenbach, 1995,

2004). This is a function shared by proximal tubules of glomerular nephrons as well, leading to the conclusion drawn by Beyenbach (2004), that '*the difference between glomerular and aglomerular urine formation is more a difference of degree than of kind*'. The very cold long-standing temperatures of Antarctica did not result in any significant change in the renal function of their fish fauna, apart from the loss of glomeruli.

Glomeruli are not the only resource adaptively dismissed in Antarctic notothenioids. These fishes also lack the heat shock response, expressing constitutively the 'normally' inducible *hsp70* gene, for housekeeping protein chaperone function during protein synthesis in a cold-denaturing environment (Place and Hofmann, 2005), but not responding with upregulation of heat shock proteins upon heat shock (Somero *et al.*, 1998; Hofmann *et al.*, 2000; Buckley *et al.*, 2004). Being the most stenothermal fishes of all (Somero *et al.*, 1998; Somero, 2005), they cannot tolerate water temperature rising to 5-6°C, a result of long-time evolution in stably cold waters (Somero *et al.*, 1998; Somero, 2005). Interestingly, the stenothermy of Antarctic notothenioids is associated with a high temperature sensitivity of the K_m of their enzymes; eurythermal species display enzymes with much less temperature-sensitive K_ms (Somero *et al.*, 1998). Antarctic fish never encounter temperature fluctuations in their environment. This finding highlights the absolute contrast between the extremely stable Antarctic and the extremely variable estuarine/intertidal habitats, as will be evident below. Interestingly, as it happens with aglomerularism, the heat shock response is lost in Antarctic notothenioids, but is found in New Zealand notothenioids (Hofmann *et al.*, 2005).

There is a large amount of data on the effect of temperature on osmoregulation. Temperature strongly influences metabolic rates in ectotherms such as fish, but the direct effects on osmoregulation may be very complex (Somero, 2002). The effects of diurnal temperature cycling as in the intertidal/estuarine habitat (Somero, 2002) may be rather different from seasonal effects on temperate species subject to freezing in winter, and also different from long-term evolution under very low temperature conditions such as in Antarctica, leading to the extreme stenothermy of Antarctic notothenioids (Somero, 2005). The last two situations will be discussed in this section of the chapter, and the first situation (diurnal/tidal variation) will be discussed later in the chapter.

Temperature will likely affect ion diffusive rates across membranes, with probable increase in diffusive fluxes of ions and water upon temperature increase (Metz *et al.*, 2003), and conversely, an expected decrease in ion diffusive fluxes upon temperature reduction, including, in Antarctic fish, a putative special role for the antifreeze peptides, in blocking ion channels, leading to reduced ion leakage (Pörtner *et al.*, 1998). Differences in acclimation temperatures are often related to differences in lipid (fatty acid, cholesterol) composition of membranes of osmoregulatory tissues of fish. In rainbow trout, for example, not only the degree of unsaturation of fatty acids from membrane lipids, but also their chain length are temperature-dependent, and cold exposure increases the incorporation of polyunsaturated fatty acids into phosphatidylserine (Hazel, 1984; Maffia *et al.*, 1998). These changes in chemical composition of the membrane lipids upon cold acclimation, in order to ensure proper barrier function for the membrane have been termed 'homeoviscous adaptation' (Hazel, 1984), a phenomenon actually first described for *Escherichia coli*, under temperature increase (Sinensky, 1974). On their turn, changes in the lipid microenvironment can affect the activity of membrane-bound transport enzymes such as the Na,K-ATPase, as indicated for the intestinal Na,K-ATPase of the gilthead seabream *Sparus aurata* (Almansa *et al.*, 2003). Temperature will also directly affect the activity of the Na,K-ATPase, and while pumps such as the Na,K-ATPase display a Q_{10} of 2-4, leak processes are much less temperature sensitive, with a Q_{10} of ~1 (Pörtner *et al.*, 1998). In some studies, a direct relationship between temperature of acclimation and specific activity of the Na,K-ATPase is observed. For example, acclimation of the common carp *Cyprinus carpio* to a lower temperature (15°C) resulted in lower activities of the Na,K-ATPase. However, this lower activity was compensated by increased expression of the enzyme (Metz *et al.*, 2003). The expression of different isoforms resulting in different enzyme activities after long periods of acclimation has also been proposed (Metz *et al.*, 2003). A direct relationship was also observed when Atlantic salmon (*Salmo salar*) smolts were transferred to seawater at different temperatures (2, 4, and 6°C); the activities of the branchial Na,K-ATPase (and their rate of increase along days in seawater) were directly related to the temperature of acclimation (Arnesen *et al.*, 1998). A positive correlation between temperature of acclimation and gill Na,K-ATPase activity was also found for the juvenile turbot *Scophthalmus maximus* (Imsland *et al.*, 2003). It is important to add that another parameter associated to temperature may actually directly interfere with fish osmoregulation: photoperiod. In the arctic charr (*Salvelinus alpinus*), it has been shown that seasonal variation in seawater tolerance by this species is related to the photoperiod, with increased seawater tolerance occurring concomitant with extended periods of daylight in late spring, when the diadromous forms of the species migrate downstream to feed in coastal waters (Johnsen *et al.*, 2000; Jørgensen and Arnesen, 2002). The same pattern was observed in the congener *Salvelinus fontinalis* (brook charr, Claireaux and Audet, 2000). In any case, independently of the exact mechanisms mediating the interaction temperature-osmoregulation, it is clear that mostly for relatively stenothermal species, either too cold or too hot temperatures, such as, respectively, 2 and 17°C for a salmon smolt (*Salmo salar*), will impair osmoregulation (Staurnes *et al.*, 2001).

It should not be forgotten that long-time (over 10 million years) evolution under constant low temperatures surely led to optimization of cellular function, 'cold compensation', under these conditions (Kunzmann, 1990; Lucassen *et al.*, 2003). To cite a few examples: (1) a higher amount of unsaturated phospholipids is observed to maintain membranes fluid next to freezing temperatures (Wöhrmann, 1998); (2) cold-adapted protein synthesis machineries display low activation energies and high RNA translational capacities at similar RNA:protein ratios (Storch *et al.*, 2005); (3) partial temperature compensation of brain ATP-generating capacities (Somero *et al.*, 1998); and (4) cold-adapted enzymes with amino acid substitutions leading to increased flexibility in regions of the molecule that affect the mobility of structures next to active sites, causing higher K_m values and higher K_{cat} values at a common temperature of assay (example, lactate dehydrogenase A_4, Fields and Somero, 1998).

The 'final product' of osmoregulatory work comprises the extracellular concentrations, blood osmolality and ion levels. Thorough investigations by Arthur DeVries and collaborators in the seventies and eighties have clearly demonstrated that Antarctic notothenioids display higher blood osmolality due to elevated NaCl concentrations, when compared to arctic, temperate/subtropical or tropical fish (Dobbs and DeVries, 1975; O´Grady and DeVries, 1982). This pattern has been confirmed in later studies (Gonzalez-Cabrera *et al.*, 1995; Romão *et al.*, 2001), and is summarized in Figure 9.1. Interestingly, arctic seawater fishes (mean ± standard error, Na 202 ± 3.1 mM, Cl 185 ± 3.9 mM) display an intermediary position between the Antarctic (Na 259 ± 4.2 mM,

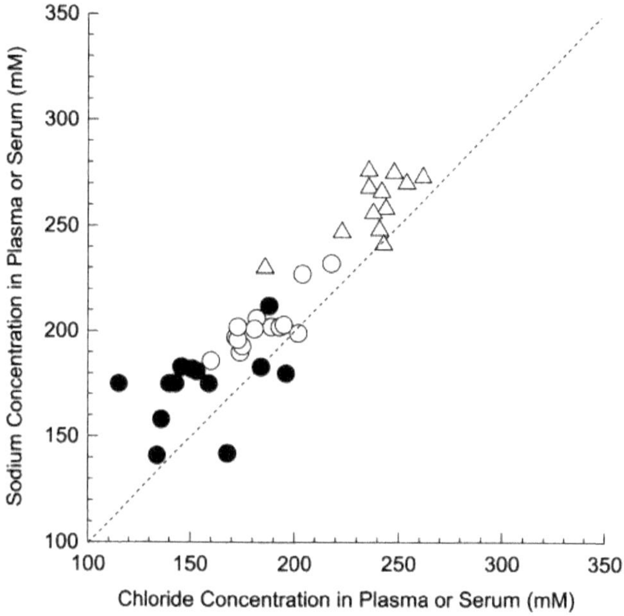

Fig. 9.1 Relationship between sodium and chloride concentrations in plasma or serum in marine temperate/subtropical/tropical (●), artic (○), and Antarctic (△) teleost fish. Values plotted for estuarine fish when in full-strength seawater. Diagonal traced line represents iso-ionic levels of sodium and chloride. Sodium concentrations are higher than respective chloride concentrations in the same fish, for all environments ($P < 0.0001$ for arctic and Antarctic, and $P < 0.05$ for temperate/subtropical/tropical fishes, paired Student´s t-test). Sodium and chloride values for Antarctic fish are higher than values for either arctic or temperate/subtropical/tropical fish ($P < 0.0001$ for all comparisons, unpaired Student´s t-test), and values for arctic fish are also higher than values for temperate/subtropical/tropical fish ($P < 0.0001$ for sodium, and $P < 0.001$ for chloride, unpaired Student´s t-test). Values for temperate and tropical species plotted are from: *Eugerres plumieri* (Plaza-Yglesias *et al.*, 1988), *Lophius americanus* (Beyenbach and Baustian, 1989), *Lophius piscatorius* and *Pleuronectes flesus* (Evans, 1993), *Fundulus* sp. and *Muraena* sp. (Bone *et al.*, 1995), *Scophthalmus maximus* (Gaumet *et al.*, 1995), *Opsanus tau* (Baustian *et al.*, 1997), *Sparus sarba* (Kelly and Woo, 1999), *Mylio macrocephalus* (Kelly *et al.*, 1999), *Sphoeroides testudineus* (Prodocimo and Freire, 2001), *Sparus aurata* (Laiz-Carrión *et al.*, 2005). Values for arctic species plotted are from: *Agonus acipenserinus, Anoplopoma fimbria, Atherestes stomias, Daisycottus setiger, Gadus macrocephalus, Hemilepidotus hemilepidotus, Hemitripterus villosus, Hippoglossus stenolepis, Lepidosetta bilineata, Lycodes polaris, Malacocottus zonurus, Myoxocephalus polyacanthocephalus, M. scorpius, Pleurogrammus monopterygius, Theragra chalcogrammus* (O´Grady and DeVries, 1982). Due to a wide latitudinal distribution, some species may also be referred as temperate, instead of polar/arctic. However, they were considered polar, as they have been collected in the Bering Sea, 56°N (O´Grady and DeVries, 1982). Values for Antarctic species plotted are from: *Trematomus dendronotus, T. lepidorhinus, T. loenabergii, Rhigophila dearborni* (Dobbs and DeVries, 1975), *Dissostichus mawsoni, Gymnodraco acuticeps, Trematomus borchgrevinki, T. hansoni, T. nicolai* (O´Grady and DeVries, 1982), *Trematomus bernacchii, T. newnesi* (Gonzalez-Cabrera *et al.*, 1995), *Notothenia neglecta* (Romão *et al.*, 2001). The names of the species are as cited in the original references.

Cl 237 ± 5.4 mM) and temperate/subtropical/tropical fish (Na 174 ± 5.1 mM, Cl 154 ± 6.5 mM), as already noted (O´Grady and DeVries, 1982), clearly apparent in Figure 9.1, and further confirmed by direct two-sample pair wise comparisons (Fig. 9.1). Higher plasma Na than Cl is apparent in all groups, being a general feature of teleosts; irrespective of latitude. Why is there an elevated NaCl in blood in Antarctic fish, and why do arctic fish display this intermediary condition? The reason is obviously related to water temperatures, as would be expected. In fact, arctic waters have been cold for a shorter period than Antarctic waters, and are warmer in summer (O´Grady and DeVries, 1982). And coherently, the Labrador species *Myoxocephalus scorpius* and *Gadus ogac* have displayed increased blood osmolalities in spring, when water temperatures were of –1.7°C, than in summer, when temperatures were of 4-7°C (Gordon *et al.*, 1962). This is also consistent with the presence of antifreeze peptides in the skin, but not in the plasma of boreal fish (Eastman, 1993). In addition, many of the arctic fish studied by O´Grady and DeVries (1982) reach lower latitudes in their distribution range, being also considered as temperate fish (www.fishbase.org).

There are at least two possible explanations for greater salt concentrations in polar than in temperate/subtropical/tropical fish. The first explanation involves the role of additional osmolytes in freezing temperature depression of the fluid. It could, however, be argued that Antarctic fish already have the antifreeze glycopeptides efficiently protecting their body fluids from freezing (Eastman, 1993). On the other hand, in fact blood NaCl in Antarctic fish contributes with 40% of the total freezing point depression (O´Grady and DeVries, 1982; Eastman, 1993). The second explanation would be to 'save energy' displaying, for example, a lower branchial Na,K-ATPase activity, resulting in less salt secretion and a smaller osmotic gradient to sea water than that maintained by fishes of milder temperatures. This possibility has been discussed by Gonzalez-Cabrera and collaborators (1995), who studied the Antarctic *Trematomus bernacchii* and *T. newnesi*, and have observed an enhancement of the activity of this enzyme in warmer temperatures (4°C), when compared to the activity measured in the normal temperature of their habitat, –1.5°C. This finding and suggestion is consistent with the report of Pörtner and collaborators (Pörtner *et al.*, 1998), that evolutionary adaptation to cold temperatures leads to reduction in the activity of the Na,K-ATPase: tropical fish display higher activities than temperate fish, which on their turn display higher activities than polar fish. On the other hand, Ventrella and collaborators (1993) have found enhancement of Na,K-ATPase activity in osmoregulatory tissues in the long-term cold-acclimated sea bass (*Dicentrarchus labrax*) (Ventrella *et al.*, 1993), and this result has been reported for other species as well (Pörtner *et al.*, 1998), which could be due to an increased number of enzyme pumps or catalytic activity of individual transporters. Most probably, both explanations are complementary. Elevated NaCl leading to a higher blood osmolality in polar fish indeed aids in the prevention of freezing, and also means energy savings for the fish, as the maintenance of a lower gradient with respect to external sea water is indeed less costly, demanding less energy input into Na,K-ATPase function (O´Grady and DeVries, 1982; Eastman, 1993; Gonzalez-Cabrera *et al.*, 1995; Pörtner *et al.*, 1998). However, one must also consider, as already mentioned above, that the enzymes of Antarctic fish evolved along millions of years in the very cold and stable temperatures, having been selected, for instance, for changes in amino acid sequence, leading to increase in the flexibility of regions involved in their catalytic conformational changes, supposedly to reduce the enthalpic energy barriers to conformational changes, increasing catalytic rates in the cold environment (Fields and Somero, 1998). Furthermore, *in vitro* values of Na,K-ATPase specific activity are strongly dependent on the purity of the membrane microsomal preparation. Crude homogenates lead to much lower specific activity. Further, *in vitro* enzyme activities quantified under optimal conditions reflect the number of functional molecules in the membrane, and not the actual *in vivo* activity of the enzyme (Kültz and Somero, 1995), and may also reflect the isoform(s) expressed under different acclimation regimes (Gonzalez-Cabrera *et al.*, 1995), as well as the protein turnover and biosynthetic capacity of the tissue (Pörtner *et al.*, 1998). In addition, all enzymes display a temperature-dependent activity curve. A comparison of activity as a function of temperature only has meaning under the same isolation/preparation procedure and with all other assay parameters kept constant. Some of the parameters have been considered in the analysis of Pörtner and collaborators (1998), but they have included another variable, mode of life (active/sluggish species) in their study, which may lead to confusion in the detection of trends associated to latitudinal temperature; active and sluggish species show very different metabolic rates under the same conditions (Somero *et al.*, 1998). Gathering data for a few species, and partially interconverting data of gill Na,K-ATPase activities obtained in different temperatures, Pörtner and colleagues (1998) found evidence that evolutionary adaptations to cold leads to reduced activities, with temperate species being intermediate between tropical and polar species (Pörtner *et al.*, 1998). In conclusion, gill Na,K-ATPase activities of polar fish may have a trend towards lower values even at their optimum temperature, when compared to their temperate/tropical orthologs (Gonzalez-Cabrera *et al.*, 1995; Pörtner *et al.*, 1998). On the other hand, the intestinal Na,K-ATPase specific activity measured at 1°C for the Antarctic *Trematomus bernacchii* was the same as that measured in the eel *Anguilla anguilla* at 18°C (Maffia *et al.*, 1998). In addition, temperate fish have been claimed to display lower Na,K-ATPase activities than tropical fish (Pörtner *et al.*, 1998), but their extracellular NaCl levels, to our knowledge, have never been shown to be distinguishable (Fig. 9.1). In addition, if the branchial Na,K-ATPase activity of Antarctic fish increases upon fish acclimation to 4°C when compared to –1.5°C (Gonzalez-Cabrera *et al.*, 1995), that may mean that the enzymes of Antarctic fish do not work at their optimum/maximum activity in the temperature of the natural habitat of the fish. But, do the enzymes of temperate/tropical fish work at their maximum in their natural environmental temperatures? If not, if raising the temperature of the assay or of the acclimation temperature of the fish raises the activity of their branchial Na,K-ATPases, does it mean that they are also saving energy in their normal habitat temperatures? Temperature directly affecting metabolism and activity is actually a general feature of ectotherms. Additionally, higher than expected metabolic rates have been measured for Antarctic fish, again indicating long-term evolutionary adaptation to this environment (Montgomery and Wells, 1993). In conclusion, biochemical and physiological cold-adaptation and their effects on osmoregulation of teleosts is a complex and partially unsettled matter which can still profit from controlled and broad comparative surveys.

The deep sea is also mostly an extremely stable environment, except perhaps in the neighborhood of thermal vents, thus with its fish fauna showing many convergent features with respect to Antarctic fish (Montgomery and Wells, 1993; Kelly and Yancey, 1999; Yancey *et al.*, 2002). Bathyal and abyssal fish display elevated levels of trimethylamine oxide (TMAO) in their tissues, when compared to fish that live in shallow depths. Teleosts of shallow waters display 10-70 mmol/kg wet weight of TMAO, while, for example, bathyal teleosts have 103 ± 9, and abyssal species have 197 ± 2 (Kelly and Yancey, 1999; Yancey *et al.*, 2002). The high content of TMAO is hypothesized to aid in osmoregulation both directly, by reducing the energetic costs of osmoregulation, and indirectly, by counteracting deleterious effects of pressure on protein stability (Kelly and Yancey, 1999; Yancey *et al.*, 2002). Again, the similarity to polar fishes arises; some of the latter have high glycerol or TMAO (~150 mmol/kg muscle) concentrations in their tissues (Kelly and Yancey, 1999; Yancey *et al.*, 2002). Elevated TMAO in polar teleosts may also exert a protective function, only this time not against high pressure, but against the high blood NaCl concentrations (Kelly and Yancey, 1999).

EXTREMELY HIGH SALINITIES

What happens in fish that face extreme values of the most important environmental variable for osmoregulation: salinity? Are there special mechanisms involved in hyporegulation in hypersaline waters? Or do the gills of these fish possess a especially powerful salt secretion system, involving the same mechanism of hyporegulation acting when in full-strength seawater, but able to establish high salt secretory fluxes and even steeper salt gradients with respect to ambient salt water? The latter possibility seems rather more probable. Some euryhaline fish are extremely tolerant of hypersaline seawater (Fig. 9.2). Amazingly, some incredibly euryhaline fish that tolerate hypersaline waters are originally freshwater fish, such as the sailfin molly *Poecilia latipinna* (Nordlie *et al.*, 1992) (Fig. 9.2), keeping huge osmotic gradients with respect to ambient saline water. Some species display tolerance to hypersaline media, but actually are not normally found in hypersaline habitats, an example of which is the Mozambique tilapia *Oreochromis mossambicus*. This species has been a model for strong hyporegulatory capacity in hypersaline medium, but it naturally inhabits freshwater and brackish water environments in Southeast Africa. The species has been introduced in other continents and is now found also in America and Asia, where it also occurs in seawater (Kültz and Onken, 1993). Morphological studies, using this species in hypersaline seawater, have demonstrated hyperplasia and hypertrophia of chloride cells, increasing the number of apical crypts, both in the gill filament and in the opercular epithelium (Kültz *et al.*, 1995), even forming chloride cell complexes (Uchida *et al.*, 2000). The same basic result was found for the euryhaline sea bass *Dicentrarchus labrax* in 70 ppt seawater, with the additional record of increased mitochondrial volume and Na,K-ATPase sites in doubly concentrated seawater (Varsamos *et al.*, 2002). An

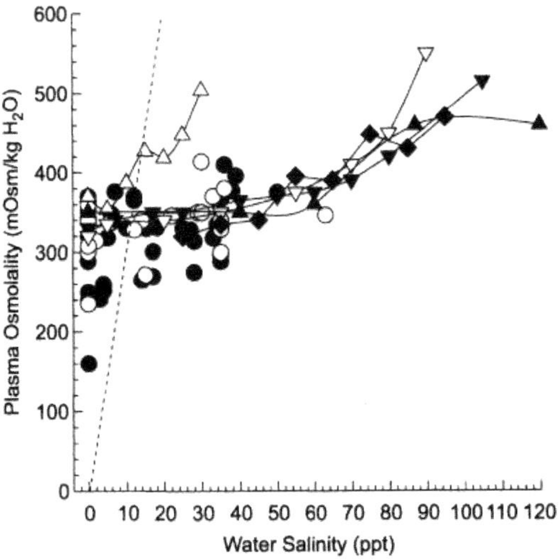

Fig. 9.2 Osmoregulatory homeostasis in euryhaline, freshwater/estuarine (○), seawater/estuarine (●), or diadromous (□, hidden by other symbols) teleost fishes when exposed/acclimated to variable salinities. Several species were gathered in those categories above; they may show widely different habits in what concerns their occupation of estuaries. When more than one time of exposure/acclimation was available in the source study, the most stable plasma osmolality value, normally for the longest time was chosen. Extremely euryhaline species for which many salinity values were available, also reaching hypersaline salinities, were plotted separately. Their symbols are connected by a line, and they are identified below. Diagonal traced line represents isosmotic values. Values for euryhaline freshwater/estuarine species plotted are from: *Fundulus kansae* (Fleming and Stanley, 1965), *Eugerres plumieri* (Plaza-Yglesias *et al.*, 1988), *Salaria fluviatilis* (Plaut, 1998), *Oreochromis mossambicus* (Uchida *et al.*, 2000), *Stizostedion lucioperca* (Brown *et al.*, 2001), *Tetraodon nigroviridis* (Lin *et al.*, 2004), *Oreochromis mossambicus* (Cataldi *et al.*, 2005). Values for euryhaline seawater/estuarine species plotted are from: *Blennius pholis* (Davenport and Vahl, 1979), *Fundulus heteroclitus* (Jacob and Taylor, 1983), *Boleophthalmus boddaerti* and *Periophthalmus chrysospilos* (Chew and Ip, 1990), *Centropomus undecimalis* (Pérez-Pinzón and Lutz, 1991), *Pleuronectes flesus* (Evans, 1993), *Sparus aurata* (Mancera *et al.*, 1993), *Scophthalmus maximus* (Gaumet *et al.*, 1995), *Fundulus heteroclitus* (Zadunaisky *et al.*, 1995), *Opsanus tau* (Baustian *et al.*, 1997), *Dicentrarchus labrax* (Jensen *et al.*, 1998), *Salaria pavo* (Plaut, 1998), *Fundulus heteroclitus* (Marshall *et al.*, 1999), *Scophthalmus maximus* (Imsland *et al.*, 2003), *Sparus aurata* (Laiz-Carrión *et al.*, 2005). Values for diadromous species plotted are from: *Anguilla anguilla* (Holmes and Donaldson, 1969), *Salmo salar* (Arnesen *et al.*, 1998), and *Salvelinus alpinus* (Jørgensen and Arnesen, 2002). Data with lines connecting the symbols are presented, for the freshwater/estuarine fishes *Poecilia latipinna* (-▽-, Nordlie *et al.*, 1992) and *Gambusia holbrooki* (-△-, Nordlie and Mirandi, 1996), the salt marshes inhabitant *Cyprinodon variegatus* (-▼-, Haney and Nordlie, 1997), the tilapia hybrid 'california mozambique tilapia' from the hypersaline Salton Sea in California, cross between *Oreochromis mossambicus* and *O. urolepis hornorum* (-◆-, Sardella *et al.*, 2004b), and the California killifish *Fundulus parvipinnis*, also found in hypersaline lagoons (-▲-, Feldmeth and Waggoner III, 1972). The species *Oreochromis mossambicus, Fundulus heteroclitus, Scophthalmus maximus*, and *Sparus aurata* appear more than once in this figure, as different studies had different protocols of exposure to different salinities. The names of the species are as cited in the original references.

electrophysiological study using the isolated opercular epithelium of *O. mossambicus* acclimated to hypersaline media led to compatible results, of increased capacity for salt secretion, with a decrease in the leak conductance resulting from less permeable tight junctions (Kültz and Onken, 1993).

Although tolerating hypersaline media, some fish display signs of a high energy cost of osmoregulation under these high salinities, such as 55 ppt for the milkfish *Chanos chanos* (Swanson, 1998), which actually is a species naturally found in hypersaline lagoons (of up to 158 ppt salinity) (Swanson, 1998). A similar result has been observed in the extremely tolerant gobiid from the Australian desert, which inhabits mound springs and temporary water bodies around Lake Eyre drainage basin, *Chlamydogobius eremius* (Thompson and Withers, 2002). These authors found increased metabolic rates in gobies maintained in double-strength seawater (70 ppt) when compared to full-strength seawater (35 ppt). Freshwater values were the lowest, i.e., there was a direct relationship between metabolic rate and salinity (Thompson and Withers, 2002), ascribed to increased energetic cost of osmoregulation in higher salinities. On the contrary, the euryhaline Mozambique tilapia *Oreochromis mossambicus*, caught in brackish water, after 1 month of acclimation to different salinities, displayed oxygen consumption rates in 1.6 × seawater similar to rates in freshwater, both rates above those measured in full strength seawater, the experimental salinity closest to the natural salinity of their environment (Iwama *et al.*, 1997). This 'U-shaped' curves of oxygen consumption conform to a pattern frequently observed for the activity of the branchial Na,K-ATPase in euryhaline species (Imsland *et al.*, 2003). A high Na,K-ATPase activity in either freshwater or seawater and hypersaline seawater, when compared to brackish water (6-12 ppt) was observed, for example, in the black seabream *Mylio macrocephalus*, also accompanied by an increase in chloride cell numbers in hypersaline media (50 ppt) (Kelly *et al.*, 1999).

The 'California' Mozambique tilapia is a hybrid of *O. mossambicus* and *O. urolepis hornorum*, and inhabits the hypersaline (43 ppt) Salton Sea in California. It has been shown to be of similar salinity tolerance as *O. mossambicus* (Fig. 9.2), with signs of osmoregulatory stress such as the increase in apoptotic gill chloride cells, appearing at salinities higher than 55-65 ppt, accompanied by a decrease in oxygen consumption rates (Sardella *et al.*, 2004b), in a pattern different from the 'high cost of osmoregulation', cited above for *Chanos chanos* (Swanson, 1998). Temperature influence is noted, as both low (15°C) and high (35°C) temperatures decrease the capacity of the California Mozambique tilapia to tolerate hypersaline media (Sardella *et al.*, 2004a).

The sheepshead minnow *Cyprinodon variegatus* is another species that occurs in hypersaline media (up to 142 ppt) for at least some time (hours to a few days), inhabiting brackish water coastal salt marshes subject to intense salinity fluctuations (Haney and Nordlie, 1997). This minnow shows increased plasma osmolality (Fig. 9.2) and, as the tilapia cited above, decreased metabolic rates when in higher salinities, above 40 ppt, a suggestion of reduced energy expenses in hypersaline media, in order to increase its tolerance time (Haney and Nordlie, 1997). The California killifish *Fundulus parvipinnis* is also extremely tolerant of hypersaline media, naturally occurring in hypersaline lagoons of 55, 87, or even 128 ppt, where its plasma osmolality increased from 350 to 450 mOsm/kg H_2O (Fig. 9.2), with a concomitant decrease in tissue water content (Feldmeth and Waggoner III, 1972). In conclusion, although metabolic strategies to deal with extreme high salinity may vary among those rare species that successfully tolerate and even thrive in such waters, all of them must be able to produce large outwardly directed salt fluxes through their branchial epithelia, and possibly reduce inwardly directed diffusive fluxes (Kültz and Onken, 1993; Kültz *et al.*, 1995; Uchida *et al.*, 2000; Varsamos *et al.*, 2002). It must be added that transport processes in the gastrointestinal tract are also potentially upregulated in those species, so that water gain is in effect, even with such enormous osmotic gradients.

UNSTABLE HABITATS: ESTUARINE AND INTERTIDAL

Rocky intertidal coasts are termed 'natural laboratories' for investigations on the effects of a changing environment on the physiology of their inhabitants (Hofmann *et al.*, 2002; Tomanek, 2002; Tomanek and Helmuth, 2002). This could be extended fairly to estuarine habitats. The large body of data available on estuarine/intertidal fish physiology in general, and particularly on fish osmoregulation, can be divided into two main approaches. The first, more classical approach, detected and revealed the remarkable euryhalinity and osmoregulatory capacity of these fish (House, 1963; Evans, 1967; Gordon *et al.*, 1970; Davenport and Vahl, 1979; Bridges, 1993; Prodocimo and Freire, 2001), as well as their eurythermicity and euryoxic capabilities (Bridges, 1993; Gracey *et al.*,

2001), and even tolerance to emersion, in intertidal amphibious fishes such as clingfishes (Gordon *et al.*, 1970) and mudskippers (Gordon *et al.*, 1969, 1978). Regular aerial exposure may lead to a shift from ammonotelism to ureotelism (Gordon *et al.*, 1970, 1978). A second, more recent approach, involves the study of their physiological response to the stress imposed by steep and fast variations in salinity, temperature, and dissolved oxygen, to highlight the three most important variables (Dahlhoff, 2004; Hofmann, 2005). These variables interact strongly, e.g., salinity and temperature directly affect fish respiration, temperature affects the metabolic rate with a Q_{10} of 2-3, salinity affects oxygen solubility and the diffusion conductance of the gills (Jensen *et al.*, 1993).

The recognition of the natural stress to which the inhabitants of these ecosystems are regularly exposed led to their appreciation as a 'natural laboratory' (Hofmann *et al.*, 2002; Tomanek, 2002). Investigating how these intertidal/estuarine animals respond physiologically to the stress of their natural habitat makes strong sense, as it integrates the variations in salinity, temperature, dissolved oxygen, pH, etc. The stress response in fish has traditionally involved the assessment of blood glucose and cortisol levels, but more recently, of the expression of regulated heat shock proteins, the heat shock response (HSR) (reviews in Dahlhoff, 2004; Hofmann, 2005). The HSR is the regulated expression of the heat-shock proteins, molecular chaperones whose function is to recognize and bind to unfolded proteins, preventing their aggregation, and sometimes allowing their correct refolding (Place and Hofmann, 2001, 2005; Dahlhoff, 2004). These proteins appear at a higher rate upon temperature, salinity, anoxia, or chemical stress (Dietz, 1994; Place and Hofmann, 2001; Deane *et al.*, 2002; Dahlhoff, 2004; Hofmann, 2005). There are many more studies on temperature activation of heat-shock genes than on salinity activation (Hofmann, 2005). With salinity, most reports refer to salinity increase, hyperosmotic shock (Pan *et al.*, 2000), although not exclusively. As would be expected, salinity reduction is even more common in intertidal or estuarine habitats, also representing an osmoregulatory challenge to the animal. However, it may well be that salinity increase more likely leads to protein denaturation than salinity reduction. Acclimation to both hyper- (50 ppt) and hypo-osmotic (6 ppt) salinities for 8 months led to increased levels of hepatic hsps upon heat shock, in the black seabream *Mylio macrocephalus* (Deane *et al.*, 2002).

Diadromous species face environmental variations, especially in salinity, although in a different time scale when compared to estuarine/ intertidal species. Species with this habit have been examined with respect to the HSR as elicited by osmotic/saline stress to isolated tissues: hsp70 was significantly induced in branchial lamellae and hepatic tissue upon hyperosmotic shock (Smith *et al.*, 1999). A specific connection between hsps and osmoregulation seems to exist, with complementary parts played by hsp70 and hsp90 at the onset of seawater adaptation of the salmon (Pan *et al.*, 2000); hsp70 has more of a chaperone-like function, while hsp90 is involved in signal transduction regulating salt secretion by the gills (Pan *et al.*, 2000). A direct linkage between hyperosmotic stress and hsps has been shown for other systems, and will probably be more clearly revealed for estuarine and intertidal fish in the near future. Indeed, in the mammalian kidney, cells of the inner medulla respond to hyperosmotic stress via TonEBP, the transcription factor 'tonicity response element binding protein', which is then translocated to the nucleus, where it can bind to osmoprotective target genes, such as that of hsp70, and genes involved with the organic osmolyte response (Kültz, 2005).

Temperature has been largely the most studied variable correlated with zonation, i.e., vertical distribution of populations in intertidal rocky shores (review in Somero, 2002). Intertidal fish have been a special focus of investigation concerning the HSR or activity of heat shock proteins, as related to the normal temperature fluctuation cycles of their habitat. A special model has been the marine goby *Gillichthys mirabilis*, of extraordinary tolerance to environmental variation (Gracey *et al.*, 2001), whose major cytosolic molecular chaperone Hsc70 has shown ATPase activity associated to its chaperoning cycles along the range of temperatures encountered by the fish in its habitat: 10-35°C (Place and Hofmann, 2001). The plasticity of the HSR response has also been demonstrated in this gobiid, in that the threshold induction temperature for the 70-kDa and 90-kDa heat shock proteins (Dietz, 1994), and the DNA-binding activity of heat-shock factor 1 (HSF1), the transcriptional regulator of all inducible hsp genes (Buckley and Hofmann, 2002) are not genetically 'fixed', but rather respond to the acclimation temperature. A comparative study has been conducted with sculpins of the genus *Oligocottus*: the intertidal sculpin *O. maculosus* found in upper and lower tidepools, and its congener, the fluffy sculpin *O. snyderi*, found only in lower tidepools during low tide (Nakano and Iwama, 2002). The tidepool sculpin was found to display higher lethal and threshold induction temperatures for hsp70 and higher constitutive hsp70 levels when compared to the fluffy sculpin, in accordance with its wider thermal tolerance (Nakano and Iwama, 2002). Again, the study has been conducted with respect to temperature fluctuations, but future work should increase the volume of data on variations in salinity and dissolved oxygen. Different stressors will not necessarily follow the same pathways, as has been demonstrated by Kültz (1996) in gill cells of the gobiid G. *mirabilis*. The isoforms of the inducible hsp70 were only upregulated by heat shock, and not by hyperosmotic shock, evidencing the stressor specificity of the HSR (Kültz, 1996).

Another cellular response intensely activated in these ecosystems is the antioxidant protection system, largely required after a cycle of hypoxia during low tide in estuaries and tide pools, followed by re-oxygenation resulting from the return of seawater with the flow tide (Ross *et al.*, 2001). It should be pointed out that this recent approach has been mostly applied to the intertidal habitat, not as much to the estuarine habitat (Ross *et al.*, 2001), and also, not as much for fish, but mainly for invertebrates and algae, a fact which coherently reflects the abundance and diversity of the latter in the intertidal habitat (Abele *et al.*, 1998; Collén and Davison, 1999). Characterization of these responses in intertidal and estuarine fishes, as a result of variation in salinity, thus in direct relationship to osmoregulation, is bound to significantly increase in the upcoming years.

Strong euryhalinity and osmoregulatory capacity of intertidal fish has been shown by Bridges (1993), revising previous data available in the literature. Euryhaline teleosts, diadromous or with some transit between estuaries and either fresh- or seawater, or even resident estuarine species, display very stable plasma osmolalities. In salinities ranging between fresh water and full strength seawater (35 ppt), the diadromous species represented in Figure 9.2 had their plasma osmolalities strictly controlled, between 322 and 377 mOsm/kg H_2O, with a mean (± SEM) of 341 ± 9.0 mOsm/kg H_2O (Fig. 9.2). Plasma osmolalities of euryhaline species of marine origin varied between 160 and 410 mOsm/kg H_2O (mean ± SEM: 318 ± 8.5 mOsm/kg H_2O), when the fish were exposed to salinities between fresh water and 50 ppt. The values were vary similar to those measured for euryhaline species of freshwater origin, in salinities ranging between fresh water and 63 ppt, and osmolalities between 235 and 414 mOsm/kg H_2O (mean ± SEM: 333 ± 8.2 mOsm/kg H_2O). It is noticeable that some species have been exposed to salinities representing extreme of their tolerance range, leading to the breakdown of their osmotic homeostatic capability. For example, the marine intertidal blenny *Salaria pavo*, when in freshwater for 3 months, displays 160 mOsm/kg H_2O of blood osmolality (Plaut, 1998) (Fig. 9.2, marine/estuarine group). On the other side, the mosquitofish *Gambusia holbrooki*, after 14 days in 30 ppt seawater, displays 504 mOsm/kg H_2O of blood osmolality (Nordlie and Mirandi, 1996) (Fig. 9.2). Apart from the mosquitofish, the other species that show tolerance to hypersaline seawater have yielded very coincident curves (Fig. 9.2, symbols with lines).

Interestingly, it seems evolutionarily easier for marine fish to move into freshwater than for freshwater fish to move into the oceans (Nelson, 1994). Marine fish enter the estuary for variable periods of their life cycles much more often than freshwater fishes. In a very extensive study conducted on estuarine salt marshes of eastern North America, Nordlie (2003) has summarized data on 237 species of fishes, and classified them into one of the following categories: permanent residents, marine nursery species, diadromous fishes, marine transients, and freshwater transients (Nordlie, 2003). Marine species that use or live in the estuary for variable periods add up to 70.0% of those 237 species studied; 52.3% were marine transients, and 17.7% were marine nursery species. In contrast, freshwater transients accounted only for 15.2% of the species. A little above nine percent were permanent residents (9.3%), and only 5.5% were diadromous (Nordlie, 2003). This fact may have a much more complex explanation than simply osmoregulation difficulties. But if osmoregulation is at least part of the reason, this may imply that in general it is less stressful to deal with hypoosmotic stress than with higher salt, hyperosmotic stress. Water and salt diffusive fluxes leading at least transiently to extracellular fluid alterations result in cell volume challenges. And *in vitro* studies with animal cells or tissues (not only from fish) show that the capacity for regulatory volume decrease after hypoosmotic shock is more frequently found than regulatory volume increase after hyperosmotic shock (Kévers *et al.*, 1979; Hoffmann and Dunham, 1995). However, there are many examples of euryhaline fish that can readily switch from salt absorption to salt secretion, including estuarine species such as the killifish, or diadromous species (Zadunaisky *et al.*, 1995; Arnesen *et al.*, 1998; Marshall *et al.*, 1999; Claireaux and Audet, 2000; Johnsen *et al.*, 2000; Lin *et al.*, 2004). The mechanisms underlying this plasticity have been discussed in other chapters of this volume. Speculatively, the superior number of marine than freshwater species using estuaries (Nordlie, 2003) may be related to migratory movements between seawater and fresh water, along bony fish evolution (Griffith, 1985). Although still a matter of debate, ostracoderms had a long period of evolution in fresh water; in late Silurian ostracoderms and early jawed fishes were abundant both in freshwater and seawater (Pough *et al.*, 1996). Evidences widely accepted as favoring the hypothesis of a period of evolution in freshwater (or diluted seawater) for a common vertebrate lineage are the facts that extant fish and vertebrates in general display low (when compared to nowadays seawater) NaCl concentrations in their extracellular fluids, and that their kidneys possess the blood filtering glomeruli (Griffith, 1985; Evans, 1993; Pough *et al.*, 1996; Vize, 2004). Reduced extracellular salt means less osmotic water entry in a dilute environment, and glomeruli are ideal structures to eliminate the water that will have entered osmotically anyway, as freshwater animals are necessarily hyperosmotic (Vize, 2004). The contrary view of early vertebrate origin, with evolution and diversification in the sea, remarks that even the Myxinoidea (hagfishes), a group of exclusively marine jawless parasite vertebrates (Agnatha), and whose osmoregulatory strategy is similar to that of marine invertebrates—isosmotic and isoionic (NaCl) with respect to seawater—display glomeruli in their kidneys (Evans, 1993; Pough *et al.*, 1996). The Myxinoidea have no features that could be indicative of a freshwater ancestor (Pough *et al.*, 1996). However, hagfishes could still descend from this putative common freshwater ostracoderm ancestor lineage, having returned early to the sea and with a long evolutionary history in the stable and salty depths in the ocean, having been selected for an osmoregulatory strategy such as that of invertebrates that never left the sea (Griffith, 1985). Elasmobranchs are intermediary in the sense that they also display lower NaCl than seawater, but have developed the strategy of organic osmolyte accumulation (mostly urea) to be slightly hyperosmotic to seawater, and also always display glomeruli in their kidneys. Interestingly, there are no examples of aglomerular elasmobranchs (Vize, 2004). As their strategy promotes osmotic water entry in seawater, the presence of glomeruli would be consistent with the availability of sufficient water to promote filtration. As teleosts face desiccation in seawater, they show the trend for partial or total morphological loss of glomeruli, or else 'physiological loss', that is, glomerular intermittency (Beyenbach, 2004; Vize, 2004). Data on the rainbow trout can illustrate this variability in filtering nephrons between teleosts in freshwater or in seawater. The kidney of the seawater rainbow trout has only 5% of its total number of glomeruli perfused and filtering, causing the rainbow trout to be almost functionally aglomerular in seawater. In contrast, 45% of all glomeruli are perfused and functional in filtration when the trout is in freshwater (Beyenbach, 2004). Independent of how concentrated was the water where the first vertebrate evolved, it is still possible (and tempting to speculate) that the extant lineages of fishes and other vertebrates derive from a lineage of freshwater ostracoderms. So, admitting this hypothesis, one may consider that evolutionarily, marine bony fish descend from a lineage that entered fresh water from the sea, evolved in diluted waters, and then sometime later returned to seawater. That is, they have ancestors performing the up- and downstream migrations, or conversion from hypo- to hyper-regulation of body fluids and vice-versa. On the other hand, extant freshwater bony fish (at least some groups) may descend from the first vertebrate migration that entered fresh water from the sea, possibly ostracoderms. Evolutionarily, they may have no freshwater to seawater migration in their past, or evolutionary history. And this ancestry may genetically influence their ability to turn off salt absorption when the salt content of the water is increased. This question is fascinating and would be well addressed by a very broad and comparative analysis of hypo- or hyper-osmoregulatory capacities among the different orders of bony fish.

CONCLUSIONS

This chapter has reviewed data on teleost fish osmoregulation as challenged by environments with extreme cold temperatures, high pressures, or high salinities, or either not as much extreme, but actually steeply variable coastal habitats such as estuaries and intertidal rocky shores. Long-time evolution in freezing cold seawater led to increased blood salt and osmolality, which aids in freezing avoidance, and, as in deep waters, a protective tissue accumulation of TMAO. Both strategies may lessen hypoosmoregulatory costs. Very strong hyporegulation in hypersaline waters is relatively rare, and most likely is the result of a quantitative (extremely high salt secretory fluxes), rather than a qualitative difference in the branchial epithelium with respect to 'regular' full-strength seawater marine hyporegulators. Estuarine and intertidal habitats challenge teleost fish osmoregulation not only from salinities varying on a short time scale, but also by interactions between variations in salinity with that of other abiotic parameters such as temperature, dissolved oxygen, and pH. Estuarine and intertidal teleosts are necessarily euryhaline, and the effect of the fluctuation in these environmental parameters is well translated by the stress response pattern of the species that inhabit such ecosystems.

References

Abele, D., H. Groβpietsch and H.O. Pörtner. 1998. Temporal fluctuations and spatial gradients of environmental PO_2, temperature, H_2O_2 and H_2S in its intertidal habitat trigger enzymatic antioxidant protection in the capitellid worm *Heteromastus filiformis*. *Marine Ecology Progress Series* 163: 179–191.

Almansa, E., J.J Sánchez, S. Cozzi, C. Rodríguez and M. Díaz. 2003. Temperature-activity relationship for the intestinal Na^+-K^+-ATPase of *Sparus aurata*. A role for the phospholipid microenvironment? *Journal of Comparative Physiology* B173: 231–237.

Arnesen, A.M., H.K. Johnsen, A. Mortensen and M. Jobling. 1998. Acclimation of Atlantic salmon (*Salmo salar* L.) smolts to 'cold' seawater following direct transfer from fresh water. *Aquaculture* 168: 351–367.

Baustian, M.D., S.Q. Wang and K.W. Beyenbach. 1997. Adaptive responses of aglomerular toadfish to dilute sea water. *Journal of Comparative Physiology* B167: 61–70.

Beyenbach, K.W. 1995. Secretory electrolyte transport in renal proximal tubules of fish. In: *Cellular and Molecular Approaches to Fish Ionic Regulation*, C.M. Wood and T.J. Shuttleworth (eds.). Academic Press, San Diego, pp. 85–105.

Beyenbach, K.W. 2004. Kidneys sans glomeruli. *American Journal of Physiology: Renal Physiology* 286: F811–F827.

Beyenbach, K.W. and M.D. Baustian. 1989. Comparative physiology of the proximal tubule. From the perspective of aglomerular urine formation. In: *Structure and Function of the Kidney*, Comparative Physiology, R.K.H. Kinne (ed.). S. Karger, Basel, Vol. 1, pp. 103–142.

Bone, Q., N.B. Marshall and J.H.S. Blaxter. 1995. *Biology of Fishes*. 2nd Edition. Chapman & Hall, London.

Bridges, C.R. 1993. Ecophysiology of intertidal fish. In: *Fish Ecophysiology*, J.C. Rankin and F.B. Jensen (eds.). Chapman & Hall, London, pp. 375–400.

Brown, J.A., W.M. Moore and E.S. Quabius. 2001. Physiological effects of saline waters on zander. *Journal of Fish Biology* 59: 1544–1555.

Buckley, B.A. and G.E. Hofmann. 2002. Thermal acclimation changes DNA-binding activity of heat shock factor 1 (HSF1) in the goby *Gillichthys mirabilis*: Implications for plasticity in the heat-shock response in natural populations. *Journal of Experimental Biology* 205: 3231–3240.

Buckley, B.A., S.P. Place and G.E. Hofmann. 2004. Regulation of heat shock genes in isolated hepatocytes from an Antarctic fish, *Trematomus bernacchii*. *Journal of Experimental Biology* 207: 3649–3656.

Cataldi, E., A. Mandich, A. Ozzimo and S. Cataudella. 2005. The interrelationships between stress and osmoregulation in a euryhaline fish, *Oreochromis mossambicus*. *Journal of Applied Ichthyology* 21: 229–231.

Chew, S.F. and Y.K. Ip. 1990. Differences in the responses of two mudskippers, *Boleophthalmus boddaerti* and *Periophthalmus chrysospilos* to changes in salinity. *Journal of Experimental Zoology* 256: 227–231.

Claireaux, G. and C. Audet. 2000. Seasonal changes in the hypo-osmoregulatory ability of brook charr: The role of environmental factors. *Journal of Fish Biology* 56: 347–373.

Collén, J. and I.R. Davison. 1999. Stress tolerance and reactive oxygen metabolism in the intertidal red seaweeds *Mastocarpus stellatus* and *Chondrus crispus*. *Plant, Cell and Environment* 22: 1143–1151.

Dahlhoff, E.P. 2004. Biochemical indicators of stress and metabolism: applications for marine ecological studies. *Annual Reviews in Physiology* 66: 183–207.

Davenport, J. and O. Vahl. 1979. Responses of the fish *Blennius pholis* to fluctuating salinities. *Marine Ecology Progress Series* 1: 101–107.

Deane, E.E., S.P. Kelly, J.C.Y. Luk and N.Y.S. Woo. 2002. Chronic salinity adaptation modulates hepatic heat shock protein and insulin-like growth factor I expression in black seabream. *Marine Biotechnology* 4: 193–205.

Dietz, T.J. 1994. Acclimation of the threshold induction temperatures for 70-kDa and 90-kDa heat shock proteins in the fish *Gillichthys mirabilis*. *Journal of Experimental Biology* 188: 333–338.

Dobbs, G.H. and A.L. DeVries. 1975. Renal function in Antarctic teleost fishes: Serum and urine composition. *Marine Biology* 29: 59–70.

Eastman, J.T. 1993. *Antarctic Fish Biology: Evolution in a Unique Environment*. Academic Press, San Diego.

Eastman, J.T. and A.L. DeVries. 1986. Renal glomerular evolution in Antarctic notothenioid fishes. *Journal of Fish Biology* 29: 649–662.

Evans, D.H. 1967. Sodium, chloride and water balance of the intertidal teleost, *Xiphister atropurpureus*. I. Regulation of plasma concentration and body water content. *Journal of Experimental Biology* 47: 513–517.

Evans, D.H. 1993. Osmotic and ionic regulation. In: *The Physiology of Fishes*, D.H. Evans (ed.). CRC Press, Boca Raton, pp. 315–341.

Feldmeth, C.R. and J.P. Waggoner III. 1972. Field measurements of tolerance to extreme hypersalinity in the California killifish, *Fundulus parvipinnis*. *Copeia* 1972: 592–594.

Fields, P.A. and G.N. Somero. 1998. Hot spots in cold adaptation: Localized increases in conformational flexibility in lactate dehydrogenase A4 orthologs of Antarctic notothenioid fishes. *Proceedings of National Academy of Sciences of the United States of America* 95: 11476–11481.

Fleming, W.R. and J.G. Stanley. 1965. Effects of rapid changes in salinity on the renal function of a euryhaline teleost. *American Journal of Physiology* 209: 1025–1030.

Gaumet, F., G. Boeuf, A. Severe, A. LeRoux and N. Mayer-Gostan. 1995. Effects of salinity on the ionic balance and growth of juvenile turbot. *Journal of Fish Biology* 47: 865–876.

Gilbert, C.R. 1993. Evolution and Phylogeny. In: *The Physiology of Fishes*, D.H. Evans (ed.). CRC Press, Boca Raton, pp. 1–45.

Gonzalez-Cabrera, P.J., F. Dowd, V.K. Pedibhotla, R. Rosario, D. Stanley-Samuelson and D. Petzel. 1995. Enhanced hypo-osmoregulation induced by warm-acclimation in Antarctic fish is mediated by increased gill and kidney Na^{+}/K^{+}-ATPase activities. *Journal of Experimental Biology* 198: 2279–2291.

Gordon, M.S., B.H. Amdur and P.F. Scholander. 1962. Freezing resistance in some northern fishes. *Biological Bulletin* 122: 52–62.

Gordon, M.S., I. Boëtius, D.H. Evans, R. McCarthy and L.C. Oglesby. 1969. Aspects of the physiology of terrestrial life in amphibious fishes. I. The mudskipper, *Periophthalmus sobrinus*. *Journal of Experimental Biology* 50: 141–149.

Gordon, M.S., S. Fischer and E. Tarifeño. 1970. Aspects of the physiology of terrestrial life in amphibious fishes. II. The Chilean clingfish, *Sicyases sanguineus*. *Journal of Experimental Biology* 53: 559–572.

Gordon, M.S., W.W.-s. Ng and A.Y.-w. Yip. 1978. Aspects of the physiology of terrestrial life in amphibious fishes. III. The Chinese mudskipper *Periophthalmus cantonensis*. *Journal of Experimental Biology* 72: 57–75.

Gracey, A.Y., J.V. Troll and G.N. Somero. 2001. Hypoxia-induced gene expression profiling in the euryoxic fish *Gillichthys mirabilis*. *Proceedings of the National Academy of Sciences of the United States of America* 98: 1993–1998.

Griffith, R.W. 1985. Habitat, phylogeny and the evolution of osmoregulatory strategies in primitive fishes. In: *Evolutionary Biology of Primitive Fishes*, R.E. Foreman, A. Gorbman, J.M. Dodd and R. Olsson (eds.). Plenum Press, New York, pp. 69–80.

Haney, D.C. and F.G. Nordlie. 1997. Influence of environmental salinity on routine metabolic rate and critical oxygen tension *of Cyprinodon variegatus*. *Physiological Zoology* 70: 511–518.

Hazel, J.R. 1984. Effects of temperature on the structure and metabolism of cell membranes in fish. *American Journal of Physiology: Regulatory, Integrative and Comparative Physiology* 246: R460–R470.

Hoffmann, E.K. and P.B. Dunham. 1995. Membrane mechanisms and intracellular signalling in cell volume regulation. *International Review of Cytology* 161: 172–262.

Hofmann, G.E. 2005. Patterns of Hsp gene expression in ectothermic marine organisms on small to large biogeographic scales. *Integrative and Comparative Biology* 45: 247–255.

Hofmann, G.E., B.A. Buckley, S. Airaksinen, J.E. Keen and G.N. Somero. 2000. Heat-shock protein expression is absent in the Antarctic fish *Trematomus bernacchii* (Family Nototheniidae). *Journal of Experimental Biology* 203: 2331–2339.

Hofmann, G.E., B.A. Buckley, S.P. Place and M.L. Zippay. 2002. Molecular chaperones in ectothermic marine animals: Biochemical function and gene expression. *Integrative and Comparative Biology* 42: 808–814.

Hofmann, G.E., S.G. Lund, S.P. Place and A.C. Whitmer. 2005. Some like it hot, some like it cold: the heat shock response is found in New Zealand but not Antarctic notothenioid fishes. *Journal of Experimental Marine Biology and Ecology* 316: 79–89.

Holmes, W.N. and E.M. Donaldson. 1969. The body compartments and the distribution of electrolytes. In: *Fish Physiology*, W.S. Hoar and D.J. Randall (eds.). Academic Press, New York, Vol. 1, pp. 1–89.

House, C.R. 1963. Osmotic regulation in the brackish water teleost, *Blennius pholis*. *Journal of Experimental Biology* 40: 87–104.

Imsland, A.K., S. Gunnarsson, A. Foss and S.O. Stefansson. 2003. Gill Na^+,K^+-ATPase activity, plasma chloride and osmolality in juvenile turbot (*Scophthalmus maximus*) reared at different temperatures and salinities. *Aquaculture* 218: 671–683.

Iwama, G.K., A. Takemura and K. Takano. 1997. Oxygen consumption rates of tilapia in freshwater, seawater, and hypersaline seawater. *Journal of Fish Biology* 51: 886–894.

Jacob, W.F. and M.H. Taylor. 1983. The time course of seawater acclimation in *Fundulus heteroclitus* L. *Journal of Experimental Zoology* 228: 33–39.

Jensen, F.B., M. Nikinmaa and R.E. Weber. 1993. Environmental perturbations of oxygen transport in teleost fishes: Causes, consequences and compensations. In: *Fish Ecophysiology*, J.C. Rankin and F.B. Jensen (eds.). Chapman & Hall, London, Vol. 9, pp. 161–179.

Jensen, M.K., S.S. Madsen and K. Kristiansen. 1998. Osmoregulation and salinity effects on the expression and activity of Na^+,K^+-ATPase in the gills of European sea bass, *Dicentrarchus labrax* (L.). *Journal of Experimental Zoology* 282: 290–300.

Johnsen, H.K., R.A. Eliassen, B.-S. Sæther and J.S. Larsen. 2000. Effects of photoperiod manipulation on development of seawater tolerance in Arctic charr. *Aquaculture* 189: 177–188.

Jørgensen, E.H. and A.M. Arnesen. 2002. Seasonal changes in osmotic and ionic regulation in Artic charr, *Salvelinus alpinus*, from a high- and a sub-arctic anadromous population. *Environmental Biology of Fishes* 64: 185–193.

Kelly, R.H. and P.H. Yancey. 1999. High contents of trimethylamine oxide correlating with depth in deep-sea teleost fishes, skates, and decapod crustaceans. *Biological Bulletin* 196: 18–25.

Kelly, S.P. and N.Y.S. Woo. 1999. The response of sea bream following abrupt hypo-osmotic exposure. *Journal of Fish Biology* 55: 732–750.

Kelly, S.P., I.N.K. Chow and N.Y.S. Woo. 1999. Haloplasticity of black seabream (*Mylio macrocephalus*): Hypersaline to freshwater acclimation. *Journal of Experimental Zoology* 283: 226–241.

Kévers, C., A. Péqueux and R. Gilles. 1979. Effects of hypo- and hyperosmotic shocks on the volume and ions content of *Carcinus maenas* isolated axons. *Comparative Biochemistry and Physiology* A64: 427–431.

Kültz, D. 1996. Plasticity and stressor specificity of osmotic and heat shock responses of *Gillichthys mirabilis* gill cells. *American Journal of Physiology: Cell Physiology* 271: C1181–C1193.

Kültz, D. 2005. DNA damage signals facilitate osmotic stress adaptation. *American Journal of Physiology: Renal Physiology* 289: F504–F505.

Kültz, D. and H. Onken. 1993. Long-term acclimation of the teleost *Oreochromis mossambicus* to various salinities: Two different strategies in mastering hypertonic stress. *Marine Biology* 117: 527–533.

Kültz, D. and G.N. Somero. 1995. Ion transport in gills of the euryhaline fish *Gillichthys mirabilis* is facilitated by a phosphocreatine circuit. *American Journal of Physiology Regulatory: Integrative and Comparative Physiology* 268: R1003–R1012.

Kültz, D., K. Jürss and L. Jonas. 1995. Cellular and epithelial adjustments to altered salinity in the gill and opercular epithelium of a cichlid fish (*Oreochromis mossambicus*). *Cell and Tissue Research* 279: 65–73.

Kunzmann, A. 1990. Gill morphometrics of two Antarctic fish species *Pleuragramma antarcticum* and *Notothenia gibberifrons*. *Polar Biology* 11: 9–18.

Laiz-Carrión, R., S. Sangiao-Alvarellos, J.M. Guzmán, M.P.M. del Río, J.L. Soengas and J.M. Mancera. 2005. Growth performance of gilthead sea bream *Sparus aurata* in different osmotic conditions: implications for osmoregulation and energy metabolism. *Aquaculture* (In press).

Lin, C.H., R.S. Tsai and T.H. Lee. 2004. Expression and distribution of Na,K-ATPase in gill and kidney of the spotted green pufferfish, *Tetraodon nigroviridis*, in response to salinity challenge. *Comparative Biochemistry and Physiology* A138: 287–295.

Lucassen, M., A. Schmidt, L.G. Eckerle and H.-O. Pörtner. 2003. Mitochondrial proliferation in the permanent vs. temporary cold: enzyme activities and mRNA levels in Antarctic and temperate zoarcid fish. *American Journal of Physiology: Regulatory, Integrative and Comparative Physiology* 285: R1410–R1420.

Maffia, M., R. Acierno, M. Rollo and C. Storelli. 1998. Ion and metabolite transport through the intestinal luminal membranes of the Antarctic fish *Trematomus bernacchii*. In: *Fishes of Antarctica. A Biological Overview*, G. di Prisco, E. Pisano and A. Clarke (eds.). Springer-Verlag, Milano, pp. 237–246.

Mancera, J.M., J.M. Perez-Figares and P. Fernandez-Llebrez. 1993. Osmoregulatory responses to abrupt salinity changes in the euryhaline gilthead sea bream (*Sparus aurata* L.). *Comparative Biochemistry and Physiology* A106: 245–250.

Marshall, W.S., T.R. Emberley, T.D. Singer, S.E. Bryson and S.D. McCormick. 1999. Time course of salinity adaptation in a strongly euryhaline estuarine teleost, *Fundulus heteroclitus*: A multivariable approach. *Journal of Experimental Biology* 202: 1535–1544.

Metz, J.R., E.H. van den Burg, S.E. Wendelaar Bonga and G. Flik. 2003. Regulation of branchial Na^+/K^+-ATPase in common carp *Cyprinus carpio* L. acclimated to different temperatures. *Journal of Experimental Biology* 206: 2273–2280.

Montgomery, J.C. and R.M.G. Wells. 1993. Recent advances in the ecophysiology of Antarctic fishes: metabolic capacity and sensory performance. In: *Fish Ecophysiology*, J.C. Rankin and F.B. Jensen (eds.). Chapman & Hall, London, Vol. 9, pp. 341–374.

Nakano, K. and G.K. Iwama. 2002. The 70-kDa heat shock protein response in two intertidal sculpins, *Oligocottus maculosus* and *O. snyderi*: relationship of hsp70 and thermal tolerance. *Comparative Biochemistry and Physiology* A133: 79–94.

Nelson, J.S. 1994. *Fishes of the World*. 3rd Edition. John Wiley & Sons, New York.

Nordlie, F.G. 2003. Fish communities of estuarine salt marshes of eastern North America, and comparisons with temperate estuaries of other continents. *Reviews in Fish Biology and Fisheries* 13: 281–325.

Nordlie, F.G. and A. Mirandi. 1996. Salinity relationships in a freshwater population of eastern mosquitofish. *Journal of Fish Biology* 49: 1226–1232.

Nordlie, F.G., D.C. Haney and S.J. Walsh. 1992. Comparisons of salinity tolerances and osmotic regulatory capabilities in populations of sailfin molly (*Poecilia latipinna*) from brackish and fresh waters. *Copeia* 1992: 741–746.

O'Grady, S.M. and A.L. DeVries. 1982. Osmotic and ionic regulation in polar fishes. *Journal of Experimental Marine Biology and Ecology* 57: 219–228.

Pan, F., J.M. Zarate, G.C. Tremblay and T.M. Bradley. 2000. Cloning and characterization of salmon hsp90 cDNA: upregulation by thermal and hyperosmotic stress. *Journal of Experimental Zoology* 287: 199–212.

Pérez-Pinzón, M.A. and P.L. Lutz. 1991. Activity related cost of osmoregulation in the juvenile snook (*Centropomus undecimalis*). *Bulletin of Marine Science* 48: 58–66.

Place, S.P. and G.E. Hofmann. 2001. Temperature interactions of the molecular chaperone Hsc70 from the eurythermal marine goby *Gillichthys mirabilis*. *Journal of Experimental Biology* 204: 2675–2682.

Place, S.P. and G.E. Hofmann. 2005. Constitutive expression of a stress-inducible heat shock protein gene, *hsp70*, in phylogenetically distant Antarctic fish. *Polar Biology* 28: 261–267.

Plaut, I. 1998. Comparison of salinity tolerance and osmoregulation in two closely related species of blennies from different habitats. *Fish Physiology and Biochemistry* 19: 181–188.

Plaza-Yglesias, M., M. Laufer and F.C. Herrera. 1988. Ionic and osmotic regulation in blood, aqueous humor, gills and retina in the euryhaline fish, *Eugerres plumieri*. *Comparative Biochemistry and Physiology* A89: 377–382.

Pörtner, H.O., I. Hardewig, F.J. Sartoris and P.L.M. Van Dijk. 1998. Energic aspects of cold adaptation: critical temperatures in metabolic, ionic and acid-base regulation? In: *Cold Ocean Physiology*, H.O. Pörtner and R.C. Playle (eds.). Cambridge University Press, Cambridge, pp. 80–120.

Pough, F.H., J.B. Heiser and W.N. McFarland. 1996. *Vertebrate Life*. 4th Edition. Prentice Hall, Upper Saddle River, USA.

Prodocimo, V. and C.A. Freire. 2001. Ionic regulation in aglomerular tropical estuarine pufferfishes submitted to sea water dilution. *Journal of Experimental Marine Biology and Ecology* 262: 243–253.

Romão, S., C.A. Freire and E. Fanta. 2001. Ionic regulation and Na^+,K^+-ATPase activity in gill and kidney of the Antarctic aglomerular cod icefish exposed to dilute sea water. *Journal of Fish Biology* 59: 463–468.

Ross, S.W., D.A. Dalton, S. Kramer and B.L. Christensen. 2001. Physiological (antioxidant) responses of estuarine fishes to variability in dissolved oxygen. *Comparative Biochemistry and Physiology* C130: 289–303.

Sardella, B.A., J. Cooper, R.J. Gonzalez and C.J. Brauner. 2004a. The effect of temperature on juvenile Mozambique tilapia hybrids (*Oreochromis mossambicus* × *O. urolepis hornorum*) exposed to full-strength and hypersaline seawater. *Comparative Biochemistry and Physiology* A137: 621–629.

Sardella, B.A., V. Matey, J. Cooper, R.J. Gonzalez and C.J. Brauner. 2004b. Physiological, biochemical and morphological indicators of osmoregulatory stress in 'California' Mozambique tilapia (*Oreochromis mossambicus* × *O. urolepis hornorum*) exposed to hypersaline water. *Journal of Experimental Biology* 207: 1399–1413.

Sinensky, M. 1974. Homeoviscous adaptation—A homeostatic process that regulates the viscosity of membrane lipids in *Escherichia coli*. *Proceedings of the National Academy of Sciences of the United States of America* 71: 522–525.

Smith, T.R., G.C. Tremblay and T.M. Bradley. 1999. Hsp70 and a 54kDa protein (Osp54) are induced in salmon (*Salmo salar*) in response to hyperosmotic stress. *Journal of Experimental Zoology* 284: 286–298.

Somero, G.N. 2002. Thermal physiology and vertical zonation of intertidal animals: optima, limits, and costs of living. *Integrative and Comparative Biology* 42: 780–789.

Somero, G.N. 2005. Linking biogeography to physiology: evolutionary and acclimatory adjustments of thermal limits. *Frontiers in Zoology* 2: 1–9.

Somero, G.N., P.A. Fields, G.E. Hofmann, R.B. Weinstein and H. Kawall. 1998. Cold adaptation and stenothermy in Antarctic notothenioid fishes: what has been gained and what has been lost? In: *Fishes of Antarctica. A Biological Overview*, G. di Prisco, E. Pisano and A. Clarke (eds.). Springer-Verlag, Milano, pp. 97–109.

Staurnes, M., T. Sigholt, T. Åsgård and G. Baeverfjord. 2001. Effects of a temperature shift on seawater challenge test performance in Atlantic salmon (*Salmo salar*) smolt. *Aquaculture* 201: 153–159.

Storch, D., G. Lannig and H.O. Pörtner. 2005. Temperature-dependent protein synthesis capacities in Antarctic and temperate (North Sea) fish (Zoarcidae). *Journal of Experimental Biology* 208: 2409–2420.

Swanson, C. 1998. Interactive effects of salinity on metabolic rate, activity, growth and osmoregulation in the euryhaline milkfish (*Chanos chanos*). *Journal of Experimental Biology* 201: 3355–3366.

Thompson, G.G. and P.C. Withers. 2002. Aerial and aquatic respiration of the Australian desert goby, *Chlamydogobius eremius*. *Comparative Biochemistry and Physiology* A131: 871–879.

Tomanek, L. 2002. The heat-shock response: its variation, regulation and ecological importance in intertidal gastropods (genus *Tegula*). *Integrative and Comparative Biology* 42: 797–807.

Tomanek, L. and B. Helmuth. 2002. Physiological ecology of rocky intertidal organisms: A synergy of concepts. *Integrative and Comparative Biology* 42: 771–775.

Uchida, K., T. Kaneko, H. Miyazaki, S. Hasegawa and T. Hirano. 2000. Excellent salinity tolerance of Mozambique tilapia (*Oreochromis mossambicus*): Elevated chloride cell activity in the branchial and opercular epithelia of the fish adapted to concentrated seawater. *Zoological Science* 17: 149–160.

Varsamos, S., J.P. Diaz, G. Charmantier, G. Flik, C. Blasco and R. Connes. 2002. Branchial chloride cells in sea bass *(Dicentrarchus labrax*) adapted to fresh water, seawater, and doubly concentrated seawater. *Journal of Experimental Zoology* 293: 12–26.

Ventrella, V., A. Pagliarani, M. Pirini, G. Trigari, F. Trombetti and A.R. Borgatti. 1993. Lipid composition and microsomal ATPase activities in gills and kidneys of warm- and cold-acclimated sea bass (*Dicentrarchus labrax* L.). *Fish Physiology and Biochemistry* 12: 293–304.

Vize, P.D. 2004. A Homeric view of kidney evolution: A reprint of H.W. Smith's classic essay with a new introduction. *Anatomical Record* A277: 344–354.

Wöhrmann, A.P.A. 1998. Aspects of eco-physiological adaptations in antarctic fish. In: *Fishes of Antarctica. A Biological Overview*, G. diPrisco, E. Pisano and A. Clarke (eds.). Springer-Verlag, Milano, pp. 119–128.

Yancey, P.H., W.R. Blake and J. Conley. 2002. Unusual organic osmolytes in deep-sea animals: Adaptations to hydrostatic pressure and other perturbants. *Comparative Biochemistry and Physiology* A133: 667–676.

Zadunaisky, J.A., S. Cardona, L. Au, D.M. Roberts, E. Fisher, B. Lowenstein, E.J. Cragoe Jr and K.R. Spring. 1995. Chloride transport activation by plasma osmolarity during rapid adaptation to high salinity of *Fundulus heteroclitus*. *Journal of Membrane Biology* 143: 207–217.

CHAPTER

10

Energy Metabolism and Osmotic Acclimation in Teleost Fish

José L. Soengas[1,*], Susana Sangiao-Alvarellos[1], Raúl Laiz-Carrión[2] and Juan M. Mancera[2]

INTRODUCTION

Euryhaline fish present the capacity to live in different environmental salinities. The osmolality of the internal milieu in marine teleosts is equivalent to 10 ± 2% of the environmental salinity (Holmes and Donaldson, 1969; Maetz, 1974). Most researchers agree on the fact that water with different ionic and osmotic concentration, with respect to internal milieu, must impose energetic regulatory costs for active ion transport: intake in hypotonic environment and uptake in hypertonic environment. However, there is less agreement concerning the magnitude of these costs and very little information on the related energetic and

Authors' addresses: [1]Laboratorio de Fisioloxía Animal, Facultade de Bioloxía, Edificio de Ciencias Experimentais, Universidade de Vigo, E-36310, Vigo, Spain.

[2]Departamento de Biología, Facultad de Ciencias del Mar y Ambientales, Universidad de Cádiz, 11510 Puerto Real, Cádiz, Spain.

**Corresponding author:* E-mail: jsoengas@uvigo.es

physiological consequences of life in different salinities (Swanson, 1998; Boeuf and Payan, 2001). In this way, many studies support the hypothesis that the energetic cost of ion regulation is lowest in an isotonic with respect to hypotonic or hypertonic environment, while others did not support this idea (Morgan and Iwama, 1991; Maxime, 2002). The energy saved from osmoregulatory processes could be derived to other physiological processes, such as growth. Thus, a relationship between environmental salinity and growth has been reported in different euryhaline fish, but the optimal salinity to obtain the best growth depends on different factors such as species, temperature, dietary intake, sex, etc. (Kirschner, 1995; Altinok and Grizzle, 2001; Boeuf and Payan, 2001).

Fish spend large amounts of energy, particularly in the osmoregulatory organs (i.e., gill, intestine, and kidney), to compensate for these salinity changes. However, the energetics of these responses to salinity change are not yet fully understood (Febry and Lutz, 1987; Morgan and Iwama, 1991;Boeuf and Payan, 2001; Brill *et al.*, 2001). This energetic cost can be estimated based on: (1) the changes observed in the whole body metabolism using respirometer measurements (McCormick *et al.*, 1989; Morgan and Iwama, 1991; Kirschner, 1993), or (2) the changes in the levels of energy substrates and/or in the activities of enzymes controlling their metabolism. This latter approach has been the one less used (McCormick and Saunders, 1987) despite of being more advantageous than the preceding (Febry and Lutz, 1987).

METABOLIC RATES

The cost of osmoregulation in freshwater (FW) and seawater (SW), according to changes in oxygen consumption, provide very different estimates ranging from 1% to 20% of the total energy cost of the fish (Febry and Lutz, 1987; Nordlie *et al.*, 1991; Morgan *et al.*, 1997; Sparks *et al.*, 2003; Wuenschel *et al.*, 2005). Moreover, those energy costs changed in a very different way when comparing fish acclimated to different environmental salinities. Apparently, those different responses are mainly related to the species assessed.

Accordingly to these data, five patterns of metabolic response to altered environmental salinities have been suggested by Morgan and Iwama (1991) in fish including: (1) no change in metabolic rate; (2) metabolic rate is minimum in isotonic salinity and increases at lower and higher salinities; (3) metabolic rate increases linearly with salinity;

(4) metabolic rate is higher in freshwater and decreases in isotonic media (not tolerate seawater); and (5) metabolic rate is highest in seawater decreasing in other salinities.

CHANGES IN ENERGY METABOLISM DURING OSMOTIC ACCLIMATION

The capacity of adaptation of euryhaline teleost to changes in environmental salinity depends on several factors, including the energy supply and the energy demand necessary for electrolyte shift between intracellular and extracellular space or the external medium. Successful salinity acclimation may require a metabolic reorganization in order to meet the increased energetic demands associated with the exposure to the new environmental salinity. The metabolic response of teleosts to different osmotic conditions undoubtedly includes both stress and osmoregulatory components, but the relative energetic demands of these processes cannot be discerned from whole animal oxygen consumption (Morgan and Iwama, 1991). Thus, not unexpectedly, alterations in intermediary metabolism related to osmoregulation are poorly studied in teleosts (Morgan *et al.*, 1997; Nakano *et al.*, 1998).

In different euryhaline species (Holmes and Donaldson, 1969; Maetz, 1974), including gilthead seabream a species used for our research group as a model to study the relationship between osmoregulation and metabolism (Mancera *et al.*, 1993; Laiz-Carrión *et al.*, 2005a; Sangiao-Alvarellos *et al.*, 2005), osmoregulatory changes occurring during acclimation to different osmotic conditions behave normally in two different stages: (1) an adaptive period with changes in osmotic parameters; and (2) a chronic regulatory period, where these parameters again reach homeostasis. However, very few studies have assessed whether or not the metabolic changes occurring in different tissues during osmotic acclimation display time courses related to changes in osmoregulatory parameters.

Therefore, in the following sections, we shall provide a general picture of the conclusions arrived at from available literature regarding metabolic changes during acclimation of teleost fish to hyperosmotic and hypo-osmotic environments. We will focus these paragraphs in different types of tissues, such as: (1) tissues involved directly in osmoregulatory work such as gills, kidney and intestine; (2) tissues involved indirectly in osmoregulatory work, providing fuels to other tissues such as the case of plasma and liver; and (3) tissues theoretically not involved in metabolic changes occurring during osmotic acclimation such as brain or heart. In addition, we will also provide information regarding the few cases in which the time courses of those metabolic changes have been assessed.

Plasma

Osmotic acclimation is known to induce changes in plasma levels of metabolites such as glucose, lactate, triglycerides or protein (see below). The assessment of those changes can provide information regarding metabolic adjustments in the fish for the osmotic acclimation to the new environment since those metabolites can be used as fuels in osmoregulatory epithelia.

Increased plasma glucose levels had been observed during acclimation from FW or brackish water (BW) to SW or hypersaline water (HSW) in several species of fish such as sea bass (Roche *et al.*, 1989), gilthead seabream (Sangiao-Alvarellos *et al.*, 2003b; Laiz-Carrión *et al.*, 2005a, b); rainbow trout (Soengas *et al.*, 1995a), cut-throat trout (Morgan and Iwama, 1996), cod (Nelson *et al.*, 1996), tilapia (Morgan *et al.*, 1997; Nakano *et al.*, 1998) or carp (DeBoeck *et al.*, 2000; Yavuzcan-Jildiz and Kirkagaç-Uzbilek, 2001), suggesting a mobilization of glucose to peripheral tissues in order to satisfy the increased energetic demand observed in osmoregulatory organs during osmotic acclimation. Moreover, in other studies using different (coho salmon: Morgan and Iwama, 1998; Atlantic salmon: Maxime, 2002) or even the same species (tilapia: Morgan *et al.* 1997, Vijayan *et al.*, 2001), no changes were observed. When euryhaline fish usually living in SW are acclimated to BW or FW, both decreases such as in red seabream (Woo and Murat, 1981) or increases such as in tilapia (Assem and Hanke, 1979), silver seabream (Kelly *et al.*, 1999), and gilthead seabream (Mancera *et al.*, 1993) have been reported in plasma glucose levels.

The time course of changes in plasma glucose levels has been assessed in different species transferred to higher salinity water, showing different patterns of variation. In rainbow trout transferred from FW to SW, the increase in plasma glucose was essentially the same throughout the study (Soengas *et al.*, 1993a). A similar time course has been reported for carp acclimated from FW to BW (DeBoeck *et al.*, 2000). In contrast, in tilapia transferred from FW to SW (Nakano *et al.*, 1998) and gilthead seabream transferred form SW to HSW (Sangiao-Alvarellos *et al.*, 2005), a two-stage behaviour was observed with a sharp increase 1 day after salinity transfer related to stress followed by a slow decline in the following days until they reached the previous values.

On the other hand, the time course of changes in plasma glucose is also available for acclimation to lower salinity water. A two-stage behaviour has been also reported for sea bass transferred from SW to FW (Roche *et al.*, 1989) and for gilthead seabream acclimated from SW to BW (Mancera *et al.*, 1993). Thus, these species showed an increase in glucose levels in the first stages of acclimation followed by a decline until the values observed in SW were recovered. However, in another study, gilthead seabream under the same osmotic transfer displayed two different peaks (Sangiao-Alvarellos *et al.*, 2005), the first on day 1 and the second one on days 7 and 14 of the experiment, suggesting an increased production of glucose.

Plasma lactate levels presented a lineal relationship with environmental salinity in gilthead seabream (Sangiao-Alvarellos *et al.*, 2003b; Laiz-Carrión *et al.*, 2005a). This result is comparable with the decrease observed in red seabream after transfer from SW to diluted SW (Woo and Murat, 1981), whereas on the other hand, increased plasma lactate levels were also reported in tilapia after acclimation to SW (Vijayan *et al.*, 1996). Considering the fact that lactate can be used in tissues like gills, kidney and brain for their energy requirements (Mommsen, 1984; Mommsen *et al.*, 1985; Soengas *et al.*, 1998), the rise of plasma lactate levels observed in parallel with increased salinity suggests that this metabolite become more important in hyperosmotic conditions, presumably related to its metabolic use in those organs.

Levels of plasma tryglyceride (TG) have been assessed in several studies during osmotic acclimation. In gilthead seabream, two clear phases were distinguished during acclimation to HSW, the first having no differences with respect to SW-acclimated fish on the first day of acclimation and the second from day 3 to day 14 with a linear increase in the levels of this metabolite (Sangiao-Alvarellos *et al.*, 2005). This result may be in agreement with the enhanced production of TG already reported in plasma of Atlantic salmon during smoltification (Nordgarden *et al.*, 2002). Plasma TG levels also increased sharply in the first days of acclimation to BW of gilthead seabream (Sangiao-Alvarellos *et al.*, 2005) in agreement with the increase in the amount of plasma lipids in angelfish transferred to hypoosmotic salinities (Woo and Chung, 1995).

Plasma protein levels showed different responses to increases in environmental salinity: (1) increase in cod (Nelson *et al.*, 1996) gilthead seabream (Sangiao-Alvarellos *et al.*, 2003b; Laiz-Carrión *et al.*, 2005a) and (2) no changes in red seabream (Woo and Murat, 1981); or (3) decreases in silver seabream (Kelly and Woo, 1999a). During acclimation of gilthead seabream to HSW, plasma protein levels described two different phases with higher levels on the first days and a lower amount at the end of the acclimation period (Sangiao-Alvarellos *et al.*, 2005). In contrast, Kelly *et al.* (1999) did not find any change in this parameter throughout the time period assessed in black seabream.

Osmotic acclimation may also produce changes in the concentration of particular amino acids. Thus, changes in osmotic conditions elicited in tilapia decreased plasma concentration of taurine and glycine (Assem and Hanke, 1983). A decrease in the concentration of total amino acids in plasma was observed when fish were acclimated to hyperosmotic conditions either for angelfish (Woo and Chung, 1995) or tilapia (Vijayan *et al.*, 1996). However, to our knowledge, the role of amino acid during osmotic adaptation is not well determined and further studies are required.

Altogether, changes observed in the levels of plasma metabolites suggested an increased availability of fuels (glucose, lactate and TG), especially during the first days of acclimation to different osmotic conditions. These enhanced levels of metabolites would be available to be used in different tissues. Since osmoregulatory tissues increase their energy requirements during that process (see below), at least part of those metabolites could be used to cover their metabolic requirements.

Gills

The gills probably constitute the osmoregulatory organ that consumes the most energy since they must ensure isosmotic regulation of intracellular fluid and also anisosmotic regulation of extracellular fluid. Thus, on transfer of teleost fish into different salinities, sodium and chloride transport across the gill epithelium switches from ion uptake in hypo-osmotic water to ion excretion into hyperosmotic water. Those ion transport mechanisms involve cotransporters, ion conductive channels, and ion transport proteins driven by ATP located in the apical and basolateral membranes of chloride cells and pavement cells (Wilson *et al.*, 2002). Whereas transport systems and associated ATPases activities have been described in ample detail (Marshall, 2002; Wilson and Laurent,

2002), the biochemical aspects of the mechanisms supplying the required ATP have been only slightly studied to date (Mommsen, 1984; Perry and Walsh, 1989). Table 10.1 summarizes the most important changes reported in literature regarding changes in gill metabolic pathways during osmotic acclimation in fish.

Fish gills are highly oxidative tissues even in FW, and its oxygen requirement increases even further when fish are transferred to SW

Table 10.1 Responses in selected metabolic pathways of gills energy metabolism after acclimation of teleost fish to different osmotic conditions. ↑, increase; ↓, decrease; ⇔ no changes.

Pathway	*Fish*	*Acclimation*	*Response*	*Reference*
Use of exogenous glucose	Rainbow trout	FW→SW	↑	Soengas *et al.* (1995b)
	Gilthead seabream	BW→SW SW→HSW	↑ ↑	Sangiao-Alvarellos *et al.* (2003b, 2005)
Glycogenolysis	Gilthead seabream	BW→SW SW→HSW	⇔ ⇔	Sangiao-Alvarellos *et al.* (2003b, 2005) Laiz-Carrión *et al.* (2005b)
Glycolysis	Rainbow trout	FW→SW	↑	Soengas *et al.* (1995b)
	Brook charr	FW→SW	↑	LeFrançois and Blier (2004)
	Gilthead seabream	SW→HSW SW→BW	↑ ↑	Sangiao-Alvarellos *et al.* (2003b, 2005) Laiz-Carrión *et al.* (2005b)
	Black seabream	SW→BW SW→HSW	↓ ⇔	Kelly *et al.* (1999) Kelly *et al.* (1999)
	Silver seabream	SW→BW	⇔	Kelly and Woo (1999b)
	Tilapia	FW→SW	⇔	Vijayan *et al.* (2001)
	Wolfish	BW→SW	⇔	LeFrançois *et al.* (2004)
Pentose phosphate	Silver seabream	SW→BW	↑	Kelly and Woo (1999b)
	Black seabream	SW→HSW SW→BW	↓ ↓	Kelly *et al.* (1999) Kelly *et al.* (1999)
	Gilthead seabream	BW→SW SW→HSW	↓ ↓	Sangiao-Alvarellos *et al.* (2003b, 2005)
	Tilapia	FW→SW	↓	Kültz and Jürss (1993)
Aerobic capacity	Atlantic salmon	FW→SW	⇔	McCormick *et al.* (1989)
	Brook charr	FW→SW	⇔	LeFrançois and Blier (2004)
	Rainbow trout	FW→SW	⇔	Leray *et al.* (1981)
Amino acid catabolism	Tilapia	FW→SW	↓	Kültz and Jürss (1993)
	Rainbow trout	FW→SW	⇔	Jürss *et al.* (1983)
	Gilthead seabream	SW→HSW SW→BW	⇔ ↓	Polakof *et al.* (2006)
Lipid catabolism	Masu salmon	FW→SW	↑	Li and Yamada (1992)
	Gilthead seabream	SW→HSW SW→BW	⇔ ⇔	Polakof *et al.* (2006)
Lactate metabolism	Gilthead seabream	SW→HSW SW→BW	⇔ ⇔	Polakof *et al.* (2006)

(Vijayan *et al.*, 1996). The energy requirement of the gills is thought to be maintained by oxidation of glucose and lactate obtained from the circulation (Mommsen, 1984). Although gill hexokinase (HK) activity—an activity involved in the use of exogenous glucose—is usually low in teleosts, it is apparently active enough to pace the CO_2 release values observed in that tissue (Mommsen, 1984). This activity increased linearly with environmental salinity in rainbow trout (Soengas *et al.*, 1995b) and gilthead seabream (Sangiao-Alvarellos *et al.*, 2003b). Specimens of this last species acclimated to HSW (Sangiao-Alvarellos *et al.*, 2005) showed higher gill Na^+,K^+-ATPase activity and the suspected increased energy demand of the gills under this osmotic conditions appear to be fuelled at least by one exogenous fuel such as glucose (as judged by changes in HK activity). The enhanced levels of plasma glucose at the same time also suggest this increased use of exogenous glucose by gills (Laiz-Carrión *et al.*, 2005a; Sangiao-Alvarellos *et al.*, 2005). This indicates an increased energy demand of gill in the first days after transfer, which is being progressively declined. From an osmoregulatory point of view, an increased osmoregulatory work of gills during the first days of acclimation (adaptive period) seems logical because in this period, the osmoregulatory system is overactive in order to reach osmotic and ionic homeostasis of internal milieu (Holmes and Donaldson, 1969; Maetz, 1974).

An enhancement of the glycolytic capacity of gill tissue based on changes observed in PK activity has been reported in several species transferred to high salinity environment (Soengas *et al.*, 1995b; Kelly and Woo 1999a; LeFrançois and Blier, 2004; Sangiao-Alvarellos *et al.*, 2005). Moreover, a decrease of ATP levels in gills along with salinity has been reported in rainbow trout (Leray *et al.*, 1981), sea bass (Roche *et al.*, 1989) and gilthead seabream (Sangiao-Alvarellos *et al.*, 2003b). These data suggest a higher use of ATP in gill of fish adapted to high salinity environment where an enhancement of gill Na^+,K^+-ATPase activity is observed (Laiz-Carrión *et al.*, 2005a; Sangiao-Alvarellos *et al.*, 2005).

On the other hand, the glycolytic capacity of gill tissue has also been measured during acclimation to BW in several species showing increases in gilthead seabream (Laiz-Carrión *et al.*, 2005b; Sangiao-Alvarellos *et al.*, 2005), decreases in black seabream (Kelly *et al.*, 1999) or no changes in silver seabream (Kelly and Woo, 1999a) when comparing them with SW-acclimated fish. This contradictory picture could indicate the existence of different metabolic pathways for supplying ATP requirements under hypo-osmotic conditions in different species.

Considering the low amount of glycogen accumulated in gill tissue it is logical to see no changes in the glycogenolytic potential of this tissue during osmotic acclimation (Sangiao-Alvarellos *et al.*, 2003b, 2005), since a raised metabolic demand in gills would not be sufficiently covered by glycogen mobilization. Since gill tissue is able to oxidize lactate at rates comparable to those of glucose (Mommsen, 1984; Perry and Walsh, 1989), and also considering the raise observed in plasma lactate levels in parallel with increased salinity in several species (see above), an enhanced use of exogenous lactate through lactate dehydrogenase working in the oxidative direction cannot be excluded. In this way, preliminary studies carried out in gilthead seabream also addressed the increase capacity for oxidation of lactate during acclimation to extreme salinities (Polakof *et al.*, unpublished). Finally, another evidence for an increased gill energy demand in high salinities comes from glucose 6-phosphate dehydrogenase (G6PDH) activity, which showed a clear decrease in parallel with increased salinity in tilapia (Kültz and Jürss, 1993) and gilthead seabream (Sangiao-Alvarellos *et al.*, 2003b). This activity displayed a time course in which transient increases were observed on the first days after transfer to HSW or BW. These initial increases coincided with similar rises observed in tilapia (Kültz and Jürss, 1993), black seabream (Kelly *et al.*, 1999), and silver seabream (Kelly and Woo, 1999a). Those transient elevations suggest a process of tissue reorganization consistent with an intermediary phase of acclimation and coincide in time with the adaptive period observed in fish after salinity transfer.

To our knowledge there are few studies regarding the impact of protein metabolism in gills during osmotic acclimation (Jürss *et al.*, 1983; Kultz and Jurss, 1993) addressing a decreased importance of these metabolites for fuelling purposes in fish during osmotic acclimation.

Several studies have shown that acclimation from FW to SW produce an increase in the amount of polyunsaturated fatty acids (PUFA) in gills of several fish species like rainbow trout (Hansen *et al.*, 1992), masu salmon (Li and Yamada, 1992), Atlantic salmon (Tocher *et al.*, 2000), sea bass (Cordier *et al.*, 2002), and eel (Hansen and Grosell, 2004). These changes cause increased fluidity of the membranes at the time of hyperosmotic acclimation. Apart from these changes in composition, only a few studies have assessed changes in lipolytic capacity of gills during osmotic acclimation addressing an increased capacity (Li and Yamada, 1992).

Altogether, the results obtained in gills of different fish species lend support to an increased energy demand in parallel with changes in environmental salinity as suggested by enhanced Na^+,K^+-ATPase activity and glycolytic capacity. This increase in demand appears to be fuelled by an enhanced use of exogenous glucose as suggested by higher plasma glucose levels, gills HK activity and levels of glucose and glycogen in gills, though the possibility of lactate also being used cannot be excluded (Sangiao-Alvarellos *et al.*, 2003; LeFrançois *et al.*, 2004). The enhancement in energy demand appears to behave in two different stages: (1) a first one characterized by decreased levels of lactate and glucose; and (2) a posterior with enhanced use of glucose through glycolysis, pentose phosphate and glycogenesis on subsequent days of acclimation. These data suggested an increased energy demand of the tissue in the first days after transfer, which is progressively declining. From an osmoregulatory point of view, an enhanced osmoregulatory work of gills during the adaptive period of acclimation to extreme environmental salinities, thus needing extra fuel seems reasonable and agrees with the two phases of salinity acclimation observed in several teleosts (Holmes and Donaldson, 1969; Maetz, 1974; Mancera *et al.*, 1993, 2002; Laiz-Carrión *et al.*, 2005a; Sangiao-Alvarellos *et al.*, 2005). Moreover, in those cases where a full range of salinities were assessed (Kelly and Woo, 1999a; Sangiao-Alvarellos *et al.*, 2003b, 2005; Laiz-Carrión *et al.*, 2005b), changes appear to be more important during acclimation to HSW rather than to BW, which may be related to the higher growth observed in several species when acclimated to BW compared with those acclimated to SW and HSW.

Kidney

Kidney, in addition to gills, is other important osmoregulatory organ that plays an active role in the extrusion of divalent ions and elimination of excess of water in hyperosmotic and hypoosmotic environments, respectively. Several changes occur in this organ during osmotic adaptation, including changes in morphology, excretion of divalent ions, glomerular filtration rate, and urine production (Beyenbach, 1995; Renfro, 1995). Most of these changes need energy in the form of ATP, and may be associated with an altered energetic demand that would lead to changes in kidney intermediary metabolism (McCormick *et al.*, 1989; Soengas *et al.*, 1994; Kelly and Woo, 1999b). Table 10.2 summarizes the most important changes reported in literature regarding modifications in kidney metabolic pathways during osmotic acclimation in fish.

Table 10.2 Responses in selected metabolic pathways of kidney energy metabolism after acclimation of teleost fish to different osmotic conditions. ↑, increase; ↓, decrease; ⇔ no changes.

Pathway	*Fish*	*Acclimation*	*Response*	*Reference*
Use of exogenous glucose	Gilthead seabream	SW→HSW BW→SW	↑ ↑	Sangiao-Alvarellos *et al.* (2003b, 2005)
	Rainbow trout	FW→SW	⇔	Soengas *et al.* (1994)
Glycogenolysis	Rainbow trout	FW→SW	↑	Soengas *et al.* (1994)
	Gilthead seabream	SW>HSW SW→BW	⇔ ⇔	Sangiao-Alvarellos *et al.* (2003b, 2005)
	Silver seabream	SW→BW	⇔	Kelly and Woo (1999b)
Glycolysis	Gilthead seabream	SW→BW SW→HSW	↑ ↑	Sangiao-Alvarellos *et al.* (2003b, 2005)
	Black seabream	SW→HSW SW→BW	⇔ ⇔	Kelly *et al.* (1999) Kelly *et al.* (1999)
Glucose export capacity	Silver seabream	SW→BW	↑	Kelly and Woo (1999b)
	Black seabream	SW→HSW SW→BW	↑ ↑	Kelly *et al.* (1999) Kelly *et al.* (1999)
	Gilthead seabream	SW→HSW BW→SW	⇔ ⇔	Sangiao-Alvarellos *et al.* (2003b, 2005)
Gluconeogenesis	Rainbow trout	FW→SW	↑	Soengas *et al.* (1994)
	Gilthead seabream	SW→HSW BW→SW	⇔ ⇔	Sangiao-Alvarellos *et al.* (2003b, 2005)
Pentose phosphate	Gilthead seabream	SW→HSW	↓	Sangiao-Alvarellos *et al.* (2003b, 2005)
	Silver seabream	SW→BW	⇔	Kelly and Woo (1999b)
	Black seabream	SW→HSW SW→BW	⇔ ⇔	Kelly *et al.* (1999) Kelly *et al.* (1999)
	Rainbow trout	FW→SW	⇔	Soengas *et al.* (1994)
Aerobic capacity	Atlantic salmon	FW→SW	↑	McCormick *et al.* (1989)
Amino acid catabolism	Rainbow trout	FW→SW	⇔	Jürss *et al.* (1983)
	Gilthead seabream	SW→HSW SW→BW	⇔ ⇔	Polakof *et al.* (2006)
Lipid metabolism	Gilthead seabream	SW→HSW SW→BW	↓ ⇔	Polakof *et al.* (2006)
Lactate metabolism	Gilthead seabream	SW→HSW SW→BW	↑ ⇔	Polakof *et al.* (2006)

Glucose appears to be an important substrate for the kidney in teleosts based on: (1) the high glycolytic and pentose phosphate shunt capacities measured in that tissue (Mommsen *et al.*, 1985); and (2) its rates of glucose use, which are similar to those observed in tissues with important rates such as brain and gills (Blasco *et al.*, 1996). In Atlantic salmon, the acclimation to SW decreased citrate synthase and cytochrome oxidase activities. It led McCormick *et al.* (1989) to suggest a diminution of activity of this tissue in higher salinities due to the fact that following SW adaptation, the teleost kidney produces smaller volumes of a more concentrated urine than is produced in FW.

Kidney HK activity showed a sharp decrease in BW- and HSW-acclimated gilthead seabream compared with SW-acclimated fish (Sangiao-Alvarellos *et al.*, 2003b), suggesting that the necessity of kidney for exogenous glucose was lower in salinities different than usual. The time course of kidney HK activity has been assessed in this species during acclimation to HSW, displaying an increase on the first days after transfer with levels being the highest on day 7 and then returning to normal values on day 14 (Sangiao-Alvarellos *et al.*, 2005). In contrast, no changes were noticed in the activity of this enzyme in kidney of rainbow trout during acclimation to SW (Soengas *et al.*, 1994). On the other hand, during acclimation of gilthead seabream to BW, a continuous decline from day 1 of experiment onwards was observed (Sangiao-Alvarellos *et al.*, 2005) suggesting a decreased potential of kidney for using exogenous glucose. In BW-acclimated fish, a reduction in the activity of the kidney could be expected because the osmotic and ionic gradients between fish body and environment are minimal. However, in HSW-acclimated fish, an increased excretion of ions by kidney could be necessary (Beyenbach, 1995; Renfro, 1995), and thus these metabolic results would lend support for an enhanced use of another fuel instead of glucose to support the higher osmoregulatory work of kidney in HSW.

Similar to gill, lactate could also be a good candidate for fuel the osmoregulatory work of kidney in hyperosmotic and hypoosmotic environments. Thus, lactate levels increase in this tissue during the adaptive period of acclimation to HSW in gilthead seabream (Sangiao-Alvarellos *et al.*, 2005). On the other hand, a sharp decrease in kidney lactate levels in BW-acclimated gilthead seabream occurred from day 1 of acclimation onwards (Sangiao-Alvarellos *et al.*, 2005). This decline, together with the increased plasma glucose, may suggest that lactate is increasingly being used as a fuel for kidney in fish adapted to hypoosmotic environments. In fact, lactate metabolization may be so high that part of the lactate molecules can be used through gluconeogenesis to enhance glycogen synthesis, which could match with the increased glycogen levels observed in kidney of gilthead seabream on days 4 and 7 of acclimation of gilthead seabream to BW (Sangiao-Alvarellos *et al.*, 2005). In fact, increased glucose levels can be related to increased conversion rates into glycogen in kidney and other tissues (Blasco *et al.*, 2001). Moreover, results obtained by Polakof *et al.* (2006) also show an increased capacity for lactate oxidation in kidney of gilthead seabream acclimated to extreme salinities (BW and HSW).

Considering the small amount of endogenous glycogen generally addressed in fish kidney, this metabolite does not seem important for fuelling purposes. Accordingly, no changes in glycogen metabolism were observed during osmotic acclimation in gilthead seabream (Sangiao-Alvarellos *et al.*, 2003b, 2005) and silver seabream (Kelly and Woo, 1999b). However, kidney glycogen levels decreased in rainbow trout acclimated to SW (Soengas *et al.*, 1994). A time course study in gilthead seabream transferred to higher environmental salinity showed increased kidney Na^+,K^+-ATPase activity on the first days after transfer that coincide with adaptive period described for osmoregulatory system (Sangiao-Alvarellos *et al.*, 2005). At the same time, changes observed in the metabolite levels suggest the existence of an increased energy demand of this tissue on the first days of acclimation in gilthead seabream (Sangiao-Alvarellos *et al.*, 2005). This energy demand appears to be reduced on the following days when the fish is in the regulative period. In HSW-acclimated fish, higher excretion of ions by kidney could be necessary, and thus metabolic changes observed would lend support for an increased use of different fuels to support the increased osmoregulatory work of kidney during the first days of acclimation to HSW.

On the other hand, gilthead seabream showed an enhancement of glycolysis at the end of the acclimation period to BW in which glucose levels also decreased and returned to normality (Sangiao-Alvarellos *et al.*, 2005; Laiz-Carrión *et al.*, 2005b). This would suggest that in the case of this acclimation, the higher energy requirements are taking place at the end of the acclimation period. The absence of important changes in the metabolic parameters assessed may also lend support to the lower activation of kidney metabolism during the first stages of hypoosmotic acclimation followed by an increase at the end of the acclimation period. Further studies are necessary to determine whether or not this pattern of metabolic variation is a common feature to other euryhaline fish or is particular to gilthead seabream.

Gastrointestinal Tract

The gastrointestinal tract (GIT) plays an essential role in regulating the water and electrolyte status of fish (Buddington and Krogdahl, 2004). In fact, acclimation from FW to SW induces an active absorption of sodium, chloride at GIT level necessary to drive water passively lost from the body (Fuentes and Eddy, 1997; Lionetto *et al.*, 2001). In the GIT, there are regional differences in the magnitude of electrolyte and water flux due to differences in the densities and proportions of transporters, ion channels and permeability. All those processes are energy expensive and, therefore, changes in energy metabolism of GIT are expected during osmotic acclimation. However, to our knowdledge, there are very few studies carried out regarding this issue in literature. The only studies available report an increased use of amino acids as fuel in rainbow trout acclimated to SW (Auerswald *et al.*, 1997), decreased triglyceride levels in masu salmon acclimated from FW to SW (Li and Yamada, 1992), and the fastest glucose absorption in intestine of rainbow trout acclimated to SW (Brauge *et al.*, 1994). Moreover, as in in the case of gills, an increased amount of PUFA has been observed in GIT of different fish during acclimation to hyperosmotic environments (Hansen *et al.*, 1992; Li and Yamada, 1992).

Liver

The liver is the main site of glycogen/glucose turnover, ammoniogenesis, fatty acid synthesis, and gluconeogenesis in teleosts (Peragón *et al.*, 1998). Thus, liver metabolism may be enhanced during osmotic adaptation in order to make fuels available for metabolic and osmoregulatory processes, especially in osmoregulatory tissues like gills and kidney (Vijayan *et al.*, 1996). Table 10.3 summarizes the most important changes reported in literature regarding metabolic pathways in liver during osmotic acclimation in fish.

Increased liver glycogenolysis has been usually observed in fish transferred to hyperosmotic salinities (Hanke, 1991), either in euryhaline (Assem and Hanke, 1979; Soengas *et al.*, 1995a; Vijayan *et al.*, 1996; Nakano *et al.*, 1997; Kelly and Woo, 1999b; Sangiao-Alvarellos *et al.*, 2003b) or in stenohaline species (DeBoeck *et al.*, 2000). However, in other cases, there were no changes in liver glycogen occurred such as in red seabream (Woo and Murat, 1981) and rainbow trout (Kroghdahl *et al.*, 2004). Changes in glycogen levels are generally in agreement with changes in glycogen phosphorylase (GPase) activity in rainbow trout (Soengas *et al.*, 1995a), tilapia (Nakano *et al.*, 1997), and gilthead seabream (Sangiao-Alvarellos *et al.*, 2003b). In contrast, in tilapia acclimated to HSW, Nakano *et al.* (1997) failed to address any significant difference in GPase activity compared with SW-acclimated fish. The mobilization of liver glycogen would provide glycosyl units ready to be used to fuel

Table 10.3 Responses in selected metabolic pathways of liver energy metabolism after acclimation of teleost fish to different osmotic conditions. ↑, increase; ↓, decrease; ⇔ no changes.

Pathway	*Fish*	*Acclimation*	*Response*	*Reference*
Glycogenolysis	Tilapia	FW→SW	↑	Vijayan *et al.* (1996)
		FW→SW	⇔	Nakano *et al.* (1998)
		SW→FW	↑	Assem and Hanke (1979)
		SW→HSW	↑	Nakano *et al.* (1997)
	Black seabream	SW→HSW	↑	Kelly *et al.* (1999)
		SW→BW	⇔	Kelly *et al.* (1999)
	Rainbow trout	FW→SW	↑	Soengas *et al.* (1993a, 1995a)
		FW→SW	⇔	Kroghdahl *et al.* (2004)
	Gilthead seabream	SW→HSW	↑	Sangiao-Alvarellos *et al.*
		SW→BW	⇔	(2003b, 2005)
				Laiz-Carrión *et al.* (2005b)
	Carp	FW→SW	↑	DeBoeck *et al.* (2000)
	Atlantic salmon	FW→SW	↑	Plisetskaya *et al.* (1994)
	Red seabream	SW→BW	⇔	Woo and Murat (1981)
	Angelfish	SW→BW	⇔	Woo and Chung (1995)
Glycolysis	Tilapia	FW→SW	↑	Nakano *et al.* (1997, 1998)
		FW→SW	↑	Vijayan *et al.* (1996, 2001)
	Rainbow trout	FW→SW	↑	Soengas *et al.* (1995a)
	Gilthead seabream	SW→HSW	↑	Sangiao-Alvarellos *et al.*
		SW→BW	↓	(2003b, 2005)
	Atlantic salmon	FW→SW	↑	Plisetskaya *et al.* (1994)
	Black seabream	SW→BW	↓	Kelly *et al.* (1999)
		SW→HSW	⇔	Kelly *et al.* (1999)
Glucose export capacity	Gilthead seabream	SW→HSW	↑	Sangiao-Alvarellos *et al.*
		SW→BW	↓	(2003b, 2005)
	Angelfish	SW→BW	↓	Woo and Chung (1995)
	Black seabream	SW→HSW	⇔	Kelly *et al.* (1999)
Gluconeogenesis	Red seabream	SW→BW	↑	Woo and Fung (1981)
	Tilapia	FW→SW	⇔	Vijayan *et al.* (1996, 2001)
	Gilthead seabream	SW→HSW	⇔	Sangiao-Alvarellos
		SW→BW	⇔	*et al.* (2003b, 2005)
				Laiz-Carrión *et al.* (2005b)
Pentose phosphate	Rainbow trout	FW→SW	↑	Jürss *et al.* (1986)
		FW→SW	⇔	Soengas *et al.* (1993a)
	Black seabream	SW→BW	↑	Kelly *et al.* (1999)
	Gilthead seabream	SW→BW	↓	Sangiao-Alvarellos *et al.*
		SW→HSW	⇔	(2003b, 2005)
	Tilapia	FW→SW	⇔	Nakano *et al.* (1997, 1998)
		FW→SW	⇔	Vijayan *et al.* (2001)
Glucose use	Gilthead seabream	SW→HSW	⇔	Polakof *et al.* (2006)
		SW→BW	⇔	

(*Table 10.3 contd.*)

(Table 10.3 contd.)

	Red seabream	SW→BW	⇔	Woo and Murat (1981)
	Angelfish	SW→BW	⇔	Woo and Chung (1995)
Amino acid catabolism	Tilapia	FW→SW FW→SW	↑ ⇔	Vijayan *et al.* (1996) Vijayan *et al.* (2001)
	Atlantic salmon	FW→SW	↓	Plisetskaya *et al.* (1994)
	Rainbow trout	FW→SW	⇔	Jürss *et al.* (1985)
	Gilthead seabream	SW→HSW SW→BW	⇔ ⇔	Polakof *et al.* (2006)
Lipid catabolism	Coho salmon	FW→SW FW→SW FW→SW	↑ ↑ ↑	Woo *et al.* (1978) Sheridan *et al.* (1985) Sheridan (1988)
	Rainbow trout	FW→SW	↑	Brauge *et al.* (1995)
	Tilapia	FW→SW	↓	Vijayan *et al.* (1996)
	Gilthead seabream	SW→HSW SW→BW	⇔ ⇔	Polakof *et al.* (2006)

endogenous pathways such as glycolysis or to be exported to other tissues that need it (i.e., osmoregulatory organs in fish submitted to salinity transfer).

The time course of changes in liver glycogen levels in fish transferred to high salinity environment present different patterns. In gilthead seabream, a first stage of a sharp decline in glycogen levels followed by a recovery of levels producing at the end of the acclimation period higher levels of glycogen in HSW- than in SW-acclimated fish has been observed (Sangiao-Alvarellos *et al.*, 2005). In other studies, different patterns were observed, such as: (1) a single phase of continuous decline in rainbow trout transferred from FW to SW (Soengas *et al.*, 1993a); (2) two phases in tilapia transferred from FW to SW with no changes at the beginning and a decline after 2 days (Nakano *et al.*, 1998); and (3) no changes in carp acclimated from FW to BW (DeBoeck *et al.*, 2000). The enhanced mobilization of glycogen levels only in the first stage of acclimation to HSW strongly suggests an increased energy demand from other organs at that time. Accordingly, the enhancement of glycogenolysis in livers of BW- or HSW-acclimated gilthead seabream is accompanied by changes in liver ATP levels that decreased in parallel with increased salinity (Sangiao-Alvarellos *et al.*, 2003b).

On the other hand, during acclimation of gilthead seabream to BW, liver glycogen levels decreased throughout the experiment suggesting an enhancement of the energy requirements of this organ through increased glycogenolysis (Sangiao-Alvarellos *et al.*, 2005). In other species, acclimation to BW indicated no changes in liver glycogen levels in angelfish (Woo and Chung, 1995) and black seabream (Kelly *et al.*, 1999).

Changes in the glycolytic capacity of the liver were evaluated when fish were transferred to increased salinities, displaying an enhancement in that capacity in several species (Soengas *et al.*, 1995a; Vijayan *et al.*, 1996, 2001; Nakano *et al.*, 1997; Kelly and Woo, 1999b; Sangiao-Alvarellos *et al.*, 2003b). The study of the time course of glycolytic potential in gilthead seabream transferred from SW to HSW showed an increase of this capacity only in the first days of acclimation suggesting that was under those salinity conditions at that time (3 days) the highest energy requirements of the liver were taking place (Sangiao-Alvarellos *et al.*, 2003b). Others studies showed an increase in PK activity on the first day of acclimation of rainbow trout from FW to SW (Soengas *et al.*, 1993a), and an increase in PFK activity also in the first day of acclimation of tilapia from FW to SW (Nakano *et al.*, 1998).

On the other hand, acclimation to BW induced in black seabream (Kelly *et al.*, 1999) and gilthead seabream (Laiz-Carrión *et al.*, 2005b; Sangiao-Alvarellos *et al.*, 2003b, 2005) a decrease in liver glycolytic potential. These changes suggest that the mobilization of glycosyl units is directed towards an increased endogenous use on the first days of acclimation to BW, whereas on subsequent days, a reduced use is apparent based on decreased enzyme activities.

The capacity of liver to synthesize glucose through gluconeogenesis do not appear to be affected by osmotic acclimation in tilapia (Vijayan *et al.*, 2001) or gilthead seabream (Sangiao-Alvarellos *et al.*, 2003b; Laiz-Carrión *et al.*, 2005b) in contrast with the increase observed in BW- compared with SW-acclimated red seabream (Woo and Fung, 1981). However, changes displayed by glucose 6-phosphatase (G6Pase) activity in gilthead seabream suggest an increased capacity of liver for exporting glucose on days 1 and 14 in HSW- compared with SW-acclimated fish (Sangiao-Alvarellos *et al.*, 2003b). No other time courses are available in literature, and the only studies performed after transfer from SW to HSW address an absence of changes in liver G6Pase activity in black seabream (Kelly *et al.*, 1999). Thus, the liver of SW-acclimated fish may have a lower capacity to export glucose to plasma, which would agree with the finding of decreased liver G6Pase activity already reported in BW-acclimated fish compared with SW- and HSW-acclimated fish (Woo and Chung, 1995; Sangiao-Alvarellos *et al.*, 2003b, 2005). So, the portion of glucose obtained from liver mobilization and, therefore, capable of being used in other tissues is probably higher in HSW- than in SW-acclimated fish in the first stages of acclimation. These two stages can be also observed when considering that the liver of BW-acclimated gilthead seabream may also have a lower capacity to export glucose to plasma from day 7 of acclimation onwards, considering the decrease observed in G6Pase activity in gilthead seabream (Sangiao-Alvarellos *et al.*, 2005). This decrease would agree with the decrease found at end point in the same enzyme activity in angelfish acclimated to BW (Woo and Chung, 1995).

Liver glucose 6-phosphate dehydrogenase (G6PDH) activity (an indicator of potential of the pentose phosphate pathway) is not affected by hyperosmotic adaptation in most species (Soengas *et al.*, 1993a; Nakano *et al.*, 1997, 1998; Kelly *et al.*, 1999; Sangiao-Alvarellos *et al.*, 2003b, 2005; Laiz-Carrión *et al.*, 2005b), whereas increased activity was observed in rainbow tout transferred from FW to SW (Jürss *et al.*, 1986). In addition, transfer to hypoosmotic environment induced different behaviour in this activity: increases in black seabream (Kelly *et al.*, 1999), decreases in gilthead seabream (Sangiao-Alvarellos *et al.*, 2003b, 2005) or no changes in red seabream and angelfish (Woo and Murat, 1981; Woo and Chung, 1995).

With respect to amino acid metabolism, no changes have been observed in hepatic potential for amino acid catabolism during acclimation of rainbow trout (Jürss *et al.*, 1986), tilapia (Vijayan *et al.*, 2001) and gilthead seabream (Sangiao-Alvarellos *et al.*, 2003b), suggesting a reduced importance of these metabolites as fuel in liver during osmotic acclimation.

In contrast, osmotic acclimation is known to produce mobilization of liver lipids (Woo *et al.*, 1978) resulting from decreased lipogenesis (Sheridan *et al.*, 1985; Brauge *et al.*, 1995) and increased lipolysis (Sheridan, 1988; Li and Yamada, 1992). These results suggest an increased importance of lipid metabolism in liver during osmotic acclimation, probably related to the use of those metabolites as fuels to support the enhanced metabolic work of liver and other tissues observed during osmotic acclimation. Moreover, fatty acid composition in liver change during acclimation to SW developed the so-called 'marine pattern', which is relatively rich in long-chain PUFA (Li and Yamada, 1992; Tocher *et al.*, 2000; Cordier *et al.*, 2002), and attributed to the increased capacity of liver for fatty acid desaturation/elongation observed during osmotic acclimation (Tocher *et al.*, 2000).

Muscle

Fish present two main muscle types: white and red, which generally behave under anaerobic and aerobic conditions, respectively (Johnston, 1982). White muscle is the main protein accumulating tissue and is generally used in burst swimming, being principally a glycogen-burning tissue. In contrast, red muscle is used in sustained swimming, with lipids being its main fuel (Driedzic and Hochachka, 1978). In addition to the metabolic changes that take place in muscle directly or indirectly related to locomotion (Driedzic and Hochachka, 1978), other changes have been described in processes such as: spawning (Weatherley and Gill, 1987), feeding (Pérez *et al.*, 1988), starvation (Kiessling *et al.*, 1990), stress (Schwalme and MacKay, 1991) or migration (Leonard and McCormick, 2001). However, changes of muscle metabolism in seawater adaptation—a process which produce an increased energetic demand (Morgan and Iwama, 1991)—have received comparatively little attention.

Only a few studies have assessed metabolic changes in red muscle of teleost fish during osmotic acclimation. Thus, elevated activities of enzymes from both the respiratory chain and tricarboxylic acid cycle were observed in rainbow trout acclimated to SW compared with fish in FW (Kiessling *et al.*, 1991). Moreover, an increase in the capacity of both glycogenolysis and glycolysis have been observed in rainbow trout transferred from FW to SW (Soengas *et al.*, 1995c). The amount of studies is too low to make clear conclusions but apparently an increased energy demand arises in this tissue during osmotic acclimation.

In contrast, there are considerably more studies available in literature regarding metabolic changes in white muscle during osmotic acclimation. Table 10.4 summarizes the most important results obtained in those studies. Thus, glycogen levels of fish white muscle present different patterns of changes during acclimation to increased environmental salinities: (1) a decrease of glycogen levels (Assem and Hanke, 1979; Woo and Murat 1981; Claireaux and Dutil, 1992; Soengas *et al.*, 1993b, 1995c); (2) no changes (Woo *et al.*, 1978; Woo and Fung, 1981); and (3) contradictory results, i.e., decreases or absence of changes, in muscle glycogen levels of the same species (Bashamohideen and Parvatheswararao, 1972; Assem and Hanke, 1979). The production of glucose from glycogen was enhanced during acclimation of rainbow trout to SW (Kiessling *et al.*, 1991; Soengas *et al.*, 1993b, 1995c) and during

Table 10.4 Responses in selected metabolic pathways of white muscle energy metabolism after acclimation of teleost fish to different osmotic conditions. ↑, increase; ↓, decrease; ⇔ no changes.

Pathway	*Fish*	*Acclimation*	*Response*	*Reference*
Glycogenolysis	Red seabream	SW→BW	↑	Woo and Murat (1981)
		BW→SW	⇔	Woo and Fung (1981)
	Tilapia	SW→FW	↑	Assem and Hanke (1979)
	Cod	SW→BW	↑	Claureaux and Dutil (1992)
	Rainbow trout	FW→SW	↑	Soengas *et al.* (1993b, 1995c)
	Coho salmon	FW→SW	⇔	Woo *et al.* (1978)
Glycolysis	Rainbow trout	FW→SW	↑	Soengas *et al.* (1993b, 1995c)
Amino acid catabolism	Red seabream	BW→SW	⇔	Woo and Fung (1981)
	Sea bass	FW→BW	⇔	Roche *et al.* (1989)
	Rainbow trout	FW→SW	⇔	Jürss *et al.* (1985)
	Carp	FW→SW	⇔	DeBoeck *et al.* (2000)
Lipid catabolism	Red seabream	SW→BW	↑	Woo and Murat (1981)
		BW→SW	⇔	Woo and Fung (1981)
	Atlantic salmon	FW→SW	↑	Sheridan (1988)
	Rainbow trout	FW→SW	↑	Brauge *et al.* (1995)
	Sea bass	FW→SW	⇔	Roche *et al.* (1989)
	Carp	FW→SW	⇔	DeBoeck *et al.* (2000)

acclimation of carp to BW (DeBoeck *et al.*, 2000). The exogenous source of glucose was not important, as judged by the absence of an increase in HK activity in the only study in which this enzyme was measured (Soengas *et al.*, 1995c).

An increased glycolytic ability related to the increased salinity—as shown by the increased activities of 6-phosphofructo1-kinase, pyruvate kinase and lactate dehydrogenase—was clearly observed in white muscle of rainbow trout (Tang and Boutilier, 1991; Soengas *et al.*, 1995c) during adaptation to seawater and in Atlantic salmon smolts *versus* parrs (Leonard and McCormick, 2001). Altogether, these changes suggest an increased mobilization of glucose obtained from glycogen stores to be increasingly used through glycolysis in white muscle during osmotic acclimation.

Of the energy stored as glycogen in white muscle, a very large fraction can be provided in the form of lactate that besides being reconverted to glycogen *in situ* (Schulte *et al.*, 1992), it can also be provided to oxidative tissues via the blood stream (Weber, 1992). Moreover, lactate levels are known to increase in white muscle and plasma of SW-adapted rainbow trout (Tang and Boutilier, 1991). Furthermore, the activity of the Cori cycle (a metabolic pathway related to the transformation of lactate produced in muscle into glucose in liver) in fish is of minor importance (Moyes *et al.*, 1992; Schulte *et al.*, 1992), resulting in a poor uptake and oxidation of lactate in liver. Considering all the above data, we can hypothesize that an increase in lactate oxidation rates by those tissues able to use lactate as fuel and involved in osmotic work, such as the gills (Mommsen, 1984; Perry and Walsh, 1989), may take place during osmotic acclimation.

As for lipid metabolism, few studies report an enhanced lipolysis in white muscle during osmotic acclimation of euryhaline fish (Woo and Murat, 1981; Sheridan, 1988; Brauge *et al.*, 1995), whereas no changes were observed in many others (Woo and Fung, 1981; Roche *et al.*, 1989; DeBoeck *et al.*, 2000). Considering the nature of lipid stores in muscle, it is not surprising to find few changes in this metabolism associated with osmotic acclimation. Furthermore, several studies addressed an increased importance of PUFA in the composition of white muscle during osmotic acclimation (Cordier *et al.*, 2002).

The importance of amino acids for fuelling purposes in muscle during osmotic acclimation is even lower since in all the studies available in existing literature, no changes were noticed in amino acid metabolism during osmotic acclimation (Woo and Fung, 1981; Jürss *et al.*, 1986; Roche *et al.*, 1989; DeBoeck *et al.*, 2000).

Brain

Brain energy metabolism in fish changes under stress conditions (Soengas and Aldegunde, 2002), and also after treatment with stress hormones such as catecholamines (Sangiao-Alvarellos *et al.*, 2003a) and cortisol (Laiz-Carrión *et al.*, 2002, 2003). In this way, a stressful situation such as adaptation to different osmotic conditions should produce effects similar to those of other stressors already evaluated in fish brain. Nevertheless, this possibility has received little attention to date (Weng *et al.*, 2002). Weng *et al.* (2002) addressed for the first time the metabolic changes in fish brain associated with osmotic adaptation describing changes in ATP levels and creatine kinase activity in tilapia during the first hours of transfer from

FW to SW. Besides this study, only a group of studies carried out in gilthead seabream (Sangiao-Alvarellos *et al.*, 2003b; Laiz-Carrión *et al.*, 2005b; Polakof *et al.*, 2006) have assessed the metabolic changes in brain during osmotic acclimation. Those studies demonstrate that acclimation of this species to salinities—either lower or higher than normal—produces a mobilization of brain glycogen levels, which constitutes the major energy store of fish brain (Soengas and Aldegunde, 2002). The role of this mobilization is not known but could be related to a stress effect of salinity on brain metabolism that lead this tissue to activate processes involved indirectly in the osmoregulatory work.

In addition, the important increase observed in HK activity in gilthead seabream brain acclimated to extreme salinities compared with those acclimated to SW (Sangiao-Alvarellos *et al.*, 2003b) reflects the fact that the necessity of glucose, the main fuel of brain energy metabolism in teleosts (Soengas *et al.*, 1998), increases under the stress conditions imposed by the acclimation to extreme salinity environments. The time course of this increase during acclimation of gilthead seabream to HSW from day 4 of acclimation onwards suggests that at least part of the increased glucose within the brain is coming directly from the blood stream (Sangiao-Alvarellos *et al.*, 2005). Furthermore, HK activity increased in fish acclimated to BW only at the end of the experimental time. This is again suggestive of the existence of two stages in the metabolic changes occurring in this case in the brain after salinity transfer: (1) a first one of reduced glycogen mobilization and reduced use of exogenous glucose, followed by (2) a second period of an increased mobilization of glycogen and potential use of glucose at the end. The enhanced availability of glucose in this final stage would help to explain the sharp increase in brain free glucose levels also observed in BW-acclimated fish (Sangiao-Alvarellos *et al.*, 2003b, 2005; Laiz-Carrión *et al.*, 2005a).

These changes in glucose phosphorylating capacity in gilthead seabream are reflected in an increased energy demand based on the high glycolytic potential observed in BW- and HSW-acclimated fish, as suggested by changes displayed by PFK activity (Sangiao-Alvarellos *et al.*, 2003b, 2005). This increased glycolysis may be related to the increase described by Weng *et al.* (2002) in both Na^+,K^+-ATPase and creatine kinase activities in the brain of tilapia transferred from FW to SW. Another interesting finding in gilthead seabream was the increase in brain ATP and lactate levels in parallel with salinity, suggesting that the increased use of carbohydrates is higher than the energy demand of the brain, resulting in the production of an accumulation of both lactate and ATP (Sangiao-Alvarellos *et al.*, 2003b). This accumulation also probably reflects the decreased energy demand of brain at lower salinities since BW appears to be of a less stressor capacity than HSW. This is in agreement with the higher growth displayed by gilthead seabream in intermediate salinity (12 ppt) (Laiz-Carrión *et al.*, 2005b).

The increased availability of glucose within the brain in the first stage of acclimation in gilthead seabream is apparently not used through glycolysis or pentose phosphate in BW-transferred fish, since no important changes were noticed in the activity of selected enzymes from those pathways, whereas an increased use through glycolysis is apparent in HSW-transferred fish at that stage (Sangiao-Alvarellos *et al.*, 2005). Also considering that fish brain appears to use glucose and lactate as fuels (Soengas *et al.*, 1998), a profound reorganization of brain energy metabolism is apparently taking place under this situation. Interestingly, the direction of changes in metabolic parameters in brain of gilthead seabream is in most cases the same when comparing HSW and BW acclimation in contrast to the situation described in other tissues (Sangiao-Alvarellos *et al.*, 2005). This may suggest that the changes occurring in brain are mainly reflecting a stress salinity response irrespective of the direction of changes in salinity.

Heart

There are very few studies assessing the changes in heart energy metabolism during osmotic acclimation. The results provided in those studies address no changes in glycogen levels of red seabream during acclimation from SW to BW (Woo and Murat, 1981), a fall in plasma protein levels in tilapia during acclimation from FW to SW (Venkatachari, 1974), and an increase in the activity of glycolytic and lipolytic enzymes in Atlantic salmon smolts compared with (Leonard and McCormick, 2001).

Acknowledgements

The author´s research has been supported in recent years by grants BOS2001-4031-C02-01 and VEM2003-20062 (Ministerio de Ciencia y Tecnología and FEDER, Spain) AGL2004-08137-C04-03/ACU (Ministerio de Educación y Ciencia and FEDER, Spain),

PGIDT04PXIC31208PN and PGIDIT05PXIC31202PN (Xunta de Galicia, Spain) to J.L. Soengas, and grant BOS2001-4031-C02-02 and BFU2004-04439-CO2-01B (Ministerio de Ciencia y Tecnología and FEDER, Spain) to J.M. Mancera.

References

Altinok, I. and J.M. Grizzle. 2001. Effects of brackish water on growth, feed conversion and energy absorption efficiency by juvenile euryhaline and freshwater stenohaline fishes. *Journal of Fish Biology* 59: 1142–1152.

Assem, H. and W. Hanke. 1979. Concentrations of carbohydrates during osmotic adjustment of the euryhaline teleost, *Tilapia mossambica*. *Comparative Biochemistry and Physiology* A64: 5–16.

Auerswald, L., K. Jürss, D. Schiedek and R. Bastrop. 1997. The influence of salinity acclimatation on free aminoacids and enzyme activities in the intestinal mucosa of rainbow trout, *Oncorhynchus mykiss* (Walbaum). *Comparative Biochemistry and Physiology* A116: 149–155.

Bashamohideen, M. and V. Parvatheswararao. 1972. Adaptations to osmotic stress in the freshwater euryhaline teleost *Tilapia mossambica*. IV. Changes in blood glucose, liver glycogen and muscle glycogen levels. *Marine Biology* 16: 68–74.

Beyenbach, KW. 1995. Secretory electrolyte transport in renal proximal tubules of fish. In: *Fish Physiology*, C.M. Wood and T.J. Shuttlewoth (eds). Academic Press, New York, Vol. 14, pp. 85–106.

Blasco, J., J. Fernández-Borrás, I. Marimon and A. Requena. 1996. Plasma glucose kinetics and tissue uptake in brown trout *in vivo*: Effect of an intravascular glucose load. *Journal of Comparative Physiology* B165: 534–541.

Blasco, J., I. Marimon, I. Viaplana and J. Fernández-Borrás. 2001. Fate of plasma glucose in tissues of brown trout *in vivo*: Effects of fasting and glucose loading. *Fish Physiology and Biochemistry* 24: 247–258.

Boeuf, G. and P. Payan. 2001. How should salinity influence fish growth? *Comparative Biochemistry and Physiology* 130: 411–423.

Brauge, C., F. Medale and G. Corraze. 1994. Effect of dietary carbohydrate levels on growth, body composition and glycaemia in rainbow trout, *Oncorhynchus mykiss*, reared in seawater. *Aquaculture* 123: 109–120.

Brauge, C., G. Corraze and F. Médale. 1995. Effects of dietary levels of carbohydrate and lipid on glucose oxidation and lipogenesis from glucose in rainbow trout, *Oncorhynchus mykiss*, reared in freshwater or in seawater. *Comparative Biochemistry and Physiology* A111: 117–124.

Brill, R., Y. Swimmer, C. Taxboel, K. Cousins and T. Lowe. 2001. Gill and intestinal Na^+-K^+ ATPase activity, and estimated maximal osmoregulatory costs, in three high-energy-demand teleosts: yellowfin tuna (*Thunnus albacares*), skipjack tuna (*Katsuwonus pelamis*), and dolphin fish (*Coryphaena hippurus*). *Marine Biology* 138: 935–944.

Buddington, R.K. and A. Krogdahl. 2004. Hormonal regulation of the fish gastrointestinal tract. *Comparative Biochemistry and Physiology* A139: 261–271.

Claireaux, G. and J.-D. Dutil. 1992. Physiological response of the Atlantic cod (*Gadus morhua*) to hypoxia at various environmental salinities. *Journal of Experimental Biology* 163: 97–118.

Cordier, M., G. Brichon, J.-M. Weber and G. Zwingelstein. 2002. Changes in the fatty acid composition of phospholipids in tissues of farmed sea bass (*Dicentrarchus labrax*) during and annual cycle. Roles of environmental temperature and salinity. *Comparative Biochemistry and Physiology* B133: 281–288.

De Boeck, G., A. Vlaeminck, A. Van der Linden and R. Blust. 2000. The energy metabolism of common carp (*Cyprinus carpio*) when exposed to salt stress: an increase in energy expenditure or effects of starvation? *Physiological and Biochemical Zoology* 73: 102–111.

Driedzic, W.R. and P.W. Hochachka. 1978. Metabolism in fish during exercise. In: *Fish Physiology*, W. S. Hoar and D. J. Randall (eds.). Academic Press, New York, Vol. 10, pp. 503–543.

Febry, R. and P. Lutz. 1987. Energy partitioning in fish: the activity related cost of osmoregulation in a euryhaline cichlid. *Journal of Experimental Biology* 128: 63–85.

Fuentes, J. and F.B. Eddy. 1997a. Drinking in marine, euryhaline and freshwater teleost fish. In: *Ionic Regulation in Animals*, N. Hazon, F.B. Eddy and G. Flik (eds.). Springer-Verlag, Heidelberg, pp. 135–149.

Fuentes, J. and F.B. Eddy. 1997b. Drinking in Atlantic salmon presmolts and smolts in response to growth hormone and salinity. *Comparative Biochemistry and Physiology* A117: 487–491.

Hanke, W. 1991. Mechanism of osmotic adaptation in fresh water teleost. *Fischerei Forschung* 29: 15–19.

Hansen, H.J.M. and M. Grosell. 2004. Are membrane lipids involved in osmoregulation? Studies *in vivo* on the European eel, *Anguilla anguilla*, after reduced ambient salinity. *Environmental Biology of Fishes* 70: 57–65.

Hansen, H.J.M., A.G. Olsen and P. Rosenkilde. 1992. Comparative studies on lipid metabolism in salt-transporting organs of the rainbow trout (*Oncorhynchus mykiss* W.). Further evidence of monounsaturated phosphatidylethanolamine as a key substance. *Comparative Biochemistry and Physiology* B103: 81–87.

Holmes, W.N. and E.M. Donaldson. 1969. The body compartments and the distribution of electrolytes. In: *Fish Physiology*, W.S. Hoar and D.J. Randall (eds.). Academic Press, San Diego. Vol 1, pp. 1–89

Johnston, I.A. 1982. Physiology of muscle in hatchery raised fish. *Comparative Biochemistry and Physiology* B73: 105–124.

Jürss, K., T. Bittorf, T. Vökler and R. Wacke. 1983. Influence of nutrition on biochemical seawater adaptation of the rainbow trout (*Salmo gairdneri* Richardson). *Comparative Biochemistry and Physiology* B75: 713–717.

Jürss, K., T. Bittorf and T. Vökler. 1986. Influence of salinity and food deprivation on growth, RNA/DNA ratio and certain enzyme activities in rainbow trout (*Salmo gairdneri* Richardson). *Comparative Biochemistry and Physiology* B83: 425–433.

Kelly, S.P. and N.Y.S. Woo. 1999a. The response of sea bream following abrupt hypo-osmotic exposure. *Journal of Fish Biology* 55: 732–750.

Kelly, S.P. and N.Y.S. Woo. 1999b. Cellular and biochemical characterization of hyposmotic adaptation in a marine teleost, *Sparus sarba*. *Zoological Science* 16: 505–514.

Kelly, S.P., N.K. Chow and N.Y.S. Woo. 1999. Haloplasticity of black seabream (*Mylio macrocephalus*): Hypersaline to freshwater acclimation. *Journal of Experimental Zoology* 283: 226–241.

Kiessling, A., L. Johansson and K.-H. Kiessling. 1990. Effects of starvation on rainbow trout muscle. I. Histochemistry, metabolism and composition of white and red muscle in mature and immature fish. *Acta Agricola Scandinavica* 40: 309–324.

Kiessling, A., K.-H. Kiessling, T. Storebakken and T. Asgard. 1991. Changes in the structure and function of the epaxial muscle of rainbow trout (*Oncorhynchus mykiss*) in relation to ration and age II. Activity of key enzymes in energy metabolism. *Aquaculture* 93: 357–372.

Kirschner, L.B. 1993. The energetics of osmotic regulation in ureotelic and hypoosmotic fishes. *Journal of Experimental Zoology* 267: 19–26.

Kirschner, L.B. 1995. Energetics of osmoregulation in fresh water vertebrates. *Journal of Experimental Zoology* 271: 243–252.

Krogdahl, A., A. Sundby and J.J. Olli. 2004. Atlantic salmon (*Salmo salar*) and rainbow trout (*Oncorhynchus mykiss*) digest and metabolize nutrients differently. Effects of water salinity and dietary starch level. *Aquaculture* 229: 335–360.

Kültz, D. and K. Jürss. 1993. Biochemical characterization of isolated branchial mitochondria-rich cells of *Oreochromis mossambicus* acclimated to fresh water or hypersaline seawater. *Journal of Comparative Physiology* B163: 406–412.

Laiz-Carrión, R., S. Sangiao-Alvarellos, J.M. Guzmán, M.P. Martin del Rio, J.M. Míguez, J.L. Soengas and J.M. Mancera. 2002. Energy metabolism in fish tissues related to osmoregulation and cortisol action. *Fish Physiology and Biochemistry* 27: 179–188.

Laiz-Carrión, R., M.P. Martín del Rio, J.M. Míguez, J.M. Mancera and J.L. Soengas. 2003. Influence of cortisol on osmoregulation and energy metabolism in gilthead seabream *Sparus aurata*. *Journal of Experimental Zoology* A289: 105–118.

Laiz-Carrión, R., P.M. Guerreiro, J. Fuentes, A.V.M. Canario, M.P. Martin del Rio and J.M. Mancera. 2005a. Branchial osmoregulatory response to salinity in the gilthead sea bream, *Sparus aurata*. *Journal of Experimental Zoology* A303: 563–576.

Laiz-Carrión, R., S. Sangiao-Alvarellos, J.M. Guzmán, M.P. Martín del Río, J.L. Soengas and J.M. Mancera. 2005b. Growth performance of gilthead sea bream *Sparus aurata* in different osmotic conditions: Implications for osmoregulation and energy metabolism. *Aquaculture* 250: 849–861.

Le François, N.R. and P. Blier. 2004. Reproductive events and associated reduction in the seawater adaptability of brook charr (*Salvelinus fontinalis*): Evaluation of gill metabolic adjustments. *Aquatic Living Resources* 16: 69–76.

Le François, N.R., S.G. Lamarre and P.U. Blier. 2004. Tolerance, growth and haloplasticity of the Atlantic wolfish (*Anarhichas lupus*) exposed to various salinities. *Aquaculture* 236: 659–675.

Leonard, J.B.K. and S.D. McCormick. 2001. Metabolic enzyme activity during smolting in stream- and hatchery-reared Atlantic salmon (*Salmo salar*). *Canadian Journal of Fisheries and Aquatic Sciences* 58: 1585–1593.

Leray, C., D.A. Colin and A. Florentz. 1981. Time course of osmotic adaptation and gill energetics of rainbow trout (*Salmo gairdneri* R.) following abrupt changes in external salinity. *Journal of Comparative Physiology* 144: 175–181.

Li, H.O. and J. Yamada. 1992. Changes of the fatty acid composition in smolts of masu salmon (*Oncorhynchus masou*), associated with desmoltification and sea-water transfer. *Comparative Biochemistry and Physiology* A103: 221–226.

Lionetto, M.G., M.E. Giordano, G. Nicolardi and T. Schettino. 2001. Hypertonicity stimulates Cl^- transport in the intestine of fresh water acclimated eel, *Anguilla anguilla*. *Cellular Physiology and Biochemistry* 11: 41–54.

Maetz, J. 1974. Aspects of adaptation to hypo-osmotic and hyper-osmotic environments. In: *Biochemical and Biophysical Perspectives in Marine Biology*, D.C. Malins and J.R. Sargent (eds.). Academic Press, New York, Vol. 1, pp. 1–167.

Mancera, J.M., J.M. Perez-Figares and P. Fernandez-Llebrez. 1993. Osmoregulatory responses to abrupt salinity changes in the euryhaline gilthead seabream (*Sparus aurata* L.). *Comparative Biochemistry and Physiology* A106: 245–250.

Mancera, J.M., R. Laiz-Carrión and M.P. Martín del Rio. 2002. Osmoregulatory action of PRL, GH, and cortisol in the gilthead seabream (*Sparus aurata* L.). *General and Comparative Endocrinology* 129: 95–103.

Marshall, W.S. 2002. Na^+, Cl^-, Ca^{2+} and Zn^{2+} transport by fish gills: retrospective review and prospective synthesis. *Journal of Experimental Zoology* 293: 264–283.

Maxime, V. 2002. Effects of transfer to sea water on standard and routine metabolic rates in smolting Atlantic salmon at different stages of seawater adaptability. *Journal of Fish Biology* 61: 1423–1432.

McCormick, S.D. and R.L. Saunders. 1987. Preparatory physiological adaptations for marine life salmonids: Osmorregulation, growth, and metabolism. *American Fisheries Society Symposium* 1: 211–229.

McCormick, S.D., C.D. Moyes and J.S. Ballantyne. 1989. Influence of salinity on the energetics of gill and kidney of Atlantic salmon (*Salmo salar*). *Fish Physiology and Biochemistry* 6: 243–254.

Mommsen, T.P. 1984. Biochemical characterization of the rainbow trout gill. *Journal of Comparative Physiology* 154: 191–198.

Mommsen, T.P., P.J. Walsh, and T.W. Moon. 1985. Gluconeogenesis in hepatocytes and kidney of Atlantic salmon. *Molecular Physiology* 8: 89–100.

Morgan, J.D. and G.K. Iwama. 1991. Effects of salinity on growth, metabolism, and ion regulation in juvenile rainbow trout (*Oncorhynchus mykiss*) and fall Chinook salmon (*Oncorhynchus tshawytscha*). *Canadian Journal of Fisheries and Aquatic Sciences* 48: 2083–2094.

Morgan, J.D. and G.K. Iwama. 1996. Cortisol induced changes in oxygen consumption and ionic regulation in coastal cut-throat trout parr. *Fish Physiology and Biochemistry* 15: 385–394.

Morgan, J.D. and G.K. Iwama. 1998. Salinity effects on oxygen consumption, gill Na^+, K^+-ATPase and ion regulation in juvenile coho salmon. *Journal of Fish Biology* 53: 1110–1119.

Morgan, J.D., T. Sakamoto, E.G. Grau and G.K. Iwama. 1997. Physiological and respiratory responses of the Mozambique tilapia (*Oreochromis mossambicus*) to salinity acclimation. *Comparative Biochemistry and Physiology* A117: 391–398.

Moyes, C.D., P.M. Schulte and P.W. Hochachka. 1992. Recovery metabolism of trout white muscle: role of mitochondria. *American Journal of Physiology* 262: R295–R304.

Nakano, K., M. Tagawa, A. Takemura and T. Hirano. 1997. Effects of ambient salinities on carbohydrate metabolism in two species of Tilapia: *Oreochromis mossambicus* and *O. niloticus*. *Fisheries Science* 63: 338–343.

Nakano, K., M. Tagawa, A. Takemura and T. Hirano. 1998. Temporal changes in liver carbohydrate metabolism associated with seawater transfer in *Oreochromis mossambicus*. *Comparative Biochemistry and Physiology* B119: 721–728.

Nelson, J.A., Y. Tang and R.G. Boutilier. 1996. The effects of salinity change on the exercise performance of two Atlantic cod (*Gadus morhua*) populations inhabiting different environments. *Journal of Experimental Biology* 199: 1295–1309.

Nordgarden, U., G.-I. Hemre and T. Hansen. 2002. Growth and body composition of Atlantic salmon (*Salmo salar* L.) parr and smolt fed diets varying in protein and lipid contents. *Aquaculture* 207: 65–78.

Nordlie, F.G., S.J. Walsh, D.C. Haney and T.F. Nordlie. 1991. The influence of ambient salinity on routine metabolism in the teleost *Cyprinodon variegatus* Lacepède. *Journal of Fish Biology* 38: 115–122.

Peragón, J., J.B. Barroso, M. de la Higuera and J.A. Lupiáñez. 1998. Relationship between growth and protein turnover rates and nucleic acids in the liver of rainbow trout (*Oncorhynchus mykiss*) during development. *Canadian Journal of Fisheries and Aquatic Sciences* 55: 649–657.

Pérez, J., S. Zanuy and M. Carrillo. 1988. Effects of diet and feeding time on daily variations in plasma insulin, hepatic c-AMP and other metabolites in a teleost fish, *Dicentrarchus labrax*. *Fish Physiology and Biochemistry* 5: 191–197.

Perry, S.F. and P.J. Walsh. 1989. Metabolism of isolated fish gill cells: Contribution of epithelial chloride cells. *Journal of Experimental Biology* 144: 507–520.

Plisetskaya, E.M., T.W. Moon, D.A. Larsen, G.D. Foster and W.W. Dickhoff. 1994. Liver glycogen, enzyme activities, and pancreatic hormones in juvenile Atlantic salmon (*Salmo salar*) during their first summer in seawater. *Canadian Journal of Fisheries and Aquatic Sciences* 51: 567–576.

Polakof, S., F.J. Arjona, S. Sangiao-Alvarellos, M.P. Martin del Rio, J.M. Mancera and J.L. Soengas. 2006. Food deprivation alters osmoregulatory and metabolic responses to salinity acclimation in gilthead seabream *sparus auratus*. *Journal of Comparative Physiology* B176: 441–452.

Renfro, J.L. 1995. Solute transport by flounder renal cells in primary culture. In: *Fish Physiology*, C.M. Wood and T.J. Shuttlewoth (eds.). Academic Press, New York. Vol. 14, pp. 147–173.

Roche, H., K. Chaar and G. Pérès. 1989. The effect of a gradual decrease in salinity on the significant constituents of tissue in the sea bass (*Dicentrarchus labrax* Pisces). *Comparative Biochemistry and Physiology* A93: 785–789.

Sangiao-Alvarellos, S., P. Bouça, J.M. Miguez and J.L. Soengas. 2003a. Intracerebroventricular injections of noradrenaline affect brain energy metabolism of rainbow trout. *Physiological and Biochemical Zoology* 76: 663–671.

Sangiao-Alvarellos, S., R. Laiz-Carrión, J.M. Guzmán, M.P. Martin del Rio, J.M. Miguez, J.M. Mancera and J.L. Soengas. 2003b. Acclimation of *S. aurata* to various salinities alters energy metabolism of osmoregulatory and monosmoregulatory organs. *American Journal of Physiology* 285: R897–R907.

Sangiao-Alvarellos, S., F.J. Arjona, M.P. Martín del Río, J.M. Míguez, J.M. Mancera and J.L. Soengas. 2005. Time course of osmoregulatory and metabolic changes during osmotic acclimation in *Sparus auratus*. *Journal of Experimental Biology* (In press).

Schulte, P. M., C.D. Moyes and P.W. Hochachka. 1992. Integrating metabolic pathways in post-exercise recovery of white muscle. *Journal of Experimental Biology* 166: 181–195.

Schwalme, K. and W.C. Mckay. 1991. Mechanisms that elevate the glucose concentration of muscle and liver in yellow perch (*Perca flavescens* Mitchill) after exercise-handing stress. *Canadian Journal of Zoology* 69: 456–461.

Sheridan, M.A. 1988. Exposure to seawater stimulates lipid mobilization from depot tissues of juvenile coho (*Oncorhynchus kisutch*) and chinook (*O. tschawytscha*) salmon. *Fish Physiology and Biochemistry* 5:173–180.

Sheridan, M.A., N.Y.S. Woo and H.A. Bern. 1985. Changes in the rates of glycogenesis, glycogenolysis, lipogenesis, and lipolysis in selected tissues of the coho salmon (*Oncorhynchus kisutch*) associated with parr-smolt transformation. *Journal of Experimental Zoology* 236: 35–44.

Soengas, J.L. and M. Aldegunde. 2002. Energy metabolism of fish brain. *Comparative Biochemistry and Physiology* B131: 271–296.

Soengas, J.L., P. Barciela, J. Fuentes, J. Otero, M.D. Andrés and M. Aldegunde. 1993b. Changes in muscle carbohydrate metabolism in domesticated rainbow trout (*Oncorhynchus mykiss*) after transfer to seawater. *Comparative Biochemistry and Physiology* B104: 173–179.

Soengas, J.L., P. Barciela, J. Fuentes, J. Otero, M.D. Andrés and M. Aldegunde. 1993a. The effect of seawater transfer in liver carbohydrate metabolism of domesticated rainbow trout (*Oncorhynchus mykiss*). *Comparative Biochemistry and Physiology* B105: 337–343.

Soengas, J.L., J. Fuentes, M.D. Andres and M. Aldegunde. 1994. Direct transfer of rainbow trout to seawater induces several changes in kidney carbohydrate metabolism. *Journal of Physiology and Biochemistry* 50: 219–228.

Soengas, J.L., M. Aldegunde and M.D. Andrés. 1995a. Gradual transfer to seawater of rainbow trout: Effects on liver carbohydrate metabolism. *Journal of Fish Biology* 47: 466–478.

Soengas, J.L., P. Barciela, M. Aldegunde and M.D. Andrés. 1995b. Gill carbohydrate metabolism of rainbow trout is modified during gradual adaptation to seawater. *Journal of Fish Biology* 46: 845–856.

Soengas, J.L., J. Fuentes, M.D. Andrés and M. Aldegunde. 1995c. The effect of gradual transfer to sea water on muscle carbohydrate metabolism of rainbow trout. *Journal of Fish Biology* 46: 509–523.

Soengas, J.L., E.F. Strong and M.D. Andrés. 1998. Glucose, lactate, and β-hydroxybutyrate utilization by rainbow trout brain: changes during food deprivation. *Physiological Zoology* 71: 285–293.

Sparks, R., B.S. Shepherd, B. Ron, N.H. Richman III, L.G. Riley, G.K. Iwama, T. Hirano and E. G. Grau. 2003. Effects of environmental salinity and 17α-methyltestosterone on growth and oxygen consumption in the tilapia, *Oreochromis mossambicus*. *Comparative Biochemistry and Physiology* B136: 657–665.

Swanson, C. 1998. Interactive effects of salinity on metabolic rate, activity, growth and osmoregulation in the euryhaline milkfish (*Chanos chanos*). *Journal of Experimental Biology* 201: 3355–3366.

Tang, J. and R.G. Boutilier. 1991. White muscle intracellular acid-base and lactate status following exhaustive exercise: A comparison between freshwater- and seawater adapted rainbow trout. *Journal of Experimental Biology* 156: 153–171.

Tocher, D.R., J.G. Bell, J.R. Dick, R.J. Henderson, F. McGhee, D. Michell and P.C. Morris. 2000. Polyunsaturated fatty acid metabolism in Atlantic salmon (*Salmo salar*) undergoing parr-smolt transformation and the effects of dietary linseed and rapeseed oils. *Fish Physiology and Biochemistry* 23: 59–73.

Venkatachari, S.A.T. 1974. Effect of salinity adaptation on nitrogen metabolism in the freshwater fish *Tilapia mossambica*. I. Tissue protein and amino acid levels. *Marine Biology* 24: 57–63.

Vijayan, M.M., J.D. Morgan, T. Sakamoto, E.G. Grau and G.K. Iwama. 1996. Food-deprivation affects seawater acclimation in tilapia: hormonal and metabolic changes. *Journal of Experimental Biology* 199: 2467–2475.

Vijayan, M.M., A. Takemura and T.P. Mommsen. 2001. Estradiol impairs hypoosmoregulatory capacity in the euryhaline tilapia, *Oreochromis mossambicus*. *American Journal of Physiology* 281: R1161–R1168.

Weatherley, A.H. and H.S. Gill. 1987. *The Biology of Fish Growth*. Academic Press, New York.

Weber, J.M. 1992. Pathways for oxidative fuel provision to working muscles: ecological consequences of maximal supply limitations. *Experientia* 48: 557–564.

Weng, C.F., C.C. Chiang, H.Y. Gong, M.H.C. Chen, C.J.F. Lin and W.T. Huang. 2002. Acute changes in gill Na^+-K^+-ATPase and creatine kinase in response to salinity changes in the euryhaline teleost, tilapia (*Oreochromis mossambicus*). *Physiological Zoology* 75: 29–36.

Wilson, J.M. and P. Laurent. 2002. Fish gill morphology: Inside out. *Journal of Experimental Zoology* 293: 92–213.

Wilson, J.M., N.M. Whiteley and D.J. Randall. 2002. Ionoregulatory changes in the gill epithelia of coho salmon during seawater acclimation. *Physiological and Biochemical Zoology* 75: 237–249.

Woo, N.Y.S. and A.C. Fung. 1981. Studies on the biology of the red sea bream, *Chrysophrys major*. II Salinity adaptation. *Comparative Biochemistry and Physiology* A69: 237–242.

Woo, N.Y.S. and J.C. Murat. 1981. Studies on the biology of the red seabream *Chrysophrys major* III. Metabolic response to starvation in different salinities. *Marine Biology* 61: 255–260.

Woo, N.Y.S. and K.C. Chung. 1995. Tolerance of *Pomacanthus imperator* to hypoosmotic salinities: Changes in body composition and hepatic enzyme activities. *Journal of Fish Biology* 47: 70–81.

Woo, N.Y.S., H.A. Bern and R.S. Nishioka. 1978. Changes in body composition associated with smoltification and premature transfer to seawater in Coho salmon (*Oncorhynchus kisutch*) and King salmon (*O. tschawytscha*). *Journal of Fish Biology* 13: 421–428.

Wuenschel, M.J., A.R. Jugovich and J.A. Hare. 2005. Metabolic response of juvenile gray snapper (*Lutjanus griseus*) to temperature and salinity: Physiological cost of different environments. *Journal of Experimental Marine Biology and Ecology* 321: 145–154.

Yavuzcan-Yildiz, H. and M. Kirkagaç-Uzbilek. 2001. The evaluation of secondary stress response of grass carp (*Ctenopharyngodon idella*, Val. 1844) after exposing to saline water. *Fish Physiology and Biochemistry* 25: 287–290.

CHAPTER

11

The Renal Contribution to Salt and Water Balance

M. Danielle McDonald

INTRODUCTION

This chapter will examine what is currently known about the kidney as it pertains to salt and water balance within the agnathans (hagfish and lamprey), elasmobranchs (sharks, rays and skates) and teleosts.

Hagfish are slightly hyperosmotic (1035 mOsm) to their marine environment (~1000 mOsm) and have achieved this status by retaining body fluid levels of Na^+ and Cl^- that are similar to concentrations found in seawater (Hickman and Trump, 1969). For this reason, hagfish are generally considered to be the only vertebrate group that, like marine invertebrates, are potentially free from any need for osmoregulation (Hardisty, 1979; Evans, 1993). However, despite the plasma being at most 2% hyperosmotic to seawater, hagfish extracellular fluid differs from seawater in almost all the major ions (Hickman and Trump, 1969).

Author's address: Rosenstiel School of Marine and Atmospheric Science, University of Miami, Miami, Florida, 33149-1098, USA.
E-mail: mcdonald@rsmas.miami.edu

Specifically, K^+ and all of the divalent ions, with the possible exception of Ca^{2+}, are concentrated in the urine and found in concentrations slightly below that of seawater in the plasma (Hickman and Trump, 1969; Evans, 1979, 1993; Hardisty, 1979). Na^+ appears to be reabsorbed by the kidney, maintaining plasma levels above seawater. Cl^- is about the same concentration in plasma as in urine. The individual regulation of the plasma ions appears to be the principal function of the hagfish kidney. However, the hagfish is believed to have the most ineffective of all vertebrate kidneys; fortunately, its regulatory task is small compared to osmoregulating organisms (Hickman and Trump, 1969). In contrast, lampreys, the other surviving agnathan, are osmoregulators and are often euryhaline as a consequence of their life cycle. In contrast to hagfish, lampreys have developed the same suite of osmoregulatory mechanisms that are present in teleost kidneys (see below).

The body fluids of elasmobranchs are similar to agnathans in the sense that they are almost isoosmotic if not slightly hyperosmotic (~1100 mOsm) to the marine environment (~1000 mOsm; Hickman and Trump, 1969). However, unlike agnathans, their high plasma osmolality is achieved by combining levels of electrolytes less than that of seawater with high levels of urea and trimethylamine oxide (TMAO) retained at concentrations far above those measured in any other vertebrate group. Consequently, water enters the body by osmosis as it does in freshwater teleosts and electrolytes enter by diffusion, similar to marine teleosts. The kidney functions to retain urea and TMAO and eliminate divalent ions such as Mg^{2+}, SO_4^{2-} and PO_4^{2-}. Excess Na^+ and Cl^- are eliminated by the rectal gland, which forms a colorless fluid that contains NaCl at nearly twice its plasma concentration.

Freshwater teleost fish live in an environment that is hypoosmotic (< 1 mOsm) to their body fluids (280-300 mOsm) and are, consequently, plagued with a continuous osmotic influx of water and depletion of salts by diffusion. Active uptake of electrolytes occurs at the gill and the main responsibilities of the freshwater kidney and urinary bladder are to rid the body of surplus water while at the same time conserve valuable salts. In contrast, marine teleosts live in an environment that is hyperosmotic (~1000 mOsm) to their body fluids (300-320 mOsm). In this environment, fish constantly gain salt by diffusion and lose water via osmosis. To compensate, they drink seawater and must desalinize the water and further modify intestinal fluid so as to promote water absorption.

Marine fish actively excrete most of their monovalent ions via the gill and while the gill extrudes Na^+ and Cl^-, the kidney and urinary bladder serve to conserve water but excrete Mg^{2+}, SO_4^{2-} and other divalent ions.

Kidney Morphology

The hagfish kidney is quite simple, with segmentally arranged glomeruli draining into short neck segments and then into paired common archinephric ducts that have some of the structural and functional attributes of proximal tubule I of other fishes and tetrapods (Fig. 11.1A; Evans, 1993). The glomeruli in hagfish are supplied by branches of segmental arteries arising from the dorsal aorta (Hickman and Trump, 1969). The capillary network of the ureters is served by the postglomerular circulation and by arteries directly from the aorta; unlike teleosts, there is no portal circulation (Hickman and Trump, 1969). In comparsion, lampreys have a much more developed kidney, with distinct glomeruli, proximal tubule I, distal tubule and collecting duct (Fig. 11.1B; Hentschel and Elger, 1989; Evans, 1993). The proximal and distal tubules are arranged in a loop, reminiscent of the loop of Henle of the mammalian kidney and the elasmobranch kidney (see below). Lampreys also lack a renal portal system which, in teleosts, enables tubular secretion to continue in the absence of glomerular perfusion (Brown and Rankin, 1999). Because of this difference in renal circulation, agnathans and teleosts regulate their GFR differently.

The internal anatomy of the elasmobranch kidney is complex but highly organized. In general, the marine elasmobranch kidney consists of a large glomerulus, proximal tubule I and II, a countercurrent loop system connecting the proximal and distal tubule and the collecting duct system (Fig. 11.1C). In general, the kidneys are divided into two regions, a sinus (ventral) zone and the bundle (dorsal) zone that is enclosed by a peritubular sheath (Lacy *et al.*, 1985; Lacy and Reale, 1986; Friedman and Hebert, 1990). A single nephron has two highly coiled loops that enter the sinus zone, which is richly endowed with blood vessels and is a region of a high rate of blood flow. There appears to be no organized pattern to the loops in the sinus zone. In contrast, five lengths of the same nephron are arranged to lie in parallel within the bundle zone. The segments pass through this region and form convoluted loops within the bundle zone and dip into the sinus zone, allowing the tubule fluid to pass twice through each of the two zones. The physiological significance of this behavior is not

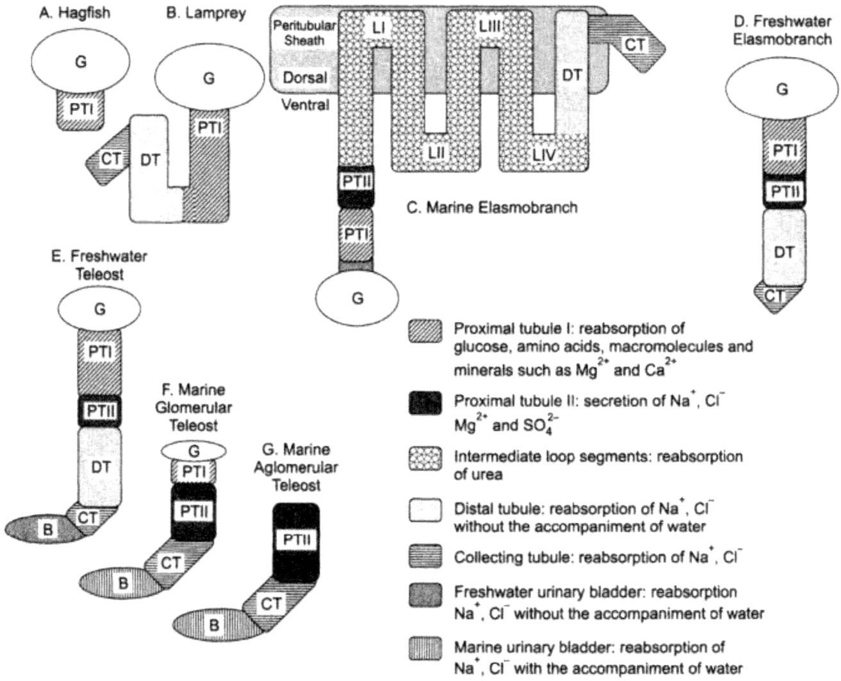

Fig. 11.1 Schematic representation of nephron structure and its relative importance to osmoregulation within (A) hagfish (B) lamprey, (C) marine elasmobranchs (D) freshwater elasmobranchs (E) freshwater teleosts, (F) marine glomerular teleosts and (G) marine aglomerular teleosts. Adapted from Hickman and Trump (1969) and Evans (1993).

entirely clear but it very likely facilitates the conservation of organic solutes in the plasma, namely urea (Braun and Dantzler, 1997). Interestingly, the freshwater elasmobranch kidney (Fig. 11.1D) lacks the nephron loops observed in marine elasmobranchs.

The teleost kidney is much less complex than that of the elasmobranch and, in some ways, the lamprey. Freshwater teleost fish have a kidney nephron that includes a glomerulus, proximal tubule I and II, distal tubule and a collecting tubule and duct (Fig. 11.1E; Nishimura and Imai, 1982; Hentschel and Elger, 1989; Evans, 1993). The kidney of marine teleosts is morphologically reduced in comparison. In marine teleosts, various stages of glomerular degeneration is observed, an adaptation that reaches its pinnacle in approximately 30 fish species that are aglomerular (i.e., without a glomerulus; Fig. 11.1F, G). Proximal tubule I is also missing in aglomerular species and the distal tubule is usually lacking in both glomerular and aglomerular marine fish. Unlike freshwater teleosts, the proximal segment(s) in marine teleosts connect directly to the collecting duct via the collecting tubule. In contrast to elasmobranchs and lamprey, the teleost kidney lacks any loop formations within the tubule. In general, arterial blood to the kidney in teleosts is supplied by renal arteries arising from the dorsal aorta or by renal branches from segmental arteries (Forster, 1953; Hickman and Trump, 1969; Nishimura and Imai, 1982; Braun and Dantzler, 1997). In glomerular forms, these arteries give rise to afferent arterioles which supply the glomerular capillaries and then drain into efferent arterioles, which break up into a network of sinusoids and peritubular capillaries (Forster, 1953; Hickman and Trump, 1969). Thus, in most teleost fish, the glomerular blood supply is arterial while the renal tubules have a double supply of blood, the first from the efferent glomerular arterioles and a second from the venous renal-portal system. In marine aglomerular teleosts, the latter is the only means of blood flow to the kidney and in marine fish in general, the venous supply becomes more and more important, depending on the extent of glomerular degeneration. Essentially, in the absence of glomeruli and in the presence of low filtration rates in glomerular marine fish, renal functions rely largely on the venous perfusion that delivers blood to the peritubular sinuses in the kidney (Beyenbach, 2004).

Osmoregulatory Processes of the Kidney

i. Glomerular filtration

The first step in renal osmoregulation is the filtration of the plasma by the glomerulus. This process, some believe, came about as a means for animals inhabiting freshwater environments to regulate the composition of their body fluids during a constant influx of water. Essentially, the GFR in fish depends primarily on its hydration state and/or the availability of freshwater in the environment in which it lives (Braun and Dantzler, 1997). Thus, the more freshwater that is available to the organism, the more filtration occurs across the glomerulus, allowing for the elimination of excess water and metabolic wastes. In times, when water is less available, there is a tendency to conserve body water by not filtering it for once it is filtered, water is potentially lost from the body (Braun and Dantzler, 1997). In hagfish, the total filtering surface of the glomeruli is larger than in most marine teleosts and as large as that of many elasmobranchs and freshwater teleosts (Hickman and Trump, 1969). However, their low dorsal aortic blood pressure and reduced water influx due to their osmoconformity is reflected in their low rates of glomerular filtration, averaging about 0.25 ml kg^{-1} h^{-1} in *Eptatretus stouti* and *Myxine glutinosa*, and similar to rates measured in marine teleosts (Johansen, 1960; Hickman and Trump, 1969; Riegel, 1978, 1986). In contrast, the GFRs of freshwater lamprey, averaging 28 ml kg^{-1} h^{-1}, are almost 100 times higher than its fellow agnathan and are 5-7 times higher than that in freshwater teleosts (Moriarty *et al.*, 1978; Evans, 1979). In half-strength seawater, the GFR of lampreys is 80% lower than that in freshwater (McVicar and Rankin, 1985; Evans, 1993).

The rates of glomerular filtration in marine elasmobranchs average about 3.5 ml kg^{-1} h^{-1}, which are much higher than marine teleosts and approach those of freshwater teleosts (Hickman and Trump, 1969; Evans, 1979, 1993). Their open glomeruli, with a very large filtering surface, suggest that even these observed rates of filtration are lower than the potentially maximal rates (Hickman and Trump, 1969). As one might expect, upon exposure to more dilute conditions, the GFRs of euryhaline elasmobranchs increase (Evans, 1993). Glomerular filtration rates in freshwater Potamotrygonidae average approximately 8.3 ml kg^{-1} h^{-1}, significantly higher than marine elasmobranchs and euryhaline elasmobranchs in dilute conditions (Goldstein and Forster, 1971).

One of the major differences between the freshwater and marine kidney is the role of glomerular filtration in urine formation. In freshwater teleosts, the rate of glomerular filtration is high (~ 4 ml kg^{-1} h^{-1}), contributing to the excretion of a dilute, hypoosmotic urine (Nishimura and Imai, 1982). In a marine environment, water is at a premium and a glomerulus, which aids in the elimination of body water, essentially becomes a liability. Thus, GFR in these fish is low (~ 0.5 ml kg^{-1} h^{-1}) and various stages of glomerular degeneration is observed, the end result being fish that are completely aglomerular (Lahlou *et al.*, 1969; Nishimura and Imai, 1982; Beyenbach, 1986; Baustian *et al.*, 1997).

a. Glomerular intermittency

When euryhaline fish—be it agnathans, elasmobranchs or teleosts—are transferred from one salinity to another, the change is reflected in the rate of glomerular filtration. There are two important ways that body water can be regulated through GFR; by changes in the amount of blood flowing into the glomerulus (the renal perfusion pressure), or by changing the number of glomeruli that are filtering (glomerular intermittency). Partly as a consequence of differences in GFR, the urine flow rates in marine teleosts with glomerular (0.03 to 0.89 ml kg^{-1} h^{-1}) or aglomerular kidneys (0.03 to 0.45 ml kg^{-1} h^{-1}) are greatly reduced as compared to those of freshwater fish (4 ml kg^{-1} h^{-1}) (Hickman and Trump, 1969). What may be apparent is that despite rather large differences in the urine flow rates of freshwater versus marine teleosts, there is no substantial difference between glomerular and aglomerular marine fish. This is due in part to glomerular intermittency, where the number of operational glomeruli changes depending on environmental conditions and results in variations in GFR.

Glomerular intermittency has been well described in hagfish, the river lamprey, *Lampetra fluviatilis*, the lesser spotted dogfish shark, *Scyliorhinus canicula* as well as the rainbow trout, *Oncorhynchus mykiss* (Brown, 1980; Brown and Green, 1987). In all species, three functional glomerular types have been found; glomeruli that are perfused with blood and filtering, those that are non-perfused and non-filtering and those that are perfused but non-filtering. Changes observed in GFRs are essentially reflections of alterations in filtration by single nephrons (single nephron glomerular filtration rate; SNGFR) which can occur in at least two ways. First, the same number of glomeruli filter but all are filtering at a different rate. Second, a different number of glomeruli are filtering, i.e., some of the glomeruli could stop filtering entirely while others continue to filter at a relatively normal rate. The end result in both cases is the same but the mechanism by which it is accomplished is different.

In fish, regulating the number of filtering glomeruli appears to be the most important mechanism for changing the rate at which the whole kidney filters. An exception is the river lamprey (*Lampetra fluviatilis*). Under freshwater conditions, almost all the glomeruli are filtering but when exposed to brackish or seawater conditions, decreases in GFR are due to a reduction in the filtration rate of all the nephrons, likely due to changes in the blood pressure of the renal arteries (Moriarty *et al.*, 1978; Logan *et al.*, 1980a; Rankin *et al.*, 1980; McVicar and Rankin, 1985; Brown and Rankin, 1999). Since lamprey do not have a renal portal system, which would allow the continuation of tubular function when glomerular blood flow is reduced, it is believed that they keep all glomeruli perfused because this is the only way blood gets to the nephron tubule. Similar to freshwater lamprey, almost all of the glomeruli are filtering in dogfish, despite being in a marine environment (Brown and Green, 1987). However, one must keep in mind that the osmoregulatory strategy of elasmobranchs results in an osmotic influx of water, making their rates of glomerular filtration and urinary excretion more comparable to freshwater fish. Infusion of adrenaline into dogfish reduces the number of filtering glomeruli from 94 to 70%, in contrast to observations in lamprey in response to seawater in which the number of filtering glomeruli never changed. Interestingly, adrenaline also increased SNGFR in dogfish, resulting in an overall diuresis (Brown and Green, 1987).

In contrast to both lamprey and dogfish, only about 45% of the glomeruli are perfused and actively filtering in rainbow trout under freshwater conditions and approximately 5% of glomeruli are filtering in the kidneys of seawater-adapted trout (Brown *et al.*, 1978, 1980; Amer and Brown, 1995). Interestingly, there appears to be a large glomerular reserve in both freshwater and marine, there being a large proportion of perfused but non-filtering glomeruli in trout found in either environment. The bottom line is that over 90% of the glomeruli are out of commission in seawater, essentially rendering a glomerular trout aglomerular in seawater and explaining why urine flow rates are similar in marine teleosts regardless of the presence of glomeruli. Another example of this is the Prussian carp, *Carassius auratius gibelio*, that decreases the number of perfused glomeruli upon exposure to seawater by half and > 90% of its glomeruli disappear over the next three months in seawater (Elger and Hentschel, 1981; Elger *et al.*, 1984, reviewed by Beyenbach, 1986). Since Mg^{2+}, SO_4^{2-}, Na^+ and Cl^- are the major urine electrolytes and osmolytes irrespective of glomeruli, urine formation is highly dependent on tubular activities in both glomerular and aglomerular fishes (see below; reviewed by Beyenbach, 2004).

ii. Tubular modification: Reabsorption and secretion

In most fish, the renal filtrate passes through the glomerulus and is then further modified by the renal tubule system, making tubular modification the second step in renal osmoregulation. Of course, this is not the case in aglomerular fish, in which urine formation starts at the level of the renal tubule, making the urine flow rate significantly greater than their 'non-existent' GFR. However, for the most part, urine flow rates are significantly less than GFRs due to tubular water reabsorption that is driven by the reabsorption of ions. The only glomerular exception here is hagfish, in which urine flow rates are almost equal to GFRs, indicating little tubular reabsorption of water as might be expected when in a nearly isoosmotic solution (Evans, 1979). In freshwater lamprey, urine flows are approximately 50% of GFR, indicating that water is reabsorbed by the kidney tubule (Moriarty *et al.*, 1978; Logan *et al.*, 1980a). When in half strength seawater, lampreys show a decrease in GFR and an increase in the tubular reabsorption of water resulting in an 80% reduction in urine flow rate (Rankin *et al.*, 1980; McVicar and Rankin, 1985). In marine elasmobranchs, 70-85% of the filtered water is reabsorbed by the tubules. Urine flows increase when euryhaline elasmobranchs are exposed to more dilute conditions, in part due to a decrease in tubular reabsorption (Evans, 1993). In freshwater teleosts, only 45% of the filtered water is reabsorbed by the kidney tubules (reviewed by Wood and Patrick, 1994). However, tubular reabsorption increases to approximately 75% in marine teleosts.

In most vertebrates, the renal tubules normally reabsorb between 35% and 97% of the filtered Na^+ and Cl^-, the highest fractional reabsorption being found in some freshwater species that have little access to these ions (reviewed by Braun and Dantzler, 1997; Dantzler, 2003). The fractional reabsorption in marine and possibly freshwater elasmobranchs appears to be even lower and in the primitive marine hagfishes, which conform osmotically to the surrounding seawater. There is little to no reabsorption of filtered sodium along the archinephric ducts (Stolte and Schmidt-Nielsen, 1978; Evans, 1979; Dantzler, 2003). However, it should be kept in mind that most data on fractional reabsorption for the entire kidney come from clearance studies that give only the net difference between the amount filtered and amount excreted in the ureteral urine (Braun and Dantzler, 1997). Within the different parts of the tubule system there is both net reabsorption and net secretion of Na^+, Cl^- as well as other solutes. The extent to which water and ions are reabsorbed or secreted is variable amongst fishes and is highly dependent on the osmolality of the environment and kidney morphology, with each segment of the kidney tubule contributing differently to salt and water balance. In the following paragraphs, the transport properties and the osmoregulatory relevance of each kidney tubule segment will be described in detail.

a. Proximal tubule I

The primary urine leaves the glomerulus and in most fish enters the first segment of the proximal tubule (proximal tubule I). In tetrapods and birds, the proximal tubule is a major site of Na^+, Cl^- and Mg^{2+} reabsorption. Due to their osmoconformity, there is some debate on whether the archinephric ducts of hagfish, which are morphologically similar to the proximal tubule of tetrapods, reabsorb a small amount of Na^+ (Evans, 1979; Braun and Dantzler, 1997). However, there is clear evidence for net Na^+ reabsorption along the proximal tubule of the freshwater lamprey, albeit minor (Logan *et al.*, 1980a). Micropuncture analysis of samples taken from various points along the lamprey proximal segment indicate that no more than 10% of filtered water is reabsorbed by these segments and fluid within the proximal tubule remains isoosmotic to blood (Logan *et al.*, 1980a). Thus, the ion transport by the proximal tubule in freshwater lampreys is insignificant when compared to other segments (see below), consistent with evidence indicating that Na^+, K^+-ATPase activity in the lamprey proximal segment is much lower than in the distal segment (Natochin, 1972; Logan *et al.*, 1980a). Upon acclimation to 50% seawater, tubular water reabsorption in lamprey increases by 40% compared to freshwater conditions contributing to a dramatic reduction in urine flow rates (Rankin *et al.*, 1980). But, similar to the freshwater condition, no more than 9% of the filtered water was reabsorbed by the proximal segment, suggesting that the increased tubular water reabsorption likely occurs further down the tubule (Logan *et al.*, 1980b). Rankin and coworkers (1980) suggested that the main function of the proximal segment in seawater lampreys is to secrete Mg^{2+} and SO_4^{2-}, however, this is yet to be determined conclusively.

Based on the tubular fluid to plasma ratios, there appeared to be some minor reabsorption in proximal tubule I of the little skate, *Raja erinacea* (Stolte *et al.*, 1977). However, no clear quantitative information is available concerning the magnitude of proximal tubule I NaCl reabsorption in elasmobranchs or teleosts. If proximal tubule I has even an average water permeability, then some NaCl reabsorption in this tubule segment may occur in marine glomerular fish to take advantage of the fact that the reabsorption of solutes would be accompanied by an osmotically equivalent quantity of water. In this way, proximal tubule I NaCl reabsorption would allow a marine fish to recover filtered fluids, the excess NaCl could then be secreted further down the nephron (see below). One must keep in mind, however, that in glomerular marine teleosts, glomerular filtration is reduced dramatically. Therefore, little NaCl reabsorption is needed to recover fluids lost by filtration. In theory, very little NaCl reabsorption should occur across proximal tubule I in freshwater glomerular fish, since nearly all (96%) the filtered NaCl is eventually reabsorbed but only 47% of filtered water is recovered (Beyenbach, 2000). In freshwater fish, most of the NaCl would have to be actively reabsorbed across more water impermeable regions of the nephron (i.e., the distal tubule) or by the urinary bladder.

Based on the observation that marine aglomerular teleosts lack proximal tubule I, Hickman and Trump (1969) suggested that proximal tubule I in freshwater teleosts was likely to be responsible for the reabsorption of valuable solutes that had entered the kidney tubule by filtration, i.e., glucose, amino acids and macromolecules as well minerals such as Mg^{2+} and Ca^{2+} (Bijvelds *et al.*, 1998; Beyenbach, 2000). Since aglomerular teleosts lack the ability to filter these substances, they have no need to reabsorb them, explaining their proximal tubule I deficiency. Interestingly, 80% of the filtered glucose in hagfish is reabsorbed across their proximal tubule I-like archinephric ducts (Munz and McFarland, 1964; Hickman and Trump, 1969), suggesting that marine glomerular fish may also use proximal tubule I for the recovery of valuable solutes. Furthermore, the structural similarity of lamprey proximal cells to those of the proximal segments of glomerular teleosts and higher vertebrates suggests that its main function is also the reabsorption of filtered macromolecules (Logan *et al.*, 1980a). However, this theory was called into question recently when evidence for net tubular secretion and subsequent phlorizin-sensitive reabsorption of glucose in the aglomerular, *Lophius americanus*, a fish believed to be lacking proximal tubule I, had surfaced (Malvin *et al.*, 1965; Braun and Dantzler, 1997). This was contradicted by a study using kidney brush border membrane vesicles made from the aglomerular toadfish, *Opsanus tau* that provided no evidence of a phlorizin-sensitive glucose reabsorptive system (Wolff *et al.*, 1987). However, it is not known whether toadfish secrete glucose to be reabsorbed, as is believed to be the case for goosefish.

Regardless, proximal tubule I likely does not play a major role in net NaCl balance in freshwater teleosts nor in hagfish, which are practically isoosmotic to their environment. However, this tubule segment contributes to NaCl and water balance to a certain extent in lamprey, which are the only fish, to date, to clearly exhibit net reabsorption across the proximal tubule in its entirety, as well as marine elasmobranchs and glomerular marine teleosts. Detailed models on the cellular transepithelial

NaCl reabsorptive processes in proximal tubule I are available only for amphibians. However, it is believed that the fractional reabsorption of Cl^- is essentially the same as that of Na^+ along proximal tubule I of lampreys, elasmobranchs and teleosts based on micropuncture or microperfusion studies (Braun and Dantzler, 1997). In addition, lumen-negative transepithelial potential and net transepithelial Na^+ reabsorption occurs against an electrochemical gradient in the proximal tubule of freshwater adapted teleosts, suggesting active Na^+ transport and passive Cl^- transport (Braun and Dantzler, 1997).

b. Proximal tubule II

The second segment of the proximal tubule (proximal tubule II) appears to be responsible for the secretion of many different osmolytes in both elasmobranchs and marine teleosts (Beyenbach and Frömter, 1985; Sawyer and Beyenbach, 1985; Beyenbach *et al.*, 1986; Cliff *et al.*, 1986; Cliff and Beyenbach, 1992; Evans, 1993). Aside from the collecting tubule and duct, proximal tubule II is the only tubular segment found in aglomerular marine teleosts. In contrast, neither hagfish nor lamprey are believed to possess this segment (Hickman and Trump, 1969; Hentschel and Elger, 1989; Evans, 1993).

Overall, most fish show a net secretion of divalent ions and both Na^+ and Cl^- along the entire length of the proximal tubule (not distinguishing between proximal tubule I and II) and there is even evidence of net secretion in freshwater (Hickman and Trump, 1969; Braun and Dantzler, 1997; reviewed by Dantzler, 2003). The first direct observation of NaCl and fluid secretion in proximal tubules occurred serendipitously. In his 1986 review, Klaus Beyenbach explains how he made his discovery while using an isolated tubule approach to study Mg^{2+} secretion in the proximal tubules of the glomerular winter flounder, *Pseudopleuronectes americanus* (Beyenbach, 1982, 1986). This approach requires the isolated tubule to be filled with oil in order to collect the luminal contents for analysis. No sooner had the tubule lumen been filled with oil than the oil column broke up at several places along the length of the tubule as a consequence of fluid being secreted into the tubule. Over time, the fluid spaces between oil droplets grew and typically expelled all the oil from the tubule lumen within an hour. Beyenbach (1982) went on to discover that the dominant electrolytes in the secreted fluid were surprisingly Na^+ and Cl^-, in nearly the same concentration as the peritubular bath. The tubule segments also secreted Mg^{2+} and SO_4^{2-} and the secretion of these divalent ions was found to increase the total rate of fluid secretion. However, the basic mechanisms of fluid excretion appeared to be driven by the secretion of NaCl, as fluid secretion continued, albeit at lower rates, with the removal of $MgSO_4$ but was inhibited by the replacement of Na^+ or Cl^- with large organic ions.

Like the flounder, the proximal tubule II of the dogfish shark, *Squalus acanthias*, secretes predominantly NaCl in concentrations not significantly different from those in the bath. The secreted fluid becomes slightly hyperosmotic to the bath due to the secretion of significant concentrations of Mg^{2+} and SO_4^{2-} (Sawyer and Beyenbach, 1985). Significant Mg^{2+}, PO_4^{2-} and SO_4^{2-} secretion, as indicated by high proximal tubule fluid to plasma ratios, has also been shown to occur in proximal tubule II of the little skate, *Raja erinacea* (Stolte *et al.*, 1977). Evidence indicates that the mechanism of NaCl secretion in dogfish proximal tubule II includes an apical, cAMP sensitive Cl^- channel (Beyenbach and Frömter, 1985). On the basolateral side of the tubule, the intracellular voltage was sensitive to ouabain (an inhibitor of Na^+, K^+ ATPase) and furosemide (an inhibitor of Na^+, Cl^- co-transport). The effects of cAMP, furosemide and ouabain on NaCl secretion and tubule electrophysiology are consistent with secondary active transport of NaCl, where the basolateral membrane Na^+, K^+ ATPase generates the primary driving force. In brief, the generation of low intracellular Na^+ concentrations via the Na^+, K^+ ATPase provides a large electrochemical gradient for Na^+ entry from the peritubular bath. The energy of this Na^+ gradient is coupled to the entry of Cl^- by way of the furosemide sensitive Na^+, Cl^- co-transporter. Na^+ arriving in the cell is returned to the bath via the Na^+, K^+ ATPase, but Cl^- diffuses out of the cell through apical membrane Cl^- channels, the conductivity of which is regulated by cAMP. The cellular secretion of Cl^- into the tubule lumen appears to be electrically balanced by the paracellular movement of Na^+ from bath to lumen. There are slight differences in NaCl secretion in teleosts but for the most part the mechanism remains the same.

Net NaCl and water secretion has also been observed in the killifish, *Fundulus heteroclitus*, a glomerular, euryhaline teleost (Beyenbach and Frömter, 1985; Cliff and Beyenbach, 1992). Net secretion of NaCl and water may be particularly important in euryhaline teleosts for the maintenance of renal excretion during periods of glomerular intermittency when the number of filtering glomeruli is reduced during adaptation to seawater (Beyenbach and Frömter, 1985). Along the same lines, tubular secretion is important in producing urine in aglomerular fish. In both cases, the cessation or lack of glomerular filtration must be compensated by tubular changes; in this case, primary urine formation depends mainly on tubular secretion, at the same time tubular reabsorption is reduced (Beyenbach and Frömter, 1985). Here the tubular secretion of NaCl could serve to excrete excess Na^+ and Cl^-, which seawater fish must do, and to excrete organic solutes via cotransport with Na^+ or Cl^- (Beyenbach and Frömter, 1985). It is not yet known whether net secretion along the proximal tubule II changes to net reabsorption when euryhaline teleosts are adapted to freshwater. However, preliminary data on killifish adapted to freshwater suggest that some, if not all, nephrons still secrete Na^+ and Cl^- (Cliff and Beyenbach, 1988).

In theory, the secretion of Na^+, Cl^-, Mg^{2+} and SO_4^{2-} along proximal tubule II without further modification could lead to a final urine osmolality that exceeds that of plasma. In fact, data reported for the killifish, *Fundulus kansae* indicates that a urine hyperosmotic to plasma is possible in fish (Stanley and Fleming, 1964; Fleming and Stanley, 1965). Furthermore, a recent study by McDonald and Grosell (2006) on the aglomerular gulf toadfish, *Opsanus beta*, measured urine osmolalities that were at times greater than those in plasma at a range of environmental salinities.

c. Nephron loop

Only lamprey and elasmobranchs have a loop within their nephron structure. In lamprey, the proximal and distal tubule themselves are arranged in a loop and in freshwater, the loop arrangement itself does not appear to play a role in the concentration of urine or the tubular handling of particular solutes. However, when in marine conditions, lamprey have the ability to produce a urine that is hyperosmotic to plasma and it is believed that the lamprey nephron loop may contribute to its concentrating ability under these conditions (Youson and McMillan, 1971; Natochin, 1977; Logan *et al.*, 1980b).

In comparison, the nephron structure in the elasmobranch is more complicated than that of the lamprey. Instead of just one nephron loop, the elasmobranch kidney has five loops arranged in countercurrent fashion, weaving in and out of the sinus zone and the enclosed bundle zone (Lacy and Reale, 1986; Friedman and Hebert, 1990). This countercurrent system is unique to marine and euryhaline elasmobranchs and is absent in the freshwater stingrays of the family Potamotrygonidae. Interestingly, freshwater elasmobranchs have also lost the ability to retain urea, suggesting that the nephron loops in the marine elasmobranch kidney may be involved in recovering the 90-96% of the urea present in the glomerular filtrate (Forster, 1967; Goldstein and Forster, 1971; Payan *et al.*, 1973). Furthermore, the fluid entering the marine elasmobranch collecting duct has been shown to have already reached its final low concentration with respect to urea (Thurau and Acquisto, 1969; Deetjen *et al.*, 1972; Stolte *et al.*, 1977).

In mammals, the countercurrent organization of nephron segments and capillaries is involved in the recycling and concentration of urea in the inner medulla (Hogg and Kokko, 1979). The energy for this process in mammals is provided by NaCl absorption from the water-impermeable thick ascending limb of Henle, also known as the diluting segment. In elasmobranchs, there is also a direct correlation between reabsorption of Na^+ and urea (Schmidt-Nielsen *et al.*, 1972). Furthermore, a nephron segment having morphological homology with the mammalian TAL has been localized to the bundle zone in elasmobranch kidney that shows a low permeability to water and active NaCl transport (Hentschel and Elger, 1987; Friedman and Hebert, 1990). With respect to urea, the elasmobranch countercurrent model of Boylan (1972) predicted a zone of low urea concentration around the terminal segment (i.e., collecting duct) of the elasmobranch nephron that allows for the passive reabsorption of urea. This was elaborated by Hentschel and coworkers (1986), who predicted that solute gradients would exist between the dorsal and ventral kidney sections that would support a passive reabsorption of urea in the dorsal region. However, there is no evidence of urea zonation between the two major kidney sections as measured urea and water content within these sections were identical (Hentschel *et al.*, 1986; Morgan *et al.*, 2003a). Friedman and Hebert (1990) determined that the permeabilities of the tubule segments running through the peritubular sheath to water, urea and Na^+ differ, a general requirement for countercurrent exchange (Friedman and Hebert, 1990). It is now believed that the peritubular sheath segregates the dorsal tubule bundle of a single nephron and creates a urea gradient from the tip to the end of the bundle, forming a countercurrent exchange 'microenvironment' (Lacy *et al.*, 1985; Friedman and Hebert, 1990; Hentschel *et al.*, 1998). The complexity of the elasmobranch kidney makes this hypothesis extremely difficult to test. However, consistent with a localized countercurrent exchange system within the dorsal bundle is the presence of a facilitative diffusion urea transporter, skUT, discovered on the brush border membrane on the dorsal bundle tubule segments (Morgan *et al.*, 2003b). Furthermore, both skUT expression and a Na^+-linked urea transport have been measured in the ventral kidney section, suggesting a role for this section in urea reabsorption as well (Morgan *et al.*, 2003a, b). When adapting to environments of lower salinity, marine elasmobranchs reduce their plasma urea concentrations. This regulatory adjustment is accomplished largely by a decrease in renal urea reabsorption (Goldstein *et al.*, 1968; Goldstein and Forster, 1971; Forster *et al.*, 1972; Schmidt-Nielsen *et al.*, 1972; Piermarini and Evans, 1998; Cooper and Morris, 2004).

d. Distal tubule and collecting duct

Found in lamprey, elasmobranchs as well as freshwater and euryhaline teleosts, the distal tubule enables the reabsorption of osmolytes without the accompaniment of water, due to its low water permeability. In this way, urine essentially becomes diluted as a consequence of salt recovery and the excess water is trapped within the tubule and excreted. Because of its function, the distal tubule is not found in stenohaline marine glomerular and aglomerular teleosts as its presence would be detrimental due to their need to retain water and eliminate excess salt. The distal tubule is also absent in hagfish.

In freshwater lamprey, most of the NaCl reabsorption occurs between the beginning of the distal tubule—which is the ascending limb of the lamprey nephron loop—and the end of the collecting duct. Greater than 95% of the filtered NaCl is reabsorbed across both tubular segments but mainly across the distal tubule, resulting in a significant dilution of the tubular fluid, despite substantial water reabsorption (Logan *et al.*, 1980a). In fact, approximately 45% of the filtered water is reabsorbed by the freshwater lamprey kidney, most of it within the distal and collecting tubules. However, one must keep in mind that the lamprey GFR is 5-7 times higher than a freshwater teleost; correspondingly, tubular reabsorption might be higher to compensate. In contrast, almost 90% of the filtered water is reabsorbed by the lamprey kidney in the marine environment and most of this is believed to occur in the collecting ducts (Logan *et al.*, 1980b). As a result, urine flow rate in marine lamprey is 10% that in their freshwater counterparts.

There are few studies outlining the micropuncture analysis of elasmobranch distal tubule fluid, specifically due to its complicated kidney morphology. However, fluid from the end collecting duct of the little skate, *Raja erinacea*, showed much higher concentrations of Mg^{2+}, PO_4^{2+} and SO_4^{2-}, than that those measured in plasma (Stolte *et al.*, 1977). In contrast, Na^+ and Cl^- concentrations were slightly lower than plasma concentrations, indicating some reabsorption. While the urine remained isoosmotic in the proximal tubule, along the collecting ducts and in the final urine the osmolality of the tubular fluid was below plasma values (Stolte *et al.*, 1977). This dilution became more pronounced when fish were kept in 75% seawater and was believed to be due to an increased reabsorption of Na^+ (Stolte *et al.*, 1977).

In freshwater teleosts, more that 90% of the filtered NaCl is reabsorbed across the distal tubule, which has little to no permeability to water (Nishimura and Imai, 1982; reviewed by Braun and Dantzler, 1997; Dantzler, 2003). In the absence of the distal tubule in stenohaline marine teleosts, it appears that some reabsorption occurs in the collecting ducts and urinary bladder (see below) (reviewed by Braun and Dantzler, 1997; Dantzler, 2003). Microperfusion studies of the distal tubule segments amongst vertebrates reveal a net Cl^- reabsorption and a lumen-positive transepithelial potential, which are dependent on the presence of both Na^+ and Cl^- in the lumen (Dantzler, 2003). Both Cl^- reabsorption and the lumen-positive TEP are reduced or eliminated by the addition of furosemide or bumetanide to the lumen or ouabain to the peritubular fluid. This supports Cl^- reabsorption occurring by a Na^+-coupled mechanism, the energy for which is derived from Na^+, K^+, ATPase at the basolateral membrane that maintains a low intracellular sodium activity. The gradient established by the Na^+, K^+, ATPase drives the coupled, electroneutral entry of Na^+, 2 Cl^- and K^+ into the cell across the luminal membrane via a furosemide or bumetanide sensitive carrier (NKCC). Cl^- then exits the cell across the peritubular membrane down an electrochemical gradient via a Cl^- conductive pathway and a coupled electroneutral carrier with K^+.

The Role of the Urinary Bladder

There is very little research conducted on the role or even the existence of the urinary bladder in agnathans or elasmobranchs. However, in most teleosts, paired mesonephric ducts leave the kidney, unite and open into a widened duct commonly termed the urinary bladder (Hickman and Trump, 1969; Nishimura and Imai, 1982). One or more sphincters guard the outlet through the urogenital papilla, which enables the urine to be stored in the bladder for some time prior to natural discharge (Curtis and Wood, 1991). Freshwater rainbow trout appear to urinate intermittently in bursts at 20-30 minute intervals, and natural rates of urine flow and NaCl excretion are at least 20 and 40% lower, respectively, than rates determined by internal catheters which bypasses bladder function (Curtis and Wood, 1991). Thus, at least in freshwater teleosts, urine appears to be stored in the bladder for some time prior to discharge and significant reabsorption occurs during the period (Curtis and Wood, 1991). Like the distal tubule, the urinary bladder of freshwater teleosts is relatively impermeable to water as compared to the bladders of marine teleosts but is much more permeable to Na^+ and Cl^- (Hirano *et al.*, 1973; Fossat and Lahlou, 1977; Demarest, 1984). In general, its main function is to reduce renal ion losses and contribute to water excretion.

In contrast, urinary bladder water permeability is variable among seawater fishes but is generally higher than the permeability of the freshwater bladder (Hirano *et al.*, 1973). In the euryhaline flounder, *Platichthys stelatus*, the rate of fluid reabsorption by bladders in seawater-acclimated fish are double reabsorption rates in freshwater acclimated fish (Demarest, 1984). While the urinary bladder performs a storage function within some marine teleosts, for the most part its main function is the reabsorption of water, participates significantly in the osmoregulation of marine fish (Lahlou *et al.*, 1969; Howe and Gutknecht, 1978; McDonald *et al.*, 2002; McDonald and Grosell, 2006). In the aglomerular gulf toadfish, there is clear evidence that the reabsorption of Na^+ across the urinary bladder is dependent on water requirements, with reduced Na^+ reabsorption across the bladder occurring in hyposaline environments (Hirano *et al.*, 1973; McDonald and Grosell, 2006). The reabsorption of Na^+ drives the reabsorption of water in the marine teleost bladder and its low permeability to Mg^{2+} and SO_4^{2-} leaves these divalent ions to concentrate. This selective permeability of the marine urinary bladder to monovalent over divalent ions fueled the belief that tubule Na^+ reabsorption was coupled to Mg^{2+} secretion in marine fishes (Beyenbach and Kirschner, 1975; Howe and Gutknecht, 1978).

CONCLUSIONS

In freshwater lamprey, it is the distal tubule and collecting duct that are most important in terms of osmoregulation, with the nephron loop playing a larger role in the urine-concentrating ability in seawater. Aside from the complex kidney morphology involved in the reabsorption of urea, urine formation and osmoregulatory processes in marine elasmobranchs are similar to teleost fish. These similarities become more apparent in the freshwater stingray, Potamotrygonidae, whose kidney morphology and function are very similar to freshwater teleosts. In freshwater teleosts, the proximal tubule plays a minor role in osmoregulation whereas the glomerulus, the distal tubule and urinary bladder play an important role in the elimination of water and the recovery of osmolytes. In marine teleosts, the proximal tubule, mainly proximal tubule II, is important for urine formation, namely the secretion of Na^+, Cl^-, Mg^{2+} and SO_4^{2-}, which cooperates with the collecting duct and urinary bladder to recover Na^+ and Cl^- and water, leaving the divalent ions. The kidney of euryhaline teleosts must shift from water excretion and monovalent ion conservation in freshwater to water conservation in seawater.

Acknowledgements

MDM is funded by an NSERC (Canada) postdoctoral fellowship.

References

Amer, S. and J.A. Brown. 1995. Glomerular actions of arginine vasotocin in the *in situ* perfused kidney. *American Journal of Physiology* 269: R775–R780.

Baustian, M.D., S.Q. Wang and K.W. Beyenbach. 1997. Adaptive responses of aglomerular toadfish to dilute sea water. *Journal of Comparative Physiology* B167: 61–70.

Beyenbach, K.W. 1982. Direct demonstration of fluid secretion by glomerular renal tubules in a marine teleost. *Nature (Lond.)* 299: 54–56.

Beyenbach, K.W. 1986. Secretory NaCl and volume flow in renal tubules. *American Journal of Physiology* 250: R753–R763.

Beyenbach, K.W. 2000. Renal handling of magnesium in fish: From whole animal to brush border membrane vesicles. *Frontiers in Bioscience* 5: D712–D719.

Beyenbach, K.W. 2004. Kidneys sans glomeruli. *American Journal of Physiology* 286: F811–F827.

Beyenbach, K.W. and E. Frömter. 1985. Electrophysiological evidence for Cl secretion in shark renal proximal tubules. *American Journal of Physiology* 248: F282–F295.

Beyenbach, K.W. and L.B. Kirschner. 1975. Kidney and urinary bladder functions of the rainbow trout in Mg and Na excretion. *American Journal of Physiology* 229: 389–393.

Beyenbach, K.W., D.H. Petzel and W.H. Cliff. 1986. Renal proximal tubule of flounder I. Physiological properties. *American Journal of Physiology* 250: R608–R615.

Bijvelds, M.J.C., J.A. Van Der Velden, Z.I. Kolar and G. Flik. 1998. Magnesium transport in freshwater teleosts. *Journal of Experimental Biology* 201: 1981–1990.

Boylan, J.W. 1972. A model for passive urea reabsorption in the elasmobranch kidney. *Comparative Biochemistry and Physiology* A42: 27–30.

Braun, E.J. and W.H. Dantzler. 1997. Vertebrate renal system. In: *Comparative Physiology*, W. H. Dantzler (ed.). Oxford University Press, New York, pp. 481–576.

Brown, D. 1980. Similarities of membrane structure in freeze-fractured *Xenopus laevis* kidney collecting tubule and urinary bladder. *Journal of Cell Science* 44: 353–363.

Brown, J.A. and C. Green. 1987. Single nephron function of the lesser spotted dogfish, *Scyliorhinus canicula*, and the effects of adrenaline. *Journal of Experimental Biology* 129: 265–278.

Brown, J.A. and J.C. Rankin. 1999. Lack of glomerular intermittency in the river lamprey *Lampetra fluviatilis* acclimated to sea water and following acute transfer to iso-osmotic brackish water. *Journal of Experimental Biology* 202: 939–946.

Brown, J.A., B.A. Jackson, J.A. Oliver and I.W. Henderson. 1978. Single nephron filtration rates (SNGFR) in the trout, *Salmo gairdneri*. *Pflugers Archives* 377: 101–108.

Brown, J.A., J.A. Oliver, I.W. Henderson and B.A. Jackson. 1980. Angiotensin and single nephron glomerular function in the trout, *Salmo gairdneri*. *American Journal of Physiology* 239: R509–R514.

Cliff, W.H. and K.W. Beyenbach. 1988. Fluid secretion in glomerular renal proximal tubules of freshwater-adapted fish. *American Journal of Physiology* 254: R154–R158.

Cliff, W.H. and K.W. Beyenbach. 1992. Secretory renal proximal tubles in seawater- and freshwater-adapted killifish. *American Journal of Physiology* 262: F108–F116.

Cliff, W.H., D.B. Sawyer and K.W. Beyenbach. 1986. Renal proximal tubule of flounder II. Transepithelial Mg secretion. *American Journal of Physiology* 250: R616–R624.

Cooper, A.R. and S. Morris. 2004. Osmotic, sodium, carbon dioxide and acid-base state of the Port Jackson shark, *Heterodontus portusjacksoni*, in response to lowered salinity. *Journal of Comparative Physiology* B174: 211–222.

Curtis, J.B. and C.M. Wood. 1991. The function of the urinary bladder *in vivo* in the freshwater rainbow trout. *Journal of Experimental Biology* 155: 567–583.

Dantzler, W.H. 2003. Regulation of renal proximal and distal tubule transport: sodium chloride and organic anions. *Comparative Biochemistry and Physiology* A136: 453–478.

Deetjen, P., D. Antkowiak and J.W. Boylan. 1972. The nephron of the skate, *Raja erinacea*. *Bulletin of Mount Desert Island Biological Laboratories* 12: 28–29.

Demarest, J.R. 1984. Ion and water transport by the flounder urinary bladder: salinity dependence. *American Journal of Physiology* 246: F395–F401.

Elger, M. and H. Hentschel. 1981. The glomerulus of stenohaline freshwater teleost, *Carassius auratius gibelio*, adapted to saline water. *Cell and Tissue Research* 220: 13–85.

Elger, M., R. Kaune and H. Hentschel. 1984. Glomerular intermittency in a freshwater teleost, *Carassius auratius gibelio*, after transfer to salt water. *Journal of Comparative Physiology* B154: 225–231.

Evans, D.H. 1979. Ionic and osmotic regulation in fish. In: *Comparative Physiology of Osmoregulation in Animals*, G.M.O. Maloiy (ed.). Academic Press, Orlando, Vol. 1, pp. 305–390.

Evans, D.H. 1993. Osmotic and ionic regulation. In: *The Physiology of Fishes*, D.H. Evans (ed.). CRC Press, Boca Raton, pp. 315–342.

Fleming, W.R. and J.G. Stanley. 1965. Effects of rapid changes in salinity on the renal function of a euryhaline teleost. *American Journal of Physiology* 209: 1025–1030.

Forster, R.P. 1953. A comparative study of renal function in marine teleosts. *Journal of Cellular and Comparative Physiology* 42: 487–509.

Forster, R.P. 1967. Osmoregulatory role of the kidney in cartilaginous fishes (Chondrichthys). In: *Sharks, Skates and Rays*, P.W. Gilbert, R.F. Mathewson and D.P. Rall (eds.). John Hopkins Press, Baltimore, pp. 187–195.

Forster, R.P., L. Goldstein and J.K. Rosen. 1972. Intrarenal control of urea reabsorption by renal tubules of the marine elasmobranch, *Squalus acanthias*. *Comparative Biochemistry and Physiology* A42: 3–12.

Fossat, B. and B. Lahlou. 1977. Osmotic and solute permeabilities of isolated urinary bladder of the trout. *American Journal of Physiology* 233: F525–F531.

Friedman, P.A. and S.C. Hebert. 1990. Diluting segment in the kidney of dogfish shark. I. Localization and characterization of chloride absorption. *American Journal of Physiology* 258: R398–R408.

Goldstein, L. and R.P. Forster. 1971. Urea biosynthesis and excretion in freshwater and marine elasmobranchs. *Comparative Biochemistry and Physiology* B39: 415–421.

Goldstein, L., W.W. Oppelt and T.H. Maren. 1968. Osmotic regulation and urea metabolism in the lemon shark *Negaprion brevirostris*. *American Journal of Physiology* 215: 1493–1497.

Hardisty, M.W. 1979. *Biology of the Cyclostomes*. Chapman & Hall, London.

Hentschel, H. and M. Elger. 1987. The distal nephron in the kidney of fishes. *Advances in Anatomy, Embryology and Cell Biology* 108: 1–151.

Hentschel, H. and M. Elger. 1989. Morphology of glomerular and aglomerular kidneys. In: *Structure and Function of the Kidney*, R.K.H. Kinne (ed.). S. Karger, Basel, pp. 1–72.

Hentschel, H., M. Elger and B. Schmidt-Nielsen. 1986. Chemical and morphological differences in the kidney zones of the elasmobranch, *Raja erinacea*. *Comparative Biochemistry and Physiology* A84: 553–557.

Hentschel, H., U. Storb, L. Teckhaus and M. Elger. 1998. The central vessel of the renal countercurrent bundles of two marine elasmobranchs—dogfish (*Scyliorhinus caniculus*) and skate (*Raja erinacea*)—as revealed by light and electron microscopy with computer-assisted reconstruction. *Anatomy and Embryology (Berlin)* 198: 73–89.

Hickman, C.P. Jr. and B.F. Trump. 1969. The kidney. In: *Fish Physiology*, W.S. Hoar and D.J. Randall (eds.). Academic Press, New York, Vol. 1, pp. 91–239.

Hirano, T., D.W. Johnson, H.A. Bern and S. Utida. 1973. Studies on water and ion movements in the isolated urinary bladder of selected freshwater, marine and euryhaline teleosts. *Comparative Biochemistry and Physiology* A45: 529–540.

Hogg, R.J. and J.P. Kokko. 1979. Renal countercurrent multiplication system. *Reviews of Physiology, Biochemistry and Pharmacology* 86: 95–135.

Howe, D. and J. Gutknecht. 1978. Role of the urinary bladder in osmoregulation in the marine teleost, *Opsanus tau*. *American Journal of Physiology* 235: R48–R54.

Johansen, K. 1960. Circulation in the hagfish, *Myxine glutinosa* L. *Biological Bulletin* 118: 289–295.

Lacy, E.R. and E. Reale. 1986. The elasmobranch kidney. III. Fine structure of the peritubular sheath. *Anatomy and Embryology (Berlin)* 173: 299–305.

Lacy, E.R., E. Reale, D.S. Schlusselberg, W.K. Smith and D.J. Woodward. 1985. A renal countercurrent system in Maine elasmobranch fish: A computer-aided reconstruction. *Science* 227: 1351–1354.

Lahlou, B., I.W. Henderson and W.H. Sawyer. 1969. Renal adaptations by *Opsanus tau*, a euryhaline aglomerular teleost, to dilute media. *American Journal of Physiology* 216: 1266–1272.

Logan, A.G., R.J. Moriarty and J.C. Rankin. 1980a. A micropuncture study of kidney function in the river lamprey, *Lampetra fluviatilis*, adapted to fresh water. *Journal of Experimental Biology* 85: 137–147.

Logan, A.G., R. Morris and J.C. Rankin. 1980b. A micropuncture study of kidney function in the river lamprey *Lampetra fluviatilis* adapted to sea water. *Journal of Experimental Biology* 88: 239–247.

Malvin, R.L., E.J. Cafruny and H. Kutchai. 1965. Renal transport of glucose by aglomerular fish *Lophius americanus*. *Journal of Cellular Comparative Physiology* 65: 381–384.

McDonald, M.D. and M. Grosell. 2006. Maintaining osmotic balance with an aglomerular kidney. *Comparative Biochemistry and Physiology* (In Press).

McDonald, M.D., P.J. Walsh and C.M. Wood. 2002. Transport physiology of the urinary bladder in teleosts: A suitable model for renal urea handling? *Journal of Experimental Zoology* 292: 604–617.

McVicar, A.J. and J.C. Rankin. 1985. Dynamics of glomerular filtration in the river lamprey, *Lampetra fluviatilis* L. *American Journal of Physiology* 249: 132–138.

Morgan, R.L., J.S. Ballantyne and P.A. Wright. 2003a. Regulation of a renal urea transporter with reduces salinity in a marine elasmobranch, *Raja erinacea*. *Journal of Experimental Biology* 206: 3285–3292

Morgan, R.L., P.A. Wright and J.S. Ballantyne. 2003b. Urea transport in kidney brush-border membrane vesicles from an elasmobranch, *Raja erinacea*. *Journal of Experimental Biology* 206: 3293–3302.

Moriarty, R.J., A.G. Logan and J.C. Rankin. 1978. Measurement of single nephron filtration rate in the kidney of the river lamprey, *Lampetra fluviatilis* L. *Journal of Experimental Biology* 77: 57–69.

Munz, F.W. and W.N. McFarland. 1964. Regulatory function of a primitive vertebrate kidney. *Comparative Biochemistry and Physiology* 13: 381–400.

Natochin, Y.V. 1972. Glomerular filtration and proximal reabsorption in the evolution of the vertebrate kidney. *Journal of Evolutionary Biochemistry and Physiology* 8: 256–264.

Natochin, Y.V. 1977. Filtration, reabsorption and secretion in the evolution of the renal function. *Journal of Evolutionary Biochemistry and Physiology* 13: 424–429.

Nishimura, H. and M. Imai. 1982. Control of renal function in freshwater and marine teleosts. *Federal Proceedings* 41: 2355–2360.

Payan, P., L. Goldstein and R.P. Forster. 1973. Gills and kidneys in ureosmotic regulation in euryhaline skates. *American Journal of Physiology* 224: 367–372.

Piermarini, P.M. and D.H. Evans. 1998. Osmoregulation of the Atlantic stingray (*Dasyatis sabina)* from the freshwater Lake Jesup of the St. Johns River, Florida. *Physiological Zoology* 71: 553–560.

Rankin, J.C., A.G. Logan and R.J. Moriarty. 1980. Changes in kidney function in the river lamprey, *Lampetera fluviatilis* L. in response to changes in external salinity. In: *Epithelial Transport in the Lower Vertebrates*, B. Lahlou (ed.). Cambridge University Press, Cambridge, pp. 50–69.

Riegel, J.A. 1978. Factors affecting glomerular function in the Pacific hagfish *Eptatretus stouti* (Lockington). *Journal of Experimental Biology* 73: 261–277.

Riegel, J.A. 1986. The absence of an arterial pressure effect on filtration by perfused glomeruli of the hagfish, *Eptatretus stouti* (Lockington). *Journal of Experimental Biology* 126: 361–374.

Sawyer, D.B. and K.W. Beyenbach. 1985. Mechanism of fluid secretion in isolated shark renal proximal tubules. *American Journal of Physiology* 249: F884–F890.

Schmidt-Nielsen, B., B. Truniger and L. Rabinowitz. 1972. Sodium-linked urea transport by the renal tubule of the spiny dogfish *Squalus acanthias*. *Comparative Biochemistry and Physiology* A42: 13–25.

Stanley, J.G. and W.R. Fleming. 1964. Excretion of hypertonic urine by a teleost. *Science* 144: 63–64.

Stolte, H. and B. Schmidt-Nielsen. 1978. Comparative aspects of fluid and electrolyte regulation by the cyclostome, elasmobranch and lizard kidney. In: *Osmotic and Volume Regulation*, C.B. Jorgenson and E. Skadhauge (eds.). Munksgaard, Copenhagen, pp. 209–222.

Stolte, H., R.G. Galaske, R.G. Eisenbach, C. Lechene and B. Schmidt-Nielsen. 1977. Renal tubule ion transport and collecting duct function in the elasmobranch little skate, *Raja erinacea*. *Journal of Experimental Zoology* 199: 403–410.

Thurau, K. and P. Acquisto. 1969. Localization of the diluting segment in the dogfish nephron: A micropuncture study. *Bulletin of Mount Desert Island Biological Laboratories* 9: 60–63.

Wolff, N.A., R. Kinne, B. Elger and L. Goldstein. 1987. Renal handling of taurine, L-alanine, L-glutamate and D-glucose in *Opsanus tau*: Studies on isolated brush border membrane vesicles. *Journal of Comparative Physiology* B157: 573–581.

Wood, C.M. and M.L. Patrick. 1994. Methods for assessing kidney and urinary bladder function in fish. In: *Biochemistry and Molecular Biology of Fishes*, P.W. Hochachka and T.P. Mommsen (eds.). Elsevier, Amsterdam, pp. 127–143.

Youson, J.H. and D.B. McMillan. 1971. Intertubular circulation in the opisthonephric kidneys of adult and larval sea lamprey, *Petromyzon marinus* L. *Anatomical Record* 170: 401–412.

CHAPTER

12

Intestinal Transport Processes in Marine Fish Osmoregulation

Martin Grosell

OSMOREGULATION IN MARINE FISH

The gastrointestinal tract (GIT) was recognized early to play a vital role in marine teleost osmoregulation by absorbing water (Smith, 1930). The extracellular fluids in marine teleosts, lampreys and chondrichthyan fishes are hypotonic (~ 300 mOsm) (Rankin *et al.*, 2001; Marshall and Grosell, 2005; Taylor and Grosell, 2006a) with respect to the surrounding seawater (~ 1000 mOsm), which results in diffusive water loss and salt gain. To offset the diffusive water loss, marine teleost fish, marine lampreys and seawater-acclimated sturgeon drink seawater at rates of 2-5 ml kg^{-1} h^{-1} (Rankin *et al.*, 2001; Rodriguez *et al.*, 2002; Marshall and Grosell, 2005), of which 60-85% is absorbed by the intestine (Smith, 1930; Shehadeh and Gordon, 1969; Sleet and Weber, 1982; Wilson *et al.*, 1996;

Author's address: Rosenstiel School of Marine and Atmospheric Sciences, Division of Marine Biology and Fisheries, University of Miami, 4600 Rickenbacker Causeway, 33145 Miami, Florida, USA.
E-mail: mgrosell@rsmas.miami.edu

Grosell *et al.*, 1999). Intestinal water absorption is ultimately linked to salt absorption and is the focus of the present chapter. For detailed discussions of integrative aspects of whole animal salt and water balance in marine fish, including renal and branchial contributions, the reader is referred to other chapters in the present book and also recent reviews (Evans *et al.*, 2005; Marshall and Grosell, 2005; McDonald, 2007).

The esophagus and the intestine of marine fish contribute in functionally distinct ways to marine teleost osmoregulation. The esophagus displays high rates of NaCl absorption and is largely impermeable to water, resulting in much lower NaCl concentrations and osmotic pressure in stomach fluids compared to the ingested seawater (Smith, 1930; Hirano and Mayer-Gostan, 1976; Kirsch and Meister, 1982; Parmelee and Renfro, 1983; Wilson *et al.*, 1996). The selective absorption of NaCl, leaving other ions and water behind, is the function of both passive and active transport with the latter ultimately driven by basolateral $Na^{+}K^{+}$-ATPase (Hirano and Mayer-Gostan, 1976; Parmelee and Renfro, 1983). The apical transport processes involved in esophageal NaCl absorption need to be fully characterized (Parmelee and Renfro, 1983). In contrast to the largely water-impermeable esophagus, the intestine displays water absorption in the order of 2-6 $\mu l\ cm^{-1}\ h^{-1}$ (Grosell and Jensen, 1999; Grosell *et al.*, 1999, 2001, 2005; Wilson *et al.*, 2002).

INTEGRATIVE SALT AND WATER MASS BALANCE

Water

Water loss to the marine environment is inevitable through renal excretion and non-renal loss, most of which is likely to occur across the gill surface. This diffusive water loss is compensated for by ingestion of seawater as first recognized by Smith more than three-quarters of a century ago (Smith, 1930). The control of seawater ingestion is complicated and beyond the scope of the present text and has been discussed recently in detail in several excellent reviews (Takei, 2000; Loretz, 2001; Ando *et al.*, 2003). Typical drinking rates of marine fish are around 2 ml $kg^{-1}\ h^{-1}$, although multiple studies reports higher rates (Marshall and Grosell, 2005) which may, at least in some cases, be elevated due to handling stress. As mentioned, 60-85% of the ingested seawater is absorbed by the gastrointestinal tract (Smith, 1930; Shehadeh and Gordon, 1969; Sleet and Weber, 1982; Wilson *et al.*, 1996; Grosell *et al.*,

1999) almost exclusively in the intestine as the esophagus is largely water-impermeable (Hirano and Mayer-Gostan, 1976; Parmelee and Renfro, 1983). The remaining water is voided in the form of rectal fluids with chemical compositions much different than the ingested seawater due to fractional ion absorption, water absorption and secretions by the gastrointestinal tract (Wilson *et al.*, 1996, 2002; Grosell *et al.*, 2001, 2004).

Monovalent Ions

Water absorption is ultimately linked to salt absorption (Skadhauge, 1974; Mackay and Lahlou, 1980; Ando *et al.*, 1986; Usher *et al.*, 1991), so it is perhaps not surprising that 98, 96 and 75% of ingested Na^+, Cl^- and K^+, respectively, is absorbed along the gastrointestinal tract (Fig. 12.1). While the esophageal and intestinal absorption of salt is necessary for water absorption, it presents an additional challenge for marine fish by adding to the diffusive salt gain from the concentrated environment. The majority of the Na^+, Cl^- and K^+ gained from the gastrointestinal tract is eliminated across the gill surface with <1, 3 and 7%, respectively, excreted along with the urine. For detailed discussions of renal function in marine teleosts and branchial transport processes, the reader is referred to recent reviews and chapters within this volume (Evans *et al.*, 2005; Marshall and Grosell, 2005; McDonald, 2007).

Divalent Ions

The dominant electrolytes in marine teleost intestinal fluids are Mg^{2+} and SO_4^{2-} reaching concentrations in some cases > 100 mM (Smith, 1930; Grosell *et al.*, 2001, 2004; Taylor and Grosell, 2006a, b), which is well above corresponding seawater levels of approximately 50 and 30 mM, respectively. From Figure 12.1 it is evident that fractional absorption of Mg^{2+} and SO_4^{2-} is low (20 and 67 %, respectively) compared to Na^+, Cl^- and K^+ despite substantial gradients from the high concentrations in the intestinal lumen to the low extracellular fluid concentrations (< 1 mM). Even with the relatively low absorption of Mg^{2+} and SO_4^{2-} marine fish experience a net gain of these divalent electrolytes, mainly from the intestine. Homeostasis is maintained despite this continuous gain by renal excretion, as discussed elsewhere (McDonald, 2007). While it is clear that SO_4^{2-} extrusion from intestinal epithelial cells (discussed below) contributes to the limited SO_4^{2-} gain across the intestine, it is unknown as

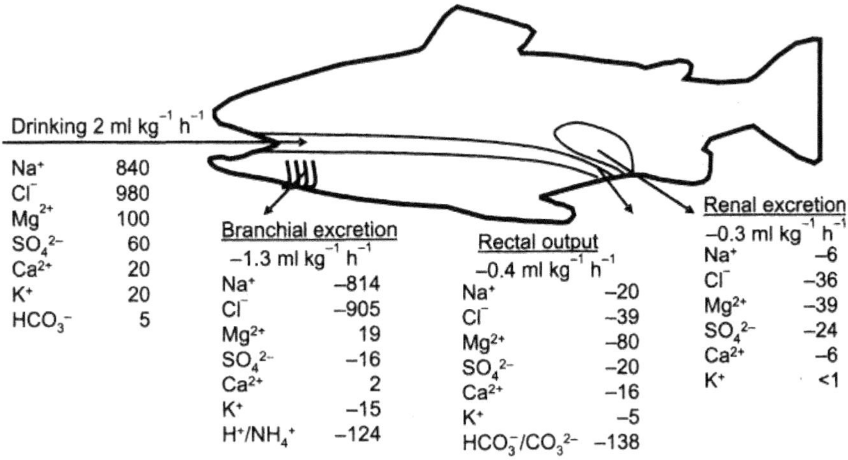

Fig. 12.1. Whole animal balance of water (ml kg^{-1} h^{-1}) and main electrolytes (μmol kg^{-1} h^{-1}) in a typical marine teleost fish. Positive values indicate net uptake while negative values indicate loss and excretion. Mass balance of water is based on assumed rates of drinking, voided rectal fluids and urine flow rates of 2.0, 0.4 and 0.3 ml kg^{-1} h^{-1}, respectively. Electrolytes ingested with drinking are calculated from typical seawater composition while rectal loss of Na^+, Cl^-, SO_4^{2-} and K^+ are calculated from average concentrations of these ions in rectal fluids (see Fig. 6.1 in Marshall and Grosell, 2005) and assumed an intestinal fractional fluid absorption of 80% (Grosell *et al.*, 1999; Shehadeh and Gordon, 1969; Wilson *et al.*, 1996). The rectal loss of Mg^{2+} and Ca^{2+} is estimated from fractional Mg^{2+} loss and absolute Ca^{2+} loss which includes ions voided as CO_3^{2-} precipitates (Wilson and Grosell, 2003). Rectal fluids contain high amounts of HCO_3^- and CO_3^{2-} in solution and considerable quantities of $CaCO_3$ precipitates formed within the intestine (Grosell *et al.*, 1999; Wilson *et al.*, 1996; Wilson and Grosell, 2003). The value for rectal HCO_3^-/CO_3^{2-} excretion depicted in this figure was chosen to yield charge balance in the rectal fluids but represents a conservative estimate since it is lower than the three empirical values reported to date (Wilson *et al.*, 1996; Grosell *et al.*, 1999; Wilson and Grosell, 2003).

Renal excretion rates of electrolytes are calculated from average concentrations of these ions in marine teleost urine (Hickman and Trump, 1969; Beyenbach, 2004) and a urine flow rate of 0.3 ml kg^{-1} h^{-1}. Net branchial rates of electrolyte exchange with the environment were determined from the difference between net gastrointestinal contributions and renal contributions to consist mainly of excretion of NaCl and acidic equivalents. Note here that "branchial" transport rates may include transport across other non-gastrointestinal and non-renal surfaces. The net excretion rate of H^+ and NH_4^+ was chosen to yield charge balance at the gill and is well within reported excretion rates and has recently been reported to reflect intestinal HCO_3^-/CO_3^{2-} excretion rates (Wilson and Grosell, 2003).

to how Mg^{2+} uptake is largely avoided despite a strongly favorable gradient for uptake.

Intestinal fluid Ca^{2+} concentrations are substantially lower than corresponding seawater levels (Smith, 1930; Grosell *et al.*, 2001, 2004; Taylor and Grosell, 2006a, b). The low Ca^{2+} concentrations in intestinal fluids has led to numerous suggestions of substantial intestinal Ca^{2+} uptake in marine teleost (Hickman, 1968; Evans, 1993; Karnaky, 1998). However, more recent investigations have revealed that low intestinal fluid Ca^{2+} concentrations are the product of alkaline precipitation rather than absorption (Wilson *et al.*, 2002; Wilson and Grosell, 2003). High luminal HCO_3^- concentrations and alkaline conditions in the intestinal fluids facilitate $CaCO_3$ formation, which is often evident as white mucus-coated structures within the intestinal lumen and holding tanks of marine fish. These precipitates can account for 30-65% of the ingested Ca^{2+}, presumably explaining, at least, in part the relatively low intestinal fractional Ca^{2+} absorption (20%; Fig. 12.1). Excess Ca^{2+} gained largely from the intestine is eliminated by the kidney (Hickman, 1965).

Luminal Alkalinity

A highly alkaline intestinal lumen containing high concentrations of HCO_3^- and CO_3^{2-} is a feature unique to marine fish that ingest seawater. This phenomenon was first reported some three-quarters of a century ago (Smith, 1930), but only recently was its functional significance for osmoregulation unraveled (Grosell *et al.*, 1999, 2001; Wilson *et al.*, 2002; Wilson and Grosell, 2003; Marshall and Grosell, 2005; Grosell, 2006; Grosell and Genz, 2006; Taylor and Grosell, 2006a, b). Considering the drinking rates and seawater concentrations of HCO_3^-, it became apparent that the source of HCO_3^- and CO_3^{2-} in the intestinal fluids was the intestinal epithelium itself (Walsh *et al.*, 1991) and that the rectal fluid represents a substantial base secretion (Wilson *et al.*, 1996; Wilson and Grosell, 2003). Although intestinal base secretion is not directly involved in the dynamic regulation of acid-base balance, it contributes significantly to the entire animal acid-base exchange with the environment. However, variations in intestinal base secretion appear to be perfectly compensated for by adjustments in branchial acid excretion (Wilson and Grosell, 2003).

Intestinal Fluid Composition

The differential absorption of Na^+, Cl^-, K^+ and water and secretion of HCO_3^- result in a unique chemical composition of fluids residing in the intestinal lumen of teleost fish. Marine teleost intestinal fluids are typically high in Mg^{2+} and SO_4^{2-}, often 3-4 fold higher than corresponding concentrations in seawater, due to Na^+, Cl^- and water absorption (Wilson *et al.*, 1996, 2002; Wilson, 1999; Bury *et al.*, 2001; Grosell *et al.*, 2001,

2004, 2005; Grosell and Wood, 2001; Grosell, 2006). In contrast, Na^+, Cl^- and K^+ concentrations in intestinal fluids are considerably lower than in seawater as a result of the active absorption of these ions. Luminal Ca^{2+} concentrations are also consistently lower than in seawater but this is due mainly to $CaCO_3$ precipitate formation within the intestinal lumen rather than Ca^{2+} absorption by the intestine (Wilson *et al.*, 2002; Wilson and Grosell, 2003).

Salinity

The data in Table 12.1 represents the first report of detailed intestinal fluid chemistry in freshwater-acclimated euryhaline teleosts (*Tilapia aureus*) and show a comparison to seawater-acclimated individuals. The most substantial differences in chemical composition between freshwater and seawater acclimated fish are Mg^{2+} and SO_4^{2-} concentrations, which are both very low in freshwater- and high in seawater-acclimated individuals. The high luminal concentrations of these divalent ions are related to drinking and differential absorption in seawater acclimated fish. Despite seawater ingestion, seawater-acclimated tilapia have similar or lower Na^+, Cl^- and K^+ concentrations in intestinal fluids compared to freshwater-acclimated fish, which demonstrates the substantial absorption of Na^+ and Cl^- in particular by the intestine of seawater-acclimated fish. Another marked difference between intestinal fluids is evident from the total CO_2 concentrations that are greatly elevated in seawater-acclimated compared to freshwater-acclimated fish. This difference is in agreement with earlier reports from euryhaline fish acclimated to different salinities (Wilson, 1999) and reflects intestinal Cl^-/HCO_3^- exchange in seawater acclimated fish. Alkaline intestinal fluids associated with the high total CO_2 concentrations in seawater acclimated fish very likely explain the relatively low Ca^{2+} concentrations in seawater-acclimated tilapia. As discussed above, alkaline precipitation of $CaCO_3$ accounts for the majority of Ca^{2+} ingested with seawater and is released from the gastrointestinal tract with the rectal fluids as clearly visible white pellets (Walsh *et al.*, 1991; Wilson *et al.*, 2002; Wilson and Grosell, 2003).

A more detailed study of the influence of salinity on intestinal fluid composition in the marine teleost *Opsanus beta*—a fish that tolerates salinity fluctuations well—revealed that osmotic pressure in the intestinal lumen remains relatively constant over a wide range of salinities (Fig. 12.2) and that plasma osmolality and intestinal fluid osmolality are strongly

Table 12.1

	Anterior		*Mid*		*Posterior*		*Rectal*	
	FW	*SW*	*FW*	*SW*	*FW*	*SW*	*FW*	*SW*
Na^+	165.4 ± 5.2	101.2 ± 14.3	131.8 ± 11.4	103.4 ± 16.7	129.6 ± 4.8	112.8 ± 10.0	122.6 ± 7.8	73.4 ± 26.1
Mg^{2+}	3.1 ± 0.9	154.2 ± 36.2	0.9 ± 0.3	155.3 ± 33.3	1.0 ± 0.3	146.2 ± 37.5	1.6 ± 0.6	179.8 ± 24.5
K^+	12.5 ± 1.3	4.5 ± 0.8	8.1 ± 0.8	3.6 ± 0.6	9.4 ± 1.0	5.1 ± 0.4	7.2 ± 0.6	8.5 ± 2.3
Ca^{2+}	6.4 ± 2.1	4.4 ± 0.5	1.2 ± 0.4	3.5 ± 0.6	2.1 ± 0.6	2.8 ± 0.7	2.4 ± 0.8	4.3 ± 1.5
Cl^-	73.2 ± 9.0	112.7 ± 17.7	87.1 ± 4.8	81.2 ± 5.5	98.8 ± 6.9	93.6 ± 7.0	96.4 ± 8.9	74.8 ± 15.5
SO_4^{2-}	2.3 ± 1.2	139.3 ± 30.5	1.2 ± 0.8	105.4 ± 23.7	0.6 ± 0.4	81.9 ± 12.7	2.1 ± 1.3	121.3 ± 17.2
TCO_2	13.5 ± 2.9	81.3 ± 16.8	36.8 ± 4.3	97.4 ± 6.8	20.4 ± 3.9	99.0 ± 13.4	10.9 ± 3.6	110.5 ± 30.2
Osmolality	336.4 ± 6.9	504.6 ± 31.6	321.1 ± 4.9	497.6 ± 36.9	325.1 ± 8.7	502.3 ± 36.7	288.0 ± 13.6	469.3 ± 26.6

Composition of fluids obtained from the anterior, mid and posterior segments of the intestine as well as from the rectum of seawater- and freshwater-acclimated *Tilapia aureus* (plasma osmolality of 303.1 ± 3.5 and 485.7 ± 34.2, respectively). Values are means ± SEM (n=6-9) with concentrations expressed in mM and osmolality expressed in mOsm. Adult tilapia were maintained in 500 L flow through tanks in Virginia Key dechlorinated tap water at 22°C and were fed commercial tilapia food pellets daily. Following gradual acclimation to increasing salinity over 7 days, fish were maintained in flow-through natural seawater (22°C) for 2 weeks prior to sampling. Food was withheld from both freshwater and seawater tilapia 48-h prior to sampling of intestinal fluids. Intestinal fluid samples were obtained and analyzed as described in detail elsewhere (McDonald and Grosell, 2005).

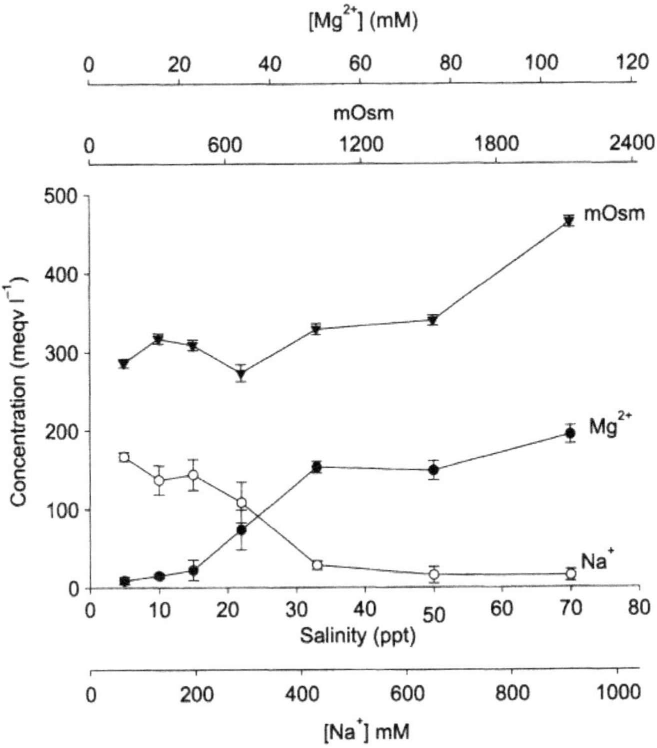

Fig. 12.2 Osmolality (mOsm), Na^+ and Mg^{2+} concentrations (Mean ± SEM) in fluids obtained from the posterior intestine in the gulf toadfish (*Opsanus beta*) at salinities ranging from 5-70 ppt. Note multiple x-axes showing ambient osmolality and Mg^{2+} concentrations (top) and salinity and Na^+ concentrations (bottom). Data taken from a recent study of Gulf toadfish salinity tolerance (McDonald and Grosell, 2005).

correlated and similar (McDonald and Grosell, 2005). Even though plasma osmolality in seawater-acclimated tilapia was elevated compared to freshwater individuals, this correlation between intestinal fluid and plasma osmolality was also observed for tilapia (Table 12.1). The data presented in Figure 12.2 demonstrate that at salinities above 15 ppt, the intestinal fluid Mg^{2+} concentrations increase gradually with salinity reflecting increased drinking rates and water absorption by the intestine. In contrast, Na^+ concentrations gradually decrease at salinities above 15 ppt despite increasing ambient Na^+ concentrations and drinking rates. The remarkable ability of the intestine to absorb Na^+ (and Cl^-, data not shown) is illustrated by luminal concentrations of less than 20 mM in fish exposed to salinities of ≥ 50 ppt (equivalent to ≥ 650 mM Na^+).

Intestinal Transport Processes—NaCl Absorption

The intestinal epithelium provides for osmoregulation by absorbing water. The mechanisms of water transport across leaky epithelia is a highly controversial topic (Larsen *et al.*, 2002; Spring, 2002), but regardless of the mechanism, water transport is tightly linked to the active absorption of Cl^- and Na^+ (Skadhauge, 1974; Mackay and Lahlou, 1980; Ando *et al.*, 1986; Usher *et al.*, 1991).

Na^+K^+-ATPase and Na^+:Cl^- Co-transporters

The active absorption of both Cl^- and Na^+ is ultimately fueled by the basolateral Na^+ K^+-ATPase (NKA) which extrudes $2K^+$ in exchange for $3Na^+$ (Skou, 1990; Skou and Esmann, 1992). The activity of this enzyme, which is highly abundant in the lateral membranes of the marine teleost intestine (Fig. 12.3), maintains low intracellular Na^+ concentrations and sustains a highly cytosol-negative membrane potential. Increased NKA mRNA expression and enzymatic activity in the intestine of euryhaline fish following transfer to seawater illustrate the importance of this enzyme for successful seawater osmoregulation (Mackay and Janicki, 1978; Jensen *et al.*, 1998; Seidelin *et al.*, 2000; Cutler and Cramb, 2002b).

The electrochemical Na^+ gradient energizes not only transport of Na^+ but also Cl^- and K^+ across the apical membrane by the Na^+:Cl^- (NC) and Na^+:K^+:$2Cl^-$ (NKCC) co-transport systems (Field *et al.*, 1978; Frizzell *et al.*, 1979; Musch *et al.*, 1982; Halm *et al.*, 1985b) (Fig. 12.4) and has recently been demonstrated to fuel HCO_3^- secretion and Cl^- absorption by apical anion exchange (see below) (Grosell and Genz, 2006). Consistent with co-transporter activity, uptake of Na^+ is abolished in the absence of luminal Cl^- while some Cl^- absorption persists even in the absence of Na^+ in the mucosal fluids (Mackay and Lahlou, 1980; Grosell *et al.*, 2001) demonstrating additional Cl^- uptake pathways. As is the case for NKA, NKCC mRNA expression in the intestinal epithelium of euryhaline teleosts increases following transfer from freshwater to seawater (Cutler and Cramb, 2002b), illustrating the importance of NKCC for marine osmoregulation.

All studies reporting simultaneous measurements of net Na^+ and Cl^- fluxes show higher net absorption rates for Cl^- than for Na^+ *in vivo*, *in situ* and *in vitro* (Hickman, 1968; Smith *et al.*, 1975; Field *et al.*, 1980; Mackay and Lahlou, 1980; Gibson *et al.*, 1987; Musch *et al.*, 1990; Marvao *et al.*,

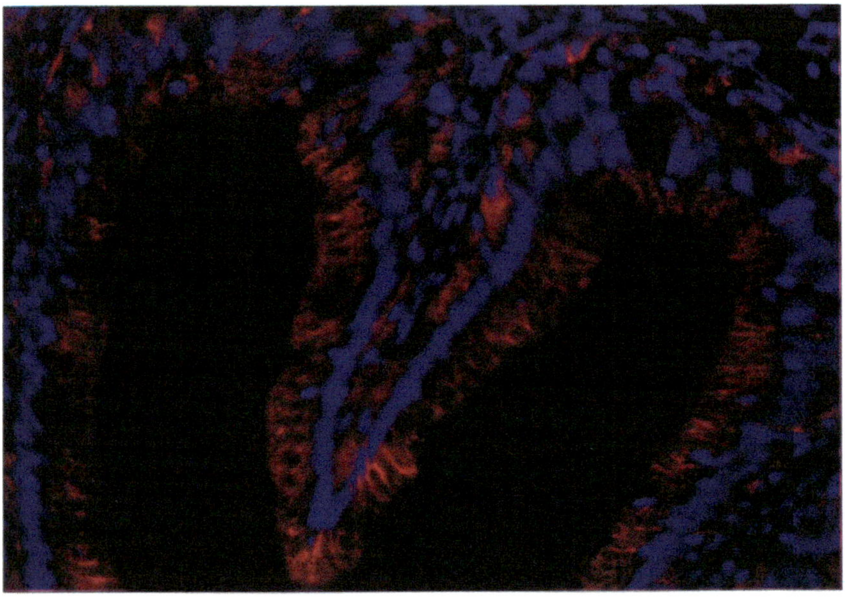

Fig. 12.3 Section of rainbow trout (65% SW) pyloric cecae immunolabelled for NKA with alpha-5 monoclonal antibody (red) (Perry, SF, Grosell, M and Gilmour, KM, unpublished). Sections were counterstained with DAPI nuclear stain (blue). Note the columnar cells with basal nucleus typical of absorbing epithelia and the clear lateral staining for NKA. Objective 40×.

1994; Loretz, 1996; Grosell and Jensen, 1999; Grosell *et al.*, 1999, 2001, 2005), as reviewed in detail by Grosell (Grosell, 2006), which may be explained partly by the stoichiometry of NKCC. However, since K^+ concentrations are relatively low in seawater and intestinal fluids, NKCC itself cannot explain the substantial excess net Cl^- absorption seen in many cases.

Intestinal Anion Exchange and Cl^- Absorption

In addition to net Na^+ and Cl^- absorption rates reported for eight species in the thirteen studies listed above, five reports on a total of four species of marine teleosts also documented net HCO_3^- secretion rates which agree well with the difference between net Cl^- and Na^+ absorption rates (Grosell and Jensen, 1999; Grosell *et al.*, 1999, 2001, 2005; Grosell, 2005, 2006). These observations strongly suggest that Cl^-/HCO_3^- exchange contributes significantly (10-71%; Grosell, 2006) to overall net Cl^- uptake in concert with the Na^+:Cl^- co-transporters discussed above. A consequence of the

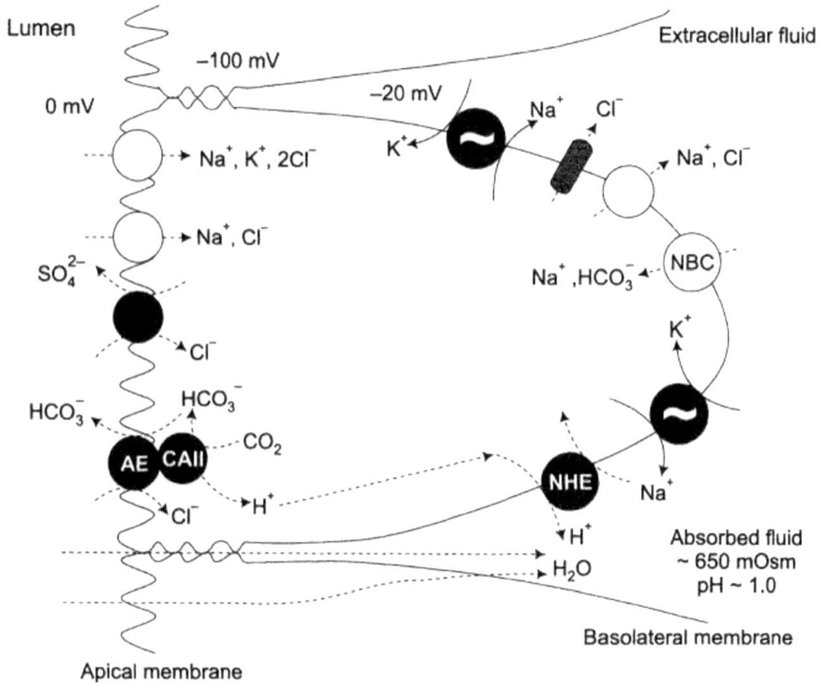

Fig. 12.4 Schematic cellular model of transport processes in the intestinal epithelium of marine teleost fish. Transcellular and/or paracellular fluid absorption is driven by active NaCl transport fueled primarily by the basolateral Na^+-K^+-ATPase (~) which provides the electrochemical Na^+ gradient allowing for Na^+, Cl^- and K^+ entry across the apical membrane. Two parallel systems, the Na^+,Cl^- and the Na^+, K^+, $2Cl^-$ co-transporters account for Na^+, K^+ and a portion of Cl^- absorption, with the remaining Cl^- uptake occurring via anion exchange (AE). The apical AE performs active transport of not only Cl^- but also HCO_3^- resulting in high luminal HCO_3^- concentrations and highly alkaline intestinal fluids. Further, anion exchange is involved in Cl^-/SO_4^{2-} exchange and likely maintains the substantial transepithelial SO_4^{2-} gradient (Pelis and Renfro, 2003). Endogenous metabolic CO_2 provides cellular HCO_3^- via carbonic anhydrase for the apical anion exchange process with the resulting H^+ being extruded across the basolateral membrane via an NHE-like transporter. The H^+ extrusion across the basolateral membrane is critical for apical HCO_3^- secretion and ultimately relies on the activity of the basolateral Na^+-K^+-ATPase (Grosell and Genz, 2006). A physical association of AE and carbonic anhydrase II (CAII) might explain how local HCO_3^- concentrations on the luminal side of the apical membrane can reach levels satisfying the thermodynamic conditions necessary for anion exchange (Grosell, 2006). Exchange of a metabolic waste product (CO_2), which exerts limited osmotic pressure, in exchange for an electrolyte provides an osmotic driving force for cellular water uptake. Basolateral import of HCO_3^- from extracellular fluids appears to contribute to luminal HCO_3^- secretion as well and may occur via Na^+: HCO_3^- co-transport (NBC). Based on previous studies summarized by Grosell (2006), fluid absorbed by the intestinal epithelium is hyperosmotic and highly acidic (Grosell *et al.*, 2005; Grosell and Genz, 2006). See text for further details.

intestinal anion exchange is high concentrations of HCO_3^- in intestinal fluids of marine teleosts. While this was reported for the first time in the first part of the previous century (Smith, 1930) and since has been confirmed for a large number of marine teleosts (Cordier and Maurice, 1956; Shehadeh and Gordon, 1969; Walsh *et al.*, 1991; Wilson, 1999; Grosell *et al.*, 2001; Wilson *et al.*, 2002), it was not until recently that attempts were made to understand the functional significance of intestinal anion exchange. The first two attempts (Wilson *et al.*, 1996; Wilson, 1999) to implicate intestinal base secretion in dynamic acid-base balance regulation were negative (Wilson *et al.*, 2002) and a role for intestinal anion exchange in osmoregulation became evident from *in situ* perfusion studies of the lemon sole intestine (Grosell *et al.*, 1999) and transport studies on isolated intestinal segments from the Pacific sanddab and the European flounder (Grosell *et al.*, 2001, 2005).

The Cl^- uptake and HCO_3^- excretion arising from the intestinal anion exchange are the function of secondary active transport and appear to occur via an apical anion exchange protein (Dixon and Loretz, 1986; Ando and Subramanyam, 1990; Wilson *et al.*, 1996; Grosell and Jensen, 1999; Grosell *et al.*, 2001, 2005; Wilson *et al.*, 2002). The cellular substrate for apical anion exchange seems to be a mix of HCO_3^- from extracellular fluids imported across the basolateral membrane and cellular hydration of endogenous CO_2 from epithelial respiration. Studies to date suggest that 30-60% of the secreted HCO_3^- is derived from cellular CO_2 hydration (Dixon and Loretz, 1986; Ando and Subramanyam, 1990; Grosell *et al.*, 2005; Grosell and Genz, 2006) but species-specific differences are quite likely to exist.

Serosal HCO_3^- appears to provide part of the substrate of apical anion exchange (Ando and Subramanyam, 1990; Grosell *et al.*, 2005; Grosell and Genz, 2006) and observations of luminal HCO_3^- secretions being dependent on serosal Na^+ and sensitive to serosal DIDS suggest the involvement of a basolateral Na^+: HCO_3^- co-transporter (NBC). A basolateral NBC may allow for HCO_3^- import by the intestinal epithelium and, thus, provide cellular substrate for the apical anion exchange. The HCO_3^- uptake across the basolateral membrane via NBC would be driven by the favorable Na^+ gradient established by the basolateral NKA (Fig. 12.4). It should be noted, however, that a role for basolateral NBC in intestinal HCO_3^- secretion remains to be conclusively demonstrated.

Cellular CO_2 hydration is partly mediated by carbonic anhydrase (Dixon and Loretz, 1986; Wilson *et al.*, 1996; Grosell and Genz, 2006) and yields not only HCO_3^- but also H^+, which must be excreted from the intestinal cells in order to avoid reversal of the CO_2 hydration. A close association between the apical anion exchanger and carbonic anhydrase has been recently suggested to facilitate Cl^-/HCO_3^- exchange by providing high local HCO_3^- concentrations (Grosell, 2006), as seen in mammalian systems (Sterling *et al.*, 2001a, b, 2002). Furthermore, since the intestinal epithelium exhibits net base secretion, these H^+ are extruded across the basolateral membrane displaying polarization of HCO_3^- and H^+ excretion from the intestinal epithelial cells (Grosell and Genz, 2006). Basolateral H^+ extrusion seems to occur via a Na^+: H^+ exchange mechanism (i.e., Gulf toadfish) and, thus, ultimately relies on the Na^+ gradient established by the activity of the basolateral NKA (Grosell and Genz, 2006) (Fig. 12.4).

Regardless of the mode of Cl^- entry across the apical membrane, cellular Cl^- concentrations (Duffey, 1979; Smith *et al.*, 1980) are above the electrochemical Cl^- equilibrium and Cl^- readily leaves the epithelial cells across the basolateral membrane via Cl^- channels (Loretz and Fourtner, 1988) or K^+:Cl^- co-transporters (Halm *et al.*, 1985a).

Anion exchange is also likely to be involved in sustaining the large SO_4^{2-} gradient across the intestinal epithelium since active extrusion of SO_4^{2-} occurs across the apical membrane in exchange for Cl^- in the winter flounder (Pelis and Renfro, 2003).

Intestinal Transport Processes—Water Absorption

While it is unequivocally clear that water absorption relies on net NaCl uptake, the exact mechanism of water absorption remains unknown and may involve both transcellular and paracellular pathways. Although water channels (aquaporins) are present in the intestinal tissue, there is no direct evidence that these proteins are involved in transepithelial water movement (Alves *et al.*, 1999; Cutler and Cramb, 2002a; Lignot *et al.*, 2002). However, several co-transport proteins, including K^+:Cl^-, Na^+:glucose and NKCC transporters, are capable of water transport in other vertebrate systems (Loo *et al.*, 2002; Hamann *et al.*, 2005) and may contribute to transcellular movement across the marine teleost intestine.

Composition of the Fluid Absorbed by the Intestine

Hypertonicity

A number of studies report simultaneous measurements of net Na^+, Cl^- and water absorption rates (Hickman, 1968; Shehadeh and Gordon, 1969; Pickering and Morris, 1973; Skadhauge, 1974; Grosell and Jensen, 1999; Grosell *et al.*, 2001, 2005; Marshall *et al.*, 2002) which allow the estimation of the Na^+ and Cl^- concentration in the water absorbed by the intestine. The tonicity of the absorbed fluid has also be estimated from these measurements (Grosell, 2006) to be hyperosmotic with respect to both the intestinal fluids and the blood plasma. These estimates of osmotic pressure yield an overall mean value of 647 mOsm, which takes into account that Na^+ concentrations in the absorbed fluids are substantially lower than corresponding Cl^- concentrations and so additional cation(s) must be absorbed along with the Na^+ and Cl^-. The absorbed fluid osmolality estimate is based on an assumed osmotic coefficient of 1 and the conservative assumption that no other anions are transported across the epithelium as discussed in detail elsewhere (Grosell, 2006).

The Missing Cation—Acidic Absorbate

From parallel and direct, although not simultaneous, measurements of water transport and serosal acid secretion in the gulf toadfish intestine, the H^+ concentration in absorbed fluids was estimated to be 77 mM, corresponding to a pH of 1.1 (Grosell and Genz, 2006). These findings verify the earlier suggestions of a highly acidic intestinal absorbate in the European flounder (Grosell *et al.*, 2005) and demonstrate that in the gulf toadfish, the majority of the 'missing' cationic charge can be attributed to H^+ secreted across the basolateral membrane. Higher net absorption rates of Cl^- rather than Na^+ seem ubiquitous to intestinal fluid absorption by marine teleosts (Grosell, 2006), suggesting that H^+ secretion across the basolateral membrane and thus acidic intestinal absorbate may be a general feature common to this group.

Dual Role of the Intestine: Feeding vs Osmoregulation

The intestine of marine teleosts, in particular, serves for both digestion and water absorption (as an important component of osmoregulation) but interactions between these two processes have not been considered extensively. Early reports demonstrated that intestinal fluid chemistry and, very likely, osmoregulation was influenced by the presence of partly digested food in the intestine (Smith, 1930). In a recent study, the influence of the ingestion of two different natural food sources on the chemical composition of intestinal fluid and the osmoregulatory status of the marine gulf toadfish was examined (Taylor and Grosell, 2006b). Feeding transiently alters the intestinal fluid chemistry, dramatically decreasing Mg^{2+} and SO_4^{2-} concentrations immediately following feeding, indicating reduced drinking and/or reduced fluid absorption. In support of this interpretation are elevated Na^+ concentrations in the anterior segments of the intestine which also indicate limited water absorption immediately following feeding (Taylor and Grosell, 2006b). In contrast, Cl^- levels are greatly reduced in all segments of the intestine for up to 48 hours post-feeding. This decrease in intestinal fluid Cl^- concentrations is seen despite of titration of the acidic stomach content (HCl) moving into the intestine and is followed by highly elevated intestinal fluid HCO_3^- concentrations. The latter suggests that intestinal anion exchange (see above) is stimulated following feeding and that this process may play an important role in reestablishing the alkaline intestinal fluids following feeding (Taylor and Grosell, 2006b). Despite the obvious influence of feeding on intestinal transport processes associated with osmoregulation, only a minor disturbance of salt and water balance was evident from plasma osmolality and ionic composition (Taylor and Grosell, 2006b).

Endocrine Control of Intestinal Salt and Water Transport

Recently, an excellent review of the endocrinology of osmoregulation (Takei and Loretz, 2005) distinguished between two classes of osmoregulatory hormones, as defined earlier (Takei and Hirose, 2002). The first class consists of fast, short-acting hormones exhibiting rapid stimulation and transient secretion and includes angiotensin II, arginine vasotocin (AVT), natriuretic peptides, vasointestinal peptide, urotensins and perhaps guanylins. The second class of slow, long-acting hormones includes cortisol, 1α-hydroxy-corticosterone, growth hormone (GH) and prolactin (PRL), with the latter likely being implicated in freshwater adaptation only. While the fast-acting hormones control the already existing transport proteins, modulating activity in many cases via phosphorylation/dephosphorylation, the slow and long-lasting hormones typically involve transcriptional modifications (see below) and *de novo* synthesis of ion transporting proteins (Takei and Loretz, 2005).

The osmoregulatory role of the gastrointestinal tract in marine fishes is intimately linked to the regulated drinking reflex which is controlled by a diverse and complicated multitude of parameters. The present text is limited to a discussion of the direct endocrine (and paracrine) regulation of intestinal transport and the reader is referred to a number of recent reviews for detailed descriptions of drinking rate regulation (Takei, 2000; Takei and Tsuchida, 2000; Ando *et al.*, 2003; Marshall and Grosell, 2005; Takei and Loretz, 2005).

Only a subset of the candidates for the endocrine control of osmoregulation in marine fish have been tested for potential action on the salt and water transport across the intestine, leaving a great deal of research yet to be done. The following paragraphs provide a brief review of the most frequently studied endocrine reagents that act on intestinal transport processes related to osmoregulation.

Guanylins

Guanylins form an intestinal peptide family which is involved in controlling salt and water absorption in mammals (Takei and Loretz, 2005) and their transcription appears to be upregulated upon seawater transfer in European eel (Comrie *et al.*, 2001a, b), suggesting a role in seawater osmoregulation. However, the impact of this peptide family on intestinal salt and water transport in marine teleosts awaits characterization.

Vasointestinal Peptide (VIP)

Vasoactive intestinal polypeptide (VIP) inhibits intestinal Na^+ and water transport in both freshwater- and seawater-acclimated tilapia (Takei and Loretz, 2005) and appears to inhibit net Cl^- absorption both in eel (Uesaka *et al.*, 1995) and winter flounder intestine; in the latter case by increasing the unidirectional Cl^- secretion (O'Grady, 1989; O'Grady and Wolters, 1990). The stimulation of luminal Cl^- secretion could be related to activation of a CFTR-like Cl^- channel in the apical membrane (Marshall *et al.*, 2002).

Neuropeptide Y

Neuropeptide Y (NPY), isolated from the Japanese eel intestine potently increases net salt absorption, as is evident from the increased serosa negative transepithelial potential and the enhanced short circuit current. NPY appears to act on the apical NKCC cotransporter since its effects can be blocked by bumetanide (Uesaka *et al.*, 1996).

Natriuretic Peptides

Atrial natriuretic peptide (ANP) is synthesized by the heart but also by non-cardiac tissues including the intestine (Loretz *et al.*, 1997). Further, the presence of ANP-ergic neurons in the intestinal tissue indicates paracrine intestinal action (Loretz, 1995). The effect of ANP on the marine teleost intestine is inhibition of net salt absorption (measured by short circuit current in Ussing chambers during radio-isotopic fluxes) (O'Grady *et al.*, 1985; O'Grady, 1989; Ando *et al.*, 1992; Loretz, 1996; Loretz and Takei, 1997) presumably by inhibiting NKCC. Release of ANP in marine teleosts is stimulated by elevated plasma Na^+ concentrations and hypervolemia. It acts to correct Na^+ concentrations and extracellular fluid volume by reducing the intestinal salt and water absorption and by stimulating branchial salt extrusion (Loretz and Pollina, 2000).

Urotensin II and Somatostatin

Urotensin II is secreted from the urophysis at rates responding to external salinities with circulating levels being higher in seawater- than in freshwater-adapted fish, suggesting a role in water retention (Bond *et al.*, 2002). In agreement with this putative role for urotensin II are observations of increased short circuit current (depending on the dose) in marine teleost intestinal epithelia conducted by apical Na^+:Cl^- co-transport systems (Loretz *et al.*, 1985; Baldisserotto and Mimura, 1997). This stimulation of net salt absorption is very likely to facilitate water absorption across the intestine epithelium. Interestingly, the partial structural analog of urotensin II somatostatin alone has no effect on Na^+ and Cl^- absorption (Loretz, 1990).

Cortisol

Cortisol is clearly involved in seawater adaptation by teleosts, as is evident from increased plasma cortisol levels following transfer to seawater (Mommsen *et al.*, 1999) and pronounced effects on the gill chloride cells and the mRNA expression of a number of ion-transporting proteins (McCormick, 2001). The effects of cortisol enhance the ability of euryhaline fish to tolerate seawater exposure (Hwang and Wu, 1993) likely via glucocorticoid receptors (Hirano, 1967; Hirano and Utida, 1968; Gaitskell and Chester Jones, 1970; Epstein *et al.*, 1971; Pickford *et al.*, 1979; Cornell *et al.*, 1994; Marshall *et al.*, 2005). Among its effects, cortisol acts on the intestinal epithelium stimulating greater water absorption and NKA activity (Cornell *et al.*, 1994; Veillette *et al.*, 1995). Thus, in contrast to ANP, NPY and Urotensin II all of which act on apical transport proteins in a rapid manner, cortisol promotes salt and water absorption by increasing the NKA activity (presumably by increasing the gene transcription and thereby amount of functional protein) in the basolateral membrane.

Prolactin

Prolactin is critical for freshwater adaptation in the euryhaline killifish (Pickford and Phillips, 1959) but, apparently, this is not a general phenomenon for euryhaline fish (Takei and Loretz, 2005). The role of prolactin (if any) in intestinal osmoregulatory function in seawater fish remains unclear but prolactin appears to stimulate intestinal NKA activity in fish held in both hypo-and hyper-osmotic media under certain conditions (Kelly *et al.*, 1999; Manzon, 2002).

Growth Hormone and Insulin-like Growth Factor

Growth hormone (GH) appears to play an important role in seawater acclimation in teleost fish (Mancera and McCormick, 1998) and plasma GH displays a transient increase following seawater transfer and protects against increased plasma osmolality during seawater exposure (Sakamoto *et al.*, 1993). GH may act directly on osmoregulatory organs or by locally secreted insulin like growth factor (IGF-1) but direct action of GH or IGF-I on the osmoregulatory function of marine teleost intestinal epithelia remains to be demonstrated.

References

Alves, P., G. Soveral, R.L. Macey and T.F. Moura. 1999. Kinetics of water transport in eel intestinal vesicles. *Journal of Membrane Biology* 171: 177–182.

Ando, M. and M.V.V. Subramanyam. 1990. Bicarbonate transport systems in the intestine of the seawater eel. *Journal of Experimental Biology* 150: 381–394.

Ando, M., H. Sasaki and K.C. Huang. 1986. A new technique for measuring water transport across the seawater eel intestine. *Journal of Experimental Biology* 122: 257–268.

Ando, M., K. Kondo and Y. Takei. 1992. Effects of eel atrial natriuretic peptide on NaCl and water transport across the intestine of the seawater eel. *Journal of Comparative Physiology* B162: 436–439.

Ando, M., T. Mukuda and T. Kozaka. 2003. Water metabolism in the eel acclimated to seawater: From mouth to intestine. *Comparative Biochemistry and Physiology* B136: 621–633.

Baldisserotto, B. and O.M. Mimura. 1997. Changes in the electrophysiological parameters of the posterior intestine of *Anguilla anguilla* (Pisces) induced by oxytocin, urotensin II and aldosterone. *Brazilian Journal of Medical and Biological Research* 30: 35–39.

Beyenbach, K.W. 2004. Kidneys sans glomeruli. *American Journal of Physiology: Renal Physiology* 286: F811–F827.

Bond, H., M.J. Winter, J.M. Warne, C.R. McCrohan and R.J. Balment 2002. Plasma concentrations of arginine vasotocin and urotensin II are reduced following transfer of the euryhaline flounder (*Platichthys flesus*) from seawater to freshwater. *General and Comparative Endocrinology* 125: 113–120.

Bury, N.R., M. Grosell, C.M. Wood, C. Hogstrand, R.W. Wilson, J.C. Rankin, M. Busk, T. Lecklin and F.B. Jensen. 2001. Intestinal iron uptake in the European flounder (*Platichthys flesus*). *Journal of Experimental Biology* 204: 3779–3787.

Comrie, M.M., C.P. Cutler and G. Cramb. 2001a. Cloning and expression of guanylin from the European eel (*Anguilla anguilla*). *Biochemistry and Biophysics Research Communication* 281: 1078–1085.

Comrie, M.M., C.P. Cutler and G. Cramb. 2001b. Cloning and expression of two isoforms of guanylate cyclase C(GC-C) from the European eel (*Anguilla anguilla*). *Comparative Biochemistry and Physiology* A129: 575–586.

Cordier, D. and A. Maurice. 1956. Etude sur l'absorption intestinale des sucres chez l'angiulle (*Anguilla vulgaris* L.) vivant dans l'eau de mer ou dans l'eau douce. *Comptes Rendes Seances Societe Biologique* 150: 1957–1959.

Cornell, S.C., D.M. Portesi, P.A. Veillette, K. Sundell and J.L. Specker. 1994. Cortisol stimulates intestinal fluid uptake in Atlantic salmon (*Salmo salar*) in the post-smolt stage. *Fish Physiology and Biochemistry* 13: 183–190.

Cutler, C.P. and G. Cramb. 2002a. Branchial expression of an aquaporin 3 (AQP-3) homologue is down-regulated in the European eel *Anguilla anguilla* following seawater acclimation. *Journal of Experimental Biology* 205: 2643–2651.

Cutler, C.P. and G. Cramb. 2002b. Two isoforms of the $Na^+/K^+/2Cl(^-)$ cotransporter are expressed in the European eel (*Anguilla anguilla*). *Biochimica et Biophysica Acta-Biomembranes* 1566: 92–103.

Dixon, J.M. and C.A. Loretz. 1986. Luminal alkalinization in the intestine of the goby. *Journal of Comparative Physiology* 156: 803–811.

Duffey, M.E. 1979. Intracellular chloride activities and active chloride absorption in the intestinal epithelium of the winter flounder. *Journal of Membrane Biology* 50: 331–341.

Epstein, F.H., M. Cynamon and W. McKay. 1971. Endocrine control of Na-K-ATPase and seawater adaptation in *Anguilla rostrata*. *General and Comparative Endocrinology* 16: 323–328.

Evans, D.H. 1993. Osmotic and Ionic regulation. In: *The Physiology of Fishes*, D.H. Evans (ed.). CRC Press, Boca Raton, pp. 315–341.

Evans, D.H., P.M. Piermarini and K.P. Choe. 2005. The multifunctional fish gill: Dominant site of gas exchange, osmoregulation, acid-base regulation, and excretion of nitrogenous waste. *Physiological Reviews* 85: 97–177.

Field, M., P.L. Smith and J.E. Bolton. 1980. Ion transport across the isolated intestinal mucosa of the winter flounder *Pseudopleuronectes americanus*: II. Effects of cyclic AMP. *Journal of Membrane Biology* 53: 157–163.

Field, M., K.J. Karnaky, W.B. Kinter, J.E. Bolton and P.L. Smith. 1978. Ion transport across the isolated intestinal mucosa of the winter flounder, *Pseudopleuronectes americanus*. *Journal of Membrane Biology* 41: 265–293.

Frizzell, R.A., P.L. Smith, M. Field and E. Vosburgh. 1979. Coupled sodium-chloride influx across brush border of flounder intestine. *Journal of Membrane Biology* 46: 27–39.

Gaitskell, R.E. and I. Chester Jones. 1970. Effects of adrenalectomy and cortisol injection on the *in vitro* movement of water by the intestine of the freshwater European eel (*Anguilla anguilla* L.). *General and Comparative Endocrinology* 15: 491–493.

Gibson, J.S., J.C. Ellory and B. Lahlau. 1987. Salinity acclimation and intestinal salt transport in the flounder: The role of the basolateral cell membrane. *Journal of Experimental Biology* 128: 371–382.

Grosell, M. 2005. Ion transport across the gulf toadfish, *Opsanus beta*, intestine. (Unpublished).

Grosell, M. 2006. Intestinal anion exchange in marine fish osmoregulation. *Journal of Experimental Biology* 209: 2813–2827.

Grosell, M. and J. Genz. 2006. Ouabain sensitive bicarbonate secretion and acidic fluid absorption by the marine fish intestine play a role in osmoregulation. *American Journal of Physiology* 291: R1145–R1156.

Grosell, M. and F.B. Jensen. 1999. NO_2^- uptake and HCO_3^- excretion in the intestine of the European flounder (*Platichthys flesus*). *Journal of Experimental Biology* 202: 2103–2110.

Grosell, M. and C.M. Wood. 2001. Branchial versus intestinal silver toxicity and uptake in the marine teleost *Parophrys vetulus*. *Journal of Comparative Physiology* B171: 585–594.

Grosell, M., G. De Boeck, O. Johannsson and C.M. Wood. 1999. The effects of silver on intestinal ion and acid-base regulation in the marine teleost fish, *Papophrys vetulus*. *Comparative Biochemistry and Physiology* C124: 259–270.

Grosell, M., C.N. Laliberte, S. Wood, F.B. Jensen and C.M. Wood. 2001. Intestinal HCO_3^- secretion in marine teleost fish: Evidence for an apical rather than a basolateral Cl^-/HCO_3^- exchanger. *Fish Physiology and Biochemistry* 24: 81–95.

Grosell, M., M.D. McDonald, C.M. Wood and P.J. Walsh. 2004. Effects of prolonged copper exposure in the marine gulf toadfish (*Opsanus beta*). I. Hydromineral balance and plasma nitrogenous waste products. *Aquatic Toxicology* 68: 249–262.

Grosell, M., C.M. Wood, R.W. Wilson, N.R. Bury, C. Hogstrand, J.C. Rankin and F.B. Jensen. 2005. Active bicarbonate secretion plays a role in chloride and water absorption of the European flounder intestine. *American Journal of Physiology* 288: R936–R946.

Halm, D.R., E.J. Krasny and R.A. Frizzell. 1985a. Electrophysiology of flounder intestinal mucosa. I. Conductance of cellular and paracellular pathways. *Journal of General Physiology* 85: 843–864.

Halm, D.R., E.J. Krasny and R.A. Frizzell. 1985b. Electrophysiology of flounder intestinal mucosa. II. Relation of the electrical potential profile to coupled NaCl absorption. *Journal of General Physiology* 85: 865–883.

Hamann, S., J.J. Herrera-Perez, M. Bundgaard, F.J. Alvarez-Leefmans and T. Zeuthen. 2005. Water permeability of Na^+-K^+-Cl^- cotransporters in mammalian epithelial cells. *Journal of Physiology* 568: 123–135.

Hickman, C.P. 1965. Studies on renal function in freshwater teleost fish. *Transactions of the Royal Society of Canada* 3: 213–236.

Hickman, C.P. 1968. Ingestion, intestinal absorption, and elimination of seawater and salts in the southern flounder, *Paralichthys lethostigma*. *Canadian Journal of Zoology* 46: 457–466.

Hickman, C.P. Jr. and B.F. Trump. 1969. The Kidney. In: *Fish Physiology*, W.S. Hoar and D.J. Randall (eds.). Academic Press, New York, Vol. 1, pp. 91–239.

Hirano, T. 1967. Effects of hypophysectomy on water transport in isolated intestine of the eel, *Anguilla japonica*. *Proceedings of the Japanese Academy of Science* 43: 793–796.

Hirano, T. and N. Mayer-Gostan. 1976. Eel esophagus as an osmoregulatory organ. *Proceedings of the National Academy of Sciences of the United States of America* 73: 1348–1350.

Hirano, T. and S. Utida. 1968. Effects of ACTH and cortisol on water movement insolated intestine of the eel, *Anguilla japonica*. *General and Comparative Endocrinology* 11: 373–343.

Hwang, P.P. and S.M. Wu. 1993. Role of cortisol in hypoosmoregulatgion in larvae of the tilapia (*Oreochromis mossambicus*). *General and Comparative Endocrinology* 92: 318–324.

Jensen, M.K., S.S. Madsen and K. Kristiansen. 1998. Osmoregulation and salinity effects on the expression and activity of Na^+,K^+-ATPase in the gills of European sea bass, *Dicentrarchus labrax* (L.). *Journal of Experimental Zoology* 282: 290–300.

Karnaky, K.J. 1998. Osmotic and ionic regulation. In: *The Physiology of Fishes*, D.H. Evans (ed.). CRC Press, Boca Raton, pp. 157–176.

Kelly, S.P., I.N.K. Chow and N.Y.S. Woo. 1999. Effects of prolactin and growth hormone on strategies of hypoosmotic adaptation in a marine teleost, *Sparus sarba*. *General and Comparative Endocrinology* 113: 9–22.

Kirsch, R. and M.F. Meister. 1982. Progressive processing of ingested water in the gut of seawater teleost. *Journal of Experimental Biology* 98: 67–81.

Larsen, E.H., J.B. Sørensen and J.N. Sørensen. 2002. Analysis of the sodium recirculation theory of solute-coupled water transport in small intestine. *Journal of Physiology* 542: 33–50.

Lignot, J.H., C.P. Cutler, N. Hazon and G. Cramb. 2002. Immunolocalisation of aquaporin 3 in the gill and the gastrointestinal tract of the European eel *Anguilla anguilla* (L.). *Journal of Experimental Biology* 205: 2653–2663.

Loo, D.D. F., E.M. Wright and T. Zeuthen. 2002. Water pumps. *Journal of Physiology-London* 542: 53–60.

Loretz, C.A. 1990. Recognition by gobi intestine of a somatostatin analog, SMS 201-995. *Journal of Experimental Zoology* 4 (Supplement): 31–36.

Loretz, C.A. 1995. Atrial natriuretic peptide regulation of vertebrate intestinal ion transport. *American Zoologist* 35: 490–502.

Loretz, C.A. 1996. Inhibition of goby posterior intestinal NaCl absorption by natriuretic peptides and by cardiac extracts. *Journal of Comparative Physiology* B166: 484–491.

Loretz, C.A. 2001. Drinking and alimentary transport in teleost osmoregulation. *Proceedings of the 14th International Congress of Comparative Endocrinology* 723–732.

Loretz, C.A. and C.R. Fourtner. 1988. Functional-characterization of a voltage-gated anion channel from teleost fish intestinal epithelium. *Journal of Experimental Biology* 136: 383–403.

Loretz, C.A. and C. Pollina. 2000. Natriuretic peptides in fish physiology. *Comparative Biochemistry and Physiology* A125: 169–187.

Loretz, C.A. and Y. Takei. 1997. Natriuretic peptide inhibition of intestinal salt absorption in the Japanese eel: Physiological significance. *Fish Physiology and Biochemistry* 17: 319–324.

Loretz, C.A., C. Pollina, H. Kaiya, H. Sakagushi and Y. Takei. 1997. Local synthesis of natriuretic peptides in the eel intestine. *Biochemical and Biophysical Research Communications* 238: 817–822.

Loretz, C.A., M.E. Howard and A.J. Siegel. 1985. Ion transport in goby intestine: Cellular mechanism of urotensin II stimulation. *American Physiological Society* G284–G293.

Mackay, W.C. and R. Janicki. 1978. Changes in the eel intestine during seawater adaptation. *Comparative Biochemistry and Physiology* A62: 757–761.

Mackay, W.C. and B. Lahlou. 1980. Relationships between Na^+ and Cl^- fluxes in the intestine of the European flounder, *Platichthys flesus*. *Epithelial Transport in Lower Vertebrates*, B. Lahou (ed.). Cambridge University Press, Cambridge, pp. 151–162.

Mancera, J.M. and S.D. McCormick. 1998. Osmoregulatory actions of the GH/IGF axis in non-salmonid teleost. *Comparative Biochemistry and Physiology* B121: 43–48.

Manzon, L.A. 2002. The role of prolactin in fish osmoregulation: A review. *General and Comparative Endocrinology* 125: 291–310.

Marshall, W.S. and M. Grosell. 2005. Ion transport, osmoregulation and acid-base balance. In: *Physiology of Fishes*, D.H. Evans and J.B. Claiborne (eds.). 3rd Edition. CRC Press, Boca Raton, pp. 177–230.

Marshall, W.S., J.A. Howard, R.R.F. Cozzi and E.M. Lynch. 2002. NaCl and fluid secretion by the intestine of the teleost *Fundulus heteroclitus*: Involvement of CFTR. *Journal of Experimental Biology* 205: 745–758.

Marshall, W.S., R.R.F. Cozzi, R.M. Pelis and S.D. McCormick. 2005. Cortisol receptor blockade and seawater adaptation in the euryhaline teleost *Fundulus heteroclitus*. *Journal of Experimental Zoology* A303: 132–142.

Marvao, P., M.G. Emilio, K.G. Ferreira, P.L. Fernandes and H.G. Ferreira. 1994. Ion-transport in the intestine of *Anguilla anguilla*—Gradients and translocators. *Journal of Experimental Biology* 193: 97–117.

McCormick, S.D. 2001. Endocrine control of osmoregulation in teleost fish. *American Zoologist* 41: 781–794.

McDonald, M.D. 2005. Renal contribution to teleost fish osmoregulation. *Fish Osmoregulation.*

McDonald, M.D. and M. Grosell. 2007. Maintaining osmotic balance with an aglomerular kidney. *Comparative Biochemistry and Physiology* (Submitted).

Mommsen, T.P., M.M. Vijayan and T.W. Moon. 1999. Cortisol in teleosts: Dynamics, mechanisms of action and metabolic regulation. *Reviews in Fish Biology and Fisheries* 9: 211–268.

Musch, M.W., S.A. Orellana, L.S. Kimberg, M. Field, D.R. Halm, E.J. Krasny and R.A. Frizzell. 1982. Na^+-K^+-$2Cl^-$ co-transport in the intestine of a marine teleost. *Nature (London)* 300: 351–353.

Musch, M.W., S.M. O'Grady and M. Field. 1990. Ion transport of marine teleost intestine. *Methods in Enzymology* 192: 746–753.

O'Grady, S.M. 1989. Cyclic nucleotide-mediated effects of ANF and VIP on flounder intestinal ion transport. *American Journal of Physiology* 256: C142–C146.

O'Grady, S.M. and P.J. Wolters. 1990. Evidence for chloride secretion in the intestine of the winter flounder. *American Journal of Physiology* 258: C243–C247.

O'Grady, S.M., M. Field, N.T. Nash and M.C. Rao. 1985. Atrial natriuretic factor inhibits Na-K-Cl cotransport in the teleost intestine. *American Journal of Physiology* 249: C531–C534.

Parmelee, J.T. and J.L. Renfro. 1983. Esophageal desalination of seawater in flounder: role of active sodium transport. *American Journal of Physiology* 245: R888–R893.

Pelis, R.M. and J.L. Renfro. 2003. Active sulfate secretion by the intestine of winter flounder is through exchange for luminal chloride. *American Journal of Physiology-Regulatory Integrative and Comparative Physiology* 284: R380–R388.

Pickering, A.D. and R. Morris. 1973. Localization of ion-transport in the intestine of the migrating river lamprey, *Lametra fluviatilis* L. *Journal of Experimental Biology* 58: 165–176.

Pickford, G.E. and J.G. Phillips. 1959. Prolactin, a factor in promoting survival of hypophysectomised killifish in freshwater. *Science* 130: 454–455.

Pickford, G.E., P.K. T. Pang, E. Weinstein, J. Torretti, E. Hendler and F.H. Epstein. 1979. The response of the hypophysectomized cyprinodont, *Fundulus heteroclitus*, to replacement therapy with cortisol: Effects on blood serum and sodium-potassium activated adenosine triphosphate in the gills, kidney and intestinal mucosa. *General and Comparative Endocrinology* 14: 524–534.

Rankin, J.C., C.S. Cobb, S.C. Franklin and J.A. Brown. 2001. Circulating angiotensins in the river lamprey, *Lampetra fluviatilis*, acclimated to freshwater and seawater: possible involvement in the regulation of drinking. *Comparative Biochemistry and Physiology* B129: 311–318.

Rodriguez, A., M.A. Gallardo, E. Gisbert, S. Santilari, A. Ibarz, J. Sanchez and F. Castello-Orvay. 2002. Osmoregulation in juvenile Siberian sturgeon (*Acipenser baerii*). *Fish Physiology and Biochemistry* 26: 345–354.

Sakamoto, T., S.D. McCormick and T. Hirano. 1993. Osmoregulatory actions of growth hormone and its mode of action in salmonids: A review. *Fish Physiology and Biochemistry* 11: 155–164.

Seidelin, M., S.S. Madsen, H. Blenstrup and C.K. Tipsmark. 2000. Time-course changes in the expression of the Na^+, K^+-ATPase in gills and pyloric caeca of brown trout (*Salmo trutta*) during acclimation to seawater. *Physiological and Biochemical Zoology* 73: 446–453.

Shehadeh, Z.H. and M.S. Gordon. 1969. The role of the intestine in salinity adaptation of the rainbow trout, *Salmo gairdneri*. *Comparative Biochemistry and Physiology* 30: 397–418.

Skadhauge, E. 1974. Coupling of transmural flows of NaCl and water in the intestine of the eel (*Anguilla anguilla*). *Journal of Experimental Biology* 60: 535–546.

Skou, J.C. 1990. The energy coupled exchange of Na^+ for K^+ across the cell membrane. The Na^+, K^+-pump. *FEBS Letters* 268: 314–324.

Skou, J.C. and M. Esmann. 1992. The Na,K-ATPase. *Journal of Bioenergetics and Biomembranes* 24: 249–261.

Sleet, R.B. and L.J. Weber. 1982. The rate and manner of seawater ingestion by a marine teleost and corresponding seawater modification by the gut. *Comparative Biochemistry and Physiology* A72: 469–475.

Smith, C.P., P.L. Smith, M.J. Welsh, R.A. Frizzell, S.A. Orellana and M. Field. 1980. Potassium transport by the intestine of the winter flounder *Pseudopleuronectes americanus*: Evidence for KCl co-transport. *Bulletin of the Mount Desert Island Biological Laboratory* 20: 92–96.

Smith, H.W. 1930. The absorption and excretion of water and salts by marine teleosts. *American Journal of Physiology* 93: 480–505.

Smith, M.W., J.C. Ellory and B. Lahlou. 1975. Sodium and chloride transport by the intestine of the European flounder *Platichthys flesus* adapted to fresh or sea water. *Pflügers Archives* 357: 303–312.

Spring, K.R. 2002. Solute recirculation. *Journal of Physiology* 542: 51–51.

Sterling, D., R.A.F. Reitmeier and J.R. Casey. 2001a. A transport metabolon. *Journal of Biological Chemistry* 276: 47886–47894.

Sterling, D., R.A.F. Reitmeier and J.R. Casey. 2001b. Carbonic anhydrase: In the driver's seat for bicarbonate transport. *JOP Journal of Pancreas* 2: 165–170.

Sterling, D., N.J.D. Brown, C.T. Supuran and J.R. Casey. 2002. The functional and physical relationship between the DRA bicarbonate transporter and carbonic anhydrase II. *American Journal of Physiology* 283: C1522–C1529.

Takei, Y. 2000. Comparative physiology of body fluid regulation in vertebrates with special reference to thirst regulation. *Japanese Journal of Physiology* 50: 171–186.

Takei, Y. and S. Hirose. 2002. The natriuretic peptide system in eels: a key endocrine system for euryhalinity? *American Journal of Physiology* 282: R940–R951.

Takei, Y. and C.A. Loretz. 2005. Endocrinology. In: *Physiology of Fishes*, D.H. Evans and J.B. Claiborne (eds.). 3rd Edition. CRC Press, Boca Raton, pp. 271–318.

Takei, Y. and T. Tsuchida. 2000. Role of the renin-angiotensin system in drinking of the seawater-adapted eels *Anguilla japonica*: Re-evaluation. *American Journal of Physiology* 279: R1105–R1111.

Taylor, J.R. and M. Grosell. 2006a. Evolutionary aspects of intestinal bicarbonate secretion in fish. *Journal of Comparative Biochemistry* A143: 523–529.

Taylor, J.R. and M. Grosell. 2006b. Feeding and osmoregulation: dual function of the marine teleost intestine. *Journal of Experimental Biology* 209: 2939–2951.

Uesaka, T., K. Yano, M. Yamasaki and M. Ando. 1995. Somatostatin-, vasoactive intestinal peptide-, and granulin-like peptides isolated from intestinal extracts of goldfish, *Carrasius auratus*. *General and Comparative Endocrinology* 99: 298–306.

Uesaka, T., K. Yano, S. Sugimoto and M. Ando. 1996. Effects of eel neuropeptide Y on ion transport across the seawater eel intestine. *Zoological Science* 13: 341–346.

Usher, M.L., C. Talbot and F.B. Eddy. 1991. Intestinal water transport in juvenile Atlantic salmon (*Salmo salar* L.) during smolting and following transfer to seawater. *Comparative Biochemistry and Physiology* A100: 813–818.

Veillette, P.A., K. Sundell and J.L. Specker. 1995. Cortisol mediates the increase in intestinal fluid absorption in atlantic salmon during parr smolt transformation. *General and Comparative Endocrinology* 97: 250–258.

Walsh, P.J., P. Blackwelder, K.A. Gill, E. Danulat and T.P. Mommsen. 1991. Carbonate deposits in marine fish intestines: a new source of biomineralization. *Limnology and Oceanography* 36: 1227–1232.

Wilson, R.W. 1999. A novel role for the gut of seawater teleosts in acid-base balance. *Regulation of Acid-Base Status in Animals and Plants, SEB Seminar Series*, Cambridge University Press, Cambridge, 68: 257–274.

Wilson, R.W. and M. Grosell. 2003. Intestinal bicarbonate secretion in marine teleost fish—Source of bicarbonate, pH sensitivity, and consequence for whole animal acid-base and divalent cation homeostasis. *Biochimique and Biophysique Acta* 1618: 163–193.

Wilson, R.W., K. Gilmour, R. Henry and C. Wood. 1996. Intestinal base excretion in the seawater-adapted rainbow trout: A role in acid-base balance? *Journal of Experimental Biology* 199: 2331–2343.

Wilson, R.W., J.M. Wilson and M. Grosell. 2002. Intestinal bicarbonate secretion by marine teleost fish—why and how? *Biochimique and Biophysique Acta* 1566: 182–193.

CHAPTER

13

The Use of Immunochemistry in the Study of Branchial Ion Transport Mechanisms

Jonathan Mark Wilson

INTRODUCTION

There are a number of recent reviews both in this volume and elsewhere that address the role of branchial ion transport proteins in fish ionoregulation (e.g., Cutler and Cramb 2001; Claiborne *et al.*, 2002; Marshall 2002; Hirose *et al.*, 2003; Perry *et al.*, 2003; Evans *et al.*, 2005; Galvez *et al.*, 2007). Therefore, I intend to focus this short review on the contribution of the immunological approach to our present state of understanding. The advantages as well as shortcomings of the technique and its contributions to our field will be discussed, in addition some technical consideration will be reviewed. In Table 13.1, I have listed the antibodies used in fish gill ionoregulatory research with their relevant

Author's address: Laboratório de Ecofisiologia, Centro Interdisciplinar de Investigação Marinha e Ambiental, Rua dos Bragas 289, 4050-123, Porto, Portugal.
E-mail: wilson_jm@cimar.org

Table 13.1 List of homologous and non-homologous antibodies used in the study of ion transport protein expression in the fish gill. The studies cited are for first use only although many have gain widespread use (notably the α5, NAK121, T4 and 24-1 antibodies (see text for additional citations)). The Table is organized into columns indicating the antibody target protein, subunit, or isoform specificity, the antibody name with the host species (rb: rabbit, rt: rat, mo: mouse, or sh: sheep) and whether the antibody is monoclonal (mo) or polyclonal (pc), the antigen the antibody was raised against (oligopeptide, purified or recombinant protein with origin), the format the antibody has been used in: immunocytochemistry (ICC) and western blot (WB), the fish species studied, and the citation of the first fish study. If appropriate the original source of the antibody is cited after in brackets.

Protein (subunit or isoform)		*Antibody*	*Antigen*	*Format*	*Fish species*	*Reference*
Na^+/K^+-ATPase	α	α5$^{(mo\ mc)}$	Purified kidney α subunit (chicken)	ICC	*Oncorhynchus mykiss*	1, (2)
		NAK121$^{(rb\ pc)*}$	CVTGVEEGRLIFDNLKKS	ICC, WB	*Anguilla japonica* *Oncorhynchus masou*	3
		eα NKA$^{(rb\ pc)}$	aa 469-773 recombinant fragment (eel)	ICC, WB	*A. japonica*	4
		eα NKA$^{(rt\ pc)}$	aa 469-773 recombinant fragment (eel)	ICC	*Tribolodon hakonensis*	5
		TG3$^{(rb\ pc)}$	125 aa fusion protein (tilapia)	WB	*Oreochromis mossambicus*	6
		a6F$^{(rb\ pc)}$	Purified kidney α subunit (chicken)	ICC, WB	*O. mossambicus*	7, (2)
		Ax2$^{(mo\ mc)}$	Purified axolemma α3 (rat)	ICC, WB		7, (8)
	β	β233$^{(sh\ pc)}$	DAGKLQEVKYFGIGD aa 222-236 (eel β1)	ICC, WB	*Anguilla anguilla*	9
Na^+:K^+:$2Cl^-$ cotransporter (NKCC)		T4$^{(mo\ mc)}$	310 aa C-term NKCC1 (human)	ICC, WB	*Periophthalmodon schlosseri*	10, (11)
Cystic fibrosis transmembrane regulator (CFTR)		TAM18$^{(mo\ mc)}$	Full length CFTR recombinant prot. (human)	ICC, WB	*P. schlosseri*	10, (12)
		24-1$^{(mo\ mc)}$	aa 1377-1480 C-term fusion prot. (human)	ICC, WB	*F. heteroclitus*	13, (14)

(Table 13.1 contd.)

(Table 13.1 contd.)

Potassium Channel (Kir)		eKir$^{(rb\ pc)}$	130 aa fusion prot. (aa 124-254) (eel)	ICC, WB	*A. japonica*	15
V-ATPase	A(70 kDa)	A$^{(rb\ pc)}$	CSHITGGDIYGIVNEN (bovine)	ICC, WB	*O. mykiss*	16, (17)
		kf A$^{(rb\ pc)}$	CAEMPADSGYPAYLGAR (killifish)	ICC, WB	*Fundulus heteroclitus*	18
	B (56kDa)	B$^{(rb\ pc)}$	279aa recombinant prot. (aa 79-357) (mosquito)	ICC, WB	*Dasyatis sabina*	19, (20)
		dB$^{(rb\ pc)}$	160aa recombinant protein (partial) (dace)	WB	*T. hakonensis*	5
		1034$^{(rb\ pc)}$	[LPDGTKRSG]-MAP4 (eel B1/B2)	WB	*Danio rerio*	21, (22)
		B2/BvA1$^{(rb\ pc)}$	RKDHADVSNQLYACYA (eel B1/B2)	ICC	*A. anguilla* *Polypterus senegalus* *Acipenser transmontanus*	23
	B1	1035$^{(rb\ pc)}$	CNRNVELNGPEAAARQHAQA (eel B1)	WB	*D. rerio*	21, (22)
		E$^{(rb\ pc)}$	CGANANRLFLD (bovine, Hemken *et al.* 1992)**	ICC, WB	*O. mykiss*	24
	E(31 kDa)	E11$^{(mo\ mc)}$	CGANANRLFLD (bovine)	ICC, WB	*Geotria australis*	25, (26)
Na^+/H^+ exchanger (NHE)	1	4E9$^{(mo\ mc)}$	aa 514-818 C-term fusion prot. (pig)	WB	*F. heteroclitus*, *Myoxocephalus octodecimspinosus*	27, (28)
	2	597$^{(rb\ pc)}$	87aa C-term fusion prot. (rabbit)	ICC, WB	*P. schlosseri*	10, (29)
		2M5$^{(rb\ pc)}$	aa 260-280 fusion prot. (rat)	WB	*F. heteroclitus*	30, (31)
	3	1380$^{(rb\ pc)}$	85aa C-term fusion prot. (rabbit)	ICC, WB	*O. mykiss*, *Pseudolabrus tetrious*	32, (33)

(Table 13.1 contd.)

(Table 13.1 contd.)

	666 $^{(rb\ pc)}$	aa 528-648 fusion prot. (rat)	WB	*F. heteroclitus* *Raja erinacea*	34, (35)
	dNHE3 $^{(rb\ pc)}$	DASVDEEASEEKPGKNHTRL (dace)	ICC	*T. hakonensis*	5
	R1B2 $^{(rt\ pc)}$	212aa C-term recombinant fragment (ray)	ICC, WB	*D. sabina*	36
Epithelial Sodium Channel (ENaC) α	1190-2 $^{(rb\ pc)}$	Full length fusion protein (bovine)	WB	*O. mossambicus*	37, (38)
β	55-1 $^{(rb\ pc)}$	GEKYCNNRDF aa 411-420 (human)	ICC, WB	*O. mykiss*, *O. mossambicus*	37, (39)
	Hillary$^{(rb\ pc)}$	Amiloride-sensitive Na^+ channel (bovine)	WB	*O. mossambicus*	37, (40)
Anion Exchanger 1	AE1t $^{(rb\ pc)}$	Purified erythrocyte AE1 (trout)	ICC, WB	*O. mossambicus*	37, (41)
Pendrin	h630-643 $^{(rb\ pc)}$	PTKEIEIQVDWNSE aa630-643 (human/rat)	ICC, WB	*D. sabina*	42, (43)
Na^+: HCO_3^- cotransporter (NBC)	dNBC1 $^{(rb\ pc)}$	42 aa COOH term fragment (dace)	ICC	*T. hakonensis*	5
H^+/K^+-ATPase	HKα1 (C2)$^{(rb\ pc)}$	GVRCCPGSWWDQELYY (pig)	ICC, WB	*D. sabina*	44, (45)
Ca^{2+}-ATPase (PMCA)	5F10 $^{(mo\ mc)}$	Purified erythrocyte PMCA (human)	ICC	*Protopterus annectens*	46, (47)
			ICC, WB	*Oncorhynchus neta*	48, (47)
Epithelial Calcium Channel (ECaC)	rtECaC$^{(rb\ pc)}$	SQFRFRLQNRKGWKEMLD (aa 18-36) (trout)	ICC, WB	*O. mykiss*	49

(Table 13.1 contd.)

(Table 13.1 contd.)

Carbonic anhydrase II	CAB[(rb pc)]	Purified gill CA (trout)	ICC, WB	*O. mykiss*	50
	dCAII[(rb pc)]	260aa recombinant protein (full length) (dace)	ICC	*T. hakonensis*	5
	tCAc[(rb pc)]	WNTKYPSFGDAASKSDGLA (trout)	ICC, WB	*O. mykiss*	51
Aquaporin 3	eAQP-3[(rb pc)]	DERIKLSNVATKDAA (eel)	ICC, WB	*A. anguilla*	52
	dAQP-3[(rb pc)]	DKTNKDMEESLKLNDVTGKN (dace)	ICC	*T. hakonensis*	5
	tilAQP-3[(rb pc)]	HTEGEARDKKQGT (tilapia)	ICC, WB	*O. mossambicus*	53
Urea transporter (UT)	eUT1[(rb pc)]	208aa recombinant fragment N-term (eel)	ICC, WB	*A. japonica*	54

* the peptide originally designed by Ura has also been used by Uchida *et al.* (2000) to make a anti-peptide polyclonal antibody they call NAK121 that is cross reactive with the denatured protein in Western blots. Ura *et al.* (1996) reported cross reactivity only with the native protein in Western blots. Katoh *et al.* (2000) have also affinity purified and directly labeled this antibody with FITC. I have also generated rabbit polyclonal antibodies using the same peptide sequences and refer to this antibody as αRbNKA.

** citation in parentheses refers to origin of peptide sequence (non-fish)

1) Witters *et al.* 1996; 2) Takeyasu *et al.* 1988; 3) Ura *et al.* 1996; [+] 4) Mistry *et al.* 2001; 5) Hirata *et al.* 2003; 6) Hwang *et al.* 1998; 7) Lee *et al.* 1998; 8) Sweadner and Gilkeson 1985; 9) Cutler *et al.* 2000; 10) Wilson *et al.* 2000b; 11) Lytle *et al.* 1995; 12) Neomarkers; 13) Marshall *et al.* 2002; 14) R&D Systems; 15) Suzuki *et al.* 1999; 16) Lin *et al.* 1994; 17) Sudhof *et al.* 1989; 18) Katoh *et al.* 2003; 19) Piermarini and Evans 2001; 20) Filippova *et al.* 1998; 21) Boesch *et al.* 2003b; 22) Boesch *et al.* 2003a; 23) Wilson (this article); 24) Sullivan *et al.* 1996; 25) Choe *et al.* 2004; 26) Hemken *et al.* 1992; 27) Claiborne *et al.* 1999; 28) Rutherford *et al.* 1997; 29) Tse *et al.* 1994; 30) Edwards *et al.* 2005; 31) Bookstein *et al.* 1997; 32) Edwards *et al.* 1999; 33) Hoogerwerf *et al.* 1996; 34) Chloe *et al.* 2002; 35) Soleimani *et al.* 1994; 36) Choe *et al.* 2005; 37) Wilson *et al.* 2000a; 38) Ismailov *et al.* 1996; 39) D. Benos, unpublished; 40) Sorscher *et al.* 1988; 41) Cameron *et al.* 1996; 42) Piermarini *et al.* 2002; 43) Royaux *et al.* 2000; 44) Choe *et al.* 2004; 45) Smolka and Swinger 1992; 46) Sturla *et al.* 2001; 47) Borke *et al.* 1989; 48) Uchida *et al.* 2002; 49) Shahsavarani *et al.* (in press); 50) Rahim *et al.* 1988; 51) Georgalis *et al.* 2006; 52) Lignot *et al.* 2002; 53) Watanabe *et al.* 2005; 54) Mistry *et al.* 2001.

information (target protein, antibody name, antigen, species, and references). From this Table, the wide use of this technique in our field is quite evident. This short review will also focus on immunocytochemistry (ICC) and will only make reference to immunoblotting, although in itself, it has proven to be a very useful technique.

The anatomical complexity of the fish gill (Wilson and Laurent, 2002) has hampered our progress in the elucidation of the molecular mechanisms involved in ion regulation (see the reviews listed above). This has been largely due to the unsuitability of the gills for use in ussing chamber type experiments that allow a detailed study of ion movements but are suited to flat two-dimensional epithelia (see Wood *et al.*, 2002). The study of freshwater ion uptake mechanisms, in particular, has lagged behind due to the lack of suitable surrogate models that can be studied with such a system. In the case of the seawater fish, the opercular epithelium preparation has served this role (see Marshall, 2002).

The immunological approach to the study of ion transport protein expression has allowed us to identify the location of many ion transport proteins and infer their function from this localization. This is a powerful means of conveying such information as the adage goes 'a picture speaks a thousand words'. For a verbally handicapped person such as myself, this technique has been a particular godsend. It is essential, however, that proper controls are run for the correct interpretation of the results (see below).

Although antibodies are well suited as probes for determining the location of proteins of interest, they do have certain obvious limitations. The most notable is that they cannot tell us whether the protein is indeed functional (active). The nature of the specificity is immunologic and not biologic (Petrusz *et al.*, 1976). The activities of many proteins are regulated by a number of different factors that would not be reflected in the antigenicity of that protein. There are, however, some antibodies that do recognize epitopes in key regulatory sites and only bind when the protein is activated (Flemmer *et al.*, 1999). One simple way to overcome this shortcoming is to use ICC in conjunction with physiological techniques from which the function can be ascertained.

Another problem relates to the use of antibodies in comparative studies. The major difficulty comes in explaining the significance of a negative result since there are a number of explanations for differences in strength of immunoreactivity between different species. These could include difference in expression levels and/or distribution, to varying levels of antibody specificity due to differences in the antigenic sites which affect antibody binding and, consequently, signal strength. In general, positive control tissue should be used when assessing antibody labeling with a new species, organ or antibody in order to rule out faults with the protocol. The α5 monoclonal antibody, which is generally considered pan-specific (recognizes a very conserved epitope of the Na^+/K^+-ATPase α subunit), has been reported to work in a wide variety of organisms, including numerous fishes. However, in the small spotted cat shark, *Scyliorhinus canicula*, I was unable to get cross-reactivity with this antibody in gill, kidney or rectal gland with either ICC or Western blotting. The presence of Na^+/K^+-ATPase was confirmed by measuring *in vitro* ouabain-sensitive activity and by using the αRbNKA antibody (which also recognized a highly conserved region of Na^+/K^+-ATPase α subunit) with both ICC and Western blots. Care must, therefore, be taken with the use of this technique in comparative studies and alternative techniques used for confirmation. Generally, it is advisable to use more than one antibody to the protein of interest, if possible (Nigg *et al.*, 1982).

ADDING TO THE SEAWATER MODEL OF ION EXCRETION

The components of secondary active chloride secretion identified using electrophysiological and pharmacological techniques have all been identified and localized by immunochemistry (Marshall, 2002). The basolateral Na^+/K^+-ATPase creates a Na^+ gradient to drive Cl^- uptake via the Na^+:K^+:$2Cl^-$ cotransporter (basolateral isoform NKCC1) into the cell, which accumulates and exits the cell apically via a cystic fibrosis transmembrane regulator (CFTR) Cl^- channel down its electrochemical gradient. These transporters have all been localized to the mitochondria-rich 'chloride' cell in the gill (e.g., Wilson *et al.*, 2000b; McCormick *et al.*, 2003), opercular epithelium (Marshall *et al.*, 2002) and yolk sac epithelium (Hiroi *et al.*, 2005). Figure 13.1 illustrates the colocalization of NKCC and Na^+/K^+-ATPase to the basolateral tubular system and the localization of CFTR to the apical crypts of these cells in gill sections of seawater pipefish (*Syngnathus*). In addition to confirming the presence of ion transport proteins in the seawater chloride cells, more recently, it has been demonstrated that the cellular distribution of CFTR and NKCC change

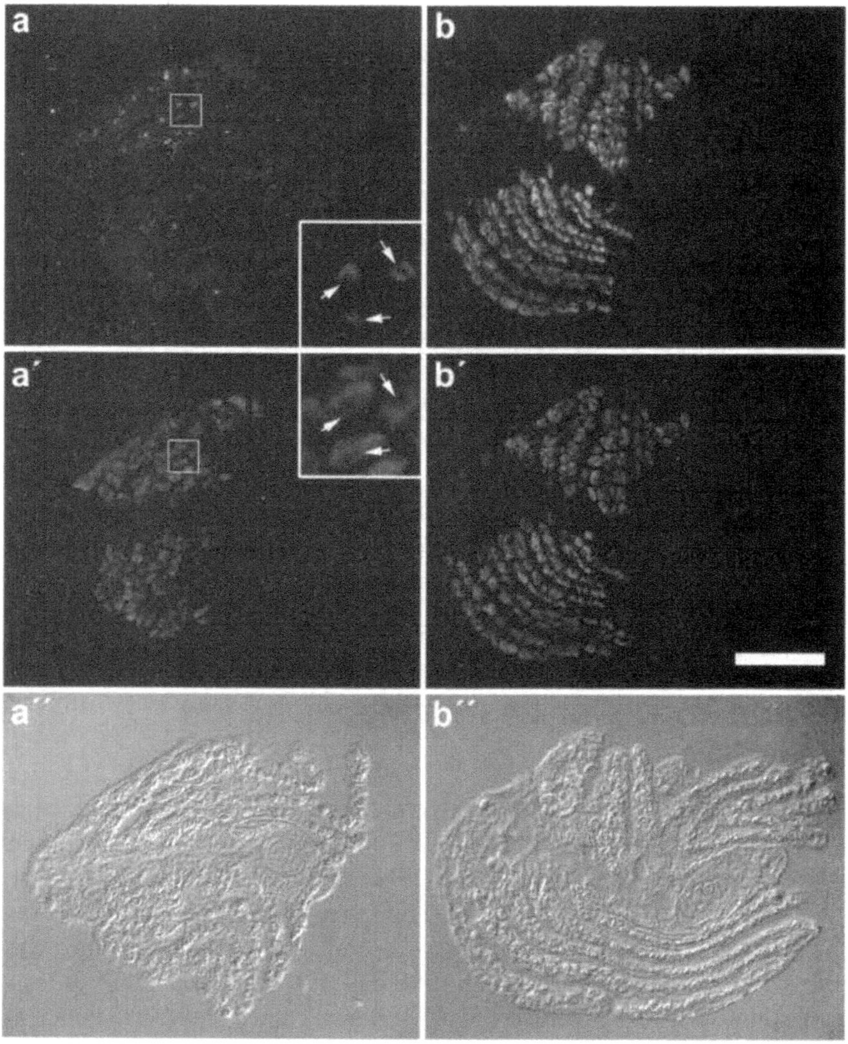

Fig. 13.1 Indirect immunofluorescent labeling of pipefish (Syngnathoidei) gill filament transverse sections for the ion transport proteins involved in secondary active Cl^- secretion. Sections are double labeled for (**a**) CFTR and (**a′**) Na^+/K^+-ATPase or (**b**) NKCC and (**b′**) Na^+/K^+-ATPase, with corresponding DIC images for tissue orientation (**a′′**, and **b′′** respectively). The insets provide higher magnification (4 ×) of the CFTR apical crypt staining of Na^+/K^+-ATPase immunoreactive cells as indicated by the arrows. CFTR and NKCC were immunolocalized with monoclonal antibodies antibody 24-1 (R&D Systems), and T4 (DSHB), respectively and Na^+/K^+-ATPase with the rabbit anti-peptide polyclonal antibody αRbNKA. Scale bar = 100 μm (J.M. Wilson and P. Laurent, unpublished observations).

with salinity (Marshall *et al.*, 2002; Hiroi *et al.*, 2005). Hiroi and co-workers working with tilapia larvae yolk sac epithelium (2005) have employed direct triple labeling of these three transporters to identify four subtypes of mitochondria-rich cell (MRC) [type I: basolateral Na^+/K^+-ATPase only; type II: basolateral Na^+/K^+-ATPase and apical NKCC (apical isoform NKCC2); type III: apical CFTR and basolateral Na^+/K^+-ATPase and NKCC1 and type IV: basolateral Na^+/K^+-ATPase and NKCC1]. The relative abundance of these MRC sub-types changes with both freshwater and seawater adaptation, indicating a role of the type III cell in seawater and the type II in freshwater ion regulation. The apical NKCC2 will be discussed further in the following sections on freshwater ion uptake mechanisms.

The accessory cell, which is found to be closely associated with the chloride cell, is an integral part of the seawater fish ion regulatory model because the tight junctions it forms with the chloride cell are leaky (see Wilson and Laurent, 2002). These leaky junctions provide the paracellular pathway for Na^+ to pass down its electrochemical gradient out of the fish. In the coho salmon, the NHE2 antibody 597 (Tse *et al.*, 1994) specifically labelled cells that, based on their morphology and appearance, appear to be the accessory mitochondria-rich cell (Wilson *et al.*, 2002b). These cells are slender and juxtaposed to Na^+/K^+-ATPase rich chloride cells and extend apically into the crypt. A similar staining pattern is observed in tilapia (*O. mossambicus*) acclimated to seawater (Fig. 13.2). In freshwater-acclimated tilapia, the frequency of these cells is reduced and there is an absence of CC-AC apical crypts. Although it is unclear why NHE-2 immunoreact with AC in this way (see Wilson *et al.*, 2002b), the antibody 597 may prove to be a useful tool for studying ACs in much the same way that some antibodies are used as marker to identify different cell types.

FRESHWATER ION UPTAKE MECHANISMS

Sodium Uptake Mechanism

One on-going debate that immunochemistry seems well suited to address is the relative importance of direct versus indirect coupling of Na^+ and H^+ exchange in acid-base regulation and sodium uptake. The first proposed mechanism involves direct coupling via a sodium-proton exchange (NHE) protein which utilizes either a sodium or proton gradient to drive the exchange. The sodium driven exchange would be predicted to function in

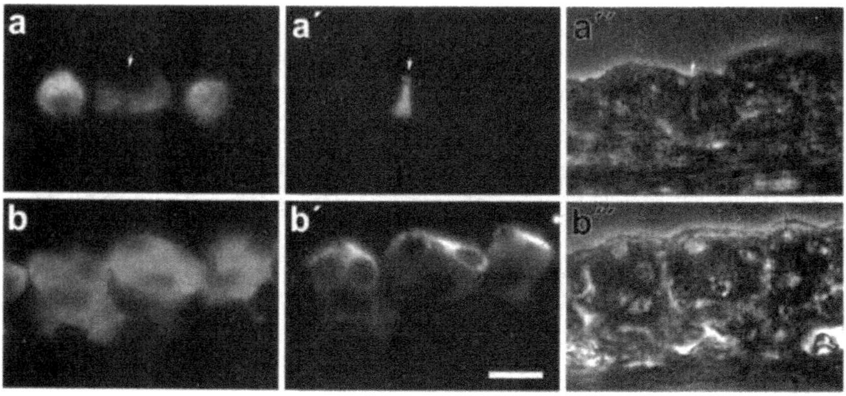

Fig. 13.2 Photomicrograph of NHE-2 (Ab597) colabeling with Na^+/K^+-ATPase (α5) in freshwater (**a** and **a**′, respectively) and seawater (**b** and **b**′, respectively) adapted tilapia (*O. mossambicus*) gill. The corresponding phase contrast images (**a**″ and **b**″). Arrows indicate location of NHE-2 immunoreactive cells in freshwater tilapia. Scale bar = 10 μm (J.M. Wilson and T.J. Lam, unpublished observations).

acid excretion, while the H^+ driven exchange would facilitate Na^+ uptake. The second proposed mechanism involves indirect coupling via an apical proton pump (vacuolar type proton ATPase) and a sodium channel, whereby the proton pump creates the favourable electrochemical gradient for sodium uptake against its concentration gradient. It should be noted that these two mechanisms need not be mutually exclusive, as is evident in the mammalian kidney (NHE in the proximal tubule and V-ATPase-ENaC in the collecting duct).

The emerging picture in fishes appears to be that the dominant mechanism is, in part, determined by the origins of the species with marine euryhaline fishes favouring NHE (sculpin, killifish and elasmobranchs) and euryhaline freshwater fishes favouring the V-ATPase (predominantly the salmonids). The acid-tolerant dace would be an exception since the NHE is responsive to acid exposure, while the V-ATPase is not. This may be a result of its adaptation to a very unusual environment (Hirata *et al.*, 2003). In the case of the NHE, there are also differences in expression of isoforms with NHE2 being upregulated in the freshwater-adapted killifish (*F. heteroclitus*, Edwards *et al.*, 2005), while the NHE3 is more important in the freshwater-adapted Atlantic stingray (Choe *et al.*, 2005), acid-tolerant dace (Hirata *et al.*, 2003) and seawater-adapted killifish (Edwards *et al.*, 2005). The killifish NHEs are responsive to environmental hypercapnia

(1% CO_2) (Edwards *et al.*, 2005). However, in the Atlantic stingray, they are not (Choe *et al.*, 2005). It would, thus, appear that the physiological roles of the NHEs are not uniform in fishes.

In dace, V-ATPase expression reported using Western analysis indicates that it is unresponsive to environmental pH, while NHE3 is strongly upregulated under acidic conditions (Northern analysis). The apical immunolocalization of proton-ATPase has been demonstrated in the gills of a number of different species of fish (salmonids (trout and salmon), tilapia, zebrafish) (e.g., Lin *et al.*, 1994; Sullivan *et al.*, 1995, Wilson *et al.*, 2002b). In the grey bichir, *Polypterus senegalus senegalus*, V-ATPase is found apically in a cell type distinct from those rich in Na^+/K^+-ATPase (Fig. 13.3). In the freshwater loach (*Misgurnus anguillicaudatus*), a similar staining pattern has been observed (J.M. Wilson, unpublished observations).

There is only one report of the presence of a sodium channel (ENaC, epithelial sodium channel) in trout and tilapia gill using a battery of non-homologous antibodies directed against mammalian ENaC (Wilson *et al.*, 2000a). This cross-reactivity is supported by the finding of phenamil (ENaC specific inhibitor) sensitive Na^+ uptake but not by molecular genetic analyses (Perry *et al.*, 2003). It is, thus, uncertain based on immunochemical results whether the V-ATPases localized in fish gill are driving Na^+ uptake via an epithelial type Na^+ channel.

Recently, the NKCC has entered the debate after being immunolocalized to the apical membrane of freshwater Na^+/K^+-ATPase immunoreactive cells in tilapia (*O. mossambicus*) using the T4 antibody (Wu *et al.*, 2003; Hiroi *et al.*, 2005). To support a possible role for the NKCC2 (apical isoform) in Na and Cl uptake is the finding of furosamide sensitivity in the goldfish (Preest *et al.*, 2005). However, I have been unsuccessful in immunolocalizing an apical NKCC in goldfish (J.M. Wilson, unpublished observations). Although in the climbing perch (*A. testudineus*) gill, I have been able to find apical NKCC immunoreactivity associated with Na^+/K^+-ATPase immunopositive cells (Fig. 13.4). In the case of the goldfish, it has long been known that Na and Cl uptake can operate independently (Maetz and Garcia-Romeu, 1964), which would indicate that the NKCC is not the only mechanism present. Hiroi *et al.* (2005) have made preliminary reports on the presence of both NKCC1 and NKCC2 mRNA in tilapia gill. In killifish, the NKCC2 mRNA expression has been reported as low but detectable in gill although not

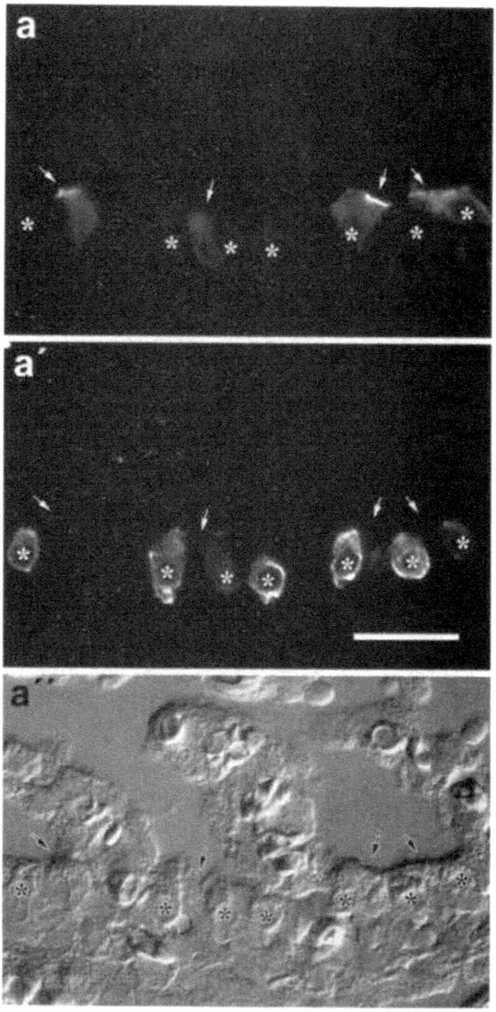

Fig. 13.3 Immunolocalization of (**a**) Na^+/K^+-ATPase (α5 antibody) and (**a′**) V-ATPase (B2 antibody) in the grey bichir, *Polypterus senegalus senegalus*, gill filament sagittal section by double immunofluorescence microscopy. (**a″**) The corresponding differential interference contrast (DIC) image. Arrows indicate the apical V-ATPase immunolabeling and asterisks are placed over the nuclei of Na^+/K^+-ATPase immunoreactive cells. Scale bar = 25 µm (J.M. Wilson and Y.K. Ip, unpublished observations).

modulated by salinity (Scott *et al.*, 2005). The NKCC2's relative importance in Na, K and Cl uptake remains to be determined as well as the gradients responsible for driving cotransport.

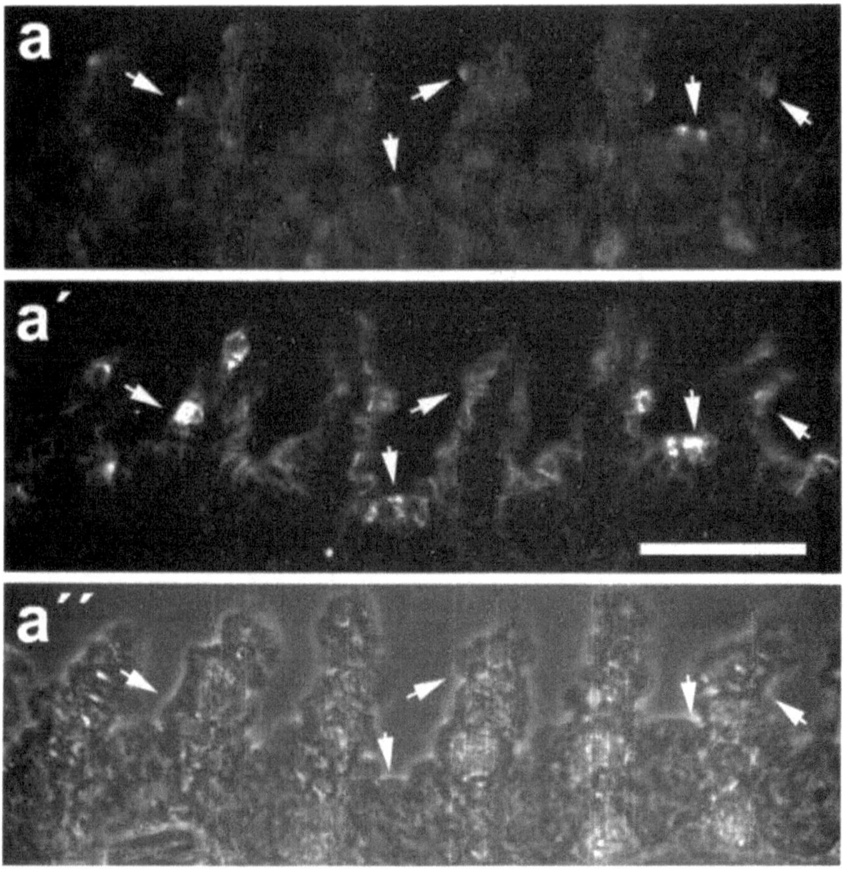

Fig. 13.4 Immunolocalization of NKCC (T4 antibody) to the apical region of Na^+/K^+-ATPase immunoreactive cells (αRbNKA antibody) in the gills of the climbing perch (*Anabas testudinus*). Arrows indicate the apical NKCC labeling. No basolateral labeling was observed. Scale bar = 50 μm (J.M. Wilson and Y.K. Ip, unpublished observations).

CHLORIDE UPTAKE

Chloride uptake in freshwater fishes is largely believed to be facilitated by a SITS sensitive electroneutral Cl^-/HCO_3^- anion exchanger (AE; e.g., reviewed by Evans *et al.*, 2005). The molecular identification of the AE in teleost fishes has been elusive. Apical AE1 (band 3) immunoreactivity has been reported in the tilapia and coho salmon associated with a subpopulation of Na^+/K^+-ATPase immunopositive cells (Wilson *et al.*, 2000a, 2002b) using an antibody developed against trout band 3 protein

(Cameron *et al.*, 1996). In coho salmon, this apical immunoreactivity is not present in seawater-adapted fish. However, the only supporting data for this type of exchanger is an *in situ* hybridization using an oligonucleotide probe in trout gill (Sullivan *et al.*, 1996). In the euryhaline Atlantic stingray (*Dasyatis sabina*), apical immunoreactivity to pendrin-like protein (a Cl^-/HCO_3^- anion exchanger) in cells with basolateral V-ATPase expression (Piermarini *et al.*, 2002). The basolateral V-ATPase helps drive the apical exchange by removing H^+ from the cell in order to maintain the intracellular HCO_3^- which directly drives the exchange. Another proposed AE in the gill is the putative anion transporter 1 (PAT1) that has been identified in trout gill. However, additional characterization is still awaited (G.G. Goss U. Alberta, unpublished observations). In the killifish, V-ATPase (A subunit) has been immunolocalized to gill Na^+/K^+-ATPase immunoreactive cells (mitochondria-rich cells) in a basolateral location and was suggested to play a similar role as in the Atlantic stingray (Katoh *et al.*, 2003). However, the killifish Cl^-/HCO_3^- exchange has not been found to be of physiological significance and, thus, the role of basolateral V-ATPase in the killifish is somewhat unclear (Kirschner, 2004). The NKCC2 also represents another pathway for Cl^- uptake to be further explored (Hiroi *et al.*, 2005; Preest *et al.*, 2005).

OTHER TRANSPORTERS IN THE GILL

In addition to the ion transport proteins involved in Na^+ and Cl^- regulation, a number of other transporters have been recognized in the gills of fishes (see Table 13.1). Of note are the urea transporter (UT; Mistry *et al.*, 2001) and aquaporin 3 (water channel proteins; Lignot *et al.*, 2002; Hirata *et al.*, 2003; Watanabe *et al.*, 2005) which are both found localized to the mitochondria-rich cell's tubular system (basolateral membrane). The plasma membrane calcium ATPase (PMCA) is also found associated with the tubular system (e.g., Sturla *et al.*, 2001; Uchida *et al.*, 2002) and an epithelial calcium channel (ECaC) with the apical membrane (Shahsavarani *et al.*, in press).

IDENTIFICATION OF NA^+/K^+-ATPASE IMMUNOREACTIVE CELLS. CHLORIDE CELLS?

Immunocytochemistry has become almost a standard technique to study the branchial mitochondria-rich 'chloride' cell. This is largely due to the high abundance of Na^+/K^+-ATPase associated with either the tubular system or basal labyrinth or these cells and, most importantly, a source of very specific and inexpensive anti-Na^+/K^+-ATPase antibody. The cellular immunolabeling pattern is characteristic of the MRC type, thus making it readily identifiable for enumeration and characterization (size, shape factor, location). The bulk of these studies have made use of the α5 mouse monoclonal antibody first developed by Douglas Fambrough and available though the Developmental Studies Hybridoma Bank at a very reasonable price (currently 25$ per ml culture supernatant). The α5 antibody recognizes an epitope in a conserved region of the α subunit and is not isoform specific (5 α subunit isoforms have been characterized so far). Another antibody that has also been widely used—although is not so readily available—is the anti-peptide antibody initially developed by Ura *et al.* (1996) and used more extensively by the group of Kareko (e.g. Uchida *et al.*, 2000; Katon *et al.*, 2000) and given the name NAK121. I have raised antibodies against the same peptide (αRbNKA).

Na^+/K^+-ATPase immunoreactive cells have been identified in a wide range of fish, including hagfish (*Myxine glutinosa* Choe *et al.*, 1999), lamprey (*Geotria australis* Choe *et al.*, 2005), ray (*Dasyatis sabina* Piermarini and Evans, 2000), shark (*Squalus acanthias* Wilson *et al.*, 2002) as well as a plethora of teleost fishes (e.g., *Anguilla japonica, Oncorhynchus masou* Ura *et al.*, 1996; *O. mykiss* Witters *et al.*, 1996; *Fundulus heteroclitus* Katoh *et al.*, 2000; *Oreochromis mossambicus* Dang *et al.*, 2000; *Periophthalmodon schlosseri* Wilson *et al.*, 2000b; *Stenogobius hawaiiensis* McCormick *et al.*, 2003). Lee *et al.* (1998) have also used other antibodies against specific α-subunit isoforms that have been found to localize to MRCs in the tilapia (*O. mossambicus*) while Hwang *et al.* (1998), Cutler *et al.* (2000) and Mistry *et al.* (2001) have developed homologous antibodies in *O. mossambicus* (tilapia), *A. anguilla* (European eel) and *A. japonica* (Japanese eel), respectively.

The immunological approach has largely surpassed other techniques that target Na^+/K^+-ATPase (^{3}[H] ouabain autoradiography and anthroylouabain fluorescence, histochemistry), or the membrane systems (Champy-Malliet stain) of MRCs (see Wilson and Laurent, 2002). The main advantage to this system of MRC identification over mitochondrial vital stains such as DASPMI (dimethylaminostyryl-methylpyridiniumiodide, emission wavelength 609 nm) and DASPEI (dimethylaminostyrylethylpyridiniumiodide, emission wavelength 530nm) (e.g., Li *et al.*, 1995; Witters *et al.*, 1996; Rombough, 1999) and Mitotracker (Molecular Probes, Galvez *et al.*, 2002; Marshall *et al.*, 2002) is that fixed and processed tissue sections can be used. Both types of mitochondrial stain require live mitochondria, although in the case of Mitotracker, the cells can be post-fixed. There have only been a few studies using immunohistochemical techniques to detect mitochondria using antibodies directed against mitochondria specific proteins (HSP 60, Mickle *et al.*, 2000; Ab-AMA, Sturla *et al.*, 2001).

In addition to counting the immunopositive cells, the signal strength (intensity), area of immunolabeling, and cell shape factor, can be measured and cellular expression calculated (area × intensity) using commercially available image analysis software or freeware (NIH image). Commonly, counting of cells is complicated by the fact that MRCs are not evenly distributed across the filament being more abundant towards the afferent side. Counting is, therefore, usually done at the afferent end. Counts are also commonly separated into lamellar and filament epithelial populations of cells (e.g., Pelis *et al.*, 2001).

Abundance of Na^+/K^+-ATPase has most commonly been measured biochemically using ouabain as the specific inhibitor of Na^+/K^+-ATPase activity *in vitro* under optimal conditions (Vmax; e.g., McCormick *et al.*, 1993). However, more recently, Western blotting has also been used to quantify Na^+/K^+-ATPase subunit abundance. Generally, there is a good correlation with the two approaches (Piermarini and Evans, 2000; Wilson *et al.*, 2004).

TECHNICAL CONSIDERATION

In the second half of this review, I will cover some technical aspects of immunochemistry and also analyse the application of methods in fish gill research.

THE ANTIBODY

The use of antibodies as a tool to localize molecules of interest in microscopy was first demonstrated in the 1940s (Coons *et al.*, 1940), with immunohistochemcial techniques coming into broader use in the 1970s (Taylor *et al.*, 1974). It is widely appreciated that antibodies can make excellent probes for detecting proteins of interest. The acquired immune response is hijacked to produce specific antibodies to proteins (or other antigens) (whole or in part; biochemically purified or synthesized) injected into a host animal. With subsequent boost injects, the levels of the polyclonal antibodies in the blood increases and can easily be collected by bleeding the host animal and separating the serum. The antibodies can then be further purified. Alternatively, hybridoma techniques allow the immortalization of B cells in order to produce monoclonal antibodies either *in vivo* or *in vitro*. It is beyond the scope of this review to elaborate further as this topic is dealt with in textbooks. One important point to keep in mind is that a monoclonal antibody recognizes a single epitope, while a polyclonal antibody recognizes multiple epitopes.

To my knowledge, the first such study to apply an immunological approach in the study of the fish gill was by Rahim and co-workers (1988) in the study of trout carbonic anhydrase. Having found that commercially available antibodies raised against mammalian erythrocyte CA were not cross reactive in the trout, they proceeded to biochemically purify gill and erythrocyte CA and generate homologous antibodies (P. Laurent, pers. comm.). Their findings have remained significant as CA which reversibly catalyzed CO_2 hydration is important in providing the intracellular counter ions, H^+ and HCO_3^-, for Na^+ and Cl^- exchange, respectively. Research of fish gill ion transporters has, however, relied on the use of non-homologous mammalian antibodies largely due to their availability and species cross-reactivity (see Table 13.1). This indicates the conserved nature of some of these proteins. However, more recently, there has been a greater tendency to developing homologous antibodies for fish studies and approximately half of the antibodies listed in Table 13.1 are homologous for the species being studied.

In many recent reviews, it is common to see that future work should involve the development of homologous antibodies. Although, this is ultimately desirable it is not, in my opinion, always necessary. The most notable examples are the $\alpha 5$ and Ura antibodies that can be considered pan-specific for the α subunit of Na^+/K^+-ATPáse as the epitopes for these antibodies are conserved in (most) vertebrates and invertebrates (the notable exception being *S. canicula* as discussed earlier). They have consequently enjoyed widespread use with over 50 publications in the fish gill literature. They have proved invaluable in characterizing the dynamics of mitochondria-rich (Na^+/K^+-ATPase immunoreactive) cells in response to environmental change as well as for determining the presence of

Na^+/K^+-ATPase-IR cells in less studied species. The T4 monoclonal antibody directed against NKCC (isoforms 1 & 2) is also widely used although some immunostaining results require further evaluation. However, the study of Na^+/K^+-ATPase will need to progress beyond the use of isoform non-specific antibodies as recent studies have indicated the importance of isoform switching in adaptation (α1a and α1b; Richards *et al.*, 2004).

In addition, there are other commercially available monoclonal antibodies that recognize conserved epitopes of plasma membrane Ca^{2+} ATPase (PMCA; 5F10 (Sturla *et al.*, 2001; Uchida *et al.*, 2002) and CFTR (24-1) (Marshall *et al.*, 2002). In the case of antibodies generated against short peptides, it is possible in many cases to compare the peptide sequence to those available on-line (genebank) using a Blast-p search. Examples of this form of validation can be found in the work of Edwards *et al.* (2005) using mammalian NHE2 antibodies and Wilson *et al.* (2000a) for V-ATPase A subunit. From this information, it is possible to determine the conserved nature of the peptide sequence and also the likelihood that the cross reactivity is indeed specific. In the cases where sequence information for the antigen of non-homologous antibodies is not known, some caution should be taken in interpreting the results in the absence of strong supporting evidence.

With advances in PCR-based cloning techniques and with sequence information readily available through the Internet, it is now reasonable for most laboratories using this information to deduce the amino acid sequences in order to have antigens for antibody production produced either through solid state peptide synthesis or recombinant protein synthesis. An impressive example is the recent publication by Hirata *et al.* (2003) on the acid-tolerant dace in which four ion transport proteins are cloned and homologous antibodies generated (V-ATPase B, NHE 3, AQP 3, CAII).

FIXATION

Fixation serves to preserve the structural integrity of the tissue and to insure that cellular components remain in place. Good fixation generally comes at a cost to tissue antigenicity and a compromise must be made.

In the existing fish gill literature, a number of different fixatives have been used successfully for ICC. The most common fixatives are paraformaldehyde (PFA) based (3-4%) in a phosphate-buffered saline (PBS; pH 7.3-74) (e.g., Uchida *et al.*, 1996; Edwards *et al.*, 1999; Seidelin and Madsen, 1999). Glutaraldehyde is added to improve fixation although generally at low concentrations 0.05% to 0.1% (e.g., Rahim *et al.*, 1988; Sullivan *et al.*, 1995; Piermarini and Evans, 2000) with the exception of Lee *et al.* (1998; 1%). Generally, tissue fixation is adequate with PFA at the light microscopy level and glutaraldehyde, which is a better crosslinking agent, may be associated with epitope masking so is best avoided. However, for electron microscopy the inclusion of glutaraldehyde in the fixative may become necessary for preservation of subcellular detail. Picric acid, which is good at preserving membrane, is commonly used either as Bouin's solution (Wilson *et al.*, 2002b) or as a 0.05% additive to PFA/PBS (Sullivan *et al.*, 1995; Piermarini and Evans, 2000).

Paraformaldehyde works by reacting with primary amines of lysine to establish a cross linking. Fixation may prevent antibody binding to its epitope by denaturing the epitope through its cross linking action or by blocking access to the epitope by crosslinking neighbouring proteins. However, it should be noted that not all epitopes are adversely affected by fixation.

Tissue can also been fixed by protein precipitation with alcohols at low temperature (ice cold EtOH: Witters *et al.*, 1996; 20% DMSO in MeOH at –20°C: Pelis *et al.*, 2001; Marshall *et al.*, 2002; Wilson *et al.*, 2004). The DMSO is added to prevent ice crystal formation that would damage the tissue. This technique avoids chemical cross linking but may still denature the epitope and block access of the antibody. I have also noticed that with 20% DMSO in MeOH in some tissue preparations fixation artifact results from the cytoplasmic content of cells being washed out. This artifact, however, is easily identifiable and tissue preservation is generally good since in the case of the gill, the diffusion distances are short. In direct comparison with PFA fixed tissue, tissue antigenicity is markedly stronger.

EMBEDDING

Embedding of fixed tissue in paraffin or cryo-embedding (OCT compound) for sectioning or whole mounting are commonly used. Both cryosectioning and whole mounting tissue avoids additional harsh treatment of the tissue that may further compromise tissue antigenicity. Generally, immunoreactivity is better in tissue that has seen the least processing with cryosections giving superior signal strength in direct comparisons with paraffin embedded tissue. However, paraffin-embedded tissue has the benefit of being easier to handle and archive as no special conditions are required (room temperature). There are a number of antigen retrieval or unmasking techniques that can sometimes be used to recover tissue antigenicity in order to make results with paraffin embedded tissues comparable to cryosections (see below).

DETECTION

A direct or indirect detection method is necessary in order to visualize the antibody bound to the section through a marker molecule. The markers commonly used are fluorochromes (e.g., FITC, TRITC, Cy3, Alexa Fluor dyes), an enzyme that catalyzes a chromogenic reaction (horseradish peroxidase (HRP) or alkaline phosphatase (AP)), or a gold particle. Direct labeling involves the conjugation of the marker directly to the antibody which results in lower background staining but requires that the antibody be purified. The tissue staining pattern must also be well characterized (by indirect detection) beforehand since it is difficult running negative controls with direct labeling. Consequently, there have been very few studies using direct labeling of antibodies with the exception of the group of Kaneko (e.g., Katoh *et al.*, 2000; Katoh and Kaneko, 2003; Hiroi *et al.*, 2005).

Practically all studies in the fish gill literature make use of indirect detection, whereby the marker molecule is conjugated to a secondary reagent that will recognize the primary antibody. This reagent is most commonly a labeled species-specific anti-immunoglobulin antibody (for example, when using a rabbit polyclonal primary antibody, a goat anti-rabbit IgG HRP conjugated secondary antibody is used). Protein A and G (bacterial cell wall proteins) can also be used as secondary reagents although their use is less common (Laurent, unpubl.). In addition to being more sensitive, indirect labeling also allows for additional signal amplification (ABC, BSA, PAP techniques see below).

The choice of marker molecule is determined in part by equipment availability and the degree of resolution required. The use of fluorochrome conjugates requires a fluorescence microscope and appropriate filter sets but offers high resolution and the ability to use multiple labels (2-5) on the same section. There are many examples of the use of fluorescence for single antibody labeling (e.g., Lin *et al.*, 1994; Sullivan *et al.*, 1995; Lignot *et al.*, 2002). Double labeling of sections using indirect detection has been done using a combination of mouse monoclonal and rabbit polyclonal primary antibodies [α5 with V-ATPase A, tAE1, 597, or 1380 (Wilson *et al.*, 2000a, b, 2002b), NAK121, and T4 (Pelis *et al.*, 2001); NAK121, and kfA (Katoh *et al.*, 2003); NAK121 and CFTR 24-1 (Katoh and Kaneko, 2003); NAK121 and CFTR 24-1 or T4 (McCormick *et al.*, 2003), and eαNKA and NBC1 with the addition of a nuclear stain (Hoechst 33342; triple staining Hirata *et al.*, 2003). The study of Hiroi *et al.* (2005) is the first to use more than two antibodies at a time with direct labeling [αNKA (NAK121) Alexa Fluor 546; NKCC (T4) Alexa Fluor 647 and CFTR (24-1) Alexa Fluor 488]. Although this study is not on gill epithelium, it does nicely illustrate the usefulness of direct labeling in multiple labeling experiments.

The most commonly used enzyme conjugate is horseradish peroxidase (HRP). It gives a stronger signal in a shorter period of time than alkaline phosphatase, another enzyme conjugate which is not widely used in fish ICC. Another advantage of using HRP is that the section can also be permanently mounted. The most common chromagenic substrate used is DAB (3,3′-diaminobenzidine), which gives a brown-black reaction product. In some double labeling experiments with HRP, Vector VIP (purple; Vector Lab USA) or Vector SG (blue) have been used (Piermarini *et al.*, 2002; Choe *et al.*, 2005). Although HRP can be used as a conjugate to the secondary antibody (Wilson *et al.*, 2000b; Hirata *et al.*, 2003), it is more commonly employed after a biotin amplification step to improve sensitivity (biotin labeled species-specific secondary anti-immunoglobulin). Avidin and streptavidin both bind to biotin with high affinity. Some examples of the use of the ABC (avidin biotin conjugate) method include Uchida *et al.* (1996), Witters *et al.* (1996), Hiroi *et al.* (1998), Suzuki *et al.* (1999), Katoh *et al.* (2000), while the BSA (biotin streptavidin) method has been used by Cutler *et al.* (2000) and Piermarini and Evans (2000). Another signal amplification technique employs the use of an anti-peroxidase antibody (PAP: peroxidase anti-peroxidase) but I can find no examples of its use in the existing fish gill literature. When using HRP, endogenous peroxidases commonly associated with erythrocytes and macrophages should be eliminated by pre-incubating the sections with H_2O_2.

Immunogold labeling is generally used for determination of cellular distribution at the subcellular level with electron microscopy although DAB staining can also be used for this purpose. In the case of immunogold a colloid gold particle of known size is conjugated to either the primary or secondary antibody. Both the gold particle and DAB product are electron dense and thus readily detectable with electron microscopy. Immunoelectron microscopy has been used to localize Na^+/K^+-ATPase to the tubular system of chloride cells (Dang *et al.*, 2000; Wilson *et al.*, 2000b) and V-ATPase to pavement cells and MRCs (Sullivan *et al.*, 1996; Wilson *et al.*, 2000a; Katoh *et al.*, 2002). Silver enhancement which enables detection of the gold particles at the light microscopic level is a very sensitive technique but is not commonly employed (Wilson *et al.*, 2002a; Tresguerres *et al.*, 2006). During the enhancement reaction silver preferentially builds up on the surface of the gold particle until it becomes visible with a light microscope.

CONTROLS

Controls should always be run when testing a new antibody or a new species in order to evaluate the specificity of the staining observed. Generally, the more controls you can run the more confidence you will have in the specificity of your staining. However, in many cases, the best control is not always available especially in the case of antibodies obtained from outside the lab (commercial or by donation).

For polyclonal antibodies, the pre-immune serum (serum collected from the same animal before immunization) serves as a good control although normal or non-immune serum from the same species is acceptable. Pre-immune and non-immune sera contain the normal antibody complement that may also cross react with your tissue and can account for non-specific labeling that you observe. An irrelevant antibody from the same host species (and isotype) can also be used since it will be present in super-abundant amounts similar to your own antibody which may result in false positive labeling. Monoclonal antibody controls should be source matched (i.e., ascites fluid, culture supernatant, or purified immunoglobulin). In the case purified antibodies are used, then a control antibody of the same isotype should be used (e.g., IgG2). Non-specific labelling of the secondary antibody can be assessed by substituting the primary antibody with dilution buffer in the labelling protocol (null control). If an enzyme conjugate is used then endogenous enzyme activity should be assessed by incubating the section with substrate. It is also standard practice to pretreat sections with hydrogen peroxide or lavaminsol to inhibit endogenous peroxidases and alkaline phosphatases, respectively (see Petrusz *et al.*, 1976; Hutson *et al.*, 1979; Harlow and Lane, 1999)

If the antigen is available, then a blocking experiments (pre-incubation of the antibody with the antigen: peptide or protein) can also be conducted. This should eliminate specific staining as the specific antibodies will have bound to antigens in solution rather than on the section.

In addition, when using amplification steps for immunodetection involving biotin-avidin or streptavidin, endogenous biotin must be taken into account in the interpretation of results. Biotin is a cofactor that is associated with the mitochondrial enzyme pyruvate carboxylase. Thus, cells with high mitochondrial densities would be expected to have more abundant biotin. This biotin artifact should be apparent on negative control slides and can be blocked if necessary (Miller *et al.*, 2004). Heat-based antigen retrieval techniques are also capable of unmasking endogenous biotin (Rodriguez-Soto *et al.*, 1997).

Finally, running Western blots is advisable as the dominant immunoreactive product can be confirmed from its predicted size. Extra bands may be observed and the band at the predicted size may even be absent which would be important in interpreting the ICC results. Factors such as purity of the preparation will influence how clean the blot is since membrane proteins only represent a small proportion of cellular proteins. However, it is not always possible to have an antibody that is cross reactive in both formats because the presentation of the antigen is different. An example is the CFTR antibody 24-1 that works very well in tissue sections, but is more difficult to get to work in Western blots.

One rather disconcerting observation that I have made is that some antibodies cross-react with mucus either covering the gills or within goblet cell mucin granules (Wilson *et al.*, 2000a). In some cases, this distribution is supported by other observations such as in the case of carbonic anhydrase (Rahim *et al.*, 1988). However, there other proteins in which there is no logical reason for them to be localized to mucin granules. One possible explanation in the case of anti-peptide antibodies is that mucin immunoreactivity results from anti-carrier antibodies which are also produced in the immune response. Peptides are usually too small to elicite an immune response so they are conjugated to immunogenic carrier molecules such as KLH or BSA. Keyhole limpet haemocyanin (KLH) is commonly used as a carrier for peptides since it is very immunogenic and it is also known to cross react with anti-carbohydrate antibodies (see Thors and Linder, 2003). In this case, affinity purification of the anti-peptide antibodies is usually effective. Alternatively, glycoprotein and glycolipid oligosaccharides are also known to present a vast array of epitopes (Feizi, 1989) and thus may be presenting an epitope similar to the target protein. In this case, the immunologic staining is specific but not for the target protein.

ANTIGEN-RETRIEVAL TECHNIQUES

In clinical histopathology, specimens are routinely fixed in 10% NBF (neutral buffered formalin) and paraffin embedded for the morphological analysis of section for disease diagnosis. Immunohistochemistry also plays an important part in disease diagnosis. However, this form of tissue preparation frequently leads to the reduction or loss of tissue antigenicity particularly with monoclonal antibodies (in comparison with cryosection which are considered the gold standard). It is, therefore, not surprising that a great deal of attention has been focused on antigen retrieval techniques that can recover antigenicity. There is now a vast literature aimed at improving, and standardizing methods and understanding the mechanisms involved (Taylor *et al.*, 1996; Werner *et al.*, 1996; Shia *et al.*, 2001). The most commonly practiced technique is heat-induced epitope retrieval (HIER; Shia *et al.*, 1991) with its many variants [Heating source (microwave, water bath, autoclave, steam), temperature, time (duration, constant, cyclic) and buffer properties (buffer, ionic composition, pH, metals)].

Although these techniques are commonly used in histopathology and, to a lesser extent, in basis research, there are very few examples of its use in the fish gill literature. In my experience, antigen retrieval has in the vast majority of cases been ineffective in resurrecting non-cross-reactive antibodies. However, I suspect that the fault most likely lies with the non-homologous antibody rather than the antigen. This situation is not the same as in its clinical applications, as the antibodies are already known to be cross-reactive in cryosections. Having said that, there are examples in which antigen retrieval has worked. With the HIER technique V-ATPase A subunit labeling in the guppy was greatly enhanced (J.M. Wilson, unpublished results). In the dogfish (*Squalus acanthias*), the same antibody required prior enzymatic digestion of sections with trypsin to unmask V-ATPase A subunit epitope(s) (Wilson *et al.*, 1997). Enzymatic digestion, unfortunately, is more difficult to standardize and control than HIER (Taylor *et al.*, 1996). SDS (sodium dodecyl sulfate) is a strong anion detergent that will denature proteins. A SDS pretreatment (1% SDS/PBS for 5 min; Brown *et al.*, 1996, Robinson and Vandré, 2001) of sections has been shown to improve α5 immunoreactivity (rat kidney; Sabolic *et al.*, 1999) and it is my general experience with fish as well. SDS pretreatment was also necessary for apical tAE1 labeling in chloride cells in fresh water tilapia (Wilson *et al.*, 2000a) and coho salmon (Wilson *et al.*, 2002b). In both the case of α5 and tAE1, the antigen used to raise the antibodies had been denatured at some point in their purification and SDS denaturation may alter the respective epitopes and improve antibody binding.

Citraconic anhydride has been recently demonstrated to be a very effective antigen retrieval agent at high temperature and physiological pH (Naminatsu *et al.*, 2005). Citraconic anhydride acts by reversing the chemical cross linking by formaldehyde fixation. So far, I have found it to be a useful method for retrieving antigenicity in a number of cases where results were initially negative or weak. In larval lamprey, Na^+/K^+-ATPase activity and protein level expression in Western blots were low but detectable in gill tissue compared to juvenile lamprey. With a standard immunolabeling protocol, I was unable to demonstrate immunoreacitivity in sections but with 0.05% citraconic annhyride treatment of sections for 30 min at 98°C labeling with either the α5 or αRbNKA antibodies was dramatically improved.

In Figures 13.5 and 13.6, I have attempted to illustrate the effects of various antigen retrieval techniques on antigen detection. In sturgeon gill, Na^+/K^+-ATPase (α5 and αRbNKA), V-ATPase (B subunit) and NKCC (T4) staining were weak under standard conditions using an indirect labeling protocol[(i)]. Antigen retrieval with 1% SDS /PBS for 5 min at room

[(i)] Gill tissue was fixed in 3% PFA/PBS overnight, dehydrated through a graded EtOH series, cleared with ClearRite (Richard-Allen) and infiltrated and embedded in paraffin (Type 6, Richard-Allen). Paraffin sections (5 μm) were collected onto APS (3-aminopropyltriethoxysilane; Sigma MO) coated slides, completely air dried, and dewaxed in Clear-Rite (Richard Allen). Sections were circled with a hydrophobic barrier (DakoPen, Dako), and rehydrated with 5% normal goat serum in 0.1% BSA/ TPBS (0.05%Tween-20/ Phosphate Buffered Saline, pH 7.4) for 20 min. Sections were then incubated with primary antibody diluted 1:200 – 1:500 in BSA/TPBS for 1 h at 37°C. Slides were rinsed in TPBS (5, 10, 15 min in Coplin jars), and incubated with goat anti-mouse Alexa Fluoro 488 and goat anti-rabbit Alexa Fluoro 594 conjugated secondary antibodies both diluted 1:200 (Molecular Probes Inc) in BSA/TPBS for 1 h at 37°C. Following a second round of rinses in TPBS, coverslips were mounted using a glycerol based fluorescence mounting media (10% Mowiol, 40% glycerol, 0.1% DABCO, 0.1 M TRIS (pH 8.5). Sections were viewed on a Leitz Ortholux 2 epi-fluorescence microscope and images captured using a digital camera (Leica DFC300).

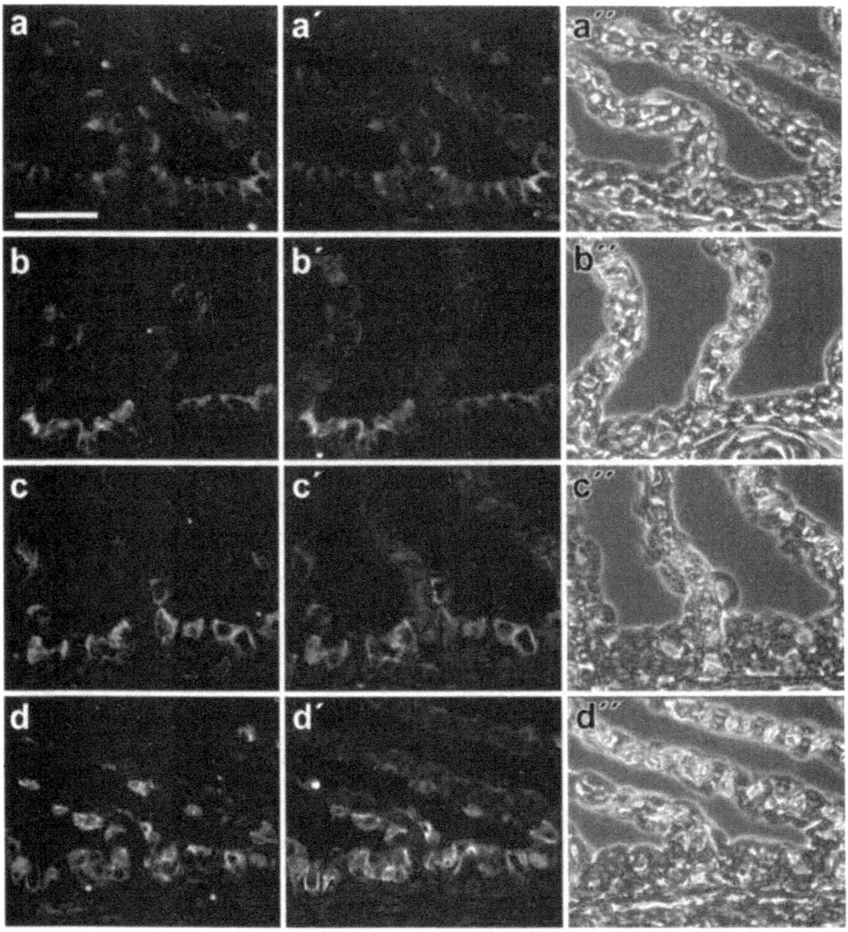

Fig. 13.5 Demonstration of the effects of different antigen retrieval techniques on immunoreactivity of sturgeon gill filament sagittal sections double labeled with antibodies against Na^+/K^+-ATPase (αRbNKA) and NKCC (T4) with phase contrast images for orientation. Section pre-treatments: (**a, a′,a′′**, respectively) Control (no pre-treatment), (**b, b′, b′′**, respectively) 1% SDS/PBS pH 7.3 for 5 min at room temperature, or heat induced epitope retrieval (HIER; (**c, c′, c′′**, respectively) 10 mM Sodium Citrate pH 6 or (**d, d′, d′′**, respectively) 0.05% citraconic anhydride pH 7.4 for 30 min at 98°C (J.M. Wilson, D. Baker and C.J. Brauner, unpublished observations). Scale bar = 50 μm.

temperature improved Na^+/K^+-ATPase labeling with both antibodies although background staining was much higher with the α5 antibody. With the NKCC T4 antibody results were only slightly improved and with the V-ATPase B2/BvA1 antibody only background staining was increased. Heat-induced epitope retrieval (HIER) improved the immunoreactivity of

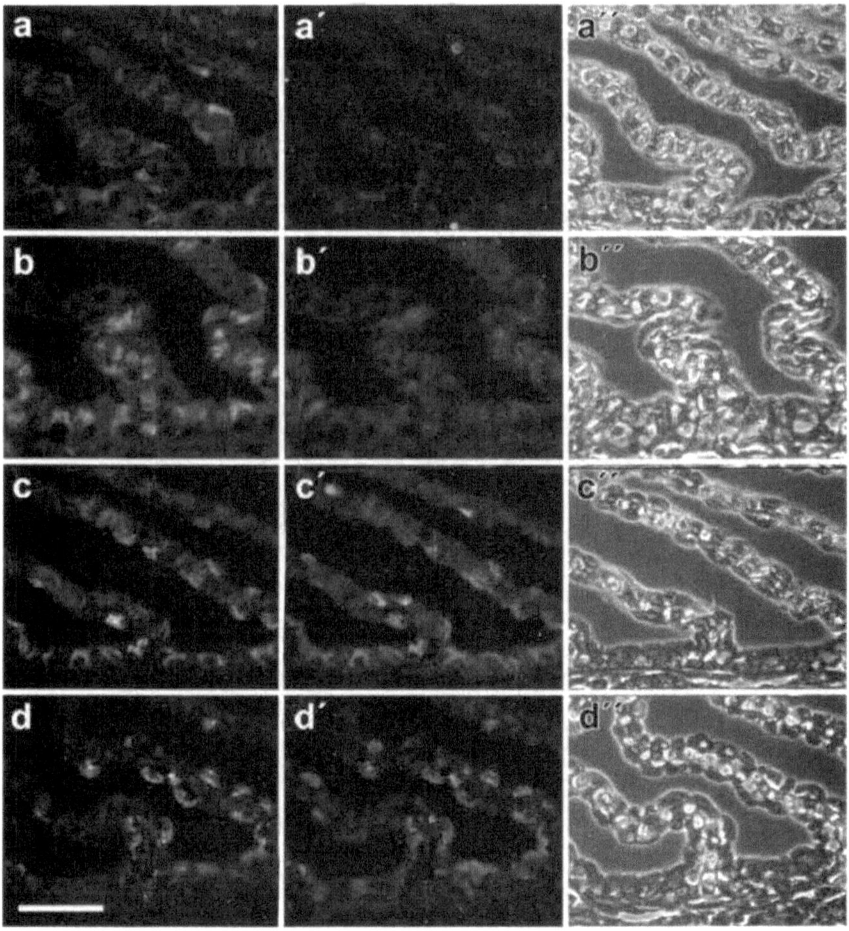

Fig. 13.6 Antigen retrieval of sturgeon gill sections probed with antibodies against Na^+/K^+-ATPase (α5) and V-ATPase B subunit (B2/BvA1, unpublished). Section pre-treatments: Section pre-treatments: (**a, a´,a´´**) Control (no pretreatment), (**b, b´, b´´**) 1% SDS/PBS pH 7.3 for 5 min at room temperature, or heat induced epitope retrieval (HIER; (**c, c´, c´´**) 10 mM Sodium Citrate pH 6 or (**d, d´, d´´**) 0.05% citraconic anhydride pH 7.4 for 30 min at 98°C (J.M. Wilson, D. Baker and C.J. Brauner, unpublished observations). Scale bar = 50 µm.

all antibodies used in this example. Heating the slides in either 10 mM sodium citrate (pH 6) or 0.05% citraconic anhydride at pH 7.4 for 30 min at 98°C using a conventional heat source (hotplate) gave similar results for the most part. Generally, the staining was more consistent throughout the sections. The citraconic anhydride treatment did however give superior results with the T4 antibody.

Thus, despite the lack of attention so far given to antigen retrieval techniques, they stand to provide a second life to some antibodies long thought useless. Since, signal strength is generally increased with these methods it is essential that negative controls are run in parallel in order to properly evaluate the results.

ADDITIONAL RESOURCES

As this review is far from complete, the reader is urged to acquire Lane and Harlow's (2000) book *Using Antibodies* for more substantial explanations and detailed protocols. An excellent online discussion group for asking questions is Histonet and the archives are searchable (Histosearch). For those interested in testing primarily non-homologous antibodies,the following web-based search engines are useful for finding appropriate commercial antibody: www.abcam.com and www.exactantigen.com. The monoclonal antibodies available through the Developmental Studies Hybridoma Bank can be found at www.uiowa.edu/~dshbwww.

Acknowledgements

The published and unpublished data of JMW included in the review were made possible by the Portuguese Foundation for Science and Technology (FCT) grants POCTI/ BSE/ 34164/ 2000, POCTI/ BSE/ 47585/ 2002 and POCTI/ MAR/ 60365/ 2004. JMW was supported by an FCT auxiliary investigator position to CIMAR.

References

Biemesderfer, D., P.A. Rutherford, T. Nagy, J.H. Pizzonia, A.K. Abu-Alfa and P.S. Aronson. 1997. Monoclonal antibodies for high-resolution localization of NHE3 in adult and neonatal rat kidney. *American Journal of Physiology* 273: F289–F299.

Boesch, S.T., B. Eller and B. Pelster. 2003. Expression of two isoforms of the vacuolar-type ATPase subunit B in the zebrafish *Danio rerio*. *Journal of Experimental Biology* 206: 1907–1915.

Boesch, S.T., H. Niederstätter and B. Pelster. 2003. Localization of the vacuolar-type ATPase in the swimbladder gas gland cells of the European eel (*Anguilla anguilla*). *Journal of Experimental Biology* 206: 475.

Bookstein, C., Y. Xie, K. Rabenau, M.W. Musch, R.L. McSwine, M.C. Rao and E.B. Chang. 1997. Tissue distribution of Na^+ /H^+ exchanger isoforms NHE2 and NHE4 in rat intestine and kidney. *American Journal of Physiology* 273: C1496–C1505.

Borke, J.L., A. Caride, A.K. Verma, J.T. Penniston and R. Kumar. 1989. Plasma membrane calcium pump and 28-kDa calcium binding protein in cells of rat kidney distal tubules. *American Journal of Physiology* 257: F842–F849.

Brown, D., J. Lydon, M. McLaughlin, A. Stuart-Tilley, R. Tyszkowski and S.L. Alper. 1996. Antigen retrieval in cryostat tissue sections and cultured cells treatment with sodium dodecyl sulfate (SDS). *Histochemistry and Cell Biology* 105: 261–267.

Cameron, B.A., S.F. Perry, C. Wu, K. Ko and B.L. Tufts. 1996. Bicarbonate permeability and immunological evidence for an anion exchanger-like protein in the red blood cells of the sea lamprey, *Petromyzon marinus*. *Journal of Comparative Physiology* 166: 197–204.

Choe, K.P., S. Edwards, A.I. Morrison-Shetlar, T. Toop and J.B. Claiborne. 1999. Immunolocalization of Na^+/K^+-ATPase in mitochondria-rich cells of the Atlantic hagfish (*Myxine glutinosa*) gill. *Comparative Biochemistry and Physiology* A124: 161–168.

Choe, K.P., A.I. Morrison-Shetlar, B.P. Wall and J.B. Claiborne. 2002. Immunological detection of Na^+/H^+ exchangers in the gills of a hagfish, *Myxine glutinosa*, and elasmobranch, *Raja erinacea*, and a teleost, *Fundulus heteroclitus*. *Comparative Biochemistry and Physiology* A131: 375–385.

Choe, K.P., J.W. Verlander, C.S. Wingo and D.H. Evans. 2004. A putative H^+-K^+-ATPase in the Atlantic stringray, *Dasyatis sabina*: Primary sequence and expression in gills. *American Journal of Physiology* 287: R981–R991.

Choe, K.P., S. O'Brian, D.H. Evans, T. Toop and S.L. Edwards. 2004. Immunolocalization of Na^+/K^+-ATPase, carbonic anhydrase II and vacuolar H^+-ATPase in the gills of freshwater adult lampreys, *Geotria australis*. *Journal of Experimental Zoology* A301: 654–665.

Choe, K.P., A. Kato, S. Hirose, C. Plata, A. Sindié, M.F. Romero, J.B. Claiborne and D.H. Evans. 2005. NHE3 in an ancestral vertebrate: Primary sequence, distribution, localization, and function in gills. *American Journal of Physiology* 289: R1520–R1534.

Claiborne, J.B., C.R. Blackston, K.P. Choe, D.C. Dawson, S.P. Harris, L.A. MacKenzie and A.I. Morrison-Shetlar. 1999. A mechanism for branchial acid excretion in marine fish: Identification of multiple Na^+/H^+ antiporter (NHE) isoforms in gills of two seawater teleosts. *Journal of Experimental Biology* 202: 315–324.

Claiborne, J.B., S.L. Edwards and A.I. Morrison-Shetlar. 2002. Acid-base regulation in fishes: Cellular and molecular mechanisms. *Journal of Experimental Zoology* 293: 302–319.

Coons, A.H., H.J. Creech and R.N. Jones. 1941. Immunological properties of an antibody containing a fluorescent group. *Proceedings of the Society for Experimental Biology and Medicine* 47: 200–202.

Curran, R.C. and J. Gregory. 1977. The unmasking of antigens in paraffin sections of tissue by trypsin. *Experientia* 33: 1400–1401.

Cutler, C.P. and G. Cramb. 2001. Molecular physiology of osmoregulation in eels and other teleosts: The role of transporter isoforms and gene duplication. *Comparative Biochemistry and Physiology* A130: 551–564.

Cutler, C.P., S. Brezillion, S. Bekir, I.L. Sanders, N. Hazon and G. Cramb. 2000. Expression of a duplicate Na,K-ATPase β_1-isoform in the European eel (*Anguilla anguilla*). *American Journal of Physiology* 279: R222–R229.

Dang, Z.-C., P.H.M. Balm, G. Flik, S.E. Wendelaar Bonga and R.A. Lock. 2000. Cortisol increases Na^+/K^+-ATPase density in the plasma membranes of gill chloride cells in the freshwater tilapia *Oreochromis mossambicus*. *Journal of Experimental Biology* 203: 2349–2355.

Edwards, S.L., B.P. Wall, A.I. Morrison-Shetlar, S. Sligh, J. Weakly and J.B. Claiborne. 2005. The effect of environmental hypercapnia and salinity on the expression of NHE-like isoforms in the gills of a euryhaline fish (*Fundulus heteroclitus*). *Journal of Experimental Zoology* A303: 464–475.

Edwards, S.L., C.M. Tse and T. Toop. 1999. Immunolocalization of NHE3-like immunoreactivity in the gills of the rainbow trout (*Oncorhynchus mykiss*) and the blue-throated wrasse (*Pseudolabrus tetrious*). *Journal of Anatomy* 195: 465–469.

Evans, D.H., P.M. Piermarini and K.P. Choe. 2005. The multifunctional fish gill: Dominant site of gas exchange, osmoregulation, acid-base regulation and excretion of nitrogenous waste. *Physiological Reviews* 85: 97–177.

Ezaki, T. 2000. Antigen retrieval on formaldehyde-fixed paraffin sections: its potential drawbacks and optimization for double immunostaining. *Micron* 31: 639–649.

Feizi, T. 1989. Oligosaccharides as recognition structures. In: *Mucus and Related Topics: Symposium Proceedings (Society for Experimental Biology)* Eric N. Chantler and N.A. Ratcliffe (eds.), Company of Biologists, Cambridge, No. 43, pp. 139–147.

Filippova, M., L.S. Ross and S.S. Gill. 1998. Cloning of the V-ATPase B subunit cDNA from *Culex quinquefasciatus* and expression of the B and C subunits in mosquitoes. *Insect Molecular Biology* 7: 223–232.

Flemmer, A.W., R. Behnke and B. Forbush III. 1999. Changes in phosphorylation state of the Na-K-Cl cotransporter (NKCC) in chloride cells of the gill of *Fundulus heteroclitus* (killifish) during salt adaptation. *Bulletin of the Mount Desert Island Biological Laboratory* 38: 80–82.

Galvez, F., S.D. Reid, G. Hawkings and G.G. Goss. 2002. Isolation and characterization of mitochondria-rich cell types from the gill of freshwater rainbow trout. *American Journal of Physiology* 282: R658–R668.

Georgalis, T., S.F. Perry and K. Gilmour. 2006. The role of branchial carbonic anhydrase in acid-base regulation in rainbow trout (*Oncorhynchus mykiss*). *Journal of experimental Biology* 209: 518–530.

Gown, A.M. 2004. Unmasking the mysteries of antigen or epitope retrieval and formaline fixation. *American Journal of Clinical Pathology* 121: 172–174.

Harlow, E. and D. Lane. 1999. *Using Antibodies: A Laboratory Manual*, Cold Spring Harbor Laboratory Press, Cold Spring Harbor, New York, pp. 495.

Hemken, P., X.L. Guo, Z.Q. Wang, K. Zhang and S. Gluck. 1992. Immunological evidence that vacuolar H^+ ATPases with heterogeneous forms of Mr = 31,000 subunit have different membrane distributions in mammalian kidney. *Journal of Biological Chemistry* 267: 9948–9957.

Hirata, T., T. Kaneko, T. Ono, T. Nakazato, N. Furukawa, S. Hasegawa, S. Wakabayashi, M. Shigekawa, M.-H. Chang, M.F. Romero and S. Hirose. 2003. Mechanism of acid adaptation of a fish living in pH 3.5 lake. *American Journal of Physiology* 284: R1199–R1212.

Hiroi, J., S.D. McCormick, R. Ohtani-Kaneko and T. Kaneko. 2005. Functional classification of mitochondria-rich cells in euryhaline Mozambique tilapia (*Oreochromis mossambicus*) embryos, by means of triple immunofluorescence staining for Na^+/K^+-ATPase, $Na^+/K^+/Cl^-$ cotransporter and CFTR anion channel. *Journal of Experimental Biology* 208: 2023–2036.

Hirose, S., T. Kaneko, N. Naito and Y. Takei. 2003. Molecular biology of major components of chloride cells. *Comparative Biochemistry and Physiology* B136: 593–620.

Hoogerwerf, W.A., S.C. Tsao, O. Devuyst, S.A. Levine, C.H. Yun, J.W. Yip, M.E. Cohen, P.D. Wilson, A.J. Lazenby, C.M. Tse and M. Donowitz. 1996. NHE2 and NHE3 are human and rabbit intestinal brush-border proteins. *American Journal of Physiology* 270: G29–G41.

Hutson, J.C., G.V. Childs and P.J. Gardner. 1979. Considerations for establishing the validity of immunocytological studies. *Journal of Histochemistry and Cytochemistry* 27: 1201–1202.

Hwang, P.-P., M.J. Fang, J.-C. Tsai, C.J. Huang and S.T. Chen. 1998. Expression of mRNA and protein of Na^+-K^+-ATPase α subunit in gills of tilapia (*Oreochromis mossambicus*). *Fish Physiology and Biochemistry* 18: 363–373.

Ismailov, I.I., B.K. Berdiev, A.L. Bradford, M.S. Awayda, C.M. Fuller and D.J. Benos. 1996. Associated proteins and renal epithelial Na^+ channel function. *Journal of Membrane Biology* 149: 123–132.

Katoh, F., A. Shimizu, K. Uchida and T. Kaneko. 2000. Shift of chloride cell distribution during early life stages in seawater-adapted killifish, *Fundulus heteroclitus*. *Zoological Science* 17: 11–18.

Katoh, F., S. Hyodo and T. Kaneko. 2003. Vacuolar-type proton pump in the basolateral plasma membrane energizes ion uptake in branchial mitochondria-rich cells of killifish *Fundulus heteroclitus*, adapted to a low ion environment. *Journal of Experimental Biology* 206: 793–803.

Kim, S.H., M.C. Kook, Y.K. Shin, S.H. Park and H.G. Song. 2004. Evaluation of antigen retrieval buffer systems. *Journal of Molecular Histology* 35: 409–416.

Kirschner, L.B. 2004. The mechanism of sodium chloride uptake in hyper-regulating aquatic animals. *Journal of Experimental Biology* 207: 1439–1452.

Lee, T.H., J.C. Tsai, M.J. Fang, M.J. Yu and P.P. Hwang. 1998. Isoform expression of Na^+-K^+-ATPase α-subunit in gills of the teleost *Oreochromis mossambicus*. *American Journal of Physiology* 275: R926–R932.

Li, J., J. Eygensteyn, R.A. Lock, P.M. Verbost, A.J.H. van der Heijden, S.E. Wendelaar Bonga and G. Flik. 1995. Branchial chloride cells in larvae and juveniles of freshwater tilapia *Oreochromis mossambicus*. *Journal of Experimental Biology* 198: 2177–2184.

Lignot, J.H., C.P. Cutler, N. Hazon and G. Cramb. 2002. Immunolocalization of aquaporin 3 in the gill and gastrointestinal tract of the European eel (*Anguilla anguilla* L.). *Journal of Experimental Biology* 205: 2653–2663.

Lin, H., D.C. Pfeiffer, A.W. Vogl, J. Pan and D.J. Randall. 1994. Immunolocalization of proton-ATPase in the gill epithelia of rainbow trout. *Journal of Experimental Biology* 195: 169–183.

Maetz, J. and F. Garcia-Romeu. 1964. The mechanism of sodium and chloride uptake by the gills of a fresh-water fish, *Carassius auratus*. II Evidence for NH_4^+/Na^+ and HCO_3^-/Cl^- exchanges. *Journal of General Physiology* 47: 1209–1227.

Marshall, W.S. 2002. Na^+,Cl^-,Ca^{2+} and Zn^{2+} transport by fish gill: retrospective review and prospective synthesis. *Journal of Experimental Zoology* 293: 264–283.

Marshall, W.S., E.M. Lynch and R.R.F. Cozzi. 2002. Redistribution of immunofluorescence of CFTR anion channel and NKCC cotransporter in chloride cells during adaptation of killifish *Fundulus heteroclitus* to sea water. *Journal of Experimental Biology* 205: 1265–1273.

McCormick, S.D. 1993. Methods for non-lethal gill biopsy and measurement of Na^+,K^+-ATPase activity. *Canadian Journal of Fisheries and Aquatic Sciences* 50: 656–658.

McCormick, S.D., K. Sundell, B.T. Björnsson, C.L. Brown and J. Hiroi. 2003. Influence of salinity on the localization of Na^+/K^+-ATPase, $Na^+/K^+/2Cl^-$ cotransporter (NKCC) and CFTR anion channel in chloride cells of the Hawaiian goby (*Stenogobius hawaiiensis*). *Journal of Experimental Biology* 206: 4575–4583.

McNicol, A.M. and J.A. Richmond. 1998. Optimizing immunohistochemistry: antigen retrieval and signal amplification. *Histopathology* 32: 97–103.

Mickle, J., K.J. Karnaky, T. Jensen, D.S. Miller, S. Terlouw, A. Gross, B. Corrigan, J.R. Riordan and G.R. Cutting. 2000. Processing and localization of the cystic fibrosis transmembrane regulator in gill and operculum from *Fundulus heteroclitus*. *Bulletin of the Mount Desert Island Biological Laboratory* 39: 75–77.

Miller, R.T. 2004. Endogenous biotin artifact. www.propathlab.com

Mistry, A.C., S. Honda, T. Hirata, A. Kato and S. Hirose. 2001. Eel urea transpoter is localized to chloride cells and is salinity dependent. *American Journal of Physiology* 281: R1594–R1604.

Montero, C. 2003. The antigen-antibody reaction in immunohistochemistry. *Journal of Histochemistry and Cytochemistry* 51: 1–4.

Namimatsu, S., M. Ghazizadeh and Y. Sugisaki. 2005. Reversing the effects of formalin fixation with citraconic anhydride and heat: A universal antigen retrieval method. *Journal of Histochemistry and Cytochemistry* 53: 3–11.

Nigg, E.A., G. Walter and S.J. Singer. 1982. On the nature of crossreactions observed with antibodies directed to defined epitopes. *Proceedings of the National Academy of Science of the United States of America* 79: 5939–5943.

Pan, J., D.C. Pfeiffer, C.L. Reimer, B.J. Crawford, A.W. Vogl and N. Auersperg. 1993. Lack of inter-species reactivity between antigens and antibodies is overcome by protease treatment of Western blots. *Electrophoresis* 14: 892–898.

Pelis, R.M., J. Zydlewski and S.D. McCormick. 2001. Gill Na^+-K^+-$2Cl^-$ cotransporter abundance and localization in Atlantic salmon: Effects of seawater and smolting. *American Journal of Physiology* 280: R1844–R1852.

Perry, S.F., A. Shahsavarani, T. Georgalis, M. Bayaa, M. Furimsky and S.L.Y. Thomas. 2003. Channels, pumps, and exchangers in the gill and kidney of freshwater fishes: Their role in ionic and acid-base regulation. *Journal of Experimental Zoology* A300: 53–62.

Petrusz, P., M. Sar, P. Ordroneau and P. DiMeo. 1976. Specificity in immunocytochemical staining. *Journal of Histochemistry and Cytochemistry* 24: 1110–1112.

Piermarini, P.M. and D.H. Evans. 2000. Effects of environmental salinity of Na^+/K^+-ATPase in the gills and rectal gland of a euryhaline elasmobranch (*Dasyatis sabina*). *Journal of Experimental Biology* 203: 2957–2966.

Piermarini, P.M. and D.H. Evans. 2001. Immunochemical analysis of the vacuolar proton-ATPase B-subunit in the gills of a eurhyaline stingray (*Dasyatis sabina*): Effects of salinity and relation to Na^+/K^+-ATPase. *Journal of Experimental Biology* 204: 3251–3259.

Piermarini, P.M., J.W. Verlander, I.E. Royaux and D.H. Evans. 2002. Pendrin immunoreactivity in the gill epithelium of a euryhaline elasmobranch. *American Journal of Physiology* 283: R983–R992.

Preest, M.R., R.J. Gonzalez and R.W. Wilson. 2005. A pharmacological examination of Na^+ and Cl^- transport in two species of freshwater fish. *Physiological and Biochemical Zoology* 78: 259–272.

Rahim, S.M., J.P. Delaunoy and P. Laurent. 1988. Identification and immunocytochemical localization of two different carbonic anhydrase isoenzymes in teleostean fish erythrocytes and gill epithelia. *Histochemistry* 89: 451–459.

Richards, J.G., J.W. Semple, J.S. Bystriansky and P.M. Schulte. 2003. Na^+/K^+-ATPase α-isoform switching in gills of rainbow trout (*Oncorhynchus mykiss*) during salinity transfer. *Journal of Experimental Biology* 260: 4475-4486.

Robinson, J.M. and D.D. Vandré. 2001. Antigen retrieval in cells and tissues: Enhancement with sodium dodecyl sulfate. *Histochemistry and Cell Biology* 116: 119–130.

Rodriquez-Soto, J., R. Warnke and R.V. Rouse. 1997. Endogenous avidin-binding activity in paraffin-embedded tissue revealed after microwave treatment. *Applied Immunohistochemistry* 5: 59–62.

Rombough, P.J. 1999. The gill of fish larvae. Is it primarily a respiratory or an ionoregulatory structure? *Journal of Fish Biology* 55: 186–204.

Royaux, I.E., K. Suzuki, A. Mori, R. Katoh, L.A. Everett, L.D. Kohn and E.D. Green. 2000. Pendrin, the protein encoded by the Pendred Syndrome gene (*PDS*), is an apical porter of iodide in the thyroid and is regulated by thryoglobulin in FRTL-5 cells. *Endocrinology* 141: 839–845.

Sabolic, I., C.M. Herak-Kramberger, S. Breton and D. Brown. 1999. Na/K-ATPase in intercalated cells along the rat nephron revealed by antigen retrieval. *Journal of the American Society of Nephrology* 10: 913–922.

Scott, G.R., J.B. Claiborne, S.L. Edwards, P.M. Schulte and C.M. Wood. 2005. Gene expression after freshwater transfer in gills and opercular epithelia of killifish: Insight into divergent mechanisms of ion transport. *Journal of Experimental Biology* 208: 2719–2729.

Seidelin, M. and S.S. Madsen. 1999. Endocrine control of Na^+, K^+-ATPase and chloride cell development in brown trout (*Salmo trutta*): Interaction of insulin-like growth factor-I with prolactin and growth hormone. *Journal of Endocrinology* 162: 127–135.

Shahsavarani, A., B. McNeill, F. Galvez, C.M. Wood, G.G. Goss, P.-P. Hwang and S.F. Perry. 2006. Characterization of a branchial epithelial calcium channel (ECaC) in freshwater rainbow trout (*Oncorhynchus mykiss*). *Journal of experimental Biology* 209: 1928–1943.

Shi, S.-R., M.E. Key and K.L. Kalra. 1991. Antigen retrieval in formalin-fixed, paraffin embedded tissues: An enhancement method for immunohistochemical staining based on microwave oven heating of tissue sections. *Journal of Histochemistry and Cytochemistry* 39: 741–748.

Shi, S.-R., R.J. Cote and C.R. Taylor. 1997. Antigen retrieval immunohistochemistry: Past, present, and future. *Journal of Histochemistry and Cytochemistry* 45: 327–343.

Shi, S.-R., R.J. Cote and C.R. Taylor. 2001. Antigen retrieval techniques: Current perspectives. *Journal of Histochemistry and Cytochemistry* 49: 931–937.

Soleimani, M., G. Singh, G.L. Bizal, S.R. Gullans and J.A. McAteer. 1994. Na^+/H^+ exchanger isoforms NHE-2 and NHE-1 in inner medullary collecting duct cells. *Journal of Biological Chemistry* 269: 27973–27978.

Sompuram, S.R., K. Vani, E. Messana and S.A. Bogen. 2004. A molecular mechanism of formalin fixation and antigen retrieval. *American Journal of Clinical Pathology* 121: 190–199.

Sorscher, E.J., M.A. Accavitti, D. Keeton, E. Steadman, R.A. Frizzell and D.J. Benos. 1988. Antibodies against purified epithelial sodium channel protein from bovine renal papilla. *American Journal of Physiology* 255: C835–C843.

Sturla, M., M.A. Masini, P. Prato, C. Grattarola and B. Uva. 2001. Mitochondria-rich cells in gills and skin of an African lungfish, *Protopterus annectens*. *Cell and Tissue Research* 303: 351–358.

Sullivan, G.V., J.N. Fryer and S.F. Perry. 1995. Immunolocalization of proton pumps (H^+-ATPase) in pavement cells of rainbow trout gill. *Journal of Experimental Biology* 198: 2619–2629.

Sullivan, G.V., J.N. Fryer and S.F. Perry. 1996. Localization of mRNA for the proton pump (H^+-ATPase) and Cl^-/HCO_3^- exchanger in the rainbow trout gill. *Canadian Journal of Zoology* 74: 2095–2103.

Suzuki, Y., M. Itakura, M. Kashwagi, N. Nakamura, T. Matsuki, H. Sakuta, N. Naito, K. Takano, T. Fujita and S. Hirose. 1999. Identification by differential display of a hypertonicity-inducible inward rectifier potassium channel highly expressed in chloride cells. *Journal of Biological Chemistry* 274: 11376–11382.

Sweadner, K.J. and R.C. Gilkeson. 1985. Two isozymes of the Na,K-ATPase have distinct antigenic determinants. *Journal of Biological Chemistry* 260: 9016–9022.

Takeyasu, K., M.M. Tamkun, K.J. Renaud and D.M. Fambrough. 1988. Ouabain-sensitive ($Na^+ + K^+$)-ATPase activity expressed in mouse L cells by transfection with DNA encoding the α-subunit of an avian sodium pump. *Journal of Biological Chemistry* 263: 4347–4354.

Taylor, C.R. and J. Burns. 1974. The demostration of plasma cells and other immunoglobulin containing cells in formalin-fixed, paraffin-embedded tissues using peroxidase labelled antibody. *Journal of Clinical Pathology* 27: 14–20.

Taylor, C.R., S.-R. Shi and R.J. Cote. 1996. Antigen retrieval for immunohistochemistry. Status and need for greater standardization. *Applied Immunohistochemistry* 4: 144–166.

Thors, C. and E. Linder. 2003. Localization and identification of *Schistosoma mansoni*/ KLH-crossreactive component in infected mice. *Journal of Histochemistry and Cytochemistry* 51: 1367–1373.

Tresguerres, M., S.K. Parks, F. Katoh and G.G. Goss. 2006. Microtubule-dependent relocation of branchial V-H^+-ATPase to the basolateral membrane in the Pacific spiny dogfish (*Squalus acanthias*): A role in base secretion. *Journal of Experimental Biology* 209: 599–609.

Tse, C.M., S.A. Levine, C.H. Yun, S. Khurana and M. Donowitz. 1994. Na^+/H^+ exchanger-2 is an O-linked but not an N-linked sialoglycoprotein. *Biochemistry* 33: 12954–12961.

Uchida, K., T. Kaneko, K. Yamauchi and T. Hirano. 1996. Morphometrical analysis of chloride cell activity in the gill filaments and lamellae and changes in Na^+,K^+-ATPase activity during seawater adaptation in chum salmon fry. *Journal of Experimental Zoology* 276: 193–200.

Uchida, K., T. Kaneko, H. Miyazaki, S. Hasegawa and T. Hirano. 2000. Excellent salinity tolerance of Mozambique tilapia (*Oreochromis mossambicus*): Elevated chloride cell activity in the branchial and opercular epithelia of the fish adapted to concentrated seawater. *Zoological Science* 17: 149–160.

Uchida, K., S. Hasegawa and T. Kaneko. 2002. Effects of a low-Ca^{2+} environment on branchial chloride cell morphology in chum salmon fry and immunolocalization of Ca^{2+}-ATPase in chloride cells. *Canadian Journal of Zoology* 80: 1100–1108.

Ura, K., K. Soyano, N. Omoto, S. Adachi and K. Yamauchi. 1996. Localization of Na-K-ATPase in tissues of rabbit and teleosts using an antiserum directed against a partial sequence of the α-subunit. *Zoological Science* 13: 219–227.

Watanabe, S., T. Kaneko and K. Aida. 2005. Aquaporin-3 expressed in the basolateral membrane of gill chloride cells in Mozambique tilapia *Oreochromis mossambicus* adapted to freshwater and seawater. *Journal of Experimental Biology* 208: 2673–2682.

Werner, M., R. von Wasielewski and P. Komminoth. 1996. Antigen retrieval, signal amplification and intensification in immunohistochemistry. *Histochemistry and Cell Biology* 105: 253–260.

Wilson, J.M. and P. Laurent. 2002. Fish gill morphology: Inside out. *Journal of Experimental Zoology* 293: 192–213.

Wilson, J.M., D.J. Randall, A.W. Vogl and G.K. Iwama. 1997. Immunolocalization of proton-ATPase in the gills of the elasmobranch, *Squalus acanthias*. *Journal of Experimental Zoology* 278: 78–86.

Wilson, J.M., D.J. Randall, M. Donowitz, A.W. Vogl and Y.K. Ip. 2000a. Immunolocalization of ion-transport proteins to branchial epithelium mitochondria-rich cells in the mudskipper (*Periophthalmodon schlosseri*). *Journal of Experimental Biology* 203: 2297–2310.

Wilson, J.M., J.D. Morgan, A.W. Vogl and D.J. Randall. 2002a. Branchial mitochondria-rich cells in the dogfish *Squalus acanthias*. *Comparative Biochemistry and Physiology* A132: 365–374.

Wilson, J.M., N.M. Whiteley and D.J. Randall. 2002b. Ionoregulatory changes in the gill epithelia of coho salmon during seawater acclimation. *Physiological and Biochemical Zoology* 75: 237–249.

Wilson, J.M., P. Laurent, B.L. Tufts, D.J. Benos, M. Donowitz, A.W. Vogl and D.J. Randall. 2000b. NaCl uptake by the branchial epithelium in freshwater teleost fish: An immunological approach to ion-transport protein localization. *Journal of Experimental Biology* 203: 2279–2296.

Wilson, J.M., A. Leitão, A. Gonçalves, C. Ferreira, P.N. Reis-Santos, A.-V. Fonseca, J.C. Antunes, C.M. Pereira and J. Coimbra. 2007. Modulation of branchial ion transport protein expression by salinity in the glass eels (*Anguilla anguilla*). *Marine Biology* DOI 10.1007/s00227-006-0579-7.

Wilson, J.M., J.C. Antunes, P.D. Bouça and J. Coimbra. 2004. Osmoregulatory plasticity of the glass eel of *Anguilla anguilla*: Freshwater entry and changes in branchial ion-transport protein expression. *Canadian Journal of Fisheries and Aquatic Sciences* 61: 432–442.

Witters, H., P. Berckmans and C. Vangenechten. 1996. Immunolocalization of Na^+,K^+-ATPase in the gill epithelium of rainbow trout, *Oncorhynchus mykiss*. *Cell and Tissue Research* 283: 461–468.

Wood, C.M., S.P. Kelly, B.S. Zhou, M. Fletcher, M.J. O'Donnell, B. Eletti and P. Pärt. 2002. Cultured gill epithelia as models for the freshwater fish gill. *Biochimica et Biophysica Acta* 1566: 72–83.

Wu, Y.C., L.Y. Lin and T.H. Lee. 2003. Na^+, K^+, $2Cl^-$ cotransporter: a novel marker for identifying freshwater- and seawater-type mitochondria rich cells in gills of the euryhaline tilapia, *Oreochromis mossambicus*. *Zoological Studies* 42: 186–192.

Yamashita, S. and Y. Okada. 2005. Mechanisms of heat-induced antigen retrieval: Analyses *in vitro* employing SDS-PAGE and immunohistochemistry. *Journal of Histochemistry and Cytochemistry* 53: 13–21.

CHAPTER

14

Rapid Regulation of Ion Transport in Mitochondrion-rich Cells

William S. Marshall

INTRODUCTION

Mitochondrion-rich Chloride Secreting Cells

Mitochondrion-rich (MR) cells are present in large numbers in the gill and opercular epithelia of estuarine teleosts. The cell density is augmented in hypersaline conditions and moderates slightly in brackish water, but in the wild these epithelia are capable of rapid chloride secretion. Typically, the cells appear with a smaller accessory cell entwined with the larger cell and between the accessory cells and MR cells is a leaky paracellular pathway (Sardet *et al.*, 1979) that allows transepithelial transport of sodium down its electrochemical gradient (Degnan and Zadunaisky, 1980). The opercular membrane of killifish (*Fundulus heteroclitus*) (Degnan *et al.*, 1977; Karnaky and Kinter, 1977; Degnan and Zadunaisky, 1980), of tilapia

Author's address: Department of Biology, St. Francis Xavier University, P.O. Box 5000, Antigonish, Nova Scotia, Canada B2G 2W5.
E-mail: bmarshal@stfx.ca

(*Oreochromis mossambicus*) and the skin of the euryhaline goby (*Gillichthys mirabilis*) (Marshall, 1977; Marshall and Bern, 1980) provide convenient flat preparations of gill-like epithelia that actively secrete Cl^- at high rates. The marine MR cell model (Marshall and Bryson, 1998; Evans *et al.*, 1999, 2005; Marshall, 2002; Marshall *et al.*, 2002; Fig. 14.1) includes a basolateral Na^+,K^+,$2Cl^-$ symport (NKCC) or cotransporter of the NKCC1 type that transports Cl^- into the cytosol, driven by the transmembrane Na^+ gradient. The Na^+ gradient is maintained by Na^+,K^+-ATPase that is located on the tubular system that is made up of infoldings of the

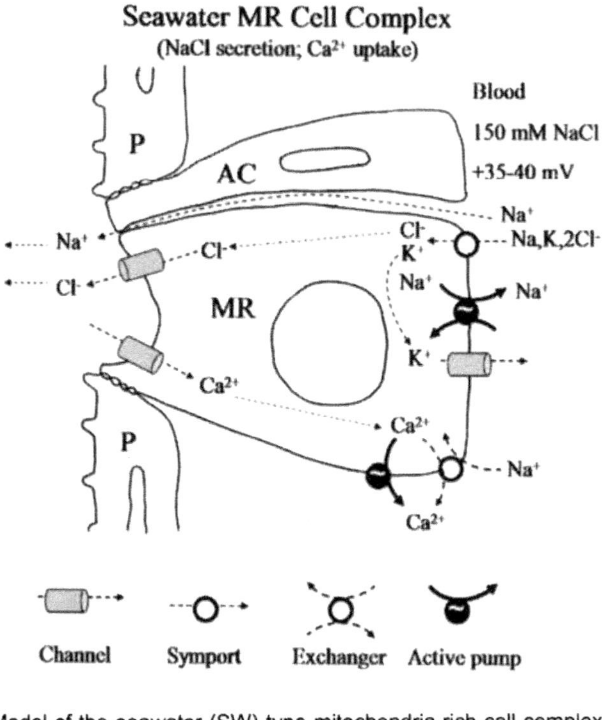

Fig. 14.1 Model of the seawater (SW) type mitochondria-rich cell complex with Na^+,K^+-ATPase, NKCC1 and K^+ channels on the basolateral membrane that accumulate Cl^- in the cytosol driven indirectly by the pump. Solid arrows and closed symbols indicate active transport; dashed lines and open symbols, passive processes. Cl^- exits through CFTR like anion channels in the apical membrane that has a typical cup shaped 'apical crypt'. Intercellular junctions between mitochondria rich (MR) cells and pavement cells (P) are many stranded and have low permeability, while junctions between accessory cells (AC) and MR cells are single stranded and provide a high conductance paracellular pathway that is cation (Na^+) selective. Also shown is the proposed uptake pathway for transepithelial Ca^{2+} uptake, via Ca^{2+} channels at the apical membrane and via Na/Ca^{2+} and Ca^{2+}-ATPase at the basolateral membrane. Acid/base and other solute transport pathways are not shown.

basolateral membrane. Cl^- secretion can be blocked by basolateral furosemide and bumetanide, loop diuretics that block NKCC type cotransporters, but not by thiazide type diuretics that block Na^+-Cl^- symports (Marshall and Bryson, 1998). Intracellularly, the Cl^- diffuses to the apical membrane where anion channels conduct Cl^- across the epithelium and into the environment. The anion channel in the apical membrane is a homolog of Cystic Fibrosis Transmembrane Conductance Regulator (CFTR), having pharmacological and single channel properties similar to h<FTR (Marshall *et al.*, 1995) and cloned CFTR from killifish, when expressed in amphibian oocytes, imparts a cAMP stimulated anion conductance (Singer *et al.*, 1998). Cl^- exit across the apical membrane generates a blood side-positive electrical potential that is sufficient to drive Na^+ out via a cation-selective paracellular pathway. Epithelial current and ion conductance are localized to MR cells, measured using the vibrating microprobe (Foskett and Scheffey, 1982). The anion channel has a low conductance, 8 pS, and deactivates on excision of an inside out membrane patch. Addition of Protein Kinase A (PKA) and ATP activates the channels in quiescent patches, demonstrating that phosphorylation activates the channel. The Cl^- transepithelial current is sensitive to the anion channel inhibitors diphenyl amine-2-carboxylate (DPC) and 5-nitro-2-(3-phenylpropylamino) benzoic acid (NPPB) but insensitive to 4,4'-diisothiocyanostilbene-2,2'-disulfonic acid (DIDS; Marshall *et al.*, 1995; Marshall and Bryson, 1998). The channel has been cloned, sequenced and found to be homologous to hCFTR (Singer *et al.*, 1998). Further, the expressed gene in *Xenopus* oocytes also produces a cAMP activated anion current (Singer *et al.*, 1998).

Since NKCC1 also carries K^+ into the cell, K^+ must be recycled across the basolateral membrane and transepithelial Cl^- secretion thus is blocked by K^+-free solutions and by barium. The K^+ channel in the basolateral membrane of gill epithelial cells has tentatively been identified (Duranton *et al.*, 2000a, b) and found to be a large conductance (122 pS) channel activated by hypotonicity and by ionomycin, while being inhibited by the addition of quinidine and charybdotoxin. There appear to be separate basolateral and apical membrane K^+ channel populations (Duranton *et al.*, 2000a).

CONTROL OF Cl^- SECRETION

Overview

The hormonal control of osmoregulation in fish has been recently reviewed. McCormick (2001) and Evans (2002) specifically studied cell signalling in fish gills. Many factors are known to affect chloride secretion in teleost fish pharmacologically (Table 14.1), including hormones and neurotransmitters that inhibit secretion: catecholamines, via α-2 adrenoceptors, acetylcholine via muscarinic receptors, somatostatin and urotensin II. The second messenger pathway for adrenergic and cholinergic receptors appears to be calcium. From mammalian research on urotensin II action, there is also an association with intracellular calcium (Conlon *et al.*, 1997). There are also many hormones, neurotransmitters and drugs that stimulate chloride secretion: vasoactive intestinal polypeptide (VIP), glucagon, arginine vasotocin (AVT) and urotensin I (Table 14.1) and in these cases the common pathway appears to be an augmentation of intracellular cAMP, activation of PKA. For instance, the β-adrenergic agonist isoproterenol at 1.0 μM causes rapid accumulation of cAMP in killifish opercular epithelia.

Beyond the second messengers, the termination of the phosphorylation and dephosphorylation cascades are of interest, as cotransporters and channels are now known to have tightly associated kinases and phosphatases including complexes involving scaffolding proteins. In spite of the rapid action, there exists a complex series of steps that follow the second messenger changes and provide the unique regulation of these ion transport systems. Elucidation of the regulatory pathways terminating at channels and transporters will be important to understanding ion transport regulation in health and disease.

Catecholamines

α-adrenoceptors

The most likely physiologically relevant response is the inhibition of Cl^- secretion by epinephrine and norepinephrine. The response is rapid, dose-dependent and occurs at catecholamine levels that are physiologically realistic, with ED_{50} of approximately 50 nM for epinephrine and 500 nM for norepinephrine. The effect is blocked by yohimbine, indicating an α-receptor type (May *et al.*, 1984; Marshall *et al.*, 1993; Marshall and

Table 14.1 Summary of potential rapid secretagogues and inhibitory agents affecting Cl^- secretion by marine teleost mitochondrion-rich (MR) cells.

Agent added	*Effect on Cl^- secretion*	*2nd Messenger*	*References*
Hormones			
Arginine vasotocin	↑	cAMP	Avella *et al.* (1999) Marshall (2002)
Glucagon	↑	cAMP	Foskett *et al.* (1982)
Urotensin I	↑	cAMP	Marshall and Bern (1980)
Atrial natriuretic peptide	0?	0	Scheide and Zadunaisky (1988) Evans (2002)
Epinephrine	↓	Ca^{2+}	Degnan *et al.* (1977) Marshall and Bern (1980) Foskett *et al.* (1982) Avella *et al.* (1999)
Urotensin II	↓	Ca^{2+}	Marshall and Bern (1979, 1980)
Neurotransmitters and Parahormones			
Vasoactive intestinal peptide	↑	cAMP	Foskett *et al.* (1982)
Serotonin	0	0	Marshall *et al.* (1993) Evans (2002)
Norepinephrine	↓	Ca^{2+}	Degnan *et al.* (1977) Marshall *et al.* (1998)
Acetylcholine	↓	Ca^{2+}	May *et al.* (1984) May and Degnan (1985)
Endothelin	↓	?	Evans (2002)
NO (sodium nitroprusside)	↓	?	Evans (2002)
Prostaglandin (PGE_2)	↓	?	van Praag *et al.* (1987)
Somatostatin	↓	↓cAMP	Davis and Shuttleworth (1986)
Agents			
Calyculin A	↓	-	Hoffmann *et al.* (2002)
IBMX/db-cAMP	↓	↓cAMP	Mendelsohn *et al.* (1981) Foskett *et al.* (1982) May *et al.* (1984) May and Degnan (1985)
A23187	0	-	Degnan *et al.* (1977) Marshall *et al.* (1993)
ionomycin	↓	Ca^{2+}	Marshall *et al.* (1993, 2000)
Furosemide, Bumetanide	↓	-	Foskett *et al.* (1982) Eriksson *et al.* (1985)
DPC, NPPB (Apical)	↓	-	Marshall *et al.* (1995) Hoffmann *et al.* (2002)
Glibenclamide	↓	-	Hoffmann *et al.* (2002)
Genistein	↓	-	Marshall *et al.* (2000)

(Table 14.1 contd.)

(Table 14.1 contd.)

Treatments			
Hypertonic shock	↑	?	Zadunaisky *et al.* (1995) Hoffmann *et al.* (2002)
Hypotonic shock	↓	Ca^{2+}?	Marshall *et al.* (2000, 2005) Daborn *et al.* (2001) Leguen and Prunet (2004)

+: stimulation, -: inhibition, 0: No effect.

Bryson, 1998). The α_2-receptor subtype was determined by comparing affinity for norepinephrine and epinephrine (Marshall *et al.*, 1993) by application of specific α_2-receptor agonist clonidine (May *et al.*, 1984; Marshall *et al.*, 1993), by comparing α_1-adrenergic agonist phenylephrine with clonidine (Marshall *et al.*, 1993) and by use of α_2-receptor blockers to inhibit epinephrine action (May *et al.*, 1984).

Physiological inhibition is via α_2-adrenoceptors

However marked the effect of β-adrenoceptor agonist isoproterenol might be in stimulating chloride secretion (see below), the application of the natural non-specific agonists (epinephrine and norepinephrine) invariably causes strong inhibition of the chloride secretion in tilapia opercular membranes (Foskett *et al.*, 1982; Foskett and Machen 1985), *Fundulus* opercular membranes (Degnan *et al.*, 1977; May and Degnan, 1985; Marshall *et al.*, 1993) and goby (*Gillichthys mirabilis*) skin (Marshall and Bern, 1980). The β-adrenergic response, therefore, appears to be pharmacological. The ED_{50} for epinephrine is between 0.1 and 1.0 μM for opercular membrane of tilapia (Foskett *et al.*, 1982) and *Fundulus* (Marshall *et al.*, 1993) and *Gillichthys* skin (Marshall and Bern, 1980). Circulating catecholamine levels in resting, unstressed teleosts are below this range (0.05 μM); thus, Cl^- secretion is not likely under tonic inhibition by the interrenal. In stressed teleosts, catecholamine levels are much higher (0.5 – 1.0 μM) and would inhibit Cl^- secretion by the gills. Spontaneous recovery of Cl^- secretion after epinephrine addition is observed in killifish and tilapia membranes and was initially ascribed to β-adrenergic activity, but was prevented by addition of antioxidants (ascorbate) (Foskett *et al.*, 1982), so oxidation of the catecholamine would appear to be the cause of the observed recovery.

The opercular epithelium was dissected with its nerve supply intact, mounted in an Ussing chamber and the nerve directly stimulated while measuring transepithelial Cl^- secretion (Marshall *et al.*, 1998). Stimulation of the branch of the glossopharyngeal nerve that innervates the opercular epithelium in isolated epithelia illustrates that the natural sympathetic reflex response in the opercular epithelium (and likely also the gill MR cells) is inhibition of chloride secretion. Autonomic nerves track alongside cranial nerves leading to the gill and head region of teleosts (Sundin and Nilsson, 2002), so stimulation of the bundle would activate autonomic and motor neurons. Nerve stimulation induced rapid inhibition that was unaffected by cholinergic blockers (atropine and tubocurarine) but was blocked effectively by the α-adrenergic inhibitor yohimbine (Marshall *et al.*, 1998), demonstrating α-adrenergic mediation. Thus, physiological activation of autonomic neurons innervating the opercular epithelium and gill produce rapid inhibition of Cl^- secretion.

Catecholamine effects on the gill vasculature are consistent with the above-mentioned inhibition of epithelial Cl^- secretion. Epinephrine induces a reduction in perfusion of the ion transporting cells, in that epinephrine dilates the arterio-arterial perfusion of flounder gill lamellae at the expense of the arterio-venous flow that supplies the gill filament epithelium where the MR cells reside (Stagg and Shuttleworth, 1984). In rainbow trout (*Oncorhynchus mykiss*), epinephrine increases functional surface area for gas exchange, a combination of vasodilation via β-receptors on the afferent side and a balanced vasoconstriction on the efferent side mediated by α-receptors, reviewed by Sundin and Nilsson (2002). These changes are consistent with a shift away from perfusion of ion transporting cells more toward perfusion of the lamellae.

Calcium as second messenger

In many epithelial transport systems, intracellular calcium is associated with stimulation of transport. This is true for avian salt gland, reptilian salt gland, mammalian airway epithelium (Liedtke, 1990), mammalian sweat gland and elasmobranch rectal gland (Evans, 2002). Thus, teleost MR cells are somewhat exceptional in having a calcium-mediated *inhibition* of NaCl secretion. Calcium is apparently the second messenger for many inhibitory hormones and neurotransmitters in gill MR cells (Table 14.1) and may also be involved in hypotonic shock (Leguen and Prunet, 2004; also see below).

β-adrenoceptors are also present

Isoproterenol stimulates Cl^- secretion in killifish opercular epithelium (May *et al.*, 1984; Marshall *et al.*, 1998) and *Gillichthys* skin (Marshall and Bern, 1980). While specific β_1 and β_2 agonists were ineffective and thus did not identify the β receptor subtype, β_1 antagonists (e.g., timolol) blocked the response to isoproterenol, whereas the β_2 antagonists were ineffective, indicating the presence of a β_1 receptor subtype (May and Degnan, 1985). The β-adrenoceptor second messenger is cAMP, as isoproterenol (Stagg and Shuttleworth, 1984; Marshall *et al.*, 2000) and norepinephrine (Stagg and Shuttleworth, 1984) significantly increase opercular epithelium tissue content of cAMP. It has been suggested that the β-adrenergic response might enhance recovery from an initial α-adrenergic inhibition.

Acetylcholine, Serotonin, Dopamine

Acetylcholine decreases Cl^- secretion rate in killifish opercular membranes via muscarinic receptors, based on blockade of the effect by homatropine (Mendelsohn *et al.*, 1981). The response is not mediated by cAMP decrease, as there was no change in cAMP levels after acetylcholine treatment, neither did acetylcholine decrease cAMP levels that had been stimulated by isoproterenol (May and Degnan, 1985). Because the α-adrenergic and cholinergic responses mirrored each other in a variety of pharmacological treatments, May and Degnan (1985) proposed that the two responses converged on a common inhibitory pathway not involving cAMP lowering.

Serotonin is known to stimulate epithelial Cl^- secretion in dog tracheal epithelium (Tamaoki *et al.*, 1997) and rabbit corneal epithelium (Marshall and Klyce, 1984) and serotoninergic neurons have been identified in teleost gills (Bailly *et al.*, 1992; Sundin and Nilsson, 2002) particularly in proximity to smooth muscle. However, serotonin at low doses (1 and 10 μM) added to the killifish opercular epithelium has no effect; at high doses (50 μM) it is modestly inhibitory and appears to be acting through adrenergic receptors because the effect is blocked by yohimbine (Marshall *et al.*, 1993). Dopamine has not been tested in the opercular membrane system to see if this substance could potentially regulate Cl^- secretion.

Urotensin I and Urotensin II

Urotensin I is a peptide resembling mammalian Corticotrophin Releasing Factor (CRF) and amphibian sauvagine that is secreted by the caudal neurosecretory system of teleosts (Bern *et al.*, 1985; Pohl *et al.*, 2001). Urotensin I stimulates Cl^- secretion in *Gillichthys* skin MR cells (Marshall and Bern, 1979, 1981) and the effect is more apparent in epithelia that were previously inhibited by epinephrine. However, thus far, urotensin I receptors have not been localized in the gills of teleosts, while three different CRF type receptors have been localized in catfish to brain, heart, pituitary and urophysis (Arai *et al.*, 2001). The potent vasodilatory effects of urotensin I would suggest primarily a role in these responses (Olson, 2002).

Urotensin II is a Somatostatin-like peptide that is secreted by the caudal neurosecretory system of teleost fish (Bern *et al.*, 1985). Urotensin II is most prevalent in the caudal neurosecretory system (Bern *et al.*, 1985) but has now been identified in the brain of many vertebrates, suggesting an ancient neurosecretory role in vertebrates (Conlon *et al.*, 1997). Urotensin II significantly inhibits Cl^- secretion by the goby skin (Marshall and Bern, 1979, 1981) and tilapia opercular membrane (Foskett and Hubbard 1981; Foskett *et al.*, 1982), an effect reversed by agents that augment tissue cAMP. Plasma levels measured by homologous RIA indicate that seawater (SW) adapted flounder (*Platichthys flesus*) had higher plasma urotensin II than their freshwater counterparts (Winter *et al.*, 1999). While the urotensins are associated with ion balance in fish, their release into the renal portal system by neurosecretory cells of the urophysis points toward a renal and intestinal responses (Bern *et al.*, 1985) rather than a more systemic effect on more remote osmoregulatory structures such as the gill and opercular membranes. Urotensin II levels in plasma of flounder transferred from SW to FW were significantly lower than SW-SW controls at 24 and 72 h after transfer (Bond *et al.*, 2002), suggesting downregulation of urotensin II during FW acclimation. In both transfer groups, the disturbance appeared to inhibit urotensin II release compared to long term acclimated FW or SW animals that had 20-25 fmol/ml urotensin II (Bond *et al.*, 2002).

VIP, Glucagon and Atrial Natriuretic Peptide

Vasoactive Intestinal Polypeptide (VIP) rapidly increases Cl^- secretion by the tilapia opercular epithelium if the Cl^- secretion had been previously inhibited by epinephrine (Foskett *et al.*, 1982), while VIP alone applied to uninhibited epithelia had only minimal response. The response is consistent with a cAMP mediated effect because addition of small amounts of the phosphodiesterase inhibitor 3-isobutyl-1-methylxanthine augmented the VIP response. Similar results to VIP were obtained by Foskett *et al.* (1982) using glucagon. In both cases, the doses necessary to produce marked stimulation of the Cl^- secretion were rather high (10^{-5} M). VIP and glucagon at physiological concentrations (10^{-13} - 10^{-11} M) stimulate adenylate cyclase in SW acclimated rainbow trout gill membranes, more so than in freshwater trout (Guibbolini and Lahlou, 1987, 1990). VIP (26-28 amino acids) belongs to the glucagon like peptides and teleost VIP shares high similarity to human VIP (85 and 88% at the protein level for *Takifugu rubripes*, fugu and *Danio rerio*, zebrafish, respectively), while zebrafish and fugu VIP share 77% amino acid similarity. However, the SW teleosts (fugu and Atlantic cod, *Gadus morhua*) have a phenylalanine in position 13 rather than leucine (Mammal) and variations in the amino terminal amino acids (Table 14.2). These variations may explain the apparent insensitivity of the tilapia receptors for mammalian VIP (Foskett *et al.*, 1982). The variation in VIP sequence suggests that human VIP could activate teleost VIP receptors but that glucagon could also be acting through the VIP receptors. VIP receptors have been identified in gill and other tissues of goldfish, *Carassius auratus* (Chow *et al.*, 1997) and the EC_{50} for cod VIP on goldfish VIP receptor is 1 nM but that the receptor will also bind glucagon at a lower affinity (Chow *et al.*, 1997). The high doses of heterologous glucagon and VIP required in earlier work suggest that native VIP will be needed to examine truly physiological responses. In nature, VIP responses would most likely be mediated by VIP neurons rather than by circulating VIP. In goldfish, there exist VIP immunoreactive cells on the gills (Chow *et al.*, 1997) and in the goldfish gill there are immunoreactive VIP receptors in nerve cell bodies in the subepithelial space (de Girolamo *et al.*, 1997). The position of these VIP neurons in the gill suggests a possible physiological response to VIP. VIP is vasodilatory in brown trout (*Salmo trutta*) gill, an effect blocked by indomethacin, hence suggesting a vasoactive response mediated by prostaglandins (Bolis *et al.*, 1984). While it is not known specifically that neurally derived VIP stimulates Cl^- secretion by the gill, these results collectively point toward such a response.

Atriopeptin II (10^{-9}-10^{-7} M), an Atrial Natriuretic Peptide (ANP), is reported to have a modest stimulatory effect on Cl^- secretion by the

Table 14.2 Transporter regulation.

Transporter	*Stimulus (Second messenger)*	*Downstream events*	*Reference*
Apical Membrane			
CFTR Anion Channel	(cAMP)	↑PKA	Marshall *et al.* (1995, 2005a)
		FAK	Marshall *et al.* (2005b)
Basolateral membrane			
NKCC1 cotransporter	Hypotonicity	↑p38 MAPK	Marshall *et al.* (2005a)
	(Ca^{2+})	↑JNK	Kültz and Avila (2001) Marshall *et al.* (2005a)
		↑ERK1	Kültz and Avila (2001)
		↑SRC?	
		↓pY^{407} FAK	Marshall *et al.* (2005b)
		↑PP1?	
NKCC1 cotransporter	Hypertonicity	↑PKC	Hoffmann *et al.* (2002) Marshall *et al.* (2005a)
		↑p38 MAPK	Marshall *et al.* (2005)
		↑JNK	Kültz and Avila (2001) Marshall *et al.* (2005a)
		↑MLCK	Hoffmann *et al.* (2002)
		↑pY^{407} FAK?	Marshall *et al.* (2005b)
		↑OSR1/SPAK?	Marshall *et al.* (2005b)
NKCC1 cotransporter	cAMP	↑PKA?	
Na^+,K^+-ATPase	?	?	
K^+-channel	?	?	

↑: activation/phosphorylation, ↓ deactivation/dephosphorylation

PK protein kinase, ERK1 extracellular signal regulated protein kinase, MAPK mitogen-activated protein kinase, JNK J-N-terminal kinase, FAK focal adhesion kinase (pY^{407}, phosphorylation at tyrosine position 407), MLCK myosin light chain kinase, OSR1 oxidative stress response protein kinase, SPAK ste20-like proline-rich protein kinase, PP protein phosphatase.

killifish opercular epithelium from SW and freshwater adapted animals (Scheide and Zadunaisky, 1988). The effect was not mediated by neural activity, as the effect was not blocked by tetrodotoxin; likewise the effect was not mediated by adrenergic receptors, as blockade of β adrenoceptors with propranalol did not affect the response. ANP receptors in European eel (*Anguilla anguilla*) gills activate cGMP pathways and in SW eels branchial cells more than in freshwater branchial cells (Broadhead *et al.*, 1992). Eel gills rapidly remove ANP from circulation and ANP and CNP have clear effects on drinking rates and intestinal salt transport in eels especially during the early phase of SW adaptation (Takei and Hirose, 2002). Eel ANP, eel CNP and the NPR-C-specific C-ANF inhibited the forskolin-stimulated production of cyclic AMP in dispersed eel gill cells (Callahan *et al.*, 2004) and this effect was abolished by pre-treatment of cells with pertussis toxin. Whereas ANP likely has some function in teleost gill cells, the re-examination of ANP for an ion transport effect has failed to confirm the ANP response in this system (Evans, 2002).

Arginine Vasotocin

There is a detailed case for arginine vasotocin (AVT) regulation of Cl^- secretion and related to SW adaptation. AVT specific binding to eel gill cells is higher in SW adapted animals than FW eels (Guibbolini *et al.*, 1988), strong evidence for the existence of AVT receptors and a function in SW osmoregulation. The AVT receptor has been isolated from flounder gill and from white sucker, *Catostomous commersoni* (Mahlmann *et al.*, 1994; Pierson *et al.*, 1995; Warne 2001; Genbank AF184966). From Northern analysis, the AVT receptor is expressed in gill as well as brain, kidney and liver. Expression of AVT receptors in *Xenopus* oocytes results in dose-response stimulation of an inward current, likely anion current, using native AVT and the mammalian V1 AVT agonist. In rainbow trout, there appears to be an increase in plasma AVT when the animals are moved to higher salinity (Pierson *et al.*, 1995). Measurement of plasma AVT in flounder transferred from SW to FW revealed marked inhibition of AVT release compared to control fish that were transferred from SW to SW (Bond *et al.*, 2002), indicating a possible role for AVT in SW. More directly, AVT plasma levels are correlated with plasma tonicity and rise significantly in flounder infused with NaCl (Warne and Balment, 1995), consistent with AVT release being stimulated by increases in plasma osmolality.

Whereas AVT has not been shown to stimulate Cl^- secretion by the opercular epithelium, the cultured gill cell epithelium from sea bass (*Dicentrarchus labrax*) shows a modest increase in Cl^- secretion with 50 nM AVT (Avella *et al.*, 1999). Although the demonstrated effective AVT dose *in vitro* is much higher than measured plasma levels *in vivo*, it remains a possible upregulator that could restore SW level ion secretion. While glucagon increases adenylate cyclase activity in trout gills, AVT and isotocin are inhibitory (Guibbolini and Lahlou, 1992).

Nitric Oxide and Endothelin

The recent observation that nitric oxide synthase co-localizes with mitochondria rich cells in killifish opercular epithelium (Evans, 2002) invites consideration of this effector in regulation of Cl^- secretion. NO donors such as sodium nitroprusside produce a dose-dependent inhibit Cl; secretion by killifish opercular membranes. NO donors also inhibit Na^+,K^+-ATPase in gill homogenates from brown trout (Towle *et al.*, 1977; Tipsmark and Madsen, 2003). In Atlantic salmon (*Salmo salar*) smolts, NOS colocalizes with Na^+,K^+-ATPase in gill MR cells and there is an increase in NADPHD in the gill during smolting suggests a regulatory role of NO in the attenuation of the smolting-related increase in Na^+,K^+-ATPase activity prior to entering SW (Ebbesson *et al.*, 2005). It is possible that other agents may be acting through NO, here potentially acting in a paracrine/autocrine manner.

Big endothelin immunoreactivity is identified in mitochondria rich cells of Atlantic stingray (*Dasyatis sabina*), suggesting presence of the peptide in gill ion transporting cells (Evans, 2002). Addition of endothelin agonists to killifish opercular epithelia shows a concentration dependent inhibition of Cl^- secretion with a threshold of 10^{-10} M (Evans, 2002). There is growing evidence that NO and endothelin may interact to downregulate Cl^- secretion.

Prostaglandins and Eicosanoids

The opercular epithelium and gills of killifish metabolize eicosanoids and produce prostaglandins and leukotrienes (van Praag *et al.*, 1987). PGE_2 rapidly inhibits Cl^- secretion by the opercular epithelium (Eriksson *et al.*, 1985; van Praag *et al.*, 1987), whereas several leukotrienes appear to stimulate Cl^- secretion (van Praag *et al.*, 1987). In cultured respiratory cells from sea bass gills, PGE_2 has a small stimulatory effect, apparently on Cl^- secretion (Avella *et al.*, 1999). In gill filaments and kidneys of both eel and trout, PGE_2 is present but PGD_2 and 6-keto $PGF_{1\alpha}$ were major prostaglandins and concentrations of these were significantly lower in the eel after SW adaptation (Brown *et al.*, 1991). Rainbow trout gill filaments synthesize a wide range of eicosanoid products following calcium ionophore challenge, indicating potential functions in freshwater gill function (Holland *et al.*, 1999). Prostaglandins are synthesized via COX-2, as the specific COX 2 inhibitor NS-398 inhibitor blocks the endothelin response (Evans *et al.*, 1999). $PGF_{2\alpha}$, PGD2, PGI2 and thromoboxane A2 are all ineffective (Evans *et al.*, 1999). As has been mentioned above, prostaglandins apparently mediate the VIP responses in gill vasculature, but thus far a prostaglandin that stimulates Cl^- secretion has not been identified, although such may exist because application of arachidonic acid alone produces a biphasic response where the first phase is stimulatory (van Praag *et al.*, 1987).

Mediation of Cl^- secretion stimulation

One locus for cAMP mediated stimulation of Cl^- secretion appears to be an apically located CFTR-like anion channel. The channel was described in patch clamp experiments on killifish MR cells to be a low conductance (8 pS) anion selective channel that is activated by cAMP (Marshall *et al.*, 1995). The killifish CFTR gene, cloned and expressed in amphibian oocytes imparts a cAMP activated current (Singer *et al.*, 1998). In excized patches from MR cells, protein kinase A and ATP addition activate CFTR-like channels in quiescent patches (Marshall and Singer, 2002). The β adrenergic agonist isoproterenol (May and Degnan, 1985; Marshall *et al.*, 2000) and forskolin (May and Degnan, 1985) rapidly increase killifish opercular membrane cAMP levels accompanied by large increases in epithelial short-circuit current. It is thus clear that cAMP activates the apical membrane chloride conductance and transepithelial chloride secretion.

It is still possible for participation of NKCC activation at the basolateral membrane as well, but this has not been demonstrated in teleost preparations. Indeed NKCC phosphorylation is a common yet complex regulatory mechanism in NKCC containing systems (Flatman, 2002) and full activation of transepithelial transport would require activation of both components in a coordinated fashion.

In the dogfish shark (*Squalas acanthias*) rectal gland, that secretes Cl^- in a manner similar to teleost MR cells, isoproterenol activates secretion via cAMP and increases phosphorylation of NKCC threonine residues in the N terminus (Flemmer *et al.*, 2002). This mode of NKCC activation seems widely distributed, as similar responses were observed in at least two other NKCC secretory systems: rat parotid gland and tracheal epithelium (Flatman, 2002). The protein serine/threonine phosphatase (PP) PP1 and PP2A inhibitor calyculin A stimulates NKCC1 in several systems (Flatman, 2002) and stimulates Cl^- secretion by the killifish opercular membrane (Hoffmann *et al.*, 2002), hence the model includes PP1 and PP2A as possible downregulators of NKCC1 and activation of NKCC1 via cAMP and PKA.

Short-term activation of Na^+,K^+-ATPase and basolateral K^+ channels, could feasibly also contribute to upregulation of Cl^- secretion through enhancement of the Na^+ transmembrane gradient that drives NKCC1. Rapid activation of Na^+,K^+-ATPase, within hours of transfer to SW, was observed in killifish gill (Towle *et al.*, 1977; Mancera and McCormick, 2000). Short-term activation of the enzyme in killifish gills can be as rapid 0.5 h (Towle *et al.*, 1977) to 3.0 h (Mancera and McCormick, 2000) after transfer to SW and the response is dependent on protein synthesis (Mancera and McCormick, 2000). Upregulation of K^+ channels, in turn could also increase turnover rate of Na^+,K^+-ATPase by enhancement of K^+ recycling across the basolateral membrane and supplying NKCC with K^+ extracellularly. Blockade of K^+ channels by barium (Degnan, 1985; Zadunaisky *et al.*, 1995) rapidly inhibits Cl^- secretion by the opercular epithelium, demonstrating the dependence on operational basolateral K^+ conductance.

Mediation of Cl^- transport inhibition

Much less is known regarding the downregulation of Cl^- secretion and the subsequent covering of inactive Cl^- cells by pavement cells. The calcium ionophore ionomycin reduced Cl^- secretion with a similar time course to that of clonidine (Marshall *et al.*, 1993) but the less efficient calcium ionophore A23187 is ineffective (Degnan *et al.*, 1977; Stagg and Shuttleworth, 1984; Marshall *et al.*, 1993), presumably because in some poikilothermic systems A23187 is not sufficiently active. The second messenger for the α_2-adrenergic response appears to be calcium, based on a blunting of the clonidine response in calcium (Marshall *et al.*, 1993) depleted media and the fact that clonidine did not decrease tissue levels of cAMP in killifish opercular epithelium (Marshall *et al.*, 1993). Whereas calcium activation of Cl^- secretion is the norm for many otherwise similar systems from diverse origins, including dogfish shark rectal gland, colon, airway and avian salt gland, the effect of calcium in teleost MR cells is inhibitory. The source of this regulatory "inversion" is assumed to be a terminal phosphatase instead of a kinase in a calcium-regulated regulatory pathway.

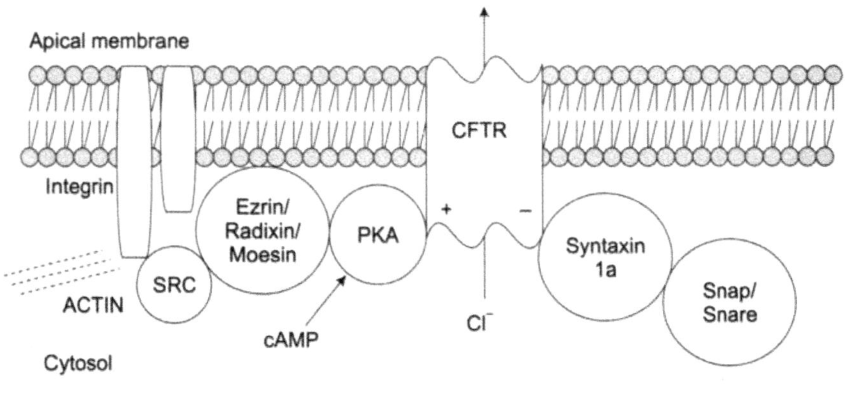

Fig. 14.2 Hypothetical model of apical membrane transport in rapid control of Cl^- secretion by seawater type mitochondria rich cells of marine and estuarine teleosts. Apical membrane regulation of CFTR anion channels includes cAMP activation via protein kinase A (PKA) and phosphorylation of the channel. Inferred from mammalian results is the involvement of EZRIN, RADIXIN, MOESIN linkage to the PDZ domain at the carboxy terminus of CFTR and the downregulation of CFTR by syntaxin 1a and SNAP/SNARE proteins.

HYPOTONIC AND HYPERTONIC STRESS AND Cl^- SECRETION

Hypotonic Shock

Hypotonic shock and cell swelling inhibits Cl^- secretion by the killifish opercular epithelium in a dose-dependent fashion. The reduction in Cl^- secretion was approximately 60% for a 40 mOsm/kg reduction in basolateral osmolality (Marshall *et al.*, 2000). The effect is freely reversible by reintroduction of iso-osmotic solutions, usually with an overshoot. Reduction in NaCl with added mannitol to maintain constant osmolality was a control that had no effect on Cl^- secretion rate, therefore the hypotonic effect is purely osmotic. Hypotonic shock on the basolateral side produced the same effect as bilateral reductions in osmolality. Hypotonic shock had no effect on tissue cAMP levels and was not impeded by pre-treatment of the tissue with the Ca^{2+} ionophore ionomycin or the calcium store depleting agent, thapsigargin (Marshall *et al.*, 2000). The protein tyrosine kinase (PTK) inhibitor genistein mimicked the response and was not additive, suggesting that PTK inhibition may be involved. Daidzein is the inactive control molecule for genistein. Daidzein was ineffective by itself and did not affect the subsequent response to hypotonic shock. Taken together, these results imply that hypotonic shock inhibits Cl^-

secretion in MR cells by a mechanism involving PTK inhibition. One PTK that has been shown to be present by immunocytochemistry is focal adhesion kinase (FAK) (Marshall *et al.*, 2005a); this is of particular interest because of the association of FAK with integrin-signalling processes and association of FAK with the cytoskeleton, two possible linkages with volume sensing mechanisms (Fig. 14.3).

Recently evidence was provided to support involvement of p38MAPK, n-terminal kinase (JNK), protein kinase C (PKC), oxidative stress response kinase (OSR1) and ste20 proline rich stress kinase (SPAK). The drug SB203580 significantly enhanced the decrease in current by hypotonic shock in killifish opercular membranes suggesting an inhibitory role of p38 MAPK in the hypotonic inhibition (Marshall *et al.*, 2005a). Hypotonic stress rapidly and transiently increased phosphorylated p38 MAPK, measured by Western analysis, by eight-fold at 5 minutes, then more slowly again to seven fold at 60 minutes. Hypertonic shock slowly increased pp38 by seven fold at 60 minutes. Phosphorylated JNK kinase was increased by 40-50% by hypotonic shock. The stress protein kinase SPAK and oxidation stress response kinase OSR1 were present in salt and fresh water (FW) acclimated gill and opercular epithelium with higher expression in FW (Marshall *et al.*, 2005a). Using immunocytochemistry, SPAK, OSR1 and phosphorylated focal adhesion kinase (pFAK-tyrosine, position 407) colocalizes with NKCC at the basolateral membrane (Marshall *et al.*, 2005a). Hypotonic shock in a rapid and stable manner dephosphorylates FAK at position 407 but not at other tyrosines (Y397, Y576, Y577, Y861), as detected by immunocytochemistry and Western immunoblots (Marshall *et al.* 2005b). The protein tyrosine kinase inhibitor genistein (100 μM) inhibits Cl^- secretion that was high, increases Cl^- secretion that was low and reduces immunocytochemical staining for phosphorylated FAK. The association of FAK with NKCC regulation is an interesting new role for this kinase in epithelial transport that deserves more in-depth investigation (see Fig. 14.3 and Table 14.2).

Cultured gill epithelial cells provide a model for gill cell volume responses. In cultured rainbow trout pavement cells, hypotonic shock, applied as two-third strength dilution of Ringer solution produced rapid cell swelling followed by a slow regulatory volume decrease (RVD) (Leguen and Prunet, 2004). Hypotonic shock induced a biphasic increase in intracellular Ca^{2+} comprising an initial peak followed by a sustained plateau. Calcium free Ringer had no effect on isotonic cell volume, but

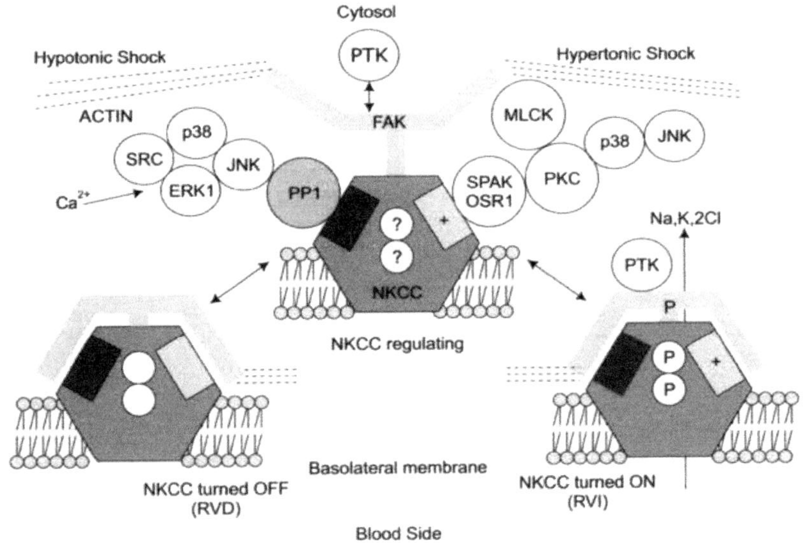

Fig. 14.3 Hypothetical model of basolateral membrane transport in rapid control of Cl^- secretion. Basolateral regulation of NKCC (Na^+,K^+,$2Cl^-$ cotransport, secretory type, NKCC1) is complex and appears to involve a stable activated state, a stable deactivated state and an unstable regulating state. NKCC may be activated by cell shrinkage (hypertonic shock, *right side*) via a cascade involving many kinases (p38, JNK, MLCK and PKC) linked in some fashion to a final kinase in the chain, SPAK or OSR1 that ultimately phosphorylate NKCC at two serines that, in turn, activate the transporter. Focal adhesion kinase (FAK) is phosphorylated at tyrosine 407 when NKCC is activated (lower right) and this state is consistent with regulatory volume increase (RVI). Not shown is the common hormonal pathway for activation via adenylate cyclase, cAMP and PKA. Hypotonic shock and elevation of intracellular calcium (*left side*) deactivate the transporter apparently via a cascade involving SRC, ERK1, p38, JNK and an ultimate protein phosphatase that dephosphorylates NKCC and FAK and turns off the transport, consistent with regulatory volume decrease (RVD). The order of enzymes in the cascades and the scaffolding role for FAK are hypothetical. The involvement of SRC is inferred from pathways involving integrin, FAK and SRC in cell volume regulation from mammalian systems. Norepinephrine acts via α_2-adrenergic receptors that are mediated through inositol tris phosphate (IP_3) from phosphoinositolbisphosphate (PIP_2) and deactivate NKCC via intracellular calcium. It is not clear what stabilizes the deactivated state of NKCC during regulatory volume decrease.

inhibited RVD after hypotonic shock. Ca^{2+} (Leguen and Prunet, 2004). Similar results were obtained in cultured cells from freshwater or SW adapted animals and from freshly isolated gill cells. Thus pavement cells regulate cell volume, associated with intracellular calcium signalling. Although low calcium solutions and thapsigargin (that depletes intracellular stores of calcium) did not diminish the hypotonic response in opercular membranes (Marshall *et al.*, 2000), it is still likely, considering the similar origins of MR cells and pavement cells, that cell volume responses in MR cells is also calcium mediated (see Fig. 14.3).

Hypotonic shock applied to the basolateral side of killifish opercular epithelia induces also a secondary decrease in transepithelial conductance that develops over an hour or so. Examination by SEM revealed significantly fewer exposed apical crypts in hypotonically stressed tissues and that pavement cells apparently close over MR cells (Daborn *et al.*, 2001). A similar reversible change in MR cell density occurs in estuarine mudskippers, *Periophthalmus modestus* (Sakamoto *et al.*, 2000). It would seem that regulation of the paracellular leak pathway, localized to leaky junctions between accessory cells and MR cells (Sardet *et al.*, 1979), is accomplished by retraction of Cl^- secreting cells and their covering over by pavement cells that have tight intercellular junctions (Sardet *et al.*, 1979; Daborn *et al.*, 2001). Furthermore, the robust actin ring surrounding the apical crypt of MR cells may be the active motor for the retraction (Daborn *et al.*, 2001).

Hypertonic Shock

Hypertonic shock has the opposite effect to hypotonic shock and can increase Cl^- secretion by 100% or more when NaCl or mannitol 50 – 100 mOsm/kg is added to the basolateral bathing medium (Zadunaisky *et al.*, 1995; Hoffmann *et al.*, 2002). In nature, blood plasma osmolality rises after SW transfer to a peak at 8-16 h and the average increase for killifish is approximately 40 mOsm/kg (Zadunaisky *et al.*, 1995). Accompanying a smaller increase in Cl^- secretion, hypertonic stress of isolated opercular epithelia from FW acclimated killifish significantly increases the density of exposed apical crypts (Daborn *et al.*, 2001), suggesting the emergence of Cl^- secreting cells. The increase in Cl^- secretion in SW adapted opercular membranes could be blocked by basolaterally added bumetanide and by basolaterally added anion channel blockers, DPC (Zadunaisky *et al.*, 1995), NPPB and glibenclamide (Hoffmann *et al.*, 2002). Hyperosmolarity stimulates the heart and enhances NaCl extrusion functions at the gill (Ando *et al.*, 2003). In the teleost intestine, hypertonic shock also activates NKCC family cotransporters, but in this case to aid NaCl uptake from the posterior intestine of eels (Ando *et al.*, 2003). Hypertonicity in SW transferred killifish produces differential gene expression and augmentation of NaCl secretion in protection of plasma ion levels and expression patterns are quantitatively different for Northern and Southern populations of killifish (Scott and Schulte, 2005). Because glibenclamide blocks ATP sensitive potassium channels in cardiac muscle of teleosts (Paajanen and Vornanen, 2002), it is possible that a secondary action via potassium channels in the tissue may impede the response to hypertonic shock. Curiously, large doses of amiloride (2 mM) on the basolateral side blocked the Cl^- transport stimulation by hypertonic shock (Zadunaisky *et al.*, 1995), suggesting that the Na^+-H^+ exchange and/or intracellular pH may affect the hypertonic response.

In killifish opercular membranes, hypertonic shock increases phosphorylation level of p38 MAPK by seven fold after 60 minutes of hypertonic shock (Marshall *et al.*, 2005a). Phosphorylated JNK kinase was also increased rapidly by 40-50% after hypertonic shock and remained elevated after 30 minutes in hypertonic medium. The PKC inhibitor chelerythrine has no effect on hypotonic inhibition but blocks the recovery (Hoffmann *et al.*, 2002; Marshall *et al.*, 2005a), indicating PKC involvement in stimulation and presumably also in NKCC activation by hypertonicity (Marshall *et al.*, 2005a). Further, myosin light chain kinase (MLCK) is involved in transport activation in the opercular membrane (Hoffmann *et al.*, 2002; see Fig. 14.3 and Table 14.2).

The second messenger for the hypertonicity response is unknown (Table 14.2). The increase in secretion is dependent on calcium in the bathing solutions (Zadunaisky *et al.*, 1995) but calcium ionophore ionomycin causes inhibition (Marshall *et al.*, 1993) so calcium instead may play some permissive role. The increase also is also not dependent on cAMP and PKA (Zadunaisky *et al.*, 1995; Hoffmann *et al.*, 2002). It is possible that a diffusible second messenger is not involved and instead the response is sensed and transporter activation takes place by a direct stimulation of a kinase cascade by a volume sensing molecular complex such as that depicted in Figure 14.3.

There is a rapid elevation of Na^+, K^+-ATPase activity within three hours of transfer of killifish to SW that also would augment C secretion rate (McCormick, 2001). This increase in transport enzyme activity is dependent on translational and transcriptional processes (McCormick, 2001) but it is not known whether the response is hormone mediated or dependent on increased plasma osmolality.

Role of Intracellular pH

Measurement of intracellular pH with cytoplasmically trapped fluorophore BCECF indicated no change in intracellular pH of MR cells even after profound inhibition of Cl^- secretion by clonidine (Marshall *et al.*, 1993), suggesting chloride secretion is not physiologically regulated by intracellular pH. However, pharmacological manipulation of intracellular pH has noticeable transport effects. Acidification of the cytosolic compartment in 5% CO_2 and addition of NH_4Cl to the basolateral but not the apical side of the killifish opercular membrane inhibits Cl^- secretion, apparently as a secondary effect of acidification of the cytosolic compartment by ammonium loading (Zadunaisky *et al.*, 1995). The recovery from this acidification is enhanced by the phorbol ester, phorbol myristate acetate (PMA), and inhibited partially by basolateral amiloride, suggesting that PKC activation may stimulate Na^+/H^+ exchange, raise intracellular pH and allow Cl^- secretion to return to normal rates. This implies intracellular pH may indirectly affect Cl^- secretion, with acidic stress being inhibitory. Meanwhile, blockage of the anion exchanger with DIDS has no effect on resting Cl^- secretion rate (Marshall *et al.*, 1995) and did not block the stimulation of Cl^- secretion by hypertonic shock (Zadunaisky *et al.*, 1995), suggesting that Cl^-/HCO_3^- exchange is not involved in Cl^- secretion. However, 0.1 mM DIDS blocks the inhibition of Cl^- secretion in response to hypotonic shock (Zadunaisky *et al.*, 1997), while amiloride was ineffective, again implicating the anion exchanger in mediation of the osmotic response. More direct evidence is needed to establish whether intracellular pH is involved in physiological regulation of Cl^- transport.

Cytoskeleton and Transporters

Hypotonic and hypertonic shock may be mediated via the cytoskeleton in stretch activated or inactivated membrane elements. The possible connection of the CFTR type anion channel and actin cytoskeleton arose from an association of cAMP dependent phosphorylation by PKA blocked by the actin disrupter cytochalasin D (Prat *et al.*, 1995). The F-actin stabilizing agent phalloidin also inhibits generation of sustained Cl^- current in colonic T84 cells (Matthews *et al.*, 1994). Partial disruption of F-actin activates, while more complete disruption inhibits CFTR function and renders CFTR refractory to cAMP/PKA activation (Cantiello, 2001). Now, it is known that CFTR is a nucleus for developing F-actin fibers, observed by confocal microscopy, and that a direct connection between the channel protein and actin exists (Chasan *et al.*, 2002). Endocytic recycling and trafficking of CFTR seems to involve Golgi PDZ binding proteins (Gentzsch *et al.*, 2003) and the well known PDZ binding site (Swiatecka-Urban *et al.*, 2002) on the carboxy terminus (-TRL) that is common to human and killifish (Singer *et al.*, 1998) CFTR proteins. However, rapid inhibition of CFTR involves syntaxin 1A expressed in *Xenopus* oocytes (Peters *et al.*, 1999) or in airway epithelia (Naren *et al.*, 2000). Current thinking has expanded to include syntaxin 1A-SNARE complex (Ganeshan *et al.*, 2003) or syntaxin 1A-SNAP-23 complex bound to the amino terminus of CFTR (Cormet-Boyaka *et al.*, 2002) where syntaxin 1A binds to both CFTR and the SNAP/SNARE proteins (Chang *et al.*, 2002; Cormet-Boyaka *et al.*, 2002; Ganeshan *et al.*, 2003). In this complex, binding of syntaxin 1A directly decreases channel open probability (Chang *et al.*, 2002). Meanwhile, activation of CFTR can occur via a complex at the carboxy terminus with the β-adrenoceptor, and ezrin/radixin/moesin binding phosphoprotein regulated by PKA (Naren *et al.*, 2003; see Fig 14.2). It is tempting to extrapolate to the physiological downregulation of Cl^- secretion in teleost MR cells by syntaxin 1A. Clearly, a demonstration of syntaxin 1A localization to MR cells and binding to kfCFTR would be a first step to establishing this relationship. Important also is the identification by (Kültz *et al.*, 2001) of a 14-3-3 gene in killifish that encodes for 14-3-3, a protein that is known to regulate ion channels, transporters and cytoskeleton for cells in changing environments. Protein 14-3-3 is also upregulated in killifish gills on transfer of killifish from SW to FW (Kültz *et al.*, 2001), implying the involvement of this protein in long term FW adaptation. Finally, transfer of killifish to FW produces physiologically relevant decrease in blood osmolality of approx. 60 mOsm/kg (Marshall *et al.*, 2000) also activates three identified MAP-kinases: ERK1, SAPK1 (= Jun N-terminal kinase, JNK) and SAPK2 (= p38) kinase, as all three show enhanced phosphorylation by immunoblotting with antibodies specific for the phosphorylated form of the enzymes (Kültz and Avila, 2001). Thus osmosensing in gill cells initiates MAP kinase transduced changes in cell physiology (see Fig. 14.2; Table 14.2).

Whereas mammalian epithelial cells will not normally experience changes in basolateral osmolality of the magnitudes tested in cell volume research, gill epithelial cells of estuarine fish actually are osmotically stressed when these animals change salinity. Hence, the teleost epithelial cell responses could be more sensitive and amplified than in mammals, hence easier to study.

Role of PKC Isozymes

The association of PKCα and PKCε with actin cytoskeleton is demonstrated by the elegant work of Song *et al.* (2002) where the PKC non-specific agonist (PMA) in T84 epithelial cells was shown to activate first PKCε in actin disruption and stimulation of basolateral endocytosis, followed by a late effect involving activation of PKCα actin stabilization and inhibition of endocytosis. The isozymes of PKC that may be involved in Cl^- transport regulation in teleosts are unknown. Clearly, the actin-PKC relationship should be clarified for the teleost system before conclusive relationships between PKC, F-actin and intracellular Ca^{2+} can be realized. A parallel between the teleost system and T84 cells is the inhibitory phase of Cl^- transport regulation by acetylcholine in killifish (Mendelsohn *et al.*, 1981; May and Degnan, 1985) and in T84 cells (Song *et al.*, 2002). The selective activation of PKCε by carbachol and the stimulation of endocytosis in the late inhibitory phase (Song *et al.*, 2002) invites speculation that the inhibitory action of acetylcholine and α-adrenoceptor agonists may be activation of a PKCε and endocytosis or dephosphorylation of NKCC1. Further, PKCα and PKCε are differentially translocated to the membrane fraction of cells subjected to hypotonic shock (Liu *et al.*, 2003), implying that hypotonic shock effects could also be mediated via PKCε.

NON-GENOMIC EFFECT OF CORTISOL

One possibility that has not been investigated for the gill and opercular epithelium is that of non-genomic responses to corticosteroids. There are well-established inhibitory roles for cortisol (distinct from other steroids) at physiological levels to depress prolactin release by cultured cells from the rostral pars distalis of tilapia (Borski *et al.*, 2002). These effects are not mediated by gene transcription or protein synthesis. The effects develop rapidly, in 10-20 minutes, and are reversible by removal of the hormone. Similar to somatostatin actions on prolactin cells, intracellular Ca^{2+} is decreased and adenylate cyclase and cAMP are reduced by cortisol (Borski *et al.*, 2002). Interestingly, the intracellular Ca^{2+} elevation produced by hypotonic shock in prolactin cells is reversed by 200 nM cortisol. Given that cortisol may be elevated in fish during salinity change, the possible downregulation by decrease in adenylate cyclase or reduction of Ca^{2+} mediated hypotonic response could either aid or exacerbate (respectively) ion regulation on entry into FW.

PROPOSED MODEL FOR RAPID REGULATION OF CHLORIDE TRANSPORT

Some of the mechanisms that enter into the hypothetical model here (Figs. 14.2, 14.3) have been established in mammalian systems and appear with question marks and some others have been demonstrated to be present in teleosts. Killifish CFTR, because it shares the carboxy terminus PDZ binding domain in common with the human, may share some aspects of PDZ binding and regulation. All vertebrate CFTR proteins share the first two amino acids at the amino terminus and killifish shares the first six of ten with the human, suggesting the syntaxin 1A binding is also feasible. Whereas intracellular Ca^{2+} is normally associated with activation of Cl^- secretion, teleost Cl^- secreting gill and gill-like systems all have Ca^{2+} mediated inhibition. The point of inversion in the regulatory pathway could be activation of a phosphatase such as Ca^{2+} activated protein phosphatase (PP2B, calcineurin) but the PP2B antagonist FK506 is not effective in blocking clonidine action in the killifish opercular epithelium (Marshall, unpublished). Another possibility is the activation of a different PKC isozyme, such as PKCε that is linked downregulation of membrane transport.

Rapid regulation of NaCl transport at the gill is necessary in fish that move often between SW and FW, as is true for many estuarine species. Estuarine animals entering FW may respond in a graded fashion over a few hours to this stress. Within minutes of exposure to FW, reflex catecholamine release from the autonomic system could rapidly inhibit Cl^- secretion via α-adrenoceptors, to conserve NaCl. Within 1-3 hours as a result of passive NaCl loss, plasma hypotonicity (hypotonic shock) would develop in turn producing sustained inhibition of Cl^- secretion and retraction of Cl^- cells, thus reducing both active and passive NaCl loss. When estuarine teleosts return to SW after a few hours, NaCl secretion may be reinitiated, perhaps stimulated by hypertonicity and by peptide hormones (AVT, glucagon or urotensin I) and neurotransmitters (VIP and epinephrine operating through β adrenoceptors). In this way, estuarine fish may rely mostly on rapid acting means of adjusting NaCl secretion and passive loss, rather than immediately invoking the more energetically expensive permanent changes to the gill epithelium.

CONCLUSION

This article examines the mechanism of ion transport by mitochondrion-rich MR cells and its rapid regulation in teleost fish. Small estuarine fish experience rapid shifts in salinity in intertidal microenvironments. Immediate stress responses mediated by neurotransmitters and the adrenal medullary homolog rapidly change ion transport rates. There are also direct effects on MR cells by changes in blood osmolality; hypotonic shock reduces and hypertonic shock increases NaCl secretion rates. Each reaction involves cascades of phosphorylation or dephosphorylation that terminate at the ion transport protein. All these are important immediate responses that help maintain ion balance in early stages of adaptation before gene expression produces new proteins or newly differentiated cells.

References

Ando, M., T. Mukuda and T. Kozaka. 2003. Water metabolism in the eel acclimated to sea water: from mouth to intestine. *Comparative Biochemistry and Physiology* B 136: 621–633.

Arai, M., I.Q. Assil and A.B. Abou-Samra. 2001. Characterization of three corticotropin-releasing factor receptors in catfish: A novel third receptor is predominantly expressed in pituitary and urophysis. *Endocrinology* 142: 446–454.

Avella, M., P. Pärt and J. Ehrenfeld. 1999. Regulation of Cl^- secretion in seawater fish (*Dicentrarchus labrax*) gill respiratory cells in primary culture. *Journal of Physiology (London)* 516: 353–363.

Bailly, Y., S. Dunel-Erb and P. Laurent. 1992. The neuroepithelial cells of the fish gill filament: indolamine-immunocytochemistry and innervation. *Anatomical Record* 233: 143–161.

Bern, H.A., D. Pearson, B.A. Larson and R.S. Nishioka. 1985. Neurohormones from fish tails: The caudal neurosecretory system I. 'Urophysiology' and the caudal neurosecretory system of fishes. *Hormone Research* 41: 533–552.

Bolis, L., M. Mandolfino, D. Marino and J.C. Rankin. 1984. Vascular actions of vasoactive intestinal polypeptide and β-endorphin in isolated perfused gills of the rainbow trout, *Salmo trutta* L. *Molecular Physiology* 5: 221–226.

Bond, H., M.J. Winter, J.M. Warne, C.R. McCrohan and R.I. Balment. 2002. Plasma concentrations of arginine vasotocin and Urotensin II are reduced following transfer of the euryhaline flounder (*Platichthys flesus*) from seawater to fresh water, *General and Comparative Endocrinology* 125: 113–120.

Borski, R.J., N.H. Gregory and S. Fruchtman. 2002. Signal transduction mechanisms mediating rapid, nongenomic effects of cortisol on prolactin release. *Steroids* 67: 539–548.

Broadhead, C.L., U.T. O'Sullivan, C.F. Deacon and I.W. Henderson. 1992. Atrial natriuretic peptide in the eel, *Anguilla anguilla* L.: Its cardiac distribution, receptors and actions on isolated branchial cells. *Journal of Molecular Endocrinology* 9: 103–114.

Brown, J.A., C.J. Gray, G. Hattersley and J. Robinson. 1991. Prostaglandins in the kidney, urinary bladder and gills of the rainbow trout and European eel adapted to fresh water and seawater. *General and Comparative Endocrinology* 84: 328–335.

Callahan, W., S. Nankervis and T. Toop. 2004. Natriuretic peptides inhibit adenylyl cyclase activity in dispersed eel gill cells. *Journal of Comparative Physiology* B 174: 275–280.

Cantiello, H.F. 2001. Role of actin filament organization in CFTR activation. *Pflügers Archives* 443: S75–S80.

Chang, S.Y., A. Di, A.P. Naren, H.C. Palfrey, K.L. Kirk and D.J. Nelson. 2002. Mechanisms of CFTR regulation by syntaxin 1A and PKA. *Journal of Cell Science* 115: 783–791.

Chasan, B., N.A. Geisse, K. Pedatella, D.G. Wooster, M.Teintze, M.D. Carattino, W.H. Goldmann and H.F. Cantiello. 2002. Evidence for direct interaction between actin and the cystic fibrosis transmembrane conductance regulator. *European Biophysical Journal* 30: 617–624.

Chow, B.K., T.T. Yuen and K.W. Chan. 1997. Molecular evolution of vertebrate VIP receptor and functional characterization of a VIP receptor from goldfish *Carassius auratus*. *General and Comparative Endocrinology* 105: 176–185.

Conlon, J.M., H. Tostivint and H. Vaudry. 1997. Somatostatin- and Urotensin II-related peptides: molecular diversity and evolutionary perspectives. *Regulatory Peptides* 69: 95–103.

Cormet-Boyaka, E., A. Di, S.Y. Chang, A.P. Naren, A. Tousson, D.J. Nelson and K.L. Kirk. 2002. CFTR chloride channels are regulated by a SNAP-23/syntaxin 1A complex. *Proceedings of the National Academy of Sciences of the United States of America* 99: 12477–12482.

Daborn, K., R.R.F. Cozzi and W.S. Marshall. 2001. Dynamics of pavement cell-chloride cell interactions during abrupt salinity change in *Fundulus heteroclitus*. *Journal of Experimental Biology* 204: 889–1899.

Davis, M.S. and T.J. Shuttleworth. 1986. Mode of adrenergic and peptidergic inhibition of ion transport in flounder gill. *American Journal of Physiology* 251: R1064–R1070.

de Girolamo, P., N. Arcamone, A. Rosica and G. Gargiulo. 1997. Vasoactive intestinal polypeptide immunoreactive nerves in the gill arch of teleost fish, *Carassius auratus* L. *Acta Histochemica* 99: 13–22.

Degnan, K.J. 1985. The role of K^+ and Cl^- conductances in chloride secretion by the opercular epithelium. *Journal of Experimental Zoology* 236: 19–25.

Degnan, K.J., K.J. Karnaky Jr. and J.A. Zadunaisky. 1977. Active chloride transport in the *in vitro* opercular skin of a teleost (*Fundulus heteroclitus*), a gill-like epithelium rich in chloride cells. *Journal of Physiology (London)* 271: 155–191.

Degnan, K.J. and J.A. Zadunaisky. 1980. Passive sodium movements across the opercular epithelium: The paracellular shunt pathway and ionic conductance. *Journal of Membrane Biology* 55: 175–185.

Duranton, C., E. Mikulovic, M.Tauc, M. Avella and P. Poujeol. 2000a. Potassium channels in primary cultures of seawater fish gill cells. I. Stretch-activated K^+ channels. *American Journal of Physiology* 279: R1647–R1658.

Duranton, C., E. Mikulovic, M. Tauc, M. Avella and P. Poujeol. 2000b. Potassium channels in primary cultures of seawater fish gill cells. II. Channel activation by hypotonic shock. *American Journal of Physiology* 279: R1659–1670.

Ebbesson, L.O., C.K. Tipsmark, B. Holmqvist, T. Nilsen, E.Andersson, S.O. Stefansson and S.S. Madsen. 2005. Nitric oxide synthase in the gill of Atlantic salmon: colocalization with and inhibition of Na^+, K^+-ATPase. *Journal of Experimental Biology* 208: 1011–1017.

Eriksson, Ö., N. Mayer-Gostan and P.J. Wistrand. 1985. The use of isolated fish opercular epithelium as a model tissue for studying intrinsic activities of loop diuretics. *Acta Physiologica Scandinavica* 125: 55–66.

Evans, D.H. 2002. Cell signalling and ion transport across the fish gill epithelium. *Journal of Experimental Zoology* 347: 293–336.

Evans, D.H., P.M. Piermarini and W.T.W. Potts. 1999. Ionic transport in the fish gill epithelium. *Journal of Experimental Zoology* 283: 641–652.

Evans, D.H., P.M. Piermarini and K.P. Choe. 2005. The multifunctional fish gill: Dominant site of gas exchange, osmoregulation, acid-base regulation, and excretion of nitrogenous waste. *Physiological Reviews* 85: 97–177.

Flatman, P.W. 2002. Regulation of Na-K-2Cl cotransport by phosphorylation and protein-protein interactions. *Biochimica et Biophysica Acta* 1566: 140–151.

Flemmer, A.W., I. Gimenez, B.F. Dowd, R.B. Darman and B. Forbush. 2002. Activation of the Na-K-Cl cotransporter NKCCl detected with a phospho-specific antibody. *Journal of Biological Chemistry* 277: 37551–37558.

Foskett, J.K. and G.M. Hubbard. 1981. Hormonal control of chloride secretion by teleost opercular membrane. *Annals of the New York Academy of Sciences* 372: 643.

Foskett, J.K and T.E. Machen. 1985. Vibrating probe analysis of teleost opercular epithelium: Correlation between active transport and leak pathways of individual chloride cells. *Journal of Membrane Biology* 85: 25–35.

Foskett, J.K and C. Scheffey. 1982. The chloride cell: Definitive identification as the salt-secretory cell in teleosts. *Science* 215: 164–166.

Foskett, J.K., G.M. Hubbard, T.E. Machen and H.A. Bern. 1982. Effects of epinephrine, glucagon and vasoactive intestinal polypeptide on chloride secretion by teleost opercular membrane. *Journal of Comparative Physiology* 146: 27–34.

Ganeshan, R., A. Di, D.J. Nelson, M.W. Quick and K.L. Kirk. 2003. The interaction between syntaxin 1A and cystic fibrosis transmembrane conductance regulator Cl^- channels is mechanistically distinct from syntaxin 1A-SNARE interactions. *Journal of Biological Chemistry* 278: 2876–2885.

Gentzsch, M., L. Cui, A.Mengos, X.B. Chang, J.H. Chen and J.R. Riordan. 2003. The PDZ-binding chloride channel ClC-3B localizes to the Golgi and associates with cystic fibrosis transmembrane conductance regulator-interacting PDZ proteins. *Journal of Biological Chemistry* 278: 6440–6449.

Guibbolini, M.E. and B. Lahlou. 1987. Adenylate cyclase activity in fish gills in relation to salt adaptation. *Life Sciences* 41: 71–78.

Guibbolini, M.E. and B. Lahlou. 1990. Evidence for the presence of a new type of neurohypophysial hormone receptor in fish gill epithelium. *American Journal of Physiology* 258: R3–R9.

Guibbolini, M.E. and B. Lahlou. 1992. Gi protein mediates adenylate cyclase inhibition by neurohypophyseal hormones in fish gill. *Peptides* 13: 865–871.

Guibbolini, M.E., I.W. Henderson, W. Mosley and B. Lahlou. 1988. Arginine vasotocin binding to isolated branchial cells of the eel: Effect of salinity. *Journal of Molecular Endocrinology* 1: 125–130.

Hoffmann, E.K., E. Hoffmann, F. Lang and J.A. Zadunaisky. 2002. Control of Cl^- transport in the opercular epithelium of *Fundulus heteroclitus*: Long- and short-term salinity adaptation. *Biochimica et Biophysica Acta* 1566: 129–139.

Holland, J.W., G.W. Taylor and A.F. Rowley. 1999 The eicosanoid generating capacity of isolated cell populations from the gills of the rainbow trout, *Oncorhynchus mykiss*. *Comparative Biochemistry and Physiology* 122: 297–306.

Karnaky, K.J. Jr and W.B. Kinter. 1977. Killifish opercular skin: A flat epithelia with a high density of chloride cells. *Journal of Experimental Zoology* 199: 355–364.

Kültz, D. and K. Avila. 2001. Mitogen activated protein kinases are *in vivo* transducers of osmosensory signals in fish gill cells. *Comparative Biochemistry and Physiology* B 129: 821–829.

Kültz, D., D. Chakravarty and T. Adilakshmi. 2001. A novel 14-3-3 gene is osmoregulated in gill epithelium of the euryhaline teleost *Fundulus heteroclitus*. *Journal of Experimental Biology* 204: 2975–2985.

Leguen, I. and P. Prunet. 2004. Effect of hypotonic shock on cultured pavement cells from freshwater or seawater rainbow trout gills. *Comparative Biochemistry and Physiology* A 137: 259–269.

Liedtke, C.M. 1990. Calcium and alpha-adrenergic regulation of Na-Cl(K) cotransport in rabbit tracheal epithelial cells. *American Journal of Physiology* 259: L66–L72.

Liu, X., M.I. Zhang, L.B. Peterson and R.G. O'Neil. 2003. Osmomechanical stress selectively regulates translocation of protein kinase C isoforms. *Federation of European Biological Societies Letters* 538: 101–106.

Mahlmann, S., W. Meyerhof, H. Hausmann, J. Heierhorst, C. Schonrock, H. Zweirs, K. Lederis and D. Richter. 1994. Structure, function and phylogeny of Arg8 vasotocin receptors from fish and toad. *Proceedings of the National Academy of Sciences of the United States of America* 91: 1342–1345.

Mancera, J.M. and S.D. McCormick. 2000. Rapid activation of gill Na^+,K^+-ATPase in the euryhaline teleost *Fundulus heteroclitus*. *Journal of Experimental Zoology* 287: 263–274.

Marshall, W.S. 1977. Transepithelial potential and short-circuit current across the isolated skin of *Gillichthys mirabilis* (Teleostei: Gobiidae), acclimated to 5% and 100% seawater. *Journal of Comparative Physiology* 114: 157–165.

Marshall, W.S. 2002. Na^+, Cl^-, Ca^{2+} transport by fish gills: Retrospective review and prospective synthesis. *Journal of Experimental Zoology* 293: 264–283.

Marshall, W.S. and H.A. Bern. 1979. Teleostean urophysis: Urotensin II and ion transport across the isolated skin of a marine teleost. *Science* 204: 519–521.

Marshall, W.S. and H.A. Bern. 1980. Ion transport across the isolated skin of the teleost, *Gillichthys mirabilis*. In: *Epithelial Transport in the Lower Vertebrates*, B. Lahlou (ed.). Cambridge University Press, Cambridge, pp. 337–350.

Marshall, W.S. and H.A. Bern 1981. Active chloride transport by the skin of a marine teleost is stimulated by Urotensin I and inhibited by Urotensin II. *General and Comparative Endocrinology* 43: 484–491.

Marshall, W.S. and S.E. Bryson. 1998. Transport mechanisms of seawater chloride cells: An inclusive model of a multifunctional cell. *Comparative Biochemistry and Physiology* A119: 97–106.

Marshall, W.S. and S.D. Klyce. 1984. Cellular mode of serotonin action on Cl^- transport in the rabbit corneal epithelium. *Biochimica et Biophysica Acta* 778: 139–143.

Marshall, W.S. and T.D. Singer. 2002. Cystic fibrosis transmembrane conductance regulator in teleost fish. *Biochimica et Biophysica Acta* 1566: 16–27.

Marshall, W.S., S.E. Bryson and D. Garg. 1993. α_2-adrenergic inhibition of chloride transport by opercular epithelium is mediated by intracellular Ca^{2+}. *Proceedings of the National Academy of Sciences of the United States of America* 90: 5504–5508.

Marshall, W.S., S.E. Bryson, A. Midelfart and W.F. Hamilton. 1995. Low conductance anion channel activated by cyclic AMP in teleost Cl^--secreting cells. *American Journal of Physiology* 268: R963–R969.

Marshall, W.S., R.M. DuQuesnay, J.M. Gillis, S.E. Bryson and C.M. Liedtke. 1998. Neural modulation of salt secretion in teleost opercular epithelium by α_2-adrenergic receptors and inositol 1,4,5-trisphosphate. *Journal of Experimental Biology* 201: 1959–1965.

Marshall, W.S., S.E. Bryson and T. Luby. 2000. Control of epithelial Cl^- secretion by basolateral osmolality in euryhaline teleost *Fundulus heteroclitus*. *Journal of Experimental Biology* 203: 1897–1905.

Marshall, W.S., E.M. Lynch and R.R.F. Cozzi. 2002. Redistribution of immunofluorescence of CFTR anion channel and NKCC cotransporter in chloride cells during adaptation of the killifish *Fundulus heteroclitus* to seawater. *Journal of Experimental Biology* 205: 1265–1273.

Marshall, W.S., C.G. Ossum and E.K. Hoffmann. 2005a. Hypotonic shock mediation by p38 MAPK, JNK, PKC, FAK, OSR1 and SPAK in osmosensing chloride secreting cells of killifish opercular epithelium. *Journal of Experimental Biology* 208: 1063–1077.

Marshall, W.S., E.K. Hoffmann, C.G. Ossum and R.R.F. Cozzi. 2005b. Osmosensing chloride cells in *Fundulus heteroclitus* rapidly regulate ion transport using p38 MAPK, JNK, PKC, SPAK, OSR1 and FAK. In: *Volume Regulation in Health and Disease*, Copenhagen (Abstract).

Matthews, J.B., C.S. Awtrey, K.J. Tally and J.A. Smith. 1994. Dynamic role of microfilaments in intestinal chloride secretion. *American Journal of Surgery* 167: 21–26.

May, S.A. and K.J. Degnan. 1985. Converging adrenergic and cholinergic mechanisms in the inhibition of Cl^- secretion in fish opercular epithelium. *Journal of Comparative Physiology* B 156: 183–189.

May, S.A., K.H. Baratz, S.A. Key and K.J. Degnan. 1984. Characterization of the adrenergic receptors regulating chloride secretion by the opercular epithelium. *Journal of Comparative Physiology* B 154: 343–348.

McCormick, S.D. 2001. Endocrine control of osmoregulation in teleost fish. *American Zoologist* 41: 781–794.

Mendelsohn, S.A., B.D. Cherksey and K.J. Degnan. 1981. Adrenergic regulation of chloride secretion across the opercular epithelium: The role of cyclic AMP. *Journal of Comparative Physiology* 145: 29–35.

Naren, A.P., A. Di, E. Cormet-Boyaka, P.N. Boyaka, J.R. McGhee, W. Zhou, K. Akagawa, T. Fujiwara, U. Thome, J.F. Engelhardt, D.J. Nelson and K.L. Kirk. 2000. Syntaxin 1A is expressed in airway epithelial cells, where it modulates CFTR Cl^- currents. *Journal of Clinical Investigation* 105: 377–386.

Naren, A.P., B. Cobb, C. Li, K. Roy, D.J. Nelson, G.D. Heda, J. Liao, K.L. Kirk, E.J. Sorscher, J.R. Hanrahan and J.P. Clancy. 2003. A macromolecular complex of β-2 adrenergic receptor, CFTR, and ezrin/radixin/moesin-binding phosphoprotein 50 is regulated by PKA. *Proceedings of the National Academy of Sciences of the United States of America* 100: 342–346.

Olson, K.R. 2002. Gill circulation: regulation of perfusion distribution and metabolism of regulatory molecules. *Journal of Experimental Zoology* 293: 320–335.

Paajanen, V. and M. Vornanen. 2002. The induction of an ATP-sensitive K(+) current in cardiac myocytes of air- and water-breathing vertebrates. *Pflügers Archives* 444: 760–770.

Peters, K.W., J. Qi, R.A. Watkins and R.A. Frizzell. 1999. Syntaxin 1A inhibits regulated CFTR trafficking in *Xenopus* oocytes. *American Journal of Physiology* 277: C174–C180.

Pierson, P.M., M.E. Guibbolini, N. Mayer-Gostan and B. Lahlou. 1995. ELISA measurements of vasotocin and isotocin in plasma and pituitary of the rainbow trout: effect of salinity. *Peptides* 16: 859–865.

Pohl, S., M.G. Darlison, W.C. Clarke, K. Lederis and D. Richter. 2001. Cloning and functional pharmacology of two corticotropin-releasing factor receptors from a teleost fish. *European Journal of Pharmacology* 430: 193–202.

Prat, A.G., Y.-F. Xiao, D.A. Ausiello and H.F. Cantiello. 1995. cAMP-independent regulation of CFTR by the actin cytoskeleton. *American Journal of Physiology* 268: C1552–C1561.

Sakamoto, T., S. Yokota and M. Ando. 2000. Rapid morphological oscillation of mitochondrion-rich cell in estuarine mudskipper following salinity changes. *Journal of Experimental Zoology* 286: 666–669.

Sardet, C., M. Pisam and J. Maetz. 1979. The surface epithelium of teleostean fish gills, cellular and junctional adaptations of the chloride cell in relation to salt adaptation. *Journal of Cell Biology* 80: 96–117.

Scheide, J.I. and J.A. Zadunaisky. 1988. Effect of Atriopeptin II on isolated opercular epithelium of *Fundulus heteroclitus*. *American Journal of Physiology* 254: R27–R32.

Scott, G.R. and P.M. Schulte. 2005. Intraspecific variation in gene expression after seawater transfer in gills of the euryhaline Killifish *Fundulus heteroclitus*. *Comparative Biochemistry and Physiology* A 141: 176–182.

Singer, T.D., S.J. Tucker, W.S. Marshall and C.F. Higgins. 1998. A divergent CFTR homologue: highly regulated salt transport in the euryhaline teleost *Fundulus heteroclitus*. *American Journal of Physiology* 274: C715–C723.

Song, J.C., P.K. Rangachari and J.B. Matthews. 2002. Opposing effects of PKCα and PKCε on basolateral membrane dynamics in intestinal epithelia. *American Journal of Physiology* 283: C1548–C1556.

Stagg, R.M. and T.J. Shuttleworth. 1984. Hemodynamics and potentials in isolated flounder gills: Effects of catecholamines. *American Journal of Physiology* 246: R211–R220.

Sundin, L. and S. Nilsson. 2002. Branchial innervation. *Journal of Experimental Zoology* 293: 232–248.

Swiatecka-Urban, A., M. Duhaime, B. Coutermarsh, K.H. Karlson, J. Collawn, M. Milewski, G.R. Cutting, W.B. Guggino, G. Langford and B.A. Stanton. 2002. PDZ domain interaction controls the endocytic recycling of the cystic fibrosis transmembrane conductance regulator. *Journal of Biological Chemistry* 277: 40099–40105.

Takei, Y. and S. Hirose. 2002. The natriuretic peptide system in eels: a key endocrine system for euryhalinity? *American Journal of Physiology* 282: R940–R951.

Tamaoki, J., A. Chiyotani, H. Takemura and K. Konno. 1997. 5-Hydroxytryptamine inhibits Na^+ absorption and stimulates Cl^- secretion across canine tracheal epithelial sheets. *Clinical and Experimental Allergy* 27: 972–977.

Tipsmark, C.K. and S.S. Madsen. 2003. Regulation of Na^+, K^+-ATPase activity by nitric oxide in the kidney and gill of the brown trout. *Journal of Experimental Biology* 203: 1503–1510.

Towle, D.W., M.E. Gilman and J.D. Hempel. 1977. Rapid modulation of gill Na^+,K^+-dependent ATPase during rapid acclimation of the killifish *Fundulus heteroclitus* to salinity change. *Journal of Experimental Zoology* 202: 179–186.

van Praag, D., S.J. Farber, E. Minkin and N. Primor. 1987. Production of eicosanoids by the killifish gills and opercular epithelia and their effect on active transport of ions. *General and Comparative Endocrinology* 67: 50–57.

Warne, J.M. 2001. Cloning and characterization of an arginine vasotocin receptor from the euryhaline flounder *Platichthys flesus*. *General and Comparative Endocrinology* 122: 312–319.

Warne, J.M. and R.J. Balment. 1995. Effect of acute manipulation of blood volume and osmolality on plasma [AVT] in seawater flounder. *American Journal of Physiology* 269: R1107–R1112.

Winter, M.J., P.C. Hubbard, C.R. McCrohan and R.J. Balment. 1999. A homologous radioimmunoassay for the measurement of Urotensin II in the euryhaline flounder, *Platichthys flesus*. *General and Comparative Endocrinology* 114: 249–256.

Zadunaisky, J.A., S. Cardona, L. Au, D.M. Roberts, E. Fisher, B. Lowenstein, E.J. Cragoe Jr. and K.R. Spring. 1995. Chloride transport activation by plasma osmolarity during rapid adaptation to high salinity of *Fundulus heteroclitus*. *Journal of Membrane Biology* 143: 207–217.

Zadunaisky, J.A., M. Balla and D.E. Colon. 1997. A reduction in chloride secretion by lowered osmolality in chloride cells of *Fundulus heteroclitus*. *Bulletin of the Mount Desert Island Biological laboratory* 24: 52. (Abstract).

CHAPTER

15

Control of Calcium Balance in Fish

Pedro M. Guerreiro* and Juan Fuentes

INTRODUCTION

Perhaps the most evident function for calcium (Ca^{2+}) in animals is the formation of a hard body shaping and protective structure, which may either be external (shell of invertebrates) or internal (skeleton of vertebrates), and which consists primarily of Ca^{2+} carbonate and/or Ca^{2+} phosphate complexes. In teleost fish about 99% of the total body Ca^{2+} is incorporated into bone, scales, teeth and otoliths (Flik *et al.*, 1986a; Wendelaar Bonga and Pang, 1991). However, Ca^{2+}, either ionic or protein-bound, plays a crucial role in numerous other physiological and biochemical processes such as muscular contraction, vision, blood coagulation, regulation of enzymatic reactions, modulation of permeability

Authors' address: Molecular and Comparative Endocrinology, Centre of Marine Sciences, CCMAR, CIMAR Laboratório Associado, University of Algarve, Campus de Gambelas, 8005-139 Faro, Portugal.

**Corresponding author:* E-mail: pmgg@ualg.pt

and excitability of plasma membranes, neural and intercellular communication and intracellular signalling (Riccardi, 1999, 2000). It is also important in the reproductive cycle of many animals, namely fish, where it is an essential component of vitellogenin, the egg yolk precursor protein (Yeo and Mugiya, 1997) and large amounts of Ca^{2+} are also necessary for the production of the eggshells of many species. The triggering and maintenance of sperm mobility and function and post-fertilization events also require the presence of this ion (Stricker, 1999).

The preservation of such vital functions requires the regulation of Ca^{2+}, within narrow limits, both in the intracellular media (usually around 100 nM) and in the extracellular fluids (about 1-1.5 mM), since deviations from the physiological range will rapidly disturb the neural, muscular and cardiovascular functions, leading eventually to either death or intermediate stages of tetany (due to lack of Ca^{2+}) or lethargy (caused by Ca^{2+} excess) (Mundy, 1990). Fish also have to comply with such limits, and many of the disturbances caused by Ca^{2+} disequilibria in tetrapods are also observable in fish, as is the case of lethargy caused in eels by the removal of the Corpuscles of Stannius (Flik *et al.*, 1995). Nonetheless, from the data of several studies (Hanssen *et al.*, 1989; Guerreiro *et al.*, 2002), it seems clear that fish—in relation to mammals—are capable to tolerate a wider variation between minimum and maximum blood Ca^{2+} concentration, being able to survive for long periods in hypercalcemic conditions that would be critical to terrestrial vertebrates (Mundy, 1990), which may be a reflection of the diversity of habitats they live and have evolved in.

Teleosts represent approximately half of the extant vertebrate species. They inhabit an immense variety of freshwater (FW) environments, from the extremely soft, ion-poor Amazonian rivers (where Ca^{2+} levels are usually less than 10 μM; Gonzalez *et al.*, 2002) to the very hard waters that drain calcareous regions. And obviously, fish are not only found in the marine environment, where Ca^{2+} concentration in seawater (SW) is constant at approximately 10 mM, but also in coastal waters with a wide variety of salinities that range from mixing brackish waters in river estuaries to hypersaline situations in evaporating enclosed pools, comprising a large variation in Ca^{2+} concentration. Some fish are even able to thrive in waters of extremely high salinity and Ca^{2+} levels, and have been observed swimming in the Dead Sea estuary, in areas where

Ca^{2+} concentration is close to 90 mM (Skadhauge and Lotan, 1974). Not only have some species adapted to life in these extreme conditions but several can live in environments where Ca^{2+} concentration fluctuates greatly, either seasonally or on a daily basis, and have the capability to adjust their mechanisms to maintain Ca^{2+} homeostasis in such potentially challenging conditions.

Salinity or water softness/hardness are not the only factors relevant in fish adaptation. Fish species also inhabit either in quite acidic or in rather alkaline environments, such as the rivers of tropical forest (Gonzalez *et al.*, 1998, 2002) and several lakes of the African rift region (Wood *et al.*, 1989, 2002; Bergman *et al.*, 2003), respectively, conditions that modulate speciation and ion availability and change the affinity of transporting mechanisms and the toxicity of certain ions (Spry and Wiener, 1991).

On land, food and drinking water are the only available source of Ca^{2+}, making the gastrointestinal tract of tetrapods the major site for Ca^{2+} uptake. Furthermore, due to the varying amount of Ca^{2+} in the diet and the episodic nature of feeding, terrestrial vertebrates have to rely on factors and mechanisms that by default conserve Ca^{2+} and have a complex system of cell types and hormonal factors that allows them to use their internal skeleton as a Ca^{2+} reservoir. Fish, however—even those living in soft FW—are surrounded by a virtually infinite supply of Ca^{2+}, readily obtainable from the environment. Therefore, in addition to the Ca^{2+} acquired via the food, aquatic animals have access to the ionic Ca^{2+} occurring in water, and the mechanisms for Ca^{2+} uptake and control reflect these important differences in availability and accessibility.

Either in land or in water, several control mechanisms are at play to maintain Ca^{2+} homeostasis, including a range of tissues; organs and cell types, transporting mechanisms, sensory systems and endocrine factors, which monitor and influence the regulation of cellular and circulating Ca^{2+} concentration. This chapter will comprehensively review the existent literature to illustrate the particularities of Ca^{2+} balance in teleost fish: the tissues and mechanisms involved in uptake and excretion in FW and marine species and the interaction with ambient Ca^{2+}. The focus will largely be on the role of endocrine factors that control Ca^{2+} homeostasis in fish and the relevant developments that arose in recent years in this field.

CA^{2+} EXCHANGE TISSUES AND SENSING MECHANISMS

Given the close contact of fish with the Ca^{2+} source, water exposed tissues with large areas and thin epithelia are adequate and energetically favorable for Ca^{2+} exchange. Therefore, in fish the gills, the gut and the integument, in early larval stages, are of crucial importance for Ca^{2+} uptake.

Gills

The relevance of the branchial tissue for Ca^{2+} uptake has been demonstrated in a large variety of fish species, and the gill epithelia is probably the most important site for Ca^{2+} uptake in fish, being those freshwater, marine or euryhaline species (Fenwick, 1989; Flik *et al.*, 1995, 1996; Perry, 1997; Marshall, 2002; Evans *et al.*, 2005).

The branchial epithelia are in direct contact with the Ca^{2+} present in the environment, which in SW is present in concentrations well above those of the Ca^{2+} in blood plasma (total Ca^{2+}, including bound and free fractions, 2-3 mM, ionic Ca^{2+} around 1-1.5 mM), and in FW in concentrations similar or lower than these internal levels. In both cases, fish extract Ca^{2+} from the water. Evidence from electrochemical studies (Potts and Hedges, 1991; Witters *et al.*, 1992; Verbost *et al.*, 1993b, 1994; Marshall *et al.*, 1995) demonstrated that the transepithelial potential in fish gill is always more positive than the equilibrium potential for Ca^{2+} across the integument and, therefore, the driving force for passive Ca^{2+} movement across the gills is directed outwards both in FW and in SW fish (Flik *et al.*, 1995, 1996). This indicates that the overall branchial Ca^{2+} uptake does not occur by diffusion via paracellular routes but primarily due to active transport mediated by a transcellular sequence of events, including a passive entry step, a cytoplasmatic transport phase and an active extrusion step (Fig. 15.1).

Branchial localization of Ca^{2+} uptake. The major sites for Ca^{2+} transport in gills are the mitochondria-rich (MR) cells, also called chloride cells or ionocytes (e.g., Zia and McDonald, 1994; Marshall *et al.*, 1995; Marshall and Bryson, 1998; Marshall, 2002). These specialized epithelial cells occupy a small fraction of the total gill surface area, largely confined to the filament epithelium and usually concentrated in the intra-lamellar regions. In general, only a few MR cells are located in the secondary lamellae. Such cells are also found abundantly in the opercular membrane,

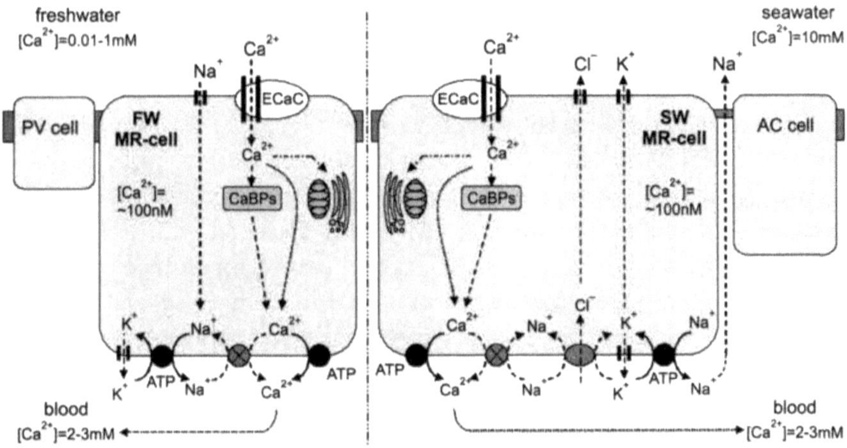

Fig. 15.1 Working model of Ca^{2+} transport in the FW MR-cell (left) and the SW MR-cell (right). Solid lines indicate active transport; dashed lines indicate secondary transport or diffusion across channels or across the cell. Solid circles represent ATPases, grey circles stand for co-transporters and exchangers, and parallel lines represent ion channels. The mechanisms for transcellular transport of Ca^{2+} are similar in both FW and SW. Ca^{2+} entry occurs down an electrochemical gradient through an apical voltage-independent Ca^{2+} channel (ECaC). Within the cell, Ca^{2+} can be sequestered by cell organelles. Ca^{2+} en route binds to Ca^{2+}-binding proteins and is transported across the cytosol. Only residual Ca^{2+} remains free. Extrusion in the basolateral membrane occurs via a Ca^{2+}-ATPase or a Na^{+}/Ca^{2+}-exchanger. Secondary transport by the Na^{+}/Ca^{2+}-exchanger depends directly on the Na^{+} gradient produced by the Na^{+}/K^{+}-ATPase. In SW, the same gradient can be used by the $Na^{+}/K^{+}/2Cl^{-}$. (Adapted from Marshall *et al.*, 1998; Marshall, 2002).

a thin epithelia exposed to the water in the fish posterior bucal cavity and that was also shown to be an important site for Ca^{2+} uptake (Marshall *et al.*, 1992, 1995; McCormick *et al.*, 1992; Verbost *et al.*, 1997). In this sense, the opercular epithelia is often regarded as an extension of the branchial ion-exchanging tissue, and it has been shown that the operculum contains 15% of the total branchial Na^{+}/K^{+}-ATPase activity in tilapia, *Oreochromis mossambicus* (Wendelaar Bonga *et al.*, 1990), and accounts for 46% of the total Ca^{2+} uptake in the killifish, *Fundulus heteroclitus* (Marshall *et al.*, 1995). Several other non-branchial areas of the skin contain MR cells and may be involved in Ca^{2+} uptake, but the relative contribution is most probably proportional to the number of these cells and, therefore, rather small. In larval stages, though, abundant MR cells can be found in the head and trunk skin, and in fin membranes, and the contribution for Ca^{2+} uptake is most probably considerable (see Early Life Stages).

MR cells have a distinct apical surface, an array of sub-apical vesicles and extensive tubular systems emanating from the basolateral membrane, but the most distinctive feature is the abundance of mitochondria, indicative of an high level of cell activity (Perry, 1997; Marshall and Bryson, 1998; Marshall, 2002; Evans *et al.*, 2005). The rates of Ca^{2+} uptake, *in vivo* and *in vitro* are well correlated with the numbers and surface area of MR cells (Marshall *et al.*, 1992; Perry *et al.*, 1992), and modifications in environmental Ca^{2+} trigger significant and rapid changes in the shape, number and distribution of MR cells in several species, regardless of the initial ion-composition of the water (Mayer-Gostan *et al.*, 1983; Katoh and Kaneko, 2002; Sakamoto and Ando, 2002; Moron *et al.*, 2003). Histological data has showed the presence of oxalate-induced Ca^{2+} precipitates in the surroundings the MR cell apical membrane in goldfish, *Carassius auratus* (Ishihara and Mugiya, 1987), and the presence of $^{45}Ca^{2+}$ was specifically localized by autoradiography in rainbow trout (*Oncorhynchus mykiss*) MR cells (Zia and McDonald, 1994). In addition, studies in isolated MR cells have demonstrated the importance of the Na^+/Ca^{2+}-exchanger activity, a key player in Ca^{2+} transport (Li *et al.*, 1997), and recently, Galvez *et al.* (2006), have demonstrated differential Ca^{2+} transport rates in two functionally distinct populations of rainbow trout MR-cells and also in gill pavement (PV)-cells *in vitro*. They showed that $^{45}Ca^{2+}$ uptake was most pronounced in PNA^+ (peanut lectin agglutinin positive) MR cells, with levels over threefold higher than those found in either PNA^- MR or in PV cells, suggesting that the PNA^+ MR cell type is a high-affinity and high-capacity site for apical entry of Ca^{2+} in the gill epithelium.

Ca^{2+} uptake in the apical membrane. Intracellular Ca^{2+} concentration is quite low (in the submicromolar range (Schoenmakers *et al.*, 1993; Larsson *et al.*, 1998)) and Ca^{2+} entry across the apical membrane of the MR cells is believed to be a passive process, following the concentration gradient, that occurs even in very soft waters. Due to the detrimental effects of high intracellular Ca^{2+} concentrations that could arise from this gradient difference, especially in SW, it is also expected that this step is subject to both short- and long-term hormonal regulation. The apical entry step occurs probably through voltage-insensitive Ca^{2+} channels, and their numbers and distribution may make this the rate-limiting step in the overall Ca^{2+} uptake process (Flik *et al.*, 1993b; Perry *et al.*, 2003). The initial studies showed that lanthanum, La^{3+}, a voltage-independent Ca^{2+}-channel inhibitor, blocked the entry of Ca^{2+} in branchial transporting cells

(Perry and Flik, 1988) and data gathered by the use of specific L-type Ca^{2+} channel antagonists verapamil and diltiazem demonstrated that these were effective in reducing Ca^{2+} uptake but that nifedipine was without effect (Bjornsson *et al.*, 1999), making inconclusive the existence or importance of voltage-sensitive Ca^{2+} channels in the uptake process. More recently Rogers and Wood (2004) re-addressed this issue using both voltage sensitive and voltage insensitive apical channel blockers. In their study, Ca^{2+} uptake by rainbow trout was 55% reduced in the presence of 1 μM La^{3+} and 74% by a similar concentration of cadmium (Cd^{2+}), while the voltage-dependent, L-type Ca^{2+}-channel blockers nifedipine and verapamil did not affect Ca^{2+} influx. Although more data is necessary to fully characterize the Ca^{2+} entry step in the brachial epithelium, this initial process seems to be chiefly mediated by voltage-independent channels.

Recently, Qiu and Hogstrand (2004), Pan and colleagues (2005) and Shahsavarani and co-workers (2006) isolated and cloned a branchial epithelial Ca^{2+} channel (ECaC) in the pufferfish *Fugu rubripes*, the zebrafish (*Danio rerio*) and in the rainbow trout, respectively, which has a sequence similarity to the mammalian non-voltage gated epithelial Ca^{2+} channels TRPV5 and TRPV6 responsible for Ca^{2+} reabsorption in the kidney (Nijenhuis *et al.*, 2005). Functional expression of the *Fugu* ECaC in Madin–Darby canine kidney (MDCK) cells confirmed that the channel mediates Ca^{2+} influx and was also permeable to zinc (Zn^{2+}) (Qiu and Hogstrand, 2004). Furthermore, the rainbow trout ECaC mRNA and protein levels respond to treatments known to increase or decrease Ca^{2+} uptake capacity (Shahsavarani and Perry, 2006). Exposure of fish to soft water caused a significant increase in ECaC mRNA levels and an increase in ECaC protein expression, while infusion with $CaCl_2$ was associated with a significant decrease in ECaC mRNA levels. ECaC mRNA and protein expression were also increased in fish treated with cortisol, known to produce hypercalcemic effects. Interestingly, the results of these studies also demonstrated that the ECaC expression is not confined to the MR cells but also exists in PV cells that could, thus, potentially contribute to Ca^{2+} uptake (see Shahsavarani *et al.*, 2006). Although more functional characterization and species comparative analysis of the ECaC is needed, there is evidence that this may be the mechanism responsible for the first step in branchial Ca^{2+} uptake.

Transcelullar Ca^{2+} transport. Once inside the cell, most Ca^{2+} is either sequestered in organelles or diffuses through the cytoplasm probably bound to Ca^{2+}-binding proteins (CaBPs) (Perry, 1997; Bjornsson *et al.*, 1999). To date, little focus has been given to this step. Several CaBPs have been identified in the gills of the channel catfish *Heteropneustes fossilis*, but immunolocalization studies indicate that with the exception of parvalbumin, all seem to be concentrated in neuroendocrine cells (Fasulo *et al.*, 1998). Fish-specific CaBPs with characteristics similar to calbindin and s-100, found in Ca^{2+} transporting tissues of higher vertebrates, were found in the branchial tissue of the European eel *Anguilla anguilla* (Hearn *et al.*, 1978) and the catfish *Ictalurus punctatus* (Porta *et al.*, 1996), but there is no functional evidence linking to Ca^{2+} transport.

Basolateral Ca^{2+} extrusion. The exit of Ca^{2+} across the basolateral membrane is an active, energy-consuming step that occurs against an electrochemical gradient. This step involves membrane transporters, such as the Ca^{2+}-ATPase and the Na^+/Ca^{2+}-exchanger (Flik and Verbost, 1993; Flik *et al.*, 1996; Perry, 1997; Marshall and Bryson, 1998; Marshall, 2002; Perry *et al.*, 2003). The relative importance of these two-membrane carriers remains to be fully determined. Earlier studies in eel and tilapia branchial membrane vesicles isolated a Ca^{2+}-ATPase with high affinity for Ca^{2+} (Flik *et al.*, 1983, 1984a, b, 1985a, b) and showed that it was up-regulated by hormonal factors that also lead to hypercalcemia (Flik *et al.*, 1984b, c, 1989; Flik and Perry, 1989). These observations led to the suggestion that Ca^{2+}-ATPase was responsible for most of the transcellular Ca^{2+} uptake (Flik *et al.*, 1985a, b), and comprehensive reviews on the characterization and kinetics of these mechanisms exist (Flik and Verbost, 1994, 1995; Flik *et al.*, 1995, 1996). However, more recent studies showed that there is also an important role played by the Na^+/Ca^{2+}-exchanger (Flik *et al.*, 1997). In the isolated opercular membrane of killifish, the Na^+/Ca^{2+}-exchanger accounted for approximately 80-85% of Ca^{2+} transport (Verbost *et al.*, 1997), which was dramatically reduced when Na^+ was removed from the bathing solution, or when ouabain was added. On the other hand, treatment with vanadate, a Ca^{2+}-ATPase inhibitor had little effect on Ca^{2+} uptake rate. Furthermore, the affinity and the number of these carriers, taken with the Na^+/K^+-ATPase, would suffice to account for Ca^{2+} and sodium homeostasis in FW and SW (Verbost *et al.*, 1994). However, acclimation of rainbow trout to 70% SW evokes an 8-fold increase in Ca^{2+}-ATPase and a 5-fold increase in Na^+/Ca^{2+} exchange, which seem to be far beyond the requirements for transepithelial Ca^{2+}

transport (Flik *et al.*, 1997), thus implying the need for still other Ca^{2+} transport mechanisms. Clearly, more studies are necessary to determine the importance of these two transporting mechanisms and, in the case of Na^{+}/Ca^{2+}-exchanger, the full coupled machinery involved in the gradient production and maintenance. It seems apparent though that different species, with different adaptation processes, may rely differentially on either transporter.

Branchial Ca^{2+} excretion. Although the renal tubule is considered the primary site for divalent ion excretion, disparate results indicate a possible branchial role for Ca^{2+} excretion in fish. Ca^{2+} loss via the gills happens by paracellular routes driven by the outward electrochemical force for Ca^{2+} both in FW and in SW (Perry and Flik, 1988; Verbost *et al.*, 1994; Flik *et al.*, 1995; Marshall *et al.*, 1995) and is further influenced by the Ca^{2+} permeability of the intercellular junctions. In the epithelium of the gills of FW fish, the electrochemical gradient for Ca^{2+} is generally directed outwardly and a substantial passive efflux of Ca^{2+} has been reported (Flik *et al.*, 1995). In marine fish, despite the transepithelial potential that prevents passive Ca^{2+} entry, Ca^{2+} loss is at first sight unlikely due to the high concentration in water. However, the ingestion of SW and subsequent intestinal absorption may present an important Ca^{2+} load which can also be excreted by branchial mechanisms. Hickman (1968) calculated that in the southern flounder (*Paralichthys lethostigma*), 68% of the ingested Ca^{2+} was absorbed, and only 11% of that could be accounted for by renal loss, and thus the remaining 89% of the Ca^{2+} taken up would have to be excreted by the gills. Evans and colleagues (2005) calculated the Nernst equilibrium electrical potential for Ca^{2+} in the order of 10–15 mV (inside positive), which is probably below that across the marine fish gill epithelium, leaving a margin for passive branchial excretion of Ca^{2+} More information is needed to substantiate this hypothesis.

Disruption of branchial uptake by metals: the example of cadmium, lead and zinc. Ca^{2+} uptake can be disrupted by the presence of other metals in water and diet. This interference can be acute or chronic, moderate or extreme and lead to imbalance in Ca^{2+} homeostasis, growth arrest, deformities and death (Spry and Wiener, 1991; Croke and McDonald, 2002). Deleterious effects of increasingly common heavy metals found in water basins such as Cd^{2+}, Zn^{2+}, lead (Pb^{2+}), and others, in Ca^{2+} uptake are associated with similarities in ionic size and charge. It is thought that at least some of these metals use the apical Ca^{2+}-channel to gain entry in epithelial cells and establish a competitive interaction with the Ca^{2+}-binding sites in basolateral membrane transporters (Verbost *et al.*, 1987, 1992; Schoenmakers and Flik, 1992; Qiu and Hogstrand, 2004; Franklin *et al.*, 2005; Morgan *et al.*, 2005; Galvez *et al.*, 2006; Niyogi and Wood, 2006). Increases in water Cd^{2+} significantly decrease branchial Ca^{2+} uptake and increase Cd^{2+} accumulation (Verbost *et al.*, 1987; Chang *et al.*, 1998; Niyogi and Wood, 2006). On the other hand, branchial Cd^{2+} transport was partially reduced in fish fed with high-Ca^{2+} diets, which caused downregulation of the branchial Ca^{2+} transporting mechanisms used by Cd^{2+} (Baldisserotto *et al.*, 2005; Franklin *et al.*, 2005). Since the effect is based in competitive kinetics, the toxicity of such metals is most severe in soft waters because of the low availability of Ca^{2+}, as was observed in the Amazonian teleost, *Colossoma macropomum*. In this species, exposure to water-borne Cd^{2+} induced an average of 42% inhibition in whole body Ca^{2+} uptake relative to controls within 3 hours of exposure, which increased to 91% after 24 hours (Matsuo *et al.*, 2005). Previous acclimation to elevated Ca^{2+} concentrations protected fish against acute Cd^{2+} influence. There are no known effects of Cd^{2+} on Ca^{2+} efflux.

Rogers and Wood (2004) addressed the interactions between water-borne Pb^{2+} and Ca^{2+} uptake in FW rainbow trout. Representative doses of Pb^{2+} significantly reduce Ca^{2+} influx, although its effects were not as pronounced as those of Cd^{2+}. Apical entry in gill cells is greatly inhibited by addition of La^{3+} and Cd^{2+} to the water but not by blockers of voltage-gated Ca^{2+} channels, and Pb^{2+} accumulation in branchial cells is reduced by increasing waterborne Ca^{2+} concentrations. This indicates that Pb^{2+} enters via the ECaC. In this study, the high-affinity Ca^{2+}-ATPase activity was not acutely affected but long-term exposure to dissolved Pb^{2+} significantly reduced the pump activity, indicating a possible non-competitive component to Pb^{2+}-induced Ca^{2+} disruption.

Ca^{2+} uptake is also inhibited when elevated concentrations of Zn^{2+} are present. As for the other two metals dealt above, the apical entry of Zn^{2+} is inhibited by La^{3+}. The putative presence of an active basolateral transporter for Zn^{2+} was investigated *in vitro* on isolated basolateral membranes from rainbow trout gill cells (Hogstrand *et al.*, 1996). In this study, there was no evidence of Na^{+}-gradient driven Zn^{2+} transport but this metal was found to be an effective blocker of the Ca^{2+}-ATPase. This inhibition was both of a competitive and a non-competitive nature, although the competitive component prevailed. Injection of $CaCl_2$, increased water Ca^{2+} concentrations and elevated dietary Ca^{2+} all greatly reduced Ca^{2+} and Zn^{2+} uptake (Barron and Albeke, 2000; Hogstrand *et al.*, 1996; Niyogi and Wood, 2006).

Intestine

In fish, the role of the intestine in Ca^{2+} uptake is poorly understood, and differences in Ca^{2+} uptake at this site occur between FW and SW fish (see Fig. 15.2). In FW, the amount of Ca^{2+} that gains entry through drinking is rather small and the intestinal contribution to total Ca^{2+} uptake is frequently considered accessory (Flik *et al.*, 1996), since the estimated branchial influx suffices for growth and homeostasis (Flik *et al.*, 1995). Seawater fish drink copiously to compensate for the water loss and substantial amounts of Ca^{2+} enter the gut. From this an important part can be absorbed, and the intestinal contribution for total Ca^{2+} uptake in marine fish is relevant.

Studies using different *in vivo* and *in vitro* techniques demonstrated a net absorption of Ca^{2+} across the marine teleosts' intestine, which has been estimated to be 20% in tilapia (Schoenmakers *et al.*, 1993), 40% in cod, *Gadus morhua* (Sundell and Bjornsson, 1988), 70% in the southern flounder (Hickman, 1968), and 90% and 40% in seabream *Sparus auratus*

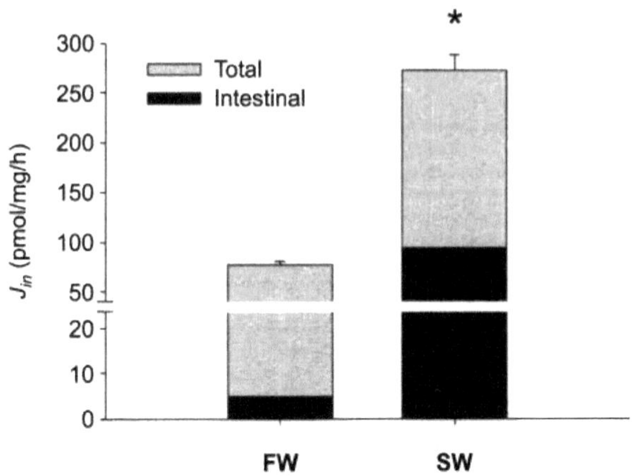

Fig. 15.2 Total and intestinal Ca^{2+} uptake by tilapia larvae in FW and SW. In SW larvae, Ca^{2+} influx is 3.5-fold higher than in FW and the calculated contribution of the intestinal route amounts to as much as 35% of the total uptake. In FW, the maximum contribution of the intestine for total Ca^{2+} uptake is only close to 5%. * indicates significant difference from the FW group ($P<0.05$, *t*-test). See text for details.

juveniles (Guerreiro *et al.*, 2002) and larvae, respectively. In SW-adapted tilapia larvae, the intestinal contribution amounts to about 35% of the total Ca^{2+} uptake (Fig. 15.2). Overall, these percentages account for up to 30-40% of the total Ca^{2+} uptake in most studied fish (Flik *et al.*, 1985b, 1990b; Sundell and Bjornsson, 1988; Vonck *et al.*, 1998) and up to 60-70% of the total Ca^{2+} uptake in SW adapted seabream juveniles (Guerreiro *et al.*, 2002). Furthermore, the intestinal contribution may increase in times of extra need for Ca^{2+}, such as during gonadal maturation (Mugiya and Watabe, 1977; Sundell and Bjornsson, 1988; Guerreiro *et al.*, 2002).

The cellular routes for Ca^{2+} uptake in intestine are similar to those in the gills, and Ca^{2+} uptake occurring in the intestinal tract is via active transport mechanisms (Flik *et al.*, 1990b; Schoenmakers *et al.*, 1993). Nevertheless, considering the leaky nature of the intestinal epithelium and the particularly high electrochemical gradients, significant paracellular transport takes place, and this fraction has been estimated as 40% of total Ca^{2+} absorption in the intestine in the marine cod (Sundell and Bjornsson, 1988).

The enterocyte cytoplasm is negatively charged when compared to the intestinal lumen, and intracellular Ca^{2+} concentrations are in the nanomolar range (Schoenmakers *et al.*, 1993; Larsson *et al.*, 1998). The combination of these two factors allows the Ca^{2+} to move across the brush-border membrane and into the enterocyte interior down an electrochemical gradient, and studies in brush border membrane vesicles isolated from FW fish indicated that uptake is composed by a saturable and a non-saturable component (Klaren *et al.*, 1993) and that the entry in these vesicles could be ATP-dependent, mediated by a P_2 purinoceptor-controlled Ca^{2+} channel (Klaren *et al.*, 1997). Studies in isolated cod enterocytes indicated that unlike the MR cell, the enterocyte apical Ca^{2+} channel is more likely a L-type voltage-dependent Ca^{2+} channel (Larsson *et al.*, 1998, 2002). As for MR cells, transport across the cytoplasma may require CaBPs (Bjornsson *et al.*, 1999). Extrusion from the enterocyte to the extracellular fluid occurs through the action of basolateral transporters which include Ca^{2+}-ATPases, but it is mainly Na^{+}-dependent (Schoenmakers and Flik, 1992; Flik and Verbost, 1993; Schoenmakers *et al.*, 1993; Larsson *et al.*, 2002). In studies with membrane vesicles, the simultaneous operation of the Ca^{2+}-ATPase and Na^{+}/Ca^{2+}-exchanger indicates that the extrusion activity of the exchanger exceeds that of the

ATPase (Flik *et al.*, 1990b). In agreement with the importance of the Na^+/Ca^{2+}-exchanger in the intestine, tilapia enterocytes display a much higher Na^+/K^+-ATPase activity than tilapia gill cells (Schoenmakers *et al.*, 1993). Na^+/K^+-ATPase activity in seabream enterocytes is also considerably higher than that measured in gills (Diaz *et al.*, 1998; Almansa *et al.*, 2001). Almansa *et al.* (2001) also showed heterogeneity in the biochemical properties of the Na^+-K^+-ATPase along the intestine of this marine species. This agrees with the importance and regional distribution of several of Na^+-driven transporters for nutrient uptake, including Na^+/Ca^{2+}-exchanger. Recently Fuentes *et al.* (2006) using intestinal segments from distinct regions of the seabream gastrointestinal tract mounted in Ussing chambers, showed the existence of differences in Ca^{2+} uptake along the intestinal tract that reflects the complexity of the calcium-transporting processes of fish intestinal epithelia.

Thus, the importance of intestinal Ca^{2+} uptake for total Ca^{2+} accumulation in fish is variable, due to Ca^{2+} concentration in water, the presence of chelating mechanisms, and the regional distribution and activity of Na^+/Ca^{2+}-exchanger and Ca^{2+}-ATPase.

The relevance of dietary Ca^{2+} is variable, depending on the content of Ca^{2+} in the food and/or on the environmental Ca^{2+} levels, and fish compensate accordingly: when dietary Ca^{2+} is low or absent, Ca^{2+} requirements can be compensated by uptake via the gills from the surrounding medium (Mugiya and Watabe, 1977) and in contrast, if dietary Ca^{2+} content becomes high there is a substantial decrease in branchial Ca^{2+} uptake (Baldisserotto *et al.*, 2005; Franklin *et al.*, 2005). Still, the contribution of dietary Ca^{2+} to fish Ca^{2+} uptake is uncertain and several studies suggest fish can live and grow normally on a Ca^{2+}-poor diet (see Flik *et al.*, 1995). However, in some cichlid and catfish species, growth was impaired or reduced when the Ca^{2+} content in the rearing water was low, and reduced growth has been shown in a variety of species in full-strength SW when reared on a Ca^{2+}-deficient food (e.g., Hossain and Furuichi, 2000). Seawater-adapted seabream juveniles fed Ca^{2+}-deficient for 9 weeks show growth rates comparable to siblings fed on a control diet, and the same was true for those fed a Ca^{2+}-sufficient diet in low ambient Ca^{2+}. When both diet and ambient Ca^{2+} were limiting, seabream experienced growth arrest, reduced Ca^{2+} accumulation and had lower circulating Ca^{2+} levels (Abbink *et al.*, 2004).

The intestine of SW fish is an important osmoregulatory organ and takes up large amounts of (sea) water (c.a. 10 mM Ca^{2+}) to compensate for the losses that occur via the integument. Most of the water is absorbed in the anterior gastrointestinal tract, which leads to increased Ca^{2+} availability in the intestine and concentrations as high as 25 mM have been measured in SW-adapted tilapia (Klaren *et al.*, 1993) and in the winter flounder *Pleuronectes americanus* (Parmelee and Renfro, 1983). The intestinal epithelium is leaky and such luminal concentrations can greatly increase diffusional Ca^{2+} transport. In an apparent measure to prevent extra Ca^{2+} loading of the fish, *in vitro* studies showed that the net Ca^{2+} transport is reduced in the proximal region in tilapia adapted to SW compared to those in FW (Schoenmakers *et al.*, 1993). An important adaptation to reduce Ca^{2+} activity and availability in the intestine of SW fish has been proposed: the chelation of ingested Ca^{2+} by bicarbonate (HCO_3^-) to produce deposits of calcium carbonate formed by the favouring high pH and calcium concentration in the intestinal fluid. According to recent studies (Wilson *et al.*, 2002; Wilson and Grosell, 2003; Taylor and Grosell, 2006), HCO_3^- secretion may play a role in preventing excess Ca^{2+} absorption by precipitating a large fraction of the imbibed Ca^{2+} as $CaCO_3$. In European flounder (*Platichthys flesus*), the increase of intestinal Ca^{2+} to 20 mM induced a 57% increase of net HCO_3^- secretion, and higher Ca^{2+} concentrations (40 and 70 mM) in ambient SW resulted in reduced plasma total CO_2, probably used to fuel intestinal HCO_3^- secretion (Wilson and Grosell, 2003). As a result, it is proposed that up to 75% of the Ca^{2+} in the intestinal fluid is precipitated and then excreted as faeces, thus reducing disproportionate branchial and/or renal excretion and risk of renal stone formation. Intestinal Ca^{2+} excretion in the cod was estimated to represent 20% of total Ca^{2+} excretion and 50% of the extrarenal excretion (Sundell and Bjornsson, 1988), but little is known in the case of other species.

Kidney

Ca^{2+} excretion in fish takes place mainly via the urine, and the amount of renally excreted Ca^{2+} varies with the Ca^{2+} concentration in water. In the American eel (*Anguilla rostrata*), Schmidt-Nielsen and Renfro (1975) observed plasma values of 2.41 ± 0.06 mM Ca^{2+} in SW (9.8 mM Ca^{2+}) and 2.63 ± 0.07 mM in FW (0.7 mM Ca^{2+}), while the concentrations in urine and the Ca^{2+} excretion rates were approximately 5 mM and

3.8 μmol/kg/h for SW fish and 1mM and 1.8 μmol/kg/h for FW fish, respectively. Similar results were obtained for the SW and brackish water adapted winter flounder (Elger *et al.*, 1987) and illustrate the fact that in the kidney of FW fish, Ca^{2+} losses are minimized by means of active tubular reabsorption of ultrafiltered Ca^{2+} (calculated by Schmidt-Nielsen and Renfro (1975) to be up to 80% of the amount in plasma in the American eel), which is demonstrated by the fact that they have a urine/ultrafiltered plasma ratio of less than 1 (Foster, 1976; Fenwick, 1981). Despite the lower concentration of Ca^{2+} in urine due to the highly dilute urine produced to counteract the osmotic inflow of water, the large volume produced makes the renal excretion an important fraction (up to 50% according to Hobe *et al.* (1984)) of the total Ca^{2+} excretion. In SW-fish, the kidney has a Ca^{2+}-excretory function, yet urine flow is reduced to a minimum. Tubular reabsorption of Ca^{2+} is reduced compared to FW animals and this ion is also secreted (Schmidt-Nielsen and Renfro, 1975; Renfro, 1978; Renfro *et al.*, 1982; Elger *et al.*, 1987). In fact, the importance of secretory processes in the SW kidney can be illustrated by studies in SW and FW-acclimated rainbow trout in which the percentage of all glomeruli that were perfusing and filtering was only 5% in SW, and 45% in FW (Marshall and Grosell, 2006). Despite the low glomerular filtration rate and urine production, renal Ca^{2+} excretion was determined to account for up to 65% of the total Ca^{2+} excretion in the marine Atlantic cod (Bjornsson and Nilsson, 1985).

The mechanisms of renal Ca^{2+} transport have not received enough attention and little is known about the pathways used for Ca^{2+} movement in the entire fish nephron and across the renal tubule epithelium. Transepithelial potential (TEP) measurements in the isolated nephron segments of FW-fish produced variable results, but the epithelium seems to be impermeable to passive paracellular Ca^{2+} movements (Cliff and Beyenbach, 1988), indicating that the processes of reabsorption must occur transcellularly. In SW fish, negative TEP in the lumen (Beyenbach *et al.*, 1986; Cliff and Beyenbach, 1988) seems to exclude the possibility for paracellular excretion. Consequently, renal Ca^{2+} reabsorption and secretion must be active processes.

In isolated vesicles from the euryhaline goby *Gillichthys mirabilis* and tilapia, Ca^{2+} transport across the basolateral membrane was found to be an energy dependent process mediated by the activity of high-affinity Ca^{2+}-ATPases (Doneen, 1993; Bijvelds *et al.*, 1995). In *G. mirabilis* activity of these pumps was increased when fish were transferred from SW to FW, and were higher in FW-acclimated fish and lower in fish exposed to 200% SW (Doneen, 1993), while tilapia plasma membrane Ca^{2+} pump activity was 56% lower in preparations from SW-adapted fish than in preparations from FW-adapted fish (Bijvelds *et al.*, 1995), which is in keeping with the differences observed in urine Ca^{2+} concentration, and indicate an environmental regulation of renal reabsorption. Furthermore, the use of specific Ca^{2+}-ATPase inhibitors and Ca ionophore A23187 showed that this pump accounted for 55% of the total Ca^{2+} transport by the vesicles and that most of the uptake is ATP-dependent (Bijvelds *et al.*, 1995). These data establish the importance of basolateral Ca^{2+}-ATPase in the reabsorption process in fish. On the other hand, and in contrast to the intestinal mechanism, the involvement of the Na^+/Ca^{2+}-exchanger in renal Ca^{2+} uptake is unclear. Neither Bijvelds *et al.* (1995), using vesicles isolated from the kidneys of FW and SW tilapia, nor Renfro *et al.* (1982), in vesicles prepared from the marine winter flounder renal tubules could find evidence of Na^+-dependent Ca^{2+} transport. It seems then that Ca^{2+} transport in renal epithelium, either in the reabsorptive or in the secretory direction, may be fully dependent on ATP-driven pumps.

Studying the secretory process in isolated and sealed renal tubules from winter flounder, Renfro (1978) showed that they readily accumulated $^{45}Ca^{2+}$ when exposed to increased bath activities of this ion, and that the kinetics of the Ca^{2+} uptake process indicate that this accumulation was not due to the gradient across the tubule wall. In a later study, inhibition of Ca^{2+} uptake by La^{3+} and stimulation by Ca ionophore A23187 suggested that ATP-altered plasma membrane Ca^{2+} transport, while experiments using different Na^+ concentrations in the incubation media showed that Ca^{2+} uptake is not directly related to the bath-to-tissue Na^+ gradient (Renfro *et al.*, 1982). Therefore, Ca^{2+} accumulation in the tubule lumen (e.g., secretion) is ATP dependent but not Na^+ dependent. Furthermore, the inhibition by La^{3+} also indicates that Ca^{2+} entry in the cell during the secretory process is mediated by a Ca^{2+}-channel.

The mechanism for Ca^{2+} entry in the ion-transporting kidney cell is not known. In mammals, Ca^{2+} enters the renal cell via a highly Ca^{2+}-selective channel TRPV5 because of a steep inward electrochemical gradient across the apical membrane (Nijenhuis *et al.*, 2005). A similar situation may occur in fish. Due to the high Ca^{2+} glomerular filtration rate, a comparable Ca^{2+} gradient exists, both in FW and SW fish. The fish kidney express the mRNA for an ECaC (Pan *et al.*, 2005) that has similarity to the TRPV5-like channel recently found in the branchial cells

(Shahsavarani *et al.*, 2006). The mammalian kidney is not capable of Ca^{2+} secretion and only the apical membrane possesses TRPV5. The fish renal tubule actively secretes Ca^{2+}, and is likely that a similar channel may also occur in the basolateral membrane. In fact, studies with aglomerular species would greatly increase our knowledge on the cellular processes involved in renal Ca^{2+} balance.

Due to the extreme importance of Ca^{2+} as a regulator of intracellular enzyme activity, all the proposed mechanisms for transepithelial transport involve the sequestration of Ca^{2+} en route through the cytosol. A 28 kDa high-affinity CaBP was found in the kidney cells of the American eel (Hearn *et al.*, 1978) and an immunoreactivity for an S-100-like protein was detected in the juxtaglomerular cells of several SW fish but apparently not in other tracts of the nephron (De Girolamo *et al.*, 2003).

The urine produced by the kidney is retained in the urinary bladder and its ionic composition is further modified there through transport of water and/or monovalent ions. A few studies involving the possible participation of the urinary bladder in Ca^{2+} balance have been carried out and suggest that Ca^{2+} is not transported across the bladder epithelia (Foster, 1975, 1976; Renfro, 1975). Reabsorption of water in this tissue greatly concentrates Ca^{2+} in the urine, and favours the production of bladder stones containing Ca^{2+} salts (Marshall, 1995; Marshall and Grosell, 2006).

Calcified Structures

Calcified structures make up for the major fraction of Ca^{2+} in fish (Flik *et al.*, 1986a; Wendelaar Bonga and Pang, 1991) and life-long growth is accompanied by long-term net accumulation of Ca^{2+} into bone and scales. Although there is still considerable debate on the implications of the cellular or acellular nature of fish bone (Herrmann-Erlee and Flik, 1989), and whether it functions as a Ca^{2+} reservoir to buffer changes in plasma Ca^{2+} as it happens in mammals, it has been demonstrated that in periods of extra demand fish can mobilize Ca^{2+} from scales and bone (Persson *et al.*, 1997, 1998; Witten and Villwock, 1997).

Osteoblast- and osteoclast-like cells, the cell types responsible for Ca^{2+} deposition and mobilization in calcified tissue are found in both cellular and acellular bone, and also in scales (Herrmann-Erlee and Flik, 1989), where they are denominated by scleroblasts and scleroclasts, respectively. However, very little is known about the cellular and biochemical processes involved in bone mineralization and regeneration and the manner in which they relate to Ca^{2+} balance. Recently, several studies have addressed these questions using the scales as a model. In scales, markers for the activity of these cell types, such as the alkaline phosphatase (ALP), for osteoblast, and tartrate-resistant acid phosphatase (TRAP) for osteoclast, and the genes involved in extracellular matrix calcification were shown to vary according to the salinity or Ca^{2+} contents of the environment, and/or are under the control of calcitropic endocrine factors, such as calcitonin, estradiol and parathyroid hormone-related protein (Persson *et al.*, 1994, 1995, 1997, 1999; Suzuki *et al.*, 2000; Suzuki and Hattori, 2003; Redruello *et al.*, 2005; Rotllant *et al.*, 2005; Yoshikubo *et al.*, 2005).

When environmental Ca^{2+} becomes scarce, mobilization of the mineral from the skeleton may provide enough Ca^{2+} for fundamental physiological processes. Fast-growing juvenile tilapia transferred to low-Ca^{2+} FW showed a decrease in Ca^{2+} density of bone when compared to fish kept in normal FW (Flik *et al.*, 1986a). This difference in Ca^{2+} contents was more evident in skeletal bone (vertebrae and scales; 12% less) than in dermal bone (operculum, 6-7%). In this study, the authors calculated the pool of readily exchangeable Ca^{2+} in the bone of FW fish to be about 7% of the total hard tissue Ca^{2+} and about 15% in fish exposed to low-Ca^{2+}. When Ca^{2+} uptake in Atlantic salmon was disrupted by excess of dietary Cd^{2+}, the Ca^{2+} content of scales was significantly reduced, and apparently this mobilization was sufficient to maintain Ca^{2+} circulating levels undisturbed (Berntssen *et al.*, 2003).

Sexual maturation requires enormous amounts of Ca^{2+} for gonadal growth and may result in substantial transient resorption from these tissues (Bjornsson *et al.*, 1999). In fact, mobilization of Ca^{2+} from scales and the vertebral skeleton has been demonstrated in salmonids during sexual maturation and spawning migration from Ca^{2+}-rich SW to FW (Persson *et al.*, 1998; Kacem *et al.*, 2000). Scale osteoclast activity is increased during vitellogenesis inducing demineralization of these structures (Persson *et al.*, 1995). So, when the Ca^{2+} demand exceeds the capacity of the Ca^{2+} uptake mechanisms, or when it is impossible or energetically unfavourable to extract Ca^{2+} from the environment, some fish mobilize Ca^{2+} from internal stores (Bjornsson *et al.*, 1999). The opposite process, Ca^{2+} deposition, has received less attention. Whether fish use the formation of Ca^{2+} aggregates in bones and scales as a means to reduce Ca^{2+} activity in plasma by precipitation is uncertain.

The overall contribution of the mineralized structures for Ca^{2+} homeostasis is debatable, but it is clear that in some fish they represent an important source of Ca^{2+} during critical periods of their life history.

Early Life Stages

Due to the specific characteristics of fish development, a word is due to Ca^{2+} uptake during this phase. In early life stages, an adequate supply of Ca^{2+} is essential for rapid development of mineralized structures such as the vertebral column, the fins and the cranium (Faustino and Power, 1998; 1999), which are essential for survival. The relatively few studies on Ca^{2+} uptake at the larval phase show that this process probably differs slightly from that of adults (Hwang *et al.*, 1994, 1996; Chang *et al.*, 1997, 1998; Hwang and Yang, 1997; Guerreiro *et al.*, 2001, 2004b; Chou *et al.*, 2002; Chen *et al.*, 2003). At these early developmental stages, the classical organs responsible for Ca^{2+} transport are not yet completely developed and in early larval stages alternative and more efficient routes for Ca^{2+} entry probably exist. Branchial tissue, which in adults accounts for the great majority of the body surface, is still poorly developed and structures such as the gill lamellae are only developing secondarily (Sarasquete *et al.*, 1998). Despite the possibility that gills, with their MR cells, are needed for ion uptake before they are needed for gas exchange (Rombough, 1999, 2002), their contribution to Ca^{2+} uptake may not be sufficient for the requirements of growth; in particular since the vascular system may not be sufficiently developed to provide developing peripheral bone structures with Ca^{2+}. It seems likely that in fish larvae MR cells initially found over the entire body surface (Fig. 15.3) contribute greatly to Ca^{2+} uptake (Wales and Tytler, 1996; Wales, 1997; Hiroi *et al.*, 1998, 1999; Van Der Heijden *et al.*, 1999a; Chang *et al.*, 2001; Kaneko *et al.*, 2002).

The contribution of the developing intestinal tract to Ca^{2+} uptake in fish larvae is largely unstudied. However, it has been proposed that it may be important in early stages (Tytler *et al.*, 1990) by increasing the relative surface available for ion exchange. The most relevant fraction of Ca^{2+} uptake in the intestine is most probably related to drinking rather than to diet, as drinking rates are significantly higher in both FW and SW larvae compared to juveniles and adults (Perrot *et al.*, 1992; Fuentes and Eddy,

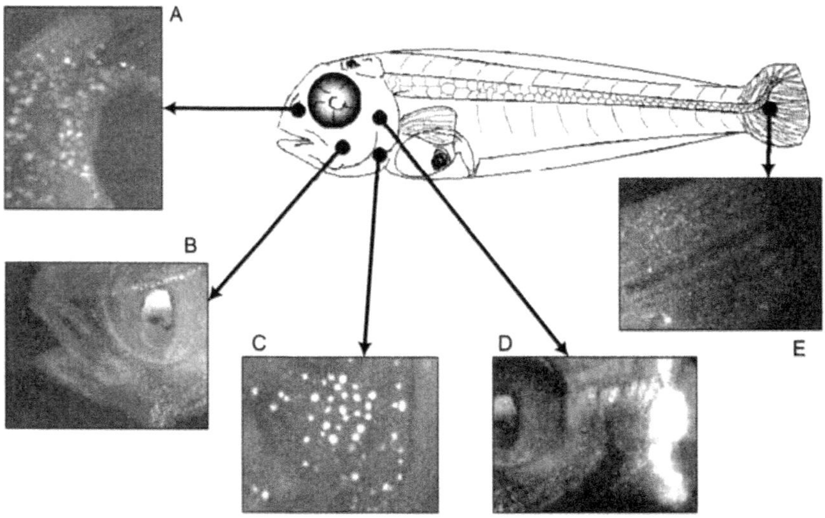

Fig. 15.3 Confocal laser-scanning micrographs showing fluorescence emitted by the mitochondria selective probe MitoTracker. It is possible to appreciate distribution and relative quantity of mitochondria-rich cells in different areas of the body of 30-day-old seabream larvae: (**A**) high magnification of the rostral area showing concentration of MR-cells; (**B**) the lower part of the head featuring MR-cells around the eye, jaw and operculum; (**C**) detail of the interior epithelia of the opercular membrane showing abundance or MR-cells; (**D**) detail of the posterior area of the head where it is possible to notice high concentration of fluorescent signal produced by the heart; and (**E**) detail of the caudal fin featuring MR-cells in the membranes in between the fin rays.

1997). Although the developing kidney seems to be at least partially functional from early stages (Nebel *et al.*, 2005), there are no studies on the contribution of renal transport for Ca^{2+} balance in fish larvae.

We have estimated the contribution of the extra-intestinal uptake in seabream larvae to be increasingly important in lower salinities, mostly due to reduction in drinking and water Ca^{2+} concentrations (Guerreiro *et al.*, 2004b). Extra-intestinal uptake was at least 60% of the total Ca^{2+} entry measured in full-strength SW more than 85% in 25% SW. Further reduction in water salinity and Ca^{2+} content to one-tenth of the original SW resulted in a decrease in the extra-intestinal influx to 55% of the original but this route still contributes with at least 94% of the overall uptake. The contribution from intestinal uptake is difficult to quantify since absorption rates are unknown. In full-strength seabream larvae, the drinking rates were reasonably high and a great amount of water enters the intestine. Roughly 40% to 50% of the $^{45}Ca^{2+}$ measured in these larvae at any given time was contained in the intestinal tract. In tilapia larvae, the contribution of the intestinal fraction was calculated as 35% of the total uptake, while in FW-adapted fish the proportion resulting from intestinal absorption was considerably low, averaging only 5% of the overall uptake (Fig. 15.2).

In previous studies, the effects of environmental Ca^{2+} concentrations on the survival, growth, body Ca^{2+} content and Ca^{2+} uptake kinetics in developing tilapia larvae were studied by Hwang and colleagues (1994, 1996) and Chou *et al.* (2002). They have shown that hatching rates and growth were similar in larvae from fertilized eggs exposed to either 0.90 or 0.02 mM Ca^{2+}. Body Ca^{2+} content in low-Ca^{2+} groups was about 90-95% that of high-Ca^{2+} group and exposure to low Ca^{2+} induced an averaged 1.2-fold stimulation of Ca^{2+} uptake rates. Ca^{2+} efflux was also altered, and after acclimation for 8 days, the effluxes of the low-Ca^{2+} group were 43% of the high-Ca^{2+} group. These studies also show that responses to low-Ca^{2+} environments are age-dependent and when newly hatched-larvae were transferred to low Ca^{2+}, the net uptake increased (from 5% to 69%) within 64 hours, while 3-day post-hatched larvae managed to reach the levels of the control within 38 hours. Declining Ca^{2+} efflux in 3-day-post-hatched larvae occurred 14 hours after exposure, much faster than those in newly hatched larvae (38 hours). These data indicate that tilapia larvae are able to modulate their Ca^{2+} uptake mechanism to maintain normal body Ca^{2+} content and rapid growth in environments with different levels of Ca^{2+}, despite having still quite under-developed Ca^{2+} exchanging structures. In fact, these larva are able to increase their Ca^{2+} content to about 60 times higher then that of the embryo within ten days after hatching (Hwang *et al.*, 1994).

Comparison of larva from different fish species also showed differences in the strategies for Ca^{2+} balance in response to environmental Ca^{2+} that may be associated with different development patterns and environments in which these fish naturally occur. Seabream larvae, although adapting to lower Ca^{2+} environments, show high dependence on environmental Ca^{2+} for normal uptake (Fig. 15.4) (Guerreiro *et al.*, 2004b), while the studies conducted on tilapia indicate that these fish easily equilibrate their Ca^{2+} balance even in low Ca^{2+} conditions. This was also demonstrated by Chen *et al.* (2003) using three FW species, goldfish, zebrafish, and ayu (*Plecoglossus altivelis*). While the goldfish larvae were able to maintain Ca^{2+} contents in low-, mid-, and high-Ca^{2+} environments by increasing

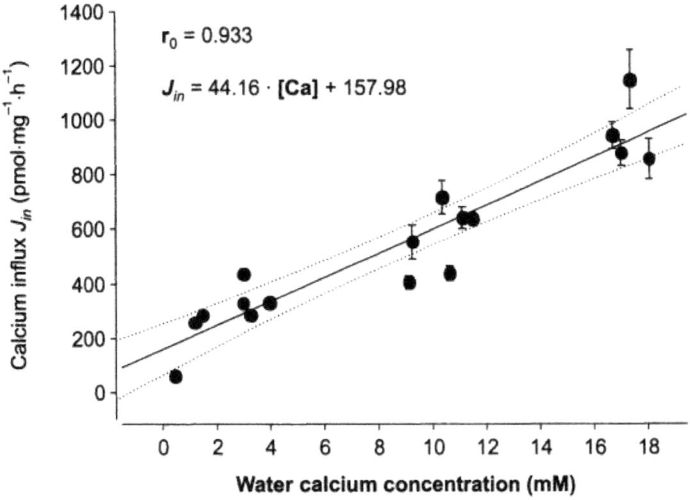

Fig. 15.4 Relationship between whole body calcium uptake and environmental calcium concentration in seabream larvae transferred from a 100% SW situation. Each point represents the average and standard error for at least 15 fish exposed to the different Ca^{2+} concentrations. A highly significant correlation was found. (Adapted from Guerreiro *et al.*, 2004b).

uptake capability in low-Ca^{2+} situations, the Ca^{2+} content of both zebrafish and ayu larva acclimated to low-Ca^{2+} were significantly lower than those acclimated to mid- or high-Ca^{2+}, due to insufficient Ca^{2+} influx rates.

Sensing Mechanisms—The Extracellular Ca^{2+}-Sensing Receptor (CaSR)

To effectively control Ca^{2+} balance, vertebrates rely on sensing systems that monitor extracellular Ca^{2+} levels. In tetrapods, the modulation of renal Ca^{2+} transporting systems and the regulatory calcitropic hormonal factors are associated with a Ca^{2+}-sensing receptor (CaSR), present in the cell membrane that binds and is activated by Ca^{2+} ions in the extracellular fluids (Tfelt-Hansen and Brown, 2005). This CaSR enables extracellular Ca^{2+} concentrations to modulate internal Ca^{2+} and other secondary messengers without being taken up by the cell (Riccardi, 2000) and in tetrapods it is involved in the regulation of parathyroid hormone (PTH) secretion by the parathyroid gland, responding to minute changes in blood Ca^{2+} levels (Tfelt-Hansen and Brown, 2005).

Fish also possess such receptors, showing remarkable amino acid conservation in most domains across species (Flanagan *et al.*, 2002), most probably due to the specificity of their function. However, the intracellular domain of the piscine molecule is considerably smaller than that of higher vertebrates, suggesting that the range of its functions and intracellular transmission mechanisms may be more limited.

The CaSR mRNA is expressed in several tissues associated to Ca^{2+} transport, osmoregulation and endocrine secretion. Expression has been found in the brain, branchial epithelium, kidney, intestine, bone, pituitary, corpuscles of Stannius of seabream and pufferfish (Ingleton *et al.*, 1999, 2002; Flanagan *et al.*, 2002; Hang *et al.*, 2005; Abbink *et al.*, 2006), and in addition to these also in the pancreas, muscle and spleen of the rainbow trout (Radman *et al.*, 2002), and in the caudal neurosecretory system of the flounder (Ingleton *et al.*, 2002). It was also found in kidney, urinary bladder, brain, pituitary, gill, stomach and intestine of the tilapia (Loretz *et al.*, 2004) and in winter flounder and Atlantic salmon (*Salmo salar*), there is expression for this receptor in kidney, stomach, intestine, gill, and brain (Nearing *et al.*, 2002). These expression studies have shown that at least in intestine and kidney expression levels are affected by environmental salinity (Nearing *et al.*, 2002; Loretz *et al.*, 2004), indicating that they may be involved in regulating the response to osmotic and ionic challenge in fish. Functional work showed that exposure of the lumen of winter flounder urinary bladder to known CaSR agonists, Gd^{3+} and neomycin, reversibly inhibit volume transport, which is important for euryhaline teleost survival in SW (Nearing *et al.*, 2002) and human cells transfected with the tilapia CaSR show important intracellular response to elevations in the extracellular Ca^{2+} concentration that was dependent on the ionic strength of the bathing medium, also supporting a role in salinity sensing (Loretz *et al.*, 2004).

The role of CaSR in the modulation of calciotropic hormones (see Endocrine Control of Ca^{2+} Balance) is still quite unknown, but it seems to be at least involved in the control of stanniocalcin secretion. This factor serves as an anti-hypercalcemic hormone such that a rise in extracellular ionic Ca^{2+} above a physiological set-point evokes an immediate secretory response (Wagner *et al.*, 1998a). Using pharmacological agents that increase the sensitivity of the CaSR to Ca^{2+}, Radman *et al.* (2002) showed that these had time and dose-dependent stimulatory effects on STC secretion that were indistinguishable from those of Ca^{2+} loading. The downstream effects of the increased secretion were the inhibition of gill Ca^{2+} transport. It is, therefore, predictable that similar modulatory actions may occur over other calciotropic factors in fish.

The CaSR is also abundantly expressed in the olfactory epithelia of both FW and SW species (Cao *et al.*, 1998; Flanagan *et al.*, 2002; Hubbard *et al.*, 2002; Nearing *et al.*, 2002). The presence in this tissue has prompted the question on its possible function as a proxy for direct determination of environmental salinity or Ca^{2+} concentration, that could serve to forewarn the animal that is reaching the limit of its osmotic tolerance (Hubbard *et al.*, 2000) and thus select between suitable or less favourable habitats, or be used to detect landmarks, features of a specific water body, as earlier suggested by Bodznik (1978) in relation to migrating salmon returning to their original rivers.

Electrophysiology studies have now shown that the fish olfactory system is highly sensitive to changes in environmental Ca^{2+} concentration. Extracellular recordings made from the olfactory nerve of seabream while the Ca^{2+} concentration of artificial seawater flowing over the olfactory epithelium was varied from 10 to 0 mM showed that reductions in Ca^{2+} caused a large increase in the firing rate of the olfactory nerve and further experiments demonstrated that the goldfish olfactory system is also acutely sensitive to changes in external Ca^{2+} within the range that this species is likely to encounter in the wild (0.05-3 mM) (Hubbard *et al.*, 2000, 2002). In both cases, the response was not due to the concomitant reduction in osmolality and was specific for Ca^{2+}, and the apparent EC50 for the response of each species was close to the levels of their circulating ionic Ca^{2+} in the plasma.

Given the specificity of the response to Ca^{2+} and its fitting to environmental concentrations in the range usually experienced by each species, a stenohaline non-migratory and a partially euryhaline, it seems that this olfactory sensing may be a general phenomenon in teleosts and is likely involved in internal Ca^{2+} homeostasis. The existence of Ca^{2+} sensors linked to the MR cells, that could respond to variations in ambient Ca^{2+} had already been predicted (Flik and Verbost, 1995), based on the existence of Ca^{2+}-responsive receptor cells in the skin of fish, and the proliferation of gill CaBPs in low ambient Ca^{2+}.

Although it seems clear that fish have internal and external CaSRs, it still needs to be determined if sensing of external Ca^{2+} levels can trigger changes in the activity of Ca^{2+}-handling tissues and/or in the secretion of calciotropic factors.

ENDOCRINE CONTROL OF Ca^{2+} BALANCE

The endocrine system of fish is responsible for maintaining circulating Ca^{2+} levels within tight limits, acting to up-regulate or down-regulate the exchange of Ca^{2+} between the blood and the environment and between the blood and the internal stores (see Fig. 15.5). In terrestrial vertebrates, the parathyroid hormone (PTH) and calcitonin (CT), together with 1,25-dihydroxy vitamin D_3 (1,25$(OH)_2D_3$; the most effective vitamin D_3 metabolite) seem to be the main factors controlling extracellular Ca^{2+} levels, Ca^{2+} uptake and excretion and bone mobilization (Mundy, 1990; Mundy and Guise, 1999). PTH is a hypercalcemic factor that acts directly by stimulating Ca^{2+} resorption in the bone and reabsorption of Ca^{2+} in the kidney. It also stimulates the synthesis of 1,25$(OH)_2D_3$ which promotes intestinal Ca^{2+} absorption, allowing enough Ca^{2+} to be taken up by the organism. 1,25$(OH)_2D_3$ is also involved in the up-regulation of plasma

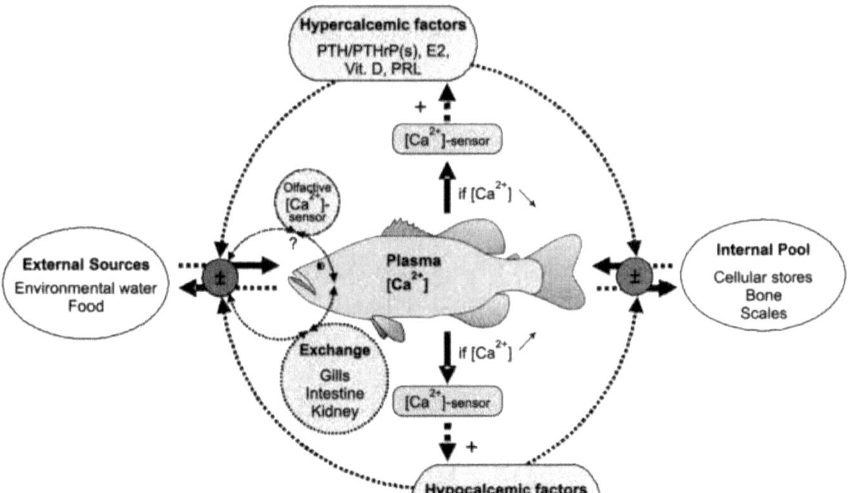

Fig. 15.5 Calcium balance in fish. Illustrative model of the overall interaction between control factors, transport tissues, sensing mechanisms and external and internal Ca^{2+} pools. Changes in extracellular Ca^{2+} are monitored by internal CaSRs that modulate the secretion and circulating levels hyper- and hypocalcemic factors. These act over the transporting tissues to promote either and increase in Ca^{2+} uptake and/or Ca^{2+} retention or a reduction in Ca^{2+} uptake and/or stimulation of Ca^{2+} excretion. Actions on mineralized tissues can also be evoked to induce Ca^{2+} resorption or deposition. Whether the exchange mechanisms can be triggered directly by changes in environmental Ca^{2+} via an olfactory CaSR is uncertain.

Ca^{2+} due to its actions in the bone and kidney. CT counteracts the action of the other two, rapidly preventing Ca^{2+} release from bone tissue by promoting osteoblastic and osteocytic activities. It also stimulates—to a lesser extent—Ca^{2+} excretion via the gut and kidney (Mundy, 1990; Mundy and Guise, 1999).

The establishment of an endocrine model for the control of Ca^{2+} balance in fish has been a difficult task, mostly due to the fact that the nature of the actions of many known factors is still inconclusive and to the constant appearance of possible new players. The range of factors reported to have direct or indirect influence in Ca^{2+} metabolism is increasing, and recently, the arrival of an entire family of heavy weight contenders opens space for new questions. Hormones with direct effects on Ca^{2+} balance include stanniocalcin, calcitonin, the metabolites of the vitamin D_3, estradiol in specific stages of the life cycle, and at least some members of the recently uncovered piscine PTH/PTHrP family of proteins (Canario *et al.*, 2006). Other factors promote changes in osmoregulatory, acid-base and metabolic parameters, and have indirect actions over Ca^{2+} balance.

Pituitary Hormones and Cortisol

Pituitary hormones are involved in a multitude of functions, and several reports indicated, with more or less certainty, that the members of the prolactin (PRL) gene family have calciotropic actions in fish.

Prolactin. The true role of PRL in fish Ca^{2+} balance is not completely clear. If on one hand it has direct actions in plasma Ca^{2+} and Ca^{2+} transporting mechanisms, on the other the disturbances in Ca^{2+} balance *per se* do not affect its expression and secretion, which seem to be regulated by osmolality or extracellular Na^+ and Cl^- levels rather than Ca^{2+} levels (Wendelaar Bonga and Pang, 1991; Arakawa *et al.*, 1993; Kaneko and Hirano, 1993; Seale *et al.*, 2003). Moreover, despite early studies showed that removal of the pituitary leads to hypocalcemia, in most cases the hypocalcemia is accompanied by decreases in the concentration of other electrolytes. In these cases, prolactin replacement restores Ca^{2+} levels but also those of the remaining electrolytes. In this context, PRL can be regarded as an osmoregulatory factor that has hypercalcemic actions in fish.

Ovine PRL administration induces an increase in plasma Ca^{2+} in tilapia (Flik *et al.*, 1986b) and carp *Cyprinus carpio* (Chakraborti and Mukherjee, 1995), confirming identical observations in other teleosts, including sticklebacks (*Gasterosteus aculeatus*), American eels, rainbow trout and the killifish (Flik *et al.*, 1986b). It also induced significant Ca^{2+} influx and reduction of Ca^{2+} efflux in tilapia, resulting in net Ca^{2+} accumulation. The increase in Ca^{2+} uptake was associated with enhanced activity of the branchial high-affinity plasma membrane Ca^{2+}-ATPase (Flik *et al.*, 1984b, 1989). These results were confirmed by the use of homologous recombinant tilapia prolactins, PRL-I and PRL-II that differ in size by 11 amino acids (Specker *et al.*, 1985). Both PRLs increased the activity of the Ca^{2+}-ATPase in a plasma membrane preparation of the branchial epithelium but dose-response studies demonstrated that PRL-I was twice as potent as PRL-II in inducing hypercalcemia (Flik *et al.*, 1994). The ratio between Ca^{2+}-ATPase and Na^{+}/K^{+}-ATPase increases upon PRL treatment, as does the proliferation of opercular MR cells, but this increase in cell density may or may not relate specifically to Ca^{2+} transport. The possible effects of PRL on intestinal Ca^{2+} balance have not yet been addressed, although the PRL receptor exists in the fish intestinal and renal cells (Tse *et al.*, 2000; Sandra *et al.*, 2001). In the intestine as in gill, prolactin induces proliferation of ionoregulatory cells, upregulation of specific ion transporters and changes in membrane permeability, which seems to be the hormone's main functions in FW adaptation (Sakamoto and McCormick, 2006). In isolated preparations of winter flounder kidney cells, ovine PRL had stimulatory effects on phosphate uptake but the salmon protein was without effect, and the actions on renal Ca^{2+} transport were not tested (Lu *et al.*, 1995).

The action of PRL in Ca^{2+} balance is slow, and the effects on plasma take between two to four days to achieve, which contrasts with the immediate responses to PTH in terrestrial vertebrates. This also supports the role of this hormone as a FW-adapting factor (McCormick and Bradshaw, 2006; Sakamoto and McCormick, 2006), rather than a true hypercalcemic factor. PRL secretion is also regulated by factors such as estradiol (Brinca *et al.*, 2003) that has reported actions in Ca^{2+} homeostasis (see ahead: Estradiol 17β), suggesting that it may be a mediator or an agonist of the actions of this steroid. However, cortisol, which may have hypercalcemic action, decreases the production of PRL (Borski *et al.*, 2001), which may be related to their opposite roles in osmoregulation (McCormick and Bradshaw, 2006; Sakamoto and McCormick, 2006).

Somatolactin. Somatolactin (SL) is another pituitary hormone with arguably hypercalcemic effects in fish (Kaneko and Hirano, 1993). The structure of this hormone is closely related to both PRL and growth hormone (GH), and its primary function(s) remain mostly unknown but seems to be involved in reproduction, acid-base regulation, and stress responses in fishes (Kaneko, 1996). Initial indication for its putative calciotropic effects came from the fact that low environmental Ca^{2+} levels—but not osmolality—are associated with activation of the SL-producing cells in the pituitary of the rainbow trout leading to increased mRNA levels as well as plasma SL levels (Kakizawa *et al.*, 1993). Plasma SL levels were also elevated in association with plasma Ca^{2+} and phosphate during acute stress and during exhaustive exercise (Kakizawa *et al.*, 1995) and the number of SL-immunoreactive cells and the secretory and synthetic activity increased in parallel with maturation and vitellogenesis, in an apparent association with increasing Ca^{2+} plasma levels (Mousa and Mousa, 2000). These studies only provide indirect evidence, and the effects of low Ca^{2+} on SL secretion are slow, not compatible with a homeostatic factor. In flounder renal cells administration of SL had no effect on Ca^{2+} transport, but induced significant reabsorption of phosphate, an ion closely associated to Ca^{2+} metabolism (Lu *et al.*, 1995). Clearly, further studies are needed before SL can be considered a calciotropic factor.

Growth hormone. GH is a factor involved in fish osmoregulation, favouring the adaptation to SW due to the capacity of this hormone to increase the number and size of gill MR cells and ion transporters involved in salt secretion (Sakamoto and McCormick, 2006), which may also reflect in increased Ca^{2+} transport. Very few studies have addressed this issue. Flik *et al.* (1993a) tested the effects of homologous recombinant tilapia GH on growth, Ca^{2+} accumulation and Ca^{2+} fluxes. GH had no significant effect on Ca^{2+} influx from the water, but stimulated Ca^{2+} accumulation in parallel to increased growth rates, which was achieved by reduction in Ca^{2+} efflux. The Ca^{2+} plasma levels were unchanged and the analysis of the Ca^{2+} contents of mineralized structures, normalized to weight, did not show difference between control and treated fish. It appears than that the effects of GH in Ca^{2+} balance are not direct but the consequence of accelerated growth rate, and that other factors may have mediated the reduction in Ca^{2+} efflux.

In an interesting approach, the involvement of PRL and GH in otolith and scale calcification was examined using hypophysectomized goldfish (Shinobu and Mugiya, 1995). As predicted by the previous data, hypophysectomy resulted in hypocalcemia, which was corrected by PRL replacement therapy, but not by GH, which induced a further reduction in plasma Ca^{2+}. Ca^{2+} incorporation into otoliths and scales was also markedly reduced after hypophysectomy due to low Ca^{2+} availability. Interestingly, PRL was ineffective in stopping this reduction but treatment with GH completely counteracted the reduction in otoliths and scales and even exceeded the respective sham levels, suggesting that GH may be relevant for Ca^{2+} deposition.

Cortisol. This steroid hormone produced in the inter-renal tissue is related to stress response and also to SW adaptation, being involved in the control of proliferation and apoptosis of MR cells (Wendelaar Bonga, 1997; McCormick and Bradshaw, 2006). In some situations it has been associated with hypercalcemic effects, but the true nature and length of this association are unclear. Cortisol administration was shown to produce hypercalcemia in coho salmon, *Oncorhynchus kisutch* (Björnsson *et al.*, 1987) and rainbow trout (Flik and Perry, 1989). Plasma concentrations of cortisol increased in rainbow trout acclimated to low-Ca^{2+} FW and remained elevated for 8 days. Branchial basolateral membrane vesicles prepared from these fish showed increased Ca^{2+}-transport capacity due to stimulation of Ca^{2+}-ATPase activity, probably as a result of MR cell proliferation, according to the authors. Fish in normal FW were treated with cortisol for 7 days. After this period these fish displayed stimulated whole body Ca^{2+} uptake and an increase of the branchial Ca^{2+} transport capacity; the combination of these changes resulted in hypercalcemia. Recently, it has been shown that cortisol is capable of rapidly stimulate gill Ca^{2+}-ATPase, through non-genomic pathways (Sunny and Oommen, 2001). Treatment with cortisol resulted also in increased mRNA expression and protein levels of the apical ECaC in rainbow trout (Shahsavarani and Perry, 2006).

Estradiol 17β

Estradiol 17β (E_2) has a marked impact on Ca^{2+} balance in fish. However, despite its clear calciotropic effects, E_2 can hardly be regarded as true Ca^{2+} regulating factor, given the intermittency and the specificity of its effects, the time it takes to act, and the fact that it does not respond to fulfil housekeeping needs or to re-establish Ca^{2+} homeostasis. To date, to the best of our knowledge, there is no evidence that E_2 synthesis is regulated by decreases in either extracellular or environmental Ca^{2+} concentrations.

This hormone is the main factor involved in gonadal maturation in (female) fish and the stimulus for the production of vitellogenin, a Ca^{2+}-binding protein essential for egg production. The E_2-mediated vitellogenin induction requires elevated extracellular Ca^{2+} concentrations (Yeo and Mugiya, 1997) and significant increases in circulating Ca^{2+} are associated with rises in plasma E_2 levels (Mugiya and Watabe, 1977; Pang and Balbontin, 1978; Pandey, 1993; Persson *et al.*, 1997; Srivastav and Srivastav, 1998; Guerreiro *et al.*, 2002; Guzman *et al.*, 2004).

Although the E_2-driven actions occur only seasonally for most species, they have great impact in Ca^{2+} balance, which is more pronounced in species such as the tilapia, with short reproductive cycles. In fact, serum Ca^{2+} levels significantly decreased after gonadectomy in females but were not altered in males, and there was no difference in serum Ca^{2+} levels between gonadectomized males and females (Tsai and Wang, 2000). Treatments with E_2 produced hypercalcemia, which was more pronounced in castrated males than in females, suggesting a differential sensitivity in estrogen receptors between males and females.

The pathways and targets for the hypercalcemic action of E_2 seem to be two-fold, depending on the species: the rise in circulating E_2 leads to an increase of Ca^{2+} uptake from the water and Ca^{2+} mobilization from mineralized structures such as the scales and skeleton. The relative importance of these two pathways seems to be species specific, and linked to the life-strategy and environment they inhabit. In salmonids, E_2 induces Ca^{2+} resorption from scales and bone (Persson *et al.*, 1994, 1998; Kacem *et al.*, 1998, 2000; Witten and Hall, 2003), in a process associated with the spawning migration. Using TRAP as a marker, Persson *et al.* (1998) demonstrated that scale osteoclast activity increased throughout sexual maturation in the Atlantic salmon and that Ca^{2+} was simultaneously accumulated in the female gonads, and proposed that the scales are reabsorbed in order to provide Ca^{2+} for the growing ovaries. Rainbow trout injected with E_2 (10 mg/kg) showed significantly high plasma Ca^{2+} values when compared to sham-injected fish. Twenty days after the initial injection, plasma Ca^{2+} reached approximately 16 mM. This rise was accompanied by an increase in scale scleroclast activity and a reduction in scale Ca^{2+} contents (Persson *et al.*, 1997). The Ca^{2+}-resorbing actions of E_2 in scales as measured by increased TRAP activity have also been demonstrated *in vitro* in other FW and SW species (Suzuki *et al.*, 2000; Rotllant *et al.*, 2005) and this steroid down-regulates the expression of osteonectin in the goldfish, an extracellular-matrix protein responsible for Ca^{2+} mineralization in scales (Lehane *et al.*, 1999).

E_2 treatment also stimulates Ca^{2+} exchange with the water. In rainbow trout juveniles, Ca^{2+} influxes were significantly increased by a factor of 1.2 while the effluxes remained unchanged (Persson *et al.*, 1994). In the marine seabream, injection of 10 mg/kg E_2 in coconut-oil implants resulted in a remarkable increase in plasma Ca^{2+}, a 2.5-fold increase in free Ca^{2+} and a 10-fold increase in plasma total Ca^{2+}, which reached almost 30mM 15 days after the injection (Guerreiro *et al.*, 2002). In these fish, we have not observed any effects on scale TRAP activity nor in scale Ca^{2+} contents, but the whole body Ca^{2+} uptake increased significantly in response to E_2. Both the intestinal and branchial Ca^{2+} uptake routes were stimulated by E_2, while there were no observable effects over Ca^{2+} efflux, resulting in a significant 31% increase in net Ca^{2+} uptake.

Despite these evident effects, the way E_2 interacts with the transporting mechanisms or with the bone reabsorbing cells is not clear. Estrogen receptors are found in scales, bone and intestine (Armour *et al.*, 1997; Persson *et al.*, 2000; Socorro *et al.*, 2000; Filby and Tyler, 2005; Pinto *et al.*, 2006), but its presence in gills is uncertain (Persson *et al.*, 2000; Socorro *et al.*, 2000; Filby and Tyler, 2005; Pinto *et al.*, 2006), and this suggests that at least part of the action of E_2 on Ca^{2+} homeostasis may be mediated indirectly via other endocrine factors.

The increasing importance of xenoestrogens in the water, as disrupters of the normal endocrine function, osmoregulatory capabilities (McCormick *et al.*, 2005), Ca^{2+} metabolism (Suzuki and Hattori, 2003; Suzuki *et al.*, 2003) and reproductive cycle (Christensen *et al.*, 1999; Kirk *et al.*, 2003) has stimulated several studies on the effects of these substances that use Ca^{2+} levels as a reliable indicator of vitellogenesis (Nagler *et al.*, 1987; Gillespie and de Peyster, 2004).

Calcitonin

Calcitonin (CT) is a protein produced in the ultimobranchial glands of fish, analogous to the parafollicular 'C' cells in the thyroid gland of mammals, where it has an established hypocalcemic role (Mundy and Guise, 1999). The nature of CT role in fish is ambiguous, as a consequence of the discrepancy of several physiological studies which yielded inconsistent results upon CT administration, showing hypocalcemia

(Wendelaar Bonga, 1981; Wales and Barrett, 1983; Fenwick and Lam, 1988; Chakrabarti and Mukherjee, 1993; Mukherjee *et al.*, 2004a), hypercalcemia (Fouchereau-Peron *et al.*, 1987; Oughterson *et al.*, 1995) or no effects on blood Ca^{2+} levels (Yamauchi *et al.*, 1978a; Wendelaar Bonga, 1980; Hirano *et al.*, 1981; Bjornsson and Nilsson, 1985).

Despite the diverse action of CT on circulating Ca^{2+} levels, there is additional evidence to support a role consistent with a hypocalcemic factor. CT had potent inhibitory effect on Ca^{2+} uptake in the salmon gills perfused under simulated *in vivo* condition (Milhaud *et al.*, 1977), and increased efflux from isolated gills of CT-treated salmon, an effect accompanied by reduction of plasma Ca^{2+} (Milhaud *et al.*, 1980). Milet *et al.* (1979) reported that Ca^{2+} influx of ultimobranchialectomized eels was lower than of controls, with unchanged efflux, and that CT perfusion induced important decreases in influx and increases in efflux. In addition to these results, CT injection caused an inhibitory action over whole-body Ca^{2+} uptake in young rainbow trout (Wagner *et al.*, 1997b) but whether it also influenced plasma Ca^{2+} levels was not reported. Significant inhibitory effects on Ca^{2+} uptake were also observed *in vivo* in the snakehead *Channa punctatus* and carp in response to salmon calcitonin, either in normal tap water or low-Ca^{2+} water, but not in high-Ca^{2+} water (Mukherjee *et al.*, 2004b). In these fish, plasma Ca^{2+} levels were considerably reduced by the hormone injection (Mukherjee *et al.*, 2004a). These results are consistent with the presence of CT receptors in the gill (Sasayama, 1999; Suzuki *et al.*, 2001). The manner in which these receptors interact with the Ca^{2+}-transporting mechanisms is unknown. The presence of CT mRNA and CT immunoreactivity (Martial *et al.*, 1994; Clark *et al.*, 2002; Hidaka *et al.*, 2004) in branchial cells was also found, indicating that it may be a local regulator of Ca^{2+} transport, activated by yet another factor.

CT secretion has also been shown to be regulated by the extracellular Ca^{2+} concentration, although results from different studies are conflicting (Wendelaar Bonga and Pang, 1991). For instance, infusion of Ca^{2+} into the Japanese eel was associated with an increase in plasma CT immunoreactivity (Sasayama *et al.*, 2002), but transfer from FW to SW, which also modified plasma Ca^{2+} concentrations was without effect on CT levels (Suzuki *et al.*, 1999). Changes of Ca^{2+} concentrations in the environment were also unable to trigger any change in CT secretion in cod (Bjornsson and Deftos, 1985) but the maintenance of immature brown trout (*Salmo trutta*) in water containing 20 and 50 mM Ca^{2+} resulted in a significant reduction of plasma calcitonin and the extent of the decrease was related to higher levels of environmental Ca^{2+} (Oughterson *et al.*, 1995). Srivastav and colleagues (2002) demonstrated—using histological techniques—that the CT cells in the ultimobranchial tissue of the mud eel (*Amphipnous cuchia*) became hyperactive after a 15-day vitamin D_3-induced hypercalcemia.

In tetrapods, CT has an important role in protecting the skeleton from demineralization. In fish, this function is also present. Prolonged administration of salmon CT to SW eels, provoked a decrease of plasma Ca^{2+} and an increase of both the osteoblastic apposition and of the degree of mineralization of the intercellular matrix in the vertebral bone (Lopez *et al.*, 1976). Ultimobranchialectomy resulted in a rise in serum Ca^{2+} levels and completely stopped the process of vertebral bone osteoblastic apposition and caused a significant decrease in the degree of mineralization of the bone organic matrix (Lopez *et al.*, 1976). Similar actions were observed in the goldfish, where CT increased Ca^{2+} deposition in the rib bone, pharyngeal bone, and scales of starved but not of fed animals (Shinozaki and Mugiya, 2000) and CT treatment was also effective in reducing the plasma activity of enzymes related to bone turnover and caused a significant increase in skeletal bone Ca^{2+} concentration in the snakehead (Mukherjee *et al.*, 2004a). In *in vitro* bioassays CT suppressed the basal osteoclastic activity in scales of the nibblerfish (*Girella punctata*) and goldfish, and when administered in combination with E_2 it reduced or abolished the osteoclastic activity induced by the steroid (Suzuki *et al.*, 2000).

There is evidence to suggest an interplay between CT and E_2 in the regulation of Ca^{2+} metabolism. Injections of E_2 significantly increase circulating CT in rainbow trout, Atlantic salmon, and coho salmon (Bjornsson *et al.*, 1989). Furthermore, plasma CT levels show seasonal variations, increasing before ovulation in salmonids (Bjornsson *et al.*, 1986; Norberg *et al.*, 1989) and eels (Yamauchi *et al.*, 1978b), and the levels of CT in plasma of sockeye salmon, *Oncorhynchus nerka* (Watts *et al.*, 1975) and rainbow trout (Fouchereau-Peron *et al.*, 1990) are higher in females rather than in males during the spawning season. A recent report demonstrated that E_2 has a direct up-regulatory action in CT secretion acting on estrogen receptors of the ultimobranchial gland in the goldfish (Suzuki *et al.*, 2004). Given the opposite actions of these two factors on bone and scale metabolism, it is possible that the rise in CT may reflect a preventive role, protecting the skeleton from demineralization.

Vitamin D_3

Despite some ambiguous results, there is significant evidence for physiological actions of vitamin D_3 metabolites in Ca^{2+} balance in fish. A comprehensive and timely review of the piscine vitamin D_3 system was produced by Sundell and colleagues (1996).

The differential roles of the several related metabolites, vitamin D_3, $1{,}25(OH)_2D_3$, $25(OH)D_3$ and $24{,}25(OH)_2D_3$, were tested in FW *H. fossilis* in a series of Ca^{2+} concentrations (Srivastav and Singh, 1992; Srivastav *et al.*, 1997). Vitamin D_3 and $1{,}25(OH)_2D_3$, injected daily, consistently and gradually increased serum Ca^{2+} over a period of 5 days, which culminated in 30% increase in total plasma regardless of the water Ca^{2+} concentration. The other metabolites had little ($25(OH)D_3$) or no effect ($24{,}25(OH)_2D_3$). Daily injections of $1{,}25(OH)_2D_3$ also increased the total plasma Ca^{2+} (but not ionic Ca^{2+}) by approximately 30% in the male tilapia at 3 and 5 days after the initial injection (Srivastav *et al.*, 1998). However, in another study, Rao and Raghuramulu (1999) reported that none of these metabolites displayed calciotropic activity in this species.

In the SW cod, daily injections of vitamin D_3, $1{,}25(OH)_2D_3$, $25(OH)D_3$ and $24{,}25(OH)_2D_3$ did not produce measurable effects on total plasma Ca^{2+}, but $1{,}25(OH)_2D_3$ treatment induced hypercalcemia by specifically elevating the free ionic Ca^{2+}, the most physiologically relevant fraction. The effect was detectable within 24 hours and lasted for 5 days (Sundell *et al.*, 1993). In another marine species, the Antarctic fish *Pagothenia bernacchii*, a single injection of vitamin D_3 increased both ionic and total Ca^{2+} levels in plasma, whereas treatment of $1{,}25(OH)_2D_3$ evoked a decrease of only the ionized Ca^{2+}, and $25(OH)D_3$ had no effects on either fraction of plasma Ca^{2+} (Fenwick *et al.*, 1994). These results, showing differences in the modulation of the bound and free Ca^{2+} fractions in relation to $1{,}25(OH)_2D_3$ treatment, suggest a possible role of the steroid in the recruitment of Ca^{2+}-binding proteins.

Similar increases in plasma Ca^{2+} occurred in American eel injected daily with either vitamin D_3 or $1{,}25(OH)_2D_3$ for 7 days but only in fed and not in unfed fish. Intestinal Ca^{2+} absorption was measured in these animals using perfused averted gut sacs, and fed fish showed higher Ca^{2+}

uptake rates than unfed animals, differences that were further increased by vitamin D_3 or 1,25$(OH)_2D_3$ (Fenwick *et al.*, 1984). As for *H. fossilis*, treatment with either 24,25$(OH)_2D_3$ or vitamin D_2 had no effect on plasma Ca^{2+} levels nor in intestinal Ca^{2+} absorption. In a similar study using goldfish, the hypercalcemic action of daily vitamin D_3 injections was also dependent on the feeding status of the animals, increasing plasma Ca^{2+} in animals fed after a long starvation period, but not in those fed daily or unfed (Fenwick, 1984). Intestinal Ca^{2+} absorption was approximately 25% higher in vitamin D_3-treated fish than in those injected with the vehicle alone. Identical effects of vitamin D_3 and 1,25$(OH)_2D_3$ on intestinal Ca^{2+} uptake were observed in the FW-adapted European eel and in tilapia (Chartier *et al.*, 1980; Flik *et al.*, 1982).

In the marine cod, perfusion with 25$(OH)D_3$ stimulated the intestinal Ca^{2+} uptake by 65%, whereas 24,25$(OH)_2D_3$ reduced it by 36%. In this bioassay, vitamin D_3 and 1,25$(OH)_2D_3$ did not affect the Ca^{2+} flux across the intestinal mucosa (Sundell and Bjornsson, 1990). In a similar experiment Larsson *et al.* (1995) found a rapid (within 10 to 25 min) dose-related down-regulatory effect of 24,25$(OH)_2D_3$ in Ca^{2+} uptake, but no effects of increasing doses of 1,25$(OH)_2D_3$. The outcome of these experiments lead the authors to suggest that an environmental adaptation could explain the need for a hypercalcemic metabolite (i.e., 1,25$(OH)_2D_3$) in FW fish and an anti-hypercalcemic (i.e. 24,25$(OH)_2D_3$) in SW fish.

The enterocytes of both FW and SW fish have cytoplasmatic or nuclear receptors for 1,25$(OH)_2D_3$ (Bjornsson *et al.*, 1999) but specific binding to enterocyte basolateral membrane has been demonstrated. The SW cod basolateral membrane only binds 24,25$(OH)_2D_3$, whereas in FW carp specific binding for both 1,25$(OH)_2D_3$ and 24,25$(OH)_2D_3$ exists (Larsson *et al.*, 2001). Furthermore, in FW-adapted rainbow trout, enterocytes display receptors for 1,25$(OH)_2D_3$ in the basolateral membrane, and respond to specific treatment with increased intracellular Ca^{2+} concentrations, but no receptors are found for 24,25$(OH)_2D_3$ and no intracellular response is observed when the cells are incubated with 24,25$(OH)_2D_3$. After acclimation to SW, 1,25$(OH)_2D_3$ receptors are down-regulated while specific binding for 24,25$(OH)_2D_3$ appears (Larsson *et al.*, 2003). There is also evidence that 24,25$(OH)_2D_3$ prevents the Ca^{2+} entry in isolated cod enterocytes, decreasing intestinal Ca^{2+} uptake via inactivation of L-type Ca^{2+}-channels, whereas 25$(OH)D_3$ but not 1,25$(OH)_2D_3$, is responsible for increasing enterocyte Ca^{2+} transport via activation of Na^+/Ca^{2+} exchangers, concurrent with activation of L-type Ca^{2+} channels (Larsson *et al.*, 2002). Taken together, these data greatly support a role of the vitamin D3 system in controlling disturbances in Ca^{2+} balance by maintaining adequate intestinal Ca^{2+} absorption.

The actions of this hormonal system in renal Ca^{2+} balance are unknown and although vitamin D_3 stimulated renal phosphate reabsorption in eel (see Sundell *et al.*, 1996, for reference), neither $1,25(OH)_2D_3$ nor $1,25(OH)_2D_3$ produced any effect in phosphate transport across winter flounder tubule cells (Lu *et al.*, 1994).

Finally, the administration of vitamin D_3 metabolites to fish also produces (ambiguous) effects on the metabolism of calcified structures. Injections of $1,25(OH)_2D_3$ decreased bone mineralization in FW-adapted eel (Lopez *et al.*, 1977, 1980) and tilapia (Wendelaar Bonga *et al.*, 1983), but in sexually mature SW-adapted female eels, injections of $1,25(OH)_2D_3$ stimulated osteoblast activity and prevented demineralization, acting in a similar way to $24,25(OH)_2D_3$, reported to increase osteoblastic activity in tilapia (Wendelaar Bonga *et al.*, 1983). Since mature female fish display higher demineralization rates due to the Ca^{2+} demand during vitellogenesis, this may be perceived as a protective action of the skeleton, although more studies are clearly necessary to confirm this possibility.

Overall, it seems clear that the vitamin D_3 family has important functions in Ca^{2+} metabolism in fish. Two metabolites, vitamin D_3 and $1,25(OH)_2D_3$ appear to have a hypercalcemic function mainly by stimulating intestinal Ca^{2+} uptake, a process which is blocked or reversed by $24,25(OH)_2D_3$. Measurable levels of vitamin D_3 and $1,25(OH)_2D_3$ have been detected in plasma of FW and SW teleosts and $24,25(OH)_2D_3$ was identified in blood of the marine bluefin tuna (Sundell *et al.*, 1996). To date, there are no studies indicating whether the regulation of these metabolites is dependent on extracellular Ca^{2+} levels but changes in salinity can alter the cellular response to specific metabolites in the intestine. Whether purely sensory-mechanisms are responsible for the transition from 'the FW'- to 'the SW'-vitamin D_3 system in the intestine or there is involvement of other factors is unknown.

Stanniocalcin

Stanniocalcin (STC) is a glycoprotein secreted primarily by the corpuscles of Stannius (CS), which are associated with renal tissue in teleost and holostean fishes only (Wendelaar Bonga and Pang, 1991). However, it is

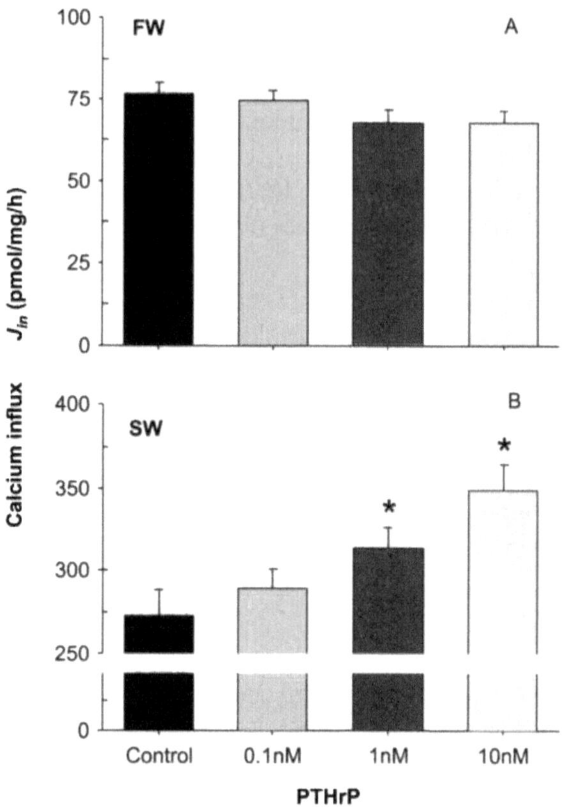

Fig. 15.6 Whole body Ca^{2+} influx rates in response to different concentrations of water-borne parathyroid hormone-related protein (1–34) PTHrP measured over a 4-h period in tilapia larvae adapted to freshwater (**A**) and seawater (**B**). Each group represents the means and SE of 15–25 fish. * indicates significant difference from the respective control ($P<0.05$, one-way ANOVA). Note that scale in freshwater is one-fourth that in seawater.

now evident that several forms of STC exist both in fish and in mammals, expressed in tissues that are not directly related to the CS, such as the ovary, kidney, gut and gills, and that it has roles in mineral metabolism, neural differentiation, reproduction, and even in cancer development (Wagner and Dimattia, 2006). Genes bearing high homology for STC variants have also been identified in aquatic invertebrates, which suggest an ancient lineage for the protein (Gerritsen and Wagner, 2005).

Initial evidence for the possible role of CS in Ca^{2+} control was obtained by observations that removal of the corpuscles of Stannius, or stanniectomy (STX), resulted in an increase of plasma Ca^{2+} in eels and killifish (Fontaine, 1964, 1967; Pang, 1971; Fenwick, 1974), while extracts of the excised glands were able to restore normocalcemia or induce hypocalcemia (Pang *et al.*, 1973; Fenwick, 1974; Bailey and Fenwick, 1975; Dubewar and Suryawanshi, 1978). Further studies have shown that the cell morphology and secretory capability of the CS were determined by alterations in environmental Ca^{2+} (Pang and Pang, 1974; Meats *et al.*, 1978; Aida *et al.*, 1980; Wendelaar Bonga, 1980; reviewed by Wendelaar Bonga and Pang, 1991).

STC was isolated and characterized from the Atlantic salmon (Wagner *et al.*, 1986) and the Australian eel, *Anguilla australis* (Butkus *et al.*, 1987), and has since been cloned or isolated from several other salmonids, white sucker (*Catostomus commersoni*), garpike (*Lepisosteus platyrhynchus*), bowfin (*Amia calva*) and arawana (*Osteoglossum bicirrhosum*) among others (see Wagner and Dimattia (2006) for references). In most fish species, the translated preproSTC is a dimer of identical polypeptide chains. The prepro-STC monomer in salmonids is 256 aa long and is processed into a mature monomer of 223 residues (Flik *et al.*, 1990a; Wagner *et al.*, 1998b).

Production of STC mRNA and release of the protein are stimulated by increased Ca^{2+} levels either *in vitro* (Wagner *et al.*, 1989, 1998a; Wagner and Jaworski, 1994; Ellis and Wagner, 1995) and *in vivo* (Hanssen *et al.*, 1991, 1992; Wagner *et al.*, 1991, 1998a). Administration of $CaCl_2$ either by injection or perfusion leads to an elevation of both ionic and total plasma Ca^{2+} resulting in the release of STC from the CS into the blood. The secretion of stored STC from the incubated CS is also stimulated by elevated levels of ionic Ca^{2+} (2.5 mM) but not by concentrations that fall within or below the normal circulating levels (1.0–1.5 mM).

Transfer of SW-adapted eels to FW or distilled water showed a significant 50% drop in plasma STC levels (Hanssen *et al.*, 1992), but in fully acclimated fish, the STC and extracellular Ca^{2+} levels were not different between SW and FW. STC secretion and clearance were 70-75% higher in SW than in FW, and STC injection-induced hypocalcemia was more pronounced in SW (Hanssen *et al.*, 1993). A posterior study with FW and SW salmon has revealed no difference in the sensitivity of STC cells to Ca^{2+} and also suggested that the increased demand for STC in marine fishes is met instead by increased rates of hormone synthesis and secretion and perhaps by a redistribution of STC receptors (Wagner *et al.*, 1998a). Taken together, these data indicate that STC metabolism responds to changes in internal and external milieus, to compensate alterations in Ca^{2+} levels. The hormone seems to be more active in fish transferred to SW, probably in order to prevent hypercalcemia.

In the tilapia, enlargement of the CS was observed in sexually mature females, and the size of the CS increased in parallel with the growth of the ovaries and the elevation of total and ionic Ca^{2+} levels. Ovariectomy is followed by a reduction in the size of the CS and by a reduction in plasma Ca^{2+} to levels typical for males, in which gonadectomy does not affect size or ultrastructure of the CS, or plasma Ca^{2+} levels (Urasa and Wendelaar Bonga, 1985). Although there are no indications on the STC levels in these fish, these observations clearly agree with an anti-hypercalcemic role.

Secretion of STC is modulated by the CaSR. Radman *et al.* (2002) showed that Ca^{2+}-stimulated STC secretion in salmon is mimicked by calcimimetics that increase the sensitivity of the CaR to Ca^{2+}. The intraperitoneal administration of NPS R-467, a molecule that acts as a positive allosteric modulator of the CaR, produced time- and dose-dependent stimulatory effects on STC secretion that were virtually indistinguishable from those of Ca^{2+} loading experiments. Whether this secretion led to reductions in circulating Ca^{2+} was not tested, but the immediate downstream result was a reduction in branchial Ca^{2+} uptake. Previously, cholinergic stimulation of STC had been demonstrated (Cano *et al.*, 1994; Fenwick *et al.*, 1995), which indicates the existence of two regulatory pathways for STC secretion: a direct regulation by Ca^{2+} levels signalling through CaSRs on the CS STC-secreting cells and a cholinergic route that may be driven by CaSRs located elsewhere.

The hypocalcemic action of STC is mostly produced by rapid reductions in Ca^{2+} influx, and also increases in Ca^{2+} efflux. Single- and cumulative-injections induce up to 60% reduction in gill Ca^{2+} transport in rainbow trout (Wagner *et al.*, 1997a), and a similar effect was observed in tilapia. In this fish native STC and a synthetic N-terminal (1-20 amino acids) fragment were equipotent in reducing Ca^{2+} uptake (Verbost *et al.*, 1993a), indicating that this region comprises bioactivity and that these first amino acids are required for receptor binding and activation.

Increases in Ca^{2+} uptake after STX were observed in several species, either in isolated gills or in *in vivo* experiments (e.g., So and Fenwick, 1977, 1979; Milet *et al.*, 1979; Lafeber *et al.*, 1988; Verbost *et al.*, 1993a, b; Van Der Heijden *et al.*, 1999b). In eel, removal of the CS induced a 4-fold increase in plasma Ca^{2+}, reaching close to 8 mM, the result of a 7-fold increase in branchial Ca^{2+} uptake (Van Der Heijden *et al.*, 1999b). Branchial Ca^{2+} influx following STX increased during the first four days and stayed elevated for at least 40 days. Removal of the CS had no effect over the transepithelial potential across the gill cell, indicating that the increase of Ca^{2+} transport occurs across the cells and not in a paracellular fashion (Verbost *et al.*, 1993b). In this study, the increased Ca^{2+}-influx after STX was not correlated with changes in ATP-dependent Ca^{2+}-extrusion across the basolateral membrane, and STC did not affect the Ca^{2+}-ATPse in isolated basolateral membranes.

This suggests that in the gill cell STC acts by modulating the permeability of the apical Ca^{2+}-channel. A signal transduction system between the two membranes must exist (Wendelaar Bonga and Pang, 1991; Flik and Verbost, 1995) but receptors for STC have not been described in fish. Assuming that the main target for STC is the apical Ca^{2+}-channel, inhibition of Ca^{2+} entry in enterocytes and reduced reabsorption in renal tubule is to be expected. Given the amount of Ca^{2+} in SW, and the high drinking rates of marine fish, the intestine of SW fish is a likely target for a hypocalcemic factor. Stanniectomy increased drinking rates in SW-adapted eels, thus increasing the amount of Ca^{2+} in the gastrointestinal tract and inducing a steeper Ca^{2+} gradient across the luminal wall. Conversely, STC decreased Ca^{2+} uptake in the intestinal membrane in isolated cod gut preparations in a dose-dependent manner, reaching a maximum of 23% reduction (Sundell *et al.*, 1992). Unfortunately, very little information exists on the action of STC in the intestine and the same holds true for the effects in renal function. Incubation with the hormone failed to produce any effect on transepithelial Ca^{2+} transport across winter flounder proximal tubule cells, but increased phosphate reabsorption, which could likely depress plasma Ca^{2+} levels by promoting deposition (Lu *et al.*, 1994). Measurements of urine Ca^{2+} concentration and Ca^{2+} fractional glomerular filtration rates prior and upon STC treatment are not available.

A stimulatory action on bone and scale mineralization would also be in line with the effects of a hypocalcemic factor, but the few report available on the effects of STC in mammalian bone assays show contradictory results, either causing Ca^{2+} resorption in a PTH-like fashion (Lafeber *et al.*, 1986, 1989) or inhibiting the effects of PTH (Stern *et al.*, 1991; Yoshiko *et al.*, 1996).

PTHrP and PTH-like Peptides

Fish apparently lack an equivalent of the parathyroid gland, and were for long thought to be devoid of parathyroid hormone (PTH), the major hypercalcemic factor in terrestrial vertebrates. Nevertheless, several studies, using heterologous antisera, have historically indicated the presence of PTH-like immunoreactivity in fish pituitary glands, brain and plasma (Parsons *et al.*, 1978; Harvey *et al.*, 1987; Kaneko and Pang, 1987; Pang *et al.*, 1988; Fraser *et al.*, 1991), but only in recent years has the existence of PTH-like peptides and their receptors in fish been firmly established, initially with the isolation of PTH/PTH-related protein receptors in zebrafish (Rubin and Juppner, 1999), the cloning of the PTHrP cDNA in seabream and the characterization of the gene in pufferfish (Flanagan *et al.*, 2000; Power *et al.*, 2000). It is now clear that fish possess PTH and PTHrP like genes and 3 PTH/PTHrP receptors (Rubin and Juppner, 1999; Canario *et al.*, 2006). PTHrP has recently arisen as a likely hypercalcemic factor in fish.

PTHrP is a multifunctional factor in mammals (see Clemens *et al.*, 2001, for a comprehensive review of its functions), first identified in relation to malignant neoplastic tissues and associated to a disease, the humoral hypercalcemia of malignancy (Moseley *et al.*, 1987). In fish, the protein has been characterized in pufferfish, seabream and European flounder. The piscine mature protein is 125 to 129 amino acids long according to the species, and the N-terminal 34 amino acids, the known bioactive region in mammals, share 100% similarity due to conserved amino acid substitutions. PTHrP and PTH/PTHrP receptors type 1 and 3 (PTH1R and PTH3R) are expressed in many tissues related to Ca^{2+} transport, such as the branchial, intestinal and renal epithelia, Ca^{2+} deposition, such as bone, scales and cartilage and also in organs involved in the endocrine response to ion and osmoregulation, e.g., the pituitary, the inter-renal and the CS (reviewed in Guerreiro *et al.*, 2007).

In terrestrial vertebrates, PTHrP acts as a paracrine or autocrine regulator, but its presence in high concentration in fish blood, determined after the development and validation of an homologous radioimmunoassay (Rotllant *et al.*, 2003), predicts a classical endocrine function, most possibly related to Ca^{2+} metabolism, but to date no producing gland has been unambiguously identified in fish. A likely candidate, however, is the pituitary gland, since tissue extracts of seabream pituitary contain significant levels of the peptide (Rotllant *et al.*, 2003) and PTHrP is also released constitutively from European flounder pituitary primary cultures over a 24-hour period (Worthington *et al.*, 2004b). However, mRNA for PTHrP and PTH1R were down regulated in the pituitary of seabream transferred directly from SW to dilute SW or acclimated to low salinities for a week. The inverse relationship was observed in the gill (Abbink *et al.*, 2006).

In fully acclimated fish, plasma circulating levels seem to be closely related to both total and ionic blood Ca^{2+} (Abbink *et al.*, 2004, 2006). In juvenile seabream fed on a Ca^{2+}-deficient diet, plasma PTHrP was significantly increased when compared to fish on a normal diet and fish fed a normal diet but exposed to lower salinity (from 10.5 mM to 0.7 mM Ca^{2+}) also showed elevated PTHrP levels in relation to salinity and diet controls (Abbink *et al.*, 2004). In a series of preliminary studies with the European flounder, Worthington and colleagues (2003, 2004a) observed that chronically acclimated SW- and FW-animals had similar plasma levels of PTHrP and Ca^{2+}, but when kept in deionized water (DIW) for 2 weeks the levels of PTHrP increased significantly. Additionally, injection of the Ca^{2+}-chelating agent EGTA lowered plasma Ca^{2+} and significantly elevated plasma PTHrP 6-8 hours later. Furthermore, a transfer from SW to FW induced a rise in PTHrP levels (when compared to fish transferred from SW to SW) within 24 hours, while the reciprocal change (FW to SW) slightly decreased the circulating protein at both 4 and 8 hours after the transfer. Increases in PTHrP immunohistochemistry staining intensity were observed in flounder kidney, gill and pituitary from animals acclimated to FW compared to those in SW (Danks *et al.*, 1998). On the whole, these results suggest that Ca^{2+} levels—both in the plasma and in external sources—either in the diet or in the environment, are a significant factor regulating PTHrP secretion in fish, although the mechanisms involved are not completely clear. The data also indicates that the hormone may be used for rapid compensations in changing environments, but also necessary to reduce hypocalcemia in long-term hypocalcemic conditions. Whether the CaSR, the main modulator of the PTH response to lowered Ca^{2+} levels in tetrapods, is involved in the regulation of PTHrP in fish is not known, but changes in CaSR mRNA expression parallel those of the PTHrP and the PTH1R (Abbink *et al.*, 2006).

PTHrP seems to be involved with other factors known to evoke hypercalcemia. In addition to the rise in Ca^{2+} influx and plasma Ca^{2+}

levels upon E_2 administration, we have also observed an increase in PTHrP levels, that preceded the rise in Ca^{2+} (Fuentes *et al.*, in press). The estradiol-induced hypercalcemia was blocked by co-implantation of the (7-34)PTHrP antagonist, which strongly suggests that the rise in Ca^{2+} levels is due to PTHrP.

In sturgeon (*Acipenser naccarii*), single injections of piscine N-terminal (1-34)PTHrP significantly increased whole Ca^{2+} uptake while decreasing Ca^{2+} efflux and, consequently, plasma Ca^{2+} levels were up-regulated within 4 hours and sustained for 24 hours (Fuentes *et al.*, 2007). Exposure of seabream larva to (1-34)PTHrP in SW also caused a significant and dose-dependent increase in whole body Ca^{2+} uptake measured 4 hours after the addition of the peptide (Guerreiro *et al.*, 2001). This effect was achieved both by stimulated Ca^{2+} influx and by a reduction in Ca^{2+} efflux. Similar results were obtained with SW-, but not with FW-acclimated tilapia (*Oreochromis mossambicus*) larvae (Fig. 15.6A, B), indicating a differential regulation that may result from variations in endogenous levels of PTHrP in the two environments. The Ca^{2+} release rate from SW tilapia isolated gill cells pre-loaded with $^{45}Ca^{2+}$ was accelerated by addition of (1-34)PTHrP to the culture medium, an effect that occurs within seconds to a few minutes from the initial exposure, and suggests rapid regulation of basolateral transporting mechanisms (Fig. 15.7A). In opercular membranes of seabream mounted in Ussing chambers, the exposure to (1-34)PTHrP also stimulated $^{45}Ca^{2+}$ transport from the apical to the basolateral side (Fig. 15.7B).

The effects of physiological levels (1-34)PTHrP on intestinal Ca^{2+} uptake were measured in seabream duodenum, hindgut and rectum preparations mounted in Ussing chambers under symmetric (saline in both hemichambers) and asymmetric (saline in basolateral, saline similar to composition of the intestinal fluid in apical) (Fuentes *et al.*, 2006). When in symmetric conditions, administration of PTHrP resulted in an increase Ca^{2+} uptake in duodenum and hindgut and a reduction in efflux in the rectum, indicating that different mechanisms are responsive to PTHrP along the intestine. In control asymmetric conditions, there was a basal positive net uptake in Ca^{2+} uptake in all regions. Addition of (1-34)PTHrP in these conditions increased net Ca^{2+} uptake 2 to 3 fold in all regions and decreased slightly the epithelial resistance of the intestinal epithelium, which may favour an increase of paracellular Ca^{2+} absorption.

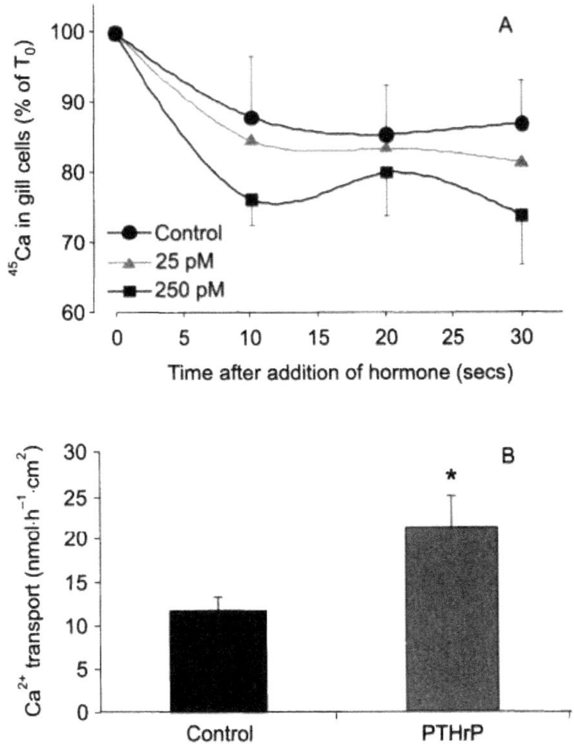

Fig. 15.7 Effects of PTHrP on calcium transport in branchial tissues. **A**—fast action of different PTHrP dose on $^{45}Ca^{2+}$ release from previously loaded gill cells. Isolated tilapia gill cells were incubated in culture medium with $^{45}Ca^{2+}$. Upon PTHrP treatment Ca^{2+} extrusion is enhanced in treated cells in relation to control cells. **B**—action of PTHrP on mucosal-to-serosal Ca^{2+} transport across the isolated opercular membrane of seabream mounted in using type chambers. Basal measurements were made in membranes exposed to saline for 60 minutes. PTHrP was added to the serosal side at this point and the effects evaluated after 60 minutes more. * indicates significant difference from the control group ($P<0.05$, *t*-test).

Which mechanisms are used by PTHrP to evoke these actions on branchial and intestinal tissue is unknown. PTH1R was identified in the gill cells by RT-PCR. In enterocytes, the existence of PTH3R was determined by RT-PCR and its affinity and transactivational characteristics determined in radioligand binding and intracellular signalling studies (Rotllant *et al.*, 2006). The receptor K_d is within the levels of PTHrP in plasma, indicating that increases in plasma hormone concentrations can have direct effects in the function of this receptor.

PTHrP may be an important regulatory factor in fish kidney. In Ussing chamber studies (1-34)PTHrP induced a slight increase in Ca^{2+} reabsorption in winter flounder proximal tubule cells, but the major effect was observed in phosphate secretion, which was significantly stimulated (Guerreiro *et al.*, 2004a). Removal of phosphate from the blood reduces Ca^{2+} deposition and increases the activity of free Ca^{2+}, as expected from a hypercalcemic factor.

There is also increasing evidence concerning the possible function of PTHrP as a factor involved in the resorption of the mineralized tissue. When treated with (1-34)PTHrP, the seabream scales had reduced mRNA expression of osteonectin, an important extracellular matrix protein (Redruello *et al.*, 2005). In addition, 10 nM (1-34)PTHrP, acting via the PTH1R, was equipotent to 1000 nM E_2 in stimulating TRAP activity in scales (Rotllant *et al.*, 2005). Taken together, these results strongly suggest a Ca^{2+} mobilizing action of PTHrP in fish scales.

The data reviewed support the view that PTHrP is a factor with hypercalcemic actions in fish that may counteract the hypocalcemic action of STC. The possible calciotropic actions of the novel PTH-like peptides are still to determine, but given the similarity of the N-terminal, interaction with the receptors targeted by PTHrP would not be surprising. At least one of the PTH-like proteins is equipotent to PTHrP in stimulating Ca^{2+} uptake in larvae (Canario *et al.*, 2006). So, after a long search for a hypercalcemic hormone, it is possible that a whole family of peptides may have such functions.

Perspectives

Given the particular environment that fish inhabit, calcium transport takes places mainly in three different epithelial tissues the gills, the intestine and the kidney, although we may also include the skin by the presence of mitochondria-rich cells. Calcium transport across epithelia shares a general model that is mostly based on the interaction of epithelial calcium channels, Ca^{2+}-ATPase and Na^+/Ca^{2+}-exchangers. With the aid of primarily vesicle studies, some of the transporters have been characterized in several species, and the relative importance in calcium transport defined. However, with the exception of a few studies, little is known about the regulation of these mechanisms at the molecular and endocrine levels. This would be of particular interest in the case of stanniocalcin that has been a well-established calcitropic factor for nearly two decades. However, the particular mechanisms responsible for its biological action in calcium transport remain elusive.

The fact that calcium availability in the water is high has been a good reason to propose a model for the endocrine control of calcium balance in fish based in the action of a single factor with hypocalcemic or anti-hypercalcemic nature: Stanniocalcin. However, the recent discovery of the PTH/PTHrP family of peptides in fish and their contribution to calcium balance with all the characteristics of hypercalcemic factors will probably mean a reassessment of this model for the endocrine control of calcium balance in fish. In addition, the interaction and putative cross-regulation of hypercalcemic and hypocalcemic factors in fish in short term and long-term calcium regulation is an issue that has obviously received little attention, but ensures new achievements on the understanding of how the endocrine control of calcium regulation takes place in fish.

Acknowledgements

The authors would like to express their appreciation to Professor A.V.M. Canario and Professor D.M. Power from CCMar (Faro), and Professor G. Flik from Radboud University (Nijmegen), for their constant support and encouragement. Unpublished results presented in this chapter have been funded by projects from the Commission of the European Union, Quality of Life and Management of Living Resources specific RTD programme (Q5RS-2001-02904) and Fundação para a Ciência e a Tecnologia (FCT), Ministry of Science, Portugal (POCTI/ CVT/48946/2002). P. G. is funded by the Ministry of Science, Portugal through grant SFRH/BPD/9464/02.

References

Abbink, W., G.S. Bevelander, J. Rotllant, A.V.M. Canarioand and G. Flik. 2004. Calcium handling in *Sparus aurata*: Effects of water and dietary calcium levels on mineral composition, cortisol and PTHrP levels. *Journal of Experimental Biology* 207: 4077–4084.

Abbink, W., G.S. Bevelander, X. Hang, W. Lu, P.M. Guerreiro, T. Spanings, A.V. Canario and G. Flik. 2006. PTHrP regulation and calcium balance in seabream (*Sparus aurata* L.) under calcium constraint. *Journal of Experimental Biolology* 209: 3550–3557.

Aida, K., R.S. Nishioka and H.A. Bern. 1980. Degranulation of the Stannius corpuscles of coho salmon (*Oncorhynchus kisutch*) in response to ionic changes *in vitro*. *General and Comparative Endocrinology* 41: 305–313.

Almansa, E., J.J. Sanchez, S. Cozzi, M. Casariego, J. Cejas and M. Diaz. 2001. Segmental heterogeneity in the biochemical properties of the Na^+-K^+-ATPase along the intestine of the gilthead seabream (*Sparus aurata* L.). *Journal of Comparative Physiology* B171: 557–567.

Arakawa, E., S. Hasegawa, T. Kaneko and T. Hirano. 1993. Effects of changes in environmental calcium on prolactin secretion in Japanese eel, *Anguilla japonica*. *Journal of Comparative Physiology* B163: 99–106.

Armour, K.J., D.B. Lehane, F. Pakdel, Y. Valotaire, R. Graham, R.G. Russell and I.W. Henderson. 1997. Estrogen receptor mRNA in mineralized tissues of rainbow trout: Calcium mobilization by estrogen. *FEBS Letters* 411: 145–148.

Bailey, J.R. and J.C. Fenwick. 1975. Effect of angiotensin II and corpuscle of Stannius extract on total and ionic plasma calcium levels and blood pressure in intact eels (*Anguilla rostrata* Lesueur). *Canadian Journal of Zoology* 53: 630–633.

Baldisserotto, B., M.J. Chowdhury and C. A. Wood. 2005. Effects of dietary calcium and cadmium on cadmium accumulation, calcium and cadmium uptake from the water, and their interactions in juvenile rainbow trout. *Aquatic Toxicology* 72: 99–117.

Barron, M.G. and S. Albeke. 2000. Calcium control of zinc uptake in rainbow trout. *Aquatic Toxicology* 50: 257–264.

Bergman, A.N., P. Laurent, G. Otiang'a-Owiti, H.L. Bergman, P.J. Walsh, P. Wilson and C.M. Wood. 2003. Physiological adaptations of the gut in the Lake Magadi tilapia, *Alcolapia grahami*, an alkaline- and saline-adapted teleost fish. *Comparative Biochemistry and Physiology* A136: 701–715.

Berntssen, M.H.G., R. Waagbo, H. Toften and A.K. Lundebye. 2003. Effects of dietary cadmium on calcium homeostasis, Ca mobilization and bone deformities in Atlantic salmon (*Salmo salar* L.) parr. *Aquaculture Nutrition* 9: 175–183.

Beyenbach, K.W., D.H. Petzel and W.H. Cliff. 1986. Renal proximal tubule of flounder. 1. Physiological properties. *American Journal of Physiology* 250: R608–R615.

Bijvelds, M.J.C., A.J.H. Vanderheijden, G. Flik, P.M. Verbost, Z.I. Kolar and S.E.W. Bonga. 1995. Calcium-pump activities in the kidneys of *Oreochromis mossambicus*. *Journal of Experimental Biology* 198: 1351–1357.

Björnsson, B.Th. and L.J. Deftos. 1985. Plasma calcium and calcitonin in the marine teleost, *Gadus morhua*. *Comparative Biochemistry and Physiology* A81: 593–596.

Björnsson, B.Th. and S. Nilsson. 1985. Renal and extra-renal excretion of calcium in the marine teleost, *Gadus morhua*. *American Journal of Physiology* 248: R18–R22.

Björnsson, B.Th., C. Haux, L. Forlin and L.J. Deftos. 1986. The involvement of calcitonin in the reproductive physiology of the rainbow trout. *Journal of Endocrinology* 108: 17–23.

Björnsson, B.Th., K. Yamauchi, R.S. Nishioka, L.J. Deftos and H.A. Bern. 1987. Effects of hypophysectomy and subsequent hormonal replacement therapy on hormonal and osmoregulatory status of coho salmon, *Oncorhynchus kisutch*. *General and Comparative Endocrinology* 68: 421–430.

Björnsson, B.Th., C. Haux, H.A. Bern and L.J. Deftos. 1989. 17β estradiol increases plasma calcitonin levels in salmonid fish. *Endocrinology* 125: 1754–1760.

Björnsson, B.Th., P. Persson, D. Larsson, S.H. Johannsson and K. Sundell. 1999. Calcium balance in teleost fish: Transport and endocrine control metabolism. In: *Calcium*

Metabolism: Comparative Endocrinology, J. Danks, C. Dacke, G. Flik and C. Gay (eds.). BioScientifica Ltd., Bristol, pp. 29–38.

Bodznik, D. 1978. Calcium ion: An odorant for natural water discriminations and the migratory behavior of sockeye salmon. *Journal of Comparative Physiology* A127: 157–166.

Borski, R.J., G.N. Hyde, S. Fruchtman and W.S. Tsai, 2001. Cortisol suppresses prolactin release through a non-genomic mechanism involving interactions with the plasma membrane. *Comparative Biochemistry and Physiology* B129: 533–541.

Brinca, L., J. Fuentes and D.M. Power. 2003. The regulatory action of estrogen and vasoactive intestinal peptide on prolactin secretion in seabream (*Sparus aurata*, L.). *General and Comparative Endocrinology* 131: 117–125.

Butkus, A., P.J. Roche, R.T. Fernley, J. Haralambidis, J.D. Penschow, G.B. Ryan, J.F. Trahair, G.W. Tregearand and J.P. Coghlan. 1987. Purification and cloning of a corpuscles of Stannius protein from *Anguilla australis*. *Molecular and Cellidar Endocrinology* 54: 123–133.

Canario, A.V.M., J. Rotllant, J. Fuentes, P.M. Guerreiro, H.R. Teodosio, D.M. Power and M.S. Clark. 2006. Novel bioactive parathyroid hormone and related peptides in teleost fish. *FEBS Letters* 580: 291–299.

Cano, T.M., S.F. Perry and J.C. Fenwick. 1994. Cholinergic control of stanniocalcin release in the rainbow trout, *Oncorhynchus mykiss*, and the American eel, *Anguilla rostrata*. *General and Comparative Endocrinology* 94: 1–10.

Cao, Y.X., B.C. Oh and L. Stryer. 1998. Cloning and localization of two multigene receptor families in goldfish olfactory epithelium. *Proceedings of the National Academy of Sciences of the United States of America* 95: 11987–11992.

Chakrabarti, P. and D. Mukherjee. 1993. Studies on the hypocalcemic actions of salmon calcitonin and ultimobranchial gland extracts in the freshwater teleost *Cyprinus carpio*. *General and Comparative Endocrinology* 90: 267–273.

Chakraborti, P. and D. Mukherjee. 1995. Effects of prolactin and fish pituitary extract on plasma calcium levels in common carp, *Cyprinus carpio*. *General and Comparative Endocrinology* 97: 320–326.

Chang, I.C., T.H. Lee, C.H. Yang, Y.Y. Wei, F.I. Chou and P.P. Hwang. 2001. Morphology and function of gill mitochondria-rich cells in fish acclimated to different environments. *Physiological and Biochemical Zoology* 74: 111–119.

Chang, M., H. Lin and P. Hwang. 1997. Effects of cadmium on the kinetics of calcium uptake in developing tilapia larvae, *Oreochromis mossambicus*. *Fish Physiology and Biochemistry* 16: 549–470.

Chang, M.H., H. C. Lin and P. P. Hwang. 1998. Ca^{2+} uptake and Cd^{2+} accumulation in larval tilapia (*Oreochromis mossambicus*) acclimated to waterborne Cd^{2+}. *American Journal of Physiology* 274: R1570–R1577.

Chartier, M.M., C. Milet, E. Martelly, E. Lopez and S. Warrot. 1980. Intestinal absorption of calcium in the eel (*Anguilla anguilla* L.) stimulated by vitamin-D_3 and 1,25-Dihydroxycholecalciferol. *Gastroenterologie Clinique et Biologique* 4: 929–930.

Chen, Y.Y., F.I. Lu and P.P. Hwang. 2003. Comparisons of calcium regulation in fish larvae. *Journal of Experimental Zoology* A295: 127–135.

Chou, M.Y., C.H. Yang, F.I. Lu, H.C. Lin and P.P. Hwang. 2002. Modulation of calcium balance in tilapia larvae (*Oreochromis mossambicus*) acclimated to low-calcium environments. *Journal of Comparative Physiology* B172: 109–114.

Christensen, L.J., B. Korsgaard and P. Bjerregaard. 1999. The effect of 4-nonylphenol on the synthesis of vitellogenin in the flounder *Platichthys flesus*. *Aquatic Toxicology* 46: 211–219.

Clark, M.S., L. Bendell, D.M. Power, S. Warner, G. Elgar and P.M. Ingleton. 2002. Calcitonin: Characterisation and expression in a teleost fish, *Fugu rubripes*. *Journal of Molecular Endocrinology* 28: 111–123.

Clemens, T.L., S. Cormier, A. Eichinger, K. Endlich, N. Fiaschi-Taesch, E. Fischer, P.A. Friedman, A.C. Karaplis, T. Massfelder and J. Rossert. 2001. Parathyroid hormone-related protein and its receptors: Nuclear functions and roles in the renal and cardiovascular systems, the placental trophoblasts and the pancreatic islets. *British Journal of Pharmacology* 134: 1113–1136.

Cliff, W.H. and K.W. Beyenbach. 1988. Fluid secretion in glomerular renal proximal tubules of freshwater-adapted fish. *American Journal of Physiology* 254: R154–R158.

Croke, S.J. and D.G. McDonald. 2002. The further development of ionoregulatory measures as biomarkers of sensitivity and effect in fish species. *Environmental Toxicology and Chemistry* 21: 1683–1691.

Danks, J.A., P.C. Hubbard, R.J. Balment, P.M. Ingleton and T.J. Martin. 1998. Parathyroid hormone-related protein localization in tissues of freshwater and saltwater-acclimatized flounder. *Annals of the New York Academy of Sciences* 839: 503–505.

De Girolamo, P., N. Arcamone, G.V. Pelagalli and G. Gargiulo 2003. Immunohistochemical localization of S100-like protein in non-mammalian kidney. *Microscopy Research and Technique* 60: 652–657.

Diaz, M., S. Cozzi, E. Almansa, M. Casariego, A. Bolanos, J. Cejas and A. Lorenzo. 1998. Characterization of intestinal Na^+-K^+-ATPase in the gilthead seabream (*Sparus aurata* L.). Evidence for a tissue-specific heterogeneity. *Comparative Biochemistry and Physiology* B121: 65–76.

Doneen, B.A. 1993. High-affinity Ca^{2+}-Mg^{2+}-ATPase in kidney of euryhaline *Gillichthys mirabilis*: Kinetics, subcellular distribution and effects of salinity. *Comparative Biochemistry and Physiology* B106: 719–728.

Dubewar, D.M. and S. A. Suryawanshi. 1978. Evidence of hypocalcemic factor from corpuscles of Stannius of the teleost *Heteropneustes fossilis* (Bloch). *Endokrinologie* 71: 210–213.

Elger, E., B. Elger, H. Hentschel and H. Stolte. 1987. Adaptation of renal function to hypotonic medium in the winter flounder (*Pseudopleuronectes americanus*). *Journal of Comparative Physiology* B157: 21–30.

Ellis, T.J. and G.F. Wagner. 1995. Post-transcriptional regulation of the stanniocalcin gene by calcium. *Journal of Biological Chemistry* 270: 1960–1965.

Evans, D.H., P.M. Piermarini and K.P. Choe. 2005. The multifunctional fish gill: Dominant site of gas exchange, osmoregulation, acid-base regulation, and excretion of nitrogenous waste. *Physiological Reviews* 85: 97–177.

Fasulo, S., A. Mauceri, G. Tagliafierro, M.B. Ricca, P. Lo Cascio and L. Ainis. 1998. Immunoreactivity to calcium-binding proteins (CaBPs) in the epithelia of skin and gill of the catfish, *Heteropneustes fossilis*. *Italian Journal of Zoology* 65: 149–153.

Faustino, M. and D. Power. 1998. Development of osteological structures in the seabream: Vertebral column and caudal fin complex. *Journal of Fish Biology* 52: 11–22.

Faustino, M. and D. Power. 1999. Development of the pectoral, pelvic, dorsal and anal fins in cultured seabream. *Journal of Fish Biology* 54: 1094–1110.

Fenwick, J.C. 1974. The corpuscles of Stannius and calcium regulation in the North American eel (*Anguilla rostrata* LeSueur). *General and Comparative Endocrinology* 23: 127–135.

Fenwick, J.C. 1981. The renal handling of calcium and renal Ca^{2+} (Mg^{2+})-activated adenosinetriphosphatase activity in freshwater and seawater acclimated North American eels (*Anguilla rostrata* LeSueur). *Canadian Journal of Zoology* 59: 478–485.

Fenwick, J.C. 1984. Effect of vitamin D_3 (cholecalciferol) on plasma calcium and intestinal ^{45}Ca absorption in goldfish, *Carassius auratus* L. *Canadian Journal of Zoology* 62: 34–36.

Fenwick, J.C. 1989. Calcium exchange across fish gills. In: *Vertebrate Endocrinology: Fundamental and Biomedical Implications*, P.K.T. Pang and M.P. Schreibman (eds.). Academic Press, San Diego, Vol. 3, pp. 319–342.

Fenwick, J.C. and T.J. Lam. 1988. Effects of calcitonin on plasma calcium and phosphate in the mudskipper, *Periophthalmodon schlosseri* (Teleostei), in water and during exposure to air. *General and Comparative Endocrinology* 70: 224–230.

Fenwick, J.C., K. Smith, J. Smith and G. Flik. 1984. Effect of various vitamin D analogs on plasma calcium and phosphorus and intestinal calcium absorption in fed and unfed American eels, *Anguilla rostrata*. *General and Comparative Endocrinology* 55: 398–404.

Fenwick, J.C., W. Davison and M.E. Forster. 1994. *In vivo* calcitropic effect of some vitamin D compounds in the marine Antarctic teleost, *Pagothenia bernacchii*. *Fish Physiology and Biochemistry* 12: 479–484.

Fenwick, J.C., G. Flik and P.M. Verbost. 1995. A passive immunization technique against the teleost hypocalcemic hormone stanniocalcin provides evidence for the cholinergic control of stanniocalcin release and the conserved nature of the molecule. *General and Comparative Endocrinology* 98: 202–210.

Filby, A.L. and C.R. Tyler. 2005. Molecular characterization of estrogen receptors 1, 2a, and 2b and their tissue and ontogenic expression profiles in fathead minnow (*Pimephales promelas*). *Biology of Reproduction* 73: 648–662.

Flanagan, J.A., L.A. Bendell, P.M. Guerreiro, M.S. Clark, D.M. Power, A.V. Canario, L. Brown and P.M. Ingleton. 2002. Cloning of the cDNA for the putative calcium-sensing receptor and its tissue distribution in seabream (*Sparus aurata*). *General and Comparative Endocrinology* 127: 117–127.

Flanagan, J.A., D.M. Power, L.A. Bendell, P.M. Guerreiro, J. Fuentes, M.S. Clark, A.V. Canario, J.A. Danks, B.L. Brown and P.M. Ingleton. 2000. Cloning of the cDNA for seabream (*Sparus aurata*) parathyroid hormone-related protein. *General and Comparative Endocrinology* 118: 373–382.

Flik, G. and S.F. Perry. 1989. Cortisol stimulates whole body calcium uptake and the branchial calcium pump in freshwater rainbow trout. *Journal of Endocrinology* 120: 75–82.

Flik, G. and P. Verbost. 1993. Calcium transport in fish gills and intestine. *Journal of Experimental Biology* 184: 17–29.

Flik, G. and P. Verbost. 1994. Ca^{2+} transport across plasma membranes. In: *Biochemistry and Molecular Biology of Fishes*, P.W. Hochachka and T.P. Mommsen (eds.), Elsevier, Amsterdam, Vol. 3, pp. 625–637.

Flik, G. and P. Verbost. 1995. Cellular mechanisms in calcium transport and homeostasis in fish. In: *Biochemistry and Molecular Biology of Fishes*, P.W. Hochachka and T.P. Mommsen (eds.), Elsevier, Amsterdam, Vol. 5, pp. 251–263.

Flik, G., P.H.M. Klaren, T.J.M. Schoenmakers, M.J.C. Bijvelds, P.M. Verbost and S.E.W. Bonga. 1996. Cellular calcium transport in fish: Unique and universal mechanisms. *Physiological Zoology* 69: 403–417.

Flik, G., F.M.J. Reijntjens, J. Stikkelbroeck and J.C. Fenwick. 1982. 1,25-Vitamin D_3 and calcium transport in the gut of tilapia (*Saretherodon mossambicus*). *Journal of Endocrinology* 94: 40.

Flik, G., S.E. Wendelaar Bonga and J.C. Fenwick. 1983. Ca^{2+}-dependent phosphatase and ATPase activities in eel gill plasma membranes. 1. Identification of Ca^{2+}-activated ATPase activities with non-specific phosphatase-activities. *Comparative Biochemistry and Physiology* B76: 745–754.

Flik, G., S.E. Wendelaar Bonga and J.C. Fenwick. 1984a. Ca^{2+}-dependent phosphatase and Ca^{2+}-dependent ATPase activities in plasma membranes of eel gill epithelium. 2. Evidence for transport by high-affinity Ca^{2+}-ATPase. *Comparative Biochemistry and Physiology* B79: 9–16.

Flik, G., S.E. Wendelaar Bonga and J.C. Fenwick. 1984b. Ca^{2+}-dependent phosphatase and Ca^{2+}-dependent ATPase activities in plasma membranes of eel gill epithelium. 3. Stimulation of branchial high-affinity Ca^{2+}-ATPase activity during prolactin-induced hypercalcemia in American eels. *Comparative Biochemistry and Physiology* B79: 521–524.

Flik, G., J.H. Vanrijs, J.C. Fenwick and S.E. Wendelaar Bonga. 1984c. Effects of ambient Ca^{2+} and prolactin on high-affinity (Ca^{2+}/Mg^{2+})-ATPase and nonspecific phosphatase-activity in gills of American eel (*Anguilla rostrata*). *General and Comparative Endocrinology* 53: 494–495.

Flik, G., S.E. Wendelaar Bonga and J.C. Fenwick. 1985a. Active Ca^{2+} transport in plasma membranes of branchial epithelium of the North American eel, *Anguilla rostrata* LeSueur. *Biology of the Cell* 55: 265–272.

Flik, G., J. Rijs and S.E. Wendelaar Bonga. 1985b. Evidence for high-affinity Ca^{2+}-ATPase activity and ATP-driven Ca^{2+}-transport in membrane preparations of the gill epithelium of the cichlid fish *Oreochromis mossambicus*. *Journal of Experimental Biology* 119: 335–347.

Flik, G., J.C. Fenwick, Z. Kolar, N. Mayer-Gostan and S.E. Wendelaar Bonga 1986a. Effects of low ambient calcium levels on whole-body Ca^{2+} flux rates and internal calcium pools in the freshwater cichlid teleost *Oreochromis mossambicus*. *Journal of Experimental Biology* 120: 249–264.

Flik, G., J.C. Fenwick, Z. Kolar, N. Mayer-Gostan and S.E. Wendelaar Bonga. 1986b. Effects of ovine prolactin on calcium uptake and distribution in *Oreochromis mossambicus*. *American Journal of Physiology* 250: R161–R166.

Flik, G., J.C. Fenwick and S.E. Wendelaar Bonga. 1989. Calcitropic actions of prolactin in freshwater North American eel (*Anguilla rostrata* LeSueur). *American Journal of Physiology* 257: R74–R79.

Flik, G., T. Labedz, J.A. Neelissen, R.G. Hanssen, S.E. Wendelaar Bonga and P.K.T. Pang. 1990a. Rainbow trout corpuscles of Stannius: stanniocalcin synthesis *in vitro*. *American Journal of Physiology* 258: R1157–R1164.

Flik, G., T. Schoenmakers, J. Groot, C. Os and S.E. Wendelaar Bonga. 1990b. Calcium absorption by fish intestine: the involvement of ATP- and sodium-dependent calcium extrusion mechanisms. *Journal of Membrane Biology* 113: 13–22.

Flik, G., W. Atsma, J.C. Fenwick, F. Renter-Delrue, J. Smal and S.E. Wendelaar Bonga. 1993a. Homologous recombinant growth hormone and calcium metabolism in the tilapia, *Oreochromis mossambicus*, adapted to fresh water. *Journal of Experimental Biology* 185: 107–119.

Flik, G., J. Velden, K. Dechering, P. Verbost, T. Schoenmakers, Z. Kolar and S.E. Wendelaar Bonga. 1993b. Ca^{2+} and Mg^{2+} transport in gills and gut of tilapia, *Oreochromis mossambicus*: A review. *Journal of Experimental Zoology* 265: 356–365.

Flik, G., F. Rentier-Delrue and S.E. Wendelaar Bonga. 1994. Calcitropic effects of recombinant prolactins in *Oreochromis mossambicus*. *American Journal of Physiology* R1302–R1308.

Flik, G., P. Verbost and S.E. Wendelaar Bonga. 1995. Calcium transport processes in fishes. In: *Fish Physiology*, C. Wood and T. Shuttleworth (eds.). Academic Press, Vol. 14, 317–342.

Flik, G., T. Kaneko, A.M. Greco, J. Li and J.C. Fenwick. 1997. Sodium dependent ion transporters in trout gills. *Fish Physiology and Biochemistry* 17: 385–396.

Fontaine, M. 1964. [Stannius' Corpuscles and Ionic (Ca, K, Na) of the Interior Environment of the Eel (*Anguilla Anguilla* L.)]. *Comptes rendus hebdomadaires des séances de l'Académie des sciences. Série D: Sciences naturelles* 259: 875–878.

Fontaine, M. 1967. Intervention of the corpuscula of Stannius in phosphocalcic balance of the internal milieu of a teleost fish, the eel. *Comptes rendus hebdomadaires des séances de l'Académie des sciences. Série D: Sciences naturelles* 264: 736–737.

Foster, R.C. 1975. Changes in urinary bladder and kidney function in the starry flounder (*Platichthys stellatus*) in response to prolactin and to freshwater transfer. *General and Comparative Endocrinology* 27: 153–161.

Foster, R.C. 1976. Renal hydromineral metabolism in starry flounder, *Platichthys stellatus*. *Comparative Biochemistry and Physiology* A55: 135–140.

Fouchereau-Peron, M., Y. Arlot-Bonnemains, M.S. Moukhtar and G. Milhaud. 1987. Calcitonin induces hypercalcemia in grey mullet and immature freshwater and sea-water adapted rainbow trout. *Comparative Biochemistry and Physiology* A87: 1051–1053.

Fouchereau-Peron, M., Y. Arlot-Bonnemains, L. Maubras, G. Milhaud and M.S. Moukhtar. 1990. Calcitonin variations in male and female trout, *Salmo gairdneri*, during the annual cycle. *General and Comparative Endocrinology* 78: 159–163.

Franklin, N.M., C.N. Glover, N.A. Nicol and C.M. Wood. 2005. Calcium/cadmium interactions at uptake surfaces in rainbow trout: Waterborne versus dietary routes of exposure. *Environmental Toxicology and Chemistry* 24: 2954–2964.

Fraser, R.A., T. Kaneko, P.K.T. Pang and S. Harvey. 1991. Hypo- and hypercalcemic peptides in fish pituitary glands. *American Journal of Physiology* 260: R622–R626.

Fuentes, J. and F. Eddy. 1997. Drinking in freshwater, euryhaline and marine teleosts. In: *Ionic Regulation in Animals*, N. Hazon, F. Eddy and G. Flik (eds.). Springer-Verlag, Berlin, pp. 135–149.

Fuentes, J., J. Figueiredo, D.M. Power and A.V.M. Canario. 2006. Parathyroid hormone-related protein regulates intestinal calcium transport in the seabream (*Sparus aurata*). *American Journal of Physiology* 299: R1499–1506.

Fuentes, J., C. Haond, P.M. Guerreiro, N. Silva, D.M. Power and A.V.M. Canario. 2007. Regulation of calcium balance in the sturgeon *Acipenser naccarii*: A role for PTHrP. *American Journal of Physiology* doi: 10.1152/ajpregu.00203.2007.

Fuentes, J., P.M. Guerreiro, T. Modesto, J. Rotllant, A.V.M. Canario and D.M. Power. A PTH/PTHrP receptor antagonist blocks the hypercalcemic response to estradiol-17β. *American Journal of Physiology*. (In Press).

Galvez, F., D. Wong and C.M. Wood. 2006. Cadmium and calcium uptake in isolated mitochondria-rich cell populations from the gills of the freshwater rainbow trout. *American Journal of Physiology* 291: R170–R176.

Gerritsen, M.E. and G.F. Wagner. 2005. Stanniocalcin: No longer just a fish tale. *Vitamines and Hormones* 70: 105–135.

Gillespie, D.K. and A. de Peyster. 2004. Plasma calcium as a surrogate measure for vitellogenin in fathead minnows (*Pimephales promelas*). *Ecotoxicology and Environmental Safety* 58: 90–95.

Gonzalez, R.J., C.M. Wood, R.W. Wilson, M.L. Patrick, H.L. Bergman, A. Narahara and A.L. Val. 1998. Effects of water pH and calcium concentration on ion balance in fish of the Rio Negro, Amazon. *Physiological Zoology* 71: 15–22.

Gonzalez, R.J., R.W. Wilson, C.M. Wood, M.L. Patrick and A.L. Val. 2002. Diverse strategies for ion regulation in fish collected from the ion-poor, acidic Rio Negro. *Physiological and Biochemical Zoology* 75: 37–47.

Guerreiro, P.M., J. Fuentes, D.M. Power, P.M. Ingleton, G. Flik and A.V. Canario. 2001. Parathyroid hormone-related protein: a calcium regulatory factor in seabream (*Sparus aurata* L.) larvae. *American Journal of Physiology* 281: R855–R860.

Guerreiro, P.M., J. Fuentes, A.V. Canario and D.M. Power. 2002. Calcium balance in seabream (*Sparus aurata*): The effect of oestradiol-17β. *Journal of Endocrinology* 173: 377–385.

Guerreiro, P.M., A.V.M. Canario, D.M. Power and J.L. Renfro. 2004a. Possible actions of PTHrP on calcium and phosphate transport by winter flounder (*Pseudopleuronectes americanus*) renal proximal tubule cells. In: *Avances en Endocrinologia Comparada*, J.P. Castaño, M.M. Malagón and S. García Navarro (eds.), Universidad de Córdoba. Córdoba, Spain, Vol. 2, pp. 113–117.

Guerreiro, P.M., J. Fuentes, G. Flik, J. Rotllant, D.M. Power and A.V.M. Canario. 2004b. Water calcium concentration modifies whole-body calcium uptake in seabream larvae during short-term adaptation to altered salinities. *Journal of Experimental Biology* 207: 645–653.

Guerreiro, P.M., J.L. Renfro, D.M. Power and A.V.M. Canario. 2007. The parathyroid hormone family of peptides: Structure, tissue distribution, regulation and potential functional roles in calcium and phosphate balance in fish. *American Journal of Physiology* 292: R679–R696.

Guzman, J.M., S. Sangiao-Alvarellos, R. Laiz-Carrion, J.M. Miguez, P. Martin del Rio Mdel, J. L. Soengas and J. M. Mancera. 2004. Osmoregulatory action of 17β estradiol in the gilthead seabream *Sparus aurata*. *Journal of Experimental Zoology* A301: 828–836.

Hang, X.M., D. Power, G. Flik and R.J. Balment. 2005. Measurement of PTHrP, PTHR1, and CaSR expression levels in tissues of seabream (*Sparus aurata*) using quantitative PCR. *Annals of the New York Academy of Sciences* 1040: 340–344.

Hanssen, R.G., F.P. Lafeber, G. Flik and S.E. Wendelaar Bonga. 1989. Ionic and total calcium levels in the blood of the European eel (*Anguilla anguilla*): Effects of stanniectomy and hypocalcin replacement therapy. *Journal of Experimental Biology* 141: 177–186.

Hanssen, R., E. Aarden, W. Venne, P.K.T. Pang and S.E. Wendelaar Bonga. 1991. Regulation of secretion of the teleost fish hormone stanniocalcin: Effects of extracellular calcium. *General and Comparative Endocrinology* 84: 155–163.

Hanssen, R., N. Mayer-Gostan, G. Flik and S.E. Wendelaar Bonga. 1992. Influence of ambient calcium levels on stanniocalcin secretion in the European eel (*Anguilla anguilla*). *Journal of Experimental Biology* 162: 197–208.

Hanssen, R., N. Mayer-Gostan, G. Flik and S.E. Wendelaar Bonga. 1993. Stanniocalcin kinetics in freshwater and seawater European eel (*Anguilla anguilla*). *Fish Physiology and Biochemistry* 10: 491–496.

Harvey, S., Y.Y. Zeng and P.K.T. Pang. 1987. Parathyroid hormone-like immunoreactivity in fish plasma and tissues. *General and Comparative Endocrinology* 68: 136–146.

Hearn, P.R., C.J. Kenyon, R.G.G. Russell, H. Mellersh, C.J. Preston and S. Tomlinson. 1978. Low-molecular weight calcium-binding protein from kidney and gill of freshwater eel (*Anguilla anguilla*). *Journal of Endocrinology* 79: P36–P37.

Herrmann-Erlee, M. and G. Flik. 1989. Bone: comparative studies on endocrine involvement in bone metabolism. In: *Vertebrate Endocrinology: Fundamentals and Biomedical Implications—Regulation of Calcium and Phosphate*, P.K.T. Pang and M. Schreibman (eds.). Academic Press, San Diego, Vol. 3, pp. 211–352.

Hickman, C.P., Jr. 1968. Ingestion, intestinal absorption, and elimination of seawater and salts in the southern flounder, *Paralichthys lethostigma*. *Canadian Journal of Zoology* 46: 457–466.

Hidaka, Y., S. Tanaka and M. Suzuki. 2004. Analysis of salmon calcitonin I in the ultimobranchial gland and gill filaments during development of rainbow trout, *Oncorhynchus mykiss*, by *in situ* hybridization and immunohistochemical staining. *Zoological Science* 21: 629–637.

Hirano, T., S. Hasegawa, H. Yamauchi and H. Orimo. 1981. Further studies on the absence of hypocalcemic effects of eel calcitonin in the eel, *Anguilla japonica*. *General and Comparative Endocrinology* 43: 42–50.

Hiroi, J., T. Kaneko, T. Seikai and M. Tanaka. 1998. Developmental sequence of chloride cells in the body skin and gills of Japanese flounder (*Paralichthys olivaceus*) larvae. *Zoological Science* 15: 455–460.

Hiroi, J., T. Kaneko and M. Tanaka. 1999. *In vivo* sequential changes in chloride cell morphology in the yolk-sac membrane of mozambique tilapia (*Oreochromis mossambicus*) embryos and larvae during seawater adaptation. *Journal of Experimental Biology* 202: 3485–3495.

Hobe, H., P. Laurent and B.R. McMahon. 1984. Whole body calcium flux rates in freshwater teleosts as a function of ambient calcium and pH Levels: A comparison between the euryhaline trout, *Salmo gairdneri* and stenohaline bullhead, *Ictalurus nebulosus*. *Journal of Experimental Biology* 113: 237–252.

Hogstrand, C., P.M. Verbost, S.E. Wendelaar Bonga and C.M. Wood. 1996. Mechanisms of zinc uptake in gills of freshwater rainbow trout: Interplay with calcium transport. *American Journal of Physiology* 270: R1141–R1147.

Hossain, M.A. and M. Furuichi. 2000. Necessity of calcium supplement to the diet of Japanese flounder. *Fisheries Science* 66: 660–664.

Hubbard, P.C., E.N. Barata and A.V. Canario. 2000. Olfactory sensitivity to changes in environmental [Ca^{2+}] in the marine teleost *Sparus aurata*. *Journal of Experimental Biology* 203: 3821–3829.

Hubbard, P.C., P.M. Ingleton, L.A. Bendell, E.N. Barata and A.V.M. Canario. 2002. Olfactory sensitivity to changes in environmental [Ca^{2+}] in the freshwater teleost *Carassius auratus*: An olfactory role for the Ca^{2+}-sensing receptor? *Journal of Experimental Biology* 205: 2755–2764.

Hwang, P.P. and C.H. Yang. 1997. Modulation of calcium uptake in cadmium-pretreated tilapia (*Oreochromis mossambicus*) larvae. *Fish Physiology and Biochemistry* 16: 403–410.

Hwang, P., Y. Tsai and Y. Tung. 1994. Calcium balance in embryos and larvae of the freshwater-adapted teleost, *Oreochromis mossambicus*. *Fish Physiology and Biochemistry* 13: 325–333.

Hwang, P., Y. Tung and M. Chang. 1996. Effect of environmental calcium levels on calcium uptake in tilapia larvae (*Oreochromis mossambicus*). *Fish Physiology and Biochemistry* 15: 363–370.

Ingleton, P.M., P.A. Hubbard, J.A. Danks, G. Elgar, R.A. Sandford and R.J. Balment. 1999. The calcium-sensing receptor in fishes. In: *Calcium Metabolism: Comparative Endocrinology*, J. Danks, C. Dacke, G. Flik and C. Gay (eds.). BioScientifica, Bristol, pp. 1–4.

Ingleton, P.M., L.A. Bendell, J.A. Flanagan, C. Teitsma and R. J. Balment. 2002. Calcium-sensing receptors and parathyroid hormone-related protein in the caudal neurosecretory system of the flounder (*Platichthys flesus*). *Journal of Anatomy* 200: 487–497.

Ishihara, A. and Y. Mugiya. 1987. Ultrastructural evidence of calcium uptake by chloride cells in the gills of goldfish, *Carassius auratus*. *Journal of Experimental Zoology* 242: 121–129.

Kacem, A., F.J. Meunier and J.L. Baglinière. 1998. A quantitative study of morphological and histological changes in the skeleton of *Salmo salar* during its anadromous migration. *Journal of Fish Biology* 53: 1096–1109.

Kacem, A., S. Gustafsson and F.J. Meunier. 2000. Demineralization of the vertebral skeleton in Atlantic salmon *Salmo salar* L. during spawning migration. *Comparative Biochemistry and Physiology* A125: 479–484.

Kakizawa, S., T. Kaneko, S. Hasegawa and T. Hirano. 1993. Activation of somatolactin cells in the pituitary of the rainbow trout *Oncorhynchus mykiss* by low environmental calcium. *General and Comparative Endocrinology* 91: 298–306.

Kakizawa, S., T. Kaneko, S. Hasegawa and T. Hirano. 1995. Effects of feeding, fasting, background adaptation, acute stress, and exhaustive exercise on the plasma somatolactin concentrations in rainbow trout. *General and Comparative Endocrinology* 98: 137–146.

Kaneko, T. and P.K.T. Pang. 1987. Immunocytochemical detection of parathyroid hormone-like substance in the goldfish brain and pituitary gland. *General and Comparative Endocrinology* 68: 147–152.

Kaneko, T. and T. Hirano. 1993. Role of prolactin and somatolactin in calcium regulation in fish. *Journal of Experimental Biology* 184: 31–45.

Kaneko, T., K. Shiraishi, F. Katoh, S. Hasegawa and J. Hiroi. 2002. Chloride cells during early life stages of fish and their functional differentiation. *Fisheries Science* 68: 1–9.

Kaneko, T. 1996. Cell biology of somatolactin. *International Review of Cytology* 169: 1–24.

Katoh, F. and T. Kaneko. 2002. Effects of environmental Ca^{2+} levels on branchial chloride cell morphology in freshwater-adapted killifish *Fundulus heteroclitus*. *Fisheries Science* 68: 347–355.

Kirk, C.J., L. Bottomley, N. Minican, H. Carpenter, S. Shaw, N. Kohli, M. Winter, E. W. Taylor, R. H. Waring, F. Michelangeli *et al.* 2003. Environmental endocrine disrupters dysregulate estrogen metabolism and Ca^{2+} homeostasis in fish and mammals via receptor-independent mechanisms. *Comparative Biochemistry and Physiology* A135: 1–8.

Klaren, P.H.M., G. Flik, R.A.C. Lock and S.E.W. Bonga. 1993. Ca^{2+} transport across intestinal brush-border membranes of the cichlid teleost *Oreochromis mossambicus*. *Journal of Membrane Biology* 132: 157–166.

Klaren, P.H.M., S.E.W. Bonga and G. Flik. 1997. Evidence for P_2 purinoceptor-mediated uptake of Ca^{2+} across a fish (*Oreochromis mossambicus*) intestinal brush border membrane. *Biochemical Journal* 322: 129–134.

Lafeber, F.P., H.I. Schaefer, M.P. Herrmann-Erlee and S.E. Wendelaar Bonga. 1986. Parathyroid hormone-like effects of rainbow trout Stannius products on bone resorption of embryonic mouse calvaria *in vitro*. *Endocrinology* 119: 2249–2255.

Lafeber, F.P., G. Flik, S.E. Wendelaar Bonga and S.F. Perry. 1988. Hypocalcin from Stannius corpuscles inhibits gill calcium uptake in trout. *American Journal of Physiology* 254: R891–896.

Lafeber, F.P., M.P. Herrmann-Erlee, G. Flik and S.E. Wendelaar Bonga. 1989. Rainbow trout hypocalcin stimulates bone resorption in embryonic mouse calvaria *in vitro* in a PTH-like fashion. *Journal of Experimental Biology* 143: 165–175.

Larsson, D., B.T. Bjornsson and K. Sundell. 1995. Physiological concentrations of 24,25-dihydroxyvitamin D_3 rapidly decrease the *in vitro* intestinal calcium uptake in the Atlantic cod, *Gadus morhua*. *General and Comparative Endocrinology* 100: 211–217.

Larsson, D., T. Lundgren and K. Sundell, K. 1998. Ca^{2+} uptake through voltage-gated L-type Ca^{2+} channels by polarized enterocytes from Atlantic cod *Gadus morhua*. *Journal of Membrane Biology* 164: 229–237.

Larsson, D., I. Nemere and K. Sundell. 2001. Putative basal lateral membrane receptors for 24,25-dihydroxyvitamin D_3 in carp and Atlantic cod enterocytes: Characterization of binding and effects on intracellular calcium regulation. *Journal of Cellular Biochemistry* 83: 171–186.

Larsson, D., L. Aksnes, T.B. Björnsson, B. Larsson, T. Lundgren and K. Sundell. 2002. Antagonistic effects of 24R,25-dihydroxyvitamin D_3 and 25-hydroxyvitamin D_3 on L-type Ca^{2+} channels and Na^+/Ca^{2+} exchange in enterocytes from Atlantic cod (*Gadus morhua*). *Journal of Molecular Endocrinology* 28: 53–68.

Larsson, D., I. Nemere, L. Aksnes and K. Sundell. 2003. Environmental salinity regulates receptor expression, cellular effects, and circulating levels of two antagonizing hormones, 1,25-dihydroxyvitamin D_3 and 24,25-dihydroxyvitamin D_3, in rainbow trout. *Endocrinology* 144: 559–566.

Lehane, D.B., N. McKie, R.G.G. Russell and I.W. Henderson. 1999. Cloning of a fragment of the osteonectin gene from goldfish, *Carassius auratus*: Its expression and potential regulation by estrogen. *General and Comparative Endocrinology* 114: 80–87.

Li, J., J. Eygensteyn, R.A.C. Lock, S.E.W. Bonga and G. Flik 1997. Na^+ and Ca^{2+} homeostatic mechanisms in isolated chloride cells of the teleost *Oreochromis mossambicus* analysed by confocal laser scanning microscopy. *Journal of Experimental Biology* 200: 1499–1508.

Lopez, E., J. Peignoux-Deville, F. Lallier, E. Martelly and C. Milet. 1976. Effects of calcitonin and ultimobranchialectomy (UBX) on calcium and bone metabolism in the eel, *Anguilla anguilla* L. *Calcified Tissue Research* 20: 173–186.

Lopez, E., J. Peignouxdeville, F. Lallier, K.W. Colston and I. Macintyre. 1977. Responses of bone metabolism in eel (*Anguilla anguilla*) to injections of 1,25-dihydroxyvitamin D_3. *Calcified Tissue Research* 22: 19–23.

Lopez, E., I. Macintyre, E. Martelly, F. Lallier and B. Vidal. 1980. Paradoxical effect of 1,25-dihydroxycholecalciferol on osteoblastic and osteoclastic activity in the skeleton of the eel *Anguilla anguilla* L. *Calcified Tissue International* 32: 83–87.

Loretz, C.A., C. Pollina, S. Hyodo, Y. Takei, W.H. Chang and D. Shoback. 2004. CDNA cloning and functional expression of a Ca^{2+}-sensing receptor with truncated C-terminal tail from the Mozambique tilapia (*Oreochromis mossambicus*). *Journal of Biological Chemistry* 279: 53288–53297.

Lu, M., P. Swanson and J.L. Renfro. 1995. Effect of somatolactin and related hormones on phosphate transport by flounder renal tubule primary cultures. *American Journal of Physiology* 268: R577–R582.

Lu, M.Q., G.F. Wagner and J.L. Renfro. 1994. Stanniocalcin stimulates phosphate reabsorption by flounder renal proximal tubule in primary culture. *American Journal of Physiology* 36: R1356–R1362.

Marshall, W.S. 1995. Transport processes in isolated teleost epithelia: Opercular epithelium and urinary bladder. In: *Cellular and Molecular Approaches to Fish Ionic Regulation*, C.M. Wood and T.J. Shuttleworth (eds.). Academic Press, New York, Vol. 14, pp. 1–23.

Marshall, W.S. 2002. Na^+, Cl^-, Ca^{2+} and Zn^{2+} transport by fish gills: retrospective review and prospective synthesis. *Journal of Experimental Zoology* A293: 264–283.

Marshall, W. and S. Bryson. 1998. Transport mechanisms of seawater teleost chloride cells: An inclusive model of a multifunctional cell. *Comparative Biochemistry and Physiology* A119: 97–106.

Marshall, W.S. and M. Grosell. 2006. Ion transport, osmoregulation, and acid-base balance. In: *The Physiology of Fishes*, D.H. Evans and J.B. Claiborne (eds.) 3rd Edition, CRC Press, Boca Raton, pp.170–230.

Marshall, W., S. Bryson and C. Wood. 1992. Calcium transport by isolated skin of rainbow trout. *Journal of Experimental Biology* 166: 297–316.

Marshall, W., S. Bryson, J. Burghardt and P. Verbost. 1995. Ca^{2+} transport by opercular epithelium of the fresh water adapted euryhaline teleost, *Fundulus heteroclitus*. *Journal of Comparative Physiology* B165: 268–277.

Martial, K., L. Maubras, J. Taboulet, A. Jullienne, M. Berry G. Milhaud, A.A. Benson, M.S. Moukhtar and M. Cressent. 1994. The calcitonin gene is expressed in salmon gills. *Proceedings of the National Academy of Sciences of the United States of America* 91: 4912–4914.

Matsuo, A.Y., C.M. Wood and A.L. Val. 2005. Effects of copper and cadmium on ion transport and gill metal binding in the Amazonian teleost tambaqui (*Colossoma macropomum*) in extremely soft water. *Aquatic Toxicology* 74: 351–364.

Mayer-Gostan, N., M. Bornancin, G. DeRenzis, R. Naon, J.A. Yee, R.L. Shew and P.K.T. Pang. 1983. Extraintestinal calcium uptake in the killifish, *Fundulus heteroclitus*. *Journal of Experimental Zoology* 227: 329–338.

McCormick, S.D. and D. Bradshaw. 2006. Hormonal control of salt and water balance in vertebrates. *General and Comparative Endocrinology* 147: 3–8.

McCormick, S., S. Hasegawa and T. Hirano. 1992. Calcium uptake in the skin of a freshwater teleost. *Proceedings of the National Academy of Sciences of the United States of America* 89: 3635–3638.

McCormick, S.D., M.F. O'Dea, A.M. Moeckel, D.T. Lerner and B.T. Björnsson. 2005. Endocrine disruption of parr-smolt transformation and seawater tolerance of Atlantic salmon by 4-nonylphenol and 17β estradiol. *General and Comparative Endocrinology* 142: 280–288.

Meats, M., P.M. Ingleton, I.C. Jones, H.O. Garland and C.J. Kenyon. 1978. Fine structure of the corpuscles of Stannius of the trout, *Salmo gairdneri*: Structural changes in response to increased environmental salinity and calcium ions. *General and Comparative Endocrinology* 36: 451–461.

Milet, C., J. Peignoux-Deville and E. Martelly. 1979. Gill calcium fluxes in the eel, *Anguilla anguilla* (L.). Effects of Stannius corpuscles and ultimobranchial body. *Comparative Biochemistry and Physiology* A63: 63–70.

Milhaud, G., J.C. Rankin, L. Bolis and A.A. Benson. 1977. Calcitonin: Its hormonal action on the gill. *Proceedings of the National Academy of Sciences of the United States of America* 74: 4693–4696.

Milhaud, G., L. Bolis and A.A. Benson. 1980. Calcitonin, a major gill hormone. *Proceedings of the National Academy of Sciences of the United States of America* 77: 6935–6936.

Morgan, T.P., C.M. Guadagnolo, M. Grosell and C.M. Wood. 2005. Effects of water hardness on the physiological responses to chronic waterborne silver exposure in early life stages of rainbow trout (*Oncorhynchus mykiss*). *Aquatic Toxicology* 74: 333–350.

Moron, S.E., E.T. Oba, C.A. De Andrade and M.N. Fernandes. 2003. Chloride cell responses to ion challenge in two tropical freshwater fish, the erythrinids *Hoplias malabaricus* and *Hoplerythrinus unitaeniatus*. *Journal of Experimental Zoology* A298: 93–104.

Moseley, J.M., M. Kubota, H. Diefenbach-Jagger, R.E. Wettenhall, B.E. Kemp, L.J. Suva, C.P. Rodda, P.R. Ebeling, P.J. Hudson, J.D. Zajac, *et al.* 1987. Parathyroid hormone-related protein purified from a human lung cancer cell line. *Proceedings of the National Academy of Sciences of the United States of America* 84: 5048–5052.

Mousa, M.A. and S.A. Mousa. 2000. Implication of somatolactin in the regulation of sexual maturation and spawning of *Mugil cephalus*. *Journal of Experimental Zoology* 287: 62–73.

Mugiya, Y. and N. Watabe. 1977. Studies on fish scale formation and resorption .2. Effect of estradiol on calcium homeostasis and skeletal tissue resorption in goldfish, *Carassius auratus*, and killifish, *Fundulus heteroclitus*. *Comparative Biochemistry and Physiology* A57: 197–202.

Mukherjee, D., U. Sen, S.P. Bhattacharyya and D. Mukherjee. 2004a. The effects of calcitonin on plasma calcium levels and bone metabolism in the fresh water teleost *Channa punctatus*. *Comparative Biochemistry and Physiology* A138: 417–426.

Mukherjee, D., U. Sen, S.P. Bhattacharyya and D. Mukherjee. 2004b. Inhibition of whole body Ca^{2+} uptake in fresh water teleosts, *Channa punctatus* and *Cyprinus carpio* in response to salmon calcitonin. *Journal of Experimental Zoology* A301: 882–890.

Mundy, G.R. 1990. *Calcium Homeostasis: Hypercalcemia and Hypocalcemia*. Martin Dunitz Publishers, London, UK.

Mundy, G.R. and T.A. Guise. 1999. Hormonal control of calcium homeostasis. *Clinical Chemistry* 45: 1347–1352.

Nagler, J.J., S.M. Ruby, D.R. Idler and Y.P. So. 1987. Serum phosphoprotein, phosphorus and calcium levels as reproductive indicators of vitellogenin in highly vitellogenic mature female and estradiol-injected immature rainbow trout (*Salmo gairdneri*). *Canadian Journal of Zoology* 65: 2421–2425.

Nearing, J., M. Betka, S. Quinn, H. Hentschel, M. Elger, M. Baum, M. Bai, N. Chattopadyhay, E.M. Brown, S.C. Hebert, *et al.* 2002. Polyvalent cation receptor proteins (CaRs) are salinity sensors in fish. *Proceedings of the National Academy of Sciences of the United States of America* 99: 9231–9236.

Nebel, C., G. Negre-Sadargues, C. Blasco and G. Charmantier. 2005. Morphofunctional ontogeny of the urinary system of the European sea bass *Dicentrarchus labrax*. *Anatomy and Embryology* 209: 193–206.

Nijenhuis, T., J.G. Hoenderop and R.J. Bindels. 2005. TRPV5 and TRPV6 in Ca^{2+} (re)absorption: Regulating Ca^{2+} entry at the gate. *Pflugers Archives* 451: 181–192.

Niyogi, S. and C.M. Wood. 2006. Interaction between dietary calcium supplementation and chronic waterborne zinc exposure in juvenile rainbow trout, *Oncorhynchus mykiss*. *Comparative Biochemistry and Physiology* C143: 94–102.

Norberg, B., B.T. Björnsson, C.L. Brown, U.P. Wichardt, L.J. Deftos and C. Haux. 1989. Changes in plasma vitellogenin, sex steroids, calcitonin, and thyroid hormones related to sexual maturation in female brown trout (*Salmo trutta*). *General and Comparative Endocrinology* 75: 316–326.

Oughterson, S.M., R. Munoz-Chapuli, V. De Andres, R. Lawson, S. Heath and D.H. Davies. 1995. The effects of calcitonin on serum calcium levels in immature brown trout, *Salmo trutta*. *General and Comparative Endocrinology* 97: 42–48.

Pan, T.C., B.K. Liao, C.J. Huang, L.Y. Lin and P.P. Hwang. 2005. Epithelial Ca^{2+} channel expression and Ca^{2+} uptake in developing zebrafish. *American Journal of Physiology* 289: R1202–1211.

Pandey, A.C. 1993. Correlative changes in the corpuscles of Stannius (CS), and total plasma calcium, phosphorus and protein levels in *Ompok bimaculatus* (Bloch) in response to bilateral ovariectomy and estradiol-17β treatment. *National Academy Science Letters-India* 16: 149–152.

Pang, P.K.T. 1971. The relationship between corpuscles of stannius and serum electrolyte regulation in killifish, *Fundulus heteroclitus*. *Journal of Experimental Biology* 178: 1–8.

Pang, P.K.T. and F. Balbontin. 1978. Effects of sex steroids on plasma calcium levels in male killifish, *Fundulus heteroclitus*. *General and Comparative Endocrinology* 36: 317–320.

Pang, P.K.T. and R.K. Pang. 1974. Environmental calcium and hypocalcin activity in the Stannius corpuscles of the channel catfish, *Ictalurus punctatus* (Rafinesque). *General and Comparative Endocrinology* 23: 239–241.

Pang, P.K.T., R.K. Pang and W.H. Sawyer. 1973. Effects of environmental calcium and replacement therapy on the killifish, *Fundulus heteroclitus*, after the surgical removal of the corpuscles of Stannius. *Endocrinology* 93: 705–710.

Pang, P.K.T., T. Kaneko and S. Harvey. 1988. Immunocytochemical distribution of PTH immunoreactivity in vertebrate brains. *American Journal of Physiology* 255: R643–R647.

Parmelee, J.T. and J.L. Renfro. 1983. Esophageal desalination of seawater in flounder—Role of active sodium-transport. *American Journal of Physiology* 245: R888–R893.

Parsons, J., D. Gray, B. Rafferty and J. Zanelli. 1978. Evidence for a hypercalcaemic factor in the fish pituitary immunologically related to mammalian parathyroid hormone. In: *Endocrinology of Calcium Metabolism*, D. Copp and R. Talmage (eds.). Excerpta Medica, Amsterdam, Vol. 421, pp. 111–114.

Perrot, M.N., C.E. Grierson, N. Hazon and R.J. Balment. 1992. Drinking behaviour in sea water and fresh water teleosts, the role of the renin-angiotensin system. *Fish Physiology and Biochemistry* 10: 161–168.

Perry, S.F. 1997. The chloride cell: Structure and function in the gills of freshwater fishes. *Annual Review of Physiology* 59: 325–347.

Perry, S.F. and G. Flik. 1988. Characterization of branchial transepithelial calcium fluxes in freshwater trout, *Salmo gairdneri*. *American Journal of Physiology* 254: R491–R498.

Perry, S.F., G.G. Goss and J.C. Fenwick. 1992. Interrelationships between gill chloride cell morphology and calcium uptake in freshwater teleosts. *Fish Physiology and Biochemistry* 10: 327–337.

Perry, S.F., A. Shahsavarani, T. Georgalis, M. Bayaa, M. Furimsky and S.L.Y. Thomas. 2003. Channels, pumps, and exchangers in the gill and kidney of freshwater fishes: Their role in ionic and acid-base regulation. *Journal of Experimental Zoology* A300: 53–62.

Persson, P., K. Sundell and B.T. Björnsson. 1994. Estradiol-17β induced calcium uptake and resorption in juvenile rainbow trout, *Oncorhynchus mykiss*. *Fish Physiology and Biochemistry* 13: 379–386.

Persson, P., Y. Takagi and B. Björnsson. 1995. Tartrate resitant acid phosphatase as a marker for scale resorption in rainbow trout, *Oncorhynchus mykiss*: Effects of estradiol-17β treatment and refeeding. *Fish Physiology and Biochemistry* 14: 329–339.

Persson, P., S. Johannsson, Y. Takagi and B. Bjornsson. 1997. Estradiol-17β and nutritional status affect calcium balance, scale and bone resorption, and bone formation in rainbow trout, *Oncorhynchus mykiss*. *Journal of Comparative Physiology* B167: 468–473.

Persson, P., K. Sundell, B. Björnsson and H. Lundqvist. 1998. Calcium metabolism and osmoregulation during sexual maturation of river running Atlantic salmon. *Journal of Fish Biology* 52: 334–349.

Persson, P., B.T. Björnsson and Y. Takagi. 1999. Characterization of morphology and physiological actions of scale osteoclasts in the rainbow trout. *Journal of Fish Biology* 54: 669–684.

Persson, P., J.M. Shrimpton, S.D. McCormick and B.T. Björnsson. 2000. The presence of high-affinity, low-capacity estradiol-17β binding in rainbow trout scale indicates a possible endocrine route for the regulation of scale resorption. *General and Comparative Endocrinology* 120: 35–43.

Pinto, P.I., A.L. Passos, R.S. Martins, D.M. Power and A.V.M. Canario. 2006. Characterization of estrogen receptor betab in seabream (*Sparus aurata*): Phylogeny, ligand-binding, and comparative analysis of expression. *General and Comparative Endocrinology* 145: 197–207.

Porta, A.R., E. Bettini, O.I. Buiakova, H. Baker, W. Danho and F.L. Margolis. 1996. Molecular cloning of ictacalcin: A novel calcium-binding protein from the channel catfish, *Ictalurus punctatus*. *Molecular Brain Research* 41: 81–89.

Potts, W.T.W. and A.J. Hedges. 1991. Gill potentials in marine teleosts. *Journal of Comparative Physiology* B 161: 401–405.

Power, D.M., P.M. Ingleton, J. Flanagan, A.V. Canario, J. Danks, G. Elgar and M.S. Clark. 2000. Genomic structure and expression of parathyroid hormone-related protein gene (PTHrP) in a teleost, *Fugu rubripes*. *Gene* 250: 67–76.

Qiu, A.D. and C. Hogstrand. 2004. Functional characterisation and genomic analysis of an epithelial calcium channel (ECaC) from pufferfish, *Fugu rubripes*. *Gene* 342: 113–123.

Radman, D.P., C. McCudden, K. James, E.M. Nemeth and G.F. Wagner. 2002. Evidence for calcium-sensing receptor mediated stanniocalcin secretion in fish. *Molecular and Cellular Endocrinology* 186: 111–119.

Rao, D.S. and N. Raghuramulu. 1999. Vitamin D_3 and its metabolites have no role in calcium and phosphorus metabolism in *Tilapia mossambica*. *Journal of Nutritional Science and Vitaminology* 45: 9–19.

Redruello, B., M.D. Estevao, J. Rotllant, P.M. Guerreiro, L.I. Anjos, A.V.M. Canario and D.M. Power. 2005. Isolation and characterization of piscine osteonectin and downregulation of its expression by PTH-related protein. *Journal of Bone and Mineral Research* 20: 682–692.

Renfro, J.L. 1975. Water and ion-transport by urinary bladder of teleost *Pseudopleuronectes americanus*. *American Journal of Physiology* 228: 52–61.

Renfro, J.L. 1978. Calcium transport across peritubular surface of the marine teleost renal tubule. *American Journal of Physiology* 234: F522–F531.

Renfro, J.L., K.G. Dickman and D.S. Miller. 1982. Effect of Na^+ and ATP on peritubular Ca transport by the marine teleost renal tubule. *American Journal of Physiology* 243: R34–R41.

Riccardi, D. 1999. Cell surface, Ca^{2+}(cation)-sensing receptor(s): One or many? *Cell Calcium* 26: 77–83.

Riccardi, D. 2000. Calcium ions as extracellular, first messengers. *Zeitschrift für Kardiologie* 89 Supplement 2: 9–14.

Rogers, J.T. and C.M. Wood. 2004. Characterization of branchial lead-calcium interaction in the freshwater rainbow trout *Oncorhynchus mykiss*. *Journal of Experimental Biology* 207: 813–825.

Rombough, P.J. 1999. The gill of fish larvae. Is it primarily a respiratory or an ionoregulatory structure. *Journal of Fish Biology* 55: 186–204.

Rombough, P.J. 2002. Gills are needed for ionoregulation before they are needed for O_2 uptake in developing zebrafish, *Danio rerio*. *Journal of Experimental Biology* 205: 1787–1794.

Rotllant, J., G.P. Worthington, J. Fuentes, P.M. Guerreiro, C.A. Teitsma, P.M. Ingleton, R.J. Balment, A.V. Canario and D.M. Power. 2003. Determination of tissue and plasma concentrations of PTHrP in fish: development and validation of a radioimmunoassay using a teleost 1-34 N-terminal peptide. *General and Comparative Endocrinology* 133: 146–153.

Rotllant, J., B. Redruello, P.M. Guerreiro, H. Fernandes, A.V.M. Canario and D.M. Power. 2005. Calcium mobilization from fish scales is mediated by parathyroid hormone-related protein via the parathyroid hormone type 1 receptor. *Regulatory Peptides* 132: 33–40.

Rotllant, J., P.M. Guerreiro, B. Redruello, H. Fernandes, L. Apolónia, L. Anjos, A.V.M. Canario and D.M. Power. 2006. Ligand binding and signalling pathways of PTH receptors in seabream (*Sparus aurata*) enterocytes. *Cell and Tissue Research* 323: 333–341.

Rubin, D.A. and H. Juppner. 1999. Zebrafish express the common parathyroid hormone/parathyroid hormone-related peptide receptor (PTH1R) and a novel receptor (PTH3R) that is preferentially activated by mammalian and fugufish parathyroid hormone-related peptide. *Journal of Biological Chemistry* 274: 28185–28190.

Sakamoto, T. and M. Ando. 2002. Calcium ion triggers rapid morphological oscillation of chloride cells in the mudskipper, *Periophthalmus modestus*. *Journal of Comparative Physiology* B 172: 435–439.

Sakamoto, T. and S.D. McCormick. 2006. Prolactin and growth hormone in fish osmoregulation. *General and Comparative Endocrinology* 147: 24–30.

Sandra, O., P. Le Rouzic, F.O. Rentier-Delrue and P. Prunet. 2001. Transfer of tilapia (*Oreochromis niloticus*) to a hyperosmotic environment is associated with sustained expression of prolactin receptor in intestine, gill, and kidney. *General and Comparative Endocrinology* 123: 295–307.

Sarasquete, C., M.L.G. de Canales, J. Arellano, J.A.M. Cueto, L. Ribeiro and M.T. Dinis. 1998. Histochemical study of skin and gills of Senegal sole, *Solea senegalensis* larvae and adults. *Histology and Histopathology* 13: 727–735.

Sasayama, Y. 1999. Hormonal control of Ca homeostasis in lower vertebrates: Considering the evolution. *Zoological Science* 16: 857–869.

Sasayama, Y., Y. Takei, S. Hasegawa and D. Suzuki. 2002. Direct raises in blood Ca levels by infusing a high-Ca solution into the blood stream accelerate the secretion of calcitonin from the ultimobranchial gland in eels. *Zoological Science* 19: 1039–1043.

Schmidt-Nielsen, B. and J.L. Renfro. 1975. Kidney function of the American eel *Anguilla rostrata*. *American Journal of Physiology* 228: 420–431.

Schoenmakers, T. and G. Flik. 1992. Sodium-extruding and calcium-extruding sodium/calcium exchangers display similar calcium affinities. *Journal of Experimental Biology* 168: 151–159.

Schoenmakers, T., P. Verbost, G. Flik and S.E. Wendelaar Bonga. 1993. Transcelullar intestinal calcium transport in freshwater and seawater fish and its dependence on sodium/calcium exchange. *Journal of Experimental Zoology* 176: 195–206.

Seale, A.P., N.H. Richman, T. Hirano, I. Cooke and E.G. Grau. 2003. Cell volume increase and extracellular Ca^{2+} are needed for hyposmotically induced prolactin release in tilapia. *American Journal of Physiology* 284: C1280–1289.

Shahsavarani, A. and S.F. Perry. 2006. Hormonal and environmental regulation of the epithelial calcium channel (ECaC) in the gill of rainbow trout (*Oncorhynchus mykiss*). *American Journal of Physiology* (In press).

Shahsavarani, A., B. McNeill, F. Galvez, C.M. Wood, G.G. Goss, P.P. Hwang and S.F. Perry. 2006. Characterization of a branchial epithelial calcium channel (ECaC) in freshwater rainbow trout (*Oncorhynchus mykiss*). *Journal of Experimental Biology* 209: 1928–1943.

Shinobu, N. and Y. Mugiya. 1995. Effects of ovine prolactin, bovine growth hormone and triiodothyronine on the calcification of otoliths and scales in the hypophysectomized goldfish *Carassius auratus*. *Fisheries Science* 61: 960–963.

Shinozaki, F. and Y. Mugiya. 2000. Effects of salmon calcitonin on calcium deposition on and release from calcified tissues in fed and starved goldfish *Carassius auratus*. *Fisheries Science* 66: 695–700.

Skadhauge, E. and R. Lotan. 1974. Drinking rate and oxygen consumption in the euryhaline teleost *Aphanius dispar* in waters of high salinity. *Journal of Experimental Biology* 60: 547–556.

So, Y.P. and J.C. Fenwick. 1977. Relationship between net ^{45}calcium influx across a perfused isolated eel gill and the development of post-stanniectomy hypercalcemia. *Journal of Experimental Zoology* 200: 259–264.

So, Y.P. and J.C. Fenwick. 1979. *In vivo* and *in vitro* effects of Stannius corpuscle extract on the branchial uptake of ^{45}Ca in stanniectomized North American eels (*Anguilla rostrata*). *General and Comparative Endocrinology* 37: 143–149.

Socorro, S., D.M. Power, P.E. Olsson and A.V. Canario. 2000. Two estrogen receptors expressed in the teleost fish, *Sparus aurata*: cDNA cloning, characterization and tissue distribution. *Journal of Endocrinology* 166: 293–306.

Specker, J., D. King, R. Nishioka, K. Shirahata, K. Yamagushi and H. Bern. 1985. Isolation and partial characterization of a pair of prolactins released *in vitro* by the pituitary of a cichlid fish, *Oreochromis mossambicus*. *Proceedings of the National Academy of Sciences of the United States of America* 82: 7490–7494.

Spry, D.J. and J.G. Wiener. 1991. Metal bioavailability and toxicity to fish in low-alkalinity lakes: A critical review. *Environmental Pollution* 71: 243–304.

Srivastav, A. and S. Singh. 1992. Effect of vitamin D_3 administration on the serum calcium and inorganic phosphate levels of the fresh water catfish, *Heteropneustes fossilis*, maintained in artificial fresh water, calcium-rich fresh water, and calcium-deficient fresh water. *General and Comparative Endocrinology* 87: 63–70.

Srivastav, A., S. Srivastav, Y. Sasayama, N. Suzuki and A. Norman. 1997. Vitamin D metabolites affect serum calcium and phosphate in freshwater catfish, *Heteropneustes fossilis*. *Zoological Science* 14: 743–746.

Srivastav, A.K., G. Flik and S.E. Wendelaar Bonga. 1998. Plasma calcium and stanniocalcin levels of male tilapia, *Oreochromis mossambicus*, fed calcium-deficient food and treated with 1,25 dihydroxyvitamin D_3. *General and Comparative Endocrinology* 110: 290–294.

Srivastav, A.K., P.R. Tiwari, S.K. Srivastav and N. Suzuki. 2002. Responses of the ultimobranchial gland to vitamin D_3 treatment in freshwater mud eel, *Amphipnous cuchia*, kept in different calcium environments. *Anatomy, Histology and Embryology* 31: 257–261.

Srivastav, S.K. and A.K. Srivastav. 1998. Annual changes in serum calcium and inorganic phosphate levels and correlation with gonadal status of a freshwater murrel, *Channa punctatus* (Bloch). *Brazilian Journal of Medical and Biological Research* 31: 1069–1073.

Stern, P.H., G. Shankar, R.C. Fargher, D.H. Copp, C.E. Milliken, K.J. Sato, D. Goltzman and M.P. Herrmann-Erlee. 1991. Salmon stanniocalcin and bovine parathyroid hormone have dissimilar actions on mammalian bone. *Journal of Bone and Mineral Research* 6: 1153–1159.

Stricker, S.A. 1999. Comparative biology of calcium signaling during fertilization and egg activation in animals. *Developmental Biology* 211: 157–176.

Sundell, K. and B.T. Björnsson. 1988. Kinetics of calcium fluxes across the intestinal mucosa of the marine teleost, *Gadus morhua*, measured using an *in vitro* perfusion method. *Journal of Experimental Biology* 140: 171–186.

Sundell, K. and B.T. Björnsson. 1990. Effects of vitamin D_3, 25(OH) vitamin D_3, 24,25$(OH)_2$ vitamin D_3, and 1,25$(OH)_2$ vitamin D_3 on the *in vitro* intestinal calcium absorption in the marine teleost, Atlantic cod (*Gadus morhua*). *General and Comparative Endocrinology* 78: 74–79.

Sundell, K., A.W. Norman and B.Th. Björnsson. 1993. 1,25$(OH)_2$ vitamin D_3 increases ionized plasma calcium concentrations in the immature Atlantic cod *Gadus morhua*. *General and Comparative Endocrinology* 91: 344–351.

Sundell, K., D. Larsson and B.Th. Björnsson. 1996. Calcium regulation by the vitamin D_3 system in teleosts, with special emphasis on the intestine. In: *The Comparative Endocrinology of Calcium Regulation*, C. Dacke, J. Danks, I. Caple and G. Flik (eds.). Journal of Endocrinology Ltd., Bristol, pp. 75–84.

Sundell, K., B.Th. Björnsson, H. Itoh and H. Kawauchi. 1992. Chum salmon (*Oncorhynchus keta*) stanniocalcin inhibits *in vitro* intestinal calcium uptake in Atlantic cod (*Gadus morhua*). *Journal of Comparative Physiology B* 162: 489–495.

Sunny, F. and O.V. Oommen. 2001. Rapid action of glucocorticoids on branchial ATPase activity in *Oreochromis mossambicus*: An *in vivo* and *in vitro* study. *Comparative Biochemistry and Physiology* A130: 323–330.

Suzuki, N. and A. Hattori. 2003. Bisphenol A suppresses osteoclastic and osteoblastic activities in the cultured scales of goldfish. *Life Sciences* 73: 2237–2247.

Suzuki, N., D. Suzuki, Y. Sasayama, A.K. Srivastav, A. Kambegawa and K. Asahina. 1999. Plasma calcium and calcitonin levels in eels fed a high calcium solution or transferred to seawater. *General and Comparative Endocrinology* 114: 324–329.

Suzuki, N., T. Suzuki and T. Kurokawa. 2000. Suppression of osteoclastic activities by calcitonin in the scales of goldfish (freshwater teleost) and nibbler fish (seawater teleost). *Peptides* 21: 115–124.

Suzuki, N., T. Suzuki and T. Kurokawa. 2001. Cloning of a calcitonin gene-related peptide from genomic DNA and its mRNA expression in flounder, *Paralichthys olivaceus*. *Peptides* 22: 1435–1438.

Suzuki, N., A. Kambegawa and A. Hattori. 2003. Bisphenol A influences the plasma calcium level and inhibits calcitonin secretion in goldfish. *Zoological Science* 20: 745–748.

Suzuki, N., K. Yamamoto, Y. Sasayama, T. Suzuki, T. Kurokawa, A. Kambegawa, A.K. Srivastav, S. Hayashi and S. Kikuyama. 2004. Possible direct induction by estrogen of calcitonin secretion from ultimobranchial cells in the goldfish. *General and Comparative Endocrinology* 138: 121–127.

Taylor, J.R. and M. Grosell. 2006. Evolutionary aspects of intestinal bicarbonate secretion in fish. *Comparative Biochemistry and Physiology* A 143: 523–529.

Tfelt-Hansen, J. and E.M. Brown. 2005. The calcium-sensing receptor in normal physiology and pathophysiology: A review. *Critical Reviews in Clinical Laboratory Sciences* 42: 35–70.

Tsai, C.L. and L.H. Wang. 2000. Sex differences in the responses of serum calcium concentrations to temperature and estrogen in tilapia, *Oreochromis mossambicus*. *Zoological Studies* 39: 55–60.

Tse, D.L.Y., B.K.C. Chow, C.B. Chan, L.T.O. Lee and C.H.K. Cheng. 2000. Molecular cloning and expression studies of a prolactin receptor in goldfish (*Carassius auratus*). *Life Sciences* 66: 593–605.

Tytler, P., M. Tatner and C. Findlay. 1990. The ontogeny of drinking in the rainbow trout, *Oncorhynchus mykiss* (Walbaum). *Journal of Fish Biology* 36: 867–875.

Urasa, F.M. and S.E. Wendelaar Bonga. 1985. Stannius corpuscles and plasma calcium levels during the reproductive cycle in the cichlid teleost fish, *Oreochromis mossambicus*. *Cell and Tissue Research* 241: 219–227.

Van Der Heijden, A.J., J.C. Van Der Meij, G. Flik and S.E. Wendelaar Bonga. 1999a. Ultrastructure and distribution dynamics of chloride cells in tilapia larvae in fresh water and sea water. *Cell and Tissue Research* 297: 119–130.

Van Der Heijden, A.J., P.M. Verbost, M.J. Bijvelds, W. Atsma, S. E. Wendelaar Bonga and G. Flik. 1999b. Effects of sea water and stanniectomy on branchial Ca^{2+} handling and drinking rate in eel (*Anguilla anguilla* L.). *Journal of Experimental Biology* 202: 2505–2511.

Verbost, P. M., G. Flik, R.A. Lock and S.E. Wendelaar Bonga. 1987. Cadmium inhibition of Ca^{2+} uptake in rainbow trout gills. *American Journal of Physiology* 253: R216–R221.

Verbost, P.M., F.P. Lafeber, F.A. Spanings, E.M. Aarden and S.E. Wendelaar Bonga. 1992. Inhibition of Ca^{2+} uptake in freshwater carp, *Cyprinus carpio*, during short-term exposure to aluminum. *Journal of Experimental Zoology* 262: 247–254.

Verbost, P.M., A. Butkus, W. Atsma, P. Willems, G. Flik and S.E. Wendelaar Bonga. 1993a. Studies on stanniocalcin: Characterization of bioactive and antigenic domains of the hormone. *Molecular and Cellular Endocrinology* 93: 11–16.

Verbost, P.M., G. Flik, J.C. Fenwick, A. Greco, P. Pang and S.E. Wendelaar Bonga. 1993b. Branchial calcium uptake: Possible mechanisms of control by stanniocalcin. *Fish Physiology and Biochemistry* 11: 205–215.

Verbost, P.M., T.J. Schoenmakers, G. Flik and S.E. Wendelaar Bonga. 1994. Kinetics of ATP-driven and Na^{+}-gradient driven Ca^{2+} transport in basolateral membranes from gills of freshwater- and seawater-adapted tilapia. *Journal of Experimental Biology* 186: 95–108.

Verbost, P.M., S.E. Bryson, S.E. Wendelaar Bonga and W.S. Marshall. 1997. Na^{+}-dependent Ca^{2+} uptake in isolated opercular epithelium of *Fundulus heteroclitus*. *Journal of Comparative Physiology* B167: 205–212.

Vonck, A., S.E. Wendelaar Bonga and G. Flik. 1998. Sodium and calcium balance in Mozambique tilapia, *Oreochromis mossambicus*, raised at different salinities. *Comparative Biochemistry and Physiology* A119: 441–449.

Wagner, G.F. and E. Jaworski. 1994. Calcium regulates stanniocalcin mRNA levels in primary cultured rainbow trout corpuscles of Stannius. *Molecular and Cellular Endocrinology* 99: 315–322.

Wagner, G.F. and G.E. Dimattia. 2006. The stanniocalcin family of proteins. *Journal of Experimental Zoology* A305: 769–780.

Wagner, G.F., M. Hampong, C.M. Park and D.H. Copp. 1986. Purification, characterization, and bioassay of teleocalcin, a glycoprotein from salmon corpuscles of Stannius. *General and Comparative Endocrinology* 63: 481–491.

Wagner, G.F., B. Gellersen and H.G. Friesen. 1989. Primary culture of teleocalcin cells from rainbow trout corpuscles of Stannius: regulation of teleocalcin secretion by calcium. *Molecular and Cellular Endocrinology* 62: 31–39.

Wagner, G.F., C. Milliken, H.G. Friesen and D.H. Copp. 1991. Studies on the regulation and characterization of plasma stanniocalcin in rainbow trout. *Molecular and Cellular Endocrinology* 79: 129–138.

Wagner, G.F., P. DeNiu, E. Jaworski, D. Radman and C. Chiarot. 1997a. Development of a dose-response bioassay for stanniocalcin in fish. *Molecular and Cellular Endocrinology* 128: 19–28.

Wagner, G.F., E.M. Jaworski and D.P. Radman. 1997b. Salmon calcitonin inhibits whole body Ca^{2+} uptake in young rainbow trout. *Journal of Endocrinology* 155: 459–465.

Wagner, G.F., M. Haddad, R.C. Fargher, C. Milliken and D.H. Copp. 1998a. Calcium is an equipotent stimulator of stanniocalcin secretion in freshwater and seawater salmon. *General and Comparative Endocrinology* 109: 186–191.

Wagner, G.F., E.M. Jaworski and M. Haddad. 1998b. Stanniocalcin in the seawater salmon: Structure, function, and regulation. *American Journal of Physiology* 43: R1177–R1185.

Wales, B. 1997. Ultrstrucutral study of chloride cells in the trunk epithelium of larval herring, *Clupea harengus*. *Tissue and Cell* 29: 439–447.

Wales, N. and A. Barrett. 1983. Depression of sodium, chloride and calcium ions in the plasma of goldfish (*Carassius auratus*) and immature freshwater- and seawater-adapted eels (*Anguilla anguilla* L.) after acute administration of salmon calcitonin. *Journal of Endocrinology* 98: 257–261.

Wales, W. and P. Tytler. 1996. Changes in chloride cell distribution during early larval stages of *Clupea harengus*. *Journal of Fish Biology* 49: 801–814.

Watts, E.G., D.H. Copp and L.J. Deftos. 1975. Changes in plasma calcitonin and calcium during the migration of salmon. *Endocrinology* 96: 214–218.

Wendelaar Bonga, S.E. 1980. Effect of synthetic salmon calcitonin and low ambient calcium on plasma calcium, ultimobranchial cells, Stannius bodies, and prolactin cells in the teleost *Gasterosteus aculeatus*. *General and Comparative Endocrinology* 40: 99–108.

Wendelaar Bonga, S.E. 1981. Effect of synthetic salmon calcitonin on protein-bound and free plasma calcium in the teleost *Gasterosteus aculeatus*. *General and Comparative Endocrinology* 43: 123–126.

Wendelaar Bonga, S.E. 1997. The stress response in fish. *Physiological Reviews* 77: 591–625.

Wendelaar Bonga, S.E. and P.K.T. Pang. 1991. Control of calcium regulating hormones in the vertebrates: Parathyroid hormone, calcitonin, prolactin and stanniocalcin. *International Review of Cytology* 128.

Wendelaar Bonga, S.E., P.I. Lammers and J.C.A. Vandermeij. 1983. Effects of 1,25- and 24,25-dihydroxyvitamin-D_3 on bone formation in the cichlid teleost *Sarotherodon mossambicus*. *Cell and Tissue Research* 228: 117–126.

Wendelaar Bonga, S.E., G. Flik, P. Balm and J. Meij. 1990. The ultrastructure of chloride cells in the gills of the teleost *Oreochromis mossambicus* during exposure to acidified water. *Cell and Tissue Research* 259: 575–585.

Wilson, R.W. and M. Grosell. 2003. Intestinal bicarbonate secretion in marine teleost fish-source of bicarbonate, pH sensitivity, and consequences for whole animal acid-base and calcium homeostasis. *Biochimica et Biophysica Acta-Biomembranes* 1618: 163–174.

Wilson, R.W., J.M. Wilson and M. Grosell. 2002. Intestinal bicarbonate secretion by marine teleost fish—Why and how? *Biochimica et Biophysica Acta-Biomembranes* 1566: 182–193.

Witten, P.E. and B.K. Hall. 2003. Seasonal changes in the lower jaw skeleton in male Atlantic salmon (*Salmo salar* L.): Remodelling and regression of the kype after spawning. *Journal of Anatomy* 203: 435–450.

Witten, P.E. and W. Villwock. 1997. Growth requires bone resorption at particular skeletal elements in a teleost fish with acellular bone (*Oreochromis niloticus*, Teleostei: Cichlidae). *Journal of Applied Ichthyology* 13: 149–158.

Witters, H.E., S. Vanpuymbroeck and O.L.J. Vanderborght. 1992. Branchial and renal ion fluxes and transepithelial electrical potential differences in rainbow trout, *Oncorhynchus mykiss*: Effects of aluminum at low pH. *Environmental Biology of Fishes* 34: 197–206.

Wood, C.M., S.F. Perry, P.A. Wright, H.L. Bergman and D.J. Randall. 1989. Ammonia and urea dynamics in the Lake Magadi tilapia, a ureotelic teleost fish adapted to an extremely alkaline environment. *Respiratory Physiology* 77: 1–20.

Wood, C.M., P. Wilson, H.L. Bergman, A.N. Bergman, P. Laurent, G. Otiang'a-Owiti and P. J. Walsh. 2002. Obligatory urea production and the cost of living in the Magadi tilapia revealed by acclimation to reduced salinity and alkalinity. *Physiological and Biochemical Zoology* 75: 111–122.

Worthington, G.P. and R.J. Balment. 2003. The role of PTHrP in the calcium homeostasis of the European flounder, *Platichthys flesus* (Abstract). *Comparative Biochemistry and Physiology* A134: S87.

Worthington, G.P., L.N. Dow, C.A. Teitsma, P.M. Ingleton and R.J. Balment. 2004a. Exogenous and endogenous modulators of circulating N-terminal PTHrP 1-34 immunoreactivity in the European flounder *Platichthys flesus*: A role in calcium homeostasis (Abstract). In: *5th International Symposium in Fish Endocrinology*. Castellón, Spain.

Worthington, G.P., P.M. Ingleton and R.J. Balment. 2004b. Circulating levels of PTHrP in the European flounder *Platichthys flesus*: So much hormone, but where is it coming from? (Abstract). *Comparative Biochemistry and Physiology* A137: S41.

Yamauchi, H., M. Matsuo, A. Yoshida and H. Orimo. 1978a. Effect of eel calcitonin on serum electrolytes in the eel *Anguilla japonica*. *General and Comparative Endocrinology* 34: 343–346.

Yamauchi, H., H. Orimo, K. Yamauchi, K. Takano and H. Takahashi. 1978b. Increased calcitonin levels during ovarian development in the eel, *Anguilla japonica*. *General and Comparative Endocrinology* 36: 526–529.

Yeo, I. and Y. Mugiya. 1997. Effects of extracellular calcium concentrations and calcium antagonists on vitellogenin induction by estradiol-17β in primary hepatocyte culture in the rainbow trout *Oncorhynchus mykiss*. *General and Comparative Endocrinology* 105: 294–301.

Yoshiko, Y., T. Kosugi and Y. Koide. 1996. Effects of a synthetic N-terminal fragment of stanniocalcin on the metabolism of mammalian bone *in vitro*. *Biochimica et Biophysica Acta* 1311: 143–149.

Yoshikubo, H., N. Suzuki, K. Takemura, M. Hoso, S. Yashima, S. Iwamuro, Y. Takagi, M.J. Tabata and A. Hattori. 2005. Osteoblastic activity and estrogenic response in the regenerating scale of goldfish, a good model of osteogenesis. *Life Sciences* 76: 2699–2709.

Zia, S. and D.G. McDonald. 1994. Role of the gills and gill chloride cells in metal uptake in the freshwater-adapted rainbow trout, *Oncorhynchus mykiss*. *Canadian Journal of Fisheries and Aquatic Sciences* 51: 2482–2492.

CHAPTER

16

Role of Prolactin, Growth Hormone, Insulin-like Growth Factor I and Cortisol in Teleost Osmoregulation

Juan Miguel Mancera[1] and Stephen D. McCormick[2]

INTRODUCTION

Maintenance of constant cellular ion concentrations is a basic requirement of all life forms. The strategy evolved by teleost fish to achieve this requirement is by maintaining nearly constant levels of extracellular ions at approximately one-third the ionic strength of seawater (SW). In freshwater (FW), teleosts must counteract the passive loss of ions and gain of water by actively taking up ions (primarily through the gills), and removing excess water by excreting a dilute urine. In SW, teleosts

Authors' addresses: [1]Departamento de Biología, Facultad de Ciencias del Mar y Ambientales, Universidad de Cádiz, 11510 Puerto Real, Cádiz, Spain.
E-mail: juanmiguel.mancera@uca.es
[2]USGS, Conte Anadromous Fish Research Center, Turners Falls, MA, USA.
E-mail: steve_mccormick@usgs.gov

counteract the gain of ions and loss of water by drinking SW, absorbing water and ions through the gut, and secreting excess monovalent ions through the gills and divalent ions through the kidney. Details of these mechanisms can be found in excellent reviews published in the last several years (Marshall, 2002; Evans *et al.*, 2005).

The demands on these ion-regulatory pathways will change as a function of environmental salinity, feeding, activity, injury, reproductive state and a variety of stressors. Therefore, control of ion regulation is critical, and the neuroendocrine system is the major means for regulating these mechanisms. Several excellent reviews on various aspects of the hormonal control of osmoregulation in fish have been published previously (Foskett *et al.*, 1983; Mayer-Gostan *et al.*, 1987; Bern and Madsen, 1992; McCormick, 1995, 2001; Sakamoto *et al.*, 2001; Sakamoto and McCormick, 2006). Here, we will focus on the endocrine mechanisms that control the overall capacity of the ion regulatory mechanisms in teleost fish, focussing on the osmoregulatory actions of prolactin (PRL), the growth hormone (GH)/insulin-like growth factor I (IGF-I) axis and cortisol. We will build on existing reviews and incorporate new data to give an integrative synthesis of the role of these hormones in the osmoregulation of teleost fish.

PROLACTIN (PRL)

PRL is a pleiotropic hormone with a wide spectrum of functions in vertebrates. Many of these functions are related to osmoregulatory processes (Bole-Feysot *et al.*, 1998; Sakamoto *et al.*, 2003; Harris *et al.*, 2004). The first evidence of the hyperosmoregulatory role of PRL in fish came from the studies by Grace Pickford and her collaborators (1959, 1970). Using hypophysectomized FW-adapted killifish, *Fundulus heteroclitus*, they demonstrated that PRL treatment was essential for survival of this species in a hypoosmotic environment. Although pituitary PRL is not necessary for FW survival of all teleosts, subsequent studies have established the hyperosmoregulatory role of PRL using other species, types of studies, and experimental approaches (see Hirano, 1986; McCormick, 1995; Manzon, 2002).

PRL has been shown to regulate several aspects of the ion regulatory mechanisms that are characteristic of FW fish. Water permeability of the gill, gut, and kidney are generally lower in FW- than in SW-acclimated fish, and PRL decreased water permeability in these tissues in several teleost species (Table 16.1; see also Manzon, 2002). To date, the mechanisms and gene products responsible for the actions of PRL on water permeability have not been identified, though they are likely to include regulation of tight junctions, membrane composition, and water channels such as aquaporins.

Treatment with PRL increases the ion uptake capacity of teleosts, and it is likely that this effect is carried out through regulation of gill chloride

Table 16.1 Physiological evidence for a hyperosmoregulatory role of PRL in teleosts.

Action	*References*
Pituitary	
Higher PRL cells activity, synthesis and secretion in FW and BW relative to SW	Nishioka *et al.* (1988) Mancera *et al.* (1993) Martin *et al.* (1999)
Low osmolality stimulates pituitary PRL secretion *in vitro*	Seale *et al.* (2003)
Plasma	
Higher PRL plasma levels in FW and BW relative to SW	Manzon (2002)
Receptors	
PRL receptor mRNA levels show a negative relationship with salinity (i.e., lower in higher salinities)	Shiraishi *et al.* (1999) Sandra *et al.* (2000)
PRL receptors present in gill chloride cells and in kidney	Ng *et al.* (1991) Weng *et al.* (1997) Santos *et al.* (2001)
Gills	
Exogenous PRL reduces gill Na^+,K^+-ATPase activity and mRNA levels	Sakamoto *et al.* (1997) Kelly *et al.* (1999) Mancera *et al.* (2002)
Exogenous PRL stimulates development of chloride cells 'fresh water morphology'	Herndon *et al.* (1991) Pisam *et al.* (1993)
Kidney	
Exogenous PRL increases Na^+ reabsorption and water excretion, through stimulation of glomerular size and urine output	Clarke and Bern (1980) Braun and Dantlzler (1987)
Contradictory effects on renal Na^+,K^+-ATPase activity, with increases or no effects	Pickford *et al.* (1970) Seidelin and Madsen (1997) Kelly *et al.* (1999)
Intestine	
Exogenous PRL decreases permeability to water and ions and Na^+,K^+-ATPase activity	Collie and Hirano (1987) Manzon (2002)
Contradictory effects on intestinal Na^+,K^+-ATPase activity, with increases or no effects	Kelly *et al.* (1999) Seidelin and Madsen (1999)
Skin	
Exogenous PRL increases mucus production by stimulation of differentiation and proliferation of mucous cells	Clarke and Bern (1980) Brown and Brown (1987)

cells. Herndon *et al.* (1991) found that PRL injection in SW-acclimated tilapia resulted in decreased chloride cell size, typical of FW-acclimated tilapia. In the Nile tilapia, Pisam *et al.* (1993) found that treatment with PRL increased the number of 'β-cells' typical of FW-acclimated tilapia and decreased the number of α-cells typical of SW-acclimated tilapia. Kelly *et al.* (1999) have found that the impact of PRL on chloride cells of *Sparus sarba* is dependent on the environmental salinity; in hypoosmotic brackish water PRL reduces chloride cell number and size, whereas in SW this hormone has no effect. Sakamoto and McCormick (2006) have suggested that the control of cell turnover (apoptosis and cell proliferation) in different osmoregulatory epithelia (e.g., gill and gastrointestinal tract) is a critical feature of the control of osmoregulation by PRL.

It is also likely that PRL affects the transporters that are involved in ion uptake. However, there is still some uncertainty regarding the transporters that are most directly involved in ion uptake in teleost fish. To date, the most favored models include a chloride-bicarbonate exchanger through which chloride uptake is driven through production of carbon dioxide. Sodium is thought to be taken up through an apical sodium channel energized by an apical H^+-ATPase. Characterization and localization of these necessary transporters to fully validate these models is ongoing, and no information on the role of PRL in regulating these transporters is currently available. This hormone has variable effects on gill Na^+,K^+-ATPase activity among teleost species (see McCormick, 1995). This may stem, in part, from differences in the relative importance of the Na^+,K^+-ATPase pump in ion uptake among teleosts (in most teleosts gill Na^+,K^+-ATPase activity is higher in SW, but in others it is lower), their relative euryhalinity, and the salinity at which the studies were carried out.

The activity of PRL cells is under hypothalamic and extra-hypothalamic control. Decreases in plasma osmolality result in increased PRL synthesis and release (Seale *et al.*, 2003). In addition, other hormones such as cortisol decrease PRL release (Borski *et al.*, 2002). At the hypothalamic level, dopamine has a clear inhibitory effect on PRL cells (Nishioka *et al.*, 1988). In mammals, a specific prolactin-releasing hormone peptide (Pr-RP) has been described, and in recent years, a Pr-RP has also been identified in teleosts (see Sakamoto *et al.*, 2003, 2005; Fujimoto *et al.*, 2006). This peptide is synthesized in hypothalamic neurons, with axons ending close to PRL cells in the *rostral pars distalis* of the pituitary (Sakamoto *et al.*, 2003). This peptide can stimulate PRL cells, increasing synthesis and release of this hormone to systemic blood. In addition, in the amphibious, euryhaline mudskipper (*Periophthalmus modestus*) molecular studies have demonstrated a strong relationship between expression of Pr-RP and environmental salinity, with higher Pr-RP expression in fish acclimated to FW and terrestrial environments relative to SW conditions (Sakamoto *et al.*, 2005). The presence of Pr-RP in peripheral organs (like gut mucus cells) suggests the possibility of other actions, including effects on hormone expression outside of the pituitary, and even direct actions on osmoregulatory tissues (see Sakamoto and McCormick, 2006).

GROWTH HORMONE (GH)/INSULIN-LIKE GROWTH FACTOR I (IGF-I) AXIS

GH is a member of the GH/PRL family with a role in osmotic acclimation (McCormick, 1995) as well as growth and energy metabolism in fish (Björnsson, 1997). GH causes both local and systemic production of IGF-I, the latter being produced primarily in the liver. IGF-I carries out many of the growth-promoting actions of GH, though GH can also have direct actions on target tissues. Also, in carrying out its osmoregulatory function in fish, GH appears to work—at least in part—by increasing circulating IGF-I and production of IGF-I by the target tissue itself (Sakamoto and Hirano, 1993).

Smith (1956) was the first to demonstrate that GH treatment increased the capacity of trout to move from FW to SW. Later, Bolton *et al.* (1987) showed that these effects were relatively rapid and independent of the growth promoting actions of GH. McCormick *et al.* (1991) demonstrated that IGF-I was as potent as GH in increasing the salinity tolerance of rainbow trout. Increased salinity tolerance in response to GH treatment has also been observed in several non-salmonid teleosts, including tilapia and killifish (Mancera and McCormick, 1998a, 1999).

GH and IGF-I impacts on hypoosmoregulatory tissue are exerted in part through their influence on gill chloride cells. Many studies of salmonids have shown an effect of GH and/or IGF-I treatment on the number, size and specific ultrastructural features of gill chloride cells (see references in McCormick, 2001). Sakamoto and McCormick (2006) have hypothesized that this impact of GH and IGF-I may be through the control of cell turnover and differentiation in the gill. This effect would be consistent with the known proliferative and anti-apoptotic roles of IGF-I in many vertebrate tissues (Wood *et al.*, 2005). It should be noted, however, that such effects have yet to be demonstrated in osmoregulatory tissues of fish.

GH and IGF-I are also involved in the upregulation of transporters critical to salt secretion by the gill. Both Na^+,K^+-ATPase and the $Na^+,K^+,2Cl^-$ cotransporter (NKCC) are upregulated by GH (Pelis and McCormick, 2001). Although GH has not been shown to have direct (*in vitro*) effects on these transporters, IGF-I increased gill Na^+,K^+-ATPase, both *in vivo* and *in vitro* (Madsen and Bern, 1993; Seidelin and Madsen, 1999). These impacts on specific transporters may be part of a proliferation and differentiation pathway for the development of salt secreting chloride cells in the gill. Surprisingly, the impact of GH and IGF-I on osmoregulatory tissues other than the gill has received little attention. To date, what is known suggests that the action of GH and IGF-I on salt secretory capacity is primarily through its impact on gill physiology (Seidelin and Madsen, 1999).

IGF-I binding proteins are known to play a critical role in regulating the interaction of IGF-I with its receptor (Wood *et al.*, 2005). Recently, Shepherd *et al.* (2005) have shown that plasma levels of three IGF binding proteins (21-, 42- and 50-kDa) are higher after salinity acclimation. To our knowledge, this is the only report of the possible role of binding proteins in ion regulation. High-affinity, low-capacity IGF-I binding sites characteristic of receptors have recently been found in Atlantic salmon gill, and are most abundant in gill chloride cells (McCormick, unpublished results). Reinecke *et al.* (1997) present evidence that local production of IGF-I in the gill occurs primarily in chloride cells (Tables 16.2-16.4).

CORTISOL

Cortisol is the major corticosteroid produced by the interrenal tissue of teleost fish. This hormone has several established physiological roles related to osmoregulation, intermediary metabolism, growth, stress and immune function (Wendelaar Bonga, 1997; Mommsen *et al.*, 1999). Evidence for the osmoregulatory role of cortisol in fish has been compiled in excellent reviews (McCormick, 1995, 2001; Sakamoto *et al.*, 2001;

Table 16.2 Physiological evidence for a hyperosmoregulatory role of GH in salmonids.

Action	*References*
Pituitary	
Higher GH cell activity, synthesis and secretion in SW relative to FW	Nishioka *et al.* (1988) Sakamoto *et al.* (1993) Björnsson (1997)
Plasma	
Higher plasma GH levels and metabolic clearance rate of GH during smolting and after transfer from FW to SW	Sakamoto *et al.* (1990) Björnsson (1997)
Receptors	
GH receptors present at high levels in gill, kidney and intestine	Sakamoto and Hirano (1991)
Gills	
Exogenous GH increases gill Na^+,K^+-ATPase activity and mRNA levels	Boeuf *et al.* (1994) Madsen *et al.* (1995) McCormick (1995) Seidelin and Madsen (1999)
Exogenous GH Stimulates proliferation of chloride cells with "seawater morphology"	See McCormick (1995) Sakamoto and McCormick (2006)
Exogenous GH increases abundance of Na^+-K^+-$2Cl^-$ cotransporter	Pelis and McCormick (2001)
Kidney	
GH treatment has not effect on kidney Na^+-K^+-ATPase activity	Madsen *et al.* (1995)
Intestine	
Exogenous GH induces 'seawater morphology' in the mucosa of the middle intestine of *Salmo salar* previous to smoltification	Nonnotte *et al.* (1995)
Exogenous GH increases the drinking response in *S. salar* pre-smolts after transfer to SW	Fuentes and Eddy (1997)

Evans, 2002). However, in recent years, new aspects of the physiology of cortisol in fish have arisen, and it is on these that we will focus our attention.

This hormone is considered a classical SW-promoting hormone, and evidence has shown a hypoosmoregulatory role of cortisol in several teleosts. Cortisol decreased plasma ion levels and osmolality in SW-adapted teleosts and enhanced salinity tolerance after transfer from low-salinity water to high-salinity water. This effect is due to increases in gill chloride cell size and density induced by cortisol treatment (McCormick, 1995, 2001). In addition, this hormone enhanced expression of gill Na^+,K^+-ATPase α-subunit and gill Na^+,K^+-ATPase activity in salmonid and no-salmonid species (Madsen *et al.*, 1995; Seidelin *et al.*, 1999;

Table 16.3 Physiological evidence for an osmoregulatory role of GH in non-salmonids.

Action	*References*
Pituitary	
GH cells activation is depending on the species studied and the environmental salinity	Nishioka *et al.* (1988) Mancera and McCormick (1998b)
Plasma	
GH levels behave differently depending on the species studied and the environmental salinity	Nishioka *et al.* (1988) Mancera and McCormick (1998b)
Receptors	
GH binding found in renal tubule of gilthead sea bream	Munoz-Cueto *et al.* (1996)
Gill and operculum	
Exogenous GH increases salinity tolerance, opercular chloride cell number and gill Na^+,K^+-ATPase activity in tilapia (*O. mossambicus*) and mummichog (*Fundulus heteroclitus*)	Flik *et al.* (1993) Xu *et al.* (1998) Mancera and McCormick (1998a)
Exogenous GH does not cause any significant changes in gill Na^+,K^+-ATPase activity or α- and β-subunit mRNA levels in silver sea bream (*Sparus sarba*)	Deane *et al.* (1999) Kelly *et al.* (1999)
Exogenous GH increases gill Na^+,K^+-ATPase activity in gilthead seabream (*Sparus aurata*)	Sangiao-Alvarellos *et al.* (2006)
Kidney	
Exogenous GH reduces Na^+,K^+-ATPase activity in SW- and BW-acclimated silver seabream (*Sparus sarba*)	Kelly *et al.* (1999)

Mancera *et al.*, 2002; Laiz-Carrión *et al.*, 2003). Finally, cortisol stimulated expression and abundance of Na^+-K^+-$2Cl^-$ cotransporter in the gills of FW-acclimated *S. salar* (Pelis and McCormick, 2001).

At the intestinal level, cortisol stimulated Na^+,K^+-ATPase activity, together with ion and water absorption, thus helping adaptation to high environmental salinity (Veillette and Young, 2005). Also, an improved drinking response after transfer to SW has been observed in *Oncorhynchus mykiss* and *S. salar* treated with this hormone (Fuentes *et al.*, 1996).

In addition to the classical hypoosmoregulatory role of cortisol, and according to several evidences (see Table 16.5), a new role of this hormone either in ion uptake in FW- or BW-adapted fish has been suggested. McCormick (2001), in his excellent revision of this topic, proposed a 'dual osmoregulatory' role for cortisol: (1) a stimulatory action on ion secretion in cooperation with GH/IGF-I axis in hyperosmotic environments; and (2) an increase of ion uptake in cooperation with PRL in hypoosmotic environments.

Table 16.4 Physiological evidence for an osmoregulatory role of IGF-I in salmonids and non-salmonids.

Action	*References*
Plasma	
IGF-I levels increased during smolting and SW acclimation	Sakamoto and Hirano (1993)
IGF-I binding proteins levels are altered after SW exposure of rainbow trout	Shepherd *et al.* (2005)
Receptors	
High affinity, low capacity IGF-I binding in salmon gill	McCormick (unpublished)
IGF-I receptor immunoreactivity present in chloride cells	McCormick (unpublished)
Gill	
IGF-I mRNA levels increase after exogenous GH and transfer to SW in salmonids and tilapia (*O. mossambicus*)	Sakamoto and Hirano (1993) Weng *et al.* (2000)
Exogenous IGF-I increases salinity tolerance, gill Na^+,K^+-ATPase activity and development of chloride cells	McCormick (1995) Mancera and McCormick (1998a) Seidelin and Madsen (1999)
IGF-I immunoreactivity present in chloride cells	Reinecke *et al.* (1997)

A large number of binding studies in fish have found evidence for a single class of corticosteroid receptors (CR) (see references in Prunet *et al.*, 2006). However, in the last several years, molecular techniques have demonstrated the presence of genes in several teleost species related to the mammalian glucocorticoid (GR) and mineralcorticoid receptors (MR). Fish GR has been characterized in several species (*Oreochromis mossambicus*, *Paralichthys olivaceus*), with a second isoform present in some species (*O. mykiss*, *Haplochromis burtoni*). In addition, MR has been molecularly characterized in *O. mykiss* and *H. burtoni*. Using a transfected cell line, Sturm *et al.* (2005) found that the rainbow trout MR (rtMR) has high transactivation efficiency for both aldosterone and 11-deoxycorticosterone (DOC), similar to the mammalian MR. Prunet *et al.* (2006) suggest that DOC, present in the plasma of some teleosts at levels that could activate the rtMR, might be a mineralocorticoid in fish. It may be possible that the teleost MR is involved in the 'dual osmoregulatory' role (ion secretion and uptake) of cortisol in teleost fish. However, the physiological function of the MR in fish and the possible physiological relevance of DOC remains to be established.

Table 16.5 Physiological evidence for a hyperosmoregulatory role of cortisol.

Action	*References*
Plasma	
Transfer from SW to FW transiently increases plasma cortisol levels	Mancera *et al.* (1994) McCormick (2001)
Effects of cortisol treatment	
Restored plasma osmolality and ion levels in hypophysectomized eels, goldfish and bowfin	McCormick (2001)
Increased surface area of gill chloride cells and the influx of sodium and chloride in FW rainbow trout, tilapia, eel and catfish	Laurent and Perry (1990) Perry *et al.* (1992)
Stimulated whole-body calcium uptake and the branchial calcium pump in freshwater rainbow trout	Flik and Perry (1989)
Enhanced H^+-ATPase activity in gills of salmonids, possibly involved in sodium uptake in hypo-osmotic environments	Lin and Randall (1995) Marshall (2002)
Increased ion regulatory capacity after transfer of *Sparus aurata* to low salinity environments	Mancera *et al.* (1994)
Stimulated gill Na^+,K^+-ATPase activity, plasma osmolality and ion levels in BW-adapted *S. aurata*	Mancera *et al.* (2002)
Interactions with other hormones	
A positive interaction of cortisol with PRL for maintenance of ion balance in FW-acclimated channel catfish *Ictalurus punctatus* and stinging catfish *Heteropneustes fossilis*	Parwez and Goswami (1985) Eckert *et al.* (2001)
A positive interaction of cortisol with PRL for promoting the transepithelial resistance and potential of cultured branchial epithelia from FW rainbow trout	Zhou *et al.* (2003)

HORMONE INTERACTIONS

In addition to the independent osmoregulatory actions of PRL, GH/IGF-I axis and cortisol, there is substantial evidence indicating the existence of synergy and antagonism of these hormones with one another.

PRL and Cortisol

Consistent with its role in promoting acclimation to low environmental salinities, PRL antagonizes the salt-secretory actions of both cortisol and GH (*O. mykiss*: Madsen and Bern, 1992; *S. salar*: Boeuf *et al.*, 1994; *S. trutta*: Seidelin and Madsen, 1997). Seidelin and Madsen (1997) found that PRL could reverse all of the increases in hypoosmoregulatory ability induced by cortisol, but did not affect the capacity of cortisol to increase gill Na^+,K^+-ATPase activity. They suggested that an interaction of PRL and cortisol on salt secretory capacity may occur in non-branchial tissue such as the intestine. Cortisol has been shown to rapidly decrease the release of PRL from the tilapia pituitary (Borski *et al.*, 1991).

As outlined above, cortisol also has an apparent role in ion uptake, and there is evidence for a positive interaction of exogenous treatment with cortisol and PRL for maintenance of ion balance in FW fish (Parwez and Goswami, 1985; Eckert *et al.*, 2001). In *S. aurata*, a greater activation of pituitary PRL and ACTH cells have been shown to occur in BW-acclimated fish relative to SW-acclimated fish, suggesting a possible cooperation of PRL and cortisol in the control of osmoregulation at low salinities (Mancera *et al.*, 1993, 2002). Using an *in vitro* gill cell preparation, it has been demonstrated that PRL and cortisol act synergistically in order to promote transepithelial resistance and potential (Zhou *et al.*, 2003).

PRL and GH/IGF-I Axis

It has also been demonstrated that treatment with PRL can antagonize the hyperosmoregulatory actions of GH and IGF-I in salmonids (*O. mykiss*: Madsen and Bern, 1992; *S. salar*: Boeuf *et al.*, 1994; *S. trutta*: Seidelin and Madsen, 1997, 1999). This effect has been shown to occur at the level of specific ion transporters (gill Na^+,K^+-ATPase), and chloride cell number and morphology, resulting in differences in decreased whole animal performance (higher plasma ions) in SW. This antagonistic action of PRL, along with the lower PRL levels seen in BW, may explain the greater efficacy of GH treatment on salt-secretory capacity in BW relative to FW (Bolton *et al.*, 1987; McCormick, 1996). Currently, it is thought that this antagonism occurs primarily at target tissues, as we are not aware of any studies indicating that GH and PRL affect one another's synthesis or secretion.

GH/IGF-I Axis and Cortisol

An important synergy of the GH axis and cortisol to improve salinity tolerance and salt-secretory capacity has been demonstrated in salmonid and non-salmonid species. This cooperation is mediated by increased expression of gill Na^+,K^+-ATPase subunits, gill Na^+,K^+-ATPase activity, and abundance of Na^+-K^+-$2Cl^-$ cotransporter in gill chloride cells (Madsen, 1990; McCormick, 1996; Mancera and McCormick, 1999; Pelis and McCormick, 2001; McCormick, 2001). GH has been shown to increase the abundance of gill cortisol receptors in two species of salmonids (*O. kisutch* and *S. salar*) (Shrimpton *et al.*, 1995; Shrimpton and McCormick, 1998), and this may explain a substantial part of the interaction between GH and cortisol. Seidelin *et al.* (1999) found an additive effect of IGF-I and cortisol on gill chloride cell number and Na^{+},K^{+}-ATPase activity, but to date, no one has examined whether IGF-I can increase the number of gill cortisol receptors. Another possible mechanism of IGF-I and cortisol interaction is through a possible anti-apoptotic action of IGF-I on gill chloride cells, permitting cortisol to affect a greater number of partially or fully differentiated chloride cells.

In addition to interactions at target tissues, GH, IGF-I and cortisol are likely to interact so as to affect one another's synthesis and secretion, though surprisingly little research has been done in this area. GH has been shown to increase the sensitivity of the interrenal tissue to adrenocorticotropic hormone (ACTH) *in vitro* and *in vivo*, thus enhancing cortisol release (Young, 1988). Exogenous cortisol has been shown to decrease the circulating levels of IGF-I (Peterson and Small, 2005; McCormick, unpublished results). It is important to remember that these hormones have active roles in growth and energy mobilization, and thus their feedback mechanisms may reflect their involvement in processes other that just osmoregulation.

CONCLUSION

The control of the osmoregulatory system of teleosts involves several hypophysial and extra hypophysial hormones (PRL, GH and cortisol), which play an important role in osmotic acclimation (McCormick, 1995, 2001; McCormick and Sakamoto, 2006). It is a well-established fact that PRL has an important role in the FW acclimation of many teleosts, though the mechanisms of ion regulation controlled by this hormone have not been fully elucidated. In contrast, the osmoregulatory role of the GH/IGF-I axis appears to be more highly species-dependent. In salmonids this axis has a hypoosmoregulatory role acting clearly as a SW-adapting hormone. However, in non-salmonid species, the evidence is contradictory, with GH exhibiting an apparent hypoosmoregulatory role in some species, and no clear osmoregulatory role in others. Finally, cortisol has been shown to have a role in SW acclimation in both primitive and advanced teleost fish. However, in recent years, evidence also suggests a role for cortisol in ion uptake in low-salinity water-adapted fish. This new evidence suggests a 'dual osmoregulatory' role for cortisol, with the classic role of stimulation of ion secretion in hyperosmotic media (in cooperation with GH and IGF-I), and an additional role of increasing ion uptake in hypoosmotic environments (in cooperation with PRL).

Acknowledgements

The present publication was supported by a grant BFU2004-04439-C02-01B (Ministerio de Educación y Ciencia, Spain) to J.M.M.. The authors would like to express their gratitude to the members of the laboratory of Dr. Mancera: Dr. Raúl Laiz-Carrión, Jose María Guzman, Francisco Jesús Arjona and Luis Vargas-Chacoff for their contribution to the understanding of fish osmoregulation. We also thank Michelle Monette for her comments and corrections on the manuscript.

References

Bern, H.A. and S.S. Madsen. 1992. A selective survey of the endocrine system of the rainbow trout (*Oncorhynchus mykiss*) with emphasis on the hormonal regulation of ion balance. *Aquaculture* 100: 237–262.

Björnsson, B.Th. 1997. The biology of salmon growth hormone: from daylight to dominance. *Fish Physiology and Biochemistry* 17: 9–24.

Boeuf, G., A.M. Marc, P. Prunet, P.-Y. Le Bail and J. Smal. 1994. Stimulation of parr-smolt transformation by hormonal treatment in Atlantic salmon (*Salmo salar* L.). *Aquaculture* 121: 195–208.

Bole- Feysot, C., V. Goffin, M. Edery, N. Binart and P.A. Kelly. 1998. Prolactin (PRL) and its receptor: Actions, signal transduction pathways and phenotypes observed in PRL receptor knockout mice. *Endocrine Reviews* 19: 225–268.

Bolton, J.P., N.L. Collie, H. Kawauchi and T. Hirano. 1987. Osmoregulatory actions of growth hormone in rainbow trout (*Salmo gairdneri*). *Journal of Endocrinology* 112: 63–68.

Borski, R.J., L.M. Helms, N.H. Richman and E.G. Grau. 1991. Cortisol rapidly reduces prolactin release and cAMP and $45Ca^{2+}$ accumulation in the cichlid fish pituitary in vitro. *Proceedings of the National Academy of Sciences of the United States of America* 88: 2758–2762.

Borski, R.J., G.N. Hyde, S. Fruchtman and W.S. Tsai. 2002. Cortisol suppresses prolactin release through a non-genomic mechanism involving interactions with the plasma membrane. *Comparative Biochemistry and Physiology* B129: 533–541.

Braun, E.J. and W.H. Dantlzler. 1987. Mechanisms of hormone actions on renal function. In: *Vertebrate Endocrinology: Fundamentals and Biomedical Implications*, P.K.T. Pang and M.P. Schreibman (eds.). Academic Press, San Diego, Vol. 2, pp. 189–210.

Brown, P.S. and S.C. Brown. 1987. Osmoregulatory actions of prolactin and other adenohypophysial hormones. In: *Vertebrate Endocrinology: Fundamentals and*

Biomedical Implications, P.K.T. Pang and M.P. Schreibman (eds.). Academic Press, San Diego, Vol. 2, pp. 45–84.

Clarke, W.C. and H.A. Bern. 1980. Comparative endocrinology of prolactin. In: *Hormonal Proteins and Peptides*, C.H. Li (ed.). Academic Press, New York, Vol. 8, pp. 105–197.

Collie, N.L. and T. Hirano. 1987. Mechanisms of hormone actions on intestinal transport. In: *Vertebrate Endocrinology: Fundamentals and Biomedical Implications*, P.K.T. Pang and M.P. Schreibman (eds). Academic Press, San Diego, Vol. 2, pp. 239–270.

Deane, E.E., S.P. Kelly and N.Y.S. Woo. 1999. Hormonal modulation of branchial Na^+-K^+-ATPase subunit mRNA in a marine teleost *Sparus sarba*. *Life Science* 20: 1819-1829.

Eckert, S.M., T. Yada, B.S. Shepherd, M.H Stetson, T. Hirano and G. Grau. 2001. Hormonal control of osmoregulation in the channel catfish *Ictalurus punctatus*. *General and Comparative Endocrinology* 122: 270–286.

Evans, D.H. 2002. Cell signalling and ion transport across the fish gill epithelium. *Journal of Experimental Zoology* 293: 336–347.

Evans, D.H., P.M. Piermarini and K.P. Choe. 2005. The multifunctional fish gill: Dominant site of gas exchange, osmoregulation, acid-base regulation and excretion of nitrogenous waste. *Physiological Reviews* 85: 97–177.

Flik, G. and S.F. Perry. 1989. Cortisol stimulates whole-body calcium uptake and the branchial calcium pump in freshwater rainbow trout. *Journal of Endocrinology* 120: 75–82.

Flik, G., W. Atsma, J.C. Fenwick, F. Rentier-Delrue, J. Smal and S.E. Wendelaar Bonga. 1993. Homologous recombinant growth hormone and calcium metabolism in the tilapia, *Oreochromis mossambicus*, adapted to fresh water. *Journal of Experimental Biology* 185: 107–191.

Foskett, J.K., H.A. Bern, T.E. Machen and M. Conner. 1983. Chloride cells and the hormonal control of teleost fish osmoregulation. *Journal of Experimental Biology* 106: 255–281.

Fuentes, J. and F.B. Eddy. 1997. Drinking in Atlantic salmon presmolts and smolts in response to growth hormone and salinity. *Comparative Biochemistry and Physiology* A117: 487–491.

Fuentes, J., N.R. Bury, S. Carrol and F.B. Eddy. 1996. Drinking in Atlantic salmon presmolts (*Salmo salar* L.) and juvenile rainbow trout (*Oncorhynchus mykiss* Walbaum) in response to cortisol and seawater challenge. *Aquaculture* 141: 129–137.

Fujimoto, M., T. Sakamoto, T. Kanetoh, M. Osaka and S. Moriyama. 2006. Prolactin-releasing peptide is essential to maintain the prolactin level and osmotic balance in freshwater teleost fish. *Peptides* 27: 1104–1109.

Gaitskell, R.E. and I. Chester Jones. 1970. Effects of adrenolectomy and cortisol injection on the *in vitro* movement of water by the intestine of the freshwater European eel (*Anguilla anguilla* L.). *General and Comparative Endocrinology* 15: 491–493.

Harris, J., P.M. Stanford, S.R. Oakes and C.J. Ormandy. 2004. Prolactin and the prolactin receptor: new targets of an old hormone. *Annals in Medicine* 36: 414–425.

Herndon, T.M., S.D. McCormick and H.A. Bern. 1991. Effects of prolactin on chloride cells in opercular membrane of seawater-adapted tilapia. *General and Comparative Endocrinology* 83: 283–289.

Hirano, T. 1986. The spectrum of prolactin action in teleosts. In: *Comparative Endocrinology: Developments and Directions*, C.L. Ralph (ed.). Alan R. Liss, New York, pp. 53–74.

Kelly, S.P., I.N.K. Chow and Y.S. Woo. 1999. Effects of prolactin and growth hormone on strategies of hypoosmotic adaptation in a marine teleost, *Sparus sarba*. *General and Comparative Endocrinology* 113: 9–22.

Laiz-Carrión, R., M.P. Martín del Río, J.M. Miguez, J.M. Mancera and J.L. Soengas. 2003. Influence of cortisol on osmoregulation and energy metabolism in gilthead sea bream *Sparus aurata*. *Journal of Experimental Zoology* A298: 105–118.

Laurent, P. and S.F. Perry. 1990. Effects of cortisol on gill chloride cell morphology and ionic uptake in the freshwater trout, *Salmo gairdneri*. *Cell and Tissue Research* 259: 429–442.

Lin, H. and D. Randall. 1995. Proton pumps in fish gills. In: *Fish Physiology—Ionoregulation: Cellular and Molecular Approaches*, C.M. Wood and T.J. Shuttlerworth (eds.). Academic Press, New York, Vol. 14, pp. 22–256.

Madsen, S.S. 1990. Effect of repetitive cortisol and thyroxine injections on chloride cell number and Na^+/K^+-ATPase activity in gills of freshwater acclimated rainbow trout, *Salmo gairdneri*. *Comparative Biochemistry and Physiology* A95: 171–175.

Madsen, S.S. and H.A. Bern. 1992. Antagonism of prolactin and growth hormone: Impact on seawater adaptation in two salmonids, *Salmo trutta* and *Oncorhynchus mykiss*. *Zoological Science* 9: 775–784.

Madsen, S.S. and H.A. Bern. 1993. *In vitro* effects of insulin-like growth factor-I on gill Na^+,K^+-ATPase in coho salmon, *Oncorhynchus kisutch*. *Journal of Endocrinology* 138: 23–30.

Madsen, S.S., M.K. Jensen, J. Nohr and K. Kristiansen. 1995. Expression of Na^+-K^+-ATPase in the brown trout, *Salmo trutta*: *In vivo* modulation by hormones and seawater. *American Journal of Physiology* 269: R1339–R1345.

Mancera, J.M. and S.D. McCormick. 1998a. Evidence for growth hormone/insulin-like growth factor I axis regulation of seawater acclimation in the euryhaline teleost *Fundulus heteroclitus*. *General and Comparative Endocrinology* 101: 103–112.

Mancera, J.M. and S.D. McCormick. 1998b. Osmoregulatory actions of the GH/IGF axis in non-salmonids teleosts. *Comparative Biochemistry and Physiology* B121: 43–48.

Mancera, J.M. and S.D. McCormick. 1999. Influence of cortisol, growth hormone, insulin-like growth factor I and 3,3′,5-triiodo-L-thyronine on hypoosmoregulatory ability in the euryhaline teleost *Fundulus heteroclitus*. *Fish Physiology and Biochemistry* 21: 25–33.

Mancera, J.M., J.M. Pérez-Fígares and P. Fernández-Llebrez. 1994. Effect of cortisol on brackish water adaptation in the euryhaline gilthead seabream (*Sparus aurata* L.). *Comparative Biochemistry and Physiology* A107: 397–402.

Mancera, J.M., R. Laíz-Carrión and M.P. Martín del Río. 2002. Osmoregulatory action of PRL, GH and cortisol in the gilthead seabream *Sparus aurata*. *General and Comparative Endocrinology* 129: 95–103.

Manzon, L.A. 2002. The role of prolactin in fish osmoregulation: A review. *General and Comparative Endocrinology* 125: 291–310.

Marshall, W.S. 2002. Na^+, Cl^-, Ca^{++} and Zn^{++} transport by fish gills: Retrospective review and prospective synthesis. *Journal of Experimental Zoology* 292: 264–283.

Martin, S.A.M., A.F. Youngson and A. Ferguson. 1999. Atlantic salmon (*Salmo salar*) prolactin cDNA sequence and its mRNA expression after transfer of fish between salinities. *Fish Physiology and Biochemistry* 20: 351–359.

Mayer-Gostan, N., S.E. Wendelaar Bonga and P.H.M. Balm. 1987. Mechanisms of hormone actions on gill transport. In: *Vertebrate Endocrinology: Fundamentals and Biomedical Implications*, P.K.T. Pang and M.P. Schreibman (eds.). Academic Press, San Diego, Vol. 1, pp. 211–238.

McCormick, S.D. 1995. Hormonal control of gill Na^+,K^+-ATPase and chloride cell function. In: *Fish Physiology*, C.M. Wood and T.J. Shuttlerworth (eds.). Academic Press, New York, Vol. 14, pp. 285–315.

McCormick, S.D. 1996. Effects of growth hormone and insulin-like growth factor I on salinity tolerance and gill Na^+,K^+-ATPase in Atlantic salmon (*Salmo salar*): Interaction with cortisol. *General and Comparative Endocrinology* 101: 3–11.

McCormick, S.D. 2001. Endocrine control of osmoregulation in fish. *American Zoologist* 41: 781–794.

McCormick, S.D., T. Sakamoto, S. Hasegawa and T. Hirano. 1991. Osmoregulatory action of insulin-like growth factor I in rainbow trout (*Oncorhynchus mykiss*). *Journal of Endocrinology* 130: 87–92.

Mommsen, T.P., M.M. Vijayan and T.W. Moon. 1999. Cortisol in teleosts: dynamics, mechanisms of action, and metabolic regulation. *Reviews in Fish Biology and Fisheries* 9: 211–268.

Muñoz-Cueto, J.A., J.P. Martinez-Barbera, C. Pendon, R.B. Rodriguez and M.C. Sarasquete. 1996. Autoradiographic localization of growth hormone binding sites in *Sparus aurata* tissues using a recombinant gilthead seabream growth hormone. *Comparative Biochemistry and Physiology* C114: 17–22.

Ng, T.B., T.Y. Hui and C.H.K. Cheng. 1991. Presence of prolactin receptors in eel liver and carp kidney and growth hormone receptors in eel liver. *Comparative Biochemistry and Physiology* A99: 387–390.

Nishioka, R.S., K.M. Kelley and H.A. Bern. 1988. Control of prolactin and growth hormone secretion in teleost fishes. *Zoological Science* 5: 267–280.

Nonnotte, L., G. Boeuf and G. Nonnotte. 1995. The role of growth hormone in the adaptability of Atlantic salmon (*Salmo salar* L.) to seawater: effects on the morphology of the mucosa of the middle intestine. *Canadian Journal of Zoology* 73: 2361–2374.

Parwez, I. and S.V. Goswami. 1985. Effects of prolactin, adrenocorticotrophin, neurohypophysial peptides, cortisol and androgens on some osmoregulatory parameters of the hypophysectomized catfish, *Heteropneustes fossilis* (Bloch). *General and Comparative Endocrinology* 58: 51–68.

Pelis, R. and S.D. McCormick. 2001. Effects of growth hormone and cortisol on Na^+-K^+-$2Cl^-$ cotransporter localization and abundance in the gills of Atlantic salmon. *General and Comparative Endocrinology* 124: 134–143.

Perry, S.F., G.G. Goss and P. Laurent. 1992. The interrelationships between gill chloride cell morphology and ionic uptake in four freshwater teleosts. *Canadian Journal of Zoology* 70: 1775–1786.

Peterson, B.C. and B.C. Small. 2005. Effects of exogenous cortisol on the GH/IGF-I/IGFBP network in channel catfish. *Domestic Animal Endocrinology* 28: 391–404.

Pickford, G.E. and J.G. Phillips. 1959. Prolactin, a factor in promoting survival of hypophysectomized killifish in fresh water. *Science* 130: 454–455.

Pickford, G.E., R.W. Griffith, J. Torreti, E. Hendlez and F.H. Epstein. 1970. Branchial reduction and renal stimulation of Na^+,K^+-ATPase by prolactin in hypophysectomized killifish in fresh water. *Science* 228: 378–379.

Pisam, M., B. Auperin, P. Prunet, F. Rentier-Delrue, J. Martial and A. Rambourg. 1993. Effects of prolactin on α and β chloride cells in the gill epithelium of the saltwater adapted tilapia *Oreochromis niloticus*. *Anatomical Record* 235: 275–284.

Prunet, P., A. Sturm and S. Milla. 2006. Multiple corticosteroid receptors in fish: From old ideas to new concepts. *General and Comparative Endocrinology* 147: 17–23.

Reinecke, M., A. Schmid, R. Ermatinger and D. Loffingcueni. 1997. Insulin-like growth factor I in the teleost *Oreochromis mossambicus*, the tilapia: gene sequence, tissue expression, and cellular localization. *Endocrinology* 138: 3613–3619.

Sakamoto, T. and T. Hirano. 1991. Growth hormone receptors in the liver and osmoregulatory organs of rainbow trout: characterization and dynamics during adaptation to seawater. *Journal of Endocrinology* 130: 425–433.

Sakamoto, T. and T. Hirano. 1993. Expression of insulin-like growth factor-I gene in osmoregulatory organs during seawater adaptation of the salmonid fish: Possible mode of osmoregulatory action of growth hormone. *Proceedings of the National Academy of Sciences of the United States of America* 90: 1912–1916.

Sakamoto, T. and S.D. McCormick. 2006. Prolactin and growth hormone in fish osmoregulation. *General and Comparative Endocrinology* 147: 24–30.

Sakamoto, T., T. Ogasawara and T. Hirano. 1990. Growth hormone kinetics during adaptation to a hyperosmotic environment in rainbow trout. *Journal of Comparative Physiology* B160: 1–6.

Sakamoto, T., S.D. McCormick and T. Hirano. 1993. Osmoregulatory actions of growth hormone and its mode of action in salmonids: A review. *Fish Physiology and Biochemistry* 11: 155–164.

Sakamoto, T., K. Uchida and S. Yokota. 2001. Regulation of the ion-transporting mitochondrion-rich cell during adaptation of teleost fishes to different salinities. *Zoological Science* 18: 1163–1174.

Sakamoto, T., M. Fujimoto and M. Ando. 2003. Fishy tales of prolactin-releasing peptide. *International Review of Cytology* 225: 91–130.

Sakamoto, T., M. Amano, S. Hyodo, S. Moriyama, A. Takahashi, H. Kawauchi and M. Ando. 2005. Expression of prolactin-releasing peptide and prolactin in the euryhaline mudskippers (*Periophthalmus modestus*): Prolactin-releasing peptide as a primary regulator of prolactin. *Journal of Molecular Endocrinology* 34: 825–834.

Sandra, O., P. Le Rouzic, C. Cauty, M. Edery and P. Prunet. 2000. Expression of the prolactin receptor (tiPRL-R) gene in tilapia *Oreochromis niloticus*. Tissue distribution and cellular localization in osmoregulatory organs. *Journal of Molecular Endocrinology* 24: 215–224.

Sangiao-Alvarellos, S., F.J. Arjona, J.M. Migues, M.P. Martín del Río, J.L. Soengas and J.M. Mancera. 2006. Growth hormone and prolactin affect osmoregulation and energy metabolism of gilthead seabream (*Sparus aurata*). *Comparative Biochemistry and Physiology* 144: 491–500.

Santos, C.R., P.M. Ingleton, J.E.B. Cavaco, P.A. Kelly, M. Edery and D.M. Power. 2001. Cloning, characterization, and tissue distribution of prolactin receptor in the sea-bream (*Sparus aurata*). *General and Comparative Endocrinology* 121: 32–47.

Seale, A.P., L.G. Riley, T.A. Leedom, S. Kajimura, R.M. Dores, T. Hirano and E.G. Grau. 2003. Effects of environmental osmolality on release of prolactin, growth hormone and ACTH from the tilapia pituitary. *General and Comparative Endocrinology* 128: 91–101.

Seidelin, M. and S.S. Madsen. 1997. Prolactin antagonizes the seawater-adaptative effect of cortisol and growth hormone in anadromous brown trout (*Salmo trutta*). *Zoological Science* 14: 249–256.

Seidelin, M. and S.S. Madsen. 1999. Endocrine control of Na^+,K^+-ATPase and chloride cell development in brown trout (*Salmo trutta*): Interaction of insulin-like growth factor-I with prolactin and growth hormone. *Journal of Endocrinology* 162: 127–135.

Seidelin, M., S.S. Madsen, A. Byrialsen and K. Kristiansen. 1999. Effects of insulin-like growth factor-I and cortisol on Na^+,K^+-ATPase expression in osmoregulatory tissues of brown trout (*Salmo trutta*). *General and Comparative Endocrinology* 113: 331–342.

Shepherd, B.S., K. Drennon, J. Johnson, J. W. Nichols, R. C. Playle, T. D. Singer and M.M. Vijayan. 2005. Salinity acclimation affects the somatotropic axis in rainbow trout. *American Journal of Physiology* C288: R1385–R1395.

Shiraishi, K., M. Matsuda, T. Mori and T. Hirano. 1999. Changes in the expression of prolactin- and cortisol-receptor genes during early-life stages of euryhaline tilapia (*Oreochromis mossambicus*) in fresh water and seawater. *Zoological Science* 16: 139–146.

Shrimpton, J.M. and S.D. McCormick. 1998. Regulation of gill cytosolic corticosteroid receptors in juvenile Atlantic salmon: Interaction effects of growth hormone with prolactin and triiodothyronine. *General and Comparative Endocrinology* 112: 262–274.

Shrimpton, J.M., R.H. Devlin, E. McLean, J.C. Byatt, E.M. Donaldson and D.J. Randall. 1995. Increases in gill corticosteroid receptor abundance and saltwater tolerance in juvenile coho salmon (*Oncorhynchus kisutch*) treated with growth hormone and placental lactogen. *General and Comparative Endocrinology* 98: 1–15.

Smith, D.C.W. 1956. The role of the endocrine organs in the salinity tolerance of trout. *Memoirs of the Society of Endocrinology* 5: 83–101.

Sturn, A., N. Bury, L. Dengreville, J. Fagart, G. Flouriot, M.E. Rafestin-Oblin and P. Prunet. 2005. 11-deoxycorticosterone is a potent agonist of the rainbow trout (*Oncorhynchus mykiss*) mineralcorticoid receptor. *Endocrinology* 146: 47–55.

Veillette, P.A. and G. Young. 2005. Tissue culture of sockeye salmon intestine: functional response of Na^+-K^+-ATPase to cortisol. *American Journal of Physiology* 288: R1598–R1605.

Wendelaar Bonga, S.E. 1997. The stress response in fish. *Physiological Review* 77: 591–625.

Weng, C.F., T.H. Lee and P.P. Hwang. 1997. Immune localization of prolactin receptor in the mitochondria-rich cells of the euryhaline teleost (*Oreochromis mossambicus*) gill. *FEBS Letters* 405: 91–94.

Weng, C.F., C.C. Chiang, H.Y. Gong, M.H.C. Chen, C.H. Lin, P.P. Hwang and J.L. Wu. 2000. Salinity change induces insulin-like growth factor and heaptocyte nuclear factors on gill in euryhaline teleost (*Oreochromis mossambicus*). *Biochemical Society Transactions* 28: A341.

Wood, A.W., C.M. Duan and H.A. Bern. 2005. Insulin-like growth factor signaling in fish. *International Review of Cytology* 243: 215–285.

Young, G. 1988. Enhanced response of the interregnal of coho salmon (*Oncorhynchus kisutch*) to ACTH after growth hormone treatment *in vivo* and *in vitro*. *General and Comparative Endocrinology* 71: 85–92.

Xu, B., H. Miao, P. Zhang and P. Li. 1998. Osmoregulatory actions of growth hormone in juvenile tilapia (*Oreochromis niloticus*). *Comparative Biochemistry and Physiology* 17: 295–301.

Zhou, B.S., S.P. Kelly, J.P. Ianowski and C.M. Wood. 2003. Effects of cortisol and prolactin on Na^+ and Cl^- transport in cultured branchial epithelia from FW rainbow trout. *American Journal of Physiology* C285: R1305–R1316.

Index

B

C

R

For Product Safety Concerns and Information please contact our EU representative GPSR@taylorandfrancis.com Taylor & Francis Verlag GmbH, Kaufingerstraße 24, 80331 München, Germany

Batch number: 10726481

Printed by Printforce, the Netherlands

THE WORKS OF CHARLES BABBAGE

Volume 1. *Mathematical papers*

THE WORKS OF
CHARLES BABBAGE

EDITED BY
MARTIN CAMPBELL-KELLY

VOLUME
1

MATHEMATICAL PAPERS

Routledge
Taylor & Francis Group
LONDON

First published 1989 by Pickering & Chatto (Publishers) Limited

Published 2016 by Routledge
2 Park Square, Milton Park, Abingdon, Oxon OX14 4RN

Routledge is an imprint of the Taylor & Francis Group, an informa business

LIBRARY OF CONGRESS

Library of Congress Cataloging-in-Publication Data

Babbage, Charles, 1791–1871.
Mathematical papers.
(The works of Charles Babbage; v. 1)
1. Mathematics. I. Title. II. Series: Babbage, Charles, 1791–1871. Works. 1989; v. 1
QA3.B18 1989 vol. 1 510 s 88–32926

ISBN: 9-7811-38763-685 (hbk)

CONTENTS

GENERAL INTRODUCTION

FOREWORD BY I. BERNARD COHEN

Charles Babbage is an enigmatic figure in the history of science and of technology. Currently respected for his anticipations of many aspects of present-day computers, his two major projects of constructing calculators – a Difference Engine and an Analytical Engine – were both egregious failures. Today, so great is the emphasis on Babbage as a computer pioneer that we are apt to neglect his other intellectual contributions. Furthermore, we are led thereby to stress his visionary side rather than his practical side and thus even to belittle his actual influence or effect on his own times and on the course of history.

For a quarter of a century or more, Charles Babbage has been featured in the history of computers as a man far ahead of his time, as a scientist and inventor whose ideas contain many of the elements of the twentieth-century digital computer. Not only is Babbage esteemed because his proposed machines are seen as immediate intellectual precursors of the computers of the 1950s and early 1960s; he is especially praised because of an alleged influence on the architecture of the machines that introduced our computer age.

Any discussion of Babbage's career and influence leads at once to the issue of what historians of science now call a 'Whig interpretation' of history. The 'Whig interpretation' places value on ideas or practices of the past primarily to the degree that they resemble, and insofar as they may be said to anticipate, ideas and practices of the present. Babbage's significance could thus be as a precursor of our technology under some such rubric as 'pioneer of the computer'.

Whig history originally signified British history written from a special viewpoint of the Whig party in which certain causes were favoured as 'progressive', and in which Whigs 'saw their own beliefs, practices and institutions as the goals for all previous beliefs, practices and institutions'.[a] Thus Whig historians saw their goal to be a search 'for past developments which anticipated the present'. Most scholars today have accepted the 'more extended meaning' of this term given by Herbert Butterfield as 'the tendency in many historians . . . to praise revolutions provided they have been successful, to emphasize certain principles of progress in the past and to produce a story which is the ratification if not the glorification of the present'.[b]

Until fairly recently, such a 'Whig' interpretation infused almost all the writings on the history of science and of technology. The reason is not hard to find. Historians of science are generally committed to the thesis that science is progressive and cumulative. George Sarton, the leading historian of science of a generation or more ago, even went so far as to declare that science is the only such progressive and cumulative branch of human knowledge.[c]

The obvious corollary to this theorem is that the historian's task is to seek in the past for ideas, concepts, experiments, theories or methods that in some way resemble our own and that can be seen accordingly as stages in the progress towards our present state of knowledge. Whig historians produced chronicles of the heroes of the past, whose achievements were celebrated because they did well on a scale of values determined by the degree of accord with the present state of scientific knowledge and belief. Scant attention was paid to the 'villains of the piece', those who advanced or advocated allegedly 'wrong' or 'bad' ideas and who therefore were considered to have acted so as to slow down or even hold up the progressive march of science.

Historians of science since the 1950s have generally abandoned the older Whiggish history, largely because – primarily under the influence of the late Alexandre Koyré[d] – they have come to see the advantages of studying the scientific thought of the past in the direct terms of the problems and intellectual currents of the time under

[a] C. B. Wilde, 'Whig history', in W. F. Bynum, E. J. Brown and R. Porter (editors), *Dictionary of the history of science* (Princeton, 1981), pp. 445–6.

[b] H. Butterfield, *The Whig interpretation of history* (London, 1931).

[c] G. Sarton, *The study of the history of science* (Cambridge, Mass., 1936).

[d] Historians of science of the last four decades were profoundly influenced by Alexandre Koyré's seminal *Études Galiléennes* (Paris, 1939), translated by John Mepham as *Galilean studies* (Sussex, 1978). See also Koyré's *Metaphysics and measurement, essays in scientific revolution* (London, 1968), *Newtonian studies* (Cambridge, Mass., 1965), and *From the closed world to the infinite universe* (Baltimore, 1957).

which any work was done, rather than merely 'grading' it in a schoolmasterish way in terms of its degree of accord with the present. The new approach is more illuminating because it concentrates on the reconstruction of the way in which a scientist grappled with the problems of his or her own day, thus forcing an examination of the historical past in its own terms. It goes without saying that the new history tends to focus more directly on the nature and creative process of discovery and invention, with the qualities of individual scientific minds in their interaction with their intellectual (and even social) environments. This history aims to see the lives and achievements of scientists and inventors in their own terms and in their own scale of values, rather than ours.[a]

In his introduction to H. W. Buxton's *Memoir of the life and labours of the late Charles Babbage Esq. F.R.S.* (first published 1987), Anthony Hyman points out that even the historical setting of Babbage is often mislabelled 'Victorian'; in fact, 'Babbage was quintessentially a figure of his time whose intellectual roots reach deep into the Enlightenment'. Hyman also notes that the problem of evaluating Babbage's work is 'exacerbated because the history of computing is primarily a twentieth-century subject', yet 'it is quite essential that Babbage's work be set in its proper historical context'. The latter can be a problem for historians of computing since they 'cannot always be relied on to have an adequate knowledge of eighteenth- and nineteenth-century history'.

The treatment of Charles Babbage by computer historians and computer scientists has all too often illustrated a Whig interpretation of the history of technology. For example, B. V. Bowden introduces his classic work, *Faster than thought* (1953) with the remark that 'Babbage's ideas have only been properly appreciated in the last ten years, but we now realize that he understood clearly all the fundamental principles which are embodied in modern digital

[a] G. B. Wilde observes that ' "Whig" criteria are still generally applied by historians of science in choosing subjects for study'. In evidence, Wilde notes that the rapidly growing literature on Newton and Darwin 'is vastly greater than that on their opponents'. A somewhat different reason for the decline of Whig history of science has been a radical change in philosophy of science; notably 'the breakdown of positivist philosophies of science has made the belief that the present state of scientific knowledge is, in any absolute, epistemological sense, superior to earlier forms of knowledge, difficult to sustain' (Wilde, *op. cit.*, p. 446). Readers should be aware that few, if any, of the authors of conventional histories of science or of technology knew anything about the Whig tradition of historical writing as such.

computers'. Bowden conceived that the whole book ('a symposium on digital computing machines') was in fact 'devoted to an account of the construction and use of the machines which his [i.e. Babbage's] vision inspired'.[a]

Again, in a study of *Brain, mind and computers*, Stanley L. Jaki concludes that 'It is readily admitted today that the giant electronic computers, a chief pride of our times, are but the embodiment of principles that Babbage laid down well before the nineteenth century reached its midpoint.'[b] At the very dawn of the computer age, on 1 August, 1946, J. H. Curtis of the U.S. National Bureau of Standards observed that 'Babbage's ideas seem to have been pretty generally neglected until the last ten or fifteen years, when a tremendous revival of interest has occurred in the subject of automatic digital computing machines'.[c] Curtis held that the chief causes for this revival were to be found in 'the extensive development during the last 40 or 50 years of punched card machinery for performing tabulations and individual arithmetic operations, the success of the Bush differential analyzer (1930), and the development of fire control apparatus just before and during the war'. Even more significant, perhaps, was 'that a team consisting of one man, with ability and tenacity, and a corporation, with resources and vision, actually brought Babbage's dream of an automatic sequenced digital calculator into realization'. He was referring, of course, to 'Dr. Howard Aiken of Harvard University and the International Business Machines Corporation', which had 'constructed the first completely automatic general-purpose digital computing machine (1944)'. The computer scientist and historian René Moreau concludes his survey with a three-point summary of 'Babbage's Machine as the Precursor of the Computer'.[d]

These authors, typical of many others, dwell on elements of Babbage's machines in relation to computer technology of the 1950s,

---

[a] B. V. Bowden (editor), *Faster than thought: a symposium on digital computing machines* (London, 1953), p. 7.

[b] Stanley L. Jaki, *Brain, mind and computers* (New York, 1969), p. 43.

[c] J. H. Curtis, 'A review of government requirements and activities in the field of automatic digital computing machinery', pp. 343–65 of M. Campbell-Kelly and M. R. Williams (editors), *The Moore School lectures: theory and techniques for design of electronic digital computers, The Charles Babbage Institute Reprint Series for the History of Computing*, Vol. 9 (Cambridge, Mass., London and Los Angeles, 1985), pp. 343–4.

[d] René Moreau, *The computer comes of age: the people, the hardware, and the software* (Cambridge, Mass., and London, 1984), p. 19.

1960s and beyond, and they tend to ignore the relation of Babbage's innovations to the thought or technology of his age. Furthermore, there is no exploration of either the fate or influence of Babbage's bold inventions during the century or so between Babbage's work and the beginning of the computer age. There is usually even an attempt to make precise the degree to which the founding fathers of the modern electronic computer were actually indebted to Babbage in the matter of general architecture or any particular aspect of computer technology.

One of the chief exhibits of those who see Babbage only in relation to today's computer is the Harvard Mark I or the IBM Automatic Sequence Controlled Calculator (ASCC). Howard Aiken, responsible for the architecture of the Mark I, was one of those who, it is usually said, was directly inspired by Babbage's 'vision'. There is evidence in abundance that Aiken presented himself as Babbage's heir. Anyone who encountered him would speedily be made aware of Aiken's great admiration for his British 'counterpart' of a century earlier. Aiken – in conversation, in lectures and in published writings – often linked himself with Babbage, presenting himself as one who finally had brought Babbage's dreams to reality. Visitors to the Harvard 'Comp. Lab.' would have their attention drawn at once to a set of original computing wheels that had once been intended to form part of Babbage's great Difference Engine. They were presented to Harvard University by Major-General Henry Prevost Babbage, son of the inventor, in 1886 on the occasion of the celebration of the 250th anniversary of the founding of Harvard College.[a]

Aiken's earliest recorded statement of his admiration for Babbage occurs in his formal proposal to IBM for a calculating machine in its final version, prepared by 4 November, 1937.[b] As is well known, Aiken's concept was transformed into a functional machine by the engineers of IBM and began operation at Harvard in 1944. When Aiken published the *Manual of operation* in 1946, he publicly announced or acknowledged in a variety of different ways that

[a] This gift is described in my article on 'Babbage and Aiken', *Annals of the History of Computing*, in press. The wheel assemblage is still on public display in the lobby of the Aiken laboratory.

[b] H. H. Aiken, 'Proposed automatic calculating machine', previously unpublished memorandum, *IEEE Spectrum*, August, 1964, pp. 62–9. Reprinted in B. Randell, *The origins of digital computers*, 3rd edition (Berlin, Heidleberg and New York, 1982), pp. 195–201.

Charles Babbage had a serious claim to priority in the realm of ideas that concerned computing by machine. Facing page 1 was a plate showing Harvard's set of Babbage computing wheels. At the head of chapter one, there was prominently displayed a quotation from Babbage's *Passages from the life of a philosopher* (1864), which reads:

If, unwarned by my example, any man shall undertake and shall succeed in really constructing an engine embodying in itself the whole of the executive department of mathematical analysis upon different principles or by simpler mechanical means, I have no fear of leaving my reputation in his charge, for he alone will be fully able to appreciate the nature of my efforts and the value of their results.[a]

When Aiken first encountered this passage, according to a report by Jeremy Bernstein of an interview with Aiken,[b] he 'felt that Babbage was addressing him personally from the past'. Aiken said simply that 'If Babbage had lived seventy-five years later, I would have been out of a job'.

Yet, despite Aiken's public reiteration of the kinship of his Mark I and Babbage's machines, a close examination of the architecture of Mark I does not show a marked similarity with Babbage's designs for either a Difference or an Analytical Engine. In fact, the Mark I did not even emulate their operation. It seems clear that Aiken never even fully knew the features of Babbage's inventions.

Current interest in Babbage has been accelerated by the computer connection, but there has been a growing awareness that he was of importance in many different areas, and thus attention is being focused more on a close study of original texts and actual machines. A subject for serious scholarly research is the history of Babbage's reputation and influence – not merely in relation to mechanical calculation and, eventually, the computer – in every area in which he laboured. Karl Marx and John Stuart Mill esteemed his contributions to economics, but modern economists have largely been silent; and what was the ultimate fate of his many proposals for economic and political reform?

---

[a] As quoted in Harvard Computation Laboratory, *A manual of operation for the Automatic Sequence Controlled Calculator* (Cambridge, Mass., 1946). See *Passages from the life of a philosopher* (1864), *Works of Babbage*, Vol. 11, p. 336.

[b] J. Bernstein, *The Analytical Engine: computers – past, present, and future*, revised edition (New York, 1981), p. 62.

Furthermore, Babbage was a spirited writer; his books and articles are a joy to read if one can get hold of them. Many of his writings are long out of print and many papers have never been reprinted or assembled. The time has come to make Babbage's works easily available in a single uniform edition. The words 'easily available' have here a triple sense. First, there is the availability in print of his writings – including not only reprints of his published books, but collections of pamphlets, articles, letters to the press and other ephemera. Secondly is the need for scholarly, explanatory introductions and annotations and an index to render intelligible to us the writings of a past age. Thirdly is the careful editing of mathematical texts to eliminate simple errors and proof faults.

The variety and scope of the volumes comprising the *Works of Babbage* should help to erase the limited image of Babbage as a mere computer precursor or pioneer. We may, by perusing these volumes, gain some insight into the quality of his multi-faceted interests and their variety and scope. First and foremost a mathematician, Babbage was one of the group who got British mathematicians to give up the dot-notation of Newton for the calculus and to adopt the Continental or Leibnizian notation (to replace 'British dot-age' with 'Continental dee-ism'). In the *Works of Babbage* this activity is complemented by the presence of his contributions to pure and applied mathematics, topics not even remotely connected with calculating or computing or machines. We are reminded that Babbage was, for a time, the Lucasian Professor of Mathematics at Cambridge University.

Babbage was seriously concerned with machines, with manufacturing and with industrial processes, which he observed in factories at first hand. His book on this topic, *On the economy of machinery and manufactures* (1832) was praised for its contribution to economics and was speedily translated into French, German, Italian and other languages. He pioneered attempts to invent a form of Mechanical Notation that would be unambiguous. Today, he is lauded for his studies in the new area of man and tool interaction, usually held to be the much later creation of Frederick Taylor. Babbage is even considered to be a founder (or precursor) of the general methods of operations research. His writings show concern for a rationalized postal system, for railroad safety, for lighthouses, for engineering, for the state of science in England, and so on and so on. It is difficult for us to conceive that a single creative mind could be so intensely active in so many areas of science, technology, social science and public affairs.

All these facets of Babbage's life and interests are gathered together in this new edition of his writings. Here are accounts of government and private commissions; of science and of its organization; of manufacturing and commerce; of political economy; and even of religious philosophy. These whet our appetite for more – for an eventual edition of Babbage's correspondence, which would include exchanges with mathematicians, scientists and public and private figures from all over Europe. The present great first step of assembling Babbage's published papers in a uniform edition should help to restore to Babbage his many and mighty dimensions.

THE WRITINGS AND LIFE OF CHARLES BABBAGE

Although Babbage is well-described as the 'pioneer of the computer', he was far more than this: he was a vigorous polymath in the nineteenth-century tradition. Indeed, it is the sheer breadth of his work that presents Babbage scholarship with its most formidable problem: we do not yet have an integrated view of Babbage. In the past decade scholars have separately investigated Babbage's mathematical work, his calculating engines and his economic writings; the arrangement of the *Works of Babbage* to some extent reflects this tripartite scholarship. Although the editorial board has drawn on the best of current thinking in planning the *Works*, the last word on Babbage scholarship remains some distance away.

Babbage published a good deal during his lifetime, but his output does not truly convey a balanced representation of his work. Babbage himself, however, considered his list of printed works to be an important measure of his achievements. At various times during his life he attempted to compile a list of his publications (the last and most complete list is reproduced in Vol. 11, pp. 370–3). None of his lists were very accurate, and we have used the best bibliographies known to us – notably Van Sinderen (1980) and Hyman (1982) – in compiling the new list annexed here, which includes a number of recently discovered items. This list we believe to be the most complete produced to date; but that said, we fully expect that there will be some further small Babbage finds during the next few years.

Babbage privately printed pamphlets, published in learned journals, and wrote full-length books. His books have mostly been reprinted at one time or another, but the *Works of Babbage* represents the first attempt to bring his entire published output together. In arranging the *Works* we have grouped Babbage's short papers together, to some extent by topic, in order to make five comfortably sized volumes:

Volume 1: Mathematical papers
Volumes 2 and 3: Calculating engines, I and II
Volumes 4 and 5: Scientific and miscellaneous papers, I and II

Volumes 1, 2 and 3 form an excellent basis for the separate study of the full scope of Babbage's work in mathematics and his calculating engines. However, Volumes 4 and 5, scientific and miscellaneous papers, are frankly a *pot-pourri* that bring together Babbage's short writings on science, technology, inventions, economic and statistical topics, and some short polemical works; these volumes, more than any others, reflect Babbage's considerable output of minor writings in a wide variety of fields. Within each volume we have preserved chronological order, with the exception of Volumes 2 and 3 where an additional level of grouping has been imposed, and there are separate sections on the Differene Engine and table making, the Analytical Engine, and the Mechanical Notation.

Babbage's books have been reprinted in the *Works*, in keeping with their original format, as individual volumes:

Volume 6: *A comparative view of the various institutions for the assurance of lives* (1826)
Volume 7: *Reflections on the decline of science in England* (1830)
Volume 8: *On the economy of machinery and manufactures*, 4th edition (1835)
Volume 9: *The ninth Bridgewater treatise*, 2nd edition (1838)
Volume 10: *The Exposition of 1851*, 2nd edition (1851)
Volume 11: *Passages from the life of a philosopher* (1864)

Several of Babbage's books and pamphlets appeared in two or more editions. Babbage generally added new material to each subsequent edition of a work but invariably took nothing of significance away. In the *Works* we have therefore reprinted only the last edition of a particular work, and have broadly indicated the material added from one edition to the next.

A particular difficulty encountered by Babbage scholars is that he did not generally write up his researches in a systematic fashion. This is true particularly of the calculating engines, and references to them are scattered throughout his works. We have therefore provided a single index for the entire *Works of Babbage* (in Volume 11) which has the valuable function of enabling the scholar to track down all significant references to individual topics.

Charles Babbage was born in London on Boxing Day 1791, the son of a wealthy banker. As a child he showed mathematical promise, and in 1810 went up to Trinity College, University of Cambridge, where he studied mathematics, graduating in 1814. His early career was predominantly mathematical, and he made a number of technical contributions to mathematics, notably on the calculus of functions. His twenty-odd papers on mathematical topics are brought together in Volume 1 and are discussed below by John M. Dubbey.

In fact Babbage's mathematical contributions were less technical than political. During his undergraduate years at Cambridge, British mathematics was at its nadir and lagged far behind the continent where the calculus had flourished using the notation of Leibniz. In Britain the mathematical establishment had doggedly clung to the notation of Newton. Whilst still an undergraduate, together with John Herschel and George Peacock, Babbage founded the Analytical Society which flowered briefly during 1812–4: its great and lasting achievement was the introduction of the continental notation for the calculus that enabled a renaissance of British mathematics.

After about 1820, Babbage's interests began to branch in several different directions. Most notably he became interested in the making of mathematical tables. He formally proposed the construction of a Difference Engine for the automatic computation of tables in 1822 and constructed a small prototype engine in the same year. His *Table of logarithms* published in 1827 were the most accurate of their day and they appeared in several editions in several countries.

The construction of the Difference Engine was a task far exceeding the mechanical engineering capabilities of the day, and for this reason he made a systematic study of all kinds of machinery, particularly machinery for manufacture. He also read deeply on the subject of political economy. This later led to his classic book *On the economy of machinery and manufactures* (1832) that established his reputation as a major pundit and commentator on industry. The book was unique in its focus on manufacture and was a turning point in economic writing. Babbage was also a major force in applied statistics: he was a founder of the London Statistical Society, and was a knowledgeable critic of actuarial practice. Maxine Berg discusses Babbage's economic and statistical work in more detail below.

Babbage had established himself at an early age as a leading scientific figure, being elected to the Royal Society in 1816. In 1827–8 he made the continental tour during which he consolidated his position by establishing contact with some of the leading scientific names of Europe. He was profoundly affected by the high prestige of science on the continent compared with Britain, and the progressive views on scientific education. This experience heightened his concern for the reform of the organization of science in England in general, and of the Royal Society in particular. Babbage advocated the formation of specialist scientific societies and he was instrumental in founding the Astronomical Society and the British Association. In 1828 he was elected Lucasian Professor of Mathematics at Cambridge, a post which he held until 1839, it was the only paid position he ever accepted. Politically, Babbage was a liberal Whig: he stood for Parliament in the early 1830s, he denounced election corruption and bribery, attacked church preferments and tithes, and was a firm supporter of the Reform Bill. Babbage's social and academic context was clearly that of early nineteenth-century liberal scientific circles, and he participated in the salon culture of the day; his own Saturday evening soirées were attended by political, scientific and literary figures of the day.

In 1829, Babbage had begun to receive government support for the construction of the Difference Engine. The support was by no means ungenerous, but it proved inadequate to the scale of his undertaking. After a long and abrasive relationship with Babbage, the Government finally abandoned and withdrew all support for the project in 1842. In fact, in about 1834, Babbage had conceived of the Analytical Engine – a generalization of the Difference Engine very close in spirit to the modern digital computer. The Analytical Engine was undoubtedly the crowning achievement of Babbage's intellectual life. Babbage published little, and somewhat obscurely, on his engines, so Volumes 2 and 3 of the *Works*, reviewed below by Allan Bromley, contain a number of contemporary writings by other authors, notably Menabrea's account of the Analytical Engine, translated and annotated by the Countess of Lovelace.

The first phase of Babbage's work on the Analytical Engine lasted from about 1834 to 1846. During this period he elaborated all the essential concepts and mechanisms for the engine. After this, for a period of some years, he did no further work on his engines. He continued, however, to keep up his interest in all matters scientific and technical: geology, cryptography and inventions of various kinds. He remained an energetic scientific and political reformer.

From about 1856, Babbage resumed the work on his engines, but without any real prospect of their construction. He still produced a trickle of publications on diverse topics, including his autobiographical *Passages from the life of a philosopher* in 1864. He died in 1871 at the age of 79. There were many obituaries, although no full-length biographies were published until recent times. Probably the most substantial biographical article on Babbage published in the nineteenth century was a fascinating but little-known pamphlet written by James Jerwood, under the pseudonym 'Judex Juris' – *On the calculation and printing of mathematical tables by machinery; the inventor and his treatment*, printed in 1861. This pamphlet is reprinted as the first article of this volume. After his death, Babbage's son Henry continued the work on the engines, publishing a book *Babbage's calculating engines* in 1889, and completing an experimental fragment of the Analytical Engine in about 1908. Thereafter Babbage's computing work lay neglected for over half a century before it was seriously taken up by the present generation of computer historians.

MATHEMATICAL PAPERS: BY JOHN M. DUBBEY

The mathematical work of Charles Babbage needs to be evaluated in the context of the severe decline of mathematical quality in Britain following the domination by Isaac Newton. The century prior to Babbage had seen the barren controversy between the supporters of Newton and the supporters of Leibniz over the prior claim to the invention of the calculus, the determination of British mathematicians to follow only the methods and notation of Newton, and the unrefuted attack by Bishop Berkeley on the logical basis of the calculus. The achievement of Babbage may be observed in his lateral approach to the advance of a discipline becoming stagnant among his countrymen, his selection of new methods to tackle new problems, and a visionary ability to generalize. It is particularly instructive to observe the convergence of the range of his mathematical ideas towards his major work in the construction of the Difference and Analytical Engines.

Babbage's recollections in *Passages* indicate that during his early education, out of a broad range of interests, mathematics rapidly became his major occupation. He read widely both British and foreign authors, acquiring a depth of knowledge which he found on arrival at Cambridge to be superior to that offered by his tutors. From the age of seventeen he began original work in the 'calculus of functions', starting from a functional equation obtained through attempting to solve a geometrical problem of Pappus.

When he entered Cambridge University in 1810, it was at a time when mathematics was perceived as an excellent training of the mind for future lawyers, clergy, doctors or men of leisure, and provided virtually the whole of the curriculum. The function of the tutor appeared to be a preparation for the difficult but standard problems of the Tripos examination, and Babbage largely ignored the formal system, preferring to pursue his own studies with his friends John Herschel, George Peacock and others with whom he formed the Analytical Society.

The purpose of the Analytical Society was to rescue the deplorable state of mathematics in the University and the country in general. The first objective was to liberate the calculus from the constraints of Newtonian method and notation, introducing the more fruitful analytical approach of continental mathematicians. The group produced in 1813 the *Memoirs of the Analytical Society*, of which Babbage and Herschel wrote the preface and to which Babbage contributed a paper 'On continued products'; this latter paper, Babbage's first mathematical publication, vividly conveys his talent for mathematical manipulation.

As part of their campaign to introduce the continental notation into Great Britain, and to influence the notation used in the Tripos examinations, Babbage and his friends translated and published in 1816 S. F. Lacroix' *An elementary treatise on the differential and integral calculus*, and produced in 1820 a volume of *Examples of the differential and integral calculus*. Appropriate extracts from these books have been reprinted in the *Works*. The campaign to introduce the differential notation to Great Britain was strongly opposed in the University, but with the powerful support of William Whewell, all opposition had been scattered by 1820. The work of the Analytical

Society was now complete and the more general aim was taken on by the newly formed Cambridge Philosophical Society.

The interest shown by the Analytical Society in the importance of notation became a preoccupation of Babbage; he studied the principles of mathematical symbolism, fascinated by the power of notation to make the ideas of generality and simplicity compatible. His subsequent work on the construction of the calculating engines was underpinned by his Mechanical Notation to describe the moving parts of machinery. In his mathematical work he always sought to improve the quality of notation and demonstrated how problems of great complexity could be simplified by this means.

Babbage's major mathematical work, the calculus of functions which attempted to systematize an approach to the solution of a wide range of problems, became possible only by means of a carefully defined notation. His two major papers – *An essay towards the calculus of functions* in two parts – were published in the *Philosophical Transactions of the Royal Society* in 1815 and 1816.[a] Two subsequent

---

[a] These essays on the calculus of functions established a cumulative calculus for solving the increasingly complex functional equations:

$$\psi x = \psi \alpha x$$
$$\psi x = Ax \cdot \psi \alpha x$$
$$\psi x = Ax \cdot \psi \alpha x + Bx$$
$$\cdot$$
$$\cdot$$
$$\cdot$$
$$F(x, \psi x, \psi \alpha x, \ldots \psi v x) = 0$$

where it must be understood that brackets are unambiguously discarded, x is the variable, $\alpha, \beta, \gamma, \ldots A, B, C, \ldots$ the fixed functions and ψ, ϕ the variable functions.

These first order equations are extended by the device of writing the repeated functions $\psi[\psi(x)]$ as $\psi^2 x$ and proceeding to solve:

$$\psi^2 x = x$$
$$\psi \mathrm{n} x = x$$
$$\psi^2 x = \alpha x$$
$$\cdot$$
$$\cdot$$
$$\cdot$$
$$F(x, \psi x, \psi^2 \alpha x, \ldots \psi^n \mathrm{v} x) = 0$$
$$F(x, \psi x, \psi \alpha(x, \psi \beta x)) = 0$$

This is extended to functions of several variables with problems analogous to these above, and ultimately to differential-functional equations such as:

$$\psi \alpha x = \frac{d\psi x}{dx}$$
$$\psi \alpha x = \frac{d^n \psi x}{dx^n}$$

subsequent papers published in 1817 took the subject only a little further. Babbage believed that his new scheme would serve as a generalized calculus to include all problems capable of analytical formulation, and it is possible to see here a hint of the inspiration for his concept of the Analytical Engine. While the work on the engines and his other scientific, social and political activities caused him virtually to abandon mathematical research at the age of thirty, the calculus of functions was the area he often yearned to continue. In fact the calculus of functions was not taken up by other workers, and it is the aspect of Babbage's mathematical work that modern mathematicians find the most fascinating.

Babbage's preoccupation with notation was the subject of three mathematical papers. A semi-popular article on 'Notation' appeared in Brewster's *Edinburgh Encyclopaedia* in 1830 (which Babbage actually wrote in 1821). A quite remarkable paper 'Observations on the notation employed in the calculus of functions' appeared in the *Transactions of the Cambridge Philosophical Society* in 1822: this not only demonstrated conclusively the excellence of his notation for the calculus of functions, but there was also a series of calculations showing the conciseness of the symbolism – a 'calculus of notations', as it were. The third paper 'On the influence of signs in mathematical reasoning' (1827) did not add significantly to the *tour de force* of this earlier paper.

Of Babbage's remaining mathematical papers, one can only say that they were the workman-like output of a journeyman mathematician: none was profound or of lasting importance. There were two further papers on analysis which appeared in 1819, one on infinite series and another on a theorem relating to prime numbers. His only publication on probability 'An examination of some questions connected with games of chance' (1821), is most interesting for its demonstration of Babbage's virtuosity in mathematical manipulation. There were three papers on geometry. The first in 1816 supplied some missing proofs from Matthew Stewart's venerable *Some general theorems of considerable use in the higher parts of mathematics* (1746). There were two publications on porisms, one of which was a semi-popular article for the *Edinburgh Encyclopaedia*. The subject of

---

and

$$\psi(x, y) = \frac{d\psi(x, \alpha y)}{dx}.$$

porisms was an eccentric choice – of scarcely more interest then than today.

An intriguing paper in 1817, which attempted to give a mathematical foundation for the Knight's move in chess, is of only limited interest. But it was the germ of an idea that led to a truly remarkable mathematical analysis of the game of noughts and crosses in his unpublished manuscript the *Philosophy of analysis* (*c.* 1821). The *Philosophy of analysis* exists only as an unrevised manuscript, and demonstrates all too clearly that Babbage did not always publish his most interesting work. It contains, for example, a major essay on algebra. This essay is a first indication that the subject was capable of development when liberated from the need for arithmetical interpretation. This notion was contemporary with, but independent of, continental work to release geometry from the rules of Euclid and provided an essential link with the new algebras of quaternions, logic and matrices developed later in the century by Hamilton, Boole and Cayley.

The flow of mathematical research ceased as Babbage became absorbed in other matters, but the analytical probing of the early part of his intellectual life characterized his approach to problems whether from science, economics, philosophy or any other area. His achievement may be seen not only in his substantial mathematical contributions but also in his pioneering ability to inspire mathematical order in wide areas of a non-mathematical universe.

TABLE MAKING AND CALCULATING ENGINES: BY ALAN G. BROMLEY

The first idea of an automatic calculating machine, automatic in the sense that it could carry out a lengthy series of calculations without intervention by the operator, occurred to Babbage in about 1820. Although the various accounts in Babbage's later writings differ from one another, they concur in tracing the idea to the extensive errors found by Babbage and John Herschel in preparing mathematical tables. The first Difference Engine was a realization of the resulting aspiration that such tables might be 'executed by steam'.

By mid-1822 the idea of the Difference Engine was well established and Babbage had constructed a small demonstration model, now unfortunately lost. The machine was to employ the method of finite differences to sub-tabulate printed values for tables between pivotal values prepared by human calculators. The Difference Engine was, therefore, to be the mechanical embodiment of the third and largest group of human computers used by de Prony in his massive project for recalculating tables after the French Revolution which Babbage took as his model. This function was most readily adapted to mechanization as it required no mathematical operation other than addition. Recognizing that the printing was as critical to the preparation of tables as the calculation, Babbage proposed that the Difference Engine should directly prepare stereotype printing plates. Barring mechanical failure, there would then be no impediment to perfect accuracy in printed tables.

Babbage's open letter of 1822 to Sir Humphry Davy, President of the Royal Society, announcing the invention was his most important publication on the Difference Engine; the letter was instrumental in securing government financial support to construct a full-scale machine. For his invention of the Difference Engine, Babbage was awarded the gold medal of the Astronomical Society in 1824.

The project to build the Difference Engine continued until 1833 when it collapsed largely through the failure of the contractual and personal relationships between Babbage and Joseph Clement, the engineer entrusted with the construction of the machine. This failure seems attributable to the excessive cost of the extremely fine quality workmanship employed by Clement in the construction, for which he was paid, effectively, on a cost-plus basis. Matters were not improved by an unfortunate tendency in Babbage to treat Clement as his servant, in an era in which the role and status of engineers had expanded dramatically. Another contributing factor was the uncertain nature of the financial arrangements between Babbage and the Government: Babbage considered himself as acting as an agent for the Government, but the Government saw itself as giving grants in aid to a project which was primarily Babbage's. The project certainly did not fail, as is commonly supposed, because the Difference Engine required a degree of mechanical precision beyond the technology of its day – the continued perfect functioning of the model assembled in 1833 and now in the Science Museum, London, is proof of that.

Very little material was published by Babbage respecting the Difference Engine, and none concerning its mechanical, as distinct from mathematical, principles. There were, however, a number of publications on the engines by contemporary scientific writers; we have included in the *Works* some of the key publications on which Babbage had a direct influence. The most extensive account of the Difference Engine was published by Dionysus Lardner, the well-known popularizer of science, in 1834. From the technical detail that it contains, much of which can be directly linked to the drawings of the machine, it is evident that Babbage was directly involved with and played a substantial hand in its preparation. We have also included in the *Works* a previously unpublished manuscript of Babbage's, 'The science of number reduced to mechanism' (probably written in 1822); this deals, rather introspectively, with the design and the design process. No detailed account was ever prepared of how the Difference Engine was to have been used in preparing tables. However, it was undoubtedly the same as for the Scheutz Difference Engine constructed in 1855; Babbage again undoubtedly played a substantial hand in the account published in *Specimens of tables calculated, stereomoulded, and printed by machinery* (1857).

It is arguable that, impressive as the Difference Engine was, it was a rather limited conception because it addressed too small a proportion of the problems inherent in table making. In 1822 Babbage had conceived of a range of possible calculating machines aside from the Difference Engine. Following the collapse of the project to build the Difference Engine in 1833 he returned to these, particularly a method of enhancing the power of the Difference Engine by feeding back the tabular value to a higher order difference, and the mechanization of multiplication. Progress was very rapid, and by 1834 he had conceived of the Analytical Engine, a general purpose calculating machine unlimited in the potential scope of the calculations it could perform. Babbage worked vigorously on the design of the Analytical Engine until 1846 during which time its power was greatly expanded as the mathematical mechanisms were simplified and refined.

It is on the Analytical Engine that Babbage's modern reputation in computing is based. The Analytical Engine anticipated many of the logical design characteristics of modern computers, although embodied in a mechanical rather than electronic mechanism. First, there was the separation of the machine into a store, where numbers were retained between steps of a calculation, and the mill to which numbers were brought for the purpose of the arithmetic operations of addition, subtraction, multiplication and division. Secondly, the computation was specified by a program of elementary steps, expressed by holes in Jacquard cards. Thirdly, internal control of the machine was by a sequence of steps identical in spirit to a modern microprogram; in the Analytical Engine these were coded on barrels not unlike the cylinders of a music box. Much of the power of the Analytical Engine relied on the flexible sequencing of program steps at the user and microprogram levels, ideas traceable to the printing mechanisms of the Difference Engine. Babbage did not, however, arrive at the modern conceptions of a stored program, nor of structuring of data in the store.

Aside from the very general and political accounts in *Passages*, nothing was published by Babbage concerning the design of the Analytical Engine. An incomplete account in a manuscript 'On the mathematical powers of the calculating engine' (1837), was not published until modern times. In 1840 Babbage travelled to Turin, in Italy, where he delivered a series of seminars describing the Analytical Engine. He intended that the eminent astronomer, Plana, would prepare an account of his design and its mathematical powers, but the task fell to the young military engineer Menabrea. Menabrea's account was subsequently translated into English with extensive notes by Ada Lovelace as the 'Sketch of the Analytical Engine' (1843). This remained the definitive account of Babbage's programming ideas until modern times. In this paper Menabrea and Lovelace played the same role, of presenters of Babbage's ideas, as Lardner had earlier played for the Difference Engine. A romantic myth has grown up surrounding Ada Lovelace to the effect that she originated the programming ideas discussed in her notes. This is discredited by modern biographical work, and the fact that all of her examples are now known to have been prepared by Babbage.

Work on the design of the Analytical Engine ceased in 1846 and Babbage then prepared a design for a new Difference Engine No. 2. This is the most complete of all of Babbage's designs and was offered by him to the Government, but the offer was declined. From 1847 to 1855 Babbage seems to have done no work on the design of calculating machines although his Mechanical Notation was systematized and published, in incomplete form, during this period. In 1856 Babbage returned to the design of the Analytical Engine and attempted to so far simplify the design that it could be built within his own means. This work, which continued until his death in 1871, seems to have involved little development of the logical design of the machine. However, the mechanical details of the mechanism were most ingeniously simplified by abandoning machining of parts in favour of pressing and later die-casting. This would have markedly reduced the cost of the machine. A trial-piece of this design was nearing completion at the time of his death. Although a considerable amount of manuscript material, drawings and notes, remains from this last phase of Babbage's work nothing was or has been published.

The published record of Babbage's work on the design of automatic calculating machines is scant indeed. There is virtually nothing in Babbage's own hand, and the writing of others, in which Babbage was directly involved, contains very little technical material. It is appropriate to enquire, therefore, as to the significance of Babbage's work.

As I. B. Cohen notes in his Foreword, Babbage had no influence on the technical development of modern computers. The published material gave only very general and superficial accounts which could have contributed little beyond the mere idea of automatic calculating machines. In England the idea of a calculating engine was also preserved by the oral traditions at Cambridge and elsewhere; but the tradition may even have been counter-productive to the modern development of computing machines: for why should one support attempts to develop a field in which a person of Babbage's intellectual calibre and energy had failed? The Difference Engine tradition bore fruit in the work of L. J. Comrie in the 1930s, but Comrie in turn argued strongly against the development of specialized calculating machinery in favour of the adaptation of standard commercial machines. In America both the Babbage tradition of automatic calculation and any negative import of his failure were absent. The developments which occurred in the 1930s, particularly the development of the Harvard Mark I by Howard Aiken, were uninfluenced by Babbage's work. Indeed, some of Babbage's ideas, particularly microprogramming, went beyond the designs of the 1940s: had they been known, they might well have influenced the development of computers in the 1950s. But the ideas were too deeply hidden in Babbage's manuscripts for them to have been readily gleaned, even had somebody looked. It has only been by working retrospectively, with a knowledge of the design of electronic computers, that it has been possible to interpret the vast array of detail in Babbage's manuscript designs.

Although Babbage's work had no direct influence on the development of modern computers, it may have importance beyond the merely antiquarian. It represents a development of the idea of automatic calculation to a high degree of sophistication completely independent of modern developments. It provides an opportunity, rare in the history of science, to examine the development of an idea in two independent contexts, and using vastly different technologies. It is striking, in this light, that Babbage's designs are so remarkably similar to those of electronic computers.

ECONOMICS AND STATISTICS: BY MAXINE BERG

Since the 1950s, Babbage's renaissance as a figure in the history of scientific thought can be almost entirely attributed to the interest in his calculating engines. This fixation on the engines has resulted in an overshadowing of his economic and statistical work: yet in both he was one of the most remarkable thinkers of his day.

Babbage made a brilliant and utterly original foray into political economy, which demonstrates vividly his remarkable ability to cut across and to go beyond the social and intellectual barriers of his time. His great economic work was the book *On the economy of machinery and manufactures*, first published in 1832. The book was immensely popular: there were four editions in two and a half years, it was reprinted in the United States and translated into six continental languages. The book sprang in part from Babbage's abiding interest in practical mechanics and a fascination with contemporary industrial technology. He learned from manufacturers, large and small, mechanical engineers, and above all from the skilled artisans he never ceased to praise. Both at home and on his continental tours he lost no opportunity to visit local workshops and factories. (An account of one such visit 'Description of the boracic acid works of Tuscany' appeared in John Murray's popular *Handbook for travellers in Central Italy* (1843), reprinted in Volume 5.)

Babbage wanted to present his readers with the mechanical principles of arts and manufactures, and he also hoped to be read by the intelligent working man. To this extent the book fell within the contemporary genre of industrial–technological literature; indeed part of it had been published in 1829 as a part of the *Encyclopaedia Metropolitana*, and extracts from it were published long after his death as a pamphlet *How to invent machinery* (1899). Tracts on the steam-engine, histories of the cotton industry and industrial manuals, dictionaries and encyclopedias were all very popular. Another well-known book, Andrew Ure's *The philosophy of manufactures* (1835) – an extraordinary panegyric on the factory system and steam-powered machinery – was very much a product of this genre, but it completely lacked the analysis of Babbage's contribution. The *Economy* was much more than popular industrial observation: it was an analysis based on economic principles, especially the account of the division of labour set forth by Adam Smith in the *Wealth of nations* (1776). Adam Smith had analysed the sources of increases in labour productivity to be found in the division of labour: Babbage took this fundamental principle of economic growth and applied it to the individual firm. His obvious first-hand knowledge of a wide variety of industrial and business processes, combined with general analysis of production systems, made the work a *tour de force*. At a time of anxiety and ambiguity over the reception of new technology, he also offered authoritative policy statements on a wide range of machinery issues, including patent reform, export of machinery, crises of overproduction, and technological unemployment.

The book's intellectual position in relationship to political economy was not, however, easily apparent, and few apart from J. S. Mill and Karl Marx appreciated its significance to their discipline. The book was criticized initially for failing to confront the major policy issues of the day. And Babbage himself conceded that his first edition did not profess to examine questions of political economy. He attempted to correct this in the second edition by introducing three new chapters: 'On a new system of manufactures', 'On the effect of machinery in reducing the demand for labour' and 'On money as a medium of exchange' (this material was also issued as a separate pamphlet *On currency, on a new system of manufacturing, and on the effect of machinery on human labour* (1833)). But the general framework constructed by Babbage was foreign to the classical political economy of his time: contemporary classical economists, such as Ricardo, Malthus and others, were primarily concerned with macroeconomic issues such as value, income distribution and international trade. Going back to Smithian principles Babbage had analysed industrial organization and the microeconomics of the manufacturing firm, never losing sight of technological constraints and opportunities.

The book was also criticized for failing to give due attention to the factory system and steam-powered textile technology. But this was precisely its strength. For it analysed the factory and the workshop as parts of the more general organization of work; and it examined machinery in the context of a more general discussion of technology, including skill. Babbage made a close observation of skills and hand processes as well as machinery, and of the workshop as well as the factory. This was a more accurate perception of contemporary industrial practice than would have been a concentration solely on the outstanding and atypical phenomenon of the factory in industrial structure.

Babbage analysed what he called 'the domestic economy of the factory'. He sought to specify what arrangement of production would succeed in selling articles at a minimum price. He made a careful analysis of economies of scale in relation to the division of labour, distinguishing the dynamics of the factory from those of the workshop. He developed Smith's principle of the division of labour to a further refinement, introducing the significance of the division of skill, or the division of mental and manual labour. Following his 'Babbage Principle' was vital, he believed, to the success of any organization of work:

> that the master manufacturer by dividing the work to be executed into different processes, each requiring different degrees of skill or of force, can purchase exactly that precise quantity of both which is necessary for each ·rocess.

This analysis of workshop organization, which was the core of the book, was complemented by topical commentary on profit-sharing, technological unemployment and trade unions. An important radical departure on wages and labour was provided in the chapter 'On a new system of manufactures' which argued for a piece-rate wage system and profit-sharing as the key to overcoming the longstanding opposition of workers to machinery. This was the problem which Babbage, along with many of his contemporaries, believed to be the major brake on Britain's industrial progress. The system was a far-reaching proposal for a workers' stake in increasing productivity, and for collective decision making on hiring and firing, and the organization of the works. The lengths to which his suggestions went were probably surpassed only by radical and Owenite co-operatives.

*

Babbage had a lifelong interest in actuarial and statistical subjects. In 1824 he produced a set of life tables for 'The Protector', a projected life assurance company that did not in the event materialize. He subsequently wrote his popular book on insurance, *A comparative view of the various institutions for the assurance of lives* (1826).

Babbage's use of practical observation and statistical data, and his critique of political economy's 'closest philosophers' induced him, with Richard Jones, J. E. Drinkwater, Malthus and Quetelet, to form a Statistics Section of the British Association in 1833, followed later by the London Statistical Society (1834). His entrepreneurial efforts in participating in the formation of these were clearly more important than the papers he eventually contributed. The Societies immediately became an important area of a statistical movement which spread into the formation of many provincial statistical societies. These societies conducted the first social surveys and contributed to the empirical critique of classical economic theory, as well as collecting data and developing statistical techniques.

The British Association's Statistical Section (Section F), was confined to the presentation of statistical data, avoiding areas of political controversy. But the less restrictive London Statistical Society made its brief the connection of political economy with the statistical investigation of economic improvement. Babbage was the first President of Section F, and wrote several statistical papers. His earlier 'On the proportionate number of births of the two sexes under different circumstances' (1829) compared the demographic structure of the Kingdom of Naples, France, Prussia and Westphalia. Much later he wrote 'On the statistics of lighthouses' for the Brussels Congress of Statistics of 1853. A paper 'Analysis of the statistics of the Clearing House during the year 1839' was read to the London Statistical Society and printed in its memoirs in 1856. His last contributions to the London Statistical Society appeared in the *Proceedings of the Fourth International Statistical Congress* held in London in 1860.

REMAINING WORKS

Volumes 4 and 5, in addition to the economic and statistical works discussed above by Maxine Berg, contain Babbage's short works on scientific and technical subjects, and a number of his polemical writings.

We do not yet have the scholarship by which to judge Babbage's scientific writings, but it is fair to say that the most impressive thing about them is their breadth: although the nineteenth-century scientific mind readily crossed subject barriers, Babbage's range was wide even by the standards of his time. He published scientific papers across the spectrum of natural philosophy and science: astronomy, electrical and magnetic science, geology and physics. Yet Babbage's total output of scientific writing amounts to barely one dozen papers with not more than two or three papers in the same field, and not one of them is profound. Although Babbage was a capable experimentalist (see for example 'On electrical and magnetic rotations' (1826)) his forte was in the Baconian tradition as a master of classification. This approach is perhaps most clearly seen in 'On tables of the constants of nature and art', first published in 1832 and republished in expanded form in the 1856 *Annual Report* of the Smithsonian Institution. This analytical approach also underpinned all of Babbage's economic, statistical and actuarial work.

The *Works of Babbage* includes the several short papers that Babbage published on inventions, technology, scientific instruments and machine tools. The collection of topics covered includes a new micrometer, the diving bell, lathes, lighthouses, submarines, the occulting telegraph and gun-laying. Though impressively wide, this range is not fully representative of his interest in inventions and the mechanical arts. For example, during 1838-9 he made a major study of the operational characteristics of railways of which nothing was published (save for a brief chapter in *Passages* and a contribution to a discussion on railway accidents in the *Proceedings of the Institute of Civil Engineers* in April 1862). None of the papers that he published are of obvious significance, except perhaps for the insight they provide on Babbage himself.

Babbage took a considerable interest in the decyphering of secret writing. In 1831 a paper was published in France as 'Sur l'emploi plus ou moins fréquent des mêmes lettres dans les différentes langues' ['On the proportion of letters occurring in various languages']. Babbage also translated a letter in secret writing that appeared in Francis Bailey's *Account of the life of the Reverend John Flamsteed, the first Astronomer-Royal* (London, 1835), which we have reprinted for no other reasons than the sake of completeness and because

Babbage himself included it in his list of publications. There was, much later, a set of correspondence in *Journal of the Society of Arts* (1854–5), in which Babbage explained *inter alia* his reluctance to undertake the breaking of codes invented by amateurs. (See O. I. Franksen's *Mr Babbage's Secret* (1984) for a full discussion of Babbage's cyphering activity.)

Babbage made but one sortie into religious philosophy in *The ninth Bridgewater treatise* (1837), a teasing title referring to the eight religious treatises written under the terms of the will of the Earl of Bridgewater. It is interesting as a pre-Darwinian view of the creation, and particularly interesting for Babbage's use of his calculating engines to explain miracles.

The *Works* includes Babbage's short polemical writings, most of which he had privately printed and published. A first hint of Babbage's view of the decline of British science is given in a piece of reportage 'Account of the great congress of philosophers at Berlin on the 18th September 1828'; this was subsequently reprinted by him in the *Decline of science in England* (1830). In this latter book Babbage made a strong attack on the Royal Society, which had long ceased to be an effective instrument for the promotion of science in Britain. The final chapter of the book was published separately as *Sketch of the philosophical characters of Dr Woolaston and Sir H. Davy* (1830).

During the 1830s reform was in the air: Babbage twice stood as a Liberal candidate (1832 and 1834), and his election addresses are published here for the first time. In the pamphlet *A word to the wise* (1833) his target was the hereditary peerage, and he advocated peerages for life. His political pamphlet *Thoughts on the principles of taxation* (1848, 1851, 1852) showed a concept of moderate reform. Less liberally, in *Thoughts upon an extension of the franchise* (1865) he compared the principle of one man–one vote with the 'advance of socialism', on the grounds that the poor, being in the majority, would vote for lower taxes for themselves and higher taxes for the rich, thus destroying private enterprise.

Perhaps the most embittered of all Babbage's writing was his book on the Great Exhibition, the *Exposition of 1851*. As Babbage's notice in the *Dictionary of national biography* observed it was 'the diatribe of a disappointed man and . . . is disfigured by personal allusions, in giving utterance to which he wronged his better nature'. Babbage's pique was due to the combination of his having been excluded from the planning committee, and his Difference Engine not having been accepted as an exhibit.

The final volume of the *Works of Babbage* is his autobiography *Passages from the life of a philosopher* (1864). This is a remarkable and very readable book, affording insights both into Babbage's life and into his engines. It is, however, notably lacking in the domestic data which is the essence of a normal biography. Chapter XXVI gives an account of his vendetta against street musicians and was published as a separate pamphlet *A chapter on street nuisances* which was widely circulated. Although less ill-tempered than his earlier writings *Passages*, which was written when Babbage was in his mid-seventies, was not without a caustic overtone. This, together with the book's cynicism and the vendetta against street musicians, undoubtedly contributed to his posthumous reputation for cantankerousness.

REFERENCES

M. Berg, *The machinery question and the making of political economy 1815–1848* (Cambridge, 1980).

A. G. Bromley, 'The evolution of Babbage's calculating engines', *Annals of the History of Computing*, Vol. 9 (1987), pp. 113–36.

H. W. Buxton, *Memoirs of the life and labours of the late Charles Babbage Esq. F.R.S.*, previously unpublished manuscript, *The Charles Babbage Institute Reprint Series for the History of Computing*, Vol. 13 (Cambridge, Mass., London, and Los Angeles, 1987).

I. B. Cohen, 'Babbage and Aiken', *Annals of the History of Computing*, Vol. 10, in press.

J. M. Dubbey, *The mathematical work of Charles Babbage* (Cambridge, 1978).

O. I. Franksen, *Mr Babbage's secret: the tale of cypher – and APL* (Strandberg, 1984).

A. Hyman, *Charles Babbage: pioneer of the computer* (Oxford, 1982).

A. W. Van Sinderen, 'The printed papers of Charles Babbage', *Annals of the History of Computing*, Vol. 2 (1980), pp. 169–85.

BIBLIOGRAPHY

This bibliography lists in chronological order Babbage's articles and monographs, including those printed in the *Works of Babbage*. Because the first five volumes of the *Works* are organized by genre, and some items were reprinted by Babbage in his monographs (and have not been separately reprinted in the *Works*), volume and page numbers are provided so that any item can be quickly located.

The bibliographical data includes all English editions of each work, but not translations or reprintings containing no significant additions. In all cases of multiple editions, the edition reprinted in the *Works* is the latest. The translation and reprinting history of each item is included in its editorial annotations. Items which are newly discovered or are not listed in the established bibliographies are prefixed with an asterisk. See also A. W. Van Sinderen, 'The printed papers of Charles Babbage', *Annals of the History of Computing,* Vol. 2 (1980), pp. 169–85, which includes translation and reprinting histories.

	Volume	*Pages*
1813		
with John Frederick William Herschel, 'Preface', *Memoirs of the Analytical Society*, Cambridge, pp. i–xxii.	1	37–60
'On continued products', *Memoirs of the Analytical Society*, Cambridge, pp. 1–31.	1	61–92
1815		
'An essay towards the calculus of functions', *Philosophical Transactions of the Royal Society*, Vol. 105, pp. 389–423.	1	93–123

	Volume	Pages
1816		
'An essay towards the calculus of functions, part II', *Philosophical Transactions of the Royal Society*, Vol. 106, pp. 179–256.	1	124–93
'Demonstrations of some of Dr Matthew Stewart's general theorems; to which is added, an account of some new properties of the circle', *Journal of Science and the Arts*, Vol. 1, pp. 6–24.	1	194–212
with John Frederick William Herschel and George Peacock (translators): Silvestre Francois Lacroix, *An elementary treatise on the differential and integral calculus*, J. Deighton and Sons, Cambridge.	1	213–5
1817		
'Observations on the analogy which subsists between the calculus of functions and the other branches of analysis', *Philosophical Transactions of the Royal Society*, Vol. 107, pp. 197–216.	1	216–30
'Solutions of some problems by means of the calculus of functions', *Journal of Science and the Arts*, Vol. 2, pp. 371–9.	1	231–8
'An account of Euler's method of solving a problem, relative to the move of the knight at the game of chess', *Journal of Science and the Arts*, Vol. 3, pp. 72–7.	1	239–44
'Note respecting elimination', *Journal of Science and the Arts*, Vol. 3, pp. 355–7.	1	245–7
1819		
'On some new methods of investigating the sums of several classes of infinite series', *Philosophical Transactions of the Royal Society*, Vol. 109, pp. 249–82.	1	248–78

	Volume	*Pages*
'Demonstration of a theorem relating to prime numbers', *Edinburgh Philosophical Journal*, Vol. 1, pp. 46–9.	1	279–82
1820		
'Examples of the solutions of functional equations', Part III of Charles Babbage, John Frederick William Herschel and George Peacock, *A collection of examples of the applications of the differential and integral calculus*, J. Deighton and Sons, Cambridge.	1	283–326
1821		
'An examination of some questions connected with games of chance', *Transactions of the Royal Society of Edinburgh*, Vol. 9, pp. 153–77.	1	327–43
1822		
'Observations on the notation employed in the calculus of functions', *Transactions of the Cambridge Philosophical Society*, Vol. 1, pp. 63–76.	1	344–54
with John Frederick William Herschel, 'Barometrical observations made at the Fall of the Staubbach', *Edinburgh Philosophical Journal*, Vol. 6, pp. 224–7.	4	1–4
'A note respecting the application of machinery to the calculation of astronomical tables', *Memoirs of the Astronomical Society*, Vol. 1, p. 309.	2	3–4
A letter to Sir Humphry Davy, Bart., President of the Royal Society, on the application of machinery to the purpose of calculating and printing mathematical tables, J. Booth; Baldwin, Cradock and Joy, London.	2	5–14

	Volume	Pages
*'The science of number reduced to mechanism', Buxton MSS collection, Museum of the History of Science, Oxford.	2	15–32
'Observations on the application of machinery to the computation of mathematical tables', *Memoirs of the Astronomical Society*, Vol. 1, pp. 311–4.	2	33–7
1823		
'On the theoretical principles of the machinery for calculating tables', *Edinburgh Philosophical Journal*, Vol. 8, pp. 122–8.	2	38–43
'On the application of analysis to the discovery of local theorems and porisms', *Transactions of the Royal Society of Edinburgh*, Vol. 9, pp. 337–52.	1	355–70
with Francis Maseres (editors), *Scriptores optici; or, a collection of tracts relating to optics*, Baldwin, Cradock and Joy, London.	4	5–10
1824		
'Observations on the measurement of heights by the barometer', *Edinburgh Journal of Science*, Vol. 1, pp. 85–7.	4	11–3
**Rates of The Protector Life Assurance Society*, privately printed, London.	4	14–9
1825		
with John Frederick William Herschel, 'Account of the repetition of M. Arago's experiments on the magnetism manifested by various substances during the act of rotation', *Philosophical Transactions of the Royal Society*, Vol. 115, pp. 467–96.	4	20–41

	Volume	Pages
1826		
'On a new zenith micrometer', *Memoirs of the Astronomical Society*, Vol. 2, pp. 101–3.	4	42–5
'On a method of expressing by signs the action of machinery', *Philosophical Transactions of the Royal Society*, Vol. 116, pp. 250–65.	3	209–23
'On electrical and magnetic rotations', *Philosophical Transactions of the Royal Society*, Vol. 116, pp. 494–528.	4	46–73
'On the determination of the general term of a new class of infinite series', *Transactions of the Cambridge Philosophical Society*, Vol. 2, pp. 217–25.	1	61–8
'Diving-bell', *Encyclopaedia Metropolitana*, Vol. 18, pp. 157–67.	4	74–103
A comparative view of the various institutions for the assurance of lives, J. Mawman, London.	6	
1827		
'On the influence of signs in mathematical reasoning', *Transactions of the Cambridge Philosophical Society*, Vol. 2, pp. 325–77.	1	371–408
'Notice respecting some errors common to many tables of logarithms', *Memoirs of the Astronomical Society*, Vol. 3, pp. 65–7.	2	69–71
Table of logarithms of the natural numbers, from 1 to 108,000, J. Mawman, London.	2	72–107
1829		
'Account of the great congress of philosophers at Berlin on the 18th September, 1828', *Edinburgh Journal of Science*, Vol. 10, pp. 225–34. Reprinted in *Decline of science in England*, 1830, appendix 1.	7	111–20

	Volume	Pages
'On the proportionate number of births of the two sexes under different circumstances', *Edinburgh Journal of Science*, New Series, Vol. 1, pp. 85–104.	4	104–23
'An essay on the general mechanical principles which regulate the application of machinery to manufacture and the mechanical arts', *Encyclopaedia Metropolitana*, Vol. 8, pp. 1–84. Reprinted with additions and rearrangement in *On the economy of machinery and manufactures*, 1832.		
1830		
*with Peter Mark Roget, 'Correspondence', *Philosophical Magazine*, New Series, Vol. 7, pp. 446–8, Vol. 8, pp. 72–4, 153–4.	7	125–33
'Notation', *The Edinburgh Encyclopaedia*, Vol. 15, pp. 394–9.	1	409–24
'Porisms', *The Edinburgh Encyclopaedia*, Vol. 17, pp. 106–14.	1	425–48
Sketch of the philosophical characters of Dr Wollaston and Sir H. Davy, R. Clay, London. Extracted from, and identical to, 'Conclusion' of *Decline of science in England*, 1830.		
Reflections on the decline of science in England, and on some of its causes, B. Fellowes, London.	7	
*'Mathematical questions', *Mathematical Repository*, New Series, Vol. 5, pp. 51, 63–4, 66–7, 71–2, 178–80.	1	449–57
1831		
'Sur l'emploi plus ou moins fréquent des mêmes lettres dans les différentes langues', *Correspondance mathématique et physique*, Vol. 7, pp. 135–7.	4	124–5
New English translation, 'On the proportion of letters occurring in various languages'.	4	126–7

	Volume	Pages
Specimen of logarithmic tables printed with different coloured inks on variously coloured papers, B. Fellowes, London. Preface reprinted in *Edinburgh Journal of Science*, New Series, Vol. 6 (1832), pp. 144–50.	2	115–7
Table of logarithms of the natural numbers from 1–108,000 on different coloured papers, B. Fellowes, London.	2	72–107

1832

'On the advantage of a collection of numbers, to be entitled the constants of nature and of art', *Edinburgh Journal of Science*, New Series, Vol. 6, pp. 334–40. Reprinted with additions as 'Sur les constantes de la nature – classe des mammifères', *Compte rendu des travaux du congrès général de statistique*, Brussels, 1853, pp. 222–30, and with further additions as 'On tables of the constants of nature and art', *Annual Report of the Board of Regents of the Smithsonian Institution for 1856*, U.S. Government Printing Office, Washington, D.C., 1857, pp. 289–302.	5	138–54
**To the electors of the Borough of Finsbury*, privately printed, London.	4	128–9
**Mr Babbage's address to the electors of the Borough of Finsbury*, privately printed, London.	4	130–3
On the economy of machinery and manufactures, Charles Knight, London. Second edition, 1832. Third edition, 1833. Fourth edition, 1835.	8	

1833

On currency, on a new system of manufacturing, and on the effect of machinery on human labour, Charles Knight, London. A separate printing of three chapters of the *Economy of machinery and manufactures*, second edition, 1832.

	Volume	Pages
A word to the wise, privately printed, London.	4	134–40

1834

Abstract of a paper entitled Observations on the Temple of Serapis at Pozzuoli; with remarks on certain causes which may produce geological cycles of great extent, privately printed, London.	4	141–6
**Statement addressed to the Duke of Wellington respecting the calculating engine*, unpublished manuscript, 23 December.	3	2–8

1835

'Une lettre à M. Quetelet de M. Ch. Babbage relativement à la machine à calculer', *Académie royale des sciences, des lettres et des beaux-arts de Bruxelles, Bulletins*, Vol. 2, pp. 123–6.	3	9–11
New English version, 'A letter to M. Quetelet from Charles Babbage respecting the calculating machine'.	3	12–4
'Letter decyphered by C. Babbage', in Francis Baily, *An account of the life of the Reverend John Flamsteed, the first Astronomer-Royal*, London, pp. 348–9.	4	147–50

1837

'On the mathematical powers of the calculating engine', Buxton MSS collection, Museum of the History of Science, Oxford.	3	15–61
The ninth Bridgewater treatise: a fragment, John Murray, London. Second edition, 1838.	9	

1839

Letter from Mr Babbage to the members of the British Association for the Promotion of Science, privately printed, London.	4	151–9

	Volume	Pages
1843		
'Addition to the memoir of M. Menabrea on the Analytical Engine', *Philosophical Magazine*, Vol. 23, pp. 235–9.	3	83–8
'Description of the boracic acid works of Tuscany', in *Handbook for travellers in Central Italy*, John Murray, London, pp. 178–9.	4	160–1
1847		
'The Planet Neptune and the Royal Astronomical Society's Medal', *The Times*, 15 March, p. 5.	4	162–4
Observations on the Temple of Serapis at Pozzuoli, near Naples. With a supplement. Conjectures on the physical condition of the surface of the moon, privately printed, London.	4	165–217
'Paper on the principles of tools for turning and planing metals', in Charles Holtzapffel, *Turning and mechanical manipulation*, Vol. 2, Holtzapffel, London, pp. 984–91.	5	1–13
1848		
Thoughts on the principles of taxation, with reference to a property tax, and its exceptions, John Murray, London. Second edition, 1851. Third edition, 1852.	5	31–56
1851		
Laws of mechanical notation, privately printed, London.	3	224–8
The Exposition of 1851; or, views of the industry, the science and the government of England, John Murray, London. Second edition, 1851.	10	

	Volume	Pages
1852		
'Note respecting the pink projections from the sun's disc observed during the total solar eclipse observed in 1851', *Monthly Notices of the Royal Astronomical Society*, Vol. 12, pp. 209–10.	5	14–5
Notes respecting lighthouses, privately printed, London.	5	16–30
*'Letter to Lord Derby', 8 June, privately printed. Reprinted in *Passages from the life of a philosopher*, 1864.	11	77–81
1853		
'On the statistics of lighthouses', *Compte rendu des travaux du congrès général de statistique*, Hayes, Brussels, pp. 230–7.	5	57–65
1854		
With John H. B. Thwaites, 'Correspondence on cypher writing', *Journal of the Society of Arts*, No. 90, pp. 663–4, No. 93, pp. 707–8, No. 95, pp. 732–3, No. 98, pp. 776–7.	5	66–78
*'Letter to the biographer', in William Charles Henry, *Memoirs of the life and scientific researches of John Dalton*, London, pp. 185–9. Reprinted in *Passages from the life of a philosopher*, 1864.	11	219–22
1855		
'Submarine navigation', *The Illustrated London News*, 23 June, pp. 623–4.	5	79–84
'On the possible use of the occulting telegraph at Sebastopol', *The Times*, 16 July, p. 6.	5	85–7
'A method of laying the guns of a battery without exposing the men to the shot of the enemy', *The Illustrated London News*, 18 August, p. 210.	5	88–90

	Volume	Pages
'Note sur la machine Suédoise de MM. Scheutz pour calculer les tables mathématiques par la méthode des différences, et en inprimer les résultats sur des planches stéréotypes', *Académie des sciences, Paris, Comptes-rendus hebdomadaires des séances*, Vol. 41, pp. 557–60.	3	229–32
New English translation 'Note on the Swedish machine of Messrs. Scheutz to calculate mathematical tables by the method of differences, and to print the results on stereotype plates'.	3	233–6
'Cypher writing', *Journal of the Society of Arts*, No. 159, pp. 40–1.	5	91–2
1856		
'Analysis of the statistics of the Clearing House during the year 1839', *Journal of the Statistical Society*, Vol. 19, pp. 28–48. Reprinted with an appendix, John Murray, London, 1856.	5	93–132
'On the action of ocean-currents in the formation of the strata of the earth', *Quarterly Journal of the Geological Society*, Vol. 12, pp. 366–8.	5	133–6
Observations addressed, at the last anniversary, to the President and Fellows of the Royal Society, after the delivery of the medals, John Murray, London.	2	187–93
1857		
'Table of the relative frequency of occurrence of the causes of breaking of plate-glass windows', *Mechanics' Magazine*, Vol. 66, p. 82.	5	137
1860		
'Observations on the discovery in various localities of the remains of human art mixed with the bones of extinct races of animals', *Proceedings of the Royal Society*, Vol. 10, pp. 59–72.	5	155–68

	Volume	Pages
'Contributions to the discussions', *Proceedings of the Fourth International Statistical Congress*, London, pp. 380, 381, 383, 394.	5	169–75
'Letter to Dr Farr, on the origin of the International Statistical Congresses', *Proceedings of the Fourth International Statistical Congress*, London, pp. 505–7.	5	176–80

1864

A chapter on street nuisances, John Murray, London. A separate printing of chapter XXVI of *Passages from the life of a philosopher*, 1864. Second and third editions, 1864.		
Passages from the life of a philosopher, Longman, Green, Longman, Roberts and Green, London.	11	

1865

Thoughts upon an extension of the franchise, Longman, Green, Longman, Roberts and Green, London.	5	181–6

1868

'Observations on the parallel roads of Glen Roy', *Quarterly Journal of the Geological Society*, Vol. 24, pp. 273–7.	5	187–92

WORKS NOT WRITTEN BY BABBAGE INCLUDED IN THE WORKS OF BABBAGE

Francis Baily, 'On Mr Babbage's new machine for calculating and printing mathematical and astronomical tables', *Astronomische Nachrichten*, No. 46 (1823), col. 409–22.	2	44–56

	Volume	Pages
Henry Thomas Colebrook, 'Address on presenting the gold medal of the Astronomical Society to Charles Babbage', *Memoirs of the Astronomical Society*, Vol. 1 (1825), pp. 509–12.	2	57–60
John Frederick William Herschel, *Report of the Royal Society Babbage Engine Committee*, Clowes, London, 1829.	2	108–14
Dionysus Lardner, 'Babbage's calculating engine', *Edinburgh Review*, Vol. 59 (1834), pp. 263–327.	2	118–86
Luigi Frederico Menabrea, 'Notions sur la machine analytique de M. Charles Babbage', *Bibliothèque universelle de Genève*, Vol. 41 (1842), pp. 352–76.	3	62–82
Augusta Ada Lovelace, 'Sketch of the Analytical Engine', *Scientific Memoirs*, Vol. 3 (1843), pp. 666–731.	3	89–170
Sir Harry Nicholas, *Statement of the circumstances respecting Mr Babbage's calculating engines*, London, 1843. Reprinted in *Passages from the life of a philosopher*, 1864.	11	53–74
Charles R. Weld, *The eleventh chapter of the history of the Royal Society*, with an annexed review by Augustus De Morgan, 1849.	10	149–74
Luigi Frederico Menabrea, 'Letter to the editor of *Cosmos*', *Cosmos*, Vol. 6 (1855), pp. 421–2.	3	171–2
New English translation, 'Letter to the editor of *Cosmos*'.	3	173–4
Anon., *Specimens of tables, calculated, stereomoulded, and printed by machinery*, Longman, Brown, Green, Longmans and Roberts, London, 1857. Extracts.	2	193–221
Judex Juris, *On the calculation and printing of mathematical tables by machinery; the inventor and his treatment*, privately printed, London, 1861.	1	1–36

	Volume	*Pages*
Benjamin Herschel Babbage, *Babbage's calculating machine; or Difference Engine*, South Kensington Museum, London, 1872.	2	222–33
C. W. Merrifield, 'Report of the committee . . . appointed to consider the advisability and to estimate the expense of constructing Mr Babbage's analytical machine, and of printing tables by its means', in *Report of the British Association for the Advancement of Science, Dublin, August 1878*, John Murray, London, 1879, pp. 92–102.	3	175–89
Henry Prevost Babbage, 'On the mechanical arrangements of the Analytical Engine', in H. P. Babbage (editor), *Babbage's calculating engines*, E. and F. N. Spon, London, 1889, pp. 331–7.	3	190–204

EDITORIAL NOTE

The *Works of Babbage* comprises about one hundred articles and monographs which were issued by many different publishing houses, each of which imposed its own house-style, to a greater or lesser extent. To provide a 'uniform' edition of the *Works of Babbage* the individual items have been lightly edited to conform to the house-style of Pickering & Chatto. These changes mainly concern the italicization and capitalization of headings, the form of paragraph indentation, the layout of lists, spelling and punctuation.

Spelling and punctuation have been modernized, and some abbreviations have been extended for uniformity and readability. Arbitrary capitalization has been dropped where it serves no purpose and dates have been put in a standard form. Typographical errors and misprints have been silently corrected, although all substantive corrections have been made either by a footnote or by an editorial interpolation in angle brackets. (Angle brackets have been used instead of square brackets because Babbage himself used the latter – *all* matched angle brackets are editorial additions.) Quotations from other works have not been edited. Where Babbage quoted authors writing in French or Latin, translations have been provided – from a standard translation if one exists, or newly prepared for the *Works*. Mathematical tables have generally been reproduced in facsimile in order to avoid the possibility of introducing typesetting errors. (This is a decision of which Babbage would no doubt have approved, since the need to eliminate typesetting errors in printing mathematical tables was a major aspect of the design of the Difference Engine.)

Cross-references, where given as a page number, have been altered to conform to the pagination of the *Works*. To assist the scholar who wishes to follow up a reference in the secondary literature, however, the original pagination is included at the top of each page, and in the text the break between pages is indicated by a solidus (/).

Babbage, not unusually for the period, was casual and sometimes inaccurate in giving references. All significant references given by

Babbage have been checked for correctness against standard bibliographies and catalogues, or in some cases by going back to the original source. Where Babbage's citation is in some way faulty or incomplete we have not corrected Babbage's citation, but have instead provided either a footnote or an editorial interpolation in angle brackets giving fuller bibliographical details. In addition, there is a separate bibliography of Babbage's sources in Volume 11.

MATHEMATICAL EDITING

The editing of the mathematics in the *Works of Babbage* is much more extensive than in the ordinary text. The original mathematical typesetting in Babbage's printed works was unacceptable by today's standards – both in terms of typography and accuracy. In resetting Babbage's mathematics the aim has been to provide an attractive and consistent typography, together with an accurate interpretation for the scholar.

Probably the most unattractive and confusing aspect of the mathematical typography of Babbage's day was that the solidus was not used for division. In running text this notation led to uneven line spacing and in crudely set equations the intention was sometimes unclear. We have therefore 'broken down' formulae in running text with a solidus, and have used solidii in displayed equations where this renders the meaning clearer. Occasionally it has been necessary to introduce parentheses, but such changes have not been highlighted since the fact that Babbage never used a solidus is sufficient warning for the reader. We have substituted parentheses where Babbage used the vinculum to group terms, as this is now archaic and sometimes it is unclear which terms are being embraced. We have standardized on an elipsis of three points (. . .) where Babbage used a variety of forms, and a centred dot (·) has been used for multiplication, except where Babbage used a cross (×) for emphasis. We have dropped the archaic period after functions (log. x, sin. x, etc.), which Babbage observed 'is not merely useless, but interferes with another principle'.[a] Apart from these editorial changes, we have tried to

[a] 'Notation' (1830), in this volume of the *Works of Babbage*, p. 422.

avoid intrusive modernization. For example, Babbage did not observe today's conventions on the priority of operators, so that he would have written $n \cdot n+1 \cdot n+2 \cdot \ldots \cdot n+p$ where we would write $n(n+1)(n+2) \ldots (n+p)$. Although this notation is mildly ambiguous, we think the intention is plain and have not changed it. We have also retained Babbage's use of ε for the base of Napierian logarithms.

Babbage had definite views on the use of roman and italic founts in setting mathematics, and stated: 'all letters that denote quantity should be printed in italics, but all those which indicate operations, should be printed in a roman character'.[a] It must, however, be said that he – or perhaps his typesetter – broke this rule far more often than he observed it. Because it is not possible to interpret Babbage's intention reliably, we have used italics throughout. Surprisingly, Babbage had nothing to say on the choice of typesize: although small type was used for subscripts and indices, it was also often used simply as a convenience to squeeze formulae into a cramped space. In the *Works of Babbage*, we have used type sizes only according to modern conventions.

In editing the papers for the press, we found very many trivial typographical errors – such as a missing parenthesis or symbol – which were clearly printers' errors, or the result of the casual way Babbage prepared his manuscripts and saw them through the press. Rather than draw attention to corrections (which would merely compound the confusion) trivial errors have simply been silently put right. All substantive corrections have been indicated by footnotes.

[a] *Ibid.*, p. 419.

ACKNOWLEDGEMENTS

The editors and publishers are grateful to the following who supplied copies of papers or other material reprinted in the *Works of Babbage*: the Crawford Library of the Royal Observatory, Edinburgh; the library of Waseda University, Tokyo; Miss Elizabeth Buxton and the library of the Museum of the History of Science, Oxford; the Syndics of Cambridge University Library; the British Library; the Royal Society; A. W. Van Sinderen and the *Annals of the History of Computing*.

The following people, in addition to the consulting editors, gave significant advice on the selection of material in the *Works of Babbage*: Brian Randell, Garry J. Tee, O. I. Franksen. and Alfred W. Van Sinderen.

ON THE

CALCULATION AND PRINTING

OF

MATHEMATICAL TABLES

BY

MACHINERY;

THE INVENTOR AND HIS TREATMENT.

BY

JUDEX JURIS.

——— "LONGA EST INJURIA, LONGÆ AMBAGES."———

Virg. Æn. 1, 345.

EDITOR'S INTRODUCTION

Charles Babbage was the subject of a number of biographical notes and obituaries, but there were no full length biographies until recent times. Of the biographical articles written in the last century, Alice M. Cooke's entry in the *Dictionary of national biography* is undoubtedly the best known and most authoritative. The little-known biographical pamphlet reprinted here was written in Babbage's lifetime and, while lacking the balance of Cooke's more scholarly biographical note, it had a lively and provocative tone that evidently did not displease its subject.

The pamphlet was written by James Jerwood (1797?–1877), who held various positions in the English legal system, and it was published under the pseudonym 'Judex Juris'. Jerwood was an amateur scientist in the 19th century tradition: he was a Fellow of the Astronomical Society, a contributor to the *Mathematical Repository*, and the author of *A lecture on the new planet Neptune* (London, 1848).

Jerwood originally wrote the pamphlet intending it for publication and submitted it to two quarterly reviews (titles unknown). It was in no sense an unbiased or objective account of Babbage's activities, and both journals rejected it – one because it hit at the reviewer's friends, and the other because some of the topics raised had already been discussed at length in previous articles. On 2 February 1861, Jerwood wrote to Babbage[a] drawing attention to the article, though not at that time claiming to be the author, suggesting it would be a good pamphlet to bring out for the forthcoming 1862 International Exhibition. There followed a long exchange of letters between Babbage and Jerwood. Babbage's end of the correspondence has not survived, but in a letter dated 24 April 1862 Jerwood noted of one of Babbage's replies 'that some of its contents convince me that parts of the Pamphlet must please you'.[b] In the same letter, Jerwood proposed that he put the pamphlet 'in the hands of a *cheap printer* and let them be dispersed in *thousands* in order that your claims, and Treatment may be known to the multitude of Englishmen and Foreigners'. This was evidently never done for the pamphlet remains very rare, possibly as few as three copies being extant.

[a] British Library, Add. MSS 37198, f. 147.
[b] British Library, Add. MSS 37198, f. 310.

ON THE CALCULATION AND PRINTING OF MATHEMATICAL TABLES BY MACHINERY

JUDEX JURIS
(London, 1861)

What achievement of modern science is entitled to be placed at the head of the first rank? The question is difficult to answer justly and satisfactorily: nor could a philosopher who flourished a century ago, when slow coaches, and slower waggons, were the most expeditious kind of transit for all purposes, very easily solve it, were he again to pay us a visit in order to review all that has been effected in the interval between his time and this. Since his era, we have adopted a mode of conveyance that almost annihilates space: we employ the lightning's aid to indicate our wants and wishes to distant friends – the fathomless oceans that separate us forming no impediment to the process: we employ the sun to take the exact likeness of our relatives, and to imprint the lineaments of nearest friends with unerring correctness. Scientific publications teem with marvels of this nature – each of which would be wholly incomprehensible to our imaginary visitant. Every branch of science has its wonders, and in numerous cases, we are at a loss which most to admire, the heads that devised them, or the hands which so nicely carried them into / operation. One rare instance of this mental devising and manual contrivance and performance has recently been brought to our notice, which, in our opinion, justly claims pre-eminence for the unique combination of abstract science and mechanical devices that produced it: we mean 'SPECIMENS OF TABLES CALCULATED, STEREOMOULDED, AND PRINTED BY MACHINERY,' a singular book which may, we believe, be purchased at the Messrs. Longmans'.[a] The tables are specimens of logarithmic numbers, and of various other astronomical and mathematical tables. The machine that has calculated and printed these tables was designed and constructed by Messrs George and Edward Scheutz, father and son, of Stockholm.

We are quite aware that only persons who have devoted much time to the mental drudgery of constructing logarithmic, mathematical, and astronomical tables, can form a proper estimate of the intellectual labour and care which are

[a] Anon., *Specimens of tables calculated, stereomoulded and printed by machinery* (1857), reprinted in *Works of Babbage*, Vol. 2.

required to render such tables accurately trustworthy; but we are quite certain that every one who is conversant with the mode and labour of constructing such sets of numbers will be astonished to see tables of the description, not only *calculated*, but also *printed* by MACHINERY. Not only has a kind of intellectual power been imparted to wheels, pullies, axles, etc., which enables them to calculate numbers not easily found by expert mathematicians, but the machine does more than even *Napier* or *Briggs* could effect, for it *prints* the tables almost simultaneously with calculating them.

Hitherto the formation of scientific tables with the necessary accuracy to facilitate the labours of astronomers, navigators, engineers, and many other votaries of science, has been deemed the most rugged and mentally exhausting application of abstract science. Happily now by an easy adjusting of the calculating apparatus, and by the mere turning of a winch, many of these tables may be calculated with unerring accuracy, and at the same time be neatly printed with equal precision. / Men noted for their skill and accomplishments in forming such tables, may, perhaps, feel chagrined, that their brain-work which made them celebrated can now be performed by machinery. However, they may learn from the book mentioned above, that the machine calculates and stereomoulds without chance of error two and a half pages of figures in as little time as a skilful compositor would occupy to set up the type for a single page! In printing scientific tables the difficulty in preventing press errors, is very great, the machine in stereomoulding *cannot* make a mistake.

The introduction of the unique volume contains an historical sketch of the inventors' labours, difficulties, and final success, and it leads to the conclusion, that they richly deserve the honours which have ultimately rewarded their unflinching perseverance and surprising ingenuity. The machine was for some time exhibited in the apartments of the Royal Society, in Somerset House; and its mechanism and capabilities were explained by Mr Gravatt, the well-known scientific engineer, who has taken a warm interest in the invention, and has furnished the inventors with many mathematical formulae highly useful in explaining and working the machine.

The King of Sweden has zealously interested himself in the astonishing piece of mechanism, and proposed to the Diet to give the inventors a national recompense, which was conferred on them in the most flattering manner: they have also been rewarded by the gold medal, at the grand Exhibition, at Paris. It is truly gratifying to learn that the inventors have been supported and duly honoured by their countrymen – their sovereign right royally taking the lead. When the singular and gigantic undertaking which the inventors have begun and completed is contemplated, it must afford sincere pleasure to every friend of science to know that their genius and labour have been so properly appreciated, and so satisfactorily rewarded./

The ingenious projectors of the machine in their unpretending historical sketch of its origin and completion, again and again gratefully state a fact, which deeply concerns Englishmen, namely, that they are wholly indebted for their astonishing achievement to an article on the merits and construction of Mr Babbage's Difference Engine, published in the *Edinburgh Review*, for July, 1834.[a] The inventors, therefore, as properly as gracefully, have dedicated their book to Mr Babbage as his sincere admirers. The Messrs. Scheutz in discussing the supposed effects of their published specimens, thus advert to Mr Babbage: 'They (the authors) believe also that if this limited collection of specimens should be deemed not sufficient to prove the practical usefulness of the invention, it will soon be followed by other proofs of greater importance, printed by the astronomers of the Western World; and that the INDIVIDUAL, to whom they have with gratitude and regard dedicated this little work, will then be further known for what he truly is, namely, one of the BENEFACTORS OF MANKIND, and one amongst the NOBLEST and MOST INGENIOUS of the sons of England.'

Whilst, however, every one must admire the dauntless ingenuity of the Swedes, who, from a verbal description of so complicated a piece of mechanism, which had been designed and nearly completed, could construct a similar one, we Englishmen ought to bear in mind, that, after all, the machine so replete with powers which are nearly akin to intellectual ones is the UNQUESTIONABLE INVENTION of our countryman.

In perusing the simple and interesting narration of the difficulties that the Messrs Scheutz had to surmount, and the encouraging aid which they received in their arduous undertaking, one must feel unfeigned respect for their sovereign who so munificently interfered with the Diet on their behalf – for their legislature, that so bountifully assisted them, and for the noble FIFTEEN of their countrymen who with such splendid / disinterestedness made themselves responsible for a pecuniary consideration, which at that juncture was likely to render the whole scheme abortive. The lovers and promoters of science, no matter of what country, must entertain feelings of esteem and honour for the true patrons and generous friends of the Messrs Scheutz: it is truly gratifying to find that at last the enterprize has been completed, and that the genius, perseverance, and anxious labours of the constructors, have been crowned with merited success, and been rewarded with really deserved honour. But their honour, and their success, naturally suggest the inquiry, what has England done for Mr Babbage to whom the invention of the marvellous piece of mechanism belongs? The Messrs Scheutz have been patronized by their sovereign, and assisted by the legislature of their country, and in consequence of this, they have been honoured by other nations, and rewarded by other countries; but what, we ask again, has England done for Mr Babbage to whom

[a] D. Lardner, 'Babbage's calculating engine' (1834), reprinted in *Works of Babbage*, Vol. 2.

the ingenious Swedes are indebted for the suggestion – the very germ of their mechanism? We regret to be compelled to say – *nothing*. We are almost ashamed to place the withering ingratitude and the blighting neglect with which some of the ruling cliques in England have treated and harassed Mr Babbage side by side, with the truly noble manner in which the Swiss ⟨*sic*⟩ governing powers treated their meritorious countrymen.

Mr Babbage is one of the most distinguished men in Europe, in whatever country science is cultivated, and intellectual achievements honoured, he is held in high repute. Moreover, England is more deeply indebted to Mr Babbage for the eminent position which she now holds in mathematical and abstract science than she is to any other countryman now in existence however illustrious he may be. We undertake to prove this allegation, and to substantiate the assertion, that Mr Babbage has been ill-treated almost in proportion to his / splendid talents, and to the distinguished services which he has rendered to his country.

To accomplish this object, it will be necessary, in the first place to advert to the state of mathematical science in England in the early part of the present century, in order to contrast it with its present position, and to do proper justice to its cultivators who have brought it from the one state to the other. During a long period, especially whilst the continental war was raging, England shut out as it were the splendid discoveries, and improvements that were effected by Laplace and many other illustrious mathematicians. We were enamoured with the magnificent discoveries of our immortal Newton, and adhered with pride and prejudice to his geometrical mode of elucidating matters – with a kind of national partiality amounting almost to bigotry, we clung to the fluxional method which he had invented, and notwithstanding the accommodating notation and comprehensive calculus which the continental investigators had invented and applied with so much success to a great variety of interesting topics wholly unknown to us, English mathematicians, as a body, could not be induced, or enticed, to adopt the improvements – perhaps because they were continental, and not thoroughly home-bred: at any rate, they were not allowed to get a footing even at the University of Cambridge – then, and always, the national nursery of British mathematical science. The consequence of this isolating prejudice, was, that in the commencement of the present century, England was far behind many European nations in abstract science. In a note to Sir John Herschel's treatise on sound in the *Encyclopaedia Metopolitana* the celebrated author, having compared the *Annales de Chemie* and the Journals of *Poggendorf* and *Schweigger* with the meagre and unsatisfactory scientific publications of our own country, says 'Here whole branches of continental discovery are unstudied, and, indeed, almost unknown by name. It is in vain to conceal the / melancholy truth. We are fast dropping behind. In

MATHEMATICS we have long since drawn the rein and given over a hopeless race.'[a]

Whether the distinguished writer – feeling the degraded state of the national science, did not somewhat over-colour his picture by way of exciting his countryment to action may be questionable, but there can be no doubt that in the beginning of this century, mathematical science in England had not reached the eminence which it had attained in France and Germany.

It is generally admitted that at this time the stigma has been removed from the national character, and that in the present era, we have regained the credit, which, at the period mentioned we had lost in the hopeless race, that Sir John Herschel so feelingly lamented: in proof of this, we would have our mathematical publications in treatises and memoirs compared with the similar works of other countries – we believe England would not lose by the comparison. How then has the honour been won, and to whom are we indebted for the achievement? Who were the pioneers in the redemption of the mathematical sciences in England? An historical sketch that attempts to put this point in its true light would be neither correct nor just, if it omitted to mention with deserved respect the works of the late Professor Woodhouse. That useful mathematician, in his various publications endeavoured to show the advantages of the continental notation and methods – and his labours were rewarded with the abuse of the scientific bigots of the day, for his un-English partialities, and for his bias to *Gallic* modes of investigation. But, perhaps, the principal initiatory step that was taken towards renovating mathematical science in England was the translation of Lacroix's Differential and integral calculus which speedily led to the adoption of that notation instead of the fluxional, at Cambridge, and by English authors generally. New life seemed, at once to be breathed into our torpid science: from this source an impulse was imparted / to it, which has since progressively carried it to the lofty position that it at present occupies.

Now to whom is the country mainly indebted for thus laying the foundation of that eminence? Unquestionably, to a large extent to Mr Babbage. He, Sir John Herschel, and Dr Peacock, the late Dean of Ely – three kindred spirits fortunately happened to be students at Cambridge at the same time, and they jointly translated Lacroix;[b] and to them in due proportion belongs the merit and the national gratitude for introducing that new light upon English mathematics, and for inspiring it with such a source of health and vigour.

[a] See *Decline of science in England* (1830), *Works of Babbage*, Vol. 7, pp. viii–ix.

[b] C. Babbage, J. F. W. Herschel and G. Peacock (translators): S. F. Lacroix, *An elementary treatise on the differential and integral calculus* (1816), extracts reprinted in this volume.

Besides the translation of the work, notes were added by each translator, exceedingly well adapted to bring English students acquainted with modes and methods hitherto unknown to most of them. We shall again refer to Dr Peacock and Sir John Herschel: at present we are only concerned with Mr Babbage, and we are quite of opinion that, for the meritorious service, just mentioned performed at such a juncture, he is entitled to the gratitude of every Englishman who has any regard for the scientific credit of his country. But this initiatory work – translating Lacroix, was done *upwards of forty years ago* – 'what,' some readers may ask, 'has Mr Babbage since done to assist in the progress of mathematical science so as to deserve the enconium which you have passed upon him?' We can undertake to answer the enquiry and with entire confidence assert that very few have laboured more energetically or more successfully than he has done down to the present time. Any trustworthy catalogue of scientific publications exhibits a list of Mr Babbage's contributions to science, amounting to about SIXTY. Many of them upon topics, at the time little known, if at all, to merely English students: all of them hearing the impress of a master mind in their comprehensiveness and completeness, and containing valuable additions to the common stock of human knowledge. The *Mathematical Repository*, the *Edinburgh Transactions*, / the *Cambridge Philosophical Transactions*, and the *Transactions of the Royal Society* contain a variety of tracts tending to illustrate and show the powers of the continental methods, then, first gaining a footing, as it were, in this country; amongst a large number of valuable memoirs, may be mentioned Mr Babbage's Essays towards the '*Calculus of Functions*,' published in the *Philosophical Transactions* for 1815 and 1816.[a] These articles formed a complete treatise on that comprehensive subject which at the time to most English mathematicians was entirely new. However a knowledge of that branch of analysis is absolutely requisite as a preparation for unravelling Laplace's celebrated coefficients. It was the only key that could successfully be employed in unlocking the mysteries of Lagrange, and for opening the way to the truly magnificent performances of the continental mathematicians. We venture to assert that these essays, and their subsequent publication in Sir John Herschel's examples, have had a paramount influence in renovating and advancing mathematical science in this country. To them we are especially indebted for the important works of Murphy, O'Brien, Kelland, and a host of others, of whom England may well be proud. Sir John Herschel, then in the vigour of manhood, and laudably zealous for the scientific credit of his country, which as we have seen, he considered to be in no flattering state, it is just to remark, also worked incessantly, in order to show his countrymen the comprehensive powers of methods and systems which

[a] 'An essay towards the calculus of functions' (1815; part II, 1816), reprinted in this volume of the *Works of Babbage*.

were generally unknown in this country, and which, he and his distinguished coadjutors, were so vigorously exerting themselves to introduce. Many readers will probably recollect his set of problems published and solved in the *Mathematical Repository* – particularly that on building a bridge from one planet to another – these, and Mr Babbage's equalling interesting ones in the same publication, were not only striking, novel, and ingenious – but they were valuable as examples to / show the application of the modern calculus then coming into use, in unravelling the most intricate problems, and in overcoming difficulties hitherto insurmountable.[a] We are quite willing and anxious to do ample justice to Mr Babbage's two celebrated coadjutors – the treatment which each has received will be discussed hereafter.

Very likely the intellectual exercise in functional analysis and its application to so great a variety of matters first suggested to Mr Babbage, the notion of performing the requisite calculations MECHANICALLY. To an inventive mind the mental process of the calculus might suggest the necessary mechanism for effecting the same end. But, however, this may be; our remarks on the treatment of the inventor would hardly be intelligible unless we gave a brief sketch of the history of the calculating machine from its commencement up to the present time.

Many years ago, the government of France, for their own purposes had mathematical tables of various kinds computed under the superintendence of the celebrated M. Prony, they filled *seventeen large folio volumes.* Some members of the Board of Longitude in England perceived the value of these tables, as a national object, and overtures were made to the Board of Longitude in France, to print an abridgement of them, at the joint expense of both countries. The proposal was declined. The manual and mental labour of calculating these tables had been immense; one table only that of logarithms contained more than *eight million* figures! These tables however followed a certain law well known to mathematicians, and with which Mr Babbage's investigations above adverted to, had just then, made him peculiarly familiar. On the refusal of the French Board of Longitude it occurred to Mr Babbage that the greatest part of the manual calculation of these sets of numbers might be effected by machinery: he set to work on the subject, and the scheme of his Difference Engine was the / result. However, well stored Mr Babbage's mind might have been with mathematical lore, and however fertile his invention might have been in its application – still, no reader we apprehend, can easily conceive the bewildering difficulty of devising machinery capable of performing such intellectual operations, but such was the task which Mr Babbage undertook – and in due time produced his scheme for completing it. But perhaps the enquiry may arise, was Mr Babbage's projected machine an imaginary con-

[a] 'Mathematical questions' (1830), reprinted in this volume of the *Works of Babbage*.

trivance, appearing it may be, exceedingly satisfactory in his description – but did it bear the test of an examination of men qualified to form an opinion upon its capability? Was it, in fact, a piece of curious machinery in embryo, the utility of which existed only in the contriver's imagination – or was it the plan of a practical reality that has since undergone the *Experimentum Crucis* of actually calculating numbers in accordance with the inventor's pretensions? We shall give satisfactory proof that the machine, even in an unfinished state, has entirely fulfilled its projector's representations, with regard to its *capability* of calculating numbers, and also *printing* them at the same time with a degree of correctness which is quite unattainable by the old methods.

Mr Babbage having completed his plans, had grounds for stating that his Difference Engine in its finished state would be capable of performing the labour of *eighty-eight* calculators, in addition to its unerring accuracy both in *calculating* and *printing* – so that if the English Board of Longitude availed themselves of Mr Babbage's piece of mechanism, they might have the same tables calculated and printed by means of TWELVE or even a less number of calculators, as mere superintendents, instead of *ninety-six*, which the French Board had employed.

The general importance of forming and publishing correct mathematical tables having been abundantly proved by what / had been done in France, Mr Babbage's schemes for effecting the same object were brought to the notice of government, and in consequence, at their request, and under the advice of the Royal Society, a committee was appointed comprising the most distinguished scientific men in England, to examine the matter and to give their opinion upon the subject. It may be mentioned that the committee, by their attainments, were fully enabled to appreciate the public value and importance of the scheme, as well as to comprehend the entire feasibility of its being carried into effect. The report of the Committee was unanimously in favour of the plan. Hence the British government, advised by the Royal Society, and guaranteed by a committee of the most eminent mathematicians and practical engineers of the day, resolved on having the proposed Difference Engine constructed, at the *national expense* and for NATIONAL PURPOSES.

In pursuance of this resolution, thus carefully formed, and at the special request of the government, Mr Babbage undertook to superintend the construction of the machinery, to find plans, etc. *all gratuitously*, the government supplying raw material, and defraying the expenses of the engineer and labourers. Under these stipulations and conditions the work was commenced in 1821, and was continued for several years; the undertaking was examined at intervals by committees of scientific men of the highest class; one of these committees reported on the *first of May*, 1823,[a] 'that Mr Babbage has displayed

[a] *Parliamentary papers*, 1823, Vol. 15, pp. 9–14.

great talents and ingenuity in the construction of his machine for computation, which the committee think fully adequate to the attainment of the objects proposed by the inventor, and they consider Mr Babbage as highly deserving of public encouragement in his public undertaking'.

Another committee, similarly composed of the most distinguished scientific men in England, in 1828, having minutely examined the drawings, tools, and the parts of the engine then / executed, admitted its public utility, and the immense advantages of accurate numerical tables in all matters of calculation which it is professedly the object of the engine to calculate and print with perfect accuracy, *report* that they 'have no hesitation in giving it as their opinion that the engine is likely to fulfil all the expectations entertained of it by its inventor'.[a]

Thus, for a lengthened period was the singular enterprize carried on with every success. At last, however, for some inexplicable reason, the government began to falter in supplying the necessary funds: Mr Babbage was frequently under the necessity of furnishing them from his own resources in order that the work might proceed. Finally, about fifteen years after the commencement of the stupendous undertaking when the engine had given most convincing proofs of its capabilities, and leaving no doubt whatever that when completed, it would not only fulfil all that its inventor had proposed, but a vast deal more – the enterprize was abandoned by the government of the day. It has been remarked, that for some time before the government relinquished the work and broke their contract made with Mr Babbage at the commencement of the business, there had been much annoyance arising from their want of punctuality in supplying the funds, so that in addition to intellectual toil of the severest kind, and the rigid employment of his time and talents for *nothing*, he had to tax heavily his pecuniary resources to keep the work in motion. Here, the fact should be borne in mind, that upon the highest authority, the government had been induced to believe that the invention would be highly beneficial to the general interests of the community, and that under such belief they had directed Mr Babbage to take the necessary steps for carrying his invention into operation – this was done for a *national purpose*.

To account for the length of time occupied, it should not be forgotten, that in prosecuting the work every step that Mr / Babbage had to take was entirely an untrodden one. The mechanical contrivances necessary to effect the requisite movements were altogether novel: the inventor had to devise not only the form of the mechanic power necessary to fulfil the required object, but he had also to invent the *tools* for shaping the peculiar parts of the mechanism: moreover, the instruments that aided to overcome one difficulty could not be applied to the next which was one of a totally different description: any one accustomed

[a] J. F. W. Herschel, *Report of the Royal Society Babbage Engine Committee* (1829), reprinted in *Works of Babbage*, Vol. 2.

to deal with such involved matters must be fully aware how much time must often be occupied in *merely experimentalizing* in the formation of tools for shaping peculiar axles, cogs, etc. requiring so much nicety before one having the requisite exactness could be produced. In this maze of mechanical difficulties Mr Babbage could derive no assistance from any quarter: he had to depend exclusively upon the resources of his own invention, not only for mechanical contrivances of which nothing of the kind had hitherto been effected, but he had also at every step to invent *new tools* for forming the required mechanism.

It is difficult to imagine a more rigid piece of brain-work: the very nature of the scheme suggests that its progress towards completion must necessarily have been exceedingly slow. Mr Babbage with a courage which was truly British, and with a perseverance which no mortifying impediments could daunt, continued to work incessantly at this intense toil for the greater part of twenty years: he had not only no pecuniary emolument to soften his labours, but during the period, he expended a very heavy sum from his own resources on the enterprize – the great labour of his life, after this long term of exhausting labour and expensive toil – the engine was so far completed as to show its wonderful capabilities. The following is an account of the engine and of some of its capacities by the late Francis Baily, Esq., extracted from Professor Schumacher's '*Astronomische Nachrichten*,' No. 46.[a]

The Board of Longitude / employed Dr Hutton to calculate square numbers as far as 25,400. In computing a table of this kind by the machine, even if extended to the most remote point that could be desired, *the whole of the mental labour would be saved*; and when the numbers 1, 1, 2, are once placed in it, it will continue to produce all the square numbers in succession without interruption. This is, in fact, one of the tables which the engine is capable of computing as far as its limited number of wheels will permit. In computing tables of cubes and higher powers square roots, and cube roots; tables of the reciprocals of numbers; tables of natural sines, cosines, tangents, etc.

This eminently practical mathematician was of opinion that the labour would be reduced to the 2,000th part of the whole. In forming hyperbolic logarithms he thought the 200th part of the labour would be saved by the engine. This account of the performances of the machine is enough to make English men proud of the inventor. But the engine does more. On page 198 of the *Economy of Machinery*,[b] in a note, the author says,

Since the publication of the second edition of this work, one portion of the engine, which I have been constructing for some years has been put together. It calculates in three columns a table with its first and second differences. Each column can be expressed

[a] F. Baily, 'On Mr Babbage's new machine for calculating and printing mathematical and astronomical tables' (1824), reprinted in *Works of Babbage*, Vol. 2.

[b] *On the economy of machinery and manufactures*, 4th edition (1835), *Works of Babbage*, Vol. 8.

as far as five figures, so that these fifteen figures constitute about one-ninth part of the larger engine. The ease and precision with which it works, leave no room to doubt its success, in the more extensive form. Besides tables of squares, cubes, and portions of logarithimic tables, it possesses the power of calculating certain series whose differences are not constant; and it has already tabulated parts of series, formed from the following equations:

$$\Delta^3 u_x = \text{units figure of } \Delta u_x$$

$$\Delta^3 u = \text{nearest whole number to } \left(\frac{1}{10{,}000}\,\Delta u_x\right)$$ /

The subjoined is one amongst the series which it has calculated:

0	3,486	42,972
0	4,991	50,532
1	6,907	58,813
14	9,295	67,826
70	12,236	77,602
230	15,741	88,202
495	19,861	99,627
916	24,597	111,928
1,504	30,010	125,116
2,340	36,131	139,272

The general term of this is,

$$u_x = \frac{x \cdot x - 1 \cdot x - 2}{1 \cdot 2 \cdot 3,} + \text{the whole number in } \frac{x}{10}$$

$$+\ 10\Delta^3 \left(\text{units figure of } \frac{x \cdot x + 1}{2}\right).$$

The reader who is accustomed to mental operations such as the solution of equations of differences require, may, perhaps, be competent to form some notion of the inventive power which was requisite to work them out, *tabulate* and *print* them by machinery: but to persons unacquainted with the higher parts of mathematics and the higher mechanical contrivances – these performances by an engine are wholly incomprehensible, and to most mathematicians who are in the habit of making such calculations, in the most expert and shortest manner they must appear to be almost incredible.

It is only by deeply considering the nature of the intellectual process, and the suitable adaptation of machinery, that the feasibility of the operation becomes apparent. The Difference Engine, even in its unfinished state, when brought to a standstill was it is believed, altogether unique; at that time, there had been no achievement of genius which could be placed in comparison with its intellectual performances. When, however, the engine had satisfied men best qualified to form a correct judgment on its capabilities, and had astonished the / scientific world by its actual performances – it was unaccountably thrust aside.

The government at first, as we have stated, vouchsafed to the invention its fostering protection, but at last after lingering under that blighting influence for many years – the marvellous contrivance became defunct, or was deemed so by its patrons: at length whether paralyzed or dead, the invention was doomed to be interred in a glass coffin. Mr Weld in his history of the Royal Society, says[a] – 'In 1843 an application was made to the government by the trustees of King's College, London, to allow the engine as it existed, to be removed to the museum of that Institution. The request was complied with, and the engine enclosed in a *glass case* now stands nearly in the centre of the museum. It is capable of calculating to five figures and to two orders of differences, and performs the work with ABSOLUTE PRECISION'). There, entombed in its glass sarcophagus, the engine remains as a striking memento of its illustrious contriver's invention – but a deep stigma on the whole nation for permitting the treatment which we are about to describe to be inflicted on their celebrated countryman.

The fact deserves iteration, that Mr Babbage, at the request of the government, and for national purposes had superintended the construction of the engine for many years, and had solely invented the parts for making it. Does the reader recollect a similar instance of intellectual and inventive powers so determinedly fixed upon one object for so lengthened a period, and under such deterring circumstances? We know nothing like it, and believe it to be without a parallel in the whole history of science. The intense stretch of mental vigour be it remembered was not sweetened by any remuneration, on the contrary, the depressing chilliness of neglect was experienced as well as the inconvenience and annoyance arising from the drains which were repeatedly made upon private pecuniary resources. The engine having been brought to that degree of / completeness which has been adverted to; seeing what even in that unfinished state it could perform, and inferring from it with almost absolute certainty what it could effect when completed – the abrupt stop put to its progress by the government must have been a mortifying disappointment to Mr Babbage of the severest kind. He had set about the vast enterprize under the directions of the government of the day: during nearly twenty years he had continued to work incessantly at the undertaking – not only without pay, but he had expended about *twenty thousand pounds* of his private fortune, of course relying on the probity of his employers, and the honour of his country to reimburse the outlay. 'Put no confidence in princes,' says an old apophthegm, the advice is sound, but its practical usefulness when applied to a set of men called THE GOVERNMENT, is abundantly more obvious, and Mr Babbage's case may be adduced as proof of the allegation. To men holding high government offices,

[a] C. R. Weld, *A history of the Royal Society*... (2 vol., London, 1848); the eleventh chapter of Vol. 2 was reprinted by Babbage – see *Works of Babbage*, Vol. 10.

the gift of speech appears to be given and used merely to disguise their intentions, and to mislead and disappoint ardent and unsophisticated men who repose any confidence in their ambiguous promises. For some occult reason, which metaphysical writers have not explained, taking high office in state matters, has a marvellous effect upon men previously of high integrity and unquestionable veracity; the assumption of important office does not, perhaps, like Circe's wand, transform such officials into hogs – but it does very often indeed metamorphose gentlemen of pure minds and untarnished honour into dissembling tricksters and daily traders in deception. There have been, and there are splendid exceptions to this unprincipled class – but whoever places implicit confidence in men holding high ephemeral power, stands a great chance to be like Mr Babbage, at last grievously disappointed and irretrievably injured. In such cases, the constituent materials of a government are so fleeting and unstable, that the law can seldom lay hold of them – they / frequently vanish before it can be put in motion especially in questions of faith which they have induced their dupes to place in their delusions. Had Mr Babbage under similar circumstances been set to work by an individual, and that individual a man of honour and honesty, he would have felt bound to act justly with Mr Babbage, and if possible, enabled him to complete the enterprize without injuring himself in any manner; not to do so, he would have felt was a disreputable breach of faith, which perhaps the law would redress and punish. The eternally shifting body of men called the government, however, are bound by no scruples of the kind. Mr Babbage during twenty years had relied upon the honour of government and he found at last that his faith had rested upon mere moonshine, and that his shabby and abominable treatment admitted of no redress.

Forty years have elapsed, since the government induced him to waste his mental powers as well as his money, and to squander his intellectual gifts of the highest order in their service for nothing, and the result to be an *abortion*, which they themselves caused – and nothing has been done. In addition to all this mortifying loss, during the long interval which has since passed, Mr Babbage has been subject to imputations as groundless as they have been malignant and base, as to the expenditure of the money which the government had advanced, although the libellous insinuations have been again and again refuted.

Thus the writer of an article on Babbage's calculating machine in the *Edinburgh Review*, as far back as July, 1834, (the article before referred to as having afforded to the Messrs. Scheutz the suggestion for their machine) says,

The execution of the workmanship was confided to an engineer by whom all the subordinate workmen were employed, and who supplied for the work the requisite tools and other machinery – the latter being his own property and not that of the government. / This engineer furnished at intervals his accounts which were duly AUDITED by *proper persons appointed for that purpose*. It was thought advisable, with a view, perhaps, to invest Mr Babbage with more strict authority over the subordinate agents, that the payment of these accounts of the engineer should pass through his hands. The amount was accordingly from time to time issued to him by the treasury, and paid over to the engineer. This *circumstance has given rise to reports that he has received considerable sums of money as a remuneration for his skill and labour in inventing and constructing the machinery. Such reports are altogether destitute of truth*, on the contrary, owing to various official delays in the issues of money from the treasury for the payment of the engineer, he has frequently been obliged to advance those payments himself that the work might proceed without interruption. Had he not been able to do this from his own private resources, it would have been impossible that the machinery could have arrived at its present advanced state.[a]

Again Mr Babbage in the second edition of his ninth Bridgewater Treatise, published in 1838, adverting to the engine said, 'Here I should willingly have left the subject, but the public having erroneously imagined that the sums of money paid to the workmen for the construction of the engine were the remuneration of my own services for inventing and directing its progress; and a committee of the House of Commons having incidentally led the public to believe that a sum of money was voted to me for that purpose, I think it right to give to that report the most direct and unqualified contradiction.'[b]

Notwithstanding the above explanation given so long ago in the *Edinburgh Review*, by a writer evidently well-acquainted with the facts and details of the case, and notwithstanding the public disavowal of Mr Babbage, he has for years been subjected to the slanderous imputation of not properly accounting for the money, when in fact he had had none to account / for. Gross insinuations that he had received large sums of public money, and had rendered no satisfactory account of them, had been iterated in every form which malice could invent, and circulated with corresponding industry, although it was matter of fact and figures, which his enemies must have known, or might easily have known, that flatly contradicted them. So far then Mr Babbage's extraordinary achievement had brought him only grievous disappointment, heavy pecuniary loss, and unfounded slander – this continued for many years – but it was not all, other annoyances and insults were in store for him.

In 1851, there was a Crystal Palace raised in Hyde Park for the Exhibition of the *chef d'oeuvres* in arts and sciences of all countries. The most celebrated cultivators of art and science of all nations were invited and encouraged to send their chief performances to the diorama in Hyde Park. Now, although not completed, we venture to assert, there was at that time, nothing like Mr Babbage's machine in the world, even in its unfinished state, there was nothing in existence that approached its extraordinary performances. The machine was

[a] D. Lardner, 'Babbages calculating engine' (1834), *Works of Babbage*, Vol. 2, pp. 184–5. The emphases have been added by Judex Juris.

[b] *The ninth Bridgewater treatise*, 2nd edition (1838), Appendix Note B, *Works of Babbage*, Vol. 9.

as unique in its construction as its nearly intellectual feats were astonishing. Yet this marvellous piece of mechanism, the sole invention of an Englishman, was not permitted to be placed in the universal display. Mr Babbage's persecutors, unhappily for the national honour, exercised a baneful influence in the management of the Exhibition, and they had the desperate meanness to exclude that unique piece of mechanism so singularly endued with something like intellectuality from the diorama. England invited all countries to send their chief works to the Exhibition, and yet its managers, to their eternal shame shut out their own countryman's justly renowned achievement from the display, although there was no similar invention to compete with it!

In the commencement of this article we gave the Messrs Scheutz's own account of the royal encouragement and national / assistance which they had received during the progress of their arduous undertaking – any one who has had the perseverance to follow our narration, may contrast their praiseworthy treatment with the deceptive shabbiness which has attended Mr Babbage the sole inventor, throughout his herculean labours in an untrodden track, and the annoying persecutions which have since been perpetrated upon him. But we will give another illustration – another contrast which we trust our countrymen will duly consider. From an article which was published at the time in the *Illustrated News* under the head of '*Staffel's calculating machine*' we learn that Mr Staffel is a native of Warsaw and sent his machine to the Exhibition of that city.[a] A committee of professors of the Academy, was appointed to examine it, who expressed their high approbation of the invention, and concluded by saying that the inventor deserved the highest regard and support. For this support and appreciation Mr Staffel did not look in vain. The Prince Paskewitch Erivaasky, Lord Lieutenant of the Kingdom of Poland, having been made acquainted with this important invention, sent for the inventor to produce his machine at the viceregal lodge, and having by ocular demonstration, convinced himself of the importance of the invention, enabled Mr Staffel to present his machine at the Imperial Academy of St Peterburgh, which reported in its favour; under the same high patronage he was afterwards presented to the Czar who was so gratified with his invention that he ordered a sum of 1,500 silver rubles to be paid to him as a gratification out of his private purse.

Mr Staffel's machine was exhibited in Hyde Park, and the jury awarded him a medal for the invention. No doubt, Mr Staffel's is a very in-

[a] 'Staffel's calculating machine,' *Illustrated London News*, 20 September 1851, pp. 354–5. The Staffel machine was one of two calculating machines which were awarded prize medals at the Exhibition, the other being the much better known Thomas de Colmar Arithmometer. The Staffel machine, which measured 18 in. wide, 9 in. deep and 4 in. high, was capable of performing the four rules of arithmetic, and (somewhat laboriously) of extracting square roots. In spite of the commendations of the jury of the Exhibition, the machine achieved none of the popularity of the Arithmometer. Ernst Martin in *Die Rechenmaschinen und ihrer entwicklungsgeschichte* (Pappenheim, 1925, p. 65) states 'Sie ist bedeutungslos geblieben' – It has remained without significance.

genious contrivance, and we have not the smallest wish to detract an iota from its merits – but giving the machine the fullest credit for the exact performance of all that it pretends to do – the contrivance is a mere bagatelle / when compared with Mr Babbage's Difference Engine – if any such comparison can be justly made.

Hence then, we learn from the Messrs Scheutz that they were assisted in the pecuniary part of their arduous undertaking by the assistance and guarantee of distinguished members of the Swedish Academy – that their Sovereign applied to the Diet for a grant of public money to reward them. Their machine was placed in the Grand Exhibition, at Paris, and the gold medal unanimously awarded to the inventors, who received it publicly through the hands of Prince Charles in the Royal Palace of Stockholm: men of all ranks from the sovereign downwards vied with each other in rewarding and honouring the inventors. In Mr Staffel's case he found warm friends in the professors near him: a powerful and effective advocate in the prince who presided over his country, and he was honoured by the friendly patronage and truly royal munificence of the imperial Czar, and lastly his invention was distinguished and rewarded by the award of a medal at our Grand Exhibition of all nations. On the other hand Mr Babbage at the request of the Government, having first satisfied them by the most cogent testimony of the feasibility of the plan, undertook to construct a Difference Engine, which should print its results: he continued to work at the engine from 1822 to 1834, and, in order to engage all his energies in the work, he refused other sources of profitable occupation, one of which was an office worth £2,500 a year. To the long term of incessant labour that has been named should be added the preliminary work effected before the scheme could be submitted to the government – for all of which he not only received no recompense but in making initiatory experiments, and in payments made during the construction he actually spent £20,000, as we said before, of his own money. Yet the whole of this time, money, and mental toil, was, by the government, that occasioned it all to be incurred, made to end in an abortion – for no kind of / recompense has since been made. As a flagitious crowning point to Babbage's iniquitous treatment, we have been told that an absolute refusal was given to the applications for placing the engine in the Exhibitions at Dublin and Paris!

There was not a scientific man of any note amongst the universal gathering in Hyde Park, to whom the name of Babbage was unknown, and who was unacquainted with the celebrity of his works, nor was there a single visitor to the Grand Exhibition who had seen any piece of mechanism endued with like powers which Mr Babbage's machine possessed – for there was nothing comparable to it then known. The celebrated men who visited the palace may rationally have asked 'why do we not see the name of Babbage in the list of Royal Commissioners, and where, in this gorgeous collection of the works of science the *chef d'oeuvres* of all nations can we find the Difference Engine, which has shed a lustre over the scientific glory of the country?' How great must have been the surprise of such enquirers when they learned that Mr Babbage had had nothing to do with the *Exhibition*, and that his *engine was not in the palace*! Thus the most astonishing piece of mechanism that human ingenuity had ever conceived and carried into effect is the production of an Englishman, and the surprising invention was excluded by its patrons! We have more than once touched upon this exclusion, because we believe the engine would have added considerably to the national credit – high as it stood in the universal concourse, and that therefore the country has been deprived of its merited honours, and that common justice to English science has been perverted. On the whole, let any one fairly contrast the treatment of the Messrs Scheutz and of Mr Staffel with that of Mr Babbage, and if he be an Englishman we are convinced he will feel a kind of loathing and detestation for the usage which our highly gifted countryman has received: he will perceive that the government patronage which was / apparently vouchsafed to him, has blighted all his prospects – damaged his pecuniary resources, and induced him to sacrifice twenty years of the most valuable and vigorous part of his life: he will be struck with the remarkable fact that his countrymen have looked on with chilling indifference – their apathy is almost as extraordinary and censurable as the withering support of their rulers.

In contrasting the treatment of the Messrs Scheutz and Mr Staffel with that of Mr Babbage, we have named the royal aid of the King of Sweden and the munificence of the late Czar of Russia: but to do simple justice to our own sovereigns – it should be remembered that in each of the above cases, the recipients of the royal patronage and encouragement, were made known to the sovereigns: had Mr Babbage's surprising invention and its performances been, in a similar manner, brought to the notice of the British sovereign, there can be no doubt that he would have received quite as ample an amount of royal support. If, therefore, the munificence, and princely encouragement of other sovereigns contrast unfavourably with the seeming indifference of our own, the censure, and we consider it a very heavy one, ought to be cast on the parties whose duty it was to bring their sovereign fully acquainted with every matter that affects so deeply the national character and the highest interests of the country. England occupies a lofty position amongst the civilized nations of the world: it has gained that eminence chiefly by the intellectual powers of its people – it can maintain that place only by a mental exertion which is truly national, when therefore, it treats men like Mr Babbage – highest in intellect, and foremost in acquirements with discreditable injustice, and blighting ingratitude, it becomes a matter for serious apprehension as indicating unsoundness and symptoms of decadence.

Having discussed the subject to such length, the questions may now fairly occur. How can Mr Babbage's treatment be / explained? How is it that his singular inventions, and extraordinary labours in the cause of British science, have been treated with so much withering apathy and depressing neglect? How is it, it may be further asked, that a sort of ill-fortune has attended all his efforts? What is the reason that his long course of intellectual exertions has brought him only personal loss, and, in some measure, public reproach? Why did England invite all nations to send their chief works of arts and productions of science to the gorgeous diorama, and exclude the most astonishing work in existence – contrived and constructed by one of her own sons –from the public view? Why did Great Britain invite all the illustrious men of science of every country to come to the grand assembly and display to the world what they deemed their principal works, and keep one of the most widely known men of science, her own son, as well as the surprizing productions of his inventive powers – in the dark? What answer can be given to all these enquiries? If the reader feels any difficulty in forming a satisfactory reply, let him turn to Mr Babbage's work, entitled 'The Exhibition of 1851; or Views of the Industry, the Science, and the Government of England.'[a] In the chapter 'On the intrigues of science' will be found the author's explanation of the mystery. He there attributes all the blighting influence which has frustrated his exertions, and rendered abortive, his great labour, to a MALIGNANT CABAL, the chief members of which, in our opinion, ought to have been the encouragers of British science, instead of having been the vindictive persecutors of one of its most celebrated promotors. However, as one of the clique whom Mr Babbage considers the main concoctor of all the atrocious mischief that has been done to him, has since gone to his account – we shall simply refer to Mr Babbage's work and abstain from personalities: kindred members of the cabal say that they shall not look upon the defunct's like again – peace to his manes – we heartily trust they never may! /

An unsophisticated man of science who resides at a distance from the metropolis, and is unacquainted with the scientific bodies in London, and their general management: seeing for instance, the distinguished men who constitute the council of the Royal Society he imagines it to be the *ne plus ultra* of a scientific man's ambition to become a member of that royal corporation. He fancies that the cultivation of science is each member's chief object, and, that in the calm region of pure intellect, the dominant aim of the association is mutually to encourage and assist meritorious labourers in the vineyard of science. Should, however, any sanguine cultivator of science who entertains these notions respecting the halcyon state of this royal body – by chance or luck get initiated into its mysteries so far as to become familiar with its working, he will find that the whole machinery of the society is under the sole management of a very few members some of whom are as unscrupulous as they are

[a] *Works of Babbage*, Vol. 10.

talented: he will learn that small minded jealousies, crafty trickery, and dishonest cajolery, are quite as rife in that region of intellect, as they are among stock jobbers of the lowest class, and knaves by profession. A man of scientific note must become the willing tool of the rulers of that body if he wish to make progress: but if he have the least spirit of independence – if he does not altogether make himself over as the most submissive lackey to those dogmatic rulers – good bye to him – he is descried, calumniated, and injured in every way – there is a pack of slanderers always ready to be let loose, and halloed on, to worry and run him down. This is certain to be the fate of any man of talent and of note who refuses to become the mere instrument of the dominant set – but especially is it the case with any man of spirit whose achievements are likely to throw into the shade the performances of any member of those self-constituted rulers.

The Royal Society, as a body, comprises a large number of men deservedly celebrated for their attainments in science, / nevertheless there is a considerable number, who have become members merely for the sake of the tail piece to their names, or in order to add to their professional importance or local reputation, these dilettante philosophers are sure to become the ready tools of the ruling clique – and to join the pack for '*putting down*' any man of note who has rendered himself obnoxious by being too independent –or too conspicuous by his intellectual performances. During the presidency of old butterfly Banks,[a] the society was noted for this hunting down and disparaging of eminent men who refused to be the mere tools of a wealthy connoisseur in grubs. Whether Mr Babbage is entirely correct in attributing all his abominable usage to the malignant spite of the clique that he names, or not, one fact is obvious, that all his mental labours have been alike unsuccessful, and hence it becomes probable that his vindictive persecutors have possessed influence in different quarters. For example the Royal Society professes to give medals to the successful cultivators of different sciences. We have already adverted to the celebrated memoirs which Mr Babbage once published in the Society's transactions –they were as masterly as they were novel, and every mathematician in England competent to form a just opinion on the subject, will admit that they were highly instrumental in renovating mathematical science in England – then at a very low ebb – they were in fact *nationally* important and valuable. Did the society pay its distinguished member the deserved compliment of a medal for these celebrated essays? We cannot find that it did anything of the kind. We venture to assert that these memoirs had a greater influence in reviving abstract science in England than perhaps any other publication that can be named, and yet they were allowed to pass undistinguished by the Royal Society. Again Mr Babbage's invention of mechanical notation was as novel as it was universally

[a] Sir Joseph Banks P.R.S. (1744–1820).

serviceable – a memoir on the subject was published in the society's transactions,[a] did that extension of human / knowledge obtain the Society's chief mark of approbation? we have searched in vain to find that it did. The society's large number of volumes filled with memoirs of all kinds, does not contain a set of essays more original and of greater importance to national abstract science than Mr Babbage's, and yet for neither of them, nor for the whole together, have they awarded one of their medals to the author.

The Difference Engine which the Royal Society originally patronized, though not finished, was so far complete as to perform almost intellectual operations, and yet for this truly astonishing invention the society which assumes to be the royal supporter of British science has not had the justice to give the inventor one of its medals.

Mr Babbage's allegations are that his vindictive persecutors have a predominant influence in the Royal Society, and also sufficient weight with the government of the day to bias its conduct in scientific matters. The striking proof of this is that his treatment has been generally alike – throughout, his usage has been the same – which renders it probable that the same enemies have been instrumental in inflicting the whole on all occasions.

It may perhaps be expected that the Royal Society in consequence of its removal to Burlington House, will amend some of its rules, and improve some of its practices, so as to become more generally useful to the community in return, for the expense which the public has incurred for its accommodation – especially if it desire to retain its assumed title of royal patron of British science. To maintain that distinction is must no longer deserve the imputation of being managed by unscrupulous partisans who award medals to one another instead of justly rewarding candidates for distinction – to whom they rightfully belong. The society's medals when bestowed from partiality to tools, or not conferred in consequence of rival jealousy, are inimical to the real progress of science. The / medals when so given are deceptive – they are at once a disgrace to the unprincipled shufflers that award them, and of not much value to the pets who get them. We will give an instance of the strange manner in which the society's medal was awarded in another most singular case. Everybody remembers the splendid discovery of the planet Neptune. The fact admits of no dispute, that Mr Adams had calculated the elements of the planet and fixed the place near which it afterwards was found, NINE MONTHS BEFORE Le Verrier had entered upon the subject. It has been undeniably proved that Mr Adams, had made his calculations fixing the place of the planet and lodged the results first with Professor Challis of the Cambridge Observatory, and then with the Astronomer Royal at the Greenwich Observatory, and had the English astronomers only performed the simple duty which attaches to the posts that they fill, by using

---

[a] 'On a method of expressing by signs the action of machinery' (1826), *Works of Babbage*, Vol. 3.

the ample means under their managements in examining the region to the 325th degree of longitude, they would unquestionably have found the new planet in the place which Adams had fixed – many months before Le Verrier's attention had been called to the topic by his friend Arago. But unfortunately for English science, there is some truth – *though not the* WHOLE *truth*, in the epigram –

> When Airy was told, he would not believe it;
> When Challis saw, he couldn't perceive it.

However, for some reason or other, which *has never been* EXPLAINED, Mr Adams's elements of the unseen planets' orbit were permitted to lie by until Le Verrier had performed the feat of completing the same task, and then it appeared that Mr Adams's results, which had so long been lodged at the two principal observatories in England, agreed very nearly with the French astronomers' elements; but there was this difference in their success, Mr Adams lodged his elements with the two chief public astronomers in England, either of these officials / might have had the results tested – but both strangely omitted to do so. On the other hand, the French astronomer sent his results as soon as they were ready to M. Galle, who at once found the planet in the spot pointed out by Le Verrier, and found by Mr Adams the *year before*! Now the English mathematician did in the PREVIOUS YEAR exactly what Le Verrier did, hand over his results to the chiefs of observatories; our astronomers have not told us what they did with Adams's papers, but M. Galle immediately set to work and found the planet. When the facts became known, as they soon did, there could be no doubt as to who *first* fixed the place of the planet: Adams did – but the two English astronomers, *burked* the discovery, and Galle established it. But how did the Royal Society act in the matter when an Englishman's undoubted rights to a splendid discovery were in jeopardy? Why the same cabal which had manoeuvred to have all Mr Babbage's valuable performances passed over, gave Adams's claims the go by, and awarded their medal to Le Verrier, who at all events did not commence his investigations until many months after Adams had placed his results containing the exact elements of the planet's orbit in the two English astronomers' hands.

There is a vast difference between the tasks of entering and following up an untrodden track such as Adams so successfully travelled over, and going over the same ground when it is KNOWN that another has gone over it before. In the one case the mind has only its own strength to depend upon, and there is a kind of depressing uncertainty attending the toil; in the other, there is the cheering incitement of knowing that the object aimed at has been attained, and that what has been done may be done again.

That Adams's achievement was completed unassisted, and entirely by his own efforts, under the belief that nothing of the kind had been effected before – there is no manner of doubt or dispute. That Le Verrier performed the same feat some / months afterwards, exactly under the same circumstances, and without any hints or knowledge that the elements of the planet had been determined by means of inverse perturbations –we shall know more certainly when we can more clearly ascertain that Mr Adams's papers were kept perfectly secure and close during the long interval – and when we know more exactly that the fact, that he had solved the problem, did not transpire, or in any way, reach France – we have palpable and abusive subterfuge on the point – but no proof, and until that be vouchsafed, we can only admit that Le Verrier MAY have made the discovery so many months later than Adams, without knowing what had been done in England.

But be this as it may, the Royal Society by the bestowal of its medal, did its utmost to deprive an Englishman of his just rights to the most splendid discovery of the age. Had the search been made, when Adams sent his results to the English Observatories, the discovery would have been unquestionably his – for Arago had not then so urgently requested his friend Le Verrier to take up the subject – why Arago's request was so emphatic may raise questions: however, Adams's discovery would at once have placed him at the head of theoretical astronomers – was this desirable? Ask the self-constituted patrons of British science –the clique that awarded the medal to Le Verrier. This magnificent discovery is truly a national affair, and therefore whoever attempts to deprive Mr Adams of his just claims to it, is endeavouring to rob the British nation of an honour of which it may well be proud: it behoves the guardians of English celebrity, to defend their countryman's rights even in spite of himself. Mr Adams with all his admitted genius as a mathematician and theoretical astronomer, may be a kind of scientific Esau who would sell his birthright on the subject for a mess of postage; nevertheless every friend of his country's fame would, one might suppose, espouse the Englishman's unquestionable rights to the discovery: it is, / therefore, really mortifying to find the whole tribe of fawners on the ruling scientific cabal that has been referred to, preferring the Frenchman's pretensions to their countryman's claims. Every writer who is under the influence of the prevalent clique of the Royal Society, or who is crawling up to become another tool, tries to obtain his object by fulsomely praising Le Verrier, and by throwing only the crumbs of his laudation to Adams. These creeping scribblers prove their fitness for being partisan tools – their sense of justice and love of truth and fair play are totally other matters. The most recent production of this description was from the pen of an egotistical old gentleman[a] who has always lived and moved and had his being as the unscrupulous hack of the Royal Society cabal. He has with singular propriety

[a] William Henry Smyth (1788–1865), admiral and scientific writer.

termed his article – 'Story of the new planet Neptune.'[a] The *story* gives all the credit of the discovery to Le Verrier, and in reference to Adams, it is made up of lame apologies and shallow subterfuges. The absurdity about the *experimentum crucis* with respect to the error of Radius Vector is repeated – this is mere hoodwinking. The Astronomer Royal's question was an idle one; for the answer to it was virtually contained in Adams's first solution. Professor Challis said that errors of Radius Vector were as readily deduced from the first solution as from the other. The enquiry therefore proved that Mr Adams's first solution had received no attention, or that the interrogator was under some misapprehension. Hence, probably, Adams did not answer the question because he had done so before. The same question was put to Le Verrier, who in effect, answered, 'What are you thinking about? Why the radius vector is rectified of itself, without its being taken into consideration in a direct manner.'

The STORY, which is printed for PRIVATE CIRCULATION, does its best throughout to establish Le Verrier's assumption to the chief honour – the scribe pretends to be a great astronomer, a universal genius, and one of the / champions of English science – and yet he thus furtively tries as hard as he can to deprive England of the discovery because the unjust deprivation is in accordance with the wishes of the un-english cabal of which he has long been the tool and scribe. We believe, however, the old gentleman can hardly be deemed a thorough Briton, and we are pleased that to some extent he belongs to a distant country. We think his eagerness to puff off his own pretensions and screen the cabal, of which he is a laborious member, from a charge of disreputable carelessness, mean jealousy, or something worse, has led him to reopen a subject which some of his fellow cliquemen had much rather should pass away. The claims of British science suggest that a complete history of the discovery of Neptune, and all its attendant incidents, should be given to the public so that strict justice may be awarded to all parties. On the present occasion we have touched upon the topic because we believe that precisely the same cabal which Mr Babbage alleges blighted all his prospects, and at length rendered all his marvellous performances an abortion, has also deprived England of the honour of the discovery of Neptune, perhaps the most brilliant application of abstract science that is to be found in history. The malignant coterie have not succeeded in depriving Mr Babbage of all the honour due to him as the inventor of his engine – they have not yet found a Le Verrier as a rival – but they have so far accomplished their ends as to have the invention entombed in a glass mausoleum – and shut out from the chief works of all nations. Will their next attempt be to prove that Mr Babbage borrowed his notions from the Messrs Scheutz? Their partisan scribes who at their bidding

[a] The article was first published anonymously as 'Story of the new planet, Neptune', *United Services Magazine*, 1847, Part 2, pp. 1–14 and 161–72. It was reprinted in W. H. Smyth, *The cycle of celestial objects* ... (London, 1860), pp. 405–31.

deprive Adams of his rights, can easily show that the Messrs Scheutz are labouring under a delusion when they ascribe their calculating machine entirely to Mr Babbage's invention.

We have given an imperfect outline of Mr Babbage's indomitable / labours through a long period in bringing his invention to perfection under difficulties the most annoying and discouraging. Yet whilst thus engaged in this intellectual toil year after year – and his inventive powers necessarily kept on the rack – other schemes of far higher pretensions occurred to him, and he has devised what he terms 'THE ANALYTICAL ENGINE' which is not only capable of tabulating the results of one particular function or formulae like the Difference Engine which he had constructed, but for developing and tabulating any indefinite function of any degree of generality or complexity as $F(x, y, z, \log x, \sin y, x^m)$ a function of all other functions.

A copy of the engravings of the Analytical Engine enabled M. Menabrea, a distinguished Italian mathematician, to enter into the spirit of the great principles on which the invention rested; and to draw up an explicit and instructive memoir on the subject.[a] The memoir was published in the '*Bibliothèque Universelle de Genève*' for October, 1842. It has since been translated and published in the third volume of Taylor's scientific memoirs, with copious and valuable illustrative notes by the late Countess of Lovelace.[b]

The inventor has the drawings and plans, which, as the above memoir and notes prove, make the feasibility of the invention quite plain to any one qualified to form a correct opinion upon the subject. Probably, hereafter the drawings and explanations will induce the SCHEUTZ of some other country to carry the singular invention into operation: of course, its practicability has been made manifest. Obviously, the illustrious inventor must let the splendid scheme take its course and its chance, he has experienced quite enough of discouraging ingratitude from his country to prevent him from attempting to carry out the design in England. The loss of time, and the loss of pecuniary means have already been too great for him to bring it before other nations – not so ruled by vindicative cabals / as his own has been. We have on the present occasion adverted to the subject simply to bring our readers acquainted with the intellectual powers of the man whom the rulers of his country have permitted his malignant persecutors to injure in every way, in order '*to put him down.*'

We have before animadverted on the fact that the country is more indebted to Mr Babbage than to any man living, for his successful labours in renovating mathematical science in England – the chief foundation of all other inductive

[a] L. F. Menabrea, 'Notions sur la machine analytique de M. Charles Babbage' (1842), reprinted in *Works of Babbage*, Vol. 3.

[b] A. A. Lovelace, 'Sketch of the Analytical Engine' (1843), reprinted in *Works of Babbage*, Vol. 3.

sciences, and for raising it to the lofty eminence which it now holds. We named two distinguished authors who were his fellow labourers, but not, in our opinion, his equals, in that glorious work of renovation. But, mark the difference in the treatment of the triumvirate: first by the government, and next by the society that claims to be the rewarder and promoter of British science. The late Mr Peacock was in due time made Dean of Ely, a lucrative position in his profession, and next to the highest in point of honour. Although a laborious and very learned man, we are not aware that he contributed any essays to the Philosophical Transactions; had he done so, no doubt, they would have been rewarded by the Society's medal, for we believe he was not obnoxious to the ruling cabal.

As to Sir John Herschel, he has been made first a knight, then a baronet, and also Master of the Mint, a public post of high honour and considerable emolument: Sir John has contributed a great number of interesting and valuable memoirs to the transactions of the Royal Society, and the medal bestowers have given him one for almost every kind of topic that he has touched upon: he has had *five* at least of these royal awards. Now we desire very emphatically to disclaim every wish to detract in the smallest degree from the merits of the two illustrious authors that we have named – nor have we the smallest objection to the public rewards which they have received; on the contrary, we willingly and unreservedly admit / that they deserved them, and if they had been higher or of greater value, we should have liked them the better. We, however, have referred to them merely to contrast them with the treatment of Mr Babbage their eminent, and, at least, equally celebrated coadjutor. The government rewarded the first two with personal dignities and with high lucrative posts: at it request Mr Babbage entered upon an undertaking of national concern worked at it incessantly and successfully for many years – not only without remuneration – but actually spending a large sum of his own resources. To obey the behests of government, he refused suitable posts of considerable emolument – the result of all was that the government which set him to work and induced him to waste his time and his substance, at last made the whole an abortion, and absolutely refused to give him posts, the duties of which he was eminently qualified to discharge. Again, Sir John Herschel as we have said contributed many valuable essays to the Royal Society's Transactions, and he has received the number of medals which we have stated. But Mr Babbage has also contributed a number of memoirs, as meritorious and important as any which are to be found in the whole set of transactions, and yet, notwithstanding the Society's reports on the Difference Engine, not one medal has been awarded to him. Is not the contrast striking, and does it not thoroughly disgrace the Royal Society for permitting such gross partiality, and monstrous injustice to be exercised? Here again we do not in any manner grudge Sir John Herschel the pocket-full of medals that have been awarded to him – but we wish to brand with the stigma that they merit, the set of vindictive tricksters who have never awarded Mr Babbage one. We have a sincere respect for Sir John Herschel, and heartily admit the real and valuable services which he has rendered to British science: but we wish he had the moral courage to rebuke and silence the malicious persecutors of his old colleague and friend Mr Babbage. The Royal medals which / the scientific jobbers so lavishly bestow on him may flatter his vanity, but as a man they ought not to induce him to connive at the base practices which knaves adopt for 'PUTTING DOWN.' These remarks not only apply to Mr Babbage's disgraceful treatment, but to that of Sir James South and others – old and valued – yes, and SERVICEABLE friends of Sir John's: and yet to gratify the unprincipled pack that flatters him, he turns his back upon the friends of his youth and manhood – aye and his venerated father's friends, and allows them to be 'PUT DOWN,' with impunity, no matter how evil the means may be that are adopted to accomplish that end. Much of Sir John's celebrity as a practical astronomer was won at Sir James South's observatory, as the coadjutor of Sir James – in their investigations respecting double stars – Sir John has ignored this fact: we are pleased to see that justice has been done Sir James in this matter in Knight's English Cyclopedia [a] – the cheapest and most complete work of the kind in the language.

The government of the day had been made fully acquainted with the public importance of correct mathematical tables before they ordered Mr Babbage to commence his mechanism for accomplishing that object, and they engaged to find the material and to pay the engineer and artizans employed in the work – the end of the undertaking was to effect a public advantage, affecting the highest interest of the country, and consequently – it was to be done at the public expense. It must have been obvious at the commencement, that the cost of such a work would be considerable, though a mere trifle taken in a public point of view.

The members of the House of Commons are supposed to be the guardians of the national purse: if then British science had been duly represented in that assembly, it would have been impossible for the government – or for any cabal, having a malign influence over the government, to use a man engaged like Mr Babbage as he has been treated. Had the Common's House of Parliament / comprised any independent members capable of forming a correct opinion upon the subject, they would have thoroughly investigated the matter, and ascertained exactly how it stood, and not permitted the government to have spent so much of the public money to end in very like nothing. When the work was so near its completion, they would have insisted on having the engine finished, so that costly or cheap, useful or impracticable, the country, should at any rate, have the engine for which it had been paying its money, and inducing

[a] Charles Knight, *The English cyclopedia* (London, 1854–72).

Mr Babbage to expend his time and his energies. But, the fact is, the House of Commons did not, nor does it now, represent English science, or any other truly British interest. A large number of the members are sent there by their patrons to represent cliques and sectional interests – perhaps a still larger number purchase their seats, with the view, if possible, to gain cent. per cent. by the outlay: they are legislative traders, and like other traffickers always ready to turn a penny by voting for measures likely to bring some advantage to them or theirs. It is often quite mortifying to observe how these custodians of the country's purse vote away of the public money in lumps of millions, and then higgle over mere candle-top and cheese-paring expenditure. The enormous sums that are squandered in building gun-boats with rotten timber – ships that are never put to sea, and in stores that are jobbed from one dealer to another until the public has paid two or three times over – and generally gets nothing at last: the incredible sums that are spent in armies and navies, and which are now annually increasing –though when either is wanted it is never in an effective state – all this proves that the constitutional keepers of the national purse, for their own purposes, generally hold it open with its mouth downwards pouring out indiscriminately and unrestrainedly.

These economical custodians very likely would grudge and grumble about a few thousands paid for an undertaking of so / much importance as Mr Babbage's machine is; we have no evidence that they did so in that case – but we know that they did not interfere, and insist on the enterprize being completed – nor did they prevent an influential set of mischief makers from inducing the government to convert the whole affair into an abortion, mortifying to Mr Babbage, and disgraceful to all the parties that occasioned it.

However the great cost of the machine was fixed upon by the maligners of Mr Babbage as a huge sum of public money sacrificed by him – or appropriated to his own use. These unscrupulous slanderers, like ill-omened birds were always hovering over him. Envy fastened her fangs on his reputation and cowardly detraction for years ceaselessly calumniated him. As we have before said, it was broadly asserted that Mr Babbage had expended large sums of the public money to no purpose, and that consequently the country instead of gaining anything by his twenty years' toil for nothing, had lost a large sum by his speculation which he did not carry into any useful operation. We have given an opinion as to what was the duty of the House of Commons in such matters, if British science had had any real supporters amongst them, or if they had been in reality the custodians of the public purse, instead, as we consider them, the abetters of lavish and useless prodigality, and the ready supporters of wasteful expenditure.

At last, however, the question as to the value of Mr Babbage's labours to the general interests of the community, and the real *quid pro quo* which the nation has received for the sums which it advanced for wages and materials whilst the engine was constructing has received a decisive reply from the highest scientific authority in the kingdom. Lord Rosse was for some years the President of the Royal Society, and was thus deservedly placed at the head of science in England – we emphatically say *deservedly* –for the Earl of Rosse is not only a scientific man among lords, but he is a lord among scientific / men. Chesterfield, it is said, advised his son if he were fond of flute playing, to hire a person to blow it; just now it is fashionable with some aspiring men to become astronomers according to Chesterfield, they are rich enough to buy a large telescope, and sufficiently wealthy to pay a man for looking through it, and thus to constitute themselves astronomers. Lord Rosse is not a dilettante astronomer of this description, but he is a labourious man of science, and a practical astronomer of universal celebrity: by his own contrivance, and under his personal direction, he has had a gigantic telescope constructed and erected, which forms another wonder of the world, and greatly exceeds in penetrating power every thing of the kind that had been previously constructed. This justly celebrated nobleman, so eminently qualified to form a just opinion on the subject – thus adverted, to Mr Babbage's engine in his parting address, as President, to the Royal Society, November 30, 1854,[a]

You are all, gentlemen, no doubt aware, that in 1823, your council, at the request of the Lords of the Treasury, appointed a Committee to report upon Mr Babbage's plan for the construction of a calculating machine, which he called a Difference Engine. The committee, I need hardly say, was composed of men eminent for their theoretical and practical acquaintance with such subjects: that committee recommended the Lords of the Treasury to assist Mr Babbage in carrying out his undertaking. The Lords of the Treasury acquiesced, and the work was proceeded with, Mr Babbage exercising a constant and vigilant superintendence, furnishing the designs, making the computations, in fact, supplying all the theoretical requirements while the government supplied the manual labour and raw materials. In the then backward state of mechanical engineering great difficulties were encountered; at length, in 1828, the Royal Society was again consulted by government, and the result was a report from a committee, to the effect that satisfactory progress had been made, considering / the difficulties, and that the engine was likely to answer the expectations of the inventor. The council adopted the report, and communicated it to government with a strong recommendation in favour of the undertaking. The government acting under that recommendation, supplied further funds, on the condition that the engine was to be public property, and the work proceeded. In 1830, the Royal Society was again consulted by government, and the council, acting as on former occasions, appointed a committee. The report which was drawn up in a detailed form was satisfactory to the Treasury, and the council were informed that *funds would be supplied from time to time until the engine was completed.* Very soon a new difficulty occurred; it became necessary to change the engineer, and it was then found that by the rules of the trade, the tools, which had been

[a] Earl of Rosse P.R.S. 'Address of the President', *Proceedings of the Royal Society*, Vol. 7 (1855), pp. 248–63; part of this extract also appears in *Passages from the life of a philosopher* (1864), *Works of Babbage*, Vol. 11, pp. 76–7.

constructed at the public expense, was the private property of the engineer, there was no choice, therefore, but to sacrifice the tools, or to endeavour to effect a compromise for a large sum. The progress of the work was suspended; there was a change of government. Science was weighed against gold by a new standard, and it was resolved to proceed no further. No enterprise could have had its beginning under more auspicious circumstances: the government had taken the initiative they had called for advice, and the adviser was the highest scientific authority in this country; your council, guided by such men as Davy, Wollaston, and Herschel. By your council the undertaking was inaugurated – by your council it was watched over in its progress. That the first great effort to employ the powers of calculating mechanism in aid of the human intellect, should have been suffered in this great country to expire fruitless because there was no tangible evidence of immediate profit; as a British subject, I deeply regret, and as a Fellow, my regret is accompanied with feelings of bitter disappointment. When a question has once been disposed of, succeeding governments rarely re-open it, still I thought I / should not be doing my duty *if I did not take some opportunity of bringing the facts once more before govenment. Circumstances had changed, mechanical engineering had made much progress, the tools required and trained workmen were to be found in the workshops of the leading mechanists, the founders' art was so advanced that casting had been substituted for cutting in making the change wheels, even of screw-cutting engines, and therefore it was very probable that persons would be found willing to undertake to complete the Difference Engine for a specific sum.*

That finished, the question would then have arisen, how far it was advisable to endeavour, by the same means, to turn to account the great labour which had been expended under the guidance of inventive powers the most original, controlled by mathematics of very high order, and which had been wholly devoted for so many years to the great task of carrying the powers of calculating machinery to its utmost limits. Before I took any step I wrote to several very eminent men of science, inquiring, whether in their opinion any great scientific object would be gained if Mr Babbage's views as explained in Menabrea's little essays, were completely realized. The answers I received were strongly in the affirmative. As it was necessary the subject should be laid before government in a form as practical as possible, I wrote to one of our most eminent mechanical engineers to inquire whether I should be safe in stating to government that the EXPENSE *of the calculating engine had been more than re-paid in the improvements in mechanism*, directly referable to it – he replied, UNQUESTIONABLY. Fortified by these opinions I submitted this proposition to government: that they should call upon the president of the society of civil engineers to report whether it would be practicable to make a contract for the completion of Mr Babbage's Difference Engine, and if so, for what sum. This was in 1852, during the short administration of Lord Derby, and it led to / no result. The time was unfortunate, a great political contest was impending, and before there was a lull in politics, so that the voice of science could be heard, Lord Derby's government was at an end.

Although, in communicating with Lord Derby, I was not acting under the directions of your council, still, as my object was to induce the government to complete a work in which the society had taken so great an interest, I conceived it to be my duty to lay the facts before you, as a basis to proceed upon, should it hereafter be considered expedient to renew the subject. [In the preceding extracts the italics are ours.]

The above quotation from Lord Rosse's address has been made at such length because it substantiates the main points in the preceding narration, with respect to Mr Babbage and the government, and entirely establishes the fact that the country has been amply re-paid for the money which it disbursed on Mr Babbage's engine. Thus the annoying calumnies to which Mr Babbage has been subject for so many years, have been at length completely refuted by a scientific authority which must carry conviction. The noble chief of British science in resigning the presidency of the Royal Society did Mr Babbage a simple act of justice, as graceful as it was liberal and perfect; and although the address must have been wormwood to some of his audience, Mr Babbage may henceforward treat his slanderers on his head, whether pious or impious, with the utter contempt which they deserve. The same reply which the eminent mechanical engineer gave to Lord Rosse's enquiry with regard to the *re-payment* of the expense of the engine in the improvements of mechanism directly referrable to it – would have been given by all the chief engineers in England, whose attention had been properly directed to the subject. The value of the improvements in tools directly consequent on the construction of the Difference Engine has been long admitted to be of great importance to the / public. The Edinburgh reviewer already referred to said –

One of the consequences which has arisen in the prosecution of the invention of the calculating machinery, has been the discovery of a multitude of mechanical contrivances, which have been elicited by the engineers of the undertaking, and which are as novel in their nature as the purposes where novel which they were designed to attain. In some cases several different contrivances were devised for the attainment of some end, and that among them which was best suited for the purpose was finally selected: the rejected expedients, those overflowings or wastes of the invention, were not however always found useless. Like the waste in many manufactories they were soon converted to purposes of utility. Those rejected contrivances, *have found their way, in many cases into the mills of our manufactories*: and we now find them busily effecting purposes far different from any which their inventor dreamed of, in the spinning frames of Manchester.

Another department of mechanical art which has been enriched by this invention, has been that of TOOLS. The great variety of new forms which it was necessary to produce, created the necessity of contriving and constructing a vast number of novel and most valuable tools, by which with the aid of the lathe, and that alone, the required forms could be given to the different parts of the machinery with all the requisite accuracy.

Is it not incomprehensibly strange, and really mortifying, that such inventions, and such results, should have been treated with ingratitude so thoroughly disgraceful, and the justly celebrated author of the whole, should remain neglected and unrewarded?

We have nearly finished the task which we imposed on ourselves to perform – namely, to give our readers some notion of the capabilities and powers of machinery constructed for calculating and printing mathematical tables – but more especially to bring them acquainted with the valuable labours / of Mr Babbage in the cause of English science, and with the treatment that he has received in return. In reference to the last part, if we have in any manner, accomplished our wishes we have shown that the nation is more deeply indebted to Mr Babbage for the renovation of abstract science, and for the eminent position which mathematical science now holds in this country, than to any other man in existence: we might have put his case on this head, even on higher ground, and challenged all doubters or detractors to disprove the allegation if they could. Our sketch of this part of Mr Babbage's deserts is necessarily imperfect and so incomplete that we feel we have not done him justice.

There is a large number of highly meritorious performances which have not been mentioned. For instance, societies have been formed for the promotion of scientific objects of different kinds. Some of these institutions have obtained a European reputation, and have tended to render Great Britain celebrated in the annals of arts and sciences. It will be found that Mr Babbage was in the foremost rank in establishing these societies, some of which owe their existence and usefulness altogether to his suggestions and exertions. On this topic, England is deeply indebted to his unceasing labours in the cause of science.* We believe, and have endeavoured to prove, that Mr Babbage is one of the most celebrated men of science of the age in which he lives, and that he is indeed, by far, the worst used man in England. We have adverted to a malignant, and not uninfluential cabal, to whom Mr Babbage attributes all the ill-luck that has attended his stupendous labours, and we have adduced some reasons to prove that his conjectures have strong probability to support them. Now, as the ardent, though feeble advocates of Mr Babbage's claims to the nation's gratitude, and the hearty detesters of his persecutors and maligners – we have no wish to set up Mr Babbage as a model / of perfection. All that we know of him induces us to honour and esteem him: but he may have opinions of his own, and he may have a manner of supporting them which may not always be agreeable to the maintainers of opposite ones: he may have views of his own on political topics, and he may not always have been sufficiently compliant and submissive to personages of influence in state matters, and consequently scientific spite and political hatred may have combined to thwart all his labours, and to render abortive all his extraordinary achievements. We do not know that it is, or that it is not so, but even admitting that this conjecture has some foundation in fact, is the blighting hatred to be immortal? Is it never to be satiated? Suppose Mr Babbage (and it is mere supposition) by the maintenance of his own views, which he has a perfect right to do, or even by some infirmity of temper, has at some time given umbrage to persons of the

* See Knight's English Cyclopaedia, which does justice to Mr Babbage on this head.

description we have referred to, is it not disgraceful to the country, that such an offence, if there be one, should be permitted to deprive the nation of the honour and utility of his invetions? – That he, deserving of the highest honours, has received none; that he, justly entitled to be presented to the assembled representatives of all nations, as one of England's most illustrious sons, should be shamefully slighted and put aside: in a word, that he, admired and esteemed by the great body of his countrymen, should for so long a period be made the victim of a base clique, made up perhaps of scientific and political malignants. But will England longer tolerate such dastardly persecution? Our strongest motive for entering upon the discussion has been to induce our country to clear itself from the stigma which Mr Babbage's treatment for so many years, has cast upon it. Time passes rapidly away: some of Mr Babbage's magnificent performances in the advancement of British science, were given to the country upwards of forty years ago, and a large portion of that time has elapsed since he invented and / contrived machinery endued with almost intellectual powers of calculation, and yet whilst others less distinguished, and immeasurably less talented, have been honoured and rewarded, he remains unnoticed for his invaluable labours, and unpaid for the heavy expense which the government of his country induced him to incur in an undertaking adapted and intended to accomplish a grand national purpose. We ask again, is this usage which the late Bomba of Naples would have been ashamed to sanction – to continue longer? We have stated the encouraging assistance and Royal support which Staffel and the Messrs Scheutz received from the Imperial Czar in the one case, and from the King of Sweden in the other in completing their enterprizes – the greatest of which was founded entirely on Mr Babbage's invention. Will the government of England longer permit our most gracious and beloved Queen to stand in such damaging contrast with the sovereigns that we have named? They were made acquainted with the contrivances by their advisors, and they munificently rewarded, and duly honoured the ingenious constructors – no doubt our sovereigns, in the first instance would have done honour to themselves by similarly encouraging and rewarding Mr Babbage, had his invention in like manner, been brought to their notice, and its surprizing performances explained to them. At all events, we are certain that England's present sovereign would not permit a month to pass without making Mr Babbage ample amends as far as she could, for all that he has done, and for all that he has borne, were the whole case fairly brought to her Majesty's notice. We have made the preceding simple statement in the hope that this may be done. His Royal Highness the Prince Consort* is justly entitled to / the nation's gratitude, for his judicious assistance and encouraging patronage of arts and sciences.

* Since this part of the article was written, Her Most Gracious Majesty has sustained an irreparable loss, by the almost sudden demise of His Royal Highness the Prince Consort. Every Englishman feels the overwhelming bereavement which their Queen and her children have suffered. The

The country has been unspeakably benefited by the Exhibition of 1851, in Hyde Park. That grand scheme was altogether devised by his Royal Highness, as his sagacious mode of teaching the nation its defects, and of imparting to that suspicious and haughty pupil lessons of valuable improvement in a vast number of subjects. Practical wisdom was never more judiciously evinced – or its comprehensive objects more completely attained. We trust, and earnestly hope, that Mr Babbage's claims and treatment, will now be fully made known to his Royal Highness, and if so, we are sure he will take steps that justice be done, and that the British Sovereign shall no longer appear to be indifferent to the most astonishing invention of this age or any other – whilst other sovereigns have been munificently rewarding the ingenious constructors of mere copies of it.

Lord Palmerston, the present Premier, is allowed, even by his political opponents, to be a thorough Englishman. His lordship did himself high honour, and paid a graceful compliment to intellect and learning generally when he advised his sovereign to make Macauley a peer; by conferring that title her Majesty added a lustrous gem to the British crown, which will continue to shine as long as literature is honoured and real talent esteemed. The celebrated Premier has more recently given another proof that he knows how to add dignity to the peerage by advising the Queen to continue the title of the truly illustrious Brougham. In these royal acts our most gracious Queen has proved that she can confer honours which exalt even majesty itself. She has done herself the immortal honour of bestowing titles on these lords of creation, and the Premier deserves the nation's esteem for the part he has taken in the matter. Our readers will recollect his lordship's declaration on a memorable occasion, as to an Englishman's rights, associating them with the noted phrase 'ROMANUS SUM'; / we hope an Englishman's being in England forms no exception to that noble intimation; on the contrary, we believe if Mr Babbage's case were fairly brought to Lord Palmerston's notice, as we hope it may, that he would do himself the credit, and his sovereign the honour, to get justice done to his meritorious labours. It was rumoured not long since, that Sir Cornewall Lewis, the Home Secretary, had much interested himself, in Scheutz's machine, and had advised that one of them should be obtained for public purposes – we do not know whether the report be correct – or whether it has been carried into effect – but be that as it may, Sir Cornewall stands high in reputation as a man of large acquirements and as a gentleman of truly

Nation's sorrow for the loss of His Royal Highness is as unaffectedly real as it is universal. The country acutely feels that a true friend of its highest interests – one who sedulously guarded its immunities, and with consummate wisdom fostered its prosperity, has been taken away. England deeply bewails the premature Death of the illustrious Prince who will deservedly stand in the nation's annals as ALBERT THE GREAT and the GOOD. His most enduring monument will be the grateful remembrance of the British people.

English propensities – we hope, therefore, that having informed himself of the utility of the invention in a public point of view – that Mr Babbage – the unquestionable inventor, will have the benefit of his influence which is considerable in getting redress for protracted injustice, and for meritorious public services which have met with persecuting treatment – instead of their deserts – high honour and public munificence.

Lord Rosse has publicly stated the plan which he would have carried into operation if he could, and he has suggested one mode of partly making amends to Mr Babbage, and at the same time of removing the stigma of indifference and ingratitude from the national character. No doubt, as his lordship remarked, that in the present state of practical mechanical science the Difference Engine could be completed for a specified sum of no very large amount. Sir Cornewall Lewis could easily have the estimate made, and we have no doubt on his recommendation, if the expense be moderate, as it seems likely to be, that the House of Commons would have the engine completed without delay. But more than this, we have described another invention of Mr Babbage still more comprehensive and extraordinary of the same description, so far theoretically completed, that a celebrated mathematician has / been enabled to set out its marvellous powers, and very like mental performances. Lord Rosse's suggestion equally applies to this. Probably there are many practical engineers who from drawings and plans would undertake to construct the Analytical Engine for a certain sum. Let Mr Gravatt's opinion, for one, be taken on the subject, so that England may preserve at least one splendid invention entirely to herself.

We are shortly to have another exhibition, in Hyde Park, comprising the chief works of all nations. We do sincerely hope for the scientific credit of the English people, that this time steps will be taken to do ample justice to our distinguished countryman, for the abominable treatment that was perpetrated upon him in 1851, and since that, in reference to the exhibitions at Dublin and Paris. We shall take a deep interest in the next gathering, and vigilantly watch the committee's doings and management, and if there be again any '*putting down*' or *shutting out of* an eminent countryman's inventions, the perpetrators of the iniquity, whoever they may be, shall not escape public reprobation. But our hopes are that detestable persecution is at end, and that substantial justice will at last be done to splendid merit, and indomitable perseverance. '*Fiat Justitia*' is all we ask.

We conclude our discussion by recommending our readers to purchase and peruse the singular little book to which we referred in the commencement of this article. /

MEMOIRS

OF THE

ANALYTICAL SOCIETY

1813.

CAMBRIDGE:

Printed by J. Smith, Printer to the University;

AND SOLD BY

DEIGHTON & SONS, CAMBRIDGE; LONGMAN & CO. PATERNOSTER-ROW, LONDON;

PARKER, OXFORD; CONSTABLE & CO. EDINBURGH; AND

GILBERT & HODGES, DUBLIN.

1813.

[i]

I

[MEMOIRS OF THE ANALYTICAL SOCIETY]

PREFACE[a]

From *Memoirs of the Analytical Society* (Cambridge, 1813), pp. i–xxii

With J. F. W. Herschel

To examine the varied relations of necessary truth, and to trace through its successive developments, the simple principle to its ultimate result, is the peculiar province of mathematical analysis. Aided by that refined system, which the ingenuity of modern calculators has elicited, and to which the term analytics is now almost exclusively appropriated, it pursues trains of reasoning, which, from their length and intricacy, would resist for ever the unassisted efforts of human sagacity. To what cause are we to attribute this surprising advantage? One, undoubtedly the most obvious, consists in the nature of the ideas themselves, whose relations form the object of investigation – and the accuracy with which they are defined. This is equally indeed the property of every branch of mathematical enquiry. Three causes however chiefly appear to have given so vast a superiority to analysis, as an instrument of reason. Of these, the accurate simplicity of its language claims the first place. An arbitrary symbol can neither convey, nor excite any idea foreign to its original definition. This immutability, no less than the symmetry of its notation (which should ever be guarded with

[a] This paper includes many references which are given with incomplete or informal titles. To avoid an excessive amount of editorial interpolation, the principal works cited are listed here:

J. L. R. d'Alembert, *Réflections sur la cause générale des vents* (Berlin, 1747)
L. F. A. Arbogast, *Du calcul des dérivations* (Strasbourg, 1800)
L. Euler, *Opscula analytica* (2 vol., Petropoli, 1783–1785)
C. F. Gauss, *Disquisitiones arithmeticae* (Leipzig, 1801)
C. Kramp, *Analyse des réfractions astronomiques et terrestres* (Strasbourg, 1798)
S. F. Lacroix, *Traité du calcul différential et du calcul intégral* (3 vol., Paris, 1797–1800)
J. L. Lagrange, *Leçons sur le calcul des fonctions* (Paris, 1804)
J. L. Lagrange, *Théorie des fonctions analytiques* (Paris, 1797)
J. L. Lagrange, *Traité de la résolution des équations numériques* (Paris, 1808)
P. S. de Laplace, *Théorie analytiques des probabilitiés* (Paris, 1812)
P. S. de Laplace, *Traité de mécanique céleste* (5 vol., Paris, 1798–1827)

a jealousy commensurate to its vital importance) facilitates the translation of an expression into common language at any stage of an operation, disburdens the memory of all the load of the previous steps, and at the same time, affords it a considerable assistance in retaining the results. Another, and perhaps not less considerable cause, is to be found in the conciseness of that notation. Every train of reasoning implies an exercise of the judgement, which, being an operation of the mind, deciding on the agreement or disagreement of ideas successively presented to it, it is reasonable to presume will be more correct, in proportion as the ideas compared follow each other more closely; provided the succession be not so rapid as to cause confusion. Were an analytical operation of any complexity converted into common language, in all its detail, the mind, after acquiring a clear conception of one part of the related ideas, must suspend its decision until it could obtain an equally perspicuous one of the remainder of the proposition; and in so long an interval as this must occupy, the impression of the former ideas would necessarily have faded in some degree from the memory, unless fixed by an expense of time and attention, sufficient to deter any one from the employment of such means of discovery. It is the spirit of this symbolic language, by that mechanical tact (so much in unison with all our faculties) which carries the eye at one glance through the / most intricate modifications of quantity, to condense pages into lines, and volumes into pages; shortening the road to discovery, and preserving the mind unfatigued by continued efforts of attention to the minor parts, that it may exert its whole vigor on those which are more important.

The last cause we have occasion to notice is, that analysis, by separating the difficulties of a question, overcomes those which appear almost insuperable when combined, or at least, reducing each to its least terms, leaves them as the

---

A. M. Legendre, *Exercices de calcul intégral . . .* (3 vol., Paris, 1811–17)
A. M. Legendre, *Essai sur la théorie des nombres* (Paris, 1803)
W. Spence, *An essay on the theory of the various orders of logarithmic trancendents, with an inquiry into their applications to the integral calculus and the summation of series* (London, 1809)
Brook Taylor, *Methodus incrementorum* (London, 1717)
R. Woodhouse, *A treatise on isopermetrical problems and the calculus of variations* (Cambridge, 1810)
R. Woodhouse, *A treatise on plane and spherical trigonometry* (London, 1809)

The following are the full titles of the lesser-known journals cited in this paper:

Académie Français. Mémoires de mathématique et de physique présentés . . . pas divers savants [cited as *Savants Étrangers*] (Paris, 1750–86)
Acta eruditorum (Leipzig, 1682–1731)
Acta academiae scientiarum imperialis petropolitanae (Petropoli, 1777–1782)
Histoire de l'Académie des sciences (Paris, 1666–1790)
Mémoires de l'Académie des sciences de Turin (Turin, 1784–1814)
Mémoires de l'Académie royale des sciences de Berlin (Berlin, 1786–1804)
Mémoires de l'Institute national des sciences et arts. Sciences mathématiques et physiques (Paris, 1795–1806).

acknowledged landmarks of its progress, open to approach on all sides, should ulterior discovery furnish any rational hope of their removal. Meanwhile that progress continues unimpeded. Simple relations are found to exist between the most refractory functions, and even when the difficulties themselves prove invincible, their nature at least becomes thoroughly understood, and means of evading them almost universally pointed out.

That the preceding observations are not founded on bare speculation, the whole history of analytical science will abundantly evince. It is our intention, in the following pages of this Preface, to give a general outline of that history up to the present time. From the space allotted to it, it is evident that little else than the most prominent points in so wide a field can be selected for observation. Faint, however, as it is, the subject cannot but communicate to it some portion of its interest; as well as the reflection, that (with the exception of one branch of it*) the history of the more modern discoveries has hitherto unfortunately found little place in our language.†

Symbolic reasoning appears to have been ushered into the world under unfavourable auspices, and to have been regarded in its infancy with an eye of extreme jealousy. And, indeed, if we consider the rudeness of its first attempts, the poverty of its first resources, and the lavish want of economy in their employment, we shall find little reason to wonder, that for a long period, the new methods were looked upon as inelegant, although serviceable auxiliaries of the ancient processes, to be regularly discarded after serving their turn. To employ as many symbols of operation and as few of quantity as possible, is a precept which is now found invariably to / ensure elegance and brevity. The very reverse of this principle forms the character of symbolic analysis, up to within fifty years of the present date.

The first and most natural object of research in the algebraic calculus, was the resolution of equations, involving simply the powers of the unknown quantity. As far as the fourth degree no difficulty occurred, but beyond this, not a step has yet been made. Almost every analyst of eminence has applied his ingenuity to the accomplishment of the *general* problem, but without success; and after more than two centuries, during which every other branch of analytics has been advancing with unrivalled rapidity, no progress whatever has been made in this. This, it must be allowed, presents little prospect of success to future researches on the subject; yet ought not the difficulty to be

* The calculus of variations, the history of whose rise and progress has been ably combined with the exposition of its theory, in a late work, *On Isoperimetrical Problems* ⟨probably R. Woodhouse, *op. cit.* (1810)⟩.

† The admirable review of the *Mécanique Céleste* (*Edinburgh Review* No. 22) will still be fresh in the minds of our readers. But it should be recollected, that the author of that essay confines his attention entirely to the subject of analytical dynamics; referring to the discoveries in the integral calculus merely as connected with that subject, and that too very cursorily. *Our business is exclusively with the pure analytics.*

considered insurmountable, until opinion has been confirmed by demonstration. Delambre notices a memoir presented to the National Institute by M. Ruffini, in which he proposes a proof of the impossibility of the resolution of equations above the fourth degree. If this demonstration be correct, it will render an important service to algebraists, by diverting them from a pursuit which must necessarily be unsuccessful. The work, however, if yet published has not arrived in this country. Recent French publications are not easily procured, nor is it surprising that to obtain those of the German analysts is almost impossible, when Delambre regrets their scarcity even in France.

Although to express in finite algebraic terms, the root of any proposed equation be impracticable, yet the inverse function of any expression, such as

$$a + bx + cx^2 + dx^2 + \dots$$

may readily be exhibited in an infinite series. When the difficulty of solving equations above the fourth degree was perceived, it was natural to seek rapid and convenient approximations, and accordingly, three of our countrymen, Newton, Ralphson, and Halley, produced nearly at the same time, modes of approximation which have since received various improvements. All such researches, when symbolically conducted, and without regard to the numerical value of the symbols, lead at the bottom to series of greater or less complexity. It seems to have been following up this idea, that Lagrange was first conducted to that very general resolution of all equations in the series which bears his name; a series which has been productive of discovery wherever it has been applied, and whose fecundity appears yet far from being exhausted. It is thus that the most distant parts of analysis hang together, nor is it possible to assign the point, however remote or unexpected, in which any proposed career of research may not ultimately terminate. This series made its first appearance in the *Mém. de l'Acad. Berlin*, 1767–8,* together with / the most systematic method of approximating to the roots of numerical equations which has yet been given. The method has this considerable advantage over all others, that, in all cases where the root is an integer, the formula of approximation will give it exactly, and in many where it is a surd, the continued fraction employed will point out the rational number of which it is the root. Though the complete solutions of equations is nearly hopeless, it might perhaps be of some advantage, and certainly of little difficulty, supposing the roots of equations

* The demonstration there given is defective in rigour. A better was given in Note XI of the *Traité de la résolution des équations numériques*, Lagrange. But the most elegant is that of Laplace, to be found in the *Mém. de l'Acad. des Sciences*, Paris, 1777; in Lacroix's *Calc. Diff. et Int.*, 2nd edn, Art 107; in ⟨Laplace's⟩ *Mécanique Céleste*, tom. I, p. 172. Adopted (in principle at least) by Lagrange in the *Théorie des Fonct. Analyt.*, Art. 97, et suiv. And in our own language, in Mr Woodhouse's *Trigonometry*, 2nd edn. Arbogast has also demonstrated this theorem. See his *Calcul des Dérivations*, Art. 282, et suiv.

known, to investigate what change would take place if one or more of the coefficients were augmented or diminished by any small quantity.

To trace the history of the differential calculus through the cloud of dispute and national acrimony, which has been thrown over its origin, would answer little purpose. It is a lamentable consideration, that that discovery which has most of any done honour to the genius of man, should nevertheless bring with it a train of reflections so little to the credit of his heart.

Discovered by Fermat, concinnated and rendered analytical by Newton, and enriched by Leibnitz with a powerful and comprehensive notation, it was presently seen that the new calculus might aspire to the loftiest ends. But, as if the soil of this country were unfavourable to its cultivation, it soon drooped and almost faded into neglect, and we have now to re-import the exotic, with nearly a century of foreign improvement, and to render it once more indigenous among us.

The most prominent feature of this calculus, is the theory of the development of functions. The theorem which has immortalized the name of Brook Taylor, forms its foundation. Elicited by its author from a formula which at first sight seemed independent of it, by a method not remarkable for its rigour, it seems to have been long considered in the light of a very general formula of interpolation. Lagrange and Arbogast have, as it were, invented it anew, and established it as the true basis of the differential calculus. The theory of Lagrange is to be found in a memoir among those of the Acad. de Berlin, 1772, which contains the independent demonstration of Taylor's theorem – in the *Théorie des fonctions analytiques*, wherein he exhibits its application to the various branches of the differential calculus, independently on any consideration of limits, infinitesimals or velocities – and lastly, in the *Journal de l'École Polytechnique*, cah. XII (1802) and in the *Leçons sur le calcul des fonctions*. The ideas of Arbogast are contained in a manuscript / presented to the Academy of Sciences in the year 1789, and of which the outline is given in the preface to his celebrated work on Derivations. Such is the brief account of the greatest revolution which has yet taken place in analytical science.

The operations of the differential calculus once well understood, and rigorously demonstrated, may be employed in improving the theory which gives rise to them. The work of Arbogast just alluded to, has shown to how vast an extent this application may be carried, and how great is the assistance thus rendered. The peculiar grace of Laplace's analysis has nowhere been more beautifully exhibited, than in his improvement and extension of Lagrange's theorem already mentioned. Nor should the labours of Paoli in this field pass unnoticed. By the aid of a very remarkable series derived from reverting that of Taylor, he has been able to assign the development of any function of a quantity given by any equation whatever, in terms of a function any how composed of the remaining symbols which enter into that equation.

The development of functions has lately been made, under the name of 'Calcul des fonctions génératrices', the foundation of a most elegant theory of finite differences, of which more hereafter.

Soon after the discovery of the integral calculus, on the discussion of some problems, between Leibnitz and the Bernouillis, respecting the variation of the parameters of curves, there occurred certain equations, which, though they satisfied the conditions, were yet not contained in the complete integral of the equation whence they were derived. It is somewhat remarkable, considering the manner in which they first appeared, that their geometrical signification should have remained so long undiscovered. Brook Taylor, according to Lagrange, was the first who arrived at a particular solution by differentiating. Clairaut, in a memoir presented to the Academy of Sciences at Paris, first remarked, that the equations so found, satisfy the geometrical conditions proposed. Euler styled them analytical paradoxes, and shewed how in some cases they might be derived from the differential equations. But their theory remained unknown, till the year 1772,* when Laplace explained it in a memoir of the Academy of Sciences, and pointed out the methods of discovering all the particular solutions of which an equation admits. This subject was still farther pursued in the Berlin Memoirs, by Lagrange, who there develops, with great perspicuity the whole theory both analytical and / geometrical.† But the most complete exposition of the subject, which has yet appeared, is to be found in a paper read before the National Institute in the year 1806, by Poisson, in which the theory is extended to partial differential equations, and also to those of finite differences. He observes of certain partial differential equations, that they admit of particular solutions equally general with the complete integral. The analytical theory, in its present state, is most elegant: still it requires some farther developments when applied to equations of partial differentials, and to those of differences, and might perhaps with advantage be applied to equations of mixed differences. Its geometrical signification has been beautifully illustrated by Lagrange; but the meaning of particular solutions, when they occur in dynamical problems, which is a question of considerable importance, remains yet undecided. Poisson has shown a case, in which the particular solution and the complete integral are both required, and has produced others, in which only one is necessary.

As the integration of expressions containing one variable is a matter of considerable importance, and the number of those which are capable of

* J. Trembley, in the 5th vol. of the *Mém. de Turin*, has given a paper on the derivation of the complete integral, having given a number of particular solutions. His method consists in multiplying the equation by these solutions, each raised to an indeterminate power.

† Fontaine first considered a differential equation as the result of the elimination of a constant between an equation and its differential; thus laying the foundation of the theory of equations, both differential, and of differences, and also of their particular solutions.

integration, is small, when compared with those which do not admit of it, some attention has been bestowed on the classification of those which are similar, and on the reduction of those which are absolutely different to the least number possible. When this is accomplished, all that remains for the perfection of this branch of analysis, is to calculate tables which shall afford a value of the integral for any value of the variable. In general, all expressions which do not admit of complete integration, are denominated transcendants. Those which most frequently occur, are logarithmic and circular functions. Tables of these had been long calculated for trigonometrical purposes, and on the discovery of the integral calculus received a vast addition to their utility. It was next proposed to admit as known transcendants, all integrals which could be reduced to the rectification of the conic sections. But, besides the preposterous idea of limiting an analytical expression by the properties of a curve, no tables had been constructed for them, and of course the determination of their arcs could only be performed by the actual calculation of the integral under consideration: nor, indeed, would it have been possible to form useful tables of any moderate length, without first discussing the properties of the transcendants themselves in the fullest manner.

The theory of the transcendant

$$\int\frac{Pdx}{R}$$

where P is a rational and integral function of x, and R a quadratic radical of the form $\sqrt{(\alpha + \beta x + \gamma x^2 + \delta x^3 + \varepsilon x^4)}$ has at length by the successive labours of Fagnani, Euler, Landen, and Legendre, / been brought to great perfection. All the transcendants comprised in this extensive formula, are reducible to three species. Those comprised in the first, are susceptible of multiplication or division, in the same way as the arcs of circles, by algebraic operations only. The transcendants of the second species, are susceptible of a similar multiplication or division, not simply, but when increased or diminished by an algebraic quantity. This algebraic quantity passes in the third species, into a transcendant of the logarithmic or circular kind. Landen has shown, that an integral of the first species here enumerated may be reduced to two of the second: so that the number of distinct transcendants comprised in the formula

$$\int\frac{Pdx}{R}$$

is no more than two. A vast variety of integral formulae have, by dint of indefatigable research on all hands, been reduced to the evaluation of these functions; but to dwell longer on them, would lead us beyond our limits.

The only other species of transcendants of any considerable extent which have received much discussion, are those contained in the formulae

$$\int \varepsilon^{-x^n} \cdot dx \quad \text{and} \quad \int^n \frac{\varepsilon^x \cdot dx^n}{1 + \varepsilon^x}.$$

Kramp, at the end of his *Analyse des Réfractions*, has given a table of the values of the first of these (in the case of $n = 2$) the integral being taken between the limits $0 \cdot 00, \ldots 3 \cdot 00, \infty$. The definite integrals dependent on the general form, we shall speak of hereafter. The second formula is (ultimately) that of the logarithmic transcendants, on the various orders of which Mr Spence, in the year 1809,* published an Essay, which displays considerable ingenuity, and a depth of reading rarely to be met with among the mathematical writers of this country. A general property there given of the transcendant ${}^nL(x)$, leads to the summation of some very extraordinary series, which are now in our possession, and which we cannot forbear mentioning.

Their general (or i^{th}) terms are comprised in the formulae

$$\frac{f\{x \cdot \varepsilon^{i\theta\sqrt{(-1)}}\} + f\{x \cdot \varepsilon^{-i\theta\sqrt{(-1)}}\}}{i^{2n}} \quad \text{and} \quad \frac{f\{x \cdot \varepsilon^{i\theta\sqrt{(-1)}}\} - f\{x \cdot \varepsilon^{-i\theta\sqrt{(-1)}}\}}{\sqrt{(-1)} \cdot i^{2n+1}}$$

where f is the characteristic of any function whatever,† developable in integer powers, either positive or negative, or both. /

Mr Spence has also given tables (of some extent) of the successive values of his functions, as we have before remarked of Kramp. Too much praise cannot be bestowed on such examples, which however, there is little hope of seeing followed. The ingenious analyst who has investigated the properties of some curious function, can feel little complaisance in calculating a table of its numerical values; nor is it for the interest of science, that he should *himself* be thus employed, though perfectly familiar with the method of operating on symbols; he may not perform extensive arithmetical operations with equal facility and accuracy; and even should this not be the case, his labours will at all events meet with little remuneration.

It sometimes happens, that the arbitrary constant does not continue the

* Legendre published his *Exercices de Calcul Intégral* in 1811. After having given Landen's and Euler's values of particular cases of the function ${}^nL(1 + x)$ he adds, 'Jusqu'a présent, on n'est pas allé plus loin dans la théorie de ces sortes des transcendantes', p. 249. ⟨*Translation:* At present, the theory of this species of transcendants goes no further.⟩ It is probable therefore, that, owing to our interrupted intercourse with the Continent, he had not seen Mr Spence's work.

† One singular result of these researches is, the evaluation in terms of the transcendants ε, and π, of the function

$$(\tan\theta)^{(1/1)^{2n+1}} \cdot (\cot 3\theta)^{(1/3)^{2n+1}} \cdot (\tan 5\theta)^{(1/5)^{2n+1}} \cdot \ldots$$

same throughout the whole extent of an integral. An instance of this, in the series

$$C - \frac{\theta}{\cdot 2} = \frac{\sin\theta}{1} + \frac{\sin 2\theta}{2} + \ldots$$

was remarked by Daniel Bernouilli, in the *Act. Petrop.* Landen also notices, that this equation is false, when $\theta = 0$, but without explanation. Other instances occur in the 'Essay on logarithmic transcendants', pp. 52–3, in Lacroix's *Traité de Calcul etc* 4to, tom. III, p. 141, as well as some reflections on this difficult subject in page 483 of the same volume. The cause of these anomalies has not been satisfactorily explained. If we may hazard a conjecture, it must be looked for in the evanescent or infinite values of some of the differential coefficients of the function integrated, causing that function for an instant to change its form. Or, it may have some connection with exponentials, since all the instances which have hitherto been adduced depend on that species of function.

The discovery of partial differentials, has been generally attributed to d'Alembert. He certainly was the first who applied them to mechanical problems, and perceived their vast utility in all the more difficult applications of analysis to physics. But, if he is to be considered as the inventor, who first solved an equation of the kind, and who, when their importance was acknowledged, contributed more than any other to the improvement and progress of this calculus; the glory of their discovery will undoubtedly belong to Euler. In the 7th vol. of the *Acta Acad. Petropolitanae*, is a memoir of his, entitled, 'Methodus inveniendi aequationes pro infinitis curvis ejusdem generis' (A.D. 1735). In this paper, and more particularly in a supplement, are given the solutions of a number of partial differential equations, of which the most general is

$$q = Xz + pR$$

X being a function of x, and R a function of x and y.

In the latter part of the 'additamentum ad dissertationem', he proceeds to / integrate some equations of the second order. The *Réflexions sur la cause générale des vents* which contains d'Alembert's first application of partial differential equations, was not published till the year 1747, and it gained the prize of the Academy of Berlin in 1746. It would lead us too far to trace, successively, the various improvements which the new theory underwent in the hands of Euler, Lagrange, Laplace, Monge, Parseval, and a multitude of great men, whom the vast importance of the subject incited to its prosecution. Notwithstanding every exertion, the theory however continues to present a multitude of difficulties. The analogy which was supposed to exist between the arbitrary functions, which enter into the integrals of partial differential equations, and the arbitrary constants in equations of total differentials, is found not to hold beyond the first degree; after which, even the number of these arbitrary functions is unknown. Instances have been adduced, where, besides the arbitrary functions, arbitrary constants also must be introduced to complete the integral.* The application of definite integrals to the integration of these equations, presents a wide field of research, as well as the promise of great discoveries. A very curious memoir of Laplace is to be found in the *Mém. Acad. des Sciences*, 1779, where this subject, among many others, is discussed with considerable success.†

In applying the test of integrability to differential equations, some were found, which could not be made to satisfy the equations of condition. These were for a long time deemed absurd, until about the year 1784, when Monge perceived their connection with the theory of curve surfaces, and demonstrated that these equations admit of solutions, corresponding to the curves of double curvature, formed by the successive intersections of a curve surface, whose parameter varies; and discovered methods of transforming any equation of this kind into an equation of partial differentials, and also of solving the converse problem. From this, he proposed obtaining solutions of equations of partial differentials, which are not integrable by other methods. But the example he gives, shows, that to expect success in such an enquiry, we must be familiar with space, considered as of three dimensions, and also with a numerous collection of curves and curve surfaces situated in it; such considerations would but add intricacy to a subject already difficult: hints for the advancement of analysis may be derived from foreign sources, but must always be improved and cultivated by its own powers. Lacroix has given an excellent analytical theory of this kind of equations.

To the practice which prevailed in the infancy of analytics, of proposing to / geometers the solution of difficult problems, we owe many of its improvements. The method of variations, among others, is much indebted to this source. Obliged however to consult brevity, we must refer to the late treatise on *Isoperimetrical Problems*, for a full account of its history: a work which being undoubtedly in the hands of the majority of our readers, must render superfluous all we could say on that subject. The chief difficulty now consists in distinguishing the maxima from the minima, which depends in general on the solution of difficult differential equations. The inverse of the method of variations does not appear to have received any attention; it would consist in

* Monge, *Savants Étrangers*, vol. VII, p. 322.

† The great desideratum in the integration of an equation by definite integrals is, that whenever it is susceptible of actual resolution, these integrals should give it – a condition which does not always hold good. See Laplace, 'Mémoire sur divers points d'Analyse', *Journ. de l'École Polytechnique*, No. 15.

solving such problems as the following: 'Given a curve, to find what properties of maxima and minima it possesses.'

The method of variations was applied by its inventor to differential equations, and also to those of finite differences. Cases may occur, in which it would be necessary to apply it to equations of mixed differences; these relate to a number of difficult problems, such, for instance, as this: 'What must be the nature of a curve, such, that drawing to any point, an ordinate and also a normal, and at the foot of this normal another ordinate; the curvilinear area intercepted between the first and last ordinate, may be a maximum or a minimum?'

There are but few instances in the history of science, in which the path of the inventor has been the shortest and most direct. Thus it occurs, that the method of finite differences, which would most naturally have preceded that of the differential calculus, was not discovered until many years after ⟨its⟩ inventor, Brook Taylor, published it in a work, entitled, *Methodus incrementorum directa et inversa*, a book noted for its obscurity. Montmort, Stirling, and Emerson, made several improvements, whch may be found in the *Philosophical Transactions*, and also in their respective works. Moivre first investigated the nature of recurring series, on which, Laplace remarks, 'Sa théorie est une des choses les plus curieuses, et les plus utiles que l'on ait trouvées sur les suites.'[a] It has been well observed, by the same author, that the first who summed a geometrical or arithmetical series, had really integrated an equation of finite differences. The same remark, as is well known, applies to any recurrent series. It was not, however, till Lagrange in the *Mélanges de Turin*, vol. I applied d'Alembert's method of indeterminate coefficients, to an equation of differences of the first degree, that this truth was perceived. In the fourth volume of the same work, Laplace published a memoir, in which the two celebrated theorems of Lagrange, respecting equations of common differentials, are extended to those of differences, and to those of partial differentials of a similar description, with constant coefficients. Returning to the subject, in a memoir communicated to the Academy of Paris, he integrates a very extensive class of equations of partial differences, involving any number of variable indices – and also a singular species of equations, frequent in the theory of chances, called by him, *equations rentrantes.* /

Of equations beyond the first degree, very few have been solved, if we consider their amazing variety and importance. Monge has given a short paper on the subject, in the Memoirs of the Academy of Sciences, for 1783. Laplace also in the 15th number of the *Journal de l'École Polytechnique*, has by a most happy combination of the equations of differences, with the discover of Euler, respecting elliptic transcendants, integrated a few very difficult ones. The nature

[a] *Translation:* His theory is one of the most interesting and useful ones discovered on series.

of the integrals obtained by Charles, requires a fuller investigation. They might perhaps receive considerable extension. When the variables are mixed in the indices, thus, u_{x+y}, u_{x-y+1}, u_{xy}, ...; the subject seems to have passed altogether unnoticed. Many equations containing such expressions, are impossible or contradictory.

Euler first remarked, that the constant introduced by integrating an equation between u_x, u_{x+1}, ... may be an arbitrary function of cos $2\pi x$, a remark which afterwards in the hands of Laplace (*Savants Étrangers*, 1773) became the foundation of a very general theory of determining functions from given conditions. To notice all the applications of the theory of finite differences, or all the profound researches which have enriched it, would occupy volumes. We cannot, however, pass over the theorems relating to the analogy of differences to powers, given first by Lagrange without demonstration, in *Mém. de Berlin*, 1772. A demonstration by Laplace, appeared in the *Mém. des Savants Étrangers* for the following year, and another in the *Mém. de l'Acad.* for 1777. In 1779, appeared that noted memoir, in which the same author exhibited the principles of his powerful and elegant 'Calcul des fonctions génératrices'. In this,* he gives a far more systematic proof of the theorems, and extends them to any number of variables. Since that period, they have been demonstrated by Arbogast, in the 6th article of his, *Calcul des dérivations*, where, by a peculiarly elegant mode of separating the symbols of operation from those of quantity, and operating *upon them* as upon analytical symbols, he derives not only these, but many other much more general theorems with unparalleled conciseness. Brinkley has given a demonstration of the theorem

$$\Delta^n u_x = \{\varepsilon^{\Delta x \cdot \frac{d}{dx}} - 1\}^n \times u_x$$

of considerable elegance, and a simplicity truly elementary.† /

We are now naturally led to say a few words on the 'Calcul des fonctions génératrices'. Its object may be best stated, in the words of its inventor:‡ 'C'est de ramener au simple développement des fonctions, toutes les opérations relatives aux différences et spécialement l'intégration des équations aux

* This memoir forms the greater part of the first chapter of the *Théorie Analytique des Probabilités*. Its first principles, and the demonstration of the theorems, on the analogy of differences with powers, are given briefly in the 15th No. of the *Journal de l'École Polytechnique*. A slight sketch of the method alluded to, is also to be found in the 9th book of the *Mécan. Cél.*, tom. IV, p. 204.

† Philos. Transactions, 1807, Part I. He has extended his researches to the actual expansion of the series themselves, to which these theorems lead; such, for instance, as

$$\Sigma u_x = \frac{1}{\Delta x} \cdot \int u_x dx - \frac{u_x}{2} + \Delta x \cdot {}^1A_x + (\Delta x)^2 \cdot {}^2A_x + \ldots$$

But in point of clearness and elegance, by no means with equal success: partly owing to an unfortunate notation, and partly to the perpetual employment of the theory of combinations. In his results, he has for the most part been anticipated by Laplace, and others.

‡ *Journal de l'École Polyt.* Mém. sur divers points d'Analyse.

différences ordinaires et partielles',[a] and from the extreme facility with which all the known theorems flow from it, and its fecundity in affording new ones, it is, perhaps, in the present state of science, the best adapted of any, for explaining the general theory of differences, and the development and transformation of series. At the same time, it must be confessed, that owing to its extreme generality, and the consequent complexity of many of its operations, particularly in what regards the transformation of series, it is an instrument to be placed only in the hands of an experienced analyst. If we except the calculus of variations, it is the only method, perhaps, of any considerable importance, which has received its first and last touches from the same hand; and which first appeared in a state of perfection, very little short of what it at present possesses. The latest work which treats of this subject, is the *Théorie Analytique des Probabilités*, the first part of which is dedicated to a very full exposition of the method.

After the solution of differential equations, and those of finite differences, it was natural to consider those, in which the difference and differential of a quantity both occurred. These have been called equations of mixed differences. They were not attempted until about the year 1779, when Condorcet and Laplace obtained the integrals of some few particular cases. But problems which required their application had been proposed and resolved by several geometers in the *Acta eruditorum*, and in the *Acta Acad. Petrop.* long before this time; their solutions were obtained by certain insulated artifices, dependent on the peculiar nature of the problems.

On this subject, a wide field is extended for investigation, and one which abounds with difficulties. The little that has yet been discovered, is chiefly contained in two papers, one by Biot, in the *Mémoires de l'Institut*, and the other by Poisson, in the *Journal de l'École Polytechnique*; the former treats chiefly of that kind of equations, called *équations successives*; the latter integrates a few particular equations, by employing a substitution used by Laplace, in integrating some equations of partial differentials.

There is no branch of mathematical science which has not received improvements, from the profound and original genius of Euler. Several are indebted to him for their existence: of this latter class is the knowledge of the nature, and use of definite integrals, a subject, to which the greatest geometers of the present age look as the most probable source of future discoveries and improvements. /

Legendre, in a work lately published, entitled *Exercices de Calcul Intégral*, has collected all that has been discovered on this subject, and demonstrated the results with peculiar elegance; the greater part is extracted from the various works of Euler, and also a considerable portion from some memoirs of Laplace.

[a] *Translation:* It is to reduce to the simple expansion of functions, all operations on differentials, and especially the integration of ordinary and partial differential equations.

Its application to the solution of differential equations, from which so much is expected, does not enter into the plan of his work; this however has been well treated, by Lacroix, in the third volume of his *Traité de Calcul etc.*

But the most elegant part of the theory of definite integrals is, their application to such problems in finite differences, as involve functions of very high numbers. In many cases (particularly in the theory of chances) it has been well remarked, that the mere impracticability of the arithmetical operations requisite to obtain a result (however simple its analytical expression) must for every preclude our advancement; were it not for some mode of approximation, which, grasping the prominent terms of an expression, in a formula easily reduced to numbers, should throw the minor ones into the background, to be valued by a series converging the more rapidly the higher the numbers employed become. A more appropriate instance cannot be adduced, than the equation.

$$1 \cdot 2 \cdot 3 \ldots s = s^{s+\frac{1}{2}} \cdot \varepsilon^{-s} \cdot \sqrt{2\pi} \cdot \left\{1 + \frac{1}{12s} + \frac{1}{288s^2} + \ldots\right\}$$

in page 129 of the *Théorie Anal. des Probabilités*, or the method in which the author, in page 259 of the same work, computing the probability of a primitive cause influencing the inclinations of the cometary orbits, employs a definite integral to effect with conciseness, a calculation surpassing, without that assistance, the utmost limits of human patience and industry.

In this part of his career, Laplace stands unrivalled. Stirling indeed, and Moivre had seen, and in some cases obviated the difficulties, arising from the immensity of the numbers under consideration. The discoveries of Euler, gave a connection and unity to their results. But it was not till the labours of Lagrange, Condorcet, and Laplace had brought the theory of finite differences to considerable perfection, that the definite integrals were applied by the latter, to the solution of these equations, and a clear and strong light thrown over this most obscure part of the mathematics. The whole of this interesting theory, has been digested into one work (*Théor. de Prob.* above cited) which for comprehensive views, for depth of investigation, and the purity of its analysis, may justly be looked up to, as marking the highest point to which the science of abstract number has yet attained.

In analytical investigations, we frequently meet with a series of quantities connected together by multiplication, whose differences are constant. These have / received various names from different geometers; Vandermonde called them powers of the second order; Kramp adopted the appellation of facultés numériques; and Arbogast named them factorials. Each of these writers treats them in a different manner, and employs a peculiar notation for them: that of Vandermonde is perhaps the best adapted to the subject, as it has a considerable resemblance to that of exponents, and possesses also an advantage, which is by no means inconsiderable, that of being capable of a ready extension to powers of all orders. Lacroix has adopted it, and by its help demonstrated many properties of powers of the second order. Those of the superior orders have not as yet been examined. Kramp has deduced from his *Théorie des facultés numériques*, some contradictions which require examination. One of his most useful theorems affords a method of transforming any power of the second order, into a series which converges *ad libitum*. It appears probable, that the thory of powers of different orders may afford a useful method of classing transcendents, as they can frequently be reduced to definite integrals, and by this means their value be obtained, when the index is fractional.

Interpolations were at first considered, as a branch of the method of finite differences, and as such they were usually treated of together. Wallis appears first to have applied this name, to the determination of the intermediate term of a series, whose law of formation is known. The extraction of roots is an interpolation of powers, and may be considered as an extension of the meaning of exponents, from whole numbers to fractions. Perhaps it might not be unworthy of consideration, whether the meaning of the indices of differentiation, could not be considerably extended. Euler seems to have had the first idea* of interpolating the series

$$dy; d^2y, d^3y, \ldots$$

Laplace has extended his researches on this subject to considerable length, and has given the value of such expressions as the following, by converging series and definite integrals

$$\frac{d^n \cdot x^m}{dx^n}, \qquad \Delta^n \cdot x^m$$

so as to allow of evaluation for fractional values of n. But the indices themselves might be supposed to vary *continuously*, and such expressions as these

$$\left\{\frac{d^n\left(\frac{d^p y}{dx^p}\right)}{dp^n}\right\}, \qquad \left(\frac{d\Delta^n y_x}{dn}\right)$$

become the subject of analytical investigation. Or the index of a function might vary, as in the following instance. /

Let

$$f(x) = \sqrt{\left(\frac{2x}{1+x}\right)}, \qquad f^2(x) = \sqrt{\left(\frac{2f(x)}{1+f(x)}\right)}, \quad \text{and so on,}$$

* Leibnitz, it is true, in a letter to J. Bernouilli, mentions fractional indices of differentiation. Euler, however, first determined the value of such an expression as $d^{\frac{1}{2}}y$, y being a certain function of x.

then we shall have,

$$\frac{df^n(\sec v)}{dn} = -\frac{v \cdot \log 2}{2^n} \cdot \tan\left(\frac{v}{2^n}\right) \cdot \sec\left(\frac{v}{2^n}\right)$$

and again (making use of Arbogast's notation) if $y = a^x$

$$D^{p_n}_{p_{n-1}} \cdot D^{p_{n-1}}_{p_{n-2}} \cdot \ldots \cdot D^{p_2}_{p_1} \cdot D^{p_1}_x \cdot y = a^x \cdot (\log a)^{p_1} \cdot (\log^2 a)^{p_2} \cdot \ldots \cdot (\log^n a)^{p_n}$$

where $\log^2 a = \log \log a$, $\log^3 a = \log \log \log a$, and so on.

It has been observed by Charles, that the equation

$$\Delta y_n = b \cdot \frac{dy_n}{dx}$$

may be transformed into an integral, in which the index of integration is variable. Its solution then is[a]

$$y_{-n} = b^{-n}\, \varepsilon^{-x/b} \cdot \int^n \varepsilon^{x/b} \cdot y_0 \cdot dx^n$$

or, which comes to the same,

$$y_n = b^n\, \varepsilon^{-x/b} \cdot D^n_z\{y_0 \cdot \varepsilon^{x/b}\}$$

y_0 being any function of x. Laplace, in the *Théorie Anal. des Prob.* gives other instances of the same kind.*

That such expressions are not merely analytical curiosities, but relate to the most difficult and important theories, is confirmed by the opinions of the most / eminent analysts; and though on mathematical subjects when proof can be produced, no weight must be allowed to authority; yet when the former is deficient, our judgement may surely be influenced by the latter.

The importance of adopting a clear and comprehensive notation did not, in the early period of analytical science, meet with sufficient attention; nor were the advantages resulting from it, duly appreciated. In proportion as science

* The integral of the equation of mixed partial differences

$$u_x = A \cdot \left(\frac{d^\alpha u_{x-1}}{da^\alpha}\right) + B \cdot \left(\frac{d^\beta u_{x-2}}{db^\beta}\right) + C \cdot \left(\frac{d^\gamma u_{x-3}}{dc^\gamma}\right) + \ldots$$

may be easily shown to be

$$u_x = A^{x/1} \cdot \left(\frac{d^{\alpha x/1}\ \psi_1(a)}{da^{\alpha x/1}}\right) + B^{x/2} \cdot \left(\frac{d^{\beta x/2}\psi_2(b)}{db^{\beta x/2}}\right) + C^{x/3} \cdot \left(\frac{d^{\gamma x/3}\psi_3(c)}{dc^{\gamma x/3}}\right) + \ldots$$

$\psi_1, \psi_2, \ldots$ being the characteristics of arbitrary functions: an expression which except $\alpha, \beta, \gamma, \ldots$ are respectively multiples of 1, 2, 3, ... must necessarily involve fractional indices of differentiation for some values of x.

[a] $\int^n$ means integrate n times.

advanced, and calculations became more complex, the evil corrected itself, and each improvement in one, produced a corresponding change in the other. Perhaps no single instance of the improvement or extension of notation, better illustrates this opinion, than the happy idea of defining the result of every operation, that can be performed on quantity, by the general term of function, and expressing this generalization by a characteristic letter. It had the effect of introducing into investigations, two qualities once deemed incompatible, generality and simplicity. It now points out a calculus* perhaps more general than any hitherto discovered, and which should be called the calculus of functions, a name that more naturally belongs to it, than to that which Lagrange has so classically treated in the work which bears this name, although this latter is a branch of it.

Its object would be in general, the determination of functions from given conditions of whatever nature, whether depending on the successive terms of their developements, or on a series of indices differing by unity; or lastly, on a species of equations depending on the successive *orders* of the same function, of which the first mention we believe is made in one of the papers which compose the present volume. The necessity of this calculus was perceived soon after the discovery of equations of partial differentials, when it became requisite to determine the arbitrary functions which enter into their integrals, so as to satisfy given conditions. Euler and d'Alembert determined a few particular cases. Lagrange, in a memoir entitled, 'Solution de différents problèmes de calcul intégral' resolved the equation

$$T = \alpha \cdot \varphi\{t + a \cdot (h + kt)\} + \beta \cdot \varphi\{t + b(h + kt)\} + \ldots$$

Monge also has given several papers upon the subject, in the Memoirs of the Societies of Turin and Paris. The method of Laplace for reducing an equation of the first order, where the difference of the independent variable is any function of the variable itself, to one wherein that difference is constant, is well known. It had not, however, hitherto been shown to be possible to reduce *every* equation of the former kind, to one of the latter. This object is, however, accomplished in the following pages. Still, it appears by no means natural, to resolve these equations, by / means of finite differences, and it were much to be wished, that some independent method could be discovered, by which they might be treated.

One of the most striking advantages of the theory *of functions*, is, that it seems equally adapted to the proof of the most elementary truths, and to that of the most complicated and abstruse theorems. The latter part of this assertion,

* Quamobrem non solum in hoc negotio, sed in plurimis aliis casibus, maxime utile foret, si functionum doctrina magis perficeretur et excoleretur. Euler, *Act. Acad. Petropol.*, tom. VII. ⟨*Translation:* Why, not only in this business, but also in many other cases, it might be of considerable use, if the wise men brought the doctrine of functions to completion and perfection.⟩

no one will be inclined to deny. An example of the former may be found in Laplace's proof of the decomposition of forces in the *Mécanique Céleste.*'*

There are still many problems in the theory of functions, which analysis seems / to afford no means of attacking directly. Among which, may be enumerated the greater part of those which lead to an equation, containing a definite integral, where the unknown function enters under the integral, sign. For instance, suppose it were required to find the form of a function $\varphi(x)$, such that the integral

$$\int dx \cdot F\{a, x, \varphi(f(x))\}$$

taken between the limits $x = 0$, $x = a$, should equal any assigned function of a. F and f being given characteristics.

* This demonstration consists of two parts. In the first he proves, that the diagonal of a rectangle whose sides represent the separate forces, will on the same scale represent the *quantity* of the resultant. The second is devoted to shew, that it represents also its direction. This part seems to be generally considered, as deficient in clearness and simplicity. What we have here to remark, is, that it is *redundant*. In fact, by the combination of Laplace's three equations,

$$x = z\cdot\varphi(\theta) \qquad y = z\cdot\varphi\left(\frac{\pi}{2} - \theta\right) \qquad x^2 + y^2 = z^2$$

we obtain

$$\{\varphi(\theta)\}^2 + \left\{\varphi\left(\frac{\pi}{2} - \theta\right)\right\}^2 = 1$$

an equation which suffices for determining the nature of the function φ, and from which, by known processes, combined with the conditions of the question; it is easy to obtain $\varphi(\theta) = \cos\theta$, and $x = z\cdot\cos\theta$, which is the equation to be deduced.

Another very remarkable instance of the use of the theory of functions, in demonstrating elementary truths, may be found in the following demonstration of Euclid's 47th, which has generally been thought to admit of none, but a geometrical proof: call a, b, c, the sides, A, B, C, the opposite angles of a right-angled triangle, $C = \pi/2$. It is easy to see, that the following equations hold good:

$$b = c\cdot\varphi(A); \qquad a = c\cdot\varphi(B); \tag{1}$$

drop a perpendicular p, dividing c into two parts x, y, and we shall have, in the same manner,

$$x = b\cdot\varphi(A); \qquad y = a\cdot\varphi(B)$$

and of course,

$$x + y = c = b\cdot\varphi(A) + a\cdot\varphi(B)$$

from which, eliminating $\varphi(A)$ and $\varphi(B)$, by equations (1)

$$c^2 = a^2 + b^2, \quad \text{QED.}$$

Having proved from other principles, that $A + B + C = \pi$, we shall have the very same equation

$$1 = \{\varphi(A)\}^2 + \left\{\varphi\left(\frac{\pi}{2} - A\right)\right\}^2$$

and thus we obtain $b = c\cdot\cos A$, $a = c\cdot\cos B$. It is only by this way of proceeding, or some analogous one, that we can ever hope to see the elementary principles of trigonometry, brought under the dominion of analysis. But this is not the place to proceed farther with the subject. It may suffice to have thrown out a hint, which may be followed up at some future opportunity.

The examination of the properties and relations of numbers, constitutes a distinct branch of mathematical enquiry, almost entirely of modern origin; so abstract, and apparently so far removed from the confines of utility, that it seems to have attracted little attention from the generality of those, who have dedicated themselves to the pursuits of science. To the few, however, who have thought it worth their while to explore its more profound recesses, it has proved a mine, fertile in the most brilliant produce. Euclid, in his 7th book, has given the elements of transcendental arithmetic (a name appropriated to it by Professor Gauss). In the work of Diophantus, notwithstanding the ingenuity of the author, we discover the infancy of science in the absence of that generalization so happily adopted by modern writers. It consists of a variety of insulated problems, relating to the solution of certain indeterminate equations, rather than to the properties of numbers. Indeed, the indeterminate analysis, and the theory of numbers, form two branches of enquiry, which (however nearly connected) ought to be carefully distinguished, in any systematic arrangement of our knowledge. The former must be considered as a province of the pure analytics. Its attainment is indeed necessary for the perfection of the latter (which should rather be regarded in the light of an *application* of analysis) but is by no means limited to this one object. Lacroix* has introduced it with a very elegant effect, in that part of the theory of curves, which relates to their construction by points.

The resolution of the indeterminate equation of the first degree, is said to be due to Bachet. Euler, in the Petersburg Commentaries, exhibited a method of obtaining any number of solutions of that of the second, provided one particular one be known; but it has been remarked, that his methods do not afford *all* the possible solutions. This, however, has been effected at length by Lagrange, in the *Mém. Acad. de Berlin* (1767 and 1768). His method consists in reducing successively by a series of operations, the coefficients of the equation

$$ax^2 + by^2 = x^2$$

(to which form, every equation of the second degree may be reduced) till one of / them becomes unity; in which case, the resolution is easy. Gauss also, in his *Disquisitiones Arithmeticae*, No. 216. has, by a method entirely different, shown how to obtain all the solutions, of which an equation of the second degree admits. It is extremely remarkable, that the Hindu algebraist, Bhaskara Acharya, who flourished about the year 1188, had also succeeded so far in his attempts on this difficult problem, as to derive any number of solutions from one, previously known in the case of $ax^2 + b = y^2$.

The theorems given by Fermat without demonstration, form, without doubt, the most remarkable era in the theory of numbers. Too general, and too extra-

* *Traité de Calc. etc.* 2nd edn, vol. I, note to p. 417.

ordinary in their nature to escape notice, they seem to have been the principal cause of the advances, which have since been made in this theory, by drawing the attention of mathematicians to their demonstration. Euler, than whom none ever entered with greater ardency into this career, has proved some of the principal. Lagrange has supplied the demonstrations to others: still, however, many remain, of which no proof has been offered. It has been suggested, that Fermat was indebted to the method of induction, for the discovery of many of his theorems; an opinion rendered probable by the observation of Euler, that one of them relating to prime numbers is not true. This method is perhaps more applicable to researches in the theory of numbers, than to any other branch of abstract investigation; but it is of dangerous use, and should be supported by a large number of instances.* /

Among the later discoveries in the theory of numbers, we have to enumerate two of the most surprising, which perhaps are to be found in the whole circle of analytics. The first is a formula of Lagrange, obtained by induction, for determining the number of primes contained between given limits. It has been demonstrated by Legendre, in his *Essai sur la Théorie des Nombres*, although not very rigorously, it must be confessed; such is the difficulty of the subject. Indeed, the author has given it only as an attempt. The other is that celebrated theorem of Gauss, given in his *Disquisitiones Arithmeticae*, on the resolution of the equation $x^n - 1 = 0$, n being any prime, viz. that this equation may be reduced to α equations of the degree a, β of the degree b, and so on, where $n - 1 = a^\alpha \cdot b^\beta \ldots$ Of course, the resolution of the equation $x^n - 1 = 0$, where

* The substitution of 0, 1, 2, ... for x, in the expression $x^2 + x + 41$, gives a series of numbers, of which the 40 first terms are primes, as Euler has remarked: yet it is easy to shew, that no algebraic function of x can in all cases represent a prime. The reason of this singular circumstance, and of a variety of similar coincidences, has since been satisfactorily explained, and the property demonstrated *a priori*. Fermat, deceived by a similar induction, asserted that all the numbers contained in the formula $2^{2^n} + 1$, are primes which Euler has since (as above alluded to) shown to fail, in the case of $n = 5$.

The following theorems are derived solely by induction:

1mo. An indefinite number of integer values of x may be found, which render $(7^x - 1)/10^a$ an integer, to which we may add, that the formula $(7^{8\cdot 10^x} - 1)/10^{x+2}$ is always an integer, as are also the formulae

$$\frac{3^{2^x} - 1}{2^x}, \frac{5^{2^x} - 1}{2^x}, \text{ and in general, } \frac{(2n-1)^{2^x} - 1}{2^x},$$

which last may be easily shown, *a priori*, as well as a variety of expressions of the same description.

2do. The expression $(3^{10^x} - 1)/10^{x+1}$ is an integer, and, it is somewhat remarkable, that this integer is always of the form $100n + 22$.

3°. If A be such, that 5^n is congruous to A (modul. 10^k) then will also $5^{n+2^{k}\,{}^{2\cdot i}}$ be congruous to the same A, to the same modulus, and consequently, $5^{n+2^{k-2}\cdot i} \equiv 5^n$ (modul. 10^k) i being any integer. In other words, the formula

$$\frac{5^{n+2^x\cdot i} - 5^n}{10^{x+2}}$$

is always an integer, x and i being integers.

n is a prime of the form $2^m + 1$, requires only the application of quadratics. Thus the division of the circle into 17, 257, 65,537 parts, may be accomplished by the description only of circles and straight lines.

To enter into any account of the advances made in the mixed analytics, would far exceed our limits. There is one point, however, which we cannot forbear cursorily touching upon, on account of the great difficulty of reducing its conditions into symbolic language. We allude to the geometry of situation. Like the theory of numbers, at the first glance it seems barren and useless, but on a nearer examination is found abounding with interesting relations. Like that theory too, its cultivators have hitherto been few, but eminent, distinguished for that restless spirit of enquiry, which is ever upon the wing in search of new truths, and that invention which knows how to extract them, from the most unpromising hints. Leibnitz, appears to have found its first application, in considering the game of solitaire. A similar case (the problem of the knight's move at chess) occupied the attention of Euler, and afterwards of Vandermonde, who adapted to it a notation analogous to that, by which the position of a point in space is determined, by three rectangular co-ordinates. In a more advanced state, it might, perhaps, embrace problems of a much higher order of difficulty, such as the following: 'Given n points in space; to find the course to be pursued, so that setting off from any one, and passing at least once through all the rest, on returning to the original position, the least possible space shall have been described.' Such is the brief account of a theory yet in its first infancy. On its basis some future Leibnitz may perhaps hereafter lay the foundation of a name great as that of its original inventor.* /

The preceding pages have been devoted to a slight account of the history and present state of analytical science, that branch of human knowledge, of which Laplace has justly observed, 'C'est le guide le plus sur qui peut nous conduire dans la recherche de la vérité'.[a] But some account will naturally be expected of the source itself, from which the present work emanates. Of this however, very little need be said, but, that it consists of a few individuals, perhaps too sanguine in their hopes of promoting their favourite science, and of adding at least some trifling aid to that spirit of enquiry, which seems lately to have been awakened in the minds of our countrymen, and which will no longer suffer them to receive discoveries in science at second hand, or to be thrown behind in that career, whose first impluse they so eminently partook. The time

* In the *Journal de l'École Polytechnique*, An. 10, is a memoir on polygons and polyhedrons, by Poinsot, in which he shews, that there exist other regular polygons besides the equilateral triangle, the sum of whose angles is equal to two right angles, and also, that there are more than five regular polyhedrons. The whole memoir relates to geometry of situation, and forms the introduction to some more considerable researches, which the author promises in a future paper.

[a] *Translation:* It is the most certain guide to lead us in the search for truth.

perhaps is not far distant, when such an attempt will be regarded in an honourable light, whatever may be its success.

Meanwhile the view we have taken of the subject, appears by no means to lead to the mortifying conclusion, deduced by a foreign geometer of considerable eminence; 'que la puissance de notre analyse est a-peu pres épuisée'.[a] The golden age of mathematical literature is undoubtedly past. Another, 'less fine in carat', may however yet succeed. The motive which could draw forth so severe a sentence on the success of future exertions we will forbear to enquire, but it must surely be looked for elsewhere, than in the real interest of science which can never be promoted by repressing the ardour of research, or extinguishing the hope of reward. The foundations of a vast edifice have been laid; some of its apartments have been finished; others yet remain incomplete: but the strength and solidity of the basis will justify the expectation of large additions to the superstructure.

Attentively to observe the operations of the mind in the discovery of new truths, and to retain at the same time those fleeting links, which furnish a momentary connection with distant ideas, the knowledge of whose existence we derive from reason rather than perception, are objects in whose pursuit nothing but the most patient assiduity can expect success. Powerful indeed, must be that mind, which can simultaneously carry on two processes, each of which requires the most concentrated attention. Yet these obstacles must be surmounted, before we can hope for the discovery of a philosophical theory of invention; a science which Lord Bacon reported to be wholly deficient two centuries ago, and which has made since that time but slight advances. Probably, the era which shall produce this discovery is yet far distant. The capital of science, however, from its very nature, must continue to increase by gradual yet permanent additions; at the same time that all such additions to the common stock yield an interest in the power they afford of multiplying our combinations, and examining old difficulties in new points of view. It is this connection with fresher sources, which can restore fertility to subjects apparently / the most exhausted, and which cannot be too earnestly recommended to those who wish to enlarge the limits of analysis. The fire of improvement, however dormant, and seemingly extinct, may yet break forth at the contact of some external flame. The history of mathematics affords too many instances of the most distant principles coming into play on the most unexpected occasions, to allow of our ever despairing of success in such enquiries.

One inconvenience however, results as a necessary consequence from the continued accumulation of indestructible knowledge. The beaten field of analysis, limited as it is when compared with the almost boundless extent which

[a] *Translation:* that the power of analysis is almost exhausted.

remains to be explored, is yet so considerable with respect to the powers of human reason, and (if we may be allowed to pursue the metaphor a little farther) so intersected with the tracks of those who have traversed it in every direction, as to become bewildering and oppressive to the last degree. The labour of one life would be more than occupied in perusing those works on the subject which the labour of so many has been spent in composing. The multitude of different methods and artifices, which for the most part lead only to the same results, and whose power is limited by the same points of difficulty, is at length grown into a very serious evil. Our continental neighbours seem sensible of this, if we may judge from the number of works which have appeared within these few years, digesting various points into a systematic form. But there is still much to be done in this line. That man would render a most invaluable service to science, who would undertake the labour of reducing into a reasonable compass the whole essential part of analysis, with its applications, curtailing its superfluous luxuriance, rejecting its artificial difficulties, and giving connection and unity to its scattered members. /

II

ON CONTINUED PRODUCTS

From *Memoirs of the Analytical Society* (Cambridge, 1813), pp. 1–31

Assume the equation,

$$\psi x \cdot \varphi x = \chi x \tag{a}$$

in which the functions denoted by ψ, φ, χ, are only subject to one condition, namely, that the product of the two first is equal to the last. In other respects they are perfectly arbitrary.

Take any other arbitrary function fx. It is evident that the generality of equation (a) is neither increased nor diminished by putting φfx for χx.

Let this be done, then,

$$\psi x \cdot \varphi x = \varphi fx. \tag{b}$$

In this equation put for x successively $fx, ffx, fffx$, and $f^n x$; then multiply the resulting equations together,

$$\begin{aligned} \psi fx \cdot \varphi fx &= \varphi f^2 x \\ \psi f^2 x \cdot \varphi f^2 x &= \varphi f^3 x \\ \psi f^3 x \cdot \varphi f^3 x &= \varphi f^4 x \\ &\vdots \\ \psi f^n x \cdot \varphi f^n x &= \varphi f^{n+1} x \end{aligned}$$

therefore,

$$\frac{\varphi f^{n+1} x}{\varphi fx} = \psi fx \cdot \psi f^2 x \cdot \psi f^3 x \cdot \ldots \cdot \psi f^n x. \tag{c}$$

In (b) we may without limiting the equation put $(\psi x)^{p-1}$ for ψx, it then becomes

$$(\psi x)^{p-1} \varphi x = \varphi fx. \tag{d}$$ /

Dividing by $(\psi x)^p$, we have

$$\frac{\varphi x}{\psi x} = \frac{\varphi fx}{(\psi x)^p}. \tag{e}$$

Let us assume

$$y = \overset{1/p}{\left\{\frac{\varphi fx}{\psi fx}\right.} \overset{1/p}{\left\{\frac{\varphi f^2x}{\psi f^2x}\right.} \overset{1/p}{\left\{\frac{\varphi f^3x}{\psi f^3x}\right.} \overset{1/p}{\left\{\vphantom{\frac{1}{1}} \right.} \cdots \overset{1/p}{\left\{\frac{\varphi f^nx}{\psi f^nx}\right\}}^{*}$$

$$y = \overset{1/p}{\left\{\varphi fx\right.} \overset{1/p}{\left\{\frac{\varphi f^2x}{(\psi fx)^p}\right.} \overset{1/p}{\left\{\frac{\varphi f^3x}{(\psi f^2x)^p}\right.} \overset{1/p}{\left\{\vphantom{\frac{1}{1}} \right.} \cdots \overset{1/p}{\left\{\frac{\varphi f^nx}{(\psi f^{n-1}x)^p}\right.} \overset{1/p}{\left\{\frac{1}{(\psi f^nx)^p}\right\}}.$$

Multiplying by $(\varphi f^{n+1}x)^{1/p^{n+1}}$, we have

$$y(\varphi f^{n+1}x)^{1/p^{n+1}}$$

$$= \overset{1/p}{\left\{\varphi fx\right.} \overset{1/p}{\left\{\frac{\varphi f^2x}{(\psi fx)^p}\right.} \overset{1/p}{\left\{\frac{\varphi f^3x}{(\psi f^2x)^p}\right.} \overset{1/p}{\left\{\vphantom{\frac{1}{1}} \right.} \cdots \overset{1/p}{\left\{\frac{\varphi f^nx}{(\psi f^{n-1}x)^p}\right.} \overset{1/p}{\left\{\frac{\varphi f^{n+x}x}{(\psi f^nx)^p}\right\}}.$$

Which becomes, by applying equation (*e*),

$$(\varphi f^{n+1}x)^{1/p^{n+1}}y = \overset{1/p}{\left\{\varphi fx\right.} \overset{1/p}{\left\{\frac{\varphi fx}{\psi fx}\right.} \overset{1/p}{\left\{\frac{\varphi f^2x}{\psi f^2x}\right.} \overset{1/p}{\left\{\vphantom{\frac{1}{1}} \right.} \cdots \overset{1/p}{\left\{\frac{\varphi f^nx}{\psi f^nx}\right\}} = \overset{1/p}{\left\{\varphi fx \cdot y\right\}}.$$

Hence,

$$y^p(\varphi f^{n+1}x)^{1/p^n} = y\varphi fx$$

$$y^{p-1} = \frac{\varphi fx}{(\varphi f^{n+1}x)^{1/p^n}}$$

$$y = \left\{\frac{\varphi fx}{(\varphi f^{n+1}x)^{1/p^n}}\right\}^{1/p-1}$$

$$\left\{\frac{\varphi fx}{(\varphi f^{n+1}x)^{1/p^n}}\right\}^{1/p-1} = \overset{1/p}{\left\{\frac{\varphi fx}{\psi fx}\right.} \overset{1/p}{\left\{\frac{\varphi f^2x}{\psi f^2x}\right.} \overset{1/p}{\left\{\vphantom{\frac{1}{1}} \right.} \cdots \overset{1/p}{\left\{\frac{\varphi f^nx}{\psi f^nx}\right\}}. \tag{f}$$

* Braces with an index as $(1/p)$ over them signify the $1/p$ power of all the following part of the expression.

Putting p for $1/p$, we find

$$\left\{\frac{(\varphi f^{n+1}x)^{p^n}}{\varphi fx}\right\}^{p/(p-1)} = \left\{\frac{\varphi fx}{\psi fx}\right\}^{p}\left\{\frac{\varphi f^2x}{\psi f^2x}\right\}^{p}\left\{\ldots\right\}^{p}\left\{\frac{\varphi f^n x}{\psi f^n x}\right\}^{p}. \qquad (g)$$

And the equation determining the functions becomes,

$$\{\varphi x\}^p\{\psi x\}^{p-1} = \{\varphi fx\}^p.$$

These equations in their present general form are sufficiently concise, but as they will become rather complex by the substitution of particular cases, I shall make use of the following notation to abbreviate them. P with an index above it placed before any function of n and other quantities signifies the continued product of that function, n being successively equal to 1,2, and n; thus,

$$\overset{n}{P}\{\psi f^n x\} = \psi fx \cdot \psi f^2 x \cdot \psi f^3 x \cdot \ldots \cdot \psi f^n x.$$

If the product is to be continued to infinity, I shall make use of Euler's method of denoting the limits of integrals.

In (b) assume

$$\psi x = 1 + x + x^2 + \ldots + x^a$$

$$fx = x^{a+1}$$

$$\{1 + x + x^2 + \ldots + x^a\}\varphi x = \varphi x^{a+1}$$

or,

$$\frac{x^{a+1} - 1}{x - 1}\varphi x = \varphi x^{a+1}.$$

Let,

$$y_z = x \text{ and } y_{z+1} = x^{a+1}$$

then,

$$y_z^{a+1} = y_{z+1}.$$

By integrating

$$y_z = y_0^{(a+1)^z} = c^{(a+1)^z} = x.$$

Let

$$\varphi x = \varphi y_z = u_z$$

$$\varphi x^{a+1} = \varphi y_{z+1} = u_{z+1}$$

then,

$$\frac{c^{(a+1)^{z+1}} - 1}{c^{(a+1)^z} - 1} u_z = u_{z+1}$$

$$(c^{(a+1)^{z+1}} - 1)u_z = (c^{(a+1)^z} - 1)u_{z+1}. /$$

But this equation is of the form

$$Q_{z+1}u_z = Q_z u_{z+1}$$

whose integral is

$$u_z = bQ_z$$

therefore,

$$u_z = \varphi x = b(c^{(a+1)^z} - 1) = (x - 1)b, \quad b = 1.$$

Substituting these values of ψ, f, and φ in (c), we have

$$\frac{x^{(a+1)^{n+1}} - 1}{x^{(a+1)} - 1} = \overset{n}{P}\{1 + x^{1\cdot(a+1)^n} + x^{2\cdot(a+1)^n} + \ldots + x^{a\cdot(a+1)^n}\}. \quad (1)$$

Let $a = 1$

$$\frac{x^{2^{n+1}} - 1}{x^2 - 1} = \overset{n}{P}\{1 + x^{2^n}\} = (1 + x^2)(1 + x^4)(1 + x^8) \cdot \ldots \cdot (1 + x^{2^n}). \quad (2)$$

If $a = 2$

$$\frac{x^{3^{n+1}} - 1}{x^3 - 1} = \overset{n}{P}\{1 + x^{3^n} + x^{2\cdot 3^n}\}$$

$$= (1 + x^3 + x^6)(1 + x^9 + x^{18}) \cdot \ldots \cdot (1 + x^{3^n} + x^{2\cdot 3^n}). \quad (3)$$

In (1) for a put any even number as $2a$, and make x negative,

$$\frac{1 + x^{(2a+1)^{n+1}}}{1 + x^{(2a+1)}} = \overset{n}{P}\{1 - x^{1\cdot(2a+1)^n} + x^{2\cdot(2a+1)^n} - \ldots + x^{2a\cdot(2a+1)^n}\}. \quad (4)$$

Let $a = 1$

$$\frac{1 + x^{3^{n+1}}}{1 + x^3} = \overset{n}{P}\{1 - x^{3^n} + x^{2\cdot 3^n}\}$$

$$= (1 - x^3 + x^6)(1 - x^9 + x^{18}) \cdot \ldots \cdot (1 - x^{3^n} + x^{2\cdot 3^n}). \quad (5)$$

Assume $fx = x^2$, and

$$\psi x = 1 - x + x^2 - \ldots + x^{2a} = \frac{1 + x^{2a+1}}{1 + x}$$

$$x = y_z \qquad x^3 = y_{z+1} \qquad y_z = c^{2^z}.$$

Let

$$\varphi y_z = u_z \qquad \varphi y_{z+1} = u_{z+1}. /$$

Equation (b) becomes

$$\frac{1 + c^{(2a+1)\cdot 2^z}}{1 + c^{2^z}} u_z = u_{z+1}$$

$$u_z(1 + c^{(2a+1)\cdot 2^z}) = u_{z+1}(1 + c^{2^z}).$$

In order to reduce this to the form of

$$Q_{z+1} u_z = Q_z u_{z+1}$$

multiply each side by $(1 - c^{(2a+1)\cdot 2^z}) \cdot (1 - c^{2^z})$

$$u_z(1 + c^{(2a+1)\cdot 2^z}) \cdot (1 - c^{(2a+1)\cdot 2^z}) \cdot (1 - c^{2^z})$$
$$= u_{z+1}(1 - c^{(2a+1)\cdot 2^z}) \cdot (1 - c^{2^z}) \cdot (1 + c^{2^z})$$

$$u_z(1 - c^{(2a+1)\cdot 2^{z+1}}) \cdot (1 - c^{2^z}) = u_{z+1}(1 - c^{(2a+1)\cdot 2^z}) \cdot (1 - c^{2^{z+1}})$$

$$u_z \cdot \frac{1 - c^{(2a+1)\cdot 2^{z+1}}}{1 - c^{2^{z+1}}} = u_{z+1} \cdot \frac{1 - c^{(2a+1)\cdot 2^z}}{1 - c^{2^z}}.$$

Hence,

$$u_z = \varphi x = \frac{1 - c^{(2n+1)\cdot 2^z}}{1 - c^{2^z}} = \frac{1 - x^{(2a+1)}}{1 - x}.$$

Putting for f, ψ, and φ, their values in (b), we have

$$\frac{1 - x^{(2a+1)\cdot 2^{n+1}}}{1 - x^{2\cdot(2a+1)}} \cdot \frac{1 - x^2}{1 - x^{2^{n+1}}} = \overset{n}{P}\{1 - x^{1\cdot 2^n} + x^{2\cdot 2^n} - \ldots + x^{2a\cdot 2^n}\}. \tag{6}$$

Let $a = 1$,

$$\frac{1 - x^{3\cdot 2^{n+1}}}{1 - x^{2\cdot 3}} \cdot \frac{1 - x^2}{1 - x^{2^{n+1}}} = \frac{1 + x^{2^{n+1}} + x^{2^{n+2}}}{1 + x^2 + x^4} = \overset{n}{P}\{1 - x^{2^n} + x^{2^{n+1}}\}$$
$$= (1 - x^2 + x^4)(1 - x^4 + x^8)\ldots(1 - x^{2^n} + x^{2^{n+1}}). \tag{7}$$

Divide (5) by (3)

$$\frac{1 + x^{3^{n+1}}}{1 - x^{3^{n+1}}} \cdot \frac{1 - x^3}{1 + x^3} = \overset{n}{P}\left\{\frac{1 - x^{3^n} + x^{2\cdot 3^n}}{1 + x^{3^n} + x^{2\cdot 3^n}}\right\}$$

$$= \left(\frac{1 - x^3 + x^6}{1 + x^3 + x^6}\right) \cdot \ldots \cdot \left(\frac{1 - x^{3^n} + x^{2\cdot 3^n}}{1 + x^{3^n} + x^{2\cdot 3^n}}\right) \quad / \qquad (7)^{a}$$

In equation (1), for x put x^2, and let a be any even number as $2a$, then,

$$\frac{x^{(2a+1)\cdot 2^{n+1}} - 1}{x^{(2a+1)\cdot 2} - 1} = \overset{n}{P}\{1 + x^{1\cdot(2a+1)\cdot 2^n} + x^{2\cdot(2a+1)\cdot 2^n} + \ldots + x^{2a\cdot(2a+1)\cdot 2^n}\}.$$

Divide both sides by

$$\overset{n}{P}\{x^{2a\cdot(2a+1)^n}\} = x^{2a\{(2a+1)+(2a+1)^2+\ldots+(2a+1)^n\}} = x^{(2a+1)^{n+1}-(2a+1)}.$$

By properly arranging the right side of the equation, it becomes,

$$\frac{x^{(2a+1)^{n+1}} - x^{-(2a+1)^{n+1}}}{x^{(2a+1)} - x^{-(2a+1)}} = \overset{n}{P}\{1 + (x^{2\cdot(2a+1)^n} + x^{-2\cdot(2a+1)^n})$$
$$+ (x^{4\cdot(2a+1)^n} + x^{-4\cdot(2a+1)^n}) + \ldots$$
$$+ (x^{2a\cdot(2a+1)^n} + x^{-2a\cdot(2a+1)^n})\}.$$

For $x^{+1} + x^{-1}$ put $2 \cos \theta$.

This substitution gives,

$$\frac{\sin(2a + 1)^{n+1} \cdot \theta}{\sin(2a + 1) \cdot \theta} = \overset{n}{P}\{1 + 2 \cos 2(2a + 1)^n\theta + 2 \cos 4(2a + 1)^n\theta + \ldots$$
$$+ 2 \cos 2a(2a + 1)^n\theta\}. \qquad (8)$$

For a in (1), put any odd number as $2a - 1$, and substituting x^3 for x,

$$\frac{x^{(2a)\cdot 2^{n+1}} - 1}{x^{(2a)\cdot 2} - 1} = \overset{n}{P}\{1 + x^{1\cdot(2a)^n\cdot 2} + x^{2\cdot(2a)^n\cdot 2} + \ldots + x^{(2a-1)\cdot(2a)^n\cdot 2}\}.$$

Divide each side by

$$\overset{n}{P}\{x^{(2a-1)\cdot(2a)^n}\} = x^{(2a-1)\{(2a)^1+(2a)^2+\ldots+(2a)^n\}} = x^{(2a)^{n+1}-2a}$$

[a] *Sic.*

then,

$$\frac{x^{(2n)^{n+1}} - x^{-(2a)^{n+1}}}{x^{2a} - x^{-2a}} = \overset{n}{P}\{(x^{1\cdot(2a)^n} + x^{-1\cdot(2a)^n}) + (x^{3\cdot(2a)^n} + x^{-3\cdot(2a)^n}) + \ldots$$
$$+ (x^{(2a-1)\cdot(2a)^n} + x^{-(2a+1)\cdot(2a)^n})\}.$$

Using the substitution

$$x^{+1} + x^{-1} = 2\cos\theta$$

$$\frac{\sin(2a)^{n+1}\theta}{\sin 2a\,\theta} = \overset{n}{P}\{\cos 1(2a)^n\theta + \cos 3(2a)^n\theta + \ldots$$
$$+ \cos(2a-1)(2a)^n\theta\} \cdot \overset{n}{P}\{2\}.$$

But

$$\overset{n}{P}\{2\} = 2^n$$

$$\frac{1}{2^n}\cdot\frac{\sin(2a)^{n+1}\theta}{\sin 2a\theta} = \overset{n}{P}\{\cos 1(2a)^n\theta + \cos 3(2a)^n\theta + \ldots$$
$$+ \cos(2a-1)(2a)^n\theta\}. \tag{9}$$

If $a = 1$, it becomes the well known expression of Euler

$$\frac{1}{2^n}\cdot\frac{\sin 2^{n+1}\theta}{\sin 2\theta} = \cos 2\theta \cdot \cos 2^2\theta \cdot \ldots \cdot \cos 2^n\theta. \tag{1,1}$$

Let $a = 2$

$$\frac{1}{2^n}\cdot\frac{\sin 4^{n+1}\theta}{\sin 4\,\theta} = \overset{n}{P}\{\cos 1\cdot 4^n\theta + \cos 3\cdot 4^n\theta\}. \tag{1,2}$$

If the same operations be performed on (4) and (6) that have been made use of on (1), the results will be

$$\frac{\cos(2a+1)^{n+1}\theta}{\cos(2a+1)\theta} = \overset{n}{P}\{\pm(1 - 2\cos 2\cdot(2a+1)^n\theta + \ldots$$
$$+ 2\cos 2a\cdot(2a+1)^n\theta)\}$$

and

$$\frac{\sin 2^{n+1}\cdot(2a+1)\cdot\theta}{\sin 2\cdot(2a+1)\cdot\theta}\cdot\frac{\sin 2\theta}{\sin 2^n\theta} = \overset{n}{P}\{\pm(1 - 2\cos 2\cdot 2^n\theta + 2\cos 4\cdot 2^n\theta - \ldots + 2\cos 2a\cdot 2^n\theta)\}$$

plus or minus being used as a is an even or odd number.

To obviate the ambiguity of the signs, it will be better to put for a, $2a$, and $2a-1$, then,

$$\frac{\cos(4a+1)^{n+1}\cdot\theta}{\cos(4a+1)\theta} = \overset{n}{P}\{1 - 2\cos 2\cdot(4a+1)^n\cdot\theta + 2\cos(4a+1)^n\theta - \ldots + 2\cos 4a\cdot(4a+1)^n\theta\} \qquad (1,3)$$

/

$$\frac{\cos(4a-1)^{n+1}\theta}{\cos(4a-1\theta)} = \overset{n}{P}\{2\cos(4a-2)\cdot(4a-1)^n\theta - 2\cos(4a-4)\cdot(4a-1)^n\theta + \ldots + 2\cos 2\cdot(4a-1)^n\theta - 1\} \qquad (1,4)$$

$$\frac{\sin 2^{n+1}(4a+1)\theta}{\sin 2\,(4a+1)\theta}\cdot\frac{\sin 2\theta}{\sin 2^{n+1}\theta} = \overset{n}{P}\{1 - 2\cos 2\cdot 2^n\theta + 2\cos 4\cdot 2^n\theta - \ldots + 2\cos 4a\cdot 2^n\theta\} \qquad (1,5)$$

$$\frac{\sin 2^{n+1}(4a-1)\theta}{\sin 2\cdot(4a-1)\theta}\cdot\frac{\sin 2\theta}{\sin 2^{n+1}\theta} = \overset{n}{P}\{2\cos(4a-2)\cdot 2^n\theta - 2\cos(4a-4)\cdot 2^n\theta + \ldots + 2\cos 2\cdot 2^n\theta - 1\}. \qquad (1,6)$$

In (8) make $a = 1$, and also make $a = 1$ in (1,4). Dividing the last result by the first,

$$\frac{\cos 3^{n+1}\theta}{\cos 3\theta}\cdot\frac{\sin 3\theta}{\sin 3^{n+1}\theta} = \frac{\tan 3\theta}{\tan 3^{n+1}\theta} = \overset{n}{P}\left\{\frac{2\cos 2\cdot 3^n\theta - 1}{2\cos 2\cdot 3^n\theta + 1}\right\} = \overset{n}{P}\left\{\frac{2 - \sec 2\cdot 3^n\theta}{2 + \sec 2\cdot 3^n\theta}\right\}. \qquad (1,7)$$

Or,

$$\frac{\tan\theta}{\tan 3^{n+1}\theta}=\left\{\frac{2-\sec 2\cdot 3\theta}{2+\sec 2\cdot 3\theta}\right\}\cdot\left\{\frac{2-\sec 2\cdot 3^2\theta}{2+\sec 2\cdot 3^2\theta}\right\}\cdot\ldots\cdot\left\{\frac{2-\sec 2\cdot 3^n\theta}{2+\sec 2\cdot 3^n\theta}\right\}.$$

In (1,6) make $a = 1$, and in (1,1), put 2θ for θ; then,

$$\frac{\sin 3\cdot 2^{n+1}\theta}{\sin 2^{n+1}\theta}\cdot\frac{\sin 2\theta}{\sin 3\cdot 2\theta}=\overset{n}{P}\{2\cos 2\cdot 2^n\theta-1\}. \tag{1,8}$$

And,

$$\frac{1}{2^n}\cdot\frac{\sin 2\cdot 2^{n+1}\theta}{\sin 2\cdot 2\theta}=\overset{n}{P}\{\cos 2\cdot 2^n\theta\}.$$

The first divided by the second produces

$$\frac{2^n\sin 3\cdot 2^{n+1}\theta}{\sin 3\cdot 2\theta}\cdot\frac{\sin 2\theta\sin 2\cdot 2\theta}{\sin 2^{n+1}\theta\sin 2\cdot 2^{n+1}\theta}=\overset{n}{P}\{2-\sec 2\cdot 2^n\theta\}. \tag{1,9}$$

/

If in theorems (8) (9) (1,3) (1,4) (1,5) (1,6) for θ be substituted, in each respectively,

$$\frac{\theta}{(2a+1)^{n+1}},\quad\frac{\theta}{(2a)^{n+1}},\quad\frac{\theta}{(4a+1)^{n+1}},\quad\frac{\theta}{(4a-1)^{n+1}},\quad\frac{\theta}{2^{n+1}},\quad\frac{\theta}{2^{n+1}}.$$

The following are the results:

$$\frac{\sin\theta}{\sin\dfrac{\theta}{(2a+1)^n}}=\overset{n}{P}\left\{1+2\cos\frac{2\theta}{(2a+1)^n}+2\cos\frac{4\theta}{(2a+1)^n}+\ldots+2\cos\frac{2a\theta}{(2a+1)^n}\right\} \tag{2,1}$$

$$\frac{1}{2^n}\cdot\frac{\sin\theta}{\sin\dfrac{\theta}{(2a)^n}}=\overset{n}{P}\left\{\cos\frac{1\cdot\theta}{(2a)^n}+\cos\frac{3\cdot\theta}{(2a)^n}+\ldots+\cos\frac{(2a-1)\theta}{(2a)^n}\right\} \tag{2,2}$$

$$\frac{\cos\theta}{\cos\dfrac{\theta}{(4a+1)^n}}=\overset{n}{P}\left\{1-2\cos\frac{2\theta}{(4a+1)^n}+2\cos\frac{4\theta}{(4a+1)^n}-\ldots+2\cos\frac{4a\theta}{(4a+1)^n}\right\} \tag{2,3}$$

$$\frac{\cos\theta}{\cos\dfrac{\theta}{(4a-1)^n}} = \overset{n}{P}\left\{2\cos\frac{(4a-2)\theta}{(4a-1)^n} - 2\cos\frac{(4a-4)\theta}{(4a-1)^n} + \dots + 2\cos\frac{2\theta}{(4a-1)^n} - 1\right\} \quad (2,4)$$

$$\frac{\sin(4a-1)\theta}{\sin\dfrac{4a+1}{2^n}\theta}\cdot\frac{\sin\dfrac{\theta}{2^n}}{\sin\theta} = \overset{n}{P}\left\{1 - \cos\frac{2\theta}{2^n} + 2\cos\frac{4\theta}{2^n} - \dots + 2\cos\frac{4a\theta}{2^n}\right\} \quad (2,5)$$

$$\frac{\sin(4a-1)\theta}{\sin\dfrac{4a-1}{2^n}\theta}\cdot\frac{\sin\dfrac{\theta}{2^n}}{\sin\theta} = \overset{n}{P}\left\{2\cos\frac{(4a-2)\theta}{2^n} - \cos\frac{(4a-4)\theta}{2^n} + \dots + 2\cos\frac{2\theta}{2^n} - 1\right\}. \quad (2,6)$$

Other theorems nearly similar may be thus derived. In (1), for x put vx, and x/v; multiply together the results; and in the left side of the equation put for $v^{+1} + v^{-1}$ its value $2\cos\theta$; then,

$$\frac{x^{2(a+)^{n+1}} - 2x^{(a+1)^{n+1}}\cos(a+1)^{n+1}\theta + 1}{x^{2(a+1)} - 2x^{a+1}\cos(a+1)\theta + 1}$$
$$= \overset{n}{P}\{1 + (xv)^{1\cdot(a+1)^n} + (xv)^{2\cdot(a+1)^n} + \dots + (xv)^{a\cdot(a+1)^n}\}$$
$$\times \overset{n}{P}\left\{1 + \left(\frac{x}{v}\right)^{1\cdot(a+1)^n} + \left(\frac{x}{v}\right)^{2\cdot(a+1)^n} + \dots + \left(\frac{x}{v}\right)^{a\cdot(a+1)^n}\right\}.$$

/

In order to reduce the right side of the equation to a series of multiple arcs, let us consider the product of two functions of this form:

$$f(xv) = 1 + (xv) + (xv)^2 + \dots + (xv)^a$$

$$f\left(\frac{x}{v}\right) = 1 + \left(\frac{x}{v}\right) + \left(\frac{x}{v}\right)^2 + \dots + \left(\frac{x}{v}\right)^a.$$

Their product is of the form

$$\begin{aligned} f(xv) \times f\left(\frac{x}{v}\right) &= (1 + x^2 + \ldots + x^{2a}) \\ &\quad + x(1 + x^2 + \ldots + x^{2a-2})(v^{+1} + v^{-1}) \\ &\quad + x^2(1 + x^2 + \ldots + x^{2a-4})(v^{+2} + v^{-2}) + \ldots \\ &\quad + x^a(v^{+a} + v^{-a}) \\ &= \{1 - x^2\}^{-1} \cdot \{(1 - x^{2a+2}) + x(1 - x^{2a})(v^{+1} + v^{-1}) \\ &\quad + x^2(1 - x^{2a-2})(v^{+2} + v^{-2}) + \ldots + x^a(v^{+a} + v^{-a})\}. \end{aligned}$$

Putting $v^{+1} + v^{-1} = 2\cos\theta$, and applying this to the preceding expression,

$$\begin{aligned} &\frac{x^{2\cdot(a+1)^{n+1}} - 2x^{(a+1)^{n+1}}\cos(a+1)^{n+1}\theta + 1}{x^{2\cdot a+1} + 2x^{a+1}\cos(a+1)\theta + 1} \\ &= (1 - x^2)^{-n} \overset{n}{P}\{(1 - x[^{2a+2)\cdot(a+1)^n}) \\ &\qquad + 2(1 - x^{2a\cdot(a+1)^n})x^{1\cdot(a+1)^n}\cos 1\cdot(a+1)^n\theta + \ldots \\ &\qquad\qquad + 2x^{a\cdot(a+1)n}\cos a\cdot(a+1)^n\theta\}. \end{aligned}$$

In the case of x equal to unity, this expression becomes

$$\begin{aligned} \frac{1 - \cos(a+1)^{n+1}\theta}{1 - \cos(a+1)\theta} &= \overset{n}{P}\{(a+1) + 2(a)\cos 1\cdot(a+1)^n\theta \\ &\quad + 2(a-1)\cos 2(a+1)^n\theta + \ldots \\ &\quad + 2(1)\cos a\cdot(a+1)^n\theta\}. \end{aligned} \tag{2,6}$$

If x do not equal unity, and if $a = 1$,

$$\frac{x^{2\cdot 2^{n+1}} - 2x^{2^{n+1}}\cos 2^{n+1}\theta + 1}{x^{2\cdot 2} - 2x^2\cos 2\theta + 1} = \overset{n}{P}\{1 + 2x^{2^n}\cos 2^n\theta + x^{2^{n+1}}\} \tag{2,7}$$

If the same operations are performed on (4) and (6), we shall obtain the following results:

$$\begin{aligned} \frac{1 + \cos(2a+1)^{n+1}\cdot\theta}{1 + \cos(2a+1)\cdot\theta} &= \overset{n}{P}\{(2a+1) - 2(2a)\cos 1\cdot(2a+1)^n\theta \\ &\quad + 2\cdot(2a-1)\cos 2\cdot(2a+1)^n\theta - \ldots \\ &\quad + 2(1)\cos 2a\cdot(2a+1)^n\theta\}. \end{aligned} \tag{2,8}$$

$$\begin{aligned} &\frac{1 - \cos(2a+1)\cdot 2^{n+1}\theta}{1 - \cos(2a+1)\cdot 2\theta}\cdot\frac{1 - \cos 2\theta}{1 - \cos 2^{n+1}\theta} \\ &= \left\{\frac{\sin(2a+1)\cdot 2^n\theta}{\sin(2a+1)\cdot\theta}\cdot\frac{\sin\theta}{\sin 2^n\theta}\right\}^2 \end{aligned}$$

$$= \overset{n}{P}\{(2a+1) - 2(2a)\cos 1\cdot 2^n\theta + 2(2a-1)\cos 2\cdot 2^n\theta - \dots$$

$$+ 2(1)\cos 2a\cdot 2^n\theta\}. \quad (2,9)$$

We will now give some examples of the use of theorem (f). The chief difficulty consists in finding functions which satisfy the equation (d), for, if the two functions ψ and f be assumed, the third must be determined by the method of Laplace, and this generally leads to equations of finite differences whose solutions are above the powers of analysis in its present state.

In one case however (namely when $p = 2$), the solutions we have obtained for equation (b), are also solutions of (d), for in this case (b) and (d) are identical.

We have already determined, that if

$$\psi x = 1 + x + x^2 + \dots + x^n, \quad \text{and} \quad fx = x^{a+1}$$

that

$$\varphi x = x - 1$$

hence, substituting these values in (f),

$$\frac{x^{a-1} - 1}{\{x^{(a-1)^{n+1}} - 1\}^{1/2^n}} = \overset{n}{P}\left\{\frac{x^{(a+1)^n} - 1}{1 + x^{1\cdot(a+1)^n} + x^{2\cdot(a+1)^n} + \dots x^{a\cdot(a+1)^n}}\right\}^{1/2^n}. \quad (3,1)$$

If $a = 1$

$$\frac{x^2 - 1}{\{x^{2^{n+1}} - 1\}^{1/2^n}} = \overset{n}{P}\left\{\frac{x^{2^n} - 1}{x^{2^n} + 1}\right\}^{1/2^n}$$

$$= \left\{\frac{x^2-1}{x^2+1}\right\}^{1/2}\left\{\frac{x^4-1}{x^4+1}\right\}^{1/2}\left\{\frac{x^8-1}{x^8+1}\right\}^{1/2}\left\{\dots\right\}^{1/2}\left\{\frac{x^{2^n}-1}{x^{2^n}+1}\right\}^{1/2} \quad (3,2)$$

put xv, and xv^{-1} for x, multiply the results

$$\frac{x^2v^2 - 1}{\{(xv)^{2^{n+1}} - 1\}^{1/2^n}} \times \frac{x^2v^{-2} - 1}{\{(xv^{-1})^{2^{n+1}} - 1\}^{1/2^n}}$$

$$= \overset{n}{P}\left\{\frac{x^{2^{n+1}} - x^{2^n}(v^{2^n} + v^{-2^n}) + 1}{x^{2^{n+1}} + x^{2^n}(v^{2^n} + v^{-2^n}) + 1}\right\}^{1/2^n}$$

for $v + v^{-1}$ put $2\cos\theta$

$$\frac{x^4 - 2x^2\cos 2\theta + 1}{\{x^{2^{n+2}} - 2x^{2^{n+1}}\cos 2^{n+1}\theta + 1\}^{1/2^n}} = \overset{n}{P}\left\{\frac{x^{2^{n+1}} - 2x^{2^n}\cos 2^n\theta + 1}{x^{2^{n+1}} + 2x^{2^n}\cos 2^n\theta + 1}\right\}^{1/2^n}. \quad (3,3)$$

/

If $x = 1$, and for θ be substituted 2θ, and the necessary reductions be made,

$$\frac{2\sin 2\theta}{(2\sin 2^{n+1}\theta)^{1/2^n}} = \overset{n}{P}\{\tan 2^n\theta\}^{1/2^n}$$
$$= (\tan 2\theta)^{1/2}\cdot(\tan 4\theta)^{1/4}\cdot\ldots\cdot(\tan 2^n\theta)^{1/2^n}. \qquad (3,4)$$

For θ substitute $\theta/2^{n+1}$

$$\frac{2\sin\dfrac{\theta}{2^n}}{(2\sin\theta)^{1/2^n}} = \left(\tan\frac{\theta}{2^n}\right)^{1/2}\cdot\left(\tan\frac{\theta}{2^{n-1}}\right)^{1/4}\cdot\ldots\cdot\left(\tan\frac{\theta}{2}\right)^{1/2^n}.$$

Raise both sides to the power 2^{n+1}, then,

$$\frac{\left(2\sin\dfrac{\theta}{2^n}\right)^{2^{n+1}}}{(2\sin\theta)^2} = \overset{n}{P}\left\{\tan\frac{\theta}{2^n}\right\}^{2^n}$$
$$= \left(\tan\frac{\theta}{2}\right)^2\cdot\left(\tan\frac{\theta}{4}\right)^4\cdot\ldots\cdot\left(\tan\frac{\theta}{2^n}\right)^{2^n}. \qquad (3,5)$$

Assume

$$fx = x^{a+1} \qquad \varphi x = 1 - x$$
$$(\psi x)^{p-1} = \frac{\varphi f x}{\varphi x} = \frac{1 - x^{a+1}}{1 - x} = 1 + x + x^2 + \ldots + x^a$$
$$\psi x = \sqrt[p-1]{1 + x + x^2 + \ldots + x^a}$$

and by substituting the values in (f),

$$\left\{\frac{1 - x^{a+1}}{\{1 - x^{(a+1)^{n+1}}\}^{1/p^n}}\right\}^{1/(p-1)}$$
$$= \overset{n}{P}\left\{\frac{1 - x^{(a+1)^n}}{\sqrt[p-1]{1 + x^{1\cdot(a+1)^n} + x^{2\cdot(a+1)^n} + \ldots + x^{n\cdot(a+1)^n}}}\right\}^{1/p^n}. \qquad (3,6)$$

For a put $2a$, and for x put x^2. Divide both sides by

$$x^{\frac{1}{p-1}\cdot\left((2a+1)-\frac{(2a+1)^{n+1}}{p^n}\right)}$$

and substituting for $x^{+1} + x^{-1}$ its value $2\cos\theta$, and for $x^{+1} - x^{-1}$ its value $+2\sqrt{-1}\sin\theta$, some power of $\sqrt{-1}$ will enter both sides of the equation, and they must be divided by it, but the $\sqrt{-1}$ is on both sides of the equation multiplied by 2; and therefore the equation is also divisible by the same power of 2; having performed these operations, the result will be

$$\left\{\frac{\sin(2a+1)\theta}{\{\sin(2a+1)^{n+1}\theta\}^{1/p^n}}\right\}^{1/(p-1)}$$

$$=\overset{n}{P}\left\{\frac{\sin(2a+1)^n\theta}{\sqrt[p-1]{\begin{array}{r}1+2\cos 2\cdot(2a+)1^n\theta+2\cos 4\cdot(2a+1)^n\theta+\dots\\+2\cos 2a\cdot(2a+1)^n\theta\end{array}}}\right\}^{1/p^n}. \qquad (3,7)$$

If we had begun by putting $2a$ for a, and $-x$ for x, the result would have been

$$\left\{\frac{\cos(2a+1)\theta}{\{\cos(2a+1)^{n+1}\theta\}^{1/p^n}}\right\}^{1/(p-1)}$$

$$=\overset{n}{P}\left\{\frac{\cos(2a+1)^n\theta}{\sqrt[p-1]{\begin{array}{r}1-2\cos 2\cdot(2a+1)^n\theta+2\cos 4\cdot(2a+1)^n\theta-\dots\\+2\cos 2a\cdot(2a+1)^n\theta\end{array}}}\right\}^{1/p^n}. \qquad (3,8)$$

Dividing the first of these expressions by the second,

$$\left\{\frac{\tan(2a+1)\theta}{\{\tan(2a+1)^{n+1}\theta\}^{1/p^n}}\right\}^{1/(p-1)}$$

$$=\overset{n}{P}\left\{\frac{\sin(2a+1)^n\theta}{\cos(2a+1)^n\theta}\left\{\frac{\begin{array}{l}1+2\cos 2\cdot(2a+1)^n\theta+2\cos 4\\ \quad\cdot(2a+1)^n\theta-\dots+2\cos 2a\cdot(2a+1)^n\theta\end{array}}{\begin{array}{l}1+2\cos 2\cdot(2a+1)^n\theta+2\cos 4\\ \quad\cdot(2a+1)^n\theta-\dots+2\cos 2a\cdot(2a+1)^n\theta\end{array}}\right\}^{1/(p-1)}\right\}^{1/p^n} \qquad (3,9)$$

If $a=1$, $p=2$,

$$\frac{\tan 3\theta}{\{\tan 3^{n+1}\theta\}^{1/2^n}}=\overset{n}{P}\left\{\tan 3^n\theta\cdot\frac{\sec 2\cdot 3^n\theta-2}{\sec 2\cdot 3^n\theta+2}\right\}^{1/2^n} \qquad (4,1)$$

In (3,6) for x, substitute successively xv and xv^{-1}; multiplying the results, and making

$$x=1,\quad\text{and}\quad v^{+1}+v^{-1}=2\cos 2\theta,$$

we have

$$\left\{\frac{\sin(a+1)\theta}{\{\sin(a+1)^{n+1}\theta\}^{1/p^n}}\right\}^{2/(p-1)}$$

$$= \overset{n}{P}\left\{\frac{(\sin(a+1)^n\theta)^2}{\sqrt[p-1]{\begin{array}{r}(a+1)+2(a)\cos 2\cdot(a+1)^n\theta+2(a-1)\cos 4\\ \cdot(a+1)^n\theta+\ldots+2(1)\cos 2a\cdot(a+1)^n\theta\end{array}}}\right\}^{1/p^n}. \tag{4,2}$$

If before the preceding substitution x had been made negative, and $2a$ put for a, we should have found /

$$\left\{\frac{\cos(2a+1)\theta}{\{\cos(2a+1)^{n+1}\theta\}^{1/p^n}}\right\}^{1/(p-1)}$$

$$= \overset{n}{P}\left\{\frac{\cos(2a+1)^n\theta}{\sqrt[p-1]{\begin{array}{r}(2a+1)-2(2a)\cos 2\cdot(2a+1)^n\theta+2(2a-1)\cos 4\\ \cdot(2a+1)^n\theta-\ldots+2(1)\cos 4a\cdot(2a+1)^n\theta\end{array}}}\right\}^{1/p^n}. \tag{4,3}$$

In (4,2) make $p = 3/2$, and for a put $2a$,

$$\left\{\frac{\sin(2a+1)\theta}{\{\sin(2a+1)^{n+1}\theta\}^{(2/3)^n}}\right\}^{2}$$

$$= \overset{n}{P}\left\{\frac{\sin\cdot(2a+1)^n\theta}{\begin{array}{l}(2a+1)+2(2a)\cos 2\cdot(2a+1)^n\theta\\ +2(2a-1)\cos 4\cdot(2a+1)^n\theta+\ldots+2(1)\cos 4a\cdot(2a+1)^n\theta\end{array}}\right\}^{(2/3)^n}. \tag{4,5}$$

In (4,3) make $p = 3/2$,

$$\left\{\frac{\cos(2a+1)\theta}{\{\cos(2a+1)^{n+1}\theta\}^{(2/3)^n}}\right\}^{2}$$

$$= \overset{n}{P}\left\{\frac{\cos(2a+1)^n\theta}{\begin{array}{l}(2a+1)-2(2a)\cos 2\cdot(2a+1)^n\theta\\ +2(2a-1)\cos 4\cdot(2a+1)^n\theta-\ldots+2(1)\cos 4a\cdot(2a+1)^n\theta\end{array}}\right\}^{2/3)^n}. \tag{4,6}$$

Dividing the first by the last,

$$\left\{\frac{\tan(2a+1)^n\theta}{\{\tan(2a+1)^{n+1}\theta\}^{(2/3)^n}}\right\}^2 = \overset{n}{P}\left\{\frac{\sin(2a+1)^n\theta}{\cos(2a+1)^n\theta}\right.$$

$$\times\left.\frac{(2a+1) - 2(2a)\cos 2\cdot(2a+1)^n\theta + 2(2a-1)\cos 4\cdot(2a+1)^n\theta - \ldots + 2(1)\cos 4a\cdot(2a+1)^n\theta}{(2a+1) + 2(2a)\cos 2\cdot(2a+1)^n\theta + 2(2a-1)\cos 4\cdot(2a+1)^n\theta + \ldots + 2(1)\cos 4a\cdot(2a+1)^n\theta}\right\}^{(2/3)^n}. \quad (4,7)$$

In (4,2) let $p = 3/2$, $a = 1$

$$\left\{\frac{\sin 2\theta}{\{\sin 2^{n+1}\theta\}^{(2/3)^n}}\right\} = \overset{n}{P}\left\{\frac{\sin 2^n\theta}{2(1+\cos 2^{n+1}\theta)}\right\}^{(2/3)^n}$$

$$= \overset{n}{P}\left\{\frac{\sin 2^n\theta}{(\cos 2^n\theta)^2}\right\}^{(2/3)^n} \times \overset{n}{P}\left\{\frac{1}{4}\right\}^{(2/3)^n}$$

but

$$\overset{n}{P}\left\{\frac{1}{4}\right\}^{(2/3)^n} = \left\{\frac{1}{4}\right\}^{2/3+4/9+\ldots+2^n/3^n} = \left\{\frac{1}{4}\right\}^{((2/3)^{n+1}-2/3)/(-1/3)} = \left\{\frac{1}{4}\right\}^{2(1-(2/3)^n)}$$

$$= \left\{\frac{4^{(2/3)^n}}{4}\right\}^2$$

therefore, multiplying both sides by $\{4^{(2/3)^n}/4\}^2$

$$\left(\frac{4\sin 2\theta}{\{4\sin 2^{n+1}\theta\}^{(2/3)^n}}\right)^2 = \overset{n}{P}\{\tan 2^n\theta\sec 2^n\theta\}^{(2/3)^n}$$

$$= (\tan 2\theta\sec 2\theta)^{2/3}\cdot(\tan 4\theta\cdot\sec 4\theta)^{4/9}$$

$$\cdot(\tan 8\theta\sec 8\theta)^{8/27}\cdot\ldots\cdot(\tan 2^n\theta\cdot\sec 2^n\theta)^{(2/3)^n}. \quad (4,8)$$

For θ put $\theta/2^{n+1}$, and raise both sides to the power $(3/2)^{n+1}$, then we find

$$\left\{\frac{\left\{4\sin\frac{\theta}{2^n}\right\}^{(3/2)^{n+1}}}{\{4\sin\theta\}^{3/2}}\right\}^2 = \overset{n}{P}\left\{\tan\frac{\theta}{2^n}\sec\frac{\theta}{2^n}\right\}^{(3/2)^n}$$

$$= \left(\tan\frac{\theta}{2}\sec\frac{\theta}{2}\right)^{3/2}\cdot\left(\tan\frac{\theta}{4}\sec\frac{\theta}{4}\right)^{9/4}$$

$$\cdot\left(\tan\frac{\theta}{8}\sec\frac{\theta}{8}\right)^{27/8}\cdot\ldots\cdot\left(\tan\frac{\theta}{2^n}\sec\frac{\theta}{2^n}\right)^{(3/2)^n}. \quad (4,9)$$

In (3,6) for a substitute $2a - 1$, put x^2 for x, and $2\cos\theta = x^{+1} + x^{-1}$

$$\left\{\frac{\sin 2a\theta}{\{\sin (2a)^{n+1}\theta\}^{1/p^n}}\right\}^{1/(p-1)}$$
$$= \overset{n}{P}\left\{\frac{\sin (2a)^n\theta}{\sqrt[p-1]{\cos 1\cdot(2a)^n\theta + \cos 3\cdot(2a)^n\theta + \ldots + \cos(2a-1)\cdot(2a)^n\theta}}\right\}^{1/p^n}$$

$$\times\ \overset{n}{P}\left\{\frac{1}{(2)^{1/(p-1)}}\right\}^{1/p^n}.$$

But

$$\overset{n}{P}\left\{\frac{1}{(2)^{1/(p-1)}}\right\}^{1/p^n} = \left\{\frac{1}{2^{1/(p-1)}}\right\}^{1/p+1/p^2+\ldots+1/p^n} = \left\{\frac{1}{2^{1/(p-1)}}\right\}^{\frac{1/p^{n+1}-1/p}{1/p-1}}$$
$$= \left\{\frac{1}{2^{1/(p-1)}}\right\}^{\frac{1-1/p^n}{p-1}} = 2^{-\frac{1}{(p-1)^2}\cdot\left(1-\frac{1}{p^n}\right)}$$

/

then,

$$\left\{\frac{2^{1/p-1}\sin 2a\theta}{\{2^{1/p-1}\sin (2a)^{n+1}\theta\}^{1/p^n}}\right\}^{1/(p-1)}$$
$$= \overset{n}{P}\left\{\frac{\sin\cdot(2a)^n\theta}{\sqrt[p-1]{\cos 1(2a)^n\theta + \cos 3\cdot(2a)^n\theta + \ldots + \cos\cdot(2a-1)\cdot(2a)^n\theta}}\right\}^{1/p^n} \tag{5,0}$$

In this expression make $a = 2$, $p = 2$,

$$\cos 4^n\theta + \cos 3\cdot 4^n\theta = \cos 4^n\theta + 2\cos 4^n\theta\cos 2\cdot 4^n\theta - \cos 4^n\theta$$
$$= 2\cos 4^n\theta\cos 2\cdot 4^n\theta$$

then,[a]

$$\frac{4\sin 4\theta}{\{4\sin 4^{n+1}\theta\}^{1/2^n}} = \overset{n}{P}\{\tan 4^n\theta\sec 2\cdot 4^n\theta\}^{1/2^n}$$
$$= \{\tan 4\theta\sec 8\theta\}^{1/2}\cdot\{\tan 16\theta\sec 32\theta\}^{1/4}\cdot\ldots$$
$$\cdot\{\tan 4^n\theta\sec 2\cdot 4^n\theta\}^{1/2^n}. \tag{5,1}$$

[a] In the lefthand sides of equations (5,1) and (5,2), the first factor in the numerator and denominator should read 2 instead of 4. No corrections have been made.

For θ substitute $\theta/4^{n+1}$, and raise both sides to the power 2^{n+1}, then,

$$\frac{\left\{4\sin\frac{\theta}{4^n}\right\}^{2n+1}}{\{4\sin\theta\}^2} = \overset{n}{P}\left\{\tan\frac{\theta}{4^n}\sec\frac{2\theta}{4^n}\right\}^{1/2^n}$$
$$= \left\{\tan\frac{\theta}{4}\sec\frac{2\theta}{4}\right\}^2\cdot\left\{\tan\frac{\theta}{16}\sec\frac{2\theta}{16}\right\}^4\cdots$$
$$\cdot\left\{\tan\frac{\theta}{4^n}\sec\frac{2\theta}{4^n}\right\}^{2n}. \quad (5,2)$$

By assuming $fx = x^2$,

$$\varphi x = \frac{1-x^{2a+1}}{1-x} = 1 + x + x^2 + \ldots + x^{2a},$$

we should find

$$\psi x = {}^{p-1}\sqrt{1 - x + x^2 - \ldots + x^{2a}}$$

and by pursuing the same method so frequently repeated, we should arrive at the following results:

$$\left\{\frac{1-x^2}{1-x^{3\cdot 2a+1}}\left\{\frac{1-x^{2^{n+1}(2a+1)}}{1-x^{2^{n+1}}}\right\}^{1/p^n}\right\}^{1/(p-1)}$$
$$= \overset{n}{P}\left\{\frac{1 - x^{1\cdot 2^n} + x^{2\cdot 2^n} - \ldots + x^{2a\cdot 2^n}}{{}^{p-1}\sqrt{1 + x^{1\cdot 2^n} + x^{2\cdot 2^n} + \ldots + x^{2a\cdot 2^n}}}\right\}^{1/p^n} \quad (5,3)$$

$$\left\{\frac{\sin 2\theta}{\sin 2\cdot(2a+1)\theta}\cdot\left(\frac{\sin 2^{n+1}(2a+1)\theta}{\sin 2^{a+1}\theta}\right)^{1/p^n}\right\}^{1/(p-1)}$$
$$= \overset{n}{P}\left\{\frac{{}^{p-1}\sqrt{\pm\{1 - 2\cos 2\cdot 2^n\theta + 2\cos 4\cdot 2^n\theta - \ldots + 2\cos 2a\cdot 2^n\theta\}}}{1 + 2\cos 2\cdot 2^n\theta + 2\cos 4\cdot 2^n\theta + \ldots + 2\cos 2a\,.\,2^n\theta}\right\}^{1/p^n}$$
$$(5,4)$$

$$\left\{\frac{\sin\theta}{\sin(2a+1)\theta}\cdot\left(\frac{\sin 2^n\cdot(2a+1)\theta}{\sin 2^n\theta}\right)^{1/p^n}\right\}^{2/(p-1)}$$
$$= \overset{n}{P}\left\{\frac{{}^{p-1}\sqrt{(2a+1) - 2(2a)\cos 1\cdot 2^n\theta + 2(2a-1)\cos 2\cdot 2^n\theta - \ldots + 2(1)\cos 2a\cdot 2^n\theta}}{(2a+1) + 2(2a)\cos 1\cdot 2^n\theta + 2(2a-1)\cos 2\cdot 2^n\theta + \ldots + 2(1)\cos 2a\cdot 2^n\theta}\right\}^{1/p^n}$$
$$(5,5)$$

$\pm$ being used in (5,4) as a is an even or an odd number.

In (5,4) let $a = 1$, $p = 2$,

$$\frac{\sin 2\theta}{\sin 2\cdot 3\theta}\left\{\frac{\sin 3\cdot 2^{n+1}\theta}{\sin 2^{n+1}\theta}\right\}^{1/2^n}$$

$$=\overset{n}{P}\left\{\frac{2\cos 2^n\cdot 2\theta-1}{2\cos 2^n\cdot 2\theta+1}\right\}^{1/2^n}=\overset{n}{P}\left\{\frac{2-\sec 2^n\cdot 2\theta}{2+\sec 2^n\cdot 2\theta}\right\}^{1/2^n}$$

$$=\left\{\frac{2-\sec 4\theta}{2+\sec 4\theta}\right\}^{1/2}\cdot\left\{\frac{2-\sec 8\theta}{2+\sec 8\theta}\right\}^{1/4}\cdot\ldots\cdot\left\{\frac{2-\sec 2^n\cdot 2\theta}{2+\sec 2^n\cdot 2\theta}\right\}^{1/2^n}. \tag{5,6}$$

For θ put $\theta/2^{n+1}$, and raise both sides to the 2^{n+1} power,

$$\left\{\frac{\sin\dfrac{\theta}{2^n}}{\sin\dfrac{3\theta}{2^n}}\right\}^{2^{n+1}}\cdot\left\{\frac{\sin 3\theta}{\sin\theta}\right\}^{2}$$

$$=\overset{n}{P}\left\{\frac{2\cos\dfrac{2\theta}{2^n}-1}{2\cos\dfrac{2\theta}{2^n}+1}\right\}^{2^n}=\overset{n}{P}\left\{\frac{2-\sec\dfrac{2\theta}{2^n}}{2+\sec\dfrac{2\theta}{2^n}}\right\}^{1/2^n}$$

$$=\left\{\frac{2-\sec\theta}{2+\sec\theta}\right\}^{2}\cdot\left\{\frac{2-\sec\dfrac{\theta}{2}}{2\sec\dfrac{\theta}{2}}\right\}^{2^2}\cdot\ldots\cdot\left\{\frac{2-\sec\dfrac{\theta}{2^{n-1}}}{2+\sec\dfrac{\theta}{2^{n-1}}}\right\}^{2^n}. \tag{5,7}$$

In (3,6) make $a = 1$, and also $a = 2$, then,

$$\left\{\frac{1-x^2}{\{1-x^{2^{n+1}}\}^{1/p^n}}\right\}^{1/(p-1)}$$

$$=\left\{\frac{1-x^2}{\sqrt[p-1]{1+x^2}}\right\}^{1/p}\left\{\frac{1-x^4}{\sqrt[p-1]{1+x^4}}\right\}^{1/p}\left\{\ldots\right\}^{1/p}\left\{\frac{1-x^{2^n}}{\sqrt[p-1]{1+x^{2^n}}}\right\}^{1/p}$$

$$=\overset{n}{P}\left\{\frac{1-x^{2^n}}{\sqrt[p-1]{1+x^{2^n}}}\right\}^{1/p^n} \tag{5,8}$$

$$\sqrt{\frac{1 \pm x^3}{\{1 \pm x^{3^{n+1}}\}^{1/3^n}}}$$

$$= \overset{1/3}{\left\{\frac{1 \pm x^3}{\sqrt{1 \mp x^3 + x^6}}\right.} \overset{1/3}{\left\{\frac{1 \pm x^9}{\sqrt{1 \mp x^9 + x^{18}}}\right.} \overset{1/3}{\Big\{} \cdots \overset{1/3}{\left\{\frac{1 \pm x^{3^n}}{\sqrt{1 \mp x^{3^n} + x^{2 \cdot 3^n}}}\right\}}. \quad (5,9)$$

In (5,3) let $p = 2$, $a = 1$, and multiply the result by (3,2)

$$\frac{x^4 + x^2 + 1}{x^2 - 1}\left\{\frac{x^{2^{n+1}} - 1}{x^{2 \cdot 2^{n+1}} + x^{1 \cdot 2^{n+1}} + 1}\right\}^{1/2^n}$$

$$= \overset{1/2}{\left\{\frac{x^6 + 2x^4 + 2x^2 + 1}{x^6 - 2x^4 + 2x^2 - 1}\right.} \overset{1/2}{\left\{\frac{x^{12} + 2x^8 + 2x^4 + 1}{x^{12} - 2x^8 + 2x^4 - 1}\right.} \overset{1/2}{\Big\{} \cdots$$

$$\overset{1/2}{\left\{\frac{x^{3 \cdot 2^n} + 2x^{2 \cdot 2^n} + 2x^{1 \cdot 2^n} + 1}{x^{3 \cdot 2^n} - 2x^{2 \cdot 2^n} + 2x^{1 \cdot 2^n} + 1}\right\}}. \quad (6,0)$$

In all the preceding expressions the value of p has not been at all restricted, but may be a whole number or fraction, positive or negative.

They may therefore be considered as variable quantities, and their differential co-efficients taken relative to p.

This will be much facilitated by the aid of the following simple theorem, relating to continued products, when differentiated relative to a quantity which only enters as an exponent.

Let v_n and u_n be functions of n and p. If

$$\overset{n}{P}\{a_n^{v_n}\} = b_n^{u_n}$$

then

$$\overset{n}{P}\{a_n{}^{d^q v_n/(dp)^q}\} = b_n{}^{d^q u_n/(dp)^q}.$$

Which may be easily demonstrated, thus, by taking the logarithm of the first equation,

$$\overset{n}{S}\{v_n \log a_n\} = u_n \log b_n$$

differentiating q times relative to p,

$$\overset{n}{S}\left\{\frac{d^q v_n}{(dp)^q} \log a_n\right\} = \frac{d^q u_n}{(dp)^q} \log b_n.$$

And reverting from logarithms to numbers,

$$\overset{n}{P}\{a_n^{d^q v_n/(dp)^q}\} = b_n^{d^q u_n/(dp)^q}. \tag{6,1}$$

If we apply this to (5,8) making $q = 1$, the result will be

$$\frac{\{1 - x^2\}^p}{\{1 - x^{2^{n+1}}\}^{(n\cdot(p-1)+p)/p^n}} = \overset{n}{P}\left\{\frac{\{1 - x^{2^n}\}^{n\cdot(p-1)^2}}{\{1 + x^{2^n}\}^{n\cdot(p-1)+p}}\right\}^{1/p^n} \tag{6,2}$$

For x substitute successively xv and xv^{-1}, multiplying the results and putting $2\cos\theta$ for $u^{+1} + v^{-1}$ we have

$$\frac{\{1 - 2x^2\cos 2\theta + x^4\}^p}{\{1 - 2x^{2^{n+1}}\cos 2^{n+1}\theta + x^{2^{n+2}}\}^{(n\cdot(p-1)+p)/p^n}}$$
$$= \overset{n}{P}\left\{\frac{\{1 - 2x^{2^n}\cos 2^n\theta + x^{2^{n+1}}\}^{n\cdot(p-1)^2}}{\{1 + 2x^{2^n}\cos 2^n\theta + x^{2^{n+1}}\}^{n\cdot(p-1)+p}}\right\}^{1/p^n} \tag{6,3}$$

Let $x = 1$, for θ put 2θ,

$$\frac{\{2\sin 2\theta\}^{2p}}{\{2\sin 2^{n+1}\theta\}^{(2n\cdot(p-1)+p)/p^n}} = \overset{n}{P}\left\{\frac{\{2\sin 2^n\theta\}^{2n\cdot(p-1)^n}}{\{2\cos 2^n\theta\}^{2(n\cdot(p-1)+p)}}\right\}^{1/p^n}. \tag{6,4}$$

Extract the square root, and make $p = 2$,

$$\frac{\{2\sin 2\theta\}^2 \times 4^{1-1/2^n}}{\{2\sin 2^{n+1}\theta\}^{(n+2)/2^n}} = \overset{n}{P}\{(\tan 2^n\theta)^n(\sec 2^n\theta)^2\}^{1/2^n}$$
$$= \{(\tan 2\theta)^1(\sec 2\theta)^2\}^{1/2}\cdot\{(\tan 4\theta)^2(\sec 4\theta)^2\}^{1/4}\cdot\ldots\cdot$$
$$\{(\tan 2^n\theta)^n(\sec 2^n\theta)^2\}^{1/2^n}. \tag{6,5}$$

By continuing the differentiations according to formula (6,1), and by making the usual substitutions, we might derive other theorems which would become gradually more complex, when expressed generally, though in particular cases they afford some curious results, these and many of the above might be still further extended by putting $1/p$ for p.

Multiplying or dividing the results, and making $p^{+1} + p^{-1} = 2\cos y$ this would introduce sines and cosines, etc. into the exponents.

We may obtain some other rather curious continued products, by combining the expression

$$\cos\frac{\theta}{2} = \left\{\frac{1\ \cos\theta}{2}\right\}^{1/2}$$

with the expression of Euler's (1,1)

$$2\cos\frac{\theta}{2} = \overset{1/2}{\{} 2 + 2\cos\theta\}$$

$$2\cos\frac{\theta}{2^2} = \overset{1/2}{\left\{ 2 + 2\cos\frac{\theta}{2}\right\}} = \overset{1/2}{\{} 2 + \overset{1/2}{\{} 2 + 2\cos\theta\}$$

$$2\cos\frac{\theta}{2^3} = \overset{1/2}{\left\{ 2 + 2\cos\frac{\theta}{2^2}\right\}} = \overset{1/2}{\{} 2 + \overset{1/2}{\{} 2 + \overset{1/2}{\{} 2 + 2\cos\theta\}$$

and in general

$$2\cos\frac{\theta}{2^n} = \overset{1/2}{\{} 2 + \overset{1/2}{\{} 2 + \overset{1/2}{\{} 2 + \ldots + (n-1)\text{ terms } \overset{1/2}{\{} 2 + 2\cos\theta\}.^{*} \tag{6,6}$$

Multiply and divide both sides by 2,

$$4\cos\frac{\theta}{2^n} = \overset{1/2}{\{} 2^{2^1+1} + \overset{1/2}{\{} 2^{2^2+1} + \overset{1/2}{\{} 2^{2^3+1} + \ldots + \overset{1/2}{\{} 2^{2^n+1} + 2^{2^n+1}\cos\theta\}. \tag{6,7}$$

$$\cos\frac{\theta}{2^n} = \overset{1/2}{\left\{\frac{1}{2^{2^1-1}}\right.} + \overset{1/2}{\left\{\frac{1}{2^{2^2-1}}\right.} + \overset{1/2}{\left\{\frac{1}{2^{2^3-1}}\right.} + \ldots + \overset{1/2}{\left\{\frac{1}{2^{2^n-1}} + \frac{1}{2^{2^n-1}}\cos\theta\right\}}. \tag{6,8}$$

/

In (6,6) let n vary from 1 to n; then,

$$\overset{n}{P}\left\{2\cos\frac{\theta}{2^n}\right\} = \overset{n}{P}\{\overset{1/2}{\{} 2 + \overset{1/2}{\{} 2 + \overset{1/2}{\{} 2 + \text{to } n \text{ terms} + 2\cos\theta\}\}$$

* When $\theta = 0$, the expression

$$\overset{1/2}{\{} 2 + \overset{1/2}{\{} 2 + \overset{1/2}{\{} 2 + n \text{ terms } \overset{1/2}{\{} 2 + 2\}$$

is constantly equal to 2, whatever be the value of n. This similar equation also is always true whatever be the value of p, viz.

$$a\cdot(1+a)^{1/(p-1)} = a\overset{1/p}{\{}(a+1)^{1/(p-1)} + a\overset{1/p}{\{}(a+1)^{1/(p-1)} + a\overset{1/p}{\{}(a+1)^{1/(p-1)} + \ldots \text{to } n \text{ terms } \overset{1/p}{\{}(a+1)^{1/(p-1)} + a\cdot(a+1)^{1/(p-1)}\}$$

but

$$\overset{n}{P}\left\{2\cos\frac{\theta}{2^n}\right\} = 2^n \times \overset{n}{P}\{\cos\theta/2^n\} = 2^n\frac{\sin\theta}{2^n\sin\theta/2^n} = \frac{\sin\theta}{\sin\theta/2^n}$$

therefore,

$$\frac{\sin\theta}{\sin\frac{\theta}{2^n}} = \overset{n}{P}\{\overset{1/2}{\{}2 + \overset{1/2}{\{}2 + \overset{1/2}{\{}2 + \text{to } n \text{ terms} + 2\cos\theta\}\}. \tag{6,9}$$

and similarly, from (6,7) and (6,8), we shall have

$$\frac{2^n\sin\theta}{\sin\frac{\theta}{2^n}} = \overset{n}{P}\{\overset{1/2}{\{}2^{2^1+1} + \overset{1/2}{\{}2^{2^2+1} + \overset{1/2}{\{}2^{2^3+1} + \dots$$

$$+ \overset{1/2}{\{}2^{2^n+1} + 2^{2^n+1}\cos\theta\}\}. \tag{7,0}$$

and,

$$\frac{\sin\theta}{2^n\sin\frac{\theta}{2^n}} = \overset{n}{P}\left\{\overset{1/2}{\left\{}\frac{1}{2^{2^1-1}} + \overset{1/2}{\left\{}\frac{1}{2^{2^2+1}} + \overset{1/2}{\left\{}\frac{1}{2^{2^3+1}} + \dots\right.\right.\right.$$

$$+ \overset{1/2}{\left\{}\frac{1}{2^{2^n-1}} + \frac{1}{2^{2^n-1}}\cos\theta\right\}\Bigg\}. \tag{7,1}$$

Many of the above theorems have a finite value, when the product is continued to infinity; thus, from (2) (5),

$$\frac{1}{1-x^2} = \overset{n}{P}\{1 + x^{2^n}\} \quad \begin{bmatrix} n+1 \\ n=\infty \end{bmatrix} \tag{7,2}$$

$$\frac{1}{1-x^3} = \overset{n}{P}\{1 - x^{3^n} + x^{2\cdot 3^n}\} \quad \begin{bmatrix} n=1 \\ n=\infty \end{bmatrix} \tag{7,3}$$

If we meet with a quantity of this kind,

$$\frac{\sin\frac{a\theta}{c^n}}{\sin\frac{b\theta}{c^n}}$$

its value, when n becomes infinite, may be easily found; thus,

$$\frac{\sin\dfrac{a\theta}{c^n}}{\sin\dfrac{b\theta}{c^n}} = \frac{a\theta c^n \cdot \dfrac{1}{1} - \dfrac{(a\theta)^3}{c^{3n}} \cdot \dfrac{1}{(1\cdot 2\cdot 3)} + \ldots}{\dfrac{b\theta}{c^n} \cdot \dfrac{1}{1} - \dfrac{(b\theta)^3}{c^{3n}} \cdot \dfrac{1}{(1\cdot 2\cdot 3)} + \ldots}$$

$$= \frac{a\theta - \dfrac{(a\theta)^3}{c^{2n}} \cdot \dfrac{1}{(1\cdot 2\cdot 3)} + \ldots}{b\theta - \dfrac{(b\theta)^3}{c^{2n}} \cdot \dfrac{1}{(1\cdot 2\cdot 3)} + \ldots}$$

and when n becomes infinite, this becomes

$$\frac{\sin a\theta/c^n}{\sin b\theta/c^n} = a/b$$

therefore (2,3) and (2,4) become

$$\cos\theta = \overset{n}{P}\left\{1 - 2\cos\frac{2\theta}{(4a \pm 1)^n} + 2\cos\frac{4\theta}{(4a \pm 1)^n} - \ldots + 2\cos\frac{4a\theta}{(4a+1)^n}\right\} \quad \begin{bmatrix} n = 1 \\ n = \infty \end{bmatrix} \quad (7,4)$$

$$\cos\theta = \overset{n}{P}\left\{2\cos\frac{(4a-2)\theta}{(4a-1)^n} - 2\cos\frac{(4a-4)\theta}{(4a-1)^n} + \ldots + 2\cos\frac{2\theta}{(4a-1)^n} - 1\right\} \quad \begin{bmatrix} n = 1 \\ n = \infty \end{bmatrix} \quad (7,5)$$

(2,5) and (2,6) are, when n is infinite,

$$\frac{\sin(4a+1)\theta}{(4a+1)\sin\theta} = \overset{n}{P}\left\{1 - 2\cos\frac{2\theta}{2^n} + 2\cos\frac{4\theta}{2^n} - \ldots + 2\cos\frac{4a\theta}{2^n}\right\} \quad \begin{bmatrix} n = 1 \\ n = \infty \end{bmatrix} \quad (7,6)$$

$$\frac{\sin(4a-1)\theta}{(4a-1)\sin\theta} = \overset{n}{P}\left\{2\cos\frac{(4a-2)\theta}{2^n} - 2\cos\frac{(4a-4)\theta}{2^n} + \ldots + 2\cos\frac{2\theta}{2^n} - 1\right\} \quad \begin{bmatrix} n = 1 \\ n = \infty \end{bmatrix} \quad (7,7)$$

All the preceding continued products have been accurately determined for any given value of (n), those which follow are only determined when n is infinite, and as these depend on a principle entirely different from the others, I have chosen to place them last.

Let

$$\varphi x = A_0 + A_1 x + A_2 x^2 + \ldots + A_n x^n.$$

For x, substitute successively $x, x^2, x^3, \ldots$;

$$\begin{aligned}
\varphi x &= A_0 + A_1 x + A_2 x^2 + \ldots + A_n x^n \\
\varphi x^2 &= A_0 + A_1 x^2 + A_2 x^4 + \ldots + A_n x^{2n} \\
\varphi x^3 &= A_0 + A_1 x^3 + A_2 x^6 + \ldots + A_n x^{3n} \\
\varphi x^4 &= A_0 + A_1 x^4 + A_2 x^8 + \ldots + A_n x^{4n} \\
&\vdots
\end{aligned}$$

Add all the vertical terms, and suppose every even line negative; then,

$$(\varphi x - A_0) - (\varphi x^2 - A_0) + (\varphi x^3 - A_0) - \ldots = \frac{A_1 x}{1 + x} + \frac{A_2 x^2}{1 + x^2} + \frac{A_3 x^3}{1 + x^3} + \ldots + \frac{A_n x^n}{1 + x^n}. \quad (a,1)$$

Supposing all positive,

$$(\varphi x - A_0) + (\varphi x^2 - A_0) + (\varphi x^3 - A_0) + \ldots = \frac{A_1 x}{1 - x} + \frac{A_2 x^2}{1 - x^2} + \frac{A_3 x^3}{1 - x^3} + \ldots + \frac{A_n x^n}{1 - x^n}. \quad (a,2)$$

Multiply $(a,1)$ $(a,2)$ by dx/x, and let

$$\int (\varphi x - A_0)\frac{dx}{x} = fx$$

then,

$$\int (\varphi x^n - A_0)\frac{dx}{x} = \frac{1}{n}\int (\varphi x^n - A_0)\frac{n x^{n-1}dx}{x^n} = \frac{1}{n} f x^n$$

therefore,

$$\frac{fx}{1} - \frac{fx^2}{2} + \frac{fx^3}{3} - \frac{fx^4}{4} + \ldots = S\left\{-(-1)^n \frac{fx^n}{n}\right\}$$
$$= \log\{(1 + x)^{(1/1)A_1} \cdot (1 + x^2)^{(1/2)A_2} \cdot (1 + x^3)^{(1/3)A_3} \cdot \ldots \cdot (1 + x^n)^{(1/n)A_n}\} \quad (a,3)$$

$$\frac{fx}{1} + \frac{fx^2}{2} + \frac{fx^3}{3} + \ldots = S\left(\frac{fx^n}{n}\right) =$$

$$-\log\{(1-x)^{(1/1)A_1} \cdot (1-x^2)^{(1/2)A_2} \cdot (1-x^3)^{(1/3)A_3} \cdot \ldots \cdot (1-x^n)^{(1/n)A_n}\}. \quad (a,4)$$

The sum of these two is

$$\frac{fx}{1} + \frac{fx^3}{3} + \frac{fx^5}{5} + \ldots =$$

$$-\frac{1}{2}\log\left\{\left(\frac{1-x}{1+x}\right)^{(1/1)A_1} \cdot \left(\frac{1-x^2}{1+x^2}\right)^{(1/2)A_2} \cdot \left(\frac{1-x^3}{1+x^2}\right)^{(1/3)A_3} \cdot \ldots \cdot \left(\frac{1-x^n}{1+x^n}\right)^{(1/n)A_n}\right\}. \quad (a,5)$$

These expressions, though they apparently give the value to n terms, are in fact of very little use, except when n is infinite. In that case, the value of the series on the left hand may frequently be obtained. /

In $(a,4)$ make

$$A_0 = 0 \qquad A_1 = 1 \qquad A_2 = 2 \qquad A_3 = 3, \quad \ldots$$

$$\varphi x = \frac{x}{(1-x)^2}$$

$$fx = \int(\varphi x - A_0)\frac{dx}{x} = \int\frac{dx}{(1-x)^2} = \frac{1}{(1-x)}$$

$$S\left\{\frac{fx^n}{n}\right\} = +\left\{\frac{1}{1-x}\cdot\frac{1}{1} + \frac{1}{1-x^2}\cdot\frac{1}{2} + \frac{1}{1-x^3}\cdot\frac{1}{3} + \ldots\right\} + C$$

$$= -\log\{(1-x)\cdot(1-x^2)\cdot(1-x^3)\cdot\ldots\}.$$

If $x = 0$,

$$-C = \frac{1}{1} + \frac{1}{2} + \frac{1}{3} + \ldots$$

$$-\left\{\frac{1}{1}\left(1 - \frac{1}{1-x}\right) + \frac{1}{2}\left(1 - \frac{1}{1-x^2}\right) + \frac{1}{3}\left(\frac{1}{1-x^3}\right) + \ldots\right\}$$

$$= +\frac{x}{1-x}\cdot\frac{1}{1} + \frac{x^2}{1-x^2}\cdot\frac{1}{2} + \frac{x^3}{1-x^3}\cdot\frac{1}{3} + \ldots$$

$$= -\log\{(1-x)\cdot(1-x^2)\cdot(1-x^3)\cdot\ldots\}$$

$$-\left\{\frac{x}{1}\cdot\frac{1}{1} + \frac{x^2}{1+x}\cdot\frac{1}{2} + \frac{x^3}{1+x+x^2}\cdot\frac{1}{3} + \ldots\right\}$$

$$= \log\{(1-x)(1-x^2)(1-x^3)\ldots\}^{1-x}. \quad (a,6)$$

If x ultimately equal unity,

$$\varepsilon^{-\{1/1^2+1/2^2+1/3^2+\ldots\}} = \{(1-x)(1-x^2)(1-x^3)\ldots\}^{1-x}. \qquad (a,7)$$

$$\varepsilon^{-\pi^2/6} = \{(1-x)(1-x^2)(1-x^3)\ldots\}^{1-x} \qquad x \text{ ult.} = 1$$

$$\varepsilon^{-\pi^2/6} = \{(1)(1+x)(1+x+x^2)\ldots\}^{1-x} \times \log\{(1-x)^{1/(1-x)}\}^{1-x}$$

/

$$\frac{\varepsilon^{-\pi^2/6}}{1-x} = \{1\cdot 2\cdot 3\cdot 4\cdot\ldots\}^{(1-x)} \qquad x \text{ ult.} = 1$$

In $(a,4)$, let $A_0 = 0$, $A_1 = n'$, $A_2 = n''$, $A_3 = n'''$, ...; n' n'' n''', ... being the successive terms of the nth order of figurate numbers.

$$\varphi x = \frac{x}{(1-x)^n}$$

$$fx = \frac{1}{(n-1)(1-x)^{n-1}}$$

$$\frac{1}{m} fx^m = \frac{(n-1)^{-1}}{m\cdot(1-x)^{n-1}}$$

$$\frac{1}{n-1}\left\{\frac{1^{-1}}{(1-x)^{n-1}} + \frac{2^{-1}}{(1-x^2)^{n-1}} + \frac{3^{-1}}{(1-x^3)^{n-1}} + \cdot\ldots\right\} + C$$
$$= -\log\{(1-x)^{n'/1}\cdot(1-x^2)^{n''/2}\cdot(1-x^3)^{n'''/3}\cdot\ldots\}$$

if $x = 0$,

$$C = -\frac{1}{n-1}\left\{\frac{1}{1} + \frac{1}{2} + \frac{1}{3} + \ldots\right\}$$

$$\varepsilon^{-\frac{1}{n-1}\left\{\frac{1-(1-x)^{n-1}}{1\cdot(1)^{n-1}} + \frac{1-(1-x^2)^{n-1}}{2\cdot(1+x)^{n-1}} + \frac{1-(1-x^3)^{n-1}}{3\cdot(1+x+x^2)^{n-1}} + \ldots\right\}}$$
$$= \{(1-x)^{n'/1}\cdot(1-x^2)^{n''/2}\cdot(1-x^3)^{n'''/3}\cdot\ldots\}^{(1-x)^{n-1}}$$

Let x ultimately become unity,

$$\varepsilon^{-\frac{1}{n-1}\left\{\frac{1}{1^n}+\frac{1}{2^n}+\frac{1}{3^n}+\ldots\right\}}$$
$$= \{(1-x)^{n'/1}\cdot(1-x^2)^{n''/2}\cdot(1-x^3)^{n'''/3}\cdot\ldots\}^{(1-x)^{n-1}} \qquad (a,8)$$

or,

$$\frac{\varepsilon^{-\frac{1}{n-1}\left\{\frac{1}{1^n}+\frac{1}{2^n}+\frac{1}{3^n}+\ldots\right\}}}{\{1-x\}^{1/(n-1)}} = \{(1)^{n'/1}\cdot(2)^{n''/2}\cdot(3)^{n'''/3}\cdot\ldots\}^{(1-x)^{n-1}} \qquad (a,9)$$

/

When n is an even number,

$$\varepsilon^{-\frac{1}{n-1}\frac{2^{n-1}B_{n-1}\pi^n}{1\cdot 2\cdot 3\cdot 4\ldots n}} = \{(1-x)^{n'/1}\cdot(1-x^2)^{n''/2}\cdot(1-x^3)^{n'''/3}\cdot\ldots\}^{(1-x)^{n-1}}$$

B_n being one of the numbers of Bernouilli.

Let, $A_0 = 0,\quad A_1 = 1,\quad A_2 = 0,\quad A_3 = -1,\quad A_4 = 0,\quad A_5 = 1, \ldots$

$$\varphi x = x - x^3 + x^5 - \ldots = \frac{x}{1+x^2}.$$

$$fx = \text{arc}(\tan x) = \tan^{-1} x.\text{*}$$

Substituting these values in $(a,3)$ $(a,4)$, we shall find

$$\frac{\tan^{-1} x}{1} + \frac{\tan^{-1} x^2}{2} + \frac{\tan^{-1} x^3}{3} + \ldots$$

$$= -\log \overset{n}{P}\{(1-x^{2n-1})^{-(-1)^n/(2n-1)}\}$$

$$= -\log\{(1-x)^{1/1}\cdot(1-x)^{-1/3}\cdot(1-x^5)^{1/5}\cdot\ldots\} \qquad (a,10)$$

$$\frac{\tan^{-1} x}{1} - \frac{\tan^{-1} x^2}{2} + \frac{\tan^{-1} x^3}{3} - \ldots$$

$$= \log \overset{n}{P}\{(1+x^{2n-1})^{-(-1)^n/(2n-1)}\}$$

$$= \log\{(1+x)^{1/1}\cdot(1+x)^{-1/3}\cdot(1+x^5)^{1/5}\cdot\ldots\}. \qquad (a,11)$$

In $(a,10)$, let $x = 1$,

$$\frac{\pi}{4}\left(\frac{1}{1}+\frac{1}{2}+\frac{1}{3}+\ldots\right) = -\log(1-x)^{\pi/4}$$

$$-\log\{(1)^{1/1}\cdot(1+x+x^2)^{1/3}\cdot(1+x+x^2+x^3+x^4)^{1/5}\}$$

$$-\frac{\pi}{4}\log(1-x) + \frac{\pi}{4}\log(1-x) = -\log\{(1)^{1/1}\cdot(3)^{-1/3}\cdot(5)^{1/5}\cdot\ldots\}$$

$$1 = \{(1)^{1/1}\cdot(3)^{-1/3}\cdot(5)^{1/5}\cdot(7)^{-1/7}\cdot\ldots\}. \qquad (a,12)$$

/

Let

$$\varphi x = \frac{x}{\sqrt{1-x^2}}$$

$$fx = \sin^{-1} x$$

$$A_0 = 0,\qquad A_1 = 1,\qquad A_2 = \frac{1}{2},\qquad A_3 = \frac{1\cdot 3}{2\cdot 4}\ldots,$$

* $\tan^{-1} x$ or, $\sin^{-1} x$, signifies the arc whose tangent or sine is equal to x.

⟨then⟩

$$\frac{\sin^{-1} x}{1} + \frac{\sin^{-1} x^2}{2} + \frac{\sin^{-1} x^3}{3} + \dots$$

$$= -\log\{(1-x)^{1\cdot\frac{1}{1}} \cdot (1-x^2)^{\frac{1}{2}\cdot\frac{1}{2}} \cdot (1-x^3)^{\frac{1}{2}\cdot\frac{3}{4}\cdot\frac{1}{3}} \dots\} \quad (a,13)$$

$$\frac{\sin^{-1} x}{1} - \frac{\sin^{-1} x^2}{2} + \frac{\sin^{-1} x^3}{3} - \dots$$

$$= \log\{(1+x)^{1\cdot\frac{1}{1}} \cdot (1+x^2)^{\frac{1}{2}\cdot\frac{1}{2}} \cdot (1+x^3)^{\frac{1}{2}\cdot\frac{3}{4}\cdot\frac{1}{3}} \dots\} \quad (a,14)$$

A number of particular cases might be produced affording some remarkable results, but we will rather proceed at once to some theorems yet more general.

Assume

$$\varphi(x, \theta) = A_1\, x \cos\theta + A_2\, x^2 \cos 2\theta + \dots = \overset{n}{S}\{A_n\, x^n \cos n\theta\}$$

the limits of n being unity and infinite.

Multiply this by $u/1$; for θ put 2θ, and multiply by $u^2/2$; again, for θ put 3θ, and multiply by $u^3/3$, and so on; then we find,

$$\frac{u}{1}\varphi(x, \theta) = \frac{u}{1}\overset{n}{S}\{A_n\, x^n \cos n\theta\}$$

$$\frac{u^2}{2}\varphi(x, 2\theta) = \frac{u^2}{2}\overset{n}{S}\{A_n\, x^n \cos 2n\theta\}$$

$$\frac{u^3}{3}\varphi(x, 3\theta) = \frac{u^3}{3}\overset{n}{S}\{A_n\, x^n \cos 3n\theta\}$$

$$\vdots \qquad\qquad \vdots$$

Adding these together,

$$\overset{p}{S}\left\{\frac{u^p}{p}\varphi(x, p\theta)\right\} = \overset{n}{S}\left\{A_n x^n \cdot \overset{p}{S}\left\{\frac{u^p}{p}\cos pn\theta\right\}\right\}.$$

But,

$$\overset{p}{S}\left\{\frac{u^p}{p}\cos pn\theta\right\} = \frac{u}{1}\cos n\theta + \frac{u^2}{2}\cos 2n\theta + \frac{u^3}{3}\cos 3n\theta + \dots$$

$$= -\frac{1}{2}\log\{1 - 2u\cos n\theta + u^2\}$$

$$\overset{p}{S}\left\{\frac{u^p}{p}\varphi(x, p\theta)\right\} = -\frac{1}{2}\overset{n}{S}\{A_n x^n \log\{1 - 2u\cos n\theta + u^2\}\}$$

$$= -\frac{1}{2}\overset{n}{S}\log\{\{1 - 2u\cos n\theta + u^2\}^{A_n x^n}\}$$

$$= -\frac{1}{2}\log \overset{n}{P}\{\{1 - 2u\cos n\theta + u^2\}^{A_n x^n}\}.$$

Making u negative, and dividing by the result, we have the three expressions,

$$\varepsilon^{-2\overset{p}{S}\left\{\frac{u^p}{p}\varphi(x,p\theta)\right\}} = \overset{n}{P}\{\{1 - 2u\cos n\theta + u^2\}^{A_n x^n}\} \quad \begin{bmatrix} p=1 \\ p=\infty \\ n=1 \\ n=\infty \end{bmatrix} \qquad (a,15)$$

$$\varepsilon^{2\overset{p}{S}\left\{\frac{-(-u)^p}{p}\varphi(x,p\theta)\right\}} = \overset{n}{P}\{\{1 + 2u\cos n\theta + u^2\}^{A_n x^n}\} \quad \begin{bmatrix} p=1 \\ p=\infty \\ n=1 \\ n=\infty \end{bmatrix} \qquad (a,16)$$

$$\varepsilon^{-4\overset{p}{S}\left\{\frac{u^{2p-1}}{2p-1}\varphi(x,(2p-1)\theta)\right\}} = \overset{n}{P}\left\{\left\{\frac{1 - 2u\cos n\theta + u^2}{1 + 2u\cos n\theta + u^2}\right\}^{A_n x^n}\right\} \quad \begin{bmatrix} p=1 \\ p=\infty \\ n=1 \\ n=\infty \end{bmatrix} \qquad (a,17)$$

If $u = 1$, since p and n are between the same limits, we have by extracting the square root,

$$\varepsilon^{-\overset{n}{S}\left\{\frac{\varphi(x,n\theta)}{n}\right\}} = \overset{n}{P}\left\{2\sin n\frac{\theta}{2}\right\}^{A_n x^n} \quad \begin{bmatrix} n = 1 \\ n = \infty \end{bmatrix} \qquad (a,18)$$

$$\varepsilon^{\overset{n}{S}\left\{\frac{-(-1)^n\varphi(x,n\theta)}{n}\right\}} = \overset{n}{P}\left\{2\cos n\frac{\theta}{2}\right\}^{A_n x^n} \quad \begin{bmatrix} n = 1 \\ n = \infty \end{bmatrix} \qquad (a,19)$$

$$\varepsilon^{-\frac{1}{2}\overset{n}{S}\left\{\frac{\varphi(x,(2n-1)\theta)}{2n-1}\right\}} = \overset{n}{P}\left\{2\tan n\frac{\theta}{2}\right\}^{A_n x^n} \quad \begin{bmatrix} n = 1 \\ n = \infty \end{bmatrix} \qquad (a,20)$$

In $(a,16)$ let $x = 1$, $A_n = 1$,

$$\varphi(x, \theta) = \cos\theta + \cos 2\theta + \cos 3\theta + \ldots = -\frac{1}{2}$$

for θ put 2θ,

$$\overset{p}{S}\left\{\frac{-(-u)^p}{p}\varphi(x, \theta)\right\} = -\frac{1}{2}\log(1 + u)$$

$$\varepsilon^{-\log(1+u)} = \overset{n}{P}\{1 + 2u\cos 2n\theta + u^2\}.$$

If $u = 1$,

$$\frac{1}{\sqrt{2}} = \overset{n}{P}\{2 \cos n\theta\} = (2 \cos \theta) \cdot (2 \cos 2\theta) \cdot (2 \cos \cdot 3\theta) \cdot \ldots.$$

Let

$$A_1 = \frac{1}{1}, \qquad A_2 = \frac{1}{1 \cdot 2}, \qquad A_3 = \frac{1}{1 \cdot 2 \cdot 3}, \ldots$$

⟨then⟩

$$\varphi x = \frac{x \cos \theta}{1} + \frac{x^2 \cos 2\theta}{1 \cdot 2} + \frac{x^3 \cos 3\theta}{1 \cdot 2 \cdot 3} + \ldots = \varepsilon^{x \cos \theta} \cos (x \sin \theta) - 1.$$

Taking the logarithms of (a,15) (a,16) (a,17),

$$-2 \log (1 - u) - 2\left\{\frac{\varepsilon^{x \cos \theta} \cos (x \sin \theta)}{1} u + \frac{\varepsilon^{x \cos 2\theta} \cos (x \sin 2\theta)}{2} u^2 + \frac{\varepsilon^{x \cos 3\theta} \cos (x \sin 3\theta)}{3} u^3 + \ldots\right\}$$

/

$$= \log \overset{n}{P}\{1 - 2u \cos n\theta + u^2\}^{x^n/(1 \cdot 2 \cdot 3 \cdot \ldots \cdot n)}$$

$$= \log \overset{x/1}{\{} 1 - 2u \cos \theta + u^2 \overset{x/2}{\{} 1 - 2u \cos 2\theta + u^2 \overset{x/3}{\{} 1 - 2u \cos 3\theta + u^2 \overset{x/4}{\{} \ldots\} - 2 \log (1 + u)$$

$$+ 2\left\{\frac{\varepsilon^{x \cos \theta} \cos (x \sin \theta)}{1} u - \frac{\varepsilon^{x \cos 2\theta} \cos (x \sin 2\theta)}{2} u^2 + \frac{\varepsilon^{x \cos 3\theta} \cos (x \sin 3\theta)}{3} u^3 - \ldots\right\}$$

$$= \log \overset{n}{P}\{1 + 2u \cos n\theta + u^2\}^{x^n/(1 \cdot 2 \cdot 3 \cdot \ldots \cdot n)}$$

$$= \log \overset{x/1}{\{} 1 + 2u \cos \theta + u^2 \overset{x/2}{\{} 1 + 2u \cos 2\theta + u^2 \overset{x/3}{\{} 1 + 2u \cos 3\theta + u^2 \overset{x/4}{\{} \ldots\}.$$

Also,

$$-4\left\{\frac{\varepsilon^{x\cos\theta}\cos(x\sin\theta)}{1}u+\frac{\varepsilon^{x\cos 3\theta}\cos(x\sin 3\theta)}{3}u^3+\frac{\varepsilon^{x\cos 5\theta}\cos(x\sin 5\theta)}{5}u^5+\ldots\right\}$$

$$=\log \overset{n}{P}\left\{\frac{1-2u\cos n\theta+u^2}{1+2u\cos n\theta+u^2}\right\}^{x^n/(1\cdot 2\cdot 3\cdot\ldots\cdot n)}$$

$$=\log \overset{x/1}{\left\{\frac{1-2u\cos\theta+u^2}{1+2u\cos\theta+u^2}\right.} \overset{x/2}{\left\{\frac{1-2u\cos 2\theta+u^2}{1+2u\cos 2\theta+u^2}\right.} \overset{x/3}{\left\{\frac{1-2u\cos 3\theta+u^2}{1+2u\cos 3\theta+u^2}\right.} \overset{x/4}{\left\{\ldots\right\}}.$$

In the second of these expressions if u be made equal to unity, and also x and 2θ be put for θ, we have,

$$-\log 2+\frac{\varepsilon^{\cos 2\theta}\cos\sin 2\theta}{1}+\frac{\varepsilon^{\cos 4\theta}\cos\sin 4\theta}{2}+\frac{\varepsilon^{\cos 6\theta}\cos\sin 6\theta}{3}-\ldots$$

$$=\log \overset{1/1}{\{}\,2\cos\theta \overset{1/2}{\{}\,2\cos 2\theta \overset{1/3}{\{}\,2\cos 3\theta \overset{1/4}{\{}\,2\cos 4\theta \overset{1/5}{\{}\ldots\}. \qquad (a,21)$$

/

Several of the expressions deduced in the latter part of this paper, are purposely left in their most general form; should any one enquire into the results produced, by assigning particular values to some of the quantities concerned, he will meet with many singular and curious theorems.

The formulae in the former part, may by differentiating and by other methods be converted into infinite series, and from these and the expressions themselves, may be derived several remarkable properties of the circle.

/

III

AN ESSAY TOWARDS THE CALCULUS OF FUNCTIONS ⟨PART I⟩

From *Philosophical Transactions of the Royal Society*, Vol. 105 (1815), pp. 389–423

Communicated by W. H. Wollaston, M.D. Sec. R. S.
Read 15 June, 1815

The term function has long been introduced into analysis with great advantage, for the purpose of designating the result of every operation that can be performed on quantity. This extent of signification has rendered it of essential use, but the various applications of which it admits, and the questions to which it gives rise, do not appear to have met with sufficient attention.

I propose in the following paper to present an outline of a new calculus, which naturally results from it. It comprehends questions of the greatest generality and difficulty, and will probably require the invention of new methods for its improvement.

Many of the calculations with which we are familiar, consist of two parts, a direct, and an inverse; thus, when we consider an exponent of a quantity: to raise any number to a given power, is the direct operation: to extract a given root of any number, is the inverse method. The differential calculus, which is a direct method, naturally gave rise to the integral, which is its inverse: the same remark is applicable to finite differences. In all these cases the inverse method is by far the most difficult, and it might perhaps be added, the most useful. /

It is this inverse method with respect to functions, which I at present propose to consider.

If an unknown quantity as x, be given by means of an equation, it becomes a question how to determine its value; similarly if an unknown function as ψ, be given by means of any functional equation, it is required to assign its form. In the first case, it is quantity which is to be determined; in the second, it is the form assumed by quantity, that becomes the subject of investigation. In the one case, the various powers of the unknown quantity enter into the equation; in the other, the different orders of the function are concerned.

Before I proceed, it will be proper to explain the meaning of the order of a functional equation, and likewise to indicate the notation made use of; α, β, γ, . . . are known functional characteristics; ψ, χ, ψ, are unknown ones.

If in any function as ψx, instead of x, the original function be substituted, it becomes $\psi\psi x$ or $\psi^2 x$: this is called the second function of x. If the process be repeated, the result is $\psi^2\psi x$ or $\psi^3 x$, the third function of x; and similarly $\psi^n x$, denotes the nth function of x. Suppose

$$\psi x = a + x$$

then

$$\psi^2 x = a + a + x = 2a + x$$

and generally

$$\psi^n x = na + x.$$

A functional equation is said to be of the first order, when it contains only the first function of the unknown quantity; as, for instance,

$$\psi\alpha x + x\psi x - x^n = 0$$

$$\left(\psi x + \psi\frac{1}{x}\right)^n - ax + x^2 = 0.$$

If the second function enter, the equation rises to the second order: thus,

$$\psi^2 x = x$$

$$\psi(x + \psi x) + (\psi x - x)^2 = 0$$

$$\left(\psi^2 x + \psi\frac{1}{x}\right)^n = ax.$$

A function of two variables admits of two second functions: thus $\psi(x, y)$ becomes $\psi(\psi(x, y), y)$, and $\psi(x, \psi(x, y))$ or they might be thus expressed $\psi^{2,1}(x, y)$ and $\psi^{1,2}(x, y)$.

These express the second functions; the first taken relative to x, the other relative to y. But besides these two there is another, which arises from taking the second function simultaneously relative to x, and y; it is $\psi\{\psi x, y), \psi(x, y)\}$. This ought not to be written $\psi^{2,2}(x, y)$ for it is not the second function first taken relative to x and then to y, nor is it the converse of this. In fact, the notation is defective; some method is wanting of indicating the order in which the successive substitutions are made. I shall for the present lay aside the consideration of functional equations, involving more than one variable.

Those of the first order have long been known, but the method in which I have treated them is, I believe, entirely new. Equations of the second and higher orders have never been even mentioned; it is these which present the most interesting speculations, and which are involved in the greatest difficulties. I shall first give some account of the enquiries which led me to this subject, and shall then treat of the various orders of functional equations.

Some few years since, while considering a problem mentioned by Pappus, relating to the inscription of a number / of circles in a semicircle, the following question occurred to me.

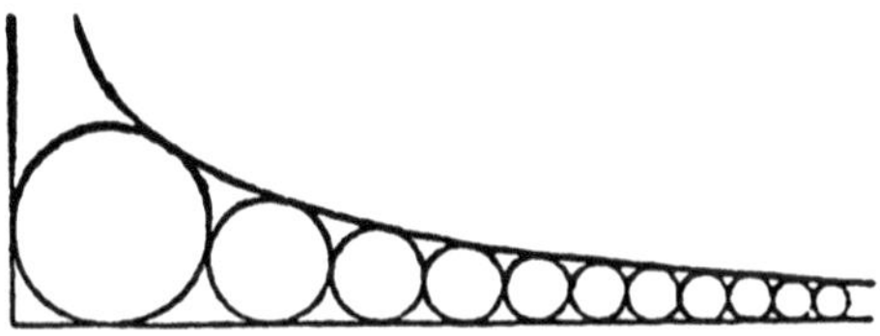

If in an hyperbola between its asymptotes a circle be inscribed touching both asymptotes and the curve, and if another circle be inscribed touching the first circle, the curve, and one asymptote, and if this be continued as represented in the figure, what ratio does the area of the circles bear to that of the figure; and conversely, if this ratio is given, what is the nature of the curve? I soon perceived the great difficulty of the subject, and that these and other problems similar to the latter of them, required the application of methods totally different from any with which I was then acquainted.

Hopeless of success, I laid aside the subject until about two years after, when the same difficulty recurred under another form. I had proposed to myself the following problem:

What must be the nature of the curve *ABC*, such that if

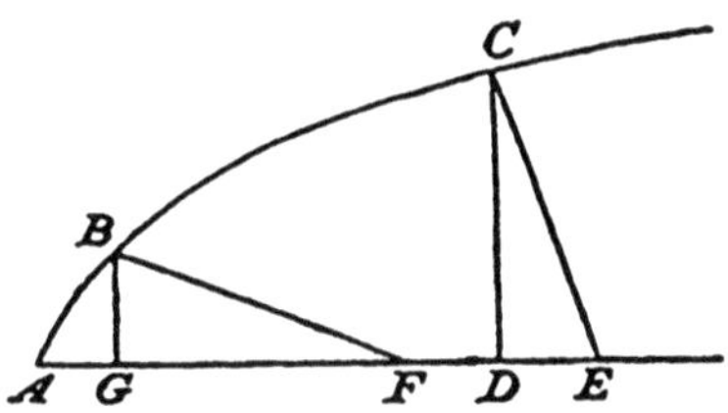

any point *C* be taken, and the ordinate *CD*, the normal *CE*, and subnormal *DE* be drawn, and if the triangle *CDE* be turned into such a position that *CD* may become the base and *DE* the perpendicular, if *DE* coincide with some new ordinate as *GB*, then the normal *CE* at the first point shall coincide with the normal *BF* at the second? /

This latter question, of far less difficulty than the former, was readily solved, but I did not at first observe that they rested on similar principles; this, however, was pointed out by Mr Herschel, to whom I had mentioned the subject. Such was the origin of the following enquiries.

The question, in its most general point of view, is the solution of functional equations of all orders.

This, however, is a generality which I do not pretend to have attained. In the first part of this paper the reader will find a new method of solving all functional equations of the first order; it depends on possessing their particular solutions. In the subsequent part, I have given various methods of solving functional equations of the second and higher orders: some of these possess considerable generality; and if we consider only those in which the *n*th and inferior orders enter simply, such as

$$F\{x, \psi x, \psi^2 x, \ldots \psi^n x\} = 0$$

I have pointed out the means of obtaining their solutions.

The determination of functions from given conditions most probably took its rise from the integration of equations of partial differentials; and we accordingly find that the authors of this calculus were soon engaged in the new problem to which it gave birth. D'Alembert was the first who occupied himself with this subject: he was soon followed by Euler and Lagrange; but it is to Monge that we are indebted for the most general view of the subject. His enquiries were directed to the determination of two functions from given conditions; they are contained in the fifth volume of the *Mélanges de Turin*, and in two papers in the seventh volume of the *Mémoires des Savants Étrangers*, 1773. /

In the first of these he explains the solution of several functional equations by means of curves of double curvature, and by curve surfaces.

In the second paper, the question is treated in a more analytical method, and he endeavours to reduce it to the solution of equations of differences. 'Je me propose,' observes Monge, 'de faire voir que la détermination des fonctions arbitraires qui se trouvent dans l'intégrale d'une équation aux différences partielles, depend en général, dans les cas que je n'ai pas encore traités, de l'intégrale d'une ou de plusieurs équations aux différences finies, dans lesquelles le rapport de la variable principale à sa différence finie est donné soit qu'il soit variable soit qu'il soit constant.'[a]

In the same volume is a paper of Laplace on this subject, which he views in the same light, and endeavours to reduce functional equations of the first order to those of finite differences. This skilful analyst first solved the functional equation $F\{x, \varphi x, \varphi \alpha x\} = 0$. The method he made use of is peculiarly elegant; he converted it into an equation of finite differences in which the difference was constant. Still, however, it appeared by no means the most direct method to make use of such an expedient, nor was it even known that all equations of the first order admitted of its application. This latter objection was, however, removed by Mr Herschel, who in an excellent paper on functional equations,

[a] *Translation:* I intend to show that, in order to determine the arbitrary functions which one obtains when integrating equations of partial differences, in general, this depends, in those cases which I have not already dealt with, on being able to integrate one or more equations of finite differences, in which the ratio of the independent variable to its finite difference has been given, whether that be variable or constant.

has extended the method made use of by Laplace to the solution of all equations of the first order. His solution is equally elegant and general; it leaves nothing to be regretted, but the narrow limits of our knowledge respecting the integration of equations of finite differences. From this and other causes, I am still / inclined to think that the solution of functional equations must be sought by methods peculiarly their own. There are some other researches on this difficult subject of which I am unable to give any account, from the impossibility of procuring the works in which they are contained; among these is the paper of Arbogast, which gained the prize of the Academy of Petersburg in the year 1790.

For the sake of convenience, I shall call any solution of a functional equation which contains one or more arbitrary functions, a general solution; but if the solution of such an equation only contains arbitrary constants, I shall call it a particular solution. With respect to the number of arbitrary functions that may enter into any solution, I shall make some observations at the conclusion of this paper.

Problem I

Required the general solution of the functional equation,

$$\psi x = \psi \alpha x$$

supposing we are acquainted with one particular solution.

Let the particular solution be $fx = f\alpha x$; then take $\psi = \varphi f$, φ being an arbitrary function. It is evident that this value of ψ will satisfy the original equation, and that

$$\varphi f x = \varphi f \alpha x$$

is identical, because $fx = f\alpha x$.

Example, let the equation be

$$\psi(x) = \psi(-x)$$

and the particular solution be

$$fx = x^2$$

the general solution is

$$\psi x = \varphi(x^2)$$

which evidently answers the conditions. /

As I shall have frequent occasion to make use of symmetrical functions of two or more quantities, I shall for the sake of brevity denote this by putting a line over the functional characteristic; thus $\bar{\varphi}(x, y)$ represents a symmetrical function of x and y, which it is well known possesses the following property,

$$\varphi\{x, y\} = \varphi\{y, x\}.$$

As we are only considering functional equations of one variable, this will be sufficient for the present purpose; it might perhaps otherwise be more advisable to put the line over the quantities relative to which the function is symmetrical; thus $\varphi\{x, y, \bar{z}, \bar{v}\}$ is symmetrical relative to z and v, but it is not so in respect to the other variables. This would possess the advantage of readily designating a function symmetrical relative to two quantities in one way, and likewise symmetrical with respect to two others, but in a different manner,* thus

$$\varphi\{\bar{x}, \bar{y}, \overset{\perp}{v}, \overset{\perp}{z}\}$$

a particular case of this is

$$\frac{v + z + axy}{v^3z^3 - ax^2y^2}$$

which is symmetrical in one sense relative to x and y, and in a different sense with respect to v and z; but these belong to other enquiries.

Problem II

Required a general solution of the equation $\psi x = \psi \alpha x$, having given a particular solution of $fx = f\alpha^2 x$. /

Assume $\psi x = \bar{\varphi}\{fx, f_1 x\}$, then it becomes

$$\bar{\varphi}\{fx, f_1 x\} = \bar{\varphi}\{f\alpha x, f_1 \alpha x\}.$$

This equation will be satisfied if we determine f and f_1, so that the following equations may be fulfilled:

$$f\alpha x = f_1 x \quad \text{and} \quad f_1 \alpha x = fx.$$

From these result

$$f\alpha^2 x = f_1 \alpha x = fx$$

but we have by hypothesis a particular solution of $f\alpha^2 x = fx$, therefore the general solution of $\psi x = \psi \alpha x$ is

$$\psi x = \bar{\varphi}\{fx, f\alpha x\}.$$

If the function α should be of such a nature that $\alpha^2 x = x$, or even if $\alpha^{2^n} x = x$, fx will then become x.

* This is not a mere imaginary refinement; I have constantly had occasion to make use of functions of many variables which were symmetrical by pairs, when investigating the nature of functional equations of more than one variable.

For example, suppose

$$\psi x = \psi\left(\frac{x}{ax-1}\right)$$

then since

$$\alpha x = \frac{x}{ax-1},$$

$$\alpha^2 x = \alpha\left(\frac{x}{ax-1}\right) = \frac{\dfrac{x}{ax-1}}{a\left(\dfrac{x}{ax-1}\right)-1} = x$$

and we have

$$\psi x = \bar{\varphi}\left\{x, \frac{x}{ax-1}\right\}.$$

A particular case of this is when $a = 0$, $\psi(x) = \psi(-x)$, its solution is

$$\psi x = \bar{\varphi}\{x, -x\} = \varphi(x^2)$$

the same as in the last Problem; as another example take

$$\psi x = \psi\frac{1}{x}$$

then since

$$\alpha x = \frac{1}{x},$$

$$\alpha^2 x = \frac{1}{1/x} = x \text{ and } \psi x = \bar{\varphi}\left\{x, \frac{1}{x}\right\}.$$

This affords a solution to the following question:

Required the nature of a curve, such that if any two abscissae, whose rectangle is equal to a given square be taken, their corresponding ordinates may also be equal. /

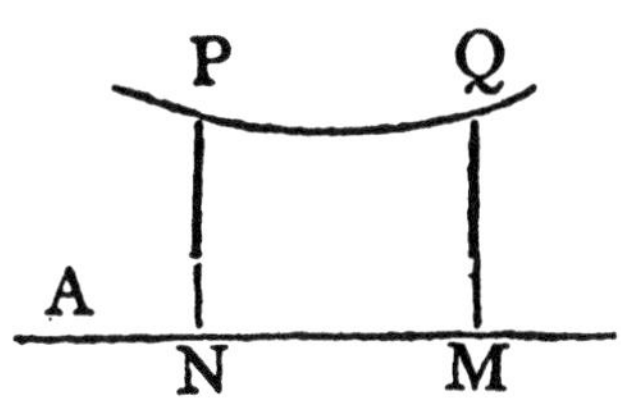

Let $y = \psi x$ be the equation of the curve, also $AN = x$, and if

$$AN \cdot AM = a^2,$$

then the property of the curve is that

$$PN = QM$$

but $PN = y = \psi x$, and $QM = \psi(AM) = \psi(a^2/x)$ consequently the equation from which ψ must be determined, is

$$\psi x = \frac{a^2}{x}.$$

Its solution is

$$y = \psi x = \bar{\varphi}\left\{x, \frac{a^2}{x^2}\right\}.$$

If we make $y = x + a^2/x$, it becomes $yx - x^2 = a^2$, an equation to the hyperbola.

Problem III

Required the general solution of

$$\psi x = Ax \cdot \psi\alpha x$$

having a particular solution, and also one of

$$\psi x = \psi\alpha^2 x.$$

Assume

$$\psi x = fx \cdot \bar{\varphi}\{f_1 x, f_2 x\}$$

making this substitution in the original equation, it becomes

$$fx \cdot \bar{\varphi}\{f_1 x, f_2 x\} = Ax \cdot f\alpha x \bar{\varphi}\{f_1 \alpha x, f_2 \alpha x\}$$

make $f_2 \alpha x = f_1 x$, and $f_1 \alpha x = f_2 x$, from this it results that $f_1 x = f_1 \alpha^2 x$ of which we possess a particular solution. Divide both sides by $\bar{\varphi}\{f_1 x, f_1 \alpha x\}$ then we have

$$fx = Ax \cdot f\alpha x$$

this is nothing more than the original equation of which, and of $f_1 x = f_1 \alpha^2 x$, we have by hypothesis particular solutions: / therefore its general solution is,

$$\psi x = fx \cdot \bar{\varphi}\{f_1 x, f_1 \alpha x\}.$$

The same equation may be solved differently, if we are acquainted with particular solutions of the equations

$$\psi x = \psi \alpha x \quad \text{and} \quad \psi x = Ax \cdot \psi \alpha x.$$

Assume $\psi x = fx \cdot \varphi x$, it becomes

$$fx \cdot \varphi x = Ax \cdot f\alpha x \cdot \varphi \alpha x$$

let fx be the particular solution of the original equation, then

$$fx = Ax \cdot f\alpha x$$

and by division

$$\varphi x = \varphi \alpha x.$$

But of this also we have given a particular solution; call it $f_1 x$. Therefore the general expression of ψx is

$$\psi x = fx \cdot \varphi f_1 x.$$

Let us take for example the equation

$$\psi x = \frac{1 - x^2}{1 - 2x} \psi \frac{x}{x - 1}.$$

The particular value of ψ is

$$fx = 1 + x$$

and the particular case of

$$\psi x = \psi \frac{x}{2x - 1} \quad \text{is} \quad f_1 x = \frac{x^2}{x - 1}.$$

From these considerations the general solution is

$$\psi x = (1 + x)\varphi \frac{x^2}{x - 1}$$

which will on trial be found to satisfy the equation.* /

* Another particular solution of

$$\psi x = \psi \frac{x}{x - 1} \qquad \text{is} \qquad \psi x = \cos \log (x - 1)$$

this gives for the general solution of

$$\psi x = \frac{1 - x^2}{1 - 2x} \psi \frac{x}{x - 1}$$

$$\psi x = (1 + x)\varphi(\cos \log (x - 1)).$$

As another example, take

$$\psi x = -\frac{1}{x}\psi\frac{1}{x}.$$

The particular solutions are, that of the equation itself

$$fx = \frac{x-1}{x}$$

and that of $\psi x = \psi(1/x)$. It is known to be $\bar{\varphi}(x, 1/x)$. Hence the general solution is

$$\psi x = \frac{x-1}{x}\bar{\varphi}\left(x, \frac{1}{x}\right)$$

which, as is readily seen, fulfils the condition.

Problem IV

Required the general solution of

$$\psi x = Ax \cdot \psi \alpha x + Bx$$

having given a particular solution fx.

Assume $\psi x = fx + \varphi x$, then $\psi \alpha x$ becomes $f\alpha x + \varphi \alpha x$, and the equation is

$$fx + \varphi x = Ax \cdot f\alpha x + Ax \cdot \varphi \alpha x + Bx$$

subtracting from this the particular solution $fx = Ax \cdot f\alpha x + Bx$, which is given by hypothesis, there remains

$$\varphi x = Ax \cdot \varphi \alpha x$$

this equation may be solved by Problem III, and we thence obtain the general solution of the given equation.

From this may readily be deduced the solution of the following equations:

$$\psi(\varepsilon^{y}) \pm (\varepsilon^{-y}) = Ay.$$

Suppose f is a particular solution, take $\psi y = fy + \chi y$, the equation becomes

$$f(\varepsilon^{y}) \pm f(\varepsilon^{-y}) + \chi(\varepsilon^{y}) \pm \chi(\varepsilon^{-y}) = Ay$$

but we have from the particular solution

$$f(\varepsilon^{y}) \pm f(\varepsilon^{-y}) = Ay$$

this subtracted from the former, leaves

$$\chi(\varepsilon^{y}) \pm \chi(\varepsilon^{-y}) = 0.$$

Let us first consider the upper sign, then a particular solution of

$$\chi(\varepsilon^{y}) = -\chi(\varepsilon^{-y})$$

is

$$\chi(y) = (\log y)^{2n+1}.$$

If we take the lower sign, then a solution of

$$\chi(\varepsilon^{y}) = \chi(\varepsilon^{-y})$$

is

$$\chi(y) = (\log y)^{2n}.$$

From these considerations it appears, that the general solutions of the given equations are

$$\psi y = fy + \varphi\{(\log y)^{2n+1}\}$$

and

$$\psi y = fy + \varphi\{(\log y)^{2n}\}$$

according as the upper or under sign is used.

The equation just solved was not constructed as an example to this particular rule, but is selected because it has actually occurred. It is used by Mr Herschel in the *Philosophical Transactions*, 1814, for the purpose of assigning the sums of several very curious series. He there observes, that when the upper sign is used ${}^{2n}L(1+y)$, and when the lower takes place ${}^{2n+3}L(1+y)$, are particular solutions, these may therefore be generalized by the introduction of an arbitrary function.

If

$$\psi\varepsilon^{y} + \psi\varepsilon^{-y} = \text{a rational function of } y \text{ containing only even powers,}$$

or if

$$\psi\varepsilon^{y} + \psi\varepsilon^{-y} = \text{a rational function of } y \text{ containing only odd powers,}$$

they admit of the following solutions: in the first case, let

$$\psi\varepsilon^{y} + \psi\varepsilon^{-y} = a_0 + a_1y^2 + a_2y^4 + \dots + a_ny^{2n}$$

/

assume

$$\psi y = A_0 + A_1(\log y)^2 + \dots + A_n(\log y)^{2n}$$

$\psi\varepsilon^y$ becomes

$$A_0 + A_1y^2 + A_2y^4 + \ldots + A_ny^{2n}$$

$\psi\varepsilon^{-y}$ becomes

$$A_0 + A_1y^2 + A_2y^4 + \ldots + A_ny^{2n}.$$

By comparing the co-efficients,

$$A_0 = \tfrac{1}{2}a_0 \quad A_1 = \tfrac{1}{2}a_1, \quad \ldots \quad A_n = \tfrac{1}{2}a_n$$

and calling the right side of the equation $F(y^2)$, if

$$\psi\varepsilon^y + \psi\varepsilon^{-y} = Fy^2$$

a particular solution is

$$\psi y = \tfrac{1}{2}F\{(\log y)^2\}$$

and similarly if

$$\psi\varepsilon^y - \psi\varepsilon^{-y} = F(y) = \text{a function containing only odd powers of } y,$$

one particular solution is

$$\psi y = \tfrac{1}{2}F\{\log y\}$$

and the general solutions may be readily deduced as above.

Problem V

To reduce the equation

$$\psi x + Ax\cdot\psi\alpha x + \ldots + Nx\cdot\psi\nu x + X = 0$$

to one in which the last term is wanting, by means of a particular solution.

Let fx be the given solution, make $\psi x = fx + \varphi x$, and substituting this value, the equation becomes

$$\left.\begin{aligned} &fx + Ax\cdot f\alpha x + Bx\cdot f\beta x + \ldots + Nx\cdot f\nu x + X \\ &+\varphi x + Ax\cdot\varphi\alpha x + Bx\cdot\varphi\beta x + \ldots + Nx\cdot\varphi\nu x \end{aligned}\right\} = 0.$$

The upper line is by hypothesis equal to nothing. Therefore the equation is reduced to this,

$$\varphi x + Ax\cdot\varphi\alpha x + \ldots + Nx\cdot\varphi\nu x = 0 \qquad (1)$$

and if we can discover the general solution of this latter equation, that of the former may be readily found. /

Supposing $\alpha, \beta, \gamma, \ldots \nu$, were to become $\alpha, \alpha^2, \alpha^3, \ldots \alpha^n$, the equation (1) would be changed into

$$\varphi x + Ax \cdot \varphi\alpha x + Bx \cdot \varphi\alpha^2 x + \ldots + Nx \cdot \varphi\alpha^{n-1} x = 0. \qquad (2)$$

After further if αx is such a function of x, that $\alpha^n x = x$, if we are acquainted with one particular solution of (2), we may easily determine the general one thus:

Assume

$$\varphi x = fx \cdot \bar{\chi}\{x, \alpha x, \alpha^2 x, \ldots \alpha^{n-1} x\}$$

here we must observe, that since χ is symmetrical relative to all the quantities contained within the brackets, it is immaterial in what order they are placed, and from the condition that $\alpha^n x = x$, it follows that if we substitute $\alpha^k x$ for x (k being successively equal to 1, 2, 3, 4, and $n - 1$), we shall always have these values $x, \alpha x, \alpha^2 x \ldots \alpha^{n-1} x$ only differently arranged, from these considerations the equation (2) will become

$$0 = (fx + Ax \cdot f\alpha x + Bx \cdot f\alpha^2 x + \ldots + Nx \cdot f\alpha^{n-1} x)\, \bar{\chi}\{x, \alpha x, \alpha^2 x, \ldots \alpha^{n-1} x\}$$

this equation may be satisfied by making the factor which multiplies $\bar{\chi}$ equal to nothing, and this is always the case when f is a particular solution, hence

$$\varphi x = fx \cdot \bar{\chi}\{x, \alpha x, \alpha^2 x, \ldots \alpha^{n-1} x\}.$$

Problem VI

To find a function of x, such that if instead of x we successively substitute $\alpha x, \beta x, \gamma x, \ldots \nu x$, the results shall all be equal to the original function; or in other words, to determine φx from the equations

$$\varphi x = \varphi\alpha x = \varphi\beta x = \ldots = \varphi\nu x.$$

/

Find f, so as to satisfy the equation

$$fx = f\alpha x. \qquad (a)$$

Take any particular value.

Find f_1, so as to satisfy the equation

$$f_1 fx = f_1 f\alpha\beta x = f_1 f\beta x \qquad (b)$$

for it is known from (a) that

$$f = f\alpha$$

Take some particular value, and determine f_2 from the equation

$$f_2 f_1 fx = f_2 f_1 f\beta\gamma x = f_2 f_1 f\gamma x$$

continue this as far as

$$f_n f_{n-1} \ldots fx = f_n f_{n-1} \ldots f\nu x \qquad (n)$$

then φ being any arbitrary function

$$\varphi\{f_n f_{n-1} \ldots f_1 fx\}$$

will satisfy the conditions of the problem.

As an example, let it be required to find a function which shall not change, when for x we substitute x, $-x$, or $x/\sqrt{x^2 - 1}$.

Here $\alpha x = -x$, and $\beta x = x/\sqrt{x^2 - 1}$.

(a) becomes

$$f(x) = f(-x)$$

hence

$$fx = \bar{\varphi}\{x, -x\}.$$

As a particular case, take

$$fx = x^2$$

then (b) becomes

$$f_1 x^2 = f_1(\beta x)^2 = f_1 \frac{x^2}{x^2 - 1}$$

whose solution is

$$f_1 x = \bar{\varphi}\left\{x, \frac{x}{x - 1}\right\}.$$

Take the case of

$$f_1 x = \frac{x^2}{x - 1}$$

then we find

$$\varphi f_1 fx = \varphi\left\{\frac{x^4}{x^2 - 1}\right\}$$

which fulfils the given conditions.

In the same manner it may be found, that the function

$$\varphi\left\{\frac{1 - ax^2 + x^4}{1 + ax^2 + x^4}\right\}$$

will remain the same, whether the variable is x, $-x$, or $1/x$.

Problem VII

Given any two series of functions $\alpha x, \beta x, \ldots \nu x$ and $\alpha_1 x, \beta_1 x, \ldots \nu_1 x$, required the form of ψ, so that the following equations may be fulfilled,

$$\begin{aligned} \psi\alpha x &= \psi\alpha_1 x \\ \psi\beta x &= \psi\beta_1 x \\ \vdots \quad & \quad \vdots \\ \psi\nu x &= \psi\nu_1 x. \end{aligned}$$

Determine f from the equation

$$f\alpha x = f_1 \alpha x \qquad (a)$$

take some particular case, and determine f_1 from the equation

$$f_1 f\alpha\beta x = f_1 f\alpha_1\beta_1 x = f_1 f\alpha\beta_1 x. \qquad (b)$$

Take a particular case and find f_2 from the equation

$$f_2 f_1 f\alpha\beta\gamma x = f_2 f_1 f\alpha\beta_1\gamma_1 x = f_2 f_1 f\alpha\beta\gamma_1 x \qquad (c)$$

and continue this to

$$f_n f_{n-1} \ldots f\alpha\beta\gamma \ldots \nu x = f_n f_{n-1} \ldots f\alpha\beta\gamma \ldots \nu_1 x. \qquad (n)$$

Then will

$$\psi x = \varphi f_n f_{n-1} \ldots f\alpha\beta\gamma \ldots \nu x$$

satisfy the conditions.

We now possess the means of solving a much more general problem than we have yet attempted; it is the general solution of any functional equation of the first order.

Problem VIII

Required the general solution of the equation[a]

$$F\{x, \psi x, \psi\alpha x, \ldots \psi\nu x\} = 0$$

having given a particular one, containing one or more arbitrary constants.

[a] Equation corrected.

Find from Problem VI an arbitrary function, such that it shall remain the same, when αx, βx, ... νx, are substituted for x, call this φx, and let the particular solution be

$$\psi x = f\{x, a, b, c, \ldots\}.$$

As these arbitrary constants are supposed not to be contained in the given equation $F\{x, \psi x, \psi\alpha x, \ldots \psi\nu x\} = 0$, when the particular solution is substituted, they must destroy each other, whatever be their value.

If therefore, instead of $a, b, c, \ldots$ we put $\varphi x, \varphi_1 x, \varphi_2 x, \ldots$ (which are arbitrary functions fulfilling this condition, viz. that they remain identical when $\alpha x, \beta x, \ldots$ are put for x), it is evident that these arbitrary functions must also destroy each other: therefore

$$\psi x = f\{x, \varphi x, \varphi_1 x, \varphi_2 x, \ldots\}$$

is a general solution containing as many arbitrary functions as the particular solution did constant quantities. /

Examples, let

$$x\psi x = \psi\frac{1}{x}$$

a particular solution is

$$\psi x = a\frac{x+1}{x}$$

and $\bar{\varphi}\{x, 1/x\}$ does not change when x becomes $1/x$, therefore the general solution is

$$\psi x = \frac{x+1}{x}\,\bar{\varphi}\left(x, \frac{1}{x}\right).$$

Take

$$\psi x - \psi\frac{1}{x} = a(1 - x^2)$$

a particular case of this equation is

$$\psi x = a + bx$$

therefore, the general solution is

$$\psi x = a + \bar{\varphi}\left\{x, \frac{1}{x}\right\}.$$

Let

$$\psi(x)\cdot\psi(-x) = \frac{1+x}{1-x}\left(\frac{1}{x}\psi\frac{1}{x}\right)^2$$

as a particular solution take

$$\psi x = \frac{1-x}{ax}$$

and ⟨as⟩

$$\varphi\left\{\frac{1-ax^2+x^4}{1+ax^2+x^4}\right\}$$

is a functional equation[a] which does not change when x becomes $-x$ or $1/x$, therefore the general solution is

$$\psi x = \frac{1-x}{x\varphi\left\{\dfrac{1-ax^2+x^4}{1+ax^2+x^4}\right\}}$$

φ being arbitrary.

Given the equation

$$2\psi\left(\frac{x}{\sqrt{x^2-1}}\right) = \psi x + \psi(-x) + \frac{\psi x - \psi(-x)}{\sqrt{x^2-1}}$$

a particular solution is

$$\psi x = a + bx$$

hence the general solution will be found to be

$$\psi x = \varphi\left(\frac{x^4}{x^2-1}\right) + x\varphi_1\left(\frac{x^4}{x^2-1}\right).$$

/

On the number of arbitrary functions introduced into the complete solution of a functional equation

When from a functional equation of the first order, we determine the form of the unknown function, one or more constant quantities are generally introduced; these as I have shown in a preceding Problem, may be changed into arbitrary functions of the unknown quantity which fulfil certain prescribed conditions.

---

[a] *Sic.*

A question naturally arises as to the number of these arbitrary functions, and how many any given equation admits of in its most general solution.

The train of reasoning usually made use of to prove, that a differential equation of the nth order, requires in its complete integral n arbitrary constants may be pursued on the present occasion, though from several reasons, it would perhaps be desirable to have a proof resting on a different principle; as I have not been successful in discovering any other, I shall give the only one I am at present possessed of.

Let

$$\psi x = F\{x, a_1, a_2, \ldots a_n\}$$

for x, put any number of known functions, as αx, $\beta x, \ldots \nu x$, the results will be

$$\psi x = F\{x, a_1, a_2, \ldots a_n\} \qquad (0)$$

$$\psi \alpha x = F\{\alpha x, a_1, a_2, \ldots a_n\} \qquad (1)$$

$$\psi \beta x = F\{\beta x, a_1, a_2, \ldots a_n\} \qquad (2)$$

$$\vdots \qquad \vdots$$

$$\psi \nu x = F\{\nu x, a_1, a_2, \ldots a_n\} \qquad (n)$$

From these $n + 1$ equations we may eliminate the n arbitrary constants, and the resulting equation will be of the form

$$0 = F_1\{x, \psi x, \psi \alpha x, \ldots \psi \nu x\}. \qquad (A)$$

In arriving at this equation, we have eliminated n arbitrary constants, and therefore it might possibly be inferred that the general solution of (A) is

$$\psi x = F\{x, a_1, a_2, \ldots a_n\}.$$

But this is too hasty a conclusion, for it is evident, that we should equally have arrived at equation (A), if each of the constant quantities in (0) had been changed into a function of x so constituted that it should not alter by the substitution of αx, $\beta x, \ldots \nu x$.

It would now appear, that putting such values for the constant quantities, the result would be the general solution of (A).

This reasoning is certainly plausible, and such a solution is undoubtedly a very general one; still, however, there are reasons which incline me to believe, that other solutions exist of a yet more general nature.

On functional equations of the second and higher orders

When we consider functional equations of an order superior to the first, new difficulties present themselves; the artifices which were used with success in the preceding part of this paper, are no longer of any avail.

Those which we have now to consider seem to possess an entirely distinct character. /

Problem IX

Required the solution of the equation

$$\psi^2 x = x. \qquad (a)$$

Subtract ψx from both sides, then we have

$$\psi^2 x - \psi x = -\psi x + x = -(\psi x - x)$$

consequently,

$$\Delta\psi x = -\Delta x \qquad (b)$$

again multiply (a) by ψx, then we have

$$\psi^2 x \cdot \psi x = \psi x \cdot x.$$

From this we learn, that (b) must be integrated on the hypothesis of $x\psi x$ being constant, hence

$$\psi x = -x + c = -x + f(x\psi x)$$

f being an arbitrary function. ψ is determined from the equation

$$\psi x + x - f(x\psi x) = 0.$$

For f we may put $f^{-1}f_1$, and it becomes

$$f(x + \psi x) - f_1(x\psi x) = 0. \qquad (c)$$

A friend to whose valuable remarks on this subject, I am much indebted, has communicated to me the following method of obtaining the same solution. Assume any symmetrical function of x and v

$$\bar{\varphi}(x, v) = 0$$

then from the nature of the equation we have the two following equations

$$x = \psi v \quad \text{and} \quad v = \psi x.$$

Consequently

$$x = \psi v = \psi\psi x = \psi x.$$

Hence ψx may be found from the equation

$$\bar{\varphi}\{x, \psi x\} = 0.$$

This at first sight appears different from (c), but it is not so / in reality, for it may easily be shown, that from the sum and product of two quantities any symmetrical function may be composed.

As particular cases of the equation

$$\psi^2 x = x$$

we may notice

$$\psi x = a - x, \quad \frac{x}{ax-1}, \quad \frac{b-x}{1-ax} \quad \text{and} \quad \frac{ax + \sqrt{c + (a^2-4)x^2}}{2}.$$

Problem X

Required the solution of

$$\psi^2 x = x.$$

Another solution may be found from the following principle.

Assume

$$\psi x = \varphi^{-1} f \varphi x$$

hence

$$\psi^2 x = \varphi^{-1} f \varphi \varphi^{-1} f \varphi x = \varphi^{-1} f^2 \varphi x$$

or

$$\varphi^{-1} f^2 \varphi x = x.$$

Now this equation can be satisfied if we are acquainted with a particular solution of $f^2 x = x$. This may be found from the last problem, and since $f^2 x = x$, the equation becomes

$$\varphi^{-1} \varphi x = x$$

which is identical; consequently

$$\psi x = \varphi^{-1} f \varphi x.$$

Some particular cases are

$$\psi x = \varphi^{-1}(a - \varphi x), \quad \varphi^{-1}\left(\frac{\varphi x}{a\varphi x - 1}\right), \quad \varphi^{-1}\left(\frac{a}{\varphi x}\right) \ldots$$

From each of these by assigning particular values to φ, new values of f may be determined, and these in their turn will furnish new forms of the function φx.

Some time ago I received from the gentleman already alluded to, the following solution of the equation

$$\psi^2 x = x$$
$$\psi x = \varphi(-\varphi x)$$

and likewise the solution $\varphi^{-1}((-1)^{1/n}\,\varphi x)$ of the equation $\psi^n x = x$; this first led me to the substitution of $\varphi^{-1}f\varphi x$, which is of such essential use in these enquiries.

Problem XI

Given the equation

$$\psi^n x = x.$$

Assume as before

$$\psi x = \varphi^{-1}f\varphi x$$

then

$$\psi^2 x = \varphi^{-1}f\varphi\varphi^{-1}f\varphi x = \varphi^{-1}f^2\varphi x$$
$$\psi^3 x = \varphi^{-1}f^2\varphi\varphi^{-1}f\varphi x = \varphi^{-1}f^3\varphi x$$

and generally

$$\psi^n x = \varphi^{-1}f^n\varphi x.$$

Hence our equation becomes

$$\varphi^{-1}f^n\varphi x = x. \qquad (a)$$

Suppose we have one particular solution of the equation. Substitute this instead of f, and the equation (a) becomes identical, whence

$$\psi x = \varphi^{-1}f\varphi x$$

and from this other values of f may be determined, and so on ad infinitum.

The equation we have just considered, affords a ready solution of the following Problem.

Required the nature of a curve, such that taking any point B in the abscissa, and drawing the ordinate BP

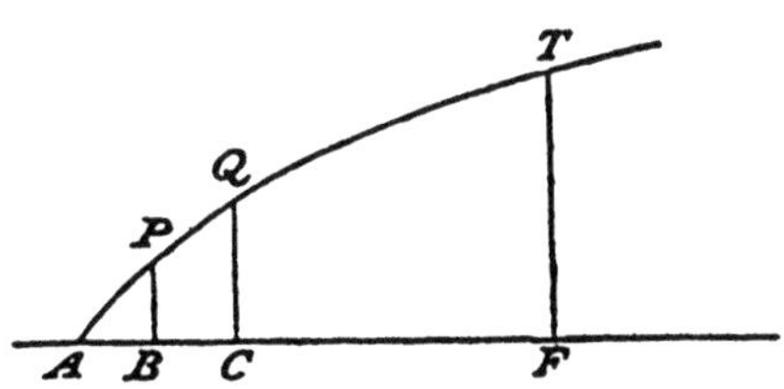

If we make AC another abscissa equal to BP the preceding ordinate, and if we continue this n times, then the nth ordinate may be equal to the first abscissa. /

If $AB = x$ and the equation of the curve be $y = \psi x$. $PB = y = \psi x$ and $AC = PB = \psi x$, and $QC = \psi\psi x = \psi^2 x$, and generally the nth ordinate TF is equal to $\psi^n x$. Hence

$$\psi^n x = x$$

which is the equation whose solution has been just found.

Problem XII

Given the equation

$$\psi^2 x = \alpha x$$

required the form of ψ.

Assume

$$\psi x = \varphi^{-1} f \varphi x$$

then

$$\psi^2 x = \varphi^{-1} f^2 \varphi x$$

and

$$\varphi^{-1} f^2 \varphi x = \alpha x$$

take the function φ on both sides, then this becomes

$$f^2 \varphi x = \varphi \alpha x.$$

This is a functional equation of the first order relative to φ, and may be solved either by the methods exhibited in the first part of this paper, or by the very elegant one of Laplace: f is a perfectly arbitrary function, except that neither fx nor $f^2 x$ must be equal to x: from not attending to this circumstance, I was at first led into several errors; the reason of these two restrictions is, that in the first case we at once determine ψx to be equal to x, and in the second, we in fact make $\alpha x = x$, neither of which are necessarily true.

Problem XIII

Given the equation

$$\psi^n x = \alpha x.$$

This admits of a solution similar to the last, by assuming ψx / equal to $\varphi^{-1} f \varphi x$.

We find

$$\varphi^{-1}f^n\varphi x = \alpha x$$

and by taking φ on both sides, it becomes

$$f^n\varphi x = \varphi\alpha x.$$

Give f some particular value and determine φ as in the last problem.

Problem XIV

Given the equation

$$\psi\alpha\psi\beta x = \gamma x.$$

Assume

$$\psi x = \varphi^{-1}f\varphi f_1 x.$$

Hence

$$\psi\alpha\psi\beta x = \varphi^{-1}f\varphi f_1\alpha\varphi^{-1}f\varphi f_1\beta x.$$

Make $f_1 = \alpha^{-1}$. Then the equation becomes

$$\varphi^{-1}f\varphi\varphi^{-1}f\varphi\alpha^{-1}\beta x = \varphi^{-1}f^2\varphi\alpha^{-1}\beta x = \gamma x$$

and by taking φ on both sides

$$f^2\varphi\alpha^{-1}\beta x = \varphi\gamma x.$$

This is an equation of the first order relative to φ.

The equation $\psi\alpha\psi\alpha\ldots(n)\psi\beta y = \gamma y$ might be solved in the same manner, but they are both reducible by a simple transformation to the form

$$\psi^n y = Fy.$$

Problem XV

Another method of solving the equation

$$\psi^2 x = x.$$

Assume $\psi x = \varphi^{-1}f\varphi x$, then $\psi^2 x = \varphi^{-1}f^2\varphi x$, and we have

$$\varphi^{-1}f^2\varphi x = x.$$

Let $y = \varphi x$, then $x = \varphi^{-1}y$ and the equation becomes

$$\varphi^{-1}f^2 y = \varphi^{-1}y$$

an equation of the first order from which y may be found. Or thus assume

$$\varphi^{-1}y = \bar{\chi}\{y, f^2 y\}$$

and make

$$f^2 y = y.$$

This method is much more extensive in its application than any of those before it.

Problem XVI

Required the solution of

$$\psi^n x = x.$$

The same method applies equally in this case.

Assume

$$\psi x = \varphi^{-1} f \varphi x$$

then

$$\varphi^{-1} f^n \varphi x = x.$$

For x put $\varphi^{-1}x$. It becomes

$$\varphi^{-1} f^n x = \varphi^{-1} x.$$

Take φ^{-1} an arbitrary symmetrical function of

$$x, f^n x, f^{2n} x \quad \text{and} \quad f^{kn} x$$

then

$$\varphi^{-1} x = \chi\{x, f^n x, f^{2n} x, \ldots f^{kn} x\}$$

and determine f such that

$$f^{(k+1)n} x = x.$$

A particular solution is sufficient, and it is evident, this value of φ^{-1} will satisfy the equation.

Problem XVII

Given the equation

$$\psi\alpha(y, \psi y) = y$$

required the form of y.

Assume

$$\psi y = \varphi^{-1} f \varphi y$$

then it becomes

$$\varphi^{-1} f \varphi \alpha(y, \varphi^{-1} f \varphi y) = y.$$

Take successively on each side the functions φ, f^{-1}, and φ^{-1}, the equation becomes

$$\alpha(y, \varphi^{-1} f \varphi y) = \varphi^{-1} f^{-1} \varphi y.$$

For y put $\varphi^{-1} y$; then

$$\alpha(\varphi^{-1} y, \varphi^{-1} f y) = \varphi^{-1} f^{-1} y$$

this is a functional equation of the first order relative to φ^{-1}; give f any determinate value and solve the equation.

From hence we may deduce the solution of the following question.

Required the nature of a curve

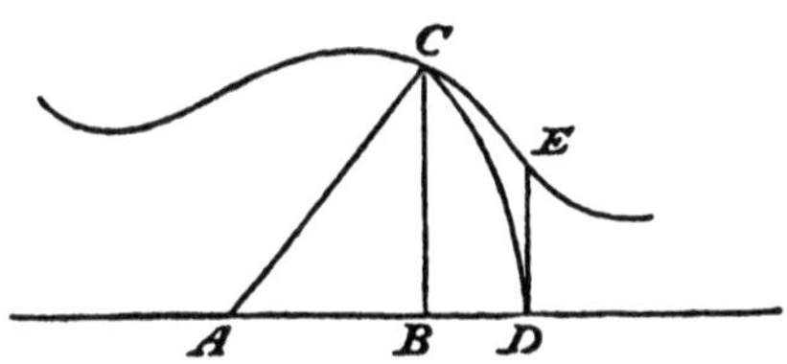

such that taking any abscissa AB and drawing the ordinate CB, if with centre A and radius AC, we describe a circle cutting the abscissa in D, the ordinate ED may be equal to the first abscissa AB.

Let

$$y = \psi x = CB$$

$$AD = CA = \sqrt{x^2 + (\psi x)^2}$$

and

$$ED = \psi(AD)$$

hence

$$x = \psi(\sqrt{x^2 + (\psi x)^2})$$

which is a particular case of the preceding problem.

Problem XVIII

To reduce the equation

$$F(x, \psi x, \psi^2 \alpha x, \ldots \psi^n \nu x) = 0$$

/

to one of the form of

$$F(x, \psi x, \psi^2 x, \ldots \psi^n x) = 0.$$

Assume $\psi x = A^{-1}\varphi A x$, the equation by this substitution becomes

$$F\{x, A^{-1}\varphi A x, A^{-1}\varphi^2 A\alpha x, \ldots A^{-1}\varphi^n A\nu x\} = 0$$

find by Problem VI such a value Ax that it shall not change by the substitution of $\alpha x, \beta x, \gamma x, \ldots \nu x$. Put for Ax the quantity y, and the equation becomes

$$F\{A^{-1}y, A^{-1}\varphi y, A^{-1}\varphi^2 y, \ldots A^{-1}\varphi^n y\} = 0$$

which is an equation of the required form.

Problem XIX

Required the solution of the equation

$$F\{x, \psi x, \psi^2 x, \ldots \psi^n x\} = 0.$$

Assume $\psi x = \varphi^{-1} f \varphi x$, then $\psi^n x = \varphi^{-1} f^n \varphi x$, and the equation becomes

$$F\{x, \varphi^{-1} f \varphi x, \varphi^{-1} f^2 \varphi x, \ldots \varphi^{-1} f^n \varphi x\} = 0.$$

For x substitute $\varphi^{-1}x$, then

$$F\{\varphi^{-1}x, \varphi^{-1}fx, \varphi^{-1}f^2 x, \ldots \varphi^{-1}f^n x\} = 0 \qquad (a)$$

which is an equation of the first order relative to φ^{-1} and may be solved by the methods in the beginning of this paper, or, by means of the method given by Mr Herschel, to which we have already alluded.

With respect to the function f it is arbitrary. There are however, some observations respecting it, which require notice; as without an attention to them we might fall into error. In the first place, it is evident, that we must not / make fx equal to x, for in this case we at once determine ψx to be equal to x, which is not always true.

The same observation may be made with respect to making $f^2 x = x$, for in this case $\psi^2 x = x$, $\psi^3 x = \psi x$, $\psi^4 x = x$, $\psi^5 x = \psi x$, and we in fact by assuming this value for f determine ψ from the equation

$$F\{x, \psi x, x, \psi x, x, \ldots\} = 0.$$

The same objection does not hold when we make $f^3x = x$, though this considerably limits the generality of the solution; apparently the most eligible mode of determining f is from the equation $f^{n+1}x = x$, for in this case supposing we are acquainted with a particular solution of (a) containing any number of arbitrary constants, such as

$$\varphi^{-1}x = A\{x, a, b, c, \ldots\}.$$

We may substitute for each of these constants an arbitrary function, such as

$$\bar{\chi}\{x, fx, f^n x, \ldots f^n x\}$$

for it is evident this function does not change when we substitute $f^m x$ for x.

But this form of f is not always correct. It may be inconsistent with the original equation. In fact, the only value we can assign to f which shall not in some measure limit the generality of the Problem, is to suppose a particular solution of the given equation.

As an example of this method take the following problem.

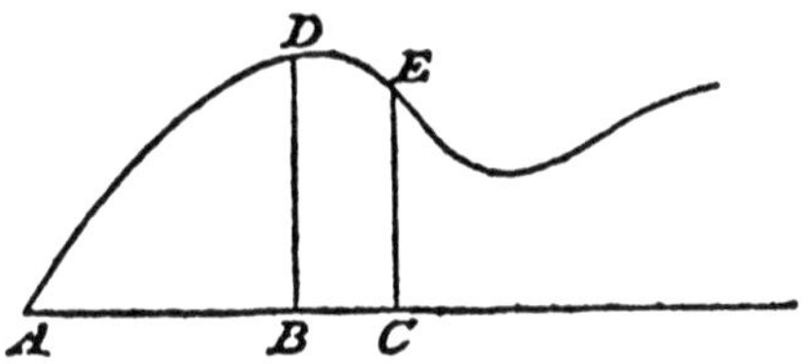

Required the nature of the curve ADE such that taking any abscissa AB and corresponding ordinate DB: if the abscissa AC be taken, equal to DB and the ordinate EC be drawn, then the rectangle under the two ordinates shall be equal to the square of the first abscissa.

Let the equation of the curve be $y = \psi x$

then

$$AB = x,\ DB = y = \psi x,\ AC = DB$$

and

$$EC = \psi(AC) = \psi(DB) = \psi y = \psi\psi x.$$

The given condition is therefore

$$\psi^2 x \cdot \psi x = x^2.$$

Making the usual substitution of $\psi x = \varphi^{-1} f \varphi x$ it becomes

$$\varphi^{-1} f^2 \varphi x \cdot \varphi^{-1} f \varphi x = x^2.$$

Putting $\varphi^{-1}x$ for x we have

$$\varphi^{-1}f^2x \cdot \varphi^{-1}fx = (\varphi^{-1}x)^2.$$

Assume

$$\varphi^{-1}x = \bar{\chi}(x, fx, f^2x)$$

the equation then becomes

$$\bar{\chi}\{f^2x, f^3x, f^4x\} \cdot \bar{\chi}\{fx, fx^2, fx^3\} = [\bar{\chi}\{x, fx, f^2x\}]^2.$$

If now f be determined from the equation

$$f^3x = x$$

we have $f^4x = fx$, and the equation becomes identical. Hence φ^{-1} and f being found, we may determine ψ.

One of the simplest curves which satisfies the equation is

$$yx^2 = a^3.$$

A line of the third order, the 69th in Newton's arrangement, its form is /

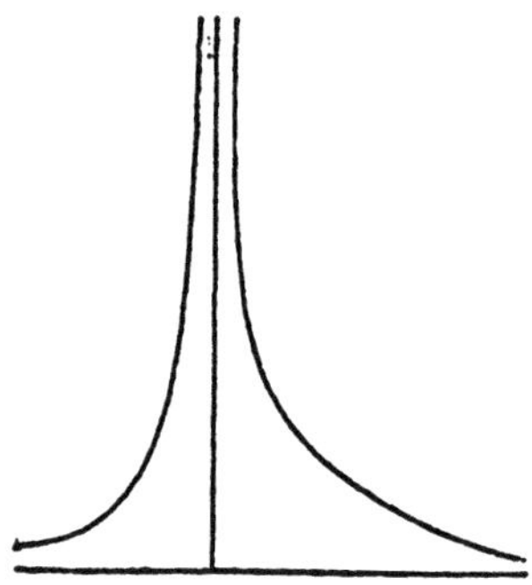

Problem XX

Required the general solution of the equation

$$F\{x, \psi x, \psi^2\alpha x, \psi^3\beta x, \ldots \psi^n\nu x\} = 0.$$

Reduce this by Problem XVIII to the form of

$$F\{x, \psi x, \psi^2x, \ldots \psi^n x\} = 0$$

in which shape it may be solved by the preceding problem.

Problem XXI

Given the equation

$$F\{x, \psi x, \psi\alpha(x, \psi\beta x)\} = 0$$

required its solution.

Assume

$$\alpha(x, \psi x) = \varphi^{-1}f\varphi x \qquad (a)$$

hence

$$\psi x = \alpha^{1,-1}(x, \varphi^{-1}f\varphi x). \qquad (b)$$

In the left side of this equation, put for x the left side of (a) and in the right side, put instead of x the right side of (a) the result is

$$\psi\alpha(x, \psi x) = \alpha^{1,-1}(\varphi^{-1}f\varphi x, \varphi^{-1}f^2\varphi x)$$

substituting this in the original equation, the result is

$$F\{x, \alpha^{1,-1}(x, \varphi^{-1}f\varphi x), \alpha^{1,-1}(\varphi^{-1}f\varphi x, \varphi^{-1}f^2\varphi x)\} = 0$$

which by putting $\varphi^{-1}x$ for x becomes

$$F\{\varphi^{-1}x, \alpha^{1,-1}(\varphi^{-1}x, \varphi^{-1}fx), \alpha^{1,-1}(\varphi^{-1}fx, \varphi^{-1}f^2x)\} = 0$$

this being an equation of the first order relative to φ^{-1} may be solved as above.

With respect to the number of arbitrary functions which enter into the complete solution of functional equations of higher orders than the first, I have little at present to offer; the difficulty of the subject, and the wide extent of the enquiries / to which it would lead, induce me to postpone it until I have more time for the consideration. The following remarks may suffice for the present to point out some of its difficulties and the mode of enquiry.

If

$$\psi x = f_1\{x, a, b, \ldots\} \qquad (1)$$

$$\left.\begin{aligned}\psi^2 x &= f_2\{x, a, b, \ldots\\ \psi^3 x &= f_3\{x, a, b, \ldots\end{aligned}\right\} \qquad \begin{aligned}(2)\\(3)\end{aligned}$$

$$\vdots \qquad \vdots$$

$$\psi^n x = f_n\{x, a, b, \ldots\} \qquad (n)$$

From this by eliminating $n - 1$ of the arbitrary constants $a, b, \ldots$ we have an equation of the form

$$F\{x, \psi x, \psi^2 x, \ldots \psi^n x\} = 0 \qquad (a)$$

and it might possibly be concluded that equation (1) containing $n - 1$ arbitrary constants is the general solution of this last equation: but this is by no means the case. In the first place between the two equations (1) and (2), more than one arbitrary constant may be eliminated. Thus let

$$\psi x = \frac{a - x}{1 - bx}$$

from which we find

$$\psi^2 x = x.$$

The two quantities a and b have been eliminated, and it is possible to select a value of ψx, between which and $\psi^2 x$ an infinite number of arbitrary constants could be eliminated.

But waiving this objection let us consider the case of (a) which is deduced from the elimination of $n - 1$ arbitrary functions.

We have seen in Problem VI that a function of the first / order may satisfy any number of conditions (which are not contradictory) simultaneously; and there appears no reason for denying this property to those of higher orders.

If now we consider the symmetrical function

$$\bar{\chi}\{x, \psi x, \psi^2 x, \ldots \psi^\nu x\}$$

and if $\psi^{\nu+1} x = x$. It is evident this function will not change by the substitution of

$$\psi x, \psi^2 x, \ldots \quad \text{or} \quad \psi^n x$$

and consequently that a different function similarly constituted may be substituted for each of the arbitrary quantities $a, b, c, \ldots$ in (1) which is the solution of the equation

$$F\{x, \psi x, \psi^2 x, \ldots \psi^n x\} = 0. \tag{a}$$

The number denoted by ν is arbitrary (it may, however, become determined from some particular circumstances of the equation (a).

Thus we have introduced into an equation of the nth order, an unlimited number of arbitrary functions, each of which contains the function whose determination was sought with all its different orders to an undefined extent.

If we take the particular case of

$$\psi x = a - x$$
$$\psi^2 x = x$$

ν must be unity, and a general solution is

$$\psi x = \bar{\chi}\{x, \psi x\} - x.$$

Taking another solution

$$\psi x = \frac{a - x}{1 - bx}, \quad \psi^2 x = x$$

and

$$\psi x = \frac{\bar{\chi}(x, \psi x) - x}{1 - x\bar{\chi}(x, \psi x)}.$$

If

$$\psi^n x = x \qquad \nu = n - 1$$

and supposing $\psi x = f(x, a, b, \ldots)$ any particular solution we have for the general one /

$$\psi x = f\{x, \bar{\chi}(x, \psi x, \ldots \psi^{n-1}x), \bar{\chi}_1(x, \psi x, \ldots \psi^{n-1}x) \ldots\}$$

from which equation of the $n - 1$th degree ψ must be found.

When we apply these considerations to functional equations of many variables, other and even greater difficulties present themselves; the first step in that direction must be an improvement in the notation.

Since the above was written, I have bestowed some attention on functional equations involving two or more variables, and I have met with considerable success: I am in possession of methods which give the general solution of equations of all orders, and even of those which contain symmetrical functions. I have also discovered a new and direct method of treating functional equations of the first order, and of any number of variables, and this new method I have applied to the solution of differential and even of partial differential functional equations. /

[179/80]

IV

AN ESSAY TOWARDS THE CALCULUS OF FUNCTIONS PART II

From *Philosophical Transactions of the Royal Society*, Vol. 106 (1816), pp. 179–256

Communicated by W. H. Wollaston, M.D. Sec. R.S.
Read 14 March, 1816

In a former paper which the Royal Society honoured with a place in the last volume of their Transactions,[a] I endeavoured to explain the nature of the calculus of functions, and I proposed means of solving a variety of functional equations containing only one variable quantity. My subsequent enquiries have produced several new methods of solving these, and much more complicated functional equations, and have convinced me of the importance of the calculus, particularly as an instrument of discovery in the more difficult branches of analysis; nor is it only in the recesses of this abstract science, that its advantages will be felt: it is peculiarly adapted to the discovery of those laws of action by which one particle of matter attracts or repels another of the same or of a different species; consequently, it may be applied to every branch of natural philosophy, where the object is to discover by calculation from the results of experiment, the laws which / regulate the action of the ultimate particles of bodies. To the accomplishment of these desirable purposes, it must be confessed that it is in its present state unequal; but should the labours of future enquirers give to it that perfection, which other methods of investigation have attained, it is not too much to hope, that its maturer age shall unveil the hidden laws which govern the phenomena of magnetic, electric, or even of chemical action.

When functional equations containing two or more variables occur, their solution presents still greater difficulties than those we have already considered; the new relations which arise, necessarily require a new notation to distinguish them. I shall endeavour, as far as I am able, to apply or extend that already in use; but, as it is almost impossible in the infancy of a calculus to foresee the

[a] 'An essay towards the calculus of functions ⟨part I⟩' (1815), immediately preceding in this volume.

extent to which it may be carried, or the new views which it may be necessary to take of it, the notation I have used should only be considered as of a temporary nature; it may be employed until some more convenient one be devised: perhaps, however, it might be more advantageous that it should not be altered until our acquaintance with this subject becomes more intimate, and until the infinitely varied and comprehensive relations displayed in the doctrine of functions, have been more minutely examined.

If ψ be the characteristic of any function of two quantities x and y, that function is thus denoted $\psi(x, y)$. Now, if instead of x in this quantity the original function be substituted, I shall call the result the second function relative to x, and I shall denote it thus

$$\psi^{2,1}(x, y) = \psi(\psi(x, y), y).$$

/

The first index 2 refers to x, and the second index 1 refers to y. Similarly if instead of y in the original function, the function itself had been substituted, the result would have been the second function relative to y; it would be thus denoted

$$\psi^{1,2}(x, y) = \psi(x, \psi(x, y)).$$

If there are more than two variables in the original function, they may be arranged in the order in which they are to be operated on, and the indices will denote the number of operations to be performed.

Thus $\psi^{2,3,1,4}(x, y, z, v)$ signifies that in the function $\psi(x, y, z, v)$ we must instead of x substitute the function itself, and in the result instead of y put the same function, this latter operation must be repeated, and finally, instead of v in the last result, put the original function; this last operation must again be repeated twice.

There are many cases which this notation does not comprehend. If, for example, in the function just proposed, we wished again to take the function relative to x or y, it would not be easy to express this. The method I propose is to have two ranks of indices, the lower one to distinguish the quantities operated on; the upper one to mark the number of operations performed. According to this method the example just chosen would be written thus:

$$\psi^{2,3,1,4}_{1,2,3,4}(x, y, z, v).$$

If only such functions as these occur, we encumber our symbol without any advantage; if, however, we now wish to perform any further operations, such, for instance, as to take the second function relative to z, and then the third relative to y, / we have a very convenient mode of doing it; these operations would be thus expressed;

$$\psi^{2,3,1,4,2,3}_{1,2,3,4,3,2}(x, y, z, v).$$

This notation may not appear sufficiently concise to those who do not consider the very complicated relation expressed by the above written symbol: it need, however, only be used in very few cases, and when the lower series of indices is omitted, it must always be understood, that the quantities themselves are arranged in the order in which they are to be operated on.

If in a function of two variables $\psi(x, y)$ we take the second function relative to x, and then the second function relative to y, we have

$$\psi^{\substack{2,2\\1,2}}(x, y) = \psi\{\psi(x, \psi(x, y)), \psi(x, y)\}.$$

If we take the second function first relative to y, and then the second relative to x we shall find

$$\psi^{\substack{2,2\\2,1}}(x, y) = \psi\{\psi(x, y), \psi(\psi(x, y), y)\}.$$

It appears from this, that the order in which these operations are performed is not immaterial, as the order in which we differentiate a function of two variables, is in the differential calculus.

The two expressions just given are the two second functions of $\psi(x, y)$, the first taken relative to x and y, and the second taken relative to y and x. But there may be another second function of $\psi(x, y)$, which will arise from substituting at the same time $\psi(x, y)$ for x, and $\psi(x, y)$ for y, it will be

$$\psi(\psi(x, y), \psi(x, y))$$

and may for the sake of distinction be called the second simultaneous function relative to x and y; it differs from the two preceding ones, and in order to denote it with brevity, I shall put a line over the two indices thus,

$$\psi^{\overline{2,2}}(x, y) = \psi(\psi(x, y), \psi(x, y)).$$

This method of distinguishing it is equally applicable when there are more variables.

There is only one other modification of the symbol denoting function to which I shall at present allude. Suppose (after any number of operations have been performed on a function of two variables for example) y becomes equal to x, and the result only is given: this will naturally be represented in a manner analogous to that in which Euler denoted the limits between which the integral of a quantity is to be taken.

Thus the equation $\psi^{\substack{2,2\\1,2}}(x, y) = fx \;\; [y = x]$ arises from the following question: What is the form of a function of x and y, such that taking the second function relative to x, and then the second relative to y, the result on making y equal to x shall be a given function of x?

It might be proposed, that after putting y equal to x, the whole should be considered merely as a function of x, and that its nth function should be taken on this hypothesis, and the result only should be given.

Such operations I would denote thus:

$$\psi^{\left.{2,2 \atop 1,2}\right\}^n}(x, y) = f(x)\,[y = x] \text{ or perhaps } \{\psi^{2,2}_{1,2}\}^n(x, y) = f(x)\,[y = x]$$

and in a similar manner all other relations of the same kind may be expressed. /

I shall give one example which will illustrate these various modifications of the original functions,

$$\psi^{\left.\left.{2,3,\overline{2,2} \atop 3,1,2,4}\right\}{3,2 \atop 3,1}\right\}^3}(x, y, z, v) = f(x) \qquad [y = \alpha x, \quad v = \beta z, \quad z = \gamma x].$$

This equation contains the analytical enunciation of the following Problem.

What must be the form of a function of four quantities $\psi(x, y, z, v)$ such that taking the second function relative to z, the third relative to x, and the second simultaneous one relative to y and v: if in the result αx be put for y and βz for v, and the whole be then considered as a function of x and z, and if on this hypothesis the third function be taken relative to z, and the second relative to x; and if γx be now put for z and the third function of the expression considered merely as a function of x be taken, then it is required that the final result shall be equal to fx a given function of x?

Symmetrical functions I shall denote as in my former paper, by putting a line over the quantities relative to which they are symmetrical, thus

$$\psi(\bar{x}, \bar{y}, \overset{\perp}{z}, \overset{\perp}{v})$$

is symmetrical relative to x and y in one sense, and relative to z and v in another.

Problem I

Required the solution of the functional equation

$$\psi(x, y) = \psi(ax, \beta y).$$

To avoid repetition, α, β, γ, ... unless otherwise mentioned, always express known functions, and φ, ψ, χ are unknown or arbitrary ones.

Put

$$\psi(x, y) = \varphi(fx, f_1 y)$$ /

then the given equation becomes

$$\varphi(fx, f_1 y) = \varphi(f\alpha x, f_1 \beta y).$$

Determine f and f_1 from the two equations

$$fx = f\alpha x \text{ and } f_1 y = f_1 \beta y$$

this may be effected by Problem I or II of my former paper, take any particular solution and φ may remain perfectly arbitrary; then the general solution of the problem is

$$\psi(x, y) = \varphi(fx, f_1 y).$$

Ex. 1. Given the equation

$$\psi(x, y) = \psi\left(-x, \frac{1}{y}\right)$$

here we have $f(x) = f(-x)$, and a particular solution is $fx = x^2$; also

$$f_1(y) = f_1\left(\frac{1}{y}\right)$$

and a particular case is

$$f_1(y) = \frac{y^2 + 1}{y}$$

hence the general solution of the equation is

$$\psi(x, y) = \varphi\left(x^2, \frac{y^2 + 1}{y}\right)$$

φ being perfectly arbitrary.

If we employ the general solutions of the equations $f(x) = f(-x)$ and $f_1(y) = f_1(1/y)$, we shall still only have one arbitrary function. In fact, the most general solution of the equation $\psi(x, y) = \psi(-x, 1/y)$ with which I am at present acquainted is

$$\psi(x, y) = \varphi\{-\overline{x}, \overline{x}, \overset{1}{\overline{y}}, \overset{1}{\overline{1/y}}\}$$

and this only involves one arbitrary function. /

Problem II

Given the same equation

$$\psi(x, y) = \psi(\alpha x, \beta y).$$

Suppose one particular solution of this equation is known, let it be $f(x, y)$, then take $\psi(x, y) = \varphi f(x, y)$, φ being perfectly arbitrary and the given equation becomes

$$\varphi f(x, y) = \varphi f(\alpha x, \beta y)$$

which is evidently satisfied since $f(x, y) = f(\alpha x, \beta y)$ by the hypothesis.

Ex. 1. Let

$$\psi(x, y) = \psi(x^n, y^{1/n})$$

one particular solution of this equation is

$$f(x, y) = x^{\log y}$$

hence the general solution is

$$\psi(x, y) = \varphi(x^{\log y}).$$

Ex. 2. Given the equation $\psi(x, y) = \psi(x^n, y^n)$ a particular case is

$$f(x, y) = \frac{\log x}{\log y}$$

hence the general solution is

$$\psi(x, y) = \varphi\left(\frac{\log x}{\log y}\right).$$

Ex. 3. Given the equation

$$\psi(x, y) = \psi(x^n, y^m).$$

In order to get a particular case let us put

$$f(x, y) = \log^2 x + a \log^2 y$$

by substituting this value we shall find that it is a particular solution of the equation, if

$$a = -\frac{\log n}{\log m}$$

hence the general solution of the equation is

$$\psi(x, y) = \varphi\left(\log^2 x - \frac{\log n}{\log m}\log^2 y\right) = \varphi\left(\log\frac{\log x}{(\log y)^{\log n/\log m}}\right)$$

or changing the value of φ it becomes

$$\psi(x, y) = \varphi\left(\frac{(\log x)^{\log m}}{(\log y)^{\log n}}\right).$$

If $n = m$ we have

$$\psi(x, y) = \varphi\left(\frac{\log x}{\log y}\right)$$

as in the last example, and if $m = 1/n$, we have the same solution as in the first.

In these equations the functions have contained the variables separated; but it may frequently happen, that they occur mixed as in the following Problems.

Problem III

Given the equation

$$\psi(x, y) = \psi(\alpha(x, y), \beta(x, y)).$$

Assume $\psi(x, y) = \varphi(f(x, y), f_1(x, y))$, and by making this substitution the equation becomes

$$\varphi\{f(x, y), f_1(x, y)\} = \varphi\{f(\alpha(x, y), \beta(x, y)), f_1(\alpha(x, y), \beta(x, y))\}.$$

In order to render this equation identical, I determine f and f_1 from the two following equations:

$$f(x, y) = f(\alpha(x, y), \beta(x, y))$$

and

$$f_1(x, y) = f_1(\alpha(x, y), \beta(x, y)).$$

From these it appears, that f_1 and f are merely two particular solutions of the original equation. If, therefore, we are acquainted with any, the general solution is

$$\psi(x, y) = \varphi(f(x, y), f_1(x, y)).$$

If only one particular solution is known, the general one is

$$\psi(x, y) = \varphi f(x, y).$$

Ex. 1. Let us examine in what cases we can find the general solution of the equation

$$\psi(x, y) = \psi(x^n y^m, x^k y^r).$$

In order to obtain a particular solution, put $\psi(x, y) = x^v y^w$ and making this substitution, we shall find the following equation of condition among the exponents

$$(1 - n)(1 - r) = km$$

hence either of the following equations may be solved generally,

$$\psi(x, y) = \psi\left\{\frac{x^n}{y^{r-1}}, \frac{y^r}{x^{n-1}}\right\} \qquad (a)$$

$$\psi(x, y) = \psi\left\{y\left(\frac{x}{y}\right)^n, x\left(\frac{y}{x}\right)^r\right\} \qquad (b)$$

the solution of the former is $\psi(x, y) = \varphi(xy)$, and that of the latter is $\psi(x, y) = \varphi(x^{r-1}y^{n-1})$.

In (a) put $n = r = \frac{1}{2}$, then the solution of $\psi(x, y) = \psi(\sqrt{xy}, \sqrt{xy})$ is $\psi(x, y) = \varphi(xy)$.

In a similar manner it may be found that the solution of

$$\psi(x, y) = \psi\left(\frac{x - y}{y}, \frac{x - y}{x}\right)$$

is

$$\psi(x, y) = \varphi\left(\frac{y}{x}\right).$$

As another example take the equation

$$\psi(x, y) = \psi\left(\frac{y}{z}\sqrt{\frac{y}{2x}}, \sqrt{2xy}\right)$$

a particular solution is $\psi(x, y) = 2xy + y^2$, hence the general solution is $\psi(x, y) = \varphi(2xy + y^2)$, but we may find another particular solution of this equation which is totally different from the former, and by combining the two we shall obtain a much more general solution. The equation (b) will coincide with the one under consideration, if we make $n = -\frac{1}{2}$ and $r = \frac{1}{2}$, then we shall have for another particular solution

$$f_1(x, y) = \frac{1}{y}\sqrt{2xy}$$

hence a very general solution of the equation

$$\psi(x, y) = \psi\left(\frac{y}{2}\sqrt{\frac{y}{2x}}, \sqrt{2xy}\right)$$

is

$$\psi(x, y) = \varphi\left\{2xy + y^2, \frac{1}{y\sqrt{2xy}}\right\}$$

φ remaining perfectly arbitrary.

In the equation of this Problem it may happen that / $\alpha(x, y)$ does not contain x nor $\beta(x, y)$ contain y, it then becomes

$$\psi(x, y) = \psi(\alpha y, \beta x).$$

This is the case when $\psi(x, y)$ is required to be a symmetrical function of x and y, the equation would then become

$$\psi(x, y) = \psi(y, x)$$

two particular solutions are $f(x, y) = xy$ and $f_1(x, y) = x + y$, hence the general solution of the equation is

$$\psi(x, y) = \varphi(xy, x + y).$$

Though these solutions may with propriety be termed general because they contain an arbitrary function, yet I am by no means inclined to think them the most general of which the questions admit, possibly we ought to except the two last equations, though I shall afterwards show that the solution of an equation of the form $\psi(x, y) = \psi(\alpha x, \beta y)$ may contain any number of known functions within the arbitrary one.

Problem IV

Given the equation

$$\psi(x, y) = \psi(\alpha(x, y), \beta(x, y)).$$

Assume as before $\psi(x, y) = \varphi(f(x, y), f_1(x, y))$, then the equation will become

$$\varphi(f(x, y), f_1(x, y)) = \varphi\{f(\alpha(x, y), \beta(x, y)) f_1(\alpha(x, y), \beta(x, y)\}.$$

In order to render this equation identical, determine f and f_1 from the two equations

$$f(x, y) = f_1(\alpha(x, y), \beta(x, y))$$

and

$$f_1(x, y) = f(\alpha(x, y), \beta(x, y))$$

putting in the first of these $\alpha(x, y)$ for x and $\beta(x, y)$ for y we find

$$f(x(x, y), \beta(x, y)) = f_1\{\alpha(\alpha(x, y), \beta(x, y)), \beta(\alpha(x, y), \beta(x, y))\} = f_1(x, y) \quad (1)$$

/

and we should find a precisely similar equation for determining f. If we are acquainted with two particular solutions of this equation, we may from them derive the general solution of the given equation. If, however, the functions α and β are of such a nature that the two following equations are fulfilled eq. (1) becomes identical without assigning any particular value to f or f_1. (The two conditions are $\alpha(\alpha(x, y), \beta(x, y)) = x$ and $\beta(\alpha(x, y), \beta(x, y)) = y$.)

It may be curious to enquire whether we can discover any forms which will satisfy these equations, for this purpose let us assume $\alpha(x, y) = a + bx + cy$, and also $\beta(x, y) = a + b_1x + c_1y$, this will only lead us to a particular solution, but I shall presently show that it may be rendered general. If the two conditions already specified are fulfilled, the arbitrary constants will be determined, and we shall have the following equations

$$\alpha(x, y) = a + bx + \frac{b^2 - 1}{b}y$$

$$\beta(x, y) = \frac{ab}{1 - b} - bx - by$$

which may be thus generalised. Let φ be any function, and let φ^{-1} be the inverse of that function, so that $\varphi\varphi^{-1}x = x$ then the conditions will be fulfilled, if

$$\alpha(x, y) = \varphi^{-1}\left\{a + b\varphi x + \frac{b^2 - 1}{b}\varphi y\right\}$$

and

$$\beta(x, y) = \varphi^{-1}\left\{\frac{ab}{1 - b} - b\varphi x - b\varphi y\right\}.$$

Some remarks, however, are necessary on the inverse function φ^{-1}. If we combine x and constant quantities by any of the direct operations, addition, multiplication, elevation of powers, etc. the result which is called a function of x admits / only of one value; let z equal the function, then we have the equation $z = \varphi x$. If from this we endeavour to discover the value of x in terms of z, the operation is an inverse one and x admits of one or more values according to the nature of the operations denoted by φ. This number may even be infinite; if φ denotes an equation of the nth degree, there are n values of x in terms of z. It may then be enquired whether in using the substitution employed in the latter part of this Problem, any of these (perhaps infinite number) may be taken, or whether only certain particular values should be used? without attending to this circumstance, our conclusions may become erroneous: all these different values will satisfy the equation $\varphi\varphi^{-1}x = x$, but only those must be used which also satisfy the equation $\varphi\varphi^{-1}x = x$: thus if $z = \varphi x = a - x^2$ we shall have $x = \varphi^{-1}z = \pm\sqrt{a - z}$ if we employ the upper sign we have

$$\varphi\varphi^{-1}x = +\sqrt{a - (a - x^2)} = +\sqrt{x^2} = +x.$$

If we use the lower one

$$\varphi^{-1}\varphi x = -\sqrt{a - (a - x^2)} = -\sqrt{x^2} = -x$$

the upper sign must therefore be taken, because in the latter part of the Problem we suppose $\varphi^{-1}\varphi x = x$ and $\varphi^{-1}\varphi y = y$. This remark, which is of some importance, extends to the conclusions in my former Paper and to the whole of the subsequent enquiries.

The equation (1) might be considered as similar to the original one, and the same transformation might be performed on this, and thus we might continue to deduce new conditions. In the first part we found that the equation $\psi x = \psi\alpha x$ always admitted of an easy solution when $\alpha^n x = x$ and by continuing the substitutions already pointed out, we should arrive at / some conclusions very analogous for functional equations of the form of those treated of in this Problem, but the length to which these enquiries would lead, render it sufficient merely to indicate them.

In the equations solved in Problem I and II it is obviously immaterial whether we first put αx instead of x, and then in the result put βy for y or conversely; but in the equation of Problems III and IV the case is different. If in the function $\psi(x, y)$ we put simultaneously $\alpha(x, y)$ for x, and $\beta(x, y)$ for y the result will be different from that which would arise from first putting $\alpha(x, y)$ for x and then in the result putting $\beta(x, y)$ for y, or from inverting this operation; the three results stand thus:

$$\psi(\alpha(x, y), \beta(x, y)) \qquad (a)$$

$$\psi(\alpha(x, \beta(x, y)), \beta(x, y)) \qquad (b)$$

$$\psi(\alpha(x, y), \beta(\alpha(x, y), y)). \qquad (c)$$

These three functions are evidently different, and in the solutions of the Problems, regard was only had to the first of them which may be called the simultaneous function. Those, however, of the second and third class might occur, and it becomes necessary to point out the means of solution which are applicable to them.

According to the notation laid down, these functions may be thus expressed

$$\psi^{\overline{1.1}}(\alpha(x, y), \beta(x, y)) \qquad (a)$$

$$\psi^{\overset{1.1}{1.2}}(\alpha(x, y), \beta(x, y)) \qquad (b)$$

$$\psi^{\overset{1.1}{2.1}}(\alpha(x, y), \beta(x, y)). \qquad (c)$$

But to avoid the trouble of indices I shall show how those of / the second and third class may be reduced to those of the first, I shall therefore always consider functions of the first order as simultaneous ones, and omit the indices, which if supplied, would be $\overline{{}^{1,1,1}_{1,2,3}}, \ldots$[a]

To transform $\psi^{1,1}_{1,2}(\alpha(x, y), \beta(x, y))$ into a function whose index is ${}^{1,1}_{1,2}$ put $\alpha(x, \beta(x, y)) = \gamma(x, y)$ then

$$\psi^{\overline{1,1}}_{1,2}(\gamma(x, y), \beta(x, y)) = \psi^{1,1}_{1,2}(\alpha(x, y), \beta(x, y))$$

and similarly if $\beta(\alpha(x, y), \beta(x, y)) = \gamma(x, y)$ we should have

$$\psi^{1,1}_{2,1}(\alpha(x, y), \beta(x, y)) = \psi^{\overline{1,1}}_{1,2}(\alpha(x, y), \gamma(x, y))$$

and generally whatever be the number of variables a similar transformation might be effected.

Problem V

Required the solution of the equation

$$\psi(x, y) = A(x, y)\psi(\alpha(x, y), \beta(x, y)).$$

Assume $\psi(x, y) = f(x, y)\varphi\{f_1(x, y), f_2(x, y)\}$ and substituting this in the given equation, we find

$$f(x, y)\varphi\{f_1(x, y), f_2(x, y)\}$$
$$= A(x, y)f(\alpha(x, y), \beta(x, y)) \cdot \varphi\{f_1(\alpha(x, y), \beta(x, y)), f_2(\alpha(x, y), \beta(x, y))\}.$$

This equation will be satisfied if we are acquainted with particular solutions of the three following equations

$$f(x, y) = A(x, y)f(\alpha(x, y), \beta(x, y))$$

$$f_1(x, y) = f_1(\alpha(x, y), \beta(x, y))$$

and

$$f_1(x, y) = f_1(\alpha(x, y), \beta(x, y))$$

the first of these is nothing more than the original equation. /

If therefore we know one particular solution of the original equation, and also one or two particular solutions of the other equation, we may deduce the general solution of the Problem.

[a] *Sic;* here and in the following paragraphs there are errors in the indices which we have not attempted to correct.

Ex. Let

$$\psi(x, y) = \left(\frac{x}{y}\right)^3 \psi(y, x)$$

in this case $f_1(x, y) = f_1(y, x)$ and two particular solutions are $f_1(x, y) = xy$ and $f_1(x, y) = x + y$ also a particular solution of the given equation is $f(x, y) = x/y^2$, hence its general solution is

$$\psi(x, y) = \frac{x}{y^2}\varphi(x + y, xy).$$

Problem VI

Given the equation

$$\psi(x, y) = A(x, y)\psi(\alpha(x, y), \beta(x, y)) + B(x, y).$$

Suppose we are acquainted with one particular solution which satisfies the equation and let it be $f(x, y)$, then assume

$$\psi(x, y) = f(x, y) + \varphi(x, y)$$

and making this substitution the equation becomes

$$f(x, y) + \varphi(x, y)$$
$$= A(x, y)f(\alpha(x, y), \beta(x, y)) + A(x, y) \cdot \varphi(\alpha(x, y), \beta(x, y)) + B(x, y).$$

Subtracting from this the particular solution

$$f(x, y) = A(x, y)f(\alpha(x, y), \beta(x, y)) + B(x, y)$$

there remains

$$\varphi(x, y) = A(x, y)\varphi(\alpha(x, y), \beta(x, y))$$

an equation which may be solved by the preceding Problem. The same substitution is applicable to the more general equation

$$0 = \psi(x, y) + A(x, y)\psi(\alpha(x, y), \beta(x, y))$$
$$+ B(x, y)\psi(\alpha_1(x, y), \beta_1(x, y)) + \dots + K(x, y).$$

/

Problem VII

Given the functions

$$\alpha(x, y), \alpha_1(x, y), \alpha_2(x, y) \dots$$
$$\beta(x, y), \beta_1(x, y), \beta_2(x, y) \dots$$

Required the nature of the function $\psi(x, y)$ such that it shall not alter its form by the simultaneous substitution of $\alpha(x, y)$, $\beta(x, y)$ for x and y, and generally that it shall remain the same when for x and y are respectively substituted any of the functions denoted by $\alpha_n(x, y)$ and $\beta_n(x, y)$. The conditions which determine ψ may be thus expressed

$$\psi(x, y) = \psi(\alpha(x, y), \beta(x, y)) = \psi(\alpha_1(x, y), \beta_1(x, y)) = \dots$$

Assume

$$\psi(x, y) = \varphi\{f(x, y), f_1(x, y)\} \tag{1}$$

then from the first condition we have

$$\varphi\{f(x, y), f_1(x, y)\} = \varphi\{f(\alpha(x, y), \beta(x, y)), f_1(\alpha(x, y), \beta(x, y))\}$$

this will be satisfied by making

$$f(x, y) = f(\alpha(x, y), \beta(x, y))$$

and

$$f_1(x, y) = f_1(\alpha(x, y), \beta(x, y))$$

these are two particular solutions of the first equation.

The second condition is

$$\psi(x, y) = \psi(\alpha_1(x, y), \beta_1(x, y))$$

which becomes

$$\varphi(f(x, y), f_1(x, y)) = \varphi(f(\alpha_1(\alpha_1(x, y), \beta_1(x, y)), f_1(\alpha_1(x, y), \beta_1(x, y)) \tag{2}$$

where f and f_1 are known functions; make

$$f(\alpha_1(x, y), \beta_1(x, y)) = K(x, y)$$

and

$$f_1(\alpha_1(x, y), \beta_1(x, y)) = {}^1K(x, y)$$

K and ${}^1K$ are therefore also known functions. /

Assume $\varphi(x, y) = \varphi_1(f_2(x, y), f_3(x, y))$, then equation (2) becomes

$$\varphi_1\{f_2(f(x, y), f_1(x, y)), f_3(f(x, y), f_1(x, y))\}$$
$$= \varphi_1\{f_2(K(x, y), {}^1K(x, y)), f_3(K(x, y), {}^1K(x, y))\}$$

This equation must be solved in the same manner as the former by means of two particular solutions, and by continuing the same method, we shall find that the form of the function ψ may be determined by means of $2n$ particular solutions of certain functional equations, when there are n pair of conditions assigned. A less general solution may, however, be found when we are only acquainted with n particular solutions.

A similar method would lead us to the form of ψ, whatever might be the number of variables. If, however, we are acquainted with any number of particular solutions which remain the same, in all the cases assigned by the conditions of the Problem, we may have the general solution by making

$$\psi = \varphi\{f_1, f_2, \dots f_i\}$$

$f_1, f_2, \dots f_i$ being i particular solutions.

Ex. Let it be required to find a symmetrical function of $x_1, x_2, \dots x_n$, the equations to be satisfied are

$$\psi(x_1, x_2, \dots x_n) = \psi(x_2, x_3, \dots x_n, x_1) = \psi(x_3, x_4 \dots x_n, x_1, x_2) = \dots$$

or the whole of the conditions may be more concisely denoted by the expression

$$\psi\{\bar{x}_1, \bar{x}_2, \bar{x}_3, \dots \bar{x}_n\}.$$

We may easily find n particular solutions which fulfill these equations: for in the first place it is evident that the sum of / any number of quantities is symmetrical with respect to them, therefore

$$f_1(x_1, x_2, \dots x_n) = x_1 + x_2 + \dots + x_n = S(x_1).$$

Again the sum of their products two by two is also symmetrical, therefore

$$f_2(x_1, x_2, \dots x_n) = x_1x_2 + x_2x_3 + x_1x_3 + \dots = S(x_1x_2)$$

and similarly the sums of their products three by three, four by four, etc. are symmetrical. We may, therefore, find n different particular solutions, and the general solution will be any arbitrary function of all these particular solutions, or

$$\psi(\bar{x}_1, \bar{x}_2, \dots \bar{x}_n) = \varphi\{S(x_1), S(x_1x_2), \dots S(x_1x_2 \dots x_n)\}.$$

Instead of taking for particular solutions the sum of all the quantities, the sum of all the products by two's, the sum of all the products by three's, etc., etc. we might have employed the sum of all the quantities, the sum of their squares, the sum of their cubes, etc. but the solution thus deduced would not be essentially different from the former.

On functional equations of the second and higher orders involving two or more variables

The notation to be employed in these enquiries has already been sufficiently explained, and the different species of second functions have been noticed. Preserving the same symbols, let it be required to solve the following Problem.

Problem VIII

Given the equation

$$\psi^{2,1}(x, y) = x.$$

This equation though apparently involving two variables / may in fact be solved by the methods of the first part; for y may be considered as a constant quantity, and if in the solution of $\varphi^2 x = x$ (Problems IX, X, and XIV, Part I) we put arbitrary functions of y instead of the constant quantities which occur, we shall have a solution of the given equation, thus a particular solution of $\varphi^2 x = x$ is $f(x) = (b - x)/(1 - cx)$; instead of c put $\chi(y)$ then a solution of the given equation is

$$\psi(x, y) = \frac{b - x}{1 - x\chi y}$$

for[a]

$$\psi^{2,1}(x, y) = \psi(\psi(x, y), y) = \frac{b - \dfrac{b - x}{1 - x\chi y}}{1 - \dfrac{b - x}{1 - x\chi y}\chi y} = \frac{x - bx\chi y}{1 - b\chi y} = x.$$

We might also instead of b put any other arbitrary function of y, and the result will be the same. The equations

$$\psi^{2,1}(x, y) = \alpha(x, y) \text{ and } \psi^{1,2}(x, y) = \alpha(x, y)$$

may be treated in a similar manner, in the first y must be considered as constant, and x must be so treated in the latter. In general, when functions are taken only relative to one of the variables, the rules delivered in my former Paper are sufficient for their solution, such is the equation

$$F\{(x, y)\psi(x, y), \psi^{2,1}(x, y), \dots \psi^{n,1}(x, y)\} = 0.$$

It might however occur, that though the order of the function does not vary relative to the other variable, yet that that variable may occur in different forms in each function. An example will render this more evident: $\alpha, \beta, \dots$ being known functions, let

$$F\{x, y, \psi(x, y), \psi^{2,1}(x, \alpha y), \psi^{3,1}(x, \beta y), \dots \psi^{n,1}(x, \upsilon y)\} = 0$$

here though the functions do not vary in *order* relative to y, / yet they do vary in a certain sense, because y is differently contained under each functional characteristic; the method of treating these kind of equations will be explained hereafter.

---

[a] The equation below has been corrected.

Problem IX

Given the functional equation

$$\psi^{\overline{2,2}}(x, y) = 0.$$

This signifies that the second simultaneous function is equal to zero. It is evident that $x - y$ or $y - x$ will be a particular solution, for if $\psi(x, y) = x - y$ we have

$$\psi^{\overline{2,2}}(x, y) = \psi(\psi(x, y), \psi(x, y)) = (x - y) - (x - y) = 0.$$

By observing the process just gone through, it appears that it would equally succeed if for x we put $f(x)$ and for y we put $f(y)$ for if $\psi(x, y) = fx - fy$, we have

$$\psi^{\overline{2,2}}(x, y) = (fx - fy) - (fx - fy) = 0.$$

This solution is considerably more general than the former, yet is by no means the complete solution; a more general one may be obtained thus: we found one particular solution to be $\psi(x, y) = x - y$, now if we multiply the right side of this equation by an arbitrary function of x and y the condition will still be fulfilled; for if $\psi(x, y) = (x - y)\varphi(x, y)$ we shall find

$$\psi^{\overline{2,2}}(x, y) = \{(x - y)\varphi(x, y) - (x - y)\varphi(x, y)\} \cdot \varphi\{(x - y)\varphi(x, y), (x - y)\varphi(x, y)\} = 0$$

provided $\varphi\{(x - y)\varphi(x, y), (x - y)\varphi(x, y)\}$ does not contain in its denominator any factor which vanishes. /

Problem X

Given the equation

$$\psi^{\overline{2,2}}(x, y) = a.$$

In this case the second simultaneous function of x and y is constant. The first solution which presents itself is $\psi(x, y) = x - y + A$, then we shall find

$$\psi^{\overline{2,2}}(x, y) = (x - y + A) - (x - y + A) + A = A = a$$

therefore $A = a$ and one particular solution is

$$\psi(x, y) = x - y + a.$$

This may be rendered more general, nearly in the same manner as the last Problem; thus let $\psi(x, y) = (x - y)\varphi(x, y) + a$ then

$$\psi^{\overline{2,2}}(x, y) = [((x - y)\varphi(x, y) + a) - ((x - y)\varphi(x, y) + a)] \cdot \varphi\{(x - y)\varphi(x, y) + a, (x - y)\varphi(x, y) + a\} + a = a.$$

Another particular solution which readily occurs is

$$\psi(x, y) = A\frac{x}{y}$$

this gives

$$\psi^{\overline{2,2}}(x, y) = A\frac{A(x/y)}{A(x/y)} = A = a$$

therefore $A = a$ and a particular solution is

$$\psi(x, y) = \frac{ax}{y} \text{ or } \psi(x, y) = \frac{ay}{x}.$$

This readily points out another general solution, let

$$\psi(x, y) = A\varphi\left(\frac{x}{y}\right) \text{ hence } \psi^{\overline{2,2}}(x, y) = A\varphi\left(\frac{A\varphi(x/y)}{A\varphi(x/y)}\right) = A\varphi(1) = a$$

make $A = a/\varphi(1)$ and the general solution is

$$\psi(x, y) = \frac{a}{\varphi(1)}\varphi\left(\frac{x}{y}\right).$$

From the combination of the two preceding solutions we / may obtain another value of ψ which will also satisfy the given equation; it will be

$$\psi(x, y) = \frac{a\chi\{(x - y)\varphi(x, y), \varphi_1(x/y)\}}{\chi\{0, \varphi_1(1)\}}.$$

This on trial will be found to agree with the condition, and χ, φ and φ_1 are arbitrary functions.

The equation we are considering will also be satisfied by making $\psi(x, y) = a\varphi(x)/\varphi(y)$ or more generally by the constant quantity a multiplied by any fraction whose numerator and denominator become equal when x is put for y: such are the following.

$$a\frac{x + xy + y^2}{y + 2x^2}, \; a\frac{x(y^2 + x^2)}{2y^3}, \; a\frac{\sqrt{x^3 + 8x^2y - 5y^3}}{2x\sqrt{x}}, \ldots$$

Problem XI[a]

Given the equation

$$\psi^{\overline{n,n}}(x, y) = ax + by.$$

Assume $\psi(x, y) = px + qy$ then we have

$$\psi^{\overline{2,2}}(x, y) = p(px + qy) + q(px + qy) = (p + q)(px + qy)$$

and

$$\psi^{\overline{3,3}}(x, y) = (p + q)^2(px + qy)$$

and generally

$$\psi^{\overline{n,n}}(x, y) = (p + q)^{n-1}(px + qy)$$

hence $p\,(p + q)^{n-1} = a$ and $q(p + q)^{n-1} = b$, which gives for the values of p and q

$$p = \frac{a}{(a + b)^{(n-1)/n}} \text{ and } q = \frac{b}{(a + b)^{(n-1)/n}}.$$

This is a very limited solution not containing even an arbitrary constant, it might easily be rendered more general, but / the problem itself would scarcely have been worth noticing had it not been for the very curious results to which it led me.

The relation between $\psi(x,y) = px + qy$ and $\psi^{\overline{n,n}}(x,y) = (p + q)^{n-1}(px + qy)$ is very remarkable, it appears from this, that in the present case in going from the nth to the $n + 1$th simultaneous function, we have only to multiply by the sum of the co-efficients of the original function. On enquiring a little more minutely into the cause of this circumstance, it will be found that it depends on the original function containing x and y of the same dimensions in all its terms, or more generally that the expression of $\psi(x, y)$ is homogeneous. Let us now assume some homogeneous function, and examine its second and higher simultaneous functions, let

$$\psi(x, y) = ax^n + by^px^{n-p} + cy^qx^{n-q} + \ldots$$

the second simultaneous function is

$$\psi^{\overline{2,2}}(x, y) = a\{\psi(x, y)\}^n + b\{\psi(x, y\}^n + c\{\psi(x, y)\}^n + \ldots$$

or

$$\psi^{\overline{2,2}}(x, y) = \{\psi(x, y)\}^n\{a + b + c + \ldots\} = \psi(1, 1)\{\psi(x, y)\}^n. \qquad (a)$$

[a] Some missing bars above the indices have been inserted in Problem XI.

If we now take the simultaneous third functions we have

$$\psi^{\overline{3,3}}(x, y) = \psi(1, 1)\{\psi^{\overline{2,2}}(x, y)\}^n = \psi(1, 1)\{\psi(1, 1)\{\psi(x, y)\}^n\}^n$$

hence

$$\psi^{\overline{3,3}}(x, y) = \{\psi(1, 1)\}^{1+n}\{\psi(x, y)\}^{n^2}.$$

Repeating the same operation we should have

$$\psi^{\overline{4,4}}(x, y) = \{\psi(1, 1)\}^{1+n+n^2} \cdot \{\psi(x, y)\}^{n^3}$$

and generally

$$\psi^{\overline{k,k}}(x, y) = \{\psi(x, y)\}^{n^{k-1}} \cdot \{\psi(1, 1)\}^{1+n+\ldots+n^{k-2}}$$

$$= \{\psi(x, y)\}^{n^{k-1}} \cdot \{\psi(1, 1)\}^{\frac{1-n^{k-1}}{1-n}}. \qquad (b)$$

/

This elegant property of homogeneous functions will assist us in solving a variety of equations.

Problem XII

Given the equation

$$\psi^{\overline{n,n}}(x, y) = a\{\psi(x, y)\}^b.$$

Determine n from the equation $b = n^{k-1}$ and also determine $\psi(1, 1)$ from the equation

$$\{\psi(1, 1)\}^{\frac{1-n^{k-1}}{1-n}} = a.$$

Or the given equation will be satisfied by any homogeneous function of the degree indicated by $b^{1/(k-1)}$ provided the sum of all its coefficients is equal to the quantity

$$a^{(1-b^{1/(k-1)})/1-b)}$$

Ex. Let

$$\psi^{3,3}(x, y) = 8(\psi(x, y))^4.$$

Here $b = 4$, $k = 3$, therefore $n^{k-1} = n^2 = b = 4$ and $n = \pm 2$; also $a = 8$ and $\psi(1, 1) = 8^{1/3} = 2$; therefore any of the following quantities will satisfy the equation

$$2xy, \quad x^2 + y^2, xy + y^2, \quad x^2 - xy + 2y^2.$$

The properties of homogeneous functions are so nearly connected with the solution of equations containing simultaneous functions, that it will be convenient to examine into them a little further, and to adopt some means of denoting them with brevity. In order to signify that a function of several variables has in each of its terms the sum of the indices of any two of them always the same, I shall make use of a line placed beneath those two variables: thus $\psi(\underline{x}, \underline{y})$ signifies an homogeneous function of x and y; and / as it may be convenient also to denote the sum of the two indices, I shall place it underneath on the outside of the parenthesis; thus then the expression $\psi(\underline{x}, \underline{y})_q$ denotes any homogeneous function of x and y of the qth degree. A function of three variables x, y, and z, may be homogeneous, with respect to two of them (x and y) in one sense, and also relative to y and z in another; but it does not from thence follow that it will be homogeneous relative to all three, such a function would be denoted thus

$$\psi(\underline{x}, \underline{\underline{y}}, \underline{z})_{p,q}$$

a particular case of this expression is

$$x^2z^2 + xyz = \psi(\underline{x}, \underline{\underline{y}}, \underline{z})_{2,2}.$$

This notation being premised, we have the following theorems relative to homogeneous functions.

$$\psi^{\overline{2,2}}(\underline{x}, \underline{y})_n = \psi(1, 1)\{\psi(\underline{x}, \underline{y})_n\}^n \tag{1}$$

$$\psi^{\overline{k,k}}(\underline{x}, \underline{y})_n = \{\psi(\underline{x}, \underline{y})_n\}^{n^{k-1}} \cdot \{\psi(1, 1)\}^{\frac{1-n^{k-1}}{1-n}} \tag{2}$$

And generally if we have any homogeneous function of the nth degree, and instead of x and y we substitute any other function whatever as $\psi_1(x, y)$, then we shall have the following equation

$$\psi\{\underline{\psi_1(x, y)}, \underline{\psi_1(x, y)}\}_n = \psi(1, 1) \cdot \{\psi(x, y)\}^k. \tag{3}$$

Assume

$$\psi(x, y) = \varphi\left\{\frac{\alpha(\underline{x}, \underline{y})_n}{\beta(\underline{x}, \underline{y})_m}\right\}$$

call the latter member, for the sake of brevity, K, and take the second simultaneous function on both sides; in this case $\alpha(\underline{x}, \underline{y})_n$ will become $\alpha(1, 1)K^n$ by equation (3), and for the same reason $\beta(\underline{x}, \underline{y})_m$ will become $\beta(1, 1)K^m$, and consequently we shall have /

$$\psi^{\overline{2,2}}(x, y) = \varphi\left\{\frac{\alpha(1, 1)}{\beta(1, 1)}K^{n-m}\right\} = \varphi\left\{\frac{\alpha(1, 1)}{\beta(1, 1)}\left(\varphi\frac{\alpha(\underline{x}, \underline{y})_n}{\beta(\underline{x}, \underline{y})_m}\right)^{n-m}\right\} \tag{4}$$

let $\alpha(1, 1) = \beta(1, 1)$ and also let $m = n$, then this equation becomes

$$\psi^{\overline{2,2}}(x, y) = \varphi(1)$$

this affords another and a more direct solution of Problem X for if $\psi^{\overline{2,2}}(x, y) = a$, a general solution is

$$\psi(x, y) = \frac{\alpha}{\varphi(1)}\varphi\left\{\frac{\alpha(\underline{x}, \underline{y})_n}{\beta(\underline{x}, \underline{y})_n}\right\}$$

$\alpha(1, 1)$ being equal to $\beta(1, 1)$, this latter condition, however, is not absolutely necessary, and if we wish to avoid it, the general solution will be

$$\psi(x, y) = \frac{a}{\varphi\left(\dfrac{\alpha(1, 1)}{\beta(1, 1)}\right)}\varphi\left\{\frac{\alpha(\underline{x}, \underline{y})_n}{\beta(\underline{x}, \underline{y})_n}\right\}.$$

The following is another solution of the same question:

Assume

$$\psi(x, y) = \varphi\left\{\frac{\begin{matrix} a + bx + cx^2 + \dots \\ b_1 y + c_1 xy \\ c_2 y^2 \end{matrix}}{\begin{matrix} {}^1a + {}^1bx + {}^1cx^2 + \dots \\ {}^1b_1 y + {}^1c_1 xy \\ {}^1c_2 y^2 \end{matrix}}\right\} = K$$

taking the second function on both sides we have

$$\psi^{\overline{2,2}}(x, y) = \varphi\left\{\frac{\left.\begin{matrix} a + b \\ b_1 \end{matrix}\right\} K + \left.\begin{matrix} c \\ c_1 \\ c_1 \end{matrix}\right\} K^2 + \dots}{\left.\begin{matrix} {}^1a + {}^1b \\ {}^1b_1 \end{matrix}\right\} K + \left.\begin{matrix} {}^1c \\ {}^1c_1 \\ {}^1c_1 \end{matrix}\right\} K^2 + \dots}\right\}$$

let

$$a = {}^1a, \quad b + b_1 = {}^1b + {}^1b_1, \quad c + c_1 + c_2 = {}^1c_1 + {}^1c_2 \dots$$

then the equation is reduced to

$$\psi^{\overline{2,2}}(x, y) = \varphi(1).$$

If therefore we assume

$$\psi(x, y) = \frac{a}{\varphi(1)}\varphi\left\{\frac{\begin{matrix} a + bx + cx^2 + \dots \\ by + c_1xy \\ c_2y^2 \end{matrix}}{\begin{matrix} {}^1a + {}^1bx + {}^1cx^2 + \dots \\ {}^1b_1y + {}^1c_1xy \\ {}^1c_2y^2 \end{matrix}}\right\} = K$$

the original equation will be satisfied.

I am inclined to think, that this solution is not the most general of which the Problem admits, even though the series were continued back, as it might, to negative powers of x and y. The two solutions which follow are possibly more general, although on this point I am not certain. It would indeed be a very important step, if we could assign the number and nature of the arbitrary functions which enter into the complete solution of functional equations.

Another solution of the equation $\psi^{\overline{2,2}}(x, y) = a$ may be thus deduced, let

$$\psi(x, y) = \varphi\left\{\frac{\alpha(\underline{x}, \underline{y})_m}{\alpha_1(\underline{x}, \underline{y})_m}, \frac{\beta(\underline{x}, \underline{y})_n}{\beta_1(\underline{x}, \underline{y})_n}, \dots\right\} = K$$

then taking the second simultaneous function on both sides, it will be perceived by the construction of the second side of the equation, that

$$\psi^{\overline{2,2}}(x, y) = \varphi\left\{\frac{\alpha(1, 1)}{\alpha_1(1, 1)}, \frac{\beta(1, 1)}{\beta_1(1, 1)}, \dots\right\}$$

call the right side of the equation A, then a very general / solution of our equation is

$$\psi(x, y) = \frac{a}{A}\varphi\left\{\frac{\alpha(\underline{x}, \underline{y})_m}{\alpha_1(\underline{x}, \underline{y})_m}, \frac{\beta(\underline{x}, \underline{y})_n}{\beta_1(\underline{x}, \underline{y})_n}, \dots\right\}$$

where the numbers $m, n, p, \dots$ are not confined to integers. Another solution may be found in the following manner: let

$$\psi(x, y) = \varphi\left(\frac{\chi(x, y)}{\chi_1(x, y)}\right) = K$$

and determine χ_1 so that $\chi_1(x, y) = \chi(x, y)$ when y is made equal to x, then taking the second simultaneous functions on both sides, we have

$$\psi^{\overline{2.2}}(x, y) = \varphi\left(\frac{\chi(K, K)}{\chi_1(K, K)}\right) = \varphi(1)$$

a general solution of the equation in question is therefore

$$\psi(x, y) = \frac{a}{\varphi(1)}\varphi\left(\frac{\chi(x, y)}{\chi_1(x, y)}\right)$$

this solution depends on that of the equation

$$\chi_1(x, y) = \chi(x, y) \qquad [y = x]$$

which belongs to a class of equations we shall speak of hereafter.

Let us now return to the consideration of equation (4) it is

$$\psi^{\overline{2.2}}(x, y) = \varphi\left\{\frac{\alpha(1, 1)}{\beta(1, 1)}\left(\varphi\left(\frac{\alpha(\underline{x}, \underline{y})_n}{\beta(\underline{x}, \underline{y})_m}\right)\right)^{n-m}\right\}$$

for n put $n + 1$ and for m put n, then it becomes

$$\psi^{\overline{2.2}}(x, y) = \varphi\left\{\frac{\alpha(1, 1)}{\beta(1, 1)}\varphi\left(\frac{\alpha(\underline{x}, \underline{y})_{n+1}}{\beta(\underline{x}, \underline{y})_n}\right)\right\}$$

take the third simultaneous function then

$$\psi^{\overline{3.3}}(x, y) = \varphi\left\{\frac{\alpha(1, 1)}{\beta(1, 1)}\varphi\left(\frac{\alpha(1, 1)}{\beta(1, 1)}\varphi\left(\frac{\alpha(\underline{x}, \underline{y})_{n+1}}{\beta(\underline{x}, \underline{y})_n}\right)\right)\right\}. \tag{5}$$

/

If we suppose $\alpha(1, 1) = \beta(1, 1)$ these questions become

$$\psi^{\overline{2.2}}(x, y) = \varphi^2\left(\frac{\alpha(\underline{x}, \underline{y})_{n+1}}{\beta(\underline{x}, \underline{y})_n}\right)$$

$$\psi^{\overline{3.3}}(x, y) = \varphi^3\left(\frac{\alpha(\underline{x}, \underline{y})_{n+1}}{\beta(\underline{x}, \underline{y})_n}\right)$$

and generally we should find

$$\psi^{\overline{p.p}}(x, y) = \varphi^p\left(\frac{\alpha(\underline{x}, \underline{y})_{n+1}}{\beta(\underline{x}, \underline{y})_n}\right) \tag{6}$$

where $\psi(x, y) = \varphi\left(\dfrac{\alpha(\underline{x}, \underline{y})_{n+1}}{\beta(\underline{x}, \underline{y})_n}\right)$ and also $\alpha(1,1) = \beta(1, 1)$.

A more general expression, and one which contains (6) as a particular case may be deduced in the following manner.

Let

$$\psi(x, y) = \left\{\frac{\beta(1, 1)}{\alpha(1, 1)}\varphi\frac{\alpha(\underline{x}, \underline{y})_n}{\beta(\underline{x}, \underline{y})_m}\right\}^{1/(n-m)} = K$$

taking the second simultaneous functions on both sides we find

$$\begin{aligned}
\psi^{\overline{2,2}}(x, y) &= \left\{\frac{\beta(1, 1)}{\alpha(1, 1)}\varphi\frac{\alpha(1, 1)K^n}{\beta(1, 1)K^m}\right\}^{1/(n-m)} \\
&= \left\{\frac{\beta(1, 1)}{\alpha(1, 1)}\varphi\frac{\alpha(1, 1)}{\beta(1, 1)}K^{n-m}\right\}^{1/(n-m)} \\
&= \left\{\frac{\beta(1, 1)}{\alpha(1, 1)}\varphi\left(\frac{\alpha(1, 1)}{\beta(1, 1)}\frac{\beta(1, 1)}{\alpha(1, 1)}\varphi\frac{\alpha(\underline{x}, \underline{y})_n}{\beta(\underline{x}, \underline{y})_m}\right)\right\}^{1/(n-m)} \\
&= \left\{\frac{\beta(1, 1)}{\alpha(1, 1)}\varphi^2\left(\frac{\alpha(\underline{x}, \underline{y})_n}{\beta(\underline{x}, \underline{y})_m}\right)\right\}^{1/(n-m)}
\end{aligned}$$

and if we continue to take the succeeding simultaneous functions we shall find generally, that when

$$\psi(x, y) = \left\{\frac{\beta(1, 1)}{\alpha(1, 1)}\varphi\left(\frac{\alpha(\underline{x}, \underline{y})_n}{\beta(\underline{x}, \underline{y})_m}\right)\right\}^{1/(n-m)}$$

$$\psi^{\overline{p,p}}(x, y) = \left\{\frac{\beta(1, 1)}{\alpha(1, 1)}\varphi^p\left(\frac{\alpha(\underline{x}, \underline{y})_n}{\beta(\underline{x}, \underline{y})_m}\right)\right\}^{1/(n-m)} \qquad (7)$$

this expression is much more general than the preceding (6) / with which it coincides when $\alpha(1, 1) = \beta(1, 1)$ and $n = m + 1$; by their assistance we may solve a variety of problems relating to simultaneous functions.

From (6) we have

$$\psi^{\overline{2.2}}(x, y) = \varphi^2\left(\frac{\alpha(\underline{x}, \underline{y})_{n+1}}{\beta(\underline{x}, \underline{y})_n}\right) = \varphi\varphi\left(\frac{\alpha(\underline{x}, \underline{y})_{n+1}}{\beta(\underline{x}, \underline{y})_n}\right)$$

putting in this for

$$\varphi\left(\frac{\alpha(\underline{x}, \underline{y})_{n+1}}{\beta(\underline{x}, \underline{y})_n}\right)$$

its value $\psi(x, y)$ we have

$$\psi^{\overline{2,2}}(x, y) = \varphi\psi(x, y)$$

from this we may deduce the solution of the following Problem.

Problem XIII

Given the equation

$$\psi^{\overline{2,2}}(x, y) = F\psi(x, y)$$

make $\varphi = F$ and take $\alpha(\underline{x}, \underline{y})_{n+1}$ any homogeneous function of the $(n + 1)$th degree, and $\beta(\underline{x}, \underline{y})_n$ a similar function of the nth, also let $\alpha(1, 1) = \beta(1, 1)$ then the equation is satisfied by making

$$\psi(x, y) = F\left\{\frac{\alpha(\underline{x}, \underline{y})_{n+1}}{\beta(\underline{x}, \underline{y})_n}\right\}.$$

Ex. Let

$$\psi^{\overline{2,2}}(x, y) = \sqrt{\psi(x, y)}.$$

Suppose $\alpha(\underline{x}, \underline{y})_{n+1} = x^2 + y^2$ and $\beta(\underline{x}, \underline{y})_n = 2x$, then one solution is

$$\psi(x, y) = \sqrt{\frac{x^2 + y^2}{2x}}$$

or let $\alpha(\underline{x}, \underline{y})_{n+1} = (x^2 + y^2)\varphi(1)$ and $\beta(\underline{x}, \underline{y})_n = 2y\varphi(x/y)$ then a more general solution is

$$\psi(x, y) = \sqrt{\frac{x^2 + y^2}{2y\varphi(x/y)}\varphi(1)}.$$

/

Problem XIV

Given the equation

$$\psi^{\overline{n,n}}(x, y) = F\psi(x, y).$$

In equation (6) we have

$$\psi^{\overline{n,n}}(x, y) = \varphi^n\left(\frac{\alpha(\underline{x}, \underline{y})_{n+1}}{\beta(\underline{x}, \underline{y})_n}\right) = \varphi^{n-1}\varphi\left(\frac{\alpha(\underline{x}, \underline{y})_{n+1}}{\beta(\underline{x}, \underline{y})_n}\right)$$

put for

$$\varphi\left(\frac{\alpha(\underline{x}, \underline{y})_{n+1}}{\beta(\underline{x}, \underline{y})_n}\right)$$

its value $\psi(x, y)$ then we have

$$\psi^{\overline{n.n}}(x, y) = \varphi^{n-1}\psi(x, y) = F\psi(x, y)$$

determine φ from the equation $\varphi^{n-1}v = Fv$ by Problem 13, Part 1, and the general solution of the equation is

$$\psi(x, y) = \varphi\left(\frac{\alpha(\underline{x}, \underline{y})_{n+1}}{\beta(\underline{x}, \underline{y})_n}\right).$$

Ex. 1. Let

$$\psi^{\overline{3,3}}(x, y) = \psi(x, y)$$

in this case $\varphi^{n-1}v = Fv$ becomes $\varphi^2 v = v$ solutions of which are

$$\varphi v = \frac{a}{v}, \quad \varphi v = a - v, \quad \varphi v = (a - v^m)^{1/m}.$$

Let

$$\alpha(\underline{x}, \underline{y})_{n+1} = x^2 + y^2 \text{ and } \beta(\underline{x}, \underline{y})_n = x + y$$

then solutions of the equation

$$\psi^{\overline{3.3}}(x, y) = \psi(x, y)$$

are

$$\psi(x, y) = \frac{a(x + y)}{x^2 + y^2},$$

$$\psi(x, y) = a - \frac{x^2 + y^2}{x + y}$$

and

$$\psi(x, y) = \left\{a - \left(\frac{x^2 + y^2}{x + y}\right)^m\right\}^{1/m}$$

Ex. 2. Let

$$\psi^{\overline{n,n}}(x, y) = \{\psi(x, y)\}^m.$$

In this example $\varphi^{n-1}v = Fv$ becomes $\varphi^{n-1}v = v^m$ and $\varphi v = v^{m^{1/(n-1)}}$; its particular solutions are therefore

$$\left(\frac{x^2 + y^2}{x + y}\right)^{m^{1/(n-1)}} \qquad \left(\frac{x^2 + 2xy - y^2}{y^2}\right)^{m^{1/(n-1)}}$$

other more general ones are

$$\psi(x, y) = \left\{\frac{(x^2 + y^2)\varphi(1)}{2x\varphi(x/y)}\right\}^{m^{1/(n-1)}} \quad \text{and} \quad \psi(x, y) = \left\{\frac{2xy\varphi(1)}{(x + y)\varphi(y/x)}\right\}^{m^{1/(n-1)}}.$$

Problem XV

Required the solution of the equation

$$\psi^{\overline{n,n}}(x, y) = F\psi^{\overline{n-p,n-p}}(x, y).$$

In (6) we have

$$\psi^{\overline{n,n}}(x, y) = \varphi^n\left(\frac{\alpha(\underline{x}, \underline{y})_{n-1}}{\beta(\underline{x}, \underline{y})_n}\right) = \varphi^p\varphi^{n-p}\left(\frac{\alpha(\underline{x}, \underline{y})_{n+1}}{\beta(\underline{x}, \underline{y})_n}\right)$$

but

$$\psi(x, y) = \varphi\left(\frac{\alpha(\underline{x}, \underline{y})_{n+1}}{\beta(\underline{x}, \underline{y})_n}\right)$$

consequently

$$\varphi^{n-p}\left(\frac{\alpha(\underline{x}, \underline{y})_{n+1}}{\beta(\underline{x}, \underline{y})_n}\right) = \psi^{\overline{n-p,n-p}}(x, y)$$

and substituting this value, we shall find

$$\psi^{\overline{n,n}}(x, y) = \varphi^p\psi^{\overline{n-p,n-p}}(x, y)$$

make $\varphi^p v = Fv$ and find the value φ, then the general solution of the equation

is $\psi(x, y) = \varphi\left(\frac{\alpha(\underline{x}, \underline{y})_{n+1}}{\beta(\underline{x}, \underline{y})_n}\right)$ if $\alpha(1, 1) = \beta(1, 1)$.

Ex. 1. Let

$$\psi^{\overline{4,4}}(x, y) = \{\psi^{\overline{2,2}}(x, y)\}^2.$$

Here

$$\varphi^2 v = Fv = v^2 \quad \text{and} \quad \varphi v = v^{\sqrt{2}}$$

therefore

$$\psi(x, y) = \left\{\frac{\alpha(\underline{x}, \underline{y})_{n+1}}{\beta(\underline{x}, \underline{y})_n}\right\}^{\sqrt{2}} \quad \text{if} \quad \alpha(1, 1) = \beta(1, 1)$$

more particular solutions are

$$\psi(x, y) = \left(\frac{x + y}{2\varphi(y/x)}\varphi(1)\right)^{\sqrt{2}} \quad \text{and} \quad \psi(x, y) = \left(\frac{x^2 + xy + y^2}{3y}\right)^{\sqrt{2}}.$$

Problem XVI

Given the equation

$$F\{\psi(x, y), \psi^{\overline{2.2}}(x, y), \ldots \psi^{\overline{p.p}}(x, y)\} = 0.$$

The same substitution as that employed in the last Problem will reduce this equation of two variables to a similar one which contains only one. Putting

$$\psi(x, y) = \varphi\left(\frac{\alpha(\underline{x}, \underline{y})_{n+1}}{\beta(\underline{x}, \underline{y})_n}\right)$$

we have

$$\psi^{\overline{2.2}}(x, y) = \varphi^2\left(\frac{\alpha(\underline{x}, \underline{y})_{n+1}}{\beta(\underline{x}, \underline{y})_n}\right)$$

and

$$\psi^{\overline{p.p}}(x, y) = \varphi^p\left(\frac{\alpha(\underline{x}, \underline{y})_{n+1}}{\beta(\underline{x}, \underline{y})_n}\right)$$

also making

$$\frac{\alpha(\underline{x}, \underline{y})_{n+1}}{\beta(\underline{x}, \underline{y})_n} = v$$

the given equation becomes

$$F\{\varphi v, \varphi^2 v, \ldots \varphi^p v\} = 0$$

an equation which contains only one variable, and may therefore be solved by the methods described in the first part.

Problem XVII

Required the solution of the equation

$$F\{\psi(x, y), \psi^{\overline{2,2}}(x, y), \ldots \psi^{\overline{p,p}}(x, y)\} = 0.$$

The following considerations lead to another mode of solution applicable to this Problem. If in the function $\psi(x, y)$ we put y equal to x it becomes $\psi(x, x)$ call this φx: then if in the second simultaneous function of $\psi(x, y)$ we put y equal x, the result will be the same as if we had taken the second function of $\psi(x, x)$ or φx relative to x, or symbolically expressed, it is

$$\psi^{\overline{2,2}}(x, y) = \psi(\psi(x, x), \psi(x, x)) = \varphi^2 x \qquad [y = x].$$

This may be rendered evident by substituting for the right side of the equation its value $\psi(\psi(x, y), \psi(x, y))$. /

In the same manner it may be shown, that if we take the pth simultaneous function, and then put x for y, the result will be the same as the pth function of φx, or expressed in symbols it is

$$\psi^{\overline{p,p}}(x, y) = \varphi^p(x) \qquad [y = x].$$

Now since this equation is identical when y is equal to x, it will remain so when any other quantity as v is put for x, if the same quantity is also put for y, therefore

$$\psi^{\overline{p,p}}(v, v) = \varphi^p(v)$$

now let $v = \psi(x, y)$ this equation becomes

$$\psi^{\overline{p,p}}(\psi(x, y), \psi(x, y)) = \varphi^p\psi(x, y)$$

but the right side of this equation is nothing more than the $(p + 1)$th simultaneous function of $\psi(x, y)$, consequently

$$\psi^{\overline{p+1,p+1}}(x, y) = \psi^p\psi(x, y).$$

If now in the equation of the Problem we substitute the several values thus formed of the simultaneous functions, we shall have

$$F\{\psi(x, y), \varphi\psi(x, y), \varphi^2\psi(x, y), \ldots \varphi^{p-1}\psi(x, y)\} = 0$$

and putting z for ψ (x, y) we have

$$F\{z, \varphi z, \varphi^2 z, \ldots \varphi^{p-1} z\} = 0$$

which is a functional equation of one variable, and may be solved by the methods of the first Part. The form of φ being thus ascertained, we have for determining $\psi(x, y)$ the equation

$$\psi(x, y) = \varphi x \qquad [y = x]$$

or expressed in words $\psi(x, y)$ may be any function of x and y which becomes equal to φx when y is equal to x. /

Problem XVIII

Given the equation

$$\psi^{\overline{p.p}}(x, y) = F(x, y)$$

$F(x, y)$ being such a function of x and y that it may be reducible to the form $F_1\gamma(\underline{x}, \underline{y})$.

Assume

$$\psi(x, y) = \left\{\frac{\beta(1, 1)}{\alpha(1, 1)}\varphi\left(\frac{\alpha(\underline{x}, \underline{y})_n}{\beta(\underline{x}, \underline{y})_m}\right)\right\}^{1/(n-m)}$$

then from equation (7) we have

$$\psi^{\overline{p.p}}(x, y) = \left\{\frac{\beta(1, 1)}{\alpha(1, 1)}\varphi^p\left(\frac{\alpha(\underline{x}, \underline{y})_n}{\beta(\underline{x}, \underline{y})_m}\right)\right\}^{1/(n-m)}$$

make

$$\frac{\alpha(\underline{x}, \underline{y})_n}{\beta(\underline{x}, \underline{y})_m} = \gamma(\underline{x}, \underline{y}) = v,$$

and since $\gamma(\underline{x}, \underline{y})$ is given and α and β are indeterminate, this equation may be easily satisfied in an infinite number of ways; put v for $\gamma(\underline{x}, \underline{y})$ our equation becomes

$$\left\{\frac{\beta(1, 1)}{\alpha(1, 1)}\varphi^p v\right\}^{1/(n-m)} = F_1 v$$

which contains only one variable and may be solved by the methods of the former part.

Ex. 1. Let

$$\psi^{\overline{p.p}}(x, y) = \frac{x+y}{2f(1)}f\left(\frac{x}{y}\right).$$

Assume

$$\alpha(\underline{x}, \underline{y}) = (x + y)f\left(\frac{x}{y}\right) \quad \text{and} \quad \beta(x, y) = 2f(1)$$

then it becomes

$$\varphi^p v = v$$

and calling f_1 any particular solution of this equation we have for the general one

$$\psi(x, y) = \varphi^{-1} f_1 \varphi\left(\frac{x + y}{2f(1)} \cdot f\left(\frac{x}{y}\right)\right).$$

/

Ex. 2. Let

$$\psi^{\overline{3,3}}(x, y) = \frac{x^2}{y}.$$

Put

$$\alpha(\underline{x}, \underline{y}) = x^2 \quad \text{and} \quad \beta(\underline{x}, \underline{y}) = y$$

then the equation to be solved is $\varphi^3 v = v$ a particular solution of which is

$$\varphi v = (1 + v)/(1 - 3v)$$

putting for v its value and using the general solution we have

$$\psi(x, y) = \varphi^{-1}\left\{\frac{1 + \varphi(x^2/y)}{1 - 3\varphi(x^2/y)}\right\}$$

as a particular case, take

$$\psi(x, y) = \left\{\frac{y^n + x^{2n}}{y^n - 3x^{2n}}\right\}^{1/n}$$

which will be found on trial to satisfy the condition.

Problem XIX

Given the equation

$$F\{\gamma(\underline{x}, \underline{y}), \psi(x, y), \psi^{\overline{2,2}}(x, y), \ldots \psi^{\overline{p,p}}(x, y)\} = 0.$$

This equation is evidently capable of solution by the same means as the last: putting

$$\psi(x, y) = \left\{\frac{\beta(1, 1)}{\alpha(1, 1)}\varphi\left(\frac{\alpha(\underline{x}, \underline{y})_n}{\beta(\underline{x}, \underline{y})_m}\right)\right\}^{1/(n-m)}$$

we have as before

$$\psi^{\overline{p,p}}(x, y) = \left\{\frac{\beta(1, 1)}{\alpha(1, 1)}\varphi^p\left(\frac{\alpha(\underline{x}, \underline{y})_n}{\beta(\underline{x}, \underline{y})_m}\right)\right\}^{1/(n-m)}$$

and assuming α and β such that

$$\frac{\alpha(\underline{x}, \underline{y})_n}{\beta(\underline{x}, \underline{y})_m} = \gamma(\underline{x}, \underline{y}) = v$$

and / making

$$\frac{\beta(1, 1)}{\alpha(1, 1)} = a \text{ our equation becomes}$$

$$F\{v, (a\varphi v)^{1/(n-m)}, (a\varphi^2 v)^{1/(n-m)}, \ldots (a\varphi^p v)^{1/(n-m)}\} = 0$$

which may be solved by Problem XIX, Part I.

Problem XX

Given the equation

$$\psi^{2,1}(x, y) = \psi^{1,2}(x, y).$$

This equation, containing no simultaneous function, is different from any we have yet solved, and requires the application of a peculiar artifice.

In my former Paper, in order to reduce the equation $\psi^2 x = \alpha x$ to one of the first order, I made use of the substitution $\varphi^{-1}f\varphi x$ for ψx: an analogous one must be employed on the present occasion; let us suppose

$$\psi(x, y) = \varphi^{-1}f(\varphi x, \varphi y)$$

the effect of this will be very similar to that of the one just alluded to, and its great utility will be evident by considering its result in the various orders of the same function, thus

$$\begin{aligned}\psi^{2,1}(x, y) &= \varphi^{-1}f(\varphi\varphi^{-1}f(\varphi x, \varphi y), \varphi y)\\ &= \varphi^{-1}f(f(\varphi x, \varphi y), \varphi y)\\ &= \varphi^{-1}f^{2,1}(\varphi x, \varphi y)\end{aligned}$$

$$\psi^{1,2}(x, y) = \varphi^{-1}f(\varphi x, \varphi\varphi^{-1}f(\varphi x, \varphi y))$$
$$= \varphi^{-1}f(\varphi x, f(\varphi x, \varphi y))$$
$$= \varphi^{-1}f^{1,2}(\varphi x, \varphi y)$$

$$\psi^{\overline{2,2}}(x, y) = \varphi^{-1}f(\varphi\varphi^{-1}f(\varphi x, \varphi y), \varphi\varphi^{-1}f(\varphi x, \varphi y))$$
$$= \varphi^{-1}f(f(\varphi x, \varphi y), f(\varphi x, \varphi y))$$
$$= \varphi^{-1}f^{\overline{2,2}}(\varphi x, \varphi y)$$

and continuing the same substitutions we shall find

$$\psi^{3,1}(x, y) = \varphi^{-1}f^{3,1}(\varphi x, \varphi y) \text{ and } \psi^{1,3}(x, y) = \varphi^{-1}f^{1,3}(\varphi x, \varphi y)$$

/

and generally

$$\psi^{n,m}(x, y) = \varphi^{-1}f^{n,m}(\varphi x, \varphi y).$$

If there are more variables than two, the proper substitution to be made is

$$\psi(x_1, x_2, \ldots x_i) = \varphi^{-1}f^{1,1,\ldots}(\varphi x_1, \varphi x_2, \ldots \varphi x_i)$$

and there would result generally

$$\psi^{n,m,p,\ldots}(x_1, x_2, \ldots x_i) = \varphi^{-1}f^{n,m,p,\ldots}(\varphi x_1, \varphi x_2, \ldots \varphi x_i).$$

By such substitutions all simple functional equations of every order and of any number of variables, may be reduced to those of the first order: but the difficulty is not then overcome, the resulting equations are by no means easy to solve, and in a variety of cases it appears, that they are contradictory or impossible.

Let us apply this substitution to the equation of this problem, then since

$$\psi(x, y) = \varphi^{-1}f(\varphi x, \varphi y)$$

we have

$$\varphi^{-1}f^{2,1}(\varphi x, \varphi y) = \varphi^{-1}f^{1,2}(\varphi x, \varphi y).$$

Performing the operation denoted by φ^{-1} on both sides it becomes

$$f^{2,1}(\varphi x, \varphi y) = f^{2,1}(\varphi x, \varphi y).$$

Put $\varphi^{-1}x$ for x and $\varphi^{-1}y$ for y then it becomes

$$f^{2,1}(x, y) = f^{1,2}(x, y).$$

which is nothing more than the original equation; from it, however we learn, that if we can find one particular solution, we can always deduce from it the general one, which supposing f a particular case, will be

$$\psi(x, y) = \varphi^{-1}f(\varphi x, \varphi y).$$

After repeated endeavours I have been unable to find any particular case which will satisfy the equation

$$\psi^{2,1}(x, y) = \psi^{1,2}(x, y).$$

I have also made some attempts at discovering particular solutions of the two following equations, and have met with no success.

$$\psi^{\overline{2,2}}(x, y) = \psi^{1,2}(x, y) \quad \text{and} \quad \psi^{\overline{2,2}}(x, y) = \psi^{2,1}(x, y).$$

Should, however, any particular case be found, their general solutions flow immediately from the method just explained.

With regard to the equation of the Problem $\psi^{2,1}(x, y) = \psi^{1,2}(x, y)$, I have little expectation of finding any particular case, as I think the following reasoning, though perhaps not quite so satisfactory as might be wished, will show the impossibility of it. First, let us suppose, that $\psi(x, y)$ is a symmetrical function of x and y, let it be $\chi(\bar{x}, \bar{y})$ then our equation becomes

$$\chi\{\overline{\chi(\bar{x}, \bar{y}), \bar{y}}\} = \chi\{\bar{x}, \overline{\chi(\bar{x}, \bar{y})}\} = \chi\{\overline{\chi(\bar{x}, \bar{y}), \bar{x}}\}.$$

Comparing the first of these expressions with the third, we may observe that in the first, wherever $\chi(\bar{x}, \bar{y})$ occurs, the same quantity $\chi(\bar{x}, \bar{y})$ also occurs in the third, consequently in this respect, the first and third are identical: but wherever y occurs in the first, x occurs similarly in the third, therefore in this respect they cannot be identical, unless y is equal to x. From this it appears, that the equation in question cannot be solved by any symmetrical function. Again, the given equation $\psi^{2,1}(x, y) = \psi^{1,2}(x, y)$ contains x and y in the same manner, and no reason can be assigned why in the solution x should be contained differently from y: this may, perhaps, be made more clear, thus. Let $f(x, y)$ be the quantity to which each side of the given equation is equal, then

$$\psi^{2,1}(x, y) = f(x, y) = \psi^{1,2}(x, y).$$

Now since $\psi(\psi(x, y), y) = f(x, y)$ and also $\psi(x, \psi(x, y)) = f(x, y)$; and since taking the second functions is a direct operation, it is evident that the original function $\psi(x, y)$ will produce the same result, whether we take the second function relative to x or relative to y; therefore it must be similarly composed of x and y; that is to say, it must be symmetrical relative to x and y: but we have before shown that no symmetrical function can satisfy the equation, consequently the equation is contradictory.

This train of reasoning I offer with considerable hesitation, well aware of the extreme difficulty of reasoning correctly on a subject so very general, and which, from its novelty, the mind has not been sufficiently habituated to consider, so as to rely with confidence on any lengthened process of reasoning. I thought it, however, right to mention this proof, that those who may seek for particular cases, might first enquire whether the equation be possible.

Problem XXI

Given the equation

$$x\psi^{1,2}(x, y) = y\psi^{2,1}(x, y).$$

Substituting $\varphi^{-1}f(\varphi x, \varphi y)$ for $\psi(x, y)$ in this equation we have

$$x\varphi^{-1}f^{1,2}(\varphi x, \varphi y) = y\varphi^{-1}f^{2,1}(\varphi x, \varphi y)$$

putting $\varphi^{-1}x$ for x and $\varphi^{-1}y$ for y it becomes

$$\varphi^{-1}x \cdot \varphi^{-1}f^{1,2}(x, y) = \varphi^{-1}y \cdot \varphi^{-1}f^{2,1}(x, y).$$

This equation will be satisfied if we could find such a form for f, that the two following equations might be fulfilled.

$$f^{1,2}(x, y) = y \quad \text{and} \quad f^{2,1}(x, y) = x$$

for in that case it would become

$$\varphi^{-1}x \cdot \varphi^{-1}y = \varphi^{-1}y \cdot \varphi^{-1}x$$

which is identical. /

Our enquiries must therefore be directed to this point, and it will be found that $f(x, y) = a - x - y$ has the required properties, and is a particular solution of the given equation: hence the general solution is

$$\psi(x, y) = \varphi^{-1}(a - \varphi x - \varphi y).$$

There are many other particular cases which fulfil the same condition, such as

$$f(x, y) = \frac{a}{xy} \quad \text{and} \quad f(x, y) = \frac{1 - x - y}{1 - bxy}$$

these give the general solutions

$$\psi(x, y) = \varphi^{-1}\frac{a}{\varphi x\varphi y} \quad \text{and} \quad \psi(x, y) = \varphi^{-1}\left(\frac{1 - \varphi x - \varphi y}{1 - b\varphi x\varphi y}\right).$$

Problem XXII

Given the equation

$$\psi^{2,1}(x, y) \cdot \psi^{1,2}(x, y) = xy.$$

Using the same substitution employed in the last Problem, this equation becomes

$$\varphi^{-1}f^{2,1}(\varphi x, \varphi y) \cdot \varphi^{-1}f^{2,1}(\varphi x, \varphi y) = xy$$

and putting $\varphi^{-1}x$ for x, and $\varphi^{-1}y$ for y we have

$$\varphi^{-1}f^{2,1}(x, y) \cdot \varphi^{-1}f^{1,2}(x, y) = \varphi^{-1}x \cdot \varphi^{-1}y$$

which becomes identical, if $f^{2,1}(x, y) = x$ and $f^{1,2}(x, y) = y$ consequently all the solutions of the last Problem also solve this.

Problem XXIII

Given the equation

$$F\{x, \psi^{1,2}(x, y)\} = F\{\psi^{2,1}(x, y), y\}$$

this equation may be solved by the same artifice as the two last, assuming $\psi(x, y) = \varphi^{-1}f(\varphi x, \varphi y)$ we have

$$F\{x, \varphi^{-1}f^{1,2}(\varphi x, \varphi y)\} = F\{\varphi^{-1}f^{2,1}(\varphi x, \varphi y), y\}$$

and putting $\varphi^{-1}x$ for x, and $\varphi^{-1}y$ for y, it becomes

$$F\{\varphi^{-1}x, \varphi^{-1}f^{1,2}(x, y)\} = F\{\varphi^{-1}f^{2,1}(x, y), \varphi^{-1}y\}$$

this is identical if we assume f so that the two conditions $f^{1,2}(x, y) = y$ and $f^{2,1}(x, y) = x$ may be fulfilled.

The same method is applicable to the equation

$$\{\psi^{2,1}(x, y) - x\}F(x, y, \psi(x, y)\ldots) = \{\psi^{1,2}(x, y) - y\}F_1(x, y, \psi(x, y), \ldots)$$

for the two factors which multiply F and F_1 vanish on account on the value of f.

Problem XXIV

Given the equation

$$x\psi^{\overline{2,2}}(x, y) = a\psi^{2,1}(x, y)$$

this equation, by means of the substitution already so frequently employed, becomes

$$x\varphi^{-1}f^{\overline{2,2}}(\varphi x, \varphi y) = a\varphi^{-1}f^{2,1}(\varphi x, \varphi y)$$

and putting $\varphi^{-1}x$ for x, and $\varphi^{-1}y$ for y we have

$$\varphi^{-1}x \cdot \varphi^{-1}f^{\overline{2,2}}(x, y) = a\varphi^{-1}f^{2,1}(x, y).$$

An artifice somewhat similar to the one already employed, will afford the solution of this equation: if we can find such a value of $f(x, y)$ that $f^{\overline{2,2}}(x, y) = c$ and also $f^{2,1}(x, y) = x$ the equation will become identical by making $c = \varphi a$. Such a value of f is $f(x, y) = c \cdot y/x$ for

$$f^{2,2}(x, y) = c\frac{c \cdot y/x}{c \cdot y/x} = c \quad \text{and} \quad f^{2,1}(x, y) = c\frac{y}{c \cdot y/x} = x$$

hence the general solution of the given equation is

$$\psi(x, y) = \psi^{-1}\left(\frac{\varphi a \varphi y}{\varphi x}\right).$$

/

Problem XXV

Given the equation

$$F\{\psi^{\overline{2,2}}(x, y), x\} = F\{a, \psi^{2,1}(x, y)\}$$

making the substitution $\varphi^{-1}f(\varphi x, \varphi y)$ for $\psi(x, y)$ and in the result putting $\varphi^{-1}x$ for x, and $\varphi^{-1}y$ for y, we shall find

$$F\{\varphi^{-1}f^{\overline{2,2}}(x, y), \varphi^{-1}x\} = F\{a, \varphi^{-1}f^{2,1}(x, y)\}$$

which will become identical if we select such a value for $f(x, y)$ that $\varphi^{-1}f^{\overline{2,2}}(x, y) = a$, and also $f^{2,1}(x, y) = x$ such a value is $f(x, y) = \varphi a + y - x$; hence the general solution of the Problem is

$$\psi(x, y) = \varphi^{-1}(\varphi a + \varphi y - \varphi x).$$

The more general equation

$$\{\psi^{2,1}(x, y) - x\}F\{x, y, \psi(x, y), \ldots\} = \psi^{\overline{2,2}}(x, y) \cdot F_1\{x, y, \psi(x, y), \ldots\}$$

may be solved nearly in the same manner, its solution will be

$$\psi(x, y) = \varphi^{-1}(\varphi y - \varphi x).$$

Problem XXVI

Given the equation

$$F\{x, y, \psi(x, y), \psi^{1,2}(x, y), \psi^{2,1}(x, y), \ldots\} = 0.$$

Assume $\psi(x, y) = \varphi^{-1}f(\varphi x, \varphi y)$ then we have $\psi^{2,1}(x, y) = \varphi^{-1}f^{2,1}(\varphi x, \varphi y)$, and $\psi^{1,2}(x, y) = \varphi^{-1}f^{1,2}(x, y)$ and generally $\psi^{n,m}(x, y) = \varphi^{-1}f^{n,m}(\varphi x, \varphi y)$.

Substituting these values the equation becomes

$$F\{x, y, \varphi^{-1}f(\varphi x, \varphi y), \varphi^{-1}f^{1,2}(\varphi x, \varphi y), \varphi^{-1}f^{2,1}(\varphi x, \varphi y), \ldots\} = 0$$

and putting $\varphi^{-1}x$ for x and $\varphi^{-1}y$ for y we have

$$F\{\varphi^{-1}x, \varphi^{-1}y, \varphi^{-1}f(x, y), \varphi^{-1}f^{1,2}(x, y), \varphi^{-1}f^{2,1}(x, y), \ldots\} = 0.$$

Some particular value of f must now be assumed, and the equation treated as one of the first order relative to φ^{-1}.

The form to be assigned to f is of some consequence; it ought to be a particular solution of the original equation: for if we assign to it any other form, this adds a limitation to the original equation which may or may not agree with it, a particular solution should therefore always be employed. This remark is applicable to several Problems in my former Paper, and with this restriction, their solutions will remain correct.

Problem XXVII

To transform the equation

$$F\{x, y, \psi(x, y), \psi^{2,1}(\alpha x, \beta y), \psi^{1,2}(\alpha_1 x, \beta_1 y), \ldots\} = 0$$

into the form of the equation of the preceding Problem.

Assume $\psi(x, y) = \varphi^{-1}f(\varphi x, \varphi y)$, then the equation becomes

$$F\{x, y, \varphi^{-1}f(\varphi x, \varphi y), \varphi^{-1}f^{2,1}(\varphi\alpha x, \varphi\beta y), \varphi^{-1}f^{1,2}(\varphi\alpha_1 x, \varphi\beta_1 y), \ldots\} = 0$$

find for φ by Problem VII, Part I, such a value that it shall not change when any of the following quantities are substituted for x

$$\begin{array}{ll} \alpha x & \beta x \\ \alpha_1 x & \beta_1 x \\ \alpha_2 x & \beta_2 x \\ \vdots & \vdots \end{array}$$

let this value be A, then the equation becomes

$$F\{x, y, A^{-1}f(Ax, Ay), A^{-1}f^{2,1}(Ax, Ay), \ldots\} = 0$$

or putting $A^{-1}x$ for x, and $A^{-1}y$ for y we have

$$F\{A^{-1}x, A^{-1}y, A^{-1}f(x, y), A^{-1}f^{2,1}(x, y), A^{-1}f^{1,2}(x, y), \ldots\} = 0$$

an equation of the required form.

One important use of this transformation is the solution of equations of the form

$$F\{x, y, \psi(x, y), \psi^{1,2}(\alpha x, y), \psi^{1,2}(\beta x, y), \ldots\} = 0$$

in which the functions are taken only relative to y, and yet x is not altogether constant. It may be transformed into

$$F_1\{x, y, \psi(x, y), \psi^{1,2}(x, y), \psi^{1,3}(x, y), \ldots\} = 0$$

which may be treated as an equation of one variable, x being constant. This is the species of equation alluded to at page (139).

After considering the various equations amongst the higher orders of functions, another question presents itself, which may be thus stated. What must be the form of a function of n variables, such that taking the functions relative to any or to all of them any number of times, and combining these quantities in any manner, the result shall (when all these variables are made equal to x) be equal to a given function of x? This question might thus be expressed when there are only two variables

$$F\{x, y, \psi(x, y), \psi^{1,2}(x, y), \psi^{2,1}(x, y), \ldots\} = f(a) \qquad [y = x].$$

This condition obviously enlarges the signification of the function ψ, and the solutions ought to be more general. We shall accordingly find that some equations, of which without this condition we cannot find even a particular solution, are capable, when it is added, of very extensive ones. When there are more than two variables, the condition may be, that making / them equal by pairs, the result shall be given: a particular case would be the equation

$$F\{x, y, z, v, w, r, \psi(x, y, z, v, w, r, \ldots)\} = F_1\{x, v, r\} \qquad \begin{bmatrix} y = x \\ z = v \\ w = r \end{bmatrix}.$$

Instead of making y equal to x, and z to v and so on, y might become a given function of x, and z a given function of v, etc. thus:

$$F\{x, y, z, v, w, r, \psi(x, y, z, v, w, r, \ldots)\} = F_1\{x, v, r\} \qquad \begin{bmatrix} y = \alpha x \\ z = \beta v \\ w = \gamma r \end{bmatrix}.$$

A few examples will sufficiently explain the method to be pursued in treating these equations.

Problem XXVIII

Given the equation

$$\psi^{2,1}(x, y) = \psi^{1,2}(x, y) \qquad [y = x].$$

This Problem, of which without the condition of y being made equal to x we could not find even a particular case, readily admits of solution in its present state. Since $\psi^{2,1}(x, y)$ is only equal to $\psi^{1,2}(x, y)$ when $y = x$, we may put for the given equation

$$\psi^{2,1}(x, y) = \varphi\{(x, y), \psi^{1,2}(x, y)\} \qquad (1)$$

provided that when y is equal to x, the latter side of the equation shall become $\psi^{1,2}(x, y)$, and this fully satisfies the condition of the Problem. If, therefore, we can find such a value of φ, the equation (1) may be treated as a common functional equation of two variables, and may be solved by the rules already given.

Nor is it at all difficult to find such a value of φ; if we make $\psi^{1,2}(x, y) = z$, φ must be such a function that

$$\varphi(x, y, z) = z \qquad [y = x].$$

/

It is evident that particular values of φ are

$$x - y + z \quad \text{and} \quad \frac{x}{y}z$$

many others might be mentioned, but it is desirable to determine φ more generally.

Since

$$\varphi(x, y, z) = z \qquad [y = x]$$

it is evident that

$$\varphi(x, x, z) = z$$

and since this is independent on any particular value of x we have

$$\varphi(v, v, z) = z$$

that is to say, that whatever quantity is substituted for x, if the same quantity is also substituted for y, the result will be equal to z. Now let $v = \varphi(x, y, z)$, it becomes

$$\varphi\{\varphi(x, y, z), \varphi(x, y, z), z\} = z$$

but this expression is nothing more than the second simultaneous function relative to x and y, and may be therefore more concisely expressed thus

$$\varphi^{\overline{2,2},1}(x, y, z) = z$$

in which equation, since it does not vary relative to z, that quantity may be considered as a constant; and the equation

$$\varphi^{\overline{2,2}}(x, y) = z = \text{constant}$$

being solved, we have only to substitute instead of the various constant quantities arbitrary functions of z: thus then the solution of the equation

$$\varphi(x, y, z) = x \qquad [y = x]$$

is reduced to that of

$$\psi^{\overline{2,2}}(x, y) = \text{constant}$$

and we have only to refer to Problem X for its general solution.

Let us apply this to the solution of the equation of this Problem

$$\psi^{2,1}(x, y) = \psi^{1,2}(x, y) \qquad [y = x]$$

take as a particular case of the equation

$$\varphi^{\overline{2,2},1}(x, y, z) = z$$

$$\varphi(x, y, z) = \frac{x}{y}z,$$

then the equation of the Problem becomes

$$\psi^{2,1}(x, y) = \frac{x}{y}\psi^{1,2}(x, y) \quad \text{or} \quad y\psi^{2,1}(x, y) = x\psi^{1,2}(x, y)$$

this is the equation solved in Problem XXI therefore all its solutions are also solutions of this equation. This however is, comparatively speaking, but a very limited answer: every different solution of the equation $\varphi^{\overline{2,2},1}(x, y, z) = z$ furnishes a new solution of our Problem, containing one or more arbitrary functions; each of these may very justly be called a general solution; but to investigate the number and nature of the arbitrary constants which enter into the complete solution, is an enquiry of considerable difficulty.

Problem XXIX

Given the equation

$$\psi^{\left.\begin{smallmatrix}2,2\\1,2\end{smallmatrix}\right\}^n}(x, y) = F(x) \qquad [y = \alpha x].$$

This signifies, that after taking the second function relative to x, and then the second relative to y; the result is considered merely as a function of x, and its nth function taken relative to that variable: lastly, the quantity to which this becomes equal, after performing these operations, is given. The manner of treating these equations is very simple; put

$$\psi^{2,2}(x, x) = \chi(x),$$

then our equation becomes

$$\psi^{\left.\substack{2,2\\1,2}\right\}^{n}}(x, y) = \chi^{n}(x) = F(x) \qquad [y = \alpha x]$$

determine χ from the equation $\chi^{n}x = F(x)$ by Problem XIII, Part I, and let its solution be $F_1(x)$, then we have

$$\psi^{2,2}(x, y) = F_1(x) \qquad [y = \alpha x].$$

This equation may be solved by nearly the same method as that employed in the last Problem.

If the function occurs in different shapes or of various orders, this method is inapplicable, as in the following Problem.

Problem XXX

Given the equation

$$F\{\psi^{2,1\}^{2}}(x, y), \psi^{1,2\}^{3}}(x, y), x, y\} = 0 \qquad [y = x].$$

The difficulties in this case appear to be much encreased from this circumstance, that the second function of $\psi^{2,1}(x, x)$ relative to x is quite different from the second function of $\psi^{1,2}(x, x)$ relative to the same quantity. The method of solution which I shall explain is equally applicable to all of this species, and consists in reducing them to a class which has been already solved.

It may be observed, that whether we take the second function of $\psi^{2,1}(x, x)$ relative to x, or whether we take the simultaneous function of $\psi^{2,1}(x, y)$ considered as a simple function, and in the result put x for y, the two expressions will be the same; the first gives

$$\psi^{2,1}(\psi^{2,1}(x, x), \psi^{2,1}(x, x))$$

and the second is

$$\psi^{2,1}(\psi^{2,1}(x, y), \psi^{2,1}(x, y))$$

which when y becomes equal to x is identical with the former; but

$$\psi^{2,1}\{\psi^{2,1}(x, y), \psi^{2,1}(x, y)\} = \psi^{2,1}\{\psi(\psi(x, y), y), \psi(\psi(x, y), y)\}$$
$$= \psi^{\substack{2,1,\overline{1,1},1\\1,2,1.2,1}}(x, y)$$

the lower line of indices denoting the quantities relative to which the operations are performed. In a similar manner it may be shown, that

$$\psi^{1,2|3}(x, y) = \psi^{1,2,\overline{2,2},2}_{1,2,1,2,2}(x, y).$$

/

Substituting these values in the original equation, we have

$$F\{\psi^{2,1,\overline{1,1},1}_{1,2,1,2,1}(x, y), \psi^{1,2,\overline{2,2},2}_{1,2,1,2,2}(x, yx, y\} = 0 \qquad [y = x].$$

This is an equation similar to that of Problem XXVIII, and may be solved by the same means.

New methods of solving functional equations of the first order, and also differential functional equations

The new methods which I now propose to explain are only applicable to equations of the form

$$F\{x, \psi x, \psi\alpha x, \psi\alpha^2 x, \ldots \psi\alpha^n x\} = 0$$

where α must be such a function that $\alpha^{n+1}x = x$. By the method of Problem VII, Part I, all functional equations of the first order may be reduced to this form; and although in many cases this reduction is very difficult, or even in the present state of analysis out of our power, yet it is theoretically possible, and we shall therefore consider all equations as so reduced. There is this remarkable difference between the former methods and the present one:

Those which I have already given always led to the general solution, and perhaps, in some cases, to the complete one; these, on the contrary, which I shall now propose, always conduct us directly to a particular solution, which does not contain even an arbitrary constant. It has, however, several advantages; it is the most direct method with which we are yet acquainted; and if by any means we could introduce into these solutions an arbitrary constant, it would afford us general ones: this is a step which is wanting to connect it with the former methods. In the case of differential functional / equations, this step is supplied by the integrations which are necessary, and we thus arrive at their general solutions.

Problem XXXI

Given the equation

$$F\{x, \psi x, \psi\alpha x\} = 0$$

and also

$$\alpha^2 x = x.$$

Find ψx in terms of x and $\psi\alpha x$; let it be

$$\psi x = F_1\{x, \psi\alpha x\}$$

put αx for x; then it becomes

$$\psi\alpha x = F_1\{\alpha x, \psi\alpha^2 x\} = F_1\{\alpha x, \psi x\}$$

put this value of $\psi\alpha x$ in the former equation, and we have

$$\psi x = F_1\{x, F_1(\alpha x, \psi x)\}$$

from which equation ψx may be found in terms of x.

If $\alpha^3 x = x$ instead of $\alpha^2 x = x$, we should find

$$\psi x = F_1\{x, F_1\{\alpha x, F_1\{\alpha^2 x, \psi x\}\}\}.$$

Ex. 1. Take the equation $(\psi x)^p \cdot \psi(a - x) = x^n$ where $\alpha x = a - x$ then

$$\psi x = \left(\frac{x^n}{\psi(a - x)}\right)^{1/p}$$

and

$$\psi(a - x) = \left\{\frac{(a - x)^n}{\psi x}\right\}^{1/p}$$

this substituted in the former gives

$$\psi x = \left\{\frac{x^n}{[(a - x)^n/\psi x]^{1/p}}\right\}^{1/p}$$

from which we find

$$\psi x = \left\{\frac{x}{(a - x)^{1/p}}\right\}^{np/(p^2 - 1)}$$

which will be found on trial to satisfy the equation. /

Ex. 2. Given the equation

$$x^n\psi x - a\psi\frac{1}{x} = bx^p$$

here $\alpha x = 1/x$ and $\alpha^2 x = x$, and we have

$$\psi x = \frac{bx^p - a\psi(1/x)}{x^n}$$

and putting $1/x$ for x it becomes

$$\psi\frac{1}{x} = \frac{bx^{-p} - a\psi x}{x^{-n}}.$$

Substituting this in the former equation we find for the value of ψx.

$$\psi x = \frac{b}{1 - a^2}\{x^{p-n} + ax^{-p}\}.$$

Ex. 3. Given the equation

$$x^n\psi x - x^m\psi\left(\frac{a - x}{1 - cx}\right) = x^p$$

by employing the same method its solution will be

$$\psi(x) = \frac{x^m\left(\frac{a - x}{1 - cx}\right)^p + x^p\left(\frac{a - x}{1 - cx}\right)^n}{\left(x\frac{a - x}{1 - cx}\right)^n - \left(x\frac{a - x}{1 - cx}\right)^m}.$$

Problem XXXII

Given the equation

$$F\{x, \psi x, \psi\alpha x, \ldots \psi\alpha^n x\} = 0$$

and also[a]

$$\alpha^{n+1}x = x$$

putting successively $x, \alpha x, \alpha^2 x, \ldots \alpha^n x$ for x, we have the following equations:

$$F\{x, \psi x, \psi\alpha x, \ldots \psi\alpha^n x\} = 0 \qquad (1)$$

$$F\{\alpha x, \psi\alpha x, \psi\alpha^2 x, \ldots \psi\alpha^n x, \psi x\} = 0 \qquad (2)$$

$$\vdots$$

$$F\{\alpha^n x, \psi\alpha^n x, \psi x, \psi\alpha x, \ldots \psi\alpha^{n-1}x\} = 0 \qquad (n + 1)$$

From these $n + 1$ equations we may eliminate the n quantities $\psi\alpha x$, $\psi\alpha^2 x$, and $\psi\alpha^n x$, and there will remain an equation of the form

$$F\{x, \alpha x, \alpha^2 x, \ldots \alpha^n x, \psi x\} = 0$$

from which ψx may readily be found.

Ex. 1. Let

$$\psi x + \psi\alpha x = fx \quad \text{and} \quad \alpha^3 x = x$$

[a] Equation corrected.

then

$$\psi\alpha x + \psi\alpha^2 x = f\alpha x$$

also

$$\psi\alpha^2 x + \psi x = f\alpha^2 x$$

hence we find

$$\psi x = \tfrac{1}{2}(fx - f\alpha x + f\alpha^2 x).$$

Ex. 2. take the equation

$$\psi x + fx\psi\alpha x = f_1 x$$

where f and f_1 are perfectly arbitrary and $\alpha^2 x = x$; then making use of process above described we find

$$\psi x = \frac{f_1 x - fxf_1\alpha x}{1 - f\alpha xf\alpha x}$$

if $\alpha^3 x = x$ we should have

$$\psi x = \frac{f_1 x - fxf_1\alpha x + fxf\alpha xf_1\alpha^2 x}{1 - fxf\alpha xf\alpha^2 x}$$

and generally when $\alpha^n x = x$ we shall have

$$\psi x = \frac{f_1 x - fxf_1\alpha x + fxf\alpha xf_1\alpha^2 x - \ldots \pm fxf\alpha x\ldots f\alpha^{n-2}xf_1\alpha^{n-1}x}{1 - fxf\alpha xf\alpha^2 x\ldots f\alpha^{n-1}x}.$$

Ex. 3. Take the equation

$$\psi x\psi\alpha x + fx\psi x = f_1 x$$

if $\alpha^2 x = x$, ψx must be deducted from the equation

$$\psi x = \cfrac{f_1 x}{fx + \cfrac{f_1\alpha x}{f\alpha x + \psi x}}$$

and generally when $\alpha^n x = x$ the form of ψx is determined by the equation.

$$\psi x = \cfrac{f_1 x}{fx + \cfrac{f_1\alpha x}{f\alpha x + \cfrac{f_1\alpha^2 x}{f\alpha^2 x + \ldots + \cfrac{f_1\alpha^{n-1}x}{f\alpha^{n-1}x + \psi x.}}}}$$

It may be observed, that this method of discovering particular solutions by elimination, will not apply when the given equation contains only the different forms of the function without the variable quantity itself: thus it is not applicable to the equation

$$F\{\psi x, \psi\alpha x, \ldots \psi\alpha^n x\} = 0$$

the reason of this is obvious; for if we eliminate from this equation (by means of the n equations which arise by changing the order in which the functions are placed), all the functions but ψx, we shall have a result containing nothing but ψx and constant quantities, and therefore, ψx is equal to a constant quantity: it is true such a value of ψx will satisfy the equation, but it scarcely deserves the name of a solution.

Another exception is, when the equation

$$F\{x, \psi x, \psi\alpha x, \ldots \psi\alpha^n x\} = 0$$

is homogeneous relative to the different forms of the unknown function; for in this case when we attempt to eliminate them, they all disappear together, leaving an equation of condition; thus given

$$\psi x = (a - x)\psi\alpha x \quad \text{and} \quad \alpha^2 x = x$$

we have

$$\psi\alpha x = (a - \alpha x)\psi\alpha^2 x = (a - \alpha x)\psi x$$

and

$$\psi x = (a - x)(a - \alpha x)\psi x \quad \text{or} \quad 1 = (a - \alpha x)(a - \alpha^2 x)$$

which equation is not necessarily true. /

Another exception is, when the given equation can be made to assume the form

$$F\{\overline{\psi x}, \overline{\psi\alpha x}, \ldots \overline{\psi\alpha^n x}\} = fx.$$

In this case the equation cannot be fulfilled unless fx is a symmetrical function of $x, \alpha x, \ldots \alpha^n x$, because the first side is such a symmetrical function: this reason, however, should be received with caution, for if the operation denoted by ψ be an inverse one, it may admit of several values, and it seems *possible*, that in such a case the condition relative to the form of f need not be fulfilled. In my former paper I explained the means of finding solutions of the equation $\psi^n x = x$. I then contented myself with explaining the theory without mentioning particular cases; as these latter may be required in our present enquiry, I shall subjoin the following particular solutions of $\psi^2 x = x$

$$\psi x = a - x \qquad \psi x = \log(a - \varepsilon^x) \qquad \psi x = (a^n - x^n)^{1/n}$$

$$\psi x = \sqrt{1 - x^2} \qquad \psi x = \frac{x - 2}{x - 1} \qquad \psi x = x - \log(\varepsilon^x - 1)$$

$$\psi x = \frac{x}{(ax^n - 1)^{1/n}} \qquad \psi x = \frac{a}{x} \qquad \psi x = \frac{1 - x}{1 + x}$$

$$\psi x = \tan^{-1}(a - \tan x) \qquad \psi x = \frac{x}{x - 1} \qquad \psi x = \frac{a - bx}{b + cx}$$

$$\psi x = \tan^{-1}\left(\frac{\sin(a - x)}{\cos a \cos x}\right) \qquad \psi x = \frac{x}{\sqrt{x^2 - 1}}.$$

Particular cases of $\psi^3 x = x$

$$\psi x = \frac{a^2}{a - x} \qquad \psi x = \frac{\sqrt{ax^2 - a^2}}{x} \qquad \psi x = \frac{a + bx}{c - \frac{b^2 + bc + c^2}{a} x}$$

$$\psi x = \frac{a^2}{ac - c^2 x} \qquad \psi x = \frac{(ax^n - a^2)^{1/n}}{x} \qquad \psi x = \log(a\varepsilon^x - a^2) - x$$

$$\psi x = \frac{ax - a^2}{x} \qquad \psi x = \left(\frac{a^2}{a - x^n}\right)^{1/n} \qquad \psi x = \log(\varepsilon^x - \varepsilon^c) - x + c$$

$$\psi x = \frac{1 + x}{1 - 3x} \qquad \psi x = \frac{1}{1 - x} \qquad \psi x = -\log(1 - \varepsilon^x).$$

/

Particular cases of $\psi^4 x = x$

$$\psi x = \frac{1}{2} \cdot \frac{1}{1 - x} \qquad \psi x = \frac{1 + x}{1 - x} \qquad \psi x = \left(\frac{2}{2 - x^n}\right)^{1/n}$$

$$\psi x = \frac{2}{2 - x} \qquad \psi x = \frac{2a^2}{2ac - c^2 x} \qquad \psi x = \frac{(2x^n - 2)^{1/n}}{x}$$

$$\psi x = 2\left(\frac{x - 1}{x}\right) \qquad \psi x = \frac{a + bx}{c - \frac{b^2 + c^2}{2a} x} \qquad \psi x = \log 2 - x + \log(\varepsilon^x - 1).$$

Particular cases of $\psi^6 x = x$

$$\psi x = \frac{1}{3(1-x)} \qquad \psi x = \frac{3x-1}{3x} \qquad \psi x = \frac{a+bx}{c - \dfrac{b^2 - bc + c^2}{3a}x}$$

$$\psi x = \frac{3}{3-x} \qquad \psi x = \frac{3a^2}{3ac - c^2 x} \qquad \psi x = \frac{1}{x}\left(x^n - \frac{1}{3}\right)^{1/n}$$

$$\psi x = 3\left(\frac{x-1}{x}\right) \qquad \psi x = \frac{3+3x}{3-x} \qquad \psi x = \log 3 - x + \log(\varepsilon^x - 1).$$

Problem XXXIII

Given the equation

$$\psi\alpha x = \frac{d\psi x}{dx}$$

α being such a function that $\alpha^2 x = x$.

For x put αx, then

$$\psi\alpha^2 x = \psi x = \frac{d\psi\alpha x}{d\alpha x}$$

by differentiating this we have

$$\frac{d\psi x}{dx} = \frac{d}{dx}\left\{\frac{d\psi\alpha x}{d\alpha x}\right\} \qquad (a)$$

but the left side of this equation is by the Problem equal to $\psi\alpha x$; therefore

$$\psi\alpha x = \frac{d}{dx}\left\{\frac{d\psi\alpha x}{d\alpha x}\right\}$$

but we also have

$$\frac{d\psi\alpha x}{dx} = \frac{d\psi\alpha x}{d\alpha x}\cdot\frac{d\alpha x}{dx}$$

consequently

$$\frac{d\psi\alpha x}{d\alpha x} = \frac{d\psi\alpha x}{dx}\left(\frac{d\alpha x}{dx}\right)^{-1}$$

/

this being substituted in (a) gives

$$\psi\alpha x = \frac{d}{dx}\left\{\frac{d\psi\alpha x}{dx}\left(\frac{d\psi\alpha x}{dx}\right)^{-1}\right\}$$

put $\psi\alpha x = z$ then

$$z = \frac{d}{dx}\left\{\frac{dz}{dx}\left(\frac{d\alpha x}{dx}\right)^{-1}\right\}$$

which is a differential equation, from whose solution z or $\psi\alpha x$ may be found.

Ex. 1. Given the equation

$$\psi(a - x) = \frac{d\psi x}{dx}$$

in this case $d\alpha x/dx = -1$, and the differential equation is

$$zdx^2 + d^2z = 0$$

its integral is

$$z = \psi(a - x) = b\cos x + c\sin x.$$

The two constant quantities which have entered by integration must be determined so as to satisfy the original equation. This condition gives

$$c = \frac{-b\cos a}{1 - \sin a}$$

the quantity b still remaining arbitrary; the solution of the equation $\psi(a - x) = d\psi x/dx$ is therefore

$$\psi x = b\cos(a - x) - \frac{b\cos a}{1 - \sin a}\sin(a - x).$$

Ex. 2. Take the equation

$$\psi\frac{1}{x} = \frac{d\psi x}{dx}.$$

In this case $\alpha x = 1/x$ and $d\alpha x/dx = -1/x^2$ and the differential equation becomes

$$zdx^2 + 2xdxdz + x^2d^2z = 0$$

whose solution is

$$z = \psi\frac{1}{x} = -\frac{b}{2p + 1}x^{-p-1} + b_1x^p, \qquad p = \frac{-1 \pm \sqrt{-3}}{2}$$

and

$$\psi x = -\frac{b}{2p+1}x^{p+1} + b_1 x^{-p}$$

/

in order to determine the constants b and b_1, substitute this expression in the given equation and it will be found that

$$-\frac{b}{2p+1} = -b_1 p$$

therefore

$$\psi x = -b_1 p x^{p+1} + b_1 x^{-p}$$

in which there still remains one arbitrary constant.

It is observable that both these solutions contain one constant. Let us suppose this to be changed into an arbitrary function of x, and let us determine what conditions it must be subject to, that it may satisfy the Problem: taking the second example we have

$$\psi x = (x^{-p} - px^{p+1})\varphi x$$

and the equation becomes[a]

$$(x^p - px^{-p-1})\varphi\frac{1}{x} = (x^{-p} - px^{p+1})\frac{d\varphi x}{dx} + \frac{d(x^{-p} - px^{p+1})}{dx}\varphi x$$

from this equation φx must be determined (the method of doing which will appear in a subsequent Problem). If this solution contains an arbitrary constant, the same process may be again repeated. We may thus continue deducing one solution from another as long as we can solve the differential equations to which they give rise, but still these will only be particular solutions.

Problem XXXIV

Given the equation

$$\psi\alpha x = \frac{d^n\psi x}{dx^n}$$

and $\alpha^p x = x$, put for x successively $\alpha x, \alpha^2 x, \ldots \alpha^{p-1}x$ then we have

$$\psi\alpha x = \frac{d^n\psi x}{dx^n}$$

/

[a] Equation corrected.

$$\psi\alpha^2 x = \frac{d^n\psi\alpha x}{(d\alpha x)^n}$$

$$\vdots$$

$$\psi\alpha^{p-1}x = \frac{d^n\psi\alpha^{p-2}x}{(d\alpha^{p-2}x)^n}$$

$$\psi\alpha^p x = \frac{d^n\psi\alpha^{p-1}x}{(dx^{p-1}x)^n}$$

but $\psi\alpha^p x = \psi x$ and combining all these equations we have

$$\psi\alpha^p x = \psi x = \frac{d^n}{(d\alpha^{p-1}x)^n}\cdot\frac{d^n}{(d\alpha^{p-2}x)^n}\cdots\frac{d^n\psi x}{(dx)^n}$$

which is a differential equation of the pnth order and putting $\psi x = z$ we have

$$z = \frac{d^{np}z}{(d\alpha^{p-1}x\cdot d\alpha^{p-2}x\cdot\ldots\cdot d\alpha x\cdot dx)^n}$$

this being integrated gives the value of z or ψx.

Problem XXXV

Given the equation

$$F\left\{x, \psi x, \psi\alpha x, \frac{d\psi x}{dx}\right\} = 0 \quad \text{also} \quad \alpha^p x = x.$$

Find the value of $\psi\alpha x$ from this equation and substitute in it $\alpha x, \alpha^2 x, \ldots$ $\alpha^{p-1}x$ for x, then we have

$$\psi\alpha x = F_1\left\{x, \psi x, \frac{d\psi x}{dx}\right\}$$

$$\psi\alpha^2 x = F_1\left\{\alpha x, \psi\alpha x, \frac{d\psi\alpha x}{d\alpha x}\right\}$$

$$\vdots \qquad \vdots$$

$$\psi\alpha^p x = F_1\left\{\alpha^{p-1}x, \psi\alpha^{p-1}x, \frac{d\psi\alpha^{p-1}x}{d\alpha^{p-1}x}\right\}$$

In each of these equations for

$$\frac{d\psi\alpha x}{d\alpha x}, \quad \frac{d\psi\alpha^2 x}{d\alpha^2 x}, \quad \ldots \quad \frac{d\psi\alpha^{p-1}x}{d\alpha^{p-1}x}$$

put their values

$$\frac{d\psi\alpha x}{dx}\left(\frac{d\alpha x}{dx}\right)^{-1}, \quad \frac{d\psi\alpha^2 x}{dx}\left(\frac{d\alpha^2 x}{dx}\right)^{-1}, \quad \ldots \quad \frac{d\psi\alpha^{p-1}x}{dx}\left(\frac{d\alpha^{p-1}x}{dx}\right)^{-1}$$

and also differentiate the results. Then we shall have the two following sets of equations

$$\psi\alpha x = F_1\left\{x, \psi x, \frac{d\psi x}{dx}\right\} \tag{1}$$

$$\psi\alpha^2 x = F_1\left\{\alpha x, \psi\alpha x, \frac{d\psi\alpha x}{dx}\left(\frac{d\alpha x}{dx}\right)^{-1}\right\} \tag{2}$$

$$\vdots \qquad \vdots$$

$$\psi\alpha^{p-1}x = F_1\left\{\alpha^{p-2}x, \psi\alpha^{p-2}x, \frac{d\psi\alpha^{p-2}x}{dx}\left(\frac{d\alpha^{p-2}x}{dx}\right)^{-1}\right\} \tag{$p-1$}$$

$$\psi\alpha^p = F_1\left\{\alpha^{p-1}x, \psi\alpha^{p-1}x, \frac{d\psi\alpha^{p-1}x}{dx}\left(\frac{d\alpha^{p-1}a}{dx}\right)^{-1}\right\} \tag{p}$$

and also

$$\frac{d\psi\alpha x}{dx} = \frac{d}{dx}F_1\left\{x, \psi x, \frac{d\psi x}{dx}\right\} \tag{$\alpha, 1$}$$

$$\frac{d\psi\alpha^2 x}{dx} = \frac{d}{dx}F_1\left\{\alpha x, \psi\alpha x, \frac{d\psi\alpha x}{dx}\left(\frac{d\alpha x}{dx}\right)^{-1}\right\} \tag{$\alpha, 2$}$$

$$\vdots \qquad \vdots$$

$$\frac{d\psi\alpha^{p-1}x}{dx} = \frac{d}{dx}F_1\left\{\alpha^{p-2}x, \psi\alpha^{p-2}x, \frac{d\psi\alpha^{p-2}x}{dx}\left(\frac{d\alpha^{p-2}x}{dx}\right)^{-1}\right\} \tag{$\alpha, p-1$}$$

$$\frac{d\psi\alpha^p x}{dx} = \frac{d}{dx}F_1\left\{\alpha^{p-1}x, \psi\alpha^{p-1}x, \frac{d\psi\alpha^{p-1}x}{dx}\left(\frac{d\alpha^{p-1}x}{dx}\right)^{-1}\right\}. \tag{α, p}$$

Since $\alpha^p x = x$ equation (p) becomes

$$\psi x = F_1\left\{\alpha^{p-1}x, \psi\alpha^{p-1}x, \frac{d\psi\alpha^{p-1}x}{dx}\left(\frac{d\alpha^{p-1}x}{dx}\right)^{-1}\right\}$$

from this by means of equations $(p-1)$ and $(\alpha, p-1)$ we may eliminate $\psi\alpha^{p-1}x$, and $d\psi\alpha^{p-1}x/dx$ the resulting equation will contain only $x, \psi x, \psi\alpha x, \ldots$ $\psi\alpha^{p-2}x$ and their differentials. From this by means of $(p-2)$ and $(\alpha, p-2)$ we may eliminate $\psi\alpha^{p-2}x$, and its differential, leaving an equation containing only $x, \psi x, \psi\alpha x, \ldots \psi\alpha^{p-3}x$ and their differentials. In the same manner $\psi\alpha^{p-3}x$ may be eliminated, and the process may be continued / until the last equation will only contain x, ψx and their differentials; this equation must be integrated, and it will determine the value of ψx in terms of x.

The same method may be employed for the solution of the much more general equation

$$F\left\{x, \psi x, \psi\alpha x, \ldots \psi\alpha^{p-1}x, \frac{d^n\psi\alpha x}{dx^n}, \frac{d^m\psi\alpha^2 x}{dx^m}, \ldots\right\} = 0$$

provided also, that $\alpha^p x = x$.

By substituting successively for x the quantities $\alpha x, \alpha^2 x, \ldots \alpha^{p-1}x$, we shall have p equations containing the functions ψx, $\psi\alpha x$, and $\psi\alpha^{p-1}x$ and their differentials.

Let each of these be differentiated as often as may be required, and we shall have two sets of equations by means of which all the quantities except x and ψx, and their differentials may be eliminated, the result is a common differential equation whose integral will afford the value of ψx in terms of x. If after satisfying the conditions of the Problem, there remain any arbitrary constants, we may suppose them functions of x, and new equations will thence arise by which they may be determined.

It might occur (when there are several arbitrary quantities) that, by assigning particular values to some of them, the others might remain in a certain degree arbitrary, should this be the case, we should obtain general solutions.

Problem XXXVI

Given the equation

$$\psi^2 x = \frac{d\psi x}{dx}.$$

Assume $\psi x = \varphi^{-1}f\varphi x$, then the equation becomes

$$\varphi^{-1}f^2\varphi x = \frac{d\varphi^{-1}f\varphi x}{dx}$$

/

putting φ^{-1} for x we have

$$\varphi^{-1}f^2 x = \frac{d\varphi^{-1}fx}{d\varphi^{-1}x} = \frac{d\varphi^{-1}fx}{dx}\left(\frac{d\varphi^{-1}x}{dx}\right)^{-1}$$

or

$$\varphi^{-1}f^2 x \cdot \frac{d\varphi^{-1}x}{dx} = \frac{d\varphi^{-1}fx}{dx}.$$

Some particular solution of the original equation must now be assumed as the value of f, and the resulting differential functional equation must be solved. The only particular case of the equation $\psi^2 x = d\psi x/dx$ with which I am at present acquainted, is

$$\psi x = \left(\frac{1 \pm \sqrt{-3}}{2}\right)^{(1 \mp \sqrt{-3})/2} \cdot x^{(1 \pm \sqrt{-3})/2}.$$

Other more complicated equations containing the various orders of functions, and their differentials may be reduced to those of the first order by the same means, but great difficulties still remain; it is by no means easy to discover particular solutions of the original equations, and even when these are found, the functional equations of the first order which remain to be solved, are of considerable difficulty. I shall therefore refrain from giving any more examples, and proceed to show how functional equations involving definite integrals may be reduced to those we have already treated. Such equations might occur in a variety of curious and interesting enquiries, few of which have yet been noticed. D'Alembert, in one of the volumes of his Opuscules,[a] has examined a question which may be referred to this class; it is the following. Suppose a sphere composed of particles of matter, what must be the law of attraction amongst these particles, so that the force of the whole sphere acting on a particle at a distance, may follow the same law? the question might be varied by supposing the law to be given, and the form of the solid to be required; / but the general solution of such questions is by no means easy.

Problem XXXVII[b]

Required the nature of the function ψ such that

$$\int dx \psi^2 x = \psi a$$

the integral being taken between the limits $x = o$ and $x = a$.

Assume $\varphi(x, v)$ such that

$$\varphi(x, v) - \varphi(o, v) = v$$

[a] J. L. R. D'Alembert, *Opuscules mathématiques; ou, mémoires sur différents subjets de géométrie, de mécanique, d'optique, d'astronomie, etc.* (8 vol., Paris, 1761–80).

[b] Babbage used both 'o' and '0' as the symbol for zero, and in the *Works of Babbage* we have normally used the latter. In Problem XXXVII, however, Babbage attached special significance to the use of 'o', which we have therefore retained. Several equations have been corrected in this Problem.

the form of φ may be ascertained from this equation by means already described. Then if we make

$$\int dx\psi^2x = \varphi(x, \psi a) \tag{1}$$

it is evident that between the two limits $x = o$ and $x = a$, the integral will be reduced to ψa, and we have therefore a differential functional equation whose mode of solution has already been pointed out. Other more complicated equations may be solved in the same way; these I shall omit. I shall, however, make some observations on this method of solution, with a view to point out some questions of considerable importance.

In equation (1) the function indicated by φ is so assumed that we may have

$$\varphi(a, \psi a) - \varphi(o, \psi a) = \psi a$$

from which, perhaps, it might be imagined, that $\varphi(x, \psi a)$ must contain only x, ψa and constant quantities, but the condition would still be fulfilled if it contained $\psi^2 a$, $\psi^3 a$, or $\psi^n a$, which though not actually variable cannot strictly be regarded as constant. To fix our ideas, let us consider the example in this Problem; one value of $\varphi(x, \psi a)$ is evidently $\varphi(x, \psi a) = (x^n/a^n)\psi a$, we have therefore

$$\int dx\psi^2x = \frac{x^n}{a^n}\psi a$$

and by differentiating

$$\psi^2x = \frac{nx^{n-1}}{a^n}\psi a$$

from which ψx may be found.

This is a solution derived from a certain form attributed to φ, but we might also give to φ the form

$$\varphi(x, \psi a) = \frac{x^n}{a^n}\psi a + x^p(x - a)^q f(a, \psi a, \psi^2 a, \ldots \psi^n a)$$

and, in that case, the equation to be solved would be

$$\psi^2x = \frac{nx^{n-1}}{a^n}\psi a + \frac{d}{dx}\{x^p(x - a)^q\} f(a, \psi a, \psi^2 a, \ldots \psi^n a)$$

this contains only the second function of the unknown quantity and must be solved as a second functional equation, considering $a, \psi a, \ldots \psi^n a$ as constant quantities; let its solution be

$$\psi x = F\{x, a, \psi a, \dots \psi^n a\} \qquad (a)$$

then we must put $x = a$ and determine ψa from the equation

$$\psi a = F\{a, a, \psi a, \dots \psi^n a\}$$

the value of ψa thus deduced, will furnish the values of $\psi^2 a, \dots \psi^n a$, and these being substituted in a, will give the value of ψx; this solution is evidently of a different nature from the former, and forms another species.

Again, the following form of φ will also agree with the conditions

$$\varphi(x, \psi a) = \frac{x^n}{a^n}\psi a + x^p(x - a)^q f\{a, \psi a, \psi^2 a, \dots \psi^n a, x, \psi x, \psi^2 x, \dots \psi^k x\}$$

which being substituted in the Problem ψx must be found from a functional equation of the kth order; must then be put equal to a, and the new functional equation of the nth order relative to a must be solved; this is a third species of solution different from either of the former. Respecting these three species of solutions, a very important question / may be proposed. What degree of generality does each possess, and how many and what sort of arbitrary functions does each solution involve? To discuss this question, and to point out the nature of other solutions yet more general, which may be found for these and other similar Problems, would far exceed the limits of a mere outline of the calculus. I shall conclude my remarks on this Problem by stating the plan to be pursued in one particular case, which may serve as a model for all similar operations. Take as the form of $\varphi(x, \psi a)$

$$\varphi(x, \psi a) = x(x - a)\frac{\psi x \psi^2 x}{\psi^2 o \psi^2 a} - \frac{a - x}{a}\frac{\psi x}{\psi o}\psi a$$

then we have

$$\psi^2 x = \frac{d}{dx}\left\{x(x - a)\frac{\psi x \psi^2 x}{\psi^2 o \psi^2 a} - \frac{a - x}{a}\frac{\psi x}{\psi o}\psi a\right\}$$

this is a differential functional equation which must be solved on the hypothesis of a, ψa, $\psi^2 a$, ψo and $\psi^2 o$, being constant quantities. Let its solution be

$$\psi x = F\{x, a, \psi a, \psi^2 a, \psi o, \psi^2 o\} \qquad (1)$$

we must now put $x = a$ and treat the resulting equation as one of the second order, considering ψo and $\psi^2 o$ as constants. Let its solution be

$$\psi a = F_1\{a, \psi o, \psi^2 o\}. \qquad (2)$$

Now substitute o for a retaining o as a letter instead of making it actually zero, there will result a new functional equation of the second order, whose solution is

$$\psi o = F_2\{o\}$$

and lastly, substituting this value of ψo, and also that of $\psi^2 o$ which may be deduced from it in (2) we have the value of ψa, from this $\psi^2 a$ may be found, and these being substituted in (1) give the value of ψx. /

Problem XXXVIII

Given the equation

$$\psi(x, y) = \frac{d\psi(x, \alpha y)}{dx}$$

where $\alpha^2 y = y$.

For y put αy and the equation becomes

$$\psi(x, \alpha y) = \frac{d\psi(x, \alpha^2 y)}{dx} = \frac{d\psi(x, y)}{dx}$$

differentiate this relative to x, then we have

$$\frac{d\psi(x, \alpha y)}{dx} = \frac{d^2\psi(x, y)}{dx^2}$$

this substituted in the original equation, gives

$$\psi(x, y) = \frac{d^2\psi(x, y)}{dx^2}$$

which is a partial differential equation, whose solution is

$$\psi(x, y) = \varepsilon^x \varphi y + \varepsilon^{-x} \varphi_1 y$$

φy and $\varphi_1 y$ being two arbitrary functions of y, so constituted as to fulfil the original equation. These may thus be determined, since

$$\psi(x, y) = \varepsilon^x \varphi y + \varepsilon^{-x} \varphi_1 y$$

we have

$$\frac{d\psi(x, \alpha y)}{dx} = \varepsilon^x \varphi \alpha y - \varepsilon^{-x} \varphi_1 \alpha y$$

and, since these two quantities must be equal, we have the following equations

$$\varphi y = \varphi \alpha y \quad \text{and} \quad \varphi_1 y = -\varphi_1 \alpha y$$

the former of these is easily satisfied by putting for φy any symmetrical function of y and αy; and a particular solution of the latter is

$$\varphi_1 y = (-y + \alpha y)c$$

and since this solution contains an arbitrary constant, it may / be changed by Problem VIII, Part I, into any arbitrary function which does not vary when y becomes αy; its general solution is therefore

$$\varphi_1 y = (-y + \alpha y)\varphi(\overline{y}, \overline{\alpha y})$$

and consequently the general solution of the equation of this Problem is

$$\psi(x, y) = \varepsilon^x \varphi(\overline{y}, \overline{\alpha y}) + \varepsilon^{-x}(\alpha y - y)\varphi_1(\overline{y}, \overline{\alpha y}).$$

Ex. 1. Take the equation

$$\psi(x, y) = \frac{d\psi(x, a - y)}{dx}$$

in this case $\alpha y = a - y$, and the general solution is

$$\psi(x, y) = \varepsilon^x \varphi(\overline{y}, \overline{a - y}) + \varepsilon^{-x}(a - 2y)\varphi_1(\overline{y}, \overline{a - y}).$$

Ex. 2. Let the given equation be

$$\psi(x, y) = \frac{d\psi(x, (1/y))}{dx}$$

here $\alpha y = 1/y$ and the general solution is

$$\psi(x, y) = \varepsilon^x \varphi(\overline{y}, \overline{1/y}) + \varepsilon^{-x}\frac{1 - y^2}{y}\varphi_1(\overline{y}, \overline{1/y}).$$

Problem XXXIX

Given the equation

$$\psi(x, y) = \frac{d\psi(x, \alpha y)}{dx}$$

supposing $\alpha^p y = y$.

By substituting successively $\alpha y, \alpha^2 y, \ldots \alpha^{p-1} y$ for y, we have the following equations

$$\psi(x, y) = \frac{d\psi(x, \alpha y)}{dx}$$

$$\psi(x, \alpha y) = \frac{d\psi(x, \alpha^2 y)}{dx}$$

$$\vdots \qquad \vdots$$

$$\psi(x, \alpha^{p-1} y) = \frac{d\psi(x, \alpha^p y)}{dx}.$$

From the first of these we may eliminate $d\psi(x, \alpha y)/dx$ by means of the differential of the second, and from the result $d\psi(x, \alpha^2 y)/dx$ may be eliminated by means of the differential of the third. And by continuing this process, observing that $\psi(x, \alpha^p y) = \psi(x, y)$ we shall find

$$\psi(x, y) = \frac{d^p \psi(x, y)}{dx^p}$$

this partial differential equation must be solved, and the arbitrary functions which enter into its integral, must be made to satisfy the conditions of the Problem.

Ex. Let $p = 4$, then $\psi(x, y) = d\psi(x, \alpha y)/dx$ and the solution of the resulting partial differential equation will be

$$\psi(x, y) = \varepsilon^x \varphi_{11} y - \varepsilon^{-x} \varphi_{12} y + \sin x \cdot \varphi_{13} y + \cos x \cdot \varphi_{14} y$$

hence

$$\frac{d\psi(x, \alpha y)}{dx} = \varepsilon^x \varphi_{11} \alpha y + \varepsilon^{-x} \varphi_{12} \alpha y + \cos x \cdot \varphi_{13} \alpha y - \sin x \cdot \varphi_{14} \alpha y$$

the first condition to be satisfied is

$$\varphi_{11} y = \varphi_{11} \alpha y$$

which is readily fulfilled by making $\varphi_{11} y = \varphi_1(\overline{y}, \overline{\alpha y}, \overline{\alpha^2 y}, \overline{\alpha^3 y})$, the next condition is

$$\varphi_{12} = -\varphi_{12} \alpha y.$$

This must be solved by Problem VIII, Part I, and we shall have

$$\varphi_{12} y = (-y + \alpha y - \alpha^2 y + \alpha^3 y)\varphi_2(\overline{y}, \overline{\alpha y}, \overline{\alpha^2 y}, \overline{\alpha^3 y})$$

the third and fourth conditions are

$$\varphi_{13}y = -\varphi_{14}\alpha y \quad \text{and} \quad \varphi_{14}y = \varphi_{13}\alpha y.$$

In the second of these put αy for y, and it becomes $\varphi_{14}\alpha y = \varphi_{13}\alpha^2 y$, / this substituted in the former, gives

$$\varphi_{13}y = -\varphi_{13}\alpha^2 y$$

whose general solution being found by the method in the first part gives

$$\varphi_{13}y = (-y + \alpha^2 y)\varphi_3(\overline{y}, \overset{1}{\overline{\alpha y}}, \overline{\alpha^2 y}, \overset{1}{\overline{\alpha^3 y}})$$

and consequently

$$\varphi_{14}y = (-\alpha y + \alpha^3 y)\varphi_3(\overline{\alpha y}, \overset{1}{\overline{\alpha^2 y}}, \overline{\alpha^3 y}, \overset{1}{\overline{y}})$$

these values being respectively substituted, we have for the general solution of the Problem in this example,

$$\begin{aligned}\psi(x, y) = {} & \varepsilon^x\varphi(\overline{y}, \overline{\alpha y}, \overline{\alpha^2 y}, \overline{\alpha^3 y}) \\ & + \varepsilon^{-x}(-y + \alpha y - \alpha^2 y + \alpha^3 y)\varphi_2(\overline{y}, \overline{\alpha y}, \overline{\alpha^2 y}, \overline{\alpha^3 y}) \\ & + (-y + \alpha^2 y)\varphi_3(\overline{y}, \overset{1}{\overline{\alpha y}}, \overline{\alpha^2 y}, \overset{1}{\overline{\alpha^3 y}}) \sin x \\ & + (-\alpha y + \alpha^3 y)\varphi_3(\overline{\alpha y}, \overset{1}{\overline{\alpha^2 y}}, \overline{\alpha^3 y}, \overset{1}{\overline{y}}) \cos x.\end{aligned}$$

If the original equation had been

$$\psi(x, y) = \frac{d^n\psi(x, \alpha y)}{dx^n}$$

the partial differential equation to be solved would have been

$$\psi(x, y) = \frac{d^{np}\psi(x, y)}{dx^{np}}.$$

This form is rather remarkable, the equation can always be integrated when np is a whole number; let us suppose n to be a fraction and p a whole number, some multiple of the denominator of n.

Ex. Let $n = \frac{1}{2}$, $p = 2$, then $np = 1$, and $\alpha^2 y = y$, and the equation to be solved is

$$\psi(x, y) = \frac{d^{1/2}(x, \alpha y)}{dx^{1/2}}$$

whose solution is $\psi(x, y) = \varepsilon^x\varphi_1 y$, or by assigning a proper form to $\varphi_1 y$ it becomes

$$\psi(x, y) = \varepsilon^x\varphi(\overline{y}, \overline{\alpha y}).$$

Not only may the index of differentiation become fractional, / but the index of the order of a function may be a fraction* or even a variable quantity, and such equations as the following might occur

$$\frac{d^{1/2}\psi^{1/2}x}{\sqrt{dx}} = \frac{d^{1/3}\psi^{n}x}{dn^{1/3}}.$$

To notice the extreme difficulty of the enquiries to which such equations would lead, might seem superfluous, though it may not be deemed equally so to support my own opinion of their utility by the authority of one well acquainted with these subjects. Lacroix, in the third volume of his Traité du Calcul, Diff. et Int.[a] speaking of fractional indices of differentiation, observes, 'L'Analyse offre foule d'expressions de ce genre, qui tiennent presque toutes aux théories les plus importantes et les plus délicates, et les réflexions que j'ai exposées dans le No. 965, me portent à croire que leur considération peut contribuer beaucoup aux progrès de la science du calcul.'[b]

Problem XL

Given the equation

$$\frac{d\psi(x, \beta y)}{dx} = \frac{d\psi(\alpha x, y)}{dy}$$

also $\alpha^2 x = x$ and $\beta^2 y = y$.

Put αx for x, and βy for y, then the equation becomes

$$\frac{d\psi(\alpha x, y)}{d\alpha x} = \frac{d\psi(x, \beta y)}{d\beta y}$$

* The difficulties which occur in treating functions with negative indices are similar to those in which they are positive; it may however be observed, that from the notation we have established, the following consequences follow:

$$\psi^{0.1}(x, y) = x \quad \text{and} \quad \psi^{1.0}(x, y) = y$$

and generally

$$\psi^{o.n}(x, y) = x \quad \text{and} \quad \psi^{n.o}(x, y) = y$$

also $\psi^{\overline{0.0}}(x, x) = x$ and if $\psi^{1.1}(x, y) = v$, then we have

$$x = \psi^{-1.1}(v, y) \quad \text{and also} \quad y = \psi^{1.-1}(x, v)$$

[a] S. F. Lacroix, *Traité du calcul différentiel et du calcul intégral* (3 vol., Paris, 1797–1800).

[b] *Translation:* Analysis offers many expressions of this type, which contain all the most important and most difficult theories, and the thoughts that I have presented in No. 965 lead me to believe that their consideration can contribute much to the progress of the science of calculus.

hence

$$\frac{d\psi(\alpha x, y)}{dx}\left(\frac{d\alpha x}{dx}\right)^{-1} = \frac{d\psi(x, \beta y)}{dy}\left(\frac{d\beta y}{dy}\right)^{-1}$$

differentiate this equation relative to y, and the original one relative to x: then the two results are

$$\frac{d^2\psi(\alpha x, y)}{dxdy} = \frac{d\alpha x}{dx}\cdot\frac{d}{dy}\left\{\frac{d\psi(x, \beta y)}{dy}\left(\frac{d\beta y}{dy}\right)^{-1}\right\}$$

/

and

$$\frac{d^2\psi(\alpha x, y)}{dxdy} = \frac{d^2\psi(x, \beta y)}{dx^2}$$

hence

$$\frac{d^2\psi(x, \beta y)}{dx^2} = \frac{d\alpha x}{dx}\cdot\frac{d}{dy}\left\{\frac{d\psi(x, \beta y)}{dy}\left(\frac{d\beta y}{dy}\right)^{-1}\right\}$$

put βy for y, observing that

$$\left(\frac{dy}{d\beta y}\right)^{-1} = \left\{\frac{dy}{dy}\left(\frac{d\beta y}{dy}\right)^{-1}\right\}^{-1} = \frac{d\beta y}{dy}$$

and also

$$\frac{d\psi(x, y)}{d\beta y} = \frac{d\psi(x, y)}{dy}\left(\frac{d\beta y}{dy}\right)^{-1},$$

then there will result the equation

$$\frac{d\beta y}{dy}\frac{d^2\psi(x, y)}{dx^2} = \frac{d\alpha x}{dx}\frac{d^2\psi(x, y)}{dy^2}.$$

This is a partial differential equation from whose solution $\psi(x, y)$ may be found.

Ex. 1. Given the equation

$$\frac{d\psi(a - x, y)}{dy} = \frac{d\psi(x, b - y)}{dx}.$$

In this case

$$\alpha x = a - x \quad \text{and} \quad \beta y = b - y,$$

and the differential equation to be solved is

$$\frac{d^2\psi(x, y)}{dx^2} = \frac{d^2\psi(x, y)}{dy^2}$$

and its solution is

$$\psi(x, y) = \varphi(x + y) + \varphi_1(x - y).$$

The two arbitrary functions φ and φ_1 must be determined so as to fulfil the given equation, for which purpose we have

$$\frac{d\psi(a - x, y)}{dy} = \varphi'(a - x + y) - \varphi'_1(a - x - y)$$

and

$$\frac{d\psi(x, b - y)}{dx} = \varphi'(b + x - y) + \varphi'_1(-b + x + y)$$

φ' and φ'_1 being respectively the differential coefficients of φ and φ_1, since these two expressions must be equal, we have

$$\varphi'(a - (x - y)) = \varphi'(b + (x - y))$$

and

$$-\varphi'_1(a - (x + y)) = \varphi'_1(-b + (x + y))$$

whose solutions are

$$\varphi'(x - y) = \chi\{\overline{x - y}, \overline{a + b - x + y}\}$$

and

$$\varphi'_1(x + y) = (a - b - 2x - 2y)\chi_1\{\overline{x + y}, \overline{a - b - x - y}\}.$$

Hence the general solution of the equation

$$\frac{d\psi(a - x, y)}{dy} = \frac{d\psi(x, b - y)}{dx}$$

is

$$\psi(x, y) = \int(dx + dy)\chi\{\overline{x + y}, \overline{a + b - x - y}\}$$
$$+ \int(dx - dy)(a - b - 2x + 2y)\chi_1\{\overline{x - y}, \overline{a - b - x + y}\}.$$

Ex. 2. Given the equation

$$\frac{(d\psi x, 1/y)}{dx} = \frac{d\psi(1/x, y)}{dy}$$

the partial differential equation to be solved is in this case

$$\frac{d\psi(x, y)}{dy^2} = \frac{x^2}{y^2}\frac{d\psi(x, y)}{dx^2}$$

and its solution is

$$\psi(x, y) = \left(\frac{x}{y}\right) + \varphi_1(xy)$$

determining φ and φ_1 so as to fulfil the conditions of the equation, we have

$$\psi(x, y) = c(x + y) + \int d(xy) \cdot (1/xy)^{1/2} \cdot \chi\{\overline{xy}, \overline{1/xy}\}.$$

Problem XLI

Given the equation

$$F\left\{x, y, \psi(x, y), \psi(\alpha x, y), \ldots \frac{d^n\psi(x, \beta y)}{dx^n}, \frac{d^k\psi(\alpha^2 x, y)}{dy^k}, \ldots\right\} = 0$$

and let

$$\alpha^p x = x \quad \text{and} \quad \beta^q y = y$$

then there may be pq different forms of the function ψ contained in the general expression $\psi(\alpha^r x, \beta^s y)$, r varying from 0 to $p - 1$, and s varying from 0 to $q - 1$.

In the first place it may be observed, that if we substitute αx for x in such a quantity as

$$\frac{d^n\psi(\alpha^3 x, \beta y)}{dx^n}$$

we shall have

$$\frac{d^n\psi(\alpha^4 x, \beta y)}{(d\alpha x)^n}$$

which may always be reduced to the form

$$f(x)\frac{d^n\psi(\alpha^4 x, \beta y)}{dx^n}.$$

If now in the original equation we substitute successively $\alpha x, \alpha^2 x, \ldots \alpha^{p-1}x$ for x, and $\beta y, \beta^2 y, \ldots \beta^{q-1}y$ for y, we shall have pq equations containing pq forms of the unknown function and their differentials. By means of these pq equations and the differentials of them, we may eliminate all the different forms of the function ψ, except one: let the one which remains be $\psi(x, y)$, then we have an equation of partial differentials containing only $x, y, \psi(x, y)$ and their differentials: and from the solution of this equation $\psi(x, y)$ may be found; a certain number of arbitrary functions will be contained in this integral; these must all be determined so as to satisfy the original equation.

Amongst the numerous questions to which the calculus of functions is applicable, I shall select a problem proposed by Euler in one of the volumes of the *Acta Acad. Petrop.* as it will offer an example of a mode of treating of functional equations of a nature yet more general than those contained in this paper.

/

Problem XLII

Required the nature of a curve such that taking any ordinate

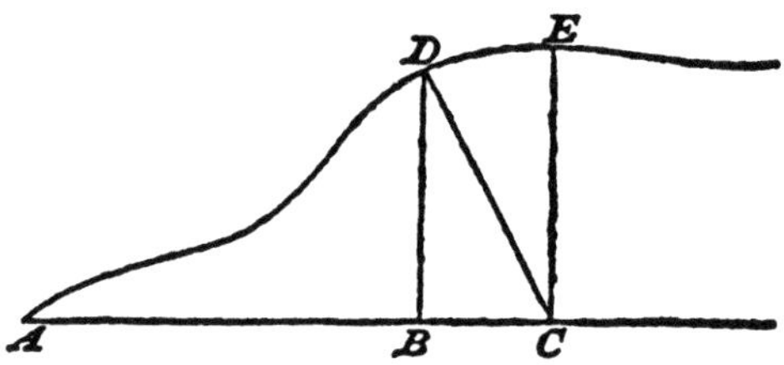

DB, and drawing a normal at the point D, the next ordinate CE raised at the foot of the normal shall be equal to that normal.

Let $AB = x$, $BD = y$ and $y = \psi x$ be the equation of the curve, then $BC = ydy/dx$, and $DC = \sqrt{y^2 + (ydy/dx)^2}$ and by the condition of the Problem we have

$$\sqrt{y^2 + \left(\frac{ydy}{dx}\right)^2} = \psi\left(x + \frac{ydy}{dx}\right)$$

hence

$$\{\psi x\}^2 + \left\{\frac{\psi x d\psi x}{dx}\right\}^2 = \left\{\psi\left(x + \frac{\psi x dx}{dx}\right)\right\}^2. \qquad (a)$$

This is apparently a very difficult functional equation, and I am not acquainted with any direct method of solving other similar ones. It is in fact only from a peculiar condition which this equation involves that any solutions have been obtained, the condition to which I allude is, that the quantity $\psi x d\psi x/dx$ does not change, when for x we substitute $x + (\psi x d\psi x/dx)$ or expressed in symbols, that /

$$\frac{\psi\left(x+\dfrac{\psi x d\psi x}{dx}\right)\cdot d\psi\left(x+\dfrac{\psi x d\psi x}{dx}\right)}{d\left(x+\dfrac{\psi x d\psi x}{dx}\right)}=\frac{\psi x d\psi x}{dx}$$

which may be thus proved: differentiate (a) which gives

$$\psi\left(x+\frac{\psi x d\psi x}{dx}\right)\cdot d\psi\left(x+\frac{\psi x d\psi x}{dx}\right)=\psi x d\psi x+\frac{\psi x d\psi x}{dx}\cdot d\left(\frac{\psi x d\psi x}{dx}\right)$$

and by dividing both sides of this equation by

$$d\left(x+\frac{\psi x d\psi x}{dx}\right)=dx+d\left(\frac{\psi x d\psi x}{dx}\right)$$

we have

$$\frac{\psi\left(x+\dfrac{\psi x d\psi x}{dx}\right)\cdot d\psi\left(x+\dfrac{\psi x d\psi x}{dx}\right)}{d\left(x+\dfrac{\psi x d\psi x}{dx}\right)}=\frac{\dfrac{\psi x d\psi x}{dx}\left(dx+d\left(\dfrac{\psi x d\psi x}{dx}\right)\right)}{dx+d\left(\dfrac{\psi x d\psi x}{dx}\right)}$$

$$=\frac{\psi x d\psi x}{dx}.$$

From this it appears, that the subnormal is constant in the same series of ordinates, but it does not follow that it must be constant in different series; this property, viz. that $\psi x d\psi x/dx$ does not change when x becomes $x+(\psi x d\psi x/dx)$ will furnish us with a solution of the equation in question; for (a) becomes by putting t for $\psi x d\psi x/dx$.

$$\{\psi(x+t)\}^2-\{\psi x\}^2=t^2$$

where t may be considered as a constant quantity, the general solution of this equation is

$$\psi x=\sqrt{xt+\varphi t}$$

φt being an arbitrary function of t, therefore the general solution of equation (a) is

$$\psi x=\sqrt{x\frac{\psi x d\psi x}{dx}+\varphi\left(\frac{\psi x d\psi x}{dx}\right)}$$

or

$$y^2=\frac{xydy}{dx}+\varphi\left(\frac{ydy}{dx}\right)$$

from which differential equation the curves which satisfy the Problem may be found. It ought, however, to be observed, / that the constant quantity introduced by integration, is not perfectly arbitrary, it must be determined so as to make the equation between x and y fulfil the equation (a). If for instance, we assume $\varphi(ydy/dx)$ to be equal to $a(ydy/dx)$, we should find the equation of the curve to be

$$y = (a + x)c$$

c being the constant introduced by integration, and on substituting this value of y in (a) we shall find $c = 0$, so that

$$y = (a + x)0$$

Let us suppose a to be infinite and equal to b/c, then we have

$$y = \left(\frac{b}{c} + x\right)c = b + cx = b, \quad \text{since } c = 0$$

which is the equation of a straight line parallel to the axis of the x's, which in fact agrees with the conditions of the Problem. If we suppose $\varphi(ydy/dx) = a^2 =$ a constant quantity, we should find

$$x = c\sqrt{y^2 - a^2}$$

this value being substituted in (a) gives for determining c the equation

$$c^2(c^2 + 1) = 0$$

whence $c = 0$ and $c = \pm\sqrt{-1}$, using this latter value we have

$$x = \sqrt{-1} \cdot \sqrt{y^2 - a^2} = \sqrt{a^2 - y^2}$$

which is the equation of the circle, and it is obvious, that this curve satisfies the conditions.

It is very necessary to attend to this mode of determining the constants, as we should otherwise meet in the solution with many curves which do not satisfy the conditions; thus in the last example, the curve is apparently an hyperbola, but owing to the constant becoming imaginary, it is in fact a circle. /

To complete the outline of this new method of calculation, it would be necessary to treat of equations involving two or more functional characteristics, and to explain methods of eliminating all but one of them; these lead to a variety of interesting and difficult enquiries, and will probably be of considerable use in completing the solutions of partial differential equations: it would also be proper to consider the maxima and minima of functions, and to apply to this subject the method of variations; these are points of considerable difficulty, and although I have made some little progress in each of them, I shall forbear for the present any further discussion on this subject. In the mean time, the sketch which I have offered, and the few applications I have given, are sufficient to point out the great importance of this method. It should however be observed, that its applications have only been noticed incidentally; my object has been to direct the attention of the analyst to a new branch of the science, and to point out the manner of treating it: the doctrine of functions is of so general a nature, that it is applicable to every part of mathematical enquiry, and seems eminently qualified to reduce into one regular and uniform system the diversified methods and scattered artifices of the modern analysis; from its comprehensive nature, it is fitted for the systematic arrangement of the science, and from the new and singular relations which it expresses, it is admirably adapted for further improvements and discoveries. /

[6]

V

DEMONSTRATIONS OF SOME OF DR MATTHEW STEWART'S GENERAL THEOREMS; TO WHICH IS ADDED, AN ACCOUNT OF SOME NEW PROPERTIES OF THE CIRCLE

From *Journal of Science and the Arts*, Vol. 1 (1816), pp. 6–24

Nearly seventy years have elapsed since Dr Matthew Stewart published under the title of General Theorems,[a] a series of geometrical propositions of singular difficulty and elegance. Many of them are capable of forming, with a slight alteration in the enunciation, the most beautiful porisms: and these, as it is well known, constitute a class of propositions invented by the ancients, and which formed a valuable part of their analysis; but the name alone remained to us until their re-discovery by Dr Simson,[b] which may rival, in point of difficulty, any of the investigations of the celebrated geometers of antiquity. The means by which Dr Stewart arrived at these theorems he never made public, nor does it appear that any traces of them have been found among his papers. Several geometers have applied themselves to this subject: Dr Small gave demonstrations of those which relate to the squares of given lines;[c] these are to be found in the second volume of the Transactions of the Royal Society of Edinburgh; and in the fifth volume of the same work are some very elegant properties of the circle by Mr Glenie,[d] from which he deduces demonstrations of many of those which relate to regular figures.

In both the papers to which I have alluded, the geometrical method of investigation is employed; and perhaps, on the whole, it may be thought best adapted to the nature of the subject. Other proofs, however, derived from other prin-

[a] M. Stewart, *Some general theorems of considerable use in the higher parts of mathematics* (Edinburgh, 1746).
[b] Robert Simson (1687–1768), Scottish mathematician.
[c] R. Small, 'Demonstrations of some of Dr M. Stewart's general theorems', *Transactions of the Royal Society of Edinburgh*, Vol. 2 (1790), pp. 112–34.
[d] J. Glenie, 'A geometrical investigation of some curious and interesting properties of the circle, etc.', *Transactions of the Royal Society of Edinburgh*, Vol. 6 (1812), pp. 21–70.

ciples, may not be uninteresting, not as adding any thing to the evidence on which these propositions rest, but as furnishing us with the means of comparing the extent and power of the two modes of reasoning made use of in their demonstration.

With this view I shall, in the following paper, deduce from the principles of common algebra and trigonometry, the proofs of a considerable number of these theorems, and I shall / at its conclusion, offer some new theorems which have occurred to me in similar investigations; thus endeavouring to prove that analysis is equally adapted for the demonstration of propositions which may be known, and for the discovery of those which are unknown, even in a branch of enquiry which has hitherto been treated by methods purely geometrical.

For the sake of convenience I shall premise two or three lemmas, but as they contain little that is new, I shall merely indicate the means by which they may be proved.

Lemma I

$$\cos\theta + \cos(\theta + b) + \cos(\theta + 2b) + \ldots + \cos(\theta + (p-1)b)$$
$$= \frac{\cos\left(\theta + \frac{b}{2}\right)\sin\frac{p}{2}b}{\sin\frac{b}{2}}$$

and

$$\sin\theta + \sin(\theta + b) + \sin(\theta + 2b) + \ldots + \sin(\theta + (p-1)b)$$
$$= \frac{\sin\left(\theta + \frac{b}{2}\right)\sin\frac{p}{2}b}{\sin\frac{b}{2}}$$

these formulae are well known and readily proved; let us now suppose a circle divided into p equal parts, and let $b = 2\pi/p$. If θ express any angle whatever, and if we constantly add to it the pth part of the circumference until we have gone round the circle, then will the sum of all the cosines of these arcs and also the sum of all the sines of the same arcs be equal to zero. This will be evident by substituting the value of b in the above expressions. We have therefore,

$$0 = \cos\theta + \cos\left(\theta + \frac{2\pi}{p}\right) + \ldots + \cos\left(\theta + (p-1)\frac{2\pi}{p}\right)$$
$$0 = \sin\theta + \sin\left(\theta + \frac{2\pi}{p}\right) + \ldots + \sin\left(\theta + (p-1)\frac{2\pi}{p}\right).$$

/

Lemma II

$$(\cos\theta)^q + \left(\cos\left(\theta + \frac{2\pi}{p}\right)\right)^q + \left(\cos\left(\theta + 2\frac{2\pi}{p}\right)\right)^q + \ldots + \left(\cos\left(\theta + (p-1)\frac{2\pi}{p}\right)\right)^q = 0$$

$$(\sin\theta)^q + \left(\sin\left(\theta + \frac{2\pi}{p}\right)\right)^q + \left(\sin\left(\theta + 2\frac{2\pi}{p}\right)\right)^q + \ldots + \left(\sin\left(\theta + (p-1)\frac{2\pi}{p}\right)\right)^q = 0$$

whenever q is an odd number.

Each term of the first series may be developed according to simple powers of the cosine of the arc and its multiples, and if those developments be arranged under each other, it will be found that the sum of each vertical column is, by the preceding lemma, equal to zero, consequently the whole series is equal to zero.

Again, each term of the second series may be developed according to the simple powers of the sine of the arc and its multiples, and these being arranged under each other, it will be found from the first lemma, that the sum of each vertical column is equal to zero.

Lemma III

$$(\cos\theta)^q + \left(\cos\left(\theta + \frac{2\pi}{p}\right)\right)^q + \left(\cos\left(\theta + 2\frac{2\pi}{p}\right)\right)^q + \ldots + \left(\cos\left(\theta + (p-1)\frac{2\pi}{p}\right)\right)^q = \frac{p}{2^q}\cdot\frac{q\cdot q-1\cdot\ldots\cdot q/2+1}{1\cdot 2\cdot\ldots\cdot q/2}$$

and

$$(\sin\theta)^q + \left(\sin\left(\theta + \frac{2\pi}{p}\right)\right)^q + \left(\sin\left(\theta + 2\frac{2\pi}{p}\right)\right)^q + \ldots + \left(\sin\left(\theta + (p-1)\frac{2\pi}{p}\right)\right)^q = \frac{p}{2^q}\cdot\frac{q\cdot q-1\cdot\ldots\cdot q/2+1}{1\cdot 2\cdot\ldots\cdot q/2}$$

when q is an even number. /

The proof for this lemma is similar to that for the last, except that in each development there will be found a constant term of the form

$$\frac{1}{2^q}\cdot\frac{q\cdot(q-1)\cdot\ldots\cdot(q/2)+1}{1\cdot 2\cdot\ldots\cdot q/2}$$

which contains neither sine nor cosine, and as there are p terms in the series, its sum will be equal to p times this term as it is given above.

By means of these two lemmas we may demonstrate many of Dr Stewart's theorems with great brevity. I shall begin with the 39th proposition, as it contains several of those which precede it; it is as follows:

'Let there be any regular figure circumscribed about a circle, and let the number of the sides be m, and let n be any number less than m; let r be the radius of the circle; and from any point in the circumference of the circle let there be drawn perpendiculars to the sides of the figure: the sum of the nth powers of the perpendiculars will be equal to

$$m \times \frac{1\cdot 3\cdot 5\cdot\ldots\cdot(2n-1)}{1\cdot 2\cdot 3\cdot\ldots\cdot n}r^n$$

Figure 1. Let AE and EF be two sides of any regular polygon circumscribed about a circle, and draw the perpendiculars CD, CF from the centre, let P be any point in the circumference of the circle, and from it draw perpendiculars to all the sides of the polygon, and also to the lines CD and CF, and let the angle PCD be called θ. Now since the figure is a regular polygon of m sides, the angle DCF is equal to $2\pi/m$ and the same angle is contained between any two perpendiculars drawn to two adjacent sides. We have for the perpendiculars

$$PA = KD = r - r\cos\theta,\quad PB = r - r\cos\left(\theta+\frac{2\pi}{m}\right),\ldots$$

or

$$PA = r(1-\cos\theta) = 2r\left(\sin\frac{\theta}{2}\right)^2,\quad PB = 2r\left(\sin\frac{1}{2}\left(\theta+\frac{2\pi}{m}\right)\right)^2,\ldots$$

consequently the sum of the nth powers of the perpendiculars is equal to /

$$2^nr^n\left\{\left(\sin\frac{\theta}{2}\right)^{2n}+\left(\sin\frac{1}{2}\left(\theta+\frac{2\pi}{m}\right)\right)^{2n}+\left(\sin\frac{1}{2}\left(\theta+2\frac{2\pi}{m}\right)\right)^{2n}\right.$$
$$\left.+\ldots+\left(\sin\frac{1}{2}\left(\theta+(m-1)\frac{2\pi}{m}\right)\right)^{2n}\right\}$$

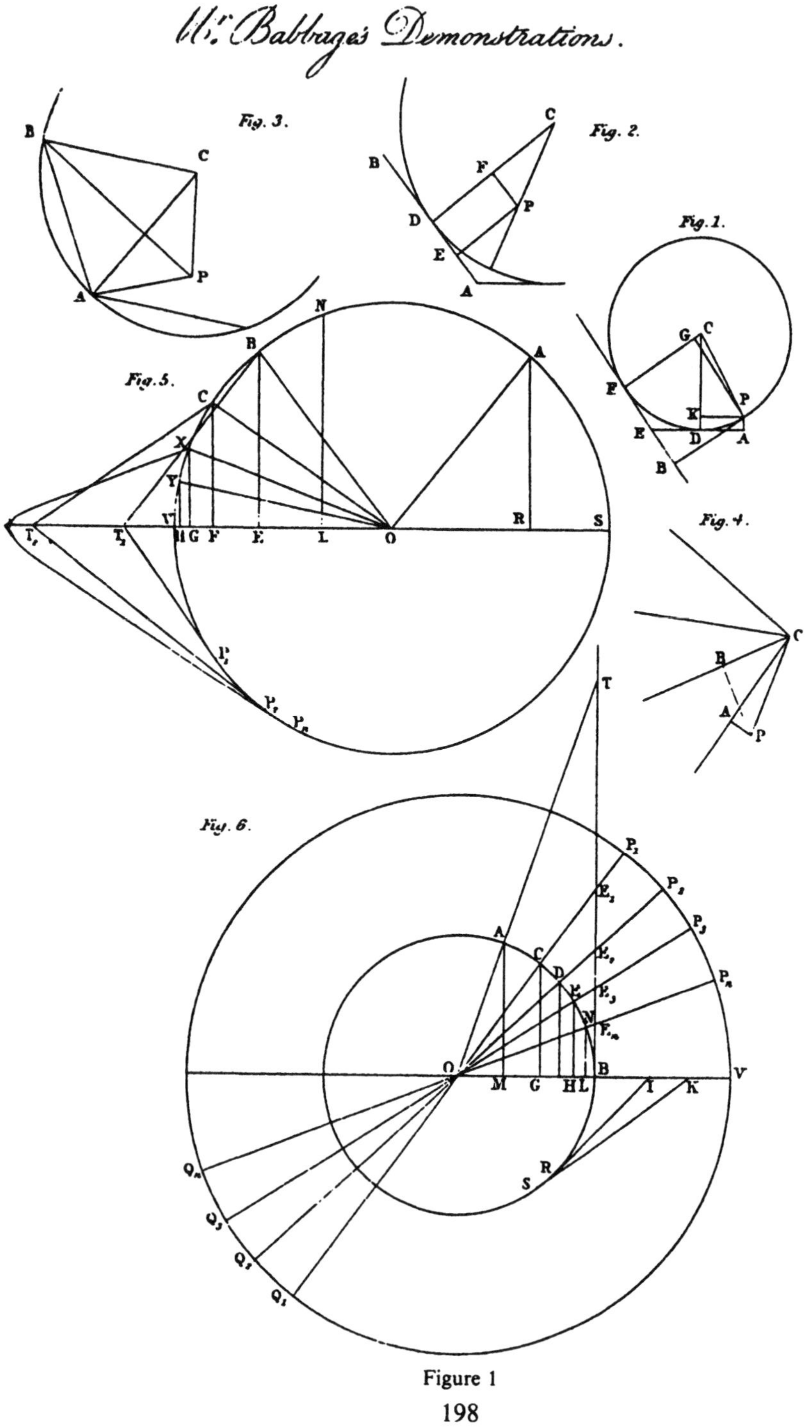

Figure 1

by combining together Lemmas I and III this expression may readily be proved to be equivalent to

$$2^n r^n\left\{\frac{m}{2^{2n}}\cdot\frac{2n\cdot 2n-1\cdot\ldots\cdot n+1}{1\cdot 2\cdot\ldots\cdot n}\right\}=\frac{mr^n}{2^n}\cdot\frac{2n\cdot 2n-1\cdot\ldots\cdot n+1}{1\cdot 2\cdot\ldots\cdot n}$$

this expression will become the same as that given by Dr Stewart if we can prove that

$$\frac{2n\cdot 2n-1\cdot\ldots\cdot n+1}{2^n}=1\cdot 3\cdot 5\cdot\ldots\cdot 2n-1$$

which may be thus effected,

$$\begin{aligned}1\cdot 2\cdot 3\cdot\ldots\cdot 2n&=1\cdot 3\cdot 5\cdot\ldots\cdot(2n-1)\times 2\cdot 4\cdot 6\cdot\ldots\cdot 2n\\&=1\cdot 3\cdot 5\cdot\ldots\cdot(2n-1)\times 1\cdot 2\cdot 3\cdot\ldots\cdot n\times 2^n\end{aligned}$$

and by dividing by $1\cdot 2\cdot 3\cdot\ldots\cdot n\times 2^n$ we have

$$\frac{n+1\cdot n+2\cdot\ldots\cdot 2n}{2^n}=1\cdot 3\cdot 5\cdot\ldots\cdot(2n-1)$$

consequently the sum of the nth powers of the perpendiculars drawn to the sides of the figure, is equal to

$$mr^n\frac{1\cdot 3\cdot 5\cdot\ldots\cdot 2n-1}{1\cdot 2\cdot 3\cdot\ldots\cdot n}.$$

If $n=3$ this becomes the 22nd proposition which is, that twice the sum of the cubes of the perpendiculars drawn from any point in the circumference of a circle to the sides of the circumscribed regular polygon, is equal five times the cube of the radius of the circle multiplied by the number of sides of the figure.

If $n=4$, it becomes proposition 28.

The next proposition I shall demonstrate is the 41st; it is as follows:

'Let there be any regular figure inscribed in a circle, and let the number of its sides be m; and let n be any number less than m, and let r be the radius of the circle, and from all the angles of the figure let there be drawn right lines to any point in the circumference; then the sum of the 2nth powers of these chords will be equal to

$$m\times\frac{1\cdot 3\cdot 5\cdot\ldots\cdot(2n-1)}{1\cdot 2\cdot 3\cdot\ldots\cdot n}\times 2^n r^{2n}.\text{'}$$

Let θ be the angular distance of any point in the circumference from the next angular point, then $\theta+2\pi/m$ will be its distance from the point beyond, and $\theta+2(2\pi/m)$ will be its distance from the next beyond and so on; and since the chord is equal to twice the sine of half the arc, the sum of the $2n$th powers of the chords will be equal to

$$(2r)^{2n}\left\{\left(\sin\frac{\theta}{2}\right)^{2n} + \left(\sin\frac{1}{2}\left(\theta + \frac{2\pi}{m}\right)\right)^{2n} + \left(\sin\frac{1}{2}\left(\theta + 2\frac{2\pi}{m}\right)\right)^{2n} + \ldots + \left(\sin\frac{1}{2}\left(\theta + (m-1)\frac{2\pi}{m}\right)\right)^{2n}\right\}$$

and his expression may by means of the first and third lemma be shown to be equal to

$$(2r)^{2n}\left\{\frac{m}{2^{2n}}\cdot\frac{2n\cdot 2n-1\cdot\ldots\cdot n+1}{1\cdot 2\cdot 3\cdot\ldots\cdot n}\right\}$$

which becomes, since $2n\cdot 2n-1\cdot\ldots\cdot n+1 = 1\cdot 3\cdot 5\cdot\ldots\cdot(2n-1)\times 2^n$

$$2^{2n}r^{2n}\left\{\frac{m}{2^{2n}}\cdot\frac{1\cdot 3\cdot 5\cdot\ldots\cdot 2n-1}{1\cdot 2\cdot 3\cdot\ldots\cdot n}2^n\right\} = m2^n r^{2n}\frac{1\cdot 3\cdot 5\cdot\ldots\cdot 2n-1}{1\cdot 2\cdot 3\cdot\ldots\cdot n}.$$

If $n = 2$ we have the following theorem, which is proposition 26, Q.E.D.

'Let there be any regular polygon inscribed in a circle, and from all the angles of the figure let lines be drawn to any point in the circumference. Then the sum of the fourth powers of the chords will be equal to six times the fourth power of the radius of the circle multiplied by the number of sides of the polygon.'

The 40th proposition of the General Theorems is as follows:

'Let there be any regular figure circumscribed about a circle, and let m be the number of sides, and let n be any number less than m, and let r be the radius of the circle; and from any point within the figure let there be drawn perpendiculars to the sides, and likewise a line to the centre of the circle: let a be the co-efficient of the third term of a binomial raised to the power n, b the co-efficient of the fifth term, c the co-efficient of the seventh, and so on: and then the sum of the nth powers of the perpendiculars will be equal, / to

$$m \times \left\{r^n + a\frac{1}{2}v^2r^{n-2} + b\frac{1\cdot 3}{2\cdot 4}v^2r^{n-4} + c\frac{1\cdot 3\cdot 5}{2\cdot 4\cdot 6}v^2r^{n-6} + \ldots\right\}$$

which may be thus proved.'

Figure 2. Let AB be one of the sides of the polygon, and P any point within it. Conceive perpendiculars drawn to each side of the polygon from the centre of the circle, and from the point P, and let the angular distance of the line $CP = v$, from the perpendicular CD be called θ, then it will appear from inspection, that the perpendicular from the point P, on the side AB, is equal to $r - v\cos\theta$, and since the angular distance of the line CP, from the perpendicular drawn from the centre to the side next to AB will be $\theta + 2\pi/m$, the perpendicular drawn from P to that side will be $r - v\cos(\theta + 2\pi/m)$ and similarly for the other perpendiculars.

If, therefore, we arrange the nth powers of these perpendiculars under each other, we have

$$r^n - \frac{n}{1}r^{n-1}v\cos\theta + \frac{n\cdot n-1}{1\cdot 2}r^{n-2}v^2(\cos\theta)^2 - \frac{n\cdot n-1\cdot n-2}{1\cdot 2\cdot 3}r^{n-3}v^3(\cos\theta)^3 + \dots$$

$$r^n - \frac{n}{1}r^{n-1}v\cos\left(\theta + \frac{2\pi}{m}\right) + \frac{n\cdot n-1}{1\cdot 2}r^{n-2}v^2\left(\cos\left(\theta + \frac{2\pi}{m}\right)\right)^2 - \frac{n\cdot n-1\cdot n-2}{1\cdot 2\cdot 3}r^{n-3}v^3\left(\cos\left(\theta + \frac{2\pi}{m}\right)\right)^3 + \dots$$

$$r^n - \frac{n}{1}r^{n-1}v\cos\left(\theta + (m-1)\frac{2\pi}{m}\right) + \frac{n\cdot n-1}{1\cdot 2}r^{n-2}v^2\left(\cos\left(\theta + (m-1)\frac{2\pi}{m}\right)\right)^2 - \frac{n\cdot n-1\cdot n-2}{1\cdot 2\cdot 3}r^{n-3}v^3\left(\cos\left(\theta + (m-1)\frac{2\pi}{m}\right)\right)^3 + \dots$$

All the even vertical columns vanish by Lemma II and if we take the sum of the odd ones by Lemma III we have[a]

$$m\left\{r^n + \frac{n\cdot n-1}{1\cdot 2}\cdot\frac{2\cdot 1}{1\cdot 2}v^2r^{n-2} + \frac{n\cdot n-1\cdot n-2\cdot n-3}{1\cdot 2\cdot 3\cdot 4}\cdot\frac{4\cdot 3}{1\cdot 2}\cdot\frac{1}{2}v^4r^{n-4} + \dots\right\}$$

/

and since

$$\frac{2p\cdot 2p-1\cdot\dots\cdot p-1}{1\cdot 2\cdot\dots\cdot p} = \frac{1\cdot 3\cdot 5\cdot\dots\cdot 2p-1}{2\cdot 4\cdot 6\cdot\dots\cdot 2p}$$

[a] The coefficients are incorrect in this formula, but this does not affect the result which uses arbitrary coefficients.

the expression becomes

$$m\left\{r^n + a\frac{1}{2}r^{n-2}v^2 + b\frac{1\cdot 3}{2\cdot 4}r^{n-4}v^4 + c\frac{1\cdot 3\cdot 5}{2\cdot 4\cdot 6}r^{n-6}v^6 + \ldots\right\}$$

which is the one given in the problem. Q.E.D.

When $n = 3$ this becomes proposition 23.

And when $n = 4$ it is proposition 29.

Proposition 42 is as follows:

'Let there be any regular figure inscribed in a circle, and let m be the number of sides, and n any number less than m; and from all the angles of the figure, and from the centre of the circle, let there be drawn lines to any point; and let r be the radius of the circle, and v the line drawn to the centre; then the sum of the $2n$th powers of the lines drawn to the angles of the figure will be equal to

$$m\left\{r^{2n} + \left(\frac{n}{1}\right)^2 v^2 r^{2n-2} + \left(\frac{n\cdot n-1}{1\cdot 2}\right)^2 v^4 r^{2n-4} + \left(\frac{n\cdot n-1\cdot n-2}{1\cdot 2\cdot 3}\right)^2 v^6 r^{2n-6} + \ldots\right\}$$

which may thus be demonstrated.

Figure 3. Let lines be drawn to the angles of the figure and to the point P from the centre of the circle, and let the line $CP = v$ make the angle θ with the first of the lines drawn from the centre to the angular points, then by inspecting the figure, it will readily appear that the square of the line drawn from P to the angle A will be equal to $r^2 + v^2 - 2vr\cos\theta$, and the angular distance of the line CP from the next angle being $\theta + 2\pi/m$ the square of the line drawn to that point will be $r^2 + v^2 - 2vr\cos(\theta + 2\pi/m)$ and so on: hence the sum of the $2n$th powers of these lines will be

$$(r^2 + v^2 - 2rv\cos\theta)^n + \left(r^2 + v^2 - 2rv\cos\left(\theta + \frac{2\pi}{m}\right)\right)^n + \ldots + \left(r^2 + v^2 - 2rv\cos\left(\theta + (m-1)\frac{2\pi}{m}\right)\right)^n$$

and if this expression be arranged in vertical columns, and if / the value of those series which contain cosines of arcs be substituted from Lemmas II and III the result will be

$$(r^2 + v^2)^n + \frac{n\cdot n-1}{1\cdot 2}(rv)^2(r^2+v^2)^{n-2}\cdot\frac{2}{1} + \frac{n\cdot n-1\cdot n-2\cdot n-3}{1\cdot 2\cdot 3\cdot 4}(rv)^4(r^2+v^2)^{n-4}\frac{4\cdot 3}{1\cdot 2} + \frac{n\cdot n-1\cdot\ldots\cdot n-5}{1\cdot 2\cdot\ldots\cdot 6}(rv)^6(r^2+v^2)^{n-6}\frac{6\cdot 5\cdot 4}{1\cdot 2\cdot 3} + \ldots.$$

The next step will be to arrange this series according to the powers of v^2 and r^2 for which purpose we will find the value of the coefficient of $v^{2q}r^{2n-2q}$ which is

$$\frac{n\cdot n-1\cdot\ldots\cdot n-q+1}{1\cdot 2\cdot\ldots\cdot q}+\frac{n\cdot n-1}{1^2}\cdot\frac{n-2\cdot n-3\cdot\ldots\cdot n-q}{1\cdot 2\cdot\ldots\cdot q-1}$$
$$+\frac{n\cdot n-1\cdot n-2\cdot n-3}{1^2\cdot 2^2}\cdot\frac{n-4\cdot\ldots\cdot n-q-1}{1\cdot 2\cdot\ldots\cdot q-2}+\ldots$$
$$=\frac{n\cdot n-1\cdot\ldots\cdot n-q+1}{1\cdot 2\cdot\ldots\cdot q}\left\{1+\frac{n-q}{1}\cdot\frac{q}{1}\right.$$
$$\left.+\frac{n-q\cdot n-q-1}{1\cdot 2}\cdot\frac{q\cdot q-1}{1\cdot 2}+\ldots\right\}$$

Now the series within the braces is equal to the coefficient of that term which does not contain x in the product of the two series

$$(1+x)^{nq}=1+\frac{n-q}{1}x+\frac{n-q\cdot n-q-1}{1\cdot 2}x^2+\ldots$$

$$\left(1+\frac{1}{x}\right)^q=1+\frac{q}{1}\frac{1}{x}+\frac{q\cdot q-1}{1\cdot 2}\frac{1}{x^2}+\ldots$$

which is equal to that term which does not contain x in the development of $(1+x)^n/x^q$ which is equal to

$$\frac{n\cdot n-1\cdot\ldots\cdot n-q+1}{1\cdot 2\cdot 3\cdot\ldots\cdot q}$$

consequently the coefficient of $v^{2q}r^{2n-2q}$ is

$$\left(\frac{n\cdot n-1\cdot\ldots\cdot n-q+1}{1\cdot 2\cdot\ldots\cdot q}\right)^2$$

and the series given as above when arranged according to the powers of v^2 and r^2 will assume the form

$$r^{2n}+\left(\frac{n}{1}\right)^2 v^2r^{2n-2}+\left(\frac{n\cdot n-1}{1\cdot 2}\right)^2 v^4r^{2n-4}$$
$$+\left(\frac{n\cdot n-1\cdot n-2}{1\cdot 2\cdot 3}\right)^2 v^6r^{2n-6}+\ldots$$

which is the same as that given by Dr Stewart. /

If $n=2$ this becomes proposition 27.

Proposition 45, is as follows:

'Let there be any number of right lines intersecting each other in a point, and making all the angles round the point of intersection equal; let m be the number of the lines, and let n be any number less than m, and from any point let there be drawn perpendiculars to the right lines, and also a line to the point of intersection; let v be the line drawn to the point of intersection, the sum of the $2n$th powers of the perpendiculars drawn to the right lines will be equal to

$$m \times \frac{1 \cdot 3 \cdot 5 \ldots 2n - 1}{1 \cdot 2 \cdot 3 \ldots n} \cdot \frac{v^{2n}}{2^n}$$

which may readily be proved thus.

Figure 4. C being the point where all the lines meet, and P the point from which the perpendiculars are drawn, let the angular distance of the line $CP = v$ from the line CA be called θ, and its angular distance from the next line will be $\theta + 2\pi/m$ and so on, consequently the perpendiculars from the point P, will be represented by $v \sin \theta$, $v \sin (\theta + 2\pi/m), \ldots$ and the sum of their $2n$th powers will be

$$v^{2n}\left\{(\sin \theta)^{2n} + \left(\sin\left(\theta + \frac{2\pi}{m}\right)\right)^{2n} + \ldots + \left(\sin\left(\theta + (m-1)\frac{2\pi}{m}\right)\right)^{2n}\right\}$$

which by Lemma III is equal to

$$v^{2n} \times \frac{m}{2^{2n}} \cdot \frac{2n \cdot 2n - 1 \cdot \ldots \cdot n + 1}{1 \cdot 2 \cdot 3 \cdot \ldots \cdot n} = v^{2n}\frac{m}{2^{2n}}\frac{1 \cdot 3 \cdot 5 \cdot \ldots \cdot 2n - 1 \cdot 2^n}{1 \cdot 2 \cdot 3 \cdot \ldots \cdot n}$$

$$= v^{2n}\frac{m}{2^n}\frac{1 \cdot 3 \cdot \ldots \cdot 2n - 1}{1 \cdot 2 \cdot 3 \cdot \ldots \cdot n}.$$

In the case of $n = 2$ this becomes proposition 34.

The general theorems may be divided into three classes: First, those which relate to regular figures, or to lines making equal angles with each other round a point; secondly, those which relate to lines drawn parallel to each other, or which all meet in a point; and, thirdly, those which relate to lines and figures irregularly drawn.

The first class, or those which relate to regular figures and lines making equal angles with each other, may, I believe, / all be demonstrated by means of the three lemmas at the commencement of this paper; and, by the same means, many other properties of regular polygons may be deduced. Of the manner of proving the second class, which relate to lines parallel to each other, or cutting each other in a point, I shall offer a few examples in the succeeding pages. The demonstrations of the third class, namely, of those which relate to irregular figures, appear to depend on other principles.

Proposition 54 is as follows:

'Let there be any number of right lines given in position, and parallel to each other; and let n be any given number; and from a point let there be drawn right lines in given angles to all the right lines given by position; and let the sum of the nth powers of the lines drawn to the given angles be invariable: and the point from which the lines are drawn will be in a line given by position.'

Let the equations of the given lines which are parallel be

$$y' = x' \tan\alpha + b, \quad y' = x' \tan\alpha + b_1, \quad y' = x' \tan\alpha + b_2, \ldots$$

and x and y being the co-ordinates of the point from which the lines are drawn, let $\theta, \theta_1, \theta_2, \ldots$ be the angles these lines form with the perpendiculars to the given lines. Now it is well known that the length of a perpendicular from the point whose co-ordinates are x and y, drawn to the line whose equation is $y' = x' \tan\alpha + b$ is equal to $y\cos\alpha - x\sin\alpha - b\cos\alpha$, and if we multiply this by $\sec\theta_1 = 1/\cos\theta_1$ we shall have the length of the line which is drawn from the point whose co-ordinates are x and y, forming a given angle with the first of the parallel lines, its nth power will therefore be

$$\left\{\frac{y\cos\alpha - x\sin\alpha}{\cos\theta_1} - b\frac{\cos\alpha}{\cos\theta_1}\right\}^n$$

and similarly the nth power of the line drawn from the same point to the next line will be

$$\left\{\frac{y\cos\alpha - x\sin\alpha}{\cos\theta_2} - b_1\frac{\cos\alpha}{\cos\theta_2}\right\}^n$$

and so on for the rest of the lines. /

Now the sum of all these quantities is to be constant; and if we put $y\cos\alpha - x\sin\alpha = v$, and if we develop these expressions, and arrange their sum according to the powers of v, we shall have a series of the form

$$Av^n - Bv^{n-1} + Cv^{n-2} - \ldots \pm Mv \mp N = P = \text{constant}$$

or

$$Av^n - Bv^{n-1} + Cv^{n-2} \ldots \mp N - P = 0.$$

Now this equation will evidently have several real roots, and if P is less than N all its roots will be real, we shall therefore have, supposing its roots to be r_1, r_2, and r_n,

$$v = y\cos\alpha - x\sin\alpha = r_1 \quad \text{or} \quad = r_2 \quad \text{or} \quad = r_n$$

hence

$$y = x\tan\alpha + (1/\cos\alpha)r_1 \quad \text{or} \quad y = x\tan\alpha + (1/\cos\alpha)r_n$$

that is to say, the locus of the point from which the lines are drawn is a straight line parallel to the given lines, and if the constant quantity P is less than N, there will always be n different lines, which will satisfy the conditions of the propositions.

Proposition 56. 'Let there be any number p of straight lines given by position, and also let there be any number q of right lines given by position: and let all the lines be either parallel to each other, or all intersecting each other in one point: and let n be any number, and from a point let there be drawn right lines in given angles to all the right lines given by position; and let the sum of the nth powers of the lines drawn in given angles to the first number of right lines given by position, be to the sum of the nth powers of the right lines drawn in given angles to the second number of lines given in position in a given ratio; the point from which the lines are drawn will be in a right line given by position.'

Case 1. Let all the lines be parallel to each other and their equations will be

$$y' = x' \tan\alpha + b_1, \quad y' = x' \tan\alpha + b_2, \quad \ldots \quad y' = x' \tan\alpha + b_p$$

$$y' = x' \tan\alpha + \beta_1, \quad y' = x' \tan\alpha + \beta_2, \quad \ldots \quad y' = x' \tan\alpha + \beta_q$$

and let x and y be the co-ordinates of the point from which / the lines are drawn in given angles, and let the given angles which these lines form with the perpendiculars to the first set of lines be $\theta_1, \theta_2, \ldots \theta_p$, and those which they form with the perpendiculars to the second set be $\eta_1, \eta_2, \ldots \eta_q$.

Now it will appear from the demonstration of the last proposition that the sum of the nth powers of the first set of lines drawn in the given angles will be

$$\left\{\frac{y\cos\alpha - x\sin\alpha}{\cos\theta_1} - b_1\frac{\cos\alpha}{\cos\theta_1}\right\}^n + \ldots + \left\{\frac{y\cos\alpha - x\sin\alpha}{\cos\theta_p} - b_p\frac{\cos\alpha}{\cos\theta_p}\right\}^n$$

and similarly the sum of the nth powers of the lines drawn in the given angles to the other set of q given lines will be

$$\left\{\frac{y\cos\alpha - x\sin\alpha}{\cos\eta_1} - \beta_1\frac{\cos\alpha}{\cos\eta_1}\right\}^n + \ldots + \left\{\frac{y\cos\alpha - x\sin\alpha}{\cos\eta_q} - \beta_q\frac{\cos\alpha}{\cos\eta_q}\right\}^n.$$

These two sums must, by the proposition, have a given ratio to each other, let the ratio be that of 1 to k, then if each term is expanded into a series, the development will assume the form (putting z for $y\cos\alpha - x\sin\alpha$).

$$Az^n + Bz^{n-1} + Cz^{n-2} + \ldots Mz + N = 0$$

and k may always be so assumed that this equation shall have one real root at least which will be of the form

$$z = y\cos\alpha - x\sin\alpha = r$$

consequently the locus of the point from which the lines are drawn is a line given by position, and parallel to the other lines. It is possible that n lines may exist, each of which may be the locus of the point, and these will all be parallel to the given lines.

Case 2. When all the lines meet in a point, the equations to the lines will in this case be /

$$y' = x' \tan \alpha_1 + b, \quad y' = x' \tan \alpha_2 + b, \quad \ldots \quad y' = x' \tan \alpha_p + b$$

$$y' = x' \tan \beta_1 + b, \quad y' = x' \tan \beta_2 + b, \quad \ldots \quad y' = x' \tan \beta_p + b$$

and it will readily be found that the sum of the nth powers of the lines drawn in the given angles to the first set will be

$$\left\{(y-b)\frac{\cos \alpha_1}{\cos \theta_1} - x\frac{\sin \alpha_1}{\cos \theta_1}\right\}^n + \ldots + \left\{(y-b)\frac{\cos \alpha_p}{\cos \theta_p} - x\frac{\sin \alpha_p}{\cos \theta_p}\right\}^n$$

and the sum of the nth powers of those drawn to the second set will be

$$\left\{(y-b)\frac{\cos \beta_1}{\cos \theta_1} - x\frac{\sin \beta_1}{\cos \theta_1}\right\}^n + \ldots + \left\{(y-b)\frac{\cos \beta_q}{\cos \theta_q} - x\frac{\sin \beta_q}{\cos \theta_q}\right\}^n$$

if the ratio of these two sums be that of one to k, and if each term be expanded, putting $y - b = z$, we shall have a series of the form

$$Az^n + Bxz^{n-1} + \ldots + Mx^{n-1}z + Nx^n = 0$$

and this by properly assigning the value of k may always have at least one real root of the form

$$z = y - b = rx$$

consequently there is always one right line which is the locus of the point from which the lines are drawn, and it appears from the equation that this line passes through the same point: it is possible that there may be n different lines, each of which may be the locus of the point, and in this case they will all meet in the common point of section of the given lines.

I shall now offer a few remarkable properties of the circle which are the geometrical translation of some trigonometrical formulae which occurred in other enquiries.

Theorem I[a]

In a circle (Figure 5) take any arc AV, bisect it in B, bisect BV in C, and so on, and let there be n bisections, and let XV be the last arc: draw sines to the points B, $C, \ldots X$, and make the angles $EBT_1, FCT_2, \ldots GXT_n$ respec-

[a] In this theorem Babbage has sometimes used T_2P_2 instead of T_nP_n, and vice versa, and sometimes neither is appropriate. No attempt has been made to correct these errors.

tively equal to / EBO, $FCO, \ldots GXO$, and let the lines BT_1, $CT_2, \ldots XT_n$ cut the radius OV produced in T_1, $T_2, \ldots T_n$; and from T_1, the first of the points, draw a tangent T_1P_1, to the circle, and from T_2, the last but one of these points, also draw a tangent T_2P_2, to the circle. Then is

$$T_1V \cdot T_2V \cdot T_3V \cdot \ldots \cdot T_nV = \left(\frac{T_1P_1}{T_2P_2}\right) OV^n$$

for the sake of brevity let the radius $OV = 1$.

Let the arc $AV = \theta$ then

$$BV = \frac{\theta}{2}, \quad CV = \frac{\theta}{2^2}, \quad \ldots \quad XV = \frac{\theta}{2^n}$$

and

$$OT_1 = 2\cos\frac{\theta}{2}, \quad \ldots \quad OT_n = 2\cos\frac{\theta}{2^n}$$

also

$$T_1V = 2\cos\frac{\theta}{2} - 1, \quad T_2V = 2\cos\frac{\theta}{2^2} - 1, \quad T_nV = 2\cos\frac{\theta}{2^n} - 1$$

and we shall find

$$T_1P_1^2 = \left(2\cos\frac{\theta}{2}\right)^2 - 1$$

and

$$T_nP_n^2 = \left(2\cos\frac{\theta}{2^{n-1}}\right)^2 - 1$$

consequently the theorem is as follows:

$$\left(2\cos\frac{\theta}{2} - 1\right)\left(2\cos\frac{\theta}{2^2} - 1\right)\ldots\left(2\cos\frac{\theta}{2^n} - 1\right)$$

$$= \frac{\left(2\cos\frac{\theta}{2}\right)^2 - 1}{\left(2\cos\frac{\theta}{2^{n-1}}\right)^2 - 1} = \frac{3 - 4\left(\sin\frac{\theta}{2}\right)^2}{3 - 4\left(\sin\frac{\theta}{2^{n-1}}\right)^2}$$

$$= \frac{3\sin\frac{\theta}{2} - 4\left(\sin\frac{\theta}{2}\right)^3}{\sin\frac{\theta}{2}} \cdot \frac{\sin\frac{\theta}{2^{n-1}}}{3\sin\frac{\theta}{2^{n-1}} - 4\left(\sin\frac{\theta}{2^{n-1}}\right)^3}$$

$$= \frac{\sin 3\frac{\theta}{2}}{\sin\frac{\theta}{2}} \cdot \frac{\sin\frac{\theta}{2^{n-1}}}{\sin 3\frac{\theta}{2^{n-1}}} \quad *$$

/

Theorem II[a]

Take any arc AV, Figure 5, and take BV equal to one third of it, take CV equal to one third of BV and so on, and let there be n trisections, and let XV be the last arc: also bisect the first arc AV and the last arc XV in the points B and Y: and let the sines NL and QH be drawn, and having drawn the lines BT_1, $CT_2, \ldots XT_n$, making with the perpendiculars BE, CF, and XG the same angles as do the lines $OB, OC, \ldots OX$, we have

$$T_1V \cdot T_2V \cdot T_3V \cdot \ldots \cdot T_nV = \frac{OL}{OH} OV^n$$

making $AV = \theta$ we have

$$BV = \frac{\theta}{3}, \quad \ldots \quad XV = \frac{\theta}{3^n},$$

$$OL = \cos\frac{\theta}{2}, \quad OH = \cos\frac{\frac{1}{2}\theta}{3^n}$$

and

$$T_1V = 2\cos\frac{\theta}{3} - 1, \quad \ldots \quad T_nV = 2\cos\frac{\theta}{3^n} - 1$$

hence the theorem becomes

$$\left(2\cos\frac{\theta}{3} - 1\right)\cdot\left(2\cos\frac{\theta}{3^2} - 1\right)\cdot\ldots\cdot\left(2\cos\frac{\theta}{3^n} - 1\right) = \frac{\cos\frac{\theta}{2}}{\cos\frac{\frac{1}{2}\theta}{3^n}}.$$

* In this and the following Theorems I shall reduce the geometrical enunciation to some trigonometrical formulae, and the demonstrations of these may be found in the Memoirs of the Analytical Society, pages 8 and 9. ⟨i.e. 'On continued products' (1813), this volume of the *Works of Babbage*, p. 68.⟩

[a] The lettering of some points in the figure and in the text of this theorem do not correspond. No attempt has been made to correct these errors.

Theorem III

In the same figure having trisected the arcs as above we have

$$ST_1 \cdot ST_2 \cdot ST_3 \cdot \ldots \cdot ST_n = \frac{NL}{YH} OV^n$$

by employing the same substitution this expression will be converted into

$$\left(2\cos\frac{\theta}{3}+1\right)\cdot\left(2\cos\frac{\theta}{3^2}+1\right)\cdot\ldots\cdot\left(2\cos\frac{\theta}{3^n}+1\right) = \frac{\sin\frac{1}{2}\theta}{\sin\frac{\frac{1}{2}\theta}{3^n}}.$$

Theorem IV

In the same figure having trisected the arcs, as above, draw from the points $T_1, T_2, \ldots T_n$ the tangents to the circle / $T_1P_1, \ldots T_nP_n$, then

$$T_1P_1 \cdot T_2P_2 \cdot T_3P_3 \cdot \ldots \cdot T_nP_n = OV^n \cdot \sqrt{\frac{AR}{XG}}$$

this may easily be deduced by multiplying the two expressions in Theorems II and III and extracting the square root.

Theorem V

With any radius OA describe a circle (Figure 6) and with the same centre and double that radius describe another circle; take any arc AB in the first circle and bisect it in C, bisect CB in D, and so on, and let there be n bisections, and let NB be the last arc; and through the extremity of each arc draw diameters to the circle, and also draw the sines AM, CG, ... NL, and let CG be the sine of the second arc in the series, and EH be the sine of the last but one: make $IG = GO$ and $KH = HO$, and from the points I and K draw the tangents IR, KS to the inner circle, and at the point B draw the tangent BT: then

$$P_1E_1 \cdot P_2E_2 \cdot P_3E_3 \cdot \ldots \cdot P_nE_n = \left(\frac{IR}{KS}\right)^2 \cdot \frac{AM}{NL}(2OB)^n.$$

Making $AB = \theta$, we have

$$CB = \frac{\theta}{2}, \quad \ldots \quad NB = \frac{\theta}{2^n}$$

$$P_1E_1 = 2\,CO - OE_1 = 2 - \sec\frac{\theta}{2}, \quad \ldots$$

$$P_nE_n = 2\,CO - OE_n = 2 - \sec\frac{\theta}{2^n}$$

$$IR^2 = \left(2\cos\frac{\theta}{2}\right)^2 - 1$$

$$KS^2 = \left(2\cos\frac{\theta}{2^{n-1}}\right)^2 - 1$$

$$AM = \sin\theta$$

and

$$NL = \sin\frac{\theta}{2}$$

consequently the theorem becomes[a]

$$\left(2 - \sec\frac{\theta}{2}\right)\cdot\left(2 - \sec\frac{\theta}{2^2}\right)\cdot\ldots\cdot\left(2 - \sec\frac{\theta}{2^n}\right)$$

$$= \frac{\left(2\cos\frac{\theta}{2}\right)^2 - 1}{\left(2\cos\frac{\theta}{2^{n-1}}\right)^2 - 1}\cdot\frac{\sin\theta}{\sin\frac{\theta}{2^n}}\;2^n$$

$$= \left(2 - \frac{1}{\cos\frac{\theta}{2}}\right)\cdot\left(2 - \frac{1}{\cos\frac{\theta}{2^2}}\right)\cdot\ldots\cdot\left(2 - \frac{1}{\cos\frac{\theta}{2^n}}\right)$$

$$= \frac{\left(2\cos\frac{\theta}{2} - 1\right)\cdot\left(2\cos\frac{\theta}{2} - 1\right)\cdot\ldots\cdot\left(2\cos\frac{\theta}{2^n} - 1\right)}{\cos\frac{\theta}{2}\cdot\cos\frac{\theta}{2^2}\cdot\ldots\cdot\cos\frac{\theta}{2^n}}$$

and multiplying by

$$\cos\frac{\theta}{2}\cos\frac{\theta}{2^2}\ldots\cos\frac{\theta}{2^n} = \frac{\sin\frac{\theta}{2^n}}{2^n\sin\theta}$$

[a] The second factor of the second line of this equation should be inverted; consequently, the statement of the theorem requires the corresponding factor to be inverted. The following equation has a similar error.

we have

$$\left(2\cos\frac{\theta}{2}-1\right)\cdot\left(2\cos\frac{\theta}{2^2}-1\right)\cdot\ldots\cdot\left(2\cos\frac{\theta}{2^n}-1\right)$$

$$=\frac{\left(2\cos\frac{\theta}{2}\right)^2-1}{\left(2\cos\frac{\theta}{2^{n-1}}\right)^2-1}$$

which is the same as the expression in Theorem I.

Theorem VI

Describe a circle (Figure 6) with any radius OA, and with the same centre and double that radius describe another circle; take any arc AB in the first circle, and let BC be one-third of it; let BD be one-third of BC and so on, and let BN be the last arc. At the point B draw the tangent BT, and through the extremity of each arc draw diameters to the circles cutting the tangent BT in the points $E_1, E_2, \ldots E_n$; then the continued product of $Q_1E_1 \cdot Q_2E_2 \cdot \ldots \cdot Q_nE_n$ is to the continued product of $P_1E_1 \cdot P_2E_2 \cdot \ldots \cdot P_nE_n$ in the ratio of the tangent of half the first arc AB, to the tangent of half the last arc BN. Making $AB = \theta$ we have

$$Q_1E_1 = 2+\sec\frac{\theta}{3}, \quad \ldots \quad Q_nE_n = 2+\sec\frac{\theta}{3^n}$$

$$P_1E_1 = 2-\sec\frac{\theta}{3}, \quad \ldots \quad P_nE_n = 2-\sec\frac{\theta}{3^n}$$

and the theorem becomes

$$\frac{\left(2+\sec\frac{\theta}{3}\right)\cdot\left(2+\sec\frac{\theta}{3^2}\right)\cdot\ldots\cdot\left(2+\sec\frac{\theta}{3^n}\right)}{\left(2-\sec\frac{\theta}{3}\right)\cdot\left(2-\sec\frac{\theta}{3^2}\right)\cdot\ldots\cdot\left(2-\sec\frac{\theta}{3^n}\right)}=\frac{\tan\frac{1}{2}\theta}{\tan\frac{1}{2}\frac{\theta}{3^n}}$$

/

$$\frac{\left(2\cos\frac{\theta}{3}+1\right)\cdot\left(2\cos\frac{\theta}{3^2}+1\right)\cdot\ldots\cdot\left(2\cos\frac{\theta}{3^n}+1\right)}{\left(2\cos\frac{\theta}{3}-1\right)\cdot\left(2\cos\frac{\theta}{3^2}-1\right)\cdot\ldots\cdot\left(2\cos\frac{\theta}{3^n}-1\right)}=\frac{\tan\frac{1}{2}\theta}{\tan\frac{1}{2}\frac{\theta}{3^n}}$$

which is nothing more than the expression in Theorem III divided by that in Theorem II. /

AN

ELEMENTARY TREATISE

ON THE

DIFFERENTIAL AND INTEGRAL

Calculus.

BY

S. F. LACROIX.

TRANSLATED FROM THE FRENCH.

WITH

AN APPENDIX AND NOTES.

Cambridge:

Printed by J. Smith, Printer to the University;

FOR J. DEIGHTON AND SONS, CAMBRIDGE;

AND LAW & WHITTAKER, NO. 13, AVE-MARIA-LANE, LONDON.

1816.

[iii]

VI

AN ELEMENTARY TREATISE ON THE DIFFERENTIAL AND INTEGRAL CALCULUS[a]

BY SILVESTRE FRANÇOIS LACROIX

With John Frederick William Herschel and George Peacock (translators)

ADVERTISEMENT

The. work of Lacroix, of which a translation is now presented to the public, forms one of a series of elementary treatises, by that distinguished author, on the different branches of the pure mathematics. It may be considered as an abridgement of his great work on the differential and integral calculus,[b] although in the demonstration of the first principles, he has substituted the method of limits of D'Alembert, in the place of the more correct and natural method of Lagrange, which was adopted in the former. The first part of this Treatise, which is devoted to the exposition of the principles of the differential calculus, was translated by Mr Babbage. The translation of the second part, which treats of the integral calculus, was executed by Mr G. Peacock, of Trinity College, and by Mr Herschel, of St John's College, in nearly equal proportions. The Appendix of Lacroix, on the calculus of differences and series, has been replaced by an original treatise, by Mr Herschel, in which many important subjects are included, which had been either entirely omitted, or very imperfectly considered in the other. /

[a] The circumstances surrounding the translation of Silvestre François Lacroix' *Traité élémentaire de calcul différentiel et de calcul intégral* (Paris, 1802) by Babbage, Peacock and Herschel are described in *Passages from the life of a philosopher*, Works of Babbage, Vol. II, pp. 20–1. In the *Works of Babbage* we have reprinted the *Advertisement*, which describes the division of the work of translation and the authorship of the additional notes. The 'sequel to this work' referred to in the *Advertisement* is *Examples to the differential and integral calculus* (1820), extracts of which are also reprinted in this volume of the *Works of Babbage*, pp. 283–326.
[b] S. F. Lacroix, *Traité du calcul différentiel et du calcul intégral* (3 vol., Paris, 1797–1800).

The first twelve of the Notes were written by Mr Peacock, and were principally designed to enable the student to make use of the principle of Lagrange, adopting those statements and examples of our author, which do not involve the theory of limits. The others were written by Mr Herschel.

It is intended to publish a sequel to this work, containing a collection of examples and results connected with the different subjects considered in it, omitting the entire operations by which they are deduced, and merely indicating such steps in the processes as cannot be expected to be discovered by an ordinary student. The work is already begun, and it is expected to be ready for publication in the course of a few months.

Cambridge,
12 December, 1816 /

[197/8]

VII

OBSERVATIONS ON THE ANALOGY WHICH SUBSISTS BETWEEN THE CALCULUS OF FUNCTIONS AND OTHER BRANCHES OF ANALYSIS

From *Philosophical Transactions of the Royal Society*, Vol. 107 (1817), pp. 197–216

Read 17 April, 1817

It is my intention in the following paper to offer to the Royal Society some remarks on the utility of analogical reasoning in mathematical subjects, and to illustrate them by some striking facts which have occurred to me, when comparing the calculus of functions with other modes of calculation with which mathematicians have been long acquainted. The employment of such an instrument may, perhaps, create surprise in those who have been accustomed to view this science as one which is founded on the most perfect demonstration, and it may be imagined that the vagueness and errors which *analogy*, when unskilfully employed, has sometimes introduced into other sciences, would be transferred to this.

It is, however, only as a guide to point out the road to discovery, that analogy should be used, and for this purpose it is admirably adapted.

It is usually more difficult to discover than to demonstrate any proposition; for the latter process we may have rules, but for the former we have none. The traces of those ideas which, in the mind of the discoverer of any new truth, connect the unknown with the known, are so faint, and his attention is so much more intensely directed to the object, than to / the means by which he attains it, that it not unfrequently happens, that while we admire the happiness of a discovery, we are totally at a loss to conceive the steps by which its author ascended to it.

From these considerations, I think it will appear, that any successful attempt to embody into language those fleeting laws by which the genius of the inventor is insensibly guided in the exercise of the most splendid privilege of intellect, would contribute more to the future progress of mathematical science than anything which has hitherto been accomplished. Amidst the total absence of all attempts of this kind, the following illustrations of one of the most obvious assistants of the inventive faculty, will not, I hope, be considered useless, even though it should have no other effect than that of directing the attention of those who are engaged in mathematical enquiries, to this most interesting and important subject.

At our first entrance into algebra, one of the most remarkable circumstances which presents itself is, that some fractions which in certain cases apparently vanish, have in fact a finite value. Such is the fraction $(a^x - b^x)/x$ which when $x = 0$ becomes $(1 - 1)/0 = 0/0$, and yet its real value is well known to be $\log(a/b)$.

Here then by assigning a certain value to a variable quantity which is capable of all degrees of *magnitude*, the expression apparently becomes illusory: let us now examine a parallel case in the Calculus of Functions. Take the expression

$$\frac{\psi x - \psi(1/x)}{\psi x}$$

and let us suppose ψ to be an arbitrary characteristic / capable of assuming all varieties of *form*; then amongst these varieties we may have $\psi x = 0$, and the expression becomes $(0 - 0)/0$.

In order to ascertain its real meaning, let us suppose

$$\psi x = v\varphi x + v^2\varphi_1 x + \ldots$$

which if we suppose $v = 0$ gives $\psi x = 0$
then

$$\psi(1/x) = v\varphi(1/x) + v^2\varphi_1(1/x) + \ldots$$

and the expression becomes

$$\frac{\psi x - \psi(1/x)}{\psi x} = \frac{(\varphi x - \varphi(1/x))v + (\varphi_1 x - \varphi_1(1/x))v^2 + \ldots}{v\varphi x + v^2\varphi_1 x + \ldots}$$

$$= \frac{(\varphi x - \varphi(1/x)) + (\varphi_1 x - \varphi_1(1/x))v + \ldots}{\varphi x + v\varphi_1 x + \ldots}$$

which becomes when $v = 0$ or $\psi x = 0$

$$\frac{\psi x - \psi(1/x)}{\psi x} = \frac{\varphi x - \varphi(1/x)}{\varphi x}$$

where φx is quite arbitrary.

If we suppose ψx to become a symmetrical function of x and $1/x$, we have $\psi x = \psi(1/x)$ and our expression becomes $0/(\psi x)$ which actually does vanish in all cases except at the same time $\psi x = 0$.

As a second example take the expression $(f_1x - fx \cdot f_1\alpha x)/(1 - fx \cdot f\alpha x)$ which becomes 0/0 if the two following equations hold true: $fx \cdot f\alpha x = 1$ and $f_1x - fx \cdot f_1\alpha x = 0$, α being any given function, such that $\alpha^2 x = x$. In order to ascertain its real value in that case, let us suppose

$$f_1x \text{ becomes } f_1x + v\varphi x$$

and

$$fx \text{ becomes } fx + v\varphi x$$

then

$$f_1\alpha x \text{ will be } f_1\alpha x + v\varphi\alpha x$$

and

$$f\alpha x \text{ will be } f\alpha x + v\varphi\alpha x$$

and these being substituted we have

$$\frac{(f_1x - fx \cdot f_1\alpha x) - (\varphi_1 x - fx \cdot \varphi_1\alpha x - f_1\alpha x \cdot \varphi x)v - \varphi x \cdot \varphi_1\alpha x \cdot v^2}{(1 - fx \cdot f\alpha x) - (fx \cdot \varphi\alpha x + f\alpha x \cdot \varphi x)v - \varphi x \cdot \varphi\alpha x \cdot v^2}.$$

Now in consequence of the two equations given above, the first term in both numerator and denominator vanishes, and dividing the remainder by v and then making $v = 0$, we have for the value of the expression when $f_1x - fx \cdot f_1\alpha x = 0$ and also $fx \cdot f\alpha x = 1$

$$\frac{f_1x - fx \cdot f_1\alpha x}{1 - fx \cdot f\alpha x} = \frac{\varphi_1 x - fx \cdot \varphi_1\alpha x - f_1\alpha x \cdot \varphi x}{fx \cdot \varphi\alpha x + f\alpha x \cdot \varphi x}$$

where φ and φ_1 are arbitrary.

If we take the expression

$$\frac{f_1x - fx \cdot f_1\alpha x + fx \cdot f\alpha x \cdot f_1\alpha^2 x}{1 + fx \cdot f\alpha x \cdot f\alpha^2 x}$$

and if fx and f_1x are so assumed that the numerator and denominator both vanish, by a similar mode of treatment to that already pointed out, we shall find its real value to be

$$\frac{\varphi_1 x - fx \cdot \varphi_1\alpha x - f_1\alpha x \cdot \varphi x - fx \cdot f_1\alpha x \cdot \varphi\alpha x - f\alpha x \cdot f_1\alpha x \cdot \varphi x - fx \cdot f\alpha x \cdot \varphi_1\alpha^2 x}{f\alpha x \cdot f\alpha^2 x \cdot \varphi x + fx \cdot f\alpha^2 x \cdot \varphi\alpha x + fx \cdot fax \cdot f\alpha^2 x}.$$

And similarly if we supposed both numerator and denominator of the fraction

$$\frac{f_1x - fx \cdot f_1\alpha x + \ldots \pm fx \cdot f\alpha x \ldots f\alpha^{n-2}x \cdot f_1\alpha^{n-1}x}{1 \pm fx \cdot f\alpha x \ldots f\alpha^{n-1}x}$$

to vanish, it would yet retain a value containing arbitrary functions.

It will be needless to multiply examples, as the mode of / treating them is sufficiently obvious from those already given. It appears then, that as in common algebra an expression may become illusory from the *variable quantity* assuming a particular *value*; so in the doctrine of functions an expression may become likewise illusory by the *variable function* assuming a particular *form*; in the one case the real value may be a *constant quantity*, in the other it may be an *arbitrary function*: nor ought this circumstance of the appearance of an arbitrary function to surprise us, when we consider that (as for instance in the second example) it is not one form only of the function fx which will satisfy the equation $fx \cdot f\alpha x = 1$, but any of the infinite variety of forms comprehended in the expression $fx = \{\chi(\overline{x}, \overline{\alpha x})\}^{x-\alpha x}$ and similarly for the values of f_1x. The circumstance of this species of vanishing fractions having an arbitrary function contained in their true value, is of considerable importance in the calculus of functions, as I shall now shortly endeavour to prove.

The Royal Society did me the honour to insert, in the last volume of their Transactions, a paper of mine,[a] in which I gave a new method of solving all functional equations of the first order, and of a certain class by means of elimination, and I there stated that all solutions so obtained were only particular cases of the general solutions, and that they did not contain any arbitrary function.

Now it may be observed, that there are certain forms which may be assigned to the coefficients which render those solutions apparently infinite, yet that on further consideration it appears, that the solution is in fact a vanishing fraction: in all such cases the process I have just pointed out will give the real solution which will contain an arbitrary function, so that it is in fact a general solution. Thus in the equation /

$$\psi x + fx \cdot \psi\alpha x = f_1x$$

whose solution is

$$\psi x = \frac{f_1x - fx \cdot f_1\alpha x}{1 - fx \cdot f\alpha x}$$

if $fx \cdot f\alpha x = 1$ it apparently becomes infinite, but by subtracting the equation

$$fx \cdot \psi\alpha x + fx \cdot f\alpha x \cdot \psi\alpha^2 x = fx \cdot f_1\alpha x$$

[a] 'An essay towards the calculus of functions, part II' (1816), in this volume of the *Works of Babbage*.

(which is deduced from the former by putting αx for x and multiplying by fx) we have since $\alpha^2 x = x$ and also $fx \cdot f\alpha x = 1$, $0 = f_1 x - fx \cdot f_1 \alpha x$, and the solution becomes a vanishing fraction whose value is

$$\psi x = \frac{\varphi_1 x - fx \cdot \varphi_1 \alpha x - f_1 \alpha x \cdot \varphi x}{fx \cdot \varphi \alpha x + f\alpha x \cdot \varphi x}.$$

Let $\alpha x = -x$ and $fx = f_1 x = 1$ then the general solution of the equation $\psi x + \psi(-x) = 1$ is

$$\psi x = \frac{\varphi_1 x - \varphi_1(-x) + \varphi x}{\varphi(-x) + \varphi x}.$$

In a similar manner the solutions of the equations $\psi x - \psi(-x) = x$, and $\psi x = fx \cdot \psi \alpha x$ if $fx \cdot f\alpha x = 1$ when $\alpha^2 x = x$ are

$$\psi x = \frac{x\varphi x - \varphi_1 x + \varphi_1(-x)}{\varphi x + \varphi(-x)} \quad \text{and} \quad \psi x = \frac{\varphi_1 x + fx \cdot \varphi_1 \alpha x}{fx \cdot \varphi \alpha x + f\alpha x \cdot \varphi x}.$$

Let $\psi x + \psi \alpha x = f_1 x$; if $\alpha^2 x = x$ this is only possible when $f_1 x = f_2(\overline{x}, \overline{\alpha x})$ then

$$\psi x = \frac{\varphi x \cdot f_2(\overline{x}, \overline{\alpha x})}{\varphi x + \varphi \alpha x}.$$

Let $\psi x - \psi \alpha x = f_1 x$ and $\alpha^2 x = x$; this is only possible when

$$f_1 x = (x - \alpha x) \cdot f_2(\overline{x}, \overline{\alpha x})$$

then

$$\psi x = \frac{(x - \alpha x) f_2(\overline{x}, \overline{\alpha x}) \cdot \varphi_1 x + \varphi x + \varphi_1 \alpha x}{\varphi x + \varphi \alpha x}.$$

The above are sufficient as examples, but the same reasoning has led me to the following curious proposition. *Whenever the method of elimination apparently fails, the real value of the vanishing fraction will give the general solution of the equation.* This principle puts us in possession of the general solutions of several classes of equations, for besides the cases / in which the solutions vanish from some particular values of the coefficients, all equations which are homogeneous relative to the different forms of the unknown function are comprehended in it, as are also *all equations which are symmetrical relative to the same quantities.*

There exists another class of equations nearly allied to those which are symmetrical relative to the different forms of the unknown functions, whose solution I shall now point out, chiefly with the view of giving another example of a method of reasoning which may frequently be employed with advantage in these inquiries, and also for the purpose of mentioning a remark respecting the elimination of variables in a certain class of algebraic equations which I do not recollect to have seen noticed. The class of functional equations to which I allude, are contained in the expression

$$F\{\psi x, \psi\alpha x, \dots \psi\alpha^{n-1}x, \overline{x}, \overline{\alpha x}, \dots \overline{\alpha^{n-1}x}\} = 0$$

which for the sake of convenience may be written thus,

$$F_1\{\psi x, \psi\alpha x, \dots \psi\alpha^{n-1}x, fx, f_1x, f_2x, \dots\} = 0$$

where $\alpha^n x = x$ and $fx, f_1x, f_2x, \dots$ are any symmetrical functions of $x, \alpha x, \dots \alpha^{n-1}x$.

In this equation none of the known functions $fx, f_1x, \dots$ are changed by the substitution of $\alpha x, \alpha^2 x, \dots \alpha^{n-1}x$ for x; and since the form of the unknown function depends on that of those which are known, it follows that the form of the unknown function will not be changed by the substitution of $\alpha x, \alpha^2 x, \dots \alpha^{n-1}x$ for x, or in other words we may suppose $\psi x = \psi\alpha x = \dots = \psi\alpha^{n-1}x$, and consequently ψx will be determined by the equation

$$F_1\{\psi x, \psi x, \dots \psi x, fx, f_1x, f_2x, \dots\} = 0$$

or from

$$F\{\psi x, \psi x, \dots \psi x, \overline{x}, \overline{\alpha x}, \overline{\alpha^2 x}, \dots \overline{\alpha^{n-1}x}\} = 0.$$

If there should exist any doubt respecting the accuracy of this reasoning, it may be confirmed by arriving at the same conclusion in rather a different manner. If in the given equation we substitute successively $\alpha x, \alpha^2 x, \dots \alpha^{n-1}x$ for x, we shall have the following equations.

$$\begin{aligned}
&F\{\psi x, \quad \psi\alpha x, \dots \psi\alpha^{n-1}x, \overline{x}, \overline{\alpha x}, \dots \overline{\alpha^{n-1}x}\} = 0 \\
&F\{\psi\alpha x, \quad \psi\alpha^2 x, \dots \psi x, \quad \overline{x}, \overline{\alpha x}, \dots \overline{\alpha^{n-1}x}\} = 0 \\
&\qquad\vdots \qquad\qquad\qquad\qquad \vdots \\
&F\{\psi\alpha^{n-1}x, \psi x, \quad \dots \psi\alpha^{n-2}x, \overline{x}, \overline{\alpha x}, \dots \overline{\alpha^{n-1}x}\} = 0
\end{aligned}$$

To eliminate $\psi\alpha x, \psi\alpha^2 x, \dots$ from these n equations would in most cases be excessively troublesome. It may however be observed, that $\psi\alpha x$ occurs in the second exactly in the same manner that ψx occurs in the first; also $\psi\alpha^2 x$ is contained in the third in the same manner as ψx is contained in the first, and similarly with the rest. From this it follows, that though no one of the equations is symmetrical relative to $\psi x, \psi\alpha x, \dots \psi\alpha^{n-1}x$, yet when all the equations are considered together, they are symmetrical relative to $\psi x, \psi\alpha x, \dots \psi\alpha^{n-1}x$; from this it follows that whether we find by elimination $\psi x, \psi\alpha x, \dots$ or $\psi\alpha^{n-1}x$ the result will be the same, therefore $\psi x = \psi \alpha x = \ldots = \psi \alpha^{n-1} x$, and we may determine ψx from the equation

$$F\{\psi x, \psi x, \ldots \psi x, \overline{x}, \overline{\alpha x}, \ldots \overline{\alpha^{n-1} x}\} = 0. \qquad (a)$$

It does not follow from hence, that this equation contains all the values of ψx: on the contrary, if the elimination between the n equations above written were actually performed, it would be found that the equation (a) would enter into the final result as one of its factors. /

If we apply this reasoning to algebraic equations containing several variables, as for instance to the two equations

$$F(x, y, a, b) = 0, \quad F(y, x, a, b) = 0$$

we shall find that one set of the value of x is always contained in the equation

$$F(x, x, a, b) = 0.$$

As an example, let us take the two equations

$$y^2 + x = a \quad \text{and} \quad x^2 + y = a.$$

One set of the values of x are contained in the equation

$$x^2 + x - a = 0$$

and this enters as a factor in the result of elimination, which gives

$$x^4 - 2ax^2 + x + a^2 - a = (x^2 + x - a)(x^2 - x - (a - 1)) = 0.$$

Another curious analogy exists between the calculus of functions and common algebra in the similarity of the relations of the roots of unity to the solutions of the functional equation $\psi^n x = x$.

In the equation $r^n = 1$ it is known that if r_1 be one of the roots then any power of r_1 will also be a root, and if n is a prime number and r_1 any root except unity, then $r_1, r_1^2, r_1^3, \ldots r_1^{n-1}$ will be all the different roots. Similarly in the functional equation $\psi^n x = x$ if αx be one form which satisfies it, $\alpha^m x$ will also fulfil it, and if n is a prime number then $\alpha x, \alpha^2 x, \ldots \alpha^{n-1}$ will all be different forms which satisfy the equation. This may be readily shown as follows, if αx is a solution, then

$$\alpha\alpha \ldots (n) x = \alpha^n x = x.$$

Suppose $m = 2$ then

$$\alpha^2\alpha^2 \ldots (n) x = \alpha^{2n} x = \alpha^n(\alpha^n x) = \alpha^n x = x$$

consequently $\alpha^2 x$ is also a solution of $\psi^n x = x$./

Again

$$\alpha^m\alpha^m \dots (n)x = \alpha^{n.m} = \alpha^n\alpha^n \dots (m)x = \alpha^n\alpha^n \dots (m-1)x = \dots$$
$$= \alpha^n\alpha^n x = \alpha^n x = x$$

consequently whenever m is a whole number $\alpha^m x$ is a solution of $\psi^n x = x$.

If we take the particular case of $n = 3$ we have $\psi^3 x = x$ and $\alpha x = 1/(1-x)$ is one of its solutions, therefore other solutions will be

$$\alpha x = \frac{1}{1-x}, \quad \alpha^2 x = \frac{x-1}{x}, \quad \alpha^3 x = x.$$

/

These expressions when generalized by the introduction of an arbitrary function, do not give solutions which are irreducible to each other, nor do they even then in my opinion contain all possible solutions; by introducing an arbitrary function they become

$$\varphi^{-1}\left(\frac{1}{1-\varphi x}\right) \quad \text{and} \quad \varphi^{-1}\left(\frac{\varphi x - 1}{\varphi x}\right)$$

the latter of which gives $1/(1-x)$ if we make $\varphi x = 1/x$.

If we consider the equation $\psi^6 x = x$ one of its values is

$$\alpha x = \frac{(3x-3)}{x}$$

this gives for the others.

$$\alpha x = \frac{3x-3}{x}, \quad \alpha^2 x = \frac{2x-3}{x-1}, \quad \alpha^3 x = \frac{3x-6}{2x-3}$$

$$\alpha^4 x = \frac{x-3}{x-2}, \quad \alpha^5 x = \frac{-3}{x-3}, \quad \alpha^6 x = x$$

All these forms will satisfy the equation $\psi^6 x = x$ and they may all be generalized by the introduction of an arbitrary function similar to that employed for the equation $\psi^3 x = x$, but I do not suppose these solutions when thus generalised would contain all possible ones.

Perhaps the following observations may throw some light on the generality of the solutions of such equations as those we are now considering.

In the first place, it is obvious that every solution of $\psi^3 x = x$ / will also be a solution of $\psi^6 x = x$, such likewise will be all the solutions of $\psi^2 x = x$. The complete solution of $\psi^6 x = x$ should therefore contain all forms of x which satisfy the equations $\psi^3 x = x$ and $\psi^2 x = x$. Again, if a function as αx satisfies the equation $\psi^3 x = -x$, it will also satisfy $\psi^6 x = x$, for since $\alpha^3 x = -x$ by putting $\alpha^3 x$ for x it becomes $\alpha^3\alpha^3 x = -\alpha^3 x = -(-x) = x$, so also any function which satisfies the equation $\psi^3 x = a/x$ will fulfil the equation $\psi^6 x = x$ which may be proved in the same manner. I shall however take the more general case, and show that if any function satisfies the equation $\psi^3 x = \beta x$ where β is any function satisfying the equation $\psi^2 x = x$, it will also satisfy the equation $\psi^6 x = x$, for since $\alpha^3 x = \beta x$, putting $\alpha^3 x$ for x, we have $\alpha^3 \alpha^3 x = \alpha^6 x = \beta \alpha^3 x = \beta\beta x = \beta^2 x = x$. In the same way it might be demonstrated, that the complete solution of $\psi^{abc\ldots}(x) = x$ must contain all functions which can satisfy any of the following conditions

$$\begin{array}{ll} \psi^a x = \alpha x, & \psi^{ab} x = \alpha_1 x, \ldots \\ \psi^b x = \beta x, & \psi^{bc} x = \beta_1 x, \ldots \\ \vdots & \end{array}$$

where

$$\begin{array}{ll} \alpha \text{ is any function satisfying the equation} & \psi^{bc\ldots}(x) = x \\ \beta & \psi^{ac\ldots}(x) = x \\ \vdots & \vdots \\ \alpha_1 & \psi^{cd\ldots}(x) = x \\ \beta_1 & \psi^{ad\ldots}(x) = x \\ \vdots & \vdots \end{array}$$

/

The comparison of the integral calculus with that of functions will supply us with several very marked analogies, some of which promise when further pursued to be of essential / service in the improvement of this latter branch of analytical science. One of the first which presents itself is the method of solving the differential equation

$$0 = y dx^n + A_x dx^{n-1} dy + B_x dx^{n-2} d^2 y + \ldots + N_x d^n y + X dx^n$$

compared with that of solving the functional equation

$$0 = \psi x + A_x \psi \alpha x + B_x \psi \alpha^2 x + \ldots + N_x \psi \alpha^{n-1} x + X.$$

Euler and D'Alembert succeeded in integrating the differential equation when all the coefficients except X are constant quantities, and Lagrange, in the Memoirs of the Academy of Berlin, 1772,[a] explained a method of treating the same when they are variable; both the processes alluded to depend in the first instance on reducing the equation to the solution of the same equation, wanting the last term; and this is effected by means of a particular integral of the given equation.

[a] J. L. Lagrange, 'Sur une nouvelle espèce de calcul rélatif à la différentiation et à l'intégration des quantités variable,' *Nouvelle Mémoires Académie de Berlin*, Vol. 3 (1772), p. 185.

The solution of the functional equation is precisely similar, it is first reduced to the solution of

$$0 = \psi x + A_x \psi \alpha x + B_x \psi \alpha^2 x + \ldots + N_x \psi \alpha^{n-1} x$$

and this is effected by knowing a particular function which satisfies the original equation: this process I have already given in a former paper.[a] The resemblance between the solutions of the two equations continues also in this respect, if fx is a particular case of the given equation, and $Kx, K_1x, K_2x, \ldots$ are particular cases of the same equation without its last term, then

$$\psi x = fx + Kx \cdot \chi(\overline{x}, \overline{\alpha x}, \ldots \overline{\alpha^{n-1} x}) + K_1 x \cdot \chi_1(\overline{x}, \overline{\alpha x}, \ldots \overline{\alpha^{n-1} x}) + \ldots$$

is a general solution of the equation. /

It may be observed that the functional equation under consideration comprehends a large class of others which may easily be reduced to it, such as

$$F\psi x + A_x F\psi \alpha x + B_x F\psi \alpha^2 x + \ldots + N_x F\psi \alpha^{n-1} x + X = 0$$

or more generally

$$F(x, \psi x) + A_x F(\alpha x, \psi \alpha x) + B_x F(\alpha^2 x, \psi \alpha^2 x) + \ldots + N_x F(\alpha^{n-1} x, \psi \alpha^{n-1} x) + X = 0$$

both of which may be reduced to the same equations wanting the last term. These transformations apply with advantage to a great variety of other equations: the solution of the latter of these equations may be reduced to that of

$$\psi_1 x + A_x \psi_1 \alpha x + \ldots + N_x \psi_1 \alpha^{n-1} x = 0$$

and if we find $\psi_1 x = Kx$ we shall have

$$\psi x = F^{1,-1}(x, Kx)$$

for the solution of the given equation.

As an instance of the utility of the former equation we may take

$$\{\psi(x)\}^2 + \{\psi(\pi/2 - x)\}^2 = 1$$

whose solution will be found by that method to be

$$\psi x = \sqrt{1/2 + (2x - \pi/2)\chi(\overline{x}, \overline{\pi/2 - x})}.$$

The same question or the more general one

$$\{\psi x\}^n + \{\psi \alpha x\}^n = 1$$

[a] See Problem V in 'An essay towards the calculus of functions ⟨part I⟩' (1815), in this volume of the *Works of Babbage*, pp. 104–105.

where $\alpha^2 x = x$ may be solved by another process which requires the aid of vanishing fractions, and we shall have

$$\psi x = \left\{\frac{\varphi x - \varphi\alpha x + \varphi_1 x}{\varphi_1 x + \varphi_1 \alpha x}\right\}^{1/n}.$$

/

The analogy between the various orders of functions which contain many variables, and that branch of the integral calculus which relates to *partial differentials*, is too apparent to require illustration. I shall therefore proceed to show that a functional equation admits of three species of solutions. 1. *The complete solution, this contains as many arbitrary functions as the nature of the given equation will admit.* 2. *The particular case, this contains all solutions which are less general than the complete solution, and which are only particular cases of it.* If they contain arbitrary functions, I have called them, for the sake of convenience, *general solutions.* 3. *The particular solution, this is a solution which satisfies the equation, and may or may not contain arbitrary functions; its peculiar property is, that it is found from one part only of the equation and independent of the rest*; thus if we have the equation

$$\{\psi^{1,2}(x, y) - y\} \cdot F\{x, y, \psi(x, y), \psi^{2,1}(x, y), \ldots\}$$
$$= \{\psi^{\overline{2,2}}(x, y) - \varphi^{-1}(0)\} \cdot F_1\{x, y, \psi(x, y), \ldots\}$$

$\psi(x, y) = \varphi^{-1}(\varphi x - \varphi y)$ is a particular solution, for it satisfies the equation by making $\psi^{2,1}(x, y) - y = 0$ and $\psi^{\overline{2,2}}(x, y) - \varphi^{-1}(0) = 0$, and is totally independent of the rest of the equation contained in the given function F and F_1, provided only that it does not make either of them infinite, whether it is contained in the complete solution, I have not yet ascertained; but I am rather of opinion, that it will be found not to be included in it. In both parts of my Essay towards the Calculus of Functions,[a] I have used the expression, particular solution, instead of particular case, this arose from not having taken a sufficiently extensive view of the calculus; it would / perhaps be desirable to confine the meaning of *particular solution* to the definition which has just been given. It is needless to add that the different species of solutions just enumerated, bear a strong resemblance to those of differential equations.

Amongst differential equations containing more than two independent variables, a large proportion do not admit of any integrals, these can only be integrated by assigning some relation between the variables, an analogous case occurs with respect to functional equations, a large number of those which contain two or more variables admit of no solution, unless we assign some

[a] 'An essay towards the calculus of functions' (1815) and 'An essay towards the calculus of functions, part II' (1816), in this volume of the *Works of Babbage*.

relation between the variable quantities; examples of such equations may be found in the second part of the Essay towards the Calculus of Functions. And here perhaps it may not be misplaced to state a difficulty of a peculiar nature with respect to functional equations which are impossible; for the sake of perspicuity I shall consider a very simple case

$$\psi x = c\psi(1/x)$$

for x substitute $1/x$

$$\psi(1/x) = c\psi x$$

and by multiplication

$$\psi x \cdot (1/x) = c^2\psi(1/x) \cdot \psi x$$

or $1 = c^2$ from which it follows that $c = \pm 1$ or in other words that the equation $\psi x = c\psi(1/x)$ is contradictory unless $c = \pm 1$. Now the functional equation $F\{x, \psi x, \psi \alpha x\} = 0$ has been reduced by Laplace by means of a very elegant artifice to an equation of finite differences, nor am I aware that this profound analyst has pointed out any restriction or any / impossible case: if we treat the equation $\psi x = c\psi x^n$ by his method, we shall find for its solution

$$\psi x = c^{\frac{-\log\log x}{\log n}}$$

and this solution satisfies the equation $\psi x = c\psi x^n$ independently of any particular value of n, and if we suppose $n = -1$ we have

$$\psi x = c^{\frac{-\log^2 x}{\log n}}$$

for the solution of the equation $\psi x = c\psi(1/x)$ whatever may be the value of c, and we have before shown that it cannot have a solution unless $c = \pm 1$. The only explanation I am at present able to offer concerning this contradiction, is one which I hinted at on a former occasion, viz. that if we suppose ψ to represent any inverse operation which admits of several values, then if throughout the whole equation we always take the same root or the same individual value of ψ, it is impossible to satisfy the equation, but if we take one value of ψ in one part, and another of the values of ψ in other parts of the equation, it is possible to fulfil it by such means. This solution may perhaps appear unsatisfactory, it is however only proposed as one which deserves examination, and I shall be happy if its insufficiency shall induce any other person to explain more clearly a very difficult subject.

One of the most extensive methods of integrating differential equations, consists in multiplying by such a factor as will render the whole equation a complete differential, the determination of this factor is, however, generally a matter of at least equal difficulty with that of solving the original equation:

analogy would lead us to suspect that some similar mode might be adopted for the solution of functional equations, / varying of course in a certain degree from the difference of the objects to be obtained: the theory which I am now to explain will show that this suspicion is not without foundation, and will at the same time unfold one of the most beautiful parallels between the integral and functional calculus which I have yet observed. It has been already shown, that when an equation is symmetrical relative to the different forms of the unknown function, as

$$F\{x, \overline{\psi x}, \overline{\psi\alpha x}, \ldots \overline{\psi\alpha^{n-1}x}\} = 0$$

the method of solution by elimination apparently becomes illusory, but that the general solution of the equation may be deduced from the vanishing fraction to which this method then leads. If therefore we can make any functional equation symmetrical, relative to the different forms of the unknown functions, we can then obtain its general solution. Now this may be effected by multiplying the equation by some factor, and determining this factor, so that the result shall be symmetrical. The discovery of this factor (as in the case of a differential equation) requires the solution of a functional equation of several variables, but fortunately the class of equations to which they belong are not of very great difficulty.

To begin with a very simple case, let us consider the equation

$$\psi x + fx \cdot \psi\alpha x = f_1 x$$

where $\alpha^2 x = x$ multiplying by φx we have

$$\varphi x \cdot \psi x + \varphi x \cdot fx \cdot \psi\alpha x = \varphi x \cdot f_1 x$$

now the first side of this equation will be symmetrical relative to ψx and $\psi\alpha x$, if we make

$$\varphi x = f\alpha x \cdot \varphi\alpha x$$

/

this equation can only be possible when $fx \cdot f\alpha x = 1$ or when

$$fx = \{\chi(\overline{x}, \overline{\alpha x})\}^{x-\alpha x}$$

and supposing $fx \cdot f\alpha x = 1$, we have $\varphi x = 1/\sqrt{fx}$ consequently the equation becomes

$$\frac{\psi x}{\sqrt{fx}} + \sqrt{fx} \cdot \psi\alpha x = \frac{\psi x}{\sqrt{fx}} + \frac{\psi\alpha x}{\sqrt{f\alpha x}} = \frac{f_1 x}{\sqrt{fx}}$$

and since the left side of this equation is symmetrical, the right side must also be so, consequently

$$\frac{f_1 x}{\sqrt{fx}} = f_2(\overline{x}, \overline{\alpha x})$$

or the given equation is impossible, unless

$$f_1 x = \sqrt{fx} \cdot f_2(\overline{x}, \overline{\alpha x})$$

and the solution of the equation will be found by taking the value of a vanishing fraction to be

$$\psi x = \frac{\sqrt{f\alpha x} \cdot f_2(\overline{x}, \overline{\alpha x}) \cdot \varphi x - \varphi_1 x + fx \cdot \varphi_1 \alpha x}{fx \cdot \varphi\alpha x + f\alpha x \cdot \varphi x}.$$

This method is applicable to all equations of the first degree, but I shall now point out a principle which extends to all equations of the first order, and which may probably be applied with some modification to those of higher orders.

Let us consider the equation

$$F\{x, \alpha x, \psi x, \psi\alpha x\} = 0$$

where $\alpha^2 x = x$ multiply this by $\varphi\{x, \alpha x, \psi x, \psi\alpha x\} = 0$ it becomes

$$F\{x, \alpha x, \psi x, \psi\alpha x\} \cdot \varphi\{x, \alpha x, \psi x, \psi\alpha x\} = 0$$

in order that this equation may be symmetrical, we must have

$$F\{x, \alpha x, \psi\alpha x\} \cdot \varphi\{x, \alpha x, \psi x, \psi\alpha x\} = F\{\alpha x, x, \psi\alpha x, \psi x\} \cdot \varphi\{\alpha x, x, \psi\alpha x, \psi x\}$$

/

from which the form of φ must be determined, one value is

$$\varphi(x, \alpha x, \psi x, \psi\alpha x) = F\{\alpha x, x, \psi\alpha x, \psi x\}$$

hence the equation

$$F\{x, \alpha x, \psi x, \psi\alpha x\} \times F\{\alpha x, x, \psi\alpha x, \psi x\} = 0$$

is symmetrical, and its general solution may therefore be found. It must however be observed, that all the solutions of this latter equation are not necessarily solutions of the former one, and it may be a matter of some difficulty to ascertain what solutions ought to be excluded. There is no part of this process which limits it to the particular case we have considered, and if the given equation were

$$F\{x, \psi x, \psi\alpha x, \ldots \psi\alpha^{n-1} x\} = 0$$

we might take the equation

$$\varphi\{\overline{F\{x, \psi x, \psi\alpha x, \ldots \psi\alpha^{n-1}x\}}, \ \overline{F\{\alpha x, \psi\alpha x, \ldots \psi x\}}, \ldots \overline{F\{\alpha^{n-1}x, \psi\alpha^{n-1}x, \ldots \psi\alpha^{n-2}x\}}\} = 0$$

which is symmetrical relative to $\psi x, \psi\alpha x, \ldots \psi\alpha^{n-1}x$.

I shall not at present enter into any further details respecting this mode of solving functional equations, as it forms the subject of some investigations on which I am now engaged, and which are as yet incomplete.

In the preceding pages I have endeavoured to point out some of the more prominent points in which the calculus of functions resembles common algebra or the integral calculus: it ought however to be observed, that several of the methods which I have applied to the solution of functional equations, directly resulted from pursuing this analogy: for when I had ascertained the remarkable similitude which exists between / the method of functions and the integral calculus, I referred to a treatise on that subject with the express purpose of endeavouring to transfer the methods and artifices employed in the latter calculus, to the cultivation and improvement of the former.

5 March, 1817 / CHARLES BABBAGE

[371/2]

VIII

SOLUTIONS OF SOME PROBLEMS BY MEANS OF THE CALCULUS OF FUNCTIONS

From *Journal of Science and the Arts*, Vol. 2 (1817), pp. 371–9

Problem I

Required the nature of a curve ABC such that if any two abscissae AD and AE be taken whose rectangle is equal to a given square, then the rectangle of their corresponding ordinates BD and CE shall be equal to another / given square.

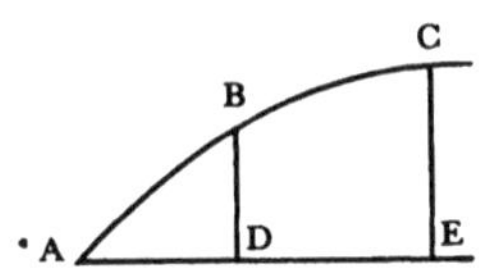

Let $AD = x$ and $BD = y = \psi x$, and let the two given squares be a^2 and c^2 then by the problem

$$AD \cdot AE = a^2 \text{ and } BD \cdot CE = c^2$$

hence

$$AE = \frac{a^2}{AD} = \frac{a^2}{x} \quad \text{and} \quad CE = \psi(AE) = \psi\left(\frac{a^2}{x}\right)$$

and the functional equation which results is

$$\psi x \cdot \psi\left(\frac{a^2}{x}\right) = c^2.$$

As this equation is not generally soluble by the direct method which I have proposed, we must endeavour to find some particular solution which may contain an arbitrary constant. If a^2 were equal to c^2 it is evident that the equation would be satisfied by supposing $\psi x = x$ or $\psi x = x^n$. Let us therefore assume $\psi x = Ax^n$ then

$$\psi\left(\frac{a^2}{x}\right) = A\left(\frac{a^2}{x}\right)^n$$

and the equation becomes

$$A^2x^n\left(\frac{a^2}{x}\right)^n = A^2a^{2n} = c^2$$

hence

$$A = ca^{-n}$$

and the solution of the given equation is

$$\psi x = c\left(\frac{x}{a}\right)^n.$$

This solution contains the arbitrary constant n which may, as I have shown in* Problem VIII be changed into any arbitrary function of x which does not vary when x becomes a^2/x; such a function is $\chi(\overline{x, a^2/x})$ which denotes any symmetrical / function of x, and a^2/x consequently the general solution of the given equation is

$$\psi x = c\left(\frac{x}{a}\right)^{\chi(\overline{x}, \overline{a^2/x})}$$

hence the equation which comprehends the species of curve which is required, is

$$y = c\left(\frac{x}{z}\right)^{\chi(\overline{x}, \overline{a^2/x})}$$

when the arbitrary function is equal to a constant quantity, we have the particular solutions already noticed

$$y = c\left(\frac{x}{a}\right)^n$$

and since n may be whole or fractional, positive or negative, it follows that all parabolas and hyperbolas of every degree, possess the required property.

The equation $\psi x \cdot \psi(a^2/x) = c^2$ may otherwise be solved in the following manner; assume $\psi x = AB^{\varphi x}$, then the equation becomes

$$Ab^{\varphi x} \cdot Ab^{\varphi(a^2/x)} = c^2$$

which will be satisfied if $A = a$ and $\varphi x = -\varphi(a^2/x)$; we must therefore endeavour to discover a particular solution of $\varphi x = -\varphi(a^2/x)$, the first which presents itself is $\varphi x = x - a^2/x$ this being substituted gives

$$\psi x = cb^{x - a^2/x}$$

* See 'An Essay towards the Calculus of Functions' in the last Volume of the Philosophical Transactions. ⟨(1815); reprinted in this volume of the *Works of Babbage.*⟩

and since b is arbitrary it may be changed into any arbitrary function which does not vary when x is changed into a^2/x, hence the general solution is

$$\psi x = c\{\chi(\overline{x}, \overline{a^2/x})\}^{x - a^2/x}.$$

/

This solution suggests a similar one of the more general problem when AE is equal to any given function α of AD, provided $\alpha^2 x = x$, in this case we have $AD = xAE = \alpha x$ and the equation to be satisfied is

$$\psi x \cdot \psi \alpha x \cdot x = c^2$$

put $\psi x = cb^{\varphi x}$ then it becomes

$$cb^{\varphi x} \cdot cb^{\varphi \alpha x} = c^2$$

which will be fulfilled if $\varphi^x = -\varphi\alpha x$, a particular solution of this equation is $\varphi x = x - \alpha x$, for then $\varphi \alpha x = \alpha x - \alpha^2 x = \alpha x - x$ hence the general solution of $\psi x \cdot \psi \alpha x = c^2$ when $\alpha^2 x = x$ is

$$\psi x = c\{\chi(\overline{x}, \overline{\alpha x})\}^{x - \alpha x}$$

A similar solution is also applicable to the same equation when $\alpha^{2n} x = x$ in that case it is

$$\psi x = c\{\chi(\overline{x}, \overline{\alpha x}, \overline{\alpha^2 x}, \ldots \overline{\alpha x^{2n-1}})\}^{x - \alpha x + \ldots - \alpha^{2n-1} x}.$$

Problem II

Required the nature of a curve ABC such that supposing AD any abscissa, if we take another abscissa AE having any given relation to the former, then the ordinate CE corresponding to the second abscissa shall have the same given relation to the ordinate BD corresponding to the first abscissa.

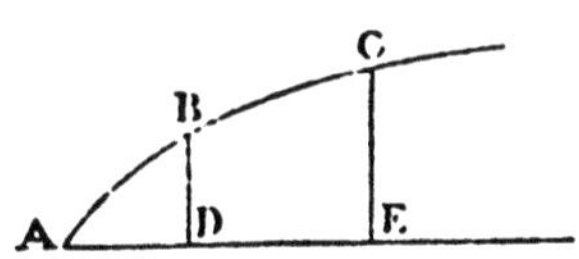

Let the first abscissa AD be denoted by x, and let the relation of the second abscissa $AE = x'$ to it be expressed by αx, α being some given function, then if $BD = y = \psi x$ we have $CE = y' = \psi \alpha x$, and by the conditions of the problem

$$\alpha \psi x = \psi \alpha x$$

this equation will evidently be satisfied if $\psi = \alpha$ or more generally if $\psi = \alpha^n$, for it then becomes

$$\alpha \alpha^n x = \alpha^n \alpha x$$

which is identical. /

If it should happen that for some particular value of n we find $\alpha^n x = x$, then the solution $\psi x = \alpha^n x$ admits of only $p - 1$ values, p being the least value of n which satisfies the equation of $\alpha^p x = x$.

Suppose it were required that the sum of the squares of the two abscissae should be equal to a constant quantity, then the sum of the squares of the corresponding ordinates must be equal to the same quantity, and the equation becomes

$$\psi(\sqrt{a^2 - x^2}) = \sqrt{a^2 - (\psi x)^2}$$

and the solution corresponding to $\psi x = \alpha x$ is

$$y = \psi x = \sqrt{a^2 - x^2}$$

which is the equation to a circle, the abscissae beginning at the centre and having for its radius the quantity a.

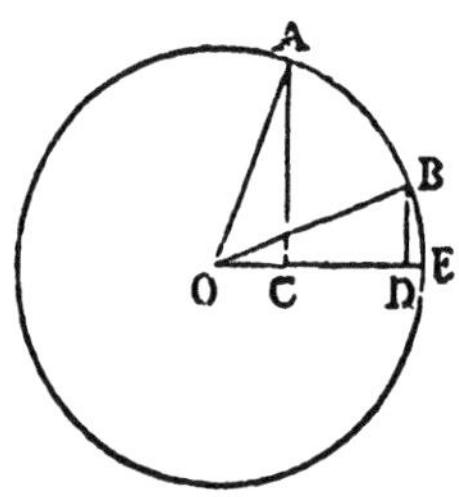

In fact, if in any circle $OABE$ we take any two abscissæ OC, and OD so related that the sum of their squares shall be equal to the square of the radius OA; it is easy to show that the two triangles OAC and OBD are similar and equal, and consequently that the sum of the squares of the two ordinates AC and BD is also equal to the square of the radius: in the case just examined, we find $\alpha^2 xc = x$, consequently we cannot from this method deduce any other solution.

The solution of the equation $\psi \alpha x = \alpha \psi x$ may be rendered more general, when $\alpha^n x = x$ by the following means, since it is satisfied by supposing $\psi x = \alpha^p x$ whatever may be the value of p, for p substitute any symmetrical function of $x, \alpha x, \ldots \alpha^{n-1} x$, and the general solution is

$$\psi x = \alpha^{\chi(\overline{x}, \overline{\alpha x}, \ldots \overline{\alpha^{n-1} x})} x.$$

Problem III

What is the relation between x and y such that if

$$x = a + by + cy^2 + \ldots$$

we may by reverting the series have

$$y = a + bx + cx^2 + \dots$$

Let $x = \psi y$ then by the condition of the problem we have also $y = \psi x$, hence $x = \psi y = \psi\psi x = \psi^2 x$, and the nature of the function ψ is determined from the equation

$$\psi^2 x = x$$

the solution of this equation is already known,* and any of the various values of ψx will when expanded, produce a series of the required form, such for example are

$$x = \psi y = \sqrt{1 - y^2} \quad \text{or} \quad x = \frac{a - by}{b + cy}.$$

In fact, if from any symmetrical function of x and y we find the value of x expressed in terms of y, it is clear that y might be expressed in terms of x in precisely a similar manner, and this is nothing else than a different mode of denoting the general solution of the equation $\psi^2 x = x$.

Problem IV

What operation must be performed on x so that if we repeat the same operation on the result, and on this result again repeat the same operation, and so on, after performing it four times, the last result may be equal to the original quantity.

Let the operation to be performed be denoted by the characteristic ψ, then the first result is ψx, and repeating the same operation four times, we have

$$\psi\psi\psi\psi x = \psi^4 x$$

and by the condition of the problem this must be equal to the original quantity; hence

$$\psi^4 x = x.$$

In order to solve this equation, let us suppose $\psi x = \varphi^{-1} f \varphi x$, φ^{-1} signifying the inverse operation of φ then / $\psi\psi x = \varphi^{-1} f \varphi\varphi^{-1} f \varphi x = \varphi^{-1} f^2 \varphi x$ and we shall find $\psi^3 x = \varphi^{-1} f \varphi\varphi^{-1} f^2 \varphi x \varphi^{-1} f^3 \varphi x$ and $\psi^4 x = \varphi^{-1} f \varphi\varphi^{-1} f^3 \varphi x = \varphi^{-1} f^4 \varphi x$ from which substitution the equation becomes

$$\varphi^{-1} f^4 \varphi x = x$$

* See Problem IX of an Essay towards the Calculus of Functions in the Philosophical Transactions for the year 1815.

and performing the operation indicated by ψ on both sides we have

$$f^4\varphi x = \varphi x$$

this equation will be identical if we assume such a value for f, that $f^4 y = y$, that is, we must make f a particular solution of the given equation.

A particular solution of the more general equation may be found from the following curious property of the function $(a + bx)/(c + dx)$, if this function be called fx, then will its nth function or $f^n x$ always be of the same form, or $f^n x = (A + Bx)/(C + Dx)$ when A, B, C, and D, are given functions of a, b, c, and d: and these latter quantities may always be so assumed that $A = 0$, $B = C$ and $D = 0$.

By employing this property, we have in the present instance as a particular solution of $f^4 x = x$,

$$fx = \frac{a + bx}{c - \dfrac{b^2 + c^2}{2a}x}$$

and by substituting this value of f we have for the general solution of the equation $\psi^4 x = x$

$$\psi x = \varphi^{-1}\left\{\frac{a + b\varphi x}{c - \dfrac{b^2 + c^2}{2a}\varphi x}\right\}.$$

This solution contains an arbitrary function φ, and by assigning various values to it we shall have an indefinite number of particular solutions; a few examples, however, will be sufficient.

Ex. 1. Let $\varphi x = x$ and $a = b = c$, then one particular solution is

$$\psi x = \frac{1 + x}{1 - x}$$

this gives $\psi^2 x = -1/x$ and $\psi^3 x = -(1 - x)/(1 + x)$ and $\psi^4 x = x$. /

Ex. 2. Let $a = -1$, $b = 1$, and $c = 0$, also $\varphi x = x$ then $\psi x = 2(x - 1)/x$ which is another solution, for $\psi^2 x = (x - 2)/(x - 1)$, $\psi^3 x = -2/(x - 2)$ and $\psi^4 x = x$.

Ex. 3. a, b, and c remaining the same, let $\varphi x = \varepsilon^x$, then $\varphi^{-1} x = \log x$, and another solution is

$$\psi x = \log(\varepsilon^x - 1) - x + \log 2.$$

Problem V

A ray of light SA falls on a surface AB which reflects it, but not according to the common law of reflection.

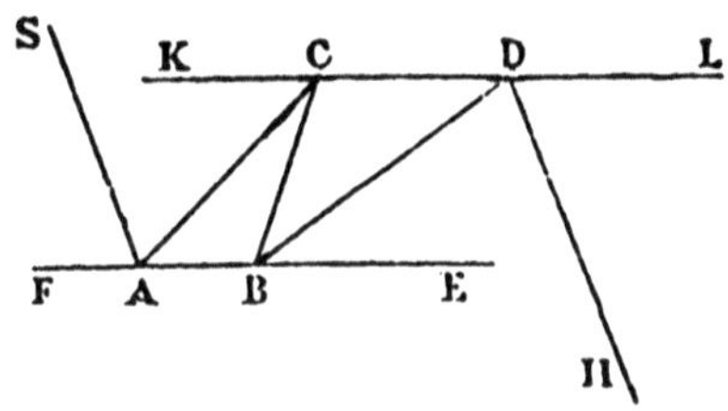

It then impinges on the surface CD which is similarly constituted and parallel to the former; after this it undergoes a second reflection at the first surface, and also a second reflection at the other surface.

Now it is found that whatever may be the angle n at which it meets the first plane, that it is always after undergoing the four reflections parallel to its first direction.

Let the angle SAF which the ray makes with the first surface be θ, and let the angle which the ray makes with that surface, after the first reflexion be denoted by $\psi\theta$: since the two surfaces are parallel, the angle KCA is equal to the angle $CAB = \psi\theta$; and since the surface KL reflects according to the same law as the surface FE, the angle LCB will have the same relation to the angle KCA as the angle CAE has to SAF, therefore $LCB = \psi(KCA) = \psi\psi\theta = \psi^2\theta$: and in a similar manner it may be shown that $DBE = \psi^3\theta$ and $LDH = \psi^4\theta$, and since by the conditions of the Problem the ray after the last reflection is parallel to ray before its first incidence we must have

$$LDH = SAF$$

which gives

$$\psi^4\theta = \theta$$

this equation is the same as that of Problem IV, and one very general solution is

$$\psi\theta = \varphi^{-1}\left(\frac{a + b\varphi\theta}{c - \dfrac{b^2 + c^2}{2a}\varphi\theta}\right)$$

φ is arbitrary, and this formula contains an infinite variety of relations between the angle of incidence and that of reflection, any of which taking place will cause the last ray to be parallel to the first.

It seems probable that this solution, although a very extensive one, does not contain all the possible answers, and if we have regard to the utmost generality, the solution must be deduced from that of the equation

$$F\{\overline{\theta}, \overline{\psi\theta}, \overline{\psi^2\theta}, \overline{\psi^3\theta}\} = 0.$$

If we suppose the ray to be parallel to itself after two reflections the equation will be

$$\psi^2\theta = \theta$$

and all the possible solutions of this equation are I believe contained in the value of $\psi\theta$ deduced from the equation

$$F\{\overline{\theta}, \overline{\psi\theta}\} = 0$$

F denoting any symmetrical function whatever. /

IX

AN ACCOUNT OF EULER'S METHOD OF SOLVING A PROBLEM, RELATIVE TO THE MOVE OF THE KNIGHT AT THE GAME OF CHESS

From *Journal of Science and the Arts*, Vol. 3 (1817), pp. 72–7

From a Correspondent

The knight being placed on any given square of the chess-board, it is required at sixty three successive moves, to cause it to move over the remaining sixty three squares.

To most of those who are familiar with the game of chess, this curious question is perhaps well known, although the method of reasoning which Euler employed in discovering its solution, is, I believe, not so generally understood.

It may be remarked, that if we have any one course of moves, by which the knight may successively move over the sixty-four squares of the chess-board, and if the square with which it terminates, is distant from the square which commenced it, by one move of the knight, then, such a course will solve the problem; the course given, Figure 1, is of this nature, and may be called a re-entering course.

Let us now suppose the knight is placed on the square marked 39, (Figure 1) then, since we are acquainted with this course, we may cause it to move on through the squares 40·41· . . . ·64, and since it can move from the square marked 64, to that marked 1, it may continue its course from 64 to 1·2·3· . . . to 38, when it will have fulfilled the conditions. This course may conveniently be represented thus:

39· . . . ·64·1· . . . ·38.

It also appears, that the knight might have moved backwards from the square 39 to 38, and so on to 1, and from thence to 64, then through 63, . . . to 40. This course would be thus represented:

39· . . . ·1·64· . . . ·40.

The arrangement just given, furnishes, therefore, two courses from whatever given square of the chess-board the knight commences his course.

As it is a much more difficult matter to discover by trial, a re-entering course, than one in which the last square does / not communicate with the first, we shall first show a simple method of transforming any course of the latter species, into several different ones of the former kind. Let us take the course (Figure 2) in which the last square, 64, does not communicate with the first, 1. It may be remarked, that the square at which the knight finishes his course, may be changed in several ways, without altering the square from which he started; thus let us notice all the squares to which the knight might move from the last one, or that marked 64: these squares are 63, 31, and 51, of which the first, or 63, contains the move already employed, to arrive at 64, and is therefore of no use. Since the knight can move from the square marked 31, to that marked 64, let him make this move on his arrival at 31, after having passed through the squares 2, 3, ... and after that let him pursue his course through the squares 64, 63, 62, ... until he arrives at 32, which will now be the last one, and his new course will be represented thus:

1·2· ... ·31·64·63· ... ·32.

In a similar manner the move from 64 to 51, gives the course

1·2· ... ·51·64·63· ... ·52

and since the last square, 52, communicates by one move with the first, 1, this course is one of the re-entering kind, and it is, in fact, the course already given. If we had not already found a course of this kind, we might make new transformations on those just given. Thus, in the course

1· ... ·31·64· ... ·32

the knight can move from the last square, 32, to the squares 43, 11, 31, 33. The number 43 gives the course

1· ... ·31·64· ... ·43·32· ... ·42.

The number 11, furnishes the course

1· ... ·11·32· ... ·64·31· ... ·12.

The third number, 31, gives the original course from whence we began, and the number 33 causes no change.

In the last course, which ends with 12, the knight may move from 12 to 59, 41, 11, and 13; these furnish the following courses:

1· ... ·11·32· ... ·59·12· ... ·31.64· ... ·60
1· ... ·11·32· ... ·41·12· ... ·31.64· ... ·42

/

and since 60 communicates with the squares 61, 59, 9, 45, 25, 27, 13, and 53, we shall have several new courses ending with 10, 26, 46, 28, 14 and 54.

We have thus an easy method of discovering a great variety of new courses, from one given one, and this number might be much increased by inverting the order of each course.

This method of discovering re-entering courses, requires us to be acquainted with one which does not possess that property; now such a course may be determined in the following manner: beginning with any square, let the knight be moved successively over as many squares as he can, marking each square with a number; those squares which remain vacant, may be marked with the letters of the alphabet, as in Figure 3. In this case there remain two squares unoccupied, which are marked *a* and *b*. The course of the knight having moved over 62 squares, may be represented thus:

1·. . . ·62

and considering 62 as the last square, let us transform this into other courses which terminate with other squares, until some of the final squares communicate by one move of the knight, with either of those marked *a* and *b*; but 62 communicates with the squares 9, 53, 59, 61, 23, 11, 55, and 21, whence we may deduce the transformations.

I. 1·. . . · 9·62· . . . ·10, which communicates with *a*.
II. 1·. . . ·53·62· . . . ·54, which communicates with *a*.
III. 1·. . . ·59·62· . . . ·60
IV. 1·. . . ·23·62· . . . ·24
V. 1·. . . ·11·62· . . . ·12
VI. 1·. . . ·55·62· . . . ·56, which communicates with *a*.
VII. 1·. . . ·21·62· . . . ·22.

The courses marked I, II and VI include the square *a*, and by applying similar transformations to either of them, we may obtain others, which also include *b*; such is the following:

1·. . . ·9·62· . . . ·58·*a*·10· . . . ·57·6

which arises from transforming

1·. . . ·9·62· . . . ·10·*a*

into

1·. . . ·9·62· . . . ·58·*a*·10· . . . ·5

where 57 communicates with *b*. /

It is obvious, that the same method is applicable, whatever be the number of squares unfilled at the beginning; and the course thus obtained, may be transformed by the process explained at the commencement of this paper, into others which have the property of returning into themselves: such are the Figures 4 and 5.

After solving the question in the manner we have just stated, Euler adds to it, the following restriction, which renders it still more curious. It is required, that the difference of the numbers which are placed in any two *opposite* squares, shall always be equal to 32. Note, each square has another which is opposite to it, and the two squares are so related, that a straight line drawn through their two centres, will always divide the chess-board into two equal parts. It is therefore required in this case, that the squares which contain the numbers 32, 33, 34, . . . 64, shall be opposite to those in which the numbers 1, 2, 3, . . . 31 occur.

In order to discover such courses we must put 1, 2, 3, . . . in the squares successively covered by the knight, and at the same time we must write the numbers 33, 34, 35, . . . in the opposite squares; continuing this as far as we are able, we shall find twelve squares remain empty, which may be filled by the letters *A*, *a*, *B*, *b*, *C*, *c*, *D*, *d*, *E*, *e*, *F*, *f*, arranged in opposite squares, as in Figure 6.

We have here two separate series of squares, which indicate the successive moves of the knight:

58· . . . ·64·1· . . . ·19
29· . . . ·51.

The square 19 communicating with 6, gives the transformations

51· . . . ·64·1· . . . ·6·19· . . . ·7·*f*·*B*·*d*·*C*
26· . . . 38·51· . . . ·39·*F*·*b*·*D*·*c*.

Since *C* communicates with the squares 8, 6, *d*, of the first series, it will furnish no new transformations; let us therefore omit the two last terms; and since it is sufficient to transform one series, because the other is determined by it, let us consider the series

58· . . . ·64·1· . . . ·6·19· . . . ·7·*f*·*B* /

where *B* communicating with 12 gives the course

58· . . . ·64·1· . . . ·6·19· . . . ·12·*B*·*f*·7· . . . ·11·*D*·*c*

but since *c* communicates with 16, we have

58· . . . ·64·1· . . . ·6·19· . . . ·16·*c*·*D*·11· . . . ·7·*f*·*B*·12· . . . ·15·*a*·*E*

and the other series will be

26· . . . ·38·51· . . . ·48·*C*·*d*·43· . . . ·39·*F*·*b*·44· . . . ·47·*A*·*c*

which comprehends all the squares.

We must now endeavour to connect these two series together, so that the end of one shall communicate with the beginning of the other; for this purpose

Fig. 1.

42	57	44	9	40	21	46	7
55	10	41	58	45	8	39	20
12	43	56	61	22	59	6	47
63	54	11	30	25	28	19	38
32	13	62	27	60	23	[illegible]	5
53	64	31	24	29	26	57	8
14	33	2	51	16	35	4	49
1	52	15	34	3	50	17	36

Fig. 2.

42	59	44	9	40	21	46	7
61	10	41	58	45	8	39	20
12	43	60	55	22	57	6	47
53	62	11	30	25	28	19	38
32	13	54	27	56	23	48	5
63	52	31	24	29	26	37	18
14	33	2	51	16	35	4	49
1	64	15	34	3	50	17	36

Fig. 3.

34	21	54	9	32	19	48	7
55	10	33	20	53	8	31	18
22	35	60	a	40	49	6	47
11	56	41	50	59	52	17	30
36	23	58	61	42	39	46	5
57	12	25	38	51	60	29	16
24	37	2	43	14	27	4	45
1	b	13	26	3	44	15	28

Fig. 4.

30	55	46	9	28	57	40	7
47	12	29	56	45	8	27	58
54	31	10	13	18	41	6	39
11	48	33	42	15	44	39	26
32	53	14	17	34	19	38	5
49	64	51	20	43	16	25	60
52	27	2	35	62	23	4	37
1	30	63	22	3	36	61	24

Fig. 5.

42	55	26	9	44	57	34	7
25	12	43	56	27	8	45	58
54	41	10	13	18	35	6	33
11	24	19	36	15	28	59	46
40	53	14	17	20	37	32	5
23	64	51	38	29	16	47	60
52	39	2	21	62	49	4	31
1	22	63	50	3	30	61	48

Fig. 6.

10	29	48	35	8	31	46	33
49	36	9	30	47	34	7	58
28	11	A	C	f	45	32	19
37	50	B	D	e	6	59	44
12	27	38	E	d	b	18	5
51	64	13	F	c	a	43	60
26	39	2	15	62	41	4	17
7	14	63	40	3	16	61	42

Fig. 7.

14	59	42	35	16	31	54	33
41	36	15	58	55	34	17	30
60	13	56	43	18	53	32	7
37	40	19	12	57	6	29	52
20	61	38	25	44	51	8	5
39	64	21	50	11	24	45	28
62	49	2	23	26	47	4	9
1	22	63	48	3	10	27	46

Fig. 8.

								33	
α									β
	1	a	b	28	7	14	19	16	
	24	27	8	c	20	17	6	13	
	9	2	25	22	11	4	15	18	
	26	23	10	3	d	21	12	5	

Fig. 9.

	37	62	43	56	35	60	41	50	
	44	55	36	61	42	49	34	59	
	63	38	53	46	57	40	51	48	
	54	45	64	39	52	47	58	33	
α									β
	1	26	15	20	7	32	13	22	
	16	19	8	25	14	21	6	31	
	27	2	17	10	29	4	23	12	
	18	9	28	3	24	11	30	5	

The Knights Moves at Chess.

we observe, that E communicates with 62, and that the first series will then end with 63, which communicates with 26, the beginning of the other series; this transformation gives

$$58\cdot\ldots\cdot 62\cdot E\cdot a\cdot 15\cdot\ldots\cdot 12\cdot B\cdot f\cdot 7\cdot\ldots\cdot 11\cdot D\cdot c\cdot 16\cdot\ldots\cdot 19\cdot 6\cdot\ldots\cdot 1\cdot 64\cdot 63$$
$$26\cdot\ldots\cdot 30\cdot e\cdot A\cdot 47\cdot\ldots\cdot 44\cdot b\cdot F\cdot 39\cdot\ldots\cdot 43\cdot d\cdot C\cdot 48\cdot\ldots\cdot 51\cdot 38\cdot\ldots\cdot 31$$

and we have a re-entering course, which possess the required property; see Figure 7.

To the condition respecting the opposite squares, we may also add the following: that the first half of the numbers 1, 2, ... 32 shall be contained in the first half of the chessboard, supposing it divided into two equal parts by a line parallel to one side passing through the centre; thus in Figure 8, all the numbers 1, 2, ... 32 must be situated below the line $\alpha\beta$, and all the other numbers above it. Let us begin by putting unity on one of the squares adjacent to the line $\alpha\beta$, then 33 must be placed on the opposite square, and causing the knight to move from 1 through various squares of the lower half of the chess-board, we arrive as far as 28; the remaining squares must now be filled up with the letters a, b, c, d, and these must, by various transformations, be included in the course: after several changes we may arrive at the following, which includes them all;

$$1\cdot\ldots\cdot 8\cdot 23\cdot\ldots\cdot 21\cdot 18\cdot\ldots\cdot 20\cdot b\cdot 24\cdot\ldots\cdot 28\cdot 17\cdot\ldots\cdot 9\cdot a\cdot c\cdot d$$

and this again may be transformed into

$$1\cdot\ldots\cdot 8\cdot 23\cdot\ldots\cdot 21\cdot 18\cdot\ldots\cdot 20\cdot b\cdot 24\cdot\ldots\cdot 28\cdot 17\cdot\ldots\cdot 15\cdot d\cdot c\cdot a\cdot 9\cdot\ldots\cdot 14$$

in which 14 communicates with 33, the beginning of the series above the line $\alpha\beta$, and the end of that series 64 also / communicates with 1; so that we have a re-entering series subject to both the restrictions.

The whole course is represented in Figure 9.

It is not difficult to find other similar ones by the same method; and this course might be transformed in several ways, some of which are,

$$7\cdot\ldots\cdot 1\cdot 8\cdot\ldots\cdot 32$$
$$7\cdot\ldots\cdot 1\cdot 8\cdot\ldots\cdot 25\cdot 32\cdot\ldots\cdot 26$$
$$15\cdot\ldots\cdot 10\cdot 7\cdot\ldots\cdot 1\cdot 8\cdot 9\cdot 16\cdot\ldots\cdot 21\cdot 24\cdot\ldots\cdot 32\cdot 23\cdot 22.$$ /

[355/6]

X

NOTE RESPECTING ELIMINATION

From *Journal of Sciences and the Arts*, Vol. 3 (1817), pp. 355–7

The following observations will frequently obviate the necessity of the tedious process of elimination, and when only a few of the roots are required, they point out a ready method of obtaining them. The class of equations to which they are applicable, comprehends the following

$$F\{x, y, z, v, \ldots\} = 0, \quad F\{y, z, v, \ldots\} = 0, \ldots$$

/

Suppose it were required to find some of the values of x and y from the equations

$$F\{x, y\} = 0 \quad \text{and} \quad F\{y, x\} = 0.$$

Since the first of these equations contains x exactly in the same manner as the second contains y, it does not appear from these two equations, that there is any reason why x should have a different value from y: we may therefore suppose it equal to y, and taking either of the two equations we have

$$F\{x, x\} = 0$$

from which we may find several values of x and also of y, since by our supposition y is equal to x.

Ex. 1. Given the equations

$$x^2 + ay = b \quad \text{and} \quad y^2 + ax = b$$

making $y = x$ we have

$$x^2 + ax = b \quad \text{or} \quad x = y = -a/2 \pm \sqrt{b + a^2/4}.$$

As a particular case take $x^2 + 4y = 5$ and $y^2 + 4x = 5$. This gives $x = y = -2 \pm 3$, and it will be found on trial that $x = y = 1$ or $x = y = -5$, satisfy the equations.

Ex. 2. As a second example, let us take the two equations $x^3 - 8x = 2y^2$ and $y^3 - 8y = 2x^2$, since these contain x and y similarly we have *by* supposing $y = x$ either $x = y = 4$ or $x = y = -2$, either of which pair of values satisfy the equations.

In a similar manner it will be found that one solution of the equations

$$x^2 + axy + by^2 = cy$$
$$y^2 + ayz + bz^2 = cz$$

and

$$z^2 + azx + bx^2 = cx$$

will be

$$x = y = z = c/(1 + a + b).$$

The same principle may be extended to another class of equations comprehended under the form

$$F\{x, f(x, y)\} = 0, \quad F\{f(x, y), x\} = 0$$

for if we substitute v for $f(x, y)$ we have $F(x, v) = 0$ and $F(v, x) = 0$, and since these equations contain x and v similarly we may suppose $v = x$, and hence

$$F(x, x) = 0$$

from which we may determine x, and since $v = x$ we have $f(x, y) = x$ or $y = f^{1, -1}(x, x)$ which gives the values of y.

Ex. 3. Let

$$x^2 + ax(a' + b'x + c'y) = b$$

and

$$(a' + b'x + c'y)^2 + ax(a' + b'x + c'y) = b$$

be the given equations, since the first contains x and $a' + b'x + c'y$ in the same manner that the second contains $a' + b'x + c'y$ and x, we may therefore suppose $a' + b'x + c'y = x$ hence

$$x^2 + ax^2 = b \quad \text{and} \quad x = \sqrt{\frac{b}{1 + a}}.$$

Also from the equation $a + b'x + c'y = x$ we have

$$y = \frac{1 - b'}{c'}\sqrt{\frac{b}{1 + a}} - \frac{a'}{c'}.$$

As a particular case take the equations

$$x^2 - 2x(y + 2x - 4) = -4$$

$$(y + 2x - 4)^2 - 2x(y + 2x - 4) = -4.$$

The two pairs of values which satisfy this equation are

$$x = 2, y = 2 \quad \text{and} \quad x = -2, y = -6.$$

By treating the equations

$$x^2 + 4(1 - y^2) = 5$$

$$(1 - y^2)^2 + 4x = 5$$

by the same method we should find

$$x = 1, y = 0 \quad \text{and} \quad x = -5, y = \sqrt{6}.$$

For another example take the equations

$$x^3 + 6x^2 - 7\frac{y}{x} = 0 \quad \text{and} \quad \frac{y^3}{x^3} + 6\frac{y^2}{x^2} - 7x = 0$$

in this case x and y/x occur similarly, we may therefore suppose $y/x = x$ or $y = x^2$ this gives

$$x = 1, y = 1 \quad \text{and} \quad x = -7, y = 49$$

both which sets of values satisfy the equations.

It is obvious that similar reasoning may be applied to equations containing any number of variables. /

[249/50]

XI

ON SOME NEW METHODS OF INVESTIGATING THE SUMS OF SEVERAL CLASSES OF INFINITE SERIES

From *Philosophical Transactions of the Royal Society*, Vol. 109, pp. 249–82

Read 1 April, 1819

The processes which it is the object of this paper to explain, were discovered several years since; but certain difficulties connected with the subject, which I was at that time unable to explain, and which were equally inexplicable to several of my friends, to whom I had communicated these methods, induced me to defer publishing them, until I could offer some satisfactory solution.

These observations refer more particularly to the second method which I have detailed in this paper, and which may not inappropriately be called the *method of expanding horizontally and summing vertically*. Some traces of this method may, perhaps, be found in former writers, and particularly in a paper by Professor Vince,[a] 'On the Summation of Series,' printed in the Philosophical Transactions for 1791; but there exists this peculiarity in that which I have employed, that after a certain number of the vertical columns are summed, all the remainder either vanish, or else have some common factor. This method, which I employed about the year 1812, gave the values of a variety of series whose sums had not hitherto been known, most of which were apparently correct, but some of the consequences which followed were evidently erroneous. About this time, Mr Herschel, to whom I had / communicated these anomalous results, by following a very different course, arrived at several general theorems, which, when applied to the series I had obtained, gave the same results. This coincidence at first increased my confidence in the values so discovered, and I continued to examine the reason why my own formulae were in some cases defective. Mr Herschel's method was published in the Philosophical Transactions for 1814;[b] and it was not until some time after that I perceived, that

[a] Samuel Vince, 'A new method of investigating the sums of infinite series' *Philosophical Transactions*, Vol. 81 (1791), pp. 295–316.
[b] J. F. W. Herschel, 'Consideration of various points of analysis' *Philosophical Transactions*, Vol. 104 (1814), pp. 440–68.

although the investigations were very different, the fundamental principle was the same in both methods. This induced me to attempt summing the same series by a direct process, and I succeeded in obtaining their sums by integration relative to finite differences, aided by certain peculiar artifices. The results obtained by this new plan, which is the first treated of in this paper, coincided with those already found, and seemed to confirm their truth, without in the least indicating the cause of the error: this cause however I now began to suspect, and, after some enquiry, I was at length able to detect. I have found that the *method of expanding horizontally and summing vertically*, will always lead to correct results, provided a certain series which I have pointed out, is finite. I have also shown how to express this series by a definite integral; and when this integral or this series has a finite value, the method may be depended on. In case this series or this definite integral is not finite, then the value of the series* multiplied by zero, must be added to / the sum given by this method. In this latter case, however, the mode of summation which I have proposed, is not well adapted for giving the sums of series; its greatest advantage is felt when the integral or series alluded to is finite: but even in this case the criterion I have pointed out is not useless, for it serves to except certain particular values of the variables, which would give incorrect results. Without this criterion, or without something equivalent to it, I am inclined to think that the principle on which this method is founded, although it will probably in many cases give accurate results, will in others produce such as are not only numerically but symbolically untrue. It is worthy of remark, that the *method of expanding horizontally and summing vertically*, in many instances, gives precisely the same formulae as the direct process of integration; yet that that method attaches limitations to them, which are necessary to their accuracy, but which are not indicated by the method last mentioned.

Before I proceed to explain these two processes, it will be convenient to prove that the values of all series of the forms

$$A_1 x \frac{(\sin\theta)^m}{(\cos\theta)^n} + A_2 x^2 \frac{(\sin 2\theta)^m}{(\cos 2\theta)^n} + A_3 x^3 \frac{(\sin 3\theta)^m}{(\cos 3\theta)^n} + \dots$$

$$A_1 x \frac{(\cos\theta)^m}{(\sin\theta)^n} + A_2 x^2 \frac{(\cos 2\theta)^m}{(\sin 2\theta)^n} + A_3 x^3 \frac{(\cos 3\theta)^m}{(\sin 3\theta)^n} + \dots$$

depend on series of the form

$$A_1 \frac{x}{(\cos\theta)^n} + A_2 \frac{x^2}{(\cos 2\theta)^n} + \dots$$

* The investigation of this series is generally a task of considerable difficulty. I have however given an example, wherein the correction thus found, added to the sum indicated by the method we are considering, gives the true value of the series, which in this case is one whose sum has been found by Euler.

$$A_1x\frac{\sin\theta}{(\cos\theta)^n}+A_2x^2\frac{\sin 2\theta}{(\cos 2\theta)^n}+\ldots$$

and

$$A_1\frac{x}{(\sin\theta)^n}+A_2\frac{x^2}{(\sin 2\theta)^n}+\ldots$$

$$A_1x\frac{\cos\theta}{(\sin\theta)^n}+A_2x^2\frac{\cos 2\theta}{(\sin 2\theta)^n}+\ldots$$

/

or else they depend partly on these and partly on other series, containing the powers of the sines or cosines of an arc in arithmetical progression in their numerators, which is a species whose sums are easily found. For the sake of brevity, I shall make use of the general term of any series with the characteristic S prefixed to it to denote that series. Beginning then with the series $SA_ix^i(\sin i\theta)^m/(\cos i\theta)^n$, we observe that when m is an even number, we have

$$SA_ix^i\frac{(\sin i\theta)^m}{(\cos i\theta)^n}=SA_ix^i\frac{\{1-(\cos i\theta)^2\}^{m/2}}{(\cos i\theta)^n}$$

$$=SA_ix^i\frac{1}{(\cos i\theta)^n}-\frac{m}{2}SA_ix^i\frac{1}{(\cos i\theta)^{n-2}}$$

$$+\frac{m\cdot m-2}{2\cdot 4}SA_ix^i\frac{1}{(\cos i\theta)^{n-4}}-\ldots \qquad (a)$$

This series will always terminate when m is an even number; and if m is greater than n, the last term will have no cosines in its denominator: if $m=n$, the last term will be SA_ix^i; and if m is less than n, the last term will be

$$SA_ix^i\frac{1}{(\cos i\theta)^{n-m}}$$

so that in all cases when m is an even number, the series in question will depend on series of the form

$$SA_ix^i\frac{1}{(\cos i\theta)^n}$$

or on others whose sums are known.

Let us now consider the case of m = an odd number; then we have

$$SA_ix^i\frac{(\sin i\theta)^m}{(\cos i\theta)^n} = SA_ix^i\frac{\sin i\theta\{1-(\cos i\theta)^2\}^{(m-1)/2}}{(\cos i\theta)^n}$$

$$= SA_ix^i\frac{(\sin i\theta)}{(\cos i\theta)^n} - \frac{m-1}{2}SA_ix^i\frac{\sin i\theta}{(\cos i\theta)^{n-2}}$$

$$+ \frac{m-1\cdot m-3}{2\cdot 4}SA_ix^i\frac{\sin i\theta}{(\cos i\theta)^{n-4}} + \ldots \qquad (b)$$

This series always terminates when m is an odd number; and in a similar manner we shall find the two following: /

$$SA_ix^i\frac{(\cos i\theta)^m}{(\sin i\theta)^n} = SA_ix^i\frac{\{1-(\sin i\theta)^2\}^{m/2}}{(\sin i\theta)^n}$$

$$= SA_ix^i\frac{1}{(\sin i\theta)^n} - \frac{m}{2}SA_ix^i\frac{1}{(\sin i\theta)^{n-2}}$$

$$+ \frac{m\cdot m-2}{2\cdot 4}SA_ix^i\frac{1}{(\sin i\theta)^{n-4}} - \ldots \qquad (c)$$

when m is an even number, and

$$SA_ix^i\frac{(\cos i\theta)^m}{(\sin i\theta)^n} = SA_ix^i\frac{\cos i\theta\{1-(\sin i\theta)^2\}^{(m-1)/2}}{(\sin i\theta)^n}$$

$$= SA_ix^i\frac{\cos i\theta}{(\sin i\theta)^n} - \frac{m-1}{2}SA_ix^i\frac{\cos i\theta}{(\sin i\theta)^{n-2}}$$

$$+ \frac{m-1\cdot m-3}{2\cdot 4}SA_ix^i\frac{\cos i\theta}{(\sin i\theta)^{n-4}} - \ldots \qquad (d)$$

when m is an odd number.

Let us now propose to investigate the sum of the series

$$\frac{A_1x}{(\sin\theta)^n} + \frac{A_2x^2}{(\sin 2\theta)^n} + \frac{A_3x^3}{(\sin 3\theta)^n}\ldots$$

Assume

$$\psi x = A_1x + A_2x^2 + A_3x^3 + \ldots$$

Put v^{2z} for x; then it becomes

$$\psi v^{2z} = A_1v^{2z} + A_2v^{4z} + A_3v^{6z} + \ldots$$

Integrate both sides, observing that $\Sigma v^{2iz} = v^{2iz}/(v^2-1)$; then we have

$$\sum\psi v^{2z} = A_1\frac{v^{2z}}{v^2-1} + A_2\frac{v^{4z}}{v^4-1} + A_3\frac{v^{6z}}{v^6-1} + \ldots$$

Integrate again, and after the nth integration we shall have

$$\overset{n}{\sum}\psi v^{2z} = A_1\frac{v^{2z}}{v^2 - 1^n} + A_2\frac{v^{4z}}{v^4 - 1^n} + A_3\frac{v^{6z}}{v^6 - 1^n} + \ldots$$

Now let $v = \cos\theta \pm \sqrt{-1}\sin\theta$; then our equation becomes

$$(2\sqrt{-1})^n\overset{n}{\sum}\psi v^{2z} = A_1\frac{v^{2z-n}}{(\sin\theta)^n} + A_2\frac{v^{4z-2n}}{(\sin 2\theta)^n} + A_3\frac{v^{6z-3n}}{(\sin 3\theta)^n} + \ldots$$

Put $z + n/2$ for z, and we have

$$(2\sqrt{-1})^n\overset{n}{\sum}\psi v^{2z+n} = A_1\frac{v^{2z}}{(\sin\theta)^n} + A_2\frac{v^{4z}}{(\sin 2\theta)^n} + A_3\frac{v^{6z}}{(\sin 3\theta)^n} + \ldots$$

$$= A_1\frac{x}{(\sin\theta)^n} + A_2\frac{x^2}{(\sin 2\theta)^n} + A_3\frac{x^3}{(\sin 3\theta)^n} + \ldots \quad (1)$$

If after the integration we had put v^{-1} instead of v, and then $v = \cos\theta \pm \sqrt{-1}\sin\theta$, we should have found

$$(-2\sqrt{-1})^n\overset{n}{\sum}\psi v^{-2z-n} = A_1\frac{v^{-2z}}{(\sin\theta)^n} + A_2\frac{v^{-2z}}{(\sin 2\theta)^n} + A_3\frac{v^{-6z}}{(\sin 3\theta)^n} + \ldots$$

$$= A_1\frac{x^{-1}}{(\sin\theta)^n} + A_2\frac{x^{-2}}{(\sin 2\theta)^n} + A_3\frac{x^{-3}}{(\sin 3\theta)^n} + \ldots \quad (2)$$

Neither of the integrals exhibited in (1) and (2) are integrable in the most simple cases, and it is only by their combination that I have been able to obtain the sums of any series. Let us suppose $A = 1$, $A_2 = -1$, $A_3 = 1$, $A_4 = -1, \ldots$ then

$$\psi v^{2z+n} = \frac{v^{2z+n}}{1 + v^{2z+n}} \quad \text{and} \quad \psi v^{-2z-n} = \frac{v^{-2z-n}}{1 + v^{-2z-n}}$$

also let $n = 1$; then the difference of the two series is

$$(2\sqrt{-1})\sum\left\{\frac{v^{2z+1}}{1 + v^{2z+1}} + \frac{v^{-2z-1}}{1 + v^{-2z-1}}\right\} = \frac{x - x^{-1}}{\sin\theta} - \frac{x^2 - x^{-2}}{\sin 2\theta} + \frac{x^3 - x^{-3}}{\sin 3\theta} \cdots$$

but the integral on the left side of this equation becomes $(2\sqrt{-1})\Sigma(1)$ which is equal to $2\sqrt{-1}(z + b)$: hence, since $z = \log x/2\log v$ we have

$$2\sqrt{-1}\left\{\frac{\log x}{2\log v} + b\right\} = \frac{x - x^{-1}}{\sin\theta} - \frac{x^2 - x^{-2}}{\sin 2\theta} + \frac{x^3 - x^{-3}}{\sin 3\theta} + \ldots$$

If $x = 1$, $b = 0$, and since $\log v = \theta\sqrt{-1}$, our series becomes

$$\frac{\log x}{\theta} = \frac{x - x^{-1}}{\sin\theta} - \frac{x^2 - x^{-2}}{\sin 2\theta} + \frac{x^3 - x^{-3}}{\sin 3\theta} - \dots \tag{3}$$

Let $x = \cos\theta' \pm \sqrt{-1}\,\sin\theta'$; then, since $\log x$ will become $\theta'\sqrt{-1}$, by dividing both sides by $2\sqrt{-1}$, we shall have

$$\frac{\theta'}{\theta} = \frac{\sin\theta'}{\sin\theta} - \frac{\sin 2\theta'}{\sin 2\theta} + \frac{\sin 3\theta'}{\sin 3\theta} - \dots \tag{4}$$

The series (3) is integrable when multiplied by dx/x, and this operation may be repeated any number of times; the first / operation produces

$$\frac{(\log x)^2}{2\theta} \pm c_1 = \frac{x + x^{-1}}{1\sin\theta} - \frac{x^2 + x^{-2}}{2\sin 2\theta} - \frac{x^3 + x^{-3}}{3\sin 3\theta} + \dots$$

the value of the constant c_1, which is equal to the series

$$c_1 = 2\left\{\frac{1}{1\sin\theta} - \frac{1}{2\sin 2\theta} + \frac{1}{3\sin 3\theta} - \dots\right\}$$

cannot be determined from this equation, but by a second multiplication by dx/x and again integrating, it may readily be found: this second operation gives

$$\frac{(\log x)^3}{2\cdot 3\cdot\theta} + \frac{\log x}{1}c_1 + c_2 = \frac{x - x^{-1}}{1^2\sin\theta} - \frac{x^2 - x^{-2}}{2^2\sin 2\theta} + \frac{x^3 - x^{-3}}{3^2\sin 3\theta} - \dots$$

If $x = 1$, $c_2 = 0$, put $x = \cos\theta + \sqrt{-1}\,\sin\theta$, then we have

$$\frac{(\theta\sqrt{-1})^3}{2\cdot 3\cdot\theta} + \frac{\theta\sqrt{-1}}{1}c_1 = 2\sqrt{-1}\left\{\frac{1}{1^2} - \frac{1}{2^2} + \frac{1}{3^2} - \dots\right\} = 2\sqrt{-1}S\frac{\pm 1}{i^2}{}^*$$

From this equation the value of c_1 may be found; it is

$$c_1 = \frac{1}{\theta}\left\{\frac{\theta^2}{6} + 2S\frac{\pm 1}{i^2}\right\}.$$

The value of c_1 thus found, we have the series

$$\frac{(\log x)^2}{2\cdot\theta} + \frac{1}{\theta}\left\{\frac{\theta^2}{6} + 2S\frac{\pm 1}{i^2}\right\} = \frac{x + x^{-1}}{1\sin\theta} - \frac{x^2 + x^{-2}}{2\sin 2\theta} + \frac{x^3 + x^{-3}}{3\sin 3\theta} - \dots \tag{5}$$

* Throughout the course of this paper I shall have continual occasion to employ the series $1/1^{2n} - 1/2^{2n} + 1/3^{2n} - \dots$; they can always be expressed by means of the numbers of Bernouilli, and the powers of π, and for the sake of brevity I shall always denote them by $S(\pm 1/i^{2n})$.

and

$$\frac{(\log x)^3}{1\cdot 2\cdot 3\cdot\theta}+\frac{\log x}{\theta}\left\{\frac{\theta^2}{6}+2S\frac{\pm 1}{i^2}\right\}$$
$$=\frac{x-x^{-1}}{1^2\sin\theta}-\frac{x^2-x^{-2}}{2^2\sin 2\theta}+\frac{x^3-x^{-3}}{3^2\sin 3\theta}-\dots \quad (6)$$

In the first of these put $x=\cos\theta+\sqrt{-1}\sin\theta$, and it becomes

$$-\frac{\theta}{6}+\frac{1}{\theta}S\frac{\pm 1}{i^2}=\frac{\cot\theta}{1}-\frac{\cot 2\theta}{2}+\frac{\cot 3\theta}{3}-\dots \quad (7)$$

/

By continuing to multiply (3) by dx/x and integrating, it is easy to perceive that we should arrive at the two following theorems:

$$\frac{(\log x)^{2k}}{1\cdot 2\cdot\ldots\cdot 2k\cdot\theta}+\frac{(\log x)^{2k-2}}{1\cdot 2\cdot\ldots\cdot 2\,k-2}c_1+\dots+c_{2k-1}$$
$$=\frac{x+x^{-1}}{1^{2k-1}\sin\theta}-\frac{x^2+x^{-2}}{2^{2k-1}\sin 2\theta}+\frac{x^2+x^{-3}}{3^{2k-1}\sin 3\theta}-\dots \quad (8)$$

$$\frac{(\log x)^{2k+1}}{1\cdot 2\cdot\ldots\cdot 2\,k+1\cdot\theta}+\frac{(\log x)^{2k-1}}{1\cdot 2\cdot\ldots\cdot 2\,k-1}c_1+\dots+\frac{\log x}{1}c_{2k-1}$$
$$=\frac{x-x^{-1}}{1^{2k}\sin\theta}-\frac{x^2-x^{-2}}{2^{2k}\sin 2\theta}+\frac{x^3-x^{-3}}{3^{2k}\sin 3\theta}-\dots \quad \langle 9\rangle$$

The constants c_1, c_3, c_5, etc. may easily be determined from each other, the value of c_1 has been already found, that of c_3 and c_5 are as follows:

$$c_3=\frac{2}{\theta}S\frac{\pm 1}{i^4}+\frac{2\theta}{6}S\frac{\mp 1}{i^2}+\frac{7\theta^3}{360}$$

$$c_5=\frac{2}{\theta}S\frac{\pm 1}{i^6}+\frac{2\theta}{6}S\frac{\pm 1}{i^4}+2\frac{7\theta^3}{360}S\frac{\pm 1}{i^2}+\theta^5\frac{31}{5040}.$$

A variety of series are deducible from those of (8); I shall only mention two of them;

$$\frac{\cot\theta}{1^{2k-1}}+\frac{\cot 2\theta}{2^{2k-1}}+\frac{\cot 3\theta}{3^{2k-1}}-\dots$$

and

$$\frac{1}{1^{2k}}\cdot\frac{\sin\theta'}{\sin\theta}-\frac{1}{2^{2k}}\cdot\frac{\sin 2\theta'}{\sin 2\theta}+\frac{1}{3^{2k}}\cdot\frac{\sin 3\theta'}{\sin 3\theta}-\dots$$

Returning to the formulae (1) and (2) by addition and subtraction, we shall have when n is even

$$(2\sqrt{-1})^n \sum^n \{\psi v^{2z+n} + (-1)^n \psi v^{-2z-n}\}$$

$$= A_1 \frac{x + x^{-1}}{(\sin \theta)^n} + A_2 \frac{x^2 + x^{-2}}{(\sin 2\theta)^n} + A_3 \frac{x^3 + x^{-3}}{(\sin 3\theta)}{}^n + \dots$$

/

and when n is an odd number,

$$(2\sqrt{-1})^n \sum^n \{\psi v^{2z+n} + (-1)^n \psi v^{-2z-n}\}$$

$$= A_1 \frac{x - x^{-1}}{(\sin \theta)^n} + A_2 \frac{x^2 - x^{-2}}{(\sin 2\theta)^n} + A_3 \frac{x^3 - x^{-3}}{(\sin 3\theta)^n} + \dots$$

From these expressions it appears, that the reason why we have succeeded in the integrations is, because we had so assumed ψ, that the sum, $\psi v^{2z+n} + \psi v^{-2z-n}$ is a constant quantity; the same success must follow whenever this condition is fulfilled: and hence, we have a method of discovering the sums of a great variety of series, containing the powers of the sines of arcs in arithmetical progression in their denominators, by solving the functional equation $\psi v^{2z+n} + \psi v^{-2z-n} = c$. This is fortunately one of a class whose general solution I have arrived at;*[a] it is

$$\psi v^{2z+n} = \frac{c\varphi v^{2z+n}}{\varphi v^{2x+n} + \varphi v^{-2z-n}} \quad \text{or} \quad \psi x = \frac{c\varphi x}{\varphi x + \varphi(1/x)}$$

In the example I have employed n was supposed equal to unity; if this is not the case, we should have found

$$(2\sqrt{-1})^n \sum^n (1) = \frac{x \pm x^{-1}}{(\sin \theta)^n} - \frac{x^2 \pm x^{-2}}{(\sin 2\theta)^n} + \frac{x^3 \pm x^{-3}}{(\sin 3\theta)^n} - \dots$$

If in the functional equation we put $c = 1$, and $\varphi x = \tan^{-1} x$, then we have $\psi x = (2/\pi) \tan^{-1} x$, and

$$(2\sqrt{-1})^n \sum^n (1) = \frac{2}{\pi} \left\{ \frac{x \pm x^{-1}}{1\,(\sin \theta)^n} - \frac{x^3 \pm x^{-3}}{3\,(\sin 3\theta)^n} - \frac{x^5 \pm x^{-5}}{5\,(\sin 5\theta)^n} - \dots \right\}$$

* See *Philosophical Transactions* for 1817, p. 202.

[a] 'Observations on the analogy which subsists between the calculus of functions and the other branches of analysis' (1817), in this volume of the *Works of Babbage*.

the upper or under sign being used as n, is even or odd; if $n = 1$, the constant is zero; and we have

$$\frac{\pi \log x}{2\theta} = \frac{x - x^{-1}}{1 \sin\theta} - \frac{x^3 - x^{-3}}{3 \sin 3\theta} + \frac{x^5 - x^{-5}}{5 \sin 5\theta} - \ldots \qquad (9)$$

/

this may be multiplied by dx/x, and integrated any number of times in the same manner as (3), and the results would be

$$\frac{\pi}{2\theta} \cdot \frac{(\log x)^{2k}}{1 \cdot 2 \cdot \ldots \cdot 2k} + \frac{(\log x)^{2k-2}}{1 \cdot 2 \cdot \ldots \cdot 2k - 2} c_1 + \ldots + c_{2k-1}$$

$$= \frac{x + x^{-1}}{1^2 k \sin\theta} - \frac{x^3 + x^{-3}}{3^{2k} \sin 3\theta} + \frac{x^5 + x^{-5}}{5^{2k} \sin 5\theta} - \ldots \qquad (1{,}1)$$

$$\frac{\pi}{2\theta} \cdot \frac{(\log x)^{2k+1}}{1 \cdot 2 \cdot \ldots \cdot 2k + 1} + \frac{(\log x)^{2k-1}}{1 \cdot 2 \cdot \ldots \cdot 2k - 1} c_1 + \ldots + \frac{\log x}{1} c_{2k-1}$$

$$= \frac{x - x^{-1}}{1^{2k+1} \sin\theta} - \frac{x^3 - x^{-3}}{3^{2k+1} \sin 3\theta} + \frac{x^5 - x^{-5}}{5^{2k+1} \sin 5\theta} - \ldots$$

and these constants may be determined one from the other in the same manner as the former. I shall only give the value of the first, in order to compare the value of the series to which it is equal, with the sum of the same series deduced in another manner

$$c = \frac{\pi\theta}{12} + \frac{2}{\theta} S \frac{\pm 1}{(2i + 1)^3} = 2\left\{\frac{1}{1^2 \sin\theta} - \frac{1}{3^2 \sin 3\theta} + \frac{1}{5^2 \sin 5\theta} - \ldots\right\} \qquad (1{,}2)$$

In order to ascertain the sums of series which contain cosines in their denominators, we must use an artifice which I shall now explain.

Assuming as before $\psi x = A_1 x + A_2 x^2 + A_3 x^3 + \ldots$, and putting v^{2z} for x, we have

$$\psi v^{2z} = A_1 v^{2z} + A_2 v^{4z} + A_3 v^{6z} + \ldots$$

If we were now to integrate this, we should introduce into the denominator of any term $v^{2i} - 1$; but we want to introduce the same expression with the signs of both terms positive. If we multiply both sides by $(-1)^z$ and then integrate, we shall have first

$$(-1)^{-z} \psi v^{2z} = A_1 v^{2z}(-1)^{-z} + A_2 v^{4z}(-1)^{-z} + A_3 v^{6z}(-1)^{-z} + \ldots$$

/

And since

$$\Delta(-1)^{-z} v^{2iz} = -(-1)^{-z} v^{2iz+2i} - (-1)^{-z} v^{2iz} = -(v^{2i} + 1)(-1)^z v^{2iz}$$

we find

$$\sum(-1)^z v^{2iz} = -\frac{v^{2iz}}{v^{2i}+1}.$$

Integrating each term separately, we have

$$\sum(-1)^{-z}\psi v^{2z} = -\left\{A_1\frac{v^{2z}(-1)^{-z}}{v^2+1} + A_2\frac{v^{4z}(-1)^{-z}}{v^4+1} + A_3\frac{v^{6z}(-1)^{-z}}{v^6+1} + \ldots\right\}.$$

Let this integration be repeated n times, it will give

$$(-1)^n\sum^n(-1)^{-z}\psi v^{2z} = A_1\frac{v^{2z}(-1)^{-z}}{(v^2+1)^n} + A_2\frac{v^{4z}(-1)^{-z}}{(v^4+1)^n} + A_3\frac{v^{6z}(-1)^{-z}}{(v^6+1)^n} + \ldots$$

Let $v = \cos\theta\sqrt{-1}\sin\theta$, and $z + n/2$ for z; this becomes

$$(-2)^n\sum^n(-1)^{z-n/2}\psi v^{2z+n}$$
$$= (-1)^{-z-n/2}\left\{A_1\frac{v^{2z}}{(\cos\theta)^n} + A_2\frac{v^{4z}}{(\cos 2\theta)^n} + A_3\frac{v^{6z}}{(\cos 3\theta)^n} + \ldots\right\}.$$

And finally,

$$(-2)^n(-1)^z\sum^n(-1)^{-z}\psi v^{2z+n}$$
$$= A_1\frac{v^{2z}}{(\cos\theta)^n} + A_2\frac{v^{4z}}{(\cos 2\theta)^n} + A_3\frac{v^{6z}}{(\cos 3\theta)^n} + \ldots$$
$$= A_1\frac{x}{(\cos\theta)^n} + A_2\frac{x^2}{(\cos 2\theta)^n} + A_3\frac{x^3}{(\cos 3\theta)^n} + \ldots \qquad (1,3)$$

The integrations here indicated will, as in a former instance, generally surpass the powers of analysis in its present state, but a contrivance similar to that which has been already stated, will in many cases elude the difficulty: the artifice consists in investigating another similar series arranged according to the descending powers of the variable, integrating it in the same manner as we have that marked (1,3), and adding these two results, we shall in many cases have a function which is integrable, and the two series become equal in the case of $x = 1$. By commencing with the descending / series

$$\psi v^{-2z} = A_1 v^{-2z} + A_2 v^{-4z} + A_3 v^{-6z} + \ldots$$

multiplying by $(-1)^{-z}$ and integrating, we shall get the expression

$$(-2)^n(-1)^z\sum^n(-1)^{-z}\psi v^{-2z-n}$$

$$= A_1\frac{v^{-2z}}{(\cos\theta)^n} + A_2\frac{v^{-4z}}{(\cos 2\theta)^n} + A_3\frac{v^{-6z}}{(\cos 3\theta)^n} - \dots$$

$$= A_1\frac{x^{-1}}{(\cos\theta)^n} + A_2\frac{x^{-2}}{(\cos 2\theta)^n} + A_3\frac{x^{-3}}{(\cos 2\theta)^n} + \dots \quad (1,4)$$

There occur very few cases in which it is possible to execute the integrations in (1,3) and (1,4); by adding the two together, we have

$$(-2)^n(-1)^z\sum^n(-1)^{-z}\{\psi v^{2z+n} + \psi v^{-2z-n}\}$$

$$= A_1\frac{x + x^{-1}}{(\cos\theta)^n} + A_2\frac{x^2 + x^{-2}}{(\cos 2\theta)^n} + A_3\frac{x^3 + x^{-3}}{(\cos 3\theta)^n} + \dots \quad (1,5)$$

Here we may observe that the new series is exactly double either of the others (1,3) or (1,4) when $x = 1$; also, that the integration on the left side can be executed any number of times, whenever $\psi v^{2x+n}\psi v^{-2z-n}$ is a constant quantity; the forms of the function ψ, which fulfil this condition, have already been given. Let $\psi x = x/(1 + x)$, then $A_1 = 1$, $A_2 = -1$, $A_3 = 1$, ... and since $\psi v^{2z+n} + \psi v^{-2z-n} = 1$, we have

$$(-2)^n(-1)^z\sum^n(-1)^{-z} = \frac{x + x^{-1}}{(\cos\theta)^n} - \frac{x^2 + x^{-2}}{(\cos 2\theta)^n} + \frac{x^3 + x^{-3}}{(\cos 3\theta)^n} + \dots$$

These integrations are easily executed; and commencing with $n = 1$, we have

$$-2\cdot(-1)^z\frac{(-1)^{-z}}{-1-1} + 2b(-1)^z$$

$$= 1 + 2b(-1)^z = \frac{x + x^{-1}}{\cos\theta} - \frac{x^2 + x^{-2}}{\cos 2\theta} + \dots$$

In order to determine the constant b, put $x = \cos\theta + \sqrt{-1}\sin\theta$; then, since z in that case becomes $1/2$, we have

$$1 + 2b\sqrt{-1} = 2 - 2 + 2 - 2 + \dots = 1$$

hence $b = 0$, and we have

$$1 = \frac{x + x^{-1}}{\cos\theta} - \frac{x^2 + x^{-2}}{\cos 2\theta} + \frac{x^3 + x^{-3}}{\cos 3\theta} - \dots \quad (1,6)$$

Continuing to integrate, it will be found that all the constants are zero, and we shall arrive at the following theorem;

$$1 = \frac{x + x^{-1}}{(\cos \theta)^n} - \frac{x + x^{-2}}{(\cos 2\theta)^n} + \frac{x^3 + x^{-3}}{(\cos 3\theta)^n} - \dots \tag{1,7}$$

Let $x = -x$, then it becomes

$$-1 = \frac{x + x^{-1}}{(\cos \theta)^n} + \frac{x^2 + x^{-2}}{(\cos 2\theta)^n} + \frac{x^3 + x^{-3}}{(\cos 3\theta)^n} + \dots \tag{1,8}$$

Putting $x = 1$ in both these, we have

$$\frac{1}{2} = \frac{1}{(\cos \theta)^n} - \frac{1}{(\cos 2\theta)^n} + \frac{1}{(\cos 3\theta)^n} - \dots \tag{1,9}$$

$$-\frac{1}{2} = \frac{1}{(\cos \theta)^n} + \frac{1}{(\cos 2\theta)^n} + \frac{1}{(\cos 3\theta)^n} + \dots \tag{2,1}$$

I propose in the next place to determine the value of the series

$$\frac{1}{1^{2k}(\cos \theta)^n} - \frac{1}{2^{2k}(\cos 2\theta)^n} + \frac{1}{3^{2k}(\cos 3\theta)^n} - \dots$$

This may be accomplished by multiplying (1,7) by dx/x, and integrating; this operation, being performed on it 2^k times, will produce the series whose sum is required; the first integration gives

$$\frac{\log x}{1} + c = \frac{x - x^{-1}}{1(\cos \theta)^n} - \frac{x^2 - x^{-2}}{2(\cos \theta)^n} + \frac{x^3 - x^{-3}}{3(\cos 3\theta)^n} - \dots$$

If $x = 1$, $c = 0$, the second operation gives

$$\frac{(\log x)^2}{1 \cdot 2} + c_2 = \frac{x + x^{-1}}{1^2(\cos \theta)^n} - \frac{x^2 + x^{-2}}{2^2(\cos 2\theta)^n} - \dots$$

If $x = 1$

$$c_{2,n} = \frac{2}{1^2(\cos \theta)^n} - \frac{2}{2^2(\cos 2\theta)^n} + \dots$$

In order to determine $c_{2,n}$ put $x = \cos \theta + \sqrt{-1} \sin \theta$, then we have /

$$-\frac{\theta^2}{2} + c_{2,n} = \frac{2}{1^2(\cos \theta)^{n-1}} - \frac{2}{2^2(\cos \theta)^{n-1}} + \dots = c_{2,n-1}.$$

The equation $c_{2,n} - c_{2,n-1} = \theta^2/2$ being integrated, gives

$$c_{2,n} = \frac{n}{2} \cdot \frac{\theta^2}{2} + b.$$

If $n = 0$

$$c_{2,0} = b = \frac{2}{1^2} - \frac{2}{2^2} + \frac{2}{3^2} - \ldots = 2S\frac{\pm 1}{i^2}.$$

Hence

$$c_{2,n} = \frac{n}{1}\cdot\frac{\theta^2}{2} + 2S\frac{\pm 1}{i^2}$$

and

$$\frac{(\log\cdot x)^2}{1\cdot 2} + \frac{n}{1}\cdot\frac{\theta^2}{2} + 2S\frac{\pm 1}{i^2} = \frac{x + x^{-1}}{1^2(\cos\theta)^n} - \frac{x^2 + x^{-2}}{2^2(\cos 2\theta)^n} + \frac{x^3 + x^{-3}}{3^2(\cos 3\theta)^n} - \ldots \quad (2,2)$$

These integrations being repeated, we shall arrive at the two following expressions:

$$\frac{(\log x)^{2k}}{1\cdot 2\cdot\ldots\cdot 2k} + \frac{(\log x)^{2k-2}}{1\cdot 2\cdot\ldots\cdot 2k-2}c_{2,n} + \ldots + \frac{(\log x)^2}{1\cdot 2}c_{2k-2,n} + c_{2k,n} = \frac{x + x^{-1}}{1^{2k}(\cos\theta)^n} - \frac{x^2 + x^{-2}}{2^{2k}(\cos 2\theta)^n} + \frac{x^3 + x^{-3}}{3^{2k}(\cos 3\theta)^n} + \ldots \quad (2,3)$$

$$\frac{(\log x)^{2k+1}}{1\cdot 2\cdot\ldots\cdot 2k+1} + \frac{(\log x)^{2k-1}}{1\cdot 2\cdot\ldots\cdot 2k-1}c_{2,n} + \ldots + c_{2k,n} = \frac{x - x^{-1}}{1^{2k+1}(\cos\theta)^n} - \frac{x^2 - x^{-2}}{2^{2k+1}(\cos 2\theta)^n} + \frac{x^3 - x^{-3}}{3^{2k+1}(\cos\theta)^n} - \ldots$$

It now becomes necessary to determine the value of $c_{2k,n}$, which is equal to twice the sum of the series we are investigating; for if $x = 1$

$$c_{2k,n} = \frac{2}{1^{2k}(\cos\theta)^n} - \frac{2}{2^{2k}(\cos 2\theta)^n} + \frac{2}{3^{2k}(\cos 3\theta)^n} - \ldots$$

For this purpose put in the first of the equations (2,3) $x = \cos\theta + \sqrt{-1}\sin\theta$, then the series on the right hand is equal to $c_{2k,n-1}$, and we have for determining $c_{2k,n}$ the equation of finite differences:

$$\frac{(\theta\sqrt{-1})^{2k}}{1\cdot 2\cdot\ldots\cdot 2k} + \frac{(\theta\sqrt{-1})^{2k-2}}{1\cdot 2\cdot\ldots\cdot 2k-2}c_{2,n} + \frac{(\theta\sqrt{-1})^{2k-4}}{1\cdot 2\cdot\ldots\cdot 2k-4}c_{4,n} + \ldots + c_{2k,n} = c_{2k,n-1}.$$

In order to integrate this equation, let us suppose $c_{2k,n}$ to represent the co-efficient of r^{2k} in the development of $f(r)/(\alpha r)^n$ where

$$\alpha(r) = 1 + Ar^2 + Br^4 + Cr^6 + \dots$$

then

$$\frac{f(r)}{(\alpha r)^n} = 1 + r^2 c_{2,n} + r^4 c_{4,n} + r^6 c_{6,n} + \dots$$

If this be multiplied by αr, it becomes

$$\begin{aligned}\frac{f(r)}{(\alpha r)^{n-1}} = 1 + r^2 c_{2,n} &+ r^4 c_{4,n} + r^6 c_{6,n} + \dots \\ + r^2 A &+ r^4 A c_{2,n} + r^6 A c_{4,n} + \dots \\ &+ r^4 B + r^6 B c_{2,n} + \dots \\ &\qquad + r^6 C + \dots\end{aligned}$$

But the co-efficient of r^{2k} in this series is equal to $c_{2k,n-1}$ hence

$$c_{2k,n} + Ac_{2k-2,n} + Bc_{2k-4,n} + \dots = c_{2k,n-1}.$$

This equation will become the one in question, if we make

$$A = \frac{(\theta\sqrt{-1})^2}{1\cdot 2} \qquad B = \frac{(\theta\sqrt{-1})^4}{1\cdot 2\cdot 3\cdot 4} \qquad C = \frac{(\theta\sqrt{-1})^6}{1\cdot 2\cdot\dots\cdot 6}\dots$$

This produces

$$c_{2k,n} + \frac{(\theta\sqrt{-1})^2}{1\cdot 2}c_{2k-2,n} + \frac{(\theta\sqrt{-1})^4}{1\cdot 2\cdot 3\cdot 4}c_{2k-4,n} + \dots = c_{2k,n-1}.$$

We have now only to determine the form of $f(r)$, and this may be easily accomplished since the values of $c_{2k,0}$ are known; for if we put $n = 0$

$$\begin{aligned}f(r) &= 1 + r^2 c_{2,0} + r^4 c_{4,0} + r^6 c_{6,0} + \dots \\ &= 1 + 2r^2 S\frac{\pm 1}{i^2} + 2r^4 S\frac{\pm 1}{i^4} + 2r^6 S\frac{\pm 1}{i^6} + \dots\end{aligned}$$

Therefore $c_{2k,n}$ is equal to the coefficient of r^{2k} in the development of

$$\left(1 + 2r^2 S\frac{\pm 1}{i^2} + 2r^4 S\frac{\pm 1}{i^4} + \dots\right)/(\cos r\theta)^n.$$

Or if

$$(\cos r\theta)^{-n} = 1 + A'_n\theta^2 r^2 + B'_n\theta^4 r^4 + C'_n\theta^6 r^6 + \dots$$

then

$$c_{2k,n} = 2S\frac{\pm 1}{i^{2k}} + 2A'_n\theta^2 S\frac{\pm 1}{i^{2k-2}} + 2B'_n\theta^4 S\frac{\pm 1}{i^{2k-4}} + 2C'_n\theta^6 S\frac{\pm 1}{i^{2k-6}}\dots$$

The quantity $c_{2k,n}$ may now be considered as completely determined, since it only depends on the coefficients of $(\cos\theta)^{-n}$, and the series marked by $S(\pm 1/i^{2k})$, both which quantities are known; the latter being given by the powers of π and numbers of Bernouilli, whilst the values of the former in functions of n are given by Legendre, in his Exercise de Calcul Integral,[a] vol. iii. art. 149, 155.

In (2, 3) let $x = 1$, and we have

$$\frac{1}{2}c_{2k,n} = \frac{1}{1^{2k}(\cos\theta)^n} - \frac{1}{2^{2k}(\cos 2\theta)^n} + \frac{1}{3^{2k}(\cos 3\theta)^n} - \dots \quad (2,4)$$

And if we put $x = v$ in the other series, it becomes

$$\frac{\theta}{2}\left\{\frac{(-\theta^2)^k}{1\cdot 2\cdot\ldots\cdot 2k+1} + \frac{(\theta^2)^{k-1}}{1\cdot 2\cdot\ldots\cdot 2k-1}c_{2,n} + \dots + \frac{(-\theta^2)}{1\cdot 2\cdot 3}c_{2k-2,n} + \frac{1}{1}c_{2k,n}\right\}$$
$$= \frac{\sin\theta}{1^{2k+1}(\cos\theta)^n} - \frac{\sin 2\theta}{2^{2k+1}(\cos 2\theta)^n} + \frac{\sin 3\theta}{3^{2k+1}(\cos 3\theta)^n} - \dots \quad (2,5)$$

If $n = 1$ this series becomes

$$\frac{\tan\theta}{1^{2k+1}} - \frac{\tan 2\theta}{2^{2k+1}} + \frac{\tan 3\theta}{3^{2k+1}} - \dots \quad (2,6)$$

The series (2,4) may be changed into another, which contains sines both in the numerator and the denominator, for it is equal to

$$\frac{1}{2}c_{2k,n} = \frac{1}{1^{2k}}\left(\frac{\sin\theta}{\sin\theta\cdot\cos\theta}\right)^n - \frac{1}{2^{2k}}\left(\frac{\sin 2\theta}{\sin 2\theta\cdot\cos 2\theta}\right)^n + \frac{1}{3^{2k}}\left(\frac{\sin 3\theta}{\sin 3\theta\cdot\cos 3\theta}\right)^n - \dots$$

But this becomes, since $\sin\theta\cdot\cos\theta = \frac{1}{2}\sin 2\theta$

$$\frac{1}{2^{n+1}}c_{2k,n} = \frac{1}{1^{2k}}\left(\frac{\sin\theta}{\sin 2\theta}\right)^n - \frac{1}{2^{2k}}\left(\frac{\sin 2\theta}{\sin 4\theta}\right)^n + \frac{1}{3^{2k}}\left(\frac{\sin 3\theta}{\sin 6\theta}\right)^n - \dots \quad (2,7)$$

By applying the theorems (*a*), (*b*), (*c*), and (*d*) to the series whose sums we have now investigated, we shall arrive at the value of many others which

[a] A. M. Legendre, *Exercices de calcul intégral*... (3 vol., Paris 1811–17).

contain the powers of tangents / and cotangents in arithmetical progression, thus (1,9) combined with (a) will produce

$$\frac{1}{2}\left\{1 - \frac{k}{1} + \frac{k \cdot k-1}{1 \cdot 2} - \dots\right\} = \frac{(\sin\theta)^{2k}}{(\cos\theta)^n} - \frac{(\sin 2\theta)^{2k}}{(\cos 2\theta)^n} + \frac{(\sin 3\theta)^{2k}}{(\cos 3\theta)^n} - \dots$$

But the left side of this equation is equal to $\frac{1}{2}(1-1)^k = 0$. Hence

$$0 = \frac{(\sin\theta)^{2k}}{(\cos\theta)^n} - \frac{(\sin 2\theta)^{2k}}{(\cos 2\theta)^n} + \frac{(\sin 3\theta)^{2k}}{(\cos 3\theta)^n} - \dots \quad (2,8)$$

And if $n = 2k$, this produces a series of tangents

$$0 = (\tan\theta)^{2k} - (\tan 2\theta)^{2k} + (\tan 3\theta)^{2k} - \dots \quad (2,9)$$

By means of the theorems already referred to, we may introduce into the numerators of each item of the series (2,4) the even powers of the sines of the same arcs whose cosines occur in the denominator: putting $l = m/2$, we shall have

$$\frac{1}{2}\left\{c_{2k,n} - \frac{l}{1}c_{2k,n-2} + \frac{l \cdot l-1}{1 \cdot 2}c_{2k,n-4} - \dots\right\}$$
$$= \frac{(\sin\theta)^{2l}}{1^{2k}(\cos\theta)^n} - \frac{(\sin 2\theta)^{2l}}{2^{2k}(\cos 2\theta)^n} + \frac{(\sin 3\theta)^{2l}}{3^{2k}(\cos 3\theta)^n} - \dots \quad (3,1)$$

And if $n = 2l$, this becomes

$$\frac{1}{2}\left\{c_{2k,2l} - \frac{l}{1}c_{2k,2l-2} + \frac{l \cdot l-1}{1 \cdot 2}c_{2k,2l-4} - \dots\right\}$$
$$= \frac{(\tan\theta)^{2l}}{1^{2k}} - \frac{(\tan 2\theta)^{2l}}{2^{2k}} + \frac{(\tan 3\theta)^{2l}}{3^{2k}} - \dots \quad (3,2)$$

If we call the sum of the series (2,5) $A_{k,n}$, and if we apply to it the formula (b), we shall have

$$A_{k,n} - \frac{l}{1}A_{k,n-2} + \frac{l \cdot l-1}{1 \cdot 2}A_{k,n-4} - \dots$$
$$= \frac{(\sin\theta)^{2l+1}}{1^{2k+1}(\cos\theta)^n} - \frac{(\sin 2\theta)^{2l+1}}{2^{2k+1}(\cos 2\theta)^n} + \frac{(\sin 3\theta)^{2l+1}}{3^{2k+1}(\cos 3\theta)^n} - \dots \quad (3,3)$$

And if $n = 2l + 1$, this becomes

$$A_{k,2l+1} - \frac{l}{1}A_{k,2l-1} + \dots$$
$$= \frac{(\tan\theta)^{2l+1}}{1^{2k+1}} - \frac{(\tan 2\theta)^{2l+1}}{2^{2k+1}} + \frac{(\tan 3\theta)^{2l+1}}{3^{2k+1}} - \dots \quad (3,4)$$

/

In the equation (1,4) if we make $\psi x = \tan^{-1}x$, we shall find

$$(-2)^n(-1)^z\sum^n(-1)^{-z}\{\tan^{-1}v^{2z+n} + \tan^{-1}v^{-2z-n}\}$$

$$= (-2)^n(-1)^z\sum^n(-1)^{-z}\cdot\frac{\pi}{4} = \frac{x + x^{-1}}{1(\cos\theta)^n} - \frac{x^3 + x^{-3}}{3(\cos 3\theta)^n} + \frac{x^5 + x^{-5}}{5(\cos 5\theta)^n} - \ldots$$

If the integrations here indicated are performed, it will be found that all the constants vanish, and ultimately that

$$\frac{\pi}{2} = \frac{x + x^{-1}}{1(\cos\theta)^n} - \frac{x^3 + x^{-3}}{3(\cos 3\theta)^n} + \frac{x^5 + x^{-5}}{5(\cos 5\theta)^n} - \ldots \qquad (3,5)$$

If $x = 1$

$$\frac{\pi}{4} = \frac{1}{1(\cos\theta)^n} - \frac{1}{3(\cos 3\theta)^n} + \frac{1}{5(\cos 5\theta)^n} - \ldots \qquad (3,6)$$

If we multiply (3,5) by dx/x and integrate, we have

$$\frac{\pi}{2}\log x + C = \frac{x - x^{-1}}{1^2(\cos\theta)^n} - \frac{x^3 - x^{-3}}{3^2(\cos 3\theta)^n} + \frac{x^5 - x^{-5}}{5^2(\cos 5\theta)^n} - \ldots \qquad (3,7)$$

If $x = 1$ $C = 0$, let $x = \cos\theta' + \sqrt{-1}\sin\theta'$, then we have

$$\frac{\pi\theta'}{2} = \frac{\sin\theta'}{1^2(\cos\theta)^n} - \frac{\sin 3\theta'}{3^2(\cos 3\theta)^n} + \frac{\sin 5\theta'}{5^2(\cos 5\theta)^n} - \ldots \qquad (3,8)$$

The equation (3,7) may be multiplied any number of times by dx/x, and integrated; and the constants thus introduced may be determined in the same manner as those of the equations (2,3); these operations will give the values of series of the following form:

$$\frac{x + x^{-1}}{1^{2k+1}(\cos\theta)^n} - \frac{x^3 + x^{-3}}{3^{2k+1}(\cos 3\theta)^n} + \frac{x^5 + x^{-5}}{5^{2k+1}(\cos 5\theta)^n} - \ldots$$

$$\frac{x - x^{-1}}{1^{2k}(\cos\theta)^n} - \frac{x^3 - x^{-3}}{3^{2k}(\cos 3\theta)^n} + \frac{x^5 - x^{-5}}{5^{2k}(\cos 5\theta)^n} - \ldots$$

Numerous other series might be found by satisfying the equation $\psi x + \psi(1/x) = 1$, whose sums would be given by this process; and if instead of putting $1/x$ for x, we had substituted αx for x, where the function α is determined by the equation $\alpha^2 x = x$, many others would be discovered. This artifice is / only a particular case of a much more general principle, which is of use in discovering certain values of the variable, in which a series admits of summation, but which generally is not expressible in finite terms. The principle is as follows: let K denote any operation, such as integration, either with respect to differential or finite differences, or any other operation, provided only $K(X + Y) = KX + KY$. Now let

$$K\psi x = A_1 x + A_2 x^2 + A_3 x^3 + \dots$$

Put $\alpha x, \alpha^2 x, \dots, \alpha^{n-1} x$ for x, and the results will be

$$\begin{aligned} K\psi\alpha x &= A_1 \alpha x + A_2 \alpha x^2 + A_3 \alpha x^3 + \dots \\ K\psi\alpha^2 x &= A_1 \alpha^2 x + A_2 \alpha^2 x^2 + A_3 \alpha^3 x^3 + \dots \\ &\vdots \\ K\psi\alpha^{n-1} x &= A_1 \alpha^{n-1} x + A_2 \alpha^{n-1} x^2 + A_3 \alpha^{n-1} x^3 + \dots \end{aligned}$$

By adding all these together, we shall have

$$K\{\psi x + \psi\alpha x + \psi\alpha^2 x + \dots + \psi\alpha^{n-1} x\}$$

equal to a series whose general term is

$$A_n\{x + \alpha x + \alpha^2 x + \dots + \alpha^{n-1} x\}.$$

Now supposing we cannot perform the operation denoted by K on the function ψx, yet if ψ is of such a form that

$$\psi x + \psi\alpha x + \dots + \psi\alpha^{n-1} x$$

is equal to a function on which the operation K can be executed, then calling this new function ψ_1, we shall have

$$K\psi_1 x = SA_i\{x^i + \alpha x^i + \dots + \alpha^{n-1} x^i\}.$$

And if α is such a function that $\alpha^n x = x$, a great variety of forms for ψ may be found, which will satisfy that condition. Now let x be determined by the equation $x = \alpha x$, and r being any root of this, we have $r = \alpha r = \alpha^2 r = \dots = \alpha^{n-1} r$.

Consequently our equation becomes

$$K\psi_1 x = nSA_i r^i = n\{A_1 r + A_2 r^2 + A_3 r^3 + A_4 r^4 + \dots\}$$

/

provided we put r for x after the operation K is executed; that is, we have found the values of the series

$$A_1 x + A_2 x^2 + A_3 x^3 + \dots$$

in the particular cases of x which satisfy the equation $\alpha x = x$.

Part II

I shall now explain another method of deducing the sums of a variety of series, which comprehend amongst them all those which are contained in the former part of this paper; it rests fundamentally on the following formulae, which have long been known:[a]

$$0 = 1^{2n} - 2^{2n} + 3^{2n} - 4^{2n} + \dots$$
$$\tfrac{1}{2} = \cos\theta - \cos 2\theta + \cos 3\theta - \dots$$
$$\tfrac{1}{2} = \cos\theta + \cos 2\theta + \cos 3\theta + \dots$$
$$0 = 1^{2n+1} - 3^{2n+1} + 5^{2n+1} - \dots$$

It is unnecessary to give proofs of these and other similar ones which have been frequently noticed, as they may be very easily demonstrated.

Let fx be any function of x developable in even powers of x, then

$$f(x) = A + Bx^2 + Cx^4 + Dx^6 + \dots$$

Divide both sides by x^{2k}, then it becomes

$$\frac{f(x)}{x^{2k}} = \frac{A}{x^{2k}} + \frac{B}{x^{2k-2}} + \dots + K + Lx^2 + Mx^4 + \dots$$

For x, put successively $1x$, $2x$, $3x$, $4x, \dots$ and let the alternate series be taken negatively; these being arranged under each other, we have

$$+\frac{f(x)}{1^{2k}x^{2k}} = +\frac{A}{1^{2k}x^{2k}} + \frac{B}{1^{2k-2}x^{2k-2}} + \dots + K + Lx^2 1^2 + Mx^4 1^4 + \dots$$

$$-\frac{f(2x)}{2^{2k}x^{2k}} = -\frac{A}{2^{2k}x^{2k}} - \frac{B}{2^{2k-2}x^{2k-2}} - \dots - K - Lx^2 2^2 - Mx^4 2^4 - \dots$$

$$+\frac{f(3x)}{3^{2k}x^{2k}} = +\frac{A}{3^{2k}x^{2k}} + \frac{B}{3^{2k-2}x^{2k-2}} + \dots + K + Lx^2 3^2 + Mx^4 3^4 + \dots$$

$$-\frac{f(4r)}{4^{2k}x^{2k}} = -\frac{A}{4^{2k}x^{2k}} - \frac{B}{4^{2k-2}x^{2k-2}} - \dots - K - Lx^2 4^2 - Mx^4 4^4 - \dots$$

$$\vdots$$

If we add the vertical columns, we shall have on the left side of the equation the series

$$\frac{1}{x^{2k}}\left\{\frac{f(x)}{1^{2k}} - \frac{f(2x)}{2^{2k}} + \frac{f(3x)}{3^{2k}} - \frac{f(4x)}{4^{2k}} + \dots\right\}$$

and the right side of the equation consists of three kinds of terms, those which contain negative powers of x, one which does not contain x, and the remaining

[a] These equations are mathematical howlers: they contain non-convergent series and are therefore nonsense. J. M. Dubbey states 'The remainder of the paper is based on these results and therefore has very little interest' – *The mathematical work of Charles Babbage* (Cambridge, 1978), p. 139.

ones which contain positive powers of x. With respect to these last, they are all of the form $Qx^{2i}\{1^{2i} - 2^{2i} + 3^{2i} - 4^{2i} + \ldots\}$; and as the series which multiplies Qx^{2i} is equal to zero, all those vertical columns which contain even powers of x will vanish: the term which is independent on x is

$$K - K + K - K + \ldots = \tfrac{1}{2}K$$

and those terms which contain negative powers of x, may be represented by the expression $S(\pm 1/i^{2k})$. All the vertical columns being summed, we shall have the equation

$$\frac{f(x)}{1^{2k}} - \frac{f(2x)}{2^{2k}} + \frac{f(3x)}{3^{2k}} - \ldots$$
$$= AS\frac{\pm 1}{i^{2k}} + BS\frac{\pm 1}{i^{2k-2}} + CS\frac{\pm 1}{i^{2k-4}} + \ldots + \tfrac{1}{2}K \quad \text{(A)}$$

As the operations by which we have arrived at this expression have been given at length, it will be unnecessary to repeat them with the slight modifications which would be required for cases nearly similar. Thus, if we suppose the function fx developable according to the odd powers of x, and if we divide both sides of the equation by x^{2k+1} and / repeat the same process we have already explained, we shall arrive at the following theorem:

$$\frac{f(x)}{1^{2k+1}} - \frac{f(2x)}{2^{2k+1}} + \frac{f(3x)}{3^{2k+1}} - \frac{f(4x)}{4^{2k+1}} + \ldots$$
$$= AxS\frac{\pm 1}{i^{2k}} + Bx^3S\frac{\pm 1}{i^{2k-2}} + \ldots + \tfrac{1}{2}K \quad \text{(B)}$$

Let $f(\theta)$ be any function of θ developable in the form

$$f(\theta) = A + B\cos\theta + C\cos 2\theta + D\cos 3\theta + \ldots$$

a very similar process to that which has been already explained will give

$$f(\theta) - f(2\theta) + f(3\theta) - \ldots = \tfrac{1}{2}f(0) \quad \text{(C)}$$

and if $f(\theta) = A\cos\theta + B\cos 2\theta + C\cos 3\theta + \ldots$ a similar course will produce the equation

$$f(\theta) + f(2\theta) + f(3\theta) + \ldots = -\tfrac{1}{2}f(0) \quad \text{(D)}$$

If a function is developable in even powers of x, then its second function is developable in the same manner, and so are all its higher functions; therefore if f_1 and f are two functions developable in even powers of x, such that

$$fx = A^1 + B^1x^2 + C^1x^4 + \ldots$$
$$f_1f^nx = A + Bx^2 + Cx^4 + \ldots$$

Then (A) will become

$$\frac{f_1f^n(x)}{1^{2k}} - \frac{f_1f^n(2x)}{2^{2k}} + \frac{f_1f^n(3x)}{3^{2k}} - \dots$$

$$= AS\frac{\pm 1}{i^{2k}} + BS\frac{\pm 1}{i^{2k-2}} + \dots + \tfrac{1}{2}K \quad \text{(E)}$$

These theorems marked (A), (B), (C), and (D), although they possess a very great degree of generality, are not entirely without restriction; it appears at first sight that they are applicable to *all* functions which have the prescribed condition of being expansible in even powers of the variable; such was my opinion of them when I first discovered them; but / several results which were evidently incorrect, soon convinced me that some limitation existed, of whose nature I was not aware: it was not until some years after, that I found out the cause of the fallacies which had perplexed me; and still more recently, I discovered that the series on whose sum their truth or falsehood depended, might be expressed by a definite integral. By applying the criterion, which I shall presently explain, we cut off a great variety of series whose sums are erroneously given by the method in question; whether this criterion does not exclude some series whose sums are correctly given, is a point which I do not consider yet completely decided; the difficulties to which the application of acknowledged principles have in this instance conducted us, appear worthy of the attention of mathematicians. A more strict method might have been pursued in determining the sum of that part of the series which is neglected; but this in general leads to such differential equations, as cannot afford us much assistance. I have, however, given one example of this method, and I have shown that when the part which had been neglected, as being apparently equal to zero (but which is in fact a finite quantity,) is added to the sum furnished by the *method of expanding horizontally and summing vertically*, the true value of the series results. This confirms the explanation I have given of the reason of the apparent failure of that method.

It will be sufficient to point out the cause which leads to error, and to determine the conditions on which its existence depends for one only of the series; suitable modifications of the reasoning will readily suggest themselves for the others. I shall therefore, at present, consider the theorem (A). If / we turn to the process employed in its investigation, we may remark, that the vertical column $Lx^2(1^2 + 2^2 + 3^2 - 4^2 + \dots)$ has been neglected, because the series which enters into it as a factor is equal to zero; so also the vertical column $Mx^4(1^4 - 2^4 + 3^4 - \dots)$ is neglected for the same reason, and similarly for all the remaining vertical columns. Now, although it would be perfectly correct to omit any one, or even any finite number of these vertical columns, as being multiplied by a factor equal to zero, yet it is not legitimate to neglect an infinite number of terms, each multiplied by zero, unless it can be proved that the sum of all the terms so multiplied is not an infinite quantity: this, then, is the latent cause of the false results at which I arrived at the commencement of these enquiries. I shall now explain how they may be obviated, or rather how to assign the condition on which the truth of the theorems just deduced depend. We have considered the series of terms

$$Lx^2(1^2 - 2^2 + 3^2 - \dots) + Mx^4(1^4 - 2^4 + 3^4 - \dots) + Mx^6(1^6 - 2^6 + 3^6 - \dots) + \dots$$

as equal to zero. Any one of the series which here multiply the powers of x, may be considered as arising from the series

$$1^{2n}y - 2^{2n}y^2 + 3^{2n}y^3 + \dots$$

When $y = 1$, call this series $K_n(y)$, and instead of making $y = 1$, let $y = 1 + o$,[a] which differs from unity by the infinitely small quantity o; then we shall have

$$K_n(1 + o) = c_{1,n} + c_{2,n}o + c_{3,n}o^2 + \dots$$

Where

$$c_{1,n} = o = 1^{2n} - 2^{2n} + 3^{2n} - \dots$$
$$c_{2,n} = 1^{2n+1} - 2^{2n+1} + 3^{2n+1} - \dots$$

by substituting this value of $K_n(1 + \text{o})$ in the series we had neglected, we shall find /

$$\begin{aligned} &\{Lx^2c_{1,1} + Mx^4c_{1,2} + Nx^6c_{1,3} + \dots\} \\ + o\ &\{Lx^2c_{2,1} + Mx^4c_{2,2} + Nx^6c_{2,3} + \dots\} \\ + o^2 &\{Lx^2c_{3,1} + Mx^4c_{3,2} + Nx^6c_{3,3} + \dots\} \\ + o^3 &\{Lx^2c_{4,1} + Mx^4c_{4,2} + Nx^6c_{4,3} + \dots\} \\ &\quad \vdots \qquad\qquad\qquad \vdots \end{aligned}$$

The first line vanishes on account of the value of $c_{1,n}$, and since o is an infinitesimal, the second line will be larger than the sum of all the rest, provided the multiplicators of the powers of o are finite; if therefore the series $Lx^2c_{2,1} + Mx^4c_{2,2} + Nx^6c_{2,3} + \dots$ is finite since it is multiplied by o, we may neglect the whole of the above expression: our next step must be to determine whether the series

$$Lx^2\{1^3 - 2^3 + 3^3 - \dots\} + Mx^4\{1^5 - 2^5 + 3^5 - \dots\} + Nx^6\{1^7 - 2^7 + 3^7 - \dots\} + \dots$$

[a] Although '0' was used here and below in the original paper, we have substituted 'o' which was probably intended.

is finite or infinite. It has been observed by Euler, that the following relations exist between the direct and reciprocal powers of the natural numbers.

$$1-2+3-4+\ldots = +2\frac{1}{\pi^2}\left\{1+\frac{1}{3^2}+\frac{1}{5^2}+\ldots\right\}$$

$$1^3-2^3+3^3-4^4+\ldots = -2\frac{1\cdot 2\cdot 3}{\pi^4}\left\{1+\frac{1}{3^4}+\frac{1}{5^4}+\ldots\right\}$$

$$1^5-2^5+3^5-4^5+\ldots = +2\frac{1\cdot\ldots\cdot 5}{\pi^6}\left\{1+\frac{1}{3^6}+\frac{1}{5^6}+\ldots\right\}$$

$$1^7-2^7+3^7-4^7+\ldots = -2\frac{1\cdot\ldots\cdot 7}{\pi^3}\left\{1+\frac{1}{3^8}+\frac{1}{5^8}+\ldots\right\}$$

$$\vdots \qquad \vdots$$

These latter being substituted for their equals, we have

$$-2Lx^2\frac{1\cdot 2\cdot 3}{\pi^4}\left(\frac{1}{1^4}+\frac{1}{3^4}+\frac{1}{5^4}+\ldots\right) + 2Mx^4\frac{1\cdot 2\cdot 3\cdot 4\cdot 5}{\pi^6}\left(\frac{1}{1^6}+\frac{1}{3^6}+\frac{1}{5^6}+\ldots\right) - 2Nx^6\frac{1\cdot\ldots\cdot 7}{\pi^8}\left(\frac{1}{1^8}+\frac{1}{3^8}+\frac{1}{5^8}+\ldots\right)+\ldots$$

The series which now multiplies each term is in all cases / less than 2, consequently this expression will be finite or infinite, according as the series.

$$-Lx^2\frac{1\cdot 2\cdot 3}{\pi^4}+Mx^4\frac{1\cdot\ldots\cdot 5}{\pi^6}-Nx^6\frac{1\cdot\ldots\cdot 7}{\pi^8}+\ldots$$

is so or the contrary: the product $1\cdot 2\ldots n$ may be expressed by means of a definite integral, thus:

$$1\cdot 2\cdot\ldots\cdot n = \int dv\left(\log\frac{1}{v}\right)^n \quad \begin{bmatrix} v=0 \\ v=1 \end{bmatrix}.$$

These products being replaced by the integral, we have

$$\frac{1}{\pi^2}\int dv\left\{-L\left(\frac{x}{\pi}\right)^2\left(\log\frac{1}{v}\right)^3+M\left(\frac{x}{\pi}\right)^4\left(\log\frac{1}{v}\right)^5-N\left(\frac{x}{\pi}\right)^6\left(\log\frac{1}{v}\right)^7+\ldots\right\}$$

$$=\frac{1}{\pi^2}\int dv\log\frac{1}{v}\cdot\left\{-L\left(\frac{x}{\pi}\log\frac{1}{v}\right)^2+M\left(\frac{x}{\pi}\log\frac{1}{v}\right)^4-N\left(\frac{x}{\pi}\log\frac{1}{v}\right)^6+\ldots\right\} \quad \begin{bmatrix} v=0 \\ v=1 \end{bmatrix}.$$

Now in order to determine the sum of this series, which evidently depends on the function $f(x)$, let us assume

$$\frac{f(x)}{x^{2k}} - \frac{A}{x^{2k}} - \frac{B}{x^{2k-2}} - \ldots - K = \chi(x)$$

then we have

$$\chi(x) = Lx^2 + Mx^4 + Nx^6 + \ldots$$

And if we put $\dfrac{x\sqrt{-1}}{\pi}\log\dfrac{1}{v}$ instead of x, it becomes

$$\chi\left(\frac{x\sqrt{-1}}{\pi}\log\frac{1}{v}\right) = -L\left(\frac{x}{\pi}\log\frac{1}{v}\right)^2 + M\left(\frac{x}{\pi}\log\frac{1}{v}\right)^4 - N\left(\frac{x}{\pi}\log\frac{1}{v}\right)^6 + \ldots$$

And the sum of the series in question is

$$\frac{1}{\pi^2}\int dv\log\frac{1}{v}\cdot\chi\left(\frac{x\sqrt{-1}}{\pi}\log\frac{1}{v}\right) \quad \begin{bmatrix} v=0 \\ v=1 \end{bmatrix}.$$

If therefore this definite integral is finite, the theorem (A) will give correct results. A more convenient form for integration may however be obtained by the following consideration; the series

$$L\frac{x^2}{\pi^2}\cdot 1\cdot 2\cdot 3 + M\frac{x^4}{\pi^4}1\cdot\ldots\cdot 5 + N\frac{x^4}{\pi^6}1\cdot\ldots\cdot 7 + \ldots$$

will always be finite, if the following series

$$A + B\left(\frac{x}{\pi}\right)^2 1\cdot 2\cdot 3 + C\left(\frac{x}{\pi}\right)^4 1\cdot 2\cdot\ldots\cdot 5 + D\left(\frac{x}{\pi}\right)^6 1\cdot 2\cdot\ldots\cdot 7 + \ldots$$

is finite, because this series when prolonged to the terms / $Mx^n + Nx^{n+1} + \ldots$ will have its terms each greater than the corresponding terms of the other series. Now this series is equal to

$$\int dv f\left(\frac{x\sqrt{-1}}{\pi}\log\frac{1}{v}\right) \quad \begin{bmatrix} v=0 \\ v=1 \end{bmatrix}. \qquad \text{(F)}$$

I shall now apply some of the theorems to the investigation of the sums of series, and then explain a method which (when the equations to which it leads can be solved) will in all cases render them correct. And first let

$$f(\theta) = (\cos\theta)^n = \left(1 - \frac{\theta^2}{1\cdot 2} + \frac{\theta^4}{1\cdot 2\cdot 3\cdot 4} - \ldots\right)^n$$

this series is capable of being expanded into another, which also proceeds according to the even powers of θ; first let $k = 0$, then comparing this with (A), we have

$$\tfrac{1}{2} = (\cos\theta)^n - (\cos 2\theta)^n + (\cos 3\theta)^n - \dots \qquad (3,9)$$

Let us now examine if the definite integral is finite, it is in this case.

$$\int dv\left\{\cos\frac{\theta\sqrt{-1}}{\pi}\log\frac{1}{v}\right\}^n$$

$$= \int dv\left\{\frac{\varepsilon^{-(\theta/\pi)\log(1/v)} + \varepsilon^{+(\theta/\pi)\log(1/v)}}{2}\right\}^n$$

$$= \int dv\left(\frac{v^{\theta/\pi} + v^{-\theta/\pi}}{2}\right)^n$$

$$= \frac{1}{2^n}\left\{\frac{1}{1 - n\frac{\theta}{\pi}} + \frac{n}{1}\cdot\frac{1}{1 + (2-n)\frac{\theta}{\pi}} + \frac{n\cdot n-1}{1\cdot 2}\cdot\frac{1}{1 + (4-n)\frac{\theta}{\pi}} + \dots\right\}\begin{bmatrix} v = 0 \\ v = 1\end{bmatrix}.$$

If n is a whole positive number we already know that the series (3,9) is correct; if n is a fraction the series which expresses the value of the definite integral is finite for all values of θ, except such as are contained in

$$\theta = -\frac{\pi}{2i - n}i$$

being any whole number; if n is negative, then the series is finite for all positive values of θ; it appears then that whatever be the value of n if θ is positive, we have /

$$\frac{1}{2} = \frac{1}{(\cos\theta)^n} - \frac{1}{(\cos 2\theta)^n} + \frac{1}{(\cos 3\theta)^n} - \dots$$

From the theorem (A) we may readily determine the value of the series

$$\frac{1}{1^{2k}(\cos\theta)^n} - \frac{1}{2^{2k}(\cos 2\theta)^n} + \frac{1}{3^{2k}(\cos 3\theta)^n} - \dots$$

For let

$$f(\theta) = 1/(\cos\theta)^n = A' + A'_n\theta^2 + B'_n\theta^4 + C'_n\theta^6 + \dots$$

and the sum of the series required will be

$$S\frac{\pm 1}{i^{2k}} + A'_n\theta^2 S\frac{\pm 1}{i^{2k-2}} + B'_n\theta^4 S\frac{\pm 1}{i^{2k-4}} + \dots + \theta^{2k}K'_n\cdot\frac{1}{2}$$

Which is precisely the same sum as we have already found in (2,4), except that we now find that it applies to fractional or surd values of n, as well as to whole numbers.

Let us next suppose

$$f(\theta) = (\tan\theta)^{2l+1} = T_1\theta^{2l+1} + T_3^{\theta 2l+3} + T_5\theta^{2l+5} + \dots$$

this give the series

$$\frac{(\tan\theta)^{2l+1}}{1^{2k+1}} - \frac{(\tan 2\theta)^{2l+1}}{2^{2k+1}} + \frac{(\tan 3\theta)^{2l+1}}{3^{2k+1}} - \dots$$

$$= T_1\theta S\frac{\pm 1}{i^{2(k-l)}} + T_3\theta^3 S\frac{\pm 1}{i^{2(k-l-1)}} + \dots$$

The definite integral in this case being

$$\int dv \left\{\tan\frac{\theta\sqrt{-1}}{\pi}\log\frac{1}{v}\right\}^{2l+1} = \frac{1}{(\sqrt{-1})^{2l+1}}\int dv\left\{\frac{-1+v^{2\theta/\pi}}{1+v^{2\theta/\pi}}\right\}^{2l+1}$$

$$= \frac{1}{(\sqrt{-1})^{2l+1}}\left\{A + \frac{B}{1+2\theta/\pi} + \frac{C}{1+4\theta/\pi} + \frac{D}{1+6\theta/\pi} + \dots\right\} \begin{bmatrix} v=0 \\ v=1 \end{bmatrix}.$$

Since

$$\frac{-1+v^{2\theta/\pi}}{1+v^{2\theta/\pi}} = A = Bv^{2\theta/\pi} + Cv^{4\theta/\pi} + \dots$$

this is always finite if θ is positive, because it is less than $A + B + C + \dots$ which is equal to zero.

In the former part we could only determine the value of the series /

$$\frac{1}{(\sin\theta)^n} - \frac{1}{(\sin 2\theta)^n} + \frac{1}{(\sin 3\theta)^n} - \dots$$

when n is an even number, nor will the method now employed enable us to find its value when n is odd; the reason of this is that

$$(\sin\theta)^{-n} = \left(\frac{\theta}{1} - \frac{\theta^3}{1\cdot 2\cdot 3} + \frac{\theta^5}{1\cdot 2\cdot 3\cdot 4\cdot 5} - \dots\right)^{-n}$$

is developable in a series proceeding according to the even powers of θ only when n is an even number, if n is odd, it proceeds according to the odd powers.

The theorems contained in this second part are applicable to a very extensive class of series which have not, I believe, yet been considered. In (*A*) let $f(\theta) = \cos^2 \theta = \cos(\cos\theta)$ and putting $k = 0$ we have

$$\frac{\cos 1}{2} = \cos^2 \theta - \cos^2 \theta + \cos^2 3\theta - \ldots \tag{4,1}$$

The definite integral which is the criterion of the truth of this value, is

$$\int dv \cos \frac{v^{\theta/\pi} + v^{-(\theta/\pi)}}{2} = A + \frac{B}{1 + 2\theta/\pi} + \frac{C}{1 + 4\theta/\pi} + \frac{D}{1 + 6\theta/\pi} + \cdots$$
$$+ \frac{B}{1 - 2\theta/\pi} + \frac{C}{1 - 2\theta/\pi} + \frac{D}{1 - 6\theta/\pi} + \cdots$$
$$\begin{bmatrix} v = 0 \\ v = 1 \end{bmatrix}.$$

And this is always finite unless θ is an even submultiple of π; if we make $k = 1$ and $k = 2$, we shall have the following theorems, which are true with the same restrictions.

$$\cos 1 \cdot S \frac{\pm 1}{i^2} + \frac{\sin 1}{2} \cdot \frac{\theta^2}{2} = \frac{\cos {}^2\theta}{1^2} - \frac{\cos {}^2 2\theta}{2^2} + \frac{\cos {}^2 3\theta}{3^2} - \cdots \tag{4,2}$$

$$\cos 1 \cdot S \frac{\pm 1}{i^4} + \theta^2 \frac{\sin 1}{2} S \frac{\pm 1}{i^2} - \frac{\theta^4}{2}\left(\frac{\sin 1}{1 \cdot 2 \cdot 3 \cdot 4} + \frac{\cos 1}{8}\right)$$
$$= \frac{\cos {}^2\theta}{1^4} - \frac{\cos {}^2 2\theta}{2^4} + \frac{\cos {}^2 3\theta}{3^4} - \cdots \tag{4,3}$$

If $f(\theta) = (\cos {}^n\theta)^m$, since this is capable of being developed / in a series proceeding according to the even powers of θ, we shall have if $k = 0$

$$\frac{(\cos^{n-1} 1)^m}{2} = (\cos {}^n\theta)^m - (\cos {}^n 2\theta)^m + (\cos {}^n 3\theta)^m - \ldots \tag{4,4}$$

and the definite integral is finite in this case, whenever θ and π are incommensurable: we may therefore in the same circumstance have the value of the series

$$\frac{(\cos {}^n\theta)^m}{1^{2k}} - \frac{(\cos {}^n 2\theta)^m}{2^{2k}} + \frac{(\cos {}^n 3\theta)^m}{3^{2k}} - \cdots$$

The theorems marked (A) and (B) in this paper correspond with that marked (12) in Mr Herschel's memoir 'On various points of Analysis', printed in the Philosophical Transactions for 1814;[a] with the first of these it coincides when *n* is

[a] J. F. W. Herschel, *op. cit.*

an even number, and with the second when it is an odd one: the theorem alluded to is

$$S\left\{\frac{(-1)^{x+1}}{x^n}\cdot\frac{f(\varepsilon^{x\theta})+(-1)^n f(\varepsilon^{-x\theta})}{2}\right\} = {}^0L(2)\cdot\alpha_n\theta^n + {}^2L(2)\cdot\alpha_{n-2}\theta^{n-2} + \dots$$

where

$${}^0L = 1 - 1 + 1 - \dots = \tfrac{1}{2}\cdot{}^{2x}L(2) = \frac{(2^{2x-1}-1)\pi^{2x}}{1\cdot 2\cdot\ldots\cdot 2x}B_{2n-1}\cdot$$

Now this latter expression is the value of those series which I have expressed by $S(\pm 1)/(i^{2k})$. Both methods give the same result, and as that result is very frequently erroneous, I shall confirm the truth of the explanation I have offered, by showing in a particular case, that if the sum of that part of the series which had been neglected as being equal to zero, is found and added to the other part, the result will no longer be erroneous: the example I shall examine is the series

$$\frac{1}{1+\theta^2} - \frac{1}{1+2^2\theta^2} + \frac{1}{1+3^2\theta^2} - \frac{1}{1+4^2\theta^2} + \dots$$

In Mr Herschel's theorem, making

$$f(\varepsilon^t) = \alpha_0 + \alpha_1 t + \alpha_2 t^2 + \dots = \frac{1+\theta t\sqrt{-1}}{1+\theta^2 t^2}$$

we have $\alpha_0 = 1$, and the equation becomes /

$$S\left\{\frac{(-1)^{x+1}\left\{\dfrac{1+x\theta\sqrt{-1}}{1+x^2\theta^2}+\dfrac{1-x\theta\sqrt{-1}}{1+x^2\theta^2}\right\}}{2}\right\} = {}^0L(2)\cdot 1$$

or

$$S(-1)^{x+1}\frac{1}{1+x^2\theta^2} = \frac{1}{2}$$

or

$$\frac{1}{2} = \frac{1}{1+\theta^2} - \frac{1}{1+2^2\theta^2} + \frac{1}{1+3^2\theta^2} - \dots \qquad (4,4)$$

The same series being summed by the theorem (A) in this paper gives the same result; but then the theorem alluded to only declares this to be the true value in case a certain series is finite, which series is

$$+\theta^2\cdot 1\cdot 2\cdot 3+\theta^4\cdot 1\cdot 2\cdot 3\cdot 4\cdot 5+\theta^6\cdot 1\cdot 2\cdot\ldots\cdot 7+\ldots$$

but this series can only be finite when θ is actually equal to zero: the method which I have explained in this paper points out that the equation

$$\frac{1}{2}=\frac{1}{1+\theta^2}-\frac{1}{1+2^2\theta^2}+\frac{1}{1+3^2\theta^2}-\ldots$$

can only be depended on when $\theta=0$, in that case it is known to be correct. I have already stated that the reason why the value of the series so found is incorrect, is that the series

$$-\theta^2(1^2-2^2+\ldots)+\theta^4(1^4-2^4+\ldots)-\theta^6(1^6-2^6+\ldots)+\ldots$$

has been neglected because the coefficient of each term is zero. I shall now proceed to investigate the sum of this series, and shall prove that it is equal to a finite function of θ: let

$$y=c^2(1^2\varepsilon^x-2^2\varepsilon^{2x}+\ldots)+c^4(1^4\varepsilon^x-2^4\varepsilon^{2x}+\ldots)+\ldots$$

then y is equal to the sum of the series whose value we are seeking; if $c=\theta\sqrt{-1}$ and $x=0$, differentiate y twice relative to x, and multiply by c^2, and we find

$$c^2\frac{d^2y}{dx^2}=c^4(1^4\varepsilon^x-2^4\varepsilon^{2x}+\ldots)+c^6(1^6\varepsilon^x-2^6\varepsilon^{2x}+\ldots)+\ldots$$

Hence the equation for determining y is

$$c^2\frac{d^2y}{dx^2}-y=-c^2(1^2\varepsilon^x-2^2\varepsilon^{2x}+\ldots)=-c^2\frac{\varepsilon^x-\varepsilon^{2x}}{(1+\varepsilon^x)^3}$$

and the value of y is

$$y=\frac{c}{2}\left\{\varepsilon^{x/c}\int\varepsilon^{-x/c}\frac{\varepsilon^{2x}-\varepsilon^x}{(1+\varepsilon^x)^3}dx-\varepsilon^{-x/c}\int\varepsilon^{x/c}\frac{\varepsilon^{2x}-\varepsilon^x}{(1+\varepsilon^x)^3}dx\right\}.$$

These integrals must be taken between the limits $x=-\infty$ and $x=0$, putting $\varepsilon^x=v$ this equation is changed into

$$y=\frac{c}{2}\left\{v^{1/c}\int\frac{v-1}{(1+v)^3}v^{-1/c}\,dv-v^{-1/c}\int\frac{v-1}{(1+v)^3}v^{1/c}\,dv\right\}$$

where the limits of v are $v = 0$ $v = 1$, in the latter integral put $v = 1/u$, and we have

$$y = \frac{c}{2}\left\{v^{1/c}\int\frac{v-1}{(1+v)^3}\,v^{-1/c}\,dv + u^{1/c}\int\frac{u-1}{(1+u)^3}\,u^{-1/c}\,du\right\}$$

$$\begin{bmatrix} v = 0 & u = 1 \\ v = 1 & u = \infty \end{bmatrix}$$

but this is equal to

$$\frac{c}{2}v^{1/c}\int\frac{v-1}{(1+v)^3}\,v^{-1/c}\,dv$$

between the limits $v = 0$ and $v = \infty$, which is equal to

$$-\pi/2c\sin(\pi/c)$$

hence

$$y = -\frac{\pi}{2c\sin(\pi/c)}$$
$$= c^2(1^2 - 2^2 + \ldots) + c^4(1^4 - 2^4 + \ldots) + c^6(1^6 - 2^6 + \ldots) + \ldots$$

If $c = \theta\sqrt{-1}$ we have

$$-\frac{\pi}{\theta\{\varepsilon^{\pi/\theta} - \varepsilon^{-\pi/\theta}\}} = -\theta^2(1^2 - 2^2 + \ldots) + \theta^4(1^4 - 2^4 + \ldots)$$
$$-\theta^6(1^6 - 2^6 + \ldots) + \ldots$$

This being added to $\frac{1}{2}$ the value given by the theorem (A) produces

$$\frac{1}{2} - \frac{\pi}{\theta\{\varepsilon^{\pi/\theta} - \varepsilon^{-\pi/\theta}\}} = \frac{1}{1+\theta^2} - \frac{1}{1+2^2\theta^2} + \frac{1}{1+3^2\theta^2} - \frac{1}{1+4^2\theta^4} + \ldots$$

which is the same value that Euler had assigned to this series.

From the value which has been found for y, or for the series

$$c^2(1^2 - 2^2 + \ldots) + c^4(1^4 - 2^4 + \ldots) + \ldots$$

I am inclined to conclude that although the series /

$$1^{2n} - 2^{2n} + 3^{2n} - \ldots$$

is equal to zero for any finite value of n, yet that when n is infinite, the sum of this series is also infinite.

It was my intention to have produced from several of the series whose sums have been found in this paper, the values of several continued products; but the length to which it has already extended will prevent me from more than merely noticing, that many very curious ones will present themselves by integrating the series whose sums have been given.

Since this paper was written, in a conversation with M. Poisson, I mentioned one of the series which it contained, and remarked, that the principle employed led to many erroneous results; that gentleman observed, that many years before he had been led to series nearly similar, in endeavouring to integrate the equations representing the planetary motions, by means of series arranged according to some other functions of the time than the usual ones of the sines and cosines: he obligingly showed me some of his papers relating to this subject, in one of which was a series which in a particular case became the one I had mentioned; the mode of investigation by which he had arrived at these series he had however laid aside, because it rested on the sums of the diverging series $1^{2n} - 2^{2n} + 3^{2n} - \dots$ on which he observed we cannot depend. To the same distinguished analyst I am indebted for some further information on this subject. M. Poisson was one of the commissioners appointed by the Institute of France to examine the manuscripts which were left by Lagrange, amongst these was one entitled 'a method of summing series', which depended on the values of the same diverging / series as those used by M. Poisson and myself; unfortunately it is very short, and its illustrious author does not appear to have resumed the subject: possibly the erroneous values which it gives for the sum of certain series might have caused him to reject it.

25 March, 1819 / C. BABBAGE

XII

DEMONSTRATION OF A THEOREM RELATING TO PRIME NUMBERS

From *Edinburgh Philosophical Journal*, Vol. 1 (1819), pp. 46–9

The singular theorem of Wilson respecting prime numbers, which was first published by Waring in his *Meditationes Analyticae*,[a] and to which neither himself nor its author could supply the demonstration, excited the attention of the most celebrated analysts of the continent, and to the labours of Lagrange and Euler we are indebted for several modes of proof; and more recently it has been considerably extended by the profound investigations contained in the *Disquisitiones Arithmeticae*.[b]

It is well known that, in the theorem in question, a certain expression is asserted to be divisible by n, whenever that number is a *prime*, but it is not divisible if n is not *prime*. In attempting to discover some analogous expression which should be divisible by n^2, whenever n is a prime, but not divisible if n is a composite number, I met with those properties of primes which form the subject of the present paper.

The theorem of Wilson asserts that

$$1 \cdot 2 \cdot 3 \cdot \ldots \cdot (n-1) + 1$$

is always divisible by n when n is a prime number, otherwise it is not. The theorem which I have arrived at is as follows, that

$$\frac{n+1 \cdot n+2 \cdot n+3 \cdot \ldots \cdot (2n-1)}{1 \cdot 2 \cdot 3 \cdot \ldots \cdot (n-1)} - 1$$

is always divisible by n^2 when n is a prime number, otherwise it is not. The demonstration is very simple. Let

$$\binom{n}{0}, \binom{n}{1}, \binom{n}{2}, \ldots$$

[a] E. Waring, *Meditationes analyticae* ... (Cambridge, 1776 etc.).

[b] C. F. Gauss, *Disquisitiones arithmeticae* (Leipzig, 1801).

represent the coefficients of the nth power of $(1 + n)$, so that

$$\binom{n}{0} = 1,\quad \binom{n}{1} = \frac{n}{1},\quad \binom{n}{2} = \frac{n \cdot n - 1}{1 \cdot 2},\quad \binom{n}{3} = \frac{n \cdot n - 1 \cdot n - 2}{1 \cdot 2 \cdot 3}, \ldots$$

then

$$\binom{n}{n-2} = \frac{n \cdot n - 1}{1 \cdot 2},\quad \binom{n}{n-1} = \frac{n}{1} \cdot \frac{n}{n} = 1$$

n being a whole number, now it is well known, that the sum of the squares of the coefficients of a binomial whose index is n, is equal to the coefficient of the middle term of another binomial whose index is $2n$, we have therefore the equation

$$\frac{2n \cdot 2n - 1 \cdot 2n - 2 \cdot \ldots \cdot n + 1}{1 \cdot 2 \cdot 3 \cdot \ldots \cdot n} = 1^2 + \binom{n}{1}^2 + \binom{n}{2}^2 + \binom{n}{3}^2 + \ldots + \binom{n}{n-1}^2 + 1^2$$

or

$$2\left\{\frac{n + 1 \cdot n + 2 \cdot \ldots \cdot 2n - 1}{1 \cdot 2 \cdot 3 \cdot \ldots \cdot n - 1} - 1\right\} = \binom{n}{1}^2 + \binom{n}{2}^2 + \binom{n}{3}^2 + \ldots + \binom{n}{n-1}^2$$

But the quantities

$$\binom{n}{1},\ \binom{n}{2},\ \binom{n}{3}, \ldots \text{ or their equals,}$$

$$\frac{n}{1},\ \frac{n \cdot n - 1}{1 \cdot 2},\ \frac{n \cdot n - 1 \cdot n - 2}{1 \cdot 2 \cdot 3}, \ldots, \frac{n \cdot n - 1 \cdot \ldots \cdot 2}{1 \cdot 2 \cdot 3 \cdot \ldots \cdot n - 1}$$

are all divisible by n when n is a prime, but they are not all divisible by n when n is not a prime; and, since the quantities on the right side consist of the sums of the squares of these, it is divisible by n^2, and consequently

$$2\left\{\frac{n + 1 \cdot n + 2 \cdot \ldots \cdot 2n - 1}{1 \cdot 2 \cdot \ldots \cdot n - 1} - 1\right\}$$

is always divisible by n in the same circumstances, as 2 cannot be divisible by n except $n = 2$ we may omit that factor. The same theorem may also be put into the following form:

$$2\left\{\frac{1\cdot 3\cdot 5\cdot\ldots\cdot 2n-1}{1\cdot 2\cdot 3\cdot\ldots\cdot n}\cdot 2^{n-1}-1\right\}$$

is always divisible by n^2 when n is a prime: This is immediately deduced from the former by the equation

$$n+1\cdot n+2\cdot\ldots\cdot 2n = 2^n\cdot 1\cdot 3\cdot 5\cdot\ldots\cdot 2n-1.$$

Several theorems of a similar kind may be deduced from the investigations of Euler, relative to the properties of the coefficients of a binomial. See the *Acta Acad. Scient. Petrop.* 1781.

Retaining the notation already employed, it is there shown, that when p, n and q are any whole number, we have /

$$\binom{p+n}{q+n} = \binom{n}{0}\binom{p}{q} + \binom{n}{1}\binom{p}{q+1} + \binom{n}{2}\binom{p}{q+2} + \ldots$$

Let $q = p - n$, and this becomes,

$$\binom{p+n}{n} = \binom{n}{0}\binom{p}{p-n} + \binom{n}{1}\binom{p}{p-n+1} + \ldots + \binom{n}{n}\binom{p}{p}.$$

All the terms except the first and last are divisible by n, and by p if both numbers are primes; therefore we have,

$$\frac{p+n\cdot p+n-1\cdot\ldots\cdot n+1}{1\cdot 2\cdot 3\cdot\ldots\cdot p} - \frac{p\cdot p-1\cdot\ldots\cdot n+1}{1\cdot 2\cdot\ldots p-n} - 1$$

always divisible by np, when the numbers n ⟨and⟩ p are primes: if $p = n$, this resolves into the former theorem.

If p is a prime, and greater than n, then since

$$\frac{p\cdot p-1\cdot\ldots\cdot n+1}{1\cdot 2\cdot\ldots\cdot p-n}$$

is divisible by p, we have,

$$\frac{p+n\cdot p+n-1\cdot\ldots\cdot n+1}{1\cdot 2\cdot 3\cdot\ldots\cdot p} - 1$$

always divisible by p, whatever n may be, if p = a prime, otherwise, it is not divisible.

This expression is also divisible by n, for the numerator is

$$n+1\cdot n+2\cdot\ldots\cdot n+p$$

and if this be arranged according to the powers of n, the term independent on

n will, when divided by the denominator, leave unity, which is destroyed by the -1, and all the remaining terms are divisible by n, so that the expression

$$\frac{p+n\cdot p+n-1\cdot\ldots\cdot n+1}{1\cdot 2\cdot\ldots\cdot p}-1$$

is always divisible by pn, if p is a prime, otherwise it is not divisible by p, but only by n.

By considering the coefficients of the cube, and other powers of

$$1+\binom{n}{1}x+\binom{n}{2}x^2+\binom{n}{3}x^3+\ldots$$

we might arrive at other theorems respecting prime numbers; but the number of the combinations which occur in all the higher powers, seem to exclude that simplicity in the expression, / which we have found in those deducted from the coefficients of the square of a binomial.

In case $n-1$ should be a prime number, we may find formulae divisible by $(n-1)^2$, thus: If instead of subtracting the two extreme terms of the coefficients, we subtract the two first and two last, we have,

$$2\left\{\frac{n+1\cdot n+2\cdot\ldots\cdot 2n-1}{\cdot 1\cdot 2\cdot\ldots\cdot n-1}-1-\frac{n^2}{1^2}\right\}=\binom{n}{2}^2+\binom{n}{3}^2+\ldots+\binom{n}{n-2}^2$$

and since, if $n-1$ is a prime, each term on the right side of the equation is divisible by $(n-1)^2$; the expression on the left side is always divisible by $(n-1)^2$ in the same circumstances; or the expression

$$\frac{n+1\cdot n+2\cdot\ldots\cdot 2n-1}{1\cdot 2\cdot\ldots\cdot n-1}-2n$$

is always divisible by $(n-1)^2$, if $n-1$ is a prime number, otherwise it is not divisible. It is obvious that similar theorems might be deduced, in which the prime divisors should be $(n-2)^2$, or $(n-3)^2$, etc. /

A

COLLECTION OF EXAMPLES

OF

THE APPLICATIONS

OF THE

DIFFERENTIAL AND INTEGRAL CALCULUS.

BY

GEORGE PEACOCK, A.M. F.R.S. F.C.P.S.

FELLOW OF TRINITY COLLEGE, CAMBRIDGE.

CAMBRIDGE:

Printed by J. Smith, Printer to the University.

AND SOLD BY J. DEIGHTON AND SONS, CAMBRIDGE; G. & W. B. WHITTAKER, AVE-MARIA LANE; J. MAWMAN, LUDGATE STREET; LONGMAN & CO. PATERNOSTER ROW, LONDON.

1820

[iii/iv]

XIII

A COLLECTION OF EXAMPLES OF THE APPLICATIONS OF THE DIFFERENTIAL AND INTEGRAL CALCULUS

With G. Peacock and J. F. W. Herschel

(1820)

PREFACE

The present work was originally intended as a supplement to a Translation of an Elementary Treatise, on the Differential and Integral Calculus, from the French of Lacroix, which was published in 1816:[a] in the course of its progress, however, it was thought proper to enlarge the plan, and to make it a collection of examples illustrative of all the more important analytical applications of the differential and integral calculus, without any particular reference to the work in question.

The principal object of this undertaking is to remedy a defect, which exists in a greater or less degree, in nearly all the foreign works on this part of analysis, in which the theory is most frequently exhibited in its most general form, without the illustration which particular examples are calculated to afford: the student is consequently bewildered in the generality of the relations which are presented to his mind, and he either rests contented with a vague and imperfect comprehension of the subject, or abandons a study which becomes repulsive and fatiguing from its obscurity and difficulty.

The works of Euler, however, form a distinguished exception to this observation, as they every where abound with the utmost fulness of illustration; his two great works on this subject, the *Institutiones Calculi Differentialis*, and the *Institutiones Calculi Integralis*,[b] are remarkable both for the / number and

[a] C. Babbage, J. F. W. Herschel and G. Peacock (translators): S. F. Lacroix, *An elementary treatise on the differential and integral calculus* (Cambridge, 1816)—extracts reprinted in this volume of the *Works of Babbage*.

[b] L. Euler, *Institutiones calculi differentialis* . . . (2 vol., Ticini, 1787), *Institutiones calculi integralis* (4 vol., Petropoli, 1768–94).

the elegance of their examples, which have been borrowed with little scruple or acknowledgement by succeeding authors: but rich and various as are the stores which they contain, they cannot be considered such, as to render the present work unnecessary; for not to mention their bulk and rarity, and the numerous additions to the science made since their publication by Lagrange and other analysts, they contain much that is unnecessary, at least, to a student, and much that is capable of being exhibited in a better form.

Whatever has been borrowed from other authors, has been generally acknowledged, and a very slight inspection of this work would show the great extent of our obligations; we say generally, for we have omitted, for the most part, any reference to examples taken from the great work of Lacroix, and in some other cases when the authors from whom they were taken, were not considered worth consulting: our object being to enable the student to refer to the original authors, whenever he may find any difficulty in the solution of a problem, or may be anxious to obtain more complete information on the subject of it.

There are several portions of this work, particularly of the first part of it, which upon more mature consideration, we should wish to see omitted or abridged, as useless or redundant; we refer more particularly to the examples connected with Maclaurin's theorem, and the development of functions.

The integrals have been taken almost entirely from the *Integraltafeln*, or *Tables of Integral Formulae*, of Meyer Hirsch, of Berlin, a work remarkable for its extent, accuracy, and / elegance.[a] The utility of this portion of the Work, might have been materially increased, if some notice of the method as well as the formulae of solution, had accompanied each class of integrals.

The same anxiety for compression, induced us to omit all notice of the important subject of definite integrals: if ever this work should be honoured by a second edition, we hope to be able to supply these defects, which we cannot but consider as very considerable.

There is one subject, for the omission of which we need make no apology, as the work has already been performed by a gentleman incomparably better calculated to do justice to it, than ourselves; we mean the Calculus of Variations: as the learned and interesting treatise of Professor Woodhouse upon this subject,[b] which has long been before the public, is accompanied by a collection of excellent examples, it would have been useless as well as impertinent in us, to have attempted to add to their number.

---

[a] Meyer Hirsch, *Integraltafeln, oder sammlung von integralformeln* (Berlin, 1810).

[b] R. Woodhouse, *A treatise on isopermetrical problems and the calculus of variations* (Cambridge, 1810).

EXAMPLES

OF THE

Solutions

OF

FUNCTIONAL EQUATIONS.

BY

CHARLES BABBAGE, A.M. F.R.S. L. & E. F.C.P.S.

AND SECRETARY TO THE ASTRONOMICAL SOCIETY OF LONDON.

EXAMPLES OF THE SOLUTIONS OF FUNCTIONAL EQUATIONS

NOTICE

The object of the following Examples of Functional Equations, is to render a subject of considerable interest, more accessible to mathematical students, than it has hitherto been. It is, perhaps, that subject of all others, which most requires the assistance of particular instances, in order fully to comprehend the meaning of its symbols, which are of the most extreme generality; that assistance is also more particularly required in this branch of science, in consequence of its never yet having found its way into an Elementary Treatise.

20 October, 1820

If a function α is of such a form, that, when it is twice performed on a quantity, the result is the quantity itself, or if $\alpha^2(x) = x$, then it is called a periodic function of the second order, if $\alpha^n(x) = x$, then it is termed a periodic function of the nth order, thus when $\alpha(x) = a - x$ the second function, or

$$\alpha(\alpha x) = \alpha(a - x) = a - (a - x) = a - a + x = x.$$

If $\alpha(x) = 1/(1 - x)$, then

$$\alpha^2 x = \alpha(\alpha x) = \frac{1}{1 - \dfrac{1}{1 - x}} = \frac{1 - x}{1 - x - 1} = \frac{x - 1}{x}$$

and

$$\alpha^3 x = \alpha^2 \alpha x = \frac{\alpha x - 1}{\alpha x} = \frac{\dfrac{1}{1 - x} - 1}{\dfrac{1}{1 - x}} = 1 - (1 - x) = x$$

the first of these examples is a periodic function of the second, the last is a periodic function of the third order.

Problem 1. To find periodic functions of the second order.

Since such functions must satisfy the equation $\psi^2 x = x$, we have

$$\psi x = \psi^{-1} x$$

or ψ must be such a function, that it shall be the same as its inverse; if therefore $y = \psi x$, we have also $x = \psi^{-1} y = \psi y$, / or if x and y are connected by some equation, it must be symmetrical relative to x and y; y or ψx must then be determined from the equation

$$* \; F\{\overline{x, \psi x}\} = 0$$

for instance, if $x + \psi x - a = 0$, $\psi x = a - x$, or if $x\psi x = a^2$, $\psi x = a^2/x$.

Another method of determining such functions is as follows: since ψx is of such a form that $\psi^2 x = x$ any symmetrical function of x and ψx remain constant when x is changed into ψx thus

$$F\{\overline{x, \psi x}\} \quad \text{becomes} \quad F\{\overline{\psi x, \psi^2 x}\} = F\{\overline{\psi x, x}\}$$

if therefore, we can find any particular solution of the equation $\psi^2 x = x$, containing an arbitrary constant we may substitute such a function for it, but $\psi x = a - x$ is a particular solution therefore

$$\psi x = F(\overline{x, \psi x}) - x$$

or

$$x + \psi x = F(\overline{x, \psi x})$$

and by changing the artitrary function into another of the same form, we find

$$F\{\overline{x, \psi x}\} = 0$$

as before.

These two methods of determining periodic functions of the second order, are not so convenient as a third process which can be extended to all orders. /

Assume $\psi x = \varphi^{-1} f \varphi x$, then

$$\psi^2 x = \varphi^{-1} f \varphi \varphi^{-1} f \varphi x = \varphi^{-1} f^2 \varphi x$$

this must be equal to x or

$$\varphi^{-1} f^2 \varphi x = x$$

this equation will be fulfilled if $f^2 v = v$, or if f is a particular solution, and if

* Bars placed above quantities under the functional sign, indicate that the function is symmetrical relative to those quantities.

also φ^{-1} is such an inverse function that $\varphi^{-1}\varphi v = v$. If therefore, φ is arbitrary, and f is a particular solution of $f^2x = x$, then the solution of $\psi^2x = x$ is

$$\psi x = \varphi^{-1}f\varphi x.$$

Example. Let

$$fx = \frac{a}{x}$$

⟨then⟩

$$\psi x = \varphi^{-1}\left(\frac{a}{\varphi x}\right)$$

if

$$f(x) = \frac{a - bx}{b + cx}$$

⟨then⟩

$$\psi x = \varphi^{-1}\left(\frac{a - b\varphi x}{b + c\varphi x}\right)$$

from these may easily be derived the following periodic functions of the second order,

$$\psi x = a - x \qquad \psi x = \frac{x}{x - 1}$$

$$\psi x = \frac{x - 2}{x - 1} \qquad \psi x = \frac{a^2}{x}$$

$$\psi x = \frac{1 - x}{1 + x} \qquad \psi x = \sqrt{1 - x^2}$$

$$\psi x = \frac{x + 1}{x - 1} \qquad \psi x = \frac{x}{\sqrt{x^2 - 1}}$$

$$\psi x = \tan^{-1}\left(\frac{\sin(a - x)}{\cos a \cdot \cos x}\right) \qquad \psi x = \log(a - \varepsilon^x)$$

$$\psi x = (a^n - x^n)^{1/n} \qquad \psi x = x - \log(\varepsilon^x - 1)$$

$$\psi x = \frac{x}{(x^n - a^n)^{1/n}} \qquad \psi x = \tan^{-1}(a - \tan x).$$

/

Problem 2. Required periodic functions of the third order, or such as fulfil the equation $\psi^3 x = x$.

Assume $\psi x = \varphi^{-1} f \varphi x$, then the equation becomes

$$\psi^3 x = \varphi^{-1} f \varphi \varphi^{-1} f \varphi \varphi^{-1} f \varphi x = \varphi^{-1} f^3 \varphi x = x$$

which will be verified if $f(v)$ is a particular solution of $f^3 v = v$, and if φ^{-1} is such an inverse value that $\varphi^{-1}\varphi v = v$, hence the solution of the equation is

$$\psi x = \varphi^{-1} f \varphi x$$

one solution is $1/(1 - x)$ and hence

$$\psi x = \varphi^{-1}\left(\frac{1}{1 - \varphi x}\right)$$

more particular cases are

$$\psi x = \frac{a^2}{a - x} \qquad \psi x = \frac{1 + x}{1 - 3x}$$

$$\psi x = \frac{a^2}{ac - c^2 x} \qquad \psi x = \frac{\sqrt{ax^2 - a^2}}{x}$$

$$\psi x = \frac{ax - a^2}{x} \qquad \psi x = \frac{1}{1 - x}$$

$$\psi x = \left(\frac{a^2}{a - x^n}\right)^{1/n} \qquad \psi x = -\log(1 - \varepsilon^x)$$

$$\psi x = \frac{(ax^n - a^2)^{1/n}}{x} \qquad \psi x = \log(a\varepsilon^x - a^2) - x$$

$$\psi x = \frac{a + bx}{c - \dfrac{a^2 + bc + c^2 x}{a}} \qquad \psi x = \log(\varepsilon^x - \varepsilon^c) - x + c.$$

Problem 3. To find periodic functions of the nth order, or to solve the equation $\psi^n x = x$.

Assume as before $\psi x = \varphi^{-1} f \varphi x$ then it becomes

$$\varphi^{-1} f \varphi \varphi^{-1} f \varphi \ldots \varphi^{-1} f \varphi x = \varphi^{-1} f^n \varphi x = x$$

which is verified if f is a particular solution of $f^n x = x$, and if φ^{-1} is such an inverse function that $\varphi^{-1}\varphi x = x$.

It now remains to find particular solutions of $\psi^n x = x$ which may be accomplished in the following manner: let fx represent $(a + bx)/(c + dx)$ then the nth function will be of the same form, or

$$f^n(x) = \frac{A_n + B_n x}{C_n + D_n x}$$

where A_n, B_n, C_n, D_n, are functions of a, b, c, d, and n, these may be so determined that $D_n = 0$, $A_n = 0$ and $B_n = C_n$ all which conditions are satisfied, if

$$d = -\frac{b^2 - 2bc\cos\frac{2k\pi}{n} + c^2}{\left(2 + 2\cos\frac{2k\pi}{n}\right)a}$$

hence

$$\varphi x = \varphi^{-1}\left\{\frac{a + b\varphi x}{c - \dfrac{b^2 - 2bc\cos\frac{2k\pi}{n} + c^2}{\left(2 + 2\cos\frac{2k\pi}{n}\right)a}\varphi x}\right\}$$

a more detailed account of this method of solution may be found in a paper by Mr Horner in the Annals of Philosophy, Nov. 1817.[a]

Instances of $\psi^4 x = x$ are

$$\psi x = \frac{1}{2}\cdot\frac{1}{1 - x} \qquad \psi x = \frac{1 + x}{1 - x}$$

$$\psi x = \frac{2}{2 - x} \qquad \psi x = \frac{2a^2}{2ac - c^2 x}$$

$$\psi x = 2\frac{x - 1}{x} \qquad \psi x = \frac{a + bx}{c - \dfrac{b^2 + c^2}{2a}x}$$

$$\psi x = \frac{\sqrt{2}}{\sqrt{2 - x^2}} \qquad \psi x = -\sqrt{\frac{1 - x^2}{1 + x^2}}$$

$$\psi x = \frac{(2x^n - 2)^{1/n}}{x}$$

$$\psi x = \log 2 - x + \log(\varepsilon^x - 1).$$

[a] W. G. Horner, 'Solution of the equation $\psi^n x = x$', *Annals of Philosophy*, Vol. 10 (1817), pp. 341–6.

All those cases which satisfy the equation $\psi^2 x = x$, also fulfil that of $\psi^4 x = x$, as well as all those which fulfil any of these equations $\psi^2 x = -x$, $\psi^2 x = 1/x$, or more generally $\psi^2 x = ax$, where ax is a particular solution of the equation $\psi^2 x = x$.

The following particular cases satisfy the equation $\psi^6 x = x$.

$$\psi x = \frac{1}{3(1-x)} \qquad \psi x = \frac{3x-1}{3x}$$

$$\psi x = \frac{3}{3-x} \qquad \psi x = \frac{3a^2}{3ac - c^2 x}$$

$$\psi x = 3\frac{x-1}{x} \qquad \psi x = \frac{3+3x}{3-x}$$

$$\psi x = \frac{a+bx}{c - \frac{b^2 - bc + c^2}{3a}x}$$

$$\psi x = \frac{1}{x}\left(x^n - \frac{1}{3}\right)^{1/n}$$

$$\psi x = \log 3 - x - \log(\varepsilon^x - 1).$$

The principle on which the solution of the functional equation $F\{x, \psi x, \psi \alpha x\} = 0$ depends, where $\alpha^2 x = x$, is that by substituting αx for x we have another equation $F\{\alpha x, \psi \alpha x, \psi x\} = 0$, between which and the given equation we may eliminate $\psi \alpha x$ and the result will be the value of ψx. A few examples will illustrate this method. /

(1). Given

$$\psi(x) + \alpha\psi(-x) = x^n$$

by putting $-x$ for x this becomes

$$\psi(-x) + \alpha\psi(x) = (-x)^n$$

and eliminating $\psi(-x)$, we have

$$\psi x - \alpha^2 \psi x = x^n - \alpha(-x)^n$$

hence

$$\psi x = \frac{1 - (-1)^n \alpha}{1 - \alpha^2} x^n.$$

(2). Given

$$\psi x - \alpha\psi\frac{1}{x} = \varepsilon^x$$

put $1/x$ for x,

$$\psi\frac{1}{x} - \alpha\psi x = \varepsilon^{1/x}$$

and

$$\psi x - \alpha\varepsilon^{1/x} - \alpha^2\psi x = \varepsilon^x$$

$$\psi x = \frac{\varepsilon^x + \alpha\varepsilon^{1/x}}{1 - \alpha^2}.$$

(3). Given

$$(\psi x)^2 \cdot \psi\frac{1 - x}{1 + x} = c^2 x$$

put $(1 - x)/(1 + x)$ for x, it becomes

$$\left(\psi\frac{1 - x}{1 + x}\right)^2 \cdot \psi x = c^2\frac{1 - x}{1 + x}$$

eliminating $\psi((1 - x)/(1 + x))$, by means of the former, we find

$$\psi x = \left(\frac{1 + x}{1 - x}c^2 x^2\right)^{1/3}.$$ /

(4). Given

$$\psi x + \frac{1}{1 - x^2}\psi\sqrt{1 - x^2} = 1 + x^2$$

putting $\sqrt{1 - x^2}$ for x, we have

$$\psi\sqrt{1 - x^2} + \frac{1}{x^2}\psi x = 2 - x^2$$

and substituting this value of $\psi\sqrt{1 - x^2}$ in the former equation

$$\psi x + \frac{2 - x^2}{1 - x^2} - \frac{1}{x^2 - x^4}\psi x = 1 + x^2$$

hence

$$\left(\frac{x^2 - x^4 - 1}{x^2 - x^4}\right)\psi x = 1 + x^2 - \frac{2 - x^2}{1 - x^2} = \frac{-1 + x^2 - x^4}{1 - x^2}$$

and $\psi x = x^2$.

(5). Given

$$\frac{\psi x}{1 + \psi x} + x\frac{\psi(-x)}{1 + \psi(-x)} = 1$$

put $\psi_1 x = \psi x/(1 + \psi x)$ thus the equation becomes

$$\psi_1 x + x\psi_1(-x) = 1$$

and changing x into $-x$ we have

$$\psi_1(-x) - x\psi_1(x) = 1$$

by eliminating $\psi(-x)$ from the former, we find

$$\psi_1 x = \frac{1 - x}{1 + x^2}$$

hence

$$\psi x = \frac{\psi_1 x}{1 - \psi_1 x} = \frac{1 - x}{x + x^2}.$$

(6). Given

$$\psi x + \frac{1 + x}{x}\psi\frac{1}{x}c$$

putting $1/x$ for x this becomes /

$$\psi\frac{1}{x} + (1 + x)\psi x = x$$

and by eliminating $\psi(1/x)$, we have

$$\psi x = \frac{1}{1 + x + x^2}c.$$

(7). Given

$$\psi x + x\psi(1 - x) = 1$$

putting $1 - x$ for x, we have

$$\psi(1 - x) + (1 - x)\psi(x) = 1$$

whence, by elimination,

$$\psi x = \frac{1 - x}{1 - x(1 - x)} = \frac{1 - x}{1 - x + x^2}.$$

(8). Given

$$\frac{\psi x}{\psi x - x} + x\frac{\psi(1 - x)}{\psi(1 - x) + x - 1} = 1$$

put $\psi_1 x = \psi x/(\psi x - x)$, then will

$$\psi_1(1 - x) = \frac{\psi(1 - x)}{\psi(1 - x) + x - 1}$$

and the equation becomes

$$\psi_1 x + x\psi_1(1 - x) = 1$$

the same as in the last example; let fx represent the solution there found, then

$$\psi_1 x = fx = \frac{\psi x}{\psi x - x}$$

whence

$$\psi x = \frac{xfx}{fx - 1}$$

if we take for fx its value $(1 - x)/(1 - x + x^2)$, we have

$$\psi x = \frac{x - 1}{x}.$$

In case the equation is symmetrical with regard to ψx and $\psi \alpha x$, the process of elimination apparently becomes illusory. By a peculiar artifice this difficulty may be overcome, and it happens rather singularly that in all these cases, the solution which is so obtained contains an arbitrary function, and in general the solution is the most extensive which the question admits of.

(9). Given

$$\psi x = \psi\frac{1}{x}.$$

If we put $1/x$ for x, this is changed into $\psi(1/x) = \psi x$, the same as the given equation; it is therefore impossible to eliminate.

Let us now suppose $\psi x = a\psi(1/x) + b$, which becomes the given equation when $a = 1$ and $b = 0$. By putting $1/x$ for x this is changed into

$$\psi\frac{1}{x} = a\psi x + b$$

and eliminating $\psi(1/x)$, we have

$$\psi x = \frac{ab + b}{1 - a^2} = \frac{b}{1 - a}$$

if $b = 0$ and $a = 1$, this becomes a vanishing fraction whose value is any constant quantity c, and we have $\psi x = c$, which fulfils the equation. This is a very limited solution, but the following plan will lead us to much more general ones.

Take the equation

$$\psi x = a\psi\frac{1}{x} + v\varphi x$$

which coincides with the given one when $v = 0$ and $a = 1$; also φx is any arbitrary function of x; putting $1/x$ for x, we have /

$$\psi\frac{1}{x} = a\psi x + v\varphi\frac{1}{x}$$

and by elimination,

$$\psi x = \frac{a\varphi\frac{1}{x} + \varphi x}{1 - a^2}v.$$

Let a become $1 + 0$ and v become 0 at the same time, then

$$\frac{0}{1 - (1 + 2\cdot 0 + 0^2)} = \frac{0}{-2\cdot 0 + 0^2} = \frac{1}{-2 + 0} = -\frac{1}{2}$$

and the solution becomes

$$\psi x = \frac{\varphi\frac{1}{x} + \varphi x}{2},$$

or changing the arbitrary function

$$\psi x = \varphi x + \varphi\frac{1}{x},$$

in which φ is indefinite.

This solution is, in fact, nothing more than an arbitrary symmetrical function of x and $1/x$, and may be expressed thus

$$\psi x = \chi\left(\overline{x, \frac{1}{x}}\right).$$

Precisely the same course of reasoning will produce the solutions of the following equations.

(10). $\psi(x) = \psi(a - x), \quad \psi x = \chi(\overline{x}, \overline{a - x}).$

(11). $\psi(x) = \psi\left(\frac{1 - x}{1 + x}\right), \quad \psi x = \chi\left\{\overline{x}, \overline{\frac{1 - x}{1 + x}}\right\}.$

(12). $\psi(x) = \psi\left(\frac{x}{\sqrt{x^2 - 1}}\right), \quad \psi x = \chi\left\{\overline{x}, \overline{\frac{x}{\sqrt{x^2 - 1}}}\right\}.$

(13). $\psi\left(\frac{x}{2 - x}\right) = \psi(1 - x), \quad \psi x = \chi\left\{\overline{x}, \overline{\frac{1 - x}{1 + x}}\right\}.$

(14). $\psi x = \psi(\alpha x), \quad \psi x = \chi(\overline{x}, \overline{\alpha x})$, where $\alpha^2 x = x$.

(15). The objection which has just been stated occurs in the equation

$$\psi x = \psi\left(\frac{x}{x - 1}\right) = c$$

and a similar mode of proceeding will obviate it. The given equation is a particular case of

$$\psi x + a\psi\left(\frac{x}{x - 1}\right) = c + v\varphi x$$

with which it coincides, if $a = 1$ and $v = 0$; putting $x/(x - 1)$ for x in this, we have

$$\psi\left(\frac{x}{x - 1}\right) + a\psi x = c + v\varphi\left(\frac{x}{x - 1}\right)$$

and elimination produces

$$\psi x = \frac{c - ac}{1 - a^2} + \left\{\varphi x - a\varphi\left(\frac{x}{x - 1}\right)\right\}\frac{v}{1 - a^2}.$$

If $v = 0$ and $a = 1$, this gives

$$\psi x = \frac{c}{2} + \varphi x - \varphi\left(\frac{x}{x - 1}\right)$$

in which the function φ has been changed into another similar one.

16. Given

$$\psi(1 + x) + \psi(1 - x) = 1 - x^2$$

put $x - 1$ for x, then /

$$\psi x + \psi(2 - x) = 1 - (x - 1)^2 = 2x - x^2$$

this is a particular case of the equation

$$\psi x + a\psi(2 - x) = 2x - x^2 + v\varphi x$$

with which it agrees if $v = 0$ and $a = 1$; changing x into $2 - x$ and eliminating $\psi(2 - x)$ from the result, we find

$$\psi x = \frac{2x - x^2 - a(2x - x^2)}{1 - a^2} + \{\varphi x - a\varphi(2 - x)\}\frac{v}{1 - a^2}$$

or

$$\psi x = \frac{2x - x^2}{1 + a} + \{\varphi x - a\varphi(2 - x)\}\frac{v}{1 - a^2}.$$

If $a = 1$ and $v = 0$, we have

$$\psi x = \frac{2x - x^2}{2} + \varphi x - \varphi(2 - x).$$

(17). Given

$$\frac{1}{x + \psi x} + \frac{x}{1 + x\psi\frac{1}{x}} = 2$$

put $1/(x + \psi x) = \psi_1 x$ then

$$\frac{1}{\frac{1}{x} + \psi\frac{1}{x}} = \psi_1\frac{1}{x} = \frac{x}{1 + x\psi\frac{1}{x}}$$

and the equation becomes

$$\psi_1 x + \psi_1\frac{1}{x} = 2$$

whose solution may be found by the method just explained to be

$$\psi_1 x = 1 + \varphi x - \varphi\frac{1}{x}$$

hence

$$\psi x = \frac{1}{1 + \varphi x - \varphi \frac{1}{x}} - x.$$

(18). Required the equation of that class of curves which possess the following property.

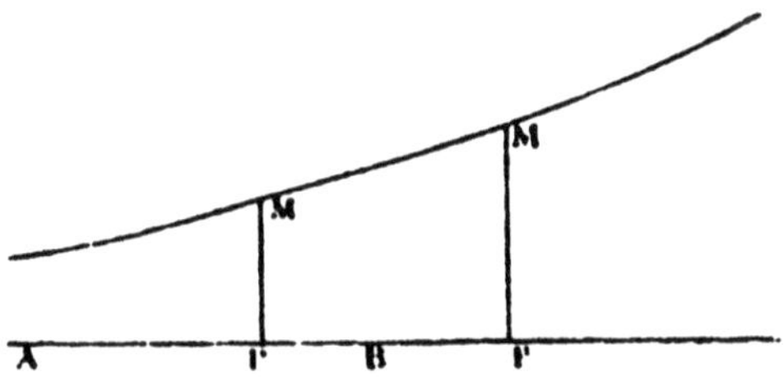

A given abscissa $AB = a$ being taken, then the product of any two ordinates at equal distances from B, shall always be equal to the square of the abscissa a. If $y = \psi x$ represent the equation of the curve, then the condition expressed analytically is

$$\psi(a - x) \cdot \psi(a + x) = a^2.$$

Putting $a - x$ for x, and then $\log \psi(x) = \psi_1 x$, we have

$$\psi_1 x + \psi_1(2a - x) = 2 \log a$$

whose solution is

$$\psi_1 x = \log a + \varphi x - \varphi(2a - x)$$

hence

$$\log \psi x = \log a + \varphi x - \varphi(2a - x)$$

and

$$\psi x = \varepsilon^{\log a} \cdot \varepsilon^{\varphi x} \cdot \varepsilon^{-\varphi(2a - x)} = \frac{a\varepsilon^{\varphi x}}{\varepsilon^{\varphi(2a - x)}}$$

and changing the arbitrary function φ into $\log \varphi$,

$$\psi x = a \frac{\varphi x}{\varphi(2a - x)}$$

and the class of curves are comprehended in the equation

$$y = \frac{a\varphi x}{\varphi(2a - x)}.$$

(19). Given the equation

$$\{\psi x\}^2 + \left\{\psi\left(\frac{\pi}{2} - x\right)\right\}^2 = 1.$$

This is the equation on which the composition of forces is made to depend in the Mecanique Celeste, ⟨Vol. 1⟩, p. 5.[a] /

Put $\psi_1 x$ for $(\psi x)^2$ then it becomes

$$\psi_1 x + \psi_1\left(\frac{\pi}{2} - x\right) = 1$$

which is a particular case of

$$\psi_1 x + a\psi_1\left(\frac{\pi}{2} - x\right) = 1 + v\varphi x.$$

Substituting $\pi/2 - x$ for x, this gives

$$\psi_1\left(\frac{\pi}{2} - x\right) + \psi_1 x = 1 + v\varphi\left(\frac{\pi}{2} - x\right)$$

and eliminating $\psi_1(\pi/2 - x)$, we have

$$\psi x = \frac{1}{1 + a} - \frac{\varphi x - a\varphi\left(\frac{\pi}{2} - x\right)v}{1 - a^2}$$

making $a = 1$ and $v = 0$, and changing φx in $2\varphi x$, we have

$$\psi_1 x = \frac{1}{2} - \varphi x + \varphi\left(\frac{\pi}{2} - x\right)$$

and therefore

$$\psi x = \sqrt{\frac{1}{2} - \varphi x + \varphi\left(\frac{\pi}{2} - x\right)}.$$

In case the coefficient of $\psi\alpha x$ in the equation $\psi x + fx\psi\alpha x = fx$ is of such a form that $fx \cdot f\alpha x = 1$, the denominator will vanish, and we must then have recourse to an artifice similar to that which has already been explained.

(20). Example 1. Let

$$\psi x + x^{2n}\psi\frac{1}{x} = x^n$$

put

$$\psi x + (x^{2n} + v\varphi x)\psi\frac{1}{x} = x^n$$

/

[a] P. S. de Laplace, *Traité de mécanique céleste* (5 vol., Paris, 1798–1827).

which coincides with the given equation if $v = 0$; then changing x into $1/x$, we have

$$\psi\frac{1}{x} + \left(x^{-2n} + v\varphi\frac{1}{x}\right)\psi x = x^{-n}$$

and by elimination

$$\psi x = \frac{x^n - (x^{2n} + v\varphi x)x^{-n}}{1 - (x^{2n} + v\varphi x)\left(x^{-2n} + v\varphi\frac{1}{x}\right)}$$

$$= \frac{x^{-n}\varphi x}{x^{2n}\varphi\frac{1}{x} + x^{-2n}\varphi x + v\varphi x \cdot \varphi\frac{1}{x}}$$

and when v vanishes,

$$\psi x = \frac{x^{-n}\varphi x}{x^{2n}\varphi\frac{1}{x} + x^{-2n}\varphi x}. \qquad (a)$$

The equation in this example may be solved differently, as follows. Multiply by x^{-n} and it becomes

$$x^{-n}\psi x + x^n\psi\frac{1}{x} = 1$$

put $\psi_1 x = x^{-n}\psi x$, then

$$\psi_1 x + \psi_1\frac{1}{x} = 1$$

whose general solution found by a process already explained is

$$\psi_1 x = \frac{1}{2} - \varphi x + \varphi\frac{1}{x},$$

hence

$$\psi x = \frac{x^n}{2} - x^n\varphi x + x^n\varphi\frac{1}{x}. \qquad /$$

This solution differs in form from that which was previously found, but it may be proved to be the same by the following substitution; since φ is quite an arbitrary function

$$\varphi x = x^{2n}\left(1 - 2\varphi_1 x + 2\varphi_1\frac{1}{x}\right)$$

this gives

$$x^{2n}\varphi\frac{1}{x} + x^{-2n}\varphi x = 2$$

and (a) becomes

$$\psi x = \frac{x^n}{2} - x^n\varphi_1 x + x^n\varphi_1\frac{1}{x}$$

exactly as the last solution.

(21). Given

$$\left(\frac{\psi x + 1}{\psi x - 1}\right)^n - \left(\frac{\psi\left(\dfrac{x}{x-1}\right) + 1}{\psi\left(\dfrac{x}{x-1}\right) - 1}\right)^n = \frac{(x-1)^2 - 1}{x - 1}$$

put[a]

$$\psi_1 x = \left(\frac{\psi x + 1}{\psi x - 1}\right)^n$$

then it becomes

$$\psi_1 x - \psi_1\left(\frac{x}{x-1}\right) = \frac{(x-1)^2 - 1}{x - 1}$$

which is a particular case of

$$\psi_1 x + (a + v\varphi x)\psi_1\left(\frac{x}{x-1}\right) = \frac{(x-1)^2 - 1}{x - 1}$$

with which it agrees, when $a = -1$ and $v = 0$.

Put $x/(x-1)$ for x, then

$$\psi_1\frac{x}{x-1} + \left(a + v\varphi\frac{x}{x-1}\right)\psi_1 x = -\frac{(x-1)^2 - 1}{x - 1}$$

whence by elimination,

$$\psi_1 x = \frac{(x-1)^2 - 1}{x - 1} \cdot \frac{1 + a + v\varphi x}{1 - (a + v\varphi x)\left(a + v\varphi\dfrac{x}{x-1}\right)}$$

[a] Equation corrected.

and when $a = -1$, and $v = 0$, this becomes

$$\psi_1 x = \frac{x^2 - 2x}{x - 1} \cdot \frac{\varphi x}{\varphi x + \varphi\left(\dfrac{x}{x-1}\right)}$$

and restoring the value of $\psi_1 x$, we find

$$\psi x = \frac{\left\{\dfrac{x^2 - 2x}{x - 1} \cdot \dfrac{\varphi x}{\varphi x + \varphi\left(\dfrac{x}{x-1}\right)}\right\}^{1/n} + 1}{\left\{\dfrac{x^2 - 2x}{x - 1} \cdot \dfrac{\varphi x}{\varphi x + \varphi\left(\dfrac{x}{x-1}\right)}\right\}^{1/n} - 1}$$

(22). $\psi x + \dfrac{x}{\sqrt{1 + x^2}} \psi\sqrt{1 - x^2} = x$, $\psi x = \dfrac{x(1 - x^2)\varphi x}{x^2 \varphi\sqrt{1 - x^2} + (1 - x^2)\varphi x}$.

This solution was found by pursuing the course so frequently pointed out: another but not a more general one may be obtained as follows: multiply by $\sqrt{1 - x^2}$; then

$$\sqrt{1 - x^2}\psi x + x\psi\sqrt{1 - x^2} = x\sqrt{1 - x^2}$$

putting $\sqrt{1 - x^2}\psi x = \psi_1 x$, we have

$$\psi_1 x + \psi_1\sqrt{1 - x} = x\sqrt{1 - x^2}$$

whose solution is

$$\psi_1 x = \frac{x\sqrt{1 - x^2}}{2} + \varphi x - \varphi\sqrt{1 - x^2}$$

hence

$$\psi x = \frac{x}{2} + \frac{\varphi x - \varphi\sqrt{1 - x^2}}{\sqrt{1 - x^2}}.$$

It would not be difficult to show the identity of these two apparently different solutions.

(23). Given the equation

$$\frac{\psi x}{\sqrt{\psi(1 - x)}} + \frac{\psi(1 - x)}{\sqrt{\psi x}} = 1$$

put

$$\psi_1 x = \frac{\psi x}{\sqrt{\psi(1 - x)}}$$

then

$$\psi_1(1 - x) = \frac{\psi(1 - x)}{\sqrt{\psi x}}$$

and the equation becomes

$$\psi_1 x + \psi_1(1 - x) = 1$$

whose general solution is

$$\psi_1 x = \frac{\varphi x}{\varphi x + \varphi(1 - x)}$$

hence

$$\frac{(\psi x)^2}{\psi(1 - x)} = \frac{(\varphi x)^2}{\{\varphi x + \varphi(1 - x)\}^2}$$

putting $1 - x$ for x, and eliminating $\psi(1 - x)$, we find

$$\psi x = \frac{\{(\varphi x)^2 \varphi(1 - x)\}^{2/3}}{\{\varphi(1 - x) + \varphi x\}^2}.$$

(24). Given

$$(\psi x)^2 + \left(\psi \frac{a^2}{x}\right)^2 = \frac{x^4 + a^4}{x^2} \psi x \cdot \psi \frac{a^2}{x}$$

divide by $\psi x \cdot \psi(a^2/x)$ then

$$\frac{\psi x}{\psi \frac{a^2}{x}} + \frac{\psi \frac{a^2}{x}}{\psi x} = \frac{x^4 + a^4}{x^2}$$

putting

$$\psi_1 x = \frac{x}{\psi \frac{a^2}{x}}$$

this becomes /

$$\psi_1 x + \psi_1 \frac{a^2}{x} = \frac{x^4 + a^4}{x^2}$$

a particular solution of which is $\psi_1 x = x^2$; hence

$$\frac{\psi x}{\psi \frac{a^2}{x}} = x^2 \text{ or } \psi x = x^2 \psi \frac{a^2}{x}$$

and the general solution of this is

$$\psi x = x\chi\left(\overline{x}, \overline{\frac{a^2}{x}}\right).$$

(25). Given

$$\frac{\psi x + x}{\psi \frac{1}{x}} + x \frac{1 + x\psi \frac{1}{x}}{\psi x} = 1 + x^2$$

put

$$\psi_1 x = \frac{\psi x + x}{\psi \frac{1}{x}},$$

then the equation becomes

$$\psi_1 x + x^2 \psi_1 \frac{1}{x} = 1 + x^2$$

whose solution is $\psi_1 x = \dfrac{(x^2 + 1)\varphi x}{\varphi x + x^4 \varphi \frac{1}{x}}$;

hence

$$\frac{x + \psi x}{\psi \frac{1}{x}} = \frac{(x^2 + 1)\varphi x}{\varphi x + x^4 \varphi \frac{1}{x}}$$

putting $1/x$ for x and eliminating $\psi(1/x)$, we find

$$\psi x = \frac{(2x + x^{-1})\varphi x + x^5 \varphi \frac{1}{x}}{(x^2 + x^{-2})\varphi x \cdot \varphi \frac{1}{x} - x^{-4}(\varphi x)^2 - x^{-4}\left(\varphi \frac{1}{x}\right)^2} \left(\varphi \frac{1}{x} + x^{-4} \varphi x\right).$$

(26). Given

$$(\psi x)^m \cdot (\psi - x)^n - (\psi x)^n \cdot (\psi - x)^m = 2x$$

putting $\psi_1 x = (\psi x)^m \cdot (\psi - x)^n$, it becomes

$$\psi_1 x - \psi_1(-x) = 2x$$

whose solution is

$$\psi_1 x = x + \chi(\overline{x}, \overline{-x})$$

hence

$$(\psi x)^m \cdot (\psi - x)^n = x + \chi(\overline{x}, \overline{-x})$$

and by the process for eliminating $\psi(-x)$, we shall find

$$\psi x = \left\{\frac{\chi(\overline{x}, \overline{-x}) + x}{\{\chi(\overline{x}, \overline{-x}) - x\}^{\frac{m}{n}}}\right\}^{\frac{m}{m^2 - n^2}}$$

(27). Given

$$\psi x + fx \cdot \psi \alpha x = f_1 x$$

where αx is such a function of x that $\alpha^2 x = x$; putting αx for x, we have

$$\psi \alpha x + f\alpha x \cdot \psi x = f_1 \alpha x$$

and by eliminating $\psi \alpha x$

$$\psi x = \frac{f_1 x - fx \cdot f_1 \alpha x}{1 - fx \cdot f\alpha x}.$$

If $fx \cdot f\alpha x = 1$ and $f_1 x - fx \cdot f_1 \alpha x = 0$, then the solution becomes a vanishing fraction; also the general value of $f_1 x$ is in that case $f_1 x = \sqrt{fx} \cdot f_2(\overline{x}, \overline{\alpha x})$ and the equation becomes

$$\psi x + fx \cdot \psi \alpha x = \sqrt{fx} \cdot f_2(\overline{x}, \overline{\alpha x})$$

dividing this by $\sqrt{fx}$ and putting instead of $1/\sqrt{fx}$ its value $\sqrt{f\alpha x}$ derived from the equation $fx \cdot f\alpha x = 1$, we have

$$\sqrt{f\alpha x} \cdot \psi x + \sqrt{fx} \cdot \psi \alpha x = f_2(\overline{x}, \overline{\alpha x})$$

which is a symmetrical equation, whose general solution is

$$\sqrt{f\alpha x} \cdot \psi x = \frac{f_2(\overline{x}, \overline{\alpha x}) \cdot \varphi x}{\varphi x + \varphi \alpha x}$$

/

hence

$$\psi x = \frac{\sqrt{fx} \cdot f_2(\bar{x}, \overline{\alpha x}) \cdot \varphi x}{\varphi x + \varphi \alpha x}.$$

(28). Given

$$a + b\psi x = \psi(a + bx)$$

⟨then⟩

$$\psi x = a\frac{1 - b^n}{1 - b} + b^n x.$$

(29). Given

$$\frac{a\psi x}{b + c\psi x} = \psi\left(\frac{ax}{b + cx}\right)$$

⟨then⟩

$$\psi x = \frac{a^n x}{b^n + c\dfrac{a^n - b^n}{a - b}x}.$$

(30). Given

$$\frac{\psi x}{\psi x - 1} = \psi\left(\frac{x}{x - 1}\right)$$

⟨then⟩

$$\psi x = \frac{x}{\dfrac{1 - (-1)^n}{2}x + (-1)^n}$$

where n is arbitrary; it may therefore be changed into any symmetrical function of x and $x/(x - 1)$.

(31). The three last examples are particular cases of the equation

$$\alpha\psi x = \psi\alpha x$$

whose general solution is $\psi x = \alpha^n x$.

(32). Given

$$\frac{\psi\left(\dfrac{x}{1 + x}\right)}{1 + \psi\left(\dfrac{x}{1 + x}\right)} = \psi\left(\frac{x}{1 + 2x}\right)$$

⟨then⟩

$$\psi x = \frac{x}{1 + nx}.$$ /

(33). Given

$$\alpha\psi\alpha x = \psi\alpha^2 x$$

⟨then⟩

$$\psi x = \alpha^n x.$$

(34). Given

$$\psi x + \alpha\psi\left(\frac{1}{1 - x}\right) = \frac{1}{x}$$

$1/(1 - x)$ is a periodic function of the third order, or of the form $\alpha^3 x = x$; putting $1/(1 - x)$ for x, we have

$$\psi\left(\frac{1}{1 - x}\right) + \alpha\psi\left(\frac{x - 1}{x}\right) = 1 - x$$

and in this again putting $1/(1 - x)$ for x, we find

$$\psi\left(\frac{x - 1}{x}\right) + \alpha\psi x = \frac{x}{x - 1}$$

$\psi(1/(x - 1))$ and $\psi((x - 1)/x)$ being eliminated between these three equations, we have

$$\psi x + \frac{1}{1 + \alpha^3}\left\{\frac{1}{x} - \alpha(1 - x) + \alpha^2\frac{x}{x - 1}\right\}.$$

(35). Given

$$\psi x - \alpha\psi\left(\frac{\sqrt{x^2 - 1}}{x}\right) = x^{2n}$$

put $\sqrt{x^2 - 1}/x$ for x, it becomes

$$\psi\left(\frac{\sqrt{x^2 - 1}}{x}\right) - \alpha\psi\left(\frac{1}{\sqrt{1 - x^2}}\right) = \frac{(x^2 - 1)^n}{x^{2n}}.$$

Again, put $\sqrt{x^2 - 1}/x$ for x, and we have

$$\psi\left(\frac{1}{\sqrt{1 - x^2}}\right) - \alpha\psi x = \frac{1}{(1 - x^2)^n}$$ /

by eliminating $\psi(\sqrt{x^2-1}/x)$ and $\psi(1/\sqrt{1-x^2})$ from these three equations, we shall find

$$\psi x = \frac{1}{1-\alpha^3}\left\{x^{2n} + \alpha\frac{(x^2-1)^n}{x^{2n}} + \alpha^2\frac{1}{(1-x^2)^n}\right\}.$$

(36). Given

$$\psi x + \psi\left(\frac{1+x}{1-3x}\right) + \psi\left(\frac{x-1}{3x+1}\right) = a$$

the function $(1+x)/(1-3x)$ is periodic of the third order, and by the process of elimination

$$\psi x = \frac{a\varphi x + \varphi_1 x - \varphi_1\left(\frac{1+x}{1-3x}\right)}{\varphi x + \varphi\left(\frac{1+x}{1-3x}\right) + \varphi\left(\frac{x-1}{3x+1}\right)}.$$

(37). Given

$$\frac{1}{\sqrt{\psi x}} + \frac{1}{\sqrt{\psi\frac{x-1}{x}}} + \frac{1}{\sqrt{\psi\frac{1}{1-x}}} = a.$$

Putting $\psi_1 x = \frac{1}{\sqrt{\psi x}}$, we have

$$\psi_1 x + \psi_1\frac{x-1}{x} + \psi_1\frac{1}{1-x} = a$$

whose solution is

$$\psi_1 x = \frac{a\varphi x + \varphi_1 x - \varphi_1\frac{1}{1-x}}{\varphi x + \varphi\frac{1}{1-x} + \varphi\frac{x-1}{x}}$$

hence

$$\psi x = \left\{\frac{\varphi x + \varphi\frac{1}{1-x} + \varphi\frac{x-1}{x}}{a\varphi x + \varphi_1 x - \varphi_1\frac{1}{1-x}}\right\}^2.$$

(38). Given

$$\psi x \cdot \psi \frac{1}{1-x} \cdot \psi \frac{x-1}{x} = c^3.$$

Putting $\psi_1 x = \log \psi x$, we find

$$\psi_1 x + \psi_1 \frac{1}{1-x} + \psi_1 \frac{x-1}{x} = \log (c^3)$$

whose solution is found in the last problem. Changing φ_1 into $\log \varphi_1$, we have

$$\psi x = \frac{\varphi_1 x}{\varphi_1 \frac{1}{1-x}} \log^{-1} \left\{ \frac{3\varphi x \cdot \log c}{\varphi x + \varphi \frac{x-1}{x} + \varphi \frac{1}{1-x}} \right\}.$$

Similarly if αx be any periodic equation of the third order.

(39). $\psi x + \psi \alpha x + \psi \alpha^2 x = a$

has for it solution

$$\psi x = \frac{a\varphi x + \varphi_1 x - \varphi_1 \alpha x}{\varphi x + \varphi \alpha x + \varphi \alpha^2 x}.$$

(40). $\psi x \cdot \psi \alpha x \cdot \psi \alpha^2 x = c^3$

has for its solution

$$\psi x = \frac{\varphi_1 x}{\varphi \alpha x} \log^{-1} \left\{ \frac{3\varphi x \cdot \log c}{\varphi x + \varphi \alpha x + \varphi \alpha^2 x} \right\}.$$

(41). Given

$$\psi x + fx \cdot \psi \alpha x = f_1 x$$

where $\alpha^3 x = x$. Putting successively αx and $\alpha^2 x$ for x, we have

$$\psi \alpha x + f\alpha x \cdot \psi \alpha^2 x = f_1 \alpha x$$
$$\psi \alpha^2 x + f\alpha^2 x \psi x = f_1 \alpha^2 x$$

and eliminating $\psi \alpha x$ and $\psi \alpha^2 x$ from these three equations, we have

$$\psi x = \frac{f_1 x - fx \cdot f_1 \alpha x + fx \cdot f\alpha x \cdot f_1 \alpha^2 x}{1 + fx \cdot f\alpha x \cdot f\alpha^2 x}$$

(42). Given

$$\psi x + fx \cdot \psi \alpha x = f_1 x$$

where $\alpha^n x = x$, a similar process of elimination will produce

$$\psi x = \frac{f_1 x - fx \cdot f_1 \alpha x + \ldots fx \cdot f\alpha x \cdot f\alpha^{n-2} x \cdot f_1 \alpha^{n-1} x}{1 - (-1)^n fx \cdot f\alpha x \ldots f\alpha^{n-1} x}.$$

(43). Given the equation

$$1 + fx \cdot (\psi x + \psi \alpha x) - \psi x \cdot \psi \alpha x = 0$$

where αx is a periodic function of the second order, and fx is any function symmetrical relative to x and αx

$$fx = \frac{\psi x \cdot \psi \alpha x - 1}{\psi x + \psi \alpha x}.$$

Consider ψx and $\psi \alpha x$ as two variables, and differentiate with respect to them, then

$$\frac{d\psi x}{1 + (\psi x)^2} + \frac{d\psi \alpha x}{(1 + \psi \alpha x)^2} = 0$$

and by integration,

$$\tan^{-1} \psi x + \tan^{-1} \psi \alpha x = C = -\frac{1}{fx}$$

whose complete solution is

$$\tan^{-1} \psi x = \frac{\varphi x}{\varphi x + \varphi \alpha x} \tan^{-1} \frac{-1}{fx}$$

hence

$$\psi x = \tan \left\{ \frac{\varphi x}{\varphi x + \varphi \alpha x} \tan^{-1} \frac{-1}{fx} \right\}$$

this process is analogous to one employed by M. Laplace, for the integration of a similar equation of differences.* /

(44). Given

$$\psi x + \psi\left(\frac{1 + x}{1 - x}\right) + \psi\left(-\frac{1}{x}\right) + \psi\left(\frac{x - 1}{x + 1}\right) = 1$$

* *Journal de l'École Polytechnique*, Cah. 15. ⟨P. S. de Laplace, 'Mémoire sur divers points d'analyse', *Journal de l'École Polytechnique*, Vol. 8 (1809), pp. 229–65, or *Oeuvres complètes de Laplace* (Paris, 1878–1912) Vol. 14, pp. 178–214.⟩

$(1 + x)/(1 - x)$ being a periodic function of the 4th order

$$\psi x = \frac{\varphi x}{\varphi x + \varphi\left(\frac{1+x}{1-x}\right) + \varphi\left(\frac{-1}{x}\right) + \varphi\left(\frac{x-1}{x+1}\right)}$$

$$+ \frac{x^2+1}{x-1}\chi\left(\overline{x}, \overline{\frac{1+x}{1-x}}, \overline{\frac{-1}{x}}, \overline{\frac{x-1}{x+1}}\right)$$

$$+ \frac{x^2+1}{x\cdot(1-x)}\chi\left\{\overline{x}, \overline{\frac{1+x}{1-x}}, \overline{\frac{-1}{x}}, \overline{\frac{x-1}{x+1}}\right\}.$$

(45). Given

$$\psi(x, y) + \psi\left(\frac{\alpha^2}{x}, \frac{\alpha^2}{y}\right) = 1$$

⟨then⟩

$$\psi(x, y) = \frac{\varphi(x, y)}{\varphi(x, y) + \varphi\left(\frac{\alpha^2}{x}, \frac{\alpha^2}{y}\right)}.$$

(46). ⟨Given⟩

$$\psi(x, y) + \psi\left(\frac{a^2}{x}, -y\right) = y^2$$

⟨then⟩

$$\psi(x, y) = \frac{y^2\varphi(x, y)}{\varphi(x, y) + \varphi\left(\frac{\alpha^2}{x}, -y\right)}.$$

(47). Given

$$\psi(x, y) + x^2\psi\left(\frac{\alpha^2}{x}, \frac{y}{y-1}\right) = \frac{xy^2}{y-1}$$

⟨then⟩

$$\psi(x, y) = \frac{(y-1)^{-1}xy^2\varphi(x, y)}{\varphi(x, y) + \varphi\left(x, \frac{y}{y-1}\right)}.$$

(48). Given

$$\psi(x, y) + f(x, y)\psi(\alpha x, \beta y) = \sqrt{f(x, y)}\cdot f_1\left(\overline{x}, \overline{\alpha x}, \overline{\frac{1}{y}}, \overline{\frac{1}{\beta y}}\right)$$

where $\alpha^2 x = x$, $\beta^2 y = y$ and $f(x, y)$ is such a function that $f(x, y) \cdot f(\alpha x, \beta y) = 1$, then

$$\psi(x, y) = \frac{\sqrt{f(\alpha x, \beta y)} \cdot f_1(\overline{x}, \overline{\alpha x}, \overline{\tfrac{1}{y}}, \overline{\tfrac{1}{\beta y}})}{f(x, y)\varphi(\alpha x, \beta y) + f(\alpha x, \beta y)\varphi(x, y)}.$$

(49). Given

$$\psi(x, y) = \psi\left(\frac{x - y}{y}, \frac{x - y}{x}\right)$$

⟨then⟩

$$\psi(x, y) = \varphi\left(\frac{x}{y}\right).$$

(50). Given

$$\psi(x, y) = \psi\left(\frac{f(x, y)}{y}, \frac{f(x, y)}{x}\right)$$

⟨then⟩

$$\psi(x, y) = \varphi\left(\frac{x}{y}\right).$$

(51). Given

$$\psi(x, y) = \psi\left(\frac{y}{2}\sqrt{\frac{y}{2x}}, \sqrt{2xy}\right)$$

$$\psi(x, y) = \chi\left(2xy + y^2, \frac{1}{y\sqrt{2xy}}\right).$$

(52). Given

$$\psi(x, y) = \psi(y, x)$$

⟨then⟩

$$\psi(x, y) = \chi(x + y, xy).$$

(53). Given

$$\psi(x, y) = \left(\frac{x}{y}\right)^3 \psi(y, x)$$

⟨then⟩

$$\psi(x, y) = \frac{x}{y^2} \cdot \varphi(\overline{x}, \overline{y}) = \frac{x}{y^2} \cdot \varphi_1(x + y, xy).$$ /

(54). Given

$$\psi(\pi - x) = \frac{d\psi x}{dx}$$

differentiating

$$\frac{d}{dx}\psi(\pi - x) = \frac{d^2\psi x}{dx^2}$$

putting $\pi - x$ for x in the given equation

$$\psi(x) = -\frac{d\psi(\pi - x)}{dx}$$

and eliminating $d\psi(\pi - x)/dx$, we have

$$-\psi x = \frac{d^2\psi x}{dx^2}$$

whence by integration

$$\psi x \doteq b\cos x + c\sin x$$

and it will be found that $c = -b$; hence

$$\psi x = b(\cos x - \sin x).$$

(55). Given

$$\psi(x, y) = \frac{d\psi(x, a - y)}{dx}$$

put $a - y$ for y, then

$$\psi(x, a - y) = \frac{d\psi(x, y)}{dx}$$

differentiate this relative to x, then

$$\frac{d\psi(x, a - y)}{dx} = \frac{d^2\psi(x, y)}{dx^2}$$

which being substituted in the given equation produces

$$\psi(x, y) = \frac{d^2\psi(x, y)}{dx^2}$$

whose solution is

$$\psi(x, y) = \varepsilon^x\varphi y + \varepsilon^{-x}\varphi_1 y$$

φ and φ_1 being two arbitrary functions so constituted as to fulfil the given equation, in order to determine them, put $a - y$ for y and differentiate relative to x, then

$$\frac{d\psi(x, a-y)}{dx} = \varepsilon^{x}\varphi(a-y) - \varepsilon^{-x}\varphi_1(a-y)$$

hence

$$\varphi y = \varphi(a-y) \quad \text{and} \quad \varphi_1 y = -\varphi_1(a-y)$$

whose solutions are

$$\varphi x = \chi(\overline{y}, \overline{a-y})$$

and

$$\varphi_1 y = (a-2y)\chi_1(\overline{y}, \overline{a-y})$$

hence the general solution of the equation is

$$\psi(x, y) = \varepsilon^{x}\chi(\overline{y}, \overline{a-y}) + \varepsilon^{-x}(a-2y)\chi_1(\overline{y}, \overline{a-y}).$$

A similar mode of solution is applicable to the three following equations.

(56). Given

$$\psi(x, y) = \frac{d}{dx}\psi\left(x, \frac{1}{y}\right)$$

⟨then⟩

$$\psi(x, y) = \varepsilon^{x}\varphi\left(\overline{y}, \overline{\frac{1}{y}}\right) + \varepsilon^{-x}\frac{1-y^2}{y}\varphi_1\left(\overline{y}, \overline{\frac{1}{y}}\right).$$

(57). Given

$$\psi(x, y) = \frac{d}{dx}\psi\left(x, \frac{y}{y-1}\right)$$

⟨then⟩

$$\psi(x, y) = \varepsilon^{x}\varphi\left(\overline{y}, \overline{\frac{y}{y-1}}\right) + \varepsilon^{-x}\frac{2y-y^2}{y-1}\varphi_1\left(\overline{y}, \overline{\frac{y}{y-1}}\right).$$

(58). Given

$$\psi(x, y) = \frac{d}{dx}\psi(x, \alpha y)$$

where $\alpha^2 y = y$

$$\psi(x, y) = \varepsilon^{x}\varphi(y, \alpha y) + \varepsilon^{-x}(\alpha y - y)\varphi_1(\overline{y}, \overline{\alpha y}).$$

(59). Given

$$\psi(x, y) = \frac{d\psi(x, \alpha y)}{dx}$$

where α is such a function that $\alpha^4 y = y$. Substituting successively αy, $\alpha^2 y$, $\alpha^3 y$ for y, we have

$$\psi(x, \alpha y) = \frac{d\psi(x, \alpha^2 y)}{dx}$$

$$\psi(x, \alpha^2 y) = \frac{d\psi(x, \alpha^3 y)}{dx}$$

$$\psi(x, \alpha^3 y) = \frac{d\psi(x, y)}{dx}.$$

From the given equation $d\psi(x, \alpha y)/dx$ may be eliminated by means of the second, and from the result $d\psi(x, \alpha^2 y)/dx$ may be eliminated by the third equation, and continuing this, we should find

$$\psi(x, y) = \frac{d^4\psi(x, y)}{dx^4}$$

the solution of this equation is

$$\psi(x, y) = \varepsilon^x \varphi y + \varepsilon^{-x} \varphi_1 y + \sin x \cdot \varphi_2 y + \cos x \cdot \varphi_3 y$$

φ, φ_1, φ_2, φ_3 must be determined so as to satisfy the given equation, taking the differential and putting αy for y, we have

$$\frac{d\psi(x, \alpha y)}{dx} = \varepsilon^x \varphi \alpha y - \varepsilon^{-x} \varphi_1 \alpha y + \cos x \cdot \varphi_2 \alpha y - \sin x \cdot \varphi_3 \alpha y$$

the first condition to satisfy is

$$\varphi y = \varphi \alpha y$$

which gives

$$\varphi y = \chi(\overline{y}, \overline{\alpha y}, \overline{\alpha^2 y}, \overline{\alpha^3 y})$$

the next condition is

$$\varphi_1 y = -\varphi_1 \alpha y$$

whose solution is

$$\varphi_1 y = (\alpha^3 y - \alpha^2 y + \alpha y - y(\chi_1(\overline{y}, \overline{\alpha y}, \overline{\alpha^2 y}, \overline{\alpha^3 y}),$$

the other two conditions are

$$\varphi_2 y = -\varphi_3 \alpha y \quad \text{and} \quad \varphi_3 y = \varphi_2 \alpha y$$

putting αy for y in the second of these it becomes $\varphi_3 \alpha y = \varphi_2 \alpha^2 y$ and this substituted in the first gives, $\varphi_2 y = -\varphi_2 \alpha^2 y$, whose solution is

$$\varphi_2 y = (\alpha^2 y - y)\chi_2\left(\overline{y}, \frac{1}{\alpha y}, \overline{\alpha^2 y}, \frac{1}{\alpha^3 y}\right)$$

hence

$$\varphi_3 y = (\alpha^3 y - \alpha y)\chi_2\left(\overline{\alpha y}, \frac{1}{\alpha^2 y}, \overline{\alpha^3 y}, \frac{1}{y}\right)$$

and the general solution of the equation is

$$\begin{aligned}\psi(x, y) = {} & \varepsilon^x \chi(\overline{y}, \overline{\alpha y}, \overline{\alpha^2 y}, \overline{\alpha^3 y}) \\ & + \varepsilon^{-x}(\alpha^3 y - \alpha^2 y + \alpha y - y)\chi_1(\overline{y}, \overline{\alpha y}, \overline{\alpha^2 y}, \overline{\alpha^3 y}) \\ & + (\alpha^2 y - y)\chi_2\left(\overline{y}, \frac{1}{\alpha y}, \overline{\alpha^2 y}, \frac{1}{\alpha^3 y}\right) \sin x \\ & + (\alpha^3 y - \alpha y)\chi_2\left(\overline{\alpha y}, \frac{1}{\alpha^2 y}, \overline{\alpha^3 y}, \frac{1}{y}\right) \cos x.\end{aligned}$$

(60). Given the equation

$$\psi(x, y) = \frac{d^n \psi(x, \alpha y)}{dx^n}$$

where α is such a function that $\alpha^n x = x$. This equation may be reduced to the solution of the partial differential equation

$$\psi(x, y) = \frac{d^{pn} \psi(x, y)}{dx^{pn}}$$

and the arbitrary functions of y which occur in its solution, must be determined by the conditions of the equation.

(61). Given the equation

$$\frac{d\psi(a - x, y)}{dy} = \frac{d\psi(x, b - y)}{dx}$$

put $a - x$ for x, also $b - y$ for y, then we have the two equations /

$$\frac{d\psi(x, y)}{dy} = -\frac{d\psi(a - x, b - y)}{dx}$$

$$-\frac{d\psi(a - x, b - y)}{dy} = \frac{d\psi(x, y)}{dx}.$$

If the first of these be differentiated relative to y, and the second relative to x; then the right side of the first resulting equation will be identical with the left side of the second, and we shall have

$$\frac{d^2\psi(x, y)}{dy^2} = \frac{d^2\psi(x, y)}{dx^2}$$

the solution of this partial differential equation is

$$\psi(x, y) = \varphi(x + y) + \varphi_1(x - y)$$

the two arbitrary functions must be determined so as to satisfy the equation; we have

$$\frac{d\psi(a - x, y)}{dy} = \varphi'(a - x + y) - \varphi_1'(a - x - y).$$

$$\frac{d\psi(x, b - y)}{dx} = \varphi'(b + x - y) + \varphi_1'(-b + x + y)$$

φ' and φ_1' being the differential coefficients of φ and φ_1 these two expressions must be identical, hence

$$\varphi'(a - (x - y)) = \varphi'(b + (x - y))$$

and

$$-\varphi_1'(a - (x + y)) = \varphi_1'(-b + (x + y))$$

the solutions of which equations are

$$\varphi'(x + y) = \chi\{\overline{x + y}, \overline{a - b - x - y}\}$$

and

$$\varphi_1'(x + y) = (a - b - 2x - 2y)\chi_1\{\overline{x + y}, \overline{a - b - x - y}\}$$

and substituting these values, we have /

$$\psi(x, y) = \int(dx + dy)\chi\{\overline{x + y}, \overline{a - b - x - y}\}$$

$$+ \int(dx - dy)(a - b - 2x - 2y)\chi_1\{\overline{x - y}, \overline{a - b - 2x + 2y}\}.$$

(63). Given the equation

$$\frac{d\psi\left(x, \frac{1}{y}\right)}{dx} = \frac{d\psi\left(\frac{1}{x}, y\right)}{dy}.$$

Put $1/y$ for y and differentiate relative to x, then

$$\frac{d^2\psi(x, y)}{dx^2} = -\frac{d^2\psi\left(\frac{1}{x}, \frac{1}{y}\right)}{dxdy}y^2.$$

Again, put $1/x$ for x, and differentiate relative to y, then

$$-\frac{d^2\psi\left(\frac{1}{x},\frac{1}{y}\right)}{dxdy}x^2 = \frac{d^2\psi(x,y)}{dy^2}$$

hence

$$\frac{d^2\psi(x,y)}{dy^2} = \frac{x^2}{y^2}\frac{d^2\psi(x,y)}{dx^2}$$

the solution of this equation of partial differentials is

$$\psi(x,y) = x\varphi\frac{x}{y} + \varphi_1(xy).$$

To determine the form of φ and φ_1, we have

$$\frac{d\psi\left(x,\frac{1}{y}\right)}{dx} = \varphi(xy) + xy\varphi'(xy) + \frac{1}{y}\varphi_1'\left(\frac{x}{y}\right)$$

$$\frac{d\psi\left(\frac{1}{x},y\right)}{dy} = -\frac{1}{x^2y}\varphi'\left(\frac{1}{xy}\right) + \frac{1}{x}\varphi_1'\left(\frac{y}{x}\right).$$

In order that these two expressions may coincide, we must have

$$\varphi(xy) + xy\varphi'(xy) = -\frac{1}{x^2y^2}\varphi\left(\frac{1}{xy}\right)$$

$$\varphi_1'\left(\frac{x}{y}\right) = \frac{y}{x}\varphi_1'\left(\frac{y}{x}\right).$$

The first of these multiplied by $d(xy)$ may be put under the form

$$d(xy)\cdot\varphi(xy) + xy\frac{d\varphi(xy)}{d(xy)}d(xy) = d\varphi\left(\frac{1}{xy}\right)$$

whose integral is

$$xy\varphi(xy) = \varphi\left(\frac{1}{xy}\right)$$

the solution of which functional equation is

$$\varphi(xy) = \frac{1}{\sqrt{xy}}\chi\left(\overline{xy},\overline{\frac{1}{xy}}\right)$$

the solution of the second equation is

$$\varphi_1'\left(\frac{x}{y}\right) = \sqrt{\frac{y}{x}} \cdot \chi_1\left(\frac{\overline{x}}{y}, \frac{\overline{y}}{x}\right)$$

employing these values of φ and φ_1, we have

$$\psi(x, y) = \sqrt{xy} \cdot \chi\left(\frac{\overline{x}}{y}, \frac{\overline{y}}{x}\right) + \int d(xy) \cdot \left(\frac{1}{xy}\right)^{1/2} \chi_1\left(\overline{xy}, \overline{\frac{1}{xy}}\right).$$

(63). Given

$$\frac{d\psi(x, \alpha y)}{dx} = \frac{d\psi(\beta x, y)}{dy}$$

where $\alpha^2 y = y$ and $\beta^2 x = x$ a process nearly similar to that / by which the two last equations were solved will lead to the partial differential equation

$$\frac{d\alpha y}{dy} \cdot \frac{d^2\psi(x, y)}{dx^2} = \frac{d\beta x}{dx} \cdot \frac{d^2\psi(x, y)}{dy^2}.$$

(64). Given

$$\psi\alpha x = \psi\psi x = \psi^2 x.$$

It is evident, that whatever be the form of α, this equation can always be satisfied by assuming $\psi x = \alpha x$, hence the solutions of the following equations,

$$\psi(-x) = \psi^2 x \qquad \psi x = -x$$

$$\psi\left(\frac{ax}{b + cx}\right) = \psi^2 x \qquad \psi x = \frac{ax}{b + cx}$$

$$\psi\sqrt{\frac{1 + x}{x}} = \psi^2 x \qquad \psi x = \sqrt{\frac{1 + x}{x}}.$$

(65). Given

$$\psi(2a - x) = \psi^3 x.$$

Put $\psi x = \varphi^{-1} f \varphi x$, then

$$\psi^2 x = \varphi^{-1} f \varphi \varphi^{-1} f \varphi x = \varphi^{-1} f^2 \varphi x$$

and

$$\psi^3 x = \varphi^{-1} f^2 \varphi \varphi^{-1} f \varphi x = \varphi^{-1} f^3 \varphi x$$

and the equation becomes

$$\varphi^{-1} f \varphi(2a - x) = \varphi^{-1} f^3 \varphi x.$$

This equation may be satisfied in the following manner: by making f a periodic function of the second order, we have $f^2v = v$, and the equation becomes

$$\varphi^{-1}f\varphi(2a - x) = \varphi^{-1}f\varphi x$$

or

$$\varphi(2a - x) = \varphi x.$$

This is satisfied by making φ any symmetrical function of x and $2a - x$. As an example take $fv = -v$, also

$$\varphi x = x \cdot (2a - x) = 2ax - x^2$$

then

$$\varphi^{-1}x = a \pm \sqrt{a^2 - x}$$

and

$$\psi x = \varphi^{-1}f\varphi x = a \pm \sqrt{a^2 - \frac{1}{x^2 - 2ax}}.$$

(66). Given

$$\psi\left(\frac{1 - x}{1 + x}\right) = \psi^3 x$$

$\psi x = \varphi^{-1}f\varphi x$, where φ and f are determined by the equations

$$\varphi x = \chi\left\{\overline{\frac{1 - x}{1 + x}}, \overline{x}\right\} \quad \text{and} \quad f^2x = x.$$

(67). Given

$$\psi\alpha x = \psi^3 x$$

where $\alpha^n v = v$; putting $\psi x = \varphi^{-1}f\varphi x$, we have

$$\varphi^{-1}f\varphi\alpha x = \varphi^{-1}f^3\varphi x.$$

Determine φ from the condition $\varphi x = \varphi\alpha x$; hence,

$$\varphi x = \chi\{\overline{x}, \overline{\alpha x}, \ldots \overline{\alpha^{n-1}x}\}$$

and let f be such a function that $f^2x = x$, then the equation is satisfied.

(68). Given

$$\psi^p\alpha x = \psi^q x$$

where $q > p$ and $\alpha^n x = x$, the substitution $\varphi^{-1}f\varphi x$ instead of ψ will give,

$$\varphi^{-1}f^p\varphi\alpha x = \varphi^{-1}f^q\varphi x$$

and this is satisfied if $\varphi x = \chi(\overline{x}, \overline{\alpha x}, \overline{\alpha^{n-1}x})$ and also $f^{q-p}v = v$, for it then becomes

$$\varphi^{-1}f^p\varphi\alpha x = \varphi^{-1}f^p\varphi x$$

where $\varphi x = \varphi\alpha x$. /

If in a function of two variables, as $\psi(x, y)$, we substitute the function itself instead of one of those quantities, the result is denoted thus,

$$\psi\{x, \psi(x, y)\} = \psi^{1,2}(x, y)$$
$$\psi\{\psi(x, y), y\} = \psi^{2,1}(x, y)$$

if the function itself is substituted simultaneously for x and y, it is denoted thus

$$\psi\{\psi(x, y), \psi(x, y)\} = \psi^{\overline{2,2}}(x, y).$$

(69). Given

$$\psi^{\overline{2,2}}(x, y) = a.$$

By means of the substitution $\varphi^{-1}f\varphi x$, for ψx, we are enabled to reduce functional equations of any order to those of the first, a substitution nearly resembling it, will be of equal value for those which contain two or more variables, by assuming

$$\psi(x, y) = \varphi^{-1}f(\varphi x, \varphi y)$$

we have

$$\psi^{\overline{2,2}}(x, y) = \varphi^{-1}f\{\varphi\varphi^{-1}f(\varphi x, \varphi y), \varphi\varphi^{-1}f(\varphi x, \varphi y)\}$$
$$= \varphi^{-1}f^{\overline{2,2}}(\varphi x, \varphi y)$$

and substituting this value in the equation

$$\varphi^{-1}f^{\overline{2,2}}(\varphi x, \varphi y) = a.$$

Put $\varphi^{-1}x$ for x, and φ^{-1} for y, also taking the function φ on both sides

$$f^{\overline{2,2}}(x, y) = \varphi a.$$

If therefore we are acquainted with a particular solution, we find the general one; let the function $A(x/y)$ be tried, then /

$$f^{\overline{2,2}}(x, y) = A\frac{A\frac{x}{y}}{A\frac{x}{y}} = \varphi a$$

hence $A = \varphi a$, and the solution is

$$\psi(x, y) = \varphi^{-1}\left(\frac{\varphi x}{\varphi y}\varphi a\right).$$

A variety of solutions may be found of different forms, such as

$$\psi(x, y) = \frac{a}{\varphi(1)}\varphi\left(\frac{x}{y}\right)$$

$$\psi(x, y) = \frac{a}{\varphi(1)}\varphi\left(\frac{\alpha(x, y)}{\beta(x, y)}\right)$$

where α and β are any two homogeneous functions of the same degree.

(70). If

$$\psi(x, y) = ax + by$$

⟨then⟩

$$\psi^{\overline{n,n}}(x, y) = (a + b)^{n-1}(ax + by).$$

(71). If $\psi(x, y)$ is any homogeneous function of x, and y of the degree n, then

$$\psi^{\overline{k,k}}(x, y) = \{\psi(x, y)\}^{n^{k-1}} \cdot \{\psi(1, 1)\}^{(1-n^{k-1})/(1-n)}.$$

(72). Given

$$\psi^{\overline{2,2}}(x, y) = \sqrt{\psi(x, y)}$$

⟨then⟩

$$\psi(x, y) = \sqrt{\frac{x^2 + y^2}{2y\varphi\left(\frac{x}{y}\right)}}\,\varphi(1).$$

(73). Given

$$\psi^{\overline{2,2}}(x, y) = \psi(x, y) + \frac{1}{\psi(x, y)}$$

⟨then⟩ /

$$\psi(x, y) = \frac{2\varphi\left(\frac{x}{y}\right)}{(x + y)\varphi(1)} + \frac{(x + y)\varphi(1)}{2\varphi\left(\frac{x}{y}\right)}.$$

(74). Given

$$\psi^{\overline{2,2}}(x, y) = \frac{1 - \psi(x, y)}{1 + \psi(x, y)}$$

⟨then⟩

$$\psi(x, y) = \frac{y\varphi(1) - x^2\varphi\left(\dfrac{x}{y}\right)}{y\varphi(1) + x^2\varphi\left(\dfrac{x}{y}\right)}.$$

(75). ⟨Given⟩

$$\psi^{\overline{2,2}}(x, y) = F\psi(x, y)$$

⟨then⟩

$$\psi(x, y) = F\left(\frac{\alpha(x, y)}{\beta(x, y)}\right)$$

provided α and β are homogeneous with respect to x and y; the first of the $n + 1$ degree, the second of the nth, and also at the same time $\alpha(1, 1) = \beta(1, 1)$.

(76). Given

$$\psi^{\overline{2,2}}(x, y) = F\psi(x, y).$$

Another solution of the same equation is

$$\psi(x, y) = F\left(\frac{\alpha(x, y)}{\beta x, y)}x\right)$$

where α and β are two such functions, that when $x = y$, we have also

$$\alpha(x, y) = \beta(x, y).$$

(77). Given

$$\psi^{\overline{3,3}}(x, y) = \psi(x, y),$$

⟨then⟩

$$\psi(x, y) = \left\{a - \left(\frac{x^2 + y^2}{x + y}\right)^n\right\}^{1/n}.$$

(78). Given

$$\psi^{\overline{n,n}}(x, y) = \{\psi(x, y)\}^m$$

⟨then⟩

$$\psi(x, y) = \left\{\frac{2xy\varphi(1)}{(x + y)\varphi\left(\dfrac{x}{y}\right)}\right\}^{m^{1/(n-1)}}.$$

(79). Given

$$\psi^{\overline{4,4}}(x, y) = \{\psi^{\overline{2,2}}(x, y)\}^2$$

⟨then⟩

$$\psi(x, y) = \left\{ \frac{(x + y)\varphi(1)}{2\varphi\left(\frac{x}{y}\right)} \right\}^{\sqrt{2}}.$$

(80). Given

$$\psi^{\overline{3,3}}(x, y) = \frac{x^2}{y}$$

⟨then⟩

$$\psi(x, y) = \varphi^{-1} \left\{ \frac{1 + \varphi\left(\frac{x^2}{y}\right)}{1 - 3\varphi\left(\frac{x^2}{y}\right)} \right\}.$$

(81). Given

$$x\psi^{1,2}(x, y) = y\psi^{2,1}(x, y)$$

put $\varphi^{-1}f(\varphi x, \varphi y)$ for ψx, then it becomes

$$x\varphi^{-1}f^{1,2}(\varphi x, \varphi y) = y\varphi^{-1}f^{2,1}(\varphi x, \varphi y)$$

putting $\varphi^{-1}x$ for x, and $\varphi^{-1}y$ instead of y, we have

$$\varphi^{-1}x \cdot \varphi^{-1}f^{1,2}(x, y) = \varphi^{-1}y \cdot \varphi^{-1}f^{2,1}(x, y).$$

If $f^{1,2}(x, y) = y$, and $f^{2,1}(x, y) = x$, this equation becomes identical; but making $f(x, y) = a - x - y$, these two equations are verified; consequently the general solution is

$$\psi(x, y) = \varphi^{-1}(a - \varphi x - \varphi y).$$

(82). Given

$$\psi^{1,2}(x, y) \cdot \psi^{2,1}(x, y) = xy$$

⟨then⟩

$$\psi(x, y) = \varphi^{-1}\left(\frac{a}{\varphi x \cdot \varphi y}\right).$$

(83). Given

$$x\psi^{\overline{2,2}}(x, y) = a\psi^{2,1}(x, y)$$

⟨then⟩

$$\psi(x, y) = \varphi^{-1}\left(\frac{\varphi a \cdot \varphi y}{\varphi x}\right).$$

/

Various methods for the solution of Functional Equations may be found in the following writings:

Speculationes Analytico Geometricae, *N. Fuss.* Mem. de l'Acad. Imp. de St. Petersburg, Vol. IV. p. 225. 1811.[a]

Memoires of the Analytical Society, p. 96. 1813.[b]

Observations on various points of Analysis, Phil. Trans. *J. F. W. Herschel.*[c]

Essay towards the Calculus of Functions, *C. Babbage.* 1815.[d]

Ditto, Part II. p. 179. 1816.[e]

Observations on the analogy which subsists between the Calculus of Functions, and other branches of Analysis, Phil. Trans. 1817. p. 197. *C. Babbage.*[f]

Spence's Essays, 1819. Note by *J. F. W. Herschel.* p. 151.[g]

Annals of Philosophy, Nov. 1817. *Mr. Horner.*[h]

Journal of the Royal Institution. *C. Babbage.*[i] /

[a] Nicolaus von Fuss, 'Speculationes analytico-geometricae', *Mémoires de l'Académie Impériale des Sciences de St. Pétersbourg*, Vol. 4 (1813), pp. 221–39.

[b] C. Babbage, 'On continued products' (1813), in this volume of the *Works of Babbage.*

[c] J. F. W. Herschel, 'Consideration of various points of analysis', *Philosophical Transactions*, (1814), pp. 440–68.

[d] C. Babbage, 'An essay towards the calculus of functions' (1815), in this volume of the *Works of Babbage.*

[e] C. Babbage, 'An essay towards the calculus of functions, part II' (1816), in this volume of the *Works of Babbage.*

[f] C. Babbage, 'Observations on the analogy which subsists between the calculus of functions and the other branches of analysis' (1817), in this volume of the *Works of Babbage.*

[g] W. Spence, *Mathematical essays ... edited by J. F. W. Herschel ...* (London, 1819).

[h] W. G. Horner, 'Solution of the equation $\psi^n x = x$', *Annals of Philosophy*, Vol. 10 (1817), pp. 341–6.

[i] C. Babbage, 'Solutions of some problems by means of the calculus of functions' (1817), in this volume of the *Works of Babbage.*

XIV

AN EXAMINATION OF SOME QUESTIONS CONNECTED WITH GAMES OF CHANCE

From *Transactions of the Royal Society of Edinburgh*, Vol. 9 (1821), pp. 153–77

Read 21 March, 1820

The questions which I propose to examine in the following paper, although not themselves dependent on chance, have arisen entirely from games in which it predominates. To determine some method of betting upon a number of successive events (and the probability of each of which is either equal to, or less than one-half) by which a profit shall be realised after a considerable number of them have been decided, is a problem which has occupied the attention, and exhausted the efforts, of one set of speculators, as completely as that of the quadrature of the circle has defeated the labours of another. The first and most simple plan, is that of doubling the stake whenever a loss occurs. This is well known, and has been so frequently practised, as to have acquired a peculiar name; it is technically called the *martingal*; it requires for its success, that the person who employs it have the power of leaving off whenever he please, and that he have the command of an unlimited capital. If the chance of the events happening is one third instead of one half the stake must be tripled. Supposing / a player adopt this system in betting on a number of independent events, the chance of occurring of each of which is one half, he would naturally desire to know, if possible, before he began, what would be his profit or loss, supposing p of the events decided in his favour, and q against him. In this particular plan of playing, it so happens, that he cannot, from the mere knowledge of the number of favourable and of unfavourable events, arrive at the conclusion he desires. I shall presently show, that these data alone are insufficient, and that, in order to determine the question, not merely the number, but the order of succession must be given.

It is probable that this difficulty, in the case of most frequent occurrence, has deterred many from attempting other similar problems. Indeed, on the first view of such questions, it is by no means apparent, that any of them can be solved without reference to the order in which the events take place. The mode of inquiry which I shall point out, will show that in many cases their mutual arrangement is not required amongst the data, and will furnish a criterion by which we may determine, in any given case, whether it is necessary.

I shall first examine the case of the martingal, which, although the results it leads to are of a negative nature, will introduce us to the method of treating these questions, and from its frequent practice is rather interesting.

Let us suppose a gamester bet a certain sum $2u$, upon an event whose chance of happening is one half; whenever he wins he repeats the same bet; but whenever he loses, he doubles his last stake. If he should win p and lose q times, it is required to ascertain how much he will have won or lost.

The first stake being $2u$, which may either be gained or lost, we may represent the gamester's profit after one event is / decided by $2u(-1)^a$, a being any whole number; for since the nature of the number a is left undecided, whether it is an even or an odd one, the expression just given will represent either a profit or a loss. First, if a is an even number, his next stake remains the same as it was before, or equal to $2u$, if it is an odd number, his second stake must be $4u$. We must therefore endeavour to find some function of a which shall be equal to $2u$, when a is an even number, and become $4u$, when a is an odd one. A great variety of functions may be found satisfying this condition; such are $2u \cdot 2^{(1-(-1)^a)/2}$, $2u((3 + \cos a\pi)/2)$, etc. and many others. Our choice amongst the infinite variety which present themselves, must be directed by the ulterior operations of the gamester; and, first, I remark, that every stake must be equal to the constant quantity u, multiplied by some power of 2. The next remark which may guide us in this choice of a proper function is, that the index of that power of 2 is determined by the number of times (in immediate succession) the event has been unfavourable, reckoning back from the event about which the stake is proposed: if, therefore, at any period, all the preceding determinations have been unfavourable, the amount of the next stake will be influenced by them all. It may also be noticed, that the successive powers of 2 can be formed by the addition of all the preceding ones, together with unity, thus,

$$1, 2^0, 2^1, 2^2, 2^3, 2^4, 2^5$$

$$1 + 2^0 = 2, \qquad 1 + 2^0 + 2^1 = 4, \qquad 1 + 2^0 + 2^1 + 2^2 = 8, \ldots$$

From these considerations it appears, that the function we require may consist of a series of functions, each multiplying the / successive powers of 2. In order to determine these, let us suppose that at some point of the game, the last event has proved favourable; then, by the conditions, the next stake is $2u$; and whatever be the course of succeeding events, $2u$ will always form part of the stake; therefore it need not be multiplied by any function of $a, b, c, \ldots$ the letters which determine the winning or losing of the subsequent events. We may therefore assume $2u$ as the constant part of every stake, without reference to any particular order in their occurrence. If the failure or happening of this event is represented by $(-1)^a$, a being an odd number in the first, and an even number in the second case, we must multiply the next power of 2, or 2^1, by some function of a which shall vanish when a is an even number, and become unity when it is an odd one. Such a function is easily found, and one of the simplest is

$$\frac{1-(-1)^a}{2}.$$

The next stake is therefore

$$u\left\{2+\frac{1-(-1)^a}{2}2\right\}$$

whatever be the form of a.

The failing or happening of this event may be represented by $(-1)^b$, and the profit of the player by this event is then represented by

$$u\left\{2+\frac{1-(-1)^a}{2}2^1\right\}(-1)^b.$$

The third stake must comprehend the second power of 2, and will be of the form

$$u(2+2^1f(a,b)+2^2f_1(a,b)).$$

If b is an even number, or the second event is favourable, in that case, the new stake would be $2u$; and therefore both $f(a, b)$ and $f_1(a, b)$ must vanish. This will take place if each has a factor of the form

$$\frac{1-(-1)^b}{2}$$

and the new stake in consequence / becomes

$$u\left\{2+2^1\frac{1-(-1)^b}{2}f(a,b)+2^2\frac{1-(-1)^b}{2}f_1(a,b)\right\}$$

f and f_1, not being the same functions as before. If a had been an even number, we may consider the first bet as not having been made, since it has no influence on the succeeding ones; and in this case the expression ought to reduce itself to

$$u\left\{2+2^1\frac{1-(-1)^b}{2}\right\}$$

$f(a, b)$ must therefore equal unity, and $f_1(a, b)$ must vanish when a is an even number. This gives

$$u\left\{2 + 2^1\frac{1-(-1)^b}{2} + 2^2\frac{1-(-1)^b}{2}f_1(a, b)\right\}.$$

This expression is reduced to $2u$ if b is an even number, and to $4u + 4uf_1(a, b)$ when b is an odd one; $f_1(a, b)$ must therefore be such a function of a, that when a is odd, it shall become unity, and when even, equal zero;

$$\frac{1-(-1)^a}{2}$$

is such a function, and we then have for the third stake

$$u\left\{2 + 2\frac{1-(-1)^b}{2} + 2^2\frac{1-(-1)^b}{2}\cdot\frac{1-(-1)^a}{2}\right\}.$$

The law by which we may represent the stake to be ventured, after the determination of any number of events, is now apparent; had it not been sufficiently so, the same reasoning which has been already explained at some length would have assigned it for the fourth stake,

$$u\left\{2 + 2\frac{1-(-1)^c}{2} + 2^2\frac{1-(-1)^b}{2}\cdot\frac{1-(-1)^c}{2}\right.$$
$$\left. + 2^3\frac{1-(-1)^a}{2}\cdot\frac{1-(-1)^b}{2}\cdot\frac{1-(-1)^c}{2}\right\}.$$

We can now represent the profit or loss on each event, without determining which of the two it is:

On the 1st he gains $u\{2\}\,(-1)^a$

2nd $u\{2 + (1-(-1)^a)\}\,(-1)^b$

3rd $u\{2 + (1-(-1)^b) + (1-(-1)^a)\cdot(1-(-1)^b)\}\,(-1)^c$

4th $u\{2 + (1-(-1)^c) + (1-(-1)^b)\cdot(1-(-1)^c)$
$+ (1-(-1)^a)\cdot(1-(-1)^b)\cdot(1-(-1)^c\})\,(-1)^d$

5th $u\{2 + (1-(-1)^d) + (1-(-1)^c)\cdot(1-(-1)^d)$
$+ ((1-(-1)^b)\cdot(1-(-1)^c)\cdot(1-(-1)^b)$
$+ (1-(-1)^a)\cdot(1-(-1)^b)\cdot(1-(-1)^c)\cdot$
$(1-(-1)^d)\}(-1)^e$

$\vdots$

The sum of all these is the object sought: with respect to that part which multiplies $2u$, it is easily found; it is

$$2u\{(-1)^a + (-1)^b + (-1)^c + \ldots\}$$

but we know that he has gained p, and lost q times; consequently p of the quantities $a, b, c, \ldots$ are even numbers, and q of them odd ones; therefore the sum of the series, multiplying $2u$, is $p - q$, and the value of the part alluded to is

$$2(p - q)u.$$

So far, then, we can proceed by means of the data furnished in the question; but they are insufficient for its complete solution, on attempting to find the sum of that part of the expression / in which (-1) has only one index, it appears to be

$$u\{1(-1)^b + 2(-1)^c + 3(-1)^d + \ldots\}$$

and the knowledge that p of the letters $a, b, c, \ldots$ represent even, and q of them odd numbers, will not assist us in determining the sum of this series; we must also be acquainted with the order of succession of the even and odd numbers. Another reason will afterwards be assigned, why it is not sufficient merely to be acquainted with the number of favourable and of unfavourable events, which may with more propriety be stated, when some other questions of a similar nature have been examined.

Let us now consider another case, in which the amount of the sum staked on each event follows a different law, we will suppose the following problem:

A gamester begins a series of bets on an event whose chance of occurring is one half, by staking the sum u. Whenever he wins he makes the next succeeding stake less than his last by the quantity v; but if he lose, he then increases his stake by the same quantity. Supposing he should win p times, and lose q times, what will he have gained or lost on the whole number $p + q$?

His first stake being u, his first profit may be expressed by $u(-1)^a$, which will be won or lost according as a is even or odd; in the first case his next stake would be $u - v$, and in the second it would be $u + v$. These two cases may be combined into one, and thus expressed $u - v(-1)^a$, and his second profit will be $\{u - v(-1)^a\}(-1)^b$; his next stake will be $u - v(-1)^a - v$ if b is even; but it will be $u - v(-1)^a + v$ / if that number is odd. These may be represented thus: $u - v\{(-1)^a + (-1)^b\}$, and his third profit will be

$$[u - v\{(-1)^a + (-1)^b\}]\,(-1)^c$$

the succeeding stakes are easily represented in the same manner, and his profits stand thus:

$$\begin{array}{l} u(-1)^a \\ u(-1)^b - v\{(-1)^a\}\,(-1)^b \\ u(-1)^c - v\{(-1)^a + (-1)^b\}\,(-1)^c \\ u(-1)^d - v\{(-1)^a + (-1)^b + (-1)^c\}\,(-1)^d \\ u(-1)^e - v\{(-1)^a + (-1)^b + (-1)^c + (-1)^d\}\,(-1)^e \\ \vdots \end{array}$$

The value of the part multiplying u, or

$$(-1)^a + (-1)^b + (-1)^c + \ldots$$

is easily found, since we know that p of the quantities $a, b, c, \ldots$ are even numbers, and q of them odd numbers. The first part of the expression is therefore

$$(p - q)u.$$

In order to determine the second part, we must observe, that the expression

$$(a')b' + (a' + b')c' + (a' + b' + c')d' + (a' + b' + c' + d')e' + \ldots$$

is equal to the sum of all the combinations, two by two of the quantities $a', b', c', \ldots$; and, moreover, that if these latter are / the roots of an equation, then the expression just written is equal to the coefficient of its third term. If we change a' into $(-1)^a$, b' into $(-1)^b$, c' into $(-1)^c, \ldots$ it becomes the same as the one whose value we are seeking. Hence then it appears, that the sum of all the quantities, multiplying v, is equal to the coefficient of the third term of an equation whose roots are $(-1)^a, (-1)^b, (-1)^c, \ldots$ or of the equation[a]

$$0 = \{x - (-1)^a\} \cdot \{x - (-1)^b\} \cdot \{x - (-1)^c\} \ldots$$

but we know that p of the quantities $a, b, c, \ldots$ are even numbers, and q of them odd ones; therefore this equation has p equal roots of the form $+1$, and q equal ones of the form -1, or the equation is

$$0 = (x - 1)^p(x + 1)^q$$

and the quantity which multiplies v, is equal to the coefficient of the third term of this expression, which is

$$\frac{p \cdot p - 1}{1 \cdot 2} - \frac{p}{1} \cdot \frac{q}{1} + \frac{q \cdot q - 1}{1 \cdot 2} = \frac{(p - q)^2 - (p + q)}{2};$$

so that the profit of the gamester is

$$W = (p - q)u - v\frac{(p - q)^2 - (p + q)}{2}. \qquad (1)$$

[a] Brackets have been inserted in the equation.

This result is entirely independent of the order in which the events occurred; and we may learn from the method that has been employed for its solution, that whenever the sum of all the winnings or losings is a symmetrical function of the quantities $(-1)^a$, $(-1)^b$, $(-1)^c$, ... the final conclusion will not depend on the order in which the events succeed each other, but on the actual number of favourable and unfavourable events. /

In the instance we are now considering, if the number of successful and of unsuccessful cases are equal, the gamester who adopts this system of play will always win, for in that case $p = q$, and $W_1 = pv$; as a numerical example of the formula just investigated. Let us suppose:

A person stakes 100 shillings on the event of a piece of money thrown into the air falling with one of its faces uppermost, in preference to the other; whenever he wins he diminishes his stake five shillings, and whenever he loses he increases it by the same sum. Supposing he makes 800 successive bets, and wins p, and loses q of them.

If p and q are equal, each being 400, his profit will be $400 \times 5 = 2000$ shillings.

From this it appears, that he will win even though a smaller number than one-half of the events prove favourable. In order to determine how many times he must win, that he may neither lose nor gain on the whole number, we must make the value of W equal to zero, putting $p + q = a$; this gives for the required value of p

$$p = +\frac{1}{2}\left(a + \frac{u}{v}\right) \pm \frac{1}{2}\sqrt{\frac{u^2}{v^2} + a}$$

in the present instance, $a = 800$, $u = 100$, $v = 5$, which gives

$$p = \tfrac{1}{2}(800 + 10) \pm \tfrac{1}{2}\sqrt{100 + 800} = 405 \pm 15;$$

the lower sign being employed, we have $p = 390$.

In fact, on substituting this number in the formula (1) we find that he neither wins nor loses money, although the number of unfavourable events has been greater by twenty than the number of favourable ones. The other root of the equation points / out another limit, beyond which if the number of favourable events increase, he will lose money.

In the problem just examined, each stake depends on that which immediately precedes it. Others may be proposed, in which it is made to depend on the result of two or more of the events which immediately precede, such is the following one:

A gamester begins by staking a given sum u, and regulates his succeeding bets in this manner; if one of the two immediately antecedent to that which he is at any time making, have been gained, and the other lost, the new stake is the same as that last made; but if both those bets have been gained, he diminishes his stake by the sum v; and if both have been lost, he increases it by the same sum v. The first stake being u, the first gain may be represented by $u(-1)^a$; and as there has only been one bet yet made, the second cannot be affected by the law prescribed; it will therefore be the same as the first, and the profit arising from it will be $u(-1)^b$. The third stake depending on the two former, we must add to the quantity u some function of a and b multiplied by v, which shall vanish if a and b are one odd and the other even, and which shall become -1 when they are both even, and $+1$ when they are both odd. Such is the function

$$-\frac{(-1)^a + (-1)^b}{2}$$

the profit from the third stake may therefore be represented by

$$\left\{u - v\left(\frac{(-1)^a + (-1)^b}{2}\right)\right\}(-1)^c$$

and in a similar manner, we may find that which arises from the fourth is expressed by

$$\left\{u - v\left(\frac{(-1)^a + (-1)^b}{2} + \frac{(-1)^b + (-1)^c}{2}\right)\right\}(-1)^d.$$

The mode of continuation is sufficiently obvious, and by arranging all these profits together, we have

$$u\{(-1)^a\}$$

$$u\{(-1)^b\}$$

$$\left\{u - v\left(\frac{(-1)^a + (-1)^b}{2}\right)\right\}(-1)^c$$

$$\left\{u - v\left(\frac{(-1)^a + (-1)^b}{2} + \frac{(-1)^b + (-1)^c}{2}\right)\right\}(-1)^d$$

$$\left\{u - v\left(\frac{(-1)^a + (-1)^b}{2} + \frac{(-1)^b + (-1)^c}{2} + \frac{(-1)^c + (-1)^d}{2}\right)\right\}(-1)^e$$

$$\left\{u - v\left(\frac{(-1)^a + (-1)^b}{2} + \frac{(-1)^b + (-1)^c}{2} + \frac{(-1)^c + (-1)^d}{2} + \frac{(-1)^d + (-1)^e}{2}\right)\right\}(-1)^f.$$

$$\vdots$$

The part which depends on u is easily determined to be $(p + q)u$, where p and q are the number of successful and of unsuccessful cases, the remaining part consists of $-v/2$ multiplied by a function of the quantities $a, b, c, \ldots$ In order to abridge as much as possible, I shall write a instead of $(-1)^a$, b instead of $(-1)^b$, and so on; so that ab will represent / $(-1)^a \times (-1)^b$, or $(-1)^{a+b}$; observing this, the factor multiplying $-v/2$ will be expressed thus:

$$\begin{aligned}&ac + b\\ &ad + 2bd + cd\\ &ae + 2be + 2ce + de\\ &af + 2bf + 2cf + 2df + ef\\ &ag + 2bg + 2cg + 2dg + 2eg + fg,\\ &\vdots\end{aligned}$$

If to these were added the two series $ab + bc + cd + \ldots$ and $a(b + c + d + e + \ldots)$ the sum would be equal to twice the sum of all the products, taken two by two of the quantities $a, b, c, \ldots$ (and since these quantities represent $(-1)^a, (-1)^b, \ldots$ we have found in the last problem that it is equal to $((p - q)^2 - (p + q))/2$. The part depending on v is therefore

$$-\frac{v}{2}\{(p - q)^2 - (p + q) - (-1)^{a+b} - (-1)^{b+c} - (-1)^{c+d} - \ldots$$
$$- (-1)^a((-1)^b + (-1)^c + (-1)^d + \ldots))\}.$$

The second of these series is evidently equal to

$$(-1)^a(p - q - (-1)^a) = (-1)^a(p - q) - 1$$

the other series

$$(-1)^{a+b} + (-1)^{b+c} + (-1)^{c+d} + (-1)^{d+e} + \ldots$$

cannot be determined merely by knowing how many of the numbers $a, b, c, \ldots$ are odd, and how many are even. /

Still, however, the knowledge of the order in which the events take place is not absolutely required. It will be sufficient to know how many changes from odd to even, and *vice versa*, occur between the quantities $a, b, d, c, e, \ldots$ for if there is no change in going from a to b, the first term will be $+1$; but if there is a change it will be -1; and generally -1 will occur as often as there is a change from odd to even, or from even to odd in the series

$$a, b, c, d, e, \ldots$$

If, therefore, k denote the number of transitions, since the number of terms is $p + q - 1$, the value of the series will be

$$p + q - 2k - 1$$

and the whole value of the part dependent on v will be

$$-\frac{v}{2}\left\{(p-q)^2-(p+q)-(p+q)+2k+1-(-1)^a(p-q)+1\right\}$$

and the profit on the whole number of bets is

$$W=(p-q)u-\frac{v}{2}\left\{(p-q)^2-2(p+q)+2k+2-(-1)^a(p-q)\right\} \quad (2)$$

If $p = q$, then

$$W_i = +v(2p-k-1) \quad (3)$$

k in the first of these expressions may vary from 0 to $p + q - 1$, and in the second from 1 to $2p - 1$.

In order to compare this with an example, let the following / series of bets be made *plus* or *minus*, denoting their success or failure:

$$+100, +100, +90, -80, +80, +80, +70.$$

The profit is 440. Here $p = 6$, $q = 1$, $k = 2$, $v = 10$, $u = 100$; and these values being substituted in (2) give $W = 440$.

The next case I shall examine is one in which the amount of each stake depends on the result of the three preceding events.

Let the law by which the stakes are regulated be such, that when the three preceding have been won, a certain sum v is taken from the last stake; if they have been lost, the sum v is added to the last; but if only two of these three have been gained, or only two lost, in the first case one third of v is subtracted, and in the second case is added to the last stake.

The first stake being u, the three first bets will be represented by $u(-1)^a$, $u(-1)^b$, $u(-1)^c$. In order to fulfil the conditions of the question, the fourth must be increased by the quantity

$$-v\frac{(-1)^a+(-1)^b+(-1)^c}{3}$$

and the profit resulting from it is represented by

$$\left\{u-v\frac{(-1)^a+(-1)^b+(-1)^c}{3}\right\}(-1)^d$$

that of the fifth is equally expressed by

$$\left\{u-v\left(\frac{(-1)^a+(-1)^b+(-1)^c}{3}+\frac{(-1)^b+(-1)^c+(-1)^d}{3}\right)\right\}(-1)^e.$$

/

The succeeding terms need not be written down, as the law is obvious. In this, as in the former questions, the coefficient of u is $p-q$; and if we employ the same abridgement of a for $(-1)^a, \ldots$ as in the last, the co-efficient of $-v/3$ is

$$\begin{aligned}&ad + bd + cd\\ &ae + be + 2ce + de\\ &af + 2bf + 3cf + 2df + ef\\ &ag + 2bg + 3cg + 3dg + 2eg + fg\\ &\vdots\end{aligned}$$

This series, which only differs from three times the sum of all the products, two by two of the quantities $a, b, c, \ldots$ by the omission of certain parts, may be called Q. On examining what parts are omitted, it will be found that

$$\begin{aligned}Q &+ 2a(S(a+b+\ldots) - a - b - c) + bd\\ &+ b(S(a+b+\ldots) - a - b - c) + (ce + df + cg + \ldots)\\ &+ 2(cd + de + ef + \ldots) + 3(ab + ac + bc)\\ &\qquad = 3S(ab + \ldots) = 3\frac{(p-q)^2 - (p+q)}{2}.\end{aligned}$$

This equation becomes, by the mutual destruction of certain terms[a]

$$\begin{aligned}Q &+ 2aS(a+b+\ldots) + bS(a+b+\ldots) + (ac + bd + cc + \ldots)\\ &\qquad + 2(bc + cd + \ldots) - 3 = 3\frac{(p-q)^2 - (p+q)}{2}.\end{aligned}$$

Hence /

$$\begin{aligned}Q = 3\frac{(p-q)^2 - (p+q)}{2} &- 2(-1)^a(p-q) - (-1)^b(p-q)\\ &+ 3 - (ac + bd + cd + \ldots) - 2(bc + cd + \ldots).\end{aligned}$$

The latter of these series is

$$\begin{aligned}\{(-1)^{a+b} + (-1)^{b+c} + (-1)^{c+d} + \ldots\} &- (-1)^{a+b}\\ &= p + q - 2k - 1 - (-1)^{a+b}\end{aligned}$$

as we found in the last question, k being the number of changes from even to odd, or *vice versa*, in the series, $a, b, c, \ldots$ The value of the series

$$(-1)^{a+c} + (-1)^{b+d} + (-1)^{c+e} + \ldots$$

is not determined without some other data. If, however, we are acquainted with the number of alterations from odd to even, and the contrary, in the series

$$a\ c \quad b\ d \quad c\ e \quad d\ f \quad \ldots$$

[a] Equation corrected.

we can assign its value; let the number be l, then the series in question is equal to $p + q - 2l - 2$. These substitutions and the necessary reductions being made, we have

$$Q = (p - q)\left\{3\frac{p - q}{2} - 2(-1)^a - (-1)^b\right\} - (p + q)\frac{q}{2} + 7 + 2l + 4k + 2(-1)^{a+b}$$

and W the profit on $p + q$ events is

$$W = (p - q)\left\{u - \frac{v}{3}\left(3\frac{p - q}{2} - 2(-1)^a - (-1)^b\right)\right\} - \frac{v}{3}\left\{7 - (p + q)\frac{q}{2} + 2l + 4k + 2(-1)^{a+b}\right\} \quad (4)$$

if $p = q$

$$W_1 = -\frac{v}{3}\left\{7 - 9p + 2l + 4k + 2(-1)^{a+b}\right\}. \quad (5)$$

/

In the first of these expressions, k may vary from 0 to $p + q - 1$, and l from 0 to $p + q - 2$; in the second k may vary from 1 to $2p - 1$, and l may vary from 0 to $2p - 2$.

The last question which I shall examine is one in which each stake depends on all those which precede it.

A gamester stakes a certain sum u on an event whose chance of happening is one half. He regulates each succeeding stake in this manner: To the constant sum u he adds the nth part of all his previous winnings; or if he has lost by the previous stakes, he subtracts the nth part of his loss, it is proposed to find his profit* at the termination of $p + q$ bets.

The first stake being u, the profit is $u(-1)^a$; the nth part of this added to u, or

$$u + (u/n)(-1)^a$$

will constitute the second, and the profit will be

$$u(-1)^b + (u/n)(-1)^a(-1)^b$$

the nth part of these two, or

$$(u/n)((-1)^a + (-1)^b) + (u/n^2)(-1)^a(-1)^b$$

* The language of analysis is so much more general than that in which we usually convey our thoughts, that it is almost impossible to make the latter keep pace with the former. This is more particularly manifest when we are treating of games of chance. The words profit, winning, gain, ... must, if we wish to avoid perpetual repetition, frequently be understood to comprehend their very opposites.

added to the quantity u, will be the third stake, and the profit on the determination of the third event will be

$$u(-1)^c + (u/n)((-1)^a + (-1)^b)(-1)^c + (u/n^2)(-1)^a(-1)^b(-1)^c \qquad /$$

the nth part of the sum of these three profits added to u, and multiplied by $(-1)^d$, gives for the profit on the fourth event

$$u(-1)^d + (u/n)((-1)^a(-1)^d + (-1)^b(-1)^d + (-1)^c(-1)^d)$$
$$+ (u/n^2)((-1)^a(-1)^b(-1)^d + (-1)^a(-1)^c(-1)^d + (-1)^b(-1)^c(-1)^d)$$
$$+ (u/n^3)(-1)^a(-1)^b(-1)^c(-1)^d.$$

From these expressions, continued a few steps further if necessary, it appears that the sum of all the profits W is equal to

u multiplied by the sum of all the quantities
$(-1)^a, (-1)^b, (-1)^c, \ldots$
$+ u/n$ multiplied by the sum of all the products, two by two of those quantities
$+ u/n^2$ multiplied by the sum of all the products, three by three of the same quantities
$+ u/n^3$ multiplied by the sum of all the products, four by four of the same quantities
$\vdots$

Hence $(1/n)W =$ the sum of all the coefficients, except the first of the terms of an equation, whose roots are $(-1)^a/n, (-1)^b/n, (-1)^c/n, \ldots$ multiplied by u; but by the conditions of the problem, / p of the quantities a, b, c, are even, and q of them odd numbers, or p of the roots of the equation are of the form $-1/n$, and q of the $+1/n$, and the equation itself is the development of

$$(x + 1/n)^p(x - 1/n)^q = 0$$

the sum of all its coefficients except the first may be found by making $x = 1$, and subtracting unity, this gives

$$\frac{1}{n}W = \left(1 + \frac{1}{n}\right)^p\left(1 - \frac{1}{n}\right)^q u - u$$

or

$$W = n\left(\frac{n+1}{n}\right)^p\left(\frac{n-1}{n}\right)^q u - nu \tag{6}$$

if $p = q$ this becomes

$$W = \frac{(n^2-1)^p}{n^{2p-1}}u - nu \tag{7}$$

if $n = 1$, whatever be the numbers p and q, the loss will be equal to u (unless at the same time $q = 0$), for in that case as soon as the first unfavourable event happens, the player loses not only all he had previously won, but also the sum u besides; and since, by the conditions of the play, he must subtract from u the nth part of all his former loss, which since $n = 1$ is $-u$, his next stake must be $u - u$, or zero; so that in fact if $n = 1$, the first unfavourable decision terminates the game.

I shall now proceed to show how a similar mode of reasoning may be applied to cases where the number of events which happen at each step are more than two.

Suppose a person draw a number of balls in succession from an urn, containing balls numbered 1, 2, 3, 4, ... k, there being many of each kind: he begins by staking the sum u, and if he / draw a ball marked one, he receives n_1 times his stake, if the ball be marked two, he receives n_2 times his stake, and so on. The next stake is thus regulated, supposing the ball last drawn to have been marked i, he adds to the last stake the sum vn_i: any of the numbers $n_1, n_2, \ldots$ may be negative, if he has drawn p balls marked[a] 1, q marked 2, r marked 3, and so on, what is the amount of his winning?

Let $\alpha, \beta, \gamma, \ldots$ be the kth roots of unity, the expression

$$S_a = \frac{\alpha^a + \beta^b + \gamma^a + \ldots}{k}$$

is always equal to zero, except when a is a multiple of k; also let

$$P_a = n_1 S_a + n_2 S_{a+1} + n_3 S_{a+2} + \ldots + n_k S_{a+k-1}$$

then P_a will in every case reduce itself to one of the quantities $n_1, n_2, n_3, \ldots n_k$. With the aid of these considerations, we can express the amount of the stake at any particular step; his first is u, and whatever be the kind of ball drawn, his profit is always expressed by uP_a, according to the form of a. In consequence of this first determination, he adds to u the quantity vP_a, which sum $u + vP_a$ forms his second stake; the number of the second ball determines the amount of his second profit, which may be expressed thus:

$$(u + vP_a)P_b$$

without at all determining the form of a, the third stake will be $u + vP_a + vP_b$ and the profit arising from it is

$$(u + vP_a + vP_b)P_c$$

that in the fourth event is

$$(u + vP_a + vP_b + vP_c)P_d.$$

/

[a] Notation corrected.

The sum of all the profits will be

$$\begin{aligned}&uP_a\\&uP_b + v(P_a)P_b\\&uP_c + v(P_a + P_b)P_c\\&uP_d + v(P_a + P_b + P_c)P_d\\&uP_e + v(P_a + P_b + P_c + P_d)P_e.\end{aligned}$$

It is easily perceived that the coefficient of u is equal to that of the second term, and the coefficient of v is equal to that of the third term of the equation

$$(x + P_a)(x + P_b)(x + P_c)(x + P_d)\ldots = 0$$

but of the quantities* $P_a, P_b, P_c, \ldots$ p are equal to n_1, q are equal to n_2, r are equal to n_3, etc.; this equation is therefore

$$(x + n_1)^p (x + n_2)^q (x + n_3)^r \ldots = 0$$

and the coefficient of the second term, in its development, is

$$pn_1 + qn_2 + rn_3 + \ldots$$

whilst that of the third term is

$$\frac{p \cdot p - 1}{2} n_1{}^2 + \frac{q \cdot q - 1}{2} n_2{}^2 + \frac{r \cdot r - 1}{2} n_3{}^2 + \ldots$$
$$+ \frac{pq}{1} n_1 n_2 + \frac{pr}{1} n_1 n_3 + \frac{qr}{1} n_2 n_3 + \ldots$$

/

so that the whole profit is

$$W = u(pn_1 + qn_2 + rn_3 + \ldots) + v\left\{\frac{p \cdot p - 1}{2} n_1{}^2 + \frac{q \cdot q - 1}{2} n_2{}^2 + \ldots + \frac{pq}{1} n_1 n_2 + \frac{pr}{1} n_1 n_3 + \ldots\right\} \quad (8)$$

In the case of $n_1 = 1, n_2 = 1, n_3 = 0$, this formula reduces itself to (1).

Supposing the urn in the last question filled with the same balls, and a person drawing out one receives $n_1, n_2, n_3, \ldots$ times the sum u, according to the number of the ball drawn; and on the second drawing he receives $n_1, n_2, n_3, \ldots$ times the sum of u plus the profit by the last drawing: and generally on extracting any ball he receives $n_1, n_2, n_3, \ldots$ times the sum of u plus the amount of the profit on all the preceding events, if the number of times each of the balls marked $1, 2, 3, \ldots$ are drawn, be respectively denoted by $p, q, r, \ldots$ what is the whole profit?

* This notation has been employed by Mr Herschel, in a paper in the *Philosophical Transactions* for the year 1818 on circulating functions. ⟨J. F. W. Herschel, 'On circulating functions ...', *Philosophical Transactions*, Vol. 108 (1818), pp. 144–68.⟩

Adopting the same notation as in the last problem,

$$P_a = n_1 S_a + n_2 S_{a+1} + n_3 S_{a+2} + \ldots$$

will represent either $n_1, n_2, n_3, \ldots$ and the first profit is uP_a, the second is $(u + uP_a)P_b$; that in the third is $(u + uP_a + uP_b + uP_aP_b)P_c$, and the sum of all the profits is

$$\begin{array}{l} uP_a \\ uP_b + uP_aP_b \\ uP_c + uP_aP_c + uP_bP_c + uP_aP_bP_c \\ \vdots \end{array}$$

/

On comparing this with the fourth problem which was solved, it appears that

$W = u \cdot$ the sum of all the coefficients except the first of the equation $(x + P_a)(x + P_b)(x + P_c) \ldots = 0$

or

$$W = u(1 + P_a)(1 + P_b)(1 + P_c) \ldots - u$$

but p of the quantities $P_a, P_b, P_c, \ldots$ are equal to n_1, q of them to n_2, r of them to $n_3, \ldots$ This equation, therefore, becomes

$$W = u(1 + n_1)^p (1 + n_2)^q (1 + n_3)^r \ldots - u \tag{9}$$

If $n_1 = 1/n$, $n_2 = -1/n$, $n_3 = 0$, and $u = nu$, this coincides with (6).

As an example, suppose an urn filled with balls of three colours, white, black, and red, and that the person who draws them out may name any sum he chose prior to each extraction; if he draw a white ball, the sum he named is paid to him; if a black, he loses one-half of it; and if a red one, he loses one-third of that sum. And suppose he regulates the sum named in the following manner, beginning with naming u whenever he has drawn a white ball, he adds the whole of his previous winnings to the sum u; but if he has drawn a black one, he adds only half his profits to the sum u; and if the ball last extracted from the urn was red, he adds one-third of all his profits to the same sum u. He has drawn out p white, q black, and r red balls, what is the amount of his profit or loss? /

In this case $n_1 = 1$, $n_2 = 1/2$, $n_3 = -1/3$, $n_4 = n_5 \ldots = 0$ and we have

$$W = u2^p \cdot \frac{1}{2^q} \cdot \frac{2^r}{3^r} - u = \frac{2^{p-q+r}}{3^r} u - u.$$

The questions examined in the preceding pages afford an instance of the immediate application of some very abstract propositions of analysis to a subject of constant occurrence, which being as far as I have been able to discover, hitherto untouched, and also requiring reasoning of rather an unusual nature,

I have preferred treating in particular instances, instead of investigating it in its most general form. The principles on which similar problems should be attempted, are first by means of some combinations of the roots of unity to represent the stake after the decision of any number of events, and then by means of any known theorems respecting the roots of equations, to ascertain the sums of the series which present themselves in the result. /

[63/4]

XV

OBSERVATIONS ON THE NOTATION EMPLOYED IN THE CALCULUS OF FUNCTIONS

From *Transactions of the Cambridge Philosophical Society*, Vol. 1 (1822), pp. 63–76

Read May 1, 1820

Amongst the various causes which combine in enabling us by the use of analytical reasoning to connect through a long succession of intermediate steps the data of a question with its solution, no one exerts a more powerful influence than the brevity and compactness which is so peculiar to the language employed. The progress of improvement in leading us from the simpler up to the most complex relations has gradually produced new modes of shortening the ancient paths, and the symbols which have thus been invented in many instances from a partial view, or for very limited purposes, have themselves given rise to questions far beyond the expectations of their authors, and which have materially contributed to the progress of the science. Few indeed have been so fortunate as at once to perceive all the bearings and foresee all the consequences which result either necessarily or analogically even from some of the simplest improvements.

The first analyst who employed the very natural abbreviation of a^2 instead of *aa* little contemplated the existence of fractional negative and imaginary exponents, at the moment when he / adopted this apparently insignificant mode of abridging his labour. So great however is the connection that subsists between all branches of pure analysis, that we cannot employ a new symbol or make a new definition, without at once introducing a whole train of consequences, and in defiance of ourselves, the very sign we have created, and on which we have bestowed a meaning, itself almost prescribes the path our future investigations are to follow.

Such being the power and influence of those symbols* by which mathe-

* Euler, to whom analysis is so much indebted, appears to have been fully aware of the power and importance of notation; 'Summa analyseos inventa maximam partem algorithmo ad certas quasdem quantitates accommodato innitantur.' ⟨*Translation*: The whole of analysis invented so far is dependent on the adaptation to a large extent of algorithms to certain fixed quantities.⟩ Specimen algorithmi singularis. Acad. Petrop. Comm. *Nov.* 1762. ⟨Vol. 9 (1764), pp. 53–69.⟩

matical reasoning is carried on, it cannot be considered as unimportant either as regards the particular branch, or with reference to the science in general, to examine some of the bearings of the notation which has been employed in the calculus of functions, and to resolve some of the unusual questions which it presents.

That the results to which such an enquiry will conduct us are of a nature purely speculative, is an objection to which every attempt to improve notation is liable; it can however never be considered an useless task to examine and strengthen that which essentially contributes to the power of an instrument, which enables us so wonderfully to trace the connection between the phenomena of nature.

When it became convenient to express without performing the repetition of an operation, whose characteristic is f, the method which first presented itself doubtless was similar to that which had been adopted with such advantage for the exponents of quantities, and $f^2(x)$ was written instead of $ff(x)$; it now followed without any other convention that $f^3(x)$ and $f^n(x)$ represented $fff(x)$, and $fff(n \text{ times})(x)$ and also that

$$f^{n+m}(x) = f^n f^m(x) \tag{A}$$

when n and m are whole numbers.

At this point of generalization a question occurred as to the meaning of f^n when n is a fractional, surd, or negative number, and in order to determine it, recourse was had to a new convention not inconsistent with, but comprehending in it the former one. The index n was now defined by means of the equation (A) and was said to indicate such a modification of the function to which it is attached that the equation shall be verified.

From this extended view of the equation (A) several curious results follow; if $n = 0$, it becomes

$$f^m(x) = f^0 f^m(x).$$

This informs us that f^0 is such an operation that when performed on any quantity it does not change it, or putting $f^m(x) = y$, it gives

$$f^0(y) = y$$

a result which is analogous to $x^0 = 1$.

Let $m = -1$, $n = 1$, we have

$$f^0 x = f^1 f^{-1}(x) \quad \text{or} \quad f(f^{-1}x) = x$$

$f^{-1}(x)$ must therefore signify such a function of x, that if we perform upon it the operation denoted by f it shall be reduced to x. The number of functions possessing this property will depend on the nature of f; thus if $f(x) = x^n$, let $r_1, r_2, \ldots, r_n$ be the roots of $v^n - 1 = 0$; then $f^{-1}(x)$ may be either of the n quantities

$$r_1 x^{1/n}, r_2 x^{1/n}, \ldots, r_n x^{1/n}$$

for if we perform the operation f upon any of them, or raise it to the nth power, the result is x.

Here then we find

$$f(f^{-1}x) = x$$

in all cases; but if the negative index is attached to the first functional sign, we have

$$f^{-1}(fx) = r_i x$$

and one only of these values gives $f^{-1}fx = x$.

It was more necessary to make this observation, because several errors have arisen from not attending to it, and because that particular form of f^{-1} which gives $f^{-1}fx = x$ possesses peculiar properties: $f^{-1}(x)$ is then the inverse function of fx; and if we have the equation $f(x) = y$, we may indicate its resolution thus

$$x = f^{-1}(y).$$

Having established the connection between positive and negative indices in functions of one variable, I shall now proceed to those containing two.

If in the function $\psi(x, y)$ we substitute at the same time $\psi(x, y)$ for x and y it becomes

$$\psi\{\psi(x, y), \psi(x, y)\}$$

and for the sake of brevity this has been denoted thus

$$\psi^{\overline{2.2}}(x, y)$$

similarly, if in this we wrote $\psi(x, y)$ for x and y we should have $\psi^{\overline{3,3}}(x, y)$, and a few steps would lead us to the equation,

$$\psi^{\overline{n,n}}\{\psi^{\overline{m,m}}(x, y), \psi^{\overline{m,m}}(x, y)\} = \psi^{\overline{n+m,n+m}}(x, y). \qquad (B)$$

The same reasons which in the former instance induced us to generalize the first definition, now point out the propriety of assuming this as the definition of simultaneous functions, and of the modifications implied by their indices. The first step is to discover the value of $\psi^{\overline{0,0}}(x, y)$; for this purpose put $n = 0$, then

$$\psi^{\overline{0,0}}\{\psi^{\overline{m,m}}(x, y), \psi^{\overline{m,m}}(x, y)\} = \psi^{\overline{m,m}}(x, y)$$

and substituting v instead of $\psi^{\overline{m,m}}(x, y)$, we have

$$\psi^{\overline{0,0}}(v, v) = v.$$

If we had put $m = 0$, $n = 1$ instead of $n = 0$, we should have found

$$\psi^{\overline{1,1}}\{\psi^{\overline{0,0}}(x, y), \psi^{\overline{0,0}}(x, y)\} = \psi^{\overline{1,1}}(x, y)$$

which shows that $\psi^{\overline{0,0}}(x, y)$ is such a function of x and y that when simultaneously substituted in $\psi(x, y)$, for x and y, it shall give either x or y: it may have different values like all other inverse functions. /

If in the function $\psi(x, y)$, we put $\psi(x, y)$ instead of x it becomes $\psi\{\psi(x, y), y\}$ which is written thus $\psi^{2,1}(x, y)$, and continuing such substitutions we arrive at

$$\psi^{n,1}\{\psi^{m,1}(x, y), y\} = \psi^{n+m,1}(x, y) \tag{C}$$

or if the nth function relative to x, and then the mth function relative to y be substituted, it would be

$$\psi^{n,1}\{x, \psi^{1,m}(x, y)\} = \psi^{n,m+1}(x, y). \tag{D}$$

These two equations are now considered as the definition by which the meaning of the indices is known, and we proceed to enquire their values in particular cases: let $n = 0$ in the first and we have

$$\psi^{0,1}\{\psi^{m,1}(x, y), y\} = \psi^{m,1}(x, y)$$

put

$$\psi^{m,1}(x, y) = v$$

and it becomes

$$\psi^{0,1}(v, y) = v, \text{ or } \psi^{0,1}(x, y) = x. \tag{E}$$

If $n = 1$ and $m = 0$ in the latter,

$$\psi^{1,1}\{x, \psi^{1,0}(x, y)\} = \psi^{1,1}(x, y)$$

whence it follows that

$$\psi^{1,0}(x, y) = y. \tag{F}$$

If $n = -m = 1$, (D) becomes

$$\psi^{1,1}\{x, \psi^{1,-1}(x, y)\} = \psi^{1,0}(x, y) = y$$

hence $\psi^{1,-1}(x, y)$ expresses such a function of x and y that when substituted in $\psi(x, y)$ for y it reduces it to y, or if

$$\psi(x, y) = v$$

then

$$y = \psi^{1,-1}(x, v) \text{ and } x = \psi^{-1,1}(v, y). \tag{G}$$

From (C) we may readily deduce as consequences,

$$\psi^{0,n}(x, y) = x, \text{ and } \psi^{n,0}(x, y) = y.$$

The same reasoning which has been employed in discovering the meaning of the indices of functions of two variables applies with / equal facility to those of many variables; without repeating it, it is sufficient to observe that

$$\psi^{0,1,1,\ldots}(x, y, \ldots) = x, \quad \psi^{1,0,\ldots}(x, y, \ldots) = y, \ldots$$

and that if

$$\psi^{1,1,1,\ldots}(x, y, z, \ldots) = v$$

then

$$y = \psi^{1,-1,1,\ldots}(x, v, z, \ldots).$$

Amongst the questions which arise from this notation there are two which are altogether different from any which have occurred in others, and although their solution is not attended with any very great difficulty they furnish an enquiry of some curiosity.

If the expression $\psi^{n,m}(x, y)$ be fully written out, we may seek, first, *how many times will either of the quantities* x *or* y *be repeated;* and secondly, *how many times will the symbol* ψ *occur.*

I shall begin with some of the more simple questions and first

Problem I

How often does the quantity x occur in the expression

$$\psi^{\overline{n,n}}(x, y)?$$

Suppose it occurs u_n times, then in

$$\psi^{\overline{n+1,n+1}}(x, y) = \psi^{\overline{n,n}}\{\psi(x, y), \psi(x, y)\}$$

it occurs twice as many times as it does in the preceding; hence

$$u_{n+1} = 2u_n, \quad \text{or} \quad u_n = C \cdot 2^n$$

if $n = 1$, $u_1 = 2C = 1$, hence $C = \frac{1}{2}$, and $u_n = 2^{n-1}$, or x occurs 2^{n-1} times in $\psi^{\overline{n,n}}(x, y)$.

Problem II

How many times is the symbol ψ repeated in $\psi^{\overline{n,n}}(x, y)$?

Let u_n be the number of times; then it is found u_{n+1} times in $\psi^{\overline{n+1,n+1}}(x, y)$; but in going from the first of these to the second we augment the number of times ψ occur by 2^n, because wherever x / occurs, we introduce ψ by putting $\psi(x, y)$ for it, and as x occurs 2^{n-1} times we by this means add 2^{n-1} times ψ, and similarly, on account of y we add 2^{n-1} times y, so that

$$u_{n+1} = u_n + 2^{n-1} + 2^{n-1} = u_n + 2^n$$

$$\Delta u_n = 2^n, \quad \text{whence} \quad u_n = 2^n + C$$

if $n = 1$, $u_1 = 2 + C = 1$, or $C = -1$, therefore ψ occurs $2^n - 1$ times in $\psi^{\overline{n,n}}(x, y)$.

Problem III

How often is x_1 repeated in $\psi^{\overline{n,n\ldots}}(x_1, x_2, \ldots x_i)$?

Suppose it to occur u_n times, then in going from the nth to the $(n + 1)$th simultaneous function wherever $x_1, x_2, \ldots x_i$ occur, we introduce x_i once, but each of these occurred u_n times, and as there are i of them, we have

$$u_{n+1} = iu_n, \quad \text{or} \quad u_n = Ci^n$$

but if $n = 1$, $u_1 = Ci = 1$, wherefore $C = 1/i$, and x occurs i^{n-1} times.

Corollary. As $x_i, \ldots x_k$ are all similarly involved, they each of them occur i^{n-1} times.

Problem IV

How often does ψ occur in $\psi^{\overline{n,n,\ldots}}(x_1, x_2, \ldots x_i)$?

Let it occur u_n times; then in the $(n + 1)$th besides the u_n times that it occurs in the nth we introduce an additional ψ wherever $x_1, x_2, \ldots x_i$ occur, or since there are i of these quantities, and each occurs i^{n-1} times, we have

$$u_{n+1} = u_n + i^n, \quad \text{whence} \quad u_n = i^n/(i - 1) + C.$$

If $n = 1$, $u_1 = i/(i - 1) + C = 1$, hence $C = -1/(i - 1)$; and ψ occurs $(i^n - 1)/(i - 1)$ times in $\psi^{\overline{n,n,\ldots}}(x_1, x_2, \ldots x_i)$. /

Problem V

How many times do x and y occur in $\psi^{n,1}(x, y)$?

In passing from the nth function to the $(n + 1)$th we change x into $\psi(x, y)$; this does not increase the number of times x occurs, and as the same reasoning applies to the $(n-1)$th, $(n-2)$th, ... down to the first it follows that x only occurs once in $\psi^{n,1}(x, y)$.

Again, in passing the nth to the $(n+1)$th we introduce y once, if therefore u_n represent the number of times it occurs, we have

$$u_{n+1} = u_n + 1, \quad \text{or} \quad u_n = n = C.$$

If $n = 1$, $u_1 = 1 + C = 1$, or $C = 0$, and y occurs n times in $\psi^{n,1}(x, y)$.

And similarly, x occurs n times, and y once in $\psi^{1,n}(x, y)$.

Problem VI

How frequently does ψ occur in $\psi^{n,1}(x, y)$?

Since $\psi^{n+1,1}(x, y) = \psi^{n,1}\{\psi(x, y), y\}$

each step introduces ψ once, it will therefore be repeated n times.

Problem VII

In $\psi^{n,m}(x, y)$ how often are x and y repeated?

$$\psi^{n,m}(x, y) = \psi^{n,1}\{x, \psi^{1,m-1}(x, y)\}.$$

In the right side of this equation x occurs once, and in $\psi^{n,m-1}(x, y)$ it is found $m-1$ times; now in each of these last x occurs $m-1$ times by Problem V, so that on the whole it appears $1 + n \cdot (m-1) = mn - n + 1$ times.

Again, y occurs once in $\psi^{1,m}(x, y)$ and this latter is found n times in $\psi^{n,m}(x, y)$, so that y occurs n times in the same expression.

Problem VIII

How many times does ψ occur in $\psi^{n,m}(x, y)$?

In $\psi^{n,1}(x, y)$, ψ is found n times by Problem VI; and y is found n / times; if for y we put $\psi^{1,m-1}(x, y)$ which contains ψ, $m-1$ times repeated, we shall have

$$\psi^{n,1}\{x, \psi^{1,m-1}(x, y)\} = \psi^{n,m}(x, y)$$

and the number of times ψ occurs is $n + n(m-1) = nm$ times.

Problem IX

How often do $x_1, x_2, \ldots x_k$ occur in $\psi^{a,b,c,\ldots}(x_1, x_2, \ldots x_k)$?

First, let us consider $\psi^{0,1,1,\ldots}(x_1, x_2, \ldots x_k)$, and let x_1 occur u_a times; then since

$$\psi^{a+1,1,\ldots}(x_1, x_2, \ldots x_k) = \psi^{1,1,\ldots}\{\psi^{a,1,\ldots}(x_1, x_2, \ldots x_k), x_2, \ldots x_k\}$$

the number of times x_1 occurs in the first side of the equation is u_{a+1}; in the second side it occurs u_a times, therefore $u_{a+1} = u_a$, or $u_a = C$; but if $a = 1$, $u_1 = C = 1$, therefore $u_a = 1$, and x occurs only once. Again, let u_a now represent the number of times x_2 occurs, then in the second side x_2 occurs u_a times in $\psi^{a,1,1,\ldots}(x_1, x_2, \ldots x_k)$, and also once besides, so that

$$u_{a+1} = u_a + 1, \quad \text{or} \quad u_a = a + C.$$

If $a = 1$, $C = 0$, and the same result will be found for any other quantity x_i except x_1, hence in $\psi^{a,1,1,\ldots}(x_1, x_2, \ldots x_k)$,

$$\begin{array}{lll} x_1 & \text{occurs} & 1 \text{ times} \\ x_2 & & a \\ \vdots & & \vdots \\ x_i & & a. \end{array} \qquad (1)$$

Let us next consider $\psi^{a,b,\ldots}(x_1, x_2, \ldots x_k\}$,

$$\psi^{a,b,1,\ldots}(x_1, x_2, \ldots x_k) = \psi^{a,1,1,\ldots}\{x_1, \psi^{1,b-1,1,\ldots}(x_1, x_2, \ldots x_k), x_3, \ldots x_k\}.$$

On the right side of this equation x_1 occurs in two places: first by itself in which state by (1) it is only found once; and secondly, it occurs in the function $\psi^{1,b-1,\ldots}(x_1, x_2, \ldots x_k)$ where by (1) it is found / to recur $b - 1$ times, but that function itself is repeated a times, therefore the whole number of times x_1 is repeated is

$$1 + a(b - 1) = ab - a + 1$$

x_2 only occurs under under the last mentioned function, and there it is found only once, but as that function is repeated a times x_2 must occur a times: x_3 occurs in two places; in the first it is repeated $a(b - 1)$ times, and in the second a times, so that on the whole x_3 is found

$$ab - a + a = ab$$

times. x_i is found the same number of times.

Therefore in $\psi^{a,b,1,\ldots}(x_1, x_2, \ldots x_k)$,

$$\begin{array}{lll} x_1 & \text{occurs} & ab - a + 1 \text{ times} \\ x_2 & & a \\ \vdots & & \vdots \\ x_i & & ab. \end{array} \qquad (2)$$

By considering the equation,

$$\psi^{a,b,c,1,\ldots}(x_1, x_2, \ldots x_k) = \psi^{a,b,1,\ldots}\{x_1, x_2, \psi^{1,1,c-1,1,\ldots}(x_1, x_2, \ldots x_k), x_4, \ldots x_k\},$$

we shall find that when there are three indices a, b, c,

$$\begin{array}{lll} x_1 & \text{occurs} & abc - a + 1 \quad \text{times} \\ x_2 & & abc - ab + a \\ x_3 & & ab \\ \vdots & & \vdots \\ x_i & & abc. \end{array} \tag{3}$$

A similar process applied to $\psi^{a,b,c,d,1,\ldots}(x_1, x_2, \ldots x_k)$ would show that in it

$$\begin{array}{lll} x_1 & \text{occurs} & abcd - a + 1 \quad \text{times} \\ x_2 & & abcd - ab + a \\ x_3 & & abcd - abc + ab \\ x_4 & & abc \\ \vdots & & \vdots \\ x_i & & abcd. \end{array} \tag{4}$$

And in $\psi^{a,b,c,d,e,1,\ldots}(x_1, x_2, \ldots x_k)$, /

$$\begin{array}{lll} x_1 & \text{occurs} & abcde - a + 1 \quad \text{times} \\ x_2 & & abcde - ab + a \\ x_3 & & abcde - abc + ab \\ x_4 & & abcde - abcd + abc \\ x_5 & & abcd \\ \vdots & & \vdots \\ x_i & & abcde. \end{array} \tag{5}$$

The law which these expressions follow is easily discovered from those which have been given, and by their assistance it may be shown that in the expression

$$\psi^{a,b,c,\ldots}(x_1, x_2, x_3, \ldots x_k)$$

ψ will be repeated $abcd\ldots$ times.

If in a function of two variables we wish after taking the $a, b, c, \ldots$[th] functions relative to the variables $x_1, x_2, \ldots$ in order, again to take the pth relative to the first, the qth relative to the second, etc. some modification of our notation becomes necessary. I have proposed in the Philosophical Transactions[a] to make use of two rows of indices the upper to signify the number of repetitions, and

[a] 'An essay towards the calculus of functions, part II' (1816), in this volume of the *Works of Babbage*.

the lower to denote the quantity for which the substitution is made, thus

$$\psi^{\substack{2,2,1\\2,1,2}}(x, y) = \psi\{\psi\{x, \psi(x, y)\}, \psi[\psi\{x, \psi(x, y)\}, \psi(x, y)]\}.$$

When these indices become numerous it is desirable, for the sake of facility in printing, to have them on a level with the rest of the symbols, and this will become almost indispensible if another or possibly two other rows of indices should be employed: the further extension of the calculus may require such additions to indicate modifications of a different kind from those yet considered. There appear to be four circumstances connected with the indices of functions whose relations it may ultimately be desirable to express by means of them: the two first of them, namely, the number of repetitions and the quantities for which substitutions are made are already pointed out by the two rows. It may become necessary to substitute / instead of one of the variables, not the original function but some modification of it such as $\psi^{1,2}(x, y)$, and this might be only substituted for one of the variables in certain particular places not universally; such changes would require two additional rows.

The alteration which I would propose in all cases where many indices are concerned is extremely simple: it consists in merely bringing the indices down to the level of the functional sign and enclosing them between two bars; the expression last-mentioned would be written thus

$$\psi\left|\substack{2,2,1\\2,1,2}\right|(x, y).$$

This would also possess some advantage in the case of a function of many variables being after the performance of given operations reduced to a function of a less number, (by means of some assigned relation between some of them) and then new functional operations being performed, or considered as a function of the reduced number of variables, thus

$$\psi\left|\substack{2,1,1\\1,2,\bar{3}}\right|\substack{2,2\\1,2}\left|(x, y, z), \quad \{z = f(x, y)\}\right.$$

which signifies that the second function of $\psi(x, y, z)$ is taken relative to x; $f(x, y)$ is then substituted for z and the second function of the result considered as a function of two variables is taken first with respect to x, then with regard to y.

The method I have pointed out for the determination of the number of times the quantities under the functional sign would be repeated if they were written out at length is perhaps the most direct and unartificial which could be proposed; I cannot however terminate this paper without pointing out another method of a character completely opposite which promises in more complicated enquiries of this nature to be of considerable use. It bears a strong resemblance to a very elegant artifice employed by M. Laplace in order to determine the numerical coefficients of Σu, and was in fact suggested by it. /

Since the number of times any variable is repeated is independent of the form* of the function, that number must be the same for all functions. If therefore we can find it for any particular function, we have it for all others. Let us then select the function $x_1 + x_2 + \ldots + x_n$ and take those functions which the index requires, we shall have the number of repetitions of each variable expressed by the coefficient attached to it.

One great advantage of this plan is that we always have it in our power to take any function however complicated of $x_1 + x_2 + \ldots + x_n$; as an example let it be required to find the number of repetitions of x and y in $\psi^{a,b}(x, y)$;

if $\psi(x, y) = vx + uy$,

$$\psi^{a,b}(x, y) = \left(v^a + uv\frac{v^a - 1}{v - 1} \cdot \frac{u^{b-1} - 1}{u - 1}\right)x + u^b \cdot \frac{v^a - 1}{v - 1}y$$

if $u = v = 1$, this becomes

$$\psi^{a,b}(x, y) = (1 + a \cdot (b - 1))x + ay = (ab - a + 1)x + ay.$$

Or x will be found repeated $ab - a + 1$ times, and y will occur a times, which coincides with what was proved in Problem VII.

The surprising condensation of meaning comprised in small space and yet exempt from even the slightest tinge of obscurity is nowhere more conspicuous than it is throughout the calculus of functions: and the solution of the problems contained in this paper enables us to give numerical results which will be viewed with surprise even by those who are best acquainted with the power of general signs.

The equation,

$$\psi^{\overline{10,10}}(x, y) = \psi(x, y)$$

is one whose solution presents no great difficulties and may be / accomplished in a few lines. To any person acquainted with the notation employed in the doctrine of functions the question is comprehended at a single glance; yet if we apply to it the rules discovered in Problems I and II, we shall find that if it were written out at length, x and y would each be repeated 512 times, and ψ would occur 1023 times; so that the whole expression would consist of 2047 letters, and it may be added, that if it were so developed it would require a much longer time merely to comprehend the enunciation of the problem than it would to understand and solve it in its contracted form.

26 February, 1820 / **CHARLES BABBAGE**

* In whatever manner the original function may involve the unknown quantities it must in this point of view be considered as containing each only once.

XVI

ON THE APPLICATION OF ANALYSIS TO THE DISCOVERY OF LOCAL THEOREMS AND PORISMS

From *Transactions of the Royal Society of Edinburgh*, Vol. 9, (1823), pp. 337–52

Read 1 May, 1820

Those who have devoted much of their attention to mathematical enquiries, must have had frequent occasion to admire the unexpected and intimate connection which subsists between branches of their science apparently the most remote and unconnected with each other; and repeated observation will have convinced them, that no researches, however recondite or abstruse, should be neglected, because they appear to stand isolated and detached from the body of the science. These reflections appear to have been felt with strongest force, by those who have most contributed to its advancement, and were particularly insisted on by Maclaurin, in the preface to his *Geometrica Organica.*[a] The present paper will add another example, to instances already numerous, of the latent affinity between departments of mathematics, usually regarded as the most opposite.

By those who have studied the writings of the illustrious restorers of the most valuable and interesting portion of the ancient / geometry, the views which are here opened may probably be regarded with surprise, not unmixed with regret, that the increasing perfection of the language of symbols should gradually cause it to usurp the hitherto exclusive domains of the higher geometry. But whatever may be the discoveries to which the geometrical interpretation of language, at once the most comprehensive and condensed which human ingenuity has devised, shall give birth, the restorer of the Porisms of Euclid,[b] and the author of the General Theorems,[c] will retain an undiminished reputation, and their works continue to be studied, by all those who wish to acquire a correct taste for the geometry of the ancients. Those propositions

[a] Colin Maclaurin, *Geometrica organica*... (London, 1740).
[b] Robert Simson, *Opera quaedam reliqua*... (Glasgow, 1776).
[c] Robert Simson, *General theorems* (Edinburgh, 1746).

which have received the appellation of Local Theorems and Porisms, may, in one point of view, be considered as differing from theorems and problems, by having something more general, or indeterminate, in their nature; by affirming that some property is possessed not merely by one individual, but by every one, of some class or species. It is on this circumstance that the algebraic investigation of porisms is founded: the arbitrary constants of an equation of two variables, are made to submit to certain conditions, which shall leave the variables themselves still indeterminate. By generalising this process, we take, instead of arbitrary constants, unknown functions of the variables themselves; and instead of the resulting algebraic equations, we find functional equations which determine the form of the functions we have assumed.

This process, which will be better understood by the subsequent enquiries, leads us at once to the highest pitch of generality, and puts us in possession of innumerable porisms, and local theorems, each comprehending whole classes of curves. By limiting and determining the form of the arbitrary functions involved in the solution, we gradually restrict the extent / of our results, and we frequently arrive at *loci* belonging to the conic sections. Sometimes, however, the simplification is still greater, and we are surprised by recognizing, amongst extensive classes of curves, possessed of some peculiar property, some well known *locus* of the straight line or circle.

The enquiries which follow are given nearly in the order in which they occurred; a different arrangement might have appeared more systematic, but it would have had the great disadvantage of concealing the means by which the results were arrived at. Several of the more restricted porisms and local theorems, might have admitted of a geometrical dress; but this would have been inconsistent with the object I had proposed to myself in the present paper. The greater part are, I believe, beyond the powers of geometry; and this opinion, if allowed to be correct, will perhaps, by the admirers of the ancient geometry, be admitted as some excuse for the present attempt to add to its stores, by means so very foreign.

There exist numerous classes of curves possessed of the following property: If we take any abscissa AE, Figure 1, and ordinate BE, and if we make $AF =$

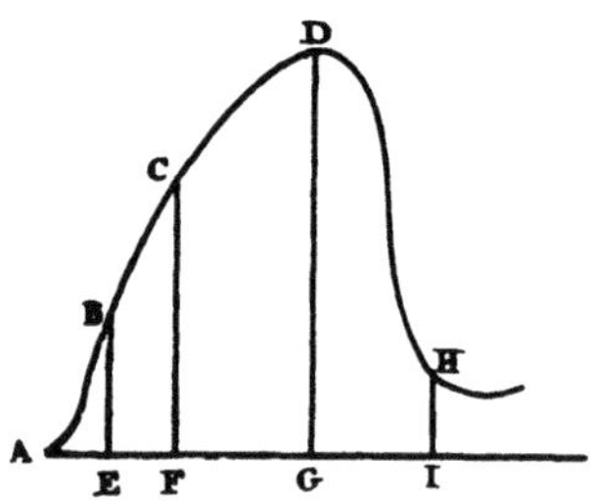

Figure 1

BE, and find the new ordinate *CF*, and repeat the same process n times, the nth ordinate *HI* shall equal the first abscissa *AE*. It is easy to perceive, that if $y = \psi x$ represent the equation of the curve, then the equation determining the form of ψ will be $\psi^n x = x$. As the classes of curves here alluded to, will frequently occur in the following inquiries, I shall venture to bestow on them the name of *periodic curves*, which was suggested by the similar name assigned by Mr Herschel to the function which satisfies the equation just given. It will also be convenient to apply / some name to the series of ordinates thus found: I shall therefore call them *corresponding ordinates*. These curves may be divided into orders, according to value of the index n; if $n = 1$, the first order consists of only one curve, namely, $y = \psi x = x$, which is the equation of a right line, making an angle of 45° with the axis.

When $n = 2$, and ψ is determined from $\psi^2 x = x$, $y = \psi x$ represents all periodic curves of the second order, and so on.

The equation $\psi^n x = x$ has been solved by me in the *Philosophical Transactions* for 1816, and numerous examples are given in another paper, published in the following year.[a] This solution has, however, been much improved by Mr Horner (see *Annals of Philosophy*, October 1817),[b] who has shown that the algebraic equation to which my method of solution leads in all cases, admits of a ready solution. The form which is thus assigned to the function ψ is

$$\psi x = \varphi^{-1}\left\{\frac{a + bx}{c - \dfrac{b^2 - 2bc\cos k\pi/n + c^2}{2(1 + \cos k\pi/n)a}x}\right\}$$

Amongst those included in the second order of periodic curves, will be found the right-angled hyperbola referred to co-ordinates parallel to its assymptotes, and also circles of all orders, where equations are $x^n + y^n = a^n$; these curves possess many singular properties. All those in which n is even are re-entering curves, without any infinite branches, whose form is nearly that of the subjoined figure, Figure 2. In this curve, if we take any abscissa, and ordinate *CD* and *BD*, and if we turn the triangle *CBD*, into such a position that *B* shall coincide with the axis *CF*, then the point *C* will coincide with *G* some / point in the curve, and also the arc *GF*, together with the arc *BF*, will be equal to the quadrant *AF* of the curve.

Let *ABC* be any periodic curve of the second order, whose equation is $y = \alpha x$, and *DB*, *EC* any two corresponding ordinates, it is required to find another curve *AFG*, such that the sum of its two ordinates, at the points *D* and *E*, may be always a constant quantity.

[a] 'An essay towards the calculus of functions' (1815; Part II, 1816), in this volume of the *Works of Babbage*.
[b] W. G. Horner, 'Solution of the equation $\psi^n = x$', *Annals of Philosophy*, Vol. 10 (1817), pp. 341–6.

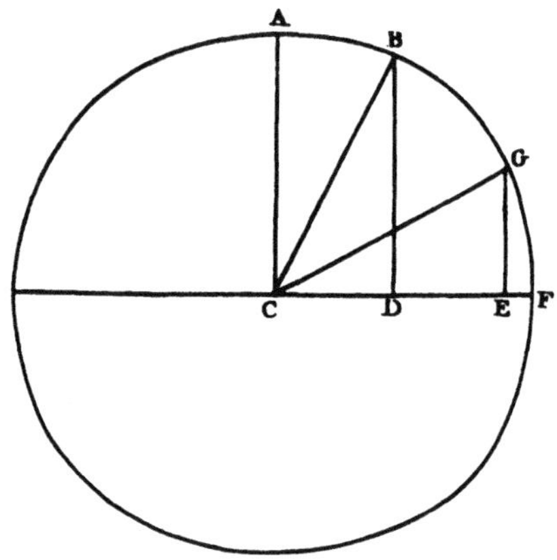

Figure 2

From the nature of the curve ABC, we have $AD = x$, and $AE = DB = \alpha x$; also $\alpha^2 x = x$.

Let $y = \psi x$ be the equation of the curve AFG, then $FD = \psi x$, and $GE = \psi(AE) = \psi\alpha x$, and by the prescribed condition, we have for the equation determining ψ

$$\psi x + \psi\alpha x = c$$

the general solution of which is,

$$\psi x = \frac{c\varphi x}{\varphi x + \varphi\alpha x}$$

the function φ is perfectly arbitrary, and of the class of curves comprehended in the equation $y = c\varphi x/(\varphi x + \varphi\alpha x)$, we may enunciate the following porisms.

Any of this family of curves being given, a periodic curve of the second order (ABC), may always be found such, that if we take any two abscissae in the curve given, respectively equal to any two corresponding ordinates of the curve found, and draw ordinates to the given curve, and if we prolong either of these ordinates (EG) above the curve, until the part above (GH) is equal to the first of the two ordinates, the extremity of the ordinate, thus increased, will always be situated in a right line given by position, Figure 3.

The line thus given by position is parallel to the abscissae, and situated at the distance denoted by c. /

If we prolong either of the two ordinates (GE) below the abscissae, by a quantity (EL) equal to the first ordinate, the extremity of the ordinate, thus prolonged, will always be situated in a curve, similar and equal to the given curve, which is also given by position.

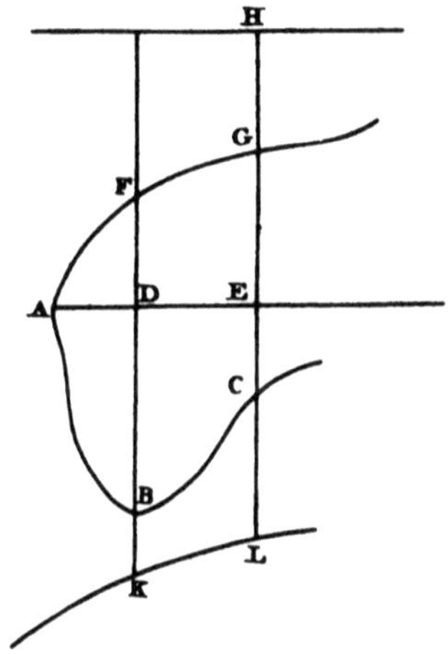

Figure 3

The position of the given curve is parallel to the original curve. The deduction of these porisms is so obvious, from the property of the class of curves, that I consider any further explanation of them unnecessary. We shall, however, by assigning particular values to some of the functions, find some very simple results. Let us suppose $\alpha x = \sqrt{a^2 - x^2}$, then, the family of curves will be contained in the equation,

$$y = \frac{c\varphi(x)}{\varphi x + \varphi(\sqrt{a^2 - x^2})}.$$

Of these curves, the following property may be stated:

Any curve of this family being given, a straight line (HK) may be found through any point, of which, if a line (HD) be drawn at right angles to the axis of the abscissae, it will cut off an abscissa (AD), and part of it will form the ordinate (CD); if an abscissa (AE) be found, whose ordinate (BE) is equal to that part of the line drawn, which is intercepted between the line found and the curve; and if both the two ordinates thus found, be prolonged below the axis, until the part of each below is equal to the abscissa belonging to the other ordinate, the two points to which these lines are prolonged, are situated in the circumference of a circle given by position. See Figure 4.

Let $\alpha x = -x$, then the curves included in the family $y = c\varphi x/(\varphi x + \varphi(-x))$ possess the following property: /

Any curve of this kind being given, if we take two equal abscissae, on opposite sides of the centre, and prolong the ordinates at both these points, until the parts

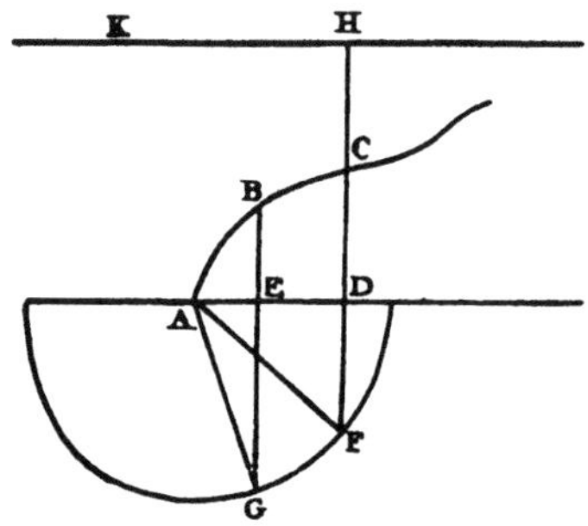

Figure 4

of each produced be equal to the ordinate at the other point, then the locus *of the points thus found is a right line given by position.*

If we make $\varphi x = a + x$, the curve becomes a right line, whose equation is $y = c/2 + (c/a)x$; and the line found is parallel to the axis, and the curve given, is a right line passing through a point in the axis of the ordinates, equally distant from each parallel.

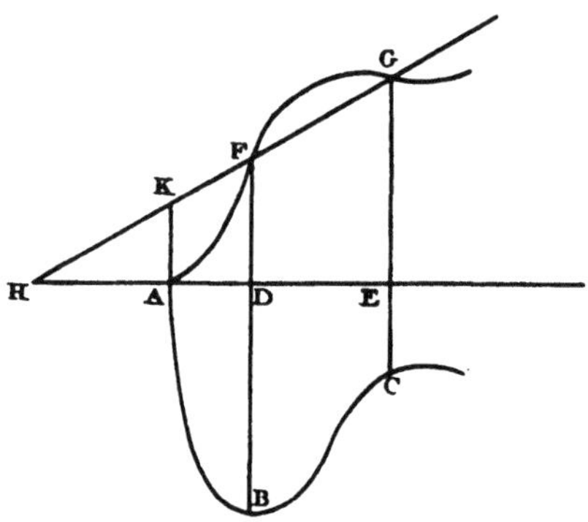

Figure 5

Let *ABC* be any periodic curve, Figure 5, and *DB*, *EC*, any two corresponding ordinates; draw the ordinates *FD*, *GC*, and draw the line *HFG*, passing through the points *F* and *G*; it is required to find the equation of this line.

w and v being the co-ordinates of any point in this line, its equation will be $v = Aw + B$; the quantities A and B being determined by the condition, that it must pass through the points F and G, whose co-ordinates are respectively x and ψx, and αx, and $\psi\alpha x$, we have

$$v = \frac{\psi x - \psi\alpha x}{x - \alpha x} w + \frac{\alpha x \cdot \psi x - x \cdot \psi\alpha x}{x - \alpha x}.$$

Let us now determine the form of the curve AFG by this condition, that the line HG shall always be parallel to itself. It is well known, that the coefficient of the abscissa w expresses the tangent of the angle of the inclination of the line to its axis; this must therefore be constant, and we have

$$\frac{\psi x - \psi\alpha x}{x - \alpha x} = c$$

or

$$\psi x - \psi\alpha x = c(x - \alpha x)$$

the solution of this equation is

$$\psi x = \frac{c(x - \alpha x)\varphi x}{\varphi x + \varphi\alpha x}.$$

This suggests the following porisms: /

Any of the family of curves contained in the equation

$$y = \frac{c(x - \alpha x)\varphi x}{\varphi x + \varphi\alpha x}$$

being given, a right line may be found, through any point of which, if a right line be drawn perpendicularly, cutting the curve in two points, and if the ordinate at each of these points is produced to a point below the axis, until the part below each is equal to the abscissa belonging to the other ordinate, the two points thus found will always be situated in a periodic curve, given in species and position.

The line to which the other lines are drawn perpendicular, is at right angles to the line HFG in the figure.

If we make $\alpha x = \sqrt{a^2 - x^2}$, the curve which is the locus of these points is a circle, whose radius is equal to a, and if we also make $\varphi x = b$, the original curve is an ellipse.

If we make $\alpha x = a^2/x$, and also $\varphi x = b$, the equation of the given curve is

$$y = \frac{c}{2\left(\dfrac{x - a^2}{x}\right)}$$

which is an hyperbola, and that of the curve found is $y = a^2/x$, a right-angle hyperbola.

If in the equation of the right line

$$v = \frac{\psi x - \psi\alpha x}{x - \alpha x} w + \frac{\alpha x \cdot \psi x - x \cdot \psi\alpha x}{x - \alpha x}$$

we make $w = 0$, we have the distance of the point K from A; let us suppose this to remain constant, then

$$\frac{\alpha x \cdot \psi x - x \cdot \psi \alpha x}{x - ax} = c$$

or

$$\alpha x \cdot \psi x - x \cdot \psi \alpha x = c(x - \alpha x)$$

the solution of which is

$$\psi x = \frac{1}{\alpha x} \cdot \frac{c(x - \alpha x)\varphi x}{\varphi x + \varphi \alpha x}.$$

/

The class of curves comprehended in the equation

$$\psi x = \frac{1}{\alpha x} \cdot \frac{c(x - \alpha x)\varphi x}{\varphi x + \varphi \alpha x}$$

are possessed of the following property:

Any curve of the kind being given, a point (K) *may be found, through which, if a right line* (KG) *be drawn in any direction, cutting the curve in two points* (F *and* G), *if the ordinate at one of these points be prolonged to a point below the axis, until the part below is equal in length to the abscissa corresponding to the other point of intersection, then the point thus found is always situated in a periodic curve of the second order, given in species and position.* Figure 5.

If in the equation of the line, we make $v = 0$, we have

$$w = \frac{(\alpha x \cdot \psi x - x \cdot \psi \alpha x)}{(\psi x - \psi \alpha x)}$$

and if we suppose this to remain constant, we in fact fix the point K in the axis of the abscissae. This gives the equation for determining ψ

$$(\alpha x - c)\psi x = (x - c)\psi \alpha x$$

whose solution is

$$\psi x = \frac{c\chi\{\overline{x}, \overline{\alpha x}\}}{\alpha x - x}.$$

Let AFG be any periodic curve, Figure 6, and AD, AE any two corresponding abscissae. Required the equation of all curves, such that the line joining the summits of the two ordinates, raised at the points D and E, shall be constant.

The equation of the periodic curve being $y = \alpha x$, let that of the family of curves sought be $y = \psi x$, then we have $AD = x$, $AE = \alpha x$, $BD = \psi x$, $CE =$

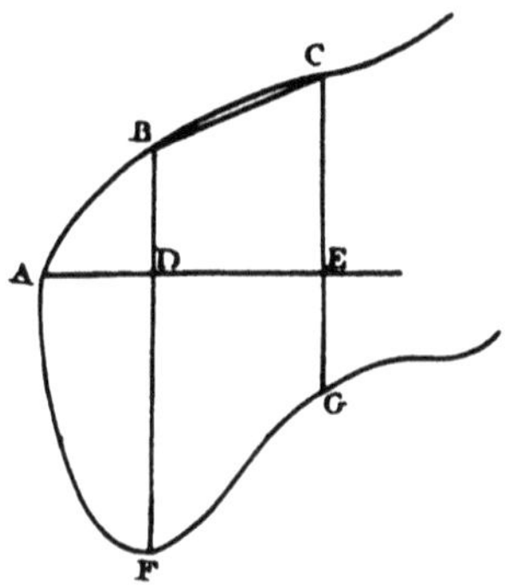

Figure 6

$\psi\alpha x$, and the equation determining the form of ψ is

$$(\psi\alpha x - \psi x)^2 + (\alpha x - x)^2 = c^2. \qquad \text{(a)}$$

/

This equation is one of those which I noticed in a paper in the *Philosophical Transactions* for 1817, p. 211.[a] I there stated, that it appeared impossible for any function to satisfy the equation, unless it contained a radicle, and unless different roots were taken in different parts of the equation. This explanation will perhaps be rendered more satisfactory by the application to geometry. The general solution of the equation in question, is

$$\psi x = \pm \frac{\varphi x \cdot \sqrt{c^2 - (\alpha x - x)^2}}{\varphi x + \varphi\alpha x} = y$$

where the upper sign must be used in one part of the equation, and the lower sign in another. From this equation we learn, that for every value of x there are two values of y, equal, but extending in opposite directions; or that the curve is symmetrical with regard to the axis of the x's: and since we must use different values of the radicle, it appears that the two points B and C cannot be in the same branch of the curve, and on the same side of the axis, but that one point being situated above the axis, the other must be placed in the corresponding branch, which exists below the axis. In fact, it seems to follow, from the very nature of the equation (*a*), that no curve possessing this property can have both the two points B and C situated in the same branch. These considerations render it necessary, in some measure, to limit the generality of the function φ; and it may be stated, that no form of φ is admissible, which takes

[a] 'Observations on the analogy which subsists between the calculus of functions and other branches of analysis' (1817), this volume of the *Works of Babbage*, p. 227.

away the double sign placed before the whole of the value of y.

Thus, if we take $\varphi x = \alpha x + \sqrt{c^2 - (\alpha x - x)^2}$, we have

$$y' = \frac{\alpha x}{2} \pm \tfrac{1}{2}\sqrt{c^2 - (\alpha x - x)^2}$$

but this does not satisfy the condition of rendering the curve symmetrical relative to its axis.

Let $\alpha x = \sqrt{a^2 - x^2}$, then we have

$$y = \frac{\sqrt{c^2 - a^2 + 2x\sqrt{a^2 - x^2}} \cdot \varphi x}{\varphi x + \varphi\sqrt{a^2 - x^2}}$$

and the following property belongs to all the curves contained in this family:

Any curve BAC of this species being given, Figure 7, *a right line BC given in*

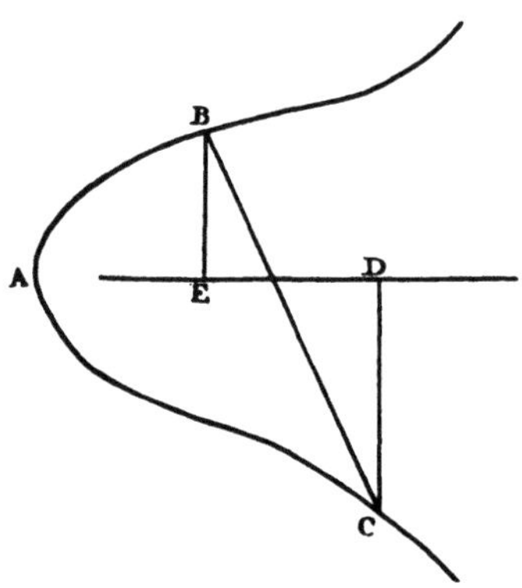

Figure 7

length may be found such, that if it be any how placed, so that its extremities B and C coincide with any two points of similar branches, but on opposite sides of the axis; then the sum of the squares described on the abscissae AE, AD, shall be equal to the square described on a right line which may be found.

If we make $\alpha x = -x$, we have the family of curves expressed by the equation

$$y = \frac{\sqrt{c^2 - 4x^2} \cdot \varphi x}{\varphi x + \varphi(-x)}$$

which possess the following property:

Any one of this class of curves being given, a right line CB may be found, which, if it be placed so that its extremities rest on two opposite branches of the

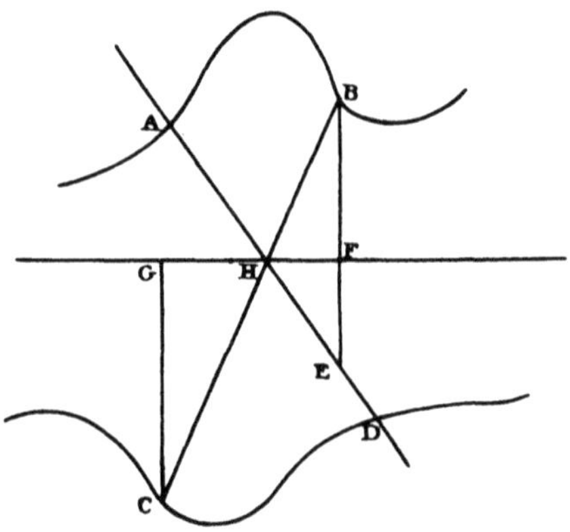

Figure 8

curve (at C and B), and if the ordinate at one of these points be prolonged on the other side the axis, until the part produced (FE) be equal to the abscissa (GH) at the other point, the extremity of the line so produced (E), will be situated in a straight line given by position.

If we make $\varphi x = b$, the equation becomes $y = \sqrt{c^2/4 - x^2}$, and the curve is a circle whose radius is $c/2$, the line to be / found is equal to twice the radius, and the line given by position passes through the centre, making an angle of $\pi/4$ with the axis.

ADE being any periodic curve of the second order, and *AF*, *AG* any two corresponding abscissae, and *ABC* being any other curve, whose equation is $y = \psi x$, required the co-ordinates of the point of intersection of its two tangents, at the points *B* and *C*.

Call

$$AF = x' \ AG = x'_1$$

$$BF = y' \ CG = y'_1$$

and let x and y be the co-ordinates of any point in either of the tangents, then

$$y = x\frac{dy'}{dx'} - \frac{x'dy'}{dx'} + y'$$

$$y = x\frac{dy'_1}{dx'_1} - \frac{x'_1 dy'_1}{dx'_1} + y'_1$$

are the equations of the tangents; and if we call v and w the co-ordinates of the point of intersection P, we have

$$v = \frac{y' - \dfrac{x'dy'}{dx'} - y'_1 + \dfrac{x'_1dy'_1}{dx'_1}}{\dfrac{dy'_1}{dx'_1} - \dfrac{dy'}{dx'}}$$

and

$$w = \frac{\dfrac{dy'_1}{dx'_1}\left(y' - \dfrac{x'dy'}{dx'}\right) - \dfrac{dy'}{dx'}\left(y'_1 - \dfrac{x'_1dy'_1}{dx'_1}\right)}{\dfrac{dy'_1}{dx'_1} - \dfrac{dy'}{dx'}}.$$

/

Let us now suppose that the point P is always situated in the right line AL, perpendicular to the axis; we have $v = 0$, and since the curve ADE is a periodic one, whose equation is $y = \alpha x$,

$$\psi x - \frac{x\alpha\psi x}{dx} - \psi\alpha x + \frac{\alpha x \cdot d\psi\alpha x}{d\alpha x} = 0.$$

Whence

$$\psi x - \frac{xd\psi x}{dx} = \chi\{\overline{x}, \overline{\alpha x}\}$$

or

$$\frac{xd\psi x - \psi x \cdot dx}{x^2} = -\frac{dx}{x^2}\chi\{\overline{x}, \overline{\alpha x}\}$$

from which

$$y = \psi x = -x\int\frac{dx}{x^2}\chi\{\overline{x}, \overline{\alpha x}\}.$$

Now, let $\alpha x = (a^4 - x^4)^{1/4}$, the family of curves are comprehended in the equation

$$y = -x\int\frac{dx}{x^2}\chi\{\overline{x}, \overline{(a^4 - x^4)^{1/4}}\}$$

and they possess the following property:

Any of this species of curve being given, if any two abscissae are taken, the sum of whose fourth powers is equal to the fourth power of a line, which may be

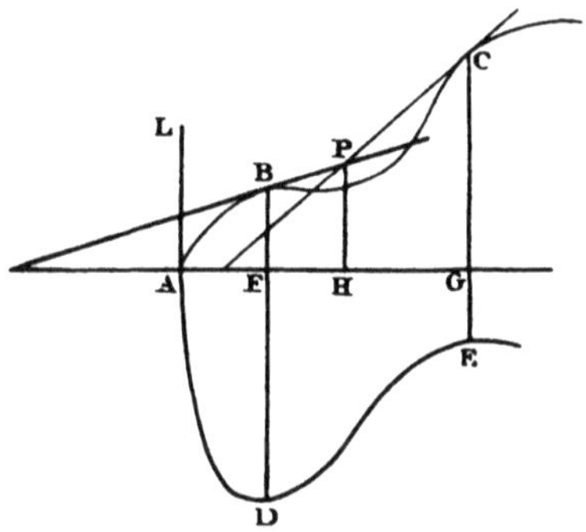

Figure 9

found, then the tangent drawn at the extremities of the ordinates corresponding to these abscissae, will intersect each other in a line given in position. This line is perpendicular to the axis.

If we suppose the point P always to be situated in the axis of the abscissae, we have $w = 0$, consequently,

$$\frac{dy'_1}{dx'_1}\left(y' - \frac{x'dy'}{dx'}\right) - \frac{dy'}{dx'}\left(y'_1 - \frac{x'_1 dy'_1}{dx'_1}\right) = 0$$

or

$$\frac{d\psi\alpha x}{d\alpha x}\left(\psi x - \frac{x d\psi x}{dx}\right) = \frac{d\psi x}{dx}\left(\psi\alpha x - \frac{\alpha x \cdot d\psi\alpha x}{d\alpha x}\right)$$

multiplying both sides by

$$\left(\frac{d\psi x}{dx} - \frac{d\psi\alpha x}{d\alpha x}\right)^{-1}$$

we have

$$\psi x\left(\frac{d\psi x}{dx}\right)^{-1} - x = \psi\alpha x\left(\frac{d\psi\alpha x}{d\alpha x}\right)^{-1} - \alpha x$$

and since both sides are symmetrical relative to x and αx, we have

$$\psi x\left(\frac{d\psi x}{dx}\right)^{-1} - x = \chi(\overline{x}, \overline{\alpha x})$$

or

$$y\frac{dx}{dy} = x + \chi(\overline{x}, \overline{\alpha x})$$

hence,

$$\frac{dy}{y} = \frac{dx}{x + \chi(\overline{x}, \overline{\alpha x})}.$$

If we suppose $\alpha x = (a^n - x^n)^{1/n}$, the class of curves comprehended in the equation

$$\log y = \int \frac{dx}{x + \chi\{\overline{x}, \overline{(a^n - x^n)^{1/n}}\}}$$

possess the following property:

If we take any two abscissae AD, AE, the sum of whose nth powers is equal to a given nth power, then the tangents BP, CP, drawn to the curve at the extremity of the ordinates corresponding to those abscissae, will always intersect each other on the axis of the abscissae. Figure 10.

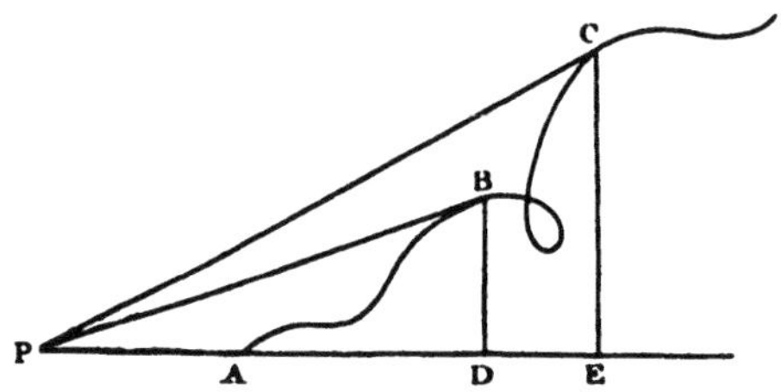

Figure 10

In the same manner, a class of curves may be found, and ordinates may be drawn, in such a manner, that the tangent / at those ordinates shall intersect each other in a line given in position.

In the curve *ABC*, taking any two ordinates, *BE*, *CD*, required its equation when the normals at these points intersect each other on the axis; let x, y, and x_1, y_1, be the ordinates, then we must have

$$AE + EP = AD + DP$$

or

$$x + \frac{ydy}{dx} = x_1 + \frac{y_1 dy_1}{dx_1}$$

if the two points E and D are so chosen, that $x_1 = \alpha x$, and if α is a periodic function of the second order, we have, supposing $y = \psi x$.

$$x + \frac{\psi x \cdot d\psi x}{dx} = \alpha x + \frac{\psi \alpha x \cdot d\psi \alpha x}{d\alpha x}$$

or

$$x + \frac{\psi x \cdot d\psi x}{dx} = \chi(\overline{x}, \overline{\alpha x})$$

hence

$$y^2 = -x^2 + 2\int dx\chi(\overline{x}, \overline{\alpha x}).$$

If we make $\chi(\overline{x}, \overline{\alpha x}) = x$, we have

$$y^2 = -x^2 + 2cx + c_1$$

which is the equation to a circle, and it is well known, that this curve possesses the property, since all its normals intersect each other in the centre.

This furnishes us with the following local theorems. *Any curve of this family being given, if from any point* (P), *in the axis of the abscissae, we draw two lines cutting the same branch of the curve perpendicularly* (*in* B *and* C), *and if we prolong the ordinate* / (CD) *at one of these points below the axis, until the part below* (DF) *is equal to the abscissa corresponding to the other point* (AE), *then the* locus *of the extremity of the ordinate* (F) *thus produced, is always a periodic curve of the second order.* Figure 11.

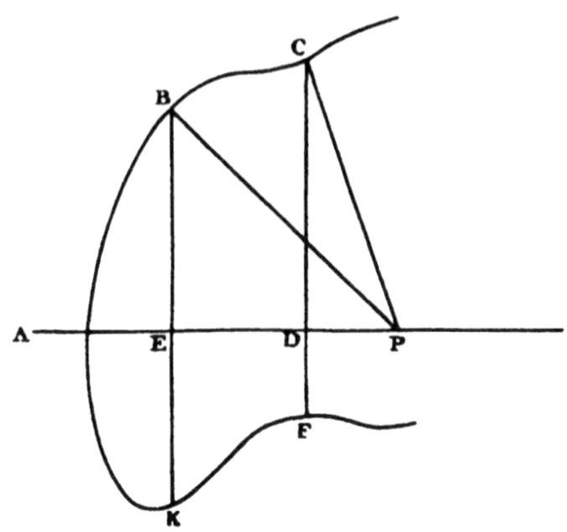

Figure 11

The questions which we have now considered, appear to me sufficient to point out the nature of that connection between the theory of functions and that of curves, which it was my object to establish. The difference between the properties thus brought to light, and those which have been hitherto known, seems to consist chiefly in two points. The first is, that the families of curves to which they relate are larger: this arises from the arbitrary functions necessarily introduced into the solution of functional equations. The other difference is, that the properties discovered relate to many points of the same curve, only connected by some given law. In the first respect, they in some measure approach to the investigations of M. Monge, in his excellent work *L'Application de l'Analyse à la Géométrie*;[a] whilst, in the second respect, they bear some analogy to the more general properties of curves, deduced from the theory of equations in the *Proprietates Curvarum* of Waring:[b] these resemblances are, however, but superficial. The nature of the questions we have considered requires, by the usual methods of analysis, the application of mixed differences; and, in most of the few instances in which any such problems have been proposed, they have been attempted by that method.

Devonshire Street, Portland Place,
1 July, 1818 /

[a] Gaspard Monge, *Application de l'analyse à la géometrie* (4th edition, Paris, 1809).
[b] Edward Waring, *Proprietates algebraicarum curvarum* (Cambridge, 1772).

XVII

ON THE INFLUENCE OF SIGNS IN MATHEMATICAL REASONING

From *Transactions of the Cambridge Philosophical Society*, Vol. 2 (1827), pp. 325–77

Read 16 December, 1821

It can scarcely excite our surprise that the earlier geometers, engaged in successfully employing the most powerful instrument of discovery which human thought has yet contrived, and seduced by the splendour of the view their science had opened to them, should press with earnestness to enlarge its boundaries by new applications, rather than exert their genius in explaining the causes which have combined to advance it to such unrivalled eminence. On the discovery of those branches which have so completely altered the face of the science, the use of the new acquisitions was too inviting to allow time for any very scrupulous enquiry into the principles on which they were founded: satisfied with the accuracy of the results at which they arrived, the desire of multiplying them naturally prevented any return on their steps for the purpose of applying themselves to the less promising task of establishing on secure foundations, principles of whose truth they felt confident.

These efforts to extend the reach rather than fix the basis of the new calculus, were undoubtedly to be admired at the period to which we refer: an acquaintance with its extensive bearings ought justly to have no inconsiderable influence on the form in which its elements should be delivered; hence the lapse of / nearly a century has been required to fix permanently the foundations on which the calculus of Newton and of Leibnitz shall rest.

Time which has at length developed the various bearings of the differential calculus, has also accumulated a mass of materials of a very heterogeneous nature, comprehending fragments of unfinished theories, contrivances adapted to peculiar purposes, views perhaps sufficiently general, enveloped in notation sufficiently obscure, a multitude of methods leading to one result, and bounded by the same difficulties, and what is worse than all, a profusion of notations (when we regard the whole science) which threaten, if not duly corrected, to multiply our difficulties instead of promoting our progress.

As a remedy to the inconveniences which must inevitably result from the continued accumulation of new materials, as well as from the various dress in which the old may be exhibited, nothing appears so likely to succeed as a revision of the language in which all the results of the science are expressed, and the establishment of general principles which shall curtail its exuberance, and regulate that which has hitherto been considered as arbitrary – the contrivance of a notation to express new relations. Previous however to this, some observations on the nature of that assistance which signs lend to our reasoning faculties, and on the causes which give such certainty to the conclusions of analysis, may render our future enquiries more intelligible.

The nature of the quantities with which the mathematical sciences are conversant, is undoubtedly one of the first of those causes: in geometry it has been well remarked* that its foundations rest on definition, and if this do not altogether hold in algebraical enquiries, at least the meaning of the symbols employed / must be regulated by definition; and here arises one of the great differences which characterise this science, the definitions themselves being exceedingly simple, comprising but few ideas, whilst in other sciences they are usually much more complicated. In geometry, definition is the beginning of an enquiry; in metaphysical science, it is frequently the result of one: thus that a triangle is a figure formed by three sides, is a convention on which many of Euclid's propositions rest, and from this, as a point of departure, numerous deductions are made: on the other hand, our idea, and consequently our definition of beauty, is only the result of considerable thought and enquiry.

In the language of analysis, it is very rare that any symbol possesses more than one meaning; in ordinary language, it is as rare to find a word having but one signification: nor is this the only difference; when an algebraical symbol has more than one meaning, they are always well defined and distinct, and should there exist several signs for the same operation, the only difference is in their external form, not the slightest in their meaning; whilst in common language, the meanings of words shade away into each other, and it is frequently difficult, even on mature consideration, to assign the precise limits of the signification of words which are nearly synonymous.

Now if this be the case when the words themselves are the especial objects of our thoughts, how open must all reasoning be to inadvertencies when the mind is compelled to occupy itself at once on the various meanings of the signs it uses and on the train of consequences which it endeavours to deduce by them.

The multitude of significations which attach to many of the words that compose our ordinary language, is a disadvantage which is completely removed from that of analysis. In our reasoning concerning any objects even of a moderately complicated nature, we are obliged to make use of the words

* Elements of the Philosophy of the Human Mind, Vol. II, p. 150 ⟨Dugald Stewart, *Elements of the philosophy of the human mind*... (London, 1792, etc.)⟩

attached to those / objects, which consequently recall to the mind the variety of particulars of which they consist, some with more, others with less vividness according to our previous habits of thought; from this cause it sometimes happens that the real ground on which our reasoning depends, is with difficulty kept in view by a laborious effort of the attention, and is in many instances very indistinctly perceived.

In the use of algebraic signs this inconvenience entirely vanishes; we can always so arrange them, that that quality on which the whole force of our reasoning turns shall be visible to the eye, whilst the numerous others which contribute to form the expression we are considering, although thrown into the background, are still by no means excluded. This species of insulation of the property whose consequences we wish to trace, enables the mind to apply that attention, which must otherwise be exerted in keeping it in view, to the more immediate purpose of tracing its connection with other properties that are the objects of our research. As an example of these ideas, I would mention the word *government*, upon which we may reason in many different directions, either as it secures domestic liberty, or protects from foreign attacks, as it discourages vice or promotes commerce: in these and in numerous other courses, our reasoning may be pursued, and the word *government* will constantly recur without the possibility of avoiding it but by the most tedious circumlocution, or of restricting the view in which it is regarded but by the most unwearied efforts of attention. The word function in analysis possesses a still more extensive signification than that which has been just mentioned:

The sign

$$\psi(\overline{x}, \overline{1/x})$$

signifies any symmetrical combination of the two quantities x /
and $1/x$. That combination is in this mode of expressing it left arbitrary and undefined. The same function ψ may at the same time be a function of

$$\frac{x^2+1}{x}, \qquad \frac{x}{x^2+1}, \qquad \frac{x^4+x^2+1}{x^2}$$

or of a thousand other quantities; all which circumstances although deducible from the original expression, are not presented to the eye, because in the consequences which it is proposed to deduce, they are entirely immaterial.

If two circumstances in the nature of the function are jointly the ground on which any of its properties depend, they may be separated from the rest and made prominent by several methods.

Thus the index n subjoined to a function

$$\psi(\overline{x}, \overline{y})_n$$

may be defined to mean that it is homogeneous with respect to x and y and of the dimensions n: these two circumstances are the causes on which the truth of the following property depends. xt being substituted for y

$$\psi(\overline{x}, \overline{xt})_n = x_n\psi(\overline{t}, \overline{1})_n.$$

In all our attempts at mathematical generalization, it is of great importance to discover and distinguish these immediate causes of successful operations; in almost all cases they lead us at once to the highest point of generality, and very frequently contribute in no inconsiderable degree to simplify the processes of the investigation. This advantage so peculiar to algebraic signs, has been remarked by M. Degerando, from whose writings I have derived much satisfaction by observing the support which many of those views that I had taken previous to my acquaintance with them, received from the reflections of that distinguished philosopher. / 'La troisième raison,' observes M. Degerando, 'est dans la propriété qu' à l'algèbre de ne saisir, dans les idées des quantités, que certains rapports généraux, de ne présentir ainsi à notre esprit que les considérations qui lui sont vraiment utiles dans les recherches auxquelles il se livre. De là il arrive que notre attention se trouve débarassée d'un grand nombre d'idées accessoires, qui étrangères au but de ses méditations, n'auroient servi qu' à la distraire.'*[a]

The quantity of meaning compressed into small space by algebraic signs, is another circumstance that facilitates the reasonings we are accustomed to carry on by their aid. The assumption of lines and figures to represent quantity and magnitude, was the method employed by the ancient geometers to present to the eye some picture by which the course of their reasonings might be traced: it was however necessary to fill up this outline by a tedious description, which in some† instances even of no peculiar difficulty became nearly unintelligible,

* Des Signes et l'art de Penser, p. 214. tom. II.

† The difficulty which many students experience in understanding the propositions relating to ratios as delivered in the fifth book of Euclid, arises entirely from this cause, and the facility of comprehending their algebraic demonstrations forms a striking contrast with the prolixity of the geometrical proofs.

A still better illustration of this fact is noticed by Lagrange and Delambre, in their report to the French Institute on the translation of the works of Archimedes by M. Peyrard.

It occurs in the ninth proposition of the 2nd book on the equilibrium of planes, on which they observe, 'La démonstration d'Archimède a trois énormes colonnes in-folio, et n'est rien moin que lumineuse.' Eutochius commence sa note 'en disant que le théorème est fort peu clair, et il promet de l'expliquer de son mieux. Il emploie quatre colonnes du même format et d'un charactère plus serré

---

[a] *Translation:* The third reason is the property of algebra which allows ideas of quantity to be expressed in certain general relationships, presenting our mind with only those considerations which are truly useful in the enquiries to which it is being applied. In this, it happens that our attention is freed from a large number of incidental ideas, which being irrelevant to our deliberations, would only serve to distract us. J. M. de Gérando, *Des signes et de l'art de penser*... (Paris, 1800).

simply from its extreme length: the invention of algebra almost entirely removed this inconvenience, and presented to the eye a picture perfect in all parts, disclosing at a glance, not merely the conclusion / in which it terminated, but every stage of its progress. At first it appeared probable that this triumph of signs over words would have limits to its extent: a time it might be feared would arrive, when oppressed by the multitude of its productions, the language of signs would sink under the obscurity produced by its own multiplication: had these expectations been realized, still its utility would have been extensive, and mankind, whilst they felt grateful for the many stages it had advanced them, must have sought some more powerful auxiliary for their ulterior progress. Fortunately however such anticipations have proved unfounded; in whatever department of analysis the number of symbols has increased to a troublesome extent, contrivances have soon occurred for diminishing it without any sacrifice of perspicuity: the inconvenience has always been temporary, the advantage permanent.

In later times the generalization and contraction introduced by the use of signs, seems even to have outstepped the discoveries which have resulted from them; and reasoning from the past course of science to its future advances, we may fairly presume that our power of condensing symbols will at least keep pace with the demands of the science.

Examples of the power of a well contrived notation to condense into small space, a meaning which would in ordinary language require several lines or even pages, can hardly have escaped the notice of most of my readers: in the calculus of functions this condensation is carried to a far greater extent than in any other branch of analysis, and yet instead of creating any obscurity, the expressions are far more readily understood than if they were written at length: the instance I shall choose as an example is the equation

$$\psi^{\overline{9,9}}(x, y) = (x, y).^{*}$$

/

To any person acquainted with the notation belonging to the calculus, it is instantly intelligible; yet if it were written out at length, the letters x and y

sans réussir d'avantage; au lieu que quatre lignes d'algèbre suffisent a M. Peyrard pour mettre la verité du théorème dans le plus grand jour.' Ouvrages d'Archimède traduites par M. Peyrard, p. 415. tom. II.

⟨*Translation:* Archimedes proof takes up three large folio columns and is anything but illuminating. Eutochius begins his note by saying that the theorem is not at all clear, and promises to do his best to explain it. He uses four columns in the same format and in a more obscure manner without succeeding in his aim; whereas four lines of algebra sufficed for M. Peyrard to bring to light the truth of the proposition. F. Peyrard, *Oeuvres d'Archimède*... (Paris, 1808).⟩

* Transactions of Cambridge Philosophical Society, Vol. I. p. 68. *Vol.* II. *Part* II. ⟨'Observations on the notation employed in the calculus of functions' (1822), this volume of the *Works of Babbage*, pp. 348–9.⟩

would be each repeated 257 times, the letter ψ would be found 512 times, whilst the expression would also contain 257 commas and 512 pairs of parentheses; thus comprising in the whole 2307 symbols; and it may be added that it would require a much longer time to understand the meaning of the equation written out at length, than it would to find its general solution.

The power which we possess by the aid of symbols of compressing into small compass the several steps of a chain of reasoning, whilst it contributes greatly to abridge the time which our enquiries would otherwise occupy, in difficult cases influences the accuracy of our conclusions: for from the distance which is sometimes interposed between the beginning and the end of a chain of reasoning, although the separate parts are sufficiently clear, the whole is often obscure. This observation furnishes another ground for the preference of algebraical over geometrical reasoning, and is one which had not escaped the notice of Lagrange. 'Chaque membre de phrase est claire et très intelligible a le considerer seul, mais le tout est si long qu'on a souvent oublié le commencement quand on arrive a l'endroit où le sens est complet.'[a]

The closer the succession between two ideas which the mind compares, provided those ideas are clearly perceived, the more accurate will be the judgement that results; and the rapidity of forming this judgement, which is a matter of great importance, inasmuch as the quantity of knowledge we can acquire in a great measure depends on it, will be proportionably increased. M. Degerando has clearly stated this advantage in comparing the decimal arithmetic with that of the Romans. 'La rapidité d'une opération intellectuelle est toujours en raison inverse des efforts qu'on demande a l'attention et à la mémoire. Cette opération / qui consiste a fixer les rapports de ses idées pour leur appliquer les même judgement s'exécutera donc autant plus promptement qu'il nous sera facile de nous rappeler et de remarquer ces rapports.'* [b]

The almost mechanical nature of many of the operations of Algebra, which certainly contributes greatly to its power, has been strangely misunderstood by some who have even regarded it as a defect. When a difficulty is divided into a number of separate ones,† each individual will in all probablity be more easily solved than that from which they spring. In many cases several of these secondary ones are well known, and methods of overcoming them have already

* Degerando, Sur les signes, tom. II. p. 196.

[a] *Translation:* Each part of the sentence is clear and quite intelligible when considered on its own, but the whole is so long that, by the time one has reached that point where the sense is complete, one has often forgotten the beginning.

[b] *Translation:* The speed of a mental operation is always inversely proportional to the amount of effort placed on the concentration and memory. This operation, which consists of determining the relations between concepts, in order to apply to them the same rule, will itself then be executed more quickly, as one finds it easy to remember and take note of these relations. J. M. de Gérando, *op. cit.*

been contrived: it is not merely useless to re-consider each of these, but it would obviously distract the attention from those which are new: something very similar to this occurs in geometry; every proposition that has been previously taught is considered as a known truth, and whenever it occurs in the course of an investigation, instead of repeating it, or even for a moment thinking on its demonstration, it is referred to as a known datum. It is this power of separating the difficulties of a question which gives peculiar force to analytical investigations, and by which the most complicated expressions are reduced to laws ⟨of⟩ comparative simplicity. One of the most elegant illustrations of this opinion I shall at present briefly allude to, as a more detailed account of it will be given in a subsequent essay. Among the papers left by the late Mr Spence,[a] is one on a method of solving certain equations of differences: elimination is the means by which he proposed to / accomplish his object, but the results soon became so complicated that little expectation could be formed of succeeding by that means. In this difficulty Mr Spence introduced into the equation to be solved an arbitrary quantity a, which is merely employed as a letter by whose powers the resulting series may be arranged: if the attempt is now made by continually eliminating, a series arises proceeding according to the powers of a, and equations are found for determining their coefficients: finally, the arbitrary quantity a having performed its office is made equal to unity, and the result is the solution of the equation. The success of this plan depends entirely on breaking into a number of separate parts a very complicated expression, each of these portions being separately reducible to known laws.

On resolving into their separate parts a vast variety of questions which have occurred, it has been found that the number of individual difficulties is by no means so large as had been originally supposed; many of very different kinds have been found to depend perhaps on the same integral, or on the solution of the same equation. In proportion to the number of questions which are reduced to these new difficulties, they themselves assume importance, and the celebrity which always attaches to those who remove obstacles regarded as insuperable by their predecessors, induces many to attempt the solution of these purely abstract questions. Perhaps these ultimate points of reference may not from their nature admit of a comparison with, or reduction to, existing transcendents: the labour and ingenuity employed in the attempt are not however thrown away; relations are discovered by which, from a certain number of particular

† Of so much importance is this maxim, that it has been adopted by Descartes as one of his principles of philosophizing. 'Diviser chacune des difficultés en autant des parcelles qu'il se pourrait et qu'il serait requises pour les résoudre.' *Discours de la Méthode.*
⟨*Translation:* Divide each of the problems into as many parts as possible, and as many as are required to resolve them. René Descartes, *Discours de la méthode* ... (1617, etc.).⟩

[a] William Spence, 1778–1815; see W. Spence, *Mathematical essays ... edited by J. F. W. Herschel* ... (London, 1819).

cases numerically given, all others may be readily calculated, approximations are discovered for determining the cases which are required as data, and, finally, they are arranged in tables and accompanied by rules for their employment, by which, as far as results in pure numbers are / required, all questions that are made to depend on them may be considered as solved.

The power which language gives us of generalizing our reasonings concerning individuals by the aid of general terms, is nowhere more eminent than in the mathematical sciences, nor is it carried to so great an extent in any other part of human knowledge. In the transition from arithmetic to algebra, when letters began to be substituted for numbers, the first step consisted rather in the circumstance of the possibility of operating on a quantity determined but unknown. Thus if it were proposed to discover such a number, that its square added to three should be equal to four times the number itself; we commence by supposing the number to be represented by x: now it is quite certain, as soon as the question is stated, that there can only exist two numbers fulfilling the condition; x therefore must in reality mean either of these two, and the rest of the process is

$$x^2 + 3 = 4x$$
$$x^2 - 4x + 4 = 1$$
$$x - 2 = \pm 1$$
$$x = 3 \text{ or } 1.$$

To point out more clearly the force of this observation, we adopt the plan which Vieta introduced into algebra, that of denoting known quantities by letters: instead of the numbers 3 and 4, let us use the letters a and b; then the process is as follows:

$$x^2 + a = bx$$
$$x^2 - bx = -a$$
$$x^2 - bx + \frac{b^2}{4} = \frac{b^2}{4} - a$$
$$x = \frac{b}{2} \pm \sqrt{\frac{b^2}{4} - a}$$

here it is true that a and b meant 3 and 4, but as no part of the reasoning employed in any manner depended on their / numerical value, the result must be independent of it, and is consequently true for all possible values. It may perhaps be contended that by the assumption of x for the number to be found, it was meant to represent number in the abstract, and that such was also the meaning of a and b; but there exists this difference, that it is not in our power to alter the value of x, but we may give to those of a and b any numerical magnitude we may please.*

The utility of the unknown quantities in algebra, arises from their capability of being operated on without reference to the determined values for which they are placed, the advantage of employing letters for the known quantities, consists in their similarity to general terms in language, and the consequent extension of the reasoning from an individual case to a numerous species. The light in which this question has been regarded, is purely arithmetical, it may however be placed in another point of view, in which without any change in the quantities concerned, it is still more general in its nature; instead of restricting the equation

$$x^2 - bx = -a$$

to number, it may be considered as indicating that x is composed of a and b in such a manner, that when its value is substituted in that equation, all the terms shall mutually destroy each other. This signification, it is true, is not contained in the original question, but arises from the equation into which it is translated: the language of signs is far more general than that of arithmetic, a circumstance which is not perhaps sufficiently attended to in the application of it to questions of pure number. In one respect this generality is not so unexpected, for if a number is required satisfying a certain condition, and if it should happen / that more numbers than one fulfil that condition, there is no reason why the answer should produce one of these numbers rather than another, it must therefore contain them all.

The reasoning which is carried on in geometry is of a general nature, and applies to a species, although it is impossible that the picture presented to the eye can be anything else than that of an individual; hence, it not unfrequently happens that some peculiarity in the figure which is actually employed, either leads us into erroneous conclusions, or when the results are correct, they are supposed to be limited by the individual nature of the figure we have employed. If a line is made use of to represent number, since some other line is the standard unit, it is impossible by such means to represent number in the abstract, but if number is denoted by a letter, there is nothing in the sign which at all indicates the magnitude of that which it represents: it is evident therefore that a property which might lead us into error in the first case, is removed from our view in the second. It will perhaps be objected that the standard unit need not be visible to the eye, since the force of the demonstration is in no way affected by its magnitude, this observation is perfectly correct, and if only one line be considered, and no unity of linear measure be stated, that line may

* There is in truth one restriction, namely, that a must always be less than $b^2/4$; but this will be removed when the question is viewed in an algebraical light, and does not in the least affect the argument.

represent length in general, and is to all purposes an arbitrary sign: but the moment any other line is introduced into the diagram, although the unit should not be mentioned, the generality of the former sign is diminished, a relation is instantly established, and whatever may be the unit of length, the ratio of those two lines is fixed and determinate.

The position of a line is another circumstance in geometry which must always remain particular, and this brings with it that of the points formed by its intersection, as well as that of the angles formed by it with other lines, and the attention which the mind must exert to perceive that no part of the reasoning it is pursuing / rests on any of these individualities, itself requires a considerable effort. The substances of these observations may be expressed in this conclusion. *The reasonings employed in geometry and in algebra are both of them general, but the signs which we use in the former, are of an* individual nature, whilst those which are employed in the latter, are as abstract as any of the terms in which the reasoning is expressed.*

The signs used in geometry, are frequently merely *individuals* of the *species* they represent; whilst those employed in algebra having a connection purely arbitrary with the species for which they stand, do not force on the attention one individual in preference to any other.

An example of the limitation which geometrical considerations introduce, we shall select from a very well known author. In determining the relation between the rectangle under the parts of two lines intersecting each other and cutting a circle, Euclid considers separately the two cases of the point of intersection being situated within and without the circle, and he shows that in the two figures

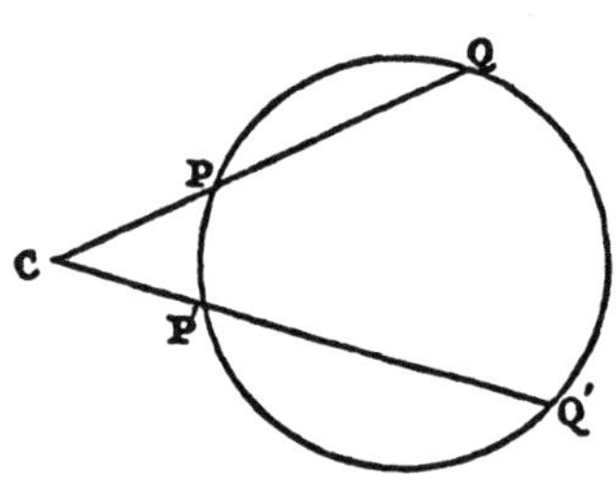

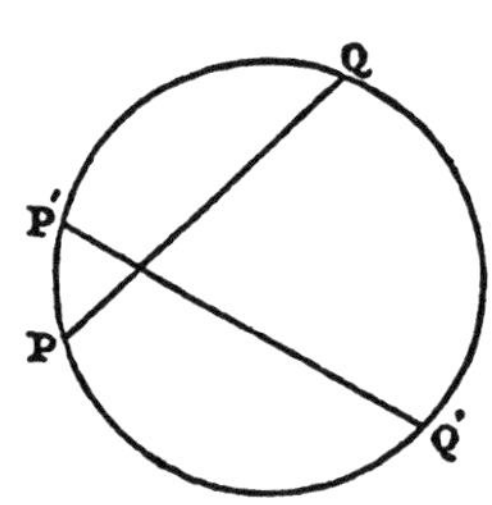

* Halley's paper on the determination of the foci of lenses, would furnish a very apposite example of this principle, and probably few of my readers will fail to recollect instances where the same identical words of a proposition, and the same letters apply to two, three, or more different geometrical figures. ⟨Edmund Halley, 'An instance of the excellence of the modern algebra, in the resolution of the problem of finding the foci of optic glasses universally', *Philosophical Transactions*, Vol. 17 (1693); reprinted in F. Maseres and C. Babbage, *Scriptores optici* . . . (1823); see *Works of Babbage*, Vol. 4).⟩

the rectangle under CP and CQ, is in both cases equal to that under CP_1 and CQ_1: the case of one of the lines becoming a / tangent is also a separate investigation. Now in the algebraic mode of treating these questions, the three cases are comprised in one formula.*

The indication of the extraction of roots by means of an appropriate sign, instead of actually performing the operation, is one of the circumstances which add generality to the conclusions of algebra, and the same principle of indicating operations, instead of executing them, when employed with judgement, contributes frequently in no small degree to the perspicuity of the result, and sometimes enables us to read in the conclusion every stage which has been passed through in the progress towards it. Any general rules to direct us in the application of this principle will be difficult to form, because they ought in a great measure to depend on the objects we have in view: it may, however, be stated generally that it is improper to adhere to it, when by an opposite course any reduction of contraction can be made in the formula; thus generally speaking it would be better to write

$$y = \sqrt{(a - x)^2 + b^2}$$

than

$$y = \sqrt{(a + x)^2 - 4ax + b^2}$$

and on the other hand, wherever in the course of any reasoning the actual execution of operations would add to the length of the formula, it is preferable merely to indicate them.

Some of the advantages which arise from the use of letters to denote known quantities, have already been adverted to; but there are others of considerable value which may now be noticed, and which relate in a great measure to the higher departments of analysis. If a player bet a certain sum of money u, he may either win it and become possessed of $+u$, or he may lose it and possess $-u$. If we now suppose that he regulates the amount of his / second stake by the result of the first, and that he makes it $u - v$ if he had won the first, but $u + v$ if he had lost it; on this second bet he will either win or lose

$$u + v, \quad \text{or} \quad u - v.$$

Supposing him to determine his third stake from his second, in the same manner as he fixed his second from his first, it is clear that according to the determination of his previous bets, he may stake on the third event either of the three sums

$$u + 2v, \quad u, \quad u - 2v.$$

* Book III. Prop. 36. and 37.

And generally on the nth event, if he proceed according to this law, there are n different bets which he may make according to the order in which the previous ones were decided. Now in any question in which such a mode of play entered, it would be exceedingly tedious to consider separately all these cases, and to repeat the same or nearly the same reasoning for each individual case. This may be avoided by rendering the events indeterminate, for we then find his first profit may be denoted by

$$u(-1)^a$$

in which the letter a represents any whole number whatever; if it is an even number he wins, and if an odd one he loses; the same artifice applied to his second stake gives for it

$$u - v(-1)^a$$

as a is still undetermined, this will represent that stake truly, whichever event has happened on the first.

The result of this second stake may be represented by

$$\{u - v(-1)^a\}(-1)^b$$

whether it is lost or gained, and this is still kept undecided by means of the letter b.

The third stake will be

$$u - v(-1)^a - v(-1)^b$$

and the profit or loss by that bet will be

$$\{u - v(-1)^a - v(-1)^b\}(-1)^c$$

a new undetermined letter being again introduced.

By these means his nth stake will be represented by

$$u - v(-1)^a - v(-1)^b - v(-1)^c - \dots v(-1)^m$$

and thus all the n cases become reduced into one expression.

The influence of this indeterminateness in contributing to the success of the solution of the following problem is very apparent.

Supposing a player to regulate his stakes in the manner just described, and that his chances of winning and losing are equal, what will be have profited after $p + q$ events, p of which have been favourable, and q have been unfavourable?

At a first view it might perhaps be considered that the data are insufficient for the solution, and that the order in which the events take place ought to be given: the result however of the investigation will prove that the profit or loss is entirely independent of that consideration.

It has already been shown that his profit on each event will be as follows:

$$u(-1)^a$$
$$u(-1)^b - v\{(-1)^a\}(-1)^b$$
$$u(-1)^c - v\{(-1)^a + (-1)^b\}(-1)^c$$
$$u(-1)^d - v\{(-1)^a + (-1)^b + (-1)^c\}(-1)^d$$
$$\vdots$$
$$u(-1)^{(n)} - v\{(-1)^a + (-1)^b + \ldots + (-1)^m\}(-1)^{(n)}.$$

On examining the composition of that collection of terms which multiply v it will appear that the sum of all the above quantities is equal to

$$u\{(-1)^a + (-1)^b + \ldots + (-1)^{(n)}\}$$
$$-v \times \left\{\begin{array}{l}\text{the sum of all the products two by two which can be}\\ \text{formed out of the quantities } (-1)^a, (-1)^b, \ldots (-1)^{(n)}\end{array}\right\}$$

and since p of the quantities $a, b, c, d, \ldots n$ are even and q are odd, this last expression is equal to

$$(p - q)u - v \times \{\text{the coefficient of the third term of } (x - 1)^p(x + 1)^q\}$$
$$= (p - q)u - v\left\{\frac{p \cdot p - 1}{2} - pq + \frac{q \cdot q - 1}{2}\right\}$$
$$= (p - q)u - \frac{p^2 - p - 2pq + q^2 - q}{2}v$$
$$= (p - q)u - \frac{(p - q)^2 - (p + q)}{2}v.$$

If the order in which the events happened, or what corresponds to it, if the quantities $a, b, c, \ldots n$ had been given, the process we have gone through could not have been completed, for the perception of the nature of the quantity, which in the sum of all the profits multiplies v, depends entirely on preserving those quantities distinct and unconnected with each other; a relation quite impossible, had each been an individual number.

The remarkable influence of signs in the successful termination of this process of reasoning, claims our particular attention: abstract number, from its very nature, admits of amalgamation when subjected to the various operations expressed by algebraic signs: hence all trace of the mode in which it originally entered is completely lost; or if this inconvenience be studiously avoided, by merely indicating instead of executing the arithmetical operations, still the individual nature of the several numbers presents so many points to which the attention is attracted, that it would be almost impossible, even for the most attentive observer, to seize that general view in which they all agree. This influence is still more remarkable in investigations, where characteristics of operation occur, and when letters are used to the exclusion of number, the relations are not merely more apparent, but the results, although attained with difficulty, are more worthy of confidence: the reason of which, is to be found / in this circumstance, that when letters only are employed, the functional characteristics convey no meaning except that on which the force of the reasoning depends; but, if numbers are used, they convey, besides this signification, a multitude of others, which distract the attention, although they are quite insigificant in producing the result.

This principle of preserving undetermined* until the conclusion, the quantities on which we are reasoning, seems to be the only one, which promises success in questions, where two parties successively make choice either of things

* The profound remark of Lagrange, that the true secret of analysis, consists in the art of seizing the various degrees of indeterminateness, of which quantity is susceptible, has been so beautifully illustrated by M. Carnot, and so well accords with my own ideas on the subject, that I should be unwilling to present it to my readers in any language but his own.

J'ai ouï dire plusieurs fois à ce profond penseur, (M. Lagrange) que le véritable secret de l'analyse consistait dans l'art de saisir les divers degrés d'indétermination dont la quantité est susceptible; idée dont je fus toujours pénétré, et qui m'a fait regarder la méthode des indéterminées de Descartes comme le plus important des corollaires de la méthode d'exhaustion.

Un nombre abstrait est moins déterminé qu'un nombre concret, parceque celui-ci spécifie non seulement le combien du nombre, mais encore la qualité de l'objet soumis au calcul; les quantités algébriques sont plus indéterminées que les nombres abstraits parce qu'elles ne spécifient pas même le combien: parmi ces dernières les variables sont plus indéterminées que les constantes, parce que celles-ci sont considérées comme fixes pendant un plus longue période de calcul; les quantités infinitésimales sont plus indéterminées que les simples variables, parce qu'elles demeurent encore susceptible de mutation, lors même qu'on est déjà convenu de considérer les autres comme fixes; enfin les variations sont plus indéterminées que les simples différentielles, parce que celles-ci sont assujéties à varier suivant une loi donnée, au lieu que la loi suivant laquelle on fait changer les autres est arbitraire. Rien ne termine cette échelle de divers degrés d'indétermination, et c'est précisément dans cet assemblage de quantités plus au moins définies, plus au moins arbitraires, qu'est le principe fécond de la méthode générale des indéterminées, dont le calcul infinitésimal n'est véritablement qu'une heureuse application. *Carnot, Réflexions sur la Métaphysique du Calcul Infinitésimal*, 2nd edit. p. 207.

⟨*Translation*: It has been remarked many times by this profound thinker, (M. Lagrange) that the real secret of analysis consists in the art of making one's self acquainted with the different degrees of indetermination, of which the quantity is capable; an idea which always struck us as forcible, and which occasioned our regarding the method of indeterminate quantities of Descartes as the most important corollary to the method of exhaustion.

... An abstract number is less determinate than a concrete, since the latter determines not only the quantity but the quality of the object submitted to the calculus. Algebraic quantities are more indeterminate than abstract numbers, because they do not specify the quantity. Amongst these last, variables are more indeterminate than constants, since the latter are considered fixed for a longer period in the calculation. Infinitesimal quantities are more indeterminate than simple variables, since they still continue susceptible of change, even when it has already been agreed to consider the others as fixed. Lastly, the variations are more indeterminate than the simple differentials, since the latter are restricted to varying according to a given law, instead of which the law according to which the others change is arbitrary. This gradation in the different degree of indetermination is endless, and it is in the union of quantities more or less defined, more or less arbitrary, that is founded the fertile principle of the general method of indeterminate quantities, of which the Infinitesimal Calculus is in truth but a fortunate application. L. N. M. Carnot, *Reflections on the metaphysical principles of the infinitesimal analysis*, translated by W. R. Browell (Oxford 1832), pp. 124–5.⟩

or of situations: of this nature are many of those questions which relate to games dependent on skill. In these cases, the number of things, out of / which the choice is to be made, is always finite, and the second player can only make his selection out of that number diminished by unity. The first player, at his second stroke, can only choose out of the same number diminished by two, and so on. Now it is evident, that if the individual actually chosen at each step, were fixed permanently, the reasoning must diverge into an immense multitude of cases; whereas, if any one indifferently is chosen by the first, and again, any one indifferently out of the $k-1$ remaining ones by the second, and so on, the things chosen by the two parties would all be comprehended in two expressions.

It is this power of representing any one of k–p things, where p have already been taken out of k (subject only to the condition that the expression for it can never represent any one of those already chosen) which enables us to delay the decision of the individuals actually selected until the conclusion; and thus by their means, to satisfy the other conditions of the problem. Several instances of such questions, will be noticed in a future paper: The only means that have hitherto been employed for this purpose, are the roots of unity, and the sines, and other similar functions of submultiples of π, and from the great length of the formulae in which they occur, I am by no means sanguine in my expectation of much success in those enquiries, until some more condensed method of indicating and to a certain extent also of executing such operations, shall have been contrived. The principle in discussion, appears to me to be the only one, by which any general and complete solutions can be arrived at: more partial views may doubtless be taken, more adapted to the present state of symbolic language, and even these become extremely valuable, in such difficult enquiries, not merely from the advances they themselves introduce, but as the ground-work of generalization to more perfect methods. /

The principle of representing any one quantity indifferently, out of a given number, has been employed on several occasions, and occurs in some mechanical questions. Lagrange has made use of it in a memoir relative to the theory of sound,* and M. Poisson has employed it on a similar occasion.† The cause, on which its successful application depends, seems to be the power which it gives of uniting together a number of cases totally distinct, and of expressing them all by the same formula, the consequence of which is, that one single investigation frequently supersedes the necessity of a multitude.

Many examples of the successful application of this principle, are to be found

* Memoirs de Turin, vol. I. ⟨J. L. Lagrange, 'Récherches sur la propagation du son', *Mémoires de l'académie de Turin*, Vol. 1 (1759), pp. i–x and 1–146.⟩

† Journal de l'École Polytechnique, cah. 18. ⟨S. D. Poisson, 'Mémoire sur la manière d'exprimer les fonctions par des séries de quantités périodiques, et sur l'usage de cette transformation dans la résolution de différens problèmes', *Journal de l'École Polytechnique*, Vol. 11 (1820), pp. 417–89.⟩

in a paper on circulating functions,* in which it is applied to the solution of a peculiar class of equations of finite differences: other instances will come under our notice, in a future communication.

An examination of the various stages, by which, from certain data, we arrive at the solution of the questions to which they belong, ought to constitute a prominent feature, in any work devoted to the philosophical explanation of analytical language. It is a subject, the consideration of which is too frequently omitted altogether, and as a correct view of it tends materially to advance the science, and also by pointing out more clearly the nature of the difficulties it has to contend with, and also by guarding us against those openings by which errors are most frequently introduced, I shall enter into the question at some length.

In the application of analysis to the various questions which are submitted to it,† there are three distinct stages, each subject / to its particular rules, and each liable to its peculiar difficulties. The order in which these succeed each other, will prescribe that which will be followed in the remarks upon them.

I. The first stage consists in translating the proposed question into the language of analysis.

II. The second, comprehends the system of operations necessary to be performed, in order to resolve that analytical question into which the first stage had transformed the proposed one.

III. The third and last stage consists in retranslating the results of the analytical process into ordinary language.

I. In the first step, which consists in translating the proposed question into the language of analysis, much caution is requisite: for unless this is correctly done, it is quite manifest that the labour bestowed on the remaining stages is useless. It is at this point that the principles applicable to the question are employed for its solution; in all the succeeding parts they are kept totally out of sight: the termination of this stage is usually marked by the circumstance of the difficulty being reduced to one or more equations,‡ which is the case in most of those questions that arise in the application of analysis to physics: but this however does not always happen, since very many questions relating to chances are reduced to the finding of the coefficient of a certain term in the development of a given function into a series, and although most of these may be reduced to equations of differences, yet it is not universally the case. It is

* Phil. Trans. p. 144. 1818. J. F. W. Herschel, Esq. ⟨J. F. W. Herschel, 'On circulating functions . . .', *Philosophical Transactions*, Vol. 108 (1818), pp. 144–68.⟩

† Of course questions already in an algebraic form are not here alluded to, such as the integration of equations, etc.

‡ The term equation in this place must be understood to comprehend such expressions as $A - B >$ or $< C$.

of some importance to observe, that those errors most difficult to detect and remove, / usually occur in this first stage; and it cannot to too strongly recommended, that every part even of the most difficult problems should be fully translated into the language of analysis, before any attempts at simplification are made. In the first stage, it is scarcely possible to see clearly in what degree the results will be affected by a proposed omission; whilst in the second, any quantity which it is conjectured will have little influence on the result, although it adds greatly to the difficulty of calculation, may be kept separate, and the operations to which it is submitted, may be indicated rather than performed. In many of the applications of analysis, and particularly in its treatment of mechanical questions, the principles which regulate the first stage of the process are completely known, and little difficulty is experienced in translating them into the language of signs, the difficulties when they occur, usually taking their rise in the solutions of the equations thus produced. A similar remark is applicable to optical questions, and indeed to by far the greater part of those which occur in the mixed sciences.

II. The second stage in the solution of any problem, generally begins with the equations into which it has been translated, and terminates with their solution. The point at which it commences is not always so well defined as that at which it ends, and this is more particularly the case when the question relates to geometrical figures, where in some instances, the first and second stages are much intermixed.

The difficulties which now occur are purely analytical, and are generally such as have been treated of in works devoted to the subject. The solution of one or more algebraic equations is frequently the object to be obtained: differential equations, or equations of finite differences are another class of analytical expressions to which physical problems are often reduced; many of these can / only be resolved approximately, but in proportion to the interest these questions have excited, the variety and accuracy of the approximations have been multiplied. This is strongly exemplified in the problem of the three bodies, as applied to determining the place of the moon: the great importance of the question has caused the approximations to be pushed to such an extent, that they have arrived at a degree of inelegance and complexity, which would long since have caused them to be rejected from any other question on the exact solution of which less important interests depended. But on this second stage in the solution of a question, it is less necessary to add many observations; the operations which are concerned in it, and the modes of effecting them, being more fully treated of in works of instruction than either of the others.

III. The last of the stages into which the resolution of a question has been distributed, has been more neglected than any other. It may perhaps appear singular that the answer to a question, which is of course the great object of research, should have been passed over without sufficient attention. It is not however of any errors in those results which are usually arrived at that I complain; but it is, that sufficient instruction is not given in elementary works, as to the full meaning of all the different circumstances which are contained in the result that analysis has presented. In those questions which lead to algebraic equations, it is not unfrequently the case, that some one or perhaps two roots are taken as the answer, whilst all the remaining ones are completely neglected. Now a question can never be said to be fully answered until every root of the equation to which it has conducted has been discovered, and its signification with reference to the data of that question been explained. It sometimes happens that superfluous roots have been / introduced by the algebraic operations that were necessary to arrive at the final equation: these ought to be pointed out, and the step at which they were introduced should be noticed, and also whether they can admit of any translation into the language of the problem considered. Imaginary roots are very frequently introduced; they sometimes imply impossibility or contradiction amongst the data; their origin ought to be carefully traced, and such a course will frequently make us acquainted with the maxima and minima which belong to the question. It is still more necessary to attend to all the real roots whether positive or negative, and to explain the various circumstances in the solution to which they refer.

It is by no means uncommon with algebraical authors, when they have led their readers through a process which terminates in an equation, to select that root which gives the answer they require, without explaining the signification of the other roots that are equally comprised in it; and this incomplete mode of solution, which is censurable from revealing only a part of the truth, has in some instances caused the most interesting circumstances attending a question to be entirely overlooked. A singular example of this occurs in several authors who have sought analytically the side of a heptagon inscribed in a circle, or the radius of a circle which would circumscribe a regular heptagon whose side is given. In neither of these questions can the equation to which we are led, be reduced below the third degree, and the three roots of the cubic are always real: the largest of the positive roots gives the answer to the latter of these questions for the common heptagon of Euclid: but no reason is stated why this root should be considered as the true answer to the question in preference to either of the others. In the Analytical Institutions of M. Agnesi,*[a] where the first problem is solved, no / notice is taken of the fact that all the three roots are possible, nor am I aware of its being noticed by any author who has treated of this question: had it been observed and enquired into, the existence of three species of heptagons answering strictly to the definition, and the

* Vol. I. p. 168. English Translation.

[a] M. G. A. M. Agnesi, *Analytical institutions* . . . (London, 1801).

knowledge of the star-shaped polygons which were discovered by M. Poinsot, could not have remained so long unknown.

If $x = OP$, denote part of the diameter intercepted between the centre, and a

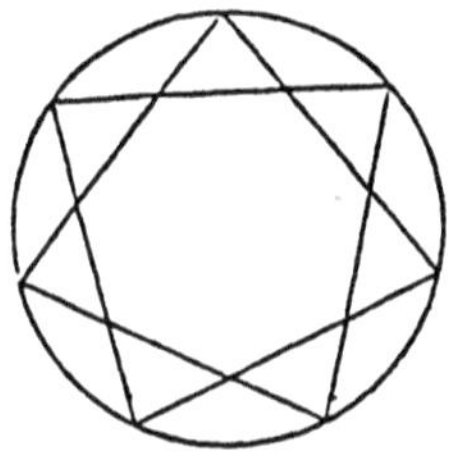

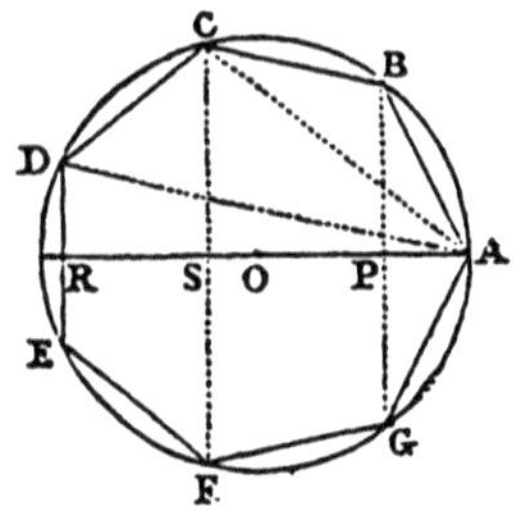

perpendicular from the extremity of the first side of the heptagon, then the usual trigonometrical formulae give

$$x^6 - \frac{5}{4}x^4 + \frac{3}{8}x^2 - \frac{1}{64} = 0$$

this contains six roots, of which the three positive are (abstracting the signs) equal to the three negative: it may be resolved into the two factors

$$\left(x^3 - \frac{1}{2}x^2 - \frac{1}{2}x + \frac{1}{8}\right)\left(x^3 + \frac{1}{2}x^2 - \frac{1}{3}x - \frac{1}{8}\right) = 0$$

the second of which is only the first with the signs of its roots changed. The three roots are real and are represented by

$$x = OP, \quad x = OS, \quad \text{and} \quad x = OR$$

the first of these gives AB for the side of the heptagon, this is the same as that which has long been known; the other two roots give AC and AD, as the sides of the polygon, and by carrying them round the circle, the two star-shaped heptagons (Figures 2 and 3) are produced which have no re-entering angles, and / which are in fact comprehended in Euclid's definition of regular polygons. The sum of the interior angles of the first of these heptagons is ten right angles, the sum of the interior angles of the second is six right angles, and that of the third species is two right angles. These new species of polygons were first noticed by M. Poinsot, in a highly interesting memoir on subjects connected with the *Geometry of Situation*, read before the Institute in 1809, and subsequently printed in the *Journal de l'École Polytechnique*, 10e Cah.[a]

Another important business which belongs to this stage of the question, is to examine carefully what changes will ensue from supposing any peculiar relations

[a] L. Poinsot, 'Mémoire sur les polygones et les polyhèdres', *Journal de l'École Polytechnique*, Vol. 4 (1810), pp. 16–48.

amongst the data; or from any of the constant quantities becoming infinite or evanescent, such circumstances frequently introduce great simplicity, and when they refer to geometrical questions, are sometimes the means of making us acquainted with general properties, by which the construction of the problem is greatly facilitated.

A careful and laborious attention to all the possible modifications of a problem which might result from any relation amongst its data, was considered by the ancient geometers as an indispensible part of its investigation, and the manner in which this was accomplished, was generally little else than a repetition of the whole process under the altered circumstances: when the data are numerous, the length of such a system of operations becomes intolerable, and if more rapid methods had not been contrived, geometry must have become stationary from the accumulation of the details with which it was thus encumbered. Many instances of the extreme length to which a full investigation of comparatively a very simple problem will lead, occur in the treatises *De Sectione Rationis*, etc. The advantage of algebraic language is in this respect very striking: all thc data of the questions are embodied in the equation in which its solution terminates, and / without repeating any part of the process by which that was produced, we can examine with ease all those modifications which any differences in the actual magnitude of the data can introduce into the question under consideration: and moreover the equation itself will suggest to us such relations amongst those quantities as will have the effect of lowering the number of its dimensions, or of rendering it the product of two or more factors.

It sometimes happens that by a peculiar relation amongst the data of a question, the number of solutions instead of being limited becomes infinite: thus, if the position of a line is determined by two points, when those points coincide, any line passing through the point in which they coalesce, will satisfy the conditions of the question which becomes to a certain extent indeterminate: this gives rise to a class of propositions in geometry which are called porisms. When the data on which questions depend are numerous, it is by no means so easy to discover by geometrical considerations that relation amongst them in which the question becomes indeterminate, as it is by an algebraical enquiry where the solution is presented in its most condensed form: one consequence of this is, that such cases have frequently escaped the notice of those who have treated the problems to which they belong in a geometrical manner. One celebrated and important oversight of this kind occurred in a problem which Newton solved in order to determine the orbit of a comet.

Having four lines given in position, it was required to draw a fifth line which should be cut by the other four into segments, having a given ratio to each other. Of this question Wren, Wallis, and Newton had given solutions, but when Zanotti, Boscovich and other astronomers made use of them, employing the observed places of a comet, the results were found greatly / erroneous. Boscovich enquired into the reason of this singular result, and having first assured himself of the accuracy of the solutions, he discovered that in a particular relation between the given lines the problem became indeterminate, and admitted of an infinite number of solutions, and that the case of a comet approached extremely near to this, and consequently that any very small error in observations must produce an extremely large one in the result.

As an instance of the curious and elegant properties to which such an examination of the relations of the data contained in the final equation sometimes leads, I shall propose the following problem.

A circle whose radius is r *being given, and also three points in one of its diameters, at what angle must three parallel chords be drawn through these points so that the sum of the squares of two of them shall be equal to a given multiple* n *of the square of the remaining one?*

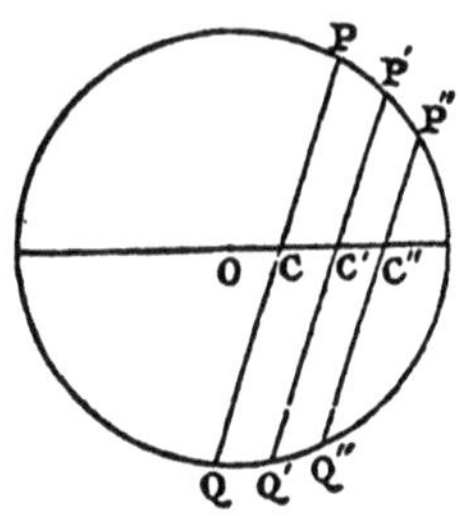

Let the distance of the three points in the diameter from the centre be v, v_1, and v_2, and calling the angle which is sought θ, we have

$$CP = -v\cos\theta + \sqrt{r^2 - v^2(\sin\theta)^2}$$

and

$$CQ = +v\cos\theta + \sqrt{r^2 - v^2(\sin\theta)^2}$$

hence

$$PQ = 2\sqrt{r^2 - v^2(\sin\theta)^2}$$

and similarly for the other two chords

$$P_1Q_1 = 2\sqrt{r^2 - {v_1}^2(\sin\theta)^2}$$

and

$$P_2Q_2 = 2\sqrt{r^2 - {v_2}^2(\sin\theta)^2}.$$

/

These values of the chords being employed give[a]

$$4(r^2 - v\,(\sin\theta)^2) + 4(r^2 - v_1{}^2\,(\sin\theta)^2) = 4n(r^2 - v_2{}^2\,(\sin\theta)^2).$$

At this step the first of the three stages which have been described terminates; the question is now translated into the language of algebra, and must be treated according to its rules: the following reductions must then be made

$$r^2 - v^2\,(\sin\theta)^2 + r^2 - v_1{}^2\,(\sin\theta)^2 = nr^2 - nv_2{}^2\,(\sin\theta)^2$$

$$(nv_2{}^2 - v_1{}^2 - v^2)\,(\sin\theta)^2 = nr^2 - 2r^2$$

$$\sin\theta = \pm r\sqrt{\frac{n-2}{nv_2{}^2 - v_1{}^2 - v^2}}.$$

The second stage is here concluded by the solution of the equation to which the first conducted us, and we have now to explain the meaning of its two roots, and the modifications which may arise from any peculiar relations amongst the data.

The two signs signify that the angle θ, may be measured either above the diameter or below it, as is apparent on inspecting the figure. As the result contains an even radical, we must enquire if in all cases a solution is possible, and if not, what are the conditions of possibility. For this purpose we observe that the numerator and denominator must be both positive or both negative, consequently

$$n - 2 > 0, \quad \text{and} \quad nv_2{}^2 - v_1{}^2 - v^2 > 0$$

or

$$n - 2 < 0, \quad \text{and} \quad nv_2{}^2 - v_1{}^2 - v^2 < 0, \quad \text{and also} \quad \sin\theta < 1$$

are the two sets of conditions; in all other cases the question is impossible.

From this, however, must be excepted the case of the numerator and denominator simultaneously vanishing in consequence of the following relations taking place amongst the data,

$$n - 2 = 0, \quad \text{and} \quad nv_2{}^2 - v_1{}^2 - v^2 = 0$$

whence

$$n = 2, \quad \text{and} \quad v_2 = \pm\sqrt{\frac{v_1{}^2 + v^2}{2}}$$

under which circumstance the value of $\sin\theta$, becomes really indeterminate, not depending even on the value of r the radius of the circle.

This indeterminate case suggests the following porism:

Any three points in a straight line being given, another point may be found

[a] The notation $(\sin\theta)^2$ has been substituted for $\sin\theta^2$ in these equations.

about which as a centre if a circle with any radius be drawn, and if through the three given points, three chords be drawn in any direction, but parallel to each other; then the sum of the squares of two of them shall be always equal to the square of the third.

It is to be observed, that the origin of the lines denoted by v, v_1 v_2, may be changed by removing it to the distance a, then the latter of the two conditions which rendered the problem indefinite becomes

$$2(v_2 + a)^2 = (v + a)^2 + (v_1 + a)^2$$

whence

$$a = \frac{2v_2{}^2 - v_1{}^2 - v^2}{2v + 2v_1 - 4v_2}.$$

Before I conclude my observations on this subject, which may perhaps be considered as a digression from that which the title prefixed to this essay would seem to imply, I shall offer one more illustration of the division of a problem into the several stages which I have pointed out.

This examination of all the circumstances attending the equation containing the solution, is still more necessary when that equation is a differential one: if it be only capable of integration by means of transcendents or by approximating series, it sometimes happens that some relation amongst the data may be assumed, by which in the one case the transcendents shall / disappear, and in the other, that the series shall terminate. Euler has taken advantage of the former of these circumstances, to discover curves whose indefinite quadrature should depend on a given species of transcendent, whilst the areas of particular portions of them are susceptible of an algebraic expression.* The integration of the equation is not always sufficient for a complete analysis of a question, for in some cases besides the general integral, there exists another not included in it, which is known by the name of a particular solution; in order to be secure of not overlooking any such, it must be observed that a change in the magnitude of an index may cause the introduction of such a solution. When the complete integral as well as all the particular solutions are found, the interpretation of them according to the circumstances of the question is not always an easy task, nor are any general rules yet established to which we can refer for information. In the theory of curves the interpretation of particular solutions is sufficiently well known: they represent the curve which touches all those formed by the complete integral when its parameter varies. In mechanical questions a considerable degree of uncertainty prevails relative to these kind of solutions, as in some instances they seem to have no reference to the problem which gave rise to them, whilst in other cases its solution can only be fully represented by their assistance; some light has been thrown on this subject by

* (Euler *Acta Acad. Petrop.*)

M. Poisson* in a memoir in which he has explained the theory of particular solutions with great perspicuity.

Of whatever kind the equation to which our question conducts us, may be, it ought to be regarded, merely in an analytical point of view; and all its various roots or solutions, should / be sought after; out of these, by means of some peculiarity in the problem, we must select that individual, which more immediately satisfies the particular view of it which we have taken, and the other solutions, must be explained if possible, by means of the data, from which we commenced the process; or should that be impossible, their entrance must be traced to some generalization in signs, to which the language of the question was incapable of adapting itself. In the demonstration of the composition of forces, given in the Mécanique Céleste,[a] which has sometimes been unjustly censured on account of its analytical nature, this does not appear to have been completely attended to. In the enquiry, to which I refer, x one of the forces is assumed equal to $z\varphi(\theta)$, where z is the resultant force, and $\varphi(\theta)$ some function of the angle between it and the force x, which function it is required to determine; by changing x into y and θ into $\pi/2 - \theta$, the two values of x and y are found to be

$$x = z\varphi(\theta), \qquad y = z\varphi(\pi/2 - \theta)$$

and the equation

$$x^2 + y^2 = z^2$$

is arrived at: this equation in fact amounts to

$$[\varphi(\theta)]^2 + [\varphi(\pi/2 - \theta)]^2 = 1$$

which results† from it, by merely substituting for x and y, their values. /

From this equation it appears to me, to be the direct course to deduce the form of φ: its general solution should first be shown, and then from the peculiar circumstances of the problem, that particular one which belongs to it should be pointed out. In the work, to which I refer, the particular form of φ has been

* *Journal de l'École Polytechnique*, Cah. 13. p. 60 ⟨S. D. Poisson, 'Sur les solution particulières des équations differentielles et des équations aux différences', *Journal de l'École Polytechnique*, Vol. 6 (1806)⟩.

† This equation is one of that class whose general solution I have ascertained, and it may be exhibited in either of the following forms

$$\varphi(\theta) = \sqrt{\varphi_1(\theta)/(\varphi_1(\theta) + \varphi_1(\pi/2 - \theta))}$$

$$\varphi(\theta) = \sqrt{\tfrac{1}{2} + (\pi/2 - 2\theta)\bar{\chi}(\theta, \pi/2 - \theta)}$$

where φ_1 is perfectly arbitrary, and $\bar{\chi}$ is any symmetrical function of θ and $\pi/2 - \theta$.

[a] P. S. de Laplace, *Traité de mécanique céleste* (5 vol., Paris, 1798–1827).

deduced at once by properties peculiar to the problem, without any reference to the general solution of the equation. A similar objection may be made to other demonstrations of this celebrated theorem: the equation to which the investigation conducts, is usually solved in a manner not sufficiently general. This is the case in a work devoted to the analytical exposition of the elements of geometry, pp. 53, 54;* the substitutions employed, although satisfying the conditions, not containing all possible solutions. M. Poisson has given an investigation of this theorem not quite so open to the objections just stated; by the introduction of two variables and the employment of one sign of function, the solution is necessarily more restricted in its extent. Equations of that class are frequently contradictory, although in the case referred to, a fortunate property leads directly to the solution. See Poisson, Mécanique,[a] p. 14. I cannot conclude this slight criticism on a detached passage of the Mécanique Céleste, without expressing that respect for its illustrious author, which is shared with all those, who are capable of appreciating the important additions he has made to mathematical science, or who have the happiness of being personally acquainted with him.

When any question leads to an algebraic equation, it is usual to resolve it generally, and then to point out amongst its roots that particular one which is sought; if the individual root required / were discovered by some artifice without the solution of the equation, we could not feel assured that it alone fulfilled all the conditions, and we should, by arriving at an equation and rejecting its use, have the semblance of generality without its reality; nor do I perceive any reasons which should induce us to change our course, when we have to consider equations of a more comprehensive nature.

Any enumeration of the causes which contribute to give such extensive power to the employment of algebraic signs, would be justly considered incomplete, if no notice were bestowed on the symmetry that ought always to prevail, where the calculations in which it is employed are in any degree complicated.

In its least restricted signification, symmetry is applied to two things, which although sometimes connected, are yet, in many instances, totally independent: it either refers to a resemblance between the systems of characters assumed to represent the data of a question, or it implies a similarity of situation, between certain of the letters, which are found in an analytical formula. An attention to it in either of these senses, has a direct and very beneficial influence in relieving

* Précis d'une nouvelle méthode pour réduire à de simples procédés analytiques la démonstration des principaux théorèmes de la géométrie. Par. I. G. C. (Paris, An. vi.) ⟨*Translation*: 'Abstract of a new method for reducing to simple analytical procedures the principal theorems of geometry.' This pamphlet does not appear in the principal national library catalogues, but it is listed in [R. Tucker], *Mathematical and scientific library of the late Charles Babbage*... (London, 1872).⟩

[a] S. D. Poisson, *Traité de mécanique* (Paris, 1811).

the memory from a considerable burthen. In the first case, its precepts would direct us to assume similar letters as the representatives of similar things: thus, if we have two series, and propose to find another, which consists of the product of the corresponding terms of the two former; if we assume for the first of two series

$$a + b + c + d + \dots$$

the terms of the second ought to be

$$A + B + C + D \dots$$

or which is still more convenient

$$a' + b' + c' + d' + \dots /$$

and the series, whose sum we wish to find is then denoted by

$$aA + bB + cC + dD + \dots$$

or by

$$aa' + bb' + cc' + dd' + \dots$$

The assumption of

$$A_1 + A_2 + A_3 + \dots$$
$$A'_1 + A'_2 + A'_3 + \dots$$

would have been equally proper, and the result equally clear: but had we assumed for the two series

$$a + b + c + d + \dots$$
$$A_1 + A_2 + A_3 + A_4 + \dots$$

the resulting series

$$aA_1 + bA_2 + cA_3 + dA_4 + \dots$$

would have been devoid of that symmetry, which forms so prominent a character in the former cases.

The plan of accenting letters, in order to represent quantities which stand in similar relations, adds, when employed with discretion, much to the perspicuity of the formulae in which it is used; but like many other innovations, whose tendency is on the whole decidedly beneficial, an attempt to extend it beyond its proper limits, has been productive of inconveniences as considerable as those which its introduction was proposed to remove. Indices in various positions have been substituted in many cases for the system of accentuation, and the admirers of this scheme, pursuing it with equal ardour, have not been more fortunate in avoiding the confusion, which a multitude of signs, differing but by the slightest shades, can scarcely fail of producing. The taste of the geometer is not less strongly tried by the choice of the letters in which he conducts his reasoning, than his skill and ingenuity are by the artifices he invents to surmount the / difficulties opposed to him; in the one case, the elegance which a judicious selection produces, carries the reader pleasantly and almost imperceptibly through an abstruse calculation, whilst the latent cause, which gives facility to his progress, is rarely appreciated, because it is spread uniformly over the whole question: in the other, whose essence often consists in some happy substitution, which is always concentrated in some point, the effect is too remarkable to escape the least attentive enquirer, and its success too striking not to command his admiration.

Unlimited variety in the use of signs, is as much to be deprecated as too great an adherence to one class of them; the one conceals the appearance of relations that really exist, whilst the other, affecting to display them too clearly, fails from its want of distinctness. It is difficult to point out models of imitation, even amongst the most eminent; the same writer, who at one time might be safely trusted as a pattern of correct judgement, has indulged at another in the most unexampled innovation; completely setting at defiance many of those principles, whose authority I have endeavoured to establish. The fate, which has attended the greatest of these proposed reformations, though sanctioned by the illustrious name of Lagrange, is no slight testimony of the validity of those views, which it is the object of several of these essays to advocate.* Having delivered a theory of the differential calculus, which rested on principles entirely new, he introduced it to the world, clothed in a language of his own creation. The value of the present was too great to allow of its utility being impeded by the garb with which it was encumbered; and the labor of acquiring the language was compensated by the truths which it revealed. Time, however, and experience, convinced even its immortal author, that the language of signs rests / on principles, which cannot be neglected without danger, or violated with impunity; – the authority of the greatest geometer of the age, failed to make converts to the language he had invented, whilst the justice of the view he had taken, was admitted, and his explanations almost universally adopted. Whilst the language, in which the theory and the calculus of functions are conveyed, is pointed out as a warning, not to be neglected by the most successful, that of the Mécanique Analytique of the same author,[a] may perhaps be held up to imitation, with fewer limitations than any other work of equal magnitude. In returning to the notation of Leibnitz, Lagrange has in this work, reduced the whole theory of mechanics to the dominion of pure analysis, and

* This relates more particularly to an Essay on the Principles of Notation, which is not yet published. ⟨Possibly 'Notation' (1830), in this volume of the *Works of Babbage.*⟩

[a] J. L. Lagrange, *Mécanique analytique* (Paris, 1788).

in the choice of his symbols has frequently displayed that happy selection, which so much facilitates the process of reading and comprehending analytical formulae.

The value of symmetrical symbols is greater in proportion to the complexity of the operations, and the number of quantities, which are concerned; but unless their selection is attended to at the outset of our studies, it is not to be expected that a correct taste can be acquired, I would therefore recommend a degree of attention to this subject, which is not usually bestowed on it by elementary writers. Some instances I shall select from the simplest applications of algebra to geometry. The equation of a right line is usually written thus:

$$y = ax + b,$$

which is sufficiently convenient. M. Biot in his Géométrie Analytique*[a] has employed this notation, as also has M. Hachette in his introduction to the admirable work of Monge:†[b] both authors in treating of lines referred to three co-ordinates have denoted it thus:

$$y = ax + \alpha, /$$

the difference is apparently trivial, but the convenience or inconvenience of notation frequently depends on differences as trifling. It may be observed, that in the first equation, a denotes the tangent of an angle, and b an absolute line; two things which have no relation to each other, and which are therefore justly represented by dissimilar signs. In the second equation, the line and the angle, are both represented by the *same letter* of different alphabets; a circumstance, which will infallibly suggest some idea of a relation that does not exist. When two straight lines enter into the question, other reasons present themselves: they may be represented in any of these four ways;

$$\begin{array}{llll} y = ax + b & y = ax + \alpha & y = ax + b & y = ax + b \\ y = a'x + b' & y = bx + \beta & y = \alpha x + \beta & y = cx + d \end{array}$$

and if we seek the ordinate of their point of intersection it will be

$$y = \frac{ab' - a'b}{a - a'}, \quad y = \frac{a\beta - b\alpha}{a - b}, \quad y = \frac{a\beta - \alpha b}{a - \alpha}, \quad y = \frac{ad - cb}{a - c}.$$

the latter of these expressions is quite devoid of all symmetry in regard to its letters, and the larger the number of lines about which we reason, the more confused will such a mode of expressing them render the result. In the first and

* p. 30.
† Application de l'Analyse à la Géométrie, 4me ed. p. 2.

[a] J. B. Biot, *Essai de géométrie analytique* ... (4th edition, Paris, 1810).
[b] G. Monge, *Application de l'analyse à la géométrie* (4th edition, Paris, 1809).

third mode, it is sufficient to remember that the letter a, under all its forms, represents the tangent of an angle, and that the letter b, in every form, always represents a particular ordinate: with this principle in our mind, we can see at a glance, however numerous the lines introduced, to what property of them each individual letter refers; whereas in the last method, we must, in order to discover the meaning of any letter, refer back for each individual one, to the original translation into algebraic language. The third plan will suffice, where only a few different lines are concerned, but its application is limited by the smallness of the number of different alphabets we can command. The second method may be defended on the ground, that the tangents are denoted by one class of letters; / namely, italics, whilst the Greek letters are reserved for lines; perhaps it might still be improved by interchanging these significations of the two alphabets.

Before I pass on to the consideration of the second species of symmetry, I shall select from the Arithmetica Universalis,[a] an example, in which the choice of the letters employed seems to have been made without any rule; and shall subjoin to it, the same problem expressed in a language consistent with the views I am illustrating. This course will render more apparent the advantages of attending even to the letters which we select to represent the quantities.

The velocities of two moving bodies A and B being given, and also their distance, and the difference of the times of the commencement of their motion, to determine the point in which they will meet.

Let A have such a velocity that it will pass over the space c in the time f; and let B have such that it will pass over the space d in the time g, and let the interval between the two bodies be e, and that of the times when they begin to move be h.

Case I. Then if both move in the same direction, and if A is further distant from the point of meeting, call that distance x, from this take away the interval e, and there will remain $x - e$ for the distance of B, from the same point. And since A passes over the space c in the time f, the time in which it will pass over the space x will be fx/c.

And so also, since B passes over the space d in the time g, the time in which it will pass over the space $x - e$, will be $(gx - ge)/d$. Now since the difference of these is supposed to be x, in order that they may become equal, add h to the smaller time; namely, to the time fx/c (if B moved first) and it will become /

$$\frac{fx}{c} + h = \frac{gx - ge}{d}$$

and by reduction

$$x = \frac{ceg + cdh}{cg - df} = \frac{ge + dh}{g - (d/c)f}$$

[a] Isaac Newton, *Arithmetica universalis*... (London, 1707, pp. 82–4). The example quoted by Babbage was translated from the Latin.

but if A began to move first, add h to the time $(gx - ge)/d$ and it will become

$$\frac{gx - ge}{d} + h = \frac{fx}{c}$$

and by reduction

$$x = \frac{cge - cdh}{cg - df}.$$

Case II. If the bodies move in opposite directions, and if x is, as before, the distance of the body A from the point of meeting, then $e - x$ will be the distance of B from the same point, and fx/c the time in which A will pass over the distance x; and $(eg - gx)/d$ will be the time in which B will perform the distance $e - x$. To the less of these times add the difference h, namely, to fx/c if B first began to move, and thus we shall have

$$\frac{fx}{c} + h = \frac{eg - gx}{d}$$

and by reduction

$$x = \frac{cge - cdh}{cg + df}.$$

If A began to move first, add h to the time $(eg - gx)/d$, and it will become

$$\frac{eg - gx}{d} + h = \frac{fx}{c}$$

and by reduction

$$x = \frac{ceg + cdh}{cg + df}.$$

This same question with the following data may be solved in nearly the same way,

v = velocity per second of A,
v' = " " " B,
s = the space one (A) is distant from the other B,
t = the time in seconds one (B) starts before the other.

Since the velocities are both positive, the bodies move in the same direction, and x being the distance of the point where they meet from the place of A, the number of seconds which A will take to pass over it, will be found from the ratio

$$v : 1'' :: x : \frac{x}{v}$$

the distance of the other body from the same point will be $x - s$, and the time it takes to arrive will be $(x - s)/v'$, but the other body began to move t seconds before B, therefore t added to the time of its motion must equal $(x - s)/v'$, or

$$\frac{x}{v} + t = \frac{x}{v'} - \frac{s}{v'}$$

hence

$$t + \frac{s}{v'} = x\left(\frac{1}{v'} - \frac{1}{v}\right)$$

and

$$x = \frac{s/v' + t}{1/v' - 1/v} = v\frac{s + tv'}{v - v'}.$$

If A move in an opposite direction, its velocity must be accounted negative; hence in that case v' becomes $-v'$, and /

$$x = v\frac{s - tv'}{v + v'}.$$

If A begin to move after B, the time t must be made negative, and then these two cases become

$$x = v\frac{s - tv'}{v - v'}$$

and

$$x = v\frac{s - tv'}{v + v'}$$

the former of these referring to the case of the bodies moving in the same direction, and the latter to that of their direction being opposite.

There are some restrictions which ought to be noticed, if the velocities of the two bodies are equal, or $v' = v$ the first and third cases show that the bodies can never meet. To this there is however an exception, if $v' = v$, and also $s = tv'$, then $x = 0/0$, an indefinite expression and x may have any value: the signification of this is that both bodies move in the same direction and with equal velocities, since $v' = v$, and that the hindermost B, which starts t seconds before the other, is situated at such a distance s from it, that it arrives at the point where the other is, exactly as it begins to move; this appears from the equation $s = tv'$, it is therefore obvious that the two bodies will be at the same point at every part of their progress and for every value of x. Whenever x is negative they can never arrive at the same point in the direction in which they move. If however we conceive that they had been moving at the same rate prior to the point of time at which we consider them, the negative value assigned to x marks a point through which they both passed at the same moment.

These two modes of translating the same question into algebra, and of re-translating the result into ordinary language, / give rise to several observations. The assumption of v to represent the velocity of one body per second, instead of the plan pursued in the first solution, was productive of two advantages: first, it substituted one letter instead of two; and secondly, it is so usual in all mechanical problems to make that letter denote velocity, that it is in such cases associated with it in the mind. The next assumption of v' for the velocity of the other body, possessed both these advantages, and tended to make the result more apparently symmetrical in case it was susceptible of that species of arrangement. The selection of the letters s and t, to represent space and time, was adopted with a similar view of making the signs recall the thing signified. In pure analysis there is but little room for taking advantage of this species of connection, but in all the mixed questions to which it is applied, it may be extensively employed. The general principle is, that either the initial or some prominent letter should be selected from the word which denotes the thing we wish to represent. The beneficial effect of this arrangement is felt a little in the first stage of the solution; it has no influence on the second; but in the last stage it saves considerable trouble by obviating the necessity of constantly referring back to enquire what particular letters represent.

In the first solution, there are in fact two distinct cases, in which the reasoning is repeated from the beginning to the end; and each of these cases has two sub-species: so that there are, in fact, four cases treated of in the Arithmetica Universalis. This defect has been remedied in the second solution, where it has been shown that the four cases are included in one formula, to which it is only necessary to give the proper interpretation, and every circumstance connected with the problem is brought to light. Two causes seem to have contributed to produce this separation into cases: in the first place, the extreme generality of the language of algebra may / not have been sufficiently noticed: the ancients were accustomed to divide their problems into cases, and the habit of treating these separately, may have produced the same cautious treatment of a question when resolved by methods of a far more general and comprehensive nature: such indeed would naturally be the case, until the degree of generalization introduced by the new method, was fully ascertained. In some instances, the action of another cause may be traced, and one that is more easily removed than that which arises from a want of confidence in the method employed: it may be referred to the imperfect manner, in which the very first elements of the application of algebra to geometry and to mechanics, have been communicated to us: in order to explain clearly the result, it is quite indispensible that we should be perfectly familiar with the principles which regulate these: without them, we may frequently put the question into an algebraic form, because we view it only in one light; and from the nature of the language made use of, that one, whichever it may be, virtually embodies the whole; but when we require to retranslate our conclusions, as they contain every possible case, we must of course be able to translate each individual. To display in a more prominent point of view, the reason why there are so few failures in expressing the conditions of a problem in algebra, and so many in giving the full account of all which the solution informs us, let us imagine a problem proposed that admits of ten cases; if we are capable of translating any one of these into an equation, such is the comprehensive nature of algebraic language, that though they are all contained in that single expression, we may be ignorant of the manner of treating nine cases, and competent to manage one of them. If this happened, we might succeed in resolving the equation, and discovering all its roots, which would answer all the ten cases; yet it is scarcely probable, with such moderate knowledge, we should succeed in explaining in common language, the meaning of many of them. /

The advantage of selecting in our signs, those which have some resemblance to, or which from some circumstance are associated in the mind with the thing signified, has scarcely been stated with sufficient force: the fatigue, from which such an arrangement saves the reader, is very advantageous to the more complete devotion of his attention to the subject examined; and the more complicated the subject, the more numerous the symbols and the less their arrangement is susceptible of symmetry, the more indispensible will such a system be found. This rule is by no means confined to the choice of the letters which represent quantity, but is meant to extend, when it is possible, to cases where new arbitrary signs are invented to denote operations.

In the formation of some of the most common algebraic signs, this maxim has been attended to; but although in many individual instances it has been admitted, it is still desirable that it should be recognized as a general principle. The sign of equality was obviously adopted from the circumstance of the same relation existing between its two parts, as that which it indicates between the two quantities which it separates, and the propriety of this selection has confirmed its use, although Girard employed $=$ to denote difference, and Descartes used $\propto$ to represent equality. In the two signs representing greater and less than

$a > b$ and $b < a$,*

the prevalence of the same motive of choice is equally apparent, for it is immediately seen, that these signs are so contrived, that the largest end is

* That this principle operated in inducing Harriot, who first used these signs, to adopt them, I have now very little doubt: I may however, remark, that on my first initiation into algebra, finding some difficulty in remembering their distinction, I formed the association above alluded to as a kind of artificial memory, a purpose, which it effectually answered.

always placed next to the largest quantity, and / consequently, the smallest end next to the smallest quantity. At a period when the more frequently occurring signs were not permanently settled, one method of denoting equality was by the following sign, 2|2, and the same author P. Hérigone employed 3|2, and 2|3 to represent respectively, greater than, and less than. In each of these, the principle we have been contending for, has undoubtedly had its influence, and the signs themselves are objectionable, on entirely different grounds. The same writer made use of the common sign of equality to denote parallelism; a purpose for which it was then well adapted; since that time, however, its long continued use in another sense, has compelled geometers to change its position; and when it is proposed to state that two lines AB and CD are parallel, instead of putting $AB = CD$ as Hérigone would have done, they merely change the position, and write $AB \parallel CB$; thus preserving the advantage, without infringing another rule, which ought never to be violated, that of avoiding the use of any sign in two senses.

In the doctrine of triangles, Lagrange has introduced a species of symmetry, which has been found productive of very advantageous results; it consists in denoting the three sides by the letters a, b, c, and the angles respectively opposite to them by the capitals A, B, C: by this arrangement, not merely the quantities themselves are indicated, but in some measure also their position, and the transition from any relation between one side and given data, to other sides and the corresponding data, are made with the greatest ease.

The more complicated the enquiries on which we enter, and the more numerous the quantities which it becomes necessary to represent symbolically, the more essentially necessary it will be found to assist the memory by contriving such signs as may immediately recall the thing which they are intended to represent. The notation which M. Carnot has contrived, for the purpose of illustrating / his view of the application of algebra to geometry, possesses in an eminent degree the qualifications we are now considering; and although I cannot altogether agree with the conclusions, which it is the object of the author of that highly ingenious work, the 'Geometry of Position,'[a] to establish, yet I am happy to acknowledge the instruction I have derived from the very original view which he has taken.

To denote a line, M. Carnot places a bar over the letters which represent it, thus, $\overline{AB}$: an arc of a circle or other curve, will naturally be represented by $\overset{\frown}{AB}$. To indicate the point where two lines meet, $\overline{AB} \cdot \overline{CD}$ is used, and similarly the points formed by the meeting of two arcs, or an arc and a circle, are denoted respectively, $\overset{\frown}{AB} \cdot \overset{\frown}{CD}$, and $\overline{AB} \cdot \overset{\frown}{CD}$; according to these principles, the line

[a] L. N. M. Carnot, *Géométrie de position* (Paris, 1803).

drawn from the point *F*, to the intersection of *AB* and *CD*, will be represented thus,

$$\overline{F\,\overline{AB}\cdot\overline{CD}}$$

and the two expressions,

$$\overline{\overline{\overline{AB}\cdot\overline{CD}}\;\overline{\overline{FG}\cdot\overline{HK}}}, \qquad \overline{\overline{\overline{AB}\cdot\overline{CD}}\;\overline{\overline{FG}\cdot\overline{HK}}}\cdot\widehat{LM}$$

denote the first, the line which joins the points, where *AB*, *CD*, and *FG*, *HK* respectively intersect each other: and the second, the point where that line cuts the arc *LM*, $\pm$ is used to signify coincidence with, as

$$\overline{AB}\cdot\overline{CD} \pm \widehat{FG}\cdot\widehat{HK}$$

indicates, that the two points formed by the intersection of the lines *AB*, *CD*, and by the arcs *FG* and *HK*, coincide.

The angle *ABC*, is indicated thus, $A\hat{B}C$, and if it is formed by the two lines *AB* and *CD*, it is thus expressed, /

$$A\hat{B}C = \overline{AB} \wedge \overline{CD}.$$

The angle formed by the intersection of arcs, will therefore stand thus,

$$\widehat{AB} \wedge \widehat{CD}.$$

Surfaces are represented by the position of two lines above the letters, which indicate them, as

$$\overline{\overline{AB\,CD}}$$

and the angles which are formed by these, with other surfaces, or with lines, can be expressed on the principles already explained; as for example,

$$\overline{AB} \wedge \overline{\overline{CDEF}}$$

is the angle formed by the line *AB*, with the surface *CDEF*. When single letters are employed to denote lines, as *a*, *b*, *c*, the angles formed by them are represented thus $a^\wedge b$, $a^\wedge c$, $b^\wedge c$.

The effect of these few abbreviations, which, when once explained, need no effort to fix them in the memory, is to render unnecessary the almost endless repetitions of the same words, and to convey the writer's meaning in the fewest terms. In the calculation of the values of annuities dependent on lives, if the question is any degree complicated, the number of independent data is very considerable, and the letters which have been employed to denote them, have in many instances been selected, without the least regard to assisting the reader's memory. The symbols made use of are far too numerous to be extracted, and the immense advantages which attend a judicious selection, can only be appreciated by those, who have had occasion to study the subject, in the writings of Price and Morgan, and have afterwards perused those of Mr Baily or Mr Milne; the former gentleman had had / the merit of explicitly stating the principle,* and the use, which he has continually made of it, has had the effect of giving great perspicuity to his formulae.

In astronomy, this principle has been adopted with much success, and signs ⊙ and ☽ for the Sun and Moon constantly occur. It is rather singular, that a principle on which the earliest and most imperfect written language rested, should be found to add so essentially to the value of the most accurate and comprehensive: yet the language of hieroglyphics, is but the next stage in the progress of the art, to the mere picture of the event recorded; and the signs it employs, in most cases, closely resemble the things they express. In algebra, although the principle has not been pushed to its extreme limits, the grounds of its observance are the same; the associations, are by its assistance, more easily and more permanently formed, and the memory most effectually assisted.†

I have entered into more detail respecting these causes, than the importance of the subject, may in the opinion of some appear / to justify: the reasons which have induced me to do so, are to be found in the incomplete manner, in which these parts of the subject are treated by the generality of our elementary writers, a circumstance which impedes the subsequent progress of the student more than is perhaps usually allowed. To those who may hereafter employ

* 'When compound quantities are represented by more simple expressions, those characters ought to be preferred, which will most readily, and with least effort of memory, bring to our recollection the original quantity intended to be expressed.' *Baily on Life Annuities*, Preface p. 39. ⟨F. Baily, *The doctrine of life annuities and assurances* ... (London, 1813).⟩

† The benefit derived from a proper choice of signs, or from a judicious mode of presenting them, is not entirely confined to mathematical studies; wherever the multiplicity of particulars renders it of importance to assist the memory, or to give quickness to the apprehension of the terms employed, such a principle, if it can be adopted, will be found of considerable value. Amongst the authors who have availed themselves of this principle, in treating other subjects, than those, in which it is so eminently useful, the name of M. Cuvier, may be mentioned. In the preface to his works, Le Règne Animal, he remarks, 'Partout les noms des divisions superieures sont en grandes majuscules; ceux des familles, des genres et de sous genres, en petit majuscules, correspondentes aux trois caractères employés dans le texte; ceux des espèces en Italiques; le nom Latin est à la suite du nom Français, mais entre deux parenthèses. – Ainsi l'œil distinguera d'avance l'importance de chaque chose et l'ordre de chaque idée, et l'imprimeur aura seconder l'auteur de tous les artifices que son art peut prêter à la mnémonique. Cuvier, *Le Règne Animal*, tom. I. Preface, p. 18. ⟨*Translation:* All the names of the principal divisions appear in upper-case, whereas those of families, genera and sub-genera are printed using lower-case, each of which corresponds to one of the three typefaces used in the text. The names of species appear in italics, the Latin name following the French, but placed in parentheses. Thus the eye is able to distinguish in advance the importance of each thing and the order of each idea, and the printer, by using all the tricks of his trade, has furthered the aims of the author in promoting mnemonics. G. L. C. F. D. Cuvier, *La règne animal* ... [*The animal kingdom*] (Paris, 1817).⟩

themselves in supplying a considerable desideratum, in English mathematics, by composing an introductory treatise on the application of algebra to geometry, I may be permitted to recommend a copious development of its very first elements, not merely a detailed account of the changes of the situation or direction of lines, by changes in the signs, or magnitudes of quantity, or by the circumstances of the radicals, in which they may be involved: but by way of giving practical illustrations of the precepts so delivered, they should be accompanied by a series of examples, in which every circumstance, attending the section of two or more right lines, should be explained with scrupulous minuteness: the influence of such a course of reading, will be sensibly felt by the student as he proceeds; and the uninviting dryness of the results will be relieved by a judicious selection, which may render them valuable points of reference, in his future reading.* How much such a system is calculated to assist all our enquiries in theoretical mechanics will be allowed by those, who have had, in some measure, to form it for themselves, at the moment when the natural difficulties of the subject were sufficient to require their undivided attention. I am indeed inclined to refer a considerable portion of those difficulties, which are so much complained of by English students, well versed in the mathematical science of their own country, when they first open the works of the continental geometers, to the want of such a previous course of instruction, as that which I have now pointed out. I would however guard myself from being supposed to imagine, that this is by any means the sole obstacle; I mention it as one, which appears to me of some weight, and which might, without much difficulty, be removed.

The other, and more general, acceptation of the word symmetry, applies to the position, as well as the choice of the letters, employed in an enquiry: in this sense, it can scarcely exist, without a previous attention to that which has just been explained, for however regularly and analogously two series of quantities may be arranged, unless the signs, by which they are represented, are so constructed, as mutually to excite the idea of their correlatives, it is impossible, that the symmetry can be apparent to the eye. By employing the first species of symmetry, we assist the memory in remembering the ideas indicated by signs; by the use of the second, we enable it more easily to retain the form in which our investigation has arranged those signs, as well as facilitate the processes, by which that final arrangement was accomplished. By the happy union of the two, our formulae acquire that wonderful property of conveying to the mind, almost at a single glance, the most complicated relations of quantity, exciting a

* These observations were written prior to the publication of the Analytical Geometry of Mr Lardner, and to the still more recent work of Mr Hamilton. I am happy that one considerable impediment to the progress of the English student is at length removed. ⟨D. Lardner, *An analytical treatise on plane and spherical trigonometry*... (London, 1826); H. P. Hamilton, *The principles of analytical geometry* (Cambridge, 1826).⟩

succession of ideas, with a rapidity and accuracy, which would baffle the powers of the most copious language.

The mode of expressing an angle of a triangle, in terms of the radii of three circles, which touch respectively each side, and the other two prolonged, will furnish the first example. Calling a, b, c those radii, and θ the angle opposite a, we have

$$\cot\frac{\theta}{2} = \sqrt{\frac{b}{a} + \frac{c}{a} + \frac{bc}{a^2}}$$

an unsymmetrical expression. This can be improved by a very trifling change, for it is equivalent to /

$$\cot\frac{\theta}{2} = \frac{\sqrt{ab + ac + bc}}{a}$$

in which the numerator is instantly perceived to be the square root of the sum of the products of the radii, two by two. In as far as regards its form, it does not admit of any further improvement; but if this expression were to be employed in any enquiry into the properties of a triangle, it would be much more convenient to use the letters A, B, C, and a, b, c, to denote respectively the angles and sides, and to employ for the radii of these circles, the letters ρ', ρ'', ρ''', thus,

$$\cot\frac{A}{2} = \frac{\sqrt{\rho'\rho'' + \rho'\rho''' + \rho''\rho'''}}{\rho'}.$$

The convenience of so employing the letters A, B, C, a, b, c, has been already noticed, and the letters ρ', ρ'', ρ''', cannot fail, after a very short use, to recall the idea of radii, as well as fix the particular side, which each circle touches.

I have now enumerated what appear to me to be the principle causes which exert an influence on the success of mathematical reasoning, and have illustrated, with examples, those which were susceptible of it. They may be recapitulated in few words. The nature of the quantities which form the subject of the science, together with the distinctness of its definitions – the power of placing in a prominent light, the particular point on which the reasoning turns – the quantity of meaning condensed into small space – the possibility of separating difficulties, and of combining innumerable cases, – together with the symmetry, which may be made to pervade the reasoning, both in the choice and in the position of the symbols, are the grounds of that pre-eminence, which has invariably been allowed to the accuracy of the conclusions deduced by mathematical reasoning. /

XVIII

NOTATION[a]

From *The Edinburgh Encyclopaedia*, Vol. 15 (1830), pp. 394–9

Notation, (in mathematics): the art of adapting arbitrary symbols to the representation of quantities, and the operations to be performed on them. The numerous symbols, which form the language of analysis in the present advanced stage of that science, have been produced by the gradually increasing wants of those who have cultivated it; new and more extensive views continually opening, have compelled them to contract modes of notation already in existence, and the improvements thus introduced into notation, have in their turn directed the attention of the mathematician to wider and more general views of the science.

The advances towards perfection which have been made in the language of analysis, and the generalization which has been introduced into its principles, alternately acting on each other, as cause and effect, have by their combined influence produced a language of unrivalled power, enabling the mind to carry on processes of deductive reasoning of almost unlimited length, with scarcely the fear of an error, and with a conviction, that, should accident have introduced one, a careful revision will not fail to eradicate it.

Brevity appears to have been the directing principle which guided the early cultivators of the algebraic art; unaware of the immense importance which, in a subsequent state of the science, would be attached to the language whose foundations they were thus unconsciously laying, they contented themselves with avoiding the tedious repetitions of the same words, by employing one or two of the initial, or, in some cases, of the final letters, to denote them. Such

---

[a] In order to convey as accurately as possible the notations used by Babbage in this paper, the original has been followed as closely as practicable, notwithstanding the mathematical typesetting conventions used elsewhere in the *Works of Babbage*. The original paper did, however, contain numerous typographical errors, and some mathematical notations that were not correctly set. In interpreting the paper, the editors have been guided by F. Cajori, *A history of mathematical notations* (2 vol., Chicago, 1928–9) and D. E. Smith, *History of mathematics* (2 vol., Boston, 1928).

Although this article was published in 1830, it was written much earlier, probably during 1821 (see J. M. Dubbey, *The mathematical work of Charles Babbage* (Cambridge, 1978), pp. 166–7).

was the case with Diophantus, the earliest author on algebra, whose writings have descended to us. The unknown quantity he denominates ἀριθμός,[a] and to avoid repeating it uses the final letter ς. He also uses the sign ⋔ or the inverted ψ to denote minus, obviously from the circumstance of its being a prominent letter in the Greek word λεῖψις.[b] This author has no sign to denote plus, but uses the word at length. He has represented the various powers thus, $\delta^{\bar{v}}$, $\kappa^{\bar{v}}$, $\delta\delta^{\bar{v}}$, ... meaning the square, cube, fourth power, etc., these letters are the same as those which commence the words square, cube, etc.

The earliest algebraical writer after the invention of printing, was Lucas Paciolus, or de Burgo, he uses *p* to signify plus and *m* for minus, and indicates the various powers by their two first letters. Such was very nearly the notation employed by his successors, Cardan, Tartaglea, and Ferrari. Stifelius, a German, who published a work, entitled *Arithmetica Integra*, Norimburg, 1544,[c] added considerably to the use of signs; according to Dr Hutton he is the first writer who employed the signs + and − and also $\surd$, to designate the root of a quantity. He had no sign to represent equality, a deficiency afterwards supplied by Robert Recorde; we also owe to him the vinculum $\overline{a + b}$ to connect compound quantities, the other mode by means of parentheses $(a + b)$ being afterwards proposed by Girarde. When Stifelius treated of several variables, he denoted them by the letters A, B, C, ...

Bombelli appears to have made a most valuable innovation in the method of denoting powers, rejecting the plans of attaching their initials to the radix, he marks them thus, $\overset{1}{\smile}, \overset{2}{\smile}, \overset{3}{\smile}, \overset{4}{\smile}, \ldots$ an improvement, perhaps more valuable than any which has yet been noticed; nearly a similar plan was adopted by Simon Stevin, who denoted the powers of the unknown quantity thus, ⓪, ①, ②, ③, and he observed that the power, whose index is zero, is equal to unity. He, however, went a step beyond his predecessors, and denoted roots by fractional indices; thus, $\textcircled{\tfrac{1}{2}}$, $\textcircled{\tfrac{1}{3}}$, with him represent the square and cube roots.

Vieta flourished several years later than Stevin, he added greatly to the science he cultivated, although he did not avail himself of all the improvements in notation, which existed previously. The most important alteration which he introduced was that of denoting *known* as well as *unknown* quantities by letters; for the former he employed the consonants, and for the latter the vowels. He appears also to have made the important remark, that negative exponents, perform the same office as positive ones, although his view seems rather to have been restricted to whole numbers.

It is a curious circumstance, that the symbol which now represents equality,

[a] *Translation*: number (*arithmos*)
[b] *Translation*: to be missing (*lei'psis*)
[c] Michael Stifel, *Arithmetica integra* (Nuremberg, 1544).

was first used to denote subtraction, in which sense it was employed by Albert Girarde, and that a word, signifying equality, was always used instead of a sign, until the time of Harriot, who at the same time proposed the signs $>$ and $<$ to denote 'greater than' and 'less than'.

Oughtred, in his 'Clavis'[a] appears first to have employed the × for multiplication, he sometimes also joins the letters without any intervening sign for the same purpose.

Such was the origin of the more common signs, which we now employ. An example of the different modes of expressing the same equation will exhibit this more clearly.[b]

Paciolus	1 *cu.* *m.* 6*ce.* p. 11*co.* eguale 6
Stifelius	1 ꝭ − 6ʒ + 11 ꝛ equantur 6
Bombelli	1$\overset{3}{\smile}$. *m.* 6$\overset{2}{\smile}$. *p.*$\overset{1}{\smile}$ eguale 6
Stevinus	1③ − 6② + 11① egale 6
Vieta	1C − 6Q + 11N egal 6
Harriot	$1.aaa - 6.aa + 11.a = 6$
Modern	$x^3 - 6x^2 + 11x = 6$

Since the time of Harriot few alterations have been made in the mode of denoting the more simple operations of analysis, although the science itself has received great improvements. The first step which necessitated the invention of new methods of denoting operations was the fluxional or differential calculus. / Newton and Leibnitz each gave a notation to this new branch of analysis, and their respective followers entering with more zeal than judgement into the unfortunate controversy respecting the claim of priority of invention, pertinaciously adhered to the notation of their masters without enquiring into their relative merits, and probably without the knowledge requisite to form a correct judgement on the question, had they turned their attention to the enquiry. In fact, the state of the science at the period at which these modes of calculation were given to the world, could not admit of the formation of a correct estimate of their relative value, and the few reasons which could then be adduced, were, on the whole, more favourable to the fluxional notation. It requires, even at the present time, an extensive acquaintance with the subsequent discoveries of the successors of Newton and Leibnitz, and an enlarged view of the present state of mathematical science, duly to appreciate the reasons which concur in rendering the notation of Leibnitz decidedly the best of any which has hitherto been proposed for accomplishing the same object.

[a] William Oughtred, *Clavis mathematicae*... (Oxford, 1652).

[b] A spurious error in the top line of the table has been corrected, but several errors due to Babbage's inaccurate interpretation of his sources have not been corrected. More reliable interpretations are given in F. Cajori, *op. cit.*, and D. E. Smith, *op. cit.*

As the proper adaptation of notation to the object of enquiry has contributed greatly to the progress of analysis, we propose offering a few general rules for the consideration of those who may have occasion to express new relations, or who may desire to abbreviate those already in use; and we shall compare several of the notations already received with these principles.

The great object of all notation is to convey to the mind, as speedily as possible, a complete idea of operations which are to be, or have been, executed; and since every thing is to be exhibited to the eye, the more compact and condensed the symbols are, the more readily will they be caught, as it were, at a glance.

The first and most obvious rule is, that *all notation should be as simple as the nature of the operations to be indicated will admit.* The reason of this is sufficiently evident, and it scarcely admits of an exception. It must, however, be remarked, that it is, in many cases, absolutely impossible to express the complicated operations required in the highest departments of analysis by formulae that can be called simple. Still, however, they may be simple with reference to the multiplied relations they express.

The next rule we shall propose is, *that we must adhere to one notation for one thing.* It is particularly unphilosophical, and completely contrary to the whole spirit of symbolic reasoning, to employ the same signs for the representation of different operations, yet instances of the infringement of this rule are occasionally met with; and as several examples are to be found in the works of an author of the very highest reputation, it is more particularly necessary to mark this deviation from the propriety of analytical language.

The notation employed by Lagrange to explain the differential calculus will be considered on another occasion. It is curious to observe, that his zeal for introducing the system of accentuation compelled him to abandon a most valuable contrivance which he had formerly employed for designating the variations of quantities which harmonized with the notation of Leibnitz, whilst that which he substituted for it (the system of dots) besides the inconveniences which it possessed in its original signification, had the additional disadvantage of having been employed in a different sense by all the followers of Newton. In the first edition of the *Théorie des Fonctions*, this notation was not employed, but in the *Leçons sur le Calcul des Fonctions*, published in 1806, the fluxional notation is used for explaining the calculus of variations.[a] The same author has also assumed the quadrant as the unity of angular measure, and represented sin. $\frac{\pi}{2}$: thus, sin. 1, which, unless previously defined, would indicate the sine of an arc equal to the radius.

Another instance of a breach of this rule is to be observed in the *Théorie des*

[a] J. L. Lagrange, *Théorie des fonctions analytiques* (Paris, 1797); *Leçons sur le calcul des fonctions* (Paris, 1804).

Nombres of Legendre,[a] who, throughout that highly valuable work, employs the sign = in two different senses; first, in its ordinary acceptation; and, secondly, he places it between two quantities, to denote, that when they are divided by the same quantity, they will leave the same remainder. This relation of quantity is of frequent occurrence in the theory of numbers; and the necessity of denoting it with brevity, induced the author[b] of the *Disquisitiones Arithmeticae* to invent the symbol ≡, and to adjoin (enclosed in brackets) the quantity which is used as the divisor. A different method has been adopted by Mr Barlow,[c] who, in his *Treatise on the Theory of Numbers*, uses the double *f* (*ff*) placed in a horizontal position. These three methods of denoting the same relation stand thus:

Legendre: $a^n = -1$ Gauss: $a^n \equiv -1 (\mathrm{mod}\, p)$ Barlow: a^n ⥲ $pv - 1$.

The first of these sins unpardonably against the rule we are endeavouring to enforce; it is much more inconvenient than the use of dots by Lagrange, because, in the instance we are considering, the same symbol = is used in the *same page*, and even in the *same line* in *two different senses*. It has also the disadvantage of requiring the divisor to be expressed in words.

This innovation in the use of a well known sign, probably arose from too strict an adherence to an admitted rule: *not to multiply the number of mathematical signs without necessity*. We may, however, here be permitted to observe, that the necessity existed; and it is acknowledged by Gauss, whose notation, although much preferable to the one we have been criticising, is decidedly inferior to that of Mr Barlow. The preference we have given to this latter is also supported by another rule too evident to require much argument. *When it is required to express new relations that are analogous to others for which signs are already contrived, we should employ a notation as nearly allied to those signs as we conveniently can.*

That analogy ought to be our guide in the formation of all new notations, is a truth, which, like many others, has been felt and acted upon, although it may not have been stated in express terms: and it was probably this feeling which induced Stifelius to inquire into the meaning of negative exponents, the consequence of which was the establishment of the connection between the direct and reciprocal powers, a deduction which, when enlarged by the consideration of fractional exponents, was no mean addition to the state of algebra at the time it was suggested.

It has long been usual to denote known quantities by the first letters of the alphabet, and to represent unknown ones by the last letters. In the sixteenth

[a] A. M. Legendre, *Essai sur la théorie des nombres* (Paris, 1811).
[b] C. F. Gauss, *Disquisitiones arithmeticae* (Leipzig, 1801).
[c] Peter Barlow, *An elementary investigation of the theory of numbers*... (London, 1811).

and seventeenth centuries, the practice used to be to employ the vowels for the unknown, and the consonants for the known quantities. This is in itself a matter perfectly arbitrary, and, provided it is adhered to, either method of distinguishing known from unknown quantities is equally proper; but whenever one of these, suppose that now in use, is fixed upon for this purpose, if we wish to consider known and unknown functions, and to treat of their relations, it is no longer a matter of indifference how / they are to be distinguished. The rule we are now illustrating must be attended to, and if we have used the first letters of the alphabet for known quantities, it compels us to employ the first letters of whatever other alphabet we may fix on to indicate functions for the known ones, and the latter letters of the same alphabet for the unknown ones. Thus, if $a, b, c, \ldots$ are known quantities, and $x, y, z, \ldots$ unknown ones, $\alpha, \beta, \gamma, \ldots$ must denote known functions, and φ, χ, ψ unknown ones. In another instance, the repetition of a letter xx has been marked thus x^2. Analogy, therefore, would direct us when we wish to repeat an operation as $\psi\psi$, to contract it by writing it thus, ψ^2, and similarly for others, as ψ^3, ψ^4.

We have already had occasion to remark that we ought not *to multiply the number of mathematical symbols without necessity*. This is a maxim of so much importance that it deserves a fuller consideration.

The natural tendency of the science is to develop new relations and new combinations of those already known. When these new relations involve complicated combinations of such as are already received, or when they are of frequent occurrence, it becomes necessary, if it were merely for the sake of brevity, that some new symbol should be employed; it is, however, a necessity that ought to be avoided as much as possible; and the rule is one which has suffered many infractions in modern times. Amongst the numberless instances may be remarked the case of a function consisting of the product of a number of terms in arithmetical progression. This species of function is of some importance, and has been frequently considered, yet scarcely any two authors have agreed in the notation they have employed. The first, in point of time, is that of Vandermonde (*Acad. des Sciences*, 1779 ⟨sic⟩).[a] He expressed the quantity, $x \cdot x - a \cdot x - 2a \cdot \ldots \cdot x - \overline{n-1}a$ by the sign $[x, a]^n$, or when $a = 1$, more shortly, thus $[x]^n$. Kramp, in his treatise on *Astronomical Refractions*,*[b] calls the functions in question *facultées numeriques*, and denotes them thus $x^{n|a} = x \cdot x + a \cdot \ldots \cdot x + \overline{n-1}a$. Arbogast has occasion to treat of the same quantities,† when he represents thus $D^n 1^x = x \cdot x - 1 \cdot \ldots \cdot x - \overline{n-1}$.

* Strasburg, 1798.
† *Calcul des Dérivations*, 1800.

[a] A. A. T. Vandemonde, 'Mémoire sur des irrationnelles des diférens ordres avec une application au cercle', *Histoire de l'Académie Royale des sciences* (1772), Part 1, pp. 489–98.
[b] C. Kramp, *Analyse des réfractions astronomiques et terrestres* (Strasbourg, 1798).

Legendre,[a] in his *Exercises du Calcul Intégral*, employs a different notation to express a function very nearly allied to that we are considering. He denotes $x \cdot x - 1 \cdot \ldots \cdot 2 \cdot 1$ by $\gamma(x)$. Euler also treated of the properties of the coefficients of a binomial, and denoted them by the characters $\left(\frac{n}{1}\right), \left(\frac{n}{2}\right), \left(\frac{n}{3}\right), \ldots$[b] Of these various methods of denoting the same thing, that of Vandermonde seems to be the most advantageous. The reasons of this preference we shall shortly state. When x is repeated n times, $x \cdot x \cdot \ldots \cdot x$, it is represented by x^n; and generally when any letter denoting either operation of quantity is repeated, it has the letter indicating the number of repetitions placed in the right hand corner above it; such is the case in the following instances, x^n, d^n, δ^n, f^n. It is therefore correct, as far as regards analogy, to place the index in that situation; so far then as relates to the position of the letter n, both the notations of Vandermonde and that of Arbogast are equally admissible; this, however, is not the case with the letter a. It is used for the difference of two successive factors, and has no regard at all to repetition; it is not therefore consistent with analogy to place it in a similar situation.

Another, and a very powerful reason for preferring the system of Vandermonde is, that it admits of unlimited extension, without in the least infringing on typographical facility. When the factors have their second difference zero, they are denoted thus:

$$[x, a]^n = x \cdot x - a \cdot \ldots \cdot x - \overline{n - 1} \cdot a$$

M. Lacroix has assigned to such expressions the appropriate name of powers of the second order. If the third differences are zero, they are called powers of the third order, and are readily denoted thus:

$$[x, a, b]^n = x \cdot \overline{x + a} \cdot \overline{x + 2a + b} \cdot \ldots \cdot \overline{x + na + \frac{n \cdot \overline{n - 1} \cdot b}{2}}$$

and similarly for the higher orders: this notation therefore admits of a very easy extension to complicated cases without increasing its difficulty of execution. Perhaps it may be urged that the notation of M. Kramp may be extended as follows:

$$x^{n|a,b} = x \cdot \overline{x - a} \cdot \overline{x - 2a - b} \cdot \ldots \cdot \overline{x - na - \frac{n \cdot \overline{n - 1}}{2} b}$$

but the difficulties of the printer will be greatly increased the moment any substitutions are made for a or b, and the other objections remain the same.

[a] A. M. Legendre *Exercises de calcul intégral*... (3 vol., Paris, 1811, 17, 16).

[b] Elsewhere in the *Works of Babbage* the form $\binom{n}{1}, \binom{n}{2}, \binom{n}{3} \ldots$ has been substituted.

These inconveniences attend the method of Arbogast in at least an equal degree. That which Euler has employed is adapted to the particular purpose he has in view; but if it be allowed that sufficient reasons have been produced to justify the preference we have assigned to that of Vandermonde, it is highly desirable that it should be employed on all future occasions. We have observed that M. Legendre uses $\gamma(x)$ to represent $x \cdot x - 1 \cdot \ldots \cdot 2 \cdot 1$, whether x is a whole number or a fraction; this can scarcely be termed a new notation; but it is certainly rather to be regretted that so able an enquirer should not have given his sanction to a notation previously established, and which agrees so well with those rules on which all notation should be founded.

The two signs which denote differentiation and integration, are, in strictness, liable to the objections of this rule; and a similar remark is applicable to the similarly related signs Δ and Σ.

In several of these instances, mathematical signs have been needlessly multiplied by not attending to this principle, that *whenever we wish to denote the inverse of any operation, we must use the same characteristic with the index* -1. This principle being acknowledged, we shall in future be delivered from one cause of the redundancies of signs. Should, for example, any future enquirer, when considering the calculus of variations, have occasion for the inverse operation of that denoted by δ, he must not, conformably to a hasty view of analogy, represent it by σ; but, directing his attention more deeply to the subject, he will perceive the necessity of indicating it by δ^{-1}.

Again, in the equation $x = \tan y$, the value of y, in terms of x is usually expressed $y = \text{arc} \tan x$; this is both long and inelegant. It should, in complying with the rule just laid down, be written thus: $y = \tan^{-1} x$, and similarly with other circular functions; this method is likewise attended with the advantage, / that all this class of functions will then be indicated by three letters.

Another principle, whose importance becomes eminent in proportion as our investigations become general, is, *that every equation ought to be capable of indicating a law.* Signs must not be employed at random, nor must new ones be introduced without grave necessity. When, however, unusual combinations, or the demonstration of new properties, render such a resource indispensable, it becomes particularly desirable that those which are first contrived, should be possessed of such propriety and power as shall effectually preclude future innovators from all temptation to change them. Analogy to those which form the established language of the science, although a precept of great importance, cannot be admitted to supersede the rigid enforcement of that we are now considering: happily however, the two principles will rarely be found at variance, for those symbols, and those inflexions of symbols (if the term may be allowed) which long experience has naturalized, generally furnish the most correct models of imitation. A precept, in some measure connected with this principle, although it may perhaps be considered of minor importance, may be introduced.

In case it is required to express a series continued to n terms, the nth term should be written, and not the words 'to n terms' be attached to a few of the first. These different methods would appear thus:

$$y = \frac{1}{1} + \frac{1}{2} + \frac{1}{3} + \ldots \text{ to } n \text{ terms.}$$

$$y = \frac{1}{1} + \frac{1}{2} + \frac{1}{3} + \ldots + \frac{1}{n}.$$

In this simple case, something is gained in point of brevity; and if deductions are to be made from it, much is gained in perspicuity. The justice of this observation will be more readily acknowledged if it is put into rather different language. Its object may be thus stated: *It is better to make any expression an apparent function of* n, *than to let it consist of operations* n *times repeated.* Thus, if it is required to express the series

$$\frac{1}{1^n} + \frac{1}{2^n} + \frac{1}{3^n} + \ldots$$

By means of a definite integral, its value would be

$$\int\frac{dx}{x}\int\frac{dx}{x}\ldots(n)\int\frac{dx}{1-x} \quad \begin{bmatrix} x=0 \\ x=1 \end{bmatrix}.$$

Where there are n integrations, or as it has been written by Mr Spence,

$$\int\frac{dx}{x}\int\frac{dx}{x}\ldots\int\frac{dx}{1-x} \quad \begin{bmatrix} x=0 \\ x=1 \end{bmatrix}.$$

Neither of these methods possess that clearness which ought instantly to convey to the mind the system of operations intended to be expressed; for it does not appear quite certain whether the n means that there are n repetitions of $\int\frac{dx}{x}$, or that the whole number of integrations is n: this latter is the signification in which it is employed by Spence. A better method of expressing it would be,

$$\left(\int\frac{1}{x}\right)^n \frac{x}{1-x}\,dx^n, \quad \text{or} \quad \left(d^{-1}\frac{1}{x}\right)^n \frac{x}{1-x}\,dx^n.$$

To any one acquainted with the integral calculus, this could not convey an incorrect idea, although presented for the first time, and without explanation. The index n denoting the repetition of the quantity to which it is attached, gives

$$\int\frac{1}{x}\int\frac{1}{x}\ldots\int\frac{1}{x}\frac{x}{1-x}\,dx^n.$$

And it is clear that the n dx's must be assigned to the n integrations. This latter expression may indeed be employed as deviating still less from the received one, and as still less liable to misinterpretation.

Another principle, which is chiefly valuable as it conduces to brevity, is, *that all notation should be so contrived as to have its parts capable of being employed separately.* By observing this rule, we avoid being encumbered with any other signs than those actually necessary to express the properties we are considering. This principle has been so universally followed, that it is difficult to produce examples of its neglect. Thus, although it has never been expressly proposed in terms, it has been virtually acted upon by analysts for the last two centuries.

It was in fact to be expected, before any general principles of notation were delivered, that it should be extended in exact proportion to the progress of the enquirer, and this progress proceeding from the simple to the more complicated, his notation would naturally increase by continued additions. Such being its origin, it will necessarily follow, that at any stage it might be used without reference to those additions with which subsequent considerations had obliged him to augment it.

It is a valuable service rendered to any science to embody into language those rules which perhaps insensibly direct or govern the minds of those who improve it; however apparently obvious some of them may be, or however well known amongst the instructed, a considerable time will always elapse before individual penetration will have formed for itself such guides; nor will the advantage be entirely confined to the student. How frequently does it happen, even to the best informed, that they prefer one thing and reject another, from some latent sense of their propriety or impropriety, without being immediately able to state the reasons on which such choice is founded; yet it cannot be doubted, when the selection appears to be the result of correct taste, that it is guided by unwritten rules, themselves the valued offspring of long experience. Any explanation of these is probably rendered unavailing at the instant, for want of having previously fixed them by language.

Another reason which appears to justify, or rather to oblige us to notice this principle, is, that although it has not hitherto been much infringed, it seems probable that, unless due care be taken in the future formation of notation, some of the rules which have been proposed, may themselves lead to its infraction. This observation more particularly applies to that which directs us 'to contrive all notation, so that it shall be capable of expressing laws.' In attempting to satisfy this condition, we are obliged to take wide and extensive views, and are therefore peculiarly liable to forget that which may appear of minor importance. The two principles are by no means incompatible; and it is very desirable that when we have occasion to employ the latter, the former may always be borne in mind.

Having discussed the general principles on which notation should be formed, it now becomes necessary to explain several rules which are proposed for the proper application of individual signs. Writers on algebra / have defined the meaning of the signs they use; but something further than these definitions is requisite in the present state of the science, or, to speak more accurately, some limitations, as well as extensions of the meaning of several of them are required.

There are in analysis two great divisions of symbols – those which denote quantity, and those which indicate operations. These latter were all of them originally arbitrary marks, such as those employed to express addition, multiplication, etc. At the present time, however, letters are frequently employed to signify operations; and hence arises occasionally a source, if not of error, at least of inconvenience. The use of letters in two senses has been objected to ⟨by⟩ the followers of Leibnitz, who employ the *d* to denote the differential of the quantity to which it is prefixed; and as differentials are much employed in the doctrine of series, it sometimes occurs that the fourth term of the series, $a + bx + cx^2 + dx^3 +$, enters into calculations in which the *d* also represents an operation performed on the letter to which it is prefixed. This inconvenience has been avoided by M. Lacroix, who uses the Roman d for differentiation, and the Italic *d* for quantity. Arbogast also, in his work on Derivations,[a] employs the Roman d for the same purpose. This completely removes the inconvenience; but in order to prevent its recurrence in any other shape, and for the purpose of affording a wider choice in the selection of letters, we would propose the following general rule: *all letters that denote quantity should be printed in italics, but all those which indicate operations, should be printed in a roman character.* Thus the letters $x, y, z, \ldots a, b, c, \ldots \alpha, \beta, \ldots \psi, \theta$, would represent quantity; and d, f, A, B, C, would denote operations. The only inconvenience which would attend this method, is of very minor importance: it would become necessary to cast new types for a few of the Greek letters in common use, because the difference between those two modes of printing does not exist in that language.

The class of letters that are termed characteristics, and which represent operations, is divided into two species; those which denote functions, as

$$\mathrm{f}, \mathrm{F}, \varphi, \psi, \chi$$

and those which signify that some alteration is to be made, which depends on the nature of the function on which it is to be executed: thus $f(1 + x^2)$ means a function of $1 + x^2$. In this case the form of the function selected for f is quite independent on $1 + x^2$; but if we have $\mathrm{d}(1 + x^2)$ the effect of the operation denoted by d, depends on the form of the function inclosed within the

[a] L. F. A. Arbogast, *Du calcul des dérivations* (Strasbourg, 1800).

parentheses: this is the case with the symbols

$$\mathrm{d}, \int, \Delta, \Sigma, S, \delta.$$

Those of the former species are called *functional characteristics*; and as it will be convenient to have some generic name for the latter, we shall appropriate to them that of *derivative characteristics*, since their own form depends on, and is derived from that which follows them.

The first of these species is subject to the following rule: *every functional characteristic affects all symbols which follow it, just as if they constituted one letter.*

Thus, if

$$\alpha x = ax + bx^2, \quad \text{and} \quad fx = 1 - x^2$$

then,

$$\alpha f x = \alpha(1 - x^2) = a(1 - x^2) + b(1 - x^2)^2$$

and

$$f\alpha x = 1 - (\alpha x)^2 = 1 - (ax + bx^2)^2.$$

Two signs have been made use of to represent multiplication: the cross (×) and the dot (·);[a] and the simple juxtaposition of two letters, if they are in italic, means the same thing: this last method is preferable in all simple cases, as

$$abc, \quad x^2y, \quad \frac{a + x}{y}a^2.$$

There exists a difference of meaning between the two former of these methods of indicating multiplication and the latter, which it is necessary to explain, and to which it is convenient to adhere. The dot and the cross imply a kind of disjunctive multiplication, or that, when they are interposed between two quantities before which a derivative characteristic is placed, this symbol only applies to the first of the two:

As,

$$\Delta\frac{x + a}{x}\cdot\frac{\varepsilon^x - 1}{x}$$

which represents

$$\frac{\varepsilon^x - 1}{x}\Delta\frac{x + a}{x}.$$

[a] A raised point has been used for multiplication here and throughout the *Works of Babbage*; in many of Babbage's papers a lowered dot was used which was ambiguous with a decimal point or an ellipsis.

If the dot or cross were not used in this case, the meaning would be quite different, for

$$\Delta \frac{x + a}{x} \frac{\varepsilon^x - 1}{x}$$

would signify the complete difference of the quantity

$$\frac{x + a}{x^2} (\varepsilon^x - 1).$$

Another common signification of the dot is, that when it is placed immediately after a derivative characteristic, it implies that the derivation extends until some other dot occurs which may separate it from any succeeding factors: this is sometimes denoted by brackets of various kinds, thus,

$$\mathrm{d}.xv.\frac{\mathrm{d}u}{\mathrm{d}x}$$

denotes the differential of u relative to x, multiplied by the complete differential of xv. If written thus,

$$\mathrm{d}.xv\frac{\mathrm{d}u}{\mathrm{d}x}$$

it would signify the complete differential of

$$xv\frac{\mathrm{d}u}{\mathrm{d}x}.$$

In the operations of the differential calculus, it is constantly required to express the second and higher powers of a differential, as $(\mathrm{d}x)^2$, $(\mathrm{d}x)^3, \ldots (\mathrm{d}x)^n$, whilst it has rarely been found necessary to express the differentials of the simple powers $\mathrm{d}(x^2)$, $\mathrm{d}(x^3), \ldots \mathrm{d}(x^n)$. These two series differ in the situation of the parentheses they contain; but as the quantities in the first recur perpetually, it was desirable that they should be contracted by omitting the parentheses. Universal usage has sanctioned this omission, and they are always written thus, $\mathrm{d}x^2$, $\mathrm{d}x^3, \ldots \mathrm{d}x^n$; and in order to remove the ambiguity which this might occasion in the few cases where the quantities mentioned in the latter series occur, they are distinguished by means of the point thus, $\mathrm{d}.x^2$, $\mathrm{d}.x^3, \ldots \mathrm{d}.x^n$.

The principle that *parentheses may be omitted, if it can be done without introducing ambiguity*, has been partly adopted in the manner of expressing circular functions; a department of analysis in which the inattention to all general principles of notation has been severely felt. Five methods have been generally employed for expressing the squares and higher powers of the sines and cosines of arcs.

$(\sin.\ \theta)^2$, $\sin.\ \theta^2$, $\sin.\ {}^2\theta$, $(\sin\ \theta)^2$, $\sin\ \theta^2$.

The first of these has nothing objectionable in it except the dot, which is not merely useless, but interferes / with another principle. The second omits the parentheses, in compliance with the principle we have just stated: like the former, it has the useless appendage of the dot. The third is by far the most objectionable of any, and is completely at variance with strong analogies. An index in the situation in which it there occurs, invariable denotes repetition of the symbol to which it is attached: in this case it would therefore mean sin. sin. θ, or the sine of the sine of the arc θ. It is true that this notation may be defined to mean the square of the sine, or any other function of the sine. Although a definition cannot be false, it may be improper; and the impropriety may arise either from its inducing ambiguity, or from its offending against received principles; both which objections occur in the present instance. Besides, if $\sin.{}^2\theta$ were allowed to signify the square of the sine of θ, how should we denote the second sine of θ? It may safely be asserted, that if this notation for the square of the sine were admitted, no convenient method could be devised, which should not also infringe some general principle. Thus one infraction would become the ground of introducing another, and instead of possessing a philosophical language founded on general laws, the science would arrive at one with no regularity to assist the memory, and devoid of those strong analogies which facilitate the acquisition of all others. The fifth ⟨*sic.*⟩ method $(\sin\ \theta)^2$, is made use of by Arbogast, in his work on the *Calculus of Derivations*;[a] an authority which is entitled to the greater weight, because, from the nature of that excellent work, and the powerful use which is made of the various notations employed in it, it is highly improbable that the author should have selected this out of the many received notations, without having well considered its propriety. It differs but little from the last method, the parentheses only being omitted, which gives $\sin\ \theta^2$ a mode of expression sufficiently clear for all simple cases. If however θ become a compound quantity, the parentheses must again be introduced. This notation has been employed by M. Lacroix, in his great work on the integral calculus.[b] M. Gauss and Delambre have also adopted it; and it is likewise used in the latest and most valuable work of Lagrange.*

In order to connect together various symbols which compose part of an expression, and which are subjected to the same operation, different species of

* *Mécanique Analytique.* ⟨J. L. Lagrange, *Mécanique analytique*, nouvelle édition, (2 vol., Paris, 1811–15.)

[a] L. F. A. Arbogast, *op. cit.*

[b] S. F. Lacroix, *Traité du calcul différential et du calcul intégral* (3 vol., Paris, 1797–1800).

vincula* are employed. From the great variety of these which are used by the printer, a considerable latitude of choice is allowed. In general, all symbols collected under a vinculum, are to be considered and operated upon, as if they formed a single symbol. The choice of parentheses, although a matter of minor importance, is not altogether arbitrary. An example, in which propriety and impropriety are apparent in different modes of writing the same expression, will be sufficient, without the aid of formal rules.

$$\left[1-\left(\tan\frac{a}{y}\right)^2\right]^2+\left\{1-\left\{\tan\frac{y}{a}\right\}^2\right\}^2$$

$$\left(1-\left[\tan\frac{a}{y}\right]^2\right)^2+\left[1-\left\{\tan\frac{y}{a}\right\}^2\right]^2$$

$$\left\{1-\left(\tan\frac{a}{y}\right)^2\right\}^2+\left\{1-\left(\tan\frac{y}{a}\right)^2\right\}^2$$

Of these three ways of writing the expression, the last is evidently to be preferred: the other two are objectionable, not from any deficiency in clearness, but from a certain want of symmetry, which is very perceptible in both of them.

Modern analysts have frequently had occasion to separate the symbols of operation from the quantities on which they act, an artifice which gives great conciseness to a variety of theorems: this separation has been marked in different ways, either by a dot, or the sign of multiplication, or by parentheses: in some instances the double dot (:) has been employed. As it is very desirable to confine each symbol to one meaning, the last plan, or the two dots, is the least objectionable for this purpose, and can create no ambiguity, since it has become entirely obsolete in the best writers in any other sense. The theorems of Lagrange would be thus written.

$$\Delta^n u = \{\varepsilon^{\frac{\mathrm{d}}{\mathrm{d}x}} - 1\}^n : u, \quad \Sigma^n u = \{\varepsilon^{\frac{\mathrm{d}}{\mathrm{d}x}} - 1\}^{-n} : u.$$

The merits of the two systems of notation employed in the differential calculus have been discussed by Euler, in the preface to his Institutiones Calculi Integralis,[a] and more fully by Mr Woodhouse, in the preface to the Analytical Calculations.[b] The superiority of the use of the d over the system of dots has become so very apparent, that the course of a few years will, in all probability, render the latter obsolete, a circumstance which induces us to forbear entering into any lengthened detail of their comparative values, and to content ourselves with merely indicating a few of the grounds for rejecting the latter: these are,

* Such as braces, parentheses, brackets, bars, etc.

[a] L. Euler, *Institutiones calculi integralis* (4 vol., Petropoli, 1792–94).

[b] Robert Woodhouse, *The principles of analytical calculation* (Cambridge, 1803).

1. The uncertainty which may arise respecting the letters to which the indices refer, and the confusion which arises from having two indices to a letter. 2. The want of analogy with other established notations, such as those relating to the symbols Δ and δ. 3. The great difficulty, if not the impossibility, of representing, by their means, theorems relating to the separation of operations from quantities.* /

* The editor has been indebted for this interesting article to Charles Babbage, Esq. F.R.S. Lond. and Edin.

XIX

PORISMS*[a]

From *The Edinburgh Encyclopaedia*, Vol. 17 (1830), pp. 106–14

Porisms are a species of proposition in geometry much employed by the ancients: they appear to have been highly valued, and to have assisted them greatly in their geometrical researches. The name is almost all that remains to us of their labours, and although, from the singularly curious nature of the subject, and the facilities they afford in the cultivation of every branch of geometrical science, it is improbable that the subject should have occupied the attention of only a single author, yet we are acquainted with but one geometer among the ancients who composed a work expressly on the subject, and, unfortunately for the science, the whole of that work is lost, if we except a small fragment preserved by Pappus, in such an imperfect manner, as almost to have rendered hopeless any divination of its meaning. The three books on porisms, by the author of the *Elements of Geometry*,[b] form the only work expressly devoted to this subject, whose title has been handed down to us; and even the meaning of the word porism was involved in considerable obscurity. Pappus Alexandrinus, through whom alone we derive any information on this subject, has given two definitions of the word porism; the first he blames as insufficient, because it might include some *loci*, and the second, which he adopts, is so general and indefinite, as to convey to us no precise knowledge of their nature. /

In the abstract he has given of the labours of those who preceded him, he has mentioned the three books on porisms by Euclid, and has given thirty-eight geometrical propositions, of no very considerable difficulty, as useful for the comprehension of the work itself. These, together with the imperfect definition, and an example of a porism which refers to a figure that is lost, and which is so remarkably confused as almost to render its reconstruction impossible, are all the data that remained. From these materials modern geometers have attempted

* The Editor has been indebted for this interesting article to Charles Babbage, Esq. F.R.S. L. and E. etc.

[a] This article, like 'Notation' which precedes it, was written about 1821.
[b] i.e. Euclid.

to restore the work of Euclid; and although the task is one of extreme difficulty, yet, when executed, a very probable estimate may be formed of its resemblance to the original.

Albert Girard, in a work on trigonometry, printed at the Hague in 1629,[a] mentions the lost books of porisms, and says that he had made a restoration of them: and again, in an edition of the works of Stevinus,*[b] he declares that he had reinvented the porisms of Euclid, and intended shortly to publish them. Unfortunately he died before this intention was accomplished. As the works of Stevinus were printed by his widow after his death, possibly the manuscript may still exist in some of the libraries in Holland. From the subsequent discoveries of Dr Simson, however, it appears that the idea which Girard had of the species of propositions to which he annexed the name of porisms, was by no means the same as that which the former writer has so ably proved to have been attached to it by the ancient geometers.

After Girard, the next attempt to explain the nature of porisms was made by Bullialdus;†[c] but this seems to have been derived from a communication with Fermat, to which distinguished mathematician we must now advert. Amongst his posthumous works‡[d] is a short paper, entitled '*Porismatum Euclidaeorum renovata doctrina*',[e] from which it appears that he had approached nearer than any of his predecessors to the true meaning of this class of propositions; and, in fact, several of those with which he illustrates his view of the subject are in reality porisms: but he did not arrive at any definition which should clearly separate porisms from local theorems, nor did he even conjecture that there existed some peculiar mode of analysis by which such propositions might be discovered, nor attempt to restore any of those of Euclid; his promised restoration of the whole of the three books never having been published.

Dr Halley, who possessed an extensive and profound acquaintance with the ancient geometry, made some attempts to decypher the enunciation of the porism given by Pappus. He had successfully restored the 8th book of the conics of Apollonius, and the two books of the same author, *De Sectione Spatii*; and had achieved a still more difficult labour, that of translating from the Arabic (a language with which he was unacquainted) the work of Apollonius *De Sectione Rationis*;[f] yet he was baffled by the obscurity which pervaded the mutilated

* Oeuvres de Simon Stevin, par A. Girard. Lugd. Bat. 1634.
† Exercitationes Geometriae. Parisiis, 1667 ⟨sic⟩.
‡ Fermat Opera Varia, p. 116. Tolosæ, 1679.

[a] Albert Girard, *Tables de sinus, tangentes*... (2nd edition; 1629).
[b] Albert Girard, *Les oeuvres mathématiques de Simon Stevin*... (1634).
[c] Ismaël Boulliau, *Exercitationes geometricae*... (Paris, 1657).
[d] Pierre de Fermat, *Varia opera mathematicae*... (Tolouse, 1679).
[e] *Translation:* The restored doctrine of Euclidean porisms.
[f] Edmund Halley, *Apollonii Pergaei conicorum libri octo*... (Oxford, 1710); *Apollonii Pergaei de sectione rationis*... *de sectione spatii*... (Oxford, 1706).

description of Pappus, and observes, 'Hactenus Porismatum descriptio nec mihi intellecta nec lectori profutura'.[a]

The failure of all who preceded in elucidating this obscure subject, as well as the high rank which Pappus assigned to these propositions, seems to have stimulated the curiosity of one whose unabated perseverance has been rewarded by complete success. Dr Robert Simson has described the progress he made in this subject, in a way which cannot fail to interest the attention of those who have devoted even a small portion of their time to geometrical enquiries. 'Postquam vero apud Pappum legeram porismata Euclidis collectionem fuisse artificiosissimam multarum rerum, quæ spectant ad analysin difficiliorum et generalium problematum, magno desiderio tenebar, aliquid de iis cognoscendi; quare sæpius et multis variisque viis tum Pappi propositionem generalem mancam et imperfectam, tum primum, lib. 1. Porisma quod solum ex omnibus in tribus libris integrum adhuc manet, intelligere et restituere conabar; frustra tamen, nihil enim proficiebam. Cumque cogitationes de hac re multum mihi temporis consumpserint, atque molestæ admodum evaserint, firmiter animum induxi hæc nunquam imposterum investigare; præsertim cum optimus geometra Halleius spem omnem de iis intelligendis abjecisset. Unde quoties mente occurrebant, toties eas arcebam. Postea tamen accidit, ut improvidum et præpositi immemorem invaserint, meque detinuerint donec tandem lux quædam effulserit, quæ spem mihi faciebat inveniendi saltem Pappi propositionem generalem, quam quidem multa investigatione tandem restitui.

'Descriptio autem quam tradit (*Pappus*) porismatum adeo brevis est et obscura, et injuria temporis aut aliter vitiata, ut nisi Deus benigne animum et vires dederit in ea pertinaciter inquirere, in perpetuum forsan geometris latuisset.' Simsoni *Opera Reliqua*, p. 513.[b]

[a] *Translation:* Thus far the definition of the term Porism neither have I understood it nor is it useful to the reader.

[b] *Translation:* It was so, having read the collection of Euclid's porisms in Pappus, which seemed to be very ingeniously contrived of many things, but which presents to the analyst difficulties and general problems, that I was seized with a great desire to get to know more about them. There were many hurdles and other problems to be overcome: Pappus' work was both incomplete and damaged, and his general proposition was wanting; foremost among the problems was that of all the three books on the porisms only Book 1 had still survived. I wanted both to research and reinstate them. However my efforts were in vain; I could make no progress. Many ideas occurred to me on this matter, and I spent much time on it. Indeed I might have broken down with anxiety had I not persuaded my reasoning never to cease its enquiry. Even the best, the mathematician Halley, gave up all hope of trying to elucidate them. From wherever and as often as ideas kept occurring to my mind I encompassed them wholly. Later, however, it happened without forethought that I allowed them to take over my memory completely, and thus they detained me, until at last a certain light shone through, which gave me hope that at least I could reconstruct Pappus' general proposition. At length and after much investigation this I succeeded in doing.

Indeed the description left by Pappus on porisms was short and obscure, moreover after either the ravages of time or through other causes it had become corrupted. If God had not been benign enough to given one mind and powers stubbornly to pursue this enquiry, mathematicians would have for evermore been deprived of the result. R. Simson, *Opera quaedam reliqua* ... [posthumous works] Glasgow, 1776).

Dr Trail, in his life of Simson, gives the following account of the discovery.

'Dr Simson maintained for some time his resolution of abstaining from all attempts at the rediscovery of porisms; but happening one day to be walking with some friends on the banks of the river Clyde at Glasgow, and by accident being left behind his company, he inadvertently fell into a reverie respecting porisms.

'Some new ideas struck his mind, and with his chalk having drawn some lines on an adjoining tree, at that moment, for the first time, he acquired a just notion of one of Euclid's porisms.'[a]

The first publication of Simson on this subject, was a paper inserted in the *Philosophical Transactions* for the year 1723;[b] it was not, however, until after his death, that the whole of his investigations were made public in the posthumous edition of his works, for which the mathematical world is indebted to the munificence of the late Earl Stanhope. Some few years after, this subject attracted the attention of Mr Playfair, who has given a most philosophical account of the origin of this class of propositions, and has removed whatever obscurity remained attached to them. The paper in which his views are explained, is indeed a model of that peculiarly beautiful style of writing for which he was so justly celebrated, and which is unfortunately so rarely met with in the literary productions of mathematicians.* [c]

In the geometrical explanation of porisms, we shall avail ourselves of the light which he has thrown on the subject, and then endeavour to supply those observations which he promised respecting their algebraical investigation.

The definition of porisms which Simson has given, / is unquestionably rather obscure; and without an example of one of these propositions, it is by no means easy to comprehend its meaning: it appears therefore preferable to postpone the explanation of the term until the reader is made acquainted with the thing. The ancient geometers examined every problem on which they bestowed their attention with the most minute scrutiny: unacquainted with the comprehensive generalization which is introduced into every geometrical problem by the application of algebra, they carefully inquired into every separate case that could cause any change in the magnitude or relative position of the data, fearful lest that mode of solution they had contrived for it in one case, might not equally apply to others. Such a laborious course of enquiry, although adverse to rapid advancement, was well calculated to make them perfectly acquainted with every thing remarkable which the solution of the problem could present;

* *Transactions of the Royals Society of Edinburgh*, 1793, vol. iii.

[a] William Trail, *Account of the life and writings of Robert Simson*... (London, 1812), pp. 19–20.
[b] Robert Simson, 'Two general propositions of Pappus, in which many of Euclid's porisms are included', *Philosophical Transactions*, Vol. 32 (1723), p. 330.
[c] John Playfair, 'On the origin and investigation of porisms', *Transactions of the Royal Society of Edinburgh*, Vol. 3 (1794), pp. 154–204.

and it must soon have occurred to them, that in many cases the general construction would fail, and no solution be obtained, in consequence of some peculiar relation between the data. Such is the case if we attempt to divide a given line into two parts, whose rectangle is equal to a given square. When the given square is greater than that described on half the given line, no solution can be obtained. In such cases, the problem became impossible, and it was always found that some two at least of the data were contradictory to each other. In the illustration we have chosen, the two conditions defining the magnitude of the line and that of the rectangle ⟨and⟩ its segments are incompatible.

When a problem contained an impossible case, another question presented itself; to determine the limits amongst the relations of the data, so that it shall just remain possible; and with respect to the problem itself, to construct it so that a certain quantity, instead of being given, shall be the greatest or least possible. The elegant constructions to which this gave rise, under the name of maxima and minima, are well known to geometers.

These circumstances would occur when the data were but few, and the problem simple; but in the consideration of questions a little less elementary, it must have been observed, that besides this method, by which the construction became useless, another of quite an opposite nature was sometimes introduced. It might happen that two lines or two circles by whose intersection the point ought to be determined, instead of cutting each other as in the general case, or not intersecting each other at all, as in the impossible one, should wholly coincide. The true interpretation of this circumstance could not long remain unnoticed. Since that point, which was common to the two lines, determined the point to be found, and in the case of two circles intersecting, there were two points in common, and therefore equally fulfilling the condition, it was natural to conclude, that when the lines or circles coincided, all points being in common, all would equally satisfy the problem. Here, then, an infinity of solutions appeared, yet they were all connected by a certain law. The reason of such a singular result, must soon have been found in the coincidence between two of the data, and thus a less number of data being given than were sufficient, the problem became indeterminate.

These curious cases would, of course, become objects of research, from the great facilities they afforded for the solutions of the problems to which they belonged, and the elegance which they introduced into them; and partaking in some measure of the nature of problems, as well as of theorems, they formed an intermediate class of propositions of great importance, to which, when enunciated in a peculiar manner, the name of porisms was attached.

As an example of the manner in which a porism might be discovered, we shall consider the following problem.

A circle *ABC*, (Figure 1) a straight line *DE*, and a point *F*, being given in

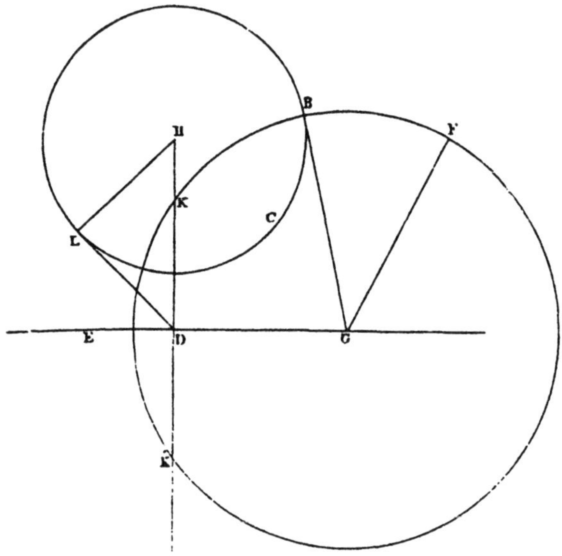

Figure 1

position, to find a point *G* in the straight line *DE*, such that *GF*, the line drawn from it to the given point, shall be equal to *GB*, the line drawn from it touching the given circle.

Suppose the point *G* to be found, and *GB* to be drawn touching the circle *ABC* in *B*; let *H* be the centre of the circle *ABC*; join *HB*, and let *HD* be perpendicular to *DE*; and from *D* draw *DL*, touching the circle *ABC* in *L*, and join *HL*. Also, from the centre *G*, with the distance *GB* or *GF*, describe the circle *BKF*, meeting *HD* in the points *K* and *K′*.

It is plain that the lines *HD* and *DL* are given in position and in magnitude. Also, because *GB* touches the circle *ABC*, *HBG* is a right angle, and *G* is the centre of the circle *BKF*; therefore *HB* touches the circle *BKF*, and consequently the square of *HB* or of *HL* is equal to the rectangle *K′HK*. But the rectangle *K′HK*, together with the square of *DK*, is equal to the square of *DH*, because *KK′* is bisected in *D*; therefore the squares of *HL* and *DK* are also equal to the square of *DH*. But the squares of *HL* and *LD* are equal to the square of *DH*; wherefore, the square of *DK* is equal to the square of *DL*, and the line *DK* to the line *DL*. But *DL* is given in magnitude, therefore *DK* is given in magnitude, and *K* is therefore a given point. For the same reason, *K′* is a given point, and the point *F* being also given by hypothesis, the circle *BKF* is given in position. The point *G*, therefore, the centre of the circle *BKF*, is given, which was to be found.

Hence this construction: Having drawn *HD* perpendicular to *DE*, and *DL* touching the circle ABC, make DK and DK' each equal to DL, and find by the centre of a circle described through the points K, F and K', that is, let FK' be joined, and bisected at right angles by the line MN, which meets DE in G; G will be the point required, or it will be such a point, that if GB be drawn from it, touching the circle ABC, and GF to the given point, GB and BF will be equal to one another.

In this instance, we have a problem which admits in general but of one solution, since only one circle can pass through three given points; yet if the point F which is given should coincide with either of the two points K or K' which are found, it is evident that an infinite number of circles can pass through two given points, and their centres will be situated on a right line perpendicular to the middle point of the line which joins them: in this case, then, the problem becomes indeterminate, since any point in that line will satisfy the conditions.

The indeterminate case is thus enunciated as a porism.

A circle ABC being given by position, and also a straight line DE, which does not cut the circle, a point K may be found such, that if G be any point whatever in the line given, the straight line drawn from G to the point K shall be equal to the straight line drawn from G touching the circle ABC. This is in fact the / 66th proposition in Dr Simson's restoration, slightly altered in its statement.

As another instance of a problem leading to a porism, we will give one which appears to have led to the invention of the second porism in the treatise of Simson.

A circle ABC, (Figure 2) and two points D and E, in a diameter of it being

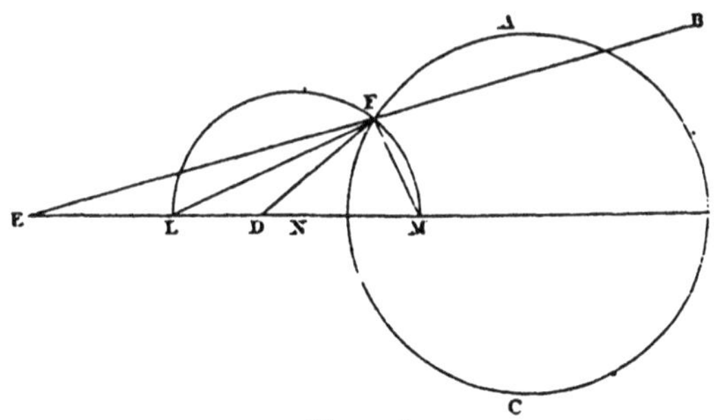

Figure 2

given, to find a point F in the circumference of the given circle, from which, if straight lines be drawn to the given points E and D, these straight lines shall have to one another the given ratio of α to β.

Suppose the problem resolved, and the F is found, so that FE has to FD the given ratio of α to β. Produce EF any how to B, bisect the angle EFD by the line FL, and the angle DFB by the line FM.

Then because the angle *EFD* is bisected by *FL*, *EL* is to *LD* as *EF* is to *FD*, that is in a given ratio; and as *ED* is given, each of the segments *EL*, *LD* is given, and also the point *L*.

Again, because the angle *DFB* is bisected by *FM*, *EM* is to *MD* as *EF* to FD, that is in a given ratio; and therefore since *ED* is given, *EM*, *MD*, are also given, and likewise the point *M*.

But because the angle *LFD* is half of the angle *EFD*, and the angle *DFM* half of the angle *DFB*, the two angles *LFD*, *DFM*, are equal to the half of two right angles, that is to a right angle. The angle *LFM* being therefore a right angle, and the points *L* and *M* being given, the point *F* is in the circumference of a circle described on the diameter *LM*, and consequently given in position.

Now, the point *F* is also in the circumference of the given circle *ABC*: it is therefore in the intersection of two given circumferences, and therefore is found.

Hence the following construction: Divide *ED* in *L* so that *EL* may be to *LD* in the given ratio of α to β, and produce *ED* also to *M*, so that *EM* may be to *MD* in the same given ratio of α to β. Bisect *LM* in *N*, and from the centre *N*, with the distance *NL*, describe the semicircle *LFM*, and the point *F* in which it intersects the circle *ABC*, is the point required, or that from which *FE* and *FD* are to be drawn.

It must, however, be remarked, that the construction fails when the circle *LFM* falls either wholly without or wholly within the circle *ABC*; so that the circumferences do not intersect; and in these cases the solution is impossible. It is plain also that in another case the construction will fail, namely, when it so happens that the circumference *LFM* wholly coincides with the circumference *ABC*. In this case it is further evident, that every point in the circumference *ABC* will answer the conditions of the problem, which therefore admits of innumerable solutions.

The indefinite case of this proposition thus enumerated becomes a porism.

A circle, *ABC*, (Figure 3) being given, and also a point *D*, a point *E* may be

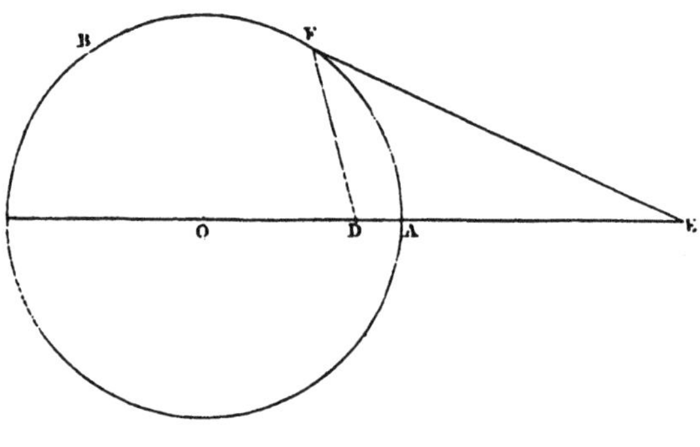

Figure 3

found, such that two lines, *DF* and *EF*, inflected from these points to any point *F* in the circumference of the circle, shall have to each other a given ratio, which ratio may be found.

From these examples, the definitions which have been given of the term porism will be better understood than by merely considering the words in which they are expressed. Dr Simson has thus described them: 'Porisma est propositio in qua proponitur demonstrare rem aliquam, vel plures datas esse, cui, vel quibus, ut et cuilibet ex rebus innumeris, non quidem datis, sed quæ ad ea quæ data sunt eandem habent relationem, convenire ostendendum est affectionem quandam communem in propositione descriptam.'[a]

The obscurity of this definition is such, that nothing but a comparison with an example can make it intelligible; that of Playfair is much happier, and is thus expressed: *A porism is a proposition, affirming the possibility of finding such conditions as will render a certain problem indeterminate.*

This latter has the advantage of indicating the course to be pursued in the discovery of porisms; for in the first case the problem was rendered indeterminate by making two out of the three points, which determined the position of a circle, coincide; and in the last example, the coincidence of two circles, whose intersections should have determined the point required in the problem, rendered it indeterminate. This mode of analysis, for the discovery of porisms, has one disadvantage, that it supposes the solution of the problem to be first found; that which was contrived by Simson is free from this objection, and when abridged by the considerations which Playfair has introduced, is admirably adapted to its object.

It may be observed, that the points or magnitudes required may generally be discovered by considering the extreme cases; but that the relation between these and the indefinite magnitudes cannot be arrived at by such limited considerations. The difference between a locus, a local theorem, and a porism, are well illustrated by Playfair in the various modes of enunciating the truth discovered in the second of the two propositions we have given.

Thus, when we say, if from two points, *E* and *D* (Figure 3) two lines, *EF*, *FD*, are inflected to a third point *F*, so as to be to one or other in a given ratio, the point *F* is in the circumference of a circle given in position: we have a locus. But when conversely, it is said, if a circle *ABC*, of which the centre is *O*, be given in position, as also a point *E*, and if *D* be taken in the line *EO*, so that the rectangle, *EO*, *OD*, be equal to the square of *AO*, the semidiameter of the circle; and if from *E* and *D*, the lines *EF* and *DF* be inflected to any point

[a] *Translation:* A Porism is a Proposition in which it is Proposed to demonstrate that one or more things are given, between which and every one of innumerable other things, not given, but assumed according to a given law, a certain relation, described in The Proposition, is to be shewn to take place. W. Trail, *Account of the life and writings of Robert Simson...* (London, 1822), p. 45.

whatever in the circumference *ABC*; the ratio of *EF* to *DF* will be a given ratio, and the same with that of *EA* to *AD*: we have a *local theorem*.

And, lastly, when it is said, if a circle *ABC* be given in position, and also a point *E*, a point *D* may be found, such that if two lines *EF* and *FD* be inflected from *E* and *D* to any point whatever *F*, in the circumference, these lines shall have a given ratio to one another: the proposition becomes a porism, and is the same we have just investigated.

The algebraical method for the investigation of porisms, may very readily be deduced from the consideration of the definition which Playfair has given, and the facilities which such a method presents in the discovery of this class of truths, is another instance of the advantages which result from a condensed method of expressing the relations of quantity. It has been stated, that *a porism is a proposition affirming the possibility of finding the indeterminate case of a problem.*

If, therefore, any problem is proposed in which the quantity sought is called x; by means of the given conditions some equation will be found between x and known quantities, which may be reduced to the form

$$A + Bx + Cx^2 + \ldots + Nx^n = 0 \qquad (a)$$

A, B,... being known functions of the constant quantities; from this equation x may be determined, or at least it cannot generally have more than a certain / determinate number of values; such values of x satisfy the conditions of the problem; but in order to discover whether any porismatic case exists, we must examine whether the data of the problem admit of such a relation amongst themselves, that we may have at the same time

$$A = B = C = \ldots = N = 0.$$

If this is the case, the equation (a) is verified independently of any particular value of x; and, instead of a limited, we have an indefinite number of solutions. This principle may be stated more generally thus, if $a, b, c, \ldots$ are given quantities, and $x, y, z, \ldots$ those which are to be found; then the solution of the problem leads to several equations of the form

$$F(x, y, \ldots a, b, \ldots) = 0.$$

If any relation which can be established amongst the constants $a, b, c, \ldots$ shall cause the equation which results from the elimination of the unknown quantities from these equations to be independent of any of them, then by supposing that relation to exist we have a porism.

As an example, let us take the following problem. Suppose a circle (Figure 4) whose radius is r, and a point C in its diameter, such that $OC = v$, also a straight line, FL, perpendicular to this diameter; let it be required to find the angle which the chord, PQ, must make with the diameter; so that if another

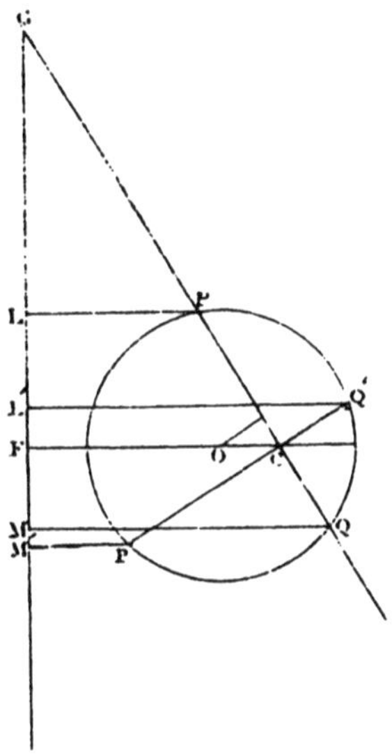

Figure 4

chord, P_1Q_1, be drawn at right angles to it, and from the extremities of these chords perpendiculars be drawn to the given line, the rectangle under those let fall from one chord on this line shall be equal to that under the perpendiculars let fall on the same line from the other chord. Let the required angle $PCF = \theta$, then[a]

$$CE = v\cos\theta \quad \text{and} \quad OE = v\sin\theta.$$

Hence

$$CP = \sqrt{r^2 - v^2(\sin\theta)^2} - v\cos\theta$$

$$CQ = \sqrt{r^2 - v^2(\sin\theta)^2} + v\cos\theta$$

also

$$CG = \frac{a - v}{\cos\theta}$$

therefore

$$PG = \frac{a - v}{\cos\theta} + v\cos\theta \pm \sqrt{r^2 - v^2(\sin\theta)^2}$$

[a] In the original, the notation $\sin\theta^2$ was used for $(\sin\theta)^2$; we have substituted the latter in preference to the modern notation $\sin^2\theta$ which Babbage considered 'by far the most objectionable of any, and is completely at variance with strong analogies' ('Notation' (1830), in this volume of the *Works of Babbage*, p. 422).

$$QG = \frac{a - v}{\cos\theta} + v\cos\theta \pm \sqrt{r^2 - v^2(\sin\theta)^2}$$

these multiplied by cos θ produce respectively

$$PL = a - v + v(\cos\theta)^2 \pm \cos\theta\sqrt{r^2 - v^2(\cos\theta)^2}$$

$$QM = a - v + v(\cos\theta)^2 \pm \cos\theta\sqrt{r^2 - v^2(\cos\theta)^2}$$

the rectangle under these two lines is

$$(a - v + v(\cos\theta)^2)^2 - (\cos\theta)^2(r^2 - v^2(\sin\theta)^2)$$

and this by proper reductions becomes

$$PL \cdot QM = (a - v)^2 + (2av - v^2 - r^2)(\cos\theta)^2.$$

In order to find the rectangle of the perpendiculars let fall from the extremities of another chord at right angles to this, we have only to change θ into $(\pi/2 + \theta)$ in this expression, and since $\cos(\pi/2 + \theta) = \sin\theta$ we must substitute $1 - (\cos\theta)^2$, instead of $(\cos\theta)^2$, which gives

$$P_1L_1 \cdot Q_1M_1 = a^2 - r^2 - (2av - v^2 - r^2)(\cos\theta)^2.$$

And since these two quantities must be equal by the conditions of the problem, we have the following equation for determining cos θ,

$$\{(a - v)^2 - a^2 + r^2\} + \{2av - v^2 - r^2 + 2av - v^2 - r^2\}(\cos\theta)^2 = 0 \quad (1)$$

The solution of this equation gives $\cos\theta = \pm 1/\sqrt{2}$ or $\theta = \pm\pi/4$, which evidently satisfies the condition; but there exists a possible relation amongst the quantities a, v, and r, which will fulfil the equation (1) without determining the value of cos θ, for that equation is the same as

$$((a - v)^2 - a^2 + r^2)^2 - ((a - v)^2 - a^2 - r^2)(\cos\theta)^2 = 0$$

and by assuming,

$$((a - v)^2 - a^2 + r^2 = 0, \quad \text{or} \quad v = a \pm \sqrt{a^2 - r^2}$$

this equation is verified independently of the value of cos θ. Here then is an indeterminate or porismatic case, and the porism so discovered may be thus enunciated.

A circle and a straight line being given in position, a point may be found within the circle, such, that if any two chords are drawn through that point at right angles to each other, the rectangle under the perpendiculars, let fall from the extremities of the first chord to the line given in position, shall always be equal to the rectangle under the perpendiculars let fall from the extremities of the other chord on the same line.

If the problem had been to find the angle which the first chord should make with the diameter, that the sum of the squares of the perpendiculars from the first chord should be equal to the sum of the squares of the perpendiculars from the extremities of the second chord, we should have found

$$PL^2 + QM^2 = 2(a - v)^2 + 2(2av - 3v^2 + r^2)(\cos\theta)^2 + 4v^2(\cos\theta)^4$$

and putting $(\pi/2) + \theta$ for θ, and making the proper reductions

$$P_1L_1{}^2 + Q_1M_1{}^2 = 2(a^2 + r^2) - 2(2av + v^2 + r^2)(\cos\theta)^2 + 4v^2(\cos\theta)^4$$

and the equation determining $\cos\theta$, is,

$$((a - v)^2 - a^2 - r^2) + (2av - 3v^2 + r^2 + 2av + v^2 + r^2)(\cos\theta)^2 = 0$$

or,

$$((a - v)^2 - a^2 - r^2) - 2((a - v)^2 - a^2 - r^2)(\cos\theta)^2 = 0$$

this, like the former, gives $\theta = \pm(\pi/4)$: but here also an indeterminate case exists, and may be found by making the coefficients vanish. The equation $((a - v)^2 - a^2 - r^2 = 0$, gives $v = a \pm \sqrt{a^2 + r^2}$; and this value of v verifies the equation independently of the magnitude of $\cos\theta$. This porism may be enunciated as follows.

A circle and a right line being given in position, a point may be found within the circle, such that drawing through it any two chords at right angles to each other, the sums of the squares of the perpendiculars from the extremities of the first chord to the given line, may be equal / to the sums of the squares of the perpendiculars drawn from the extremes of the other chord to the given line.

It is proper to observe, that in both these cases the lower sign must be employed, otherwise the point found will not be within the circle. In retranslating algebra into geometry, it frequently happens that only some one of the roots which satisfy the conditions of the problem algebraically, will fulfil the geometrical conditions.

In both examples which have been given, the equation determining the value of the unknown quantity, was multiplied by a factor independent of it; thus the first equation may be put under the form

$$\{(a - v)^2 - a^2 - r^2\} \cdot \{1 - 2(\cos\theta)^2\} = 0$$

and the second,

$$\{(a - v)^2 - a^2 - r^2\} \cdot \{1 - 2(\cos\theta)^2\} = 0$$

and as these equations contain all the conditions of the respective problems, provided they are verified, solutions will be found. If the relation amongst the quantities a, v, r is such that the first factor vanishes, it is evident that the equations are, in all cases, verified, without assigning any particular values to $\cos\theta$.

This evanescence of a factor is not, however, the only cause which produces porismatic cases, as will appear in several of the subsequent examples.

The following problem leads to a porism, which is already well known by the writings of Simson and Playfair. To show how it might have been discovered by the algebraic method, will not therefore be without interest.

A circle and a straight line being given, and also a point in that diameter of the circle which is perpendicular to the given line, it is required to find a point in the given line, such that if a line be drawn through it, and the given point cutting the circle, then the rectangle under the segments of that line contained between the points found and the circle, shall be a given multiple of the square of the line joining the two points.

Employing the same notation and letters as in the Figure 4, let G be the point required; then G will be determined by the angle θ, which the line CG makes with the diameter; and the values of the several lines will be as follows:

$$PG = \frac{a-v}{\cos\theta} + v\cos\theta \pm \sqrt{r^2 - v^2(\sin\theta)^2}$$

$$QG = \frac{a-v}{\cos\theta} + v\cos\theta \pm \sqrt{r^2 - v^2(\sin\theta)^2}$$

$$CG = \frac{a-v}{\cos\theta}.$$

Hence the equation expressing the condition is

$$\left(\frac{a-v}{\cos\theta} + v\cos\theta\right)^2 - r^2 + v^2(\sin\theta)^2 = n\left(\frac{a-v}{\cos\theta}\right)^2$$

or,

$$(1-n)\left(\frac{a-v}{\cos\theta}\right)^2 + 2v(a-v) - r^2 + v^2 = 0$$

where n is the given multiple. From this equation the angle θ may easily be found, and consequently the position of the point G may be determined. But if $n = 1$, the angle altogether disappears from the formula, and it can only be satisfied by supposing

$$2av - v^2 - r^2 = 0$$

or,

$$v = a \pm \sqrt{a^2 - r^2}.$$

If this relation take place amongst the data, any value of θ will fulfil the condition; but if it does not, no value can satisfy it.

Observing that the rectangle under the segments of the line is equal to the square of the tangent to the circle, we have the following porism.

A circle and a right line GE (Figure 5) *being given, a point* C *may be found*

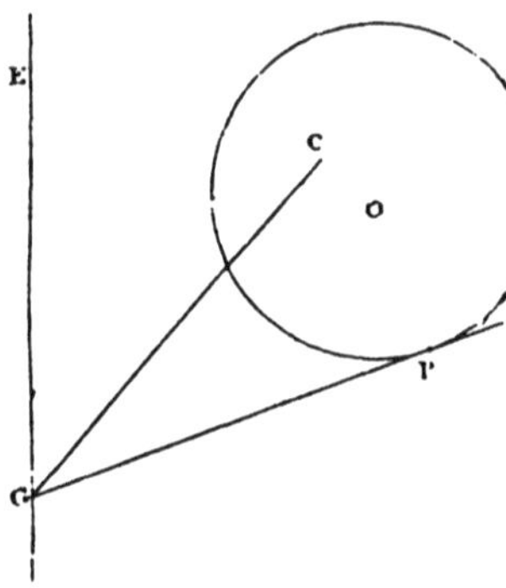

Figure 5

within the circle, such that, if from any point G *in the given line a tangent* GP *be drawn to the circle, and also if that point* G *and the point found* C *be joined, the line* CG *joining these points shall be equal to the tangent to the circle* GP.

This porism is similar to propositions 63 and 66 of Simson's restoration, and is one of the illustrations made use of in the paper of Playfair. It would be easy to investigate, by this method, many other of the porisms of Euclid; but since it is proposed as a method of invention, it will be more satisfactory to employ it in the discovery of new ones. We shall therefore proceed to investigate a few others.

A circle being given, and also a line, and a point situated in the line drawn from the centre, perpendicular to the given line, and the quantities a, v, r, ... remaining the same as in the previous questions, if a chord is drawn through the given point, and from its extremities perpendiculars are drawn to the given line, we shall have (in Figure 4)

$$PL = a - v + v(\cos\theta)^2 \pm \cos\theta\sqrt{r^2 - v^2(\cos\theta)^2}$$

$$QM = a - v + v(\cos\theta)^2 \pm \cos\theta\sqrt{r^2 - v^2(\cos\theta)^2}$$

and the sum of these two perpendiculars, or

$$PL + QM = 2(a - v + v(\cos\theta)^2)$$

which is independent of the value of r, or of the magnitude of the radius of the circle.

This circumstance allows us to enunciate this truth as a very simple porism. *Two lines making a given angle with each other* GQ and GM (Figure 6) *being*

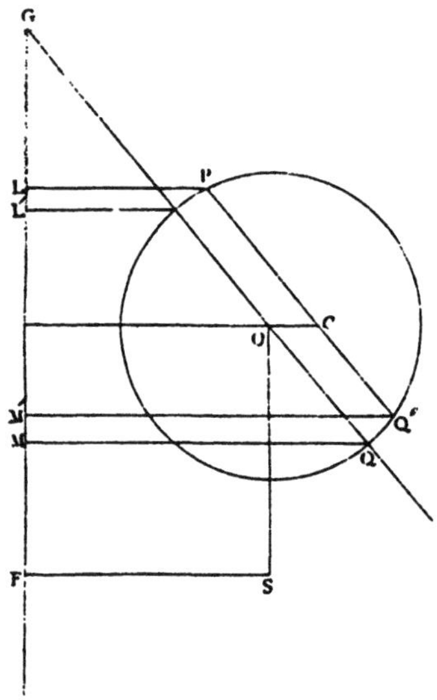

Figure 6

given, a point O *may be found in one of them, such that if about that point as a centre, a circle with any radius be described cutting one of the given lines in two points* P *and* Q, *the sum of the perpendiculars drawn from these points to the other line shall be equal to a given line.*

If FS perpendicular to GM be made equal to half the line to which the sum of the two perpendiculars is equal, and if SO be drawn perpendicular to GQ, any point O may be taken as the centre of the circles.

In Figure 7 the same notation being preserved, the sum of the perpendiculars is

$$PL + QM = 2(a - v + v(\cos\theta)^2).$$

Let some other point C be taken, $OC = v'$, and a chord be drawn parallel to the former, then the sum of the perpendiculars drawn from the extremities of this chord is

$$P_1L_1 + Q_1M_1 = 2(a - v' + v'(\cos\theta)^2).$$

Let us now determine the value of $\cos\theta$, so that the former sum shall differ from n times the latter by a constant quantity $2c$, the resulting equation is /

$$2(a - v + v(\cos\theta)^2) + 2c = 2n(a - v' + v'(\cos\theta)^2)$$

or,

$$(n - 1)a - nv' + v - c + (nv' - v)(\cos\theta)^2 = 0.$$

If we assume $v = nv'$, and $(n - 1)a - c = 0$, this equation is verified without assigning any particular value to $\cos\theta$; hence result various porisms by giving particular values to n; if $n = 2$, we have the following:

A line LM (Figure 7) *and two points* C *and* C_1 *in a perpendicular to it being*

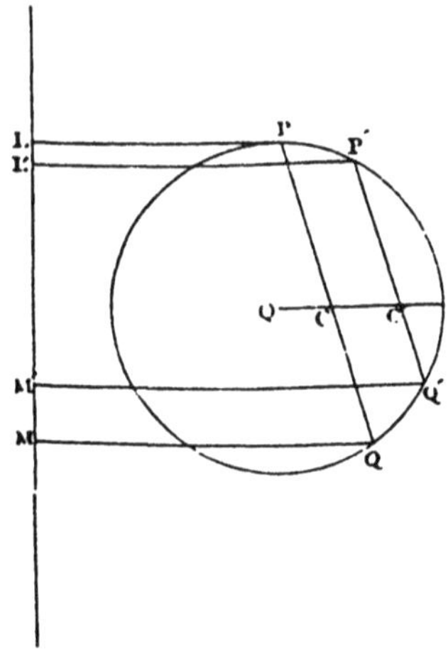

Figure 7

given, a third point may be found about which if a circle be described with any radius OR, *and if through the two given points any two parallel chords* PQ *and* PQ_1 *be drawn; the sum of the perpendiculars* PL + QM *drawn from the extremities of one of these chords, together with a line which may be found shall be equal to twice the sum of the perpendiculars* $PL_1 + QM_1$ *drawn from the extremities of the second chord to the given line.*

The demonstration is sufficiently obvious from the algebraic investigation. The centre of the circle is situated in the same line with the two given points, and its distance OC from the point C is equal to CC_1 and the line which may be found is equal to OF.

This porism is rather remarkable from the circumstance of two of the quantities being indeterminate, namely, the angle which the chords make with a given line, also the magnitude of the radius of the circle; and it may be observed that instances of such double or even triple indeterminations will occur much more frequently in the algebraic discovery of porisms than in their geometrical invention.

If in the problem from which this porism was deduced, we had supposed the chords at right angles to each other instead of being parallel, then we should have had the equation

$$2(a - v + v(\cos\theta)^2 + c) = 2n(a - v' + v'(\sin\theta)^2) = 2n(a - v'(\cos\theta)^2).$$

Hence,

$$(n - 1)a + v - c - (nv' + v)(\cos\theta)^2 = 0.$$

This equation may be satisfied without determining the value of θ, by assuming the two equations,

$$(n - 1)a + v - c = 0$$

and

$$nv' + v = 0.$$

From the latter we have

$$v' = c - \frac{v}{n}$$

and from the former

$$v = c - (n - 1)a.$$

This produces the following porism:

A point O *and a right line* LM (Figure ⟨missing⟩) *being given in position, two other points* C *and* C_1 *may be found such that if a circle be described round the first point* O *with any radius, and if two chords* PQ *and* P_1Q_1 *be drawn at right angles to each other through the latter points the sum of the perpendiculars* PL + QM *drawn from the extremities of one of those chords to the given line, together with a line which may be found, shall be equal to a given multiple* n *of the sum of the perpendiculars* $P_1L_1 + Q_1M_1$ *drawn from the extremities of the other chord to the given line.*

The radius vector from any point within the circle being

$$CP = -v\cos\theta + \sqrt{r^2 - v^2(\sin\theta)^2}$$

that which is distant from it half a revolution will be,

$$CQ = v\cos\theta + \sqrt{r^2 - v^2(\sin\theta)^2}$$

and consequently the expression for any chord passing through a point C distant from the centre by the quantity v, will be

$$PQ = 2\sqrt{r^2 - v^2(\sin\theta)^2}$$

and the value of any other chord parallel to this at a distance from the centre, denoted by v', measured on the same diameter, is

$$P'Q' = 2\sqrt{r^2 - v'^2(\sin\theta)^2}.$$

Let us now suppose three chords parallel to each other, and that the sum of the squares of the first two is equal to twice the square of the third; the equation which results is

$$4(r^2 - v^2\sin\theta) + 4(r^2 - v'^2(\sin\theta)^2) = 8(r^2 - v''^2(\sin\theta)^2)$$

or

$$2r^2 - (v^2 + v'^2)(\sin\theta)^2 = 2r^2 - 2v''^2(\sin\theta)^2$$

and this can only be satisfied by supposing

$$v'' = \sqrt{\frac{v^2 + v'^2}{2}}$$

if the origin of the co-ordinates, which is now at the centre, be removed the distance a, then v, v', and v'' become $a + v$, $a + v'$, and $a + v''$; and the last equation gives

$$2(v'' + a)^2 = (v + a)^2 + (v' + a)^2$$

if v, v', and v'' are given, a may be found, and will be

$$a = \frac{2v''^2 - v'^2 - v^2}{2v + 2v' - 4v''}.$$

This gives rise to the following porism:

Three points C, C_1, *and* C_2, *being given* (Figure ⟨8⟩) *in a right line, another point* O *may be found such that if with any radius a circle be described about* O *as a centre, and if three parallel chords be drawn through the three points, then the sum of the squares of two of them,* PQ^2 *and* ${P_1Q_1}^2$ *shall always be equal to twice the square of the remaining chord* P_2Q_2.

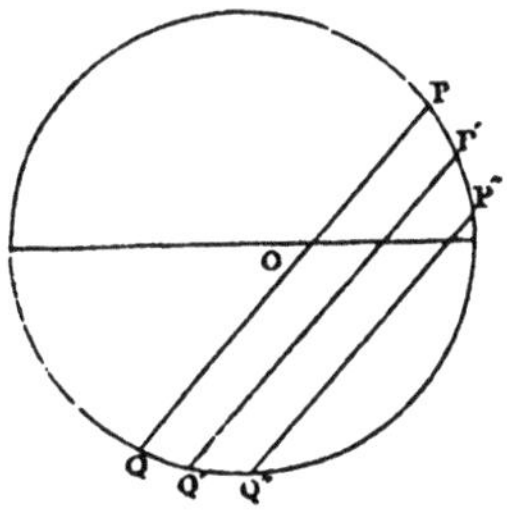

Figure 8

It has been shown that if any chord be drawn through a point C, we have

$$PQ^2 = 4r^2 - 4v^2(\sin\theta)^2.$$

If we consider another chord at right angles to the former, and drawn through the same point, we have

$$P_1Q_1{}^2 = 4r^2 - 4v^2(\cos\theta)^2$$

the sum of these two is

$$PQ^2 + P_1Q_1{}^2 = 8r^2 - 4v^2$$

which is a constant quantity; and since the angle θ is variable, this suggests the following porism:

A circle PP_1QQ_1 (Figure ⟨9⟩) *being given, another circle may be found, such*

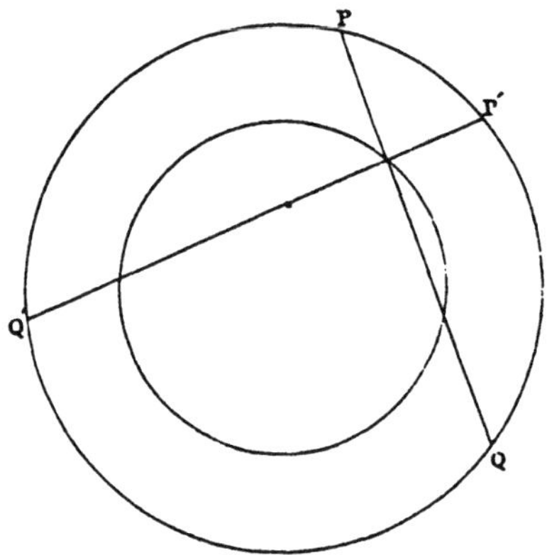

Figure 9

that, if through any point of its circumference two chords be drawn to the first circle, and perpendicular to each other, the sum of the squares of these two chords shall be equal to a given square.

If through any point in a circle n chords be drawn, making with each other the angle $2\pi/n$ these squares will be represented by

$$4(r^2 - v^2(\sin\theta)^2)$$
$$4(r^2 - v^2(\sin(\theta + 2\pi/n))^2)$$
$$4(r^2 - v^2(\sin(\theta + 4\pi/n))^2)$$
$$\vdots$$
$$4(r^2 - v^2(\sin(\theta + (2n - 2)\pi/n))^2)$$

/

but the value of the series

$$(\sin\theta)^2 + (\sin(\theta + 2\pi/n))^2 + \ldots + (\sin(\theta + (2n-2)\pi/n))^2$$

is

$$\frac{n}{2}$$

consequently the sum of the squares of all the chords is equal to

$$4nr^2 - 2nv^2$$

which is a constant quantity. This produces another porism, which comprehends the last as a particular case.

A circle being given, a point within it may be found, such that, if any number n *of chords be drawn through it, making equal angles with each other, the sums of the square of these chords shall be equal to the square described in a given line.*

A circle being given, a point within it may be found, such that, if any number n *of chords be drawn through it, making equal angles with each other, the sum of the fourth powers of those chords shall be equal to a given fourth power.*

The fourth powers of these chords are thus algebraically expressed,

$$2^4(r^4 - 2r^2v^2(\sin\theta)^2 + v^4(\sin\theta)^4)$$

$$2^4(r^4 - 2r^2v^2(\sin(\theta + 2\pi/n))^2 + v^4(\sin(\theta + 2\pi/n))^4)$$

$$\vdots$$

$$2^4(r^4 - 2r^2v^2(\sin(\theta + (2n-2)\pi/n))^2 + v^4(\sin(\theta + (2n-2)\pi/n))^4)$$

the sums of series containing the powers of the sines or cosines of arcs in arithmetical progression have long been known,* and when the common difference is the twentieth part of the circumference, and the number of terms equal to n, they receive a very remarkable simplification, for in that case

$$(\sin\theta)^{2m} + (\sin(\theta + 2\pi/n))^{2m} + \ldots + (\sin(\theta + (2n-2)\pi/n))^{2m} =$$

$$= \frac{2m \cdot 2m - 1 \cdot \ldots \cdot m + 1}{1 \cdot 2 \cdot 3 \cdot \ldots \cdot m} \cdot \frac{n}{2^{2m}}$$

and also,

$$(cos\ \theta)^{2m} + (\cos(\theta + 2\pi/n))^{2m} + \ldots + (\cos(\theta + 2n - 2)\pi/n))^{2m} =$$

$$= \frac{2m \cdot 2m - 1 \cdot \ldots \cdot m + 1}{1 \cdot 2 \cdot \ldots \cdot m} \cdot \frac{n}{2^{2m}}$$

where m must be less than n; if we apply these considerations to the several series in the above sum, which respectively multiply $2r^2v^2$ and v^4, we shall find

* *Acad. Petrop. Com. Nov.* 1773, Euler. ⟨'Summatio progressionum . . .', Vol. 18 (1774), pp. 24–36.⟩

for the value of the first $n/2$, and for that which multiplies v^4, $3n/8$; so that the sum of all the fourth powers of the chords is

$$2^4nr^4 - 2^4nr^2v^2 + 6nv^4.$$

If m is less than n, then it may be readily shown that if chords be drawn through any point within a circle, making equal angles with each other, the sum of the $2m$ powers of those chords will always be equal to a constant quantity; for the $2m$ powers of such chords are

$$2^{2m}(r^2 - v^2(\sin\theta)^2)^m$$
$$2^{2m}(r^2 - v^2(\sin(\theta + 2\pi/2))^2)^m$$
$$\vdots$$
$$2^{2m}(r^2 - v^2(\sin(\theta + (2n - 2\pi/n))^2)^m.$$

And if these expressions are expanded, the series of the powers of sines, which constitute each vertical column, are each equal to some constant quantity: the whole sum is therefore independent of θ.

A straight line and a circle being given, and also a point in that diameter of the circle which is perpendicular to the given line, it has been found that the sum of the perpendiculars to the given line drawn from the extremities of any chord passing through the given point, is expressed by

$$2(a - v + v(\cos\theta)^2).$$

If another chord pass through the same point at right angles to the former, the expression for the sum of the perpendiculars drawn from its extremities to the given line, will be

$$2(a - v + v(\sin\theta)^2).$$

The sum of the four perpendiculars is therefore

$$4(n - v) + 2v = 4a - 2v.$$

This produces the following porism:

A circle and a straight line being given, a point may be found within the circle, such that, if any two chords be drawn through it at right angles to each other, and if from the extremities of these chords perpendiculars be drawn to the given line, the sum of these four perpendiculars shall be equal to a given right line.

This property may be generalized, by supposing n chords, instead of two, to be drawn through the point found, and it will be perceived, that, if they make equal angles round that point, the sums of the perpendiculars drawn from their extremities to the given line will be a constant quantity.

In the same figure, the sums of the rectangles under the perpendiculars, let fall from each chord, will be thus expressed:

$$(a - v)^2 + (2av - v^2 - r^2)(\cos\theta)^2$$

$$(a - v)^2 + (2av - v^2 - r^2)(\cos(\theta + 2\pi/n))^2$$

$$\vdots$$

$$(a - v)^2 + (2av - v^2 - r^2)(\cos(\theta + (2n - 2/n))^2$$

And since

$$(\cos\theta)^2 + (\cos(\theta + 2\pi/n))^2 + \ldots + (\cos(\theta + (2n - 2)\pi/n))^2 = \frac{n}{2}$$

we have for the sum of all the rectangles

$$n(a - v)^2 + \frac{2av - v^2 - r^2}{2}n.$$

This suggests another porism, as follows:

A straight line and a circle being given, a point within may be found, such that a certain number n *of chords being drawn through it; the sums of all the rectangles under the perpendiculars, drawn from the extremities of each chord, shall be equal to a given square.* /

In the same figure, the sums of the squares of the perpendiculars are represented by

$$2(a - v)^2 + 2(2av - 3v^2 + r^2)(\cos\theta)^2 + 4v^2(\cos\theta)^4$$

$$2(a - v)^2 + 2(2av - 3v^2 + r^2)(\cos(\theta + 2\pi/2))^2 + 4v^2(\cos(\theta + 2\pi/n))^4$$

$$2(a - v)^2 + 2(2av - 3v^2 + r^2)(\cos(\theta + (2n - 2)\pi/n))^2 + 4v^2(\cos(\theta + (2n - 2)\pi/n))^4$$

and the sums of the vertical series being found by means of the formula already given, we have for the sum of the squares

$$2n(a - v)^2 + n(2av - 3v^2 + r^2) + \frac{3n}{2}v^2$$

which is independent of the value of θ; it may therefore be enunciated thus:

A circle and a right line being given in position, and also a square being given; a point within the circle may be found, such that if through the point found a given number n *of chords be drawn, making equal angles with each other, and from their extremities perpendiculars be drawn to the given line, the sum of the squares of all these perpendiculars shall be equal to the square which is given.*

In this, as in many of the preceding propositions, the given quantities must be contained within certain limits, otherwise they may become impossible: these limits will always be pointed out distinctly by the algebraic analysis.

The theorems relating to the sums of series of the powers of sines or cosines of arcs in arithmetical progression, which have been noticed in a former page, are of very extensive use in the discovery of porisms. In fact, whatever line (whether relating to the circle or ellipse, or to any other combination of lines or curves), can be expressed in a rational integral form, with respect to the powers of the sine or cosine, if only the angle be continually increased by some aliquot part of the circumference, until it returns into the first line; then the sum of all these lines will be constant, as will also the sums of any powers of such lines.

The explanation which has been given of the method of applying algebraic reasoning to the discovery of the most elegant class of geometrical propositions, will, it is presumed, have demonstrated that this instrument of investigation is at least as fertile in the number of the conclusions to which it leads, as that analysis which was contrived by the celebrated restorer of the lost books of Euclid. In comparing the time and attention which must be expended in employing the geometrical method with that which is requisite for the complete success of the algebraic analysis, the superior value of the latter is strikingly pre-eminent; and if, by adding to the number of these propositions, any considerable benefit would accrue to the science, that number might easily be enlarged to an unlimited extent. Such, however, is not the case; and it must be acknowledged that these truths, in their geometrical form, are useful only for the purpose of cultivating those mental habits which mathematical studies tend so strongly to promote. Signs and figure are less abstract than mere number, and are therefore more easily conceived, and their relations more calculated to excite and fix the attention: this, together with the imperfection of those instructions which are usually given to the learner at the commencement of his algebraic studies, seems to be the real ground of the asserted superiority of geometrical over algebraical reasoning, for the purpose to which we have alluded. /

XX

MATHEMATICAL QUESTIONS[a]

From the *Mathematical Repository*, New Series, Vol. 5 (1830), pp. 51, 63–4, 66–7, 71–2, 178–80.

QUESTION 423
⟨Solution required by 1 December 1819⟩

In a circle of the nth order, whose equation is $x^{2n} + y^{2n} = a^{2n}$, any point C being taken, and the abscissa and ordinate OF and CF being drawn; if the triangle COF is placed in the situation OEB, then the point B shall be situated in the curve, and the sum of the two arcs, DB and DC, shall always be constant.

SOLUTION
by C. B. the proposer

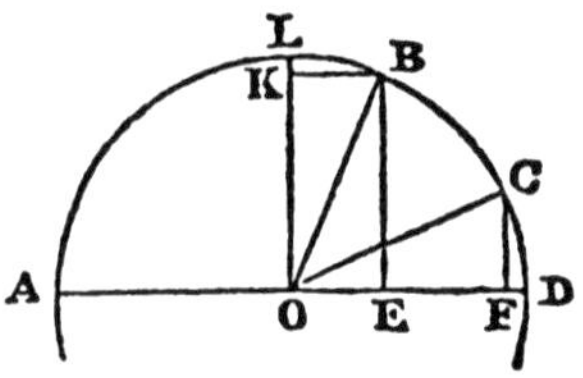

Let $OF = x$, $CF = y$; then $y = (a^{2n} - x^{2n})^{1/2n}$. Now it is first to be shown that if $OE = CF$, then BE the ordinate at E is equal to OF. Since $BE = (a^{2n} - OE^{2n})^{1/2n}$. Put for OE its value $CF = (a^{2n} - x^{2n})^{1/2n}$ then $BE = \{a^{2n} - (a^{2n} - x^{2n})^{1/2n}\}^{1/2n} = x$, which was to be proved.

Also since $KB = OE = CF$ and $KO = BE = OF$, and B and C are two points in the curve it is symmetrical relative to the two axes LO and OD; therefore the arc LB is equal to the arc CD, and the sum of the two arcs $BD + CD$ is always equal to $BD + LB$, which is constant.

[a] The third series of the *Mathematical Repository* was published by Thomas Leybourn during 1806–35. The *Mathematical Repository* is representative of a genre of British mathematical publishing which included such periodicals as the *Gentleman's Diary; or Mathematical Repository* and the *Ladies' Diary* (the three are sometimes confused). These publications regularly set mathematical

QUESTION 427
⟨Solution required by 1 December 1819⟩

Prove that $\frac{1}{2} = \frac{1}{\cos\theta} - \frac{1}{\cos 2\theta} + \frac{1}{\cos 3\theta} - \ldots$

SOLUTION
by C. B. the proposer

Put

$$\frac{v^{2z}}{1 + v^{2z}} = v^{2z} - v^{4z} + v^{6z} - v^{8z} + \ldots$$

Multiply by $(-1)^{-z}$ and integrate, observing that

$$\sum(-1)^{-1}v^{2iz} = -\frac{(-1)^{-z}v^{2iz}}{v^{2i+1}}$$

we have

$$\sum\frac{(-1)^{-z}v^{2z}}{1 + v^{2z}} = -\frac{(-1)^{-z}v^{2z}}{v^2 + 1} + \frac{(-1)^{-z}v^{4z}}{v^4 + 1} - \frac{(-1)^{-z}v^{6z}}{v^6 + 1} + \ldots \qquad (1)$$

---

problems for their readers to solve which, whilst challenging, were well removed from the mainstream of mathematical research:

In these [publications] many curious problems, not of the highest order indeed but still having a considerable degree of difficulty, and far beyond the mere elements of science, are often to be met with; and the great number of ingenious men who take a share in proposing and answering these questions, whom one has never heard of any where else, is not a little surprising. Nothing of the same kind, we believe, is to be found in any other country. (*Edinburgh Review*, Vol. 11 (1807), pp. 249–284; as quoted in J. M. Dubbey, *The mathematical work of Charles Babbage* (Cambridge, 1978), p. 14.)

The solutions to the problems in the *Mathematical Repository* were generally anonymous or pseudonymous, but it can be assumed that the problem solvers were well-educated gentlepeople. The *Mathematical Repository* also included the Cambridge problems for the degree of Batchelor of Arts, which again hints at the social class and education of the readership.

The sets of problems for the *Mathematical Repository* were published from time to time, and each set specified a latest date by which readers' solutions were to be submitted. Periodically, the problems and solutions for several years were published as a complete volume. Although Babbage's problems appeared in Volume 5 of the *Mathematical Repository*, dated 1830, they were originally issued in problem sets requiring solutions to be submitted during 1819–21. These dates indicate that the problems were set by Babbage when he was still active as a mathematician.

Again

$$\frac{v^{-2z}}{1 + v^{-2z}} = v^{-2z} - v^{-4z} + v^{-6z} - \ldots$$

Multiply by $(-1)^{-z}$ and integrate; then

$$\sum \frac{(-1)^{-z} v^{-2z}}{1 + v^{-2z}} = -\frac{(-1)^{-z} v^{-2z}}{v^{-2} + 1} + \frac{(-1)^{-z} v^{-4z}}{v^{-4} + 1} - \frac{(-1)^{-z} v^{-6z}}{v^{-6} + 1} + \ldots \quad (2)$$

Add (1) and (2) after dividing the numerator and denominator of each term by v^i in the first and by v^{-i} in the last

$$\sum \left\{ \frac{v^{2z}}{1 + v^{2z}} + \frac{v^{-2z}}{1 + v^{-2z}} \right\} (-1)^{-1} = -(-1)^{-z} \times$$

$$\left\{ \frac{v^{2z-1} + v^{-2z+1}}{v + v^{-1}} - \frac{v^{4z-2} + v^{-4z+2}}{v + v^{-2}} + \frac{v^{6z-3} + v^{-6z+3}}{v + v^{-3}} - \ldots \right\}$$

but the integral is $+\Sigma(-1)^{-z} = -(-1)^{-z}/2 + c$: put $z + \frac{1}{2}$ for z, and let $v = \cos\theta \pm \sqrt{-1}\sin\theta$, then it is /

$$-\frac{(-1)^{-z-\frac{1}{2}}}{2} + c =$$

$$-(-1)^{-z-\frac{1}{2}} \left\{ \frac{v^{2z} + v^{-2z}}{2\cos\theta} - \frac{v^{4z} + v^{-4z}}{2\cos 2\theta} + \frac{v^{6z} + v^{-6z}}{2\cos 3\theta} - \ldots \right\}.$$

If $z = \frac{1}{2}$,

$$\tfrac{1}{2} + c = \left\{ \frac{2\cos\theta}{2\cos\theta} - \frac{2\cos\theta}{2\cos\theta} + \ldots \right\} = \tfrac{1}{2}$$

hence $c = 0$; and multiplying by $-(-1)^{-z-\frac{1}{2}}/2$ we have

$$1 = \frac{v^{2z} + v^{-2z}}{\cos\theta} - \frac{v^{4z} + v^{-4z}}{\cos 2\theta} + \frac{v^{6z} + v^{-6z}}{\cos 3\theta} - \ldots$$

Now put $z = 0$, and dividing by 2, we have

$$\frac{1}{2} = \frac{1}{\cos\theta} - \frac{1}{\cos 2\theta} + \frac{1}{\cos 3\theta} - \frac{1}{\cos 4\theta} + \ldots$$

In the same way it may be proved that

$$1 = \frac{x + x^{-1}}{(\cos\theta)^n} - \frac{x^2 + x^{-2}}{(\cos 2\theta)^n} + \frac{x^3 + x^{-3}}{(\cos 3\theta)^n} - \ldots$$ /

QUESTION 428

⟨Solution required by 1 December 1819⟩

Prove that the sum of the series

$$\frac{1}{1\cdot 3\cdot\ldots\cdot 2n+1}-\frac{1}{3\cdot 5\cdot\ldots\cdot 2n+3}+\ldots \text{ is equal to}$$

$$\frac{2^{-1}}{1\cdot 2\cdot\ldots\cdot n}\left\{\frac{\pi}{2}-\frac{1}{1}-\frac{1}{1\cdot 3}-\frac{1\cdot 2}{1\cdot 3\cdot 5}-\ldots \text{ to } n \text{ terms}\right\}.$$

SOLUTION

by C. B. the proposer

Assume

$$u_n=\frac{1}{1\cdot 3\cdot\ldots\cdot 2n+1}-\frac{1}{3\cdot 5\cdot\ldots\cdot 2n+3}+\frac{1}{5\cdot 7\cdot\ldots\cdot 2n+5}-\ldots$$

then

$$u_n-\frac{1}{1\cdot 3\cdot\ldots\cdot 2n+1}=-\frac{1}{3\cdot 5\cdot\ldots\cdot 2n+3}+\frac{1}{5\cdot 7\cdot\ldots\cdot 5}-\ldots$$

adding these together, we have

$$2u_n-\frac{1}{1\cdot 3\cdot\ldots\cdot 2n+1}=$$

$$\frac{2n+3-1}{1\cdot 3\cdot\ldots\cdot 2n+3}-\frac{2n+5-3}{3\cdot 5\cdot\ldots\cdot 2n+5}+\frac{2n+7-5}{5\cdot 7\cdot\ldots\cdot 2n+7}-\ldots$$

but the series on the right hand is equal to $(2n+2)u_{n+1}$ hence the equation of first differences

$$2u_n-2(n+1)u_{n+1}=\frac{1}{1\cdot 3\cdot\ldots\cdot 2n+1}.$$

Put

$$u_n=\frac{w_n}{1\cdot 2\cdot 3\cdot\ldots\cdot n}$$

then

$$u_{n+1} = \frac{w_{n+1}}{1 \cdot 2 \cdot 3 \cdot \ldots \cdot n+1}$$

and our equation becomes

$$2\left\{w_n \frac{1}{1 \cdot 2 \cdot \ldots \cdot n} - \frac{n+1}{1 \cdot 2 \cdot \ldots \cdot n+1} w_{n+1}\right\} = \frac{1}{1 \cdot 3 \cdot \ldots \cdot 2n+1}$$

or $w_{n+1} - w_n = -\frac{1}{2}\frac{1 \cdot 2 \cdot \ldots \cdot n}{1 \cdot 3 \cdot 5 \cdot \ldots \cdot 2n+1}$,

hence $w_n = -\frac{1}{2}\left\{\frac{1}{1} + \frac{1}{1 \cdot 3} + \frac{1 \cdot 2}{1 \cdot 3 \cdot 5} + \ldots \text{ to } n \text{ terms}\right\} + c.$

If $n = 0$, $w_0 = \frac{1}{1} - \frac{1}{3} + \frac{1}{5} - \ldots = \frac{\pi}{4} = c,$

hence, the series in question or w_n is equal to

$$\frac{2^{-1}}{1 \cdot 2 \cdot \ldots \cdot n}\left\{\frac{\pi}{2} - \frac{1}{1} - \frac{1}{1 \cdot 3} - \frac{1 \cdot 2}{1 \cdot 3 \cdot 5} - \ldots \text{ to } n \text{ terms}\right\}.$$

QUESTION 429
⟨Solution required by 1 December 1819⟩

Prove that the series

$$\frac{1}{1 \cdot 2 \cdot 3 \cdot \ldots \cdot n} - \frac{1}{2 \cdot 3 \cdot \ldots \cdot n+1} + \frac{1}{3 \cdot 4 \cdot \ldots \cdot n+2} - \ldots \text{ is equal to}$$

$$\frac{1}{1 \cdot 2 \cdot \ldots \cdot n-1}\left\{2^{n-1}\log 2 - \frac{2^{n-2}}{1} - \frac{2^{n-3}}{2} - \ldots - \frac{2^0}{n-1}\right\}.$$

SOLUTION
by C. B. the proposer

Put

$$u_n = \frac{1}{1 \cdot 2 \cdot \ldots \cdot n} - \frac{1}{2 \cdot 3 \cdot \ldots \cdot n+1} + \ldots$$

Then

$$u_n - \frac{1}{1\cdot 2\cdot\ldots\cdot n} = -\frac{1}{2\cdot 3\cdot\ldots\cdot n+1} + \frac{1}{3\cdot 4\cdot\ldots\cdot n+2} - \ldots$$

$$2u_n - \frac{1}{1\cdot 2\cdot\ldots\cdot n} = \frac{n+1-1}{1\cdot 2\cdot\ldots\cdot n+1} - \frac{n+2-2}{2\cdot 3\cdot\ldots\cdot n+2} + \ldots = nu_{n+1}$$

hence the equation of first differences /

$$2u_n - nu_{n+1} = \frac{1}{1\cdot 2\cdot\ldots\cdot n}.$$

Assume

$$u_n = \frac{2^n w_n}{1\cdot 2\cdot\ldots\cdot n-1}$$

then it becomes

$$\frac{2^{n+1}w_n}{1\cdot 2\cdot\ldots\cdot n-1} - \frac{n2^{n+1}\cdot w_{n+1}}{1\cdot 2\cdot\ldots\cdot n} = \frac{1}{1\cdot 2\cdot\ldots\cdot n}$$

or

$$w_n - w_{n+1} = \frac{2^{-n-1}}{n}.$$

Therefore

$$w_n = -\sum\frac{2^{-n-1}}{n} = -\frac{2^{-2}}{1} - \frac{2^{-3}}{2} - \frac{2^{-4}}{3} - \ldots - \frac{2^{-n}}{n-1} + c$$

and

$$u_n = \frac{1}{1\cdot 2\cdot\ldots\cdot n-1}\left\{2^n c - \frac{2^{n-2}}{1} - \frac{2^{n-3}}{1} - \frac{2^{n-4}}{3} - \ldots - \frac{2^0}{n-1}\right\}.$$

If $n = 1$, then

$$2c = \frac{1}{1} - \frac{1}{2} + \frac{1}{3} - \ldots$$

or $c = \frac{1}{2}\log 2$.

Therefore the sum of the series is,

$$\frac{1}{1\cdot 2\cdot\ldots\cdot n-1}\left\{2^{n-1}\log 2 - \frac{2^{n-2}}{1} - \frac{2^{n-2}}{1} - \frac{2^{n-3}}{2} - \frac{2^{n-4}}{3} - \ldots - \frac{2^0}{n-1}\right\}.$$

PRIZE QUESTION 470
⟨Solution required by 1 December 1821⟩

Two persons A and B play at '*heads and tails*'. A has the privilege of regulating the bets at each throw, which he does thus: 'To the constant sum u he adds the nth part of all his winnings by all the previous throws, or if he have lost he subtracts the nth part of his loss from the same sum u. After $p + q$ throws, out of which p have been won and q lost, what is the amount of his profit?

SOLUTION
by C. B. the Proposer

The first stake is u and whether he is successful or otherwise his profit may be expressed thus, $u(-1)^a$ for if he lose a is an odd but if he wins it is an even number. The second stake is $u + (-1)^a\, u/n$ and his profit on this may be similarly expressed by

$$\left\{u + (-1)\frac{u}{n}\right\}(-1)^b$$

without deciding which it is: /

The third stake is

$$u + (-1)^a\frac{u}{n} + (-1)^b\frac{u}{n} + (-1)^a(-1)^b\frac{u}{n^2}$$

and the profit

$$\left\{u + (-1)^a\frac{u}{n} + (-1)^b\frac{u}{n} + (-1)^a(-1)^b\frac{u}{n^2}\right\}(-1)^c.$$

Continuing in this manner and collecting the profits on each, we have for its amount

$$W =$$

$$u(-1)^a$$

$$u(-1)^b + \frac{u}{n}(-1)^a(-1)^b$$

$$u(-1)^c + \frac{u}{n}\{(-1)^a(-1)^c + (-1)^b(-1)^c\} + \frac{u}{n^2}\{(-1)^a(-1)^b(-1)^c\}$$

$$u(-1)^d + \frac{u}{n}\{(-1)^a + (-1)^b + (-1)^c\}(-1)^d$$

$$+ \frac{u}{n^2}\{(-1)^a(-1)^b + (-1)^a(-1)^c + (-1)^b(-1)^c\}(-1)^d$$

$$\left. + \ldots + \frac{u}{n^3}\{(-1)^a(-1)^b(-1)^c(-1)^d\right\}.$$

The right side of this equation consists of u multiplied by the sum of all the quantities $(-1)^a$, $(-1)^b$, $(-1)^c, \ldots$ plus u/n multiplied by the sum of all their products two by two, plus, u/n^2 multiplied by the sum of all their products three by three, plus etc.

From this consideration it appears that if the right side were divided by n it would be equal to the sum of all the coefficients except the first of the equation

$$\left\{x + \frac{(-1)^a}{n}\right\}\left\{x + \frac{(-1)^b}{n}\right\}\left\{x + \frac{(-1)^c}{n}\right\}\ldots = 0$$

or

$$\frac{1}{n}W = u\left\{\left(1 + \frac{(-1)^a}{n}\right)\left(1 + \frac{(-1)^b}{n}\right)\left(1 + \frac{(-1)^c}{n}\right)\ldots - 1\right\}$$

but p of the quantities a, b, c, etc. are by the question even numbers and q are odd, or

$$\frac{1}{n}W = \left\{\left(1 + \frac{1}{n}\right)^p\left(1 - \frac{1}{n}\right)^q - 1\right\}u$$

hence the whole profit is /

$$W = n\left(\frac{n+1}{n}\right)^p\left(\frac{n-1}{n}\right)^q u - nu.$$

Corollary If $p = q$

$$W = \frac{(n^2 - 1)^p}{n^{2p-1}}u - nu.$$

/

For Product Safety Concerns and Information please contact our EU representative GPSR@taylorandfrancis.com Taylor & Francis Verlag GmbH, Kaufingerstraße 24, 80331 München, Germany

Batch number: 10725800

Printed by Printforce, the Netherlands